2009 European Microelectronics and Packaging Conference

(EMPC 2009)

Rimini, Italy
16 – 18 June 2009

Pages 1-407

IEEE Catalog Number: CFP0954H-PRT
ISBN: 978-1-4244-4722-0

Copyright © 2009, International Microelectronics and Packaging Society-ITALY
All Rights Reserved

***This publication is a representation of what appears in the IEEE Digital Libraries. Some format issues inherent in the e-media version may also appear in this print version.*

IEEE Catalog Number: CFP0954H-PRT
ISBN 13:   978-1-4244-4722-0

Additional Copies of This Publication Are Available From:

Curran Associates, Inc
57 Morehouse Lane
Red Hook, NY 12571 USA
Phone: (845) 758-0400
Fax: (845) 758-2633
E-mail: curran@proceedings.com

TABLE OF CONTENTS

In-mould Integration of Electronics into Mechanics and Reliability of Overmoulded Electronic and Optoelectronic Components..1
T. Alajoki, M. Koponen, E. Juntunen, J. Petaja, M. Heikkinen, J. Ollila, A. Sitomaniemi, T. Kosonen, J. Aikio, J.T. Makinen

A 3-D Packaging Concept for Cost Effective Packaging of MEMS and ASIC on Wafer Level.......................7
T. Baumgartner, M. Topper, M. Klein, B. Schmid, D. Knodler, H. Kuisma, S. Nurmi, H. Kattelus, J. Dekker, R. Schachler

Trends in IC Packaging..12
C. Beelen-Hendrikx

Temporary Adhesives for Wafer Bonding: Deep Reactive Ion Etching Application....................................20
D. Belharet, P. Dubreuil, D. Colin, L. Mazenq, H. Granier

Integrated LTCC-Glass Microreactor and µTAS with Thermal Stabilization for Biological Application..27
P. Bembnowicz, D. Nowakowska, L. Golonka

Fabrication & Characterization of S-Band Power Amplifier using GaAs Die...31
N.S. Bhatti, M. Imran

High Speed Packaging Solutions for LiNbO$_3$ Electro-Optical Modulator..35
S. Bonino, R. Galeotti, L. Gobbi, M. Belmonte, M. Bonazzoli

Packaging and Wired Interconnections for Insertion of Miniaturized Chips in Smart Fabrics..................40
J. Brun, D. Vicard, B. Mourey, B. Lepine, F. Frassati

Aluminum Ribbon on a Power Device...45
G. Cristaldi, G. Malgioglio, E. Scrofani

Advanced Modeling Techniques for System-level Power Integrity and EMC Analysis.............................52
G. Graziosi, P.J. Doriol, Y. Villavicencio, C. Forzan, M. Rotigni, D. Pandini

MEMS Pressure Sensors – New LGA Packagings..58
F. Fontana, L. Baldo, M. Azzopardi, S. Gatt

Wafer Level Packaging Fan Out Thermal Management: Is Smaller Always Hotter?..................................65
D. Gualandris, C.M. Villa

Multifunctional Coatings for Wafer-Level Chip Scale Packaging..69
R. Stapleton, D. Zoba, C. Brannen, P. Hough

Highly Conductive Adhesives via Novel Heterogeneous Structures...74
T.D. Fornes, P.W. Hough

Application of 3D Modeling Tools for Advanced Packaging on a Broad Range of Industrial Applications..78
Y. Imbs, L. Marechal, D. Auchere, G. Graziosi, J. Debono

Experimental Characterization of Thermo-Mechanical Properties of Lead-based Solders for Power Electronics Packaging Reliability Applications..87
S. Jacques, J. Roubion, N. Batut, R. Leroy, L. Gonthier

Investigation of Compliant Interconnect for Ball Grid Array (BGA)..92
R. Johannessen, F. Oldervoll, H. Kristiansen, H. Tyldum, H. Nguyen, K. Aasmundtveit

Reliability of 100 µm Bi- and In- Solder Balls..98
S. Kemethmuller, R. Dohle, J. Pohlner, T. Dunne, J. Goßler

Hydrolysis Testing of ACF Joined Flip Chip Components with Conformal Coating..................................103
K. Kokko, H. Harjunpaa, P. Heino, M. Kellomaki

Package Design for Alleviating Stress in Materials Embedded with Electronic Systems..........................109
M. Lishchynska, K. Delaney

Development of Low-firing Lead-free Thick-film Materials on Steel Alloys for Piezoresistive Sensor Applications..116
C. Jacq, T. Maeder, P. Ryser

Structuration of Zero-shrinkage LTCC Using Mineral Sacrificial Materials..122
T. Maeder, C. Jacq, Y. Fournier, W. Hraiz, P. Ryser

Process Development for a Very Precise Placement of a Lens for Micro-Optics Based Components.........128
D. Caccioli, L. Maggi

Electromagnetic Simulations for the Packaging Design of Telecommunication Component......................133
L. Maggi, G. Ticozzi

WPLGA: New Package Family for Medium Pin Count with Design Flexibility..139
P. Magni, G. Graziosi, C. Villa, R. Tiziani, R. Gacusan

Advancements in Bumping Technologies for Flip Chip and WLCSP Packaging 145
D. Manessis, R. Patzelt, A. Ostmann, H. Reichl

Assembly - Chip Interactions Leading to PPM-level Failures in Microelectronic Packages 151
A. Mavinkurve, H. Cobussen, W.D. Van Driel, L. Endrinal M. Van Dort

Small Size LTCC FlipChip-Package for RF-Power Applications 159
J. Muller, M. Noren, M. Mach, S. Brunner, C. Hoffmann

X-ray nanoCT of Interconnections in IC Packages: Visualizing of Internal 3D-Structures with Submicrometer Resolution 163
J. Luebbehuesen, H. Roth, T. Neubrand, O. Brunke

Cu Wire Bonding: Reliability Improvement for High Temperature in Plastic Packages 167
C. Passagrilli, B. Vitali, R. Tiziani, C. Azzopardi

Size and Microstructure Effects on the Stress-Strain Behaviour of Lead-Free Solder Joints 171
P. Hegde, D.C. Whalley, V.V. Silberschmidt

Experimental Study of Polymers as Encapsulating Materials for Photovoltaic Modules 180
S.A. Sala, M. Campaniello, A. Bailini

Fine Die-Attach Delamination Analysis by Scanning Acoustic Microscope 187
G. Santospirito, A. Terzoli

Materials Science Challenges in Green High Power Density Devices 1. The Ag Loaded Glues for Die Bonding 191
A. Scandurra, G.F. Indelli, R. Zafarana, A. Cavallaro, E. Scrofani, S. Russo, J.P. Giry, S. Pignataro

Modeling of Flip Chip Bump Patterns to Minimize Crosstalk on a BU-BGA Package Design 199
K. Sheach, G. Xiang, P. Brunet

Joint Project for Mechanical Qualification of Next Generation High Density Package-on-Package (PoP) with Through Mold Via Technology 203
M. Dreiza, J.S. Kim, L. Smith, D. Campos, E. Saugier, P. Jarvinen

Introduction of a Unified Equipment Platform for UV Initiated Processes in Conjunction with the Application of Electrostatic Carriers as Thin Wafer Handling Solution 211
D. Tonnies, M. Gabriel, B. Neubert, M. Hennemeyer, M. Zoberbier, R. Zoberbier

Roll-to-Roll Manufacturing of Organic Photovoltaic Modules 217
M. Tuomikoski, P. Kopola, H. Jin, M. Ylikunnari, J. Hiitola-Keinanen, M. Valimaki, M. Aikio, J. Hast

Encapsulation of the Next Generation Advanced Mems& Sensor Microsystems 221
A. Bos, L. Wang, T. Van Weelden

Next Generation Leadless RF Packages Utilizing 1st Level Low Cost Flip Chip Interconnect Technology 226
S. Walczyk, P. Dijkstra, N. Kramer, J. Verspeek

BCB-Based Wafer-Level Packaging of Integrated CMOS/SOI Piezoresistive MEMS Sensors 232
D. Weiland, A. Chaehoi, S. Ray, D. O'Connell, M. Begbie, C. Wang

New Flipchip Technology 237
R. Windemuth, T. Ishikawa

New Plasma Cleaning Technology 243
R. Windemuth, M. Nonomura, G. Dunn

A Study on High-density High-speed SerDes Design in Buildup Flip Chip Ball Grid Array Packages 248
G. Xiang, K. Sheach, P. Brunet

A Novel Methodology for Analyzing Variation Risk Introduced by the Manufacturing Process in Microsystems 252
Y. Sun, C.R. Fowkes, N. Gindy

Fracture Toughness Assessment of ACF Flip-chip Packages under High Moisture Condition with Moire Interferometry 259
J. Park, J. Jang, K. Jang, K. Paik, S. Lee

The Design and Improvement of LTCC-based Capacitive Pressure Sensors Employing Finite Element Analysis 266
C. Ionescu, P. Svasta, C. Marghescu, M.S. Zarnik, D. Belavic

Investigation of Solder Joints by Thermographical Analysis 270
P. Svasta, C. Ionescu, N.D. Codreanu, D. Bonfert

Fully Embedded Optical and Electrical Interconnections in Flexible Foils 275
E. Bosman, G. Van Steenberge, P. Gerrinck, J. Vanfleteren, P. Van Daele

Metal Trace Impact Life Prediction Model for Stress-Buffer- Enhanced Package 280
C.Y. Chou, C.J. Huang, M. Sano, K.N. Chiang

3D Packaging and Supply Chain Management 286
P. Collander

The Necessity of Corrosion Protection for Solderable Pure Tin Deposits on IC Outer Leads 290
J. Barthelmes, P. Crema, P. Kuhlkamp, O. Kurtz

Pot Life Improvement of Low Temperature and High-speed Curable Anisotrpic Conductive Adhesive (ACA) ... 294
J. Lee, J. Kim, C. Hyun

Local Hardening Behavior of Free Air Balls and Heat Affected Zones of Thermosonic Wire Bond Interconnections ... 299
C. Dresbach, G. Lorenz, M. Mittag, M. Petzold, E. Milke, T. Muller

Mechanical Properties and Microstructure of Heavy Aluminum Bonding Wires for Power Applications ... 307
C. Dresbach, M. Mittag, M. Petzold, E. Milke, T. Muller

Effect of Microstructure Design on Reliability of FBGA Lead-Free Solder Joints ... 315
F.X. Che, J.E. Luan

Thermoelastic Properties of Printed Circuit Boards: Effect of Copper Trace ... 321
G. Hu, G.K. Yong, J. Luan, L.W. Chin, X. Baraton

New Developments in High Performance Solder Products for Power Die Assemblies ... 327
M. Fenner, A. Mackie, G. Wilson

Robust LTCC/PZT Sensor-Actuator-Module for Aluminium Die Casting ... 333
M. Flossel, U. Scheithauer, S. Gebhardt, A. Schonecker, A. Michaelis

SMD Pressure and Flow Sensors for Industrial Compressed Air in LTCC Technology ... 338
Y. Fournier, A. Barras, G. Boutinard Rouelle, T. Maeder, P. Ryser

ALX Permanent Polymer Dielectrics For Microelectronic Packaging Applications ... 345
P. Garrou, A. Huffman, J. Piascik

Experimental Analysis on the Mechanism of Moisture Induced Interface Weakening in ACF Package ... 354
G. Sim, C. Chung, K. Paik, S. Lee

3D IC Products Using TSV for Mobile Phone Applications: An Industrial Perspective ... 361
Y. Guillou, A. Dutron

Process Characterisation of Dupont MXA140 Dry Film for High Resolution Microbump Application ... 367
W. Liebsch, H. Yun, C.E. Balut, A. Ahr, A. Phung, A. Huffman

Reliability Studies on High Current Power Modules with Parallel MOSFETs ... 371
G.H. Sarma, G. Nitin, Ramanan, Manivannan, K. Mehta, A. Bhattacharjee

Solder Process Optimization: Influence of Heating and Cooling Rate on the Thermo-Mechanical Stress Generated in Components ... 378
M. Hertl, D. Weidmann, J. Lecomte

Wafer Post-Processing for a Reconfigurable Wafer-Scale Circuit Board ... 383
M. Radji, A. Lakhssassi, M. Bougataya, A. Hamoui, R. Izquierdo

Evaluation of Printed Electronics Manufacturing Line with Sensor Platform Application ... 391
E. Halonen, K. Kaija, M. Mantysalo, A. Kemppainen, R. Osterbacka, N. Bjorklund

Development of Matrix Clip Assembly for Power MOSFET Packages ... 399
M. Kengen, W. Peels, D. Heyes

Electrostatic Wafer Handling for Thin Wafer Processing ... 403
C. Landesberger, R. Wieland, A. Klumpp, P. Ramm, A. Drost, U. Schaber, D. Bonfert, K. Bock

Long-Term Joint Reliability of SiC Power Devices at 330ºC ... 408
F. Lang, S. Tanimoto, H. Ohashi, H. Yamaguchi

Chip to Chip Bonding using micro-Cu Bumps with Sn Capping Layers ... 413
J.S. Lee, K.Y. Byun, Q.H. Chung, M.S. Suh, S.C. Kim, Y. Kim

Thermal Design Considerations on Wire-Bond Packages ... 418
M. Mach, J. Muller

A New Methodology for Multi-Level Thermal Characterization of Complex Electronic Systems : From Die to Board Level ... 425
O. Martins, N. Peltier, S. Guedon, S. Kaiser, Y. Marechal, Y. Avenas

Impact of Package Parasitics on the EMC Performance of Smart Power SoCs ... 437
M. Merlin, F. Fiori

Characteristics of Electrically Conductive Adhesives Filled with Copper Nanoparticles with Organic Layer ... 443
L.N. Ho, H. Nishikawa, T. Takemoto, Y. Kashiwagi, M. Yamamoto, M. Nakamoto

Design and Fabrication of Corrosion and Humidity Sensors for Performance Evaluation of Chip Scale Hermetic Packages for Biomedical Implantable Devices ... 447
N. Saeidi, A. Demosthenous, N. Donaldson, J. Alderman

Realisation of Embedded-Chip QFN Packages - Technological Challenges and Achievements ... 451
A. Ostmann, D. Manessis, L. Boettcher, S. Karaszkiewicz, H. Reichl

Far-End Maximum Crosstalk for Coupled Lines as Function of Load ... 457
A. Owzar, R. Stephan, W. Petersen, M. Helfenstein

Effects of Test Conditions on Bending Impact of Lead Free Solder..461
    J. Park

Connector Reliability Testing Using Salt Spray..466
    A. Parviainen, J. Perala, L. Frisk, S. Kuusiluoma

Accelerating the Temperature Cycling Tests of FBGA Memory Components with Lead-free Solder Joints without Changing the Damage Mechanism..474
    J. Reichelt, P. Gromala, S. Rzepka

Bisected Thermodynamic Sensor as the Power AC/DC Transmitter..482
    M. Reznicek, I. Szendiuch, Z. Reznicek Jr., Z. Reznicek

3D Integration with AC Coupling for Wafer-Level Assembly...487
    M. Scandiuzzo, L. Perugini, R. Cardu, M. Innocenti, R. Canegallo

Kirkendall Voiding in Au Ball Bond Interconnects on Al Chip Metallization in the Temperature Range from 100-200ºC After Optimized Intermetallic Coverage..491
    M. Schneider-Ramelow, S. Schmitz, B. Schuch, W. Grubl

Thermo-Mechanical Stress Analysis...497
    K. Niehoff, T. Schreier-Alt, F. Schindler-Saefkow, F. Ansorge, H. Kittel

Simulation and Experimental Analysis of Substrate Overmolding..502
    T. Schreier-Alt, C. Rebholz, F. Ansorge

Mechanical and Microstructural Properties of SiC-Mixed Sn-Bi Composite Solder Bumps by Electroplating...509
    Y. Shin, S. Lee, S. Yoo, C. Lee

Some Facts from Lead-free Solders Reliability Investigation...513
    I. Szendiuch, J. Stary, J. Sandera, M. Bursik, E. Hejatkova

Surface-Enhanced Copper Bonding Wire for LSI and Its Bond Reliability under Humid Environment.................518
    T. Uno, K. Kimura, T. Yamada

A Study of Thermal Performance for Chip-in-Substrate Package on Package..528
    T. Hung, M. Yew, C. Chou, K. Chiang

Novel Interconnection Processes for Low Cost PEN/PET Substrates..534
    J. Van Den Brand, R. Kusters, H. Fledderus, E. Rubingh, T. Podprocky, A. Dietzel

Characterization of PTC Resistor Pastes Applied in LTCC Technology...543
    J. Vanek, W. Smetana, M. Weilguni, I. Szendiuch

Processes for Integration of Microfluidic Based Devices...548
    D.P. Webb, P.P. Conway, D.A. Hutt, B.J. Knauf, C. Liu

High Linearity and Broadband WiMAX Power Amplifier Design Using Board Level Integration Technology...555
    W. Chen, K. Chin, C. Tsai, L. Chang, Y. Chang, C. Liu

Influence of the Fabrication Errors on Multilayer Thick Film Circuits..559
    W. Ali, C. Min

Correlation between Material Selection and Moisture Sensitivity Levels of Quad Flat No-lead (QFN) Packages..563
    M. Zhang, S.W.R. Lee, J. Zhang, H. Yun, D. Starkey, H. Chau

Passive Phase Change Tower Heat Sink & Pumped Coolant Technologies for Next Generation CPU Module Thermal Design..569
    M. Vogel, D. Copeland, A. Masto, S. Kang, B. Whitney, G. Upadhya, M. Connors, J. Marsala

Packaging of Silicon Photonic Devices: Grating Structures for High Efficiency Coupling and a Solution for Standard Integration...575
    J.V. Galan, A. Griol, J. Hurtado, P:. Sanchis, G.B. Preve, A. Hakansson, J. Marti

Encapsulation Challenges for Wafer Level Packaging..581
    E. Kuah, J.P. Ding, Q.F. Li, J.Y. Hao, W.L. Chan, S.C. Ho, H.M. Huang, Y.J. Jiang

Mechanical Behaviour of SAC-Lead Free Solder Alloys with Regard to the Size Effect and the Crystal Orientation..587
    J. Villain, W. Mueller, U. Saeed, C. Weippert, U. Corradi, A. Svetly

NanoBond® Assembly – A Rapid, Room Temperature Soldering Process...591
    G. Caswell

Characterization of Oxidation of Electroplated Sn for Advanced Flip-chip Bonding..............................597
    W. Zhang, W. Ruythooren

Miniaturisation of a LTCC High-Frequency Rat-Race-Ring by Using 3-Dimensional Integrated Passives and Embedded High-K Capacitors...601
    R. Perrone, P. Kapitanova, D. Kholodnyak, I. Vendik, S. Humbla, M. Hein, J. Muller

DreamPAK – Small Form Factor Package...607
    L.A. Lim, M. Ramkumar, C.J. Vath III

Fatigue Life Prediction of Plated Through Holes (PTH) Under Thermal Cycling 613
N. Park, J. Kim, C. Oh, C. Han, B. Song, W. Hong

Thin Hermetic Borosilicate Glass Layers for Highly Reliable Chip-Passivations in Wafer-Level-Packaging 617
U. Hansen, J. Leib, S. Maus, O. Gyenge, M. Topper

Modeling and Quantification of Conventional and Coax-TSVs for RF Applications 624
I. Ndip, B. Curran, S. Guttowski, H. Reichl

Stacking of Full Rebuilt Wafers For SiP and Abandoned Sensors/Applications 628
C. Val, P. Couderc, P. Lartigues

Creep Mechanism Fractography Analysis on SnPb Eutectic Solder Joint Failure 637
C. Oh, C. Han, N. Park, B. Song, W. Hong

Direct Interconnection of Chemical Mechanical Polishing (CMP)-Cu Thin Films at 150ºC in Ambient Air 641
A. Shigetou, T. Suga

Damage Risk Assessment of Under-Pad Structures in Vertical Wafer Probe Technology 646
T. Hauck, I. Schmadlak, C. Argento, W.H. Muller

Impact of Substrate Coupling Induced by 3D-IC Architecture on Advanced CMOS Technology 651
M. Rousseau, M. Jaud, P. Leduc, A. Farcy, A. Marty

Versatile MEMS and MEMS Integration Technology Platforms for Cost Effective MEMS Development 656
P. Pieters

Reliability Testing of Frequency Converters with Salt Spray and Temperature Humidity Tests 661
J. Kiilunen, L. Frisk

Screen-Printed Polymer-Based Microfluidic and Micromechanical Devices Based on Evaporable Compounds 666
N. Serra, T. Maeder, C. Jacq, Y. Fournier, P. Ryser

3D Integration of Ultra-thin Functional Devices Inside Standard Multilayer Flex Laminates 671
W. Christiaens, T. Torfs, W. Huwel, C. Van Hoof, J. Vanfleteren

3D Integration Process Flow for Set-top Box Application: Description of Technology and Electrical Results 676
S. Cheramy, J. Charbonnier, D. Henry, A. Astier, P. Chausse, M. Neyret, C. Brunet-Manquat, S. Verrun, N. Sillon, L. Bonnot, X. Gagnard, J. Vittu

Advanced Failure Analysis Methods and Microstructural Investigations of Wire Bond Contacts for Current Microelectronic System Integration 682
R. Klengel, S. Bennemann, J. Schischka, C. Grosse, M. Petzold

Electrical Modeling and Analysis of the Impact of Slits on Microstrip Lines in Thin Film Polymer Layers 688
I. Ndip, M. Topper, K. Becker, M. Hirte, I. Eidner, T. Fischer, B. Curran, J. Bauer, W. Scheel, S. Guttowski, H. Reichl

3D Integration Technologies for Ceramic Substrates in a SHM Application 692
S. Hildebrandt, K. Wolter

A New Low Cost, Elastic and Conformable Electronics Technology for Soft and Stretchable Electronic Devices by use of a Stretchable Substrate 697
F. Bossuyt, T. Vervust, F. Axisa, J. Vanfleteren

Comparison between Die Attach Film (DAF) and Film over Wire (FOW) on Stack-die CSP Application 703
C.L. Chung, C.W. Ku, H.C. Hsu, S.L. Fu

A Novel Thermo-Mechanical Test Method of Fatigue Characterization of Real Solder Joints 706
R. Metasch, M. Roellig, K.J. Wolter

Thermo Mechanical Characterization of Packaging Polymers 713
B. Boehme, K.M.B. Jansen, S. Rzepka, K. Wolter

Au–Sn SLID Bonding: Fluxless Bonding with High Temperature Stability, to Above 350ºC 723
K.E. Aasmundveit, K. Wang, N. Hoivik, J.M. Graff, A. Elfving

Optimization of Flip-chip Laser Soldering for Low Temperature Stability Substrate 729
T. Hurtony, B. Balogh, P. Gordon

Low Energy Consumption Thick-film Pressure Sensors 735
D. Belavic, M.S. Zarnik, M. Mozek, S. Kocjan, M. Hrovat, J. Holc, M. Jerlah, S. Macek

Reliability Comparison of Aluminum Redistribution based WLCSP Designs 741
U. Sharma, H. Gee, P. Holland

Miniaturization of Printed Wiring Board Assemblies into System in a Package (SiP) 746
S.G. Rosser, I. Memis, H. Von Hofen

Long Term Stability of Polymer Based Resistors Tested by Noise, Non-Linearity and Electro-Ultrasonic Spectroscopy 754
 V. Sedlakova, P. Tofel, J. Sikula

System Packaging & Integration for a Swallowable Capsule Using a Direct Access Sensor 759
 P. Jesudoss, A. Mathewson, W. Wright, C. McCaffrey, V. Ogurtsov, K. Twomey, F. Stam

Interface Resistance between Polymer Based Conducting and Resistive Layers 763
 P. Tofel, V. Sedlakova, M. Chvatal, J. Majzner

New Packaging Technology Enabling Integration of Magnetics and Semiconductors in One Component 767
 A. Pot, H. Roehm, R.V.D. Berg, T. Shanmugam, S. Ong, F. Van Der Burgt, T.P. Sidiki

Large Panel, Highly Flexible Multilayer Thin Film Boards 776
 H. Burkard, W. Kapischke, J. Link

Closing Technology Knowledge Gaps: Projects Arising from the iNEMI Technology Roadmap 782
 B. Pfahl, J. Arnold, G. O'Malley

Addressing Opportunities and Risks of Pb-Free Solder Alloy Alternatives 787
 G. Henshall, R. Healey, R.S. Pandher, K. Sweatman, K. Howell, R. Coyle, T. Sack, P. Snugovsky, S. Tisdale, F. Hua, G. O'Malley

A Comprehensive Overview on Today's Ceramic Substrate Technologies 798
 F. Bechtold

Compression Molding Solutions for Various High End Package and Cost Savings for Standard Package Applications 810
 H. Matsutani

Advanced Solutions for Ultra-Thin Wafers and Packaging 814
 G. Klug

Author Index

In-mould Integration of Electronics into Mechanics and Reliability of Overmoulded Electronic and Optoelectronic Components

T. Alajoki, M. Koponen, E. Juntunen, J. Petäjä, M. Heikkinen, J. Ollila, A. Sitomaniemi, T. Kosonen, J. Aikio, and J-T. Mäkinen

VTT Technical Research Centre of Finland, Kaitoväylä 1, FI-90571 Oulu, Finland

+358 20 722 2149, fax +358 20 722 2320, teemu.alajoki@vtt.fi

Abstract

Next generation of smart systems in different application areas such as automotive, medical and consumer electronics will utilize various electronic, optical and mechanical functions integrated in plastic product structures. In this study, in-mould integration of electronic and optoelectronic modules is examined in order to embed novel functionality into polymer matrix. Thermal load of the electronic components embedded in polymer material was briefly modelled, and a series of test structure samples realized. The test structure consisted of a flexible printed circuit (FPC) substrate with assembled electronic and optoelectronic components, which was set as an insert into injection moulding mould and thermoplastic polymer was cast on the substrate in injection moulding process. After the overmoulding of the samples, characterization of components was carried out by visual inspection and by functional testing. It was found out that with correct injection moulding process parameters, all components survived the challenging conditions of injection moulding process. The reliability of the samples was evaluated by subjecting them to constant humidity test +85 °C/ 85% RH and -40 °C...+85 °C temperature cycling test. After the tests, the samples were analyzed by visual inspection, functional testing and cross-sections. The analysis revealed that when using an appropriate silver paste, over 90% of the components remained functional, although problems were observed in the adhesion between the FPC substrate and the overmoulding material.

Key words: Injection overmoulding, in-mould integration, adhesive bonding, Moulded Interconnect Device (MID)

Introduction

The trend in electronic devices is towards thinner and thinner solutions. At the same time the number of functions must stay constant or even increase [1]. The integration of optical, electrical and mechanical functionalities into one system can greatly improve the cost efficiency of systems, due to the fact that the packaging costs and material consumption are reduced [2]. New opportunities are emerging also in mechatronic applications due to the intelligent combination of molded parts and flexible circuits. The use of FPCs can lead to very low costs when Reel-to-Reel (R2R)-processes are utilised [3].

Thermoset epoxies are the most widely used materials today in electronic packaging and printed circuit boards, but this may change with increasing technical, economical and regulatory demands. Modern halogen-free thermoplastics offer superior properties, such as environmental sustainability and precision moulding capability, and they can be manufactured with automated, high-volume processes. Injection moulded thermoplastics are already used, for instance, in MEMS packaging as well as in 3D electronic circuits (MIDs, Moulded Interconnect Devices), but future designs could combine moulded circuits and packages into single units such as RF antenna modules, filters and oscillators as well as into optoelectronic systems [4].

Concept of In-mould Integration of Electronics into Mechanics

We have studied possibilities to add functionality of the plastic products and at the same time to reduce packaging and material costs by utilizing in-mould integration technology into electronic and optoelectronic module manufacturing. The concept idea is to assemble bare dies or SMT components on a substrate or circuit board, and use thereby formed sub-assembly as an insert in injection moulding process. Mechanical and optical structures are formed and bare chips sealed in injection moulding process in such a way that mechanical and optical interconnections and required device encapsulation are provided for the module. Earlier, we have demonstrated in-mould integration concept by using conventional FR4 and low temperature co-fired ceramics (LTCC)

978-1-4244-4722-0/09 $25.00
© 2009 IMAPS-ITALY

substrates [5][6][7]. We have also studied adhesion issues between foil and overmoulding material in order to use low-cost thermoplastic materials not only as the overmoulding material, but also as the printed circuit substrate [8]. 3D integration of electronics and mechanics utilizing insert moulding concept and both thermoset and thermoplastic materials has been studied also in [1][9]. By using flexible plastic foils and cost-effective printing methods together with efficient 3D shaping of the foils and injection overmoulding technology, there is a great potential for building a new high-throughput manufacturing technology platform that can provide products in various applications fields with complex optical, electrical and mechanical functions at very low cost [8]. Process flow for in-mould integration of electronics when using plastic foil materials is shown in Figure 1.

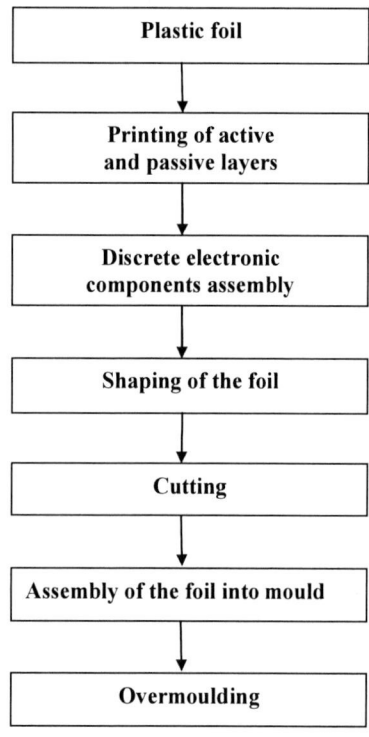

Figure 1: In-mould integration process flow.

Focus of this work was to implement test vehicles where SMT components and bare chips are assembled on thermoplastic polymer foil and ovemoulded with transparent thermoplastic polymer material. The objective was to find screen printing pastes as well as discrete component and bare chip interconnection methods that are compatible with the injection overmoulding process. Particularly interesting aspect was to study adhesive flip-chip bonding on thermoplastic foil and survival of this type of interconnection of the injection moulding process. Reliability of the overmoulded test structures was characterized with environmental tests.

Test Structure

*Polyethylene terephthalate (*PET) was used as the printed circuit substrate material and polycarbonate (PC) as the overmoulding polymer, both being transparent materials and showing a strong adhesion between each other [8]. The wirings on PET substrate were realized by screen-printing with conductive silver paste, two different types of commercially available pastes were tested. Both bare die and SMT components were assembled on PET substrate by using different adhesive bonding technologies and embedded into PC material in overmoulding. Schematic illustration of the overmoulding process is shown in Figure 2. Substrate was printed on both sides with through substrate vias which allows electrical testing of the embedded components either by using contact pads on flex "tail" or on the reverse side of the overmoulding, as shown in Figure 2. Altogether 12 test structures were realized for the overmoulding tests plus additional three reference samples which were not overmoulded.

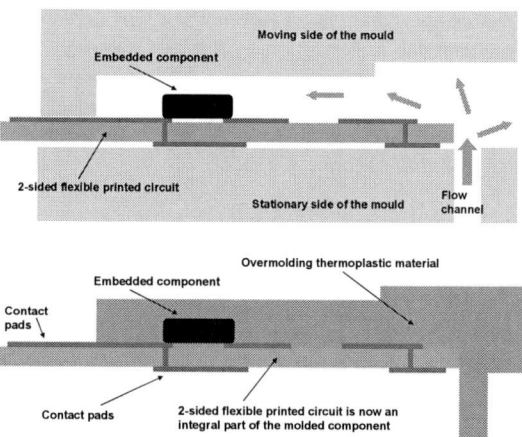

Figure 2: Schematic of the overmoulding process.

Thermal Modelling

Basic thermal model was created of the test structure in order to find out, what is the thermal load of embedded components with different heat dissipation powers. This quite simple model consisted of silicon test die bonded with gold stud bumps on 125 µm thick PET substrate with printed silver paste wirings, and the whole system covered with 1.875 mm thick layer of injection molded plastic (PC). This model represents the test structure rather well. Plane plot of the basic model simulation is shown in Figure 3.

978-1-4244-4722-0/09 $25.00
© 2009 IMAPS-ITALY

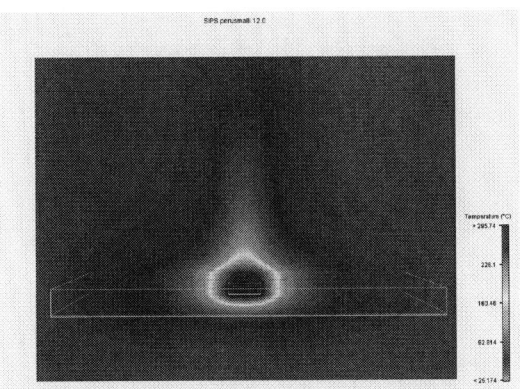

Figure 3: Thermal simulation of the basic model.

Since no heat sink is applied and the thermal conductivity of the PC material is quite poor (0.2 W/mK), the temperature of the chip rises rather quickly when the heat dissipation power is increased (Table 1). This gives constraints of which kind of electronic componets could be embedded into polymer matrix.

Table 1: Basic model with varying power levels

Power [W]	Temperature at chip [°C]
0.5	296
0.25	173
0.20	139
0.10	85

Screen-printed Flexible Printed Circuits

The simplified version of the process flow for the double sided screen-printed FPC with through substrate vias is shown in Figure 4.

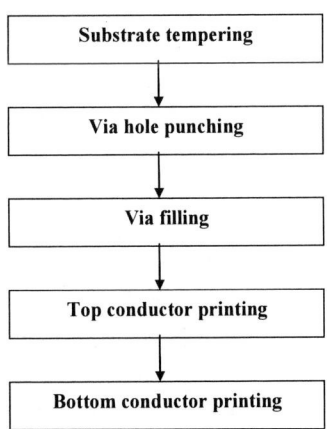

Figure 4: Process flow for double sided screen-printed FPC with through substrate vias.

To minimize the shrinkage of the substrate during the curing cycles of the pastes the substrates were tempered before use. After one tempering cycle in 150° / 60min the shrinkage stabilised. Via hole punching tool diameter was 100 µm. Via filling was done as a separate step. Collars caused by the punching tool around via holes are lower on the punch side and therefore, via filling was done from this side. 30µm thick metal stencil with 100µm hole was used for via filling process.

After preliminary printing tests, two commercially available silver pastes were chosen as the condcuctive wiring material of the test structure. More thorough printing tests were made for these two pastes. Both pastes were one component type silver pastes which use polyester resin as their binder. Therefore, the cured foil was expected to have excellent flexibility and good adhesion on the PET substrate material. Paste#1 had a bit higher solid content than paste#2, nominal resistivities were 40 and 80 µΩ·cm for paste#1 and #2, respectively. Murakami 400 mesh trampoline screen was used in printing tests. The hardness of the squeegee was 70 duro, print speed 10-20 mm/s and print pressure 20-28%. Tests revealed that both of pastes had their own characters: paste#1 does not spread much so the printing resolution was best. Small spreading leads to thick prints but also to rough surface. With paste#2 smoother surfaces were obtained, but it spreaded more leading to thinner print layers. Table 2 summarizes the average results of the exhaustive characterization of the realised conductive wirings.

Table 2: Characteristics of the realised wirings

	Paste#1	Paste#2
Line width (nominal 100 µm)	111 µm	129 µm
Line thickness (100 µm nominal line width)	12 µm	6 µm
Contact pad thickness	14.5 µm	7 µm
Contact pad surface roughness (Ra)	1.4 µm	0.8 µm
Square resistance (100 µm nominal line width)	39 [mΩ/□]	181 [mΩ/□]
Resistivity (100 µm nominal line width)	32 [µΩ·cm]	78 [µΩ·cm]
Stdev resistivity (100 µm nominal line width)	0.4 [µΩ·cm]	9 [µΩ·cm]

Component Assembly

Test structure had two pieces of SMT resistors in 0402 package and two in 0603 package; two pieces of 0603 SMT LEDs; and a flip-chipped test die equipped with Au stud bumps. All SMT components were assembled using silver particle filled epoxy, this isotropic conductive adhesive (ICA) was dispensed to the contact pads and dried at 100°C for 30 min. Flip-chipped test die was interconnected to the foil by using anisotropic conductive adhesive (ACA) in six overmoulding samples and non-conductive adhesive (NCA) in the rest six samples. Additionally, three reference samples were realised which were not overmoulded. Flip-chip bonding was done by dispensing the adhesive (ACA or NCA) to the whole chip area and then placing the chip with Finetech flip-chip bonder to the contact area and curing the adhesive at 150ºC. More information of the flip-chipping process can be found from [10].

Overmoulding of the Test Structures

Test structures with assembled components were set as an insert into injection moulding mould and thermoplastic polymer was cast on the substrate in injection moulding process. Overmouldings were made with a conventional hydraulic injection moulding machine Engel ES200 / 50 HL equipped with ISO 294-3 plate cavity mould, shown in Figure 5. Test structure was inserted to the fixed side of the mould.

Figure 5: ISO 294-3 plate cavity mould.

Table 3 shows the most important injection moulding process parameters.

Table 3: Injection moulding process parameters

Temperature of the IM material	280ºC
Temperature of the mould	80ºC
Pressure limit	50 bar
Rate of injection	110 mm/s
Packing pressure	53 bar
Dose	72 mm

Figure 6 shows the overmoulded test structure.

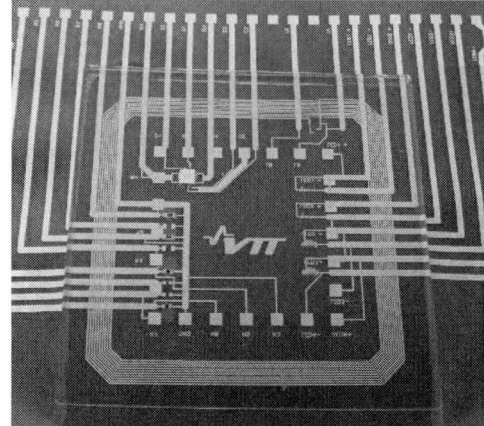

Figure 6: Overmoulded test structure.

All components in all samples were still functional after the overmoulding process. The test structures were also checked visually and no joint damages, component damages or displacement or other evident defects were observed. Figure 7 and Figure 8 show photos of the overmoulded components.

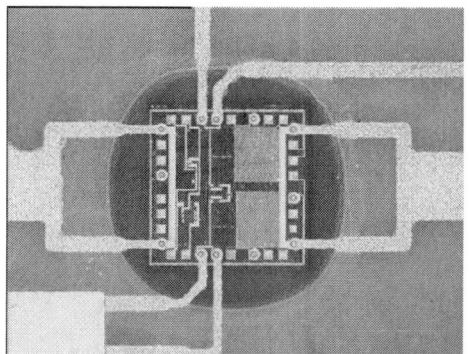

Figure 7: Flip-chipped test die (photo taken from the reverse side).

Figure 8: SMT LED.

Environmental Testing, Results and Discussion

Test structures were subjected to constant humidity and thermal cycling tests. In the constant humidity test the test structures were subjected to the temperature of +85 ºC and the relative humidity of

85% for 500 h. Thermal cycling was used to reveal reliability problems caused by CTE mismatch between different materials. Temperature range was -40 ºC…+85ºC and cycle length 1.5 h. In the tests altogether 280 cycles were carried out.

After the environmental tests, all components in all samples were electrically tested to find out, if they still remain functional. The results are shown in Table 4.

Table 4: Malfunctioned components after environmental tests

Sample#	Paste	Adhesive	Overmoulding
#1	paste#1	NCA	yes
#2	paste#2	NCA	yes
#3	paste#1	ACA	yes
#4	paste#1	ACA	yes
#5	paste#1	NCA	yes
#6	paste#2	NCA	yes
#7	paste#1	ACA	yes

Sample#	Environmental test	Malfunctioned components
#1	85/85	R2
#2	85/85	R3, R4
#3	85/85	R1, R3, DIE, LED1, LED2
#4	85/85	DIE
#5	thermal cycle	R3, R4, DIE, LED1, LED2
#6	thermal cycle	R1
#7	thermal cycle	R1, R2, R3, R4, DIE, LED2

Both 85/85 and thermal cycle test showed roughly the same amount of malfunctioned samples/components. It can be summarized that 18 of the total amount of 42 overmoulded components in samples patterned with paste#1 were malfunctioned, whereas in paste#2 samples only 3 of 42 components were not functional after the environmental tests. Reference samples (that is, the samples that weren't overmoulded but were subjected to the same environmental tests as the overmoulded samples) did not have any malfunctioned components.

The reason for the large number of malfunctioned components when using paste#1 was found out to be due to adhesion failures between PET foil and PC overmould materials causing air bubbles etc. in some samples resulting in the breakage of the wirings of the components. This phenomenon is illustrated in Figure 9. However, the adhesive interconnections (ICA, ACA and NCA) between contact pads and embedded components seem to have withstood the tests well. Adhesion failures were caused by thermomechanical stresses and moisture absorption leading to warpage of the quite large-area PET foil of the tests structure.

Paste#1 forms thicker, rougher wirings which are not as flexible as the ones made with paste#2. Even though adhesion failures were visible also in test structure samples made with paste#2, in most of these cases contacts to the embedded components remain functional. Broken paste#1 conductor is shown in Figure 10. A cross-section of an overmoulded test die is shown in Figure 11.

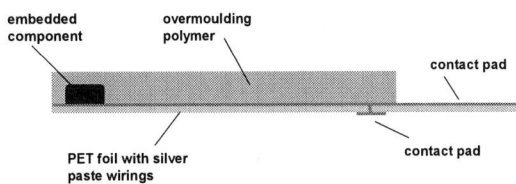

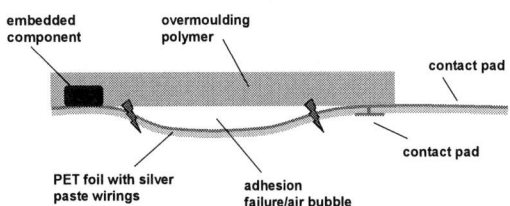

Figure 9: Adhesion failure of the PET foil and overmoulding polymer.

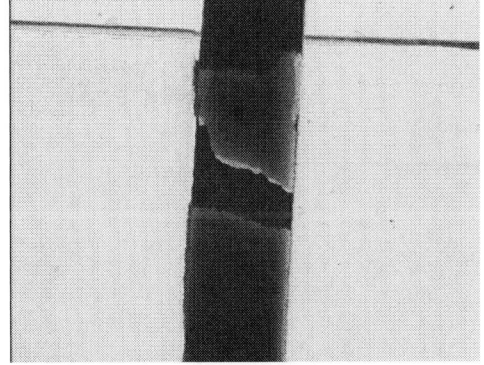

Figure 10: Broken paste#1 conductor.

Figure 11: Cross-section of an overmoulded test die.

978-1-4244-4722-0/09 $25.00
© 2009 IMAPS-ITALY

Conclusions and Future Work

Promising results were obtained that SMT conductive adhesive bonding as well as ACA and NCA flip-chip bonding to flexible polymer foils could be compatible with injection overmoulding process. Especially the short term reliability was found out to be good: with the limited number of samples (altogether 72 pcs of components) no failures were found after the overmoulding. However, survival of the componets and interconnections require careful adjusting of the injection moulding process parameters.

Problems in the long term reliability were found in the quite rough environmental tests of +85 ºC /85% RH and thermal cycle -40…+85ºC. In these tests, adhesion failures between foil and overmoulding material were perceived leading to conductor breakage in some cases. By using paste with lower solid content and thus more flexible realized conductors, the number of broken conductors could be decreased significantly.

Future work could include development of procedures to increase the adhesion between printed circuit foil and overmoulding material. 3D forming of the FPCs with assembled discrete components and overmoulding of these 3D structures will be also a topic for further study in order to fully exploit the benefits of using flexible thermoplastic circuit carriers.

Acknowledgements

Authors would like to acknowledge Airi Weissenfelt, Jyrki Ollila, Miia Aitta and Pekka Avellan for printing, assembly and injection moulding work of the test structures.

References

[1] T. Peltola, "Integration of Multilayer PWB into Plastic Covers by Injection Moulding", Electronics Systemintegration Technology Conference, Dresden, pp. 1342-1346, 2006.

[2] H. Cho, K.M. Chu, S. Kang, S. Hwang, B. Rho, W. Kim, J.S. Kim, J.J. Kim and H.H. Park, "Compact packaging of optical and electronic components for onboard optical interconnections", IEEE Trans. Adv. Packaging, pp. 114-120, February, 2005.

[3] K. Feldmann, M. Pfeffer, A. Reinhardt, "Creative Developments and Innovative Technologies for the Further Succes of MID", 7th International Congress on Molded Interconnect Devices, Fürth, Germany, September 27-28, pp. 1-15, 2006.

[4] K. Gilleo, D. Jones, G. Pham-Van-Diep, "Thermoplastic Injection Molding: New Packages and 3D Circuits", ECWC 10 Conference at IPC Printed Circuits Expo 2005.

[5] K. Keränen, M. Silvennoinen, A. Lehto, J. Ollila, T. Salmi, J.-T. Mäkinen, A. Ojapalo, M. Schorpp, P. Hoskio and P. Karioja, "Short and long term reliability of in-mould sealed bare and glob-top shielded LED devices", Proceedings of 16th European Microelectronics and Packaging Conference (EMPC), Oulu, Finland, June 17-20, pp. 280-284, 2007.

[6] J.T. Mäkinen, K. Keränen, J. Hakkarainen, M. Silvennoinen, T. Salmi, S. Syrjänen, A. Ojapalo, M Schorpp, P. Hoskio and P. Karioja, "Inmould integration of a microscope add-on system to a 1.3 Mpix camera phone", Proceedings of SPIE, Vol. 6585, 2007.

[7] K. Keränen, T. Saastamoinen, J.T. Mäkinen, M. Silvennoinen, I. Mustonen, P. Vahimaa, T. Jääskeläinen, A. Ojapalo, M. Schorpp, P. Hoskio and P. Karioja, "Injection moulding integration of a red VCSEL illuminator module for a hologram reader sensor", Proceedings of SPIE, Vol. 6585, 2007.

[8] M. Koponen, T. Alajoki, T. Kosonen, J. Petäjä, M. Heikkinen, T. Vuorinen, J-T. Mäkinen, "Adhesion of Flexible Printed Circuit Substrate to Overmoulded Polymer and Characterization of Overmoulded Electronic Components". IMAPS Nordic Annual Conference, Helsingör, Denmark, pp. 207-212, 2008.

[9] T. Peltola, P. Mansikkamäki, E.O. Ristolainen, "3D Integration of Electronics and Mechanics", Proceedings of the 10th Advanced Packaging Material Symposium Conference, Irvine, California, 2005.

[10] J. Lenkkeri, T. Jaakola, M. Lahti, M. Allen, T. Kaskiala, "Chip/antenna interconnections for contact-less smart card applications", IMAPS Nordic Annual Conference, Helsingör, Denmark, pp. 197-202, 2008.

978-1-4244-4722-0/09 $25.00
© 2009 IMAPS-ITALY

A 3-D packaging concept for cost effective packaging of MEMS and ASIC on wafer level

Tobias Baumgartner*, Michael Töpper*, Matthias Klein*, Bernhard Schmid**, Dieter Knödler**, Heikki Kuisma***, Sami Nurmi***, Hannu Kattelus+, James Dekker+ and Ralph Schachler++

*Fraunhofer IZM, Gustav-Meyer-Allee 25, 13355 Berlin, Germany Phone: +49-30-46403-602, Fax: +49-30-46403-123 and e-mail:tobias.baumgartner@izm.fraunhofer.de;

**Continental Teves AG & Co. oHG, Guerickestraße 7, 60488 Frankfurt/Main, Germany,

***VTI Technologies Oy, Myllynkivenkuja 6, Vantaa 01621, Finland,

+VTT Tietotie 3, Espoo 02044, Finland,

++AemTec GmbH Carl-Scheele-Str. 16, 12489 Berlin, Germany

Abstract

Heterogeneous integration bridges the gap between nanoelectronics and its derived applications. Currently MEMS and their signal conditioning ASICs are produced and packaged at different industry sectors (different fabs). To reduce costs and enhance yield and performance at the same time this quite expensive way of packaging has to be modified. This paper presents a different packaging concept. It uses standard redistribution layer technology (RDL) to package thinned chips on a full wafer substrate e.g. thinned ASIC chips on a MEMS wafer. For this approach no Through Silicon Vias (TSV) are needed. Standard chips can be used without redesign. Only Known Good Dies (KGDs) are packaged with the cost benefit of wafer level technology.

At the starting point for this type of packaging both ASIC and MEMS chips are still parts of full wafers. The wafer with the larger sized chips (e.g. MEMS chips) is used as a substrate for the further process steps. The wafer with the smaller sized chips (e.g. ASIC chips) is thinned down on wafer level to a thickness of 10µm to 40µm and diced. These thinned chips are glued onto the base wafer with a polymer layer (BCB from Dow Chemical). The polymer has been deposited and structured before gluing the next chip on top. After placement of the thinned chips the wafer is again coated with BCB to embed the chips. This polymer layer is photostructured to open contact pads on the base chips as well as on the embedded chips. The next step is the built-up of metal routing. Here a semi-additive process is used, which means electroplating on a sputter seed layer of TiW/Cu. This metal layer is followed by another polymer layer for passivation and acting as a solder mask. Then Under Bump Metallization (UBM) is applied again by electroplating. Finally Balling is done either by Ball Placement or by Solder Paste Printing. Now the wafer is diced and the full ASIC-MEMS package can be flip chiped onto a Printed Circuit Board (PCB). The technology will be demonstrated by the project RESTLES (Reliable System Level Integration of Stacked Chips on MEMS).

RESTLES will integrate technologies like silicon MEMS, ASIC, wafer thinning, chip stacking and flip chip to one packaged chip stack at die scale. The influence of the heterogeneous stack on performance and control mechanisms to eliminate parasitic effects will be investigated.

Key words: thin chip, embedding, wafer level, MEMS packaging,

Introduction

In IC production "Moore's Law" is a well-known proclamation that predicts the miniaturization from one generation of ICs to the next one. For MEMS a similar "Moore's MEMS law" does not exist, but there is still a tendency to shrink device size for cost reduction and the access to new applications. . Due to the heterogeneous technologies and requirements like standard CMOS chips and silicon micromachined sensors the packaging thereof is often a limiting factor. To overcome this problem the integration on silicon level was considered very intensively, but hasn't become the mainstream. Mainly yield degradation, silicon area consumption and performance treats keeps the chips individual.

Therefore, making MEMS devices smaller is not only a matter of the silicon technology itelf. The sizes of the chips (signal conditioning ASIC and MEMS element), the connection technology between ASIC and MEMS as well as the packaging are important for the size of the whole device.

The technology of Thin Chip Integration (TCI) [1] that was developed at Fraunhofer IZM enables to integrate MEMS and ASIC at chip scale by stacking and connect them with each other by the use of standard redistribution layer (RDL) technology on wafer level. There are further benefits given by the use of TCI such as well-defined and short routing appropriate for high frequency applications as well as cost reduction. The process flow of TCI is explained in the next chapter.

The process flow of TCI

TCI is used to stack at least two chips and connect them with eachother without TSV. Both types of chips are fully processed with all frontend processes. The process flow of TCI is schematically shown in figure1. The chips of the base wafer stay at wafer level until the end of the TCI process. In order to stack the thin chips on the base wafer they have to be smaller than the base chip. Before stacking the chips are thinned down on wafer level to a final thickness of 10-40μm. and are singulated during thinning before further processing

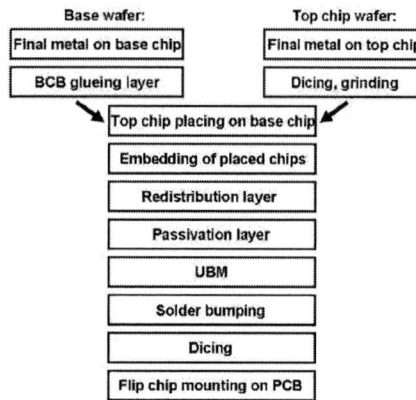

Figure 1: Process flow for TCI technology

The TCI process starts with the spin coating of a Photo-BCB layer (Benzocyclobutene - Cyclotene 4000 series from Dow Chemical Company) on the base wafer and photostructuring. This layer has two purposes: First it functions as the passivation of the final metal of the base wafer chips and secondly it works as the "glue" for the thinned top chips. After probing of top chip- and base wafer known good top dies are placed face up on top of known good base dies by a flip chip bonder. The next step is the embedding of the stack with another layer of Photo-BCB. This layer is opened at the pads

of the base chip as well as at the pads of the embedded thinned chip. In the next step a copper rerouting is built up to connect the pads in the desired routing and to from the I/O contact pads of the final device. The RDL is passivated with a 3rd layer of Photo-BCB and the I/O contact pads are opened by photolithography. Then a Ni/Au-UBM is deposited, patterned and solder balls are placed or screen printed on top of the UBM. After reflow of the solder the wafer is diced. Now the device is ready for flip chip mounting on a PCB or other substrate.

Technology demonstrator

The technology was evaluated by a demonstrator that was built up as part of the European project RESTLES (Reliable System Level Integration of Stacked Chips on MEMS). Here a MEMS wafer is used as the base wafer. Within this project the TCI technology is named SCOM (Stacked Chip on MEMS). The process flow for the RESTLES project is shown in figure2.

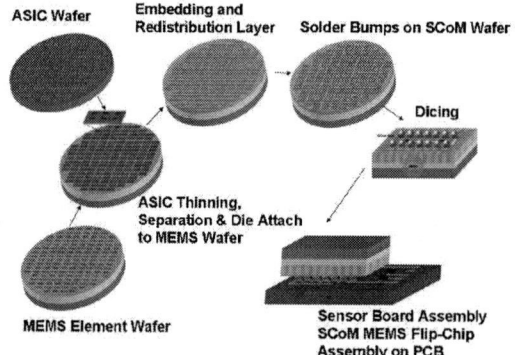

Figure 2: Process for RESTLES project

For the prototype device the base wafers contain inertial sensors and are made by project partner VTI. VTI's 3D MEMS technology consist of a triple wafer stack with a cap wafer which has high doped through wafer Si contacts embedded in glass [2]. The contacts have Al pads on top and are separated by glass.

For the technology demonstrator wafers without sensor function were used as base wafers. They do have the same thickness and mechanical geometry as the final functional MEMS wafers and the same surface metallization. Non-functional dummy ASIC chips with Al structures were used as top chips. Those were thinned down to 30μm thickness. Figure3 shows a part of a base chip seen in an optical microscope.

978-1-4244-4722-0/09 $25.00
© 2009 IMAPS-ITALY

Figure 3: Base chip seen in an optical microscope before TCI process

A structured BCB layer is used to glue the top chip onto the base chip (Figure4).

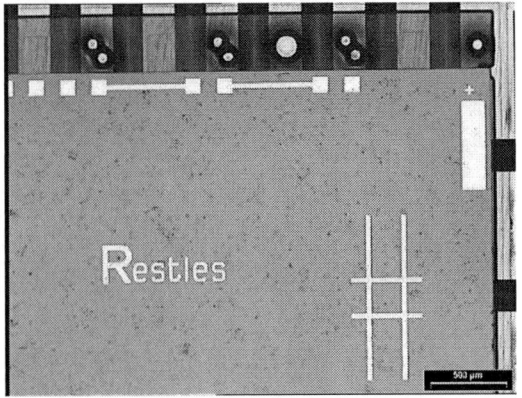

Figure 4: Thinned top chip after mounting on the base chip (optical microscope)

The top chip is embedded in another BCB layer. Subsequently the RDL layer is deposited. The RDL process is an additive process that uses a TiW/Cu seed layer, photoresist coating and structuring and electroplating in opened areas. Finally the photoresist is stripped and the seed layer etched. Figure5 shows the chip after RDL.

Figure 5: Thinned top chip on base chip after embedding and RDL seen in an optical microscope

Now the final BCB-passivation is coated and and the I/O-pads are opended. The final step is the deposition of the Ni/Au-UBM, agian by electroplating. (Figure6).

Figure 6: Thinned top chip on base chip after final passivation and Ni/Au-UBM (optical microscope)

At the project partner AEMtec Pb-free solder bumps were deposited by screen printing and fixed by a reflowed process. Figure7 shows the processed wafer after reflow of the solder bumps. A detailed picture is shown in figure8.

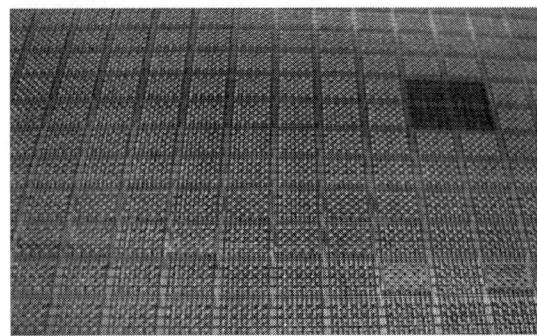

Figure 7: RESTLES wafer with embedded thinned chips after full TCI process before dicing

978-1-4244-4722-0/09 $25.00
© 2009 IMAPS-ITALY

Figure 8: RESTLES SCoM-chip with embedded thinned chip after full TCI process

Then the SCoM-wafers were diced at VTI. After that the singulated devices were flip-chip mounted on PCB at AEMtec. (Figure 9 shows the mounted chip on PCB with underfill).

Figure 9: RESTLES SCoM-device after full TCI process mounted on PCB

A cross section of the PCB-mounted chip is shown in Figure10.

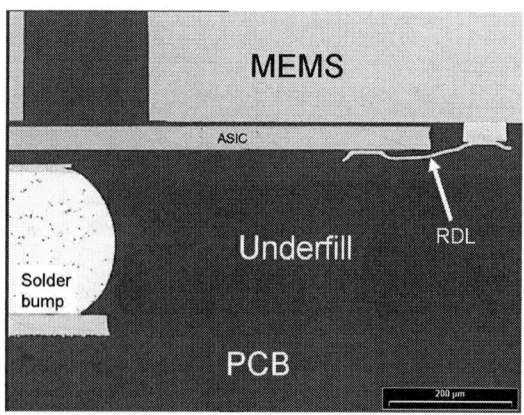

Figure 10: Cross section of a RESTLES SCOM-device stack mounted on PCB after full TCI process

Resistance Measurement Results

First daisy chain (DC) resistance measurements reveal very low parasitic resistance. The currently analysed DC's are shown in Figure 11 (DC 1 blue, DC 2 red, DC 3 green and DC 4 yellow).

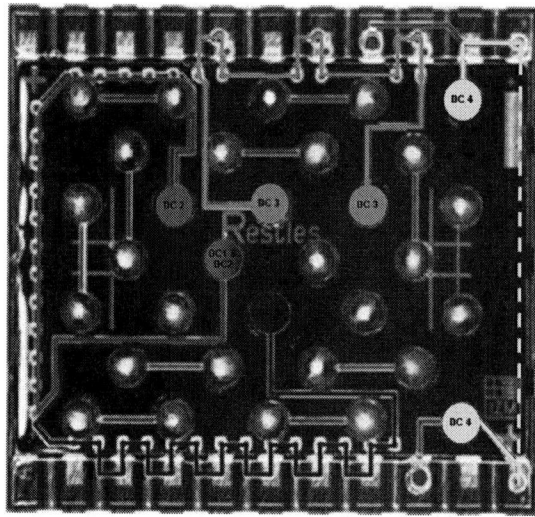

Figure 11: RESTLES SCoM-chip with daisy chain indication: DC 1 blue, DC 2 red, DC 3 green and DC 4 yellow (the dottet line is a metal runner on the MEMS layer)

DC 1 and DC 2 have only vias down to the top chip, but DC 1 has the RDL going over the top chip edge. The mean resistances of DC 1 is 2,12 Ohms and of DC 2 is 2,01 Ohms. The standard deviation of the measured 14 parts were 0,06 Ohms. Two of DC-2 have shown no connectivity. DC 3 has vias going down to MEMS and top chip pads. They have 1,95 Ohms (sigma 0,08 Ohms and 9 of 14 failing parts). DC 4 has only vias going down to MEMS pads. They have 2,00 Ohms (sigma 0,16 Ohms and 9 of 14 failing parts). The root cause of the failures is found in the embedding material when the top chip edge is to close to the dicing line of the SCoM device. Currently improvement of the embedding process is in progress to avoid these failures. In parallel these SCoM test devices are going into life time tests for reliability investigations.

Summary

A SCoM technology demonstrator that uses the same materials and surfaces as the final device was successfully built up. Via chain structures were successfully measured and show resistance of few Ohms. Currently the reliability of the demonstrators is under test. After passing these tests and further

978-1-4244-4722-0/09 $25.00
© 2009 IMAPS-ITALY

improvements the next step is to build up functional devices. This is going to take place until end of 2009.

Acknowledgements

Thanks to the Bundesministerium für Bildung und Forschung (German Federal Ministry of Education and Research, EURIPIDES project RESTLES, Project no. V3EUR015) and the Finnish Funding Agency for Technology and Innovation for funding this work as part of the project RESTLES.

References

[1] M. Töpper, K. Scherpinski, H.-P. Spörle, C. Landesberger, O. Ehrmann, H. Reichl, "Thin Chip Integration (TCI-Modules) – A Novel Technique for Manufacturing Three Dimensional IC-Packages", Proceedings IMAPS 2000, Boston, USA, September 2000

[2] Heikki Kuisma, "Wafer level packaging of MEMS", Keynote at SMART SYSTEMS INTEGRATION conference, Barcelona, Spain, 9 –10 April 2008

Trends in IC Packaging

Caroline Beelen-Hendrikx

NXP Semiconductors, Gerstweg 2, 6534 AE Nijmegen, The Netherlands

Tel: +31 (0) 24 353 3415, Fax: +31 (0) 24 353 3350, E-mail address: C.C.M.Beelen-Hendrikx@nxp.com

Abstract

Drivers for new package development are cost reduction, form factor requirements, reliability, electrical performance and function integration. Cost reduction is realized by improvement of existing packaging platforms, such as material cost reduction, and changes in processes and design methods, and by introducing new packaging platforms. Future low cost packaging platforms will be based on batch processes, large format processing, use of low cost substrates or no substrates at all, and will avoid over specification. Embedded packaging is a new technology that is either based on wafer fab technology or PCB fab technology. Wafer fab technology based embedded packages give the highest reduction in form factor but are relatively expensive. PCB fab technology based embedded packages are less expensive but have lower definition. Therefore, they will first be used for low pin count packages. Embedding in motherboards or interposers is a way to reduce package thickness; it is expected to be mainstream for simple, low pin count products. Package reliability requirements are tightened by automotive and lighting businesses, requiring, high temperature, long lifetime and zero delamination. To meet the requirements, interfacial strength and interconnect robustness need to be improved. New functions are realised by silicon based sensors and MEMS, e.g. for automotive, medical, environmental and communication purposes. Packaging challenges include wafer treatment, MEMS protection and the realization of die access for fluids and gases to be analyzed. Function integration drives the development of 3D packaging technologies. Ultimate technology is heterogeneous integration by 3DIC.This technology optimizes electrical performance and form factor. For 3DIC ultra-thin dies are stacked by fine pitch flip chip interconnect and silicon through vias. Disadvantage is cost. Therefore, for many applications, stacked die wire-bonded IC packages are still popular. They offer improved bandwidth, power consumption and size, and equal cost compared to separate packages. For 3DIC and also for products such as RFID tags and medical plasters, ultra-thin dies are needed. These require dedicated thinning and dicing technologies with use of temporary carriers.

Key words: Cost Reduction, Reliability, Sensors & MEMS, Embedding, 3D Packaging, Memory Integration, Thin Dies

Introduction

Today the main drivers for package technology development are cost reduction, form factor requirements, reliability, electrical performance, and function integration.

Cost Reduction

To keep up with the competition and ensure market share, continuous cost reduction is essential. On average, cost of existing packaging platforms is reduced with 5-10% per year. This is realized in various ways.

Obvious ways are use of lower cost packaging materials, use of less packaging materials, price negotiations with vendors, standardization of materials and formats, use of cheaper processes and larger processing formats. Examples of cost reduction activities are reduction of gold wire diameter, introduction of copper wire, use of large area transfer molding and matrix lead frames, wire length reduction (by using low loops and lead frame redesign), and introduction low cost solder balls, die

attach and molding materials, stamped lead frames and laminates with ENEPIG pad finish.

Copper wire introduction is not a simple switch from gold to copper wire. It provides huge challenges in wire-bond yield and reliability. To limit oxidation, wire bonding needs to be done in inert gas to keep the surface fresh and enable an acceptable spherical free-air ball (FAB). Because copper is harder than gold, it can damage IC bond pads and reduce the process window for the second bond. Copper wire has first been introduced for power packages. As these packages need thick wires (50-38 um), cost benefits are high and also they benefit most from the higher thermal and electrical conductivity of copper wire compared to gold wire. Gradually, use of copper wire is being expanded to packages that have thinner wires, but enough wires to keep introduction of copper wire cost beneficial, see Figure 1. Implementation of copper wire is most difficult on (ultra)lowK dies. The weakness and brittleness of the dielectric and the hardness of the copper wire make the ball bond on the IC very critical.

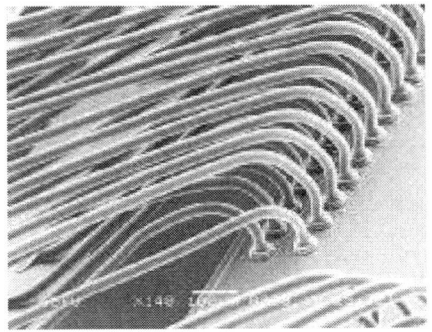

Figure 1: 25 um copper wire in LBGA

Another example of cost reduction is reducing the sawing lane width, so that the number of dies per wafer can be increased. Using standard blade sawing, sawing lane widths of 50 um can be handled, but below that, a different die separation technology needs to be used. An interesting technology is stealth laser separation. With this technology, sawing lane widths down to 15 um wide can be processed. With stealth laser dicing, the energy of the laser beam is absorbed only around the focal point of the laser beam inside the silicon wafer. Locally, the silicon structure is modified and weakened. Depending on the wafer thickness and doping level, a number of different scans need to be made across the wafer at different depth levels. After scanning, separation takes place by expanding the wafer on a foil. Advantages of this technique are very straight sawing lanes, minimum cracking, extreme cleanliness, absence of thermal damage and no use of cooling water. This makes this laser separation technique not only suitable for MEMS dies but also for saw lane reduction in conventional dies. For a 0.4x0.4 mm die, laser separation allows about 20% more dies per 8-inch wafer, see Figure 2.

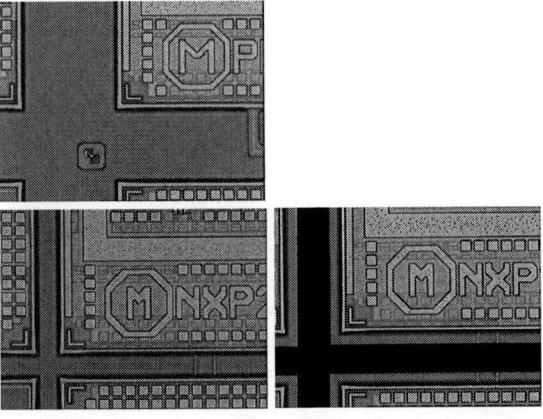

Figure 2: Product with 15 um sawing lane width (top picture: standard product with 60/80 um sawing lane width; bottom left: new product with 15/15 um sawing lane width before separation; bottom right: new product after laser separation)

A less obvious cost reduction method is chip-package codesign. Due to increasing product complexity, off-chip effects have a significant influence on electrical product performance. When designing the chip and package in an integrated way, not only electrical performance is optimized but also total system cost can be minimized. Chip I/O configuration can be made such that a package substrate with minimized number of layers and density can be used.

Besides cost reduction within an existing package platform, cost reduction can also be achieved by switching to a different package platform or by development of a new, lower cost, platform. This is illustrated in Figure 3 for lead frame based packages. Switching between package platforms enables much higher cost reduction than can be achieved by cost reduction within a platform.

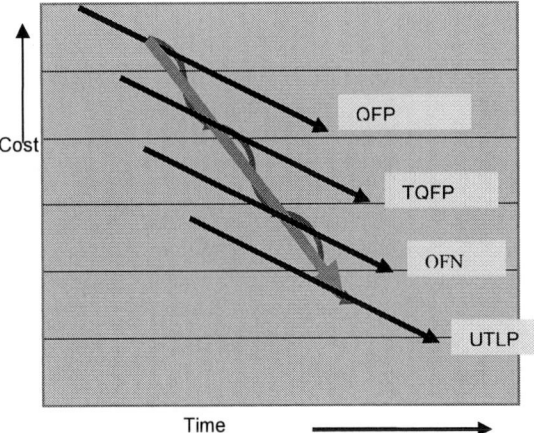

Figure 3: Cost savings between and within lead frame based package platforms

The latest package platform that was introduced for this reason is UTLP (Ultra Thin Leadless Package). UTLP is a low cost package for low pin count dies. Electrical and thermal behavior is comparable to HVQFN, however UTLP enables multiple rows of I/Os and therefore enables smaller, lower cost packages. A ULTP is also thinner than most HVQFN: presently 0.5 mm minimum but smaller in the near future. The substrate is a Cu-barrier-Cu stack with patterned NiPdAu plating on both sides. The substrate is half etched on the topside using the NiPdAu as etch mask. The packaging flow is as follows: dies are attached on the topside, wire-bonded and overmoulded. After that, the bottom side of the substrate is etched, using the NiPdAu as etch mask. After electrical test, separation (through compound only) takes place. Limited rerouting is possible. To improve electrical performance flip chip first level interconnect will be available. Figure 4 shows a range of UTLP products.

978-1-4244-4722-0/09 $25.00
© 2009 IMAPS-ITALY

Figure 5 shows a schematic cross-section after the bottom side-etching step.

Figure 4: UTLP products, maximum package size is 8x8 mm2, minimum I/O pitch is 0.35 mm, both will go down in the future

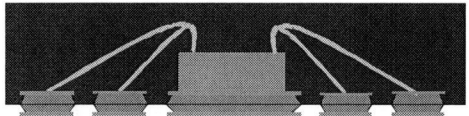

Figure 5: UTLP products after etching of the bottom side

It is expected that future low cost package platforms will as much as possible make use of batch processes for die placement and interconnect: all dies are placed in one step and all dies are interconnected in one step. Further, working areas will be large to make the batch size as large as possible and substrates will be very low cost or even non-existent. Perhaps only a temporary substrate will be used. Also important to realize low cost, is that accuracies and tolerances are not over-specified.

Embedded Packages

IC packages made by embedding technology have three characteristics: 1. First level interconnect is realized by plating (usually copper) or another type of metal deposition process. There are no flip chip interconnects or wire-bonds; 2. The packages provide a fan out of the bond pads on the die to a larger scale array of pads or bumps; 3. Thin packages of 0.5 mm or below are realized. Two embedding technologies can be distinguished: one processed with wafer fab technology and one processed with PCB fab technology.

Industry examples of embedded packages using wafer fab technology are the Embedded Wafer-Level Ball Grid Aarray (eWLB) and the Redistributed Chip Package (RCP). We will call the technology Fanout WLCSP, see Figure 6. First, a reconfigured wafer is made by placing thinned and sawn dies face down onto a carrier and applying an encapsulation material, usually a molding compound. Next, a wafer fab redistribution is applied on the side with the exposed die front sides. The last steps are ball attach and separation.

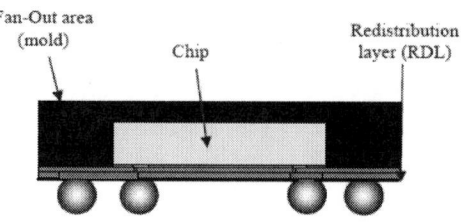

Figure 6: Schematic drawing of Fanout WLCSP

Fanout WLCSP cost is determined by the size of the reconfigured wafer and the number of wafer fab redistribution layers required. For big dies, often more than 2 layers are needed, making it an expensive technology.

Fanout WLCSP can realize a smaller and thinner package compared to BGA because the definition of the redistribution layers is higher than the pattern definition on the BGA interposer. Depending on the size reduction, cost can even be lower than the original BGA. However, this makes only sense when the customer can handle the small ball pitch that goes with it. FanoutBGA might also provide a more cost effective solution for large FC BGA that use expensive multi layer built up substrates.

A Fanout WLCSP is also seen as an expansion of the WLCSP platform; when the bumps do not fit on the die, they are placed onto the fanout area. In most cases this is less expensive than enlarging the die.

Next step in Fanout WLCSP is the addition of through mold vias and additional plating on the topside of the package. In this way the Fanout WLCSP can provide a thinner and smaller alternative for PoP. Also, more than one die will be embedded in the package together with the realization of passive functions in the redistribution layers. In this way, complete modules can be built.

Issues with Fanout WLCSP are board level reliability for bigger devices (failure during temperature-cycling test), die shift during molding and mold curing and selection of the dielectric. Standard dielectrics cannot be used, because the encapsulant does not withstand the required temperature.

Concluding, it can be said that application of Fanout WLCSP is most likely in highly miniaturized mobile products as alternative for thin BGA, PoP and WLCSP for low to medium pin count.

The second type of embedding technology is one processed with PCB fab technology. As tolerances are less tight than wafer fab technology, this is a cheaper, but less accurate process. Institutes, dedicated companies and substrate suppliers are developing different technologies. In most cases the dies are laminated face up into the substrate core and the interconnect between the substrate and the die is established by drilling microvias at the bond pad locations of the dies followed by copper plating. The accuracy of the microvia position relative to the die defines the minimum bond pad pitch that can be handled. In most cases bond pad pitch is limited to 300 um. Most accurate is the Imbera process in which, before lamination, dies are aligned and flipped onto a copper sheet relative to micro via holes, which are already present in the sheet. Bond pad pitches of 100 um can be handled. Also here first level interconnect is made during copper plating of the microvias. Figure 7 shows an example of a bare die interconnected by the copper plating in a microvia. The bond pads on the die are redistributed to enlarge the pitch.

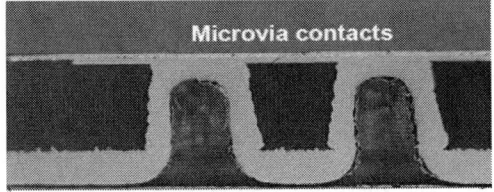

Figure 7: Bare die laminated into a substrate and interconnected by copper plating in a microvia

Usually embedding with PCB fab technology is done in substrates on which other dies and/or components are mounted. Three levels can be distinguished: embedding in motherboards, embedding in module boards and embedding in interposer boards. However, embedding with PCB fab technology can also result in a package, see the example in Figure 8.

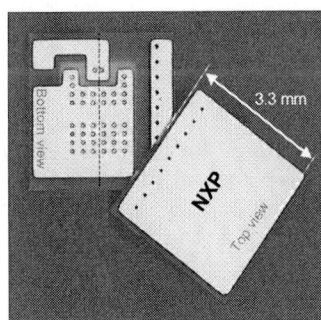

Figure 8: MOSFET Embedded Package based on PCB fab technology

Assuming a lamination process with interconnect through copper plating in microvias, the yield of the embedding process is defined by the yield of the die itself, the yield of the PCB microvia process and the yield of the interconnect process. Therefore, embedding in expensive motherboards is at present only interesting for high yielding, small, low I/O count dies, as otherwise substrate yield would be too low. Examples of dies that are being embedded in motherboards are Integrated Discrete

and RFID dies. It is expected that this will be a mainstream technology. Embedding of more complex, larger, higher I/O count dies will first be implemented in interposer boards. With only one embedded die per substrate, risk is not too high. Application for instance is a BGA containing a microprocessor with a memory die embedded in the interposer substrate, see Figure 9. A next step would be a module in which several dies are embedded in the interposer substrate.

Figure 9. A package with an additional die embedded in the interposer substrate

To enable embedding, dies need to be supplied with Au or Cu bumps. Cu bumping is preferred as this is cheaper, see Figure 10. In case dies have been designed for wire bonding, a redistribution of the I/Os into a larger pitch is required. For Integrated Discrete products, embedding enables a die size reduction. Usually, these products have a WLCSP package of which the size is determined by the bump pads and pitch. In case of embedding, the bump pad size and pitch can be reduced and therefore also the die size.

Figure 10: Example of a copper bumped Integrated Discrete die to be used for embedding in substrates

Embedding of dies in motherboards changes the supply chain, as the semiconductor supplier does not supply to the OEM anymore but to the PCB manufacturer. An issue that is especially important for large, complex dies is liability. It needs to be defined carefully who is responsible for test and field returns. Finally, the semiconductor suppliers' gross revenue is decreased because packaging is no longer needed. Therefore, embedding of complex dies in motherboards will only be a niche technology applied in those cases where miniaturization is very important, e.g. watch phones or other fancy gadgets.

PCB embedding technology to realize a package is likely to be cheaper than Fanout WLCSP. Because of pitch/accuracy limitations it will first be implemented for small dies with few I/O at large pitch. Like Fanout WLCSP, packages will be thinner

978-1-4244-4722-0/09 $25.00
© 2009 IMAPS-ITALY

than conventional packages. Also, when dies are imbedded in package interposers, this will give a thickness reduction.

Reliability

Up to now, semiconductor packages used in cars were not optimized for automotive requirements. In general, conservative leaded packages are used and changeover to new products and/or packages is slow. For product qualification, dedicated automotive grade qualifications need to be passed. However, with traffic density increasing year on year, the required safety and reliability levels in cars is also increasing. This does not only increase the number of advanced electronic systems in cars, but also puts higher demands on package reliability. Electronics in a control unit in the engine compartment need to cope with an ambient temperature of 150 degrees C, a junction temperature of 175-200 degrees C, and even higher peak temperature. Another change in automotive reliability requirements is a stricter requirement on delamination in plastic packages after temperature cycling: no delamination is allowed

An application that also puts high demands on temperature resistance is the lighting business, i.e. the drivers for CFL and SSL lamps. Here not only electronics needs to cope with a 125 degrees C ambient temperature, but also with a lifetime of 30000 to 50000 hrs.

To cope with these high temperature, zero delamination and long life time requirements, dedicated leaded plastic packages need to be developed. The interfacial strength between compound, die and lead frame needs to be improved and the wire-bond interconnects need to be more robust. Adhesion can be improved by e.g. optimized compounds, rougher lead frames and stress reduction by changing the package construction. Wire-bond interconnects can be made more robust by changing to monometallic interconnects or by applying a coating on top of the Al bond pad. All this goes together with the identification of new failure modes and development of new accelerated reliability test methods.

Sensors & MEMS

There are several factors that contribute to a growing need for sensors and MEMS.

Worldwide there is more and more attention for safety and security in vehicles, this not only results in the need for more reliable electronics, but also in the need for more sensor systems. For example MEMS sensors are needed for measuring tire pressure, electronic stability control, car roll over detection and particle emission in diesels. MEMS sensors provide relatively simple, miniaturized solutions consisting of few parts. Provided they are protected in the right way, they are very robust and can easily meet the high automotive reliability requirements.

Another trend is a change in healthcare from illness treatment using generalized medicine, towards illness prevention and monitoring and personalized medicine administration. Expensive diagnostics systems located at central laboratories testing single parameters are changing into low cost, disposable systems used at point of care or even at home, doing multiple analyses. So called, In Vitro Diagnostics (IVD) systems, used for analysis and monitoring of human body fluids are enabled by miniaturized cheap silicon based biosensors.

Third trend is more awareness for the environment. There is more and more focus on energy saving systems, environmental pollution and waste reduction. Sensors are being developed that check water, air and soil on contaminants. Smart RFID sensor tags on perishable food or other perishables could monitor the status of food and avoid people getting ill and help to optimize the supply chain and save unnecessary waist.

Finally, the still ongoing trend for function integration and miniaturization in mobile phones and other small portable devices demands small devices like MEMS microphones, camera modules and MEMS oscillators. Compared to their conventional counterparts, these enable more flexibility in product design because they are smaller and thinner and can be reflow soldered. Also other functions can easily be integrated into the package. The trend towards a more intuitive device user interface for these products needs multi touch sensitive displays and 3-axis accelerometers to change the display according to the orientation of the phone.

General challenges in MEMS and sensor packaging are wafer treatment, die protection, and the realization of die access for liquids and gasses to be analyzed. Trends are towards processes that can be done at wafer level, are low cost and use low temperature to avoid damagement of MEMS structures.

With respect to wafer treatment, MEMS separation is an issue in case this is done before capping or when capping is omitted such as for MEMS microphones and biosensors. As already explained in the beginning of the paper, laser separation brings a solution here. With laser separation there is no debris that could damage the MEMS dies. An example of a laser sawn MEMS die is shown in Figure 11.

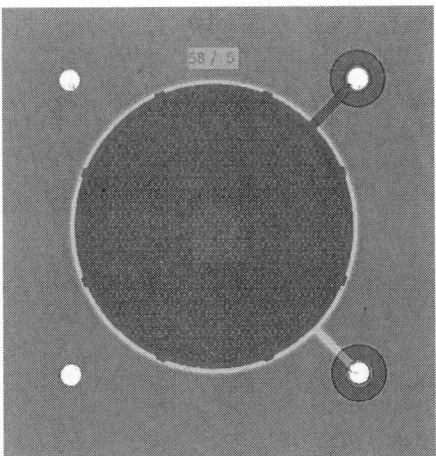

Figure 11: Laser sawn MEMS microphone die, side length 1.5 mm

MEMS devices like oscillators need hermetic capping. Capping can be done on the wafer as part of the MEMS wafer fab process or afterwards using a silicon cap wafer or individual caps. Capping as part of the MEMS wafer fab process can be done in many different ways. Cavities can be made by etching sacrificial layers (e.g. metal) or by using thermally degradable polymer sacrificial layers. The first has the disadvantage that there needs to be a hole in the cap for the etch medium. Afterwards this hole has to be filled. In some cases a metal overcoat needs to be applied over the cap to ensure hermetic capping. Capping as part of the wafer fab process gives the most miniaturized caps. Thickness can easily be lower than 50 micron and size is equal to the MEMS area. Other advantages are the fact that several wafers can be capped in one batch, and simple logistics.

Capping by means of silicon caps is a wafer-by-wafer or cap-by-cap process and results in bigger and thicker caps compared to wafer fab process integrated capping. Another disadvantage is the electrical interconnect. Tracks need to be fed through the cap seal ring, or else, silicon through vias need to be used. Different types of seal rings are used. Low cost, low temperature processes are metal soldered seal rings, e.g. AuSn or polymer seal rings.

In case the final MEMS package is an overmoulded plastic package, there is an additional requirement: the cavity needs to withstand molding pressure (about 90 bar). Depending on the cavity size, it may be necessary to build in cap supports.

Devices that analyze gases or liquids need an access hole to the silicon sensor. At the same time, the interconnects and other device parts need to protected from the material to be analyzed. One of the ways to realize this is by film assisted molding with film at the top side of the product in combination with mold inserts. The inserts are

pushed downwards on the die surface and keep part of the die area free of compound, see Figure 12.

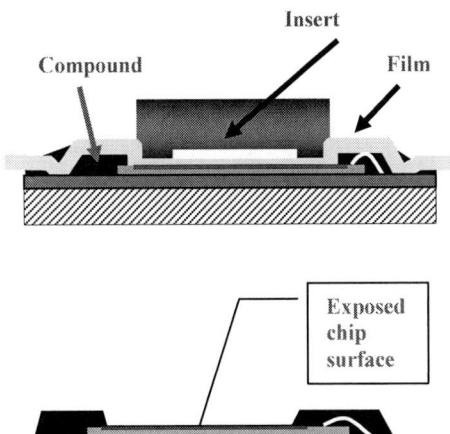

Figure 12: Film assisted molding with inserts to realize a mold compound free area on the die

3D Packaging

Function integration can either be done on die level (SoC) or on package level (SiP). Both routes are being followed where SiP is often an intermediary step towards the next SoC. Making use of existing dies provides a shorter time to market and is less risky than designing a big new SoC. The choice for SiP is more permanent when it is cheaper to use different wafer fab technologies for different functions or when interconnect to the outside world is easier and cheaper when there are multiple dies.

Standard method to include more than one die in a package is wire-bonding dies side-by-side or stacked onto the interposer. New developments in stacked die technology are aimed at reducing package thickness and including more dies. Trend is towards thinner dies, lower wire-bond loop heights, and more flexibility in die sizes, realized by use of film over wire and wire bonding on overhanging dies.

As discussed earlier, a new possibility to include more dies into a package is embedding of dies in interposers. Compared to standard stacked die packages, this would offer thinner packages.

Another emerging technology is making a die stack with fine pitch flip chip interconnects and through wafer vias (TSVs). Compared to wire-bonded stacked die packages this enables performance improvement and miniaturization, but cost would be much higher. Die stacks with a high density of TSVs and interconnects at block level could even provide an alternative for SoC.

978-1-4244-4722-0/09 $25.00
© 2009 IMAPS-ITALY

Performance would improve as interconnect length between the dies would be shorter than within a SoC. To enable this, TSV and interconnect pitch need to be at least 50 um or less. Die thickness needs to be reduced to 50 um or below as otherwise realizing these small TSVs would not be possible. This type of stacked die interconnect is now first emerging for memory die stacks. For other, so-called 3DIC applications, even finer via and interconnect pitch at cell or even transistor level are needed. There is no consensus in the world about the best method for via processing and interconnect yet. Many different technologies and flows co-exist.

Memory Integration

The trend towards more functionality per die requires a higher resolution for display processing and a higher data rate for communication functions. To realize this, a bigger memory capacity and a higher memory bandwidth are needed.

There are several possibilities to increase memory capacity: stacking a memory package on top of the IC package (PoP); including memory dies inside the IC package; or using on-chip memory. On-chip memory offers limited memory capacity only. The wafer process type cannot be optimized for cost and memory cell size because it is already fixed, and, of cause, the die size cannot be increased infinitely.

Memory bandwidth can be increased by increasing the number of interconnects between the memory die and the IC, and by decreasing interconnect length. Here, on-chip memory and off-chip memory connected to the IC by fine pitch flip chip interconnect, if needed with TSVs in case of more than one memory die, offer the best possibilities. Hence, off-chip memory connected through wire-bonds (PoP, side-by-side or stacked wire-bonded dies) offers a lower memory bandwidth.

A third discriminating factor for the memory options is power consumption. Power consumption is related to memory bandwidth. A higher power consumption results in a lower bandwidth. Power consumption is minimized by reducing interconnect length. Therefore off-chip memories connected by flip chip (with or without TSVs) consume least power. On-chip memory with possibly long on-chip interconnect lines performs a little bit worse. Next is memory in package connected by wire-bond, and least performing is PoP where the memory die is external to the package. Comparing PoP with the situation where the memory and the IC are packaged in individual packages and assembled side-by-side on the motherboard, PoP clearly offers lower power consumption and higher bandwidth but it is worse than memory in package.

Table 1 gives an overview of the different memory integration options including also form factor and cost.

Table 1: Overview of different memory integration options

	Separate package		In IC package			On chip
	Side-by-side	On top (PoP)	WB[1] side-by-side	WB stack	FC[2] Stack[3]	
Memory capacity	++	++	++	++	++	-
Memory bandwidth	--	0	+	+	++	++
Power consumption	--	0	+	+	+++	++
Form factor	--	+	0	++	+++	0
Cost	++	-	+	++	--	---

1) wire bond; 2) flip chip; 3) this includes also interconnect with TSVs

Considering all these factors, memory in package using a wire-bonded interface looks most attractive. There is no additional cost compared with individual packaged IC and memory dies assembled side-by-side on the motherboard, while bandwidth and power consumption are much better and size is smaller. Memory in package using a flip chip interface offers even better bandwidth, lower power consumption and smaller size, but it is more expensive, especially when TSVs are needed. Therefore it will first be used for high performance, miniaturized products where cost is no issue. When it becomes cheaper in the future and when the bandwidth requirements get higher, its application will be expanded.

PoP is very popular at present in mobile phones. It increases the memory bandwidth, offers power saving and miniaturization compared to side-by-side packages but less than memory in package options and it is expensive. So why is it used? PoP offers the possibility of full functional test before stacking, it is flexible, the memory supplier does not have to deliver bare dies and disclose die yield, there is no discussion about liability, and die pinning can be freely selected.

Embedding of a memory die into the interposer of an IC package is also an enabler for memory integration. The distance between the IC and the memory is small and the number of interconnects could be large, but not as large as in the case of die to die flip chip, because the interposer definition is less than the definition on a die. Therefore, for this type of technology, memory bandwidth would be slightly worse compared to die on die flip chip but still much better than stacked wire-bonded dies. Power consumption for the embedded and flip chip option would be comparable.

Thin dies

Two applications need extreme thinning of dies down to 10-30 um. The first is 3DIC. To make stacks of heterogeneously integrated dies with fine pitch TSVs and interconnects, dies need to thinned down to about 5 times the via diameter. So, in case via diameter is 5 um, wafer thickness needs to be about 25 um. Another case in which extremely thin dies are needed is when dies need to be integrated in foil or paper. Examples are RFID tags, passports or banknotes with Identification dies, and medical plasters with sensor dies or dies for drug administration purposes.

For ultra-thin dies, not only the thinning is a challenge, but also the dicing. Thinning needs to be done on a temporary carrier wafer, either silicon or glass. Standard grinding or polishing can be used or etching with SOI or epi layer as etch stop. In both cases, a wafer-to-wafer bonder and debonder are needed. Bonding medium is either adhesive tape or spin on adhesive.

Options for die separation are: 1. Standard or laser dicing after debonding the wafer from the temporary carrier wafer. Disadvantage here is that thin wafers need to be handled; 2. Dice by grind by partly etching or sawing the sawing lanes before grinding. The actual separation takes place during grinding on the carrier wafer; 3. Standard dicing after grinding but before removal of the carrier wafer. The carrier wafer is separated as well. Advantage here is that handling of the individual dies is easy, but removal of the carrier wafer now needs to be done at die level after placement and bonding. Also dicing alignment is a point of attention.

Concluding Remarks

Cost reduction is realized by improvement of existing packaging platforms, and by introducing new packaging platforms. Future low cost packaging platforms will be based on batch and large area processing, use of low cost substrates, and will avoid over specification.

Embedded packages based on PCB fab technology provide low cost miniaturised packages suitable for low to medium pin count dies. Embedded packages based on wafer fab technology provide ultra miniaturised packages that will be used in mobile equipment as alternative for FCBGA, PoP and as extention for WLCSP.

Automotive and lighting applications require packages suitable for high temperature, long lifetime and zero delamination. Dedicated technologies need to be developed.

Emerging silicon based sensors and MEMS for automotive, medical, environmental and mobile applications need dedicated wafer treatment processes, MEMS protection and the realization of die access for fluids and gases to be analyzed.

Function integration drives the development of 3D packaging technologies with 3DIC as ultimate technology for miniaturisation and performance. Cost is still a barrier for general application. For most applications standard SiP still is adequate.

For 3DIC and also for products such as RFID tags and medical plasters, ultra-thin dies are needed. These require dedicated thinning and dicing technologies with use of temporary carriers.

Acknowledgements

The author would like to thank Freek van Straten and Eef Bagerman for their useful input and discussions.

Temporary adhesives for wafer bonding: Deep reactive ion etching application

Djaffar BELHARET[1,2], Pascal DUBREUIL[1,2], David COLIN[1,2],
Laurent MAZENQ[1,2] and Hugues GRANIER[1,2]

[1] CNRS ; LAAS ; 7 avenue du colonel Roche, F-31077 Toulouse, France
[2] Université de Toulouse ; UPS, INSA, INP, ISAE ; LAAS ; F-31077 Toulouse, France

belharet@laas.fr

Abstract

Deep reactive ion etching is a critical step for Micro-ElectroMechanical Systems devices fabrication (MEMS). Some applications are bulk silicon etching with stop on oxinitride membranes, using so both side of a wafer. Usually, a first deep silicon etch is performed on the backside to form a thin silicon membrane with a thickness of a few micrometers. This membrane is then etched from the front side to delimitate active areas and to release the membrane structure. In other cases, vias are performed through the membranes. Nevertheless it's necessary to use a carrier wafer where the device wafer is bonded on.

Our approach is to bond the two wafers with an adhesive which can be considered as an intermediate layer. Temporary wafer bonding requires the adhesive to be easily removed without damaging the features on the active side of the device wafer over a short debonding time to increase throughput. It will also allow to continue the MEMS fabrication. The conventional adhesives (tapes, waxes, Fomblin...) used for bonding were studied. Only the Fomblin showed good results compared to the other adhesives. Fomblin presents the inconvenient of a removing process which takes 40 minutes to 1 hour; and a very bad surface state which makes it not re-usable for other applications.

With temporary adhesives, we achieved a better processing window. We will detail deposition process and bonding process conditions of the adhesive, demonstrating the benefits of using the wafer bonding technique which offers a good thermal transfer. We'll demonstrate that after the bonding, the wafers can move to lithography operations and be processed as single wafers to pattern the structures to be etched. After etching, we proceed to debonding in a few minutes. One other advantage is the good surface state of the wafer which can continue the flow of the process.

Key words: Adhesives, wafer bonding, Etching, MEMS

Introduction

Microelectromechanical systems (MEMS) technology encompasses an enormous variety of applications, including sensors of almost any kind, imagers, micropositioners, optical beam steering and filtering, microphones, RF tunable components and switches

Micro-Electro-Mechanical Systems (MEMS) is the integration of mechanical elements, sensors, actuators, and electronics on a common silicon substrate through microfabrication technology. While the electronics are fabricated using integrated circuit (IC) process sequences (e.g., CMOS, Bipolar, or BICMOS processes), the micromechanical components are fabricated using compatible "micromachining" processes that selectively etch away parts of the silicon wafer or add new structural layers to form the mechanical and electromechanical devices.

There are numerous possible applications for MEMS and Nanotechnology. As a breakthrough technology, allowing unparalleled synergy between previously unrelated fields such as biology and microelectronics, many new MEMS and Nanotechnology applications will emerge, expanding beyond that which is currently identified or known. Here are a few applications of current interest: Biotechnology, Communications, Power management.

No single process flow can be used to fabricate all possible MEMS. However, a handful of canonical process flows cover the basic MEMS fabrication concepts and form a basis for many other derivatives. The canonical process flows covered in the following discussion are silicon wet etching and bonding, surface micromachining, deep reactive-ion etched silicon micromachining, CMOS MEMS, and microstructural molding processes [4]. The only

978-1-4244-4722-0/09 $25.00
© 2009 IMAPS-ITALY

process which will be described in this manuscript is the dry etching process.

The fabrication of micro-electro mechanical systems (MEMS) usually requires at least one silicon etching step in the entire process. Especially for bulk micromachined devices, silicon etchings with high etch rates; high aspect ratios and high selectivity to mask materials are required for a through-wafer etching. Nowadays, etchings with such specifications are mainly achieved by deep reactive ion etching (DRIE) using high-density plasma sources.

Generally, etching silicon material thickness upper than 150µm in plasma reactor having mechanical clamp is impossible to achieve it. The solution consists in bonding the device wafer with an other wafer called generally carrier wafer.

Direct wafer bonding is a method for fabricating advanced substrates for microelectromechanical systems (MEMS) and integrated circuits (IC). Bonding in general means the joining of two pieces of the same or a different material together. Bonding can be divided into three categories: bonding with a conducting interlayer, with an insulating interlayer, or without any intermediate layer [9]. This paper discusses about the advantage of using the direct bonding wafer technique for the planarity of the stack wafer and therefore for the uniformity post-eth measurements.

1. Etch process

Plasma etching is an ion-enhanced chemical process, thus often referred to as reactive ion etching (RIE). Etch systems use RF powered plasma sources for the creation of ions and chemically reactive species. In deep silicon etching, the primary source gas used is sulfur hexafluoride (SF6), which supplies highly reactive free fluorine for high etches rates in silicon. Ions in the plasma are accelerated toward the wafer with strong directionality by a potential difference (RF Bias) between the plasma and the wafer (the electrode on which the wafer is placed). While this gives rise to enhanced removal rate in the vertical direction, additive chemistry for passivating the etched sidewalls is required for highly anisotropic profile evolution [3,6].

The way to provide sidewall passivation for deep silicon etching; is the so-called Bosch process, which uses rapid alternating steps of etching with an SF6 plasma and deposition with a polymerizing gas such as C4F8. Because of the polymer deposition and low RF Bias voltage, this process has high selectivity to photoresist and in some cases can exceed 100:1. The deep trench etch sequence is illustrated in Figure1 and is also known as deep reactive ion etching (DRIE). The mask is usually either photoresist or silicon oxide, however other mask materials can be used. In step (a), a high density inductively coupled SF6 plasma etch

achieves selectivity to the mask. The gas in the plasma chamber is then switched to C4F8 in step (b), which deposits a thin fluorocarbon polymer onto the wafer surface. The following etch step (c) uses physical ion assist to etch the polymer at the bottom of the trench, leaving some sidewall polymer. The polymer masks lateral etching and thereby maintains the vertical sidewall profile [6]. The desired trench depth is obtained by cycling etch steps (a) and deposition step (b), with an effective etch rate of around 2.5 µm/min.

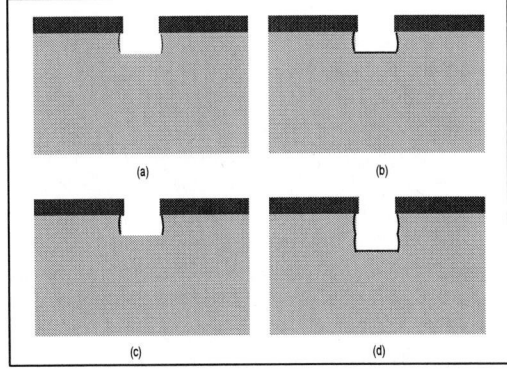

Figure1. Deep reactive etching process mechanisms

The DRIE equipment has an inductively coupled plasma (ICP) source operating at 13.56MHz. This etch system also has a mechanical wafer clamp and helium backside cooling to improve the evacuation of heat from the wafer, thus providing an accurate temperature control [1]. The DRIE equipment used in this study is STS equipment (for 4 inches wafers).

Controlling these various aspects at high silicon etch rates requires a system that has a wide process window in terms of operating conditions. Because most plasma etch reactors are radially symmetric, non uniformity typically appears as a variation in etch rate, profile shape, or profile tilt between the wafer center and edge.

2. Temporary adhesive bonding technique

Fomblin (FBL) and waxes showed their limitations on the thermal transfer. The adhesive layer between device and carrier wafer provides the mechanical strength required for thin wafer handling. Adhesives should process adequate flow properties to flow into structures on the front side of the device wafer to provide good bonding quality. The adhesives properties required to do easy and efficient bonding are : low mechanical strength, good thermal stability and chemical resistance and exhibit minimum total thickness variation across the wafer [5,7,8]. The adhesive studied is HT10-10

978-1-4244-4722-0/09 $25.00
© 2009 IMAPS-ITALY

product fabricated by Brewer science. The advantage of using HT10-10 is that there is no outgassing during bonding (at high temperature) and in other processes like DRIE etch process.

2.1. Carrier film preparation

A typical process flow for temporary bonding involves fully processing the device wafer on the front side. The fundamental operation is carrier film preparation. The carrier wafer should be cleaned before deposition step. This step is very important to ensure that there is no particles on the carrier wafer surface able to avoid homogenous deposition, the clean step can be done chemically using (H2SO4/H2O2) or physically by processing the wafer on plasma oxygen.

The carrier wafer is spin-coated with a spin-on adhesive (HT10-10 Brewer science). We studied two modes of deposition. The first one consists in deposition of the adhesive film on "open cover" and the second mode is "close cover". In the two modes, we have operated the same experimental deposition conditions (flow and rotational rate). After spin-coating deposition step, the carrier wafer is baked at high temperatures. A first bake is done at 120°C during 120 seconds, a second bake is followed at 160°C during 120 seconds.

2.2. Bonding experimental conditions

After deposition step, both wafers (device wafer and carrier wafer) are transferred to a bond chamber, the bonding equipment used in this study is AML-WB04 equipment. The wafers should be carefully centered and vacuum bonded at elevated temperatures. The process flow is illustrated in the figure2. The silicon wafers processed are 4 inch wafers.

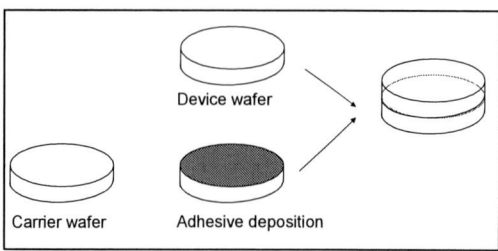

Figure2. Temporary bonding process flow

We have explored some bonding parameters like bonding force and temperature to understand their impact on the next process operations. Thinning the intermediate layer is very important in the etching process because it provides to the stack wafer a good thermal transfer.

2.3. Photolithography process

After bonding step, the stack wafers follow the next process operations among them the photolithography process. This operation process permits to define the patterns to be etched and protect the others. The first step is photoresist deposition on the front side of the stack wafer. After baking the photoresist, the insolation process consists in the insolation of this photoresist on certain parts of the wafer, for this a mask is often designed to determine exactly the regions to be insolated. The figure 3 shows the mask designed for this study. As mentioned in the figure 3.b, there is 2 types of patterns, the dimensions of patterns are : 590μm*768μm and 590μm*833μm).

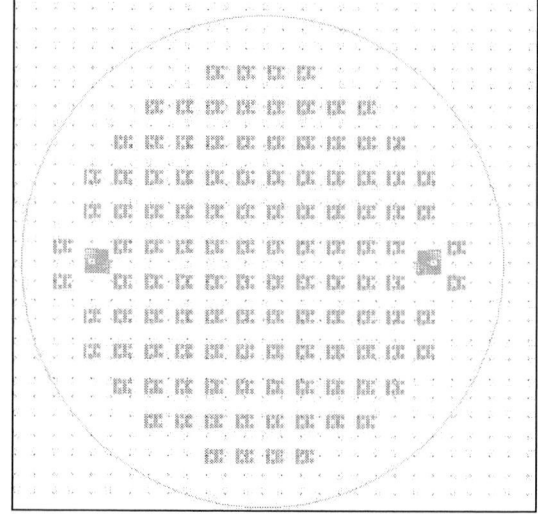

a)

b)

Figure3. a) the mask design (4 inch) b) the repeated shape (the dimensions of patterns are: 590μm*768μm and 590μm*833μm)

The next step is the development, which permits to define clearly the regions to be etched, the others are protected by the presence of the photoresist. At last an ultimate resist bake insure solvents evaporation.

2.4. DRIE etch process

All the wafers are processed with the same recipe on the plasma etch reactor (STS reactor able to etch silicon material at 2,5µm/min). The recipe used is Bosch process described previously.

2.5. Measurements step

After the process operations, the wafers are measured using an Olympus confocal microscope. A 9 points mapping was used, this mapping is illustrated in the figure 4.

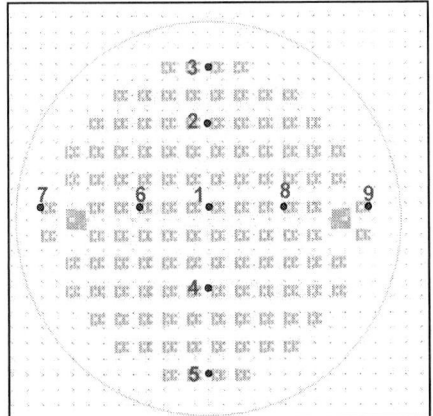

Figure4. 9 points measurements mapping

3. Experimental results

All measurement sites corresponding to the wafers processed are collected and analyzed. The following parameters were calculated : the average etch rate and the uniformity. The uniformity is calculated following this relation:

$$Uniformity(\%) = \frac{\max thickness - \min thickness}{2 * average thickness} * 100$$

Where : max thickness is the maximum thickness, min thickness is the minimum thickness and the average thickness is the average of the 9 points measurements.

These parameters are sufficient to characterize the etch process. All measurements values are compared with the experimental bonding conditions in order to identify and quantify their impact on Etch process.

3.1. Energy bonding effect

The post-etch process measurement were collected and compared with the bond energy for each wafer as illustrated in figure5.

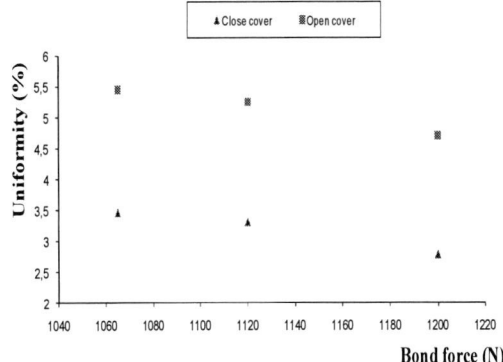

Figure5. Uniformity post etch measurements vs bond energy

The figure 5 shows that the uniformity improves as the bond energy increases. The "close cover" mode offers a good uniformity compared to the "open cover". The advantage of using the close cover mode is the condensation of the solvents inside the film surface which contributes to form homogenous deposition and thin film thickness. In the "open cover" mode, a part of the solvents are evaporated during spin-coating resulting the formation of non homogenous film deposition. Only the close cover was studied because it offers a good homogenous film deposition compared to the "open cover mode". The bonding force is a critical parameter [7] because it determines directly the film thickness value (intermediate layer), for fixed parameters and varying the bonding force from 1050N to 1200N. The film thickness obtained for 1050N is about 22µm, increasing the force to 1200N, the film thickness can be reduced to 12µm. In order to quantify the role of film thickness on the uniformity, we studied the uniformity post etch vs the film thickness as illustrated in the figure 6.

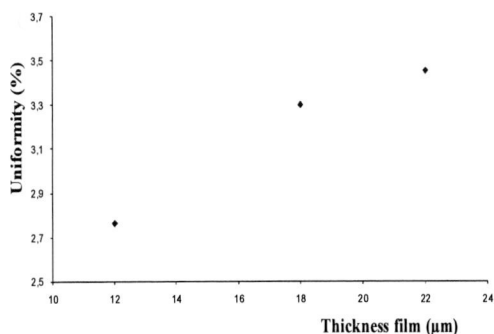

Figure6. Uniformity post etch measurement vs thickness film (close cover mode)

The uniformity degrades as the film thickness increases as illustrated in the figure 6. The

978-1-4244-4722-0/09 $25.00
© 2009 IMAPS-ITALY

thickness film value determines the thermal transfer efficiency.

3.2. Temperature effect

Optimizing the film thickness is not sufficient to ensure a good bond quality, it requires also a rigorous control of the film properties which are related to the bond conditions such as temperature and pressure. The bonding temperature is another critical parameter [7] on the bonding quality. High temperature provides a good bonding quality, but a low temperature can impact the non uniformity of the intermediate layer across the wafer. In order to define the optimal temperature value, we have explored a range of temperatures (100°, 130° and 160°). The post etch uniformity measurements were compared to the bond temperatures as illustrated in the figure7.

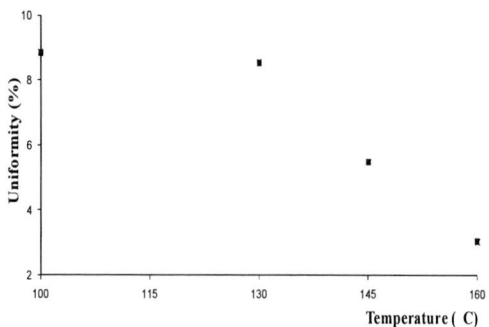

Figure7. Uniformity post etch measurements vs bond temperature (for fixed bond force 1200N).

The figure 7 shows that the uniformity post etch measurement improves as the bond temperature increases. We observe that there is two regimes, the first one corresponds to bad uniformity for temperature below 130°C, when the temperature reaches some value, the uniformity is better. Depending on the desired technological applications, the two regimes can be used, it's obvious that the second regime is the best one.

For MEMS fabrication, both sides of the device wafer are processed. The backside of the device wafer has been processed and must be protected for the next process operations. In some cases, the presence of certain thick photoresist as SU8 is very sensitive to the high temperatures. For this type of materials, the first regime will be used. It's to be noted that the use of Fomblin and waxes is not useful because the bonding is bad quality.

3.3. Comparaison between wafer bonding technique and the classical bond methods

We defined the critical process parameters and we achieved a better processing window. As the wafer bonding offers a good planar surface for the stack wafers, the intermediate layer (adhesive) plays an important role on the surface state of the stack wafers. We did experiments by using TENCOR station to measure the constraints before adhesive deposition, after adhesive deposition and after bonding process. This measurement technique consists of analyzing the radius of curvature compared to the planar axis. We will give the results obtained on one stack wafer. The radius of curvature value for the carrier wafer is -10µm before adhesive deposition, the value of the radius becomes -14µm. The radius of curvature for the device wafer is +3µm, after bonding process, this value is about -5µm. So, we can conclude that the bonding process ensures a good level of constraints. This is very important for controlling the post etch uniformity measurements across the wafer.

We studied this profile by analyzing the measurement along the axis containing the measurements sites (1,2,3,4 and 5) for two stack wafers bonded with optimal process conditions (force bond=1200N and T°=160°C), two modes were used; "close cover mode" and "open cover mode". Two other stack wafers were bonded with Fomblin and wax. All the stack wafers were processed with the same etch process recipe during 40 minutes. The etched thickness measurements were done using the 9 points mapping. In order to illustrate the profile of the thickness etched along the axis of the wafer, measurements of sites 3,2,1,4 and 5 were collected and analyzed as illustrated in the figure 8.

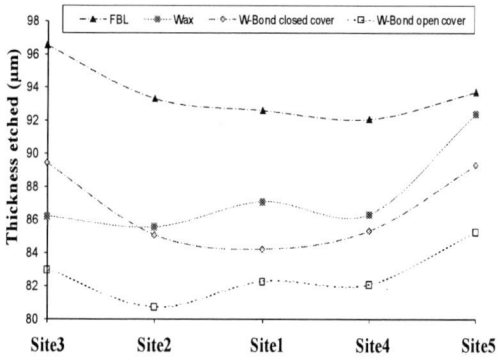

Figure8. Profile etched thickness across radius wafer

The figure 8 shows that the profile of the uniformity (wafer bonded-close cover mode) along the axis chosen is the typical profile compared to classical uniformity profile of a single wafer, where the etch rate is very important at the edge of the wafer and the etch rate is low at the center of the wafer. This is due certainly to the homogenous adhesive deposition. It's not the same situation in the case of "open cover mode". Finally, the profiles of Fomblin and waxes are not typical and not reproducible (variation on wafer to wafer). We

978-1-4244-4722-0/09 $25.00
© 2009 IMAPS-ITALY

observe that the etch rate is important when we use fomblin compared to the wafer bond technique. We can explain this by the difference of the thermal transfer coefficients.

In order to validate this approach, analyzes by scanning electron microscope (SEM) were performed on two stack wafers where the vias were etched through the device wafer. The SEM observations are illustrated in figure 9.

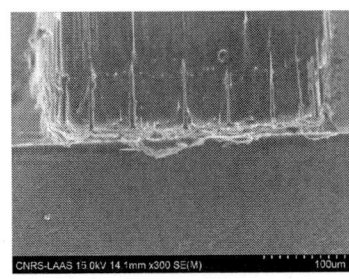

a)

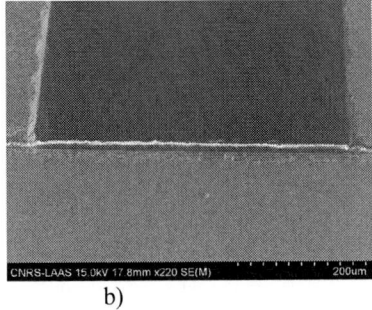

b)

Figure9. SEM observations; film interface between device wafer and carrier wafer a) using Fomblin b) using bond techniques

In figure 9a, we observe that there is high surface rugosity in the trench bottom and also in the walls trench. The Fomblin can not be used as an intermediate layer for the MEMS process flow fabrication. The figure 9b shows that there is low rugosity surface in the trench bottom and walls trench and the film adhesive is very homogenous.

3.4. Debonding operation

Two methods were used to debond the wafers, the two methods are illustrated in the figure 10.

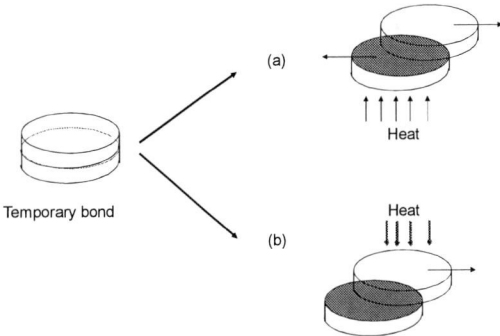

Figure10. Debonding methods a) manual method b) wafer demounting system.

3.4.1. Manual method

The first method used for debonding consists in heating the stack wafer at high temperature during 10 minutes. When the wafers are separated at the edge, we can remove the device wafer manually. The last step is to clean the wafers and especially the backside of the device wafer by wafer bond remover (Brewer science product). The carrier can continue the flow of the process due to its good surface state.

3.4.2. Wafer demounting system

The second method is quite similar to the first method described previously. Two platens are used (horizontal movable plate and heated vertical plate). This second technique is under development.

5. Conclusion

One advantage of using the adhesive is that there is no outgassing during bonding. The critical bond process parameters were identified. The impact of the energy bond on the post etch uniformity was quantified. We demonstrated that the bond energy determines directly the film thickness value which determines the thermal transfer level. We demonstrated also that the temperature is another critical process parameter because it determines the film adhesive properties and therefore the bonding quality. We determined bonding process window to achieve a good etching quality and to get a good post etch uniformity. We compared our wafer bond techniques experimental results with the classical techniques such as waxes and Fomblin. We demonstrated that the post etch uniformity is better using bond techniques. Finally, we illustrated the technique of debonding.

This intermediate layer technique is under development and it will be generalized in the future in other processes. It's seems to be very promising

978-1-4244-4722-0/09 $25.00
© 2009 IMAPS-ITALY

as the stack wafers can flow classical process flow as a single wafer.

References

[1] A.A. Ayon, R. A. Braff, R. Bayt, C.C. Lin, H. H. Sawin and M.A. Schmidt, "Characterization of a time multiplexed inductively coupled plasma etcher", J.Electrochem. Soc. 146 339. 1999

[2] A.Mehra, A.A. Ayon, I. Waitz and M.A. Schmidt, "Microfabrication of high temperature silicon devices using wafer bonding and deep reactive ion etching", J. Microelectromech. Syst. 8 152, 1999

[3] F. Laermert & al, "Method of anisotropically etching silicon", US Patent 5,501,893, March 1996.

[4] G.K. Fedder, "MEMS FABRICATION", Proceedings of the 2003 ITC INTERNATIONAL TEST CONFERENCE (ITC), Charlotte, NC, USA, 28 September - 3 October 2003, pp. 691-698, 2003.

[5] J. Lamb, B. Kim and S. Pargfrieder, "Temporary bonding/debonding for ultrathin substrates", Solid state technology, july 2008

[6] K.E. Bean, "Anisotropic etching of silicon", IEEE Transactions on Electron Devices, , Vol. 25, Issue: 10, pp. 1185- 1193, Oct 1978

[7] R. Puligadda, S.Pillalammari, W. Hong, C. Brubaker, M. Wimplinger and S. Pargfrieder, "High-performance temporary adhesive for wafer bonding applications", Mater.res. Soc.Symp.Proc.Vol.970, Materials Research Society, 2007

[8] S.Pargfrieder, P.Lindner, G. Mittendorfer, J.Weixlberger, "Ultrathin wafer processing using temporary bonding", Semiconductor International, 2006.

[9] T. Suni, "Direct wafer bonding for MEMS and microelectronics", Phd thesis, Helsinki University of Technology, 2006.

Acknowledgements

We want to acknowledge Abhi Budhwar and Amandine Jouve from Brewer Society company for all helps and providing HT1010 samples and the bond remover samples.

All these works were performed in the technology platform of LAAS, a member of the french Basic Technological Research network.

Integrated LTCC-Glass Microreactor and µTAS with Thermal Stabilization for Biological Application

Pawel BEMBNOWICZ, Dorota NOWAKOWSKA, Leszek GOLONKA

Wroclaw University of Technology, Faculty of Microsystem Electronics and Photonics,
Wybrzeze Wyspianskiego 27, 50-370 Wroclaw, Poland
phone (fax): +48-71-3554822;
e-mail: pawel.bembnowicz@pwr.wroc.pl

Abstract

Recently, analytical microdevices are frequently described in technical papers. The µTAS (Micro Total Analysis System) often implements optical read-outs which require a usage of transparent materials. There is a wide range of materials which are applied to construct devices. However, the analytical ceramics based systems are seldom described. Mechanical, thermal and chemical properties of the LTCC (Low Temperature Co-fired Ceramic) substrate make it an adequate material to build chemical and biochemical reactors. However, alumina ceramics is not transparent. Hence, the transparent material is needed to provide an optical port. In this paper sodium glass material was applied. A tight ceramic microchamber with a sodium glass window and a lateral glass waveguide is presented. Moreover, the structure is equipped with a resistor which works as a heater. The additional thermocouple sensors enable a temperature measurement. As the result, LTCC-glass µTAS with thermal stabilization is obtained. The FEM (Finite Element Method) is used to optimize thermal properties of the designed microstructure. A PC (Personal Computer) based thermal management system allows for appropriate temperature shifting.

Key words: chamber, reactor, µTAS, LTCC

Introduction

The LTCC technology was invented for the realization of multilayer circuits. The process has long been used only for the fabrication of electronic circuit components. Unexpectedly, the LTCC technology was transferred to many new different fields. A lot of new applications have appeared since it becomes evident that three dimensional structures can be easily realized [1]. The technique is used in microfluidics, packaging, MEMS and MOEMS, automotive, aerospace, military, telecommunications as well as chemistry and even genetics [2 - 4]. Chemistry and biochemistry are application fields where a chemically robustness is important. Groß et al. [5] tested chemical stability of the fired LTCC. The ceramics was stored for one year at ambient temperature in aggressive aqueous solutions. The results showed that the LTCC material underlie the same stability constrains as the standard used laboratory glass ware. It makes the material suitable for constructing the chemical/biochemical microreactors. In many microsystem's applications a

microchamber requires an optical port which enables reagents inspection during the process. These kinds of devices are the first step to construct the µTAS. Hence, the micro device, which is made of robust ceramics and equipped with optical elements, can be widely applied in the µTAS field.

Optical systems are common used in genetics to verify the reaction product. Nucleic-acid fluorescent analyzing techniques have become a significant set of tools for many different applications. Besides the frequent molecular diagnosis of diseases and assessments of therapies in clinics and hospitals, they are also broadly applied in environment surveillances, food processing industry, agricultural researches, and forensic identifications. However, the amount of nucleic-acid analyze in the test samples is usually not sufficient for direct detection. Hence, nucleic-acid amplification is an essential step to increase the amount of the target sequence [6]. In recent years, research groups have been developing PCR devices made of wide range of materials: silicon, glass, PDMS, PC, PMMA, Epoxy, SU-8, PET, Polycyclicolefin resin etc. [7]. The LTCC properties make it a suitable material to construct the DNA amplification reactor. Moreover, the additional optical ports enable real time product analyzes. The most common detection scheme for micro PCR amplification is based on fluorescence markers [8]. This technique is sensitive and widely used in PCR analyzes.

In this paper a new construction of microchamber based on the low-cost, LTCC structure with integrated a sodium glass waveguide, a transparent window and a PdAg heater is presented. The glass waveguide and the transparent window enable optical excitation and fluorescent analyze during the thermal process. The additional temperature control system is presented. The required temperature profile can be achieved. The system consists of the ceramic microchamber, additional thermocouple sensors, digital controller and PC display software. The possible applications of the device are seen in chemistry and biochemistry especially in genetic to carry-out real-time PCR.

An optical elements integration

Despite of many LTCC advantages the ceramics is a non-transparent material. This is a significant disadvantage when optical measurements are required. There have been few attempts to solve this problem in literature: glue fibers, attach polymer, the use of optical glue, or the use of high pressure lamination of sapphire window etc. However, all of them have some undesirable features which cause problems with adapting it to the LTCC technology.

The presented technology of the glass window and waveguide integration with ceramic chip is compatible with the LTCC technique.

The idea of the integration is based on usage the thin glass wafer. The softening temperature point of glass is required to be about 700°C. The wafer needs to be located over the ceramic cavity. As the firing temperature increase first the glass wafer become soft then the LTCC shrinkage start. In this way the soft glass plate does not resist the shrinkage process. Simultaneously, the viscosity in the maximum firing temperature is high enough to create a surface tension. The force is so strong that it can hold a soft glass membrane spread over the cavity. Figure 1 shows glass window in the LTCC structure.

Fluorescence methods of analyze require the laser excitation and the optical detection. Moreover, laser light should be perpendicular to the detection area. Hence, vertical and horizontal ports are required. The vertical port, which works as a waveguide, can be achieved by lamination of thin glass stripe between the LTCC layers. The glass stripe needs to be situated

inside a prepared channel to prevent deformation. Thus the glass optical waveguide was achieved.

Fig.1 Glass window made in the LTCC structure

The reactor construction

The 10x15 mm chip was made of DP951P2 LTCC tape (DuPount, USA). Fourteen layers of the ceramics were used for the chip manufacturing. The dimensions of the layers were defined in un-fired tape by Nd-YAG laser patterning with Aurel NAVS 30 laser trimming and cutting system. Figure 2 shows patterned ceramic layers.

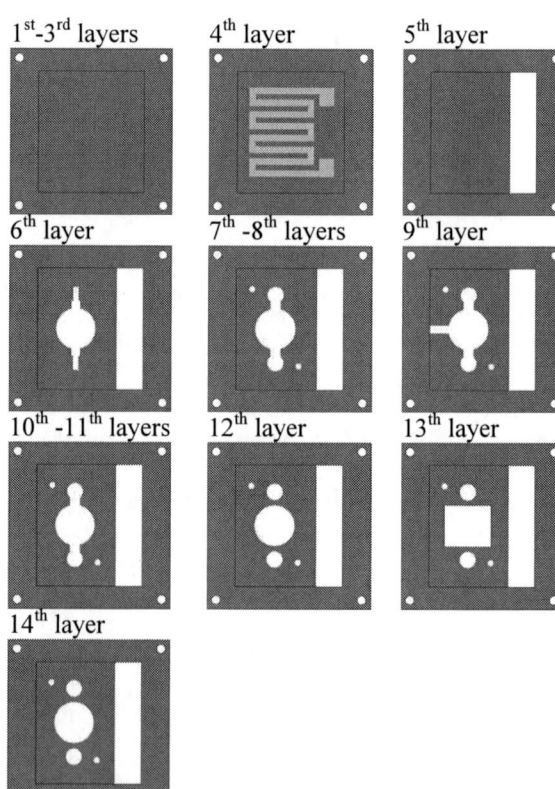

Fig. 2. Patterned ceramic layers – scheme

Both optical ports of the structure (window and waveguide) were made of precisely cut glass cover slips. The thick strips are made of extra white glass. This material is dedicated to fluorescence measurements.

The 6x6 mm wafer was inserted into square-shaped pattern in second layer in order to create transparent window. The piece of glass strip was situated in rectangular pattern in the seventh layer of LTCC. After the layers had been aligned an unconventional ceramic lamination technique was apply. In this case the standard, a high pressure lamination was inadmissible. The glass substrate was fragile at room temperature. The standard procedure would crush the glass wafer. Roosen describes non-conventional lamination methods [9]. The cold chemical lamination (CCL) was applied [10]. Afterward the structure was co-fired in 875 °C temperature profile recommended by DuPont for the DP951 tape. Eventually, flexible pipes were glued to an inlet and outlet in order to provide an easy reaction mixture injection to the chamber.

Vertical and horizontal windows, situated inside a single structure, enabled both easy laser excitation and optical detection. The LTCC-glass chip with both kinds of windows was constructed. The structure with coupled laser beam is demonstrated in figure 3.

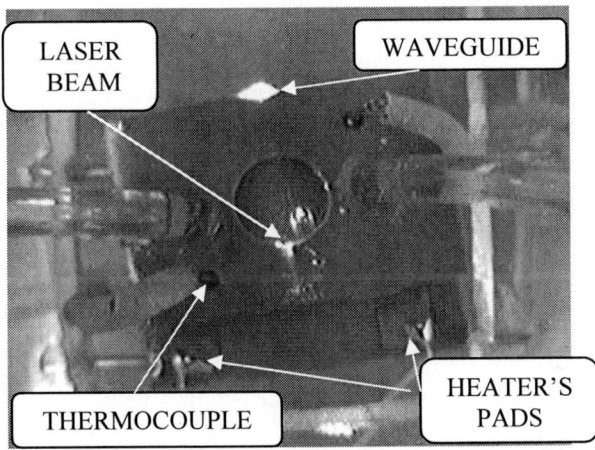

Fig. 3. Photo of the LTCC-glass structure with coupled laser beam during thermal cycle

Simulation of the temperature distribution

The biochemical reactions require precision thermal management. The FEA (Finite Element Analyze) was used to optimize thermal properties of the designed reactor. The steady – state calculations gave information which was very helpful in heating system improvement. 3-D solid elements, which had a thermal and electrical conduction capability, were used in simulations. The elements had eight nodes with two degrees of freedom, temperature and voltage, at each node. Boundary conditions were

defined. The voltage was set on the structures pads. Joule heat generated by the current flow was included in the heat balance. A convective cooling was set on all walls. Temperature distribution was investigated. The simulations (fig. 4) show that the temperature distribution at the chamber was uniform. The difference between the hottest and the coldest point of the chamber bottom was less than 1.5 K. It fulfils the thermal requirements for the biochemical reactor.

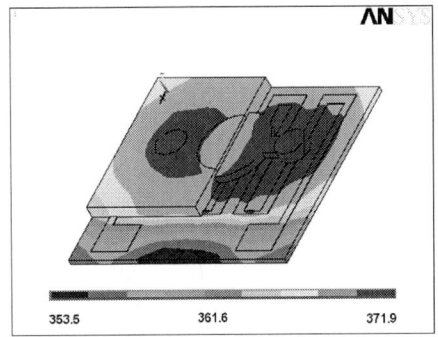

Fig. 4. Temperature (in K) distribution of designing microchamber – sectional view.

Idea of temperature management system

The ceramic chip with integrated heater was inserted into the special designed grip (fig. 5). The tool provided easy thermocouples addition to the ceramic structure. Temperature management system was based on the idea of temperature information feedback. Precise thermocouple thermometer with optical digital interface (CHY 502A, Centenary Materials) was used. The data were transmitted to the computer by RS232 protocol. A JAVA language was applied to write data management program. The software was based on the PID algorithm.

Fig. 5. The reactor inside the grip with additional thermocouples

The PC computer compares actual and set temperatures and then decided on power control. The decision data were transferred into a PWM (Pulse-Width Modulation) generating device. The peripheral device was based on ATmega8 microcontroller. Afterwards the PWM signal was amplified. Thus the chamber resistor was properly supplied. Joule heat was generated and distributed in whole ceramic reactor. Eventually, the proper temperature was applied to the reaction mixture.

The results of temperature management

The LTCC microreactor and the dedicated thermal management system was constructed. The system enabled proper thermal management of the liquid sample inside the reactor. The experiment required a change of the temperature setpoint in the following sequence: 90 °C, 60 °C, 72 °C over 60 s each. The system thermal properties were investigated. Figure 7 present a temperature versus time plot measured by system.

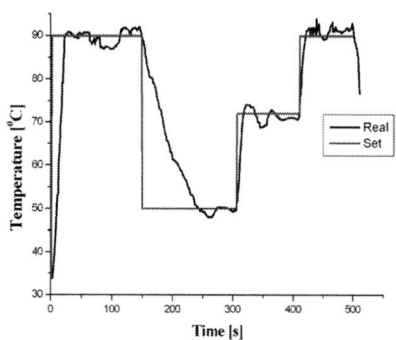

Fig. 7. Temperature versus time measured by system

Conclusions

The combination of the LTCC and glass technology was demonstrated. The glass window and the waveguide were integrated in the LTCC chamber. The screen printing was use to achieve chamber heater. The unconventional, CCL method of green tape lamination was applied. FEM simulations have been applied in order to determine temperature distribution in the chamber. Dedicated digital temperature management system was constructed and tested. Proper thermal management has been achieved. In the nearest future, the biochemical reaction will be developed as well as the influence of the ceramic components on the liquid sample will be investigated in the microreactor.

Acknowledgement

The authors whish to thank Polish Ministry of Science and High Education (grant no. R0201702) and Wroclaw University of Technology (grant no. 343578) for financial support.

References

[1] W. Smetana, B. Balluch, G. Stangl, E. Gaubitzer, M. Edetsberger, G. Köhler, "A multi-sensor biological monitoring module built up in LTCC-technology", Microelectronic Engineering , vol. 84, 2007, pp. 1240–1243

[2] L. Rebenklau, G. Schlotting, J. Uhlemann, G. Vollmer, K.J. Wolter, "LTCC Packaging for bio-application's demands", Proceedings IMAPS/AcerS, April 25-27, 2006, Denver,

[3] P. Ciosek, K. Zawadzki, J. Łopacińska, M. Skolimowski, P. Bembnowicz, L. J. Golonka, Z. Brzózka, W. Wróblewski, Monitoring of cell cultures with LTCC microelectrode array, Journal of Solid State Electrochemistry vol. 13, January, 2009, pp. 129 – 135

[4] CF Chou, R Changrani, P Roberts, D Sadler, J Burdon, F Zenhausern, A miniaturized cyclic PCR device-modelling and experiments, Microelectron Eng, vol. 61–62, 2002, pp. 921–925

[5] G.A. Groß, T. Thelemann, S. Schneider, D. Boskovic, J.M. Köhlera, Fabrication and fluidic characterization of static micromixers made of low temperature cofired ceramic (LTCC), Chemical Engineering Science vol. 63, 2008, pp. 2773 – 2784

[6] T.H. Fang, N. Ramalingam, D. Xian-Dui, T.S. Ngin, Z. Xianting, A. Tan, Lai Kuan, E. Yap, P. Huat, G. Hai-Qing, Real-time PCR microfluidic devices with concurrent electrochemical detection, Biosensors and Bioelectronics vol. 24, 2009, pp. 2131–2136

[7] C. Zhang, J. Xu, W. Ma, W. Zheng, PCR microfluidic devices for DNA amplification, Biotechnology Advances, vol. 24, 2006, pp. 243–284

[8] E.A. Ottesen, J.W. Hong, S.R. Quake, J.R. Leadbetter, Microfluidic digital PCR enables multigene analysis of individual environmental bacteria,. Science vol. 314 (5804), 2006, pp. 1464–1467.

[9] A. Roosen, New lamination technique to join ceramic green tapes for the manufacturing of multilayer devices, Journal of the European Ceramic Society vol. 21, 2001, pp. 1993–1996

[10] D. Jurkow, H. Roguszczak, L. J. Golonka, Cold chemical lamination of ceramic green tapes, Journal of the European Ceramic Society vol. 29, 2009, pp. 703–709

978-1-4244-4722-0/09 $25.00
© 2009 IMAPS-ITALY

Fabrication & Characterization of S-Band Power Amplifier using GaAs Die

[1]Nadeem Shahzad Bhatti, [2]Muhammad Imran

[1,2]National Engineering & Scientific Commission
Plot # 94, Sector H-11/4, Islamabad -Pakistan

[1]nadeemshahzad1@gmail.com, [2]imranepg @hotmail.com, Ph [1](+92) 321 5509344 Fax (+92) 51 4434092

Abstract

This paper presents the fabrication and characterization of S-band power amplifier MMIC (GaAs die) implemented in hybrid fixture, which is compatible with different microwave applications. The configuration used for device fabrication is relatively simple, low cost and easy to assemble with manual die & wedge bonder. It can be used in any microwave assembly as driver/power amplifier. Due to use of substrate material RO4003C, epoxy, and solder preform (SPF) and molytab it is useful for prototype production. The MMIC has gain as high as 21.37 dB at S-band. A number of Key methods are discussed for MMIC performance improvement.

I. Introduction

Recently several research activities have been reported for MMIC packages. Ceramic based packages were introduced with high RF performance [1, 2] but required relatively high manufacturing costs. Plastic packages can reduce the material cost but have large fluctuation in TCE (Thermal Coefficient of Expansion), which induces fabrication inaccuracy. The trend of microwave & millimeter wave ICs is now towards cost effective packaged products compatible with SMD assembly line [3].

In semiconductor industry some devices are available without hermetic package for customized application & cost reduction. Among the number of packages available in market, it is cost effective to use bare die in fixture. The Power die hereafter is called MMIC in further discussion. The fixture can be used later in microwave modular assembly. The use of multifunction MMICs (monolithic microwave integrated circuits) although reduces the cost and chip size, but on module level assembly cost is still significant. This is due to costly automatic die and wedge bonders with bonding tools and clean room facilities. In this fixture assembly manual die and wedge bonders are used for prototype production. The principal of such fixture is described in Fig-1. The Gold plated Molybdenum bar is bonded to Pedestal of fixture floor with epoxy. Pedestals of same size as MMIC are often required to match the MMIC elevation to that of the substrate [4].

The thermal management at lower cost is another key factor for MMICs assembly product. The epoxy die-attach methods are low temperature processing, easy application and ability to tolerate greater TCE (Thermal Coefficient of Expansion) Mismatches. The thin layer of epoxy results in corresponding small temperature gradients.

For solder die-attach method the alloy Au80/Sn20 with 1 mil thickness is the preferred material. A primary advantage of solder preform is its excellent thermal conductivity (TC).

The MMIC is bonded to solder preform (SPF) using eutectic bonding in a separate process, then to molybdenum bar (Moly tab) with epoxy. For thermal management at lower cost SPF & two layers of Epotek-H20 are used in this configuration.

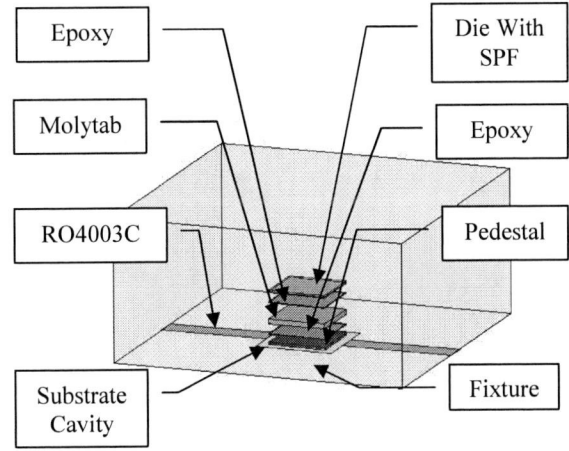

Fig. 1: Configuration of Fixture for GaAs Die

Due to use of pedestal, epoxy, molytab and SPF the exposed ground exhibits low thermal resistance which is useful for device performance. The advantages of this configuration are low cost, easy to fabricate with manual die & wire bonder. It can be used in any microwave assembly as driver or power amplifier.

This paper presents and demonstrates the integration and performance of S-band power amplifier MMIC fabricated in the hybrid fixture.

978-1-4244-4722-0/09 $25.00
© 2009 IMAPS-ITALY

II.   S-Band GaAs Power MMIC

A.   Power Amplifier MMIC

The developments in the multifunction self-aligned gate (MSAG) process has enabled to develop gallium arsenide(GaAs) high power MMIC that offers improvements in bandwidth, integration, power handling capability, noise & reliability. The MMIC used here is a two stage amplifier with on chip bias network. It is fabricated using MSAG process which is versatile for PA, LNA with no air bridges. It employs polyimide scratch protection, so easy to fabricate with manual die bonder. The MSAG is a planar, ion-implanted MESFET process.

The power amplifier device in this process is 5A FET that provides 800 mW/mm of output power at 65% power added efficiency (PAE) at 10 GHz and 10V as shown in Fig-2 [5].

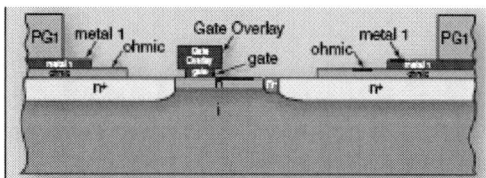

Fig-2. MSAG 5A FET cross section (not to scale)

The unique feature of the MSAG process is its intra-wafer, intra-reticle uniformity. This characteristic allows the efficient reactive combining of several tens of mm of FET gate periphery in a single stage. The M/A-COM MMIC of chip size 2.986 x 2.982 mm² is used.

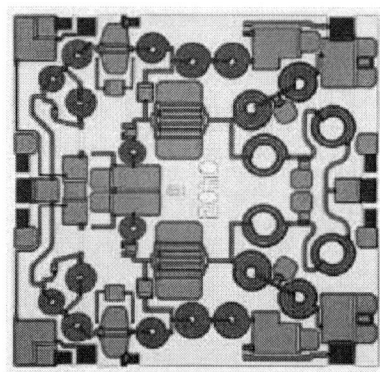

Fig-3.Chip photograph of two stage S-Band Power Amplifier MMIC [6]

Table I
Electrical Parameters of MMIC

Parameters	Typical values
Frequency range(GHz)	1.2-3.2 GHz
Linear gain(dB)	20
Out put power P_{1dB}(dBm)	30
Power added efficiency (%)	30
OTOI (dBm)	40

B.   Fabrication of MMIC in Fixture

The mechanical fixture of size 36.75 x 36.75 mm² is fabricated with pedestal of MMIC size and height 5mil. Roger's advanced packaging material RO4003C 20 mil is gold plated (Ni=1.8μm, Au=4.2 μm) for corrosion protection and bonding of gold wire. For wire bonding applications, Filtran [7] recommends a minimum of 100 micro inches (2.54 μm) of gold over a minimum of 50 micro inches (1.27 μm) of electroplated nickel.

The RO4003C high frequency laminates are glass reinforced hydrocarbon/ceramic materials. The high Glass Transition Temperature ($T_g > 400°C$) helps to maintain the expansion characteristics over wide range of temperatures [8].

Fig-4: Photograph of Die in Fixture

The 20 mil substrate with I/P,O/P 50Ω T/lines and coupling capacitors is bolted into Al fixture that have RF SMA connectors. The bypass tantalum capacitors are used on substrate to eliminate bias line instability. The DC connection points are provided with feed through capacitors on fixture. The MMIC is fabricated in the cavity of the substrate. The 1mil Au80/Sn20 solder preform (SPF) is bonded to MMIC (3mil) at 295 °C under inert N2 gas atmosphere. The Molybdenum Tab (10 mil) is bonded to pedestal (5mil) with epoxy (1-2mil). If pedestal is plated with copper / copper slug is used, it is a good choice for thermal resistance. Since Molytab is Cu/Mo/Cu alloy with gold plating when used with epoxy also exhibit low thermal resistance for device. The MMIC with SPF is bonded to Molytab using manual die bonder & epoxy (1-2mil). As a result fabricated MMIC's elevation is now aligned to substrate for wire bonding.

The Molybdenum Tab is selected due to close matching of its TCE to GaAs and good TC. Epotek's H20-E is used due to its reliability as being used in industry for at least 30 years with minimum thickness of 1-2 mils. It is specially designed for use in chip bonding of microelectronics & heat sensitive devices for thermal management applications due to its high thermal conductivity 29W/mK.For H20-E Lowest thermal resistance is obtained when using between thicknesses of 1-2 mil [9].

978-1-4244-4722-0/09 $25.00
© 2009 IMAPS-ITALY

After MMIC fabrication epoxy is cured as per data sheet. The substrate to die spacing is minimum (3-6 mil) in order to keep minimum bond length. The wire bonds for RF & DC connections with 1-mil gold wire are made using Kulicke & Soffa manual wedge bonder. To obtain minimum inductance of gold wire, multiple wire bonds on the same pad are used. The gold ribbon 0.5 x 3 mil can give better results if used. Finally fixture is hermetic sealed using Al cover for dust and humidity protection of device.

III. Measured MMIC Performance

A. Small Signal Characterization

All the measurements discussed hereafter were performed at room temperature (T_a= 25 °C). During testing the fixture is directly attached to cold heat sink to maintain the required temperature. The testing of an MMIC amplifier in die form begins with the measurement of pulsed small signal S-parameters tested over required frequency range. For this purpose Agilent's Network Analyzer with pulsed utility is used. The full 2-port calibration on required frequency range is done for correction of coaxial-to-microstrip transitions and RF cables reflections.

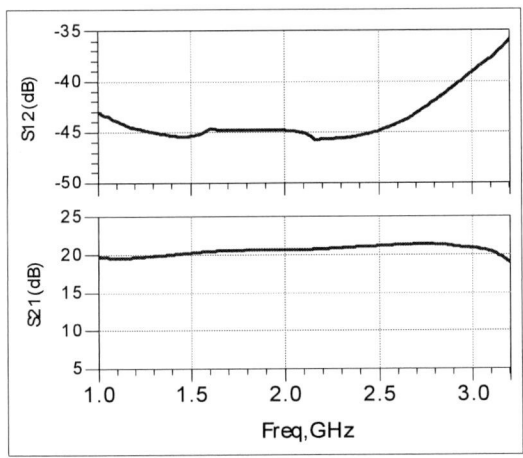

Fig.5: Measured reverse isolation (S12), Gain (S21) of the MMIC.

Fig-5 shows measured gain & isolation of MMIC in frequency range of 1-3.2 GHz. At V_{dd}=7V and bias current of I_{dq}= 680 mA, the gain (S_{21}) of amplifier is >20 dB with max. gain of 21.37dB at 2.74 GHz. For the same bias condition the reverse isolation (S_{12}) of amplifier is better than 35 dB over entire bandwidth. The input (S_{11}) and output return loss (S_{22}) not shown, are typically well below -10 dB under measurement conditions.

B. Large Signal Characterization

Power characterization is the first CW examination of the die product. The MMIC was also characterized under large signal measurement using a vector signal generator and spectrum analyzer.

Fig-6 shows output power, gain and PAE (%) versus CW input power at 2.5 GHz. At V_{dd}=7V and bias current of I_{dq}= 680 mA the output power near saturation is 32.2 dBm with gain of 20.9 dB and PAE (%) is 30.59 .The measured 1 dB compression point is 30.9dBm.

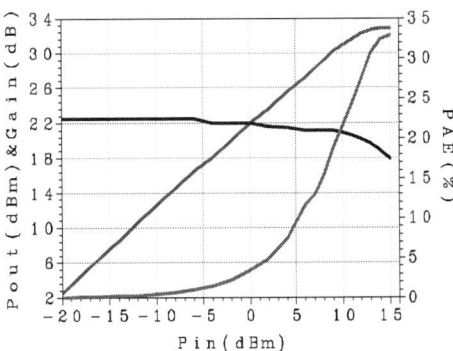

Fig-6: Measured CW output power, gain and PAE of the MMIC at 2.5GHz.

Fig-7 shows Psat versus frequency, which is greater than 30 dBm for Freq 1.3-3.2 GHz. At stop band frequencies Psat is relatively low but still > 30 dBm .

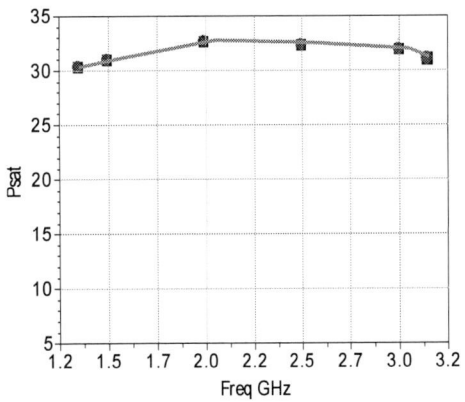

Fig-7: Measured saturated power versus Frequency.

The measured OTOI point of MMIC is shown in Fig-8 which is 39.9dBm.It is measured under the same conditions as above. For this purpose vector signal generator with two tone pulse utility is used.

978-1-4244-4722-0/09 $25.00
© 2009 IMAPS-ITALY

The level of carrier and IM3 product for a number of input power values is measured, which are plotted to obtain TOI of the MMIC [Fig-8].

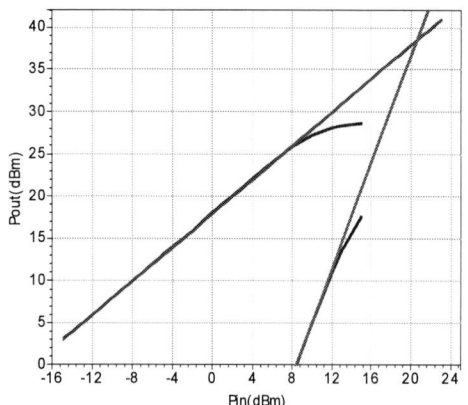

Fig-8: Measured Output third order intercept point (OTOI) of MMIC at 2.5 GHz.

IV. Conclusion

The performance of an S-band power amplifier GaAs Die fabricated in fixture has been presented. Due to the use of this configuration the MMIC performance is good for entire bandwidth with Gain greater than 20 dB, CW max. Linear power (P_{1dB}) of 30.9dBm and PAE (%) is 30.59. These all parameters are in a close match with the product datasheet. This fixture allows cost reduction with good thermal management.

References

[1] S. Koriyama, K. Kitazawa, N.Shino, H. Minamiue, "Millimeter-wave ceramic package for a surface mount" Microwave Symposium Digest, 2000 IEEE MTT-S International Volume 1,Issue 2000Page(s):61-64vol.1.

[2] H.C. Huang, A. Ezzeddine, A. Darwish, B. Hsu, J. Williams, S. Peak, "Ku-band MMIC's in low-cost SMT compatible packages" Microwave Symposium Digest 2002 IEEE MTT-S International Volume 1, Issue , 2002 Page(s):27 – 30.

[3] A. Bessemoulin, M. Parisot, P.Quentin, C.Saboureau, M. van Heijningen, J. Priday, "A 1-watt Ku-band power amplifier MMIC using cost-effective organic SMD package" 34th European Microwave Conference, 2004. Volume 1, Issue, 11-15 Oct. 2004 Page(s): 349 – 352.

[4] Application Note no.54 Rev.A, "GaAS MMIC ESD, Die Attach and Bonding Guidelines" May 2000 Agilent technologies.

[5] David. Conway, Michael. Fowler,Jack. Redus, "New process enables wideband high-power GHz amplifiers to deliver up to 20W" RFdesign Feb 2006.

[6] Product datasheet M/A Com, "MAAPGM0036-DIE".

[7]Dr.K.Ramachandran,"GoldPlating Considerations for Microwave Circuits" Filtran Microcircuits Inc.

[8] "Device attachment methods & wire bonding notes for RT Duroid and RO4000 series high frequency Laminates" Rogers Corporation.

[9]Frank W.Kulesza, "New Epoxy Systems for microelectronics" Epoxy Technology Inc. Jan. 1976.

[10] M. van Heijningen, J. Priday, "Novel organic SMD package for high power Millimeter wave MMICs" 34th European Microwave Conference, 2004. Volume 1, Issue, 11-15 Oct. 2004 Page(s): 357 – 360.

High Speed Packaging Solutions for LiNbO₃ Electro-Optical Modulator

S. Bonino, R. Galeotti, L. Gobbi, M. Belmonte, M. Bonazzoli

Avanex Corporation, Sede secondaria in Italia

Via Federico Fellini, 4, San Donato Milanese, 20097 – Milano • Italy

Phone: +39 02 518841, Fax: +39 02 51884222, E-mail Address: roberto_galeotti@avanex.com

Abstract

The increasing bandwidth demand from the telecom market, driven mainly by video application, is requiring high bit rate optical transmission systems. In this scenario the LiNbO₃ modulator is playing a key role as core optical component of the transponder both at 10Gb/s and at higher bit rates.

The competition among the component makers is driven by different factors: for 10Gb/s applications the cost and dimensions are the key features to get market share; for 40Gb/s applications RF performances with the new modulation formats (Duobinary, DPSK and DQPSK) are still the main challenge to be addressed.

In order to achieve these targets a strong development effort on both Lithium Niobate chip and packaging technology is required. In this paper we describe how the technical challenge on packaging is being addressed for the Avanex electro optical modulators.

Key words: LiNbO₃ modulator; packaging; integration; 40Gb/s

Introduction

Starting from the 2001 downturn of the Telecom market, together with the performance, the competition between the optical component manifacturers has been played on cost, form factor and integration. In this competition the packaging technologies has been a key factor to meet the customer requests. The new demand on 40Gb/s products is now asking a further challenge to the packaging technology.

In this paper we focus on describing the evolution on the 10Gb/s LiNbO₃ modulator form factor and on a successfull example of integration: the Duobinary modulator.

For the 40Gb/s applications we describe the results obtained with the DQPSK Avanex modulator.

Small Form Factor

In the last years a special effort has been required by the system integrator to reduce the components form factor. In specific for LiNbO₃ modulators this has been mainly due by the customer demand of devices that could allow moving from the old non standard format to the new MSA 4" and 3" optical transponders. This reduction requirement implies significant challenges in particular for LiNbO₃ modulators due to the chip length constraints strictly connected to the performance required.

From the packaging point of view the form factor reduction is a challenge affecting all the technologies involved in the product and process design: first to be mentioned the package body, RF interconnection and optical train design and all the packaging assembly processes.

The reduction target is becoming progressively tighter and now requires, as already mentioned, devices compliant with 3" MSA transponders.

Usually the modulator is positioned in the transponder package as shown in figure 1.

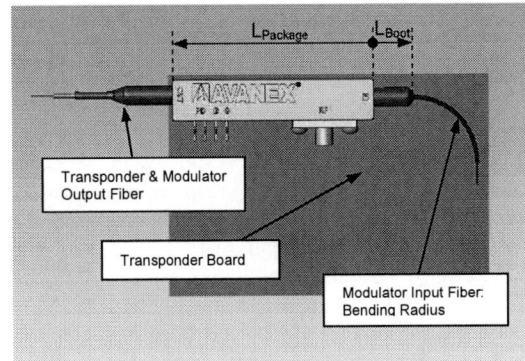

Figure 1: Scheme of modulator positioning inside the transponder

The output fiber of the modulator is typically also the output fiber of the transponder and in order to save room the modulator is aligned to the transponder wall. In the case of 3" MSA

978-1-4244-4722-0/09 $25.00
© 2009 IMAPS-ITALY

transponder, considering the transponder package wall and a fiber bending radius of 15mm the requirement on modulator dimension is:

$$L_{boot} + L_{package} = 58mm$$

Avanex corporation in 2002 (as Corning OTI) presented the first small form factor 10Gb/s modulator (IM10) and is continuing the trend of footprint reduction till the first 48mm package length ($L_{package}$) modulator commercially available for the 3" transponder (XS series modulators see figure 2, with $L_{boot}=10mm$)

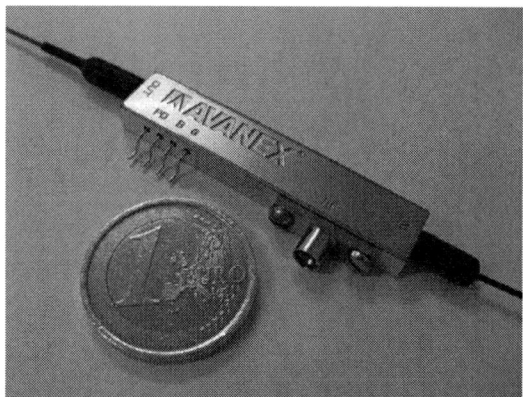

Figure 2: XS10 series modulator

XS modulator (in all versions based both on X-cut and Z-cut LiNbO$_3$ chip technology) despite of the strong dimensional reduction (17mm) maintains substantially the same performance of the standard Avanex 10G modulator designed for the 4" MSA transponder . This result has been obtained with big efforts on both chip technology and design and innovative solutions in packaging technology.

The main challenge for the packaging technology has been to reduce as much as possible the difference between the chip and the package body overall length. In order to obtain the mentioned results on the XS platform a new packaging design and a new assembly process has been developed for the optical feedthrough section.

Another important aspect in which the effort has been focused is the optimization of the pigtailing process. The optical interconnections between the input PM and output SM fibre to the LiNbO$_3$ chip have been redesigned by utilizing custom designed glass ferrule with extremely short dimension. Very complex and precise tooling has been developed in order to allow a stable and high yield pigtailing process in the manufacturing environment. This has been a complex task in particular for the PM fibre management where we have strict alignment tolerances not only in the linear axis but also in the angular ones.

These solutions allow a length difference between package and chip (see figure 3) of less than 10mm.

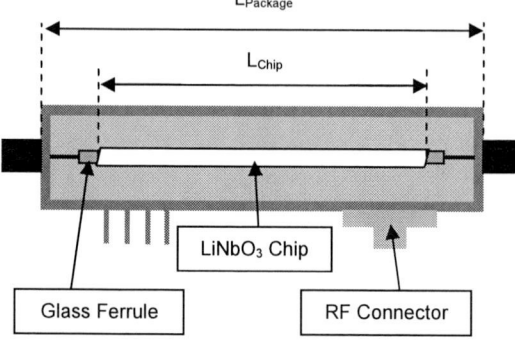

Figure 3: Difference between Package length and chip length in Avanex XS modulator

The new XS platform has been fully Telcordia qualified and passed the standard test protocol:

Telcordia GR-468-CORE		
Test	**Method**	**SS**
Mechanical Integrity [a]		
group 1 — Mechanical Shock	MIL-STD-883 2002.A	
group 1 — Vibration	MIL-STD-883 2007.A	11
group 1 — Thermal Shock [b]	MIL-STD-883 1011.A	
group 2 — Fiber Pull	Telcordia GR 326	
group 2 — Side Pull	Telcordia GR 326	11
group 2 — Twist	Telcordia GR 326	
Environmental		
group 3 — Temperature Cycling	MIL-STD-883 1010.A	11
Endurance		
group 4 — Damp Heat	Tc = 85°C R.H. = 85% operating 2000 hrs pass/fail	11
group 5 — Aging	Tc = 85°C R.H. = 85% operating 2000 hrs pass/fail	11

Table 1: Telcordia test performed on new packaging solutions for small form factor modulator

Despite of the tight dimensions and tolerances this package platform has demonstrated a strongly reliable behavior and all the developed solutions are ready to be applied to the future modulator platforms.

So, the new packaging technologies developed for the 10Gb/s market, where the form

978-1-4244-4722-0/09 $25.00
© 2009 IMAPS-ITALY

factor reduction is a key feature, are being applied on all new Avanex LiNbO₃ modulator design and also to the 40Gb/s devices.

It is important to mention that this choice gives benefit also where the form factor reduction is not a mandatory customer request; in particular these solutions as "new standard" allow cost reduction in terms of economy scale (material, purchasing, etc.) and allow improving performances through the new room available for chip design.

Integration: the case of Duobinary Application

Another strategy that could give sensible cost reduction and improve performance is the integration between the LiNbO₃ chip and other key elements.

The transfer inside the modulator package of external key elements or components give to module designers a cost saving in terms of package (1 instead of 2 or more) and interconnections (avoiding additional connectors).

In the past Avanex has obtained win-win results with the integration in the modulator package of the Driver and Bias control loop circuit board [1] (see Figure 4).

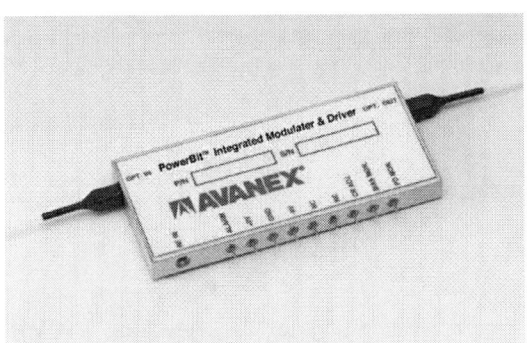

Figure 4: DM4 Modulator with integrated driver and bias control loop

More recent application of integration developed for 10Gb/s product is the LiNbO₃ modulator with integrated Bessel filter of the fifth order for duobinary modulation format.

Duobinary modulation format exhibits a high spectral efficiency, which results in enhanced tolerance to chromatic dispersion. This allows for transmission over extended distances without the need for costly dispersion management. Therefore this modulation format finds its typical application in metro/regional networks where the transmission distance can reach $150 \div 200$ km [2].

Duobinary modulation format is most frequently generated simply by filtering a NRZ stream using a low pass filter between the driver amplifier and the modulator, as in the scheme below.

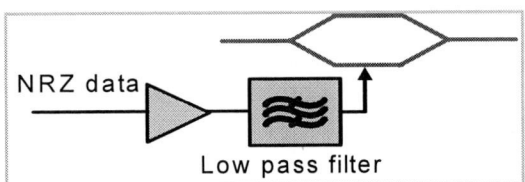

Figure 5: Duobinary modulation format generation scheme

The peculiar spectral shaping of the signal performed by the cascade of the low pass filter and the modulator frequency response is key for a good quality duobinary transmission. Therefore, on one hand it is essential to synthesize a Bessel-type low pass filter tailored for the specific modulator frequency response so as to optimize the cascaded spectral shaping; on the other, possible signal reflection at the RF interconnections should be eliminated to reduce the noise of the optical signal. Accurate filter modeling and integration within the modulator package, which reduces the number of RF interconnections, was proved to be effective in obtaining a high optical signal quality.

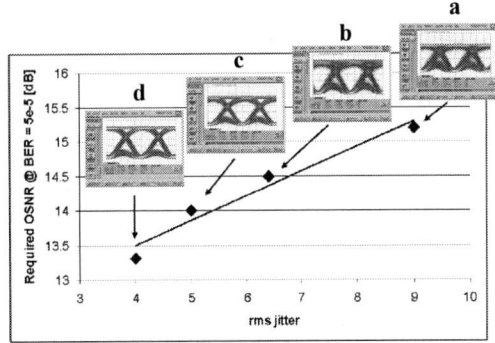

Figure 6: Eye diagram and OSNR performances of duobinary modulator – (a) external filter, (b) and (c) integrated filter, (d) integrated optimized filter

In fact, by optimizing filter design and RF interconnections, we were able to reduce the jitter of the optical eye diagram and improve the transmission performances (i.e. the required OSNR).

A standard method to realize the transition between the connector and chip inside the package is to use an Al2O3 or AlN substrate, with coplanar or microstrip structure, interconnected by soldering to the pin connector and wire bonding to the chip. Standard thickness used for these kinds of substrates is 0.508mm (20mils) and 0.254mm (10mils). Smaller thicknesses are commercially available but they are normally avoided in order to reduce the risk related to fragile parts inside the package.

In the Avanex duobinary modulator, the filter is integrated inside the package and it is designed with tight dimensional specification in order to be inserted as a normal transition in the standard Small

Form Factor modulator by using the room dedicated to the standard transition substrate. The possibility to use the same package for duobinary and non-duobinary applications means no additional cost and volumes scalability advantages for the package, which are the most expensive parts of the modulator BOM.

The filter layout dimension has been a design and process challenge. The filter is realized in distributed parameters with elements like stubs and resistors that make it possible only by using a microstrip structure. With the microstrip configuration, as the dimensions of the circuit strip are proportional to the material thickness, with 0.254mm thickness substrate the filter dimension is not compatible to the room available inside the standard package.

The design and process strategy chosen to solve this issue has been to reduce the material thickness and to focus on solving the mentioned problem related to the fragility of a thin substrate.

The main issue related to the decreased thickness is the CTE (Coefficient of Thermal Expansion) mismatch between the different elements involved (package, substrate, soldering and attaching material) that could induce mechanical stresses on the substrate during thermal variation.

To reduce this kind of effect an intermediate metal layer between package (Stainless Steel CTE=17ppm/°C) and substrate (Alumina CTE=7ppm/°C) is introduced. This metal layer, chosen with a CTE very close to Alumina one, adsorbs all the expansion coming from package without effect on the substrate.

This solution, combined with the right attach materials, gives high reliability level (the same of the standard substrate) and it works as a dimensional adapter to the standard package (designed for thicker substrate). In this way we have obtained the mentioned advantage of volume scalability for the most expensive BOM part.

The same approach has been extended to 40Gb/s modulation, where we propose a duobinary modulator version with low driving voltage and integrated Bessel filter. Very good eye diagram performances are obtained when the module is coupled to commercial drivers, both at back-to-back and after fiber propagation

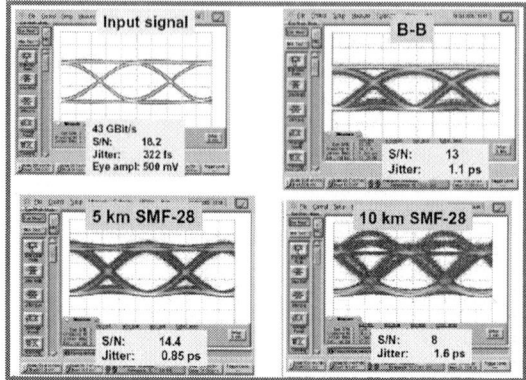

Figure 7: 40Gb/s duobinary eye diagram results

40Gb/s modulation format: DQPSK application

Among the most promising modulation formats for 40Gb/s transmission, DQPSK plays a significant role. This format combines a high spectral efficiency, which allows employing already installed 10 Gb/s infrastructure, with high tolerance to chromatic dispersion and, particularly, PMD [3]. [4]. [5].

The generation of this modulation format is achieved through a dual parallel modulator, where two high bandwidth Mach-Zehnder modulators are embedded in an outer MZI.

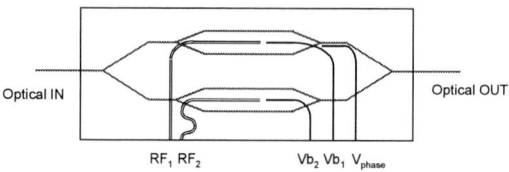

Figure 8: Modulator scheme for DQPSK modulation

Given the high level of integration at chip level, challenges arise in relation to routing all the relevant signals to the $LiNbO_3$ chip. The high frequency signal streams must be routed through large bandwidth RF lines avoiding potential crosstalk that would be detrimental to the transmission quality. Also, differential delays between the two MZI RF paths must be avoided to allow for synchronous operation of I and Q streams.

From the packaging point of view the mentioned challenge has been addressed by designing a double RF transition, with extremely tight mechanical tolerances, from high frequency SMPM connector to the RF line on the chip. A new interconnection process has been developed and qualified in order to meet the required RF performances.

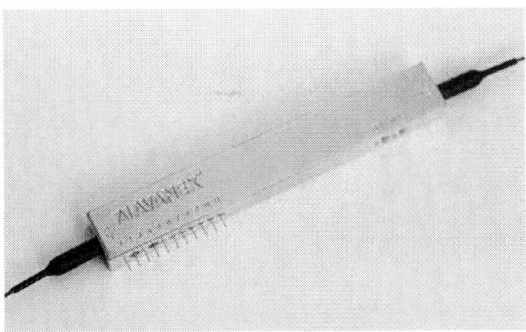

Figure 9: 40Gb/s DQPSK modulator

A typical DQPSK eye diagram obtained with our integrated dual parallel modulator is shown below.

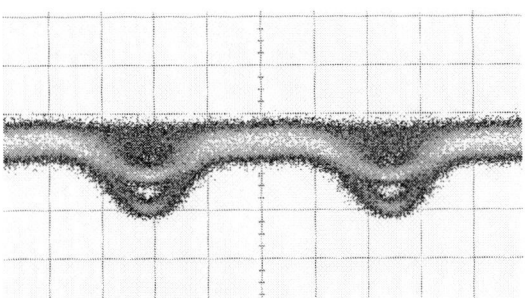

Figure 10: DQPSK modulation eye diagram

Conclusions

The LiNbO3 modulator is playing a key role in the transmission chain. To compete with other emerging technologies in the last years a strong effort is being spent to continuosly improve the performance, cost and form factor. The new 40Gb/s application demand is opening new technology challanges and market opportunities.

In this paper we have showed some examples of successfull developments of the Avanex modulators.

Acknowledgements

The authors would like to thank all the Product Development team of Avanex in Italy.

References

[1] F. Schiattone et al., "Low cost, small form factor, and integration as the key features for the optical component industry takeoff", Proceedings of the 2002 Integrated Optical Devices: Fabrication and Testing Conference, Brugge, Belgium, October 30, Proceedings of the SPIE, Volume 4944, pp. 227-241 (2003).

[2] P. Bravetti et al., "Impact of response flatness on duobinary transmission performance: An optimized transmitter with improved sensitivity," IEEE Photon. Technol. Lett., vol. 16, no. 9, pp. 2159-2161, Sep. 2004

[3] G. Kramer, A. Ashikhmin, A. J. V. Wijngaarden, and X. Wei, "Spectral efficiency of coded phase-shift keying for fiber-optic communication", J. Lightw. Technol., vol. 21, no. 10, pp. 2438–2445, Oct. 2003.

[4] R. A. Griffin and A. C. Carter, "Optical differential quadrature phase shift key (DQPSK) for high-capacity optical transmission", presented at the Optical Fiber Commun. Conf. (OFC), Anaheim, CA, 2002, Paper WX6.

[5] J. Wang and J. M. Kahn, "Impact of chromatic and polarization-mode dispersions on DPSK systems using interferometric demodulation and direct detection", J. Lightw. Technol., vol. 22, no. 2, pp. 362–371, Feb. 2004.

Packaging and wired interconnections for insertion of miniaturized chips in smart fabrics

Jean Brun, Dominique Vicard, Bruno Mourey, Benoît Lépine, François Frassati
CEA-LETI, MINATEC, F-38053 Grenoble, France
Phone, 33-4 38 78 54 48 Fax , 33-4 38 78 50 12 E-mail Address   Jean.brun@cea.fr

Abstract

Electronic devices are currently used in a complete system including mobile phone, television etc... Now, more and more applications need functionality as close as possible to the final system for in situ measurement and control (smart textile, automotive or medical applications) or for security. The components concerned are sensors, light emitting diodes or RFID for example. Concerning microsystem, being integrated in a material means several key properties, including a packaging form adapted to the material manufacturing process, autonomy in terms of energy means for being powered, neutrality versus the initial material properties and resilience, as materials usually get post-processed once manufactured. A particularly attractive support for such integration is the textile way, since textile or textile-like materials are often part of composite materials and are extremely common in the design of complex objects.

The "Diabolo" process aims at a direct connection from a chip assembly to external wires without using the traditional bonding / packaging stage. Through a very limited set of wafer scale operations, one or several chip dies can be assembled and connected to conductive wires directly from the chip surface. The result of a fully processed Diabolo assembly is a spool of chips connected to a flexible wire that can be used for incorporation into materials through taping, weaving, knitting, extrusion or more generally inclusion in a liquid phase before curing. The process is still in its early stage of development but our current studies make us confident that assembly throughput as high as 10k UPH (Unit per Hour) can be reached with a relatively simple equipment.

Introduction

One of the key issues for inclusion of microelectronic devices in materials and particularly in fabrics is the connection problem

In a regular microelectronic device, the chip itself (the die) is connected to the external world through a package including a grid of external pins (which are designed for being soldered on a PCB (Printed Circuit Board) for example. This external grid is itself linked to the die by thin bonding wires (as thin as 15μ in diameter) linking the chip pads to the grid connectors. Some assembly techniques are connecting directly the die to the PCB using bumps (flip-chip), the active side of the chip being directly in contact with the PCB and the global assembly being generally protected by a polymer cover (glop top).

1- Packaging Functions

The packaging of the chip has for main function protecting the chip from external harsh conditions (humidity, chemical aggression, shocks, ESD). Such a protection requires a certain volume of insulating material being wrapped around the die. The final volume of the packaging makes it usable by fast pick and place machines even in a dirty environment.

On the other hand, the packaging size is usually much more important than the size of the die itself, and one can reach the paradox of having an extremely small electronic die encapsulated in a quite bulky packaging (proportionally). Currently, the field of application requiring the smallest packages is the field of RFID (Radio Frequency Identification), where the chip is usually connected to a RF antenna using a flip-chip technique before being embedded in a polymer material (usually through a lamination)

2- Diabolo: Direct Die to Wire Connection

The "Diabolo" process described in this paper aims at connecting directly a die to external connection wires without going through a classical regular package. The die itself is protected by a cover glued on the chips using a wafer-scale process.

978-1-4244-4722-0/09 $25.00
© 2009 IMAPS-ITALY

The connections are established at the edge of the chip, and the cover insures the function of mechanical stability. The result is a string of chips mechanically and electrically connected to a set of wires, suited for further roll-to-roll processing and which can be used in industrial processes like weaving or extrusion.

3- The "Diabolo" Process

The Diabolo process comprises 10 stages at the wafer level operating currently in a 200mm environment. In the following paragraphs, the top wafer will be identified as the Cover Wafer, and the bottom wafer will be identified as the Active Wafer. The Cover Wafer can either be a wafer including electronic devices (like MEMs or regular electronics) or simply a mechanical cover: The Active Wafer is usually a wafer of active chips (either MEMS or regular electronics).

Process Steps

Cover wafer manufacturing

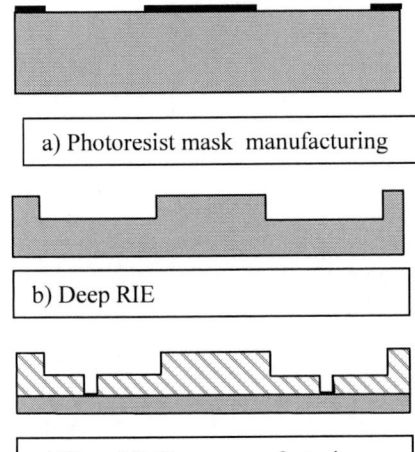

a) Photoresist mask manufacturing

b) Deep RIE

c) X an d Y Groves manfacturing

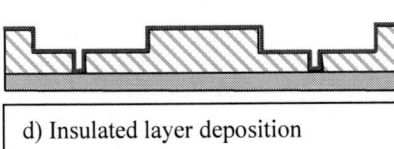

d) Insulated layer deposition

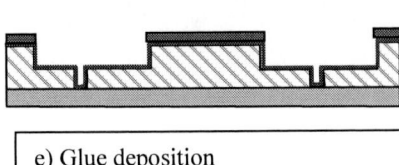

e) Glue deposition

Active wafer preparation

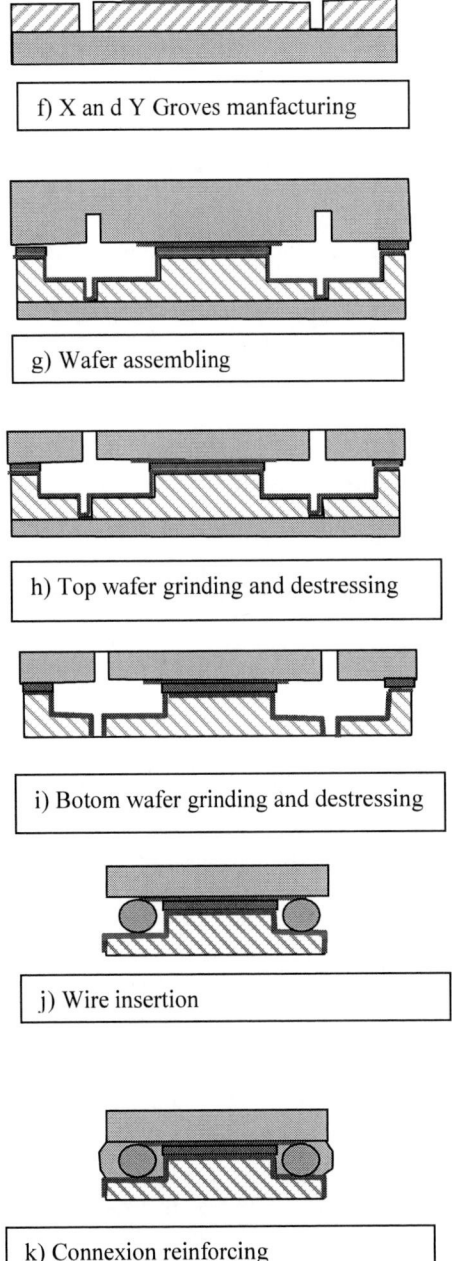

f) X an d Y Groves manfacturing

g) Wafer assembling

h) Top wafer grinding and destressing

i) Botom wafer grinding and destressing

j) Wire insertion

k) Connexion reinforcing

Figure 1 : Diabolo process Steps

The first step consists in manufacturing the main grooves in Si substrate. The depth should be

978-1-4244-4722-0/09 $25.00
© 2009 IMAPS-ITALY

related to the diameter of the future wire used in step j. It has to be noticed that the cover could be constituted by several different materials. For silicon, deep RIE is suitable.

The second step consists in pre-slicing the Cover Wafer by dicing stage. The depth of the pre-slicing etch is related to the final thickness of the assembly and to the thickness of the part removed by the thinning operation in step h and i. In case of using Polymer, one step is necessary to obtain d). Step f consists on preslicing the active wafer

Step g : aligns the Active Wafer and the Cover Wafer and performs the wafer bonding. In step h, the Cover Wafer is thinned from the top up to the pre-slicing slice etch, freeing up dices of the Cover Wafer. Diabolo dies are completely diced at step I by a second step of grinding. It has to be noticed that such process is likely inducing stess in the assembly. Due to the fact that the further stage j will consist in introducing a wire by mechanical insertion a stress release is mandatory to avoid damages. This stage is performed by plasma etching after h and i step.

4- Connection Characteristics

Mechanical: the initial mechanical cohesion of the wire is done by the snapping effect of the thinned Cover and Active wafer. The diameter of the wire has to be slightly over the height of the snap and the force generated by the strain of the top and bottom walls keeps the wire in place. Further mechanical strengthening can be obtained by a glue spot or an electrolytic contact.

Electrical: the electrical contact between the Active Wafer and the wire is intrinsically a dry contact. The use of stud bumps on the Active Wafer eases the establishment of this contact as well as generating a mechanical contact strain. Further electrical contact can be established by the use of conductive glue or by an electrolytic treatment.

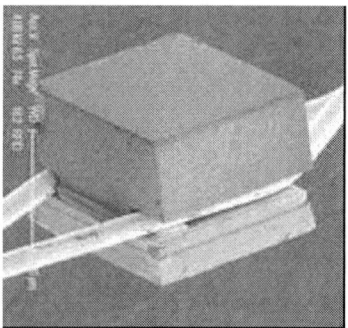

Figure 2 : SEM views of a Diabolo Assembly f an electrolytic contact (right)

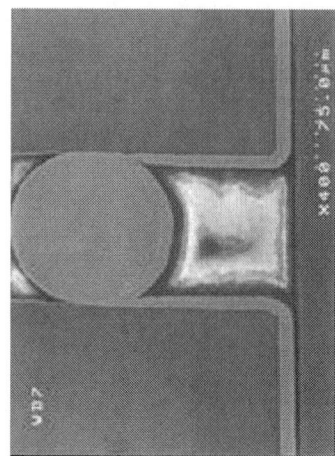

Figure 3 : Cross section of a 100µm wire embedded in its groove by electroplating layer

5 -Application Examples

Two demonstrative types of prototypes have been set-up with the Diabolo process.

5-1 LED Textile Assembly

The first one is the inclusion of a 0402 CMS LED assembled on two conductive wires by using a Diabolo glass Cover Wafer.

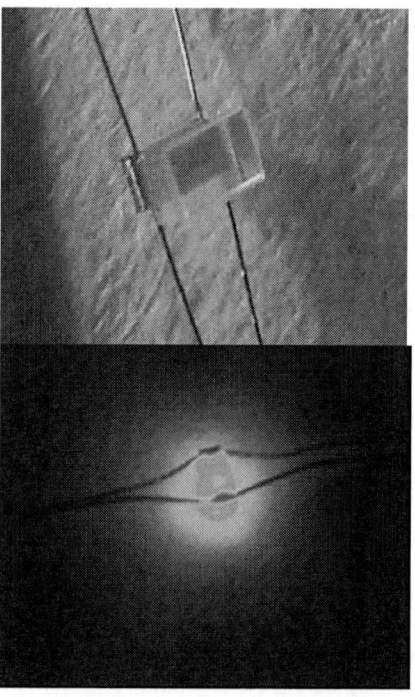

Figure 4: LED Diabolo Assembly on free wires

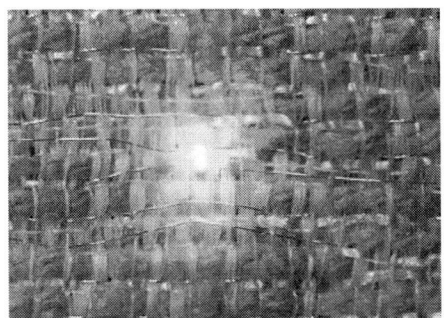

Figure 5 : LED Diabolo Assembled on pre-weaved wires

The fabric has been realized by HTH (Holding Textile Hermes) trough the METIS association. Silver wires have been weaved as chain wires using an 800µ spacing inside a viscose fabric. Diabolo LED assemblies have been inserted between two adjacent silver wires and those wires have been connected to a power supply at the edge of the fabric. The results are shown in figure 4.

Typical applications of such a LED inclusion in textile include safety garments, home decoration textiles and may be included in the elaboration of composite materials.

Step j represents the diabolo die when wires are inserted. Of course a minimum strain is necessary to introduce the wire in the groove in order to obtain a close contact with the pad (see figure 2).

RFID Assembly

The second one uses a UHF RFID chip for which the wires constitute the RF antenna.

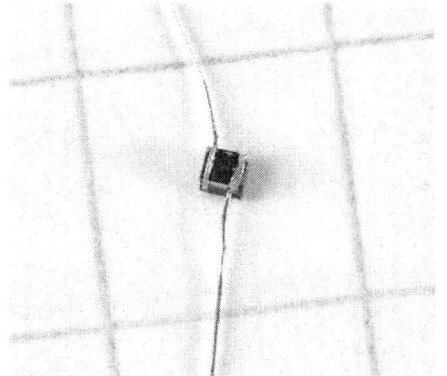

Figure 6 : UHF RFID – each square on the paper support is 5mmx5mm

The chip used is a ST Microelectronics XRAG2 chip. Its shape is roughly a square of 700µmx650µmx750µm. The contacts for the

antenna have been covered by a set of hard bump (Au-Ni). The Cover Wafer used is a 550µm thick glass wafer (which explains the transparency of the assembly on figure 5). The chip operates in the UHF range from 860MHz to 960MHz and is compliant with the EPC Global Class 1 Generation 2 RFID UHF specification (revision 1.0.9).

The antenna is a dipole antenna made of a set of Copper wires covered with Silver with a diameter of 100µ with a total length of about 15cm for air applications as shown on figure 5.

Performance tests show that the RF range complies with the specification of the chip, and certain tuned assemblies have been successfully included in various materials including polymer while remaining fully functional.

Applications of such Diabolo RFID assemblies include material identification and traceability, and all the current applications of UHF RFID tags, while providing an extremely small global size (chip + antenna).

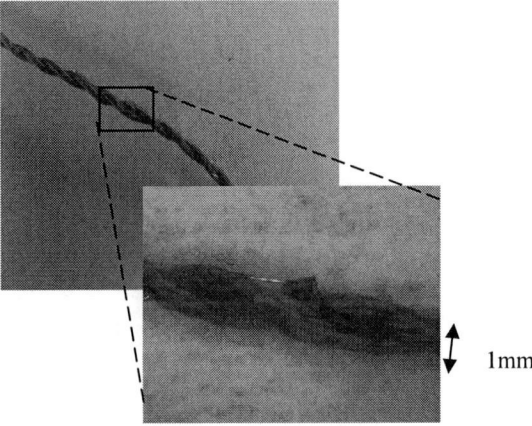

Figure 7 : 700µmx650µmx750µm RFID + antenna incorporated in cotton yarn

5-2 RFID integration

Diabolo die can be integrated to the final devices by different way. For example, it could be directly inserted in metallic thread knitted in the textile as we can see on figure 5. But we are also working on the embedment of the Diabolo RFID and its antenna in a cotton yarn (figure 7).

This yarn is likely compatible with a weaving loom but we don't know yet if it suits knitting system. Now, smaller Diabolo RFID dies are in progress to manufacture a yarn as small as possible.

6-Future Works

Work in progress at CEA includes the design and realization of an automated Diabolo wiring equipment, as well as the design of several dedicated electronic chips able to take advantage of the 2 wires (or more) Diabolo connection.

6-1Fast Assembly Equipment

CEA-LETI is currently designing with industrial partners a prototype for an automated Diabolo insertion machine. Such prototype equipment is a first step toward an industrial equipment able to insert more than 10000 Diabolo units per hour on metallic wires.

6-2Multi-wire system

Diabolo applications are huge and some of then require more than two wires. Indeed manufacturing a compact system including micro-systems, processor, energie, passives devices etc ... could be a technical issue. Connecting several serial different devices permits to overcome these difficulties. For this reason, we are studying a specific connection design allowing several connections (figure 8). The solution consists on making a toron of wires. A set of pads permits to connect sequentialy each pad with a specifique wire (figure 9).

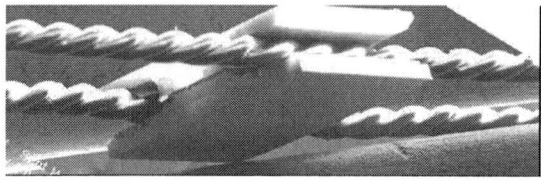

Figure 8 : Multi–wires connexions concept

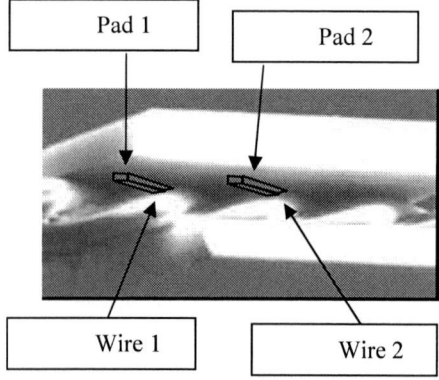

Figure 9 : Multi–wires connexions concept

Conclusion

For years, LETI has been successfully developing an original attachment system named "Diabolo" allowing an electrical connection and a strong mechanical assembly. The "Diabolo" process aims at a direct connection from a chip assembly to external wires without using the traditional bonding / packaging stage. Through a very limited set of wafer scale operations, one or several chip dies can be assembled and connected to conductive wires directly from the chip surface. Several applications have been demonstrated with RFID and LED components. Test vehicles have been already perfomed in yarns or textiles. Very promising results were obtained. A prototype for an automated Diabolo insertion machine is in progress.

Acknowledgements

This work has been sponsored by a grant of GRAVIT, a consortium of 7 institutions of the Rhône-Alpes region (www.gravit-innovation.org). Textile samples have been elaborated through the METIS association, a consortium of textile, film and paper firms located in the Rhône-Alpes region. This work is protected by several patents (issued or pending).

Aluminum Ribbon on a Power Device

Giuseppe Cristaldi, Giuseppe Malgioglio, Emanuele Scrofani
ST Microelectronics
Packaging Engineering and Development – Industrial and Multi-Segment Sector
Stradale Primosole, 50 – 95121 Catania (Italy)
Phone: +39 095 7407347 / +39 095 7407296 - E-mail: giuseppe.cristaldi@st.com, emanuele.scrofani@st.com

Abstract

In the semiconductor industry to increase the power density, improve the electrical performances and optimize the robustness in the application, more and more key roles are covered by back-end processes and in partcular by bonding technology used to connect power silicon chip and metallic leadframe that for Power devices is one of more impacting process. The performed studies, evaluations and production results show that currently the ribbon process can be considered the best compromize between multi-wires approach and clip bonding technology, normally used as interconnection methods to minimize the resistance contribution, improving the electrical performances, minimizing the global stress on the silicon well improving quality. The ribbon bonding allows to have connections resistance better than multi-wires bonding and values close to clip process, also obtaining a more robust process for power device packaging.

Keywords: power density, quality, robustness, competitiveness

Introduction

In the market of Power devices one key electrical parameter is the conductivity performance. Here following the schematic of interconnections adopted to connect the chip pads to the pin/lead of the pkg that are in a power package one of the main factor which is contributing to the global resistance:

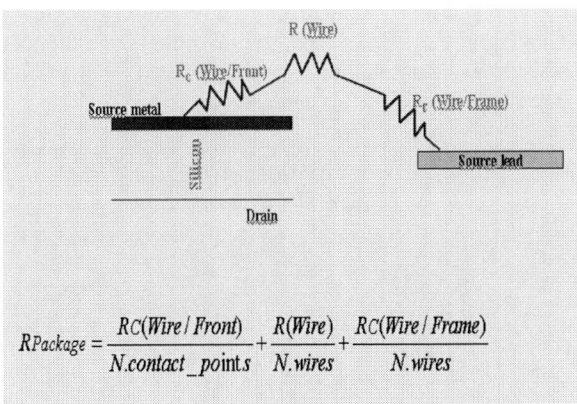

$$RPackage = \frac{RC(Wire/Front)}{N.contact\_points} + \frac{R(Wire)}{N.wires} + \frac{RC(Wire/Frame)}{N.wires}$$

In this regard, the electrical resistance given by the interconnections (RPackage) is contributing to get a lower or higher global Resistance parameter depending of connection resistance and contacts resistance (on chip and on lead). In general, a lower interconnection resistence will assure a better and lower global Resistance value minimizing the electrical power dissipation (for istance, for a Power MOSFET technology with 30V of breakdown assembled in TO220 or DPAK package the interconnection can impact on about 30% of global

RDSon parameter that is the resistance when Power MOSFET is ON).

Currently, the main semiconductor manufacturers are using to bond multiple wires (Au or Al) to reduce the interconnection resistance (fig. 1 and 2). More and more a lot of semiconductor industries are applying in pkgs like TO-220, D2pak, SO families, up to 3 or more heavy Al wires or up to 7 or more thin Au or Cu wires.

Increasing the number of wires in the front metal, will help to reduce the resistance of the interconnection. In fact, considering that:

$I=V/R$ and than $R= \rho * L / S$ where:
I = current (A)
V= tension (V)
R = interconnection resistance (Ω)
ρ = wire or interconnection material specific resistivity
L = wire or interconnection length
S = wire or interconnection section

Mantaining fixed all the others variables, to get an heavy R reduction, it is necessary to increase the number of the interconnections/wire or to increase the interconnection/wire material section. In the R parameter, the ratio L/S must be mantained smaller as much as possible.

If the multi-wire bonding is considered as method to reduce the R factor, it is easy to understand that such solution is not the best one considering both quality and productivity aspects.

978-1-4244-4722-0/09 $25.00
© 2009 IMAPS-ITALY

Infact, no in the assembly steps neither at testing level there are effective and robust methods to check the bonding process integrity and quality.

In addition, in the multi-wire methods a big limitation is given by the bonding equipment UPH (Unit Per Hour) due to the productivity reduction. Infact, bonding multiple wires, the productivity of the equipment can drop down till 30 or 40%.

Considering the multi-wire process negative aspects, today, a valid alternative to it could be the ribbon. Thanks to a larger contact area and a larger material section, it is able to easily substitute the multi-wires technology. In addition, as during the bonding operation a lower bonding "connectors" are handled, it is possible to minimize the main bonding accidents generally encountered by multi-wire process.

Fig. 1   Fig. 2

The following activity describes the usage of the ribbon process as sobstitute of the multi-wire one considering different aspects. Here following the main topics will be treated in this study:

1. Interconnection resistance

2. Mechanical stress on silicon

3. Impact on product robustness

4. Manufacturability and Productivity

1. Interconnection resistance

To see the "pure" impact of the electrical resistance introduced by the interconnection materials, different bonding tests were finalized; as test vehicles the TO-220 leadframe and a Pmos device were used [1-3].

The first test was to bond direct on the leadframe, without silicon, a certain number of Al wires having a diameter of 15 mils and Al ribbons of 80x10 and 60x8 mils sizes.

Looking at the tables 1 and 2 it is possible to see the pure resistance values (by the RDSon parameter) introduced by the interconnection materials only:

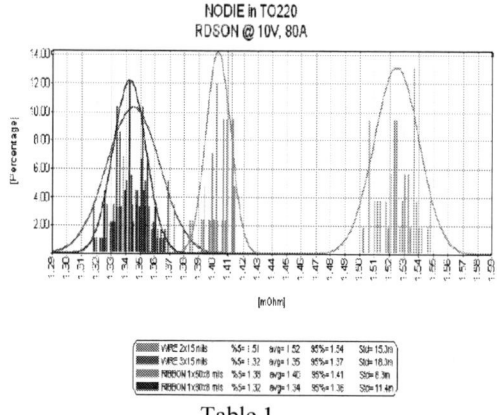

Table 1

Tesec measurement with socket

Ribbon	Rontyp @10V, 80A	Wire 4x15mils d.s. Rontyp @10V, 80A	
	[mOhm]	[mOhm]	Delta [mOhm]
1x80x10 mils cont	2.41		-0.17
2x60x8 mils cont	2.53	2.58	-0.05
60x8+80x8 mils cont	2.44		-0.14

Table 2

Using the same interconnection materials and leadframe, the second test was to bond a Power MOSFET device.

Looking at the tables 3 and 4 is possible to see the total resistance contribution introduced by both interconnection materials and silicon:

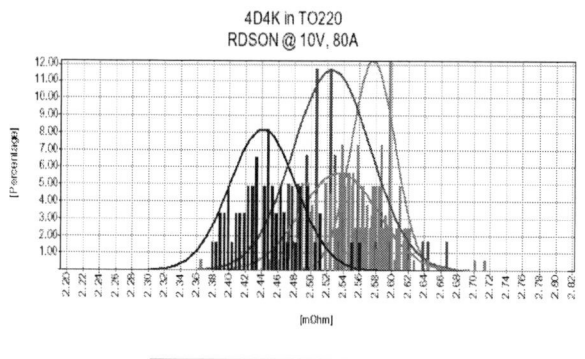

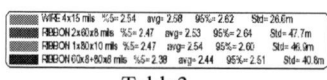

Table3

	Ribbon	Wire 2x15mils d.s. Rontyp @10V,80A		Wire 3x15mils d.s. Rontyp @10V,80A		Wire 4x15mils d.s. Rontyp @10V,80A	
Ribbon	[mOhm]	[mOhm]	Delta [mOhm]	[mOhm]	Delta [mOhm]	[mOhm]	Delta [mOhm]
1x60x8 mils cont	1.4		-0.12		0.05		0.14
1x80x8 mils cont	1.34		-0.18		-0.01		0.08
1x80x10 mils cont	1.19	1.52	-0.34	1.35	-0.17	1.26	-0.08

Table 4

To complete this exercise, an additional test was done comparing the ribbon vs. thin wires. It was realized using as test vehicle the SO8 pkg with a Power Mosfet device. The table 5 is showing the differences of the two interconnction materials.

	Ribbon 40x4 mils (lot# Y504531)		7 Cu wires 2 mils (lot # Y524880)	
	4.5A/ 4.5V	4.5A/ 10V	4.5A/ 4.5V	4.5A/ 10V
95%	10.67	9.58	15.71	13.04
AVG	10.13	9.16	14.8	12.58
Delta				
95%			5.04	3.46
AVG			4.67	3.42

Table 5

TO-220 Ribbon

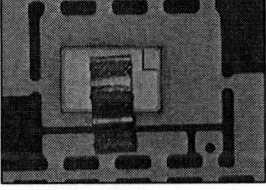
SO8 Ribbon

Many different tests were performed on all the main power packages to eliminate the most complex multi-wire configuration confirming interesting results.

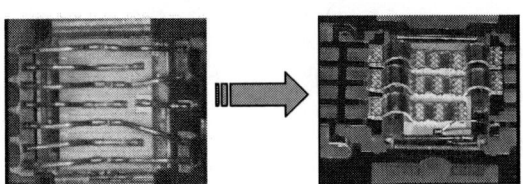
Power SO-10
Multi-Wires vs. Ribbon

Looking at the results obtained, the ribbon is a very valid technology to substitute the multi wire bonding process.

In terms of electrical performances, we can have a consistance RDSon reduction. Comparing the gain for different configurations we've found a gain that can arrive to 30%:

-1x80x10 Al ribbon vs. 2x15 mils Al wires: up to -22%

-1x80x10 Al ribbon vs. 3x15 mils Al wires: up to -12%

-1x40x4 Al ribbon vs. 7x2 mils Cu wires: up to -31%

If we consider the Cu Clip bonding, normally used as bonding alternative to multi-wires to reduce the connection resistance we generally found a mean RDSon reduction of about 25-30%:

Cu Clip vs. Multi-wires Comparison on a Power MOSFET in SO8

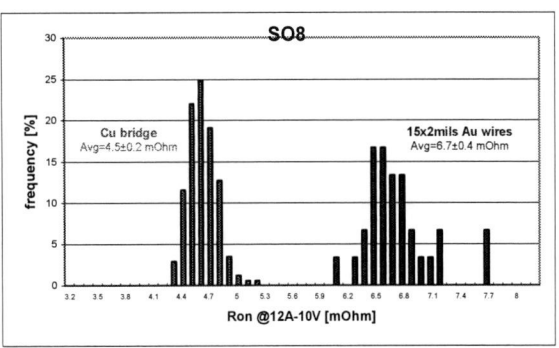

Cu Clip vs. Multi-wires Comparison on a Power MOSFET in DPAK

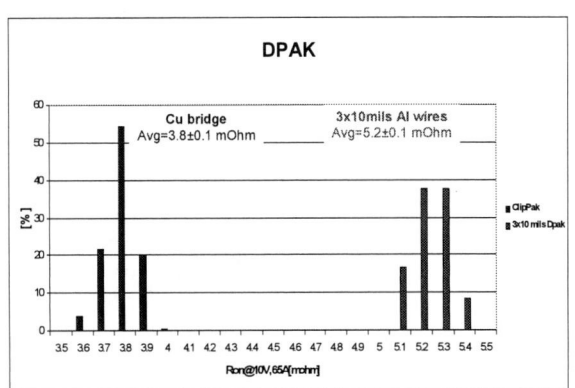

978-1-4244-4722-0/09 $25.00
© 2009 IMAPS-ITALY

So, in conclusion the gain of Ribbon in terms of resistance performance are very much interesting vs. multi-wires and close to gain obteined with Clip interconnection technology that is a more complex manufacturing process, with higher defectiviness and cost.

2. Mechanical stress on silicon

Taking into account that during the bonding operation the silicon is submitted to the mechanical stresses, it could be demonstrated that the stress applied at silicon surface by ribbon is comparable or even less than wires. Therefore well considering that the failure probability is linked to the number of bonding events, the using of one ribbon intead of multi wires (2,3 or more) improve significatively the process robustness as well proved by testing yield.

To make "visible" the mechanical stress induced by both bonding technologies (wire & ribbon) at silicon level, mechanical FEM simulations were finalized (by ANSYS-Wokbench). [4]

The package taken into account was the TO220-FP single gauge. At die attach level it was considered a 30 μm soft solder (Pb/Sn/Ag) thickness and at silicon level following features were considered:

- Die size: 6.35 x 4.57 mm
- Die thickness: 275.5 (bulk) + 4.5 μm (metal layer)
- Aluminum wire diameter: 15 mils
- Aluminum ribbon size: 80 x 10 mils

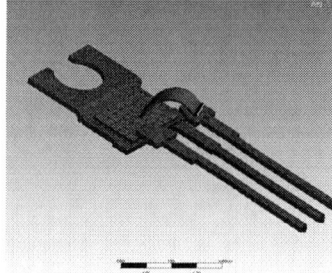

TO220, FEM model with ribbon

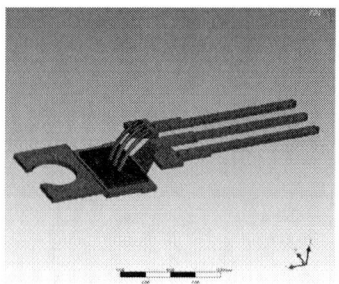

TO220, FEM model with wires

Moreover, a zoomed case was taken into account in wich it was studied the mechanical stress on the strips of polysilicon and silicon dioxide under the metal layer. For this case the following feature were taken into account:

- Die zoom size: $2.5*10^{-3}$ mm^2
- Polysilicon size: 1.600 x 0.300 μm
- Silicon dioxide (thermal) size: 1.600 x 0.047 μm
- Silicon dioxide (film deposited) size: 1.600 x 0.700 μm

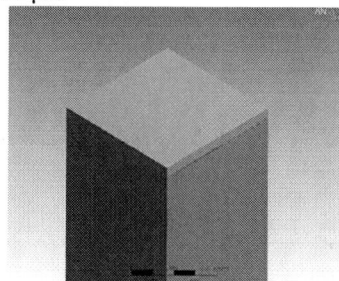

Zoom on silicon little share

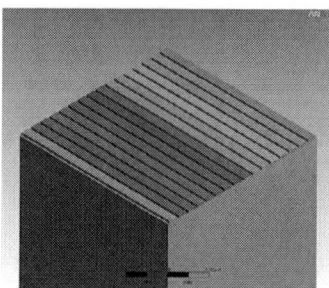

Zoom without metal layer

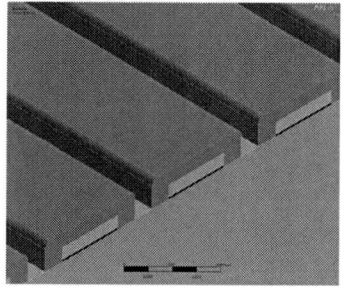

Zoom on poly (yellow)
and oxides (black and green) strips

Regarding FEM models, following materials properties were taken into account:

Material	Density (Kg/mm³) *10⁻⁶	Young's module (GPa)	Poisson's ratio
Silicon	2.34	183	0.22
Copper	8.90	137	0.37
Aluminum	2.71	70.0	0.33
Solder	11.2	20.0	0.35
Polysilicon	2.33	172	0.22
SiO₂ thermal	2.21	70.0	0.17
SiO₂ film	2.20	60.0	0.17

Next table instead resumes all the boundary conditions we referred for the FEM analysis:

Case	Force/surface Kg$_f$/mm²
Ribbon	1.300
Wires	1.790

FEM analysis boundary conditions

According to above boundary conditions and material involved properties, following results come from the analysis:

Max stress On	Bonding type	Max equivalent Stress (σ) Kg/mm²
Silicon	Ribbon	1.02
	Wires	1.45
Poly	Ribbon	2.36
	Wires	3.25

The next pictures show the stress contour plot for all the cases.

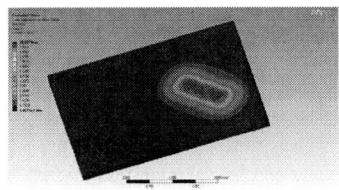

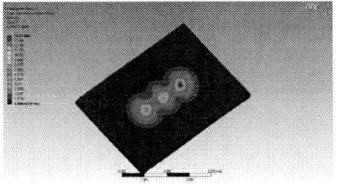

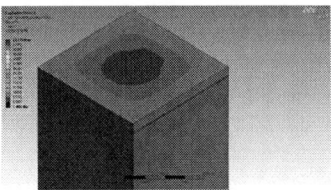

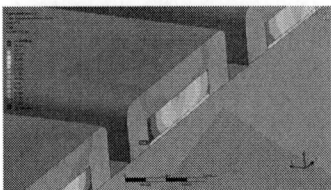

Conclusions of FEM analysis and simulations are that in any case when we use ribbon the max equivalent stress value is smaller than when we use wires. Both at package/silicon bulk level and poly/oxides substructures level we confirm this behaviour.

To pratically validate the simulation results, focused trials were performed using a particular sensitive silicon technology. It was bonded using both wires and ribbon (15 mils and 80x10 mils).

To be sure that the bonding parameters to be used for the above trial (wires and ribbon), are the rigth ones, a dedicated debuging was performed using normal (planar) silicon lines. The final parts coming from this debuging exercise were electrically tested, and the testing yields found were above 98% for both bonding tecnologies.

After that, the "sensitive" (trench) silicon was bonded using ribbon and wire, after the parts were submitted at electrical testing. The yields encounterd were 93.4% for the parts bonded with wires while 98.1% for the parts bonded with ribbon. The parts coming from ribbon are showing 4.7% less of electrical rejects.

Here following the main details of comparison trials:
- Bonding parameters
Wire   : SF450-EF500-SP80-EP90-RT30-TT150
Ribbon: SF2100-EF2200-SP180-EP190-RT50-TT150
- Materials propeties
Wire   : Al 99.99 - EL 10-30% TS 550-750g
Ribbon : Al 99.99 - EL 34% TS 1580g
-Silicon details
Thickness   : 220um
Front metal : Al 99.99 – 4.5um Thick
- Equipment type
M360C for wires - M360R for ribbon

So, in conclusion as with Ribbon the global number of bonding events is lower than multi-wires, the global impact on the silicon is well reduced with consequent improvement of testing yield, process robustness and global quality.

3. Impact on product robustness

Comparing the robustness of ribbon bonding configuration vs. wires it was found also an interesting improvement in terms of robustness vs. UIS (Unclamped Inductive Switching) a very important dynamic test performed on Power MOSFETs to verify the robustness during switching off (when the parassatic inductances of boards create an extravoltage that can destroy the device that is working at maximum voltage and current condition).[5]

It was compared in the Laboratory the same silicon power mosfet device assembled with one gate Au wire for gate and two different bonding configurations for source :

- 1 ribbon 50 x 8 mils section dimensions
- 6 wires Au of 2 mils diameter

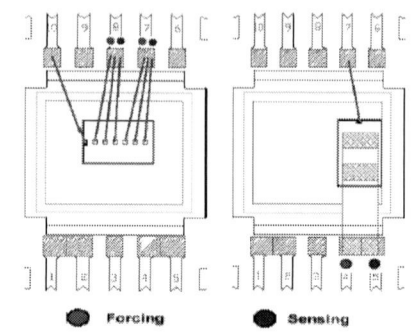

Forcing   Sensing

Unclamped Inductive Switching Test

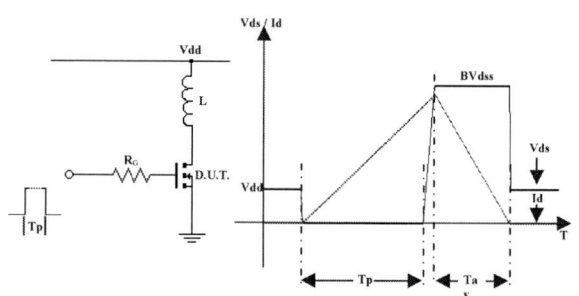

T=25°C	Ribbon + 1x1.5Au			T=150°C	Ribbon 1x1.5Au		
pz n°	Iout	V (unclamped)	Energy	pz n°	Iout	V (unclamped)	Energy
	A	V	J		A	V	J
1	43.83	39.52	1.43	1	33.96	40.06	0.84
2	45.58	39.53	1.53	2	34.88	38.62	0.9
3	45.09	39.17	1.52	3	34.22	40.23	0.85
4	44.3	39.33	1.45	4	34.12	39.92	0.86
5	44.41	38.93	1.47	5	33.73	39.93	0.83
6	44.85	39.37	1.49	6	34.53	39.16	0.88
7	44.45	38.98	1.47	7	34.34	39.02	0.87
8	43.87	39.56	1.43	8	35.05	38.41	0.91
9	44.08	39.86	1.43	9	34.91	38.79	0.9
10	44.99	38.75	1.5	10	34.09	40.13	0.85
Vbatt=13V , Vin =10V , Rg=1K , L=1mH							

T=25°C	6x2mils Au + 1x1.5 Au			T=150°C	6x2mils Au + 1x1.5 Au		
pz n°	Iout	V (unclamped)	Energy	pz n°	Iout	V (unclamped)	Energy
	A	V	J		A	V	J
1	37.66	39.64	1.05	1	29.47	38.87	0.55
2	37.49	40.59	1.04	2	29.63	38.7	0.56
3	38.59	39.25	1.12	3	29.21	39.04	0.63
4	37.06	39.68	1.02	4	29.64	40.55	0.54
5	38.59	39.12	1.12	5	29.66	39.59	0.54
6	38.21	39.3	1.09	6	29.22	39.34	0.52
7	38.35	39.43	1.1	7	29.18	40.29	0.52
8	38.49	39.9	1.1	8	30.34	39.33	0.56
9	38.15	39.69	1.08	9	29.96	38.71	0.57
10	38.18	39.56	1.08	10	30.07	38.63	0.56
Vbatt=13V , Vin =10V , Rg=1K , L=1mH							

It was found about 40% increase of sustainable energy and about 15% increase of maximum sustainable current.

This positive results are due to the reduction of connections inductances introduced from ribbon vs. wires. The Ribbon can be infact considered as a parallel of infinite wires with a significant reduction of global Inductance value.

4. Manufacturability and Productivity

At Front End the Ribbon bonding technology doesn't need any additional masks vs. standard manufacturing flow for wire bonding interconnection technology. It's a significant advantage vs. other alternative to multi-wires bonding like the clip technology that needs the wet-able metal with at least two additional masks and relevant increase of cycle time, defectiveness and cost.

At Back End level Ribbon bonding technology allows to reduce/avoid [3, 6]:

-multi-wires interconnection issues (clamp, bond off, weld off, cratering, ...)

-complex process and defects introduced by clip process (positioning, solder reflow, cleaning, …)

The assembly and testing yield show improvements vs. standard bonding technologies proving what well predicted by mechanical simulations.

The ribbon is a simple and robust bonding process for power packages. It can be realized using the same flexible equipments already used for Al wire bonding by simple adaptation.

It's also a flexible technology able to be implemented either in traditional bigger power packages (like TO220, TO220FP, D2PAK, DPAK, TO247, PoweSO-10, ..) either in smaller power packages (like SO8, PQFN, ..) allowing the easy extention of seen benefits to all the power packages.

In terms of uph (units per hour) the Ribbon has shown an interesting increasing of productivity compared to the current multiwire process:

2x15 mils double sticth wires	uph = 917
1x60x8 mils ribbon	uph = 1410
4x15 mils double.stitch wires	uph = 663
1x80x10 mils cont. ribbon	uph = 1350

Conclusion

Considering the more and more stringent requirements of Power Devices in terms of electrical performances to well improve the power density by global conductivity increase, in terms of robustness in the customers application to increase the maximum allowable energy, in terms of competitiveness to improve productivity and yield, in terms of process robustness to reduce the global stress on the silicon in particular the bonding one that is impacting very much, the Ribbon bonding technology is a simply, reliable, effective and extendable solution representing a significant step versus power packaging improvement and evolution.

Acknowledgements

The Autors wish to thank all the STMicroelectronics collegues of Divisions Product Engineering and B/E Plants Engineering have been supported the many evaluations, workabilities and qualifications activities.

References

[1] Fabio Zara, Nicola Villari (Catania STMicroelectronics), " Ribbon vs. Wires contribution resistance" internal report, 2006 September 9th

[2] Massimiliano Mauro, Nicola Villari, Fabio Zara (Catania STMicroelectronics) "TO220 Ribbon Electrical Characterization" internal report, 2006 September 27th

[3] Jaouad Benchaoui (Ain Sebaa STMicroelectronics), "Ain Sebaa Ribbon bonding activity" internal report, 2007 February 14th

[4] Giuseppe Malgioglio (Catania STMicroelectronics), "Structural Stress from Bonding on Silicon Ribbon vs. Wire Bonding" internal report, 2009 April 7th

[5] Emanuela Messina (Catania STMicroelectronics), "Assembly Comparison Ribbon vs. Wires VR23 in Power SO-10" internal report, 2008 November 24th

[6] Jaouad Benchaoui (Ain Sebaa STMicroelectronics), "SO8 Bouskoura PMOS LV Ribbon qualification lots" internal report, 2009 March 3rd

Advanced Modeling Techniques for System-level Power Integrity and EMC Analysis

Giovanni Graziosi[†], Patrice Joubert Doriol[†], Yamarita Villavicencio[‡],
Cristiano Forzan[†], Mario Rotigni[†], and Davide Pandini[†]

[†]STMicroelectronics, Agrate Brianza, 20041 Italy
[‡]Politecnico di Torino, Torino, 10129 Italy

Abstract

In modern digital ICs, the increasing demand for performance and throughput requires operating frequencies of hundreds of megahertz, and in several cases exceeding the gigahertz range. Following the technology scaling trends, this request will continue to rise, introducing new challenges to ensure the power integrity (PI) of the electronic systems, and increasing the electromagnetic interference (EMI). The enforcement of strict governmental regulations and international standards, mainly (but not only) in the automotive domain, are driving new efforts towards design solutions and modeling techniques to assess and guarantee PI and electromagnetic compatibility (EMC) across the overall system that comprises the chip, package, and printed circuit board (PCB). Hence, PI and EMC/EMI are rapidly becoming a major concern for high-speed circuit, package, and board designers. In this work we investigate the impact of the chip power rail noise on system PI and EMI, and we show that by reducing the power rail noise thus assuring the system PI, it is possible to significantly reduce the electromagnetic (EM) conducted emissions. Furthermore, we present a transistor-level lumped-element simulation model of the system power distribution network (PDN) that allows chip, package, and PCB designers to predict the power integrity and the conducted emissions at critical chip I/O pads. The experimental results obtained on an industrial microcontroller for automotive applications demonstrate the effectiveness of our approach.

1. Introduction

The continuous shrinking of the device feature sizes introduced by aggressive technology scaling trends, and the increasing complexity of digital ICs, require higher operating frequencies with faster clock rates. Therefore, power integrity has become a critical challenge that must be addressed at the system level considering the parasitic effects of package and board. Furthermore, because of high-frequency square-waves rich in harmonics and distributed throughout the die, ICs are also becoming prolific EMI generators. However, until re-

cently, circuit, package, and board designers have not considered PI from a system perspective. In contrast, various optimization techniques were focused separately on chip, package, and PCB design. Such a disjoint effort often resulted in circuits that even though meeting the design constraints before tape-out, they did not function correctly (or at all) once inserted on the PCB.

Traditionally, the problem of reducing on-chip EMI has been more an art than a science, and different engineering solutions were considered, tested, and implemented by *trial-and-error*, without any structured approach. In fact, when the on-chip EM emissions exceeded the level imposed by either the international legislation or by customers (mainly in the automotive domain), patches and workarounds could be found on the PCB by inserting off-chip decoupling capacitances (i.e., decaps) in close proximity of the die noisy I/O pads. Often, a similar practical approach was also followed to solve PI weaknesses.

Obviously, this method is no longer acceptable, since the international standards and regulations [1][2], the customer's requirements, and an increasingly aggressive competition in the automotive market dictate a deep theoretical understanding of the PI and EMC/EMI problem. Moreover, a systematic method to deploy effective and reusable solutions for a wide range of applications, an accurate and practical modeling approach for the system PDN to estimate the power integrity and EMC behavior before fabrication, and the development of an overall PI/EMC-aware design methodology, are also necessary to avoid costly design re-spins. PI and EMC have become critical objectives for first-silicon success. In fact, even if the traditional constraints such as area, timing, and power consumption are satisfied, but the chip does not meet the PI and EMC requirements, then the circuit has to be re-designed with a dramatic increase in terms of NRE costs, and subsequent delays in the product chain, thus missing critical *time-to-market* windows.

The most effective method to achieve PI and reduce on-chip EM emissions considers the switching current waveform generated on the power grid. The simultaneous toggling of logic gates and macroblocks typical of synchronous circuits results in narrow current glitches

978-1-4244-4722-0/09 $25.00
© 2009 IMAPS-ITALY

on both power supply and ground rails localized in close proximity of the clock edges. These pulsed currents are conducted off-chip through the I/O pads, thus driving the unintentional emitting antennas composed of the PCB traces and cables. Adequate power rail noise suppression is a critical requirement for the successful design of today's microelectronic systems, and it necessitates among other things the availability of low-impedance on-chip decaps to provide electric charge locally to the switching circuitry. In this work, we present a design methodology for decap insertion and power rail noise minimization, which can be seamlessly integrated into the standard design flow. This approach has been successfully exploited to guarantee the power integrity and reduce on-chip EMI on an industrial microcontroller for the automotive market.

Another essential component of a PI/EMC-aware design methodology is the availability of a simulation flow, which allows designers to efficiently and accurately evaluate the power integrity and EMC behavior of the die before tape-out considering the other system components such as the package and board. In fact, by neglecting the PCB traces loading effects when evaluating the chip power integrity and EMC performance, not only we will obtain inaccurate results with respect to post-fabrication measurements, but we will also ignore and overlook potentially critical design weaknesses, thus loosing the fidelity between the optimization techniques used during design, and their actual effectiveness measured after fabrication. In this paper, models of the IC core, package, and PCB were generated using commercial tools, and were used to predict the voltage noise at the power supply I/O pads and then estimate the PI and EM conducted emissions.

This paper is organized as follows: Section 2 overviews some relevant previous work on power rail noise suppression, while the methodology for decap characterization and insertion is introduced in Section 3. Section 4 describes our system PDN modeling technique and PI/EMC simulation flow, while Section 5 presents the experimental results validating the effectiveness of the proposed approach. Finally, Section 6 summarizes a few conclusive remarks.

2. Previous Work

In high-performance microcontrollers and cores, clock rates are steadily increasing, forcing very fast rise/fall times, and clocking to be strongly synchronous all over the chip. This means that all clock edges occur at the same time, thus causing large current pulses on the PDN and increasing the chip EMI [3]. Since fast current variations have a large harmonic content, a detrimental factor for PI and EMC is the power rail noise or *simultaneous switching noise* (SSN) caused by the

dynamic power and ground rail current fluctuations. In [4] it was demonstrated that SSN is originated mainly from package parasitic inductance, and with the increasing circuit frequencies and decreasing supply voltages, it is becoming more and more important [7][8][9]. The impact of SSN on the PI and EM emissions of digital circuits was discussed in [5], and in [6] an approach to analyze the on-chip power supply noise for high-performance microprocessors was presented. Because the dynamic SSN is a major detrimental factor for PI, it is necessary to insert decoupling capacitors in proximity of current-hungry blocks [10]. In [11] a systematic study to understand the effects of off-chip and on-chip decoupling was presented. It was demonstrated that off-chip decaps are effective at reducing the EM radiation in the low-frequency range, while they do not have any significant impact in the high-frequency region. In contrast, on-chip decoupling is a valuable technique to suppress the EM radiation in the high-frequency region, and by a careful combination of off- and on-chip decoupling it was possible to achieve a significant improvement in PI and suppression of EM emissions over the whole frequency spectrum.

3. Decap Characterization and Insertion for Power Integrity and Low-EMI Design

The fundamental issue is how many decaps should be used and where they should be placed. This matter becomes particularly pressing in today's SoC designs, especially in the automotive domain, where critical requirements such as a fast time-to-market and challenging silicon area constraints must be met by the design teams. Therefore, not only it is important to understand the effective behavior of intentional on-chip decaps, but it also mandatory to efficiently place them. It was confirmed through measurements [4][12] that on-chip decaps are efficient only when placed within a short physical distance from the switching components (standard cells, macroblocks, and I/O drivers) and noise sources. Indeed, with stringent area limits it is not possible to overfill the die with extra decaps, since there may not be enough white space available on the floorplan. Decaps are allocated either by surrounding each macroblock or by targeting specific edges of the macro bounding box. Edge targeting is useful when the switching activity is known to be highly non-uniform and confined primarily within regions near specific circuit boundaries (for example the sense amplifiers or the I/O drivers of the memories). This approach becomes much more important for those area-limited applications, where in order to effectively reduce the noise propagating through the I/O power supply pads, it is necessary to maximize the damping effect of the decoupling capacitances.

978-1-4244-4722-0/09 $25.00
© 2009 IMAPS-ITALY

Table 1. Fillercap cell characterization

Typical case (V_{DD}=1.8V, T=25C)	FILLER16	FILLER32	FILLER64
Nominal Cap Value (fF)	94.77	237.24	522.16
Cut-off Cap Value (fF)	67.01	167.75	369.22
Cut-off Frequency (GHz)	10.47	2.14	0.51

In SoC design, decaps are realized with fillercap cells (i.e., fillercaps), which are implemented by MOS transistors with long and wide channels to obtain sufficiently large capacitance values. The transistor works as a capacitor connected between the power supply (V_{DD}) and ground with the top plate corresponding to the MOS device gate and the bottom one is the inversion layer. The time response of such fillercaps is crucial because of the short switching times, where the circuit components require a large amount of charge in a very brief time interval (typically a few tens of picoseconds for more recent technologies). Therefore, in such a small period of time, fillercaps have to be considered as distributed elements, since their timing constant can be larger than the switching time. The analysis performed in [14] demonstrated that at high working frequencies, the effective capacitance value of the MOS gate decreases. This detrimental effect is much more relevant for large fillercaps, implemented with MOS transistors with a long channel. At high frequencies, only a fraction of the channel close to the source and drain electrodes follows that gate voltage variation, while most of the channel remains isolated. As a consequence, the effective capacitance decreases, and the effectiveness of the fillercap cell is reduced.

The microcontroller considered in this work is the STXX, which was taped out in 0.18μm embedded Non-Volatile Memory (eNVM) CMOS technology, with the NVM devices shrunk to 0.13μm. The area constraints were very tight, and it was critical to maximize the decoupling effect of each fillercap. Therefore, it was necessary to use only those cells that did not suffer from capacitance value degradation due to the frequency increase. A frequency characterization of the fillercaps available in the library was performed for best-, worst-, and typical-case, and the results obtained on the most relevant cells in typical-case (V_{DD}=1.8V, T=25C) are summarized in Table 1. The operating frequency of the microcontroller was 24MHz, which is at least one order of magnitude below the cut-off frequency (where the effective capacitance value is $-3dB$ with respect to the nominal value) of the fillercaps in Table 1 (this is also confirmed for best- and worst-case, whose results are not reported). Hence, even the large decaps were not subject to any effective capacitance decrease, and they

could be used to reduce the on-chip EM conducted emissions, when enough white space was available around the noise sources.

After decap placement, a full-chip transient power noise simulation and EMI analysis based on the spectral content obtained with FFT [13] was carried out to validate the effectiveness of fillercap insertion.

4. System Power Distribution Network Modeling

Power integrity requires a smooth delivery of supply currents without fast current transients that would generate noise and EMI propagating into the surrounding electronic system. Reliable PDN models become mandatory to evaluate the system PI and EMC performance prior to fabrication. Not only IC designers demand PDN models before tape-out, but also PCB designers using these ICs ask for the models to evaluate the board power integrity.

The system-level PI/EMC behavior can be evaluated with a circuit lumped-element model suitable for transistor-level simulations shown in Figure 1, which includes the voltage regulator module (VRM), the PCB planes and traces, the package lead frames and bond wires, the chip power grid rail parasitics and its switching current. The off-chip capacitors are usually represented as an ideal capacitor with an effective series resistance (ESR) and effective series inductance (ESL) modeling the parasitics of the capacitor package, where larger capacitors have larger effective series inductances.

The voltage regulator seeks to reproduce a constant output voltage independent of the load current, and in Figure 1 it is represented by an ideal voltage source in series with a small resistance and inductance of its pins. Usually near the VRM there is a large bulk capacitor (typically electrolytic or tantalum). Power ground planes on the PCB carry the supply current to the package, contributing some resistance and inductance. Typically the board designer places several small ceramic capacitors near the package critical pins. The package and its pins introduce some parasitic resistance and inductance. It is important to notice that high-frequency packages often contain small capacitors inside the package for further decoupling. Finally, the chip connects to the package through solder bumps or bond wires, with additional resistance and inductance. The chip lumped-element equivalent model includes the contribution of the I/O pads, the power grid rail parasitic resistance, and the bypass (i.e., decoupling) capacitances. The dynamic and static current demands of the chip are modeled as a variable current source. The on-chip bypass capacitance consists of the symbiotic (or intrinsic) capacitance and possibly some explicit decap

978-1-4244-4722-0/09 $25.00
© 2009 IMAPS-ITALY

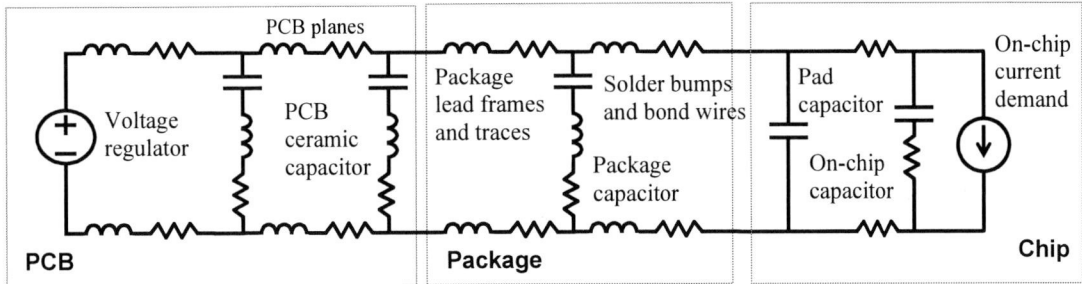

Figure 1. System power distribution network model

due to the fillercap insertion, as it was discussed in Section 3.

A critical component of the system PDN model is the chip switching activity. In our PI/EMC-aware design flow we used the Apache's tool suite (RedHawk and Sentinel-CPM), which can generate a lumped SPICE-compatible chip power model (CPM) composed of the current waveform generator along with the die equivalent impedance network seen at a given pin or pin group. RedHawk/Sentinel-CPM can simulate the dynamic current behavior of the entire SoC. The current profile of each library cell is characterized with a circuit simulator taking into account the input transition time and output load capacitance of the cell library timing characterization grid. The determination of an accurate current signature for each component is one of the most difficult aspects of power rail noise estimation. In particular, for complex and large macros such as eNVM blocks, the current profile must be derived from a specific characterization. In our approach, this current signature is determined with Apache's RedHawk-MMX, either using the memory validation stimuli, or the VCD file obtained from the logic simulation. In order to derive the SoC's CPM, a vector-based set of stimuli such as the VCD file is preferred. However, a realistic simulation can also be achieved with some behavioral specification such as the full-chip/block/cell toggle rate and power consumption.

The package model was extracted considering the quasi-static assumption and using the Ansoft's Q3D tool. The physical length of the bond wires and the signal bandwidth of the application affect the frequency range validity region for using the quasi-static approach. In fact, the ratio between the maximum length of the package interconnection (L) and the wavelength (λ) of the application plays the major role for package modeling. An iterative algorithm divides the package 3-D structure into a mesh and calculates the electromagnetic field in each element to obtain the final result. The SPICE model of the package including resistance, self inductance, and self capacitance of each connection, as well as mutual inductances and coupling capacitances, is derived from the computed field quantities. When $L < \lambda/10$, the quasi-static approach is correct and each lead frame and bond wire can be described through RLC cells (lumped-element model), and the analyzed QFP package respects the above limitations. In contrast, if $L > \lambda/10$ then the quasi-static approach is no longer acceptable as propagation phenomena appear on the interconnections. In this case, the full wave approach without any frequency limitation is needed.

Finally, the PCB model was obtained with Sigrity's PowerSI and Broadband SPICE tools. Since the board of the automotive microcontroller considered in this work included an off-chip VRM, first we built a linear model of the VRM following the approach described in [15]. Subsequently, such linear model was included into the PCB description imported in PowerSI. In this way, the PCB lumped-element model obtained with Sigrity's tools included the filtering effect of the off-chip VRM.

The system PDN model illustrated in Figure 1 can be used to simulate the power integrity and EMC performance of the microcontroller considering the loading and parasitic effects of the complete system.

5. Experimental Results

Typical microcontrollers for automotive applications include analog and mixed signal blocks, eNVM memories, I/O circuitry, and a digital core that usually takes electric energy from an independent power supply net. The need to keep the digital power grid separated from the other building blocks stems from the necessity to minimize the crosstalk between noisy circuits like the digital core and susceptible analog circuitry. In particular, the STXX is a microcontroller that contains a digital core with a separated power supply network of 1.8V, a large 128K-byte eeprom, two SRAMs, one ROM, and a few analog circuits, where the most important one for power consumption is the analog-to-digital converter.

Because the memories and the analog macros are a significant portion of the STXX in terms of power consumption, to perform an accurate full-chip evaluation of the power rail noise, their current profile must be obtained. This dynamic model should capture the current waveforms as well as parasitics for all the modes of the

978-1-4244-4722-0/09 $25.00
© 2009 IMAPS-ITALY

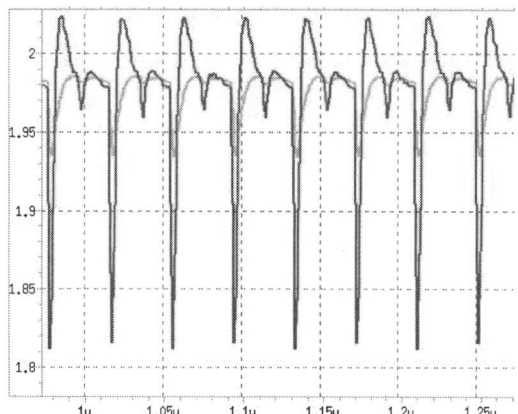

Figure 2. Power rail noise waveforms at the power supply I/O pad

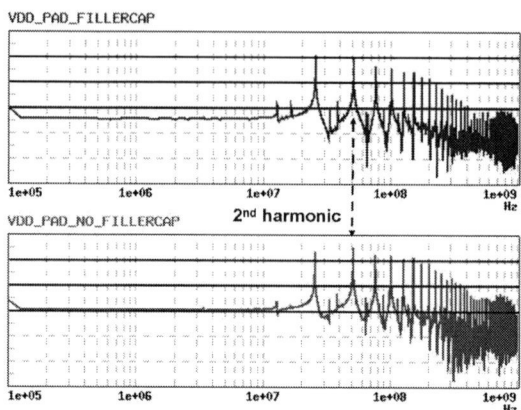

Figure 3. Power rail noise frequency spectra at the power supply I/O pad

block (e.g., different modes of operation for a memory are the read/write/test modes). In our methodology, the current signatures of the hard macros can be generated by either using the VCD input or a probabilistic vector-less analysis. Since most of the macroblocks are too large, their characterization with a traditional circuit simulator is impractical. Hence, the dynamic current waveforms at each device that is connected to power and ground nets are obtained with a fast circuit simulator based on a set of input simulation vectors. The characterization engine extracts the effective intrinsic or intentional capacitance, equivalent resistance, and time-variant and voltage-dependent current models for each transistor-level device. While for large analog and mixed-signal (AMS) blocks (like the eeprom memory) it is necessary to obtain an accurate current signature, for smaller AMS circuitry, whose power consumption is negligible with respect to the power dissipated by the larger blocks, it may not be essential to derive such an accurate current profile, given that the corresponding characterization effort will be quite significant. As such, we considered these small macros simply as "*black boxes*", and we obtained the equivalent impedance of their power grid that was connected to top-level power grid. In this way, the loading effects of these circuits were taken into account in the PDN of the overall SoC.

The fillercaps (a few of them are shown in Table 1) were inserted in the STXX floorplan for a total amount of 2.2nF, mostly in close proximity of the large memories, where the preliminary dynamic power analysis showed the power rail noise hotspots, without introducing neither area nor timing penalties. The noise reduction at the V_{DD} I/O pad of the digital core PDN was about 12mV (azure plot in Figure 2). Therefore, also achieved was a significant EM conducted emission attenuation, as illustrated in Figure 3, where the blue spectrum corresponds to the EMI analysis carried out after fillercap placement and optimization, and shows a

significant and consistent decrease in harmonic magnitude across the overall frequency range (from the fundamental frequency at 24MHz up to 1 GHz). In particular, at the second harmonic (48MHz singled out by the vertical dashed line) the harmonic amplitude reduction was 9.9dBμV. The IC lumped-element model (i.e., CPM) was used for a fast assessment of the fillercap insertion. The critical requirement of this model was its accuracy with respect to the gate-level analysis, to drive the power rail noise minimization, allowing the designer to efficiently evaluate the amount of extrinsic on-chip decoupling capacitance necessary to achieve a target noise reduction. The waveforms reported in Figure 4 demonstrate an excellent accuracy between the transistor-level simulations with CPM, and the gate-level analysis performed with RedHawk. First, with CPM we determined the amount of fillercaps necessary to significantly reduce the voltage fluctuations at the power supply I/O pad of the digital core, and then we optimally distributed the 2.2nF decoupling capacitance in close proximity of the most critical noise sources.

Lastly, the impact of on-chip decoupling was also evaluated at the PCB pin where the conducted emission measurements are performed. The results are reported in Figure 5 and summarized in Table 2. It can be observed a consistent harmonic amplitude reduction within the frequency range up to 1GHz, thus demonstrating the effectiveness of our EMC-aware design and simulation methodology.

6. Conclusions

In this work we have presented a design methodology based on decoupling capacitance insertion and optimization for power rail noise and EMI reduction. We successfully exploited this methodology to tape out an industrial microcontroller for automotive applications with tight area constraints and critical PI/EMC require-

978-1-4244-4722-0/09 $25.00
© 2009 IMAPS-ITALY

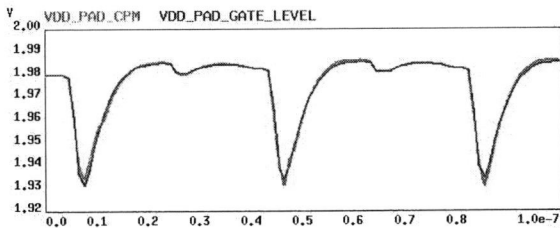

Figure 4. Gate-level analysis vs. CPM simulations

Table 2. Harmonic amplitude reduction at PCB pin for conducted emission measurements

Harmonic	2	3	4	5	6	7	8
MHz	48	72	96	120	144	168	192
Amplitude Reduction (dBμV)	9.0	11.0	13.8	14.1	14.7	14.2	14.3

ments. We integrated this methodology into the standard design flow, and demonstrated that it is possible to consider PI/EMC during the design steps of complex industrial SoCs. Moreover, we also developed a system PDN model that was exploited in the design optimization phase, to efficiently estimate the amount of decoupling capacitance necessary to significantly decrease the noise at the power supply I/O pads, and at the PCB pin where the conducted emission measurements are performed. The experimental results on the STXX confirm the effectiveness of our methodology at reducing the power rail noise and EMI.

7. References

[1] IEC 61967, 2001, *Integrated circuits – Measurements of electromagnetic emissions, 150 kHz to 1 GHz*, IEC standard; www.iec.ch.

[2] IEC 62132, 2003, *Characterization of integrated circuits electromagnetic immunity*, IEC standard; www.iec.ch.

[3] D. Pandini, G. A. Repetto, and V. Sinisi, "Clock Distribution Techniques for Low-EMI Design," in *Proc. PATMOS*, Sep. 2007, pp. 201-210.

[4] S. Pant and E. Chiprout, "Power Grid Physics and Implications for CAD," in *Proc. Design Automation Conf.*, Jul. 2006, pp. 199-204.

[5] T. Osterman, B. Deutschman, and C. Bacher, "Influence of the Power Supply on the Radiated Electromagnetic Emission of Integrated Circuits," *Microelectronics Journal*, vol. 35, pp. 525-530, Jun. 2004.

[6] H. H. Chen and D. D. Ling, "Power Supply Noise Analysis Methodology for Deep-Submicron VLSI Chip Design," in *Proc. Design Automation Conf.*, Jun. 1997, pp. 638-647.

[7] P. Larsson, "Resonance and Damping in CMOS Circuits with On-Chip Decoupling Capacitance," *IEEE Trans. on CAS-I*, vol. 45, pp. 849-858, Aug. 1998.

[8] H.-R. Cha and O.-K. Kwon, "An Analytical Model of Simultaneous Switching Noise in CMOS Systems," *IEEE Trans. on Advanced Packaging*, vol. 23, pp. 62-68, Feb. 2000.

[9] K. T. Tang and E. G. Friedman, "Simultaneous Switching Noise in On-Chip CMOS Power Distribution Networks," *IEEE Trans. on VLSI Systems*, vol. 10, pp. 487-493, Aug. 2002.

[10] S. Bobba, T. Thorp, K. Aingaran, and D. Liu, "IC Power Distribution Challenges," in *Proc. Intl. Conf. on Computer-Aided Design*, Nov. 2001, pp. 643-650.

[11] J. Kim, H. Kim, W. Ryu, J. Kim, Y.-h. Yun, S.-h. Kim, S.-h. Ham, H.-k. An, and Y.-h. Lee, "Effects of On-chip and Off-chip Decoupling Capacitors on Electromagnetic Radiated Emission," in *Proc. Electronic Components and Technology Conf.*, May 1998, pp. 610-614.

[12] N. Na, T. Budell, C. Chiu, E. Tremble, and I. Wemple, "The Effects of On-chip and Package Decoupling Capacitors and an Efficient ASIC Decoupling Methodology," in *Proc. Electronic Components and Technology Conf.*, Jun. 2004, pp. 556-567.

[13] C. R. Paul, *Introduction to Electromagnetic Compatibility*. New York N.Y.: J. Wiley and Sons, 1992.

[14] J. R. Vásquez and M. Meijer, "Modeling the Dynamic Response of On-Chip Decoupling Capacitors," in *Proc. Intl. Workshop on Signal Propagation on Interconnects*, May 2004, pp. 39-42.

[15] P. Crovetti and F. Fiori, "A Linear Voltage Regulator Model for EMC Analysis," *IEEE Transactions on Power Electronics,* vol. 22, pp. 2282-2292, Nov. 2007.

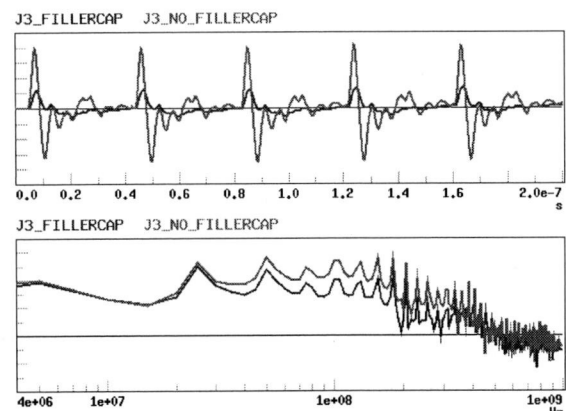

Figure 5. Noise waveforms and EMI spectra at PCB J3 pin for conducted emission measurements (w/o and w/- decaps)

978-1-4244-4722-0/09 $25.00
© 2009 IMAPS-ITALY

MEMS Pressure sensors – new LGA Packagings

Fulvio Fontana * Lorenzo Baldo**,Mark Azzopardi,***, Selma Gatt ***

* ST Microelectronics Corporate Packaging and Automation (Agrate-Italy),
** ST MEMS Division (Castelletto-Italy)
*** ST Microelectronics Corporate Packaging and Automation (Kirkop-Malta)

Tel ++39 0396032879, Fax ++390396036930,fulvio.fontana@st.com

Abstract

Cell phone, medical ,white goods and other markets are looking for absolute pressure sensors as well as differential pressure sensors . From a customer point of view these two configurations means to have one or two pressure ports, which could require a piping interface if the measured pressure is not the ambient one. For ambient measurement the MEMS LGA module needs only onean hole , which connect the device pressure port with the external environment .ST developed the so-called RHLGA packaging (Reversed Hole LGA), which is made by a flip chip placed on an holed substrates, an epoxy dam connects the pressure die port with the hole, insulating it by the molding compound.. This packaging technology allowsthe reduction of the module dimension close to the die dimensions.
Another pressure packaging is based on the integration of the HLGA (Holed HLGA packaging based on die capped and molded with a film assisted molding which leave the pressure port open). -This packaging was already presented in the past year- , with a plastic cap with piping connections (one or two) The plastic cap is assembled on the HLGA by means of an epoxy adhesive screened around the pressure holes , which attached and seals the cap.

Introduction

MEMS pressure sensors market is growing fast with many different potential applications. On cell phone where they can be used as an altimeter or to measure the barometric pressure , on car navigation system to allow them to discriminate between multi plan motorways (common in Japan), on disk drives to compensate for the altitude pressure variation to optimize the head height, on medical device to measure blood pressure.
If on some applications the driver are the packaging dimensions (mainly the thickness if we mind the cell phone) , on other one it is the need of dedicated interfaces with piping system to be the decisive factor to enter the market.

Status of art

An aleady developed packaging for pressure devices is based on the so-called HLGA (holed LGA) The MEMS sensor device is based on a thin silicon membrane with a wheatstone bridge, protected by a silicon
cap with a small pass thru hole . The silicon wafer cap is attached to the sensor wafer by means of a material "glass frit", which is applied with a stencil screening process. The capped

pressure device is glued to the LGA strip BT substrate and then wire bonded with gold wire . On the Altimeter-barometer version a Custom ASIC die is bonded by the side of the pressure sensor .
The open packaging is realised by means of a film assisted molding which leaves the pressure port hole exposed around the molded case
There are different packaging dimensions measures , the smallest stand alone sensor packaging is 3x3x1 mm

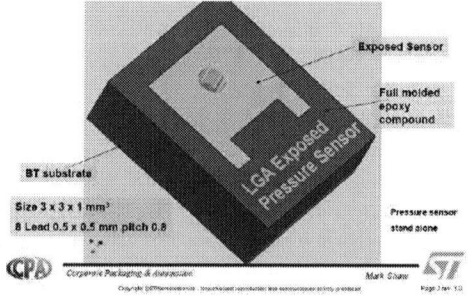

Figure 1a: HLGA packaging

Figure 1b HLGA Packaging

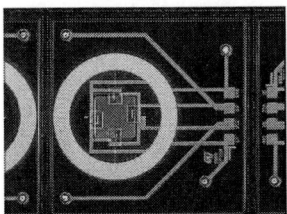

Figure 2 RHLGA Packaging

RHLGA packaging

The new packaging design RHLGA (Rear holed LGA) has been driven by the following needs:
1) Simplify and low the cost of Front End and Back End (packaging) process
2) Allow to reduce the packaging dimension below 2x2x1 mm

The basic idea is to move toward a flip chip packaging, where the electrical contacts of the pressure device are on the same side of the sensitive membrane . The electrical contacts are rerouted and bumped (with lead free alloy)

Figure 3: Die bumping rerouting

A dam must be made to connect the sensor membrane on die backside with a pressure port realized with a hole made in the BT substrate

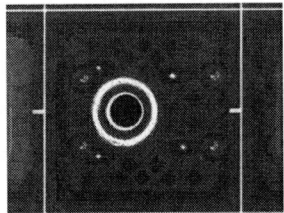

Figure 4: Hole in BT substrate

Because the bumps must be underfilled to whitstand the thermomechancal stresses due to CTE mismatch between silicon and substrate BT, the purpose of dam is also to avoid that the underfill bleeds on the sensor area impacting the sensitivity of the sensor

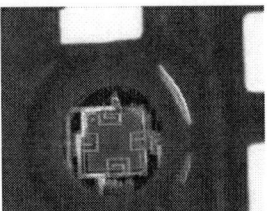

Figure 5 Vision of sensor thru substrate hole

Processes evaluations

The critical process step is related with the dam making. Four different packaging processes have been experimentally evaluated, On the first two process A, B , other sub-process have been evaluated .

Process A) is based on the differential wettability between Underfills and Chip interfaces , a hole have been made in the polyimide passivation layer exposing the silicon around the sensor . The underfill front during the dispense makes a fillet due to its surface tension along the polyimide-silicon step. This solution works with a pre-curing of the material, but it is not enough reliable for volume process because its strongly depends on UF dispense parameters like viscosity, temperature, quantity, surface cleanliness .

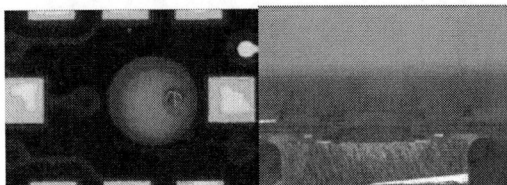

Figure 6: UF stop around passivation hole

A solution to "freeze" the Underfill before it reach the critical area has been tested, using . A

UV curable underfill and a fibre optic light shone though the central hole

Process B) A lead free solder paste is screened on a gold ring made opening the solder mask of the BT substrate around the central hole. It is not simple to calculate the right paste volume to avoid openings in the ring (low paste volume) or tilt on the chip (high paste volume compared to bump volume) . Another problem is related with the oven reflow, when some flux degass and the residuals risks to cover the critical area
A molding underfill has been tested

Figure 7: Solder paste screened /tilted chip

Process C) A solid lead free solder ring is made on wafer during the bumping process , the BT gold ring area will be optimized and the substrate bonding pad were plated with Sn to allow a correct collapse of the dam during the oven reflow .

Figure 8: chip with solder dam/Bump

Height correct/no tilt

There is still a problem when the RHLGA will be soldered on PCB . Because the lead free solder paste used for PCB soldering has the same melting temperature of the lead free solder paste used for dam , there is a problem of solder extrusion. A possible solution is to use for the dam an alloy with a lower melting temperature

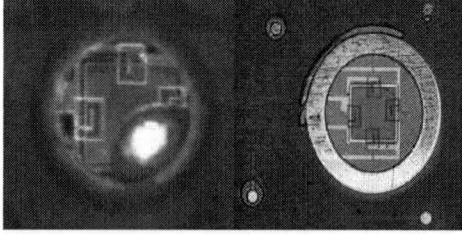

Figure 9: Solder extrusion outer and inner

Process D) A B-stage epoxy adhesive have been screened o the BT substrate strip on the solder ring , then the adhesive is procured .Next step is

Table 1: Packaging Processes

the die placement on ring . During the oven reflow the eutectic bump joins and the B-stage melt bonding on die surface

Figure 10 :B-stage Epoxy /no estrusion

Jedec Lev.3 results on Process B1 and B2

The main problem was related with the delamination occurred in the interface between the die and the molding compound , so in the process B3 an Ar/O2 plasma step have been Introduced before molding process to improve adhesion

Thermal Cycle (0-100C for 1000 Cycles) results on Process B1 and B2

No failures evidenced on bump connections

Functional Test results on Process B1 and B2

The functional test is done applying a positive pressure to the sensor and measuring the electrical answer to different temperatures
The main parameters are:
- Sensitivity (mV/V/Bar)
- TCO Thermal drift of Offset
- TCS Thermal drift of sensitivity
- NL non linearity
- Offset (mV/V)

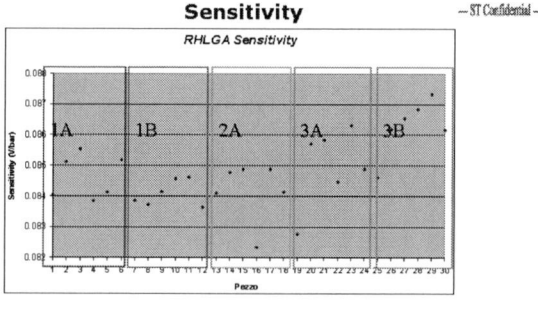

Figure 11: Sensitivity

		A1	A2	B1	B2	B3	C	D
Die Bump		SnAg	SnAg	SnAg	SnAg	SnAg	SnAg	SnAg
Bump dimension	μm	80	80	80	80	80	80	80
Die ring		Opening in Passivation	Opening in Passivation	NiAu Ring	NiAu Ring	NiAu Ring	SnAg Ring	NiAu Ring
Ring diameter	μm	950	950	950	950	950	950	950
Ring thickness	μm	na	na	125	125	165	180-190	180-190
Substrate material		BT	BT	BT	BT	BT	BT	BT
Substrate pad metallization		Cu	Cu	Cu	Cu	Cu	Cu-Sn	Cu-Sn
Substrate thickness	μm	230	230	230	230	230	230	230
Baking		2HR@150C	2HR@150C	2HR@150C	none	2HR@150C	2HR@150C	2HR@150C
Plasma		ArO2	ArO2	Ar	none	ArO2	ArO2	ArO2
Underfill		Ablestick 8828	UV curing Underfill	Ablestick 8828	None (moulding as Underfill)	Ablestick 8828	Ablestick 8828	Ablestick 8828
Ring screening Material		none	none	SnSb solder paste	SnSb solder paste	SAC solder paste	none	B-stage epoxy
Reflow peak temperature	C	250	250	250	250	250	250	250
Plasma		ArO$_2$	ArO$_2$	none	none	ArO$_2$	ArO$_2$	ArO$_2$
Molding material		Nitto GE-100 LF1-2	Nitto GE-100 LF1-2	Nitto GE-100 LF1-2	Nitto GE-100 LF1-2	Nitto GE-100 LF1-2	Nitto GE-100 LF1-2	Nitto GE-100 LF1-2
Process Results								
Ball shear		good	good	good	good	good	good	good
Chip tilt		Yes	Yes	Yes	Yes	No	No	No
Openings in dam		na	na	Yes	Yes	No	No	No
Solder ball on sensor area		na	na	Yes	Yes	No	No	No
UF bleeding		low	low	On parts with dam openings	On parts with dam openings	no	no	no
Solder extrusion After LGA soldering on PCB		yes	yes	yes	yes	yes	yes	
Delamination after Jedec 3 between die and molding		no	no	yes	yes	no	no	no

The parts with moulding underfill evidenced a higher sensitivity; the improvement could be related with the higher stiffness of the modules With moulding underfill compared to the modules with underfill .This means that the stress transmitted by pressure change is absorbed less by a stiff material

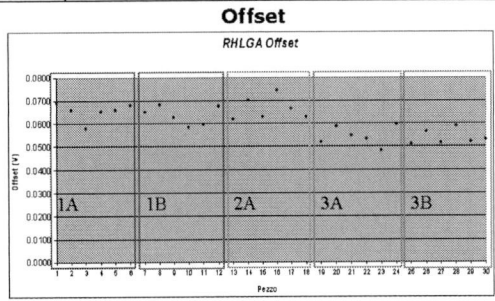

Figure 12: Offset

The parts with moulding underfill have a lower and more uniform offset

Thermal drift of the Offset: TCO

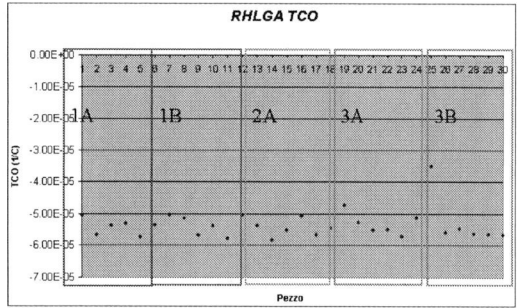

Figure 14: Thermal drift of the offset

There is not a significant difference between the two processes

Thermal drift of the Sensitivity: TCS

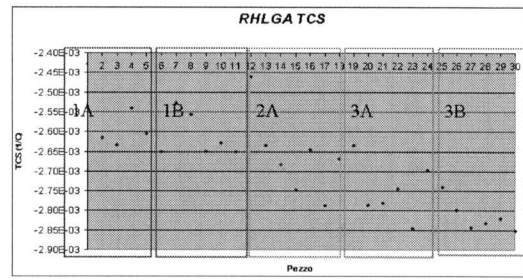

Figure 15: Thermal drift of the sensitivity

The parts with moulding underfill evidenced an higher thermal drift even though the lower CTE (=8 PPM) of the moulding compound compared to the underfill (=30) . One possible cause can be the higher moisture absorption ratio of the molding compound

Non linearity: NL

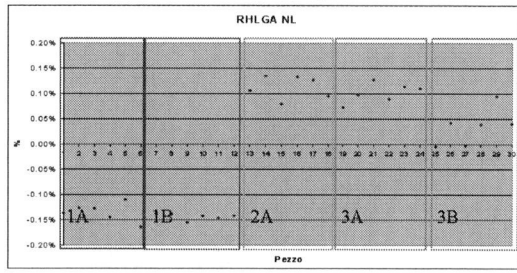

Figure 16: Non linearity

The value of non linearity is the opposite on the parts with underfill and moulding underfill.

Comparison among RHLGA331 – HLGA331 – CERAMIC DIP16

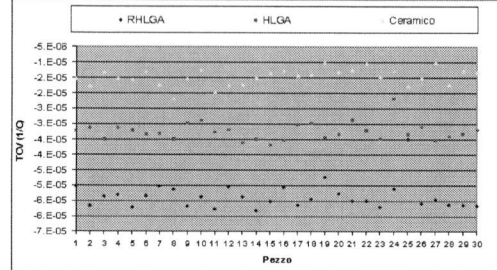

Figure 17: Comparison of Thermal Drift of Offset between different packagings

Comparison Between Ceramic Sensor, Hlga and RHLGA (process B1 and B2)

The Ceramic packaging is the better while the RHLGA is worst if compared to HLGA , one cause of this behaviour can be related with the stress transmitted from the solid dam to the sensor membrane . This is one more reason to change from solder dam to an epoxy one , before to start with experimental phase a thermo-mechanical simulation has been run to evaluate the behaviour of epoxy dam

Warpage Analysis

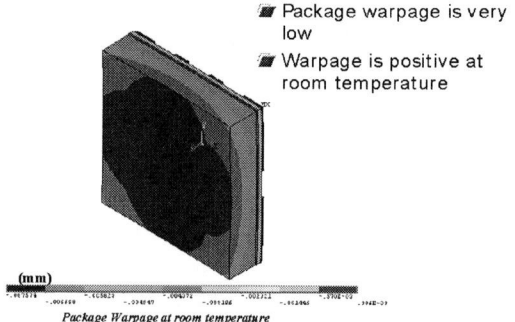

Fig 18: Thermo-mechanical model with Von Mises Stress

Thermo-mechanical simulations on Process "D" (B-stage epoxy)

The epoxy dam is able to relief the thermo-mechanical stresses transmitted to the sensor more than the solder dam. This is because the epoxy is less rigid than solder so it is more compliant with LGA warpage

978-1-4244-4722-0/09 $25.00
© 2009 IMAPS-ITALY

Membrane Stress Analysis

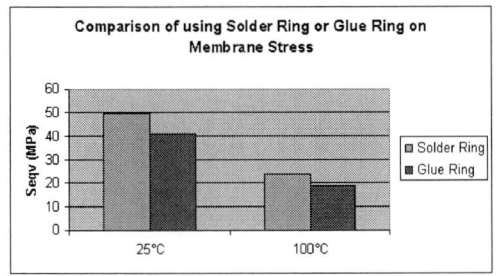

Fig:19: comparison of stress on membrane between solder and epoxy ring

The stress transmitted to the membrane is less with an epoxy ring, being less stiffer than the solder ring.

This seems to contradict the results experimentally found on sensitivity (see page 5) comparing parts with underfill and moulding underfill . But the results found on sensitivity were achieved applying only a pressure to the sensor (mechanical stress) while this simulation has been done applying a thermo-mechanical stress.

This means that to fully validate the epoxy solution it need to run also experimental test on parts

Warpage Analysis

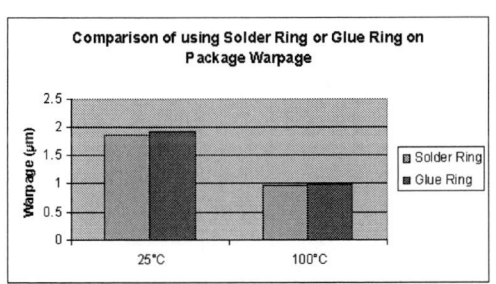

Fig:20 : comparison of LGA warpage between solder and epoxy ring

Difference between using a solder ring or an epoxy ring is marginal for normal stress SZ and von Mises stress Seqv, with the epoxy ring leading to slightly higher value.

While for shear stresses (which are more critical from a point of view of cracks) Sxy the epoxy glue leads to lower stresses

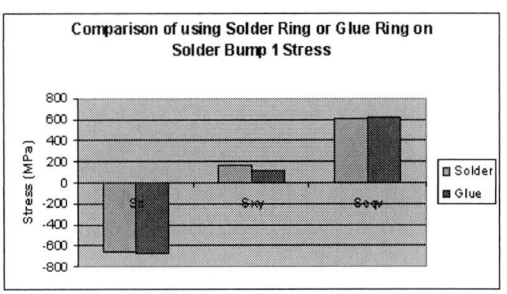

Fig:21 : comparison of Bump stress between solder and epoxy ring

HLGA Differential and Absolute pressure packaging with piping interfaces

To compare two different pressures the MEMS sensor membrane have to put in contact on the opposite sides with the pressure ports, so there are two basic configurations of die structure , the first one with the port on the opposite die side and the second one with the port on the same side . Around this last one a plastic packaging has been developed by ST. The same technology has been applied also to Absolute pressure if they need a piping connection.

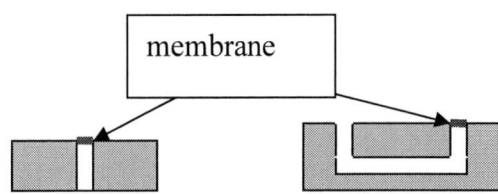

Fig 22 Differential pressure die configurations

The solution consists into using a B-stage material stencil screened above the HLGA moulded strip , leaving two openings above the two ports of each module in the strip, then the B-stage is procured and the strip is sawn to singulate the modules. Next step is to place the moulded caps with pipes above the HLGA module and glue it applying to the B-stage a temperature and a pressure controlled applied.

The plastic cap is metallised on the corner, so it can be soldered to the PCB increasing the mechanical resistance of the LGA, the material of cap (LCP) is capable to withstand SMT soldering temperatures.

This process can be applied also to absolute pressure HLGA or to an ambient pressure referred HLGA modules (with only one port to be connected with one pipe) but which need a piping connection as in white goods applications ,

978-1-4244-4722-0/09 $25.00
© 2009 IMAPS-ITALY

in this case the process can be greatly simplified because it has been demonstrated that also a cap without "shoulders" with a vertical pipe can withstand the mechanical shear stresses .

The assembly of a cap without shoulder can be done attaching a plastic panel with pipes above the HLGA strip and doing the singulation of modules as the last operation

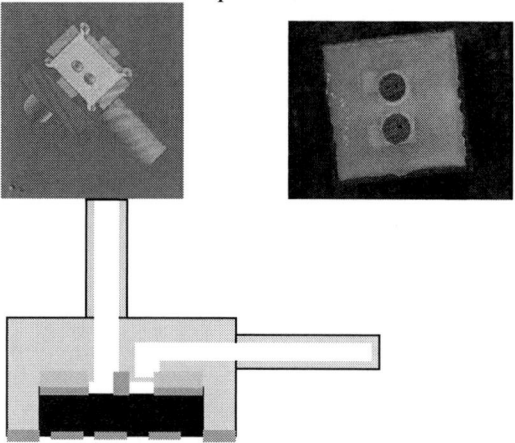

Figure 23 HLGA Differential Pressure packaging

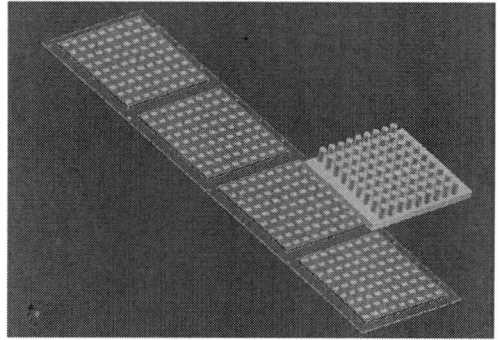

Fig 24:Cap Plastic block assembly on HLGA strip

Leakage tests

Leakage tests have been done before and after soldering simulation, putting the assembled parts in water and connecting the piping to compressed air. Results were good

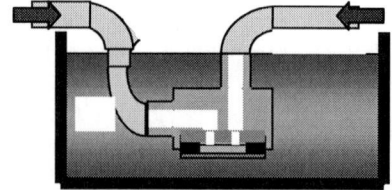

Figure 25: Leakage test

Functional test

Functional test is done connecting the piping to pressure ports and measuring the Voltage output, results were positive

Conclusions

The needs for new packaging of MEMS pressure sensor devices can be satisfied integrating the existing packaging solutions like the LGA , Flip Chip and Wire bonded with completely new solutions , modifying the substrate design with pass thru holes like in the RHLGA or making imbedded channels , changing the standard materials of the substrate (BT or FR4) with metallised plastic materials . Coupling the existing packaging with plastic moulded interface which can be modelled to satisfy any customer requirements using new materials like B-stage adhesives, adhesive pre-holed tapes to make the connections.

Thanks

Many thanks to Julien Vittu of ST (Grenoble) for his contribute to RHLGA process development, Roseanne Duca of ST (Malta) for her indispensable work on thermo-mechanical simulation, to Carlo Passagrilli and Giancarlo Santospirito of CPA ST laboratories (Agrate) for their contribution to physical analyses

References

(1) F.Campabadal et al. "Flip chip Packaging of Piezoresistive pressure sensors", Sensors and Actuators A132 (2006) pp.415-419

(2) R.Dean et al. "Micromachinned LCP for packaging MEMS sensors" , IEEE 2005 proceedings pp.2363-2367

(3) M.Shaw et al. "High Volume MEMS packaging" , Proceedings of the 16th European Microelectronics and packaging conference EMPC 2007 Procedings pp.182-187

(4) M.Shaw "Package Design of Pressure Sensors for High Volume Consumer applications" , Proceedings of the 16th European Microelectronics and packaging conference EMPC 2008 Procedings

Wafer Level Packaging Fan Out thermal management: is smaller always hotter?

Donata Gualandris, Claudio M. Villa
ST Microelectronics
Via C.Olivetti 2, Agrate Brianza- Milano Italy
Phone: +39 039. 603.5561, Fax: +39 039. 603.6930
Email: donata.gualandris@st.com; claudio-maria.villa@st.com

Abstract

Actually the electronic packaging industry is developing the Fan Out Wafer-Level Packaging (FO WLP) technology that shows an intrinsic capability for higher integration and relevant supply chain.

Size and cost reduction, better signal integrity and higher speed compared to standard laminate BGA make this new solution very attractive. In this paper we analyze from a thermal point of view the same chip assembled in standard and flip chip BGA and in Fan Out WLP in order to compare the different thermal behaviours. The main advantage of this new technology is a quite low vertical thermal resistance.

The present study shows thermal modeling analysis results achieved using FLOTHERM®.

Total thermal behaviour is due to the balance between vertical thermal resistance and lateral heat spreading. Substrate thickness is determined as the dominating factor controlling the total WLP thermal resistance. The impact of die size and hot spot presence has been investigated. Different board designs are also considered.

Keywords: Wafer Level Packaging fan out, thermal performances

1. Introduction

In the last few years the request of further and further miniaturization and integration have driven the development and industrialization of new package families. At the same time multi-functionality and higher performances request have caused a power dissipation increase leading to critical thermal behaviour. In the past, 2D BGA packaging and subsequently stacked dice and System in Package (SiP) have been adopted as standard solution mainly in mobile phone applications.

Actually the new Wafer-Level Packaging technology shows an intrinsic capability for higher integration and functionality. This new technology is basically a true Chip Scale Package, but packaging assembly steps are performed on the whole wafer before singulation using silicon-like processes. This fact causes a significant cost lowering. The packaging equipment and processes are similar to wafer fab and even testing of separated dice could be replaced by wafer level one.

Reduced size and cost, better signal integrity, higher integration density and higher speed compared to standard laminate BGA make this new solution very attractive.

Finally the capability of very fine pitch interconnection really makes this package the new frontier of package integration.

At the beginning WLP technology has been introduced using Fan In structure. In this package family the balls are redistributed on die area and this means that die and package size are the same.

Currently, WLP technology is moving to Fan Out structure evaluation. This solution shows the possibility of increasing the number of I/Os thanks to a larger organic area compared to the die one.

Obviously, both miniaturization and integration lead to a higher stress level in electrical and thermal response. Therefore it is not only interesting, but even absolutely necessary to investigate thermal performance of this new package solution.

2. WLP FO technology

This new technology shows a simplified assembly compared to conventional methods of fabrication. Briefly, assembly steps can be summarized as follow: each single die is located on tape and the entire reconstructed wafer is encapsulated. After tape removal step, a plastic wafer with dice is ready for re-routing. During fan-out redistribution one or more metal and dielectric layers are applied on the new artificial wafer using silicon-like processes. Subsequently passivation and ball placement are performed on the whole wafer. After wafer testing, the last assembly step is the singulation.

The mainly advantages of this new solution is that the entire process could be handled in-house

978-1-4244-4722-0/09 $25.00
© 2009 IMAPS-ITALY

using silicon-like processes with a significant costly reduction.

Actually the capability of assembly multiple re-distribution layers (RDL) is under development. This new trend could be useful in case of multi chip package as a better re-routing is achieved. Obviously, both cost and package robustness must be deeply investigated.

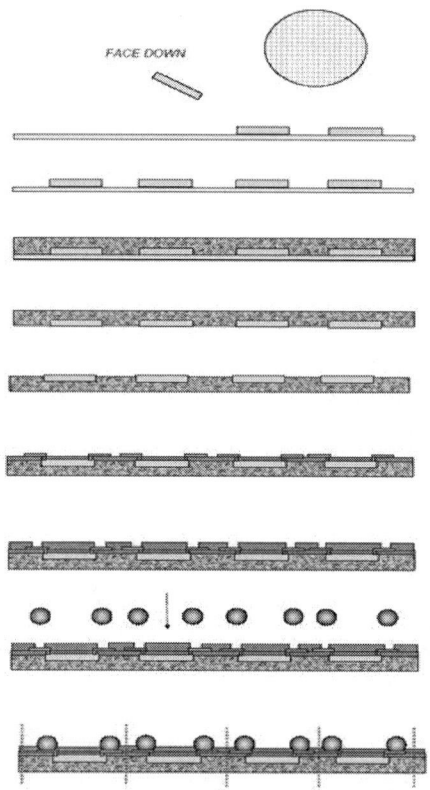

Figure 1: FO WLP assembling steps

3. Chip and packages details

In order to investigate the thermal performances of FO WLP technology, BGA 7x7 flip chip and classical wire bonding die have been compared to WLP with the same package body size.

The main differences between these package technologies involve laminate substrate thickness, materials and especially assembly.

TFBGA and VFBGA typical substrate thickness is 200 μm. Usually both 2 and 4 metal layers have been adopted as standard solution. Four metal layers substrate allows higher integration density and better thermal performance. In fact, multiple copper layer substrates lead to an increase of heat spreading. Under the same condition, four layers substrate could gain some 8-10 % in terms of thermal

resistance compared to standard two layers. Still the disadvantage of this solution is the higher price.

We have considered a standard 200 μm laminate substrate with two copper trace layers.

On the other hand, a single re-distribution layer WLP substrate have been modeled as 8 μm dielectric layer with 0.25 W/mK thermal conductivity and 6 μm re-distribution layer, modeled as averaged uniform block with different copper coverage percentage.

All packages have 148 balls with 250 μm diameter and 0.5 mm pitch.

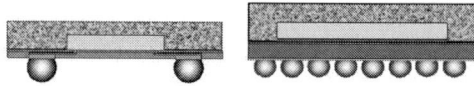

Figure 2: FO-WLP and std BGA structure.

4. Thermal behaviour

Theta j-a comparison

The first step of this study is to directly compare thermal resistance junction to ambient (Theta j-a) of the three different package technologies with same ball configuration, body and chip sizes mounted on the same board.

In standard mounting IC, the heat out balance is usually distributed as follow: 5% from side and 10% from top by convection and radiation, while the remaining 85% from bottom, i.e. via substrate and balls through board by conduction. From this point of view, we do not expect that BGA and WLP technologies show a different heat flow balance.

Simulation results show WLP thermal resistance is some 15% lower compared to BGA one.

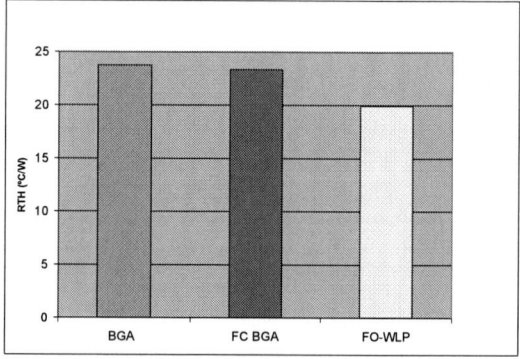

Figure 3: Theta j-a values

The different thermal behaviour observed is due to the main difference between the two packages, that is the substrate. Therefore total thermal performance is due to different compromise between lateral heat spreading and vertical thermal resistance due to package intrinsic structure.

In WLP heat spreading inside the dielectric is limited because of substrate thickness and metal trace structure. Anyway its peculiar lower vertical resistance gives rise to an optimum compromise.

RDL impact

At this point it is clear that maximising copper content in Cu layer and executing copper pad touching the first rows of inner balls leads up to some 10 % increase in BGA thermal performance with no extra cost. But considering WLP technology, re-distribution layer plays a role mainly as electrical signal path between the chip and external world. Even if we image a high planar copper content, i.e. very close and wide traces, the lower metal trace thickness compared to BGA substrate layers doesn't affect significantly total thermal performance.

This means than changing Cu percentage no relevant impact on thermal resistance has been observed. In fact the dominating factor in total thermal response is vertical thermal resistance and not lateral heat spreading.

On the other side, this means that no relevant impact is expected maximizing copper content. This enables to optimize both thermal and electrical response reducing electromagnetic interference with no drastic thermal performance worsening.

Die size and floorplan

Chip size strongly affects package thermal performance. The miniaturization trend of semiconductor industry obviously involves die size decreasing that leads to higher and higher power density. Consequently, the chip reaches a temperature that it is always closer and closer to Tj limit.

We evaluated die size impact on Theta j-a and in flip chip BGA configuration we observed an increase of some 15-20 % in Theta j-a compared to some 30 % in WLP.

Die size	Theta j-a (°C/W)	
(mm)	FC BGA	FO-WLP
6x6	23.3	19.9
5x5	25	23.5
4x4	27.4	25.8

Figure 4: die size impact on Thetaj-a in FC BGA technology vs. FO-WLP one

This means that optimizing BGA substrate heat spreading makes possible to reduce thermal gap performance between the two technologies.

No relevant difference has been observed due to hot spot presence on chip area. Both flip chip and WLP technologies show a Theta j-a increase of some 45-50% comparing 6x6 mm powered area with a 2x2 mm one.

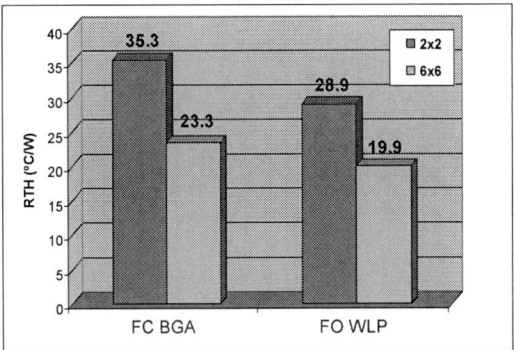

Figure 5: hot spots impact on Theta j-a

Board

The package to ambient thermal resistance is dominated by the thermal resistance to the board and to the heat spreading capabilities of the board itself.

Further analyses have also been performed in order to understand the impact of board design and a possible improvement of WLP thermal response compared to BGA one. It has been shown that the presence of vias under the package improves WLP theta j-a, but no relevant differences have been detected between the two package technologies. In fact, both technologies show some 15 % improvement with thermal vias.

Even if we model the board as a uniform block with equivalent thermal conductivity, the impact is quite the same: 20% difference between the two packages has been calculated.

Board	Theta j-a (°C/W)	
	BGA	FO-WLP
Vias Top layer ON	23.7	19.9
Vias Top layer OFF	51.8	47.1
No vias	27.1	23.1
Eq. block	23.1	19

Figure 6: board impact on Theta j-a

5. Conclusions

By the way, in actual applications it is not fully correct to directly compare the same body packages with the same device assembled with different technologies. In fact it is reasonable to assume that the same 6x6 mm device could be assembled in a 7x7 WLP fan out or in a larger BGA with more balls. This means that the advantage of low vertical WLP thermal performance will be reduced considering the decreased BGA temperature due to larger body size and consequent higher ball number. Clearly, it is possible to consider even a larger WLP body, but BGA substrate thickness and technology allows taking major advantages in heat spreading compared to WLP re-distribution layer when the whole package body increases.

Currently some preliminary works are on going to investigate side by side chips thermal performance in order to keep advantage from good vertical thermal resistance and low spreading.

Finally, FO WLP technology enables to overcome the intrinsic limit of BGA laminate substrate that not allows an increase of integration density together with reasonable costs. Actually, besides the well known advantages intrinsically due to a wafer level configuration, this new technology provides a good thermal performance. In this case really smaller is not hotter indeed.

Acknowledgments

We would like to thank Carlo Cognetti and Eric Saugier from ST Microelectronics, Yann Guillou and Eric Cirot from ST-Ericsson for their support.

References

1. Meyer T., Pressel K., Ofner G., et al. "Recent Developments in WLB and eWLB Technology".

2. The International Technology Roadmap for Semiconductor: 2007, Assembly and Packaging.

3. Plieninger R., Dittes M., Pressel K., "Modern IC Packaging Trends and their Reliability Implications" Microelectronics Reliability 46 2006, p 1868-1873

978-1-4244-4722-0/09 $25.00
© 2009 IMAPS-ITALY

Multifunctional Coatings for Wafer-Level Chip Scale Packaging

Russell Stapleton,[‡] Dave Zoba,[‡] Candice Brannen,[‡] Paul Hough[†]

[‡]110 Lord Dr., Cary, NC, USA; [†]LORD Germany GmbH, Hilden, Germany

+1-919-469-2500, +1-919-469-9688, russ_stapleton@lord.com

Abstract

For advanced wafer-level chip scale packages (WLCSP), board level solder joint reliability is a major concern, and typical stress-relieving methods such as capillary underfills and molding compounds are costly. One method of low cost reliability improvement for WLCSPs is the use of a wafer level SolderBrace™ coating, which delivers improved reliability with minimal material and capital cost. In this presentation, several new SolderBrace materials are characterized and screened for their use in two different application methods. Processing of the SolderBrace coating can be achieved by two methods. The first is similar to that of polyimides: spin coat, bake, photo-image, solvent develop, and ball drop. The second application process involves printing the material on the already-balled wafers followed by solder cleaning and cure. Unlike polyimides, these coatings are low temperature cured, have low CTE values (13-18ppm), and have minimal wafer bow. These properties make a SolderBrace coating attractive as an alternate passivation coating, while also functioning as a partial underfill SolderBrace coatings are thermally, mechanically, and chemically robust, offering a unique method to package low cost high performance WLCSPs.

Key words: Wafer, Underfill, SolderBrace™, Reliability, Package, WLCSP

Background

Advanced WLCSP designs are transitioning from 0.5mm pitch to 0.4mm pitch, while existing 0.5mm designs are increasing in die size. New WLCSP like designs such as Package-on-Package (PoP) and reconstituted wafers also appear to share the same packaging needs as WLCSPs. Flip chips on organic substrates often suffer from thermal mechanical stress induced from mismatches in thermal expansion. While larger pitch and smaller sized CSP devices have proven successful reliability performance, the continuing trends are pushing the limits of the existing designs and materials.

WLCSP package designs are frequently found in consumer products, and thus these packages are competitive in size, technology, and are yield and cost sensitive. The substrates are almost exclusively organic, lending low cost and flexible/rapid designs, but the low cost boards are limited in their ability to deliver fine pitch pads and routing. The expectation is that die stacking, such as PoP will maintain the low cost board designs, which in turn suggests wafer thinning. The combination of increased die size, higher IO, finer pitch, and lower cost is a significant challenge to the WLCSP design.

Underfilling is a common solution to addressing WLCSP reliability concerns. Capillary underfills are effective at delocalizing the stress in a package, and improving the overall lifetime of the part.[1] Capillary underfills fill the gap between the substrate and the die, creating a strong adhesive bond. This technique makes rework of die difficult, at best. The mechanical properties of capillary underfills are tuned to provide an outlet for excessive solder joint stress and to mitigate shear forces on fragile CMOS dielectrics.

Molding compounds and No-Flow materials have also proven successful in CSP protection, but their added cost in tooling is generally prohibitive in assembly environments. These solutions also suffer from their lack of convenient reworkability.

Successful low cost reliability solutions for WLCSPs have generally been adaptations of existing materials, such as modification of redistribution layer materials and solder types, but the increased performance is limited. One new approach is to reexamine the final passivation layer as more than a dielectric, but also an underfill. Wafer level underfills for flip chips have been discussed in the literature in some detail, primarily in the methods.[2-4] Few commercial wafer level protective materials are availible. PolymerCollar™ is the most notable stress reducing solution, but is actually a *cost added* underfill replacement.[5] The SolderBrace™ material not old replaces capillary underfills by also replacing the final passivation layer of the silicon process, thus an attractive low cost solution for future WLCSP designs.

Processing of SolderBrace materials has been described elsewhere.[6] One such method is outlined in Figure 1. The coating methods are either spin-coating or stencil printing, with printing being more common. With exception of the solder reflow, all of the steps are "low temperature" wafer level

978-1-4244-4722-0/09 $25.00
© 2009 IMAPS-ITALY

steps, where the maximum processing temperature is 100 °C.

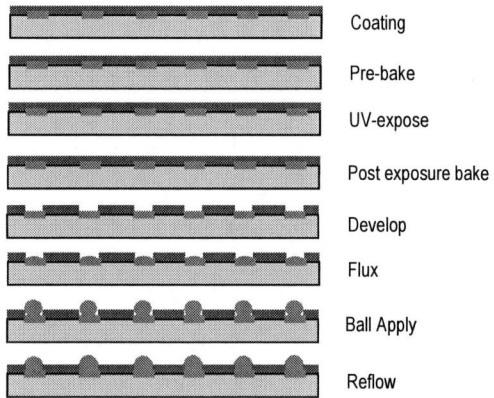

Figure 1. Application method of SolderBrace by photo definition.

Optionally there are other ways to introduce a SolderBrace type material to a WLCSP wafer, including printing on bumped wafers, as outlined in Figure 2. Although less appealing than the UV defined approach from a cost standpoint for high volume WLCSPs, the implementation of the print over ball method is mask-less and not prone to manufacturing difficulties.

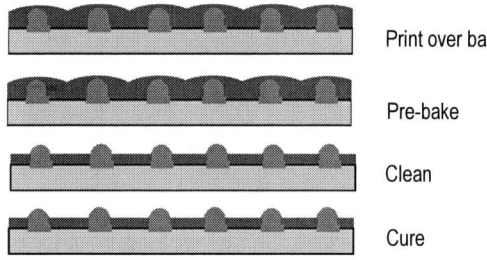

Figure 2. Application of SolderBrace material by maskless printing over balled wafer.

While both UV defined and print-over-ball methods of the SolderBrace application are used to address the solder / die junction stresses, the materials differ significantly in their function.

In this study a variety of SolderBrace type materials were screened for their process performance and material characteristics.

Test Methods and Sample Preparation

WLCSP materials are processed as thin or thick films. Traditional underfills are typically characterized by their flow behavior, and their mechanical performance gauged by their bulk properties. Because of the heating rates and special conditions (e.g. UV exposure) bulk property determination of SolderBrace type materials was inappropriate or not possible.

Samples were prepared by drawing down on a clean aluminum panel of 0.4mm in thickness. Tensile samples were masked by Kapton™ tape to dimensions of 6mm by 75mm. Multiple layers of each sample were made, as needed. Dynamic mechanical analysis (DMA) samples were similarly prepared to 12mm by ~75mm in dimension. The samples were prebaked according to Table 1 on a preheated hot plate. Prebake time and temperatures were selected based on the thickness and sample stability.

Sample	A	B (flux)	C	D
Prebake (°C)	100	0	140	140
Prebake time (min)	10	0	12	12
UV exposure (s)	60	0	0	0
Reflow peak (°C)	255	255	255	255
Viscosity (Pa*s @ 10 1/s)	7200	640	8900	12000
Filler loading (% by wt.)	75	0	50	75

Table 1. Sample preparation conditions for various thin film and formulation data.

Pre-baked film samples, typically 70um in thickness, were then reflow cured using a typical lead free ball drop reflow profile with a short dwell and peak of 255 °C. Samples were debonded from the carrier substrates by dissolution of aluminum in concentrated hydrochloric acid. Samples were rinsed with water and dried.

Thermal expansion samples of A, C, and D were prepared by stacking cut segments of used tensile samples to form approximately 5mm cube. The sample was compressed with a spring clip and passed through the reflow oven. The spring clip was removed and the samples sanded flat on all edges. The cut segments were oriented such that the constituent film plane was parallel with the thermal expansion probe. A bulk sample was cast in a 40mmx10mm mold of sample B, and cured at 200 °C for 1 hour.

Tensile tests were preformed on a The TA.XTPlus Texture Analyzer at ambient (18 °C) conditions. The pull rate was 0.03 mm/s. DMA was performed on a TA model Q800 with a single cantilever clamp, 1 Hz frequency cycle with a displacement limit of 0.1mm. The DMA was run from -60 °C to 350 °C at 2 °C/min. Thermal expansion was measured on a TA model 2940 with a flat 0.5 mm probe. The thermal expansion was observed from -50 °C to 200 °C at 2 °C/min.

Results and Discussion

The UV defined SolderBrace sample (A) is unique in that it is processed both thermally and with UV. The UV dose needed to gel the material in ~20 seconds is 500mJ, where 60s was used to ensure complete conversion of the sample. This was

978-1-4244-4722-0/09 $25.00
© 2009 IMAPS-ITALY

preformed to mimic the behavior of the SolderBrace material in a UV defined WLCSP process.

A typical tensile elongation to break sample of Sample A is shown in Figure 3, where the modulus at the ambient temperature is measured in the flat slope of the curve, and percent elongation to break was taken at the termination of the curve. The toughness was characterized by the area under the curve. Five to seven samples were run per sample set, normalized to the sample geometry, and averaged. The material properties of sample A are given in Table 2.

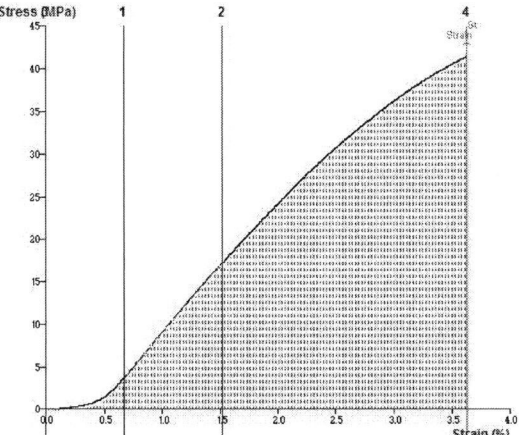

Figure 3. Typical tensile test elongation to break result of sample A.

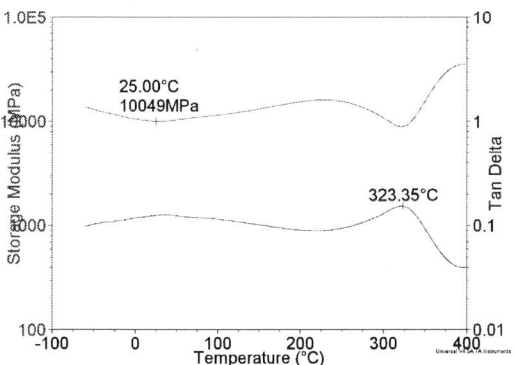

Figure 4. Storage modulus and tan δ of a thin film of sample A.

The high tensile modulus of sample A is consistent with highly filled materials, such as molding compounds, encapsulants, and underfills. And similarly the expected toughness and elongation are low, as this slightly brittle coating is designed for transitioning stress away from silicon die at thicknesses of 20-100um, which requires a high degree of inorganic nature. This is further supported by the almost imperceptible Tg in both the thermal expansion and DMA data, which only show a very slight transition in the ~130 °C range, and another at ~330°C in an otherwise flat storage and loss moduli curves, as shown in Figure 4. The lack of obvious

Tg is consistent with highly filled and highly cross linked materials. The low CTE appears to be closely matched with the solder and copper redistributions.

Sample	A (+/-σ)		B (+/-σ)	
Modulus (GPa)	3.9	0.8	1.2	0.2
Toughness (MPa)	0.04	0.03	2.3	0.6
% Elongation to break	0.5	0.1	6.3	1.6
CTE α1 (ppm/°C)	14		59	
CTE α2 (ppm/°C)	29		96	
Tg by expansion (°C)	(130)		60	
Storage modulus @ 25°C (GPa)	10		1.9	
Tg by DMA (°C)	>300		113	

Table 2. Material properties of Samples A and B.

In the implementation of the SolderBrace material, a non-residue polymerizable flux is used in the ball drop process. The use of this special no-clean flux is to avoid cleaning flux residue for more uniform solder heights, and to fill gaps where flux residue may have impeded solder wetting, is shown in Figure 5. Sample B is generally used in conjunction with sample A, and was similarly analyzed. Table 2 shows the material properties of sample B. The fluxing material has a distinct Tg at 60 °C, and a modulus consistent with an unfilled material, such as a No-Flow underfill. Relative to sample A, sample B is significantly tougher with higher elongation. The increased durability of the Sample B is offset by the high thermal expansion, which if used in large areas would likely cause wafer bow, thus making it a poor choice if used without the Sample A coating. The hybrid approach to the SolderBrace materials, having both ridged and durable materials is a unique solution to transition the stress away from both the solder interconnect and the surface of the die.

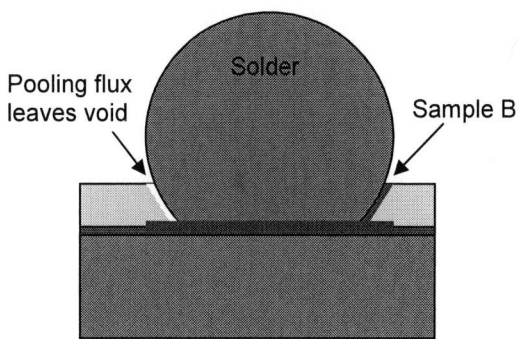

Figure 5. Depicts SolderBrace flux filling areas next to solder in WLCSP ball collapse process.

In this study the compressive modulus, adhesion, and creep properties were not evaluated on these reflow cured thin films. While the elastic properties give some indication as to the influence of the SolderBrace on package performance, the compliance is expected to play an important role

978-1-4244-4722-0/09 $25.00
© 2009 IMAPS-ITALY

with these SolderBrace materials. Furthermore the two material structure also complicate the predictive ability of the material properties, as the flux influences key interfacial forces.

To confirm sample A behaves as expected in the SolderBrace UV defined process, a spin coating was made on a bare 200 mm wafer to a thickness of 50 um. The coating was pre-baked at 100 °C for 6 min, exposed at 700 mJ/cm^2 (i-line), post exposure baked at 100 °C, and developed with propyleneglycolmethyletheracetate. The resulting imaged wafer showed good dimension uniformity, with minimal residue, as shown in Figure 6. Sample B was effectively printed by stencil into the open vias, but without a bond pad balls were not placed or reflowed. Anecdotally, when sample B was used with BGA devices we observed 100% wetting of the solder onto copper clad laminate.

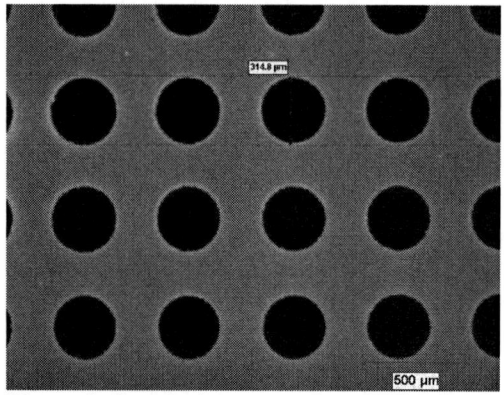

Figure 6. Sample A processed as a UV defined SolderBrace.

Compared to polyimides, which have CTEs in the 30 ppm/°C range, moduli of 1-4 GPA, with high elongation to breaks, sample A has significantly lower CTE, suggesting less overall thermal mechanical stress in package. But decrease in elongation to break suggests that sample A does not act to absorb stress. Conveniently the hybrid approach with the polymerized flux (sample B), introduces a much more elastic material at the interface, giving back the robustness of the coating, although it's unclear the necessary amount needed to optimize the low CTE / high elongation ratio for good package reliability.

An alternate approach to having a two material SolderBrace is to change the process so that only one material is required. As outlined in Figure 2, a print and cleaning process allows the ball drop process to occur independently of the SolderBrace where a special flux is not needed. Also, the printed materials do not require UV sensitivity. Despite the benefits of a print-over-ball approach, the WLCSPs wafers do require a costly solder stop as compared to the UV defined approach. In these experiments the print-over-bump process allowed us to evaluate two different material's thin film properties.

Sample C and D thin film samples were prepared and reflow cured. The mechanical properties of samples C and D are shown in Table 3. These samples represent both a durable film (sample C) with a moderate CTE, and a high modulus film with a solder matched CTE. The filler content of sample D is significantly higher than sample C, 75% by weight compared to 50% by weight. The crosslink density of sample C can also be inferred to be lower than sample D, accounting for the difference observed in the tensile and DMA results. The Tg determination in both thermal expansion and DMA were unclear, as the transitions were not significant. The DMA, like with sample A, showed a flat G' curve over an extended range. These two differing materials should allow contrast in the performance and needs, where low CTE films maybe preferred in some package designs, while others may require a more durable film. The relationship between material characteristics and reliability WLCSP devices will be given in future publications.[7]

Sample	C (+/-σ)		D (+/-σ)	
Modulus (GPa)	1.7	0.3	4.5	0.6
Toughness (MPa)	1.3	0.4	0.13	0.06
% Elongation to break	4.6	0.8	1.3	0.4
CTE α1 (ppm/°C)	32		13	
CTE α2 (ppm/°C)	110		51	
Tg by expansion (°C)	95		90	
Storage modulus @ 25°C (GPa)	6.5		14.5	
Tg by DMA (°C)	110		180	

Table 3. Mechanical data of samples C and D used as printable SolderBrace.

The key step in the print-over-ball process is the cleaning step, where the uncured, soft, organic film is partially removed from the solder along with any filler particles that may be on top of the solder balls. This process step can be accomplished in a number of high yielding methods. We have found Sample C to be particularly amenable to efficient cleaning by a simple vibratory polishing technique. As shown in Figure 7, 20x20mm panels of 250um pitch full area array test die were printed and pre-baked. The resulting coatings were not tacky and durable enough for further handling. After cleaning, the solder balls are uniformly well defined, with approximately half the solder height without residue. The cleaning process does etch the surface of the sample C coating between bumps, but this effect is minimized by cleaning for shorter periods. Cleaning does result in a matte surface finish.

978-1-4244-4722-0/09 $25.00
© 2009 IMAPS-ITALY

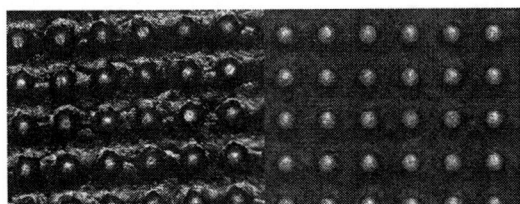

Figure 7. Cleaning results of sample A on 250um pitch flip chip devices. SolderBrace as printed on left (black/glossy) and after cleaning (black/matte) with exposed solder.

Sample D and sample A are similar in nature, having low CTEs and high moduli. The temperature dependant properties infer, like polyimides, the useful temperature range for these materials ranges from -50 to +300 °C. The lack of a distinct Tg in either the expansion or moduli data for samples A, C, and D suggest that these materials take on a significant inorganic character once processed on the wafers. We have noted that these films were processed in concentrated hydrochloric acid and organic solvents, indicating potentially good barrier properties.

Conclusion

SolderBrace™ materials are similar to polyimides in their UV processing, but have the added flexibility of being processed in other ways, notably print-over-ball. The UV defined SolderBrace combines in a hybrid approach the mechanical stability of molding compounds and polyimides with the ease of processing of a typical dielectric in a low cost solution for WLCSPs.

Acknowledgements

We would like to thank Robert Kyles, Russell Walls, John Hill, George Sears, Susan Chen at LORD Corp., Huihua Shu at Auburn University, and Chris Gregory at RTI for their contributions and support.

References

[1] U. Sharma; H. Gee; P. Holland; R. Swamy, "Thermal Reliability Performance of Large Area WLCSP Arrays", Proceedings of the 2009 International Microelectronics and Packaging Society (IMAPS), Scottsdale, Arizona, March 9-12, 2009.

[2] Gilleo, K.; Blumel, D.; "Flip Chip with Integrated Solder Mask and Underfill", US 6228678 B1 (2001).

[3] Capote, M.A.; Zhu, X.; "Semiconductor Flip-Chip Assembly With Pre-Applied Encapsulating Layers", US 6297560 B1 (2001).

[4] Sun, Y.; Zhang, Z.; Wong, C.P.; "Photo-Definable Nanocomposite for Wafer Level Packaging", Proceedings of the Electronic Components and Technology Conference (ECTC), 2005.

[5] D. Luttrull; A. Curtis; H. Pawlowski "Development and Screening of Polymer Collar WLP Candidates for Lead-Free Solder Sphere Technology to Enhance Reliability" Proceedings of the International Wafer Level Packaging Conference (IWLPC), San Jose, CA, October, 2005.

[6] R. Stapleton "Non-Capillary Protection Options for WLCSPs" Proceedings of the International Wafer Level Packaging Conference (IWLPC), San Jose, CA, October, 2008.

[7] Contact authors for manuscript

Highly Conductive Adhesives via Novel Heterogeneous Structures

Timothy D. Fornes* and Paul W. Hough**

LORD Corporation
*LORD Corporation, Cary, NC USA, +1 919-469-2500, +1 919-469-9688, tim_fornes@lord.com
** LORD Germany GmbH, Itterpark 8, 40724 Hilden, Germay, +49 2103 252 310,
paul_hough@lord.com

Abstract

There is an ever-growing need in the electronics packaging industry for high performance, polymer-based materials having increased electrical and thermal conductivity. All too often the steps taken to reach higher conductivities are accompanied by undesirable property tradeoffs such as reduced adhesion, compliance, and toughness and increased viscosity. In certain cases, fundamental physical limitations restrict the possibilities for formulating highly compliant, high strength adhesive materials. Recent technical developments at LORD Corporation have shown that many of the undesirable tradeoffs encountered in conductive adhesives can be dramatically reduced, if not eliminated. With careful control of adhesive morphology, materials that possess equivalent thermal and electrical conductivities are achievable at a fraction of the filler loadings seen in conventional technologies. Moreover, the morphology affords a broader performance space thus enabling previously unachievable property combinations.

Key words: electrical conductivity, thermal conductivity, interconnect, interface

Introduction

In recent years, electrically and thermally conductive adhesives have increasingly become an essential element in advancing electronic circuitry and device technologies. Numerous technical approaches for creating conductive materials have received considerable attention such as nanofilled polymers, isotropically conductive adhesives, and anistropically conductive adhesives [1-4]. Many of these approaches rely on heavily loading an organic matrix with metallic fillers in order to achieve high bulk conductivities. Unfortunately, this results in undesirable sacrifices in properties such as adhesion, modulus, toughness, viscosity, and reliability. Material cost may also increase, especially if expensive conductive fillers are used. In certain cases, fundamental physical limitations restrict certain property combinations from being achieved.

Recent technical advances at LORD Corporation have demonstrated that many of the undesirable tradeoffs encountered in conductive adhesives can be dramatically reduced, if not eliminated [5]. With careful control of adhesive morphology, heterogeneous materials can be produced that possess equivalent thermal and electrical conductivities at a fraction of the filler loadings seen in conventional technologies. Moreover, the unique morphology enables expansion of the performance space of conductive adhesives to levels previously unattained. This paper will highlight some the key aspects of this technology and how it addresses various challenges related to electricity and heat management in electronic packages.

Experimental

Adhesive samples were mixed using a Hauschild® mixer and heat cured using a programmable oven under a controlled ramp to a set peak temperature and time at peak temperature. Scanning electron microscopy (SEM) was performed using a FEI XL30 SEM. As mentioned later in this paper, highly conductive samples required no gold coating. Gold was applied to less conductive adhesives using a Denton Desk II sputter coater. Bulk thermal conductivity was measured via the Light Flash Method (ASTM E1461) using a Holometrix µFlash unit. The electrical resistance (or conductance) of each sample dictated the choice of resistivity instrumentation. Samples having resistances in excess of $\sim 10^{10}$ ohms were measured via ASTM D-257 using a HP 4339B High Resistance Meter equipped with a 16008B resistance cell. Samples having lower resistances, namely in the ranges of $\sim 10^2$ to 10^{10} ohms and below $\sim 10^2$ ohms were measured using a Keithley 610C Electrometer and a Keithley 580 Micro-ohmmeter, respectively. Lower resistance samples were in the form of well-defined cured strips with copper wire inserted at the ends of the sample prior to curing. The volume conductivity was calculated from the sample dimensions and the measured resistance. Modulus experiments were conducted using TA DMA2980 Dynamic Mechanical Analyzer (DMA) in a single cantilever mode at frequency of 1 Hz. Die shear strength measurements were conducted on

80 mm square silicon adhered to Ni-coated copper substrates using a Dage 4000 Series die shear tester.

Adhesive Structure – Property Relationships

In our laboratories, we have demonstrated that through appropriate selection of cure conditions and adhesive chemistry, namely that of the resin, curative, and filler, unique heterogeneous structures can be formed that possess exceptional conductivities. The SEM photomicrograph in Figure 1(a) provides an example of the type of morphology generated from an epoxy-based resin containing 33 vol% silver flake. The structure consists of discrete polymer-rich domains (very light, globular regions) distributed throughout a continuous filler-rich phase. (Note that the sample in Figure 1(a) was not coated with gold prior to SEM imaging. Thus regions that are electrically insulating will be heavily charged under the microscope and appear very bright and amorphous, i.e. polymer rich domains. Regions that are highly conductive will help dissipate the incident electron beam and appear less charged and have finer detail, i.e. silver rich domain.). Given the continuous nature of the filler-rich phase, the adhesive's conductivity is isotropic, i.e. conductive in all three orthogonal directions.

In contrast to Figure 1(a), the morphologies of traditional isotropic adhesives are typically homogeneous in appearance as shown in Figure 1(b). The SEM photomicrograph in this figure is that of an epoxy adhesive possessing the same type and amount of filler as seen in Figure 1(a). The morphology consists of an even distribution of silver throughout the polymer matrix. In this type of structure, the local concentration of silver is comparable to that of the bulk concentration, which is not the case in the heterogeneous system pictured in Figure 1 (a).

Ultimately, the differences in morphology of the cured adhesives lead to dramatic differences in thermal and electrical properties. The heterogeneous structure seen in Figure 1(a) results in nearly 20 times higher the bulk thermal conductivity (BTC) than that of the homogeneous structure shown in Figure 1(b). To gain perspective on where such heterogeneous data fall relative to data for traditional thermally conductive materials or thermal interface materials (TIMs), thermal conductivity is plotted versus vol% filler along with a plethora of TIM data based on adhesives, gels, and greases, all of which exhibit a homogenous morphology (see Figure 2). Note, the homogenous data fall on a common curve, exponential in nature, which rapidly increases in thermal conductivity at higher filler concentrations. This conductivity increase is concurrent with the concentrations at which filler-filler interactions begin to dominate.

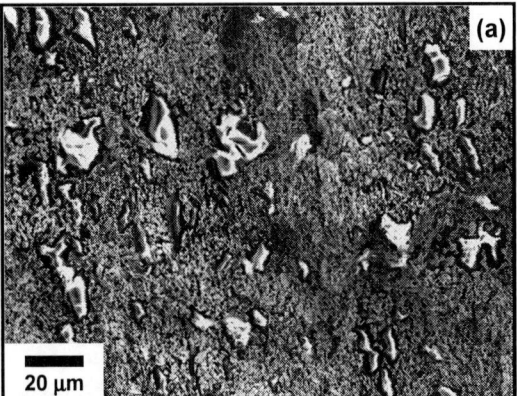

Figure 1: SEM photomicrographs of cross-sections of cured adhesives containing 33 vol% silver flake based on (a) heterogeneous morphology (BTC = 22 W/mK) and (b) homogeneous morphology (BTC = 0.7 W/mK).

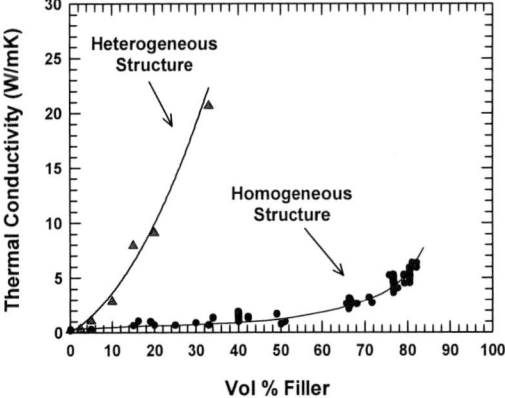

Figure 2: Influence of filler concentration on bulk thermal conductivity for composites based on heterogeneous and homogeneous structures.

As for the heterogeneous materials, thermal conductivity rapidly increases at relatively low filler loadings. Moreover, the levels of thermal conductivity are far higher than what is achievable through filler maximization strategies carried out in homogenous systems. For example, compliant (low modulus) homogenous systems are limited to BTCs of ~ 3-6 W/mK owing to the unreacted thermosetting or thermoplastic compositions becoming too viscous to be dispensed or handled [6-7]. It is for this reason that scientists have resorted to adding considerable amounts of low molecular weight species to the formulation, e.g. solvents, plasticizers, reactive diluents, or other liquid viscosity reducing species. However, a downside to this approach is that such species can cause shrinkage issues, void formation, and/or delamination which compromise thermal (and/or electrical) performance and reliability. The work presented here shows that the undesirable tradeoff in viscosity and conductivity can be greatly reduced, if not eliminated entirely, via adhesives that form heterogeneous structures upon curing.

Similar to thermal conductivity, electrical conductivity is strongly affected by the morphology of the cured adhesive. Figure 3 shows the influence of filler (silver flake) concentration on the volume electrical conductivity for cured adhesives possessing homogeneous and heterogeneous structures. The heterogeneous structure enables electrical percolation, i.e. the point at which the material abruptly changes from electrically insulating to conducting, to be achieved at a fraction of the silver concentration needed to do the same in the homogeneous system.

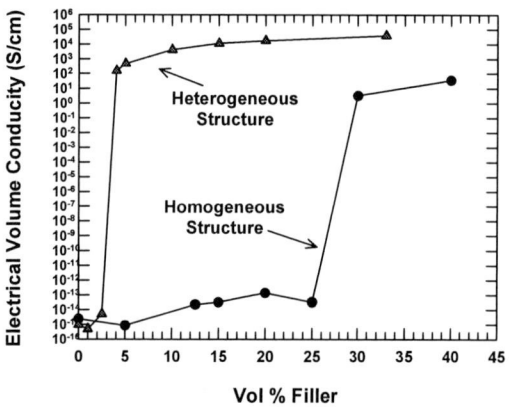

Figure 3: Electrical percolation curves for adhesives based on heterogeneous and homogeneous structures.

Specifically, the percolation threshold is only ~ 3 vol% filler for the heterogeneous composite versus ~27 vol% filler in the homogeneous composite. This much lower threshold is a result of creation of continuous, concentrated domains of filler upon curing. Figure 4 shows such morphological features observed just above the percolation threshold at 5 vol% percent filler.

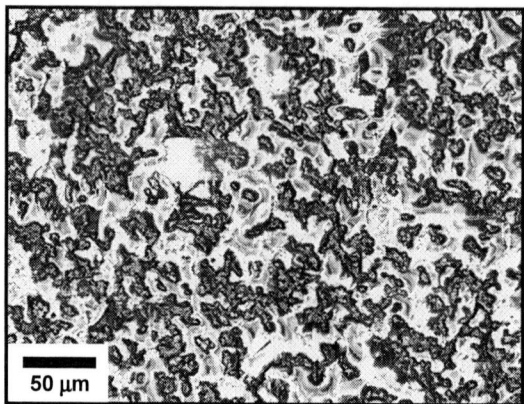

Figure 4: SEM photomicrograph of heterogeneous adhesive containing 5 vol% silver flake.

Interestingly, at high concentrations of silver flake, i.e. > ~30 vol% Ag, the heterogeneous structure affords approximately two orders of magnitude higher electrical conductivity than the homogenous adhesive, i.e. ~ 4×10^4 S/cm versus ~ 4×10^2 S/cm, respectively (See Figure 3). From a design standpoint, comparing the filler levels needed for the two systems to achieve a given level of conductivity is most compelling. For example, the concentration of silver flake (filler) needed to achieve a volume conductivity of about 100 S/cm (10^{-2} $\Omega \cdot$cm) is roughly 4 vol% for the heterogeneous composite whereas over 40 vol% silver flake is required for the homogeneous system. Such a difference is especially important with regard to minimizing tradeoffs related to such properties as viscosity, toughness, adhesion, and raw material cost. For the previously mentioned comparison, the viscosity of the 4 vol% systems is over an order of magnitude less than that of the 40 vol% Ag filled homogenous system. Lastly, it should be emphasized that more in-depth studies on the heterogeneous systems have shown that electrical (and thermal) performance can be further improved through refinement of chemical structure and cure conditions.

Addressing Packaging Needs

Development studies in LORD Corporation laboratories have recently shown that this heterogeneous technology has potential to address numerous emerging needs within the electronics community. With regard to thermal interface materials, increasing power densities coupled with larger die sizes have generated a need for materials that are both highly conductive and highly compliant (low modulus). Through appropriate selection of material chemistry and processing, adhesives possessing novel property combinations are attainable, namely low modulus combined with high conductivity and high adhesive strength, as shown in Table 1. Table 1 also shows that modulus can be

978-1-4244-4722-0/09 $25.00
© 2009 IMAPS-ITALY

controlled to range over an order of magnitude while maintaining high thermal conductivity and die shear strength with fixed filler content.

Table 1: Unique property combinations relevant to TIM applications afforded by heterogeneous adhesives.

Filler Content (vol %)	Modulus[a] (GPa)	Bulk Thermal Conductivity (W/mK)	Die Shear Adhesion (MPa)
20	0.5	10.9	26.4
20	5.1	9.1	38.0

Numerous needs exist in the area of electrical interconnects. A good example is a need for low melting, interconnect materials for replacement of lead-based solder and application involving printing circuits on temperature-sensitive substrates such as polyethelene terephthalate. Such applications require conductivities in the range of traditional solder, if not higher, and processing temperatures as low as ~120°C. Table 2 shows that electrically conductive adhesives based on the heterogeneous structure have the potential to meet these needs. High conductivies, i.e. those close to traditional solders, are achieved at temperatures as low as 120°C. Moreover, attractive levels of conductivity are maintained at low filler loadings.

Table 2: Influence of cure temperature on heterogeneous adhesives containing 10 and 33 vol% silver flake.

Cure Temperature (°C)	Electrical Volume Conductivity (S/cm) per Filler Concentration	
	10 vol% Ag	33 vol% Ag
120	2.3×10^3	2.1×10^4
160	4.2×10^3	3.8×10^4
200	5.5×10^3	4.4×10^4

Conclusions

Through appropriate selection of adhesive chemistries in concert with processing conditions, novel adhesives possessing heterogeneous morphologies were produced upon curing. The heterogeneous materials were shown to exhibit extremely high electrical and thermal conductivities at dramatically reduced filler loading relative to traditional materials based on homogenous morphologies. The high efficiency of the heterogeneous adhesives is capable of reducing, if not eliminating, undesirable tradeoffs encountered in conductive adhesives. Moreover, these materials are excellent candidates for addressing the thermal and electrical needs of existing and emerging electronic applications.

Acknowledgements

The authors would like to especially thank Nicolas Huffman for his help with adhesive formulating and conductivity measurements and Mike Owen and Sara Paisner for their technical advice.

References

[1] D. Kim and J. Moon, "Highly Conductive Ink Jet Printed Films of Nanosilver Particles for Printable Electronics", Electrochemical and Solid-State Letters,, Vol. 8, No. 11, J30-33, 2005.

[2] D.D. Lu, C. P. Wong, "Recent Advances in Developing High Performance Isotropic Conductive Adhesives", Journal of Adhesion Science and Technology, Vol. 22, No. 8-9, pp.835-852, 2008.

[3] Y. Li *et al.,* "Enhancement of Electrical Properties of Anisotropically Conductive Adhesive Joints via Low Temperature Sintering", Journal of Applied Polymer Science, Vol. 99, pp. 1665-73, 2006.

[4] X. Hu *et al.,* "Thermal Conductance Enhancement of Particle-Filled Thermal Interface Materials Using Carbon Nanotube Inclusions", Proceedings of the 2004 ITherm Conference on Thermal and Thermomechanical Phenomena in Electronic Systems, Las Vegas, Nevada, June 1-4 12-15, pp. 63-69, 2004

[5] T.D. Fornes, N.D. Huffman, "Method for producing heterogeneous composites", PCT Application WO 2008/118947, October 2, 2008.

[6] J.W. Bae *et al.,* "The properties of AlN-filled epoxy molding compounds by the effects of filler size distribution", Journal of Materials Science, Vol. 35, pp. 5907-13, 1992.

[7] T. Takahashi, K. Yamada, K. Isobe, "Thermally Conductive Grease Composition and Semiconductor Device using the same", U.S. Patent 6,372,337, April, 16, 2002.

Application of 3D modeling tools for advanced packaging on a broad range of industrial applications

Yvon IMBS[1], Laurent MARECHAL[1], David AUCHERE[1], Giovanni GRAZIOSI[2], Jason DEBONO[3]

STMicroelectronics Grenoble[1], Agrate[2], Kirkop[3]

yvon.imbs@st.com, laurent.marechal@st.com, david.auchere@st.com, giovanni.graziosi@st.com, jason.debono@st.com

Abstract

The scope of this article is concerning the electrical modeling tools used for the package parasitic extraction. Many 3D electrical modeling tools are on the market today enabling different type of electrical models to represent the electrical behavior of the packages. Semi-conductor companies are using many of them to address a wide range of applications, going from Communications, Consumer and Computer to Automotive and Industrial market segments. To get the right performances at the rights cost for all of these applications, the diversity of developed packages is considerable requesting also several strategies for the electrical modeling.

In parallel, IC interfaces speed is increasing for many of these applications fields and the new SOCs (System on Chip) are including more and more interfaces like USB, SATA etc...This is leading to an increasing impact of the package in term of electrical behavior. This trend makes electrical models mandatory for packages in a growing part of the product designs.

This paper will present several types of packages to be modeled to cover the different products. The usage and limitations of several modeling tools in regards with the applications will also be presented. Some additional limitations will appear for advanced packaging required by the most challenging products. Finally, the direction chosen to overcome these challenges through increasing collaborative work with CAD vendors and Institutions will be presented.

Key words: Packaging, Modeling, Quasi-static, Full Wave

1. Introduction

Integrated semiconductor companies are developing internally many IC packages to have custom solutions for their products in term of performances and cost. The range of applications is wide: Communications, Consumer, Computer, Automotive and Industrial markets. Therefore the diversity of developed packages is considerable. In parallel, IC interfaces speed is increasing for many of these applications fields, leading to an increasing impact of the package in term of electrical behavior. This trend makes electrical models mandatory for packages in a growing part of products.

Depending on applications, Digital or RF, Power or Analog, the needed model will be different. This disparity is also linked to the simulation flows which are in place in the different design teams. For the same type of package we can have to provide an S-parameter model in case of frequency simulations, a simple cell or a multiple cells RLCG Spice model to be used in time domain simulation.

This paper will present several types of packages and the existing commercial modeling tools used at the moment. A review of modeling capabilities, limitations and bottlenecks will be done. To overcome these limitations several actions are on going: a package electrical model validation program, some participation to European projects and collaborative developments with the modeling software providers.

2. Overview of the package portfolio

The package portfolio that we will cover is focusing on the communication, consumer, automotive and industrial markets excluding the specific packages developed for the microprocessor sector. All of these markets are subjects to pressure on cost while the package complexity is increasing.

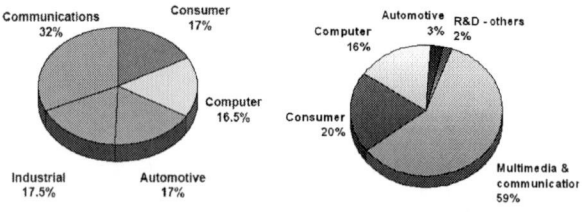

2008 sales (ST)   2008 models share (ST)
Figure 1: Package modeling repartition

978-1-4244-4722-0/09 $25.00
© 2009 IMAPS-ITALY

We consider two main mature platforms (still being developed for advanced packaging): the leadframe package platform and the laminate substrate platform (BGA). We also consider two emerging technology platforms: Sensors, MEMS & Optical and Wafer Level Packaging (WLP).

Mature platforms

The leadframe platform is mainly driven by the performance and manufacturability of the package at the lowest possible cost. There are two main families: small signal packages where electrical performances are the main drivers and power packages where power dissipation is the key element. Nowadays, the two families tend to converge to new packages where both electrical and thermal performances are optimized. This family has also the capability to integrate several chips.

Signal package: Power package: Signal & Power :
TSOP   HiQuad   MultiChip QFN
Figure 2: Lead frame packages

The laminate based package platform is driven by the miniaturization and the integration. These two trends are giving evolutions like the stacked dice packages and the System In Package (SIP). Recently a new configuration of BGA called Package over Package has been developed to overcome the limitation of the die stacking. For instance, in case of memory die stacking over an application processor, a new package development is needed each time the memory die is modified. With the PoP configuration, it is now possible to keep the bottom package unchanged while changing the top memory package.

Single chip BGA   Package over Package

 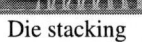 

Die stacking   System in Package
Figure 3: BGA packages

SIP enables the integration of several dice in a same package (mix of front-end technologies can be done) and permits to add surface mounted component like R,L & C elements or specific filters.

Emerging platforms

The sensor and optical package platform is driven by many applications. Main markets are phones, PCs, toys, automotive and health. Here, the challenge is the necessity of multi-disciplinary developments covering electronics, optics, mechanics, chemistry, etc… Integration, miniaturization and cost are the final goals of these developments with multi-chip packages.

Lab on Chip   Camera module
Figure 4: Emerging platform packages

The Wafer level package platform is driven by the wireless and handheld applications. This technology is targeted to get the highest level of integration. This platform is becoming reality thanks to several technological improvements like the very fine pitch bumping, through Silicon vias, wafer level encapsulation and multilayers technologies. The main difference between this platform and the BGA is that the die is included in the core material of the final product and that metal layers and dielectric layers are build on top of this core layer using silicon-like processes.

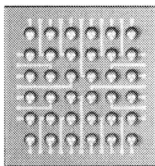 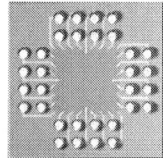
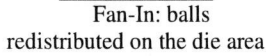

Fan-In: balls   Fan-Out: area extended
redistributed on the die area   thru organic material
Figure 5: Wafer level packages

MEMS (Micro Electro-Mechanical Systems) packaging platform is evolving to new low cost packages suitable for consumer applications. MEMS package does not provide only the electrical connection between the device and the external world; in fact the package itself is the mechanical connection to the outside world so it can impact the device performance in the same way the electrical RLCG parameters do for the electrical performance.

New applications like accelerometers (such as free fall protection for hard disk drives on laptop and cell phones, gaming interfaces, etc) require lower cost and smaller packages. The first MEMS packages were open cavity; however in standard electronic applications low cost packages are dominated by full molded solutions either on lead frames (for example SO or QFN) or increasingly using substrates.

978-1-4244-4722-0/09 $25.00
© 2009 IMAPS-ITALY

 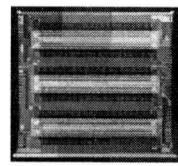

Accelerometer (LGA)   Printer heads

Figure 6: MEMS packages

The LGA (land grid array) package which is laminate based but without balls, has been recently introduced to increase the flexibility in manufacturing.

Figure 7: Stacked accelerometer

For each of the presented package families, electrical models are needed with specific targets depending on the application areas.

3. Tools

In this section highlights will be given on design tools, on conversion tools that permit to generate 3D structures of the packages and on 3D electrical modeling tools.

Package design tools

Cadence SIP and Sigrity UPD are the most common tools used in the industry for the laminate based package designs. These tools are able to cover the full design activity (routing, wire bonding, Flip-Chip, etc...) and are generating the manufacturing and documentation files (gerbers). For the leadframe design one mechanical tool is the most commonly used: Autocad from Autodesk.

Due to the type of interconnections existing on the different packages (wire bonding, flip-chip, vias, balls), 3D modeling tools are required to capture the electrical behaviors of the packages. In the coming sections we will therefore focus on these tools, not considering any planar tool.

Conversion tools

CAD design tools are basically 2D tools. A translation is therefore required to generate a 3D physical model of the package compatible with electrical modeling tools. This translation is done by different ways (Fig. 8), depending on the extraction tool and on the package design format. Tools used in our teams are well representative of the current modeling market offer for industrial applications.

The existing links between CAD tools dedicated to laminate package design and extraction tools allow a good level of automation for physical structure import (substrate stack-up, wires and balls geometries ...) and electrical setup (excitation assignments, return path definition...). For lead frame packages, the translation is currently totally manual. 2D geometry files are imported in the extraction tool and then have to be rebuilt in 3D structures. Electrical settings are also made by the user, leading to longer treatment time.

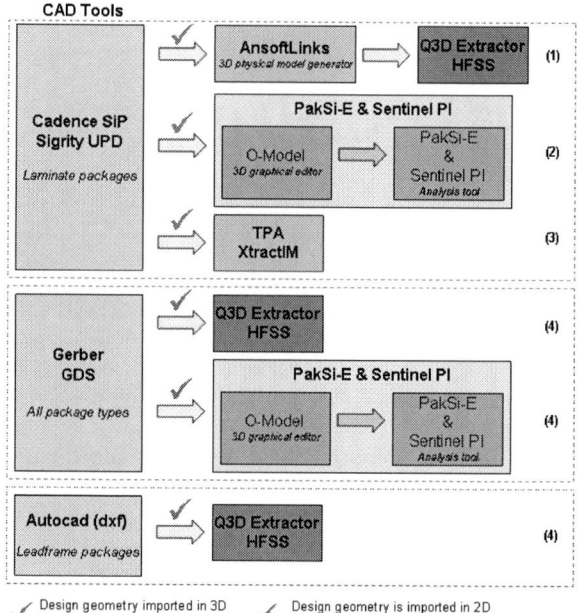

Figure 8: Conversion design/modeling tool

(1) Generation by an external tool of a 3D physical model compatible with the extraction tool.
(2) Package database import and translation in extraction tool through an embedded 3D graphical editor.
(3) Direct export from the design tool interface through a plug-in. 3D physical model can be edited in the extraction tool environment.
(4) Manual import of 2D geometry files and edition of the 3D structure.

Package electrical modeling tools

3D modeling tools are using an iterative algorithm that divides the structure into a meshing and then calculate the electromagnetic field in each element to get the final result. Two main types of modeling tools can be defined.

Full-wave tools compute electromagnetic fields inside the structure and an S-parameters matrix, defined from the ports specified in the 3D model, can be extracted. Full-wave solvers require important hardware resources and long computation times. This limits the size and complexity of the structures that

978-1-4244-4722-0/09 $25.00
© 2009 IMAPS-ITALY

can be modeled to few elements or to a single application interface (USB, SATA, DDR…). However, full-wave analysis is considered to be accurate in a broad range of frequencies, excluding the DC to ~100MHz range due to limitation of the numerical solvers.

Quasi-static tools compute the electrical fields and the magnetic fields in two separate processes. Resistance, inductance, capacitance and conductance values of the nets are derived from the computed field quantities.

Application of quasi-static approach (fig. 9) is appropriate to describe electrical behavior in a frequency range between DC and just below the first structure resonance frequency.

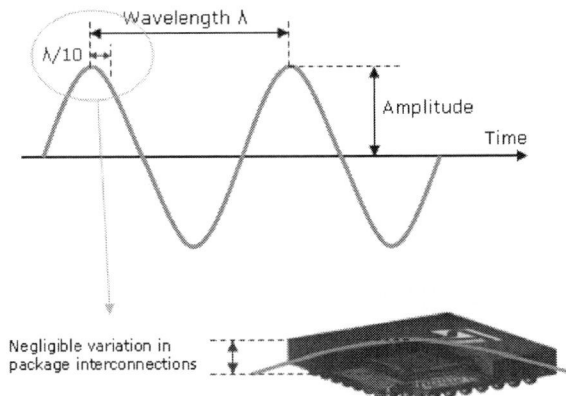

Figure 9: Quasi-static approach

The ratio between the maximum length of the package interconnection (L) and the wavelength ( ) of the application is playing a major role for the package modeling.

- If L < /10 we are in the quasi static approach validity region and the signal can be considered as only time dependent (no propagation inside the package): each connection can be described through RLCG cells (lumped model).

- If L> /10 the quasi-static approach is no more acceptable as propagation phenomena are appearing on the interconnections: the full wave approach is needed.

Quasi-static solvers can provide solutions for many nets and complex structures in a reasonable time. Some quasi-static tools are able to handle a full package extraction with several hundred pins enabling therefore coupling studies over a full interface or from an interface to another.

Figure 10: Electrical modeling tools features

Solver	Quasi-static			Full-wave
Software	TPA	Paksi-E	Q3D	HFSS
Output	RLCG model			S-Params
Extraction	Partial or full package		Partial package or user drawn structure	
Package type	Laminate package		Laminate and leadframe package	

The quasi-static approach is accurate at low frequencies but is considered as very limited for the increasing signal speed. However, there are solutions to extend the model validity while keeping the capability to take into account full interface and easy time domain simulations.

The following chart (fig. 11) is presenting the comparison of 4 modeling approaches for a USB interface:
- Full wave approach
- Quasi-static approach with a lumped model (one T-shaped cell)
- Quasi-static approach with a distributed model (3 T-shaped cells)
- Quasi-static approach with physical split of the 3D structure so that the /10 rule applied on each separate part of the nets.

Results show that the distributed model can improve the model validity and that the physical split is even better. However, this fourth solution has a drawback in term of 3D structure set-up as it is a manual approach.

The signals behaviors of the interfaces are also largely related to the quality of their return paths done mainly thru the power & ground networks. Generally power & ground networks are spread all over the package area with layouts of the planes that cannot be considered as perfect as they are including many vias, via holes and cut-outs. All these 3D structures have to be considered to get an accurate model of the loop inductance. Some solvers are therefore reaching their limits for the meshing and/or solving. This is of course depending of the size of the package. Some of the above mentioned tools (PakSI, TPA) are able to provide models either on a full design or on some specific areas that could be used for Power Distribution Network simulations. The drawback of these models is the number of ports that are produced if all Power & GND die pads and balls are considered. It can lead to very heavy models, means difficult to simulate, and make the use of the model very complicated as all these ports will have to be connected to the die and the PCB.

978-1-4244-4722-0/09 $25.00
© 2009 IMAPS-ITALY

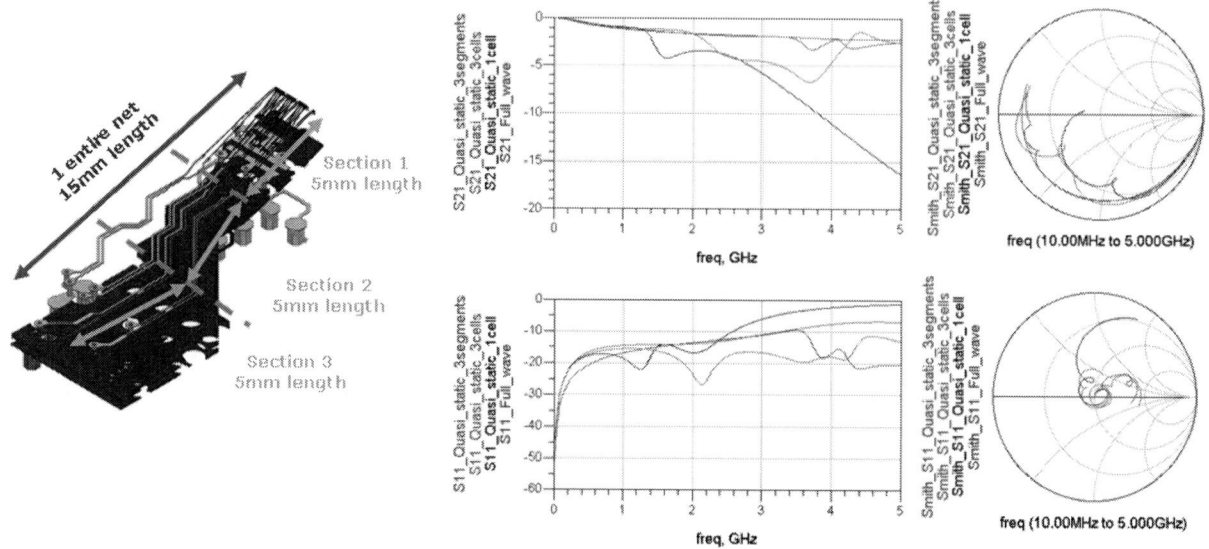

Figure 11: Electrical model validity over frequency

Moreover, due to the new silicon technologies with reduced voltages, the behavior of the power network inside the package needs more and more investigation. In many applications the package can be one of the major contributors to the noise on the power and ground nets [1].

This trend is reinforced on low cost laminate packages were the number of metal layers is reduced to the minimum and the routing density pushed to its maximum. In this case, the power network can be reduced to traces not larger than signal traces and a much reduced set of planes.

Recently, new tools (like Sigrity PowerDC or Apache Sentinel PI), able to compute the static and dynamic voltage drop, have been introduced to ease the link between the package model, the die and PCB models for complete system voltage drop estimation. These tools are also enabling some ports grouping on either die or balls side so that the produced model can be implemented more easily.

The results of these tools can also be color maps that are very useful to optimize the power network in the packages.

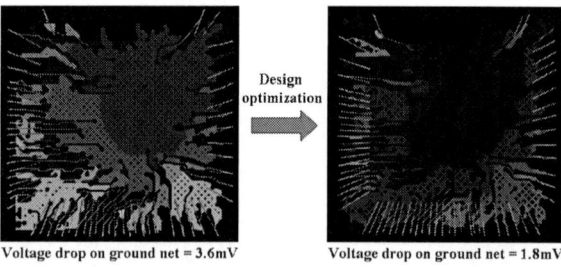

Figure 12: Voltage drop map – Before and after design optimization

Model accuracy limitations

Many questions addressed to people in charge of package modeling are linked to the accuracy of the models. Some limitations are well known, some are more complex.

Materials characteristics: Electrical properties of materials vary over the frequency (dielectric constant, loss tangent …). To get the maximum accuracy, these variations should be well known over the full frequency range covered by the model. This is not the case for all suppliers or all materials. In addition, some tools are not able to take into account the frequency dependency of materials parameters. The figure 13 curves are showing a good correlation between a low cost dielectric micro-strip structure measured and extracted with a full wave tool using a fixed description of the dielectrics till 7 GHz lowering the impact of these missing data.

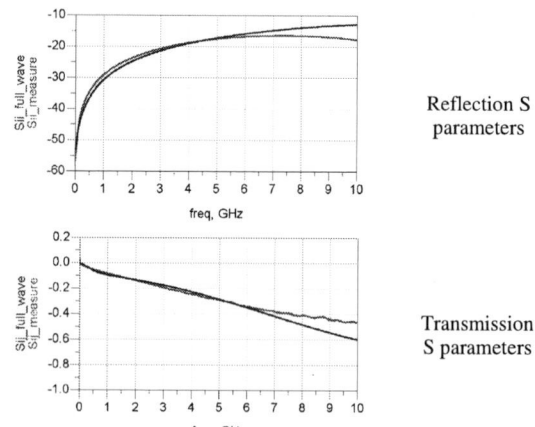

Figure 13: Correlation measure/full wave model

External Environment: For many projects, elements surrounding the package (PCB, companion chips etc...) are not defined or available when the package model is requested. This is the case when a new product is developed and that the final customer (who will assemble the board) is not yet defined. Another common case is when the final customer doesn't want to disclose any information on their board design. Consequently, in many cases, the package will be modeled as an isolated component or with generic datas on the board environment.

Manufacturing tolerances: IC packages are also subject to assembly and manufacturing process tolerances (fig. 14) affecting physical dimensions of the stack-up, of the traces and position of elements inside the package (dice, wirebonding etc...). Consideration of all parameters combinations would require a large computation time, much too long for an acceptable model delivery time.

Category	Physical	Tolerance	Electrical values
Substrate manufacturing	Material thickness	+/- 10%	Zo = 50 Ω ΔZo = +/- 6%
Substrate manufacturing	Trace width	+/- 30%	Zo = 50 Ω ΔZo = +/- 12%
Substrate manufacturing	Eps R	Data not available	-
Die attach	Positionning	+/- 50 µm	ΔRmax= 6 mΩ ΔLmax= 0.1 nH
Wire bonding	3D profil	+/- 30 µm	ΔRmax= 2 mΩ ΔLmax= 0.03 nH

Figure 14: Manufacturing tolerances impact

4. Modeling capabilities vs. applications

In this section we will review the product applications and their associated package modeling requests. The main drivers for the package modeling were historically Radio-Frequency (RF) applications (still demanding). With the increasing speed of digital interfaces this fact is changing drastically. The new Front-End technologies working at lowest voltage are also increasing the impact of package power network on the overall system voltage drop.

For each application field, the strategy for electrical modeling of the package is specific. It will depend on the package platform, on their sizes, on package database formats, on the requested frequency range (DC, HF, narrow & wide band...) and on the expected output file (Lumped, S-parameters, distributed models). All these specificities have to be covered and no single tool is currently able to fulfill these requirements entirely.

Radio-Frequency applications

Even if digital applications are becoming more and more common with the convergence trend, RF IPs are mandatory in main applications. In term of packaging the preferred packages for RF parts (transceivers, Power Amplifiers) are small, 5mm x 5mm to 10mm x 10mm usually and can be BGAs or QFN packages. The main use of models for the RF applications is linked to the impedance matching.

Depending on the frequency of the application and thanks to the small package sizes, many provided models are quasi-static based models. When several harmonics have to be considered in simulations which are mainly frequency domain simulations, S parameters are provided.

As mentioned in the packaging section, one trend for the advanced packaging on BGA is the System in Package. This type of package can integrate several dice, several type of interconnection to the laminate substrate like wire bonding, Flip-chip and can include surface mounted devices to ensure the matching inside the package implying therefore to evaluate the coupling between nets when computing the isolation.

One additional need difficult to fulfill for the moment is to get a model able to cover the coupling between package elements and some specific components at die level like VCO inductor. For PA modules having important signal levels the isolation between the several dice should also be computed but due to the difference of scale between the silicon geometries and the package geometries this computation remains difficult to do for the moment for meshing reasons.

One new area under investigation is the integration of passive elements inside the BGA laminate [2]. For these modeling requests the need of full wave tools to produce S-parameters is mandatory to compute the self values and the associated parasitics. For the embedded inductance inside BGA laminate, S-parameters models produced by HFSS are showing a good correlation between measurements and extracted models.

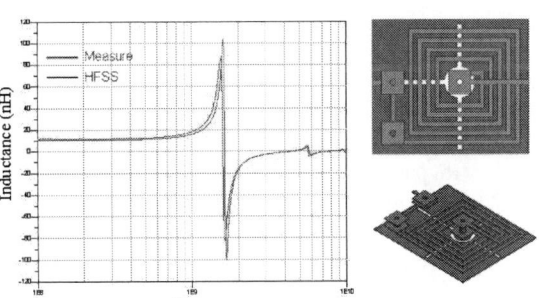

Figure 15: Laminate passives correlation

Digital applications

This family of products is moving quickly to the latest Front-End technologies enabling silicon die size reduction and new functionalities integration. Consumer, Communication (Multimedia processors) & Computer & Memory markets are following this trend. It has an impact in terms of packaging as the voltage noise margin is decreasing at each new silicon technology node (65 nm, 45 nm etc...). In addition, new generations of interfaces developed to support the customer market are also showing a speed increase.

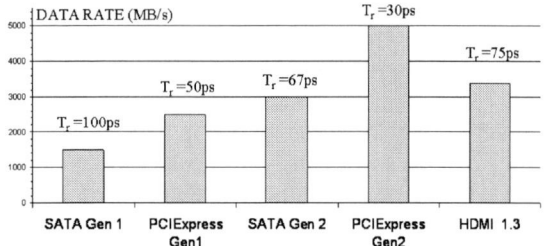

Figure 16: Data rate and rise time of high speed interfaces

These two evolutions are responsible of the number and type of models that have to be provided for each new project:
- accurate modeling for the high speed interfaces
- accurate models including coupling between the several interfaces in the same package,
- IBIS models of the entire chip (including a full package model) that will be used to design the PCB
- power models to ensure the functionality of the system and to define the decoupling strategy of the system.

The first two points are highly dependent of the interface we are considering. For instance a 75 ps rise time signal (fig. 16) requires a bandwidth from DC to at least 5 GHz. The size of the consumer packages is in the range of 23mm x 23mm to 35 mm x 35 mm. This means that the interconnection lengths can reach 15 mm, leading to a max frequency of 1 GHz according to the quasi-static approximation. As demonstrated in the previous section this means that the adapted models are either multi-section quasi-static models or S parameters models. As the interfaces frequencies are continuously increasing the need of S parameters is also growing. The trade-off has to be found between the needed accuracy, the coupling coverage, the size of the model and the computation time.

Ibis modeling is very limited in accuracy but is still requested by many companies for there PCB design due to easy model instantiation and short simulation run time.

For the Power Distribution Network the key electrical factors are the reduction of power/ground impedance to reduce the voltage drop (increasing capacitance and decreasing inductance). The dynamic voltage drop requires the reduction of the switching signals inductance to limit the signal ringing. From the modeling point of view is fundamental to choose carefully the current return path to have the correct self inductance estimation; in addition the mutual parasitic values are needed to check the level of cross-talk between aggressor signals (e.g. clocks signals) and the victims.

Analog, Mixte & MEMS applications

Applications in the Analog & Mixed signal area are mainly focusing on power management products and are demanding package modeling to evaluate the resistance of the paths. For MEMS structures the requests remain sporadic and are highly depending on the final use of the device.

Industrial applications

The current aggressive technology scaling trend and the increased complexity of digital ICs lead to higher operating frequencies with faster clock rates. In this context, the heterogeneous System on Chip (SoC) or the technological alternative of SIPs, mixing digital and analog parts with memories, are becoming relevant EMI (Electromagnetic Interference) generators.

Till now, in case of EMI emission out of the level imposed by international regulations or by the customer, the standard approach was to put on the final PCB off-chip decoupling capacitances; obviously this way of working is no longer acceptable. Package designers have now to work in close cooperation with silicon designers implementing a co-design flow to achieve EMI reduction. The electrical package modeling is part of this flow; in fact the package parasitic inductance can be a relevant EMI source that can also induce ringing effects. This is why package design for automotive applications is mainly focused on inductance optimization through the overall electrical path reduction (mainly done reducing wire length). The extraction of RLCG spice model for package power/ground and clocks connections is done with the quasi-static approach and simulated together with ICs models in time domain based tools.

5. Next steps

The previous sections showed that for each type of applications there are some specific points where the current way of generating package model will need to be revisited for the next product generation. Several actions have been started to ensure the design of next generation chips.

To ensure the validity of the models over a wider frequency range some internal validations thru measurement have been started. The problematic is to define with more accuracy the link between the announced I/O rise time and the real signal that the package has to carry. From this data we should be able to define which tool can be used for the package

978-1-4244-4722-0/09 $25.00
© 2009 IMAPS-ITALY

modeling. It will also permit to validate package models, activity which is done at the full system level at the moment.

Participation to European projects will also permit to define a common vision of the requested models between several European companies and work to solve the known limitations. Another advantage of these collaborations is to be able to deliver a consolidated specification concerning our requests to EDA companies. In parallel with the European projects, bilateral developments have started with the main CAD vendors to reach an acceptable compromise between the accuracy and the model generation time. A convergence between leadframe design, BGA design and modeling tools is targeted in near future.

6. Conclusion

This paper showed an overview of the package modeling activity in relation with the specificities of diverse products. After having reviewed the IC packaging trends, we considered the package modeling flows and tools adapted to the several types of package. We also mentioned the main limitations for the tools linked to either a full wave solver or a quasi-static approach. Finally we reviewed the main bottlenecks we are facing for different applications and the direction to overcome these new challenges.

References

[1] Digital Signal Integrity: Modeling and Simulation with Interconnects and Packages by Brian Young.

[2] RF packaging and passives: design, fabrication, measurement, and validation of package embedded inductors - Chickamenahalli, S.A.; Braunisch, H.; Srinivasan, S.; Jiangqi He; Shrivastava, U.; Sankman, B.; IEEE Transactions on Advanced Packaging.

978-1-4244-4722-0/09 $25.00
© 2009 IMAPS-ITALY

Experimental Characterization of Thermo-Mechanical Properties of Lead-based Solders for Power Electronics Packaging Reliability Applications

S. JACQUES[a,b], J. ROUBION[c], N. BATUT[b], R. LEROY[c], and L. GONTHIER[a,b]

[a] STMicroelectronics, 16 rue Pierre et Marie Curie – BP7155 – 37071 Tours Cedex 2, France
[b] "Laboratoire de Microélectronique de Puissance (LMP)", University of Tours, France
[c] "Laboratoire de Mécanique et de Rhéologie (LMR)", University of Tours, France

Tel.: +33 (0)2 47 42 83 45 / Fax: +33 (0)2 47 51 01 34 / E-mail: sebastien.jacques@st.com

Abstract

This paper deals with the thermo-mechanical properties experimental characterization of solder alloys used in power electronics packaging, and especially for TRIAC. The lead-based solder – $Pb_{92.5}Sn_{5.0}Ag_{2.5}$ – is evaluated. The Young's modulus, the yield strength, the mechanical strength, and the elongation at break are measured from 25°C up to 150°C. Additional creep measurements are performed to explain the differences between static and dynamic measurements at high temperatures. Finally, fatigue tests are presented. The fatigue lifetime could be correlated with some power cycling reliability test results. All these data could be implemented into some thermo-mechanical simulations using finite elements to predict TRIAC lifetime in real applications.

Key words: Lifetime prediction, solder joints, DTMA, creep measurements, fatigue tests.

1. Introduction

A new high-temperature TRIAC family has been introduced. This family helps to reduce the bulk of the required heatsink. These TRIACs are particularly suitable for hot or limited environments found in home appliances, such as vacuum cleaners [1].

At the present time, new functional reliability approaches are used because experimental reliability tests are expensive (devices under test vs. Test duration) and take time. Given this situation, general trend consists in developping some physical models to predict devices lifetime in real applications [2].

These new approaches are built on calculations of strain range, energy curves or damage accumulation, thanks to numerical simulations using finite elements. Then, these data are integrated into lifetime prediction models (e.g. Coffin-Manson). However, these models have to be calibrated by targeted experimental procedures.

To predict TRIACs lifetime using the methodology previously described, it is necessary to have the thermo-mechanical properties of the assembly materials, and especially for the solder joints. In many applications, data usually found in literature miss some information: temperature impact, thermal and mechanical stress validation ranges etc. That is the reason why, experimental characterization of the solder alloys physical properties has to be performed.

In this paper, dynamic thermo-mechanical analysis (DTMA) results on TRIAC interconnection materials – $Pb_{92.5}Sn_{5.0}Ag_{2.5}$ – are described. First, we explain the process to obtain the samples used for the characterization. In a second part, the Young's modulus, the mechanical strength and the elongation at break are measured between 25°C and 150°C. Additional creep measurements and fatigue tests are performed. The fatigue lifetime could be correlated with some power cycling test results.

2. Experimental environment

2.1. Context reminder

In power cycling, TRIACs are potentially vulnerable to thermal fatigue because of temperature cycling due to devices self-heating [3], [4]. The failures can mainly affect the package, and especially the solder joints. **Figure 1** shows a cross-section example for an insulated TRIAC, failed after 10000 cycles at a 155°C junction temperature swing (ΔT_j). A crack has been observed between the heatsink and the ceramic. In fact, at this location, one finds the worst CTE (Coefficient of Thermal Expansion) mismatch, the maximum temperature swing combined with the largest lateral dimensions. In these conditions, TRIACs lifetime could be impacted, and could limit the application use.

Realistic finite element modeling could be very helpful to estimate stess and strain levels to predict TRIACs reliability. However, the thermo-

978-1-4244-4722-0/09 $25.00
© 2009 IMAPS-ITALY

mechanical properties of the solder joints are necessary to build the prediction model.

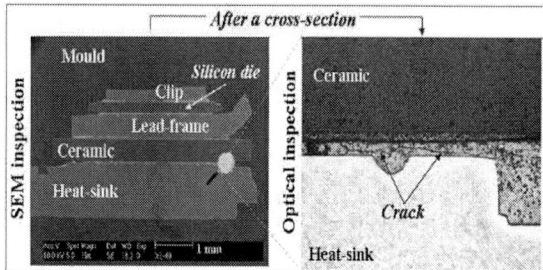

Figure 1: Example of solder joint degradation for an insulated TRIAC subjected to power cycling

2.2. Samples preparation

This part describes the process to make the "dog bones" used for the thermo-mechanical measurements.

The lead-based solder – $Pb_{92.5}Sn_{5.0}Ag_{2.5}$ – comes from solder paste. We use raw materials to be as closed on TRIAC actual solders as possible. The solder paste is inserted into a mould in graphite, not to require any chemical substance to separate the sample from the mould. The oven temperature profile is chosen so as not to have any void in the sample that could modify its mechanical properties.

The sample is then inspected using X-Rays analysis. **Figure 2** shows a "dog-bone" example with no void formation. The sample geometrical dimensions are 30mm × 2mm × 0.6mm. Its section area is equal to 1.2mm². These dimensions are measured in three key points to guarantee a 10^{-2}mm measurement precision. The targeted section is 2 × 0.6mm, according to the DTMA requirements, as explained below.

Figure 2: X-Rays analysis of a sample

2.3. Dynamic Thermo-Mechanical Analysis

The Dynamic Thermo-Mechanical Analysis (DTMA) consists in applying a sinusoidal stress (force or displacement) to a material, and measuring the sample response. The material stiffness is then calculated from this data and converted to a modulus.

Figure 3 shows the DTMA bench used for the measurements. The analyzer can perform some static and dynamic (up to 200Hz) tensile tests, with temperature-dependence. The manufacturer guarantees the following measurement precisions: ±2.2% for the forces or displacements measurements, ±4.8% for the stiffness, and ±0.01%

for the frequency [5]. We can see that the most important source of error is due to the sample geometrical dimensions, its set-up in the jaws, as well as the placement of the column (for the support).

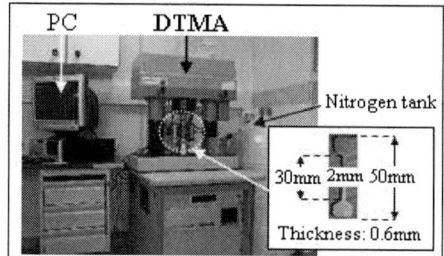

Figure 3: Experimental test environment

3. Experimental tests set-up and main results

3.1. Static measurements

The tensile tests are performed according to the NF 03-151 standard. A 100N static force has been applied with a 1N/s rate. The tests have been performed for four sample temperatures: 25°C, 70°C, 100°C, and 150°C. Five samples have been tested for each temperature.

Figure 4 and **Figure 5** show respectively the yield strength evolution and the mechanical strength versus the temperature. The yield strength is measured for a 0.2% strain, because it is a standard value for most of mechanical engineers [6]. That is the reason why, it has been chosen for this study.

Figure 4 and **Figure 5** show that the two parameters decrease about 50% when the temperature increases from 25°C to 150°C. In fact, when the temperature increases, the material is softer (due to thermal agitation). Thus, the yield strength and the mechanical strength decrease.

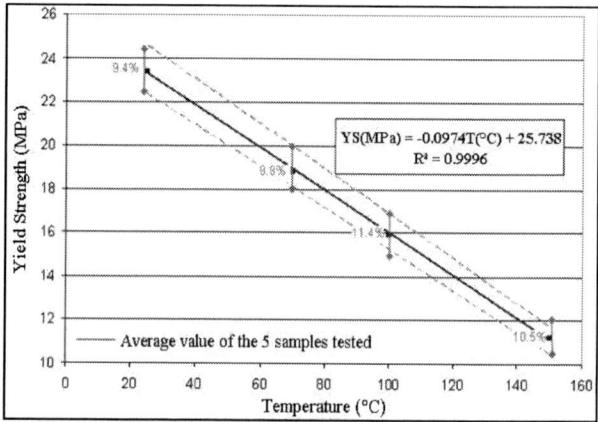

Figure 4: 0.2% Yield Strength evolution versus temperature

978-1-4244-4722-0/09 $25.00
© 2009 IMAPS-ITALY

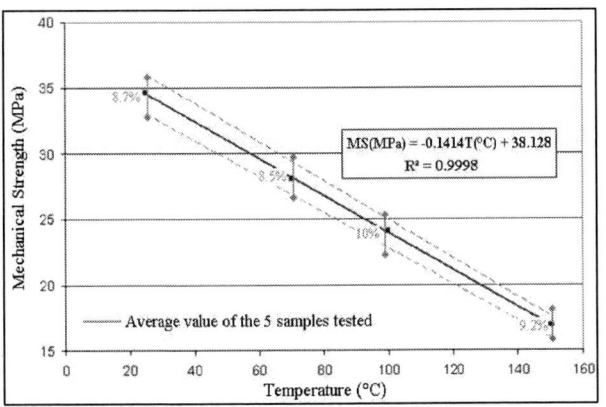

Figure 5: Mechanical Strength evolution versus temperature

3.2. Dynamic measurements

It is quite difficult to measure the Young's modulus, because it is necessary to have the slope of the Stress vs. Strain curve linear part. It is also difficult to guarantee a measurement with a good enough accuracy. That is the reason why, we define a dynamic Young's modulus. This one is deduced from the material resonance.

The dynamic analysis is performed for various frequency values, from 1Hz up to 200Hz. The tests have been realized for four sample temperatures: 25°C, 70°C, 100°C, and 150°C. Five samples have been tested for each temperature and each frequency value.

Figure 6 shows the dynamic Young's modulus evolution with the frequency. The useful dynamic Young's modulus range is between 10Hz and 80Hz. For the values lower than 10Hz, it is difficult to use this parameter value. Beyond 80Hz (particularly, between 80Hz and 130Hz), we observe a disturbance due to the sample mechanical resonance.

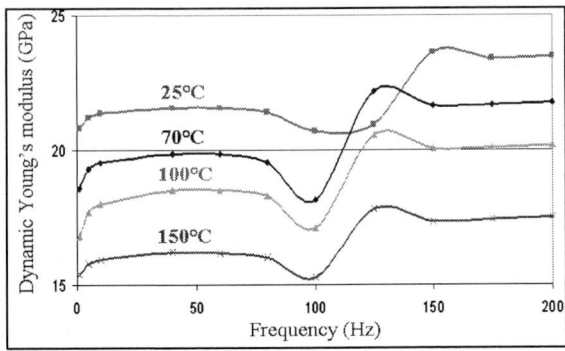

Figure 6: Dynamic Young's modulus evolution versus frequency

Figure 7 shows the dynamic Young's modulus evolution versus the sample temperature, in the useful frequency range (that is to say between 10Hz and 80Hz, because this parameter has approximately a constant value). The results show that the dynamic Young's modulus decreases about 33% when the temperature increases from 25°C to 150°C. This is due to the fact that the material expands when the temperature increases.

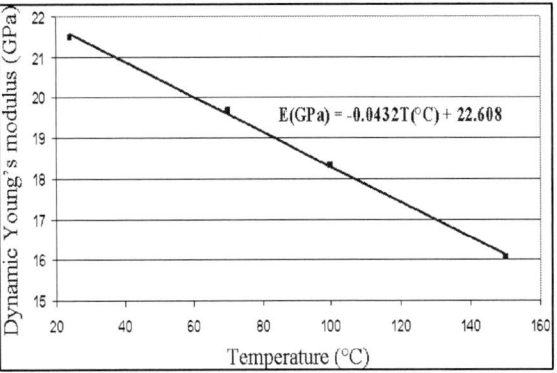

Figure 7: Dynamic Young's modulus evolution versus temperature

3.3. Creep behavior

Creep is a persistent form of deformation. For solder joints, and especially for the $Pb_{92.5}Sn_{5.0}Ag_{2.5}$ metallic alloy, this phenomenon appears even at room temperature because of the material low melting temperature (296°C). At higher temperatures, creep mechanism tends to dominate deformation processes, and could lead to solder joint failure [7].

Creep measurements are performed with various static forces from 4N to 14N, with a constant loading rate (2N/s). We keep the same temperature values as previously mentioned. Three samples have been tested for each temperature.

Figure 8 shows a creep test example for a 100°C temperature. Immediately after the load is applied, there is an instantaneous strain. At a given stress and temperature, the steady-state creep rate can be determined by calculating the slope of the strain vs. Time curves in the linear part. Knowing the slope value, it is possible to calculate the creep activation energy. The average activation energy value for this Pb-rich solder, $Q_a \approx 55kJ.mol^{-1}$, is closed to what we can find in literature, that is to say between $40kJ.mol^{-1}$ and $60kJ.mol^{-1}$ [8].

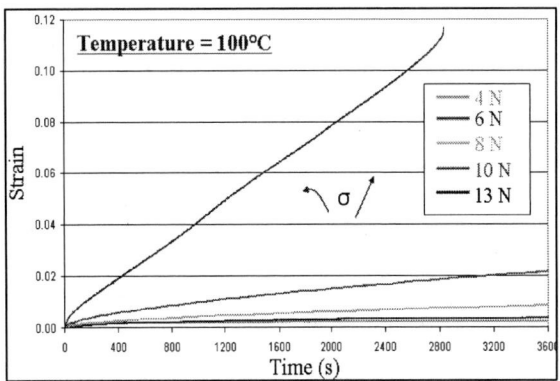

Figure 8: Strain versus time curves for various load creep tests at 100°C

Finally, **Figure 9** shows the strain rate evolution versus the stress, in logarithmic scale, for the various temperature values (25°C, 70°C, 100°C, and 150°C). For each temperature value, there is a linear dependence of the strain rate with the stress. The slope of each straight line permits to evaluate the Norton's coefficient. It should be remembered that this coefficient defines strain rate sensitivity to the stress. The Norton's coefficient, whose values are summed-up in **Table 1**, is approximately three times higher when the temperature increases from 25°C to 150°C. In fact, at room temperature, this coefficient is higher than 1, which indicates that the creep mechanism is very important at this temperature value.

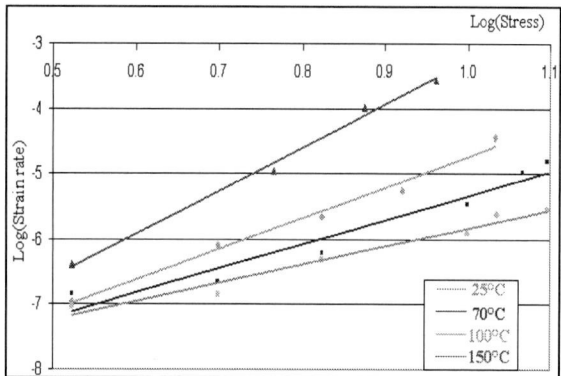

Figure 9: Strain rate versus stress for various temperature values

Temperature (°C)	Norton's coefficient
25	2.8
70	3.7
100	4.7
150	6.6

Table 1: Norton's coefficient values versus temperature

3.4. Fatigue test results

Lifetime prediction models describe the number of cycles to failure dependent on power cycling test parameters. To use some models (e.g. Coffin-Manson or Engelmeier [9]), it could be necessary to have material fatigue properties.

The easiest way to have the $Pb_{92.5}Sn_{5.0}Ag_{2.5}$ solder fatigue performances is to characterize the "S-N curve", also known as a "Wöhler's curve". This is a graph of the cyclic stress magnitude versus the number of cycles to failure (in logarithmic scale).

Fatigue tests are performed with various dynamic stress values, which are lower than the ultimate tensile stress limit, and may be below the yield stress limit. In this paper, the tests have been realized at room temperature (25°C), for dynamic stress values, from 15MPa to 30MPa, and for a 50Hz frequency. Two samples have been tested for each stress value.

Figure 10 shows maximum dynamic stress versus the number of cycles to failure (in logarithmic scale). The Wöhler's curve represents here the median curve with a 50% failure probability. Three fields can be distinguished. The first one (I) corresponds to maximum stress ranges between the yield stress and the mechanical stress, that is to say for stress values higher than 23MPa. Each cycle leads then to a macroscopic plastic strain. In the limited endurance field (II), which corresponds to stress values from 18MPa up to 23MPa, the number of cycles to failure strongly changes. This field can last up to 300000 cycles. When the stress is lower than the endurance limit value (lower than 18MPa here), the lifetime is almost never-ending. This means that there is maximum stress amplitude below which the material never fails, no matter how large the number of cycles is.

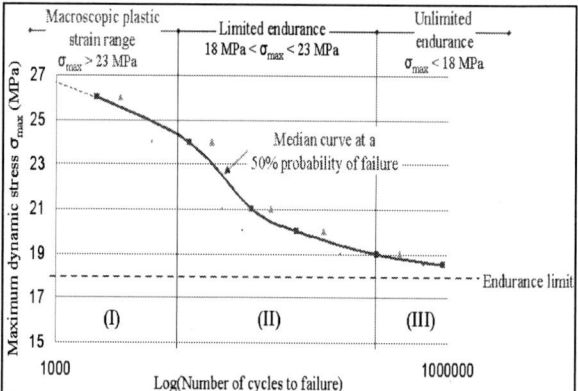

Figure 10: Wöhler's curve at 25°C and 50Hz

The fatigue constants of the lifetime prediction models can be extracted using the first part of the Wöhler's curve (see (I) in **Figure 10**). Particularly, it consists in measuring the macroscopic plastic deformation versus the number of cycles to failure. However, additional fatigue tests, with temperature-dependance (70°C, 100°C, and 150°C) have to be performed to guarantee a sufficient accuracy of the prediction model. These tests will be carried out in a next future.

4. Conclusion

In this paper, we have set up an experimental method to characterize the thermo-mechanical properties of solder alloys.

The static and dynamic measurments, performed on the $Pb_{92.5}Sn_{5.0}Ag_{2.5}$ solder alloy, have confirmed the Young's modulus, the yield strength, and the mechanical strength linear dependence to the temperature.

The $Pb_{92.5}Sn_{5.0}Ag_{2.5}$ steady-state creep behavior has been studied. The solder has a diffusion-type creep mechanism and an activation energy value very closed to what we can find in literature.

978-1-4244-4722-0/09 $25.00
© 2009 IMAPS-ITALY

Finally, fatigue tests have been performed to extract the constant values of a lifetime prediction physical model (e.g. Coffin-Manson). The fatigue lifetime could be correlated with some power cycling reliability test results. We aim to analyze this point in a next future.

The main interest to characterize the solder alloys physical properties is to obtain an experimental database. These data could be easily implemented into thermo-mechanical simulations using finite elements. The aim is to build a lifetime prediction model for TRIACs in real power cycling operation conditions.

References

[1] L. Gonthier, J.-M. Simonnet, S. Mercier, A. Passal, "New High-Temperature, High-Performance TRIACs for Optimized Vacuum Cleaner Designs", International Conference & Exhibition for Power Electronics Intelligent Motion Power Quality (PCIM), Shanghai, China, March 18-20, 2008.

[2] M. Ciappa, "Lifetime Modeling and Prediction of Power Devices", European Centre for Power Electronics (ECPE) Workshop, Toulouse, France, June 25 – 26, 2008.

[3] S. Jacques, N. Batut, R. Leroy, L. Gonthier, "Aging Test Results for High Temperature TRIACs During Power Cycling", IEEE Power Electronics Specialists Conference (PESC), Rhodes, Greece, June 15-19, 2008.

[4] J.-M. Thebaud et al., "Strategy for designing accelerated aging tests to evaluate IGBT power modules lifetime in real operation mode", IEEE transactions on components and packaging technologies, vol. 26, n° 2, pp. 429-438, June 2003.

[5] METRAVIB R.D.S., "DTMA user manual for VA815, VA2000, and VA4000 models".

[6] J. Lemaître, J.-L. Chaboche, "Mécanique des matériaux solides", Dunod Publishers, 2001.

[7] A. Schubert, H. Walter, R. Dudek, B. Michel, G. Lefranc, J. Otto, and G. Mittic, "Thermo-Mechanical Properties and Creep Deformation of Lead-Containing and Lead-Free Solders", Proceedings. International Symposium on Advanced Packaging Materials: Processes, Properties and Interfaces, pp. 129-134, 2001.

[8] S. Wiese, A. Schubert, H. Walter, R. Dudek, F. Feustel, E. Meusel, and B. Michel, "Constitutive Behaviour of Lead-Free Solders vs. Lead-Containing Solders – Experiments on Bulk Specimens and Flip-Chip Joints," Proceedings of Electronic Components and Technology Conference, pp. 890-902, 2001.

[9] W. Engelmaier, "The use environments of electronic assemblies and their impact of surface mount solder attachment reliability", IEEE Trans. Comp., Hybrids, Manufact. Technol., vol. 13, n°4, pp. 903-908, December 1990.

Investigation of Compliant Interconnect for Ball Grid Array (BGA)

Rolf Johannessen, Frøydis Oldervoll, SINTEF ICT
Helge Kristiansen, Conpart AS
Hallvard Tyldum, Norwegian University of Scienece and Technology (NTNU)
Hoang-Vu Nguyen, Knut Aasmundtveit, Vestfold University College

SINTEF ICT
Forskningsveien 1
0373 Oslo Norway
rolf.johannessen@sintef.no

Abstract

Ball grid array (BGA) carriers use solder balls as interconnects between package and underlying substrate. Stress will be introduced to the solder joints during thermal excursions when the ceramic BGA carrier is interconnected to a standard FR-4 printed circuit board (PCB) substrate due to mismatch in coefficient of thermal expansion (CTE). This has traditionally been overcome by added solder joint heights achieved by high-melting solder balls and/or solders columns. This was feasible when SnPb solder alloys were the standard in use. High density electronic systems demands fine pitch and small bump area while stand off height should be maintained. Transition to lead free alloys introduces new reliability aspects for fine pitch BGA. Lead free alloys exhibit a high Young's modulus and high solidus temperature. Sn-Ag-Cu (SAC) alloys do not mix well with other solders to form a uniform and strong solder structure, hence adequate stand off height is difficult to realize.

Introduction of polymer-core solder balls can offer compliant interconnects combined with enhanced stand-off height after reflow. Less thermal and mechanical stress will be introduced due to the improved material match thereby increasing the reliability of the interconnects compared to lead free BGAs.

In this paper, initial characterization of 240 µm polymer-core solder balls will be presented. A low temperature co-fired ceramic (LTCC) test vehicle with pad sizes ranging from 250 µm – 300 µm has been designed to investigate pad size and solder-joint geometry for BGA carriers. Result from this investigation is presented and compared to results from finite element modelling and simulation. Light microscopy has been performed on samples after assembly and on cross-sectioned samples, and will be presented together with results from destructive die shear test.

Introduction

Ceramic Ball grid array (BGA) carriers are widespread used in electronic systems as they combine high I/O density with standard mounting procedures like reflow or vapour phase soldering. The ceramic BGA is normally interconnected to a standard FR4 circuit board. Solder balls are used as interconnects between the package and substrate. When the temperature varies stress will be introduced in these solder balls due to a large mismatch in coefficient of thermal expansion (CTE) between FR4 and ceramic. This problem has traditionally been overcome by added solder joint heights achieved by a non-melting solder ball core, typically composed of a high lead solder alloy (PbSn 90/10). Lead is now banned in most electronic products and thus this solution is not available any more.

Introducing polymer-core solder balls can offer more compliant interconnects combined with enhanced stand-off height after reflow. These features can improve the interconnect reliability due to relaxation of thermal and mechanical induced stress [1]. Figure 1 shows a cross-section of this type of metal coated polymer spheres (MPS) produced by Sekisui Chemical Ltd. It consists of a 240 µm polymer sphere with 10 µm plated copper and 25 µm solder. A range of sphere sizes from 110 – 800 µm are available. Several studies during the last few years have reported improvements in the number of thermal cycles the BGAs can endure when changing from standard solder balls to MPS. Some of these results are summarized in Table 1. We see that improvement factors from 1.5 up to 4 have been obtained. Galloway [3] obtain as many as 7745 cycles to 63% failure with MPS, the first failures occurred after more than 3000 cycles.

Different ball sizes and type of solder balls (non-melting or eutectic) are used in these studies and thus the results are not directly comparable. In addition the results will also depend heavily on the pad metallization and pad size, amount of solder

978-1-4244-4722-0/09 $25.00
© 2009 IMAPS-ITALY

Table 1 Thermal cycling performance for standard solder balls vs. metal coated polymer spheres (MPS) reported in literature.

Test condition	Ball size		Cycles to 63% failure (hours)		Reference
	Solder ball [µm]	MPS [µm]	Solder ball	MPS	
-40 - +125°C	300 Eutectic solder	300	4903	7745	Galloway [3]
-40 - +125°C	760 PbSn 90/10	800	783	1378	Kangasvieri [5]
-55 - +125°C	760 Eutectic solder	800	350	1500	Okinaga [2]
0 - +100°C	635 PbSn 90/10	635	2095	6332	Movva [4]
0 - +100°C	760 PbSn 90/10	800	2660	3952	Kangasvieri [5]

paste applied and soldering process parameters. Marín et al [6] have presented finite element modelling of polymer core interconnects under cyclic shear loads. The solder fillet shape is predicted for different solder volumes (defined by stencil thickness) and pad sizes, and the FEM simulations shows that the number of cycles to failures depend to a great extent on the geometrical shape of the interconnections.

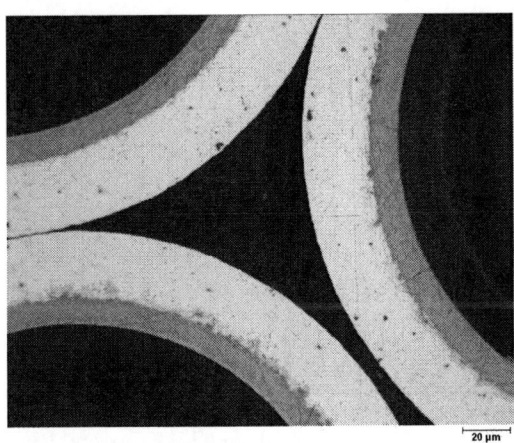

Figure 1 Cross section of metal coated polymer spheres with copper and tin.

SINTEF is currently running a research project on MPS technology together with several Norwegian industry partners. A specific industry application is focused in the project. This application requires low BGA stand-off height (preferably in the range of 250 µm) and robustness against thermal variations and mechanical vibrations.

In the present paper initial work on MPS technology will be presented. Process issues and selection of pad geometries has been the main scope of this work. The low stand off height sought

after in this project demand a minimum of solder to be supplied prior to reflow.

Assembly procedures of 240 µm polymer spheres without supply of additional solder have been investigated and characterized. Experimental results have been compared with finite element modelling.

Sample Preparation

Metal coated polymer spheres have been provided by Sekisui in Japan [10]. The MPS consist of a 240 µm polymer sphere with 10 µm plated copper and 25 µm Sn3.5%Ag solder.

BGA test modules were fabricated using a standard multilayer LTCC process by VTT in Finland [11]: Heraeus thick film screen printed conductor system [12] based on Heratape CT 800 Clad on Heralock 2000 LTCC substrate with TC 0307 silver thick film conductor for inner routing layer and LPA 405-067 for outer layer was specified for the BGA test modules. Solder pads were covered by 4 µm electroless nickel and immersion gold. The overall thickness of the LTCC substrate was 1 mm. Pad geometries of 250, 275 and 300 µm were realised for this investigation. The fabricated BGA modules displayed pad geometry accuracy within 5 % of specified size...

A standard 4 layer FR-4 PCB with copper conductor system covered by electroless nickel and immersion gold was designed and manufactured as test substrates for the BGA. Solder pads on the PCB were defined by a glass based solder mask and solder pad geometries of 250, 275 and 300 µm were realized. The accuracy of the glass matrix aperture was specified to be within ± 50 µm for all pad geometries.

MPS ball attach to the LTCC BGA carrier has in this work been investigated and compared to corresponding attach to PCB substrate. Initial work showed that MPS ball attach to the PCB prior to BGA carrier assembly showed up more promising result than vice versa.

978-1-4244-4722-0/09 $25.00
© 2009 IMAPS-ITALY

The low stand off height sought after in this project demanded a minimum amount of solder in the MPS solder joint. Maintaining low stand-off height requires no additional solder to be supplied during the assembly process.

The MPS were individually aligned to the substrate, positioned without applying any additional solder paste, and held in place by a small amount of flux paste dispensed on solder pads prior to assembly. Fully assembled substrates were transferred to an IR reflow oven and ball assembly was completed by a standard lead-free reflow process. A total of 16 MPS were attached to each substrate.

LTCC BGA carriers were aligned onto the PCB pre-soldered spheres and held in place by flux paste dispensed on solder pads. A standard lead-free reflow process completed the BGA carrier - to - PCB soldering. Figure 2 shows a cross sectional schematic view of the MPS interconnection point.

Flux paste was dispensed on solder pads by an OKI DX-200 precision dispenser. Reflow process was performed in a 3 temperature zone Madell IR reflow oven at peak temperature 260°C in ambient atmosphere.

Experimental

Visual inspection of MPS assembled on PCB prior to LTCC BGA assembly was performed by light microscopy. Solder joint geometry, MPS alignment and ball height was considered to be of interest in the initial phase of this study.

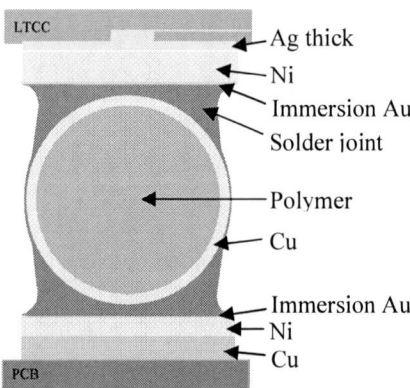

Figure 2: Schematic view of cross sections of a MPS solder joint interconnection point

Light microscopy was performed on cross sectioned units after assembly. A row of solder joints for each sample was cross sectioned. The devices were embedded under vacuum in epoxy resin. They were grinded with soap water lubrication with emphasis on precise cross sectioning through series of interconnection points and low material damage level. Finally, the devices were polished on medium hard cloth with diamond

with alcohol based lubricant, and subsequent fine polish with water based lubricant.

Light microscopy was performed with a Leica M420 on MPS ball attach samples and with a NEOPHOT 32 microscope on cross section samples after BGA assembly.

Die and ball shear testing is a simple and fast method to check bump integrity for bonded dies. Die shear test measurement was performed using a DAGE 2400A shear tester Shear strength and failure category was of interest as a reliability monitor for the bonded units.

Results

Well defined solder balls were apparent after MPS ball attach as shown in Figure 3. The flux paste dispensed on the solder pads held the MPS in place during assembly and provided adequate wetting during reflow. A ball height variation within 10 μm was measured for MPS assembly on LTCC and PCB substrate. Coplanarity for the substrate itself was measured to be within 15 μm for the LTCC substrate and 30 μm for the PCB substrate. From ball height measurements the pad diameter displayed to have only minor impact on stand-off height for the assembled BGA carrier as shown in Table 2.

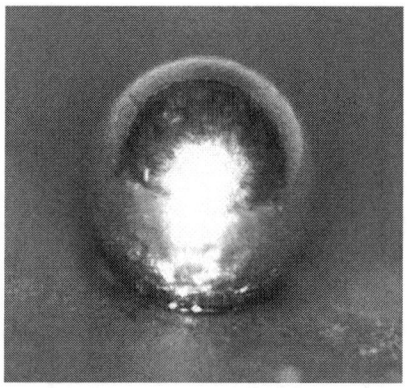

Figure 3: Microscope image of MPS assembled on 250 μm LTCC BGA solder pads

Figure 4 shows cross section images of MPS solder joints for pad diameter 250, 275 and 300 μm. The smaller pad diameters shows up a small radius curvature in the corner region of the solder joint and the PCB solder pad.

From visual inspection after reflow, some solder joints were observed not to attach to the LTCC solder pad as shown in Figure 5. Unsuccessful MPS solder joint was observed to be more frequent for the larger pad diameter compared to the smaller. Result from analysis of number of successful solder joints is presented in

Table 3. Two samples of each pad diameter were selected for this analysis. Figure 6 shows an oblique positioned MPS solder joint.

978-1-4244-4722-0/09 $25.00
© 2009 IMAPS-ITALY

a) Solder pad diameter 250 µm.

b) Solder pad diameter 275 µm

c) Solder pad diameter 275 µm

Figure 4: Light microscope image of a cross sectioned MPS solder joint. PCB interface (top) and LTC interface (bottom).

Die shear test was performed on BGA carriers assembled on PCB. Shear strength result was dependent on number of successful solder joints. For 16 successful solder joints (pad size 250 µm) shear strength of ~2.5 kg force (150 gram force/solder joint) was obtained. The predominant failure from die shear test was lift off at the LTCC BGA carrier. Failure in the silver thick film to LTCC substrate interface was apparent form analyzes performed after die shear test. One

incident of failure in the solder joint - to - solder pad interface at the PCB was found on one test sample after die shear test.

Figure 5: Light microscope image of an unsuccessful MPS solder joint. No attach to LTCC solder pad after reflow.

Figure 6: Light microscope image of a oblique positioned MPS solder joint.

Table 2: Summary of stand-off height measurements

Pad diameter	Stand-off height
250 µm	~280 µm
275 µm	~275 µm
300 µm	~275 µm

Table 3: Summary of successful solder joints for pad diameter 250, 275 and 300 µm

Pad diameter	Successful solder joint
250 µm	16 of 16
	16 of 16
275 µm	14 of 16
	13 of 16
300 µm	7 of 16
	3 of 16

Finite Element Modelling

Finite element modelling and simulation have been included in order to investigate the effect of varying the solder pad diameter for a given polymer-core solder ball diameter.

The model was simplified to a single joint imposed the shear deformation from a thermal cycling test of the BGA package. Any stiffening effect from any neighbour joint was not taken into account in this model.

The relative motion of the component and the PCB is given by [4] and this corresponds to the worst case loading condition for the interconnection point.

$$\delta = (\alpha_{PCB} - \alpha_{component}) \cdot \Delta T \cdot DNP$$

$$\delta = \left(\alpha_{PCB} - \alpha_{component}\right) \cdot \Delta T \cdot DNP$$

Where

$$\alpha_{PCB} = \text{CTE of PCB}$$
$$\alpha_{component} = \text{CTE of component}$$
$$\Delta T = \text{Temperature change}$$
$$DNP = \text{Distance from Neutral Point}$$

Simulation was performed for a target bump height of 300 µm and thermal excursion involving 90K temperature change. The interconnect shape was simplified to a MPS solder column structure as shown in

Figure 7. Simulation of a 3D model of a half interconnect with 1st order cubic elements with reduced integration was performed with ABAQUS 6.8-2.

Polymer and metal plating materials were in these models represented as linear elastic and the solder an elastic-perfect material. The polymer core properties were taken from Sekisui data sheet [7]. The copper layer Young's modulus was taken from nanoindentation measurements on electroplated copper film performed by Volinsky et al [8]. They reported the Young's modulus approaching 110 GPa for a thickness greater than 2 µm. Although solder properties are temperature, rate and aging dependent, we assumed the perfect plastic model for simplification. The specific properties were taken from tests on SAC305 by Zhao et al. [9]. Varying the dimensions in a model of a single interconnect, the effect on the crack driving forces can be observed and form a basis for optimizing the solder process. Boundary conditions and material parameters were held constant in the model while the ball diameter to bump height (d/h) and the solder column diameter were varied.

From finite element modelling stress concentrations was found to arise at the interface of the solder-pad periphery and at the top and bottom of the ball. This stress concentration is due to material stiffness mismatch and is illustrated in Figure 8, which is showing the solder material in the 280 µm ball-300 µm column configuration. The accumulated plastic strain per cycle is displayed in Figure 9

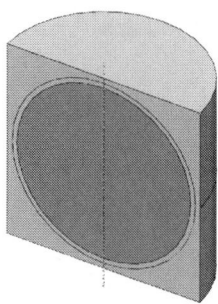

Figure 7: Cross section view of a MPS solder column structure.

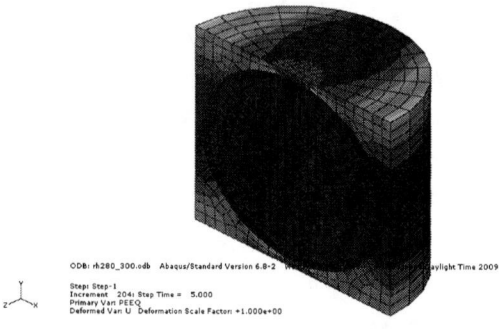

Figure 8: Equivalent plastic strain in solder for 280µm ball-300µm column after 5 complete cycles.

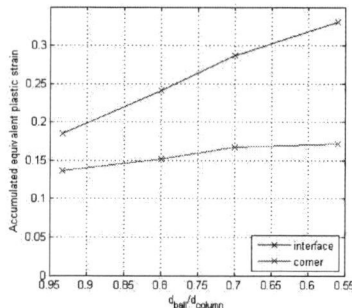

Figure 9: Accumulated plastic strain for different d_{ball}/d_{column}.

Discussion

Pad diameter of 250 µm shows up small radius solder joint curvature at the solder pad periphery at the PCB side. The PCB pad geometry is defined by a glass matrix solder stop. Nousiainen et al [13] has among others reported this region to be crack origin site when exposed to subsequent

978-1-4244-4722-0/09 $25.00
© 2009 IMAPS-ITALY

thermal excursions. Probability for crack initiation in this region is expected to increase with smaller curve radius. There is no evidence of small radius curve in the solder joint - to - solder pad region of the LTCC BGA carrier. Solder stop mask was not specified for the BGA carrier.

Low die shear strength combined with lift off at LTTC was predominant from die shear test. Failure was observed to be in the silver thick film - to - LTCC substrate interface. Material system with good adhesion is inherently important as feature sizes are reduced.

A simple geometry and material model has been developed for finite element modelling in this study. The purpose of this model was to explore the basic effects of introducing a soft polymer core. For large ball diameters the stress concentration at the top of the ball was found to be most severe. The plastic strain in this region was found to increase as the ball diameter increased. This is related to the reduced solder volume as the ball diameter increases. Although the polymer core is softer than the solder, the thickness and stiffness of the copper layer results in a stiffer core than in a pure solder bump for large ball diameters. Together this increases the plastic strain in the solder. Movva et al [4] has previously reported that the solder volume in the MPS interconnect point is a critical parameter. The high strain region was found to be near the package and the board interfaces and the MPS acted as a buffer that relived the strain in the reported study.

Previously crack has been reported to typically be initiated at the solder pad periphery [13] and grown along the pad - to - solder joint interface. The regions of high strains in our model are found to be most severe in the centre of the bump, surface defects may act as further stress concentration sites. From our model we see clear evidence that the ball to column diameter ratio has effect on plastic strain in the bump. As the pad diameter increases the plastic strain increases, and this is more evident in the centre than at the periphery of the solder bump.

One of the major benefits of the polymer core is that it provides almost no collapse and stand-off height remains after soldering; this might be the prime reason for the life-time increase previously reported and summarized in Table 1.

Conclusion

This work shows that reflow of MPS are feasible for pad geometries in the range of 250 to 300 µm without further addition of solder. In this work the smaller pad diameter is more promising with respect to successful assembly compared to the larger pads. Semi automated or automated processes together with process optimization are needed for reproducibility.

Results from FEM analysis shows that care should be taken when choosing the material combination and bump geometry for MPS BGA solder joints.

References

1. W. Engelmaier; *Achieving solder joint reliability in a lead-free world, Part 2.*, Global SMT & Packaging, August 2007

2. N. Okinaga, H. Juroda, Y. Nagai; *Excellent Reliability of Solder Ball Made of Compliant Plastic Core*, IEEE Electronic Components Technology Conference, 2001

3. J. Galloway, A. Syed, W. Kang, J. Kim, J. Cannis, Y. Ka, S. Kim, T. Kim, G. Lee, S. Ryu; *Mechanical, Thermal and Electrical Analysis of a Compliant Interconnect*, IEEE Trans. Components and Packaging Tech., Vol. 28, No. 2, 2005

4. S. Movva, G. Aguirre; *High Reliability Second Level Interconnects Using Polymer Core BGAs*, IEEE Electronics Components and Technology Conference, 2004

5. T. Kangasvieri, O. Nousiainen, J. Putaala, R. Rautioaho, J. Vähäkangas; *Reliability and RF performance of BGA solder joints with plastic-core solder balls in LTCC/PWB assemblies*, Microelectronics Reliability, 46, 2006

6. F. Marín, D. Whalley, H. Kristiansen, Z. Zhang; *Mechanical Performance of Polymer Cored BGA Interconnects*, 10th IEEE Electronics Packaging Tech. Conf., 2008

7. A. Hasegawa, *Area array interconnection material, Micropearl SOL*, Sekisui Chemical Co., Ltd., Tech. Rep., 2001.

8. A. Volinsky, J. Vella, I. Adhihetty, V. Sarihan, L. Mercado, B. Yeung, and W. Gerberich, *Microstructure and mechanical properties of electroplated Cu thin films*, in *Fundamentals of Nanoindentation and Nanotribology II*, ser. MRS Proceedings, vol. 649, 2001.

9. J. Zhao, Y. Mutoh, Y. Miyashita, and S. Mannan, *Fatigue crack-growth behavior of Sn-Ag-Cu and Sn-Ag-Cu-Bi lead-free solders Journal of Electronic Materials*, vol. 31, no. 8, pp. 879–886, Aug. 2002.

10. http://www.sekisui

11. http://www.vtt.fi

12. http://www.wc-heraeus.com

13. O.Nousiainen, J. Putaala. T. Kangasvieri, R. Rautioaho, J. Vahakangas, *Failure Mechanishm of Thermomrchanically Loaded SnAgCu / Plastic Solder Ball Composite Joints in Low-Temperature Co-Fired Ceramic / Printed Board Assemblies*, Journal of Electronic Materials. Vol 36. 3, 2007

Reliability of 100 µm Bi- and In- Solder Balls

S. Kemethmüller, R. Dohle, J. Pohlner, Th. Dünne and J. Goßler

Micro Systems Engineering GmbH, Schlegelweg 17, 95180 Berg/Ofr., Germany

Phone: +499293-78716, Fax: +499293-7841, Email: skemethmueller@mse-microelectronics.de

Abstract

Low-melting solder alloys are considered as one approach to lower assembly process temperature and therefore thermal stress on temperature sensitive component materials. In this study the reliability of 100 µm-solder bumps of low-melting Bi- and In-alloys has been investigated on a thin-film Al_2O_3 substrate (18 x 14 mm²) with more than 5000 I/Os. Two different bumping processes have been assessed:

1) Screen printing of solder paste and subsequent reflow soldering of the solder reservoirs to bumps.

2) Placement of preformed solder balls via screen in printed tacky flux followed by reflow soldering. Silicon chips with Daisy Chain structures (each >1250 I/Os) were assembled via flip-chip technology on the substrate. The hybrids were underfilled. Accelerated life tests like temperature cycling have been performed. A temperature range between -10 and +90°C has been selected according to the melting points of the alloys (between 119°C and 140°C).

The reliability and stability of the bismuth alloy and indium alloy joints themselves have been investigated thoroughly by mechanical and electrical tests. The results of the reliability tests of the 100 µm bumps have been compared with reference to their manufacturing process (solder paste or preformed solder balls) and their alloy composition.

Key words: 100 µm solder bumps, reliability, shear strength, indium alloy, bismuth alloy, bumping processes

Introduction

The continuing miniaturization in microelectronic applications requires increasing I/O-numbers which demand reduced structure geometries and decreased pitch sizes. In the industry flip-chip technologies for solder bumps with a diameter of >150 µm (produced by screen printing) and >200 µm (placed via screen) are state of the art [1].

In addition a continuously increasing demand on components with new features can be observed in the microelectronics market. These features can be achieved by the usage of new materials. Some of these materials are conductive to lower thermal stresses during assembly. The use of low-melting solder alloys can lead to these required lower assembly process temperatures.

In this study, two different bumping processes with low-melting alloys have been developed and assessed. Solder balls with 100 µm diameter were created by solder paste printing, or placing and reflow of preformed solder balls, respectively. Silicon chips with more than 5000 I/Os in total could be assembled on an Al_2O_3 thin film substrate using these low-melting solder balls with reflow temperatures below 170 °C. The reliability of these two different low-melting solder alloy joints

respectively of the two different manufacturing processes has been assessed.

Selected low-melting solder alloys

Several low-melting alloy systems have been investigated in the past. The availability of fine grained solder pastes and preformed 100 µm solder balls is strongly limited, however.

Table 1 lists the solder alloys used in this study with their Liquidus and Solidus temperatures. For both alloys, 100 µm solder balls are available.

Table 1: Liquidus and solidus temperatures of the selected low-melting solder alloys.

Solder alloy	Liquidus	Solidus
52In48Sn	118 °C	118 °C
57Bi42Sn1Ag	140 °C	139 °C

The Sn-Bi system was widely studied since as early as Glazer, 1994 [2]; a lot of other authors can be found using the Sn-Bi or Sn-Bi-Ag system in the last years [3,4,5,6]. The main concerns regarding the use of bismuth based solders fall into the categories of brittleness and reliability problems when mixed with lead. The eutectic 58Bi42Sn alloy has shown good accelerated thermal fatigue characteristics; also the elastic modulus and the shear strength are comparable to eutectic PbSn-solder alloy. McCormack et al. found, that with addition of Ag the ductility of the binary BiSn-solder can be improved significantly [3]. The melting point of the 57Bi42Sn1Ag alloy is 139 °C. The wetting behavior is worse than of PbSn but after Mei and Morris [7], still acceptable.

The other low-melting alloy system is the indium-tin system. InSn solder alloys possess a very low elastic modulus, are very ductile and possess very low shear strengths [8, 9]. Solder alloys with high indium content tend to cold shuts. The eutectic composition is 50.9In49.1Sn, but the most common solder alloy is the 52In48Sn with a melting point of 119 °C. In order to reach sufficient wetting reactive fluxes are necessary [10].

Bumping Processes

Two manufacturing processes were developed in order to create the 100 μm low-melting solder balls.

1) Placement of preformed solder balls via screen in printed tacky flux, followed by reflow.
2) Screen print of solder paste and reflow of the solder reservoirs to bumps.

The InSn-alloys with high indium contents strongly tend to cold shuts, as described above. Therefore no fine-grained solder paste type 6 or 7 is available on the world market. For this reason, the InSn solder bumps were only created by preformed 100 μm balls. Figure 1 shows the placed preformed 100 μm InSn solder balls in tacky flux before reflow.

The average height of the preformed solder balls after the reflow was 85 μm ± 5μm, independent which solder alloy was used (InSn- or SnBiAg-balls).

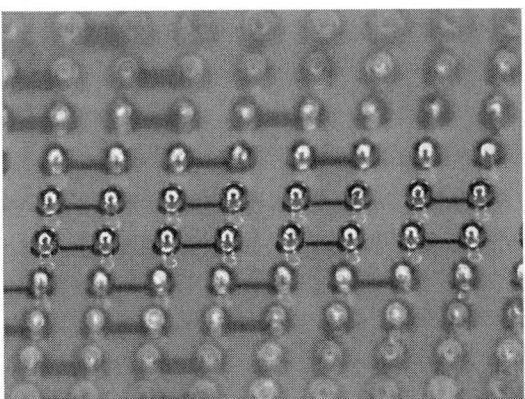

Figure 1: Placed 100 μm InSn balls before reflow at 25 x magnifications.

For the SnBiAg-alloy the solder balls could be created as well by solder printing using a type 7 solder paste from Heraeus Corp. (see figure 2).

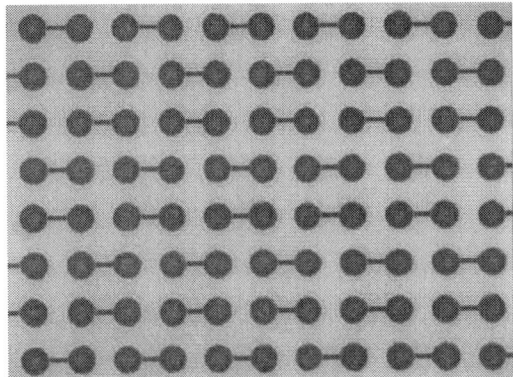

Figure 2: Printed SnBiAg solder reservoirs before reflow at 20 x magnifications.

Through the usage of special stencils and after several optimization steps of the printing parameters, SnBiAg solder balls with heights in average of 75 μm ± 10 μm after reflow could be created. Figure 3 illustrates the difference of the heights and deviation of SnBiAg solder bumps in dependency of their manufacturing process.

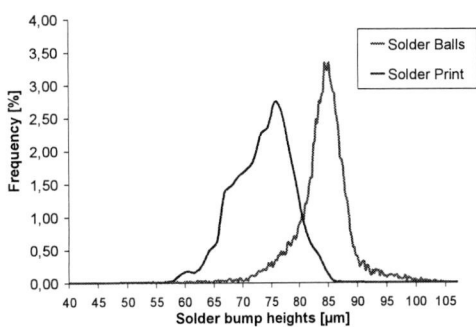

Figure 3: Comparison of solder bump heights and deviation in dependency of their manufacturing process.

The larger standard deviation of the height of the printed solder bumps did not show any significant influence on the assembly process of the silicon chips on the Al_2O_3 substrate. Anyway, the lower height of the bumps caused a smaller distance between silicon chips and substrate, of course. This lower stand-off should influence the reliability of the test vehicles manufactured via solder printed bumps.

Reliability Investigations

The influence of the differences caused by the manufacturing process (lower height and larger deviation of the printed bumps) and the various influences of the solder alloys themselves (brittleness, ductility, corrosion) on the reliability were investigated by following accelerated life tests:

1. Shear tests of the 100 μm low-melting solder balls initial, after storage at 85 °C and after storage in humidity (85 % r.h., 85 °C).
2. Electrical test via Daisy Chain structures of assembled silicon chips on Al_2O_3 substrates after storage of in humidity (85 % r.h., 85 °C).
3. Electrical test via Daisy Chain structures of assembled silicon chips on Al_2O_3 substrates after temperature cycling (-10 °C to +90 °C).

Shear strength of the 100 μm low-melting alloy balls

Figure 4 and 5 show the trend of the shear values of preformed 100 μm balls placed and reflowed on 80 μm Au/Ni pads on the Al_2O_3 thin film substrate during storage at temperature (fig. 4) and humidity (fig. 5).

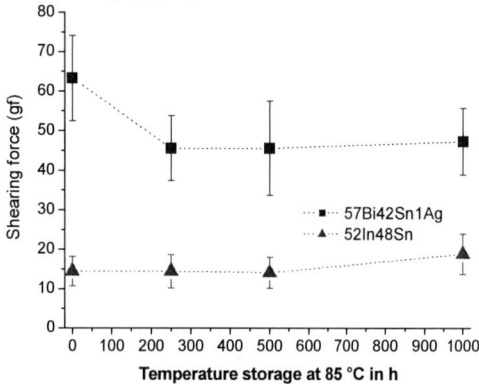

Figure 4: Trends of the shear strengths of the 100 μm InSn and SnBiAg solder balls during temperature storage. Pad sizes are 80 μm, all values represent the average of 100 single measurements and the error bars represent 3σ.

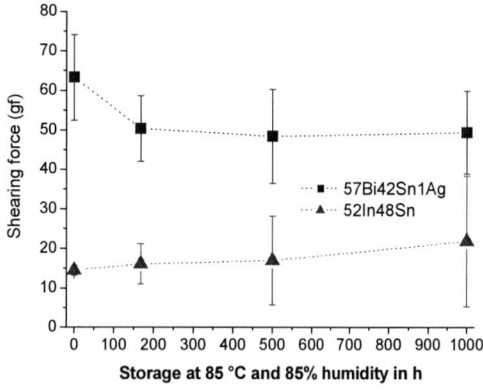

Figure 5: Trends of the shear strengths of the 100 μm InSn and SnBiAg solder balls during humidity storage. Pad sizes are 80 μm, all values represent the average of 100 single measurements and the error bars represent 3σ.

Both diagrams show similar effects. The initial shear strength of the SnBiAg balls is approximately 4 times higher (64 gf) as for the InSn-balls (15 gf). During the storage the shear values of the InSn balls remain mainly stable, whereas the shear strength values of SnBiAg balls show a slight decrease after 168 h storage to around 45 gf at temperature storage and 50 gf at humidity storage. After this decrease the shear strength values remain stable over the whole remaining time.

Figure 6 compares the distribution of the initial shear strength values of the 100 μm Balls (100 values of each). As shown in fig. 4 and 5, the InSn-balls show the lowest shear strength (15 gf in average), but with a very narrow distribution. The SnBiAg-balls show similar distributions of their shear strength values, independent of which manufacturing process was used. But the placed balls possess much higher shear strengths than the printed ones. The printed balls reach 34 gf in average, the placed preformed SnBiAg-balls 64 gf. This lower shear strength results from voids in the printed balls, which may be caused by the high flux content of the paste and therefore residual flux in the solder after reflow (see fig. 7).

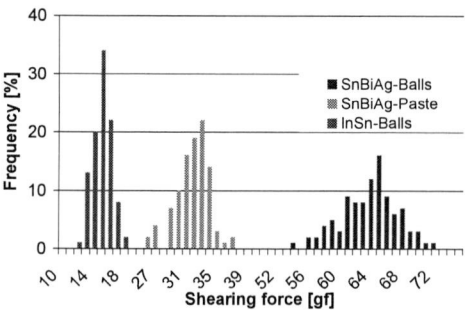

Figure 6: Comparison of the initial shear strength values of the placed InSn and SnBiAg balls and the printed SnBiAg bumps. Solder pad sizes are 80 μm.

Figure 7: Void in a printed SnBiAg solder ball after reflow.

Electrical tests after temperature cycling and humidity storage

The electrical conductivity test of the Daisy Chain structures was performed with a "bed of nail"-adapter. The resistances of the Daisy Chains were measured at 1 V. In summary 64 Daisy Chains per module with chain lengths between 100 and 400 solder ball joints were measured to verify the connectivity of the test Si-chips/Al_2O_3-modules.

The electrical tests were performed initial, after humidity storage (measurements after 168 h, 250 h, 500 h and 1000 h) and after temperature cycling between -10 and +90 °C (measurements after 10, 20, 50, 100, 250, 500, 1000 and 1500 cycles).

No significant changes could be detected after humidity storage and temperature cycling. No failures of the electrical connection could be found at any times. The connections remain stable.

Furthermore, no difference between the placed SnBiAg solder ball joints and the printed could be detected. Both show absolutely similar behavior with stable electrical connections over 1500 temperature cycles and 1000 h at humidity storage, respectively. No influence of the smaller distance between the silicon chips and the Al_2O_3 substrate with the printed balls on the reliability could be found.

Microstructure after accelerated life tests

The figures 8-11 show cross sections of the solder balls after the accelerated life tests. No corrosion has been detected at the samples with SnBiAg and InSn balls after storage over 1000 h in humidity.

Furthermore no changes in the microstructure of the InSn- and SnBiAg-balls could be detected after humidity storage and after 1500 temperature cycles (-10 to +90 °C). These results confirm the observations of Mei et al., 1999 [4] and Freer Goldstein and Morris, 1992 [8].

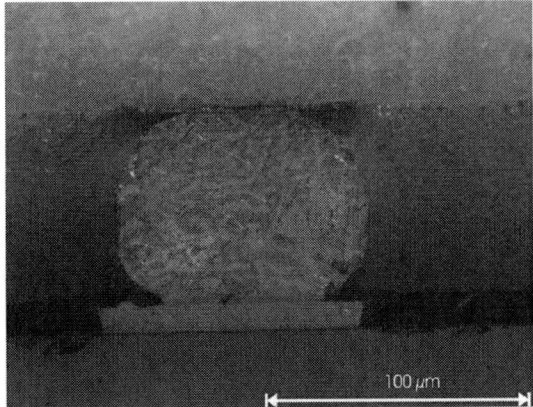

Figure 8: SnBiAg solder ball manufactured by solder print after 1000 h at 85 % r.h./ 85 °C.

Figure 9: InSn solder ball after 1000 h at 85 % r.h./ 85 °C.

Figure 10: SnBiAg solder ball manufactured by solder print after 1500 cycles between -10° and +90 °C.

Figure 11: InSn solder ball after 1500 cycles between -10° and +90 °C.

Conclusion

It has been shown, that reliable 100 µm low-melting solder joints can be achieved with both solder alloys (52In48Sn and 57Sn42Bi1Ag). There is also no influence of the manufacturing process (printing solder paste and reflow or placing of preformed balls) on the reliability on the solder joints.

978-1-4244-4722-0/09 $25.00
© 2009 IMAPS-ITALY

The InSn solder balls possess very low shear strength, but they are also very ductile, which might have had a positive influence during the accelerated life tests.

The printed SnBiAg solder balls possess lower shear strength values than the preformed SnBiAg-balls, due to remaining voids after reflow. But their shear values are still high enough for reliable solder joints. The about 10 μm lower height in comparison to the preformed balls did not show any influence during the accelerated life tests.

The fine pitch assembly process with solder printed SnBiAg balls offers a low-cost process with lower thermal stress than other lead free assembly processes. The process is less cost intensive than the placement of preformed solder balls via screen in printed tacky flux. The material costs of SnBiAg are much lower in comparison to the high prize of indium based alloys.

Acknowledgements

The authors would like to thank Mr. Bernd Burger for preparation of the cross sections and Mrs. Jana Müller for performing the shear tests.

References

[1] D. Manessis et al., "UBM Structures for Lead Free Solder Bumping using C4NP-Technology", Deutsche IMAPS Konferenz, Munich, October 2007.

[2] J. Glazer, "Microstructure and mechanical properties of Pb-free solder alloys for low-cost electronic packaging", Journal of Electronic Materials, Vol. 23, No. 8, pp. 701-7, 1994.

[3] M. McCormack, H.S. Chen, G.W. Kammlott and S. Jin, "Significantly improved mechanical properties of Bi-Sn solder alloys by Ag-doping", Journal of Electronic Materials, Vol. 26, No. 8, pp. 954-8, 1997.

[4] Z. Mei, F. Hua and J. Glazer, "Sn-Bi-X solders", SMTA Proceedings, Chicago, Il, September 1999.

[5] V. Schroeder, J. Gleason and F. Hua, "Strength and fatigue behaviour of joints made with Bi42Sn1Ag solder paste: an alternative to Sn-3.5Ag-0.7Cu for low cost consumer products", SMTA Proceedings, Chicago, Il, September 2001.

[6] J. Lau, J. Gleason, V. Schroeder, G. Henshall, W. Dauksher, B. Sullivan, „Design, materials, and assembly process of high-density packages with a low-temperature lead-free solder (SnBiAg)", Soldering & Surface Mount technology, Vol. 20, No. 2, pp. 11-20, 2008.

[7] Z. Mei and J. Morris, "Characterization of Eutectic Sn-Bi solder joints", Journal of Electronic Materials, Vol. 21, No. 6, pp. 599-607, 1992.

[8] J.L. Freer Goldstein, and J.W. Morris, "Microstructure and Creep of Eutectic In/Sn on Cu and Ni Substrates". Journal of Electronic Materials, Vol. 21, 647–652, 1992.

[9] C. Melton, "Reflow Soldering Evaluation of Lead Free Solder Alloys", Proceedings 43rd Electronic Component and Technology Conference, Piscataway, NJ, 1993

[10] K. Seelig, D. Sklarski, L. Johnson, J. Sartell, "The Advantages of Low Melting Temperature Solder when Applied to Plated-Through Hold Technology." Proceedings of the Technical Program: NEPCON East '87. Des Plaines, IL, 1987.

Hydrolysis Testing of ACF Joined Flip Chip Components with Conformal Coating

Kati Kokko*, Hanna Harjunpää**, Pekka Heino*, and Minna Kellomäki**

* Tampere University of Technology, Department of Electronics, P.O. BOX 692, FIN-33101 Tampere, Finland

** Tampere University of Technology, Department of Biomedical Engineering, P.O. BOX 589, FIN-33101 Tampere, Finland

First author Tel.: +358 40 849 0086; Fax +358 3 3115 3394. E-mail address: kati.kokko@tut.fi

Abstract

The reliability of electronics is very important, especially in implantable applications. Electronics need to be protected against moisture and body fluids. Various coating solutions are used for this. Polymer coatings offer protection against harsh environment, and even if metallic or other coatings are used, polymer conformal coatings are used underneath. Thus, their reliability issues need to be considered. In vitro testing of implantable devices is needed prior to in vivo testing and implantation.

In this study, altogether three test lots were prepared for in vitro tests. FR-4 substrate was used and an 8 × 8 mm chip with daisy chain test structure was attached using anisotropically conductive adhesive. Epoxy and parylene C conformal coatings were studied. The test lots had epoxy, parylene C, and epoxy and parylene C composite structure as conformal coatings. All the test lots were medically sterilized prior to testing using gamma rays. The hydrolysis testing was conducted using sodium phosphate buffer solution. The duration of the test was ten months. The test samples were measured once a week for their functionality and the pH value of the solution was controlled to ensure that the samples were not dissolving in the solution. The aim of the study was to compare the different conformal coatings in the hydrolysis test and to determine the failure mechanisms occurring in the ACF joints.

Key words: electronics reliability, conformal coating, hydrolysis testing, ACF, flip chip joining

Introduction

Conformal coatings protect electronics inside the coating against the environment. Since today electronics is used in very harsh environments, the need for protective coatings is increasing. One example of a harsh environment is inside the human body, where implantable electronic devices are used. This biological environment includes many aggressive compounds that may harm electronics. Body fluids contain metal cations, salts, acids, dissolved oxygen and cells [1]. These body fluids cause corrosion and degradation of the materials used in electronics packages. Coating materials need to be resistant to these mechanisms.

The electronics used inside the implantable device needs to be packaged in a space and weight saving way, since an implant may not be large or heavy. Flip chip assembly technologies offer excellent reduction in interconnection distance and area needed for the component [2], and this offers great benefits for electronics packaging inside an implant. The interconnection between flip chip and substrate can be formed using anisotropic conductive films (ACF). This technology has many advantages including lower process temperature than in solder reflow process, environmentally friendly joining, e.g. fluxless joining, low fabrication costs, and compatibility with very fine pitch applications [2, 3]. Disadvantages include higher contact resistance and lower current capability [3].

While the interconnection of the component inside an implant is selected carefully, the coating material also needs to be well suited for the application. The two key functions of coatings in medical electronic applications are environment protection, particularly moisture protection, and electrical insulation or isolation. To meet these requirements, the adhesion of the coating is critical. This has been studied elsewhere [4]. Additional challenges arise, if the coating layer must be very thin, flexible, inert, or when the device to be coated has a fragile, complex shape or operates in a high

978-1-4244-4722-0/09 $25.00
© 2009 IMAPS-ITALY

impedance circuit. When organic coatings are considered, flexibility is one advantage that can be achieved, and thus it is intended. In medical implant applications all surfaces of a device have to be biocompatible materials that do not cause excessive tissue reactions. The coatings and devices must not deliver material to the tissue or react chemically with resulting corrosive products [5].

The reliability of the electronic package is very important in such applications as implants. Before implantation, the device needs to be tested *in vitro* and after adequate testing in simulated atmosphere *in vivo* testing takes place. The importance of *in vitro* testing lies in the different behavior of the materials in ordinary reliability testing and in testing in a real body environment [6, 7, 8]. Ordinary reliability testing of electronics is not sufficient when electronics is used in implantable applications. However, ordinary reliability testing is also important, since the electronics need to function reliably throughout the life of the implant.

This paper describes a study where hydrolysis testing in Na-PBS buffer solution was performed on three different test lots. These test lots had parylene C, epoxy and epoxy – parylene C composite coating as conformal coatings. Before testing all the test lots were sterilized using gamma sterilization. During testing the daisy chain resistance was measured from the test chips. After testing cross sections were made and the interconnections and conformal coatings were studied from the cross sections using optical and scanning electron microscopy (SEM). The goal of this study was to obtain relevant information on different conformal coating materials on ACF joined flip chip components in hydrolysis testing.

Experimental

Testing was performed using test samples consisting of ACF flip chip joined test chips on FR-4 substrates. Each test sample had 3 x 5 cm² FR-4 substrate where one test chip was attached. There were also test wires soldered on the substrate to measure the daisy chain resistance during testing. The FR-4 had copper wiring with NiAu finishing. Table 1 shows the main properties of FR-4. Figure 1 shows a photo of one test sample after the chip and the test wires have been attached.

Figure 1: A photo of one test sample after chip attachment. The test wires are seen on the right.

The test chips used in this study were flip chip components specially designed for testing purposes. They had a daisy chain test structure encircling the chip. The daisy chain was used to control the functionality of the joints. The chips were 8 x 8 mm² and had 275 gold bumps.

Anisotropic conductive adhesive was used to join the test chips to the FR-4 substrates. The ACF used was commercially available epoxy-based thermoset. It was 30μm thick and had 5μm diameter Au coated polymer balls as conductive particles and SiO_2 non-conductive filler particles, 0.8μm in diameter. The most important material parameters of the ACF are shown in Table 1.

The attachment was done using a TORAY FC1000 semi-automatic flip chip bonder. The bonding conditions used were those recommended by the adhesive manufacturer. First the pre-bonding step was performed, where the adhesive was applied to the substrate. In this step the bonding conditions were 5 s at 100 °C and the pressure was 1 MPa. The final bonding conditions were 25 s at 210 °C and the pressure during final bonding was 136 MPa.

Conformal coatings used in this study were parylene C and epoxy. The material properties of these two are shown in Table 1. Parylene C coating was applied using vapour deposition polymerization, which was done at room temperature [9, 10]. The advantages of parylene C include pinhole-free conformality, low water permeability, high flexibility and mechanical strength, and chronic implantability as an ISO 10993, United States Pharmacopeia (USP) Class VI material [9]. The USP Class VI material classification verifies the biocompatibility of the material. Class VI is the highest class for biocompatibility of plastics.

Table 1: Properties of FR-4, ACF, parylene C and epoxy [10].

Property	FR-4	ACF	Parylene C	Epoxy
T_g / °C	130-140	112	87-97	60
CTE (below T_g)/ppm/°C	12-16	40	40-50	60-90
CTE (above T_g)/ppm/°C	-	561	-	-
Moisture absorption /wt%	<0.25	1.9	0.01-0.06	0.08-0.15
Water permeability/g/m² over 24h	-	-	3-4	28-37

978-1-4244-4722-0/09 $25.00
© 2009 IMAPS-ITALY

The epoxy used as conformal coating was diglycidyl ether of bisphenol A epoxy with a polyoxypropylene-diamine hardener. The test samples were moulded to the epoxy in order to ensure a thick enough layer of epoxy coating.

The test samples were medically sterilized using a gamma radiation dose of 25 kGy. Gamma sterilization is widely used to sterilize implantable medical devices, since it is an effective sterilization method [11]. Surface sterilization methods are not adequate to sterilize implantable devices that are to be implanted long-term. The effects of gamma sterilization on the reliability of ACF joined flip chip components have been studied elsewhere [12].

For this study, three test lots were prepared. Test lot HP was parylene C coated, test lot HE was epoxy coated and test lot HEP was first coated with epoxy and then coated with parylene C. Each test lot contained eight similar test samples. The test lots are summarized in Table 2. Before coating test wires were soldered to the test samples in order to measure the daisy chain resistance during testing. These test wires are seen in Figure 1.

Table 2: Test lots used in this study.

Test lot	Conformal coating
HP	Parylene C
HE	Epoxy
HEP	Epoxy and parylene C

The hydrolysis testing was carried out in Natrium phosphate buffered saline, Na-PBS. Buffer solutions are used to obtain biologically or medically relevant results with aqueous systems of proteins, lipids or cells [13]. The composition of Na-PBS was the following: 3.48 g/dm^3 Na_2HPO_4 – 0.755 g/dm^3 NaH_2PO_4 – 5.9 g/dm^3 NaCl. The pH of this buffer solution was 7.4 and was not temperature-dependent. During hydrolysis the containers were kept at +37°C in a heating chamber. The hydrolysis solution used, Na-PBS, was changed at intervals of two weeks and the pH of the solution was analyzed. The daisy chain resistance of the test samples was measured first daily for four days, then every fourth day and then once a week until week 24. After that the measurements were performed every second week. The test lasted altogether 300 days, i.e. ten months.

Results and Discussion

During hydrolysis testing the pH of the buffer solution was measured every second week. Figure 2 shows the results of these measurements. The pH was measured only from three containers from each test lot. Figure 2 shows the averages of the pH values calculated from each test lot. As can be seen from Figure 2, the pH of the buffer solution had only slight fluctuation between 7.33 and 7.43, and thus it can be stated that the buffering capability of the Na-PBS was sufficient. Furthermore, the pH from each

test lot behaved similarly. The ability of the buffer solution to maintain the pH in the specified range when some impurities are introduced into the solution is called the buffering capability [13].

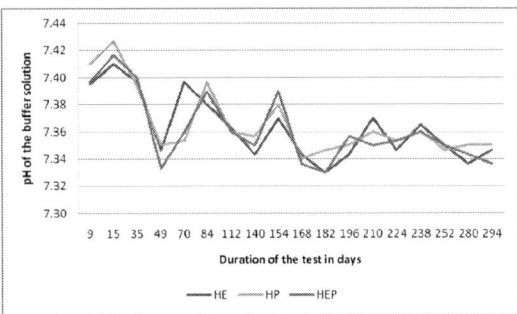

Figure 2: pH measurement of the buffer solution.

The samples were studied visually after hydrolysis testing, and it was seen that the epoxy coated test lot HE behaved differently from the other two test lots. Test lots HP and HEP, where parylene C was the topmost coating layer, had some constituents of the buffer solution condensed on the surface. This was seen in the surfaces after the samples were removed from the hydrolysis and dried out. Figure 3 shows these condensed constituents on the back of one test sample from test lot HP. These residuals were adhered to the surface of parylene C, since they did not come off when the samples were washed with water.

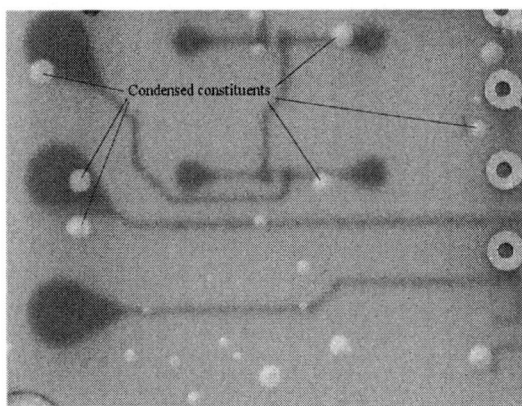

Figure 3: A test sample from test lot HP showing constituents of the buffer solution condensed on the coating.

During hydrolysis testing the daisy chain resistance was measured in order to check the functionality of the ACF joints. During testing some problems occurred with the testing wires. These wires were soldered to the test substrates and during hydrolysis these solder joints and wires suffered from corrosion most, and were first to sustain damage. To avoid this problem the wires were shielded with shrinkable plastic socks which were assembled during testing. Otherwise the test structure withstood the hydrolysis testing well. After

300 days of hydrolysis, all eight test samples in test lot HP were still functioning. Test lot HE showed two failures during testing. Test lot HEP also showed two failures. All these failures already occurred during sterilization or early in the test. The failure percentages from this functionality testing are presented in Figure 4. The verification of the failures during testing was difficult since the test wires were damaged earlier during the test, and the coating layer could not be pierced before the testing was completed. After testing the coating was drilled and the resistance was measured from the test samples. The actual results from the functionality were then collected from the measurements done after the complete test.

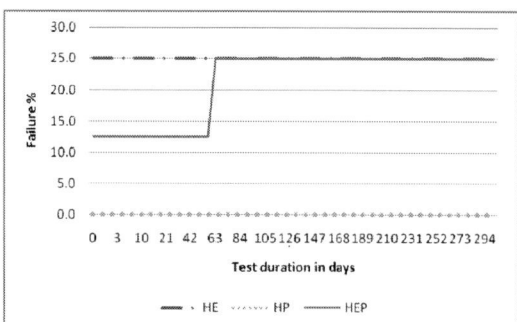

Figure 4: Failure per cents for test lots HE, HP and HEP during hydrolysis testing.

After hydrolysis testing cross sections were made to study the interconnections. Even though there were no failures during hydrolysis, the coating layer and the quality of the interconnections were examined from cross sections using optical and scanning electron microscopy (SEM). Parylene C coating forms a uniform layer, as can be seen from a SEM image taken from one test sample in test lot HP in Figure 5. The layer thickness was measured from the SEM images, and in Figure 5 the thickness measurement is also shown. The approximate thickness of the parylene C coating was 14 µm. Epoxy coating was formed by moulding, and a thick layer of epoxy then formed on the samples. This improves the water permeability properties of epoxy. On the other hand one important advantage of using organic coating materials in implantable electronic devices is to gain a flexible entity, and if a thick layer of epoxy is needed to ensure appropriate shielding properties, the coating is not flexible, and it is worth questioning whether epoxy coating is the best choice. This, of course, is the case only when flexibility is considered. There are many other advantages that may be achieved when epoxy coating is used. With parylene C the flexibility is achievable, since the coating layer is thin, as can be seen in Figure 5.

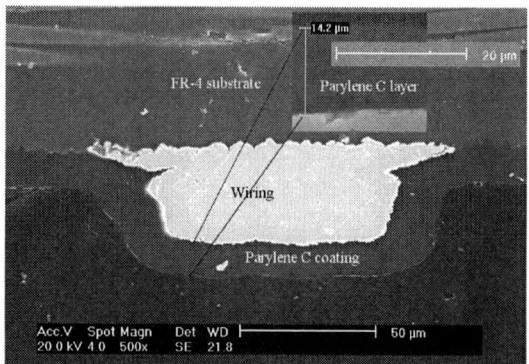

Figure 5: A micrograph showing the uniform parylene C layer formed on the samples. The picture also shows the approximate thickness of the layer.

The epoxy interface to the chip and substrate was studied from the cross sections made from test lots HE and HEP. These studies showed that from silicon chip and NiAu coated wires epoxy coating delaminated but from epoxy to FR-4 interface no delamination was found. This can be seen in Figure 6, showing these interfaces. This delamination may occur due to poor adhesion to silicon and Au. On the other hand, the delamination might easily proceed from these points and this is the start of a failure.

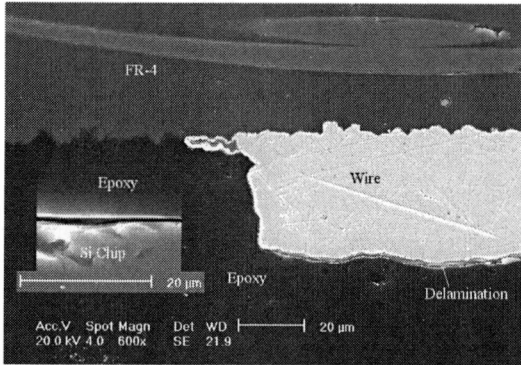

Figure 6: A picture of an epoxy coated test sample where delamination from NiAu coated Cu wiring and Si chip can be seen.

The ACF joints formed well. This was seen from the cross sections, and the daisy chain resistance measurements after joining also implied that the joints were conducting well. The alignment accuracy was good but, as can be seen in Figure 7, the pads on the substrates were over-etched. This could impair the reliability, since it reduces the contact area. However, in this study the reliability was good even with over-etching and the effects of over-etching can only be quessed at. Air bubbles were found in the adhesive and they are shown in Figure 7. These too, may cause reliability issues in the ACF joints.

978-1-4244-4722-0/09 $25.00
© 2009 IMAPS-ITALY

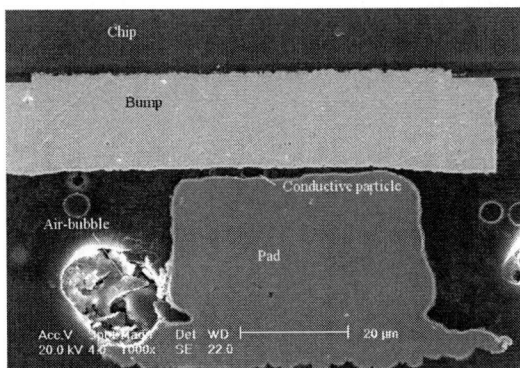

Figure 7: A micrograph of one ACF joint. The over-etching of the pads and air bubbles in ACF can be seen.

The test samples withstood the test well, which is good in the case of implant applications. However, if we want to study the failure mechanisms considering ACF joints in aqueous systems, this reliability testing was not adequate. Either longer hydrolysis testing is needed or some accelerated testing should be considered. The difficulty with accelerated testing lies in the comparability of the accelerated test results to the actual operating conditions and how *in vitro* testing can be reliably accelerated. Hydrolysis testing itself is not an accelerated test.

Conclusion

The *in vitro* testing of ACF joined flip chip components was carried out in Na-PBS buffer solution. Three different test lots were studied; one with parylene C conformal coating, one with epoxy conformal coating and one with epoxy and parylene C composite coating. The aim of this study was to gain relevant information about the coating materials during hydrolysis testing, and to ascertain how well they protect the electronics inside them. The functionality of the test samples was measured by using a test chip having a daisy chain test structure where all the junctions could be measured.

The results showed that all the coating materials were very durable in hydrolysis testing. The failures that occurred during testing were initial failures, and only one test sample failed during hydrolysis. The failure mechanism for this sample could not be worked out. The initial failures occurring with epoxy and epoxy – parylene C coatings likely already failed during sterilization. Best results were obtained with parylene C conformal coating, while no failures occurred during testing. However, according to this study, all these coating choices could be considered as coating materials for *in vivo* testing and use in implants.

Acknowledgements

The authors would like to thank the staff at the Department of Automation Science and Engineering, the Department of Biomedical Engineering and the Department of Materials Science all at Tampere University of Technology. This study was carried out as part of a technology programme funded by the Academy of Finland.

References

[1] N. Beshchasna, J. Uhlemann, K-J Wolter, "Biostability of Electronic Packaging Materials", Proceedings of the 2006 Electronics Systemintegration Technology Conference, Dresden, pp.1047-1053, 2006.

[2] C.W. Tan, Y. C. Chan, N. H. Yeung, "Effect of Autoclave Test on Anisotropic Conductive Joints", Microelectronics Reliability, Vol. 43, No. 2, pp. 279-285, February, 2003.

[3] L. Frisk, A. Cumini, "Reliability of ACA Bonded Flip Chip Joints on LCP and PI Substrates", Soldering & Surface Mount Technology, Vol. 18, No. 4, pp. 12-20, 2006.

[4] K. Kokko, H. Harjunpää, A-M. Haltia, P. Heino, M. Kellomäki, "Effects of Conformal Coating on Anisotropically Conductive Adhesive Joints; A Medical Perspective", Accepted for publication in Soldering & Surface Mount Tecnology, 2009.

[5] W. Mayr, M. Bijak, D. Rafolt, S. Sauermann, E. Unger, H. Lanmüller, "Basic Design and Construction of the Vienna FES Implants: Existing Solutions and Prospects for New Generations of Implants", Medical Engineering & Physics, Vol. 23, No. 1, pp.53-60, 2001.

[6] N.Beshchasna, J. Uhlemann, K-J Wolter, "Researching of Biochemical Degradation of Electronic Materials in Fluid Electrolytic Mediums", Proceedings of the 29th International Spring Seminar on Electronics Technology, Marienthal, May, pp. 149-155, 2006.

[7] J. Chlopek, G. Kmita, "The Study of Lifetime of Polymer and Composite Bone Joint Screws under Cyclical Loads and *in Vitro* Conditions", Journal of Materials Science: Materials in Medicine, Vol. 16, No. 11, pp. 1051-1060, November, 2005.

[8] "Guidance Document for Testing Biodegradable Polymer Implant Devices", U.S. Food and Drug Administration, Draft, April, 1996.

[9] D. Rodger, A. Fong, W. Li, H. Ameri, A. Ahuja, C. Gutierrez, I. Lavrov, H. Zhong, P. Menon, E. Meng, J. Burdick, R. Roy, R. Edgerton, J. Weiland, M. Humayun, Y-C. Tai, "Flexible Parylene-based Multielectrode Array Technology for High-Density Neural Stimulation and Recording", Sensors and Actuators B, Vol. 132, No. 2, pp. 449-460, September, 2008.

[10] J. Licari, "Coating Materials for Electronic Applications: Polymers, Processes, Reliability, Testing", William Andrew Publishing, New York, Chapter 2, pp. 65-278, 2003.

[11] W. Rogers, "Sterilisation of Polymer Healthcare Products", Rapra Technology

Limited, Shawbury, Chapter 5, pp. 147-204, 2005.

[12] K. Kokko, H. Harjunpää, P. Heino, M. Kellomäki, "Influence of Medical Sterilization on ACA Flip Chip Joints Using Conformal Coating", Microelectronics Reliability, Vol. 49, No. 1, pp. 92-98, January, 2009.

[13] Y. Park, S. H. Kim, S. Matalon, N-H. L. Wang, E. I. Franses, "Effect of Phosphate Salts Concentrations, Supporting Electrolytes, and Calsium Phosphate Salt Precipitation on the pH of Phosphate Buffer Solutions", Fluid Phase Equilibria, Vol. 278, No. 1-2, pp. 76-84, April, 2009.

Package Design for Alleviating Stress in Materials Embedded with Electronic Systems

Maryna Lishchynska, Kieran Delaney

TEC Centre, Cork Institute of Technology, Cork, Ireland

Phone: +353 21 4326668; Fax: +353 21 4326399; email: maryna.lishchynska@cit.ie

Abstract

Recent advances in system miniaturisation have led to the development of a multitude of new sophisticated multifunctional electronics. Efficient application of these devices to creating smart objects by embedding digital systems into materials is currently hampered by challenges posed by the system integration and packaging. High stresses induced by the embedded packaged systems are particularly detrimental to the host materials' integrity and cause significant degradation and failure of the host material. This in turn affects the performance and reliability of the host materials and, ultimately, smart objects. Therefore, new packaging design solutions are required to ensure an unobtrusive low-stress integration of packaged electronic systems in materials and objects. This work approaches the challenge as a package-in-a-package problem; that is a packaged electronics system encapsulated into a host material. Through simulation and experiments, it investigates effects of embedded packages on the bulk host materials and discovers package design guidelines for low-stress integration of electronics into smart objects.

Key words: electronic systems, packaging, stress, finite element method

Introduction

The concept of ubiquitous intelligence has firmly surfaced in recent years [1-5] envisaging myriads of applications in areas such as the next generation home and business automation systems and smart medical appliances. By embedding intelligence in everyday objects and environments the efficiency, safety and comfort of our lives will be taken to another level. Smart objects of the future will incorporate physically unobtrusive, distributed, networkable sensing, computing and actuating systems providing a close connection between the physical world and virtual intelligence. To realise these smart environments, robust and reliable technology platforms facilitating effective packaging and seamless integration of electronics into host materials and objects will be needed. Current approaches to creating digitally-enhanced objects usually work by developing the intelligent system and the object itself through separate processes and then integrating them at the latest stages of the development. A more effective and innovative approach would be to embed the packaged digital systems into the material before making an object from it; this has now become possible due to recent advances in materials technology, context-aware software and system miniaturisation [6]. One of the key challenges with this concept is incorporating packaged distributed, networkable computing and sensing systems into various materials without compromising the desired performance and reliability specifications of the host materials and ultimately smart objects. In this scenario, the system's packaging becomes the main interface between the electronics and the host material and subsequently needs to be functional to both, the system inside and the bulk material outside. Existing standard electronics packaging solutions, the majority in a cubical shape, have not necessarily been developed with the purpose of embedding and as a result are mostly unsuitable for use in large scale embedded systems. In fact, no dedicated technology serving the above purpose exists to date. Therefore, new robust and reliable packaging design solutions and technologies for "seamless" integration of digital systems into materials will be needed to realise smart objects and environments of the future. This will require an entirely different approach to the package design.

The approach adopted in this work is to address the system integration as an electronic packaging problem [7-8], where the challenge is to integrate packaged intelligent digital systems into a host object/material. In this context, a smart object is effectively a package-in-a-package (PiP); that is, a packaged electronic system encapsulated into a host structure or material. When embedded into a material, the packaged system perturbs the natural structural morphology in a local continuum [9-11] thus generating undesirable stresses and becoming a stress concentrator (fig. 1a). The stresses are

978-1-4244-4722-0/09 $25.00
© 2009 IMAPS-ITALY

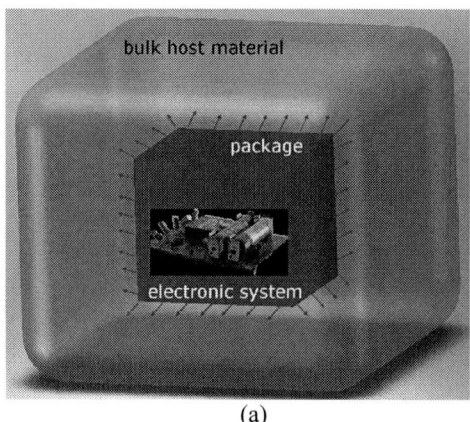 

(a)   (b)

Figure 1. Embedded node in a cubic package (a) and a "smoother" capsule shape package in a host material (b). Arrows indicate stresses exerted by the embedded package onto the host material.

Table 1. Tensile test results

Specimen type	Max load, kN	Elongation at break, mm	Tensile strength, MPa	Tensile modulus, MPa
Blank	21.6	9.7	34.56	267
With embedded insert	6.57	4.15	10.51	137.5

induced into the host material during the thermal cycle of encapsulation and during everyday use and result from mismatches in thermal/mechanical properties between the embedded package and the host material. These stresses are highly undesirable as they undermine the material's structural integrity and ultimately the smart object's functionality and reliability. Problems such as delaminating, cracking and fatigue fracture may occur and eventually render the object unusable. The magnitude of the stresses depends upon a combination of several factors, including the package design, the materials used and the encapsulation/operating conditions. In order for the smart object to be fully functioning and reliable, new package design solutions are required to mitigate these stresses and ensure an unobtrusive presence of the packaged component in the host material. Through experimental studies and employing finite element analysis (FEA), this paper investigates in detail effects of embedded packaged system on host materials and suggests package design solutions to minimise the stress levels (as in fig. 1b).

Effect of Embedded Package on Host Material Strength: Experimental Study

To experimentally study the effects of embedded packages on mechanical properties and behaviour of the bulk host material glass cubes and spheres, representing dummy packages, were embedded into a number of polyurethane blocks (fig. 2a). The inserts were ranging between 6 mm to 9 mm in plane dimensions to mimic existing miniaturised wireless sensor nodes. The samples were created using the liquid plastic moulded into a shape of a tensile test specimen and with inserts fully embedded into plastic. The liquid urethane

material cured chemically and required no thermal processing. A number of pure blank plastic samples were also made for a reference. Specimens were subjected to tensile tests and averaged results are compiled in table 1.

It was found that all specimens with embedded dummies broke at the site of the inserts as opposed to all blank specimens having broken at random points, usually off centre (fig. 2b). This indicates that an embedded component is the dominant "flaw" overriding other imperfections in the plastic responsible for the location of fracture points in blank specimens. Analysis of the results in table 1 shows that presence of an embedded object in the bulk plastic reduces the host material's tensile strength by a factor of three and causes 48.5 % decrease in tensile modulus. These indicate a substantial degradation in material properties and strength. One can speculate that if the bulk plastic required thermal curing, as most industrial polymers do, the detrimental effect of the inserts would be even bigger due to uneven shrinkage and resultant excessive residual thermal stresses in the host material. The following section addresses the effects of this through FEA simulations.

Effect of Embedded Package Design on Stress State in Host Material: Simulation Study

Commercial FEA software ANSYS [12] was used to model a block of bulk host material with an embedded dummy package subjected to thermally induced stresses. The system was modelled as a $10 \times 10 \times 10$ cm^3 PVC cube (host material) with an embedded ceramic node representing a packaged "dummy" digital system.

978-1-4244-4722-0/09 $25.00
© 2009 IMAPS-ITALY

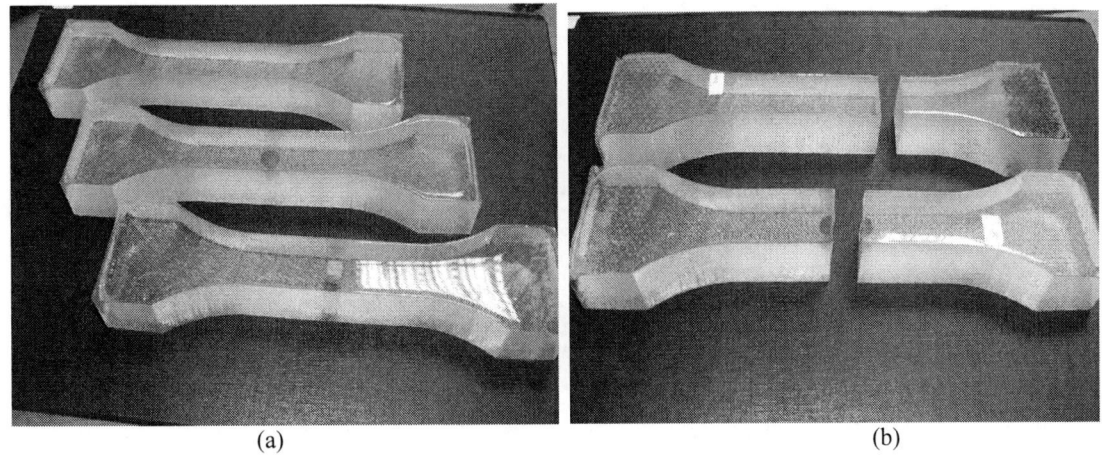

(a)   (b)

Figure 2. Specimens before (a) and after (b) tensile tests.

Table 2. Material properties

Material	Young's modulus, GPa	Poisson's ratio	CTE, 10^{-6}, 1/K
Bulk host (PVC)	3	0.3	80
Ceramic package (LFCSP)	23	0.28	10
Buffer (PDMS)	$5 \cdot 10^{-4}$	0.45	310

Due to the symmetry of the structure only a quarter was modelled with symmetry boundary conditions applied to the relevant planes. The thermomechanical simulations made use of the 20-node, linear, structural element SOLID95. To simulate the stresses induced by thermal cure during an encapsulation process the materials were assumed stress-free at the peak of the temperature cycle, 125 °C, and then the system was subjected to the temperature drop to the ambient of 25 °C. The material properties adopted in simulations are given in table 2.

Firstly, the effect of embedded package shape on stresses induced in the host material was investigated. Cubic, capsule, spherical and rounded cubic shapes were considered. Studied radiuses of rounding were 1/8, 1/4 and 1/2 of the cube edge length. For the purpose of uniform comparison dimensions of all packages were kept to 10 mm radius or edge length correspondingly. The results are presented in figure 3 showing the cubic shape to be the least favourable choice and sphere to be the best package shape for stress alleviation. In fact, choosing the spherical package shape over the cube brings about 47 % reduction in stress induced in the host material.

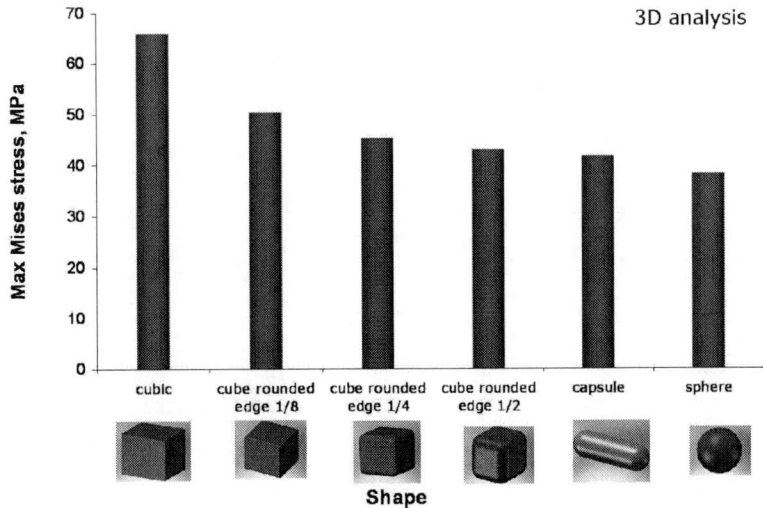

Figure 3. Effect of embedded package shape on maximum Mises stress levels induced in the bulk host material.

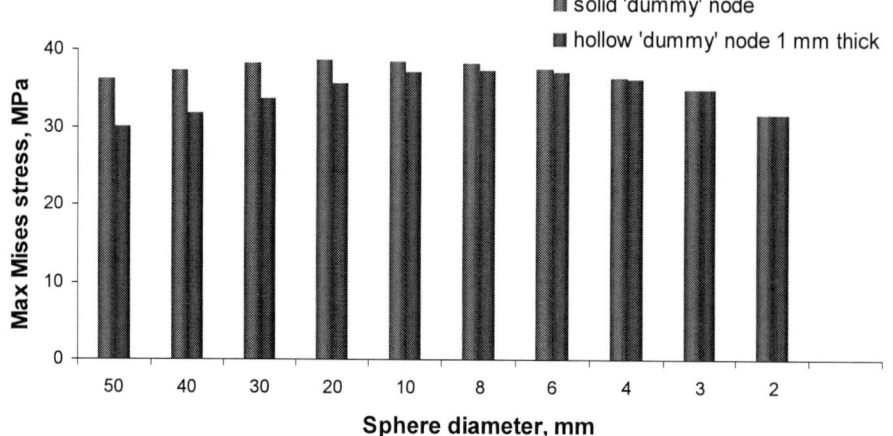

Figure 4. Simulated effect of embedded package size on maximum Mises stress levels in the host material.

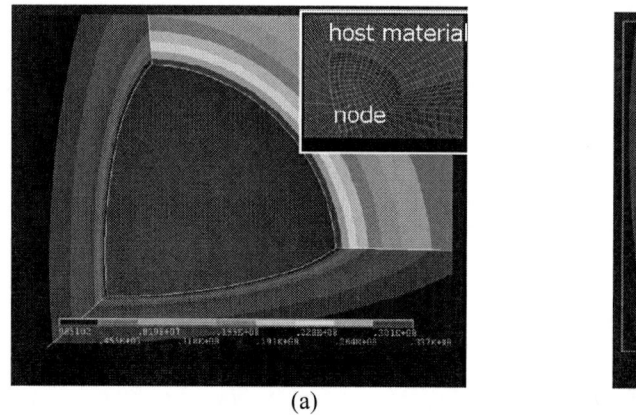

(a)

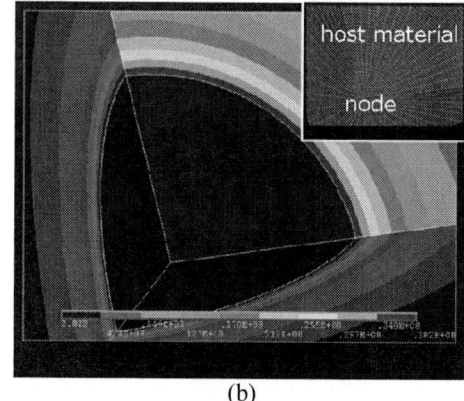

(b)

Figure 5. Simulated Mises stress distribution in the bulk host material with a hollow (a) and a solid (b) spherical packages embedded (sphere diameter is 10 mm). Insets show FEM models of the structures.

To investigate the effect of the package size two types of embedded nodes were modelled as "dummies", solid or hollow spheres, containing no real electronic components. Package dimensions varied in 2 mm to 50 mm diameter range. Simulation results show hollow package nodes to be more dependent on the size factor than the solid ones (fig. 4, 5). Some noticeable variation in stress

induced by hollow packages of various dimensions suggests a possibility of establishing optimal package dimensions for an application. Comparing maximum stress values for cubic and spherical packages (fig. 3, 4) one may conclude that a stress reduction of up to 50 % can be achieved through manipulation with the package shape and size.

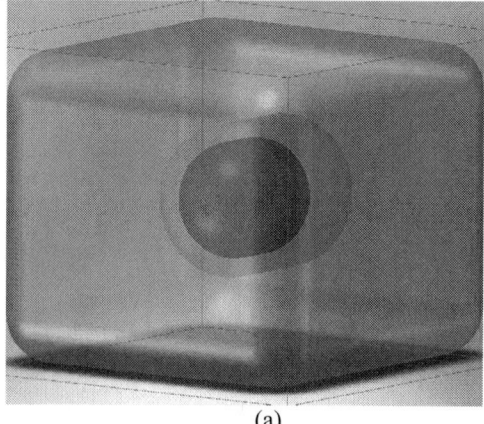

(a)

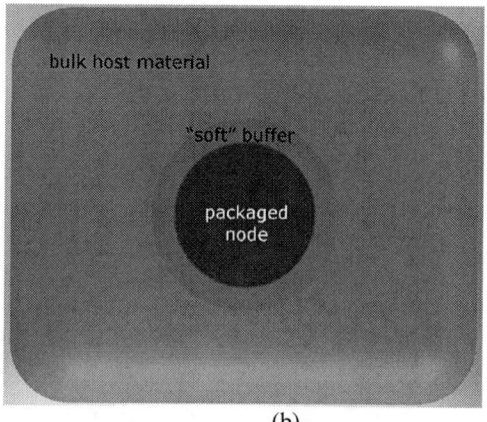

(b)

Figure 6. Schematic of a buffered node embedded in a host material: 3D view (a) and a cross-section (b).

978-1-4244-4722-0/09 $25.00
© 2009 IMAPS-ITALY

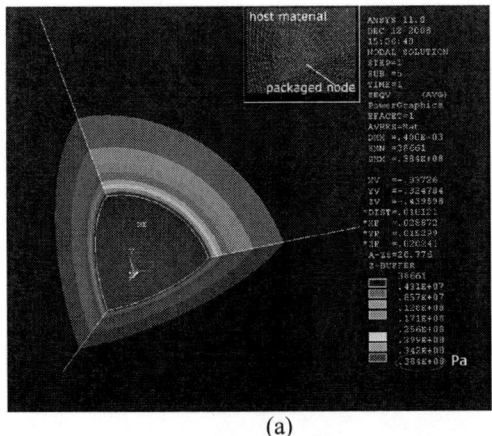
(a)

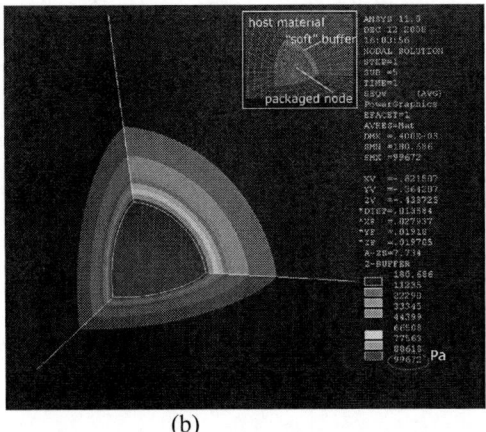
(b)

Figure 7. Simulated Mises stress distribution in the host material with a "bare" spherical node (a) and a "buffered" spherical node (b); only the host material displayed in simulated results, nodes and the buffer are hidden for better visibility. Insets show FEM models of the full structures. Node diameter is 10 mm, buffer is 1 mm thick. A drop in maximum stress level from 38.4 MPa to 99.7 kPa (by over one order of magnitude) is evident.

Using Buffer for Stress Relief

In an attempt to further reduce the undesirable stresses we propose to introduce a mechanically "soft" buffer layer between the package and the bulk host material (fig. 6). Such a buffer is intended to absorb some of the stresses and deformations thus allowing for a significant reduction of stresses induced in the host material. This concept has previously been successfully exploited in various engineering applications [e.g. 13]. In this study PDMS is chosen as a buffer material for its "soft" thermal/mechanical properties (table 2) and it is the subject of ongoing experiments.

Series of simulations with "bare" and "buffered" spherical packages embedded into the host material were performed in this work. Buffer thicknesses in the range of 40µm to 5mm were considered. The results are promising and show that

a significant reduction in stresses, by more than one order of magnitude, can be achieved through introducing a buffer layer. For example, introduction of a 1 mm thick buffer layer results in maximum Mises stress dropping from 38.4 MPa to 99.7 kPa in the case of a 10 mm spherical package (fig. 7).

Further investigation focused on the effect of varying the buffer thickness. Analysis of simulation results in figure 8 suggests that an optimal buffer thickness $t_d^{optimal}$ exists for each diameter d of the packaged component. It was found that the following linear equation accurately describes the relationship between the optimal buffer thickness and package diameter: $t_d^{optimal} = 0.04d$. The formula can be especially useful at the design stage of the system development.

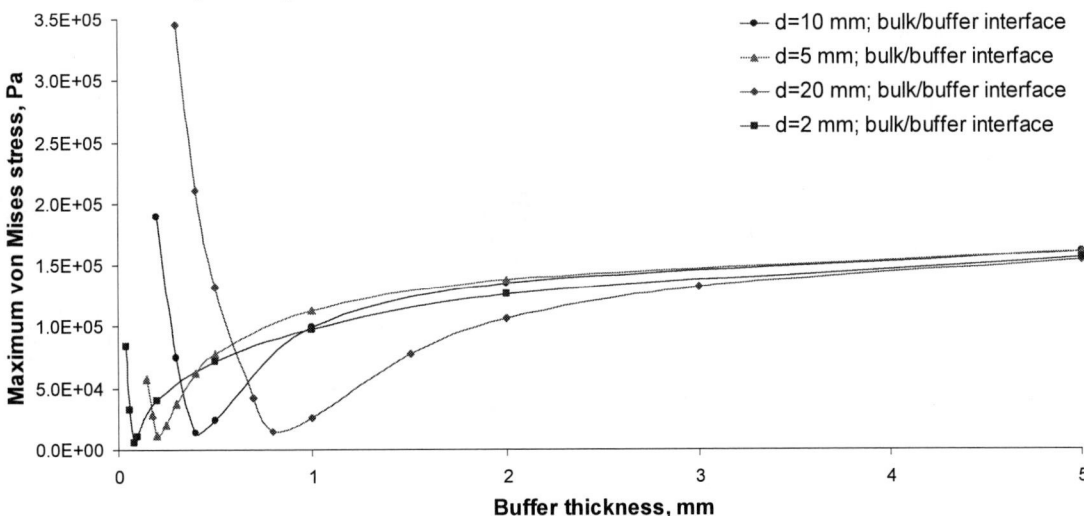

Figure 8. Maximum Mises stress in the bulk host material with 10 mm spherical node embedded.

978-1-4244-4722-0/09 $25.00
© 2009 IMAPS-ITALY

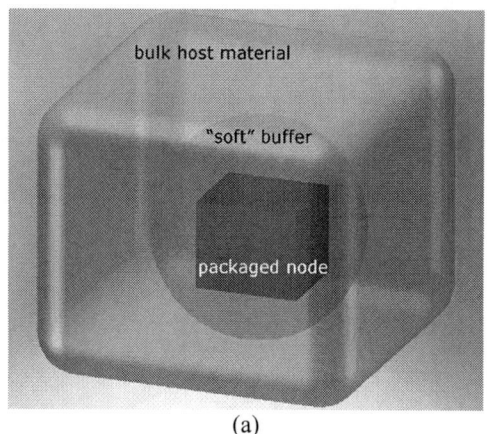

(a)

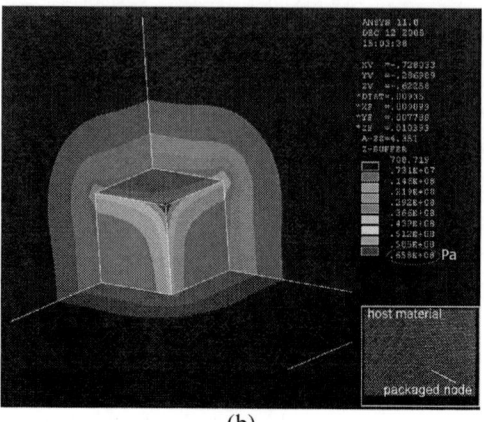

(b)

Figure 9. Schematic of a cubic node surrounded by a spherical buffer embedded in host material (a) and simulated Mises stresses induced in the host material by 10mm bufferless cubic package (b). Inset shows corresponding 3D model.

Given the successful reduction in stresses achieved using spherical package shape and buffer design we next look to use the spherical buffer to alleviate the stresses in other package shape forms. For example, if a cubic shape factor is unavoidable in an application or for certain node architecture, a spherical buffer can be considered (fig. 9a). Figures 9b and 10a compare simulated stresses induced by a "bare" (bufferless) cubic package and a cubic package enclosed in a spherical buffer and then embedded in the host material. A drop in the maximum stress level in the host material from 65.8 MPa to 167 kPa (by more than one order of magnitude) is evident.

Conclusions and Package Design Suggestions

Experimental and simulation techniques were used to investigate the detrimental effects of embedded packages on the host materials and to discover methods of alleviating the excessive stresses. Experimental studies indicated a substantial degradation in mechanical properties and strength of the plastic material embedded with inserts. Presence

of an embedded object in the bulk plastic was found to reduce the host material's tensile strength by a factor of three and causes 48.5 % decrease in tensile modulus.

Finite element modelling techniques were used to investigate the effects of the package geometry (shape and size) upon the thermal stresses occurring in the host material. It has been found that a 100°C temperature drop at the end of the encapsulation cycle generates excessively high levels of stresses up to 65 MPa. These stresses constitute 74% of the material's yield strength (88 MPa for PVC) and they are only residual post-fabrication stresses not considering the operational conditions of the object. Simulation studies undertaken in this work established that a reduction of up to 50 % in stress levels can be achieved through manipulation of the package geometry. An alternative approach to managing stresses by introducing a mechanically soft stress buffer layer was also investigated. The results are promising and show that a reduction in stresses by over one order

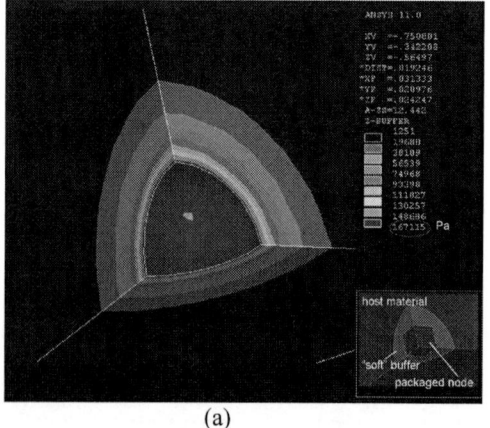

(a)

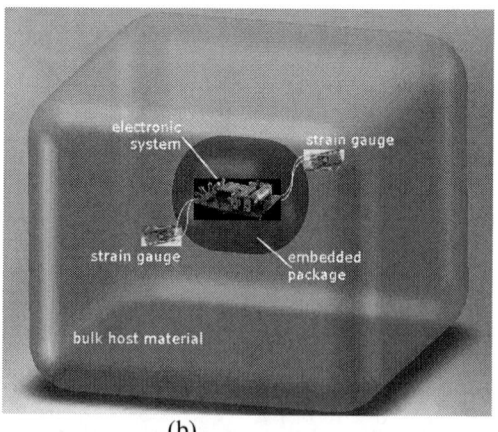

(b)

Figure 10. Simulated Mises stresses induced in the host material by 10mm cubic package surrounded by 20 mm in diameter spherical buffer (a). Inset shows corresponding 3D model. Schematic of an embedded packaged electronic system with strain gauges "wired out" of the package (b).

978-1-4244-4722-0/09 $25.00
© 2009 IMAPS-ITALY

of magnitude can be achieved through introducing a buffer layer. An optimal buffer thickness exists and can be calculated using the equation $t_d^{optimal} = 0.04d$. It is worth noting that certain types of applications may impose restrictions on the node architecture and subsequently on the package design. For instance, in applications requiring strain mapping strain gauges should be placed in alignment with the anticipated strain directions. In such cases strain gauges can be "wired out" of the package, and the buffer zone if present, and be placed "bare" in the host material (fig. 10b).

In conclusion, the following design guidelines apply to development of packaging solutions for low-stress integration of electronic systems into smart objects:

- choice of spherical or capsule-like shapes over cubic;
- optimisation of package size to minimise stresses;
- using "soft" buffer layer surrounding the package;
- optimal buffer thickness can be calculated using the equation $t_d^{optimal} = 0.04d$;
- choosing spherical buffer with packages where a cubic node shape is unavoidable;
- "wiring off" strain gauges, and other components that need to be placed within the bulk material, out of the package and the buffer, if this is present.

Acknowledgements

The authors would like to thank Padraig O'Herlihy and Fergal Cronin (CIT) for making specimens and Sean O'Leary (Mechanical Engineering, CIT) for tensile testing.

References

[1] Ambient Intelligence with Microsystems: Augmented Materials and Smart Objects (Microsystems) by Kieran Delaney (Editor), Springer, 2008.

[2] Ferscha, et al., Peer-it: Stick-on solutions for networks of things, Pervasive and Mobile Computing, 4 (2008) 448-479.

[3] K. S. J. Pister, J. M. Kahn and B. E. Boser, "Smart Dust: Wireless Networks of Millimeter-Scale Sensor Nodes", Electronics Research Laboratory Research Summary, 1999.

[4] H. Duman, H. Hagras, V. Callaghan, Intelligent association of embedded agents in intelligent inhabited environments, Pervasive and Mobile Computing, 3 (2007) 117-157.

[5] D. J. Cook, S. K. Das, How smart are our environments? An updated look at the state of the art, Pervasive and Mobile Computing, 3 (2007) 53–73.

[6] K. D. Wise, Integrated sensors, MEMS, and microsystems: Reflections on a fantastic voyage, Sensors and Actuators A 136 (2007) 39–50.

[7] C.-C. Lee, C.-T. Peng, K.-N. Chiang, Packaging effects investigation of CMOS compatible pressure sensor using flip-chip and flex board technologies, Sensors and Actuators A, 126 (2006), 48-55.

[8] R. P. Cain, et al., Packaging for a sensor platform embedded in concrete, Microelectronics and Microsystems Packaging, MRS Proceedings, Volume 682E, N. 3.9.

[9] B. Lauke, T. Schuller, Calculation of stress concentration caused by a coated particle in polymer matrix to determine adhesion strength at the interface, Composite Science and Technology, 62 (2002) 1965-1978.

[10] P. R. Marur, Estimation of effective elastic properties and interface stress concentrations in particulate composites by unit cell methods, Acta Materialia, 52 (2004) 1263-1270.

[11] Y. Fan, M. Kahrizi, Characterisation of a FBG strain gage array embedded in composite structure, Sensors and Actuators A, 121 (2005), 297-305.

[12] ANSYS 11.0. User's manual.

[13] M. Ranjan, et al., How buffer layers can provide stress management for wafer level chip-scale packaging, Solid State Technology, August 2004.

978-1-4244-4722-0/09 $25.00
© 2009 IMAPS-ITALY

Development of low-firing lead-free thick-film materials on steel alloys for piezoresistive sensor applications

Caroline Jacq, Thomas Maeder, and Peter Ryser

Laboratoire de Production Microtechnique, Ecole Polytechnique Fédérale de Lausanne (EPFL), CH-1015 Lausanne, Switzerland

+41.21.693.53.85, +41.21.693.38.91, **caroline.jacq@epfl.ch**

Abstract

Piezoresistive sensors based on steel and other metallic substrates provide higher strain response than on standard ceramic substrates and are more easily packaged. But exposing high-strength steels to the standard high-temperature 850°C thick-film firing cycle affects their mechanical properties. In previous studies, we have developed a range of low-firing thick-film materials based on lead borosilicate glass, which allows processing at low temperatures. However, it is desirable to develop alternatives to potentially toxic lead-based glasses that to not include alkali metals, which degrade high-temperature insulation characteristics of dielectrics. To this end, this work concerns investigations in essentially substituting lead for bismuth, and presents a series of low-melting Bi-B-Zn-Si-Al oxide glasses having good stability against devitrification. However, these glasses, when formulated as thick-film pastes using standard vehicles based on ethylcellulose binders, were found to be quite sensitive to incomplete binder burnout, with strong bubble generation within the layer. Therefore, a novel organic binder based on polypropylene carbonate, featuring clean low temperature burnout, had to be introduced. On this basis, thick-film dielectric compositions have then been developed and tested, aiming to optimise the mechanical strength and their expansion matching with the steel substrates. In the goal of a complete materials system, first tests on compatible conductors and resistors, using the same glasses, are presented as well.

Key words: thick-films, lead-free, bismuth glass, steel, sensor

Introduction

Piezoresistive sensors based on thick-film technology deposited on ceramic substrates have found success due to their low production cost. However, alumina is not optimal for piezoresistive sensing applications due to its brittleness, high elastic modulus and rather low strengh [1]. Additionally, the applications of this kind of sensors remain limited because their assembly requires elastomer seals. Stainless steels and other metals [2] potentially offer much better strength and toughness, hermetic assembly by welding and machinability [3,4].

However, the high firing temperatures associated with commercial thick-film processing (850°C) are not compatible with high-strength steel, owing to degradation of mechanical properties due to annealing or dimensional changes associated with martensitic transformation (which tend to destroy the thick-film layers). In order to avoid this problem, a solution resides in developing a thick-film system with a lower firing temperature, ideally below ca. 650°C, a temperature still compatible with good strength retention and that avoids phase

transformations. In previous studies [5,6,7,8,9,10], we have developed and studied several low-firing systems based on low-melting lead borosilicate glasses, where the composition was adapted to achieve compatibility with steel substrate. Such systems need also to be thermally matched to steels, which have a range of thermal coefficients of expansion (TCE) from 11 ppm/K to 17 ppm/K, compared to standard thick-film materials, which are thermally matched to alumina (7 ppm/K).

The lead content of these materials is a problem, as lead is restricted under the European Union RoHS (Restriction of Hazardous Substances) directive [11]. Although its use in glasses for electronics applications is mentioned under the list of exemptions, this may change in the future, and further restrictions are therefore likely in the medium term.

Bismuth is a potential alternative to lead, and bismuth-based frits for thick-film conductors – albeit with a somewhat high alkali content – were already patented in 1960 [12]. Inoue [13] formulated alkali-free conductor frits implicitly, by mixing Bi_2O_3 with ZnO-B_2O_3-SiO_2 glass. The resulting "modern" Bi_2O_3-ZnO-B_2O_3-SiO_2 system, with optional MgO,

BaO and Al_2O_3 additions was disclosed in several Soviet patents [14, 15, 16]. These glasses have a high Bi content, processing temperatures down to ca. 500°C, and are therefore useful as sealing glasses and frits. A limited volume of recent work specifically mentions their use or that of similar glasses in the fabrication of thick-film resistors as well [17, 18], which were found to have interesting piezoresistive properties from preliminary investigations [19].

In this work, we present the results of the development of a range of bismuth glasses and the first investigations of new low-firing lead-free thick-film system, including dielectrics, conductors and resistors. The sheet resistance (SR) and temperature coefficient (TCR) of a bismuth-glass – RuO_2 resistive composition are examined as a function of the processing conditions, the underlying dielectric, the substrate and the conductor termination.

Experimental

The following substrate materials were used: 96% pure alumina (Kyocera, Japan, A-476, TCE = 7 ppm/K) as standard thick film substrate and ferritic stainless steel 1.4016 (TCE = 11 ppm/K, comparable to the high-strength precipitation hardening martensitic stainless steels).

The thick-film materials used in this work as matrix for the dielectrics, the conductor and the resistor are based on the lead-free Bi-B-Zn-Si-Al oxide glasses.

Glass	Bi8	Bi12	Bi16	Bi9	Bi11	Bi17
Bi_2O_3 (%mol.)	45	40	50	50	45	50
B_2O_3 (%mol.)	35	35	35	30	30	25
ZnO (%mol.)	10	15	5	10	15	5
SiO_2 (%mol.)	6	6	6	6	6	6
Al_2O_3 (%mol.)	4	4	4	4	4	4
Sintering Temp. [C]	475	475	475	450	450	450

Table 1 : Bismuth glasses composition

Two dielectric types were tested: in the first one, glasses were filled with 30% mass. alumina (Alfa Aesar, aluminium oxide alpha, 99.99%, 1 μm), which is relatively inert and therefore has limited chemical reactions with the glass or piezoresistor material. However, alumina decreases the TCE of the thick-film, which could result in excessive compressive stress on steel substrates. In order to match the TCE of the thick-films to the substrate, strontium titanate ($SrTiO_3$, AlfaAesar. 99+%) was tested.

The conductor (AgBi12) is composed of Ag powder (Alfa Aesar. 99.9%, APS 0.5-1 μm) and 10% vol. Bi12 glass binder. It was fired at different

temperature before (prefired) or after the resistors (postfired). A commercial gold conductor, ESL 8837, was also used to characterise the resistive composition on bare alumina.

First tests of the dielectric as a basis for the low-firing commercial 10 kΩ resistive composition ESL 3114 were unsuccessful; excessive melting and spreading of the resistor – and unusable resistance values – were observed (Figure 1). Obviously, the reaction product between both glass types gives a very low-melting mix; these materials are incompatible.

Therefore, we extended our study to resistors as well. The studied resistive composition is based on the bismuth glass Bi12 filled with 11% by mass RuO_2 powder (Aldrich. 99.9%, 40nm). The resistors were fired for 10 min at peak (45 min total cycle) at different temperatures from 500 to 625°C.

The organic vehicle is composed of 25% mass polypropylene carbonate (binder) in 75% mass. propylene carbonate (solvent).

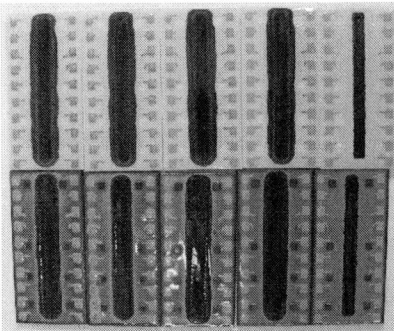

Figure 1: ESL 3114 resistors fired on bismuth dielectric alumina and steel @ 625°C

The bismuth glass – RuO_2 resistor was characterised on bare alumina (reference substrate) with three terminations schemes: a) commercial Au ESL 8837, pre-fired at 850°C; b) pre-fired AgBi12 conductor; c) post-fired AgBi12 conductor, i.e deposited onto the resistors, in light of our previous results with low-temperature lead-based glasses [8].

Following these tests, the resistor was characterised on both Al_2O_3-filled and $SrTiO_3$-filled dielectrics, deposited on alumina or stainless steel. Both (b) and (c) AgBi12 termination schemes were examined for the Al_2O_3-filled dielectrics, and only (b) was used for the $SrTiO_3$-filled ones. Samples prepared with scheme (b) were also refired at the 500°C conductor post-firing temperature used in scheme (b) for comparison of both schemes, in order for the resistor to experience the same heat treatment sequence.

Samples for electrical characterisation (sheet resistance SR and temperature coefficient of resistance TCR) were 1.5 mm wide resistors of several lengths (Figure 2), and were measured at 30°C, 65°C and 100°C.

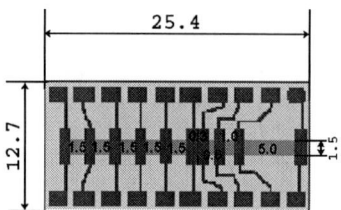

Figure 2: Layout of the test sample for measurement of electrical properties.

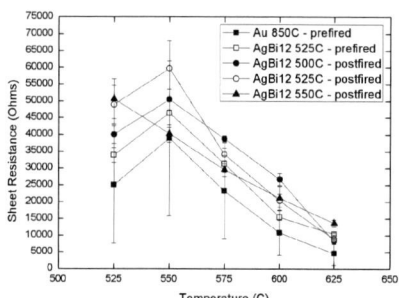

Figure 3: Sheet resistance of Bi glass – RuO$_2$ resistors as a function of the resistor firing temperature for different termination schemes and firing temperatures, on bare alumina.

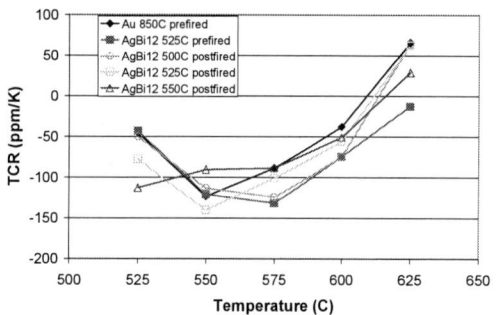

Figure 4: TCR of the resistors on bare alumina substrate.

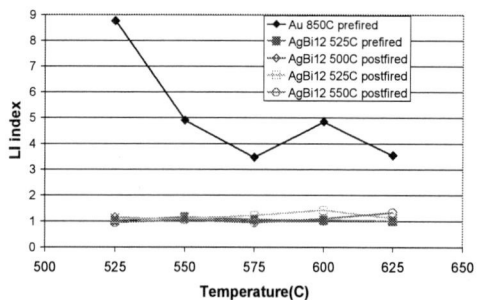

Figure 5: Length index LI the resistors on bare alumina substrate.

In order to have a quick assessment of the termination effects, we calculated a "length index" LI, defined as the ratio of the average sheet resistance of the three short resistors (0.3, 0.6 and 1.0 mm length) to the average of the 1.5 mm long ones. LI values greater than 1 imply the existence of highly resistive zones near the terminations, whereas the other case (LI < 1) corresponds to a locally decreased sheet resistance near the terminations.

Results and discussion

The results of the first characterisation of the resistor deposited directly on bare alumina substrates are depicted on Figure 3 (SR) and Figure 4 (TCR). For post-fired conductors, (scheme c), three post-firing temperature (500, 525 and 550°C) were used in order to identify the most favourable one for the next tests.

The sheet resistance is higher with the AgBi12 conductor than the gold conductor. TCR values are surprisingly moderate for a resistor series without additives, and are close to zero at firing temperatures near 600°C.

Figure 5 allows to study the terminaison effect on the resistance, and immediately show that the gold conductor is incompatible with the bismuth-glass – RuO$_2$ resistor, giving very strong inverse termination effects suggesting the occurrence of an insulating zone at the conductor-resistor interface. This problem is absent with pre-fired AgBi12 conductors. For post-fired AgBi12, the higher firing temperatures give an inverse termination effect, although it is totally overshadowed by the effect with Au; 500°C is therefore the preferrable post-firing temperature and was therefore used for the next tests.

The bismuth resistance was then charaterised on the both dielectric types (Al$_2$O$_3$ or SrTiO$_3$-filled bismuth glasses), themseleves deposited onto alumina or steel.

The consistence of the dielectrics was too glassy; when subsequently firing the conductors and resistors, extensive re-softening of the dielectrics occurs, leading to breakages in the conductor tracks. In an effort to alleviate this problem, conductor and resistor were fired at a temperature 25°C lower than the firing temperature of the dielectric, hence the shifted 500...600°C resistor firing temperature range (compared to 525...625°C on bare alumina). In spite of this measure, problems with the conductor tracks still occurred.

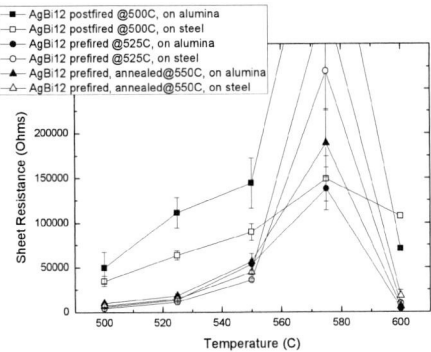

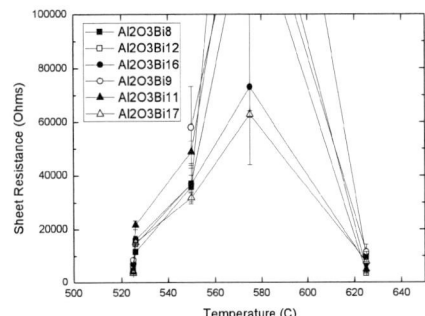

all curves exhibit a peak of SR for a firing temperature of 575°C.

Figure 6: Sheet resistance of the resistors on dielectric filled with Al₂O₃, as a function of the resistor firing temperature, for different firing temperatures and processes of the conductor.

The Figure 6 allows to determine the firing process of the conductor. The comparison has been made on the Al₂O₃-filled dielectric based on the Bi8 bismuth glass. For the post-fired AgBi12 the sheet resistance on alumina subtrate is unreliable: the resistors are apparently cracked (most likely due to cracking in the dielectric), and no meaningful measurements are possible. On the contrary, on the steel substrate, the results are very reproducible, but markedly different from those observed on bare alumina (Figure 3). While some evidence of a peak of SR for firing temperatures around 550°C is seen on bare alumina, a much stronger peak, rather centred at 575°C, is seen on dielectric. The peak of resistor value vs. firing temperature have been observed previously by Kubový et al. [20]; they attributed it to the transition between a wetting phase of the conducting powder by the glass (which moves the particles apart and therefore increase SR) and a phase where diffusion from the conductive particles into the glass progressively decreases SR. Similar effects were also observed by us in our previous studies of low-firing lead-based glass – RuO₂ resistors [21,22]. In our case, the shift to higher temperatures and increase of magnitude of the peak could be due to excess glass moving out of the dielectric into the resistor, a hypothesis that must still be considered as speculative, although the dielectrics do look more glassy at high firing temperatures. On the other hand, this peak is definitely not due to the substrate, as it is observed on both alumina and steel.

Annealing the films at 550°C results in a moderate and reproducible increase of SR for all firing temperatures.

Figure 7 depicts the sheet resistance of the resistor on the Al₂O₃-filled dielectrics based on the 6 bismuth glasses deposited on steel substrate. The conductor AgBi12 is prefired on these samples. A more glassy appearance is visible starting at ca. 550°C firing temperature for all dielectrics, and

Figure 7: Sheet resistance of bismuth resistance in function of the firing temperature of the resistance on dielectric filled with Al₂O₃ based on different bismuth glasses.

The SrTiO₃-filled dielectrics, which should have better TCE matching with the steel substrate, gave results different from the Al₂O₃-filled ones. First, only 3 bismuth glasses suited were compatible with SrTiO₃: Bi8, 12 and 16. The three other react with this filler (Figure 8) or crystallise in contact with it. It is interesting to note that there is a good correlation between glass stability and compatibility with SrTiO₃; the three compatible glass lie comfortably inside the stability zone of glass formation, while the three incompatible ones lie at the composition boundary with glasses that crystallise too rapidly to be useful as frits

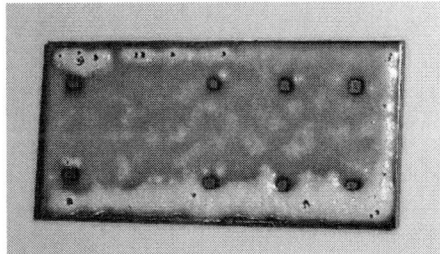

Figure 8: Stability problem with SrTiO₃-filled Bi9 dielectric

Figure 9 depicts the sheet resistance of resistors on steel substrates with AgBi12 prefired conductor. As expected from the high TCE imparted by SrTiO₃, the dielectrics cracked on alumina, and no meaningful measurements could be made on this substrate in the case of SrTiO₃-filled dielectrics.

With the compatible glasses, we obtain values similar to that on alumina-filled dielectrics for low firing temperatures (≤550°C), but the peak of SR at 575°C and subsequent drop at 600°C are much less pronounced on the SrTiO₃-filled dielectrics. The TCR (Figure 10) values are not too different from those observed on bare alumina substrates – they are also quite close to zero, given the fact that no TCR-driving additives were used.

978-1-4244-4722-0/09 $25.00
© 2009 IMAPS-ITALY

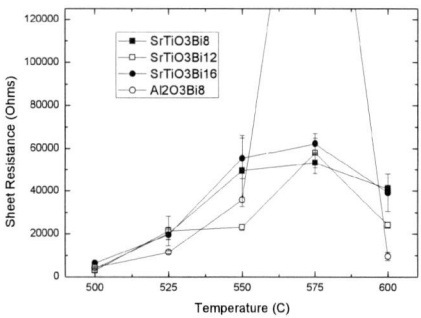

Figure 9: Sheet resistance of bismuth resistance in function of the firing temperature of the resistance on dielectric filled with SrTiO₃ on steel substrate.

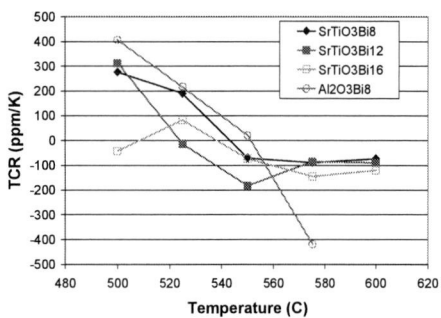

Figure 10: TCR of bismuth resistance in function of the firing temperature of the resistance on dielectric filled with SrTiO₃ on steel substrate.

Conclusion

In this first study of thick-film materials based on low-firing glasses with bismuth as a replacement for lead, functional dielectric, conductor and resistor compositions could be formulated, and the resulting resistor properties were very promising, with SR in the 10…100 kΩ range and low TCR values even without using TCR drivers. Both Al_2O_3-filled dielectrics (better chemical compatibility) and $SrTiO_3$-filled ones (better TCE matching to steel) were successfully used.

These results were achieved at firing temperatures below 650°C, and thus we can safely assert that a low-firing thick-film system compatible with high-strength steel for piezoresistive force and pressure sensing is possible (the existence of appreciable piezoresistivity was confirmed by manually bending the resistive test patterns, but no quantitative measurements were made).

The main current issue is the stabilisation of the dielectrics, in order to avoid softening upon re-firing. This may be carried out by controlled reaction and / or cristallisation of the glass in contact with the filler. More extensive studies will of course also be carried out on the resistor materials in the future, including characterisation of their piezoresistive properties.

References

[1] C. Jacq, T. Maeder and P. Ryser, "High-strain response of piezoresistive thick-film resistors on titanium alloy substrates", Journal of the European Ceramic Society, Vol. 24, No. 6, pp. 1897-1900, 2004.

[2] T. Maeder, H. Birol, C. Jacq, P. Ryser, "Strength of ceramic substrates for piezoresistive thick-film sensor applications", Proceedings, European Microelectronics and Packaging Symposium, Prague (CZ), pp. 272-276, 2004.

[3] N.M. White, "A study of the piezoresistive effect in thick-film resistors and its application to load transduction", University of Southampton, Faculty of Engineering & applied Science, 1988.

[4] L. Fraigi, D. Lupi, L. Malatto, "A thick-film pressure transducer for cars propelled by natural gas", Sensors and Actuators A, Vol. 41, pp. 429-441, 1994.

[5] C. Jacq, T. Maeder, S. Vionnet, P. Ryser, "Low-temperature thick-film dielectrics and resistors for metal substrates", Journal of the European Ceramic Society, Vol. 25 (12), pp. 2121-2124, 2005.

[6] C. Jacq, T. Maeder, S. Menot-Vionnet, H. Birol, I. Saglini, P. Ryser, "Integrated thick-film hybrid microelectronics applied on different material substrates", Proceedings of the 15th European Microelectronics and Packaging Conference (EMPC), Brugge (BE), IMAPS, pp. S13.04, 319-324, 2005.

[7] C. Jacq, T. Maeder, S. Martienerie, G. Corradini, E. Carreño-Morelli, P. Ryser, "High performance thick-film pressure sensors on steel", Proceedings, 4th European Microelectronics and Packaging Symposium, Terme Čatež (SI), IMAPS, pp. 105-109, 2006.

[8] C. Jacq, T. Maeder, N. Johner, G. Corradini, P. Ryser, "High performance low-firing temperature thick-film pressure sensors on steel", Proceedings of the 16th European Microelectronics and Packaging Conference (EMPC), Oulu (FI), IMAPS, pp. 167-170.

[9] C. Jacq, T. Maeder, N. Johner, G. Corradini, and P. Ryser. High performance low-firing temperature thick-film pressure sensors on steel. In Ceramic Interconnect and Ceramic Microsystems Technologies, pages 411-416 (P13), 2008.

978-1-4244-4722-0/09 $25.00
© 2009 IMAPS-ITALY

[10] Jacq-C Maeder-T Ryser-P, "Load sensing surgical instruments", Journal of Materials Science: Materials in Medecine, 2008.

[11] "On the restriction of the use of certain hazardous substances in electrical and electronic equipment", directive 2002/95/EC of the European Parliament and of the Council, 2002.

[12] Dumesnil-ME, "Vitrifiable inorganic ceramic binder and silver compositions containing the same", US Patent 2'942'992, 1960.

[13] Inoue-T, "Conductive paste for thick-film circuit", Japanese patent 54060497, 1979.

[14] Ermolenko-NN Manchenko-ZF Samujlova-VN Shamkalovich-V, "Легкоплавкое стекло / Low-fusible glass", Soviet Patent SU775'061, 1980.

[15] Ermolenko-NN Manchenko-ZF Saevich-NG Samujlova-VN Grigorev-VM Ivolgin-VM Titova-RA, "Легкоплавкое стекло / Low-melting glass", Soviet Patent SU923'976, 1982.

[16] Tikhonov-IA Ermolenko-NN Manchenko-ZF Saevich-NG Savelov-IN, "Легкоплавкое стекло / Low-melting glass", Soviet Patent SU1'477'706, 1989.

[17] Hormadaly-J, "Cadmium-free and lead-free thick film paste composition", United States Patent 5'491'118, 1996.

[18] Busana-MG Prudenziati-M Hormadaly-J, "Microstructure development and electrical properties of RuO_2-based lead-free thick film resistors", Journal of Materials Science: Materials in Electronics Vol. 17, 951-962, 2006.

[19] Totokawa-M Yamashita-S Morikawa-K Mitsuoka-Y Tani-T Makino-H Shimizu-H, "Microanalyses on the RuO_2 particle–glass matrix interface in thick-film resistors with piezoresistive effects", International Journal of Applied Ceramic Technology Vol. 6 (2), 195-204, 2008.

[20] Kubový-A Havlas-I, "The effect of firing temperature on the properties of model thick film resistors", Silikàty Vol. 32, 109-123, 1988.

[21] Vionnet-Menot-S Grimaldi-C Maeder-T Ryser-P Strässler-S, "Study of electrical properties of piezoresistive pastes and determination of the electrical transport ", Journal of the European Ceramic Society Vol. 25 (12), 2129-2132, 2005.

[22] Vionnet-Menot-S Maeder-T Grimaldi-C Jacq-C Ryser-P, "Properties and stability of thick-film resistors with low processing temperatures - effect of composition and processing parameters", Journal of Microelectronics and Electronic Packaging Vol. 3 (1), 37-43, 2006.

Structuration of zero-shrinkage LTCC using mineral sacrificial materials

Thomas Maeder, Caroline Jacq, Yannick Fournier, Wassim Hraiz and Peter Ryser
Laboratoire de Production Microtechnique
École Polytechnique Fédérale de Lausanne (EPFL)
Station 17, BM2.137, CH-1015 Lausanne, Switzerland
Phone: +41 21 693 58 23; fax: +41 21 693 38 91; thomas.maeder@epfl.ch; http://lpm.epfl.ch

Abstract

Recently, LTCC (low-temperature co-fired ceramic) technology has increasingly found applications beyond pure electronics, in fields such as microfluidics, sensors and actuators, due to the ease of shaping the tapes in the green (unfired) state. Accurate control of hollow structures such as channels, membranes, cavities and gaps below cantilevers has remained difficult, however, although carbon-based sacrificial materials and adhesive/solvent-assisted low-pressure lamination techniques are adequate for several uses.

Mineral sacrificial pastes (MSP), introduced by several groups including our laboratory, allow in principle much better control of open structures such as bridges and cantilevers, as they are removed only after the firing step. In practice, accurate dimensional control has been limited by deformation of the LTCC during sintering, due to shrinkage mismatch with the MSP. Attempts to eliminate this problem have met with limited success, as it is very difficult to perfectly match the shrinkage curve of the MSP (which must retain open porosity) to that of the LTCC substrate. Therefore, in this work, we endeavour to investigate MSP materials on self-constraining "zero-shrinkage" LTCC tape, which is therefore compatible with a low degree of sintering of the MSP. We present results of optimising the MSP formulation accordingly, to achieve reasonable consolidation, low deformation of LTCC and easy removal in weak acid solutions. Important topics such organic vehicle formulation and complete release processes (etching, rinsing and drying) of thin structures are also addressed.

Key words: sensors, thick-film, LTCC, 3D structuration, mineral sacrificial layers / MSP, processing.

Introduction

Thick-film technology is a versatile tool for the manufacture of sturdy and reliable devices, at relatively low cost, and has thus made important inroads into sensor applications [1], albeit with relatively simple products, a limitation due to the difficulty and cost of shaping the substrates. The relatively recent development of LTCC has heralded new possibilities in 3D structuration [2], with devices such as chemical microreactors & mixers [3], hot-plate gas sensors [4] and structured force sensors [5].

The 3D shaping of LTCC devices raises several issues regarding the lamination step, which, due to the common presence of closed cavities, channels, etc., is much less straightforward than that of "standard" electronic modules. In the simple "cut-and-laminate" method, lamination parameters are a compromise between ensuring sufficient quality and avoiding excessive deformation. To address this issue, several techniques have been successfully applied, such as solvent- or adhesive-assisted lamination to ensure a good bond between LTCC layers even at low lamination pressures and temperatures, and carbon-based sacrificial layers

that allow lamination, then burn away during firing [6].

Nevertheless, these techniques do not fully address the problem of slender structures, which tend to warp due to sagging and/or to stresses imposed by thick-film materials having different sintering curves. Therefore, there has been a continuing effort [7-13] to develop, for both thick-film and LTCC technology, mineral-based sacrificial layers (MSP, mineral sacrificial paste) that would survive during firing, supporting the structures, and then could be removed by a post-etching step in a relatively benign substance such as a weak acid.

Very recently [13], success has been obtained for thick-film technology with MgO sacrificial layers using H_3BO_3 (or a limited amount of borax) as a "glue" between the MgO grains. MgO is a well-known sacrificial material in surface micromachining [14]. In order to allow easy later etching, it is important that the material remain porous [11], which makes for a relatively low shrinkage. While this is fine for traditional thick-film technology, the low shrinkage is less suited for integration with standard LTCC materials using the free-sintering process. Previous tests resulted in important deformations [6], which could be somewhat lowered by incorporating some carbon

978-1-4244-4722-0/09 $25.00
© 2009 IMAPS-ITALY

powder to provide some shrinkage due to partial burnout.

Therefore, it is a logical idea to turn to constrained sintering, with essentially zero shrinkage in the plane. This may be accomplished by constraining the LTCC with setter tape or, much more easily, by using self-constrained tapes such as the Heraeus Heralock 2000 (HL2000) and 800 (HL800) compositions [15]. This work therefore endeavours to explore the compatibility of these tapes with the mineral sacrificial layer process. Also, a standard (shrinking) LTCC composition known to have sufficient chemical resistance to the acid etching used to remove the MSP after firing [12], DuPont (DP) 951 was used for comparison purposes.

Experimental

LTCC tapes

The nominal supplier data for the three LTCC compositions are given in table 1. We used several thicknesses of DP 951, but only a single one for HL2000 and HL800 (HL2000 is available in several thicknesses, in contrast to HL800, a newer material for which a better chemical resistance is claimed [15]).

MSP formulations – organic phase

As sacrificial layers are often printed in large amounts, there is an issue of the solvent in the printing vehicle attacking the tape, especially if the latter it thin. However, the nonaggressive formulation developed before [16] for sacrificial carbon pastes on LTCC was not suitable, as it contains some water, with which MgO (slowly) reacts. Therefore, a new vehicle was formulated (table 2), with an effort to minimise aggressivity towards LTCC.

MSP formulations – mineral phase

The mineral content of the sacrificial pastes was composed of various mixtures of MgO (magnesium oxide) and $CaB_2O_4 \cdot 2H_2O$ (Vimsite, hydrated calcium borate), both from Sigma-Aldrich, in order to adjust the degree of sintering and shrinkage. MgO, being very refractory, essentially does not sinter, which, from a dimensional standpoint, is appropriate for HL2000 and HL800, while the moderate melting point of CaB_2O_4 [17] leads us to expect significant sintering of this compound, with additional shrinkage coming from dehydration (loss of the $2 H_2O$ molecules). Five different mixes (table 3), progressively going from pure MgO (P00) to pure $CaB_2O_4 \cdot 2H_2O$ (P40) were prepared and first tested on the three tapes. From the aforementioned considerations, one would expect the MgO-rich mixes to have good compatibility with HL2000 and HL800, whereas MSPs rich in $CaB_2O_4 \cdot 2H_2O$ would fare better with DP 951.

Manufacturing of test device structures

Two kind of test structures were designed for this study: 1) bridges and cantilevers for capacitive sensors, having only conductors (DP 6146, co-firing Ag:Pd) on the thin elements, and 2) anemometric flow sensors, where the thin structures are bridges carrying a thick-film thermistor element (DP 5092D, PTC 100 Ω). Please see companion paper in the same conference [18] for more details.

Table 1. LTCC tapes
(nominal supplier data)

Tape	XY shrinkage [%]	Green thickness [µm]	Fired thickness [µm]
HL2000	0.2	133	91
HL800	0.2	130	88
DP 951	12.7	254	216
		165	140
		114	97
		50	43

Table 2. Formulation of screen-printing vehicle for MSP (chemicals: Sigma-Aldrich, reagent grade; filled with ca. 40% vol. mineral phase).

Role	Compound	Parts (by mass)
Solvent	1-Hexanol	10
	Butylene glycol	20
Binder	Ethylcellulose 46 cps 48% ethoxyl	3

Table 3. MSP mineral formulations
(as paste with vehicle from table 2).

Code	Composition (by mass)
P04	MgO only
P13	$CaB_2O_4 \cdot 2H_2O$: MgO 1 : 3
P22	$CaB_2O_4 \cdot 2H_2O$: MgO 1 : 1
P31	$CaB_2O_4 \cdot 2H_2O$: MgO 3 : 1
P40	$CaB_2O_4 \cdot 2H_2O$ only

Lamination of the structures processed with MSPs was carried out using a pseudo-isostatic process, whereby the stacks were pressed uniaxially between two flat metal plates, but with a thick rubber disk inserted between the LTCC device and the top metal plate [19]. This process combines a reasonable approximation of true isostatic lamination with the simplicity of uniaxial pressing.

978-1-4244-4722-0/09 $25.00
© 2009 IMAPS-ITALY

All the structures (the three kinds of tapes) were always fired together, with a firing profile adapted to the requirements of the HL tapes, with a sintering ramp at 3 K/min and 30 min peak dwell at 880°C.

After firing, chemical etching was carried out, at ca. 50°C or room temperature, with 10% acetic or phosphoric acid solutions.

Results and discussions

Firing of single bare tapes

Figure 1 shows the result of cutting and firing a single layer of tape (unlaminated). In line with our previous studies, DP 951 is very forgiving – in fact, good flatness is even achievable when firing a single 50 µm thick tape – and has a low process sensitivity. Given its moderate thickness, HL2000 showed little deformation as well, which means that multilayer structures should fire without problems. HL800, on the other hand, was very deformed, which hints at a larger minimal module thickness to achieve flatness.

Compatibility with the MSP vehicle

The printing vehicle selected for the MSPs did show some reactivity with the LTCC, which was principally due to the very large deposited amount of MSP (dried thicknesses • 100 µm) and only a problem when printing on single thin (ca. 100 µm) HL2000 or HL800 tapes. In practice, this required pre-laminating two such layers.

Qualification of the MSPs on the LTCC tapes

The result of the rapid qualification of MSPs on the LTCC tapes is shown in figure 2. For DP 951, a very clear decrease of deformation is seen going from pure MgO to pure $CaB_2O_4 \cdot 2H_2O$. The last sample shows very little deformation, even on a single LTCC layer. On HL2000, the apparently best composition is P13; P00 has too little cohesion and turns to powder upon firing, due to the lack of sintering of MgO, and the compositions richer in $CaB2O_4 \cdot 2H_2O$ obviously have too large shrinkage. As HL2000 is quite resistant to pull from the MSP, the latter tends to fragment instead (P31 and P40). No clear-cut trend is seen with HL800, which is expected as strong deformation was already seen on the bare tape.

The test of P40 on DP951 was repeated, this time with a cover strip (figure 3), which confirmed the low observed deformation. The MSP formulation, however, is still not ideal, as the MSP is seen to crack when fired in a thick layer.

Structure build-up and firing

Structures [18] (figures 4-6) were fabricated starting with one (DP 951) or two (HL2000 / HL800) "base" layers, followed by a cut out "channel" layer that was pre-laminated at low pressure on the base, then filled by MSP by one or several screen printing operations (up to complete filling). Then, the "structure" layer was overlaid and

the stack laminated using the pseudo-isostatic process, and, finally, the LTCC was fired.

All the LTCC structures had good firing characteristics, as shown in figures 4-6. However, even with several layers, HL800 still showed some residual deformation (intrinsic, not due to the presence of MPS), indicating that the process conditions still need to be optimised for this tape.

Chemical etching

The chemical etching phase, however, turned out to be problematic, regardless of the used acid or temperature, for different reasons.

DP 951 fared the best. It is *per se* quite resistant to attack by acid, and no alteration was visible. However, as shown in figure 6, alteration of this chemical resistance was observed when fired in contact with P40. This was not observed in previous studies with $CaCO_3$ [8], and it is therefore thought that B_2O_3 evolution from CaB_2O_4 is a probable cause of the chemical alteration of the surface. Therefore, an MSP composition between P31 and P40 would probably give better results: a lower B_2O_3 volatility and less crackling. Nevertheless, free structures could be obtained, as evidenced in figure 6.

The HL2000 and HL800 LTCC compositions, in contrast to DP 951, were directly attacked by the acids; HL2000, especially, completely lost consistency, i.e. its glass matrix was entirely degraded. HL800 suffered extensive surface attack, but did show qualitatively much better chemical resistance than HL2000.

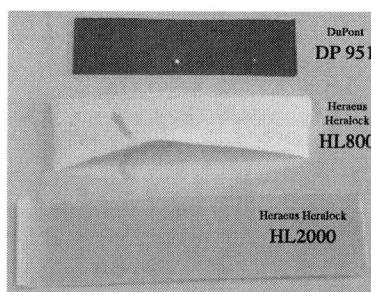

Figure 1. Single fired layers of the three LTCC compositions

978-1-4244-4722-0/09 $25.00
© 2009 IMAPS-ITALY

MSP	DP 951	HL2000	HL800
P04	DP 951 + MgO	HL2000 + MgO	HL800 + MgO
P13	DP 951 + 1:3	HL2000 + 1:3	HL800 + 1:3
P22	DP 951 + 1:1	HL2000 + 1:1	HL800 + 1:1
P31	DP 951 + 3:1	HL2000 + 3:1	HL800 + 3:1
P40	DP 951 + CaB$_2$O$_4$	HL2000 + CaB$_2$O$_4$	HL800 + CaB$_2$O$_4$

Figure 2. Qualification tests of the MSPs on LTCC (strips ca. 3 mm wide)

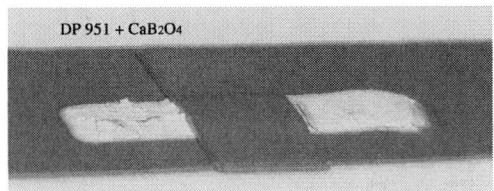

Figure 3. Cracking of P40 on DP 951.

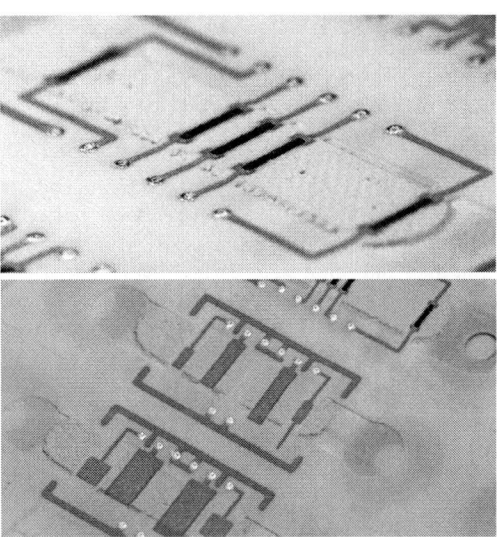

Figure 5. Flowmeter (top) and mechanical (bottom) structures – HL800 + MSP13 (channel width : ca 4. mm).

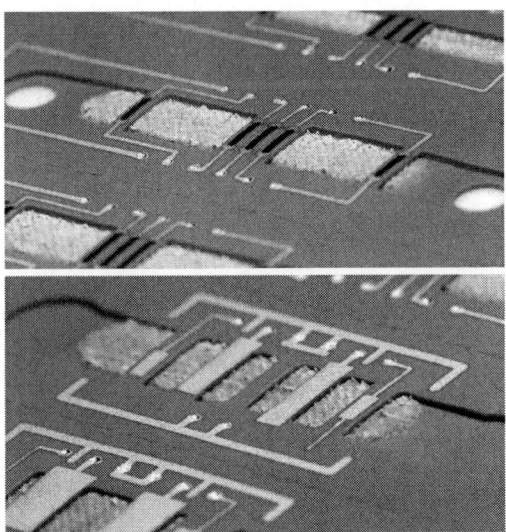

Figure 4. Flowmeter (top) and mechanical (bottom) structures – DP 951 + MSP40 (channel width : ca. 4. mm).

Figure 6. Mechanical structure after etching in 10% acetic acid, with insert to demonstrate opening of structures.

978-1-4244-4722-0/09 $25.00
© 2009 IMAPS-ITALY

This poor performance, in line with other studies [20], is not necessarily intrinsic, as we did not optimise the firing conditions, which we intend to do in later studies. Moreover, the MSP compositions, which used relatively unreactive coarse MgO powder and CaB_2O_4, still need to be optimised to etch away in more benign conditions.

Conclusions

In this work, we examined new mineral sacrificial paste (MSP) processes to achieve slender suspended structures in LTCC technology. To this end, materials based on MgO (refractory alkaline earth oxide, negligible sintering) and CaB_2O_4 (lower melting temperature, sintering) were formulated and applied to both classical (shrinking in XY direction) and "zero-shrinkage" LTCC materials: DP 951, Heraeus HL2000 and HL800.

Compared to the devices laminated between two flat metal plates, the MSP method allows a much wider process window to obtain undeformed structures; (pseudo-)isostatic lamination is possible, ensuring good interlayer bonding on the whole surface of the device.

Formulation of the MSPs to have some cohesion (rather than turning into loose powder) is advantageous for later handling and in order to avoid curling up of cantilevers [13].

The MSPs exhibited reasonably good printing and firing compatibility with the LTCC tapes. Nevertheless, a lower reactivity of the vehicle solvent system with LTCC would still be desirable, and some more fine-tuning is still needed on DP 951 to more exactly match the MSP and LTCC shrinkage. This is of course not an issue on HL2000 and HL800, which is a considerable advantage of these two tapes.

The acid etching characteristics of the structures, however, did not match our expectations. DP 951 was in general unaffected by the 10% acetic or phosphoric acid etchants, but the surface in contact with CaB_2O_4 exhibited some damage, in contrast to our earlier studies with $CaCO_3$ [8], most likely due to B_2O_3 sublimation and reaction with the LTCC surface, locally affecting its chemical resistance. Nevertheless, the structures could be etched away successfully, albeit with some residue.

Future work will therefore first concentrate on optimising the LTCC firing conditions to achieve maximal chemical resistance for the zero-shrinkage LTCC compositions (not necessarily optimal for other considerations) and characterisation of the materials in a wider range of etchants. Furthermore, the formulation of the MSPs for HL2000 and HL800 will be refined to allow more benign etching conditions. One possible path is to switch to finer, more reactive MgO powder, without any "binder" such as CaB_2O_4 or with a more soluble one, in order to increase its etching rate. On

DP 951 efforts will focus on decreasing the boron of the MSP loss while keeping the desirable sintering and shrinkage characteristics of CaB_2O_4.

Finally, the formulation of the screen-printing vehicle will be further refined to minimise solvent interactions with the LTCC tape during drying.

Acknowledgments

The Authors gratefully acknowledge Mrs. A. Kipka of Heraeus for providing the HL800 tape, and the help of Mr. M. Garcin in fabricating the samples.

References

[1] N.M. White, J.D. Turner-JD, "Thick-film sensors: past, present and future", Proceedings, 14th International Conference on Solid-State Sensors, Actuators and Microsystems - Transducers / Eurosensors'07, Lyon, France, 107-111, 2007.

[2] T. Maeder, Y. Fournier, S. Wiedmer, H. Birol, C. Jacq, P. Ryser, "3D structuration of LTCC / thick-film sensors and fluidic devices", Proceedings, 3rd International Conference on Ceramic Interconnect and Ceramic Microsystems Technologies (CICMT), Denver, USA, THA13, 2007.

[3] V. Mengeaud, O. Bagel, R. Ferrigno, H.H. Girault, A. Halder, "A ceramic electrochemical microreactor for the methoxylation of methyl-2-furoate with direct mass spectrometry coupling", Lab On A Chip (2), 9-44, 2002.

[4] R. Moos, J. Kita, "Ceramic multilayer gas sensors - an overview", Proceedings, XXXI International Conference of IMAPS Poland Chapter, Krasiczyn, Poland, 75-82, 2007.

[5] H. Birol, T. Maeder, I. Nadzeyka, M. Boers, P. Ryser, "Fabrication of a millinewton force sensor using low temperature co-fired ceramic (LTCC) technology", Sensors and Actuators A Vol. 134, 334-338, 2007.

[6] H. Birol, T. Maeder, P. Ryser, "Application of graphite-based sacrificial layers for fabrication of LTCC (low temperature co-fired ceramic) membranes and micro-channels", Journal of Micromechanics and Microengineering Vol. 17, 50-60, 2007.

[7] C.B. Sippola, C.H. Ahn, "A thick film screen-printed ceramic capacitive pressure microsensor for high temperature applications", Journal of Micromechanics and Microengineering Vol. 16, 1086-1091, 2006.

[8] H. Birol, T. Maeder, P. Ryser, "Preparation and application of minerals-based sacrificial pastes for fabrication of LTCC structures", Proceedings, 4th European Microelectronics and Packaging Symposium, Terme atež (SI), IMAPS, 57-60, 2006.

978-1-4244-4722-0/09 $25.00
© 2009 IMAPS-ITALY

[9] C. Lucat, P. Ginet, F. Ménil, "New sacrificial layer based screen-printing process for free-standing thick-films applied to MEMS", Journal of Microelectronics and Electronic Packaging Vol. 4 (3), 86-92, 2007.

[10] C. Castille, C. Lucat, P. Ginet, F. Ménil, M. Maglione, "Free-standing piezoelectric thick-films for MEMS applications", Proceedings, International Conference on Ceramic Interconnect and Ceramic Microsystems Technologies (CICMT), Munich (DE), 82-86 (TP13), 2008.

[11] Y. Fournier, S. Wiedmer, T. Maeder, P. Ryser, "Capacitive micro force sensors manufactured with mineral sacrificial layers", Proceedings, 16th IMAPS European Microelectronics & Packaging Conference (EMPC), Oulu, Finnland, 298-303, 2007.

[12] Y. Fournier, O. Triverio, T. Maeder, P. Ryser, "LTCC free-standing structures with mineral sacrificial paste", Proceedings, International Conference on Ceramic Interconnect and Ceramic Microsystems Technologies (CICMT), Munich (DE), 11-18 (TA12), 2008.

[13] T. Maeder, C. Jacq, Y. Fournier, W. Hraiz, P. Ryser, "Structuration of thin bridge and cantilever structures in thick-film technology using mineral sacrificial materials", Proceedings, 5rd IMAPS / ACerS International Conference on Ceramic Interconnect and Ceramic Microsystems Technologies (CICMT), Denver, USA, 2009.

[14] T. Kotani, T. Nakanishi, K. Nomura, "Fabrication of a new pyroelectric infrared sensor using MgO surface micromachining", Japanese Journal of Applied Physics Vol. 32, 6297-6300, 1993.

[15] A. Kipka, C. Modes, F. Gora, M. Deckelmann, "Zero shrinkage LTCC", Proceedings, XXXII International Conference of IMAPS Poland Chapter, Pułtusk, Poland, I10, 2008.

[16] T. Maeder, C. Jacq, Y. Fournier, P. Ryser, "Formulation and processing of screen-printing vehicles for sacrificial layers on thick-film and LTCC substrates", Proceedings, XXXII International Conference of IMAPS Poland Chapter, Pułtusk, Poland, B16, 2008.

[17] H. Yu, Q. Chen, Z.P. Jin, "Thermodynamic assessment of the $CaO-B_2O_3$ system", Calphad Vol. 23 (1), 101-111, 1999.

[18] Y. Fournier, A. Barras, G. Boutinard Rouelle, T. Maeder, P. Ryser, "SMD pressure and flow sensors for industrial compressed air in LTCC technology", Proceedings, 17th IMAPS European Microelectronics & Packaging Conference (EMPC), Rimini, Italy, 2009.

[19] F. Seigneur, Y. Fournier, T. Maeder, P. Ryser, J. Jacot, "Hermetic package for optical MEMS", Proceedings, International Conference on Ceramic Interconnect and Ceramic Microsystems Technologies (CICMT), Munich (DE), 627-633 (THA24), 2008.

[20] R.E. Eitel, W. Zhang, "Biostability of LTCC materials for microfluidics and biomedical devices", Proceedings, International Conference on Ceramic Interconnect and Ceramic Microsystems Technologies (CICMT), Munich (DE), 198-201 (WA23), 2008.

Process Development for a Very Precise Placement of a Lens for Micro-Optics Based Components

Danilo Caccioli, Luca Maggi

PGT Photonics s.p.a., Viale Sarca 222, 20126 Milano, Italy

Danilo.Caccioli@pgt-photonics.com, Luca.Maggi@pgt-photonics.com

Abstract

This paper describes the development of a process for placing a lens very precisely in front of a laser source, using a UV resin. The problem of placing accurately each other some optical components is crucial in micro-optics based components. A good alignment is necessary to guarantee the desired performances. Our approach combines the optical simulation and design with the experimental knowledge of the process to achieve a repeatable and cost effective alignment of a lens in front of a laser source. The lens is actively aligned and is fixed to the substrate by UV resin. Some optical simulations have been performed to calculate the right working distance of the lens and to evaluate the target beam shape. Also the substrates have been designed in order to minimize the pre alignment tolerances. We developed an experimental set up for monitoring the beam shape in real time and for precise alignment and attachment of the lens. The monitor is operating also during UV irradiation process. An optimization of all the process parameters has been carried out. The beam peak and beam centroid have been measured before and after the attachment. The post curing shift has been evaluated by the optical simulator and with analytical method. The final post curing shift is around 1 micron, which is compliant with the most stringent requirements in micro optics packaging.

Key words: Micro-optics, UV curing, Lens placement, Optoelectronic Packaging, Automatic Alignment

Introduction

One critical issue in opto-electronic packaging is the precise alignment of laser source with other optical devices [1]. Different solutions have been proposed for light coupling and for dynamic and static alignment [2-3]. The main goals are to couple as much light as possible and to reduce the cost of operation. Two main philosophies are possible. One is the optical integration using a unique technological platform, the other one is the hybridization and the use of micro-optics. Both of them have advantages and drawbacks. The choice of the adopted technology is strictly related to the final application.

In our approach a lens is actively placed in front of a laser source to collimate the light. A collimated beam is the ideal one to deal with discrete optical elements. In this way is possible to perform different operation (e.g. filtering, splitting, reshaping etc.) using very common devices. The optical performances of the system are deeply affected by the goodness of the alignment and are extremely sensible to the placement tolerances. The beam shape is probably the most important parameter when working with collimated beam [4-6] and so we monitored the beam shape during the whole process [7-8]. Finally we fixed the lens in place using a UV resin. The process parameters have been optimized to properly manage the resin shrinkage and strain [9-10]. Moreover a feedback action is possible thanks

to the real time monitoring, so that the final pot curing shift is small.

To target the right beam shape and to evaluate the effect of the post curing shift a complete optical simulation has been carried out [11].

The optical simulation of the system, the definition of target beam shape and the analysis of the tolerances are described in section II of this paper, while the experimental set up and process are described in section III. The results are discussed in section IV. Finally in section V, we provide conclusions and recommendations.

Optical model

The system to be studied consists of a laser diode, which is characterized by quite high divergence angle, and a micro lens for collimating the beam. Even if it's a simple system, due to small dimensions, the tolerances are tight and a right alignment is crucial. The rules of beam propagation are well known and valid both for collimated and gaussian divergent beams [4-5], while a good collimation is necessary for properly interacting with other optical devices [6]. So the problem is to define correctly the level of collimation that is expected and which are the discrepancies respect to the ideal case those occur in the real assembly. For evaluating the system the ZEMAX software [12] has been adopted. Using its non sequential model it's possible to simulate the physical propagation of the beam,

978-1-4244-4722-0/09 $25.00
© 2009 IMAPS-ITALY

overcoming the approximations of basic ray tracing software.

In order to reach the condition of collimated beam, the easiest way is to analyze the image on a screen, which is at infinite distance from the source. The image on the screen changes moving the lens along the optical axis. The position of collimation is reached when the image has the smallest diameter.

This is an operating condition, which can be implemented in an experimental set up.

The predictions of the model have been verified over a very large range of distances.

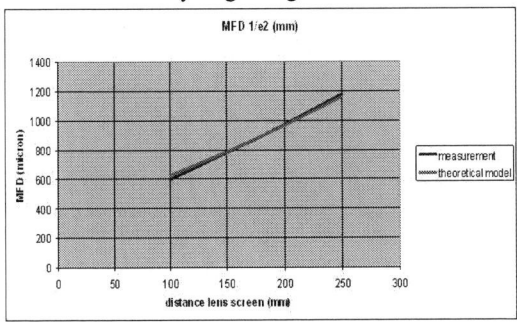

Figure 1 - Comparison between model and experimental measurement

After having verified the accuracy, the optical model has been used to define the target beam shape.

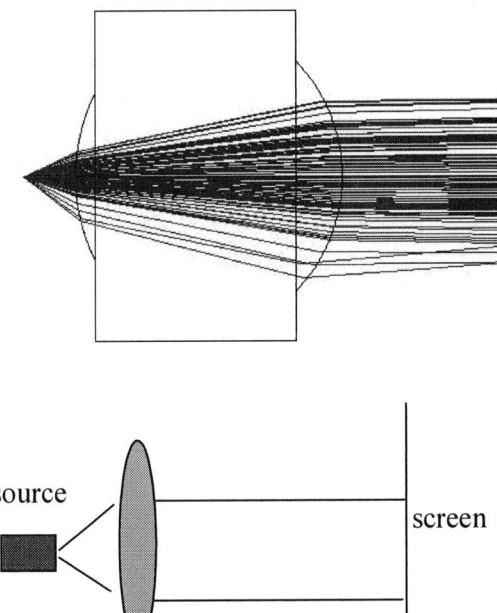

Figure 2 - Scheme of optical system

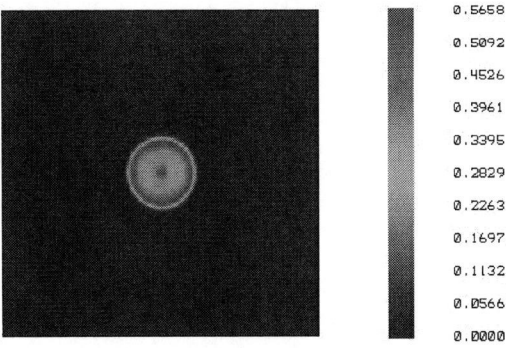

Figure 3 - Target beam shape

The tolerances of placement are more or less critical according to the axis where they occur [11]. Along the optical axis (Z axis) they are extremely tight, as a difference in working distance affects immediately the collimation. Fortunately placing the screen at very big distance, any changes in the beam diameter can be immediately detected. We placed the screen at 35 cm distance, which is infinitely bigger compared to the wavelength (λ = 1.55 micron) and to any geometrical dimension (lens dimension = 1 mm). Along the transversal axis the tolerances are less critical, even if it's more difficult to detect them.

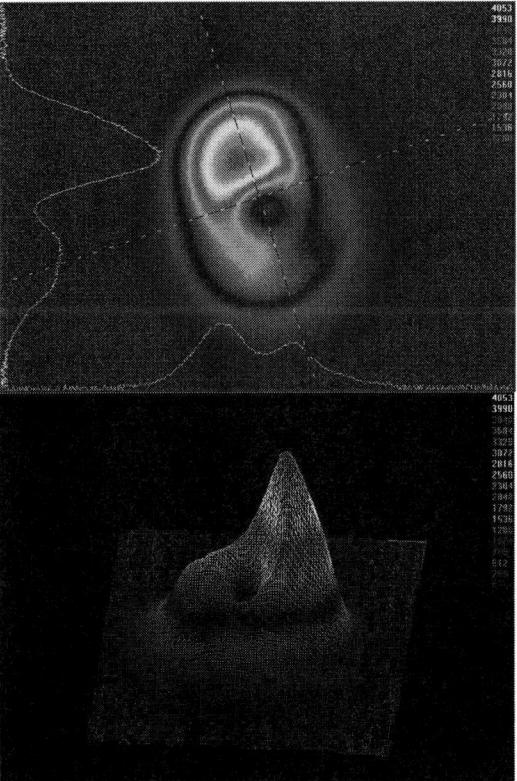

Figure 4 – Image of defocused laser beam

978-1-4244-4722-0/09 $25.00
© 2009 IMAPS-ITALY

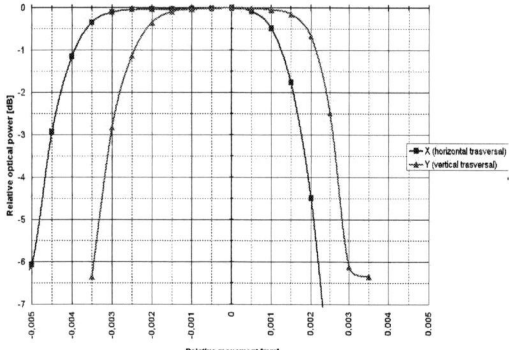

Figure 5 - tolerances along X and Y

Set up and process

The laser diode is soldered on an Optical Bed (OB), which is solder on a second Optical Bed, where the lens is placed.

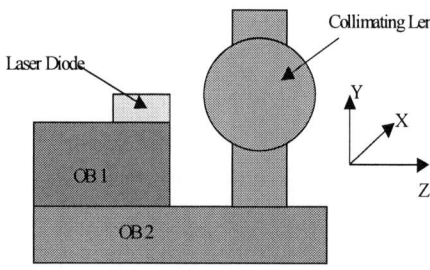

Figure 6 - Scheme of assembly

The OB 1 has been properly patterned to avoid any tilt of the laser diode. Moreover OB 1 has been sized so that, also in the worst case, no permeation of the lens with OB 2 is possible. The OB 2 has been also patterned for permitting a coarse alignment of the lens and for minimizing any momentum on the lens, due to the asymmetry in glue deposition.

Some probes supply the voltages for the laser. The lens is moved by a customized pick up tool, which is connected to a alignment system, with six degree of freedom and a very precise movement.

A very sensible IR camera was placed at 35 cm far from the laser and it was used for real time monitoring of the beam [7-8]. Analyzing the image of the camera is possible to find the position of the lens for obtain a collimated beam. The IR camera is connected to a software tool, which calculates the beam properties (beam diameter, energy peak position, energy centroid position etc.) frame by frame.

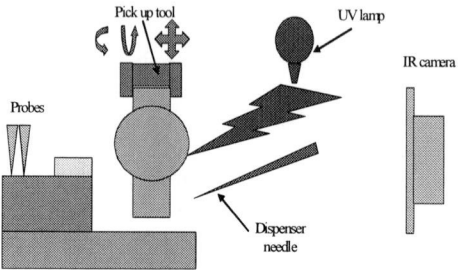

Figure 7 - Scheme of experimental set up

The glue is dispensed under the lens and some UV lamps provide the radiation for the curing

Several tests have been carried out to find the right amount of resin and the proper curing time/irradiation, for a reliable joint.

As we need to continuously monitor the beam shape during the curing process, we chose an UV resin rather then a thermal epoxy. Using an UV adhesive it's possible to bond the lens working at room temperature, without affecting the laser optical power level. In this way any degradation in optical power is due to misalignment and can be recovered applying any compensation movement if necessary. At high temperature a power drop occurs, and the energy level is below the sensitivity threshold of our IR sensor, so a thermal curing is incompatible with a real time monitor.

During the UV curing the resin suffers a shrinkage that causes a displacement of the lens toward OB 2. We applied an offset in Y direction to compensate the shrinkage of the resin. Even if any UV resin requires also a thermal step to complete the curing, most of the shrinkage appears during the UV irradiation. So if you compensate this, a very small post curing shift can be obtained.

The main steps of our process are:

- Active alignment of the lens for finding the collimating position
- Real time monitor of the beam shape
- Dispensing the glue
- Applying the Y offset
- UV curing
- Simultaneous feedback and eventual compensation movement
- Thermal curing
- Final analysis of the beam

With the final analysis we compared the beam properties after the curing with those present during the active alignment. From the differences is possible to evaluate the post curing shift.

978-1-4244-4722-0/09 $25.00
© 2009 IMAPS-ITALY

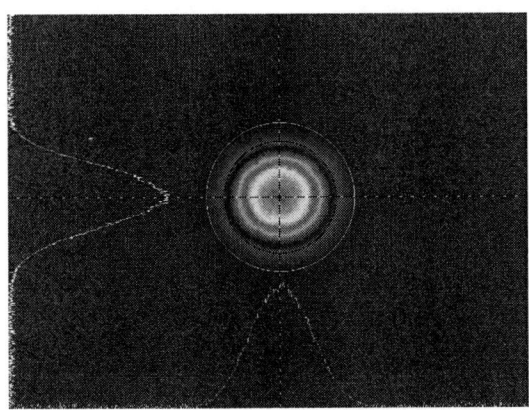

Figure 8 - Image of collimated laser beam

Results

The difference of centroid position before and after curing is summarized in the following table

Table 1 - Centroid position pre and post curing (all dimensions are in micron)

	Pre curing		Post curing		Difference	
Item	X centr.	Y centr.	X centr.	Y centr.	ΔX centr.	ΔY centr.
1	4041	2493	3999	3260	-42	767
2	4158	2516	4113	3168	-45	652
3	4040	2465	4079	3366	39	901
4	4076	2607	3964	3280	-112	673
5	4033	2472	4137	3149	104	677
6	3889	2633	3952	2885	63	252
7	4174	2656	4169	3031	-5	375
8	3943	2678	3998	2628	55	-50
9	3955	2513	3786	3116	-169	603
10	4044	2466	4170	2505	126	39
11	3640	1858	3548	1946	-92	88
12	3572	2426	3693	2586	121	160
13	3936	1899	3888	1897	-48	-2
14	4453	2190	4669	2257	216	67
15	4054	2015	3913	2212	-141	197
16	4215	1820	4331	1872	116	52
17	4109	2199	3881	2873	-228	674
Aver.	4020	2347	4017	2708	-2.5	360.3
Std. Dev.	203	291	254	515	121.8	319.7

The centroid is defined as the center point of the region which collected the 86.5% ($1/e^2$ level) of the beam energy.

For obtaining the post curing shift by the displacement of the centroid two methods have been applied. The first one use the optical model developed in ZEMAX, the second one is an analytical calculation derived by the thin lens theory and by geometrical optics.

Basically the post curing shift has the same effect of an object which is not on optical axis and which generates an image on a reference plane. The ratio between the image and object height is related to the ratio between the focal distance of the lens and the position of the image plane.

In our system we know the focal distance of the lens, the position of the image plane (that is the position of the IR camera), and the difference in the position of the centroid. The only unknown variable is the post curing shift, which can be calculated easily.

The evaluation of the post curing shift using the two methods is shown in the following table:

Table 2 - Evaluation of the post curing shift (all dimensions are in micron)

	Analytical method		Optical simulation	
Item	X shift	Y shift	X shift	Y shift
1	-0.07	1.21	-0.07	1.25
2	-0.07	1.02	-0.07	0.95
3	0.06	1.42	0.07	1.31
4	-0.18	1.06	-0.13	0.96
5	0.16	1.06	0.10	0.96
6	0.10	0.40	0.09	0.35
7	-0.01	0.59	0	0.40
8	0.09	-0.08	0.09	-0.09
9	-0.27	0.95	0.20	0.85
10	0.20	0.06	0.20	0.07
11	-0.14	0.14	0.10	0.10
12	0.19	0.25	0.19	0.20
13	-0.08	0	-0.07	0
14	0.34	0.11	0.35	0.09
15	-0.22	0.31	0.20	0.15
16	0.18	0.08	0.20	0.09
17	-0.36	1.06	-0.35	0.96
Aver.	0	0.5	0.07	0.51
Std. Dev.	0.19	0.50	0.17	0.48

The results obtained by the methods are absolutely comparable each others. The post curing shift occurs only along Y axis (the vertical axis in our coordinate system) and it's due to the shrinkage of the resin during the UV irradiation process.

The shift in the horizontal transversal direction (X direction) is absolutely negligible. This means that the pattern for avoiding the momentum works properly. Moreover the alignment process is well controlled and repeatable.

Due to controlled process, to the real time monitor and feedback is possible to minimize the post curing shift. We can estimate the shift of our process in the range of 1 micron. This value is compliant with the most stringent requirements in micro optics packaging.

Conclusion

In this paper we have presented a process for obtaining a very precise placement of a lens in front of a laser source.

Combining the simulation model with the development of custom substrates and highly precise mechanical equipment, we have obtained a very

978-1-4244-4722-0/09 $25.00
© 2009 IMAPS-ITALY

accurate process. Using a real time monitor it was possible to compensate the shrinkage of the resin during the UV curing.

Comparing the position of beam centroid before and after the curing it was possible to evaluate the post curing shift. Two methods have been used to perform the calculation of the shift. The results of both of them are comparable.

The post curing shift occurs only in vertical direction and it's due to resin shrinkage during the UV curing. The horizontal shift is absolutely negligible. The vertical shift has been evaluated in 1 micron range, which is compliant with the most stringent requirements in micro optics packaging.

References

[1] A. R. Mickelson, N. R. Basavanhally, Y. C. Lee, "Optoelectronic Packaging", John Wiley Publisher, New York, Chapter, pp. 50-51, 1997

[2] A. C. Pliska et al., "Low Cost Optoelectronic Packages: Development of a Fast alignment Technique and a Stable Bonding Process for Singlemode Optical Fiber", Proceeding of 15th European Microelectronic and Packaging Conference (EMPC), Brugge, Belgium, june 12-15, pp.128-132, 2005

[3] M. Luetzelschwab, M. P. Y. Desmulliez, D. Weiland, " MicroLens/UV-LED Array Packaging for Dynamic and Static Alignment", Proceedings of 2nd Electronics Systemintegration Technology Conference (ESTC), Greenwich, London, UK, September 1-4, pp. 1121-1126, 2008

[4] J. Alda, "Laser and Gaussian Beam Propagation and Trasnformation", Encyclopedia of Optical Engineering, Marcel Dekker Inc, New York, pp. 999-1013, 2003

[5] W. B. Joyce, B. C. DeLoach, "Alignment of Gaussian Beams", Applied Optics, Vol. 23, No. 23, pp. 4187-4196, 1984

[6] Y. Yoon et al. "Transmission Spectra of Fabry-Perot Etalon Filter for Diverged Input Beams", IEEE Photonics Technology Letters, Vol. 14, No. 9, pp. 1315-1317, 2002

[7] L. Green, "On-Line Beam Performance Monitoring to Improve Laser Reliability and Performances", Proceedings of SPIE, Laser Resonators and Beam Control V, Vol. 4629, pp. 12-23, 2002

[8] L. Green, "Pitfalls of Beam Profiling", OE Magazine, pp. 48, 2002

[9] M. J. Hodgin, "Epoxies for OptoElectronic Packaging: Applications and Material Properties", Proceedings of 36th Annual IMAPS Conference, Boston, MA, Nov 17-20, pp. 26-30, 2003

[10] J. Kuczynski, A. K. Sinha, " Strain Measurement and Numerical Analysis of an Epoxy Adhesive Subjected to Thermal Loads", IBM J. Res. & Dev., Vol. 45, No. 6, pp. 783-788, 2001

[11] A. Stockham. J. Smith, "Tolerancing Microlenses Using ZEMAX", Proceedings of SPIE, Micromachining Technology for Micro-Optics and Nano-Optics, Vol. 6110, pp. 172-183, 2006

[12] ZEMAX User Guide

978-1-4244-4722-0/09 $25.00
© 2009 IMAPS-ITALY

Electromagnetic Simulations for the Packaging Design of Telecommunication Component

Luca Maggi, Giovanni Ticozzi

PGT Photonics s.p.a., Viale Sarca 222, 20126 Milano, Italy

Luca.Maggi@pgt-photonics.com

Abstract

This paper describes the use of a fully 3D electromagnetic simulator for designing a complete packaging solution for a telecommunication component. Typically the electromagnetic simulators are used to evaluate the RF performances and for improving the electronic design. N or approach we use the software as a tool for evaluating the overall packaging solution and for improving the complete packaging architecture. Several simulations have been performed to evaluate the feed through, the connection to PCB and the PCB itself. First of all each part has been optimized separately and finally the whole RF chain has been simulated. The results have been used to compare a standard gull wing leaded package and a leadless package. In the second case the connection to PCB is realized by a flex substrate, which is more performing according to the simulation.. The electromagnetic simulator has been used to evaluate the Electro-Magnetic Interference (EMI), the influence of length and number of wires and the use of ribbon in place of wire bonding. Also a detailed analysis of the process tolerance has been carried out. As results, we have obtained a S11 parameter (reflected signal) lower than –10 dB and s21 (transmitted signal) higher than –1.5 dB over a 20 GHz bandwidth. These results are compliant with the typical requirements for HF telecommunication modules. Using the 3D simulator as a packaging tool permits to realize a virtual prototyping. In this way it's possible to analyze different configurations saving time and money.

Key words: Electromagnetic simulations, Packaging design, Virtual prototype

Introduction

In the development of telecommunication component the packaging design plays a major role as it directly impacts the final performances. Especially the RF performances can be badly affected by not optimal packaging connections. So it's crucial to design correctly the high frequency transition. Moreover this goal must be reached without time-intensive and costly iteration steps. So a complete electromagnetic simulation is required [1].

Typically the electromagnetic simulations are used to evaluate the RF performances of a device [2], for improving the electrical circuit [3] and for evaluating new substrates [4-6].

As the signal integration becomes more and more important the RF simulations extend to some package structures [7, 9] and to the Printed Circuit Board (PCB) [10].

We use the electromagnetic simulator for evaluating the overall packaging architecture and solution. Using a bottom-up approach, each single part has been optimized separately and then the whole RF chain has been simulated.

The system to be studied is a general case composed by a ceramic substrate placed inside a package, the package feed through (butterfly package used as reference) the connection to PCB

and the PCB itself. The ideal target is a telecommunication component operating at 10 Gb/s.

The desired goals are insertion loss S11<-10 dB and transmission loss S21>-3 dB in whole broadband up to 20 GHz. These are typical requirements for telecommunication components working at 10 Gb/s.

We have compared a gull wing leaded package and package connected to PCB by a flex substrate. A detailed analysis of the influence of the length and number of wires has been carried out for finding the optimized design.

The external connection, that is the comparison between leads and flex, is presented in section II. The internal connections, the influence of wires and the ceramic substrate are described in section III. The simulation of the full RF chain is described in section IV. Finally the conclusions are reported in section V of this paper.

For our purpose we have used the CST STUDIO SUITE [11] software for its feature of using a time domain solver [12] and so to deal with broadband problems without time cost.

External connections

We considered a system formed by ceramic feed through (typical of a butterfly package), the gull wing leads and a PCB. Alternatively a system using a flex substrate instead of leads has also analyzed.

978-1-4244-4722-0/09 $25.00
© 2009 IMAPS-ITALY

The transmission line for high bit-rate signals is a Co-Planar Waveguide (CPW). The reference impedance is Z = 50 ohm for single ended line or Z = 100 ohm if a differential line is considered [3].

Thanks to the electromagnetic simulations we would evaluate the effects of leads dimension, analyze the influence of PCB design and estimate the Electro-Magnetic Interference (EMI) between high frequency and bias signals.

A scheme of the leaded system is shown in the following picture

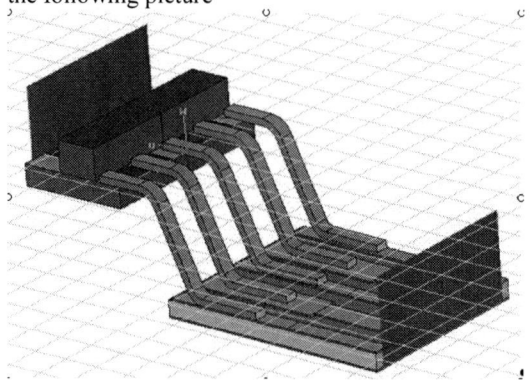

Figure 1 - Scheme of leaded system

The material of PCB is FR4, while for feed through Alumina has been used, all the metal lines are considered as Perfect Electrical Conductor (PEC).

First of all the feed through and the PCB [10] have been optimized for GSGSG differential configuration. Then the complete transition has been simulated. As expected the presence of via holes [8] both in the PCB and in the feed through deeply affected the results.

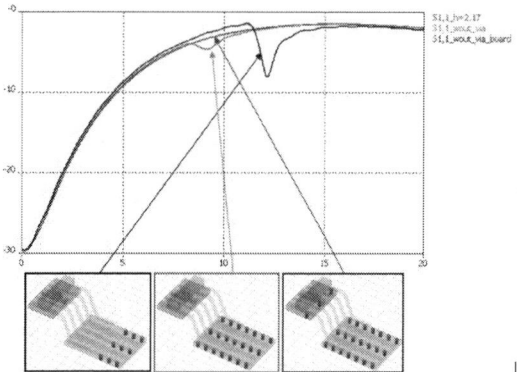

Figure 2 - Effect of the via holes

The via-holes improve the ground distribution and so the signal transmission. However the S-parameters are far from the desired goals. This is more evident carrying out a TDR (Time Domain Reflection) analysis of the signal [11-12].

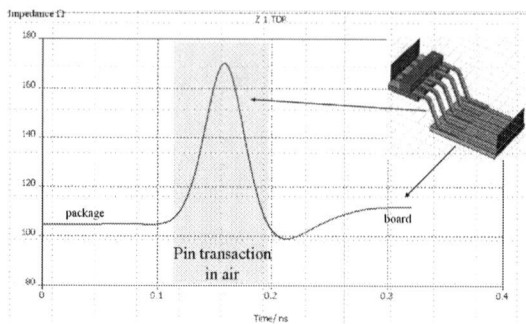

Figure 3 - TDR analysis

It's evident the inductive effect of the leads. In fact, while the impedance is well matched both at PCB and package side, a great mismatch occurs at pin transaction in air. In order to minimize this effect two possible solutions have been pursued. The first one is the compensation of inductive effect by capacitive pads; that is increasing the width of the soldering pads present on the PCB or on the package. The second solution is to limit the inductive contributor, reducing the length and height of the leads. Both the possibilities have been simulated and compared. According to the simulations the compensation of inductive effect by capacitive pads is less effective then decreasing directly the inductive mismatch.

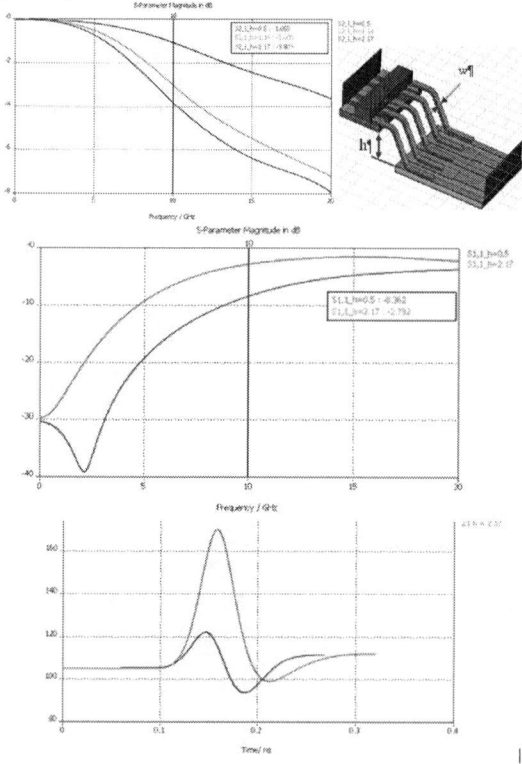

978-1-4244-4722-0/09 $25.00
© 2009 IMAPS-ITALY

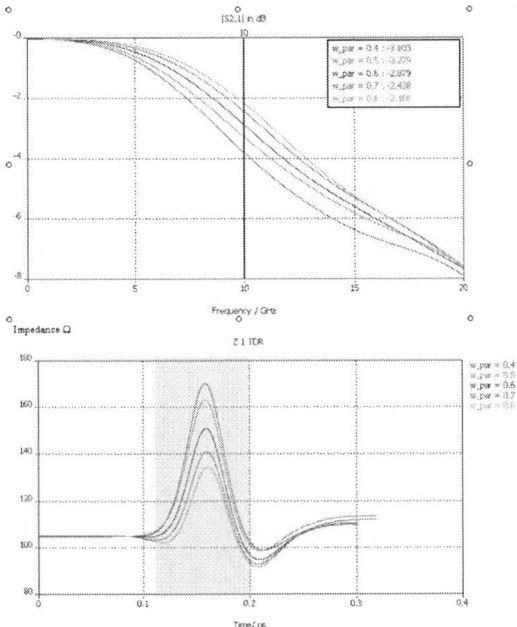

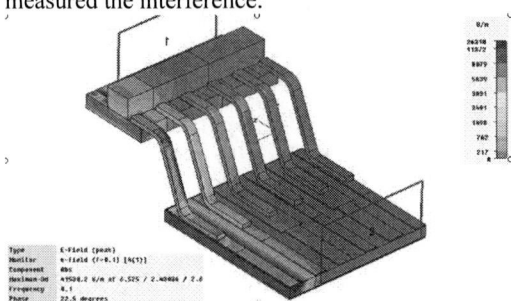

Figure 4 - Reducing the leads mismatch

Even modifying the pin dimension, the target of reflected signal seems to be unreachable, at least on the whole bandwidth. The goal on transmitted signal is achieved sizing properly the width and length of the pins.

A further investigation concerned the electromagnetic interference (EMI) between high frequency and bias signal. We have stimulated the pin near to the ground with a voltage signal and have measured the interference.

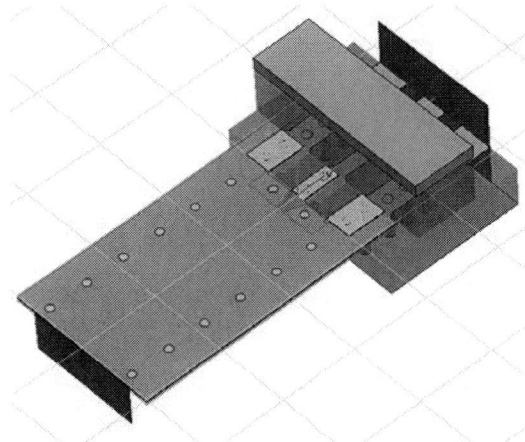

Figure 6 - System with flex connection

We have designed a proper transition on the flex, passing from CPW, in the area of soldering pads, to microstrip line in the central body of the flex. Such transmission line has been preferred in order to limit the amount of metal and so to guarantee the necessary mechanical flexibility of the substrate.

Using a flex substrate, the RF performances greatly improve and the desired targets on reflected and transmitted signals can be reached.

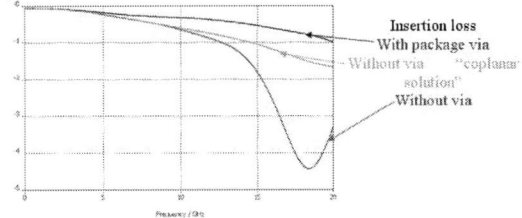

Figure 7 - S-parameters with flex

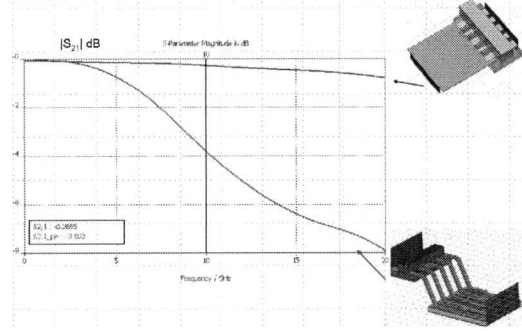

Figure 5 - EMI analysis

Thanks to the use of CPW lines, the effect of bias is completely shielded by ground pads and no interference occurs at signal lines. It's recommended using CPW transmission line when signal lines are close to bias pins. A good isolation can be reached also considering long leads with high inductive behavior. In the working range the isolation is better than –40 dB.

For overcoming the problem of poor performances of reflected signals, we have considered a different system, in which a flex substrate replaced the leads. For the flex we have used standard commercially available materials with typical $\varepsilon_r = 3.4$.

Figure 8 - Comparison of S21 with leads and flex

Adopting a flex substrate is extremely recommended for improving the RF performances, especially regarding the reflected signal.

Internal connections

We considered a system formed by ceramic feed through of the package plus a ceramic substrate placed inside the package for managing the routing

978-1-4244-4722-0/09 $25.00
© 2009 IMAPS-ITALY

of the signals. The routing substrate is made of alumina and it's connected by wire bonds to the package. Also for this substrate the selected transmission line was a differential CPW in GSGSG configuration. It was designed for routing the signals from a chip to the package I/O, so it is characterized by a basic taper structure. The pitch of the package I/O is assumed to be 1.27 mm, while the chip pads are separated by 150 micron. The layout of the stand alone substrate has been optimized, before considering the substrate in the simulations for evaluating the internal connections.

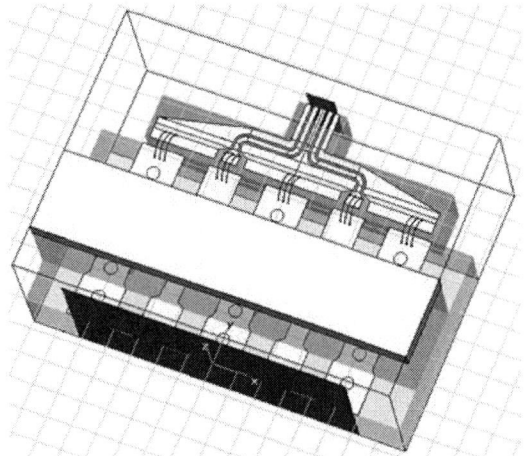

Figure 9 – Scheme of internal connections

As for the external connection case, also in this configuration the presence of the via-holes [8] deeply affected the RF performances.

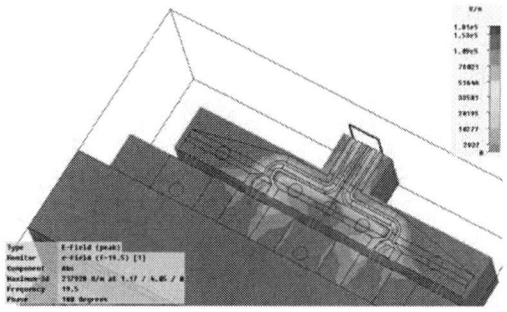

Figure 10 - Effect of the via holes on routing substrate

Due to the taper structure of the routing substrate the presence of the via-holes is necessary to confine the electric field.

Another aspect, which is important in physical realization, is the influence of the number of wires on RF performances. We have simulated the connection between routing substrate and package using different wire diameter and number. Moreover we have also simulated a ribbon bond connection in place of wire bond one. In this way it was possible to analyzing different configurations, each experimentally feasible, without realizing any physical sample. Using the virtual prototype is

possible to optimize any aspect of process implementation, reducing the time and money costs to the minimum.

The effects of different bonding parameters are summarized in the following figure.

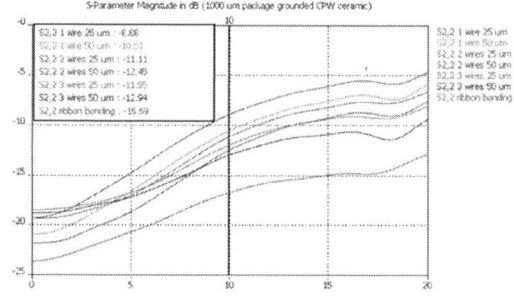

Figure 11 - Effects of the bond parameters

It's evident that the ribbon bonding is the more performing choice, but a wire connection with 3 wires is also working.

Besides the number of wire also the relative placement of the substrate and package is important. This takes into the consideration the process tolerances of placing the substrate and the mechanical tolerances those affect the parts.

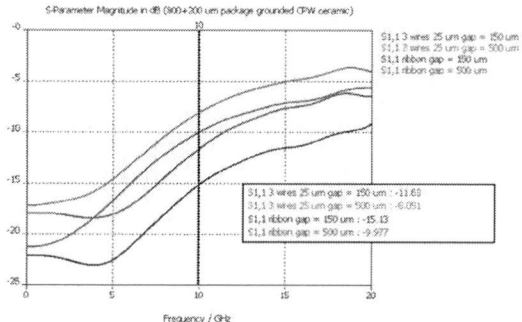

Figure 12 - Effects of relative displacement

The relative displacement plays a major role on RF performance. The placement must be as precise as possible or, equivalently, the mechanical and process tolerances must be very tight for guaranteeing the desired goals.

The gap between package and ceramic substrate should not be bigger than 500 micron.

Full RF path

After having simulated separately the internal and external connections and have found the design solution for each part, the simulation of the complete RF chain has been performed.

In this way a global optimization can be reached and it's possible to simulate the signal integrity as it will be realized in the physical sample and experimental measurement set up.

978-1-4244-4722-0/09 $25.00
© 2009 IMAPS-ITALY

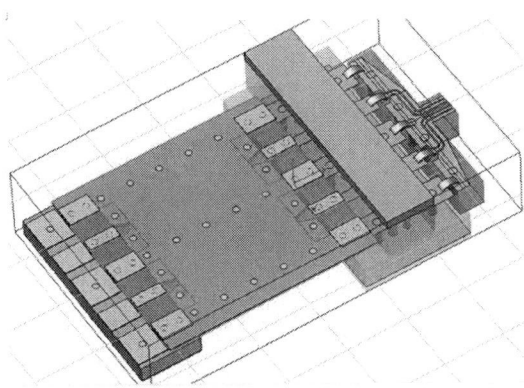

Figure 13 - Full RF path

Adjusting the position of the various via holes for guaranteeing good ground, and optimizing the pad dimension at each interface for compensating the inductive and capacitive effects, we have obtained a complete RF path, which fulfills or requirements.

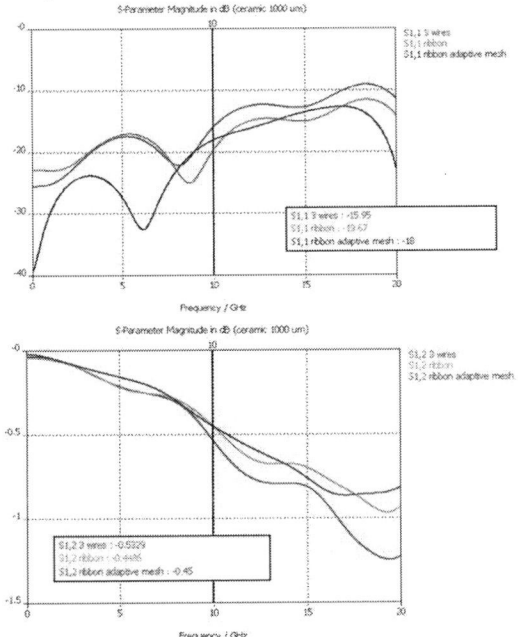

Figure 14 - S-parameter of the full RF path

As it can be seen, a good global impedance match is obtained and the goals of S11<-10 dB and S21>-3 dB overt he whole bandwidth are reached.

Conclusion

Using a 3D electromagnetic software as a tool for packaging design, it's possible to evaluate the RF performances and investigating different technological solution without the need to physical realize several sample. The virtual prototyping is an effective approach, which saves time-intensive and costly iteration steps.

With a bottom up approach the goals are reached by first optimization of each single part, followed by the global optimization of the full RF chain. In this way it's possible to analyze separately the different contributors and evaluate the global behavior.

According to our simulation the use of flex connection in place of gull wing leads is highly recommended, moreover the use of ribbon bond is suggested. If traditional wire bonds are preferred, then the wire number per pad must be increased. 3 wires act as a ribbon for improving the RF performances. A correct placement of via holes in all the structures is crucial for obtaining a good ground and so good impedance match. Finally the choice of CPW as transmission line is recommended for high bit-rate components. This type of transmission line not only guarantees good RF performances, but offers also a solution to the EMI problem.

All this suggestions can be easily implemented in the packaging design of a telecommunication component for 10 Gb/s application.

References

[1] T. Weiland, M. Timm, I. Muntenau, "A practical guide to 3-D simulation", IEEE Microwave Magazine, Vol. 9, No. 6, pp. 62-75, 2008

[2] O. Sotoudeh, T. Wittig, "Electromagnetic simulation of a mobile phone antenna performance", Microwave Journal, Vol. 51 No. 1, pp 78-82, 2008

[3] F. Xiao, Y. Kami, "Modeling and analysis of crosstalk between differential lines in high speed interconnects", Progress In Electromagnetics Research Symposium PIERS on line, Vol. 3, No. 8, pp. 1293-1297, 2007

[4] X. Ziang et al., "Development of SOP module technology based on LCP substrate for high frequency electronics applications" Proceedings of Electronics Systemintegration Technology Conference (ESTC), Dresden, Germany, September 5-7, pp. 118-125, 2006

[5] K. H. Drue et al. "LTCC multilayer technology enables very compact 20 GHz switch unit for space applications", Proceedings of 16th European Microelectronics and Packging Conference (EMPC), Oulu, Finland, June 17-20, pp. 500-504, 2007

[6] S. Pinel et al., "System-On- Package (SOP) architectures for compact and low cost RF front–end modules", Proceedings of 11th GAAS Symposium, Munich, Germany, October 6-10, pp. 433-436, 2003

[7] A. Chandrasekhar et al., "Characterization, modelling and design of bond-wire interconnects for chip-package co-design", proceedings of 11th GAAS Symposium, Munich, Germany, pp. 427-430, 2003

978-1-4244-4722-0/09 $25.00
© 2009 IMAPS-ITALY

[8]   M. Mantysalo, "Analysis of via structures in 3D package", Proceedings of 15th European Microelectronics and Packaging Conference (EMPC), Brugge, Belgium, June 12-15, pp. 298-302, 2005

[9]   M. Kunze et al., "120 GHz broadband chip-interconnects for high bit-rate communication systems", IEEE MTT-S Digest, pp. 485-488, 2004

[10]   N. Codreanu et al., "new investigations of high performance PCB structures for signal integrity compliance", Proceedings of 15th European Microelectronics and Packaging conference (EMPC), Brugge, Belgium, June 12-15, pp. 534-539, 2005

[11]   CST STUDIO SUITE release 2009, User Guide

[12]   M. Clemens, T. Weiland, "Discrete electromagnetism with the finite integration technique", Progress in Electromagnetics Research PIER, Vol. 62, pp. 65-87, 2001

WPLGA: new package family for medium pin count with design flexibility

Pierangelo Magni, Giovanni Graziosi, Claudio-Maria Villa,
Roberto Tiziani, Rodolfo Gacusan

*ST Microelectronics, Via C. Olivetti 2, 20041 Agrate Brianza, Italy
*ST Microelectronics, 9 Mountain Drive, Light Industry & Science Park II, Calamba, Phil.

Abstract

In this article we present a relatively new and innovative family of packages that is suitable for medium pin count needs and that locates itself between BGAs and "standard" QFN with single row of pads. It is characterized by the extreme design flexibility in term of I/Os size, location and shape, allowing a wide range of customization; only limited by PCB layout constraints. Nevertheless its structure looks like metal QFN, the pad-out is more similar to LGA array, being a multi-row design, with the possibility to have exposed partial/full ground and/or power rings. All this allows a relatively large footprint reduction compared with the same I/Os of a TQFP package, including a significant total package height reduction. Simplified assembly flow and easy handling leads to package total cost competitive respect to BGAs and TQFPs, allowing also testing on strip.

Results from samples will be shown demonstrating good reliability characteristics, thermally in between QFNs and BGAs, and from electrical point of view good performances due to reduction of the overall signal path length die-application board. Also solder joints show an improvement respect to standard QFNs. We will show the package structure and its design options, thermal and electrical modeling data and overall package performances and possible applications. The innovative nature of this new package along with its promising results data and cost structure is proving of great interest and this new package is not moving towards mass production.

Introduction

Needs of packages with medium pin count, relatively thin, with acceptable electrical and thermal performances and low cost are more and more present for several applications; in few words, a package with the thermal performances of exposed pad QFP, small size of BGA and the cost of QFN.
WPLGA is a package family that targets all these characteristics, including an easy Assembly and Testing&Finishing, allowing an almost infinite number of designs of pads, ground rings, ground bars and so on for a full customization.

It directly diverts from Staggered and Multi-Row QFN Outlines (JEDEC MO247) of which sheared the external look (see Figure 1). But it differs from them through the main frame: by design and manufacturing method.

Lead Frame Design

While staggered QFN leadframes look like the ones in Figure 2 with all future pads connected to the main frame, WPLGA at the end of assy and before final separation, presents all pads electrically isolated, still remaining all packages onto the original metallic frame. This could allow "strip testing" option with multiple packages contacted at

the same time (parallel testing) and an extremely easy handling of parts.

Separation of units from the leadframe is done using sawing process with the big advantage respect to QFNs of no metal cut (copper) but just molding compound.

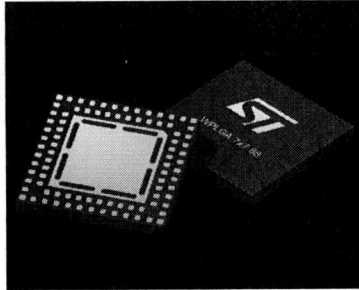

Figure 1: WPLGA 7x7x0.7 mm 88 I/O

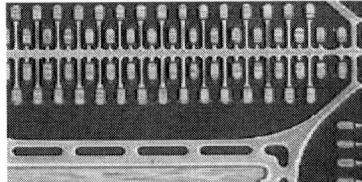

Figure 2: Standard Multi-Row QFN leadframe

978-1-4244-4722-0/09 $25.00
© 2009 IMAPS-ITALY

On Figure 3 a back-etched strip is shown and on bottom-right corner is clearly visible one device of it with all pads disconnected, ready for a possible strip testing. This option is extremely useful for ultra small packages where single device handling is extremely critical: sawing can be done on metal frame with sticky tape as well as testing. Final sorting and finishing is done on Pick and Place machines that use Die Attach basic concepts.

Size, location and shape of the pads can be designed in an infinite combination, varying the pitch of some pad rows to facilitate the inner row routing on final application board.

Ground ring, power rings or multiple pad connections are also feasible, as shown in Figure 1.

In term of package size, only BGAs can have a better ratio die/package outline if its full area can be used to distribute the ball array.

WPLGA has the possibility to reach the thinner package thickness among the packages suitable for the same application, due to its thin leadrame.

Manufacturing Process

On Figure 3 a simplified Assembly flow is shown:
1) Leadframe is received form Supplier with top etching featuring the future bottom design
2) Standard Die Attach and Wire Bonding on rigid substrate.
3) Standard molding (map type)
4) Back-etching of the excess of metal uncovered by plating
5) Possible strip testing
6) Separation by sawing

The only draw-back respect to QFNs, BGAs and QFPs is the wire bonding, being all wires forced to reach their final I/O, sometime far away from the pads or the die. This disadvantage, as we will see, will revert to an advantage for the electrical performances.

Complex bonding can be overcome by specific molding processes like "compression molding". On Figure 5 a WPLGA 13x13 312 I/O with 394 Au wires of 0.8 mils diameter is shown trough X-ray view. Molding has been done with "compression" technology that completely avoid any resin flow in the mold cavity, allowing an extremely reduced wire sweep (less than 1.5%) even on long wires (4.9 mm) and thin diameters (20 microns).

Figure 3: Back-Etched leadframe

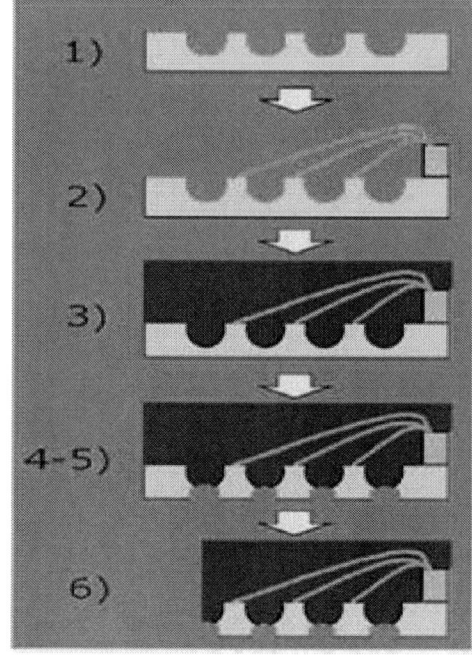

Figure 4: Assy-Testing flow

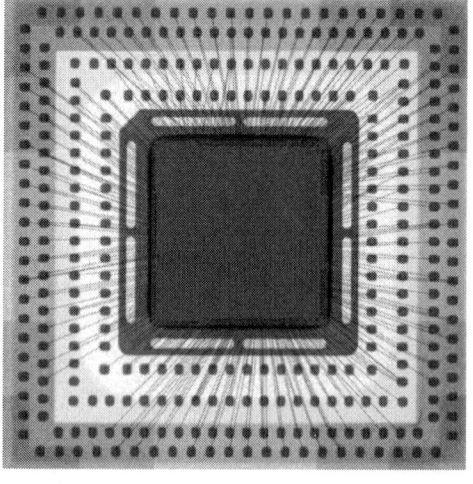

Figure 5: Complex bonding

Electrical Performance

Analyzing the WPLGA structure, the package electrical parameters are mainly depending on wire contribution; thanks to the high level of miniaturization and short electrical paths, the RLC parasitic values show better electrical performance versus other leadframe package families comparable with WPLGA for number of leads and cost.

With a medium number of I/Os (range between 100 and 200 signals) the QFP is the main alternative to WPLGA inside the leadframe based packages.

Significant data have been extrapolated from the comparison between WPLGA 11x11mm – 182 leads and the QFP20x20mm – 176 leads, both including a device dedicated to computer applications.
RLC values have been extracted in DC frequency domain where, thanks to quasi-static approach, each connection can be fully described by one RLC cell (lumped model).
In DC range, that extends his validity till 100MHz for the two packages under evaluation, the skin effect contribution is negligible and the effective cross-section for current flow is the complete conductor section; RL values can be considered constant in DC frequency region.
The pictures below show RL values related to few critical connections.

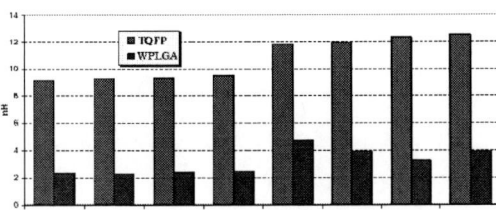

Figure 6: DC inductance

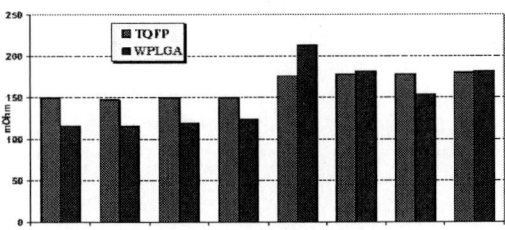

Figure 7: DC resistance

WPLGA shows better inductance values because of shorter electrical paths due to the short length of the leads; in addition mutual inductance values are about 5 times lower than QFP where there is a high parallelism of signal paths (long parallel leads).
Resistance values of the wires are the main contributors; the similar behavior is justified by the comparable wire length of the two configurations.
In terms of Capacitance, the values are not frequency dependent; WPLGA shows lower values because of smaller dimension.

Furthermore WPLGA has lower mutual capacitance values (2 times average value) than QFP; the physical mutual distance is the main contributor to mutual capacitance and it depends on bonding diagram definition as well as package shape.

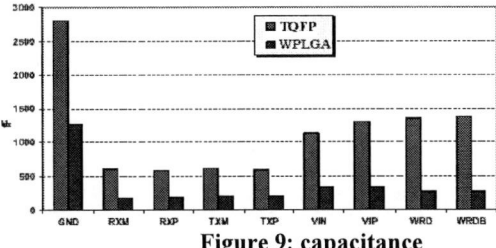

Figure 9: capacitance

The comparison has been completed with an S parameters extraction using the full wave approach (no frequency limitations) for which the electromagnetic radiation is taken into account.
Two differential pairs located on package corner area (worst case because of longer paths) have been simulated in order to compare the insertion loss till 8GHz frequency.

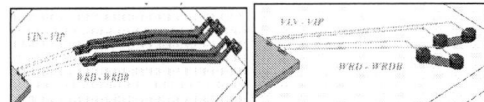

Figure 10: 3D models: designs not to scale!

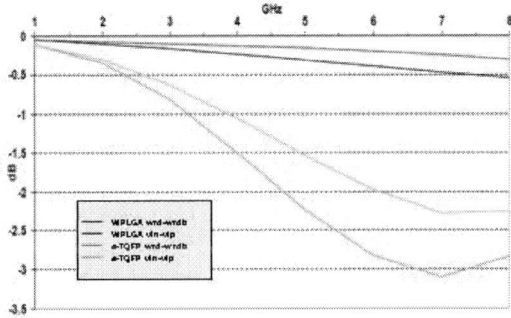

Figure 11: WPLGA vs. QFP insertion loss

Assuming -1db as acceptable lost, WPLGA working range is more than 8GHz and much more better than QFP for which the working range is between 3-4GHz only.

All above values highlight the promising electrical performance of WPLGA package.

The validation of the data through the electrical measurements is in progress; next step will be the consolidation of system level simulation approach (device + package + PCB) in a context where the system electrical design is becoming the real challenge to develop effective solutions able to satisfy the high level of electrical performance (voltage drop, EMC/EMI, cross-talk analysis, higher frequencies) requested by the applications.

978-1-4244-4722-0/09 $25.00
© 2009 IMAPS-ITALY

Thermal Performance

WPLGA package with exposed pad structure has in thermal dissipation capability one of the key features that makes it an attractive alternative to BGAs. On the other hand, having a higher I/O density compared to standard exposed pad packages, WPLGAs show a comparable thermal resistance compatible with most of the medium power applications in consumer electronics.

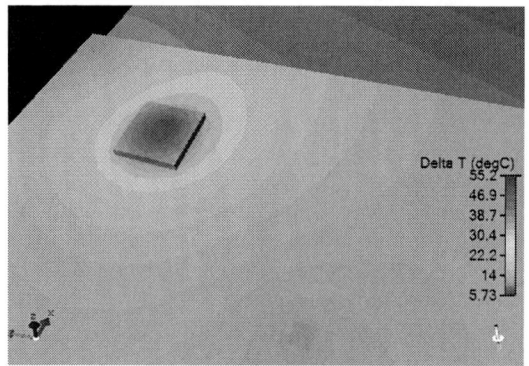

Figure 12 : WPLGA thermal simulation temperature map

While it is well known that standard plastic BGAs with body size in the range of 100 mm² can not handle more than 1 - 1.5 W, WPLGAs with similar footprint can dissipate up to 2 - 3 W. WPLGAs thermal resistance is fully determined by the actual dimension of the exposed pad area that is directly soldered to the PCB footprint. This allows achieving a very low thermal resistance between the silicon die and the board exploiting any specific thermal enhanced features in the PCB footprint itself like dedicated vias and copper areas.

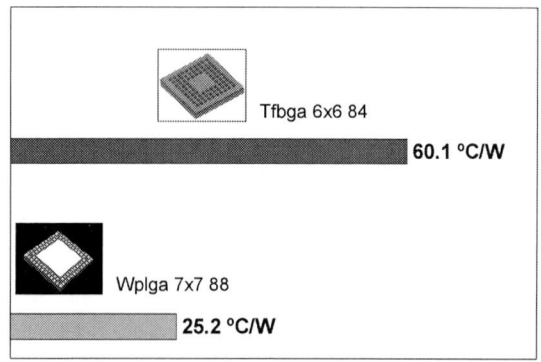

Figure 13: BGAs vs. WPLGA Theta J-A on 2s2p PCB

Comparing WPLGAs thermal dissipation capability with standard exposed package structure

like QFNs or QPFs, the main difference is the leadframe thickness. The leadframe thickness and hence the exposed pad thickness has some influence on heat flow spreading from die to PCB. Overall the thermal dissipation provided by the copper pad is determined by the vertical and the lateral thermal resistance. Thicker the copper pad, better the lateral heat spreading but higher the vertical thermal resistance. An optimum thickness exits that for a given pad area provides the minimum thermal resistance. As copper pad thickness is defined mostly by manufacturing process constraints, a certain variability in thermal dissipation is expected for various exposed pad package families produced with different copper pad thicknesses.

If we compare the most common exposed pad packages assuming to have the same die size and the components mounted on the same system, we can easily evaluate the impact of the exposed pad thickness on Theta J-A. As typical copper pad thicknesses range from 60 μm for WPLGAs to 200 μm for QFNs, the thermal resistance variation there is some 5 %.

	E-Pad (mm3)	PCB footprint (mm2)	Theta j-a (ºC/W)
TQFP 7x7 e-pad	4x4x0.125	9x9	20.2
TQFP 10x10 e-pad	4.5x4.5x0.125	12x12	19.7
TQFP 10x10 slug	6x6x0.5	12x12	18.0
QFN 9x9	4.1x4.1x0.2	9x9	22.1
WPLGA 7x7	4.6x4.6x0.06	7x7	23.2

Figure 14: E-Pad packages Theta J-A on hard disk drive

Finally, benchmarking all the possible exposed pad packages with similar pin counts, WPLGA results to be the best compromise in terms of miniaturization and thermal dissipation at a very competitive cost.

978-1-4244-4722-0/09 $25.00
© 2009 IMAPS-ITALY

Reliability performance

Being basically an LGA package, WPLGA should "suffer" even bigger problems than BGAs during surface mounting, when its planarity could change due to deformation. First step in material selection was to find a molding compound with the lowest deformation variation in the temperature range for surface mouningt. After selecting two of them, WPLGA 11x11 hasv been submitted to thermoire and a resin selected (Fig 15).

Next step was epoxy glue selection among possible candidates suitable for the specific application. WPLGA with such combinations has been submitted to MSL and results are shown into Fig 16.

WPLGA 11x11 with Daisy Chain device inside has submitted to Solder Joint Reliability cycles, assembled on standard ST board.

Cycle conditions : -40 to +125^{0}C
Soak Time: Ts (min) -40^{0}C (+5 / -10^{0}C)
Ts (min) 125^{0}C (-5 / +10/15^{0}C)
5-10 minutes 2 cycles/h
(JESD22-A104-B, G, 2, C)

Weibull Graph on Fig 17

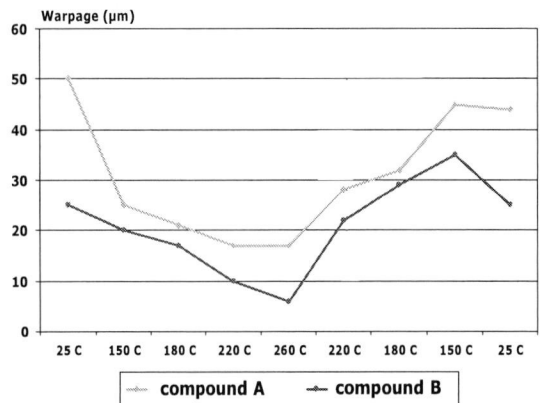

Figure 15 : warpage in temperature

WPLGA 11x11

Epoxy Glue Type	MSL2 @ 260C	MSL1 @ 260C
glue A	PASS	FAIL
glue B	PASS	PASS

WPLGA 7x7

Epoxy Glue Type	MSL2 @ 260C	MSL1 @ 260C
glue A	PASS	PASS
glue B	PASS	PASS

Figure 16: MSL Results

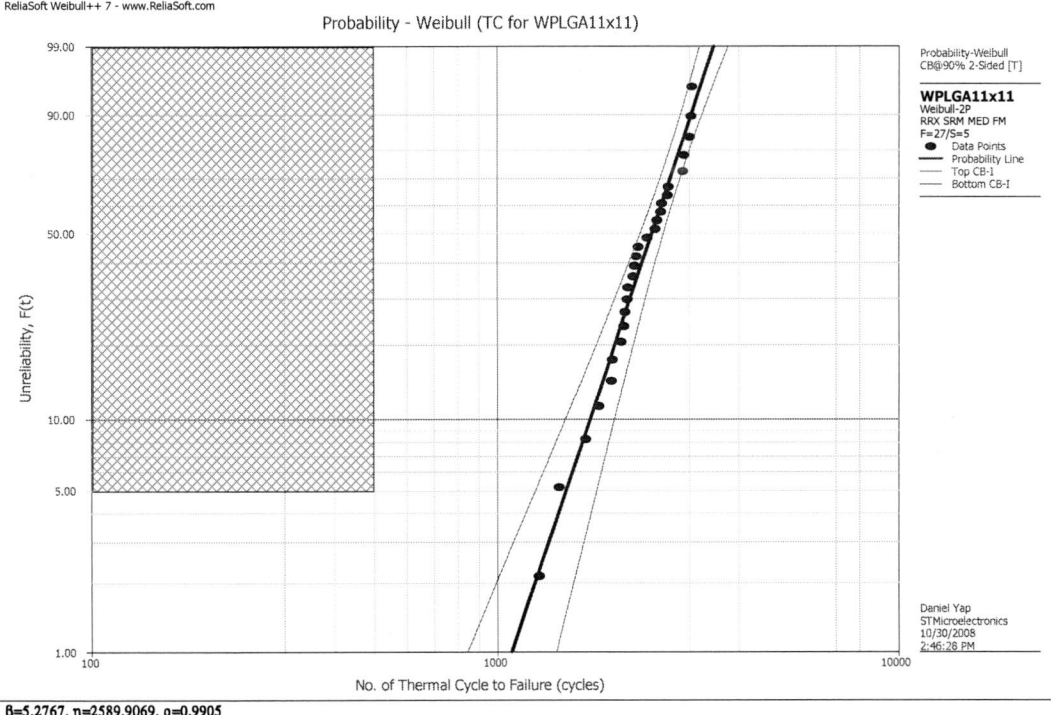

β=5.2767, η=2589.9069, ρ=0.9905

Figure 17: Weibull Graph WPLGA 11x11

978-1-4244-4722-0/09 $25.00
© 2009 IMAPS-ITALY

Package definitions and structure

WPLGA is a technical code, based on JEDEC JESD30E spec (Descriptive Designation System for Semiconductor-device Packages) where:

W = very,very thin (A1 > 0.65 and <= 0.8 mm)
P = plastic (material)
LGA = land grid array

The choice of LGA, instead of Staggered QFN as for JEDEC MO-247 (Plastic Quad no-lead Staggered Multi-row Packages) is due to the extreme flexibility in term of pad size, location and shape that this Package technology could accept. In any case, even customized designs, always based on the the above spec.

On Fig 18, a package cross section is shown. Due to thin leadframe (4 mils thick) of which only 2/3 remains inside the plastic body, the total package thickness (A1) can be contained to very small values. For other applications based on this technology, the total package height has been contained within 0.28 mm.
When die thickness cannot be maintained within 280 microns max and several loop height are necessary, package thickness increases, giving this package the name of VPLGA (1 mm max).

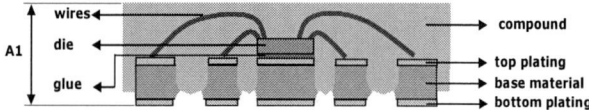

Figure 18

References

Yi-Shao Lai et al "Development and performance characterizations of a QFN/HMT package", ECTC 2008. May 2008 Page(s):964 - 967

Advancements in bumping technologies for flip chip and WLCSP packaging

Dionysios Manessis[1], Rainer Patzelt[1], Andreas Ostmann[2], and Herbert Reichl[2]

[1] Microperipheric Research Center, Technical University Berlin (TUB)
[2] Fraunhofer Institute for Reliability and Microintegration (IZM)
Gustav-Meyer-Allee 25, 13355 Berlin, Germany
E-Mail: manessis@izm.fhg.de, Tel: +49-30-46403229

Abstract

At R&D level, IZM has advanced stencil printing very close to its technological limits at pitches even down to 50μm. Innovative electroformed and laser-cut with nano-treatment stencils have been manufactured with an extreme thinness of 20μm for bumping wafers at Ultra fine pitches (UFP) of 100μm, 80μm and 60μm. Specifically, for 100μm pitch bumping, both type 7 (2-11μm) and type 6 (5-15μm) pastes of eutectic composition Sn63/Pb37 have been successfully employed. Bumping using 25 μm electroformed stencil thickness has yielded bump heights of 42.3±3.8μm and 43.6±3.5μm for type 7 and type 6 pastes, respectively. A newly prototype developed type 8 paste (2-8μm) has been used for the first time to bump chips with peripheral contacts at 80μm and 60μm pitch. Bumping at 80μm pitch with nano-treated laser-cut stencil has yielded bumps of 28μm in height. For bumping at 60μm pitch, a 20μm thick electroformed stencil was used with 35μmx80 μm oblong apertures. Printing at 60μm pitch has yielded very promising results and has proved the capability of electroformed technology to manufacture accurate and robust thin stencils. The bump height at 60μm pitch was measured to be 28 ±3 μm.

Paste-in-Resist technology has been developed as an alternative to stencils in order to overcome the manufacturing difficulties of making extremely small apertures. Paste is printed in resist apertures which have been opened by photolithographic processes. In this way, bumping has been demonstrated up to 50μm pitches. Complimentary to stencil printing processes, IZM has developed balling technologies up to 400μm pitch up to 8" wafers with a thickness of 150μm. Solder balling can be achieved either by "perform ball print" using conventional stencil printers with specially designed stencils or by "ball drop" techniques. Balling technologies have demonstrated the application of 300μm and 250μm Sn-Pb and Pb-free balls at respective area array pitches of 500μm and 400μm, the main I/O pitches for WL-CSP bumping.

Keywords: flip chip bumping, solder balling, WLCSP packaging, stencil printing, solder paste.

1. Introduction

The attractiveness of flip chip technologies lays on the superior electrical performance, higher thermal conductivity, smaller size and higher I/O counts which are mandatory requirements for advanced semiconductor applications. However, a substantive shift towards flip chip interconnection technologies will be witnessed only with accomplishment of cost reduction, reliability improvement and cost-efficient high density substrate technologies [1]. Low-cost flip chip bumping technology has become a reality with implementation of electroless nickel plating process for under bump metallization (UBM) in conjunction with stencil printing of solder pastes for the formation of solder bumps [2].

Stencil printing of solder paste for flip chip wafer bumping offers among others the advantages of cost-effectiveness and compatibility with pre-existing printing equipment in a surface mount assembly line [3-7]. The state-of-the-art in wafer bumping using the conventional stencil printing technology (laser-cut steel and Nickel-electroformed stencils) is at about 120μm pitch for peripheral arrays and 150μm for area arrays [3]. Significant technological improvements have been reported in literature regarding the capability of Ultra Fine Pitch (<120μm) (UFP) wafer bumping [9-11]. This progress is extremely significant especially in view of Moore's law prediction that the bits/chip grow by a factor of 4x every three years [8]. However, the advancements to UFP bumping can not be realised without parallel progression towards very fine pastes and overwhelming advancements in stencil manufacturing. The

978-1-4244-4722-0/09 $25.00
© 2009 IMAPS-ITALY

emergence of ultra fine type 7 pastes (2-11μm) and lately of type 8 (2-8μm) pastes along with recent developments in fabrication of very thin electroformed stencils with very small aperture dimensions and nano-treated laser-cut stencils have sparked significant work in the area of UFP wafer bumping. IZM`s printing group keeps abreast with recent developments in paste and stencil manufacturing by teaming up with renowned international companies in the semiconductor packaging field. In parallel, it presents new alternative technologies like "paste-in-resist" which still utilize economic printing processes in resist structures instead of paste printing in delicate stencil apertures. Equally important are also the developments for bumping wafers with chips having small number of I/Os by solder balling for reliable and robust mainstream WL-CSP packaging.

This paper discusses in depth all the technological issues pertinent to UFP wafer bumping and shows representative wafer bumping results at ultra fine pitches from 100μm to 60μm.

2. Technology Implementation

2.1 Wafer Designs & Chemical Metallisation

The wafers used in this study had a diameter of 6". The wafers at 100μm and 80μm pitch had a thickness of 680 μm whereas the wafer at 60μm pitch had a thickness of 320μm. The wafers at 100μm pitch consist of 540 chips with a size of 5mm x 5mm. Each chip has 176 I/O's with the pads arranged in a peripheral configuration. A total number of 95040 pads exist on the wafer. The pads have an octagonal shape with a diameter of 40 μm. The electroless Ni/Au plating technology of (TUB) was used to deposit 2 μm Ni /80 nm flash Au (over the chip passivation layer) high UBM pads on the Al metallization pads. The lateral overlappment of the Ni/Au UBM pads on the chip passivation layer is also 2 μm on both sides. The Electroless chemical metallization process finally yields 44 μm UBM pad size. Detailed description of TUB's Electroless Ni/Au technology approach (ENIG) can be found in literature [2]. The wafers at 80μm pitch have 606 chips 2.5mmx2.5mm in size and a number of 80 pads per chip. A total number of 48840 I/Os correspond to the wafer. Wafers at 80μm pitch have 60μm pad dimensions after 2μm Ni/Au processing. Correspondingly, the wafer with peripheral arrays at 60μm pitch has 577 chips with a size of 5mmx5mm with 296 pads per chip. A total number of 170792 pads exist. The passivation opening on the Al pad was 20μm and the same ENIG process was applied to deposit 4 μm Ni/80nm Au over the chip passivation layer. The UBM size of the wafer after application of Ni/Au is 28μm. Figure 1 shows the Ni/Au UBMs at 60μm pitch which proves that the ENIG process is feasible to very small wafer pitches.

Figure 1: Electroless Ni/Au UBMs at 60μm pitch. UBM size is 28μm.

2.2 State-of-the-art stencils & solder paste Materials

2.2.1 State-of-the-art stencils

The criteria for the selection of stencil manufacturing technology are always aperture wall quality, dimensional consistency, positional accuracy and stencil production cost. The two main stencil manufacturing methods for UFP bumping are laser-cutting stencils and electroformed stencils. Especially for the laser-cut stencils the quality is dependent on the technical expertise of laser stencil manufacturers, and the availability of fine-tuned laser guns in the market [9]. Recent refinements in the laser cutting process like smaller laser spot sizes and water guided laser cutting techniques have brought about large improvements in wall quality and ability to cut very small apertures. One advantage of laser-cut stencils compared to electroformed is that the material is pre-tensioned on the frame prior to cutting, which in turn reduces the stencil deformation. Laser cutting is a sequential process; therefore as the number of I/O increase on the design so does the time to manufacture the stencil. This can be a significant disadvantage for wafer bumping stencils at fine and ultra fine (<120μm) pitches with a high number of apertures because as the number of holes increases the manufacturing time rises to an almost uneconomical point. Also the heat melt interaction during the cutting process can create rough inner sidewalls which in turn produce apertures with a larger surface area and therefore have a tendency to minimize paste release. In addition, the heat generated in the cutting process must be controlled as not to warp the stencil during manufacturing or damage the fine webs required for UFP printing. Electro polished stencils are normally laser cut stencils which have been smoothed using electrochemical or mechanical polishing methods. However all metal stencil types can be treated in this manner after manufacturing. The cost factor becomes profoundly significant considering an approximate cost of 3 $/100 laser-cut apertures and the large number of apertures needed for UFP structures. The latest development in laser-cut stencils has been a nano-coating applied on the inner walls of the apertures and the bottom surface of the stencil to enhance paste release and thereby reduce stencil cleaning intervals[12]. Nano-coated laser cut stencils have started to be used extensively in SMD industry and they are still under examination for usage in UFP bumping. A nano-coated stencil has been used in the present work for bumping the 80μm pitch wafer. The nano-treatment improves the smoothness of the aperture wall and also makes the bottom surface highly hydrophobic which can be very beneficial for less flux/paste bleeding.

On the other hand, electroformed stencils with superior aperture wall quality than laser-cut apertures and reasonable manufacturing cost per wafer design; and not per aperture as is the case for laser-cut stencils, appear as a rational alternative solution especially for UFP wafer structures. In the area of UFP bumping with more than 50000 I/O`s the cost of an electroformed stencil is acceptable as compared to the high cost of a laser-cut stencil. Nevertheless, it should be mentioned that electroformed stencils are framed after aperture manufacturing and the associated stretching can lead to even slight nickel expansion and therefore misalignment problems during printing set-ups. In the present study, an electroformed stencil was selected for the bumping at 60μm pitch with extreme care from our partner to avoid stencil deformation during foil framing/stretching. The selection of electroformed stencils for 100μm and 60μm pitch was done in order to prove the capabilities of the electroformed technology to make very thin foils with good robustness for continuous printing and with good aperture smoothness and dimension accuracy. For the wafer at 100μm pitch, oblong apertures of 50μmx125μm were designed and a foil thickness of 25μm was selected. At 80μm pitch, a laser-cut stencil was manufactured with 20μm foil thickness and apertures of 49μmx108μm. For bumping at 60μm pitch, an electroformed stencil of 20μm thickness with 35μmx80μm apertures was prepared in a R&D level. Figures 2 and 3 show SEM and cross-section pictures of the stencil used at 60μm pitch and at 100μm pitch. Figure 2 evidences the superior smoothness of electroformed apertures and figure 3 shows clearly the vertical aperture walls (almost no tapering) by using special masks in the electroforming process [13] which underlines the difference with the tapered walls of laser-cut stencils. Further developments in electroformed stencil manufacturing concentrate on different nickel hardness up to 480HV. This harder nickel material will be less likely to wear from the printing forces and will also deform less from both the framing and the printing process.

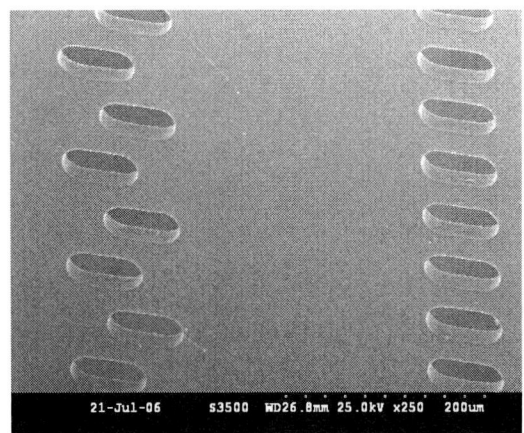

Figure 2: Electroformed nickel apertures 35μmx 80μm at 60μm pitch. Stencil thickness: 20μm

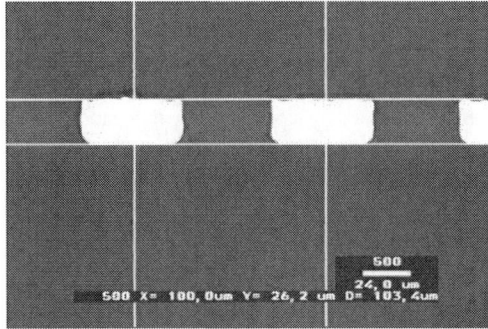

Figure 3: Cross section view of electroformed stencil at 100μm pitch. Foil thickness:~25μm.

Other changes in the plating setup target to modify the surface roughness of the stencils. Initially, it was thought that a rough stencil top surface would aid paste roll and hence aperture fill however with the very fine particle solder pastes the solder spheres actually get entrapped in the large nickel grains and therefore may cause smearing across the stencil surface [11]. This can be slightly compensated for by using a higher print pressure however this high force can speed up stencil degradation. In the case of "paste-in-resist" technology, major developments have occurred in resist technologies which can be opened photo-lithographically up to 50μm pitch and can shine even a brighter future for UFP printing since no conventional stencils are used.

The aperture design for wafer bumping always takes into consideration the existing rules for fulfillment of aspect (>1.5) and area (>0.66) ratios for efficient paste transfer on the pads but also considers to hold the minimum separation distance to avoid paste bridging. The largest challenge in bumping at UFP is to find the gold solution to satisfy the above mentioned design principles for good paste release and sufficient paste volume and on the hand to avoid paste and bump bridging [4,9,11]. The design for 100μm pitch wafer yields an aspect ratio of 1.5 and an area ratio of 0.71, respectively. For 80μm pitch the aspect and area ratios were 2.45 and 0.84, respectively whereas for 60μm pitch the respective ratios were 1.75 and 0.61. One of the main goals of the present UFP studies is to examine if state-of-the-art stencils and materials can facilitate qualitative printing even at designs with lower aspect and area ratios than 1.5 and 0.66 as well as help improve production times by avoiding frequent stencil cleaning [9-12].

Apart from stencil developments, stencil printing is very dependent on printing machine capabilities. Most of IZM`s R&D bumping work has been performed by using a DEK Horizon machine which has the capability of +/- 25μm alignment. It should be noted that alignment at UFP especially smaller than 120μm is very challenging and very close to limits of the today`s printing machines. Figure 4 shows a "real-time in-situ" very critical alignment at 80μm pitch, with a wafer of 60μm Ni/Au pads in contact with the nano-treated stencil with apertures of 49μmx108μm.

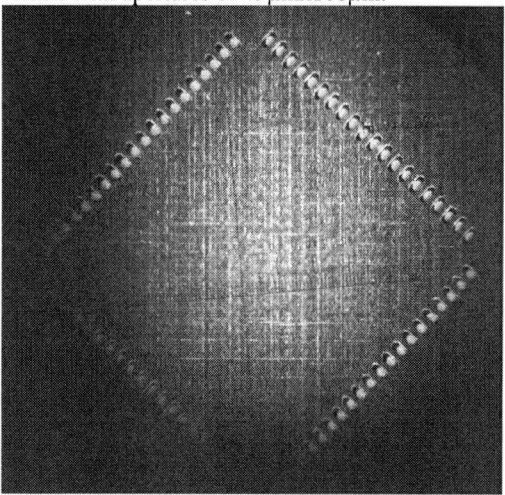

Figure 4: View of a "real-time" (at the stencil printer platform) alignment of Ni/Au pads (gold color) with stencil apertures of 49μmx108μm at 80μm pitch. (Stencil thickness:20μm)

2.2.2 State-of-the-art solder paste materials

Previous studies have shown that type 6 (5-15μm) and even finer pastes are appropriate for UFP bumping (pitches<120μm). It has been shown that type 6 may marginally be used up to 100μm pitch wafer bumping based on the aperture design [3,11]. Experimental studies have compared bumping results at 100μm pitch of type 7 (2-11μm) paste with the standard type 6 (5-15μm) paste [11]. The difference in printing and bump height distribution

was not significant and therefore the differentiation of type 7 from type 6 is not necessary. Type 6 paste which has been already established in the market for wafer bumping can be used for wafer bumping applications ranging from 300μm to 100μm pitch. For UFP applications (<120μm) and always based on the aperture design decided for each specific application, a new type 8 powder was fabricated and was provided as paste. Type 8 has a powder size from 2-8μm. Type 8 was used in the present study for bumping at 80μm and 60μm pitch (35μmx80μm apertures) whereas type 6 was employed for bumping at 100μm pitch (50μmx125μm apertures). The standard type 6 paste is commercially available whereas the type 8 paste is a developmental product. The powder size of type 8 paste has been confirmed to be 2-8μm at a percentage larger than 90%.

Our studies have shown that type 6 paste may be used up to 25μm stencil thickness and 80μm pitch provided that the stencil design is appropriate (at least 5 particles) and the stencil has a nano-treatment which improves paste release.

3. Bumping applications-Results & Discussion

3.1 Bumping at UFP using stencils

Printing at 100μm pitch was performed effectively with type 6 paste and the electroformed stencil of 25μm. For bumping at 80μm pitch, the nano-coated stencil has yielded better paste release even at very marginal aspect and area ratios. For 80μm pitch, both type 6 and type 8 were used and compared. Figure 5 shows a view of a bumped chip at 80μm pitch.

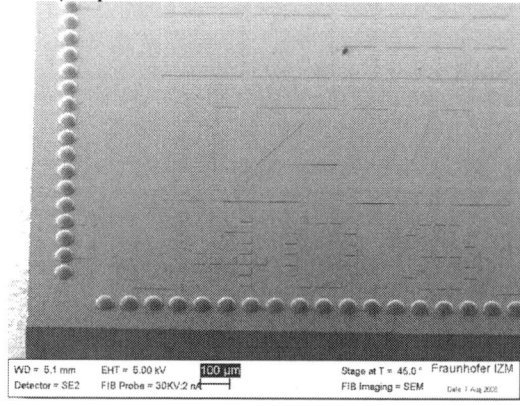

Figure 5. Bumping at 80μm pitch. Bump height: 23μm.

Printing speeds in the range of 5-15 mm/sec were applied along with a printing pressure of 2-3 Kg for a 300mm long stainless steel squeegee with 60 degrees printing angle. These parameters could ensure a clean print sweep with the minimum pressure used and proper filling of the apertures. The printing speed range is in agreement with other reported values in the literature [4]. Contact printing is used and in general is the most appropriate choice

978-1-4244-4722-0/09 $25.00
© 2009 IMAPS-ITALY

compared to snap-off printing for wafer printing at pitches smaller than 120μm. In fact, the 20μm thin foil at 80μm pitch can severely be flexed upon snap-off printing resulting in significant stencil deformation. In contrast, contact printing might protect the thin foil from repeated elastic deformations.

Printing at 60μm pitch is in fact very challenging for the alignment between wafer pads and the stencil apertures. Although 3 fiducials have been used for alignment always an offset should be always used to improve the registration. In addition to deformation due to the framing process, repeated printing may also contribute to stencil stretching. In the present study, measurements were taken to monitor deformation to ensure that the alignment of the framed stencil to the wafer was acceptable. After 20 prints and manual cleaning of the stencil-which may also deform it- good alignment was always achieved. Figure 6 shows print deposits of type 8 paste at 60μm pitch. The deposit thickness is 20μm in agreement with the foil thickness.

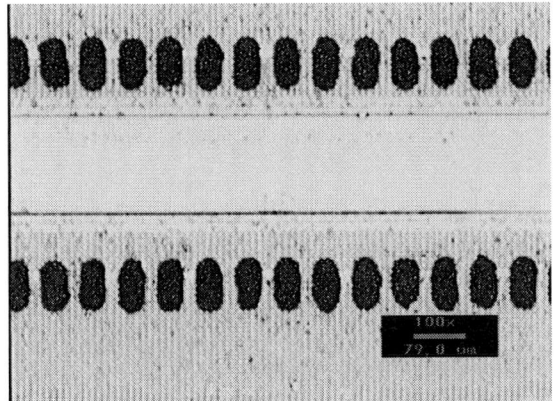

Figure 6. Wafer printing at 60μm pitch with type 8 (2-8 μm) paste.

The deposits have dimensions about 35μmx77μm on the top surface and about 41μmx84μm on the bottom surface. Based on the fact that the aperture walls are straight as shown in Figure 2, the larger deposit dimensions at the bottom may have resulted from either a slight slump of the paste under the given printing conditions or due to the small amount of undercut on the stencil at the wafer side. The reflow of type 8 paste creates many tiny solder balls. Furthermore, flux bleeding can carry particles between the deposits (see Figure 6) which can not fully coalesce to the main bump during reflow and appear as tiny balls on the bump foot. Solder balling is much more intense for bumping at 60μm and 80μm pitch than for bumping at 100μm pitch with type 6. Bumping at 60μm pitch yields bumps with a height of 28μm±3μm. Figure 7 shows a wafer bumped at 60μm pitch.

Figure 7: Wafer bumping at 60μm pitch.
Bump height: 28μm ± 3μm.

3.2 "Paste-in-Resist" technology

It is similar to stencil printing but structured resist is used instead of a metal stencils in which a solder paste is printed. Very fine paste type 8 has been printed in the apertures and subsequently reflowed in nitrogen atmospheres. One of the main developments was for the resists to withstand the lead-free temperatures. After reflow, the resist is stripped out and the bumps can be reflowed one more time for good shape formation. Figure 8 shows Sn4Ag0.5Cu bumps after resist removal at 60μm pitch. Paste in resist technology has demonstrated bumping up to 60μm pitch and it looks very promising especially for UFP wafer bumping.

3.3 Solder balling

Complimentary to stencil printing processes, IZM has developed balling technologies up to 400μm pitch up to 8" wafers with a minimum thickness of 150μm. Solder balling can be achieved either by "perform ball print" using conventional stencil printers with specially designed stencils or by "ball drop" techniques. Balling technologies have demonstrated the application of 300μm and 250μm Sn-Pb and Pb-free balls at respective area array pitches of 500μm and 400μm, the main I/O pitches for WL-CSP packaging. Figure 9 shows a 6" wafer at 500μm pitch balled by printing 300μm Sn4Ag0.5Cu balls. Yield has been measured to be 99.998%.

4. Conclusions

IZM has developed stencil design rules for any demanding wafer bumping application and offers a complete service portfolio for wafer bumping including: stencil design & selection, paste printing & reflow, fluxing/balling, wafer cleaning,

978-1-4244-4722-0/09 $25.00
© 2009 IMAPS-ITALY

Figure 8. Solder bumps at 60μm pitch using "Paste-in-Resist" technology.

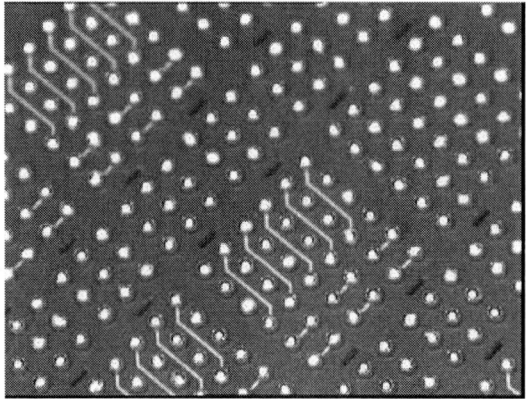

Figure 9. Printing of 300μm lead-free balls (Sn4Ag0.5Cu) at 500μm pitch (43940 balls on 6"wafer).

bump height measurements & yield statistics, microstructural characterisation of UBM/solder bump interfaces, and reliability assessment of assembled flip chips.Many investigations have devoted to UFP bumping by stencil printing of solder paste. Bumping at pitches smaller than 120μm has been demonstrated up to 60μm using developmental type 8 pastes. Specially fabricated laser-cut stencils with nano-coatings and very thin electroformed stencils have been employed through collaborative R&D work with international companies to bring printing capabilities very close to the limits of stencil manufacturing. Complimentary to stencil printing, balling technologies have been developed for WL-CSP bumping up to 400μm pitch.

Acknowledgements

The authors would like to thank the companies Heraeus, Laser Job, Microstencil Ltd, DEK Ltd for their R&D collaboration in stencil and paste materials for bumping and balling. Part of this work has been funded by EU projects "Blue Whale" and "Imecat".

References

[1]E. Vardaman.,"Growing demand for Flip Chip" Advancing Microelectronics, Jan/Feb.2003,pp.10-12.

[2] A. Ostmann et al. "Electroless Metal Deposition for Back-End Wafer Processes", Advancing Microelectronics, pp. 23-26, May/June 1999.

[3] D. Manessis et al. "Technological advancements in Lead-free Wafer Bumping using Stencil Printing Technology", Proc. In EMPC 2005, Brugge, Belgium, June 2005, pp.427-433.

[4] J. Schake, "Stencil Printing for Wafer Bumping", Semiconductor International, pp. 1-8, October 2000.

[5] P. Coskina et al. "Wafer Bumping for wafer-level CSP's and flip chips using Stencil printing technology", Proc. IMAPS Europe , Harrogate, June 7-9, 1999.

[6] P. Elenius et al., "Recent Advances in Flip Chip Wafer Bumping using Solder paste Technology", Proc. 49th ECTC, San Diego, 1999.

[7] B. Huang and N.C. Lee, "Solder Bumping via Paste Reflow for Array packages", Jour.. of SMTA, Vol. 15, Issue 1, pp. 16-31, March 2002.

[8] G. Meyer-Berg, "KGD Roadmapping", Good-Die & Europractice-HDP newsletter, No. 10, pp. 4-7.

[9] D. Manessis et al., "Technical challenges of stencil printing technology for ultra fine pitch flip chip bumping", Proc. IMAPS International, Denver, Colorado,2002,pp. 727-732.

[10] G.J.Jackson et al., "Differences in the sub-processes of ultra fine pitch stencil printing due to type-6 and type-7 Pb free solder pastes used for flip chip", Proc. ECTC, New Orleans, 2003, pp.536-543.

[11] D.Manessis et al, "Stencil Printing Technology for 100μm Flip Chip Bumping", Proc. in IMAPS 2003, November 2003, pp. 241-246.

[12] M.Roesch, D.Kozic, K.Feldmann, "Qualifizierung des Schablonendrucks unter verwendung nanobeschichterer SMT-Druckschablonen", Plus 11/2007, pp. 2175-2179.

[13] Robert Kay et al."Stencil Printing Technology for Fine Pitch Deposition of Pb-Free Flip Chip Interconnects", IMAPS Device Packaging Conference, Scottsdale, March 2006.

Assembly - Chip Interactions leading to PPM-level Failures in Microelectronic Packages

A. Mavinkurve[1], H. Cobussen[1], W.D. van Driel[1, 2], L. Endrinal[1] and M. van Dort[1]

1) NXP Semiconductors, Nijmegen, The Netherlands

2) Delft University of Technology, Delft, The Netherlands

Phone: +31-24-3533378; Fax: +31-24-3533350; Email: amar.mavinkurve@nxp.com

Abstract

It is known that molding compounds, containing irregularly shaped filler particles, used in microelectronic devices can cause damage within the diffusion stack. Electrical shorts between the top two metal layers often manifest this phenomenon. Such effects are usually seen in non-planarized diffusion processes due to their relatively larger irregularity in topography. The main factors that control this phenomenon are the properties of the molding compound (the filler size and shape), the mechanical integrity of the passivation layer, and the stresses in the chip. In this study, we focus on assembly - chip interactions that lead to a low PPM level failure. This is considered as unacceptable in today's zero PPM mindset in the automotive industry. FEM simulations are performed to estimate the stresses within the backend stack of the chip. With the help of these simulations, the most sensitive locations within the structure are identified. The scaling of this stress with filler size is analyzed and compared to that obtained with a high end low thermal expansion molding compound, containing round fillers with better size control. Using estimations and measurements of the particle size distribution, flow modeling, and statistical methods, the PPM level of the failures could be attributed to the low chance that a filler particle would land on the critical location. Modulating these effects yielded some interesting interactions between assembly and chip processing, which will be described in more detail.

Key words: PPM, filler particles, filler attack, non-planarised diffusion processes

Introduction

Understanding the interactions between a package and the chip is essential towards designing a product with a longer lifetime, in other words, a better reliability performance. The thermo-mechanical properties of the materials used to manufacture the die (e.g. silicon nitride, aluminium) and the packaging materials (e.g. the molding compound, die attach, leadframe) are significantly different. Hence, temperature changes during the manufacturing process, or during board assembly at the customer, can result in stresses within the package. If the critical strength of a particular material is exceeded, this will result in a high probability that it will fail. Failure or damage can be found not only on the die surface, for example, but also within the die [1, 2].

The so-called problem of inter-metallic shorts has been studied and reported before in the literature in microelectronic devices [3 - 7]. The failure mode is characterized by an electrical short between metal layers (usually aluminium or alloy based on aluminium), which are separated by an inter-metal insulation layer. One school of thought professes that the failure mode is usually caused by electrical discharges (ESD), either during the processing of the wafer or during assembly [3, 4, 5]. On the other hand, thermo-mechanical stress induced by the molding compound, in particular through a silica filler particle, could also lead to stresses that are sufficient to cause inter-metal shorts: this so-called filler attack mechanism is a known failure mode described in several Semiconductor Quality Handbooks and in the literature [6-8].

Conventional molding compounds, containing irregularly shaped silica filler particles, can cause damage to chip surfaces (e.g. the passivation layer) leading to cracks within the diffusion stack. In this study, we focus on assembly - chip interactions that lead to a very low failure level (< 1-2 PPM). This is unacceptable in today's zero PPM mindset in the automotive industry [9].

For this purpose, Finite Element (FEM) simulations are performed to estimate the stresses within the backend stack of the IC (especially the nitride stress and metal plasticity). With the help of these simulations, the most sensitive locations within the structure are identified. In particular one sensitive location was identified based on the layout (design-related) as also the position on the die with respect to other feastures. Special consideration for the effect of the presence of large filler particles on the passivation surface was included in the modeling

978-1-4244-4722-0/09 $25.00
© 2009 IMAPS-ITALY

work. The scaling of this stress with filler size is analyzed and compared to that obtained with a low thermal expansion molding compound, containing round fillers with better size control. Details of the modeling work will be presented in the near future elsewhere [10].

Furthermore, measurements of the particle size distribution were performed to estimate the fraction of large particles and the chance of their ending up at the most critical location. In addition to this, some interesting interactions between assembly and chip were found, which will be described in more detail. This helped us in devising an effective measure to contain the failures in the field. Finally, the effectiveness of a low thermal expansion molding compounds (one of the generally proposed solutions) to solve this issue is also given.

Problem Description

Our customers reported PPM failure rates below <1 PPM for a dual layer metal device encapsulated in small outline (SO) plastic packages. As the production volumes increased, the number of returned failures increased on a weekly basis (1-2 ppm level). Influenced by the Surface ESD theory [3,5], routine failure analyses consisting of decapping or removal of the molding compound, after which SEM / FIB were performed on the customer returned units after localisation of the short through Optical Beam Induced Resistance Change (OBIRCH). Figure 1 shows the result of such a returned sample: a metal-to-metal short. In most cases, passivation damage was also observed on the die surface, as shown in Figure 1(a). However, it soon became clear that reducing the chance of ESD (e.g. through the installation of deionisers) did not reduce the failure rate siginificantly, implying that a different failure mechanism was at play.

Subsequently, a new technique for package failure analysis was used to analyze the exact failure cause. Conventional package failure analysis usually involves mechanical polishing and cross sectioning and it poses a challenge to perform a mechanical cross-section to identify the exact location of the defect. It is highly probable that the location of the failure might be missed or that the cross-sectioning procedure causes damage to the critical location itself. Thus, a new technique of Focused Ion Beam micro-machining was developed to analyze the failing device from the backside of the silicon, in order to preserve the top plastic mould compound. The failure localization is done by backside Obirch analysis. By utilizing the FEI Strata Dual Beam FIB-SEM, a cross-section can be made at regular and controlled intervals, thus eliminating the possibility of missing the defect site [11]

Figure 2 shows an example of a returned sample analyzed by this new technique. Since the package is opened from the backside, the filler particle causing the damage is clearly visible. Six consecutive devices that were analysed in this way

(including both in-line as well as customer rejects) showed the same failure mechanism, thereby showing that this mechanism was the primary cause of the PPM level failures.

The short between the two metal layers caused by this failure resulted in an electrical short between external pins 2 and 4. To identify failing devices in line, a specific test programme was devised that could pinpoint (with 90% certainty) that this failure mode had occurred. The in-line failure rate was found to be about 30 PPM.

In order to determine the root cause of the failures, FEM simulations were performed to understand the sensitivity of the particular failing location shown above, and to identify other sensitive locations. Further investigations focused on other interactions of the chip with the assembly process, with a view to understand the PPM level of the failures, and to work towards a solution. Basically, the main interaction of the die with assembly was found to be one that increases the chance that a large filler particle from the molding compound is present at the critical location on the die.

(a)

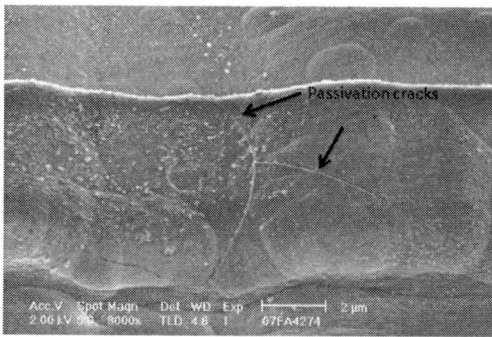

(b)

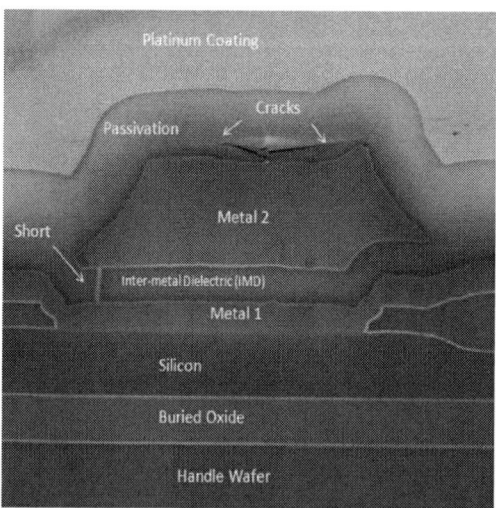

Figure 1: Example of the metal-to-metal short by conventional failure analysis (a) Top view SEM image showing nitride crack (b) SEM image of FIB cross section showing the layout.

978-1-4244-4722-0/09 $25.00
© 2009 IMAPS-ITALY

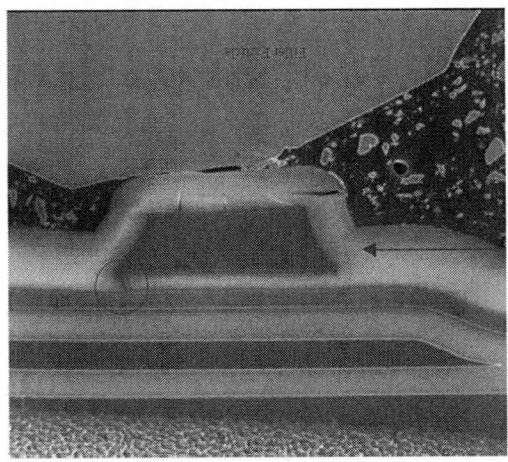

Figure 2: Example result showing the metal-to-metal short by the newly developed failure analysis technique.

Results of Finite Element simulations

Details of the approach used are given in [10]. Parametric FE models are created to predict global and local stress and strain levels. One of the important results of this simulation is shown in Figure 3, corresponding to the most critical location on the die.

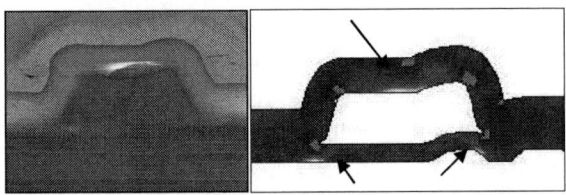

Figure 3: Stress distribution at a specific design location (right) compared with the observed cracked areas (left), at −65°C.

Intuitively, it is not hard to imagine that the higher thermal expansion coefficient of the metal (Al) will result in an asymmetric distribution of tensile stress within the nitride layer when the package-die cobnination cools down. The maximum stress was found to result at lower temperatures (e.g. −65°C, which is the lowest temperature that the product is subjected to during lifetime testing. A comparison of various locations on the die showed that the above location (overlapping metal layers, thick top metal layer, relatively thin inter-metal nitride) was the most sensitive in terms of stress. Other positions that were less critical but also sensitive were also identified, which accounted for a smaller fraction of the PPM level failures.

In the above figure, the positions of the cracks coincide well with the regions of maximum stress in the nitride layers. The maximum stress at the most critical location, based on the above model is about 300 – 400 MPa, which is too low to exceeed the critical stress needed to fracture the nitride leyer,

which is usually assocated with strength values of about 500 – 1000 MPa [2], depending on processing conditions and the presence of defects.

We have also shown [10] that when a contacting filler particle is included in the model, it is feasible that the critical stress is exceeded, even in the inter-metal nitride layer, since the hard filler particles act as stress enhancers. Further proof was furnished by comparing the calculated distance travelled by a contacting filler particle with critical displacements that the nitride layers undergo prior to fracture, in indentation tests carried out on the nitride layers [10]. In addition to this, crack modeling was also used to visualize the sequence of events starting from the filler pressing against the top nitride layer resulting eventually in a crack within the inter-metal nitride layer: the final step involving the crack getting filled with aluminium.

Another important output from the FEM simulations is the scaling of peak nitride stress with the particle size. This is shown in Figure 4 (b), for the conventional compound used, as well as a low thermal expansion biphenyl compound with a higher filler content and 100% spherical particles with a better size control. For this comparison the model shown in Figure 4 (a) was used. Differences in resin properties were calculated from the rule of mixtures and the modified Kerner equation [ref 12].

A comparison of stress levels was calculated assuming purely spherical particles for the sake of

(a)

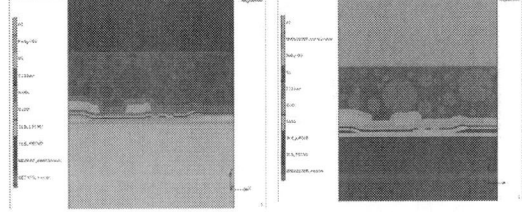

(b)

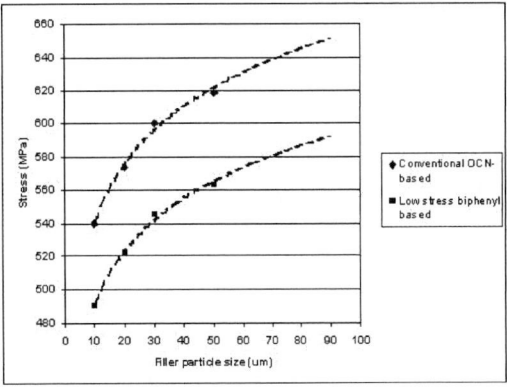

Figure 4: (a) Model used to compare stess between the conventional molding compound, and a low thermal expansion biphenyl molding compound (b) Peak stress at room temperature at the most critical location.

978-1-4244-4722-0/09 $25.00
© 2009 IMAPS-ITALY

convenience and the difficulty in assuming a precise shape of the irregular particles in the conventional compound. Taking the sharpness into account for the conventional molding compound would make the difference even higher. The diagram shows that the low thermal expansion molding compound would result in a significantly lower peak stress in the nitride layer. In this case, the value at room temperature is shown since that reflects the situation corresponding to in-line failures. Please note that it is essential for direct contact between the particle and the nitride layer to induce stresses that would exceed the critical stess needed for failure.

Assembly interactions

PPM level failures are usually a result of interactions between a few or sometimes several parameters. For this particular problem, we will restrict ourselves to the defect caused by the presence of a relatively large, irregularly shaped filler particle on the sensitive locations on the die. Other possible interactions like scratches induced during the assembly process could also occur, but form a smaller fraction of the PPM level failures in this case.

The necessary conditions for this failure are obviously the critical location on the die, and presence of a partice that is above the critical size that is directly in contact with the surface. Further investigations focus on the factors in assembly that can control the probability that a critical particle lands at the critical location. A first order estimation of that probability was a product of the volume fraction of particles above a certain critical size (e.g. area of 2000 μm^2) and the fractional occupation of each of the particles in all dimensions (x, y, z). This simple exercise, based on a number of simplifying assumptions showed that this probability is of ~PPM level.

Although intuitively satisfying, we found the need to refine this crude approach. Also, we found another strong interacting effect from the package or assembly that was strongly affecting this probability. From a mapping of the most critical locations on the die we found that the most critical location was in the die, we found that the most critical location was located in the vicinity of the bondwire connected to bondpad 3to a bondwire connected to Pin 3. Figure 5 shows the wiring scheme and the layout of the die in the package. This gave the first indication of an influence of wire at bondpad 3 on the probability that a large particle is present at the critical locations.

. Following this, the failure rate was mapped per position on the matrix leadframe. The positions are shown in Figure 6 (a). We noted that a majority of failures (~50%) were found to occur at leadframe position 3, followed by position 4 showing about 25% of the failures. The results are given in Figure 6 (b). We suspected some blocking effect of the wire

at Bondpad 3 on critical filler particles, dependent on the direction of the molding compound a it filled

(a)

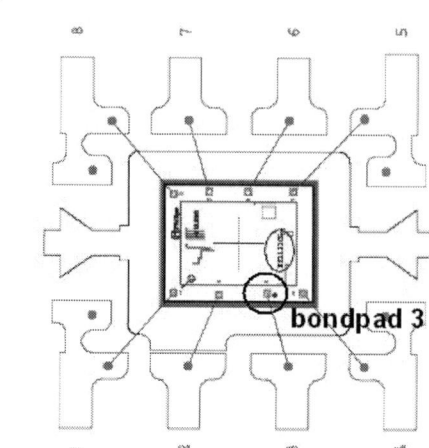

(b)

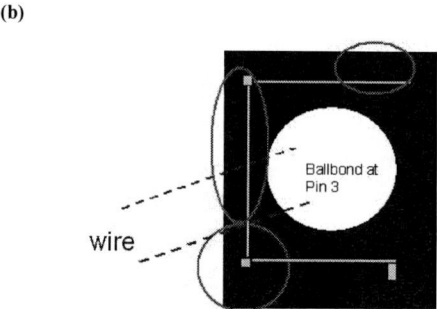

Figure 5: (a) Wiring scheme and layout of die in the package, showing bondpad 3 (black cicrle) and the most critical location (red dot) (b) Zoom in to the bondpad area of Pin 3 showing the most critical location (red circle) accounting for ~ 50% of the failures, and other less critical locations (red ellipses).

(a)

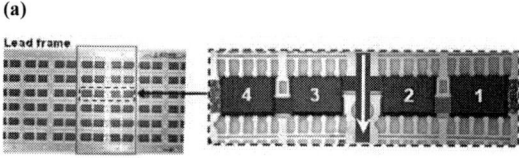

(b)

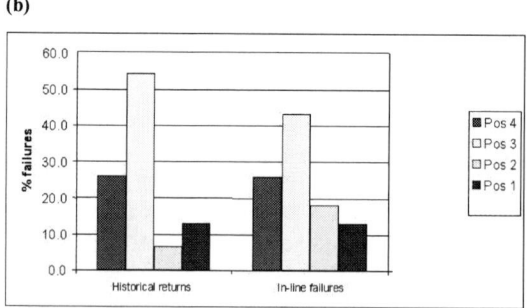

Figure 6: (a) Leadframe position for the matrix leadframe. The white arrow shows the direction of mold flow (b) Failure rate per position.

978-1-4244-4722-0/09 $25.00
© 2009 IMAPS-ITALY

the mold cavity. Why this effect is so strong at Position 3 was investigated by a combination of flow modeling, and making short shots in the matrix mold. Flow modeling was done by using *Inject*, a software tool developed at Philips, modified to account for reacting systems like molding compound [13]. The results are shown in Figure 7 (a) and (b), respectively.

The results show the direction of the flow at each of the postions with respect to the bondingwire at bondpad 3. It also shows that the wireloop at bondpad three at positions 3 and 4 could trap relatively large particles and cause them to locate themselves at or very close to the most critical location. On the other hand, the wireloop at bondpad three at leadframe positions 1 and 2 serve to block out large particles from reaching the most critical location, although they would serve the same trapping function with respect to the less sensitive locations.

To confirm this effect, planar lapping of several failing devices devices from leadframe position 3 was done, a typical photo corresponding to the wireloop at Bondpad 3 is shown in Figure 8.

(a)

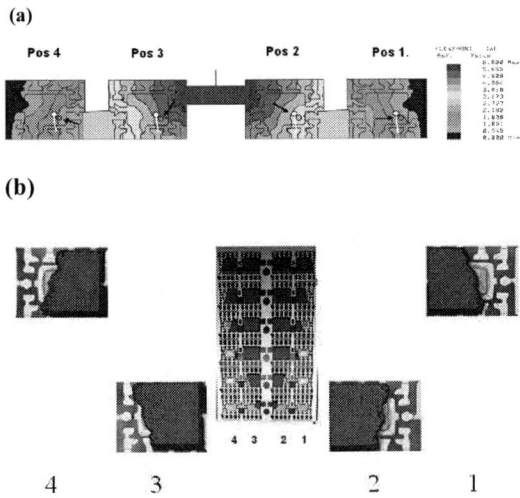

(b)

Figure 7: (a) Flow modelling to show flow fronts vs time in the mold relative to the most critical location (red circle) and bondwire at Bondpad 3 (white). The black arrows show the flow direction. (b) Short shots made to confirm the results of the flow modelling.

This confirms the suspected trapping effect of the wireloop at bondpad 3. The result can also be used to understand a blocking effect of the same wireloop at leadframe positions 1 and 2.

There are three effects that explain the preference of leadframe positions 3 and 4 compared to 1 and 2 with respect to failure rate: the described trapping and blocking effects respectively, and the volume of material (read: number of filler particles) passing the critical location. From the flow modelling results, a rough indication of total volume

passing the most sensitive location can be obtained, which is in turn directly proportional to the number of critical particles passing that location.

Even qualititatively, it is not difficult to envision that the number of filler particles passing the critical location at Position 3 is much higher than at Position 4. Table 1 summarizes the situation at

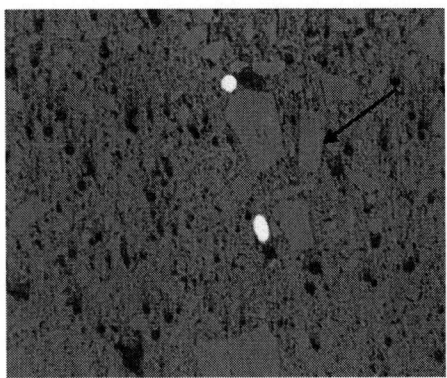

Figure 8: Planar lapping above the wireloop at bondwire 3 to show the trapping effect of the bondwire on large filler particles., at leadframe position 3 The black arrow shows the direction of the mold flow. The diameter of the cross-section of the upper two bondwire is 25 um, giving length scaling to the above micrograph.

	Pos 1	Pos 2	Pos 3	Pos 4
Blocking effect	+ -	+ +	- -	- -
Trapping effect	+ +	+ +	- -	+ -
# passing particles	+ +	+ -	- -	+ +
Preference for failures	3	3	1	2

Table 1: Overall qualitative summary of three effects at the 4 different leadframe positions, showing why a majority of the failures occur at position 3.

each leadframe position, showing why failures are preferably located at leadframe position 3. Finally, we need to refine the calculation of failure probability, and for this purpose we need to address the missing link in the equation: the filler particle size distribution, in more detail.

Particle size distribution

Even though a necessary condition for our failure mode is the presence of a relatively large filler particle in contact with the most sensitive location on the die, it does not automatically mean that when this does happen, that the chance of failure will be 100%. It will depend on actual the level of stress (particle geometry and position), as well as the presence of micro-defects within the backend stack, which would determine the strength of the nitride layers. This is known from basic Fracture Mechanics principles [14]. However, based on the results of the FE simulations that shows that the critical stress is exceeded [10], we can assume that the failure chance, in this case must be relatively high (of the order of %, say anywhere between say 10% and 90%. Hence the controlling factor for the PPM failure rate will be the probability that a critical filler particle lands on the critical location. Obviously, one of the key parameters to determine this would be the filler particle size distribution in the molding compound, which in turn would determine the number of particles exceeding the critical size.

If the filler particles were perfectly spherical, it would have been feasible to determine this quantitatively from cross-sections [15] or directly from the supplier. However, we know that the filler particles in the used conventional molding compound are irregular. A typical cross-section is shown in Figure 9 (a). Figure 9(b) shows a cross-section of the same scale of the low thermal expansion compound described earlier.

(a)

(b)

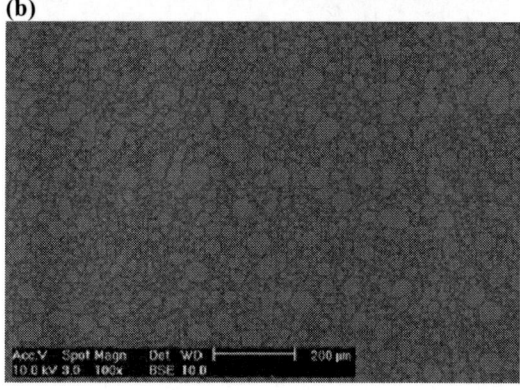

Figure 9: (a) Typical SEM cross-section (using a Back-scattering detector for better contrast) of the conventional molding compound containing ireegularly shaped filler particles (b) Corresponidng cross-sesction of the low thermal expansion molding compound showing almost 100% spherical fillers.

This would mean that it would be very difficult to exactly quantify the particle size distribution based on a characteristic dimension (e.g. a diameter). At the most we can assume that if we take a sufficiently large number of cross-sections, (or cross-sectional area), and that the chance of crossing the largest dimension is highest, we could use cross-sectional areas as a good way to get a rough estimate of particle size distribution. Cross-sections as shown in Figure 9(a) were analysed with Image analysis software (Image Pro Plus), to obtain

the corresponding particle size distribution (scanned and analysed area: 1 mm², averaged from at least two cross-sections). A cut off point of ~ 2000 µm² was used to focus on the larger particles, as also to reduce the computation time. The results are shown in Figure 10: in line with expectations, the fraction or density of particles of larger size decreases strongly.

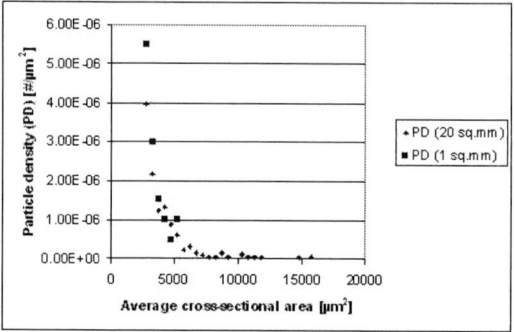

Figure 10: Particle size distribution in terms of particle density for a 1 mm2 area (Figure 9) and a 20 mm² area (not shown).

Valid questions would be: how much area does one need to measure to get a representative image so as not to miss the large particles, and how many similar areas would need to be measured to get sufficient statistics for this distribution? One must also realize that making and analyzing these cross-sections is extremely tedious (especially above a certain image size, e.g. >1-2 mm²) since the images need to be stitched together perfectly to enable accurate image analysis.

A similar analysis was done on a much larger area (~ 20 mm²), which is included in Figure 10. Now we can observe that a number of larger particles show up that were not seen in the first cross-section. Based on our experiences with this molding compound, the results from several cross-sections, as well as the size of particles usually found at the failing locations, we believe that the trend corresponding to the 20 mm² area is quite representative, keeping in mind the care we have to take when dealing with irregular particles.

One way of approaching the probability problem would be by using the Poisson's distribution, or linking this probability (P) to what is known as the defect density (i.e. density of defects per unit area), an approach often used in yield engineering in the semiconductor industry [16].

$$P = M(1 - e^{-D_0 A}) \ldots\ldots\ldots(1)$$

where M: unity, D_0: defect density (#/µm²), A_c: critical area (µm²). When $D_0 A_c \ll 1$, the above equation reduces to the simple form:

$$P = D_0 A_c \ldots\ldots\ldots\ldots\ldots(2)$$

978-1-4244-4722-0/09 $25.00
© 2009 IMAPS-ITALY

In order to derive the defect density (hence in our case the average density of critical particles per unit area) that can be trapped by the wireloop and, very likely, cause failures on the most critical locations on the die, we need to define the critical area (A_c). Since we now understand the modulating effect of the wireloop at bondpad three, we can define this area as the area covered by the molding compound as it fills the mold, but within reach of the wireloop at bondpad three. Note that by doing this we make the disputable assumption that all critical particles that pass the wireloop will be trapped exactly at the surface of the critical location. We will come to this point later. From Figure 7(a), this area can be derived as the product of the total flow length downstream of the wireloop at bondpad three, at positions 3 and 4, and a representative width of a relatively large particle, giving a value of ~ 2×10^6 μm^2.

Based on the in-line results, we know that the in-line PPM level is ~30, 70% of which are generated at positions 3 and 4. Even if we assume that all the failures at these positions are caused by the trapping effect of the loop, we would end up with about 20 PPM. Since the chance of failure due to presence of the particle is assumed to be somewhere between 10 & 90%, we end up with a range between 200 and 22 PPM, respectively (probability that a particle is present at the location). From Equation (2) we can now calculate the theoretical defect density based on filler particles needed to achieve this PPM range, which is found to be between 1×10^{-10} and 1.15×10^{-11}, respectively. To compare this with the defect density that can be estimated from the cross-section, we calculate the defect density for all particles with a cross-sectional area larger than about 10000 μm^2 from the data in Figure 10. This value is found to be ~2.5×10^{-7} particles/um^2, which is a factor of between 2500 and 22000 higher! Most assumptions made earlier would not be able to explain such a large difference.

This would mean that the disputable assumption made earlier (that all large particles passing the bondwire will be trapped, and end up at the surface of the critical location) is incorrect. The only thing that the wireloop does is that it increases the probability that a large particle is trapped at the critical location, and obviously, only a fraction of the particles actually end up exactly on the critical location (based on the ratios calculated above the chance would somewhere between 50 and 400 PPM, obtained simply by dividing 1×10^6 by the obtained chance).

Towards a solution

Based on these learnings, the first step to deal with the issue would be to devise effective measures to contain the number of failures in the field, from an assembly point of view. The facts are as follows:

- we have a very sensitive location on the die (in terms of thermo-mechanical stress)

- The most sensitive locations are present in the vicinity of the bondwire/wireloop at Bondpad 3, which increases the chance that a particle is trapped at the critical location
- Leadframe positions 3 (especially) and 4 show the highest failure rate, accounting in total for ~75% of the failure

Based on these facts, one obvious form of containment would be not to deliver products from at least leadframe position 3, to the customer. However, this would result in a large loss in production capacity (25%), as well as a large waste of material. Hence a better method would be to manipulate the wireloop at bondpad 3 in some way to reduce the chance that the particle is trapped by it. To deal with this, it was proposed to increase the wireloop height at bondpad three, from the nominal height (viz 180 μm) to 220μm, in such a way that the highest value is still within the upper control limit, viz, 250μm. This would effectively reduce the chance that large particles are trapped by the wire. A trial run on several million devices showed that the in-line PPM dropped from about 30PPM to ~ 12 PPM (hence effectiveness of this containment was shown to be about 70%), based purely by reducing the chance that a particle is trapped by the wireloop. This large difference (by reducing purely the chance that a large particle gets trapped by the wireloop) also supports the assumption that when a relatively large particle is in contact with the critical location, the chance of failure is relatively high.

For a long term containment, many options are thinkable, which we will come to subsequently, however, in this work we will only report on the corrective action that can be introduced via assembly, viz. the use of a low thermal expansion molding compound. As can be seen in Figure 4, the low thermal expansion compound exerts significantly lower stress on the nitride layer. Besides, these compounds have spherical particles of better size control, hence reducing the chance that a large, sharp particle will be in contact with the critical location.

On the other hand the spherical shape (rolling motion during flow) would also decrease the chance that a particle is trapped by the bondwire, at the critical location. Running about 3 million devices with the low thermal expansion compound after qualification, has shown that the in-line failure rate has reduced to 5 PPM (so an overall reduction of ~80%). So far only one suspected case of filler attack has been confirmed. Obviously, to reduce the PPM level further, the recommended way forward would be to combine the low thermal expansion molding compound with changes in diffusion, aimed to make the backend stack more robust (e.g. strength of the nitride layer, plasticity of the metal layer, geometrical effects like the thickness of the last metal layer, the nitride thickness, or if feasible the

978-1-4244-4722-0/09 $25.00
© 2009 IMAPS-ITALY

use of wafercoat [8]. We have shown that an interaction between assembly and the chip is responsible for this PPM level failure, hence it obvious that to reduce to failure rate to as low a value as possible, changes will need to be made at both ends, assembly as well as diffusion.

Concluding remarks

The failure mode described in this paper, called inter-metal shorts, is induced by a mechanical interaction between the die (intrinsically weak structure due to layout) and the assembly process, which controls the presence of a relatively large filler particle from the molding compound at the most critical location. This is the main contributor to the low PPM level failure rate, which is unacceptable for an automotive customer. An effective containment was devised based on reducing the trapping effect of the wireloop at bondpad 3. As a corrective action, a low thermal expansion molding compound is implemented. However, a very low PPM level of failures still seems to exist (~ 5 ppm in-line compared to ~ 30 ppm with conventional molding compound).

To reduce the failure rate further towards the zero PPM level, we propose to address changes in the layout (die design to relocate the sensitive locations), the diffusion process itself (e.g. nitride strength, plasticity of the metal layer, geometrical effects like the thickness of the last metal layer, the top nitride thickness) or if feasible the use of a poly-imide-based wafercoat as the final layer applied in the waferfab, that acts as a stress buffer not only for filler attack, but for other mechanical failure modes like scratches that can be induced anywhere between the diffusion process and the molding process.

Acknowledgements

The authors are extremely grateful to other members of the Inter-metal shorts Technical Team (Manfred van Eckendonk, Kees van Hasselt, Kees Vreeburg, Jacco Scheer, Adrie Buijsman and Piet van Kessel). Other people that deserve special mention are:
Jan van Kempen for proposing the option for higher wirelooping, Leon van Nimwegen for the particle size distribution analysis, Joop Verwijst for useful discussions regarding the defect density model, Will Balemans for the flow modelling and finally the assembly plant in Bangkok (especially Taweesak Choorat & C.M. Su for their support during the evaluation and implemention of the containment actions).

References

[1] J. Bisschop, "Failure mechanisms in plastic package IC's", Proceedings EuroSimE Conference, pp. 328-331, 2002.

[2] G.Q. Zhang, W.D. van Driel, X.J. Fan (editors), "Mechanics of microelectronics", Springer Dordrecht, The Netherlands, 2006.

[3] P. Jacob, W. Rothkirch, "Unusual defects, generated by wafer sawing: Diagnosis, mechanisms and how to distinguish from related failures", Journal of Microelectronics & Reliability 48 (8-9), pp. 1253-1257, 2008.

[4] C. Llamas, "IN/INS shorting due to surface ESD", Proceedings Electronics Packaging Technology Conference, pp. 684 – 687, 2006.

[5] P. Jacob, "Surface ESD (ESDFOS) in assembly fab machineries as a functional and reliability risk – Failure analysis, tool diagnosis and on-site-remedies", Journal of Microelectronics & Reliability 48 (8-9), pp. 1608-1612, 2008.

[6] P. Yalamanchili, V. Baltazar, "Filler Induced Metal Crush Failure Mechanism in Plastic Encapsulated Devices", 37th Annual International Reliability Physics Symposium, San Diego, California, pp. 341-346, 1999.

[7] I. De Wolf, F. Duflos, B. Vandevelde, P. Vercruysse and D. Vanderstraeten, "Impact Induced Metal-Crush Failures", Proceedings of 14th IPFA, Bangalore, India, pp. 209-213, 2007.

[8] C. Scheider & S. Chilton, "Filler Particle Induced Data Retention Failures in Plastic Encapsulated EPROM's", ISTFA '92, pp391-395, 1992.

[9] F. Kuper, "Automotive IC reliability: Elements of the battle towards zero defects", Journal of Microelectronics & Reliability 48 (8-9), pp. 1459-1463, 2008.

[10] W.D. van Driel, R.A.B. Engelen, A. Mavinkurve, H. Cobussen, M. van Dort, M. van Eckendonk and L. Endrinal, "Proceedings of EuroSimE Conference", 2009.

[11] L. Endrinal & E. Coyne, submitted to the IRPS2009.

[12] D. Yang, "Cure-Dependent Viscoelastic Behaviour of Electronics Packaging Polymers: Modelling, Characterisation, Implementation and Applications", PhD thesis, Delft Univeristy of Technology, 2007.

[13] (a) "INJECT: Thermosets User Manual MKJOB and Interpretation of Results", Version 2.2.0, Philips Applied Technologies, Apr 1995 (b) J. van der Werf & A. Boshouwers, "INJECT-3, a simulation code for the filling stage of the injection moulding process of thermoplastics", Technical Univeristy of Eindhoven, May 1988.

[14] Brian Lawn, "Structure of Brittle Solids", Cambridge Solid State Science Series, 2nd Edition, 1994.

[15] J. Serra, Image "Analysis & Mathematucal Morphology", 1982.

[16] See for example:
http://www.icyield.com/index.html.

Small Size LTCC FlipChip-Package for RF-Power Applications

Jens Müller[1], Markus Norén[2], Matthias Mach[1], Sebastian Brunner[2], Christian Hoffmann[2]

[1]TU Ilmenau, ZIK MacroNano, Institute for Micro- and Nanotechnologies, D-98693 Ilmenau, Germany,
Gustav-Kirchhoff-Str. 7

Tel. +49(0)3677 69 2606 e-mail: jens.mueller@tu-ilmenau.de

[2]EPCOS OHG Deutschlandsberg, Siemensstrasse 43, A-8530 Deutschlandsberg, Austria, P.O.Box 90

Abstract

Heat dissipation is becoming a key issue in designing packages due to growing power densities associated with IC downscaling and the introduction of new semiconductor technologies (e.g. AlGaN). Additionally, FlipChip assembly is being favoured due to small form factor and reduced interconnect parasitics which can be essential for RF and microwave applications. Compared to chip and wire technology the FlipChip approach suffers from smaller thermal interfaces since mainly the bumps contribute to the heat conduction. Minor improvements can be achieved by underfill or overmold materials.

Concepts which use the entire die surface for heat conduction are based on cavity down FlipChip packages. The direct heat path from die to the circuit board is established during package mounting (e.g. soldering or gluing). However, mechanical tolerances of cavity depth, die thickness and bump height are resulting in a gap between the die surface and the board. This gap needs to be filled by a thermal interface material which degrades the thermal performance.

The new wafer level compatible concept uses a back grinding step after FlipChip mounting and underfilling. The underfill seals the cavity and provides a smooth surface for the metal interconnect layer. Signal and thermal pads are achieved by sputtering and galvanic processes. This thermally enhanced package allows a six times higher power dissipation compared to a conventional LTCC Package with top side FlipChip assembly and thermal vias.

The paper describes the concept and the sample preparation of alternative FlipChip packages. Simulation and measurement results are compared.

Key words: LTCC, Power-Package, Flip Chip, thermal optimisation

Introduction

Mobile wireless handheld devices require small electronic components and packages. Standard system functions like W-LAN or Bluetooth interfaces and front end modules should be as small as possible to allow implementation of additional functions (e.g. further wireless standards). Typical strategies for size and volume reduction are so called System-in-Packages (SiP) which support the combination of different semiconductor technologies in one package. Low temperature cofired ceramics (LTCC), often used as the package substrate for RF-functions, further allows the integration of passives within the substrate. The bare die mounting technology plays a decisive role for package size optimization. FlipChip assembly is known for its efficient form factor and the excellent high frequency behaviour of the interconnects.

However, the power capability is rather limited for FlipChip ICs due to the relatively poor thermal path in comparison to wire bond solutions. In general, only the bumps contribute to the heat flow from the die to the substrate. Typical underfills, applied to improve first level reliability, show only a minor influence on the thermal management [1]. The substrate itself can be thermally improved with heat spreaders (metal planes) and thermal vias. Ground connections can be both used as electrical and thermal vias. The remaining signals are routed differently and contribute less to the thermal path (Fig. 1).

The previous work [1] which was focused on thermal design optimization being compatible to an existing high volume LTCC manufacturing line for zero shrink substrates has demonstrated the impact of the substrate thickness but has also shown the limitations given by the design constraints.

978-1-4244-4722-0/09 $25.00
© 2009 IMAPS-ITALY

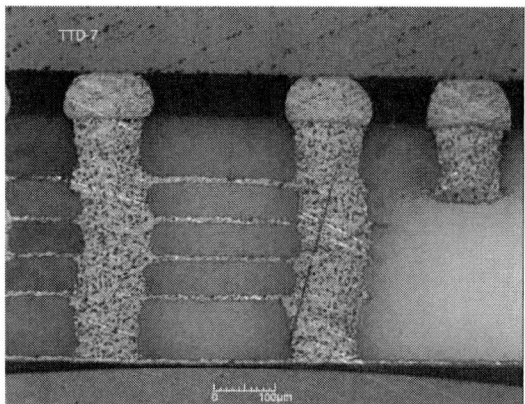

Figure 1: Cross section of FlipChip interconnects and thermal vias

Alternative FlipChip Concepts

In order to include the FlipChip back side into the thermal path four design solutions were considered. Fig. 2 shows the "standard" FlipChip with a 200µm thick copper cover soldered to it and to the substrate. The relatively thick metal provides ideal thermal conditions. The metal ring on the substrate has thermal vias underneath for optimum thermal conductivity (not visible in Fig. 2). In addition, the lid serves also as an electrical shield. On the other hand the cover increases the package thickness and requires an additional assembly step with difficult solder deposition. The tolerances of bump height, IC thickness and solder depot might become a critical issue in series production in terms of repeatability and the mismatch of thermal expansion coefficients can reduce reliability.

Figure 2: Cross section of a FlipChip with a copper lid as heat spreader

Alternatively, the copper cover can be made by thin film deposition and galvanic processes. Fig. 3 depicts the cross section of the so called FlipChip CSSP derived from the hermetic package solution for SAW devices [2]. The underfill fillet serves as deposition base. The entire process is compatible to wafer level production and addresses the tolerance limits of the previous solution.

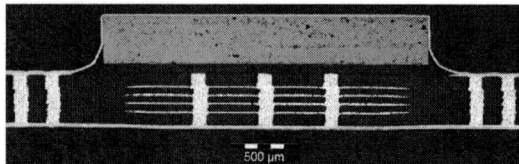

Figure 3: Cross section of a "CSSP"-Flip Chip

Free sintering of LTCC suffers from higher lateral substrate tolerances but offers various structural design options like cavities, windows or channels. A cavity can be applied to embed the chip. Fig. 4 shows a drawing of a cavity-up FlipChip package with a heat spreader. Similar to the approach with the copper cover the heat is spread to the thermal vias at the edges. If convection is possible (typically not available in hand held devices) the heat spreader serves also as cooling plate. Again, the design needs to be compensated for tolerances of die thickness, bump height and cavity depth by means of a high die-to-lid solder thickness.

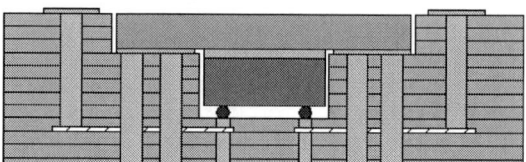

Figure 4: Schematic view of the cavity up FlipChip test package

The flipped version of the previous design was introduced by Heyer et al. [3]. In this design, the heat spreader is already on the pcb and the die rear side is glued or soldered to it. However, the repeatability of thermal performance is still suffering from the mechanical tolerances mentioned above.

An improved design of a cavity down FlipChip package which overcomes these limitations is disclosed under [4]. First, the FlipChips are assembled in the cavities of the substrate. After filling up the cavities with an appropriate filler material (underfill or glob top) the entire substrate is ground and polished to achieve a flat surface. Adhesion (NiCr) and plating base (Cu) layers are then deposited by sputtering. Finally, the Cu metal thickness is increased by electro deposition. Pads and other electrical structures can be obtained by implementing photolithography, etching or other mask processes. Fig. 5 displays a cross section of such a package. Careful grinding leads to very thin dies and therefore a very short heat transition. This option is particularly interesting for semiconductor technologies with poor thermal conductivity (e.g. GaAs).

Figure 5: Cavity down FlipChip package with deposited metal pad

978-1-4244-4722-0/09 $25.00
© 2009 IMAPS-ITALY

Simulation and Experimental

LTCC substrate preparation was done on two sites. The parts without cavity (Fig. 1 – 3) were manufactured by EPCOS in the material system MKE 100 with silver metallisation und subsequent electroless plating of NiPdAu. Cavity substrates were prepared by Ilmenau University of Technology with the tape system 951 from Du Pont with silver metallisation. These substrate were also electroless plated by EPCOS.

2.54 mm x 2.54 mm thermal test dice [5] with SnAgCu-bumps were mounted onto the substrates for thermal evaluation. Postprocesses (like underfilling and lid soldering) were carried out according to the description in the previous section. The test die applied comprises a heater and a chain of five diodes which are used to monitor the surface temperature (Fig. 6). The diode characteristic was measured in a temperature chamber and was approximated by equation (1).

$$\Theta_{jct} = \frac{V_D - 0.70085V}{-0.00208 V/_{\circ C}} \qquad (1)$$

Θ_{jct} - junction temperature [°C]

V_D - diode voltage [V]

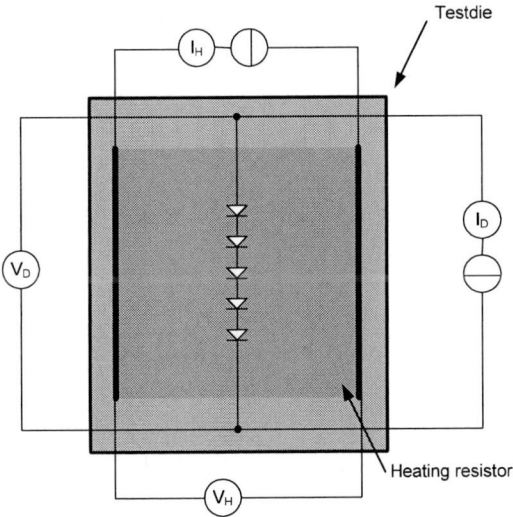

Figure 6: Schematic of the thermal test die

Each FlipChip Package was soldered to a flat 2 mm thick brass plate. This was necessary to minimise the thermal interconnect resistance betweeb package and the large thermal ground (aluminium heat sink) of the test fixture. Earlier investigation without the brass plate showed very strong measurement deviations which were attributed to punctional pad to heat sink contacts due to pad waviness. The test fixture for thermal evaluation is depicted in Fig. 7. All test specimen are pressed with a defined pressure of 50kPa onto the heat sink. Electrical contacts on the top side of the

packages are realised by pogo pins. The entire fixture was placed in a climatic chamber to assure a constant ambient temperature of 25°C (±0.4°C). The chip temperature was calculated based on the measured diode voltage under controlled heater power of one Watt and the thermal resistance was derived according to equ. (2).

$$R_{th} = \frac{\Theta_{jct} - 25°C}{1W} \qquad (2)$$

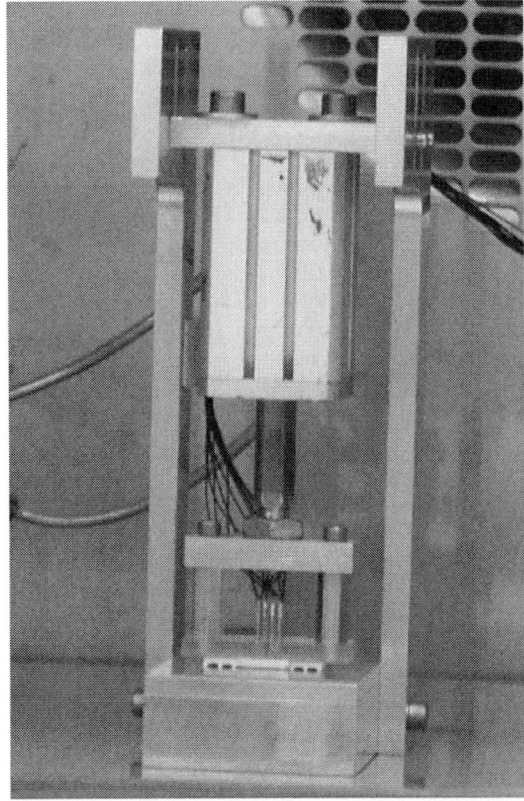

Figure 7: Test fixture for thermal evaluation

Simulations of FlipChip packages were made by finite element simulation with ANSYS workbench. Fig. 8 shows the half package model of a conventional FlipChip on a test substrate with fine meshing and common nodes.

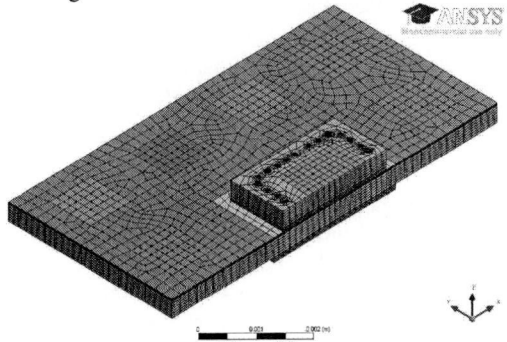

Figure 8: ANSYS meshing of the FlipChip package

978-1-4244-4722-0/09 $25.00
© 2009 IMAPS-ITALY

The material parameters which were used in the simulation are listed in Table 1.

Table 1: Thermal properties of package materials

Material	Thermal Cond. λ [W/m K]	Specific Heat Capacity c_p [J/g K]
LTCC MKE 100	3.45	0.75 @ 25°C – 0.85 @ 125°C
Silver Paste	200 – 250	0.27
Underfill	0.3	1.1
SAC-solder	50	0.23
Silicon	148	0.71

Results and Discussions

Among the different designs a large variation of results (Table 2) are observed. Simulation and measurement are in good agreement. The measurement results of the cavity up version are not included in the evaluation due to significant deviations within this test group. A statistical evaluation is not possible due to the small number of test parts in this group. X-ray analysis however, showed that some parts have large voids in the solder which might be related to the tolerance problem mentioned earlier (Fig. 9).

As expected, the cavity down FlipChip package with the shortest heat path reveals the best thermal performance. Differences between simulation and measurement might be attributed to the underestimated solder to brass plate interface resistance in the simulation.

Table 2: Package thermal resistance simulation and measurement results

FC-Package	Simulated R_{th} [K/W]	Measured R_{th} [K/W]
Standard (thermal vias, heat spreader, thickn.=8μm)	18	20
Standard + CSSP	13	15
Standard + soldered Cu-Cover	11	12
Cavity up + heat spreader	5	-
Cavity down	1	2-3

The optimised standard FlipChip package (300 μm LTCC thickness) with a thermal resistance of about 18 K/W is capable of handling approximately 2.2 W power dissipation if the circuit is working under the maximum specified ambient temperature of 85°C (junction temperature ≤ 125°C). The cavity down FlipChip package with approximately 2 K/W thermal resistance can handle

20 W power under the same conditions. These figures demonstrate the capability of the new FlipChip package. Due to the construction of the latter type, the thermal performance should be equal to or even better than a wire bond solution.

Furthermore, the chip is hermetically sealed which offers new potential applications in fields where hermeticity is required.

Fig. 9: X-ray picture of a cavity-up substrate with soldered heat sink

Acknowledgements

The authors would like to thank the German Government for the financial support. Part of the work was carried out under the project MultiSysTeM which is funded by the BMBF under the initiative "Centres for Innovation Competence".

[1] M. Norén, C. Hoffmann, W. Salz, K. Aichholzer:"Aspects on Advanced Thermal Management and Reliability for Flip Chip on LTCC", Proc. 4th International Conference and Exhibition on Ceramic Interconnect and Ceramic Microsystems Technologies, April 21-24, 2008, Munich/Germany.

[2] Gregor Feiertag, Hans Krüger, Christian Bauer, "Surface Acoustic Wave component packaging", Proceedings of the 16th EMPC, June 17-20, 2007, Oulu/Fin.

[3] Johann Heyen, Arne F. Jacob: "A Novel Package Approach for Multichip Modules based on Anisotropic Conductive Adhesives", Proc. 35th European Microwave Conference, pp. 1499-1502, October 3-7, 2005, Paris.

[4] Holger Fluhr, Christian Block, Christian Hoffmann: "Transmitter Module with Improved Heat Dissipatiion" US 2007/0108584 A1, May 17, 2007.

[5] Datasheet Delphi PST1-02 / 5PU Thermal Test Die

X-ray nanoCT of interconnections in IC packages: Visualizing of internal 3D-Structures with Submicrometer Resolution

J. Luebbehuesen, H. Roth, T. Neubrand and O. Brunke

GE Sensing & Inspection Technologies GmbH, phoenix|x-ray, Niels-Bohr-Str. 7, 31515 Wunstorf, Germany
Phone: +49 5031/172-111, Fax: +49 5031/172-299, Jens.Lubbehusen@ge.com

Abstract

Miniaturised and concealed electrical interconnections in IC packages including copper and gold wire bonds, flip chips and microvias were inspected by nanofocus X-ray tube technology and nanofocus computed tomography. The 2D and 3D X-ray imaging technology is concisely described and various analysis results and defect examples at sub-micrometer resolution are presented.

Key words: nanofocus X-ray inspection, nanoCT, high resolution Computed Tomography, 3D micro-analysis

Introduction

In backend qualification and spot checks of IC packaging, X-ray systems are customarily used for the inspection of classical features such bond wires, die bonds, moulding and seals. New challenges to X-ray inspection equipment arise from three trends in package technolgy: miniaturisation, the use of novel materials and increasing device complexity. Package miniaturisation leads to higher density and smaller size of internal structures such as microvias and FlipChip interconnections, demanding for resolution in the sub-micron range at highest magnifications as provided by novel nanofocus tube technology. The absorption of some materials like non-conductive die adhesives or copper bond wires either is to low to yield sufficient contrast in customary X-ray images or is strong enough to conceal other package features, as observed for copper-tungsten alloy caps. Highly sensitive a-Si 16-bit digital flat panel detectors and image processing techniques now have remarkably improved the detectability of such low contrast features.

The complexity of devices with 3D set-ups such as stacked or multiple die including corresponding wire connections leads to confusing overlaps in the two-dimensional X-ray images, in other words, they must be inspected slice by slice or in 3D visualisations as provided by computed tomography. However, up to now, many laboratory CT systems are using X-ray tubes of some microns focal spot size and maximum tube voltages around 100 kV. Since electronic devices contain very fine structures and strongly absorbing materials like gold or copper this results in unsatisfying image resolution and strong image artefacts, respectively. In view of this situation, a nanoCT system was designed for highest resolution CT of electronic components with an 180 kV nanofocus tube providing excellent penetration at sub-micron detail detectability. Adjacently some typical failure and quality analysis results are presented.

nanofocus X-ray - Method and Results

The following 2D investigations were performed by means of a automatic X-ray shadow microscope with 2 MPixel digital image chain for magnifications and a 180 kV nanofocus X-ray tube designed for up to 24.000x magnification and a detail detectability of 200-300 nm (phoenix|x-ray nanome|x 180 of GE Sensing & Inspection Technologies). The system can provide highly magnified images also under viewing angles up to 70° (ovhm), by tilting the detector.

One of the most challenging tasks in 2D X-ray inspection is the proper imaging of the delicate dendrites as created by electromigration: sub-micron spatial resolution at high magnification is required as well as outstanding contrast resolution. Fig. 1 sharply displays dendrites growing out of a copper conductor and causing a short circuit. The measured dendrite width is 10 microns. Another frequent inspection task is the detection of cracks and lifted up wire bonds. If the resulting gap is not only to be found but also measured and clearly displayed for further investigation of its cause, the image resolution must be much smaller than its size.

978-1-4244-4722-0/09 $25.00
© 2009 IMAPS-ITALY

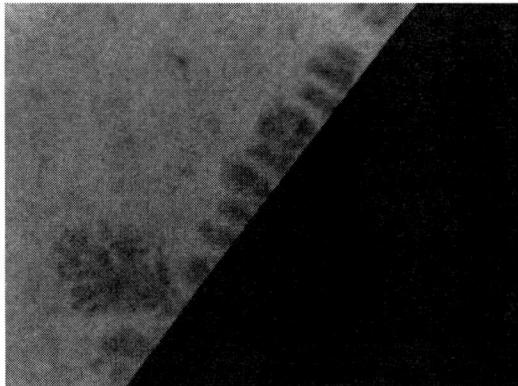

Figure 1: nanofocus X-ray image of dendrites growing out of a conductor. The measured dendrite diameter is 10 microns.

In Fig. 2a a ball bond is shown by nanofocus X-ray technology to be lifted by about 2 microns. Note that at slightly less resolution the edges of ball bond and bond pad could not be differentiated any more and the gap would be not visible. Due to nanofocus resolution the ripped of wedge bond displayed in Fig. 4b is not only detected, but also proved to be cracked a at the rim of the weld area, since the edge of the bond capillaries footprint is still visible. Note that the images in Fig. 2 a and b require precise adjustments of the viewing angle with out loss of magnification (ovhm).

Figure 2 a) nanofocus X-ray image of a lifted ball bond at highest magnification.

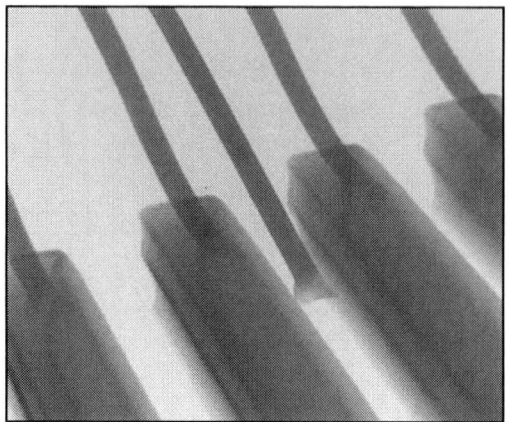

Figure 2 b) Lifted wedge bond at 55° ovhm.

In IC processor packages the internal bonds are flip-chip solder joints which show features and defects similar to area array solder joints as known from electronic assemblies. However, the flip chip interconnection may be more 10 times smaller (25 to 100 μm) and require for an appropriate image resolution to detect defects like voids, opens and shape deviations in the micrometer range, see fig 3.

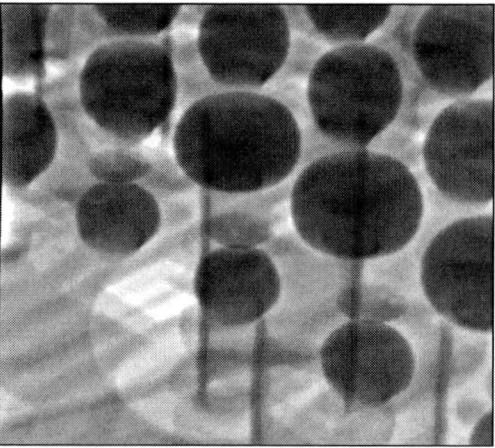

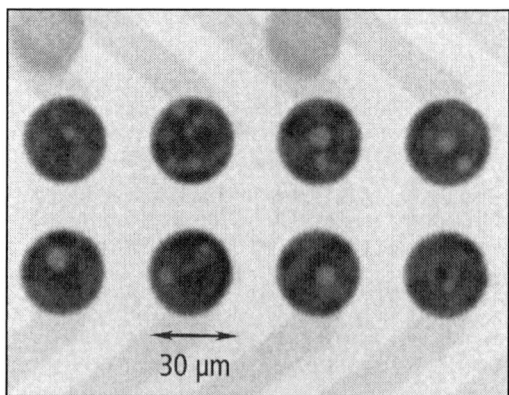

Figure 3: nanofocus X-ray images of flip chip solder joints: a) Open solder joints, b) micrometer sized voids.

Copper bond wires (when compared to the highly absorbing gold wire as shown above) yield a poor X-ray image contrast which is enhanced not only by using digital imaging but also by high magnification and sharp imaging, cf. fig. 5.

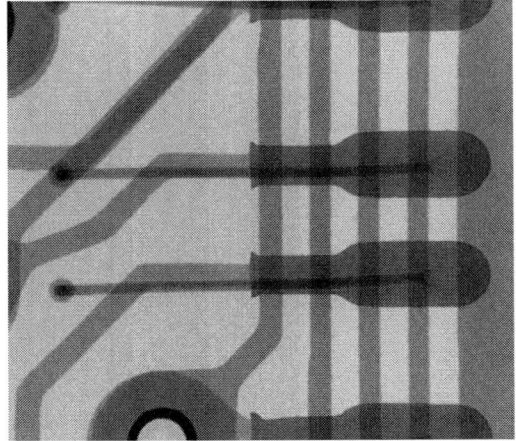

Figure 5: nanofocus X-ray image of 0.7 mil copper bond wires in an electronic device.

nanoCT® - Method and Results

The nanoCT® scans were carried out with a recently improved very compact laboratory CT system (phoenix|x-ray nanotom 180 of GE Sensing & Inspection Technologies) dedicated to the analysis of small samples at submicron voxel-resolution. The system comprises a 180 kV nanofocus tube with a focal spot size below 1 micron and a 5-megapixel flat panel detector with an active area of 120 x 120 mm (2300 x 2300 pixels, 50 μm pixel size) and a 3-position virtual detector (up to 360 mm detector width), providing a voxel resolution of 0.5 microns (500 nm) and a detail detectability of 200-300 nm. The influence of vibrations as well as thermal expansion and drift are minimised by a granite based manipulation system, vibration damping, low expansion materials and temperature stabilisation.

While 2D images of a memory cube with stacked wires and stacked dies (fig. 6 a) are not suitable for analysis due to overlaying features, 3D nanoCT virtual slices or sections allow to examine each individual die-attach for voids (fig. 6) as well as the flow of the wire bonds. In stacked die and similar devices the wires are arranged in layers so that the wires overlap in the X-ray images and short circuits cannot be clearly told from crossings. In a 3D visualisation as provided by nanoCT the spatial arrangement of the wires may be examined slice by slice and their distances may be measured at any position. Further examples for advanced failure analysis with nanoCT are defects in concealed conductors in package redistribution planes, bond wires or the spatial localisation of package voids or defects in chip components, see fig. 7.

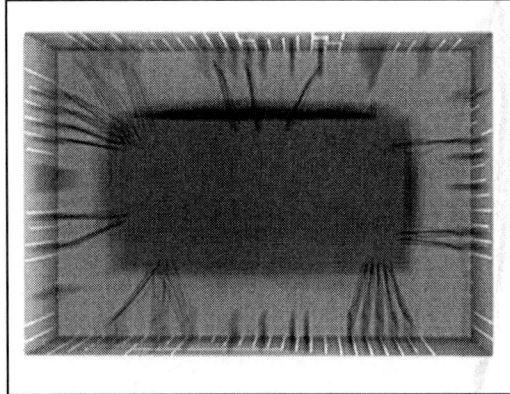

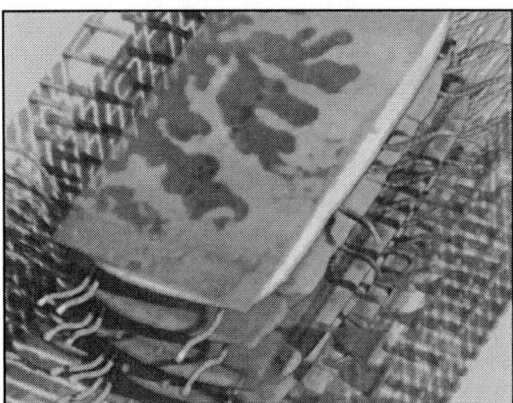

Figure 6 a: Frontal 2D X-ray image of a memory cube with stacked dies and b) tomographic section visualising the die attach (by courtesy of 3D-Plus). Size of the sample is about 15 mm x 10 mm x 10 mm.

Figure 7: Tomographic section visualisation of a ceramic IC package, showing vias and wire loop.

978-1-4244-4722-0/09 $25.00
© 2009 IMAPS-ITALY

Conclusion

Digital nanofocus X-ray inspections systems with the capability of oblique views at highest magnification are a fast and most effective tool for the analysis of electronic packages and detection of defects down to the sub-micron range. nanoCT widely expands the spectrum of detectable micro-structures in complex electronic devices and packages by 3D visualisation and slice by slice analysis. This opens a new dimension of 3D-microanalysis and will partially replace destructive methods – saving costs and time per sample inspected. nanofocus tube technology pushes computed tomography systems into application fields that very recently were exclusive to expensive synchrotron techniques.

Cu Wire Bonding: Reliability Improvement for High Temperature in Plastic Packages

C. Passagrilli, B. Vitali, R. Tiziani and C. Azzopardi

ST Microelectronics
Agrate Brianza (Italy) and Kirkop (Malta)

Abstract

The request of the electronic market to increase the working temperature of power integrated circuits up to 180-200°C has a big impact on the reliability of standard plastic packages. The joint between bonding ball and pad is mainly affected in conditions of high temperature and high current. Various solutions were tried in order to reach the target with standard interconnection solutions. If Au or Cu wires on Al pad are used, intermetallics growth takes place leading to high increase of electrical resistance and possible ball lift. The insertion of an Under Bump Metallization between wire and pad is known to be a solution for Au wire. In this paper this solution is investigated using Cu wire instead of Au wire. Cu is less expensive than Au and its use reduces the impact of the cost of the UBM.

Introduction

In the automotive electronics market the request to increase the working temperature of power ICs up to 180°C and 200°C (several thousands hours with bias) is growing, because the evolution of motor techniques involves the assembly of the components in direct contact to the object to control (engine, transmission, gear…).

Therefore the need exists to increase the maximum operating temperature of the automotive components. In particular, the electronic components in plastic packages for automotive are guaranteed for 150°C as maximum operating temperature. The new requests are 180°C and 200°C (ambient temperature) with bias.

This request is critical for plastic packages when Au wire bonding is used, because storage at high temperatures causes known phenomena of Au bonding degradation: high increase of electrical resistance due to intermetallics and possible ball lift.

Using Cu wire instead of Au wire, the intermetallics growth is much reduced. But the limitation of Cu wire at high temperature is due to the high brittleness of Cu/Al intermetallics, sometimes leading to earlier failures than Au wire [1].

The deposition of Under Bump Metallization (UBM) on Al pad as barrier between wire and pad is known to be a robust solution for high temperature applications with Au wire [2] because this barrier prevents the interdiffusion between Au and Al and the formation of intermetallics with consequent high increase of electrical resistance and possible formation of voids leading to ball lift. The problem

of the formation of voids with ball lift is more evident when the ball diameter is reduced (less than 60 μm) as in devices with fine pitch.

Since the introduction of the Under Bump Metallization on pad has an impact on the cost per wafer, the use of Cu wire instead of Au wire is desirable to compensate the increased cost per wafer by a decreased cost of assy. This work shows a reliability characterization of 2 mils Cu wire on pad with UBM at 200°C with high current, and reliability tests at 180°C using 50 μm ball diameter for fine pitch applications

Reliability Characterization of 2 Mils Cu Wire on Pad with UBM at High Temperature and High Current

This test is done with the same method and apparatus used for other evaluations of 2 mils Au and Cu wire at high temperature and high current [1],[2],[3]. A current is forced into the wires of a daisy chain and the electrical resistance of the chain is measured versus time, fixed the ambient temperature (fig.1). The device is assembled in a plastic power package (FW-25).

A "green" (Br-free) molding compound is used in order to avoid possible corrosion of copper by bromine. The packages are fixed to a big heat sink in order to minimize the self-heating due to the current flow.

In general, using this test it is possible to compare different materials or, fixed the materials, to compare different conditions of temperature and current.

In this case the comparison has been done between the following materials:

978-1-4244-4722-0/09 $25.00
© 2009 IMAPS-ITALY

- Au wire and Cu wire, both with UBM
- Cu wire with and without UBM

The test conditions are fixed: 200°C with maximum allowable current in steady state, 2.5A.

These conditions of temperature and current are indeed extreme. Molding compounds suppliers recommend to avoid higher temperatures than 200°C because of the rapid resin deterioration. The maximum allowable current in steady state is the worst condition of current stress and also causes further temperature increase. In fact, at least 20°C increase for Joule effect are measured on the wire loops.

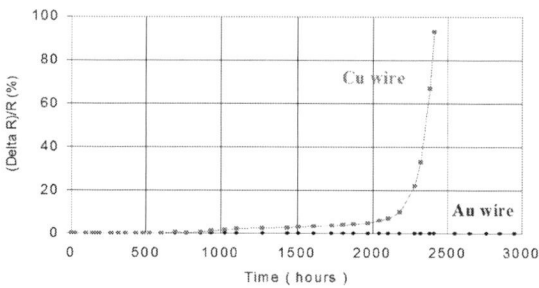

Figure 2. Increase of resistance ΔR/R versus time of daisy chains of Au and Cu wires on pads with UBM. T = 200°C, I = 2.5A, "green" resin.

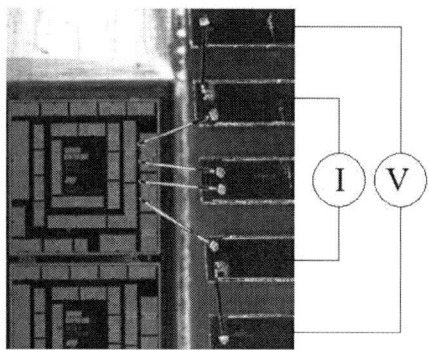

Figure 1: Daisy chain used for the test

In fig.2 the results of this test are reported up to 3000h comparing Cu wire to Au wire. The results with Au wire are a confirmation of known results [2],[3] showing the robustness of the combination Au wire + UBM + green molding compound in high temperature conditions. In fact there is not degradation of the Au daisy chain at least up to 10000h. Cu wire of course does not achieve the same result of Au wire because Cu is less chemically resistant than Au, but the results are good: no change of electrical resistance up to 1000h at 200°C with 2.5A steady state, 5% increase at 2000h, then rapid increase and open circuit between 2000h and 2500h. The origin of the increase of resistance is probably a chemical interaction at high temperature between Cu wire and molding compound leading to deterioration of wires up to macroscopic damages (fig.3), while there is not a visible degradation in the Cu/UBM joint (fig.4).

In fig.5 the results of this test are reported comparing Cu wire with UBM and without UBM. The absence of UBM allows the Cu/Al intermetallics growth. The failure in this case is due to the formation of cracks inside the brittle intermetallics [1] (fig.6), leading to sudden increase of resistance, much earlier than in the case with UBM.

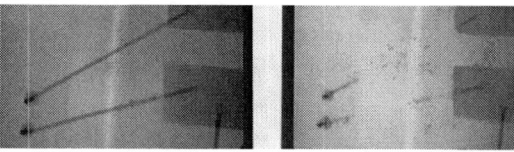

Figure 3. X-ray images of damaged Cu wires of daisy chain after 2400h at 200°C, I = 2.5A

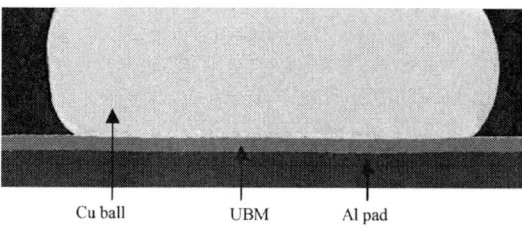

Cu ball   UBM   Al pad

Figure 4. No visible degradation of the joint between Cu wire and pad with UBM after 2400h at 200°C

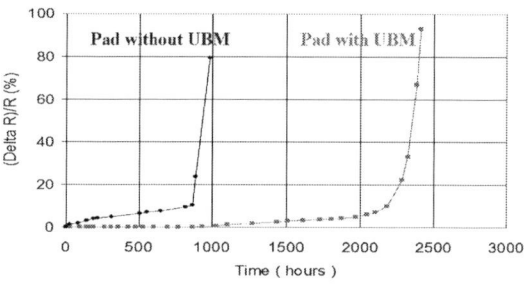

Figure 5. Increase of resistance ΔR/R versus time of daisy chains of Cu wires with and without UBM. T = 200°C, I = 2.5A, "green" resin.

978-1-4244-4722-0/09 $25.00
© 2009 IMAPS-ITALY

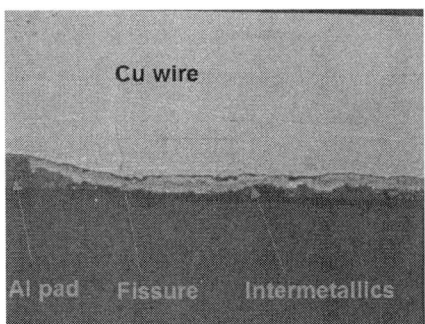

Figure 6. Failure mode of Cu wire on Al pad after 1000h at 200°C

High Temperature Reliability of Cu Wire on UBM Using 50 μm Ball Diameter for Fine Pitch Application

One of the issues which can arise using Au wires with balls diameter lower than 60 μm on Al pad for fine pitch applications is the difficulty to obtain good uniformity of intermetallics (fig.7). Poor uniformity of intermetallics leads to early ball lifts during High Temperature Storage test (HTS). Typically ball lifts start to occur after 100-200h at 180°C or after 1000h at 150°C. Introducing UBM the problem is solved because there is not more intermetallics.

In order to reduce the impact of the cost of UBM, there is the need to evaluate Cu wire. The tests are the standard ball shear and pull test, performed at time zero and after 3500h at 180°C, in LQFP packages with "green" molding compound. Cu balls diameter is 50 μm.

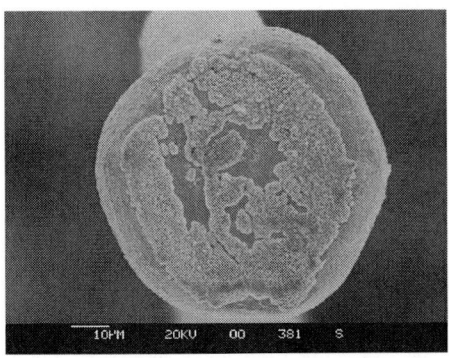

Figure 7. Example of not uniform Au/Al intermetallics, visible under the ball after pad etching

In table 1 the results of wire pull test and ball shear are shown at time zero and after 3500h at 180°C. No failures are found at pull test and there is

not decrease of the ball shear values after high temperature storage. In fig.8 a cross section of a 50 μm Cu ball on UBM after 3500h at 180°C is shown. The joint has remained unchanged, without any degradation.

Table 1. Results of wire pull test and ball shear of 50 micron Cu ball diameter on UBM at time zero and after 3500h at 180°C

	Pull test F.M.			Ball shear		
	Neck	Loop	Ball	Avg	Max	Min
Time zero	100%	0%	0%	46.9 g	58.0 g	41.2 g
3500h 180°C	100%	0%	0%	54.0 g	69.4 g	45.0 g

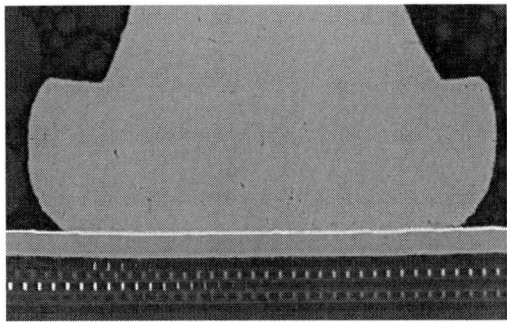

Figure 8. Cross section of a 50 μm Cu ball diameter on UBM after 3500h at 180°C

Conclusions

UBM on Al pad is known to be a robust solution for high temperature applications (180-200°C) of plastic packages using Au wires and "green" molding compounds. The reliability issues in fine pitch applications due to the difficulty to obtain uniform intermetallics are avoided using UBM on pad.

In order to reduce the impact on the cost per wafer of UBM, there is the need to evaluate also Cu wire, which is less expensive than Au wire and reduces the cost of assy. Although Cu is chemically less resistant than Au, the obtained results are good. There is not degradation of ball shear and pull test after 3500h at 180°C with balls for fine pitch. A characterization done at 200°C and high current does not show change of resistance up to 1000h and only 5% change at 2000h.

References

[1] C.Passagrilli, R.Tiziani, L.Renard, B.Vitali, "Reliability of Copper Wire Bonding in Plastic

Packages at High Temperature (200ºC) and High Current", SMTA High Temperature Electronics Workshop, 2006

[2] R.Tiziani, C.Passagrilli, L.Gobbato, "Reliability of Au/Al Bonding in Plastic Packages for High Temperature and High Current Applications", ESREF 13th European Symposium Reliability of Electronic Devices, 2002.

[3] R.Tiziani, C.Passagrilli, L.Gobbato, "Au Wire Bonding Reliability Improvements for High Temperature and High Current Application", SMTA High Temperature Electronics Workshop, 2003

Size and Microstructure Effects on the Stress-Strain Behaviour of Lead-Free Solder Joints

Pradeep Hegde, David C. Whalley and Vadim V. Silberschmidt

Wolfson School of Mechanical and Manufacturing Engineering, Loughborough University,
Loughborough, Leicestershire, LE11 3TU, UK
Email: pradeephegde127@yahoo.com

Abstract

The properties of lead-free solders such as Sn3.8Ag0.7Cu are less understood than traditional Sn-Pb solders, as well as the factors affecting these material properties. Therefore, the present paper focuses on determination of stress-strain properties for both small-scale solder joints and for bulk solder for three different strain rates. The paper also analyses the obtained experimental results, and compares material properties between the solder joint and the bulk solder. The reasons for the differences in the observed material properties are discussed, and their effects are separated through the use of finite-element analysis.

Keywords: Stress-strain, Lead-free, Sn3.8Ag0.7Cu, size effect and microstructure effect

Introduction

One of the biggest challenges for electronic device packaging is to achieve high reliability in harsh environments. Thermal fatigue of the solder joints is typically one of the most significant concerns as they form the connecting part in an assembly comprised of materials with different thermal expansion behaviour. Therefore, to ensure structural integrity of an electronic assembly, an appropriate solder material and knowledge of its material properties are essential.

Although a number of studies have determined different material properties for Sn3.8Ag0.7Cu solder using bulk solder specimens [1-3], its properties at the microscale, i.e. for joints with dimensions below 100 μm, are less well understood. Bulk specimens contain a large number of randomly oriented grains and the characteristic effective behaviours, such as elastic, plastic and creep, describe the average performance of these grains. Therefore, bulk specimens can usually be considered as isotropic and homogeneous. However, the small-scale solder joints typical in modern electronics packaging technologies such as flip-chips are different in that they may contain only one or a few grains [4-6]. Their performance is therefore expected to shift from that characteristic of a polycrystalline material to an inter- or intra-granular based one. In such cases the individual grains play an important role in the behaviour of the individual solder joints. The β-Sn matrix of SnAgCu solder has a body centred tetragonal structure, so the mechanical behaviour of a single grain is expected to have significantly anisotropic characteristics. In such a case, the grain orientations will be a key factor in determining the behaviour of the solder joint [7, 8].

In addition to the grains and their orientation, the size of the solder joints and the materials they are in contact with can also affect their mechanical behaviour. Since solder joints are usually used to join dissimilar materials, the influence of joint size on the solder material behaviour increases as the size decreases. Solder joints can also be compared with brazed joints where the tensile strength of a thin transverse brazed joint has been shown to be considerably higher than that of the bulk joint material [9-11]. Therefore, the assessment of reliability of small-scale solder joints based on the available properties data for bulk solders may not be accurate. Thus it is essential to evaluate the material properties of solder joints which are of a scale commensurate to the real application, as well as the various factors affecting their mechanical behaviour.

In the present paper, an experiment designed to evaluate the stress-strain properties of small-scale solder joint is described. A commercial solder paste containing Sn3.8Ag0.7Cu powder was used to fabricate solder joints between copper substrates using a process representative of typical industrial electronics manufacturing processes. Comparison of the properties is made with those for bulk specimens, and factors affecting them, such as the microstructure and size of the solder joints, are discussed. The paper also presents the use of finite element analysis to separate the influences of the solder joint's microstructure and its size on the observed material properties.

978-1-4244-4722-0/09 $25.00
© 2009 IMAPS-ITALY

Experiment

The objectives of the experimental work were to determine the stress-strain relationship for both bulk solder and for solder joints which are commensurate in scale with real life applications. These measured properties were used to isolate and study the effects of microstructure and joint size in the lead-free solder materials. The properties were obtained under tensile loading conditions. Since the solder joint specimens used in the tests were small, an Instron 5848 MicroTester high precision tensile testing machine was used, with displacement measurement using a highly sensitive Instron 2630-100 series extensometer.

Specimen Fabrication

For fabrication of specimens Sn3.8Ag0.7Cu solder paste was used. For the bulk solder specimen fabrication, the solder paste was placed into a ceramic container with internal dimensions of 65 mm (length) × 10 mm (width) × 5 mm (depth). This container containing the solder paste was then passed through a reflow process conforming to the IPC/JEDEC J-STD-20C standards. The cooling rate during solidification of the solder was about 1.5 K/s, which is in line with the cooling rates of 1-2 K/s typically used in surface mount solderng processes. After solidification of the solder, it was carefully ground to bring its size down to 55 mm (length) × 7.5 mm (width) × 2 mm (thickness), as shown in Fig. 1 (a). Next, the specimen was milled to a dog bone shape, as shown in Fig. 1 (b) along with its final dimensions. These dimensions conform with the ASTM standard for tensile testing of metallic materials [12].

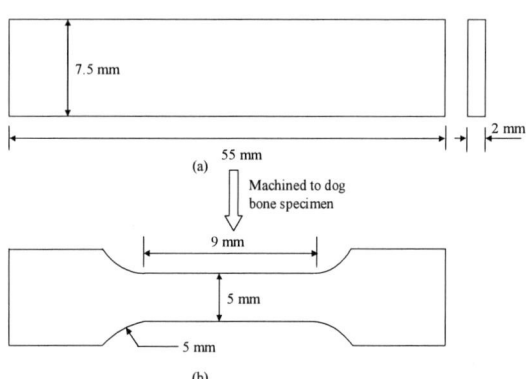

Figure 1: Fabrication of bulk specimens: (a) ground specimen after reflow; (b) final fabricated specimen

For the fabrication of small-scale solder joints, 1 mm thick copper sheet was used which was cut in to pieces 60 mm × 17 mm, as shown in Fig. 2 (a). To create a small-scale solder joint in the copper piece, a slot was cut using a low-speed diamond cutting wheel with thickness 0.3 mm, which results

in an approximate average gap of 0.34 ±0.015 mm. The slot was cut to a depth of 14 mm to avoid separation of the copper piece into two parts. With this method it is easier to maintain the parallelism of two substrates after fabrication of the specimen. Having created the slot, one side of the copper was covered with a high-temperature tape and then the gap was filled with solder paste. These copper pieces were then reflowed as per IPC/JEDEC J-STD-20C standards and with the same cooling rate of about 1.5 K/s. After the solder joints were fabricated, the sample was carefully hand ground to remove any surplus solder that had spread onto the copper surface. Then the ground specimens were cut to a width of 10.5 mm using the low-speed diamond saw to the dimensions shown in Fig. 2 (b). The specimens were finally polished using 800 grit paper on all sides to make them ready for mechanical testing. The solder joint gap length of each specimen was measured using a SIM universal optical measuring machine. Both bulk and solder joint specimens were fabricated in batches of 7 and were then stored at room temperature. Finally, these specimens were tensile tested within two days of their fabrication.

The tensile tests for both the bulk and small-scale solder joint specimens were carried out at three different displacement rates. The displacement rates were selected in such a way that the resulting strain rates do not fall into the creep regime. This was decided based on the time spent at the maximum stress level (saturation stress) during trial tests. The three displacement rates used were 0.005 mm/s, 0.05 mm/s and 0.5 mm/s, resulting in average strain rates of 0.00036 s^{-1}, 0.0037 s^{-1} and 0.0357 s^{-1} respectively for the bulk solder tensile tests, while those for the solder joint tensile tests the resulting strain rates were 0.00075 s^{-1}, 0.004 s^{-1} and 0.013 s^{-1}.

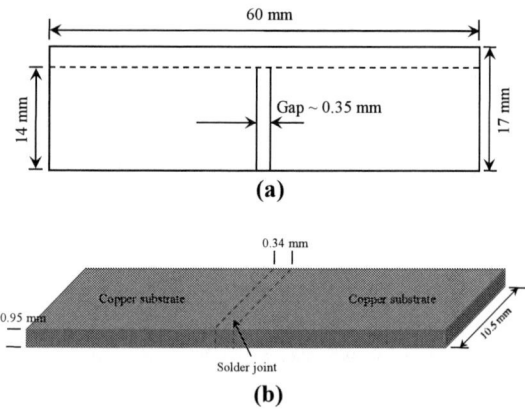

Figure 2: Fabrication of solder joint specimens: (a) dimensions of the copper pieces used; (b) final fabricated specimen

Experimental Results and Discussion

Bulk Solder

The tensile tests were carried out on 7 specimens at each strain rate. The obtained stress-strain data were converted to true stress-true strain and the average true stress-true strain data was calculated. The average true stress-true strain curves for the three strain rates are shown in Fig. 3. As the strain rate increases the maximum load carrying capacity of the solder increases, since the time available for any crystallographic deformation diminishes. The effect of the strain rate on the stress-strain behaviour of the bulk solder material is considered in terms of the Young's modulus, yield stress and ultimate strength. In experiments with the bulk solder, the displacement measurement was good enough to capture sufficient data below the proportionality limit so that the Young's modulus can be calculated. The change in the apparent Young's modulus with strain rate, calculated according to ASTM standards [13], is demonstrated by Fig. 4 (a). It increases approximately linearly in proportion to the log strain rate. Since there is no distinguishable yield point for the tested solder alloy, it is calculated based on the offset method recommended by the ASTM standard. The variation of both the yield stress (YS) and the ultimate tensile strength (UTS) is shown in Fig. 4(b). This demonstrates that both the YS and the UTS increase nearly linearly with the applied strain rate, which is a common feature in metals and alloys.

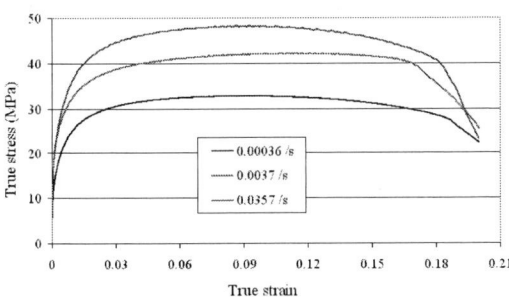

Figure 3: Average stress-strain curves for bulk solder at different strain rates

Solder Joint

As for the bulk solder samples, the tensile tests for the small-scale solder joints were carried out for 7 specimens at each strain rate. The obtained stress-strain data were again converted to true stress-true strain and average data was calculated for each strain rate condition. A comparison of the averaged true stress-true strain data was carried out for the three different strain rates and is presented in Fig. 5 (a). Since the stress-strain curves for the solder material are dependent on the strain rate, increased strain hardening is exhibited by the solder joints for the increased strain rate. Under high-rate loading, the time available for any crystallographic

deformation diminishes, resulting in an increased load at solder joint failure. The characteristics of the averaged true stress-true strain curves are very much the same, but the strain rate has an effect on the important mechanical behaviour parameters i.e. the apparent Young's modulus, yield stress and ultimate tensile strength. The deformation of the solder joint exceeds the proportionality limit at low stresses/strains, making it difficult to capture enough data for accurate Young's modulus calculation. Therefore, only the plastic parts of the true stress-true strain curves were focused on in the analysis. The yield stress and the ultimate tensile strength were calculated for each strain rate, and their variation with the strain rate was studied. The estimation of the yield strength was based on the stress offset method [13]. Figure 5 (b) shows the effect of strain rate (logarithmic scale) on both the yield stress and the ultimate tensile strength. As expected both parameters increased with the strain rate, and the variation is proportional to the log of the strain rate.

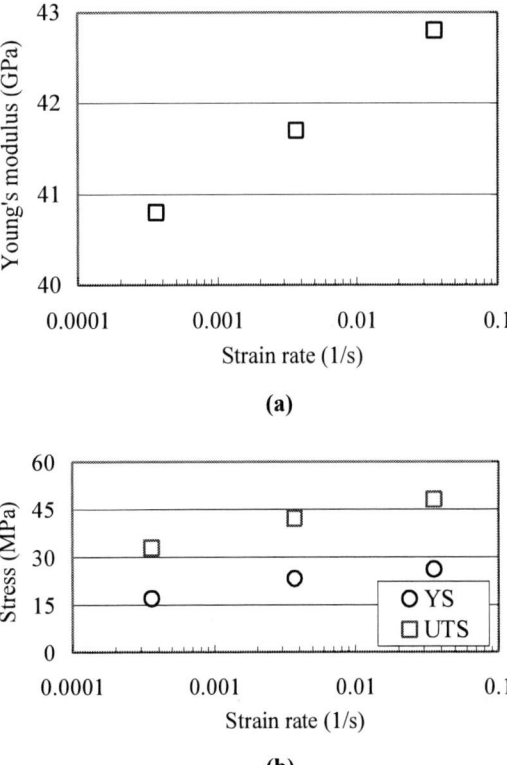

Figure 4: Effect of strain rate on: (a) Young's modulus; (b) yield stress (YS) and ultimate tensile strength (UTS) of bulk solder

978-1-4244-4722-0/09 $25.00
© 2009 IMAPS-ITALY

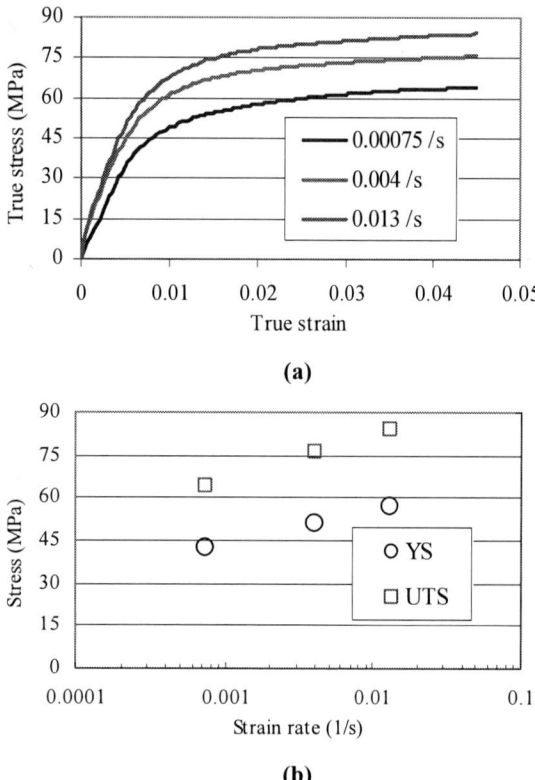

(a)

(b)

Figure 5: (a) Average stress-strain curves for bulk solder at different strain rates; (b) yield stress (YS) and ultimate tensile strength (UTS) of bulk solder

Comparative Study of Stress-Strain Properties

Having measured the stress-strain behaviour for both small-scale solder joints and bulk solder, a comparison was carried out. The comparison of the yield stress for the two types of solder specimens is illustrated by Fig. 6 (a) in semi-logarithmic coordinates. It is evident from the comparison that the apparent solder joint yield strength is more than twice that of the bulk solder. The graphs show slight divergence as the strain rate increases due to a higher strain rate sensitivity for the solder joints. Similar observations can be made for the comparison of the UTS in Fig. 6 (b), where the tensile strength in the solder joints is higher by a factor that is slightly less than 2.

It is very important to understand the contribution to these apparent increases in the material properties of the solder joint samples from the geometry of the test samples and from the differences in microstructure. The comparison of averaged stress-strain curves indicates that in the case of small-scale solder joints, strain softening is hardly present. Unlike the reflowed bulk solder, the fabricated solder joints have copper substrates on either side, which are still in the elastic state and

constrain the solder joints from plastic deformation. Therefore, during the tensile testing the solder joints failed without showing significant strain softening. The constraining effect is dependent on the size of the solder joint. So, the effect of size and constraints on the solder joints' material properties are discussed in the following section.

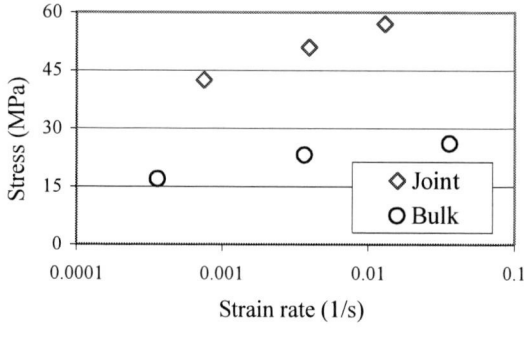

(a)

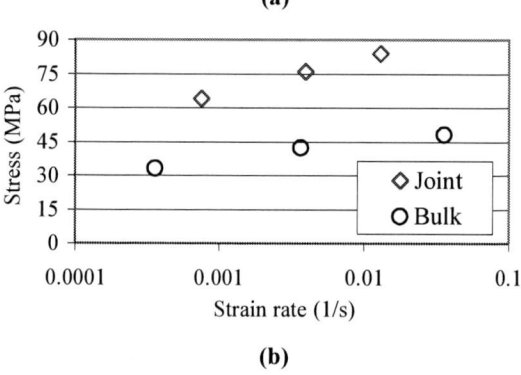

(b)

Figure 6: Comparison of average: (a) yield stress; (b) UTS for solder joints and bulk solder

Effect of Size and Constraints

The effects of joint size and the constraining effect of the copper substrates on the solder joint were studied using a finite element simulation of the tensile tests. In the finite element analysis 1/8th of the fabricated solder joint specimen was considered, as shown in Fig. 7 (a). A parametric model of the geometry of the 1/8th of the solder joint specimen was built. The gap (g), which represents the axial size of the solder joint between the two copper substrates, was varied along with the length (L) of the specimen to keep the gap to length ratio (g/L) constant for all geometries of the joint. With this approach the effect of the solder joint's size, due to a varying gap to thickness (g/t) ratio, could be studied. The gap was varied between 0.15 mm and 1.5 mm. Since the gap is small compared to the total size of the test sample, capturing the stress field in the solder joint was difficult with the full model. Therefore, the analysis was carried out in two steps. In the first step, a full-model analysis with a coarse

978-1-4244-4722-0/09 $25.00
© 2009 IMAPS-ITALY

mesh was carried out in order to obtain the displacement field for a sub-model used in the second step. In this step a sub model of the full FE model, which is shown in Fig. 7 (b), was built with a high mesh density near the solder joint, which is the area of interest. The cut plane for the sub-model was chosen so that it is sufficiently far from the solder joint and does not affect the developing stress field in it. The distance of the cut plane from the transverse symmetry plane used is illustrated in Fig. 7 (b).

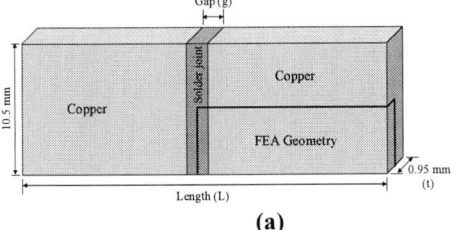

(a)

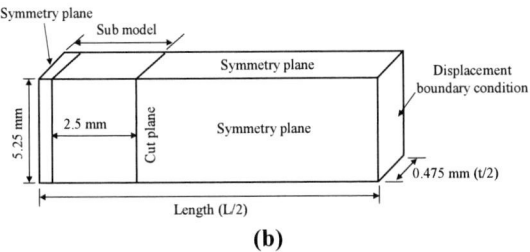

(b)

Figure 7: (a) Geometry of the solder joint specimen and its octant used in the full model analysis; (b) Full model, sub model and boundary conditions for the FEA

In the finite element analysis, the solder joint was modelled using the homogeneous bulk solder material properties as measured for the strain rate of 0.00036 s^{-1}. A comparison of the experimental and simulated true stress-true strain curves is presented in Fig. 8. The simulated stress-strain curve was created using a multi-linear isotropic hardening material model due to the monotonic uniaxial tensile loading. The copper substrate was modelled as linear elastic with a Young's modulus of 110 GPa and Poisson's ratio of 0.343. This FE analysis was carried out in ANSYS Classic version 11.

The finite-element modelling was performed using 8-noded 3D hexahedral structural solid elements. The mesh used for both full and sub models with a solder joint gap of 0.35 mm is illustrated in Fig. 9. The symmetry boundary conditions used in the full model and sub-model are shown in Fig. 7 (b). The displacement boundary condition was applied at the far end-face of the full model (Fig. 7 (b)) and the value of the applied displacement is such that the maximum total equivalent strain induced in the solder joint is about

11%. This value is selected so that the ultimate tensile stress of 32.6 MPa, as measured for the strain rate of 0.00036 s^{-1}, was achieved in the solder joint. After the full model was analysed, the displacements were extracted at the cut plane and applied to the sub-model along with the symmetry boundary conditions.

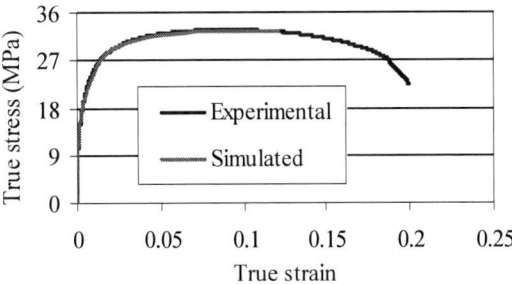

Figure 8: Experimental and simulated stress-strain curves for the solder joint

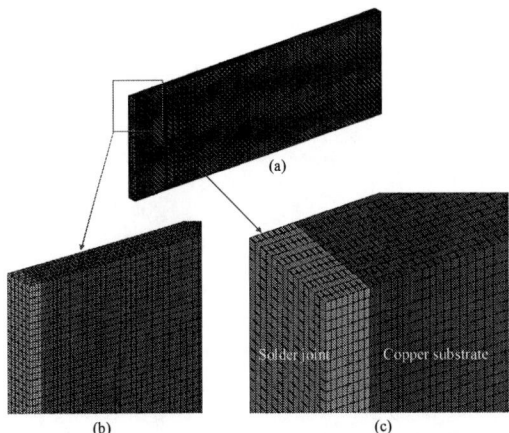

Figure 9: Meshes used in the FEA: (a) full model; (b) zoomed view of full model; (c) zoomed view for sub-model

The stress distribution in the solder joint for the two extreme gap lengths, 1.5 mm and 0.15 mm, are shown in Figures 10 and 11, respectively. For the solder joint gap of 1.5 mm, the maximum equivalent stress occurs at the centre of the solder joint (i.e. on the symmetry plane) and diminishes towards the joint interface. The equivalent stress distribution in the 0.15 mm gap solder joint is however quite different, with the maximum stress occurring at the interface between the solder joint and copper. The comparison of equivalent stresses for these two extreme gap lengths illustrates the effect of the solder joint's size. As the gap length increases, the plastic deformation taking place in the solder joint is less restricted by the copper substrate, which is in the elastic state. However, as the gap length decreases, the space available to accommodate the same amount of plastic deformation in the solder joint decreases and the copper plates constrain the transfer of this plastic deformation.

978-1-4244-4722-0/09 $25.00
© 2009 IMAPS-ITALY

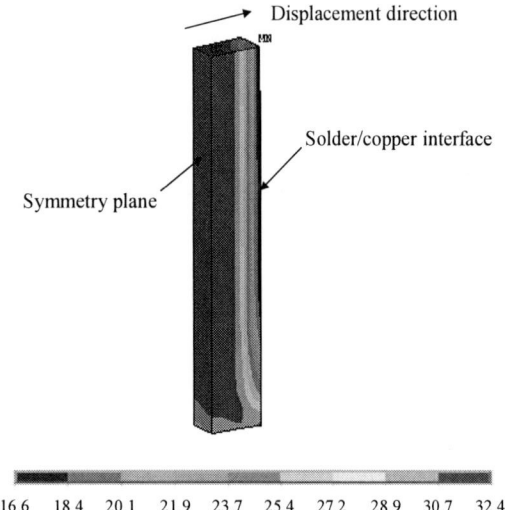

Figure 10: Equivalent stress distribution in the solder joint with a 1.5 mm gap length

The size effect on the solder joint is studied by calculating a dimensionless factor, S:

$$S = \frac{\sigma_n}{\sigma_e} \qquad (1)$$

where σ_n is the normal stress (applied load divided by the area of the solder joint) and σ_e is the equivalent stress. The size effect can be better explained by considering the variation of S with the gap to thickness ratio (g/t), as illustrated in Fig. 12. As the gap length decreases, the size effect becomes more prominent, resulting in an increased normal stress being required to induce the same strain in the solder joint. Thus, the solder joint's apparent strength is artificially increased. The size effect also causes a change in the stress state in the solder joint. During plastic or creep deformation, the material tends to keep its volume constant. Since the plastic flow in the direction of the substrate is restricted, the solder joint shrinks in the lateral direction resulting in a 3D stress state. This can be better explained by considering a triaxiality ratio, R_t, that can be given as [14]:

$$R_t = \frac{\sigma_h}{\sigma_m} \qquad (2)$$

where σ_h is the volume average of hydrostatic stress in the solder joint and, σ_m is the volume average of von Mises (equivalent) stress in the solder joint. The stress triaxiality versus gap to thickness (g/t) ratio is shown in Fig. 12. As the gap length (g) decreases, the hydrostatic stress in the solder joint increases, due to the constraining effect in the loading direction and the Poisson's effect in

the lateral direction. Therefore, even though loading is uniaxial, in the small-scale solder joints under consideration the stress field becomes triaxial due to the size effect, but as the gap length increases the triaxility decreases.

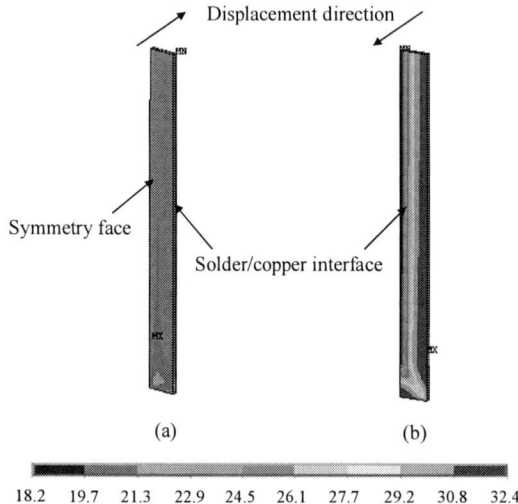

Figure 11: Distribution of equivalent stresses in the solder joint with a 0.15 mm gap length: (a) on the symmetry plane; (b) at the interface

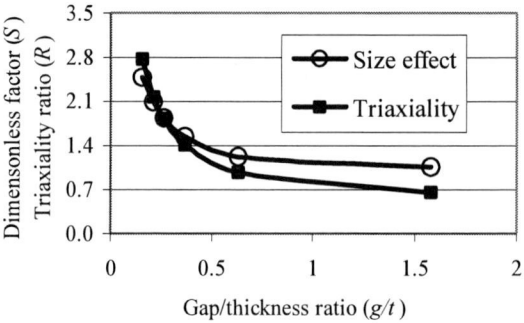

Figure 12: Size effect and triaxiality ratio as functions of gap length to thickness ratio

The size effect on the apparent solder material properties was confirmed experimentally by conducting further tensile tests with a solder joint gap of 1.1 mm. The specimen preparation, displacement measurement technique, and test conditions are exactly the same as those described for the solder joint gap of 0.35 mm. A comparison of the yield stress and the ultimate tensile strength are made for the two different solder joints (0.35 mm and 1.1 mm) and bulk solder. Figures 13 and 14 demonstrate this comparison for solder joints and bulk solder for three different strain rates. The variation of the yield stress and ultimate strength are nearly linear when plotted on a logarithmic strain rate scale. It is evident from these graphs that the stress-strain properties for solder joints with the 1.1

mm gap lie between the properties of the bulk solder and the solder joints with the gap of 0.35 mm. This is consistent for all three strain rates. Therefore, as the solder joint's gap decreases, its apparent yield stress and ultimate strength increases. The comparison study also indicates that the properties of solder joints with sufficiently large gaps converge towards the properties of the bulk solder.

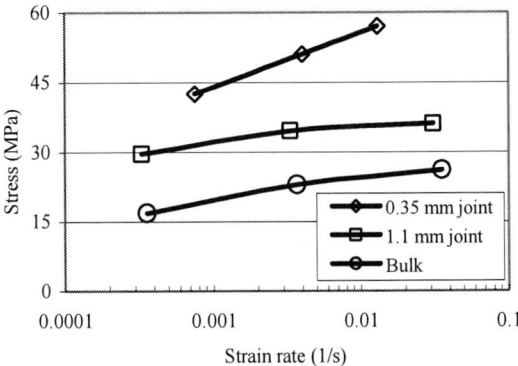

Figure 13: Comparison of yield stress for solder joints and bulk solder for various strain rates

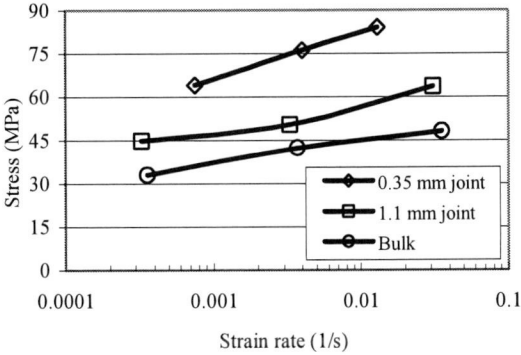

Figure 14: Comparison of ultimate tensile strength for solder joints and bulk solder for various strain rates

The contribution of the joint size effect to the increased tensile material properties of the solder compared with bulk solder is also studied. It is explained by considering the ultimate tensile strength of solder joints and the bulk solder at the strain rate of 0.00036 s^{-1}. Table 1 presents an analysis of the size effect on the solder material properties. From Fig. 12, the percentage contribution of the size effect to the increase in the solder joint's strength above that of bulk material can be determined for a specific gap to thickness (g/t) ratio. For example, for a given ratio of g/t, S is determined from the g/t vs. S graph, and $((S-1)\times 100)$ gives the percentage contribution from the size of the solder joint, which is given in the fourth column of Table 1. Table 1 shows the increase in strength cannot be explained only by the size effect. The remainder of

the increase is attributed to the different microstructure of the joints, as further discussed in the following section.

Table 1: Size and microstructure effect on solder material properties

Solder type	UTS (MPa)	Increase above bulk solder (%)	Size effect $[(S-1)\times 100]$ (%)	Microstructure effect (%)
Bulk	32.8			
1.1 mm	44.6	36	15	21
0.35 mm	59.1	80	50	30

Microstructure Effect

In the above section the size effect on the solder's material properties was discussed. The microstructure of the solder joint also has an effect on its material properties. Table 1 presents the effect of microstructure on the solder material properties. For example, for a solder joint with a gap of 0.35 mm, the increase in the UTS above bulk solder for the 0.00036 s^{-1} strain rate is 80 %, out of which 50 % can be explained due to the size effect. The remaining effect (30 %) therefore comes from the differences in the microstructure between the solder joint and bulk solder. This microstructure effect is separated using the results of the FE analysis, which is based on the assumption of homogeneous solder material properties, and does not include the properties of different constituents of the microstructure such as Sn-dendrites, the eutectic phase, and any intermetallic compounds (IMCs).

In order to study the microstructural differences between the bulk solder and solder joint, as reflowed specimens were selected. These specimens were cut into an appropriate size and cold mounted using an epoxy resin. The reason for using cold mounting is that the low curing temperature in cold mounting avoids any changes to the solder microstructure. Once mounted, the samples were ground on a series of polishing papers with increasingly finer grits of 200, 400, 800 and 1200. These ground samples were then polished using standard metallographic techniques to 0.5 μm with diamond slurry. Final polishing of the samples was done on a colloidal silica pad using 0.05 μm colloidal silica slurry for 2 minutes. This exposed the intermetallic compounds and grains present in the solder.

Figures 15 (a) and (b) present bright field images of the microstructure of an as reflowed solder joint on a copper substrate, which typically consists of a mix of Sn-dendrites, the eutectic phase, and intermetallic compounds. The microstructure of

978-1-4244-4722-0/09 $25.00
© 2009 IMAPS-ITALY

the reflowed bulk solder shown in Fig. 16 consists predominantly of Sn-dendrites and the eutectic phase. The Sn-dendrites in the bulk solder are large compared to those in the solder joints. The overall comparison of microstructures also shows that the fraction of the eutectic phase is higher in the bulk solder. Unlike the solder joint, Cu_6Sn_5 intermetallic compounds are hardly present within the bulk solder, due to the absence of dissolution of additional copper in the molten solder during reflow. In addition, there is no intermetallic layer in the bulk solder due to the absence of a substrate, which is shown in Fig. 15 (b). Ag_3Sn intermetallic compound is also more common in the solder joints than in bulk solder. Grains of Sn are observed both in the bulk solder and the solder joint, but they are larger in the bulk solder. Thus, the microstructure of the solder joint is comparatively finer than that of the bulk solder. Since the microstructure also depends on the cooling rate, the same cooling rate was used for both cases during solidification. However the observed difference could be due to the diffusion of copper into the solder from the substrate, which alters the composition of the solder joint. It is well known that a finer microstructure improves the material properties of solder materials [15-17]. Therefore, the differences in the microstructures between the solder joint and the bulk solder also contribute to the increased tensile material properties of the solder joints along with the joint size effect.

Conclusions

1. The comparison of the stress-strain data for the solder joint and the bulk solder showed significantly increased apparent tensile properties for the former.

2. The study of the reasons for this increase in comparison with the bulk solder revealed the effects of the joint's size and microstructure. As the solder joint's size decreases the plastic deformation taking place in the solder joint is constrained by the stiffer copper substrates, which remain in the elastic state.

3. This constraining effect of the copper results in a 3D stress state in the solder joint, even though the specimen was loaded uniaxially. The extent of the size effect on the solder joint's material properties was determined with the help of finite element analysis.

4. Microstructure studies of both the solder joints and the reflowed bulk solder illustrated that the solder joint has a finer microstructure than the bulk solder. It is well established that a finer microstructure improves the material properties; a similar effect was observed by comparing material properties of the bulk solder and the solder joint.

5. The effects of the joint size and microstructure were separated with the help of the finite-element analysis, and hence the size effect can be quantified for the given solder joint gap.

6. Finally, the strength of the solder joint is the result of a superposition of material properties of the bulk solder, the effect of the solder joint's size and the effects of microstructure.

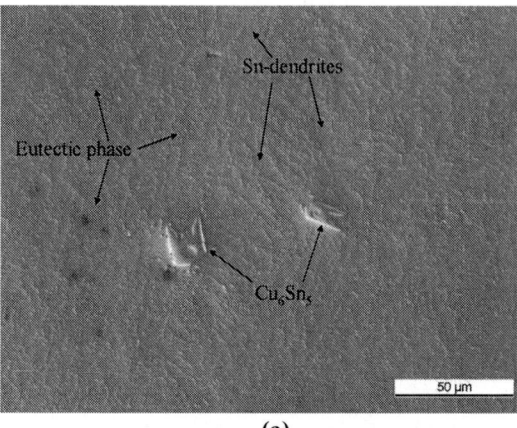

(a)

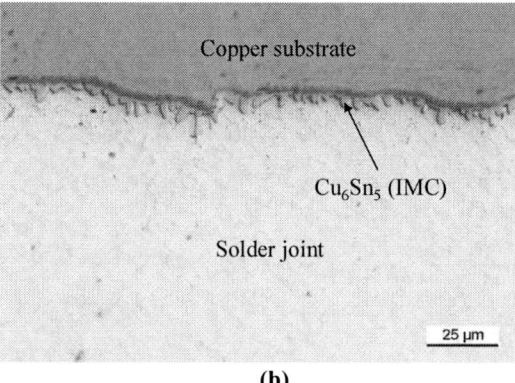

(b)

Figure 15: Microstructure of small-scale solder joint: (a) inside solder joint; (b) at the interface between substrate and solder joint

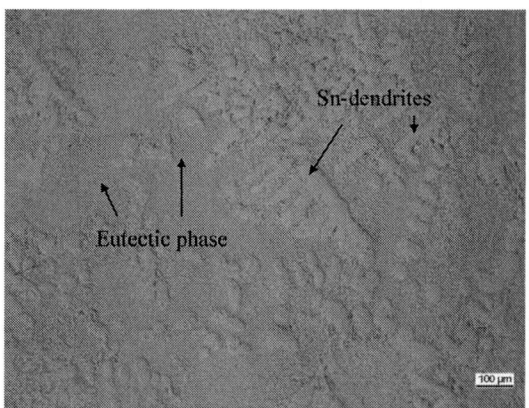

Figure 16: Microstructure of reflowed bulk solder

References

[1] J. H. L. Pang and B. S. Xiong, "Mechanical Properties for 95.5Sn-3.8Ag-0.7Cu Lead-Free Solder Alloy", IEEE Transactions on Component and Packaging Technologies, Vol. 28, pp. 830-840, 2005.

[2] W. Qiang, L. Lihua, C. Xuefan and W. Xiaohong, "Experimental Determination and Modification of Anand Model Constants for Pb-free Material 95.5Sn4.0Ag0.5Cu", Proceedings of the 8th International Conference on Thermal, Mechanical and Multiphysics Simulation and Experiments in Micro-Electronics and Micro-Systems, EuroSimE, London, UK, pp. 308-316, 2007.

[3] X. Qiang, L. Nguyen and W. D. Armstrong, "Aging and Creep Behaviour of Sn3.9Ag0.6Cu Solder Alloy", Proceedings of the 54th IEEE Electronic Components and Technology Conference, Las Vegas, USA, Vol. 2, pp. 1325-1332, 2004.

[4] A. U. Telang and T. R. Bieler, "Orientation Imaging Microscopy of Lead-Free Sn-Ag Solder Joints", Journal of the Minerals, Metals and Materials Society, Vol. 57, pp. 44-49, 2005.

[5] A. LaLonde, D. Emelander, J. Jeannette, C. Larson, W. Rietz, D. Swenson and D. W. Henderson, "Quantitative Metallography of -Sn Dendrites in Sn-3.8Ag-0.7Cu Ball Grid Array Solder Balls via Electron Backscatter Diffraction and Polarized Light Microscopy", Journal of Electronic Materials, Vol. 33, pp. 1545-1549, 2004.

[6] A. U. Telang, T. R. Bieler, S. Choi and K. N. Subramanian, "Orientation Imaging Studies of Sn-based Electronic Solder Joints", Journal of Material Research, Vol. 17, pp. 2294-2306, 2002.

[7] W. J. Plumbridge, "Long Term Mechanical Reliability with Lead-free Solders", Soldering & Surface Mount Technology, Vol. 16, pp. 13-20, 2004.

[8] T. R. Bieler and H. Jiang, "Influence of Sn Grain Size and Orientation on the Thermomechanical Response and Reliability of Pb-free Solder Joints", Proceedings of the 56th IEEE Electronic Components and Technology Conference, San Diego, CA, USA, pp. 1462-1467, 2006.

[9] H. J. Saxton, A. J. West and C. R. Barrett, "Deformation and Failure of Brazed Joints - Macroscopic Considerations", Metallurgical Transactions, Vol. 2, pp. 999-1007, 1971.

[10] P. Zimprich, A. Betzwar-Kotas, G. Khatobo, B. Weiss and H. Ipser, "Size effect in Small Scaled Lead-free Solder Joints", Journal of Material Science, Vol. 19, pp. 383-388, 2008.

[11] A. J. West, H. J. Saxton, A. S. Tetelman and C. R. Barrett, "Deformation and Failure of Thin Brazed Joints - Microscopic Considerations", Metallurgical Transactions, Vol. 2, pp. 1009-1017, 1971.

[12] "Standard Test Methods for Tension Testing of Metallic Materials", ASTM Standard E 8-04, West Conshohocken, USA, 2004.

[13] "Standard Test Methods for Young's Modulus, Tangent Modulus, and Chord Modulus", ASTM Standard E 111-04, West Conshohocken, USA, 2004.

[14] D. F. Socie and G. B. Marquis, "Multiaxial Fatigue", Society of Automotive Engineers, Warrendale, USA 2000.

[15] F. Ochoa, X. Deng and N. Chawla, "Effects of Cooling Rate on Creep Behaviour of a Sn-3.5Ag alloy", Journal of Electronic Materials, Vol. 33, pp. 1596-1607, 2004.

[16] B. Yeung and J.-W. Jang, "Correlation between Mechanical Tensile Properties and Microstructure of Eutectic Sn-3.5Ag Solder", Journal of Materials Science Letters, Vol. 21, pp. 723-726, 2002.

[17] S. Wiese, E. Meusel and K. J. Wolter, "Microstructural Dependence of Constitutive Properties of Eutectic SnAg and SnAgCu Solders", Proceeding of the 53rd IEEE Electronic Components and Technology Conference, New Orleans, USA, pp. 197-206, 2003.

978-1-4244-4722-0/09 $25.00
© 2009 IMAPS-ITALY

Experimental study of polymers as encapsulating materials for photovoltaic modules

S. A. Sala, M. Campaniello, A. Bailini

Materials Laboratory, Services for Electronic Manufacturing (SEM) S.r.l. - Solar Thin Film (STF) S.r.l.

via Lecco 61, 20059 Vimercate (MB) Italy

Abstract

Due to the introduction in the market of new generations of photovoltaic (PV) modules, there is the need to develop improved packaging materials. A reliable encapsulating layer is required to guarantee outdoor operation for more than twenty years, paying attention to factors like protection from weather agents and humidity, promotion of the adhesion between the active layer and the substrate, and electrical isolation.

The most popular encapsulant is a thermoplastic polymer called Ethylene-Vinyl Acetate (EVA), which is produced in rolls of extruded thick film. Anyway, another material, Poly-Vinyl Butyral (PVB), seems to be the best choice to satisfy the more demanding mechanical and safety requests in the field of Building Integrated PhotoVoltaic (BIPV). This is one of the reasons why PVB is nowadays widely used with thin film PV, while EVA is still the standard material for crystalline modules.

The success of the qualification campaign of a PV module can strongly depend on the performances of the packaging polymer. Many tests can be done to determine the material properties and predict its behaviour under stress conditions.

In this work, we are performing an experimental study to test different formulations of polymers as interlayer materials for PV modules. The mechanical properties are one of our focues. They strongly depend on the temperature and the loading intensity because of the thermoplastic nature of these polymers. They deeply affect the reliability of the modules and determine their application.

Key words: photovoltaic, EVA, PVB, encapsulant, reliability, glass transition temperature

Introduction

Solar modules must survive to changing and challenging weather conditions like no other high-tech products. They are also expected to have extremely long life spans, since manufacturers generally guarantee that their products will produce at least 80% of the nominal power after twenty to twenty five years of operations.

Such a long life expectancy is achieved with a durable packaging of the array of cells that constitutes the active layer of the module, which is responsible for the photocurrent generation (Fig. 1a and Fig 1b).

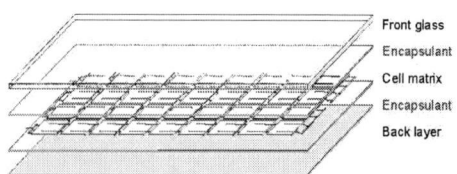

Figure 1a: stack of a traditional crystalline silicon PV module

The upper or front layer needs to be translucent, scratch-resistant and weatherproof; it must provide also mechanical rigidity. These are the main reasons why this layer normally consists of a 2- to 3- mm thick soda lime glass with low iron content to have a good light transmission. The outer layer at the non-illuminated module side (back layer) is usually a composite polymeric sheet (called backsheet) acting as a barrier for humidity and corroding species. Another glass can be used instead of the polymeric sheet, or even flexible substrate, according to the application.

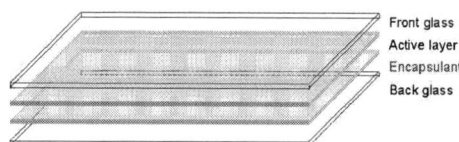

Figure 1b: stack of a thin film PV module

Finally, one (for thin film modules) or two (crystalline silicon modules) layers of encapsulant or pottant material provide the protection of the active layer and the bonding between the front and the back layer, together with the adequate mechanical compliance to accommodate stresses induced by differences in thermal expansion coefficients between glass and cells. The encapsulant material has also to electrically isolate the cells from one another, protect them from water and guarantee

978-1-4244-4722-0/09 $25.00
© 2009 IMAPS-ITALY

mechanical seal for two decades under variable temperature and humidity conditions.

The aim of this paper is to shows the results of some tests performed on the two most "popular" encapsulants for photovoltaic (PV) modules, EVA and PVB.

Key properties of encapsulant material and their applications

As already introduced, encapsulant materials must provide good performances in terms of: adhesion and cohesion strength, moisture barrier effect, corrosion protection, electrical insulation, good optical transmission and mechanical resistance. There is a huge variety of encapsulant for PV modules, since many of these materials originally were developed for application in laminated windshields and laminated architectural glass.

Ethylene-Vinyl Acetate (EVA) has the longest history in the PV encapsulant market. It is typically provided as a copolymer-based sheet; it is still used in crystalline modules and can be found in some thin film PV applications. Commercial EVA formulations differ each other mainly for the content of the two monomers (Vynil Acetate varies from 10% to 40%) and of the additives (especially curing agents and UV absorbers). In fact, EVA, like most polymers, is known to undergo photothermal degradation. The break of the molecular chains by UV radiation is the cause of the so-called "yellowing" or "browning" of EVA, which reduces its optical transmission and affects module current. Degradation can also decrease the strength of the encapsulant, leading to loss of adhesion to the cells and even detachment of the layer (delamination).

Poly-Vinyl Butyral (PVB) is a Vinyl polymer that can be classified as belonging to the Poly-Vinyl Acetates (PVAC). It was first utilized for solar PV modules in the 1970s. Even though it possesses good adhesion qualities (granted by the formation of hydrogen bonds with the glass) and the desired durability (in particular, strong UV resistance), if PVB absorbs moisture, the foil clouds up resulting in a not-negligible loss in visible light transmission. As a result, solar module manufacturers shifted for the most part to EVA. Anyway, a new generation of PVB was introduced in the market some years ago. It mitigates the cloudiness issue and seems to be very attractive to some applications like thin film PV (in particular, large area modules) or glass/glass module for Building Integrated Photovoltaic (BIPV).

EVA and PVB are both thermoplastic polymers, which means that shape changes made under heating are reversible. The two materials show a distinct viscous-elastic behaviour. This behaviour is characterized by dependence on temperature, time, kind and entity of the load. Furthermore, these materials do not spontaneously react on loads but with some time delay. They are sold in rolls of extruded films; as already said, EVA films contain also curing agents to promote cross-linking between the macro-molecules when the temperature is increased over 150 °C.

In fact, during modules assembling process EVA films are processed inside a laminator using a two-steps process (lamination and curing). During the lamination stage, temperature is raised above EVA softening point (120 °C), so that the material is free to flow and embed the cells. Vacuum is also required inside the laminator to extract air, moisture and other gases. The curing step is analogous to the vulcanization of rubber; it can take up to 40-45 minutes for standard cure EVA, while fast cure EVA requires only 10 minutes. This step is the most critical one, since cross-linking determines most of the EVA final properties.

PVB lamination is a two-steps process too. The first one is de-airing (the removal of interfacial air trapped between the various component). It is commercially achieved through mechanical or under-vacuum processes. The second one is autoclaving: the parts are subjected to high pressure and temperature (145 °C), PVB flows around obstructions filling any voids left from the de-air step and dissolving any residual air into the interlayer. Pressure and hold time are function of the quantity of material loaded in the autoclave in a single batch as well as the autoclave size and shape. Unlike EVA, PVB requires low temperature storing.

For modules using two glass panes, resin fill-in is an alternative to EVA and do not require heating to cure. Silicone resins are expensive but very stable and some modules for building integration use them. Also polyurethane resins (PU) can be used for encapsulation of glass-glass modules.

Reliability and qualification concerns

Because of the imperative to minimize module cost, the degree of protection provided by the packaging of a PV module must necessarily be imperfect, and a design trade-off between cost and protection exists. Since the late 1970s, there have been a number of PV qualification test sequences (developed by government laboratories and international standards organizations), intended to gauge the ability of module designs to protect effectively the active layer from the environment. The test sequences are relatively short in duration and are separated into several "legs" performed on different modules that can be done in parallel. At the end of the sequences, test modules must retain a specified percentage of their initial output power in order to be judged as having passed the qualification. As the PV industry has grown and become an international commodity market, the International Electrotechnical Commission (IEC) test standards are now the only test accepted by both module manufacturers and buyers. The standard test procedure for crystalline silicon modules is IEC 61215, while IEC 61646 is the reference for thin film modules.

978-1-4244-4722-0/09 $25.00
© 2009 IMAPS-ITALY

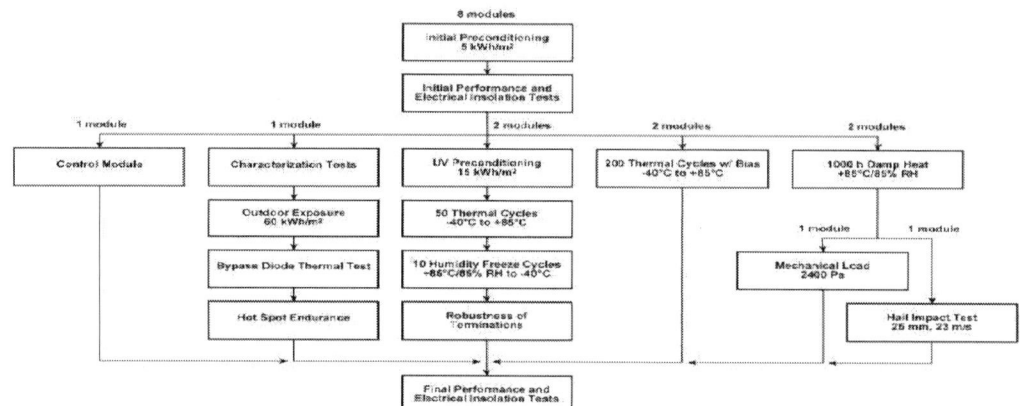

Figure 2: typical qualification testing sequence for crystalline-silicon modules

Fig. 2 shows the qualification testing sequence prescribed by IEC 61215, with major tests such as UV exposure, outdoor exposure (OE), thermal cycling (TC), humidity-freeze (HF) cycling and damp-heat (DH).

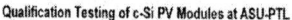

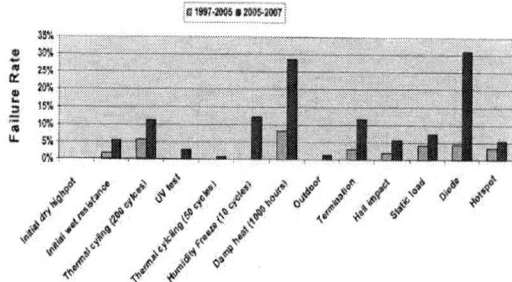

Figure 3a: failure rate comparison of crystalline silicon modules for the 1997-2005 and 2005-2007 periods.

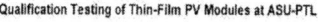

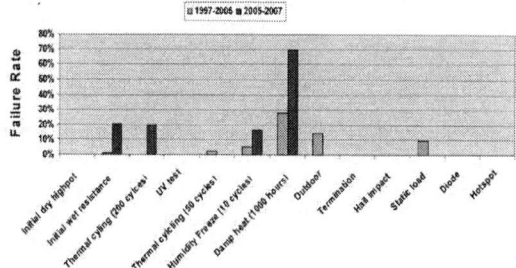

Figure 3b: failure rate comparison of thin-film modules for the 1997-2005 and 2005-2007 periods.

According to data provided by the Photovoltaic Testing Laboratory (PTL) of the Arizona State University (ASU), environmental tests have the highest failure rates among all the IEC tests (Fig 3a and Fig 3b). The last three stress tests are all accelerated environmental tests, which apply stress factors like

temperature, humidity and the combination of the two (see Table 1 for the parameters).

Test	Temperature (°C)	UR (%)	Duration
TC	−40 ÷ +85	0	50 or 200 3-h cycles
HF	−40 ÷ +85	85	10 25-h cycles
DH	+85	85	1000 h

Table 1: environmental parameters for Thermal Cycling (TC), Humidity Freeze (HF) and Damp Heat (DH) tests.

A reliable and affordable encapsulant material can be very important for the success of a qualification campaign. Nowadays, there is a strong need to develop improved packaging materials, since module designs continue to evolve and new generations of PV modules have recently entered the market. That's the reason why we have decided to test some commercial formulations of EVA and PVB evaluating their most important properties in the same environmental conditions (temperature and humidity) of the IEC tests.

Technical Approach

The test we are performing at SEM's Materials Laboratory on encapsulants can be divided in two categories:

- Characterization of material properties: verify material bulk properties comparing different EVA and PVB commercial formulations;
- Reliability testing: verify materials properties in a sample stack (glass-encapsulant-backsheet) before and after environmental tests.

The first category comprehends rheological and thermal measurements, made to determine at what temperatures the phase transitions occur and their effect on the mechanical properties, and the determination of water absorption of EVA and PVB foils. They can provide useful information for a module producer when choosing the material for the encapsulant layer.

978-1-4244-4722-0/09 $25.00
© 2009 IMAPS-ITALY

We have also performed adhesion strength study of EVA encapsulants inside a glass-backsheet laminated stack. Adhesion is a reliability issue, and it doesn't depend only on the chosen materials but also on the production process.

Rheological and Thermal Measurements - Dynamic Mechanical Analysis

Dynamic Mechanical Analysis (DMA), also known as Dynamic Mechanical Thermal Analysis (DMTA), measures the stiffness and damping properties of a material. The stiffness depends on the mechanical properties of the material and its dimensions. It is frequently converted to a modulus (for example, tensile storage modulus E') to enable sample inter-comparisons. Damping is expressed in terms of $tan\delta$, the phase lag between the force (stress), which is applied sinusoidally to the sample, and the strain, which can be calculated from the displacement of the sample after deformation. DMA is the most sensitive technique for monitoring relaxation events, such as glass transitions, since the mechanical properties change dramatically when relaxation behaviour is observed.

Tests were performed using a Rheometric Scientific DMTA VA5 Analyser equipped with a refrigeration unit. A rectangular tensional testing fixture was used with samples less than 1-mm thick, 8-mm wide, and 20-mm long with about 12-mm of the length covered by the clamps holding the sample. During the measurements, the temperature was changed between -60 °C and +90 °C, to cover the entire range of temperatures experienced by the modules during outdoor exposure. A typical response from a DMA shows both E' and $tan\delta$. As the material goes through its glass transition, the modulus reduces and $tan\delta$ goes through a peak. The data give information on the position of the glass transition temperature (T_g), sample stiffness and other viscoelastic properties.

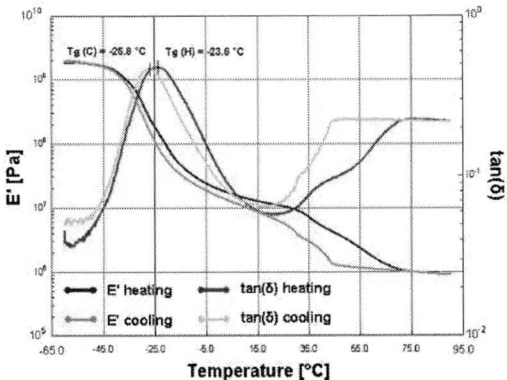

Figure 4: EVA tensile storage modulus (E') and phase angle (δ) as a function of temperature. Glass transition temperatures are also indicated. A hysteresis between the behaviour while heating and the behaviour while cooling can be appreciated.

For EVA (Fig. 4), T_g was measured between -19 °C and -27 °C. It is consistant with values reported in the literature. Another transition can be observed at higher temperatures: when cooling, it occurs quite abruptly at about 40 °C, while it covers the temperature range from 35 °C to 65 °C when heating. It should be caused by the crystallization of the crystalline part of the polymer since EVA can exibhit, depending on the Vynil Acetate content, partial-crystalline polymer like properties. Tensile storage module values confirm the semi-crystalline behaviour: 10^9 MPa below T_g, 10^7 MPa between T_g and the crystallization transition, 10^6 MPa above the crystallization transition. So, below T_g EVA is very stiff and its use may be a significant concern: high intensity winds at such low temperatures can cause a module to flex, possibly breaking some components.

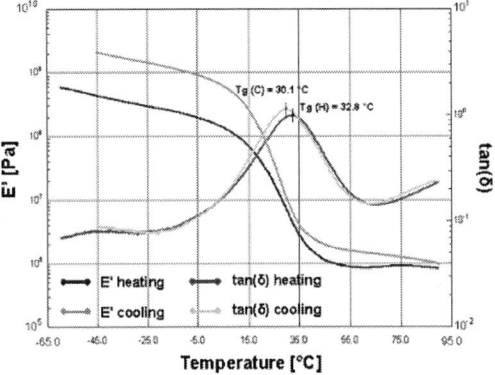

Figure 5: PVB tensile storage modulus (E') and phase angle (δ) as a function of temperature. Glass transition temperatures are also indicated. Even in this case, a hysteresis between the behaviour while heating and the behaviour while cooling can be appreciated.

PVB (Fig. 5) is an amorphous polymer, so it exhibits only the glass transition. According to our measurements, T_g of PVB is at about 30 °C. Above T_g, E' is in the range of 10^9 MPa, while below T_g it is in the range of 10^6 MPa. Since the Normal Operating Cell Temperature (NOCT) of a PV module is about 40 °C, it often works around the T_g of PVB.

Rheological and Thermal Measurements - Thermo-Mechanical Analysis

Thermo-Mechanical Analysis (TMA) can be defined as the measurement of a specimen's dimensions (length or volume) as a function of temperature whilst it is subjected to a constant mechanical stress. In this way the Coefficient of Thermal Expansion (CTE) can be determined and changes in this property with temperature monitored. Many materials will deform under the applied stress at a particular temperature that is often connected with the material melting or glass transition. So, it

978-1-4244-4722-0/09 $25.00
© 2009 IMAPS-ITALY

can be used as a means to verify T_g as it has been calculated by DMA. Tests were performed using a Perkin Elmer Pyris Diamond TMA.

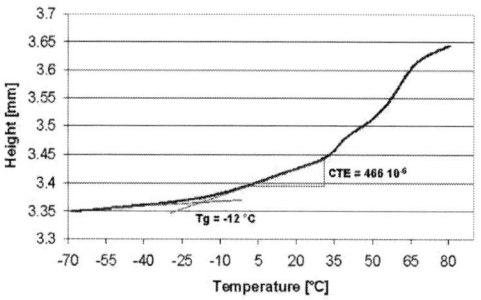

Figure 6: EVA response to TMA as a function of temperature. Glass transition temperature and thermal expansion coefficient are also indicated.

According to TMA (Fig. 6), EVA T_g ranges between -19 °C and -9°C, below the values measured by DMA. In fact, glass transission is a second order transition, so different measurement methods and techniques can give different results. TMA allows also the identification of the crystallization transition, which should be around 30 °C. The linear thermal expansion coefficient varies between $10*10^{-5}$ K^{-1} below T_g and $10*10^{-4}$ K^{-1} above the crystallization transition; between this and T_g, CTE is in the range of $20*10^{-5}$ K^{-1}. It agrees with values coming from datasheets of commercially available EVA.

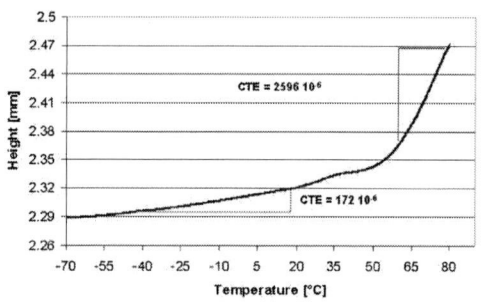

Figure 7: PVB response to TMA as a function of temperature. Thermal expansion coefficient above and below glass transition temperature is also indicated.

It is more difficult to identify the T_g of PVB from TMA analysises (Fig. 7); anyway it should be between 35 and 45 °C, as predicted by DMA. Below T_g, PVB CTE is comparable to the one of EVA ($10*10^{-5}$ K^{-1}), while above T_g it is between $20*10^{-4}$ and $20*10^{-3}$ K^{-1}. Information about the rate and the amount of thermal expansion helps design around mismatches that can cause failure in the final product. In fact, different expansion coefficients

upon diurnal thermal cycles are accompanied by shear stresses that can cause delamination together with optical and thermal degradation.

Water Absorption

Water absorption is an important lifetime-limiting factor. If the moisture content of the interlayer material increases, its adhesion to glass reduces, and also the module's high voltage leakage current characteristic is affected. Moreover, water condensing at the interface between the encapsulant and the solar cell materials will create areas of increased corrosion and the risk of encapsulant delamination.

The simplest approach to this issue is to determine the relative rate of absorpiton of water by the plastic sheet when immersed. The reference standard is ASTM D 570 - 81. We performed the 24-h immersion test and determined the percentage increase in weight (%wt) during immersion. Measurements were done immediately after immersion (t = 0) and with some time delay (t = 15 min, t = 30 min, t = 40 min).

Sample	%wt			
	t = 0 min	t = 15 min	t = 30 min	t = 40 min
EVA1	0.13	0.08	0.06	0.02
EVA2	0.14	0.10	0.07	0.02
EVA3	0.16	0.13	0.11	0.05
PVB1	4.02	3.71	3.44	3.27
PVB2	0.47	0.42	0.40	0.39

Table 2: absorption coefficient of encapsulant materials according to ASTM D 570 - 81.

EVA has a low coefficient of water absorption, comparable to that of other polymers, and moisture rapidly ri-evaporates after water immersion.

On the other hand, PVB (especially PVB1) has a huge coefficient of water absorption, because it is a hydrophilic material and it can easily gain or lose moisture depending on the environment to which it is exposed. Anyway, there are different and more recent formulations (like for example PVB2) that have a lower coefficient of water absorption (the difference can be of even one order of magnitude).

Adhesion strength

Adhesion strength of encapsulants to the glass on PV modules is an important factor that can affect critically the performance reliability and durability of modules exposed to weathering environments. The adhesion strength depends on a number of factors, which include EVA type and formulation, backfoil type and manufacturing source, glass type, and surface priming treatment on the glass surface or on the backfoil. One of the methods that can be used for its determination is peel test according to ASTM D903-98. We conducted 180-degree peel tests using a tensile-compression machine model Instron 6025 with 5500R Control System (Fig. 8). The separation rate was of 152 mm/min. The samples were prepared

978-1-4244-4722-0/09 $25.00
© 2009 IMAPS-ITALY

using a laminator, and consisted in various glasses, different commercially available EVA formulations and backing foils from different sources. Industry partners provided them. Samples were tested at ambient room temperatures before and after a Damp Heat test, with also an intermediate read-out after 500 hours (half-test).

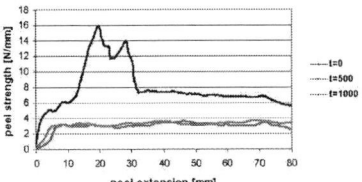

Figure 10: load-peel distance curves for a glass-EVA-Tedlar stack with a weak EVA-Tedlar interface. Measurements were done before, in the middle and after the Damp Heat test.

In both cases, the Damp Heat Test weakens the glass-EVA interface and, considering Fig. 10, it may become even weaker than the EVA-Tedlar interface. The results from these preliminary tests suggest that, as already mentioned, many are the factors that can determine the adhesion strength. Anyway, it's clear how the Damp Heat test may affect the adhesion strength, so it's very important to do the appropriate measurements in order to choose the most adeguate materials, able to withstand long-lasting outdoor exposures.

Conclusions

The preliminary results coming from our tests clearly show how many can be the limits of the two most widely used encapsulant materials (especially EVA) with respect to long-time outdoor applications. As quoted in many literature works, EVA is the dominant encapsulant not because it has the best properties possible, but because it is a very economical solution, and it is quite easy to process it into a manufacturing line. Its mechanical properties are not excellent, and they can be deeply affected by temperature changes. The Damp Heat test seems to have a tremendous impact on the performances of the material. Moreover, light transmission and UV stability are worse than those of other materials. EVA by itself does not provide adequate electrical insulation and scratch resistance.

PVB seems to be a good substitution for EVA, if moisture-related problems are solved. The long experience in the sector of automotive technology and building integrated vitrification can be carried over to the PV industry. Mechanical properties of PVB are good enough for many PV applications, like for example BIPV. As drawback, PVB storaging and processing is more demanding: the first requires low temperatures, the second the use of an autoclave together with the laminator.

Another interesting application field for PVB is thin film PV. Anyway, for thin film PV materials there is a much greater concern over moisture ingress. This is especially true of Cadmium Telluride (CdTe) and Copper, Gallium, Indium di-Selenide (CIGS) technologies; moreover, CIGS materials constructed on metal foils require flexible

Figure 8: testing equipment for peel test (sandwich sample included).

Considering the measurements performed before the Damp Heat test, two different kinds of curves can be identified, depending on which is the weakest interface: the glass-EVA one or the EVA-Tedlar one. For the first case, peel strength is never superior to 10 N/mm and the curve has a quite regular trend, with many spikes (Fig. 9).

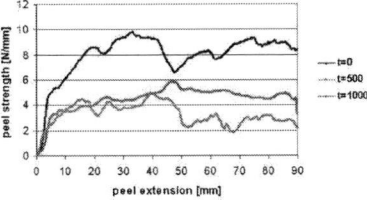

Figure 9: load-peel distance curves for a glass-EVA-Tedlar stack with a weak glass-EVA interface. Measurements were done before, in the middle and after the Damp Heat test.

In the second case, an initial peak can be observed, and it's due to the peel-off of the glass-EVA interface; afterwards, the EVA-Tedlar delamination begins with peel strength value comparable to those of the previous case. This confirms the fact that the glass-EVA interface is stronger than in the previous one (Fig. 10).

978-1-4244-4722-0/09 $25.00
© 2009 IMAPS-ITALY

encapsulants with good barrier properties to make BIPV products cost effective and easier to install.

The field of PV modules encapsulation seems to be a very open one: many are the materials and the properties involved, and many are the tests that can be done to evaluate them. The results coming from recent qualification tests indicate how critical can be the environmental tests, and polymers seem to heavily feel the effects of temperature and humidity. The introduction into the market of new products and technologies, addressing new application fields, requires more investigations and a deeper understanding of the reliability issues related to encapsulant materials. In fact, only a limited number of stress factors can be accelerated, and accelerated tests cannot provide a full prediction of product lifetime and long-lasting behaviour in field conditions.

Acknowledgements

The authors are gratefull to people from SEM's Materials Laboratory for their important help and contribution to the execution of the tests.

References

[1] C. R. Sosterwald, "History of Accelerated and Qualifications Testing of Terrestrial Photovoltaic Modules: A Literature Review", Progress in Photovoltaics: Research and Applications, 17: 11-33, August 2008

[2] A. Luque, "Handbook of Photovoltaic Science and Engineering", John Wiley & Sons Ltd., 2003

[3] M. D. Kempe, "Rheological and Mechanical Considerations for Photovoltaic Encapsulants", National Renewable Energy Laboratory publications, November 2005

[4] G. Jorgensen, "Materials Testing for PV Module Encapsulation", National Renewable Energy Laboratory publications, May 2003

[5] F. J. Pern, "Adhesion Strength Study of EVA Encapsulants on Glass Substrates", National Renewable Energy Laboratory publications, March 2003

[6] G. Jorgensen "Measurements of Backsheet Moisture Permeation and Encapsulant-Substrate Adhesion" ", National Renewable Energy Laboratory publications, October 2001

[7] B. Weller, "Interlayer Materials for Laminated Glass", Tchnische Univesitat Dresden publications

[8] G. TamizhMani, "Failure Analysis of design qualification testing: 2007 vs. 2005", Photovoltaics International, 112-116, Third Quarter 2008

[9] M. D. Kempe, "Design criteria for photovoltaic back-sheet and front-sheet materials", Photovoltaics International, 100-104, Forth Quarter 2008

[10] D. Tanner, "Large area thin film solar module lamination", pv-tech.org

978-1-4244-4722-0/09 $25.00
© 2009 IMAPS-ITALY

Fine Die-Attach Delamination Analysis by Scanning Acoustic Microscope

G. Santospirito and A. Terzoli

ST Microelectronics
Via C.Olivetti 2, Agrate Brianza- Milano Italy
Email addresses: giancarlo.santospirito@st.com; alessandro.terzoli@st.com

Abstract

Analysis of glue die-attach integrity of IC packages is typically performed by Acoustic Microscope in transmission mode (THRU-SCAN). THRU-SCAN mode can easily detect delamination at die-attach (D.A.) interface but is not able to define which is the interface delaminated.

D.A. delamination could be located at die-pad/glue or at glue/die interface.

In failure analysis it is important to fix the delaminated interface, to discover the failure mode and understand the root cause.

Up to now this analysis has been performed by cross sections with uncertain results and waste of time and money.

An analysis method using reflection mode has been developed in "CPA AGRATE stress analysis lab" to detect and fix D.A. delamination in exposed pad packages with glue D.A.

Acoustic microscope used for this trial is a SONOSCAN D9000; package test vehicle is a TQFP14x14-Ep.

Key words: Acoustic microscope, die-attach, exposed pad, cross sections, delamination

What is exposed pad TQFP

TQFP with exposed pad is made with a deep down-set leadframe such as the bottom side of the die paddle is flush with the bottom side of the mold and is exposed.

This allows the exposed pad to be soldered directly to the PCB and facilitates heat dissipation.

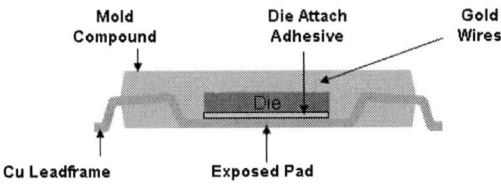

Benefits are related to thermal and electrical performance compared to full plastic packages; furthermore it offer a low cost solution if compared to packages with slug.

Die-attach acoustic microscope analysis

Standard method to perform glue D.A. inspection is THRU-SCAN analysis.
Ultrasound waves are transmitted through the entire sample thickness; defects, if present, may block the ultrasound from reaching the detector placed under the sample and will appear as dark shadows in the acoustic image. One scan reveals delamination that could be located at all the interfaces and there is no way to determine which interface is delaminated. (See Figure 1)

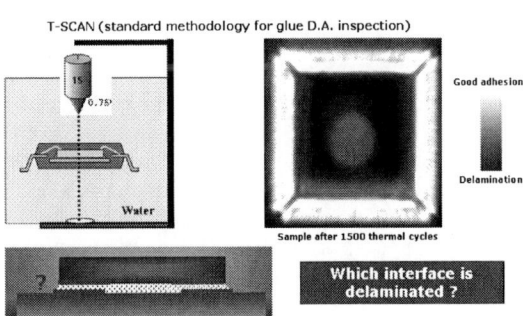

Figure 1: T-SCAN analysis

Another analysis method using acoustic microscope is the reflection mode or pulse echoes (C-SAM).
In this case ultrasound is reflected from the sample: this method guarantees more spatial resolution than THRU-SCAN mode. Delaminations are detected by phase inversion and analyzing peak amplitude.
In case of glue D.A. delamination, waveform does not show the typical phase inversion and glue detachment can be detected only accurately analyzing the peak amplitude.

978-1-4244-4722-0/09 $25.00
© 2009 IMAPS-ITALY

In order to fix D.A. delamination, previously it is necessary to set the signal amplitude to clearly identify and directly compare both good and delaminated interfaces.

Knowing acoustic materials properties it is possible to give the right interpretation of different peak amplitudes and detect delamination.

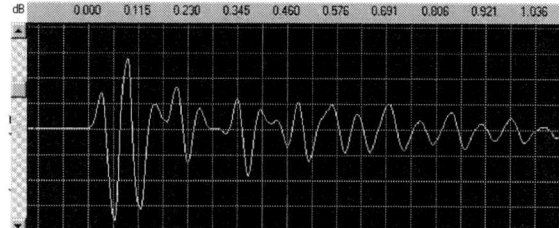

Figure 2: TQFP acoustic materials properties

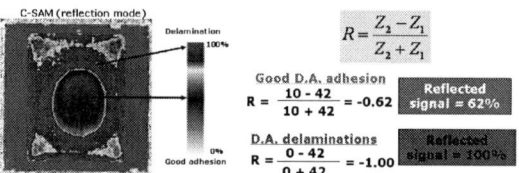

Figure 3: Typical C-SAM waveform (A-scan display)

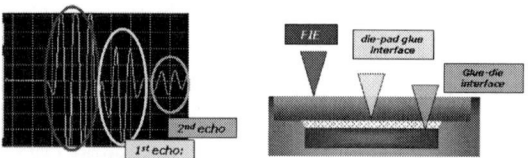

Figure 4: C-SAM analysis for D.A. integrity

In figure 4 is represented D.A.image obtained in C-SAM (reflection) mode. Delamination is detectable comparing the different quantity of reflected signal (the two echoes have the same negative phase)

But also reflection analysis performed in this way is not able to discover which interface is delaminated.

The only way to locate delamination is to identify the echoes related to both the interfaces (die-pad/glue and glue/die) in signal waveform. This is possible knowing package structure, thickness and acoustic impedance of materials and analyzing the time distance relationship in SAM waveform.

Time/depth correlation in reflection mode microscope analysis

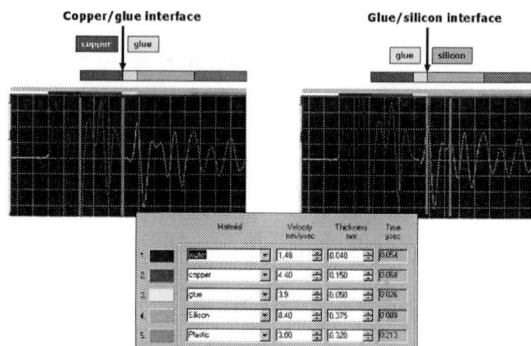

Fig 5: correlation between time on the A-Scan display and depth within the package.

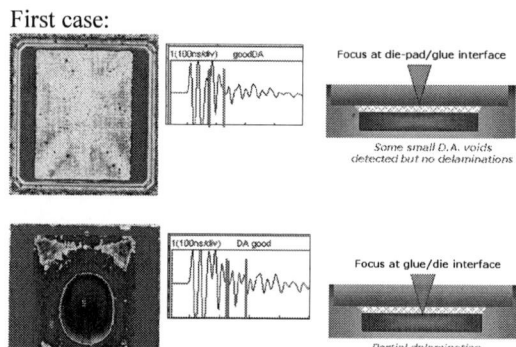

Figure 6: echo selection

Figure 6 shows how getting information about both D.A. interfaces; this is done identifying the two echoes related to that interfaces (die-pad/glue and glue/die) and restricting the electronic gate so that it includes only that peak. In case of a dual channel acoustic microscope (as the one used for this trial) analysis of both interfaces could be performed at the same time in one scan with still good resolution.

There are two possible cases:

First case:

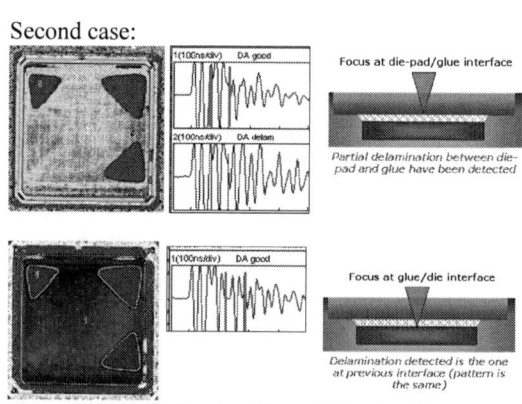

Figure 7: delamination is at glue/die interface

In sample analysed in figure 7, delamination is at glue/die level (the second interface encountered by acoustic signal). So, no delamination is detected selecting the first echo with the gate.

Delamination is only observed analyzing the echo related to glue/die interface.

Second case:

Figure 8: delamination is at die-pad/glue interface.

In the second case (figure 8) delamination is detected at both interfaces. In fact air gap present at die-pad/glue interface does not permit to the ultrasound signal to reach the interface of interest, in this case glue/die (see figure 9)

So delamination pattern is detected setting the gate on both peaks and we can easily conclude that detachment is at the first interface encountered (in this case die-pad/glue)

Figure 9: air gap does not permit signal propagation to the following interface

Cross section images (detachment confirmation)

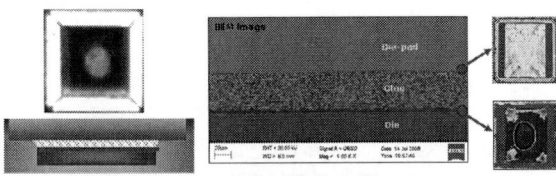

Figure 10: delamination at glue/die interface

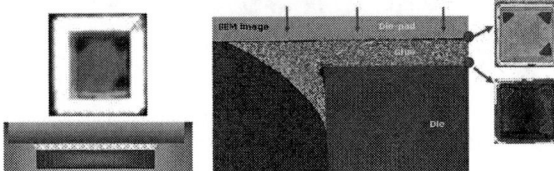

Figure 11: delamination at die-pad/glue interface

Analysis limitation

The technique described above is reliable until die-attach layer is thicker than a few wavelengths of ultrasound. Signal reflected from the two die-attach layer should be clearly separated in the time domain (oscilloscope trace or A-scan).

The wavelength of ultrasound in the adhesive glue for a 50 MHz transducer is approximately 30 μm.

When the layer thickness is less than or comparable to the wavelength of the ultrasound, the reflections from the front and back surface of the layer overlap and consequently become difficult to analyze the interface integrity.

From experimental point of view we observed some difficult starting from 25 μm of bondline thickness.

In this case it is obviously possible to clearly detect delamination but it is not easy to determine the side.

Conclusions

This non-destructive method is reliable to fix die-attach delaminated interface in IC packages with exposed pad.

As no more cross sections are needed, this analysis is useful to speed up the failure analysis and save money; in fact cross sections takes some few hours to be done and needs expensive comsumable materials.

This method could be easily extended to all packages with exposed pad, after proper equipment setting.

Acknowledgements

The authors are grateful to Claudio Maria Villa for the important technical contribution and Roberto Rossi providing the samples used in this study

References

[1] Tom Adams, Sonoscan Inc., and Rex Lee, Texas Instruments "Nondestructive Bondline Measurement in Advanced Flip Chips"

[2] Sridhar Canumalla and Bryan P. Schackmuth Sonoscan, Inc. "Metrology of Thin Layers in IC Packages Usingan Acoustic Microprobe: Bondline Thickness"
Proceedings of the 1999 International Symposium on Microelectronics. Assembled and Edited by the 1999 Technical Program Committee and the IMAPS Staff., p.264

[3] Janet E. Semmens, Sonoscan Inc., "Evaluation of Stacked Die Packages Using Acoustic Micro Imaging"

Appendix – Analysis procedure
(Equipment: SONOSCAN D9000)

1) Use the the 50 MHz transducer and map#27

2) Select the echo **r**eferred to die-pad / glue interface; it is the first one encountered beyond the FIE (front interface echo)

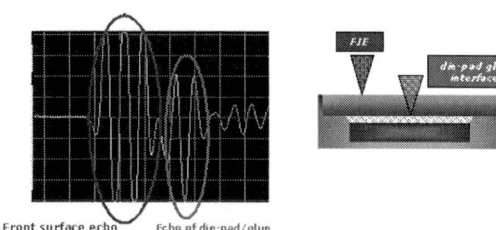

3) Use a reference sample (partial D.A. delamination between die-pad and glue) to locate the echo of die-pad / glue interface and select the appropriate signal amplitude in order to discriminate delamination area (red)

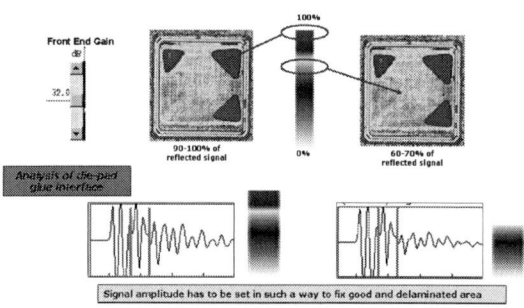

4) Remove the reference sample and put the package to be tested maintaining the same setting. Perform the analysis.

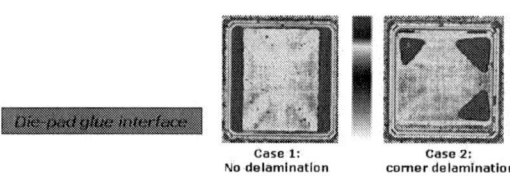

5) Select the next echo (referred to glue/die interface)

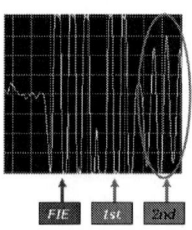 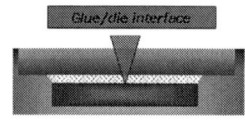

6) Use a reference sample (partial D.A. delamination between glue and die) to locate the echo of glue/die interface and select the right signal amplitude.

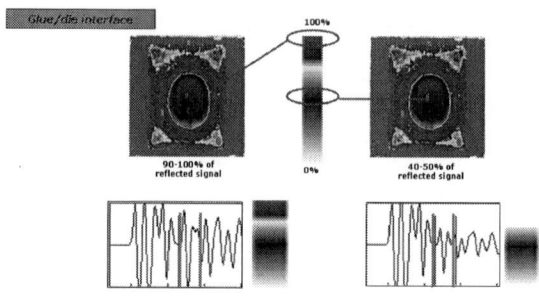

7) Remove the reference sample and put the package to be tested maintaining the same setting. Perform the analysis.

8) In case of detachment glue/die, delamination is detected with both the analysis (gate on both die-pad/glue and glue/die peak)

Focus is on glue/die interface but as delamination at die-pad /glue interface reflects the entire signal, no signal reaches the wanted interface.

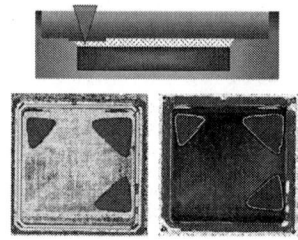

9) In conclusion there are 2 different cases:

a) <u>Detachment is at glue/die interface</u>: delamination is detected only with the second echo (no delamination detected with focus and gate on the first echo)

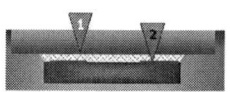  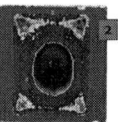

b) <u>Detachment is at die-pad/glue interface</u>: delamination is detected with both the first and second echo.

978-1-4244-4722-0/09 $25.00
© 2009 IMAPS-ITALY

Materials Science Challenges in Green High Power Density Devices
1. The Ag Loaded Glues for Die Bonding

Antonino Scandurra[a], Giuseppe Francesco Indelli[a], Roberto Zafarana[b], Angelo Cavallaro[b],
Emanuele Scrofani[b], Sebastiano Russo[b], Jean Paul Giry[b] and Salvatore Pignataro[a, c]

a) Laboratorio Superfici ed Interfasi, Consorzio Catania Ricerche, c/o STMicroelectronics, Stradale Primosole 50,
95121 Catania, Italy ++390957139148; http://www.ccr.unict.it
b) STMicroelectronics, Stradale Primosole 50, 95121 Catania, Italy; http://www.st.com
c) Dipartimento di Scienze Chimiche dell'Università di Catania, Viale Andrea Doria 6, 95124 Catania, Italy;
http://www.dipchi.unict.it

Abstract

Thermal and electrical performances of two conductive adhesive materials loaded with Ag particles suitable for power die bonding have been studied. The study has been performed using dice of Power MOSFETs having an area higher than 30 mm^2 assembled in TO220 package in comparison with Pb95.5-Sn2-Ag2.5 (wt.%) soft solder as reference; the interfaces between chip backside metal, leadframe and glues have been characterized by means of X-ray Photoelectron Spectroscopy after mechanical test and their compositions have been correlated to the thermal conductivity and electrical resistivity of the die-leadframe joints.
The results shows that interfaces between die backmetal, leadframe and glue in the die-leadframe joint play an important role in determining the final electrical and thermal conductivities. XPS analyses show the presence of silver depleted insulating regions inside the glues layers 9-10 nm thick, at least, and Ag particles electrically floating.

Key words: power green devices, conductive adhesive, XPS.

Introduction

An important innovation in high power density semiconductors technologies was represented by introduction of new advanced semiconductors materials with high electrical and reliability performances at high temperatures such as silicon carbide (SiC) and gallium nitride (GaN). One of the most important success key of high power density devices is represented by the packaging technology. Fig. 1 reports, as an example, typical high power density devices.

**Fig. 1. Example of high
power density devices.**

Die attaching of semiconductor devices to metallic leadframe or other metal plated ceramic materials that

requires a high current and/or a high heat flow through die attach material, (vertical flow current) represent one of the most critical process; traditionally die attach has been done with lead bearing soft solder alloys (typical Pb95.5-Sn2-Ag2.5 and Pb95-Sn5 wt. %) having also an high melting point, in the range of 310-312 °C, high thermal conductivity (of the order of 50 W(mK)$^{-1}$) and high thermo-mechanical fatigue resistance, being the electrical resistivity the less problematic parameter.

Environmental, health and safety regulatory, as well as consumer pressure, have pushed industry to remove lead and other dangerous elements from electronic products. Soft solders used for die bonding (die attach) is the most critical for replacing. At moment RoHS (restriction of hazardous substances) directives are accepting an exemption for high melting temperature type Pb solders, i.e. lead-based alloys containing 85% by weight or more lead [1]. However in spite of this exemption the research has been finalized to find new Pb free die attach materials having thermal conductivity, one of the most critical parameter, as much as possible higher, approaching that of soft solder or better. Some examples of Pb free die attach materials are the solders based on Bi alloys [2], on quaternary alloys such as Sn/Ag/Bi/Cu and Sn/Ag/Cu/Sb or more recently on Ag

978-1-4244-4722-0/09 $25.00
© 2009 IMAPS-ITALY

sintered nanoparticles [3]. However reliability of all these last materials has yet to be well understood, brittleness and grain growth under thermo-mechanical fatigue have to be expected, in particular when materials with large differences in the coefficients of thermal expansion, i.e. Cu vs. Si, are direct bonded. A large class of Pb free die bonding materials are based on conductive adhesive composites made of an organic polymeric matrix, e.g. acrylic, acrylic/epoxy, polyester or bismaleimide (BMI) filled with Ag particles and eventually with Ag nanoparticles [4, 5]. Actually their highest bulk thermal conductivity is of the order of 10 $W(mK)^{-1}$, in spite of some producers claims values of the order of 20-25 $W(mK)^{-1}$. The actual values are still too far from the most common requirements in power microelectronics devices. Typical Pb based soft solder values range from about 30 to 50 $W(mK)^{-1}$, depending on their composition. However it has to be pointed out that, when the materials are employed in die attach process in the package assembly, the values of bulk electrical conductivity and, particularly, the values of bulk thermal conductivity drop out furtherly to the so called "effective" thermal conductivity, just in assembled not aged devices. Consequently thermal conductivity values of the die to leadframe joints, often, are not acceptable in the applications of high power device. G.Q. Lu et Al. [3] showed that electrical resistance measurements of the Ag sintered resistor patterns gave an average electrical conductivity of 4 x $10^7 (\Omega m)^{-1}$, compared to 6.3 x $10^7 (\Omega m)^{-1}$ for bulk silver. The corresponding thermal conductivity calculated by the Wiedemann-Franz law is 290 $W(mK)^{-1}$ versus 459 $W(mK)^{-1}$ for the bulk. The nearly 40% lower conductivity values found in the sintered films are attributed to a substantial porosity (about 20%) remaining in the sintered microstructure. An Ag filled conductive materials have similar Ag contents with the difference that the porosity is filled with polymer matrix. Consequently, in principle, the electrical and thermal characteristics of conductive glues could greatly improve with respect to the actual values. This work give a contribution to explain some causes of the above inconsistencies.

Experimental

Two Ag filled glues named a) and b) respectively having different chemical and physical characteristics have been used for this study and compared with Pb95.5-Sn2-Ag2.5 (wt. %) soft solder alloy. Tab. 1 reports their main chemical and physical characteristics. The two glues have different polymer matrix. In particular a) has a lower viscosity and a much higher Ag content.

Bulk thermal conductivity λ_{bulk} of glues have been measured by laser flash method using the Netzsch LFA 447 Nanoflash® instrument. Electrical resistivity σ_{bulk} has been obtained using the 4-point probe method. Dice of n-channel Power MOSFETs with an area higher than 30 mm^2 with Au back metal have been assembled into TO220 plastic package with glues a) and b) and soft solder using Ag pre-plated copper leadframes.

Material	λ_{bulk} [W(mK)$^{-1}$]	σ_{bulk} [$\mu\Omega$cm]
Glue a)	7.2	30
Glue b)	1.8	300
Pb95.5Sn2Ag2.5 (wt.%)	53	5.5

Tab. 1. Main chemical and physical characteristics of studied die bond materials.

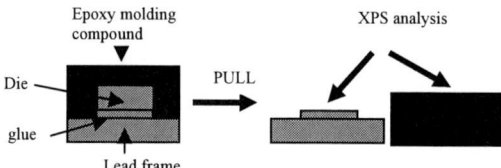

Fig. 2. Schematic representation of the procedure used for the XPS characterization of fracture surfaces obtained by mechanical test.

The choice of this package is justified by its simplicity and well known reliability behaviour. The die attach has been done by using a semi automatic glue dispenser and then the glues have been cured at the following conditions: 30 minute ramp to 175 °C followed by 30 minute hold at 175 °C in N_2 oven. Thermal resistances of junction to case θ j-c of assembled devices have been measured through the measurements of the diode forward voltage V_{SD} (that is function of junction temperature) before and after a power pulse of 50 W. Die shear test has been done by using the Shear Test Dage Series 4000 instrument. Soft solder and glues thickness layers have been measured by optical microscopy of devices cross sectioned. XPS glue-metals interfaces characterization after mechanical rupture by pull test (see Fig. 2) have been obtained by using a Kratos AXIS-HS spectrometer. Radiation MgKα$_{1,2}$ of 1253.6 eV has been used at the conditions of 10 mA and 15 keV. Areas of 2 mm × 2 mm have been analyzed. The pass energy of 40 eV has been used. The binding energy scale was referred to Au4f and Ag3d lines and to the lowest binding energy components in the C1s spectra fixed at 285 eV. During the analysis the residual pressure in the chamber was of the order of 10^{-7} Pa. Charge balance consisting in low kinetic energy electrons flux has been used to

978-1-4244-4722-0/09 $25.00
© 2009 IMAPS-ITALY

compensate the surface charging up of insulating surfaces. Spectra fitting have been done after linear background subtraction by using VISION Software (Version 1.4.0) by Kratos Analytical.

Results and discussions

The die attach process involves bonding semiconductor die to a Cu leadframe or other metallized ceramic substrates both in discrete power, power ICs or multichip modules with conductive adhesives or solder in the form of paste, solder wire or solder preforms [6]. The bond normally is between the die backside metallization and the leadframe metal surface. Die bonding workability and reliability depend strongly on the surface chemical compositions of the die backmetal and leadframe. Their full characterization is highly recommended to thoroughly understand the die attach process.

1. Physicochemical characterization of metals surfaces to be bonded
  a) Backside metal of the die

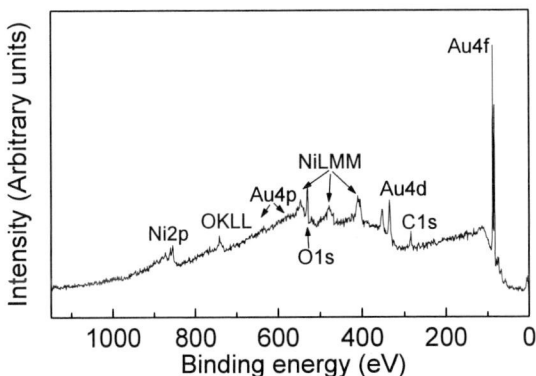

Fig. 3. XPS survey spectrum of backside metal with Au capping layer after 120 hours, 250 °C thermal treatment

In power dice with vertical flow current the backside must be coated with several layers of different metals optimized to allow soldering to the leadframe. Wafer silicon back surface is sputtered with 100 to 300 nm of Ti or Cr. Titanium and chromium form good diffusion barriers, and can remove any residual of SiO_2 on the surface by chemical reaction. Nickel or Ni-V also can be applied by sputtering. Often, 200 to 400 nm of nickel is added on top of the titanium or chromium, followed by 50 to 100 nm layer of flash gold or silver for protection and to improve solderability. Many variations are used, but most start with titanium,

include a layer of nickel, and then finish with protective layers of readily solderable flash gold, or silver. The same outermost metal layers are well suitable for the use with conductive adhesives. The back side of the die is subject to two main drawbacks that are related to the carbon contamination and Ni diffusion through the capping Au or Ag layers. Carbon contamination is ubiquitous, a consequence of wafer handling and back end processes such as die sawing and may be released by the sticky foils. Its presence represent a cause of worsening of glue adhesion to the metal surface. Moreover carbon contamination layer represents an additional insulator thin layer in series between metal and adhesive. Another important cause of adhesion loss and thermal and electrical isolation at interface is represented by the diffusion and oxidation of Ni through the Au or Ag capping layer. This diffusion becomes important at temperature higher than 200 °C. Fig. 3 reports a typical survey XPS spectra of a backmetal with Au capping layer after thermal treatment for 120 hours at 250 °C. The presence of Ni signals can be observed as consequence of its diffusion

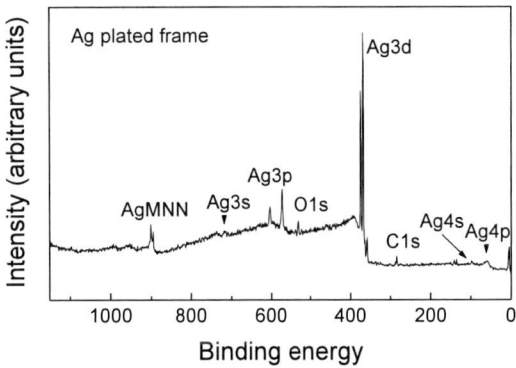

Fig. 4. XPS survey spectrum of a typical Ag plated leadframe surface.

through the Au layer. Surface segregation leads to nickel oxidation and reaction with air with final results of a layer of oxide-hydroxide-carbonate [7]. The surface employed in the bond in the general case could be formed of a mixture of Au or Ag and carbon and Ni in the above chemical states. In our case the surfaces are free of undesirable species.

  b) Leadframe

Leadframes show surface compositions that are the consequence of the plating processes and/or stamping and packaging, storage and shipment history until the assembly process. XPS spectra of leadframe surfaces usually show the presence of carbon signals and those of the metals plating (Cu, Ni, P, Ag) commonly in the

978-1-4244-4722-0/09 $25.00
© 2009 IMAPS-ITALY

form of mixture oxide-hydroxide-carbonate species. The layer of oxide-hydroxide-carbonate species often has a thickness less than 3 nm, but this is enough to worse dramatically the die attach process. Additional signals of other contaminants can be also found. In the case of Ag plating rarely contaminations of chlorine and sulphur can be found as consequence of wrong wet chemistry plating processes. A typical XPS survey spectrum of a clean Ag plated leadframe is reported in Fig. 4. The spectrum shows a little contamination of carbon species. In some cases Ar/H_2 plasma treatment of leadframe surfaces is recommended to improve the yield and reliability of die attach process.

c) Adhesion strength

Glue	25 °C	150 °C
a	163.7±33.3	54.9±14.7
b	348±66.6	193±15.7

Tab. 2. Adhesion strength (Newton) measured in shear mode.

Table 2 reports the adhesion strength measured in shear mode at 25 and 150 °C respectively for the two Ag filled glues. The comparison of values shows a worsening of adhesion strength of the high conductivity glue with respect to the low conductivity both at low and particularly at high temperature. The fracture occurs predominantly at the glue-die and leadframe glue for the glues a) and b) respectively. This finding is in agreement to the fact that high Ag content of the glue a), necessary to obtain high thermal and electrical conductivity, leads to an high viscosity material. This, in turn, for workability aspects, requires to adjust the glue viscosity by using proper polymer matrix and the overall chemical formulation that sometime penalizes the adhesion performances.

d) Thermal resistance θj-c measured on devices

Material	θ j-c K(W)$^{-1}$
Glue a)	0.4
Glue b)	1.1
PbSn2Ag2.5 (wt.%)	0.36

Tab. 3. Thermal resistance measured on devices.

Table 3 shows the average values of overall thermal resistance measured directly on assembled unstressed devices. The values corresponding to soft solder and glue a) are similar and this is explained considering that the low bulk thermal conductivity of glue is, at least partially, compensated by lower joint thickness obtained with respect to the soft solder. A similar

explanation is valid also for the glue b), but in this case the thermal resistance is about three times higher than of solder according to the lowest λ_{bulk}. However an

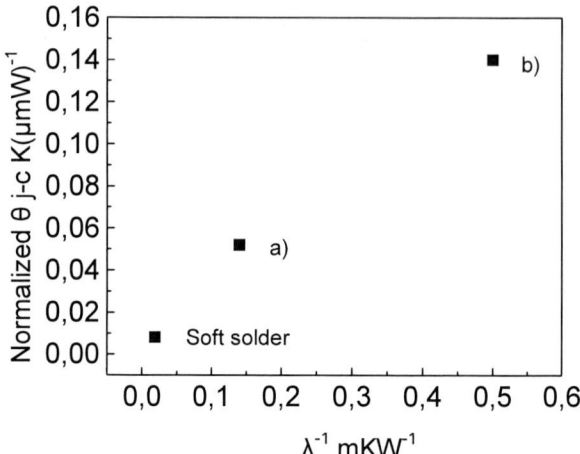

Fig. 5. normalized thermal resistance θj-c versus reciprocal of bulk thermal conductivity of die attach materials λ^{-1} for glues a) and b) and soft solder.

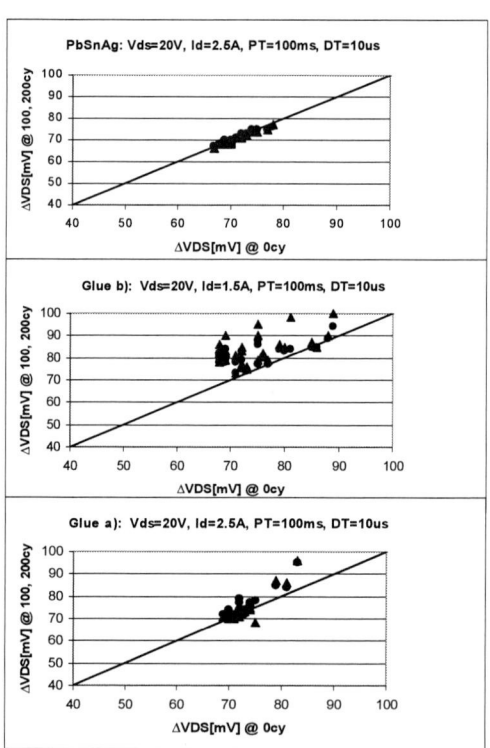

Fig. 6. ΔV_{DS} measured at 0 cycles (abscissa values) and after 100 (●) and 200 (▲) thermal cycles.

978-1-4244-4722-0/09 $25.00
© 2009 IMAPS-ITALY

excessive low die attach material thickness is cause of crack and delamination of the bond (see next section) due to the increased thermo-mechanical stress. The joint thickness normalized θj-c, as expected, is influenced by the bulk thermal conductivity of die attach materials. In particular Fig. 5 shows that the normalized θ j-c exhibits a good correlation with the reciprocal bulk thermal conductivity λ^{-1} of die attach materials, being obviously the other device components, i.e. die, Cu leadframe thickness and molding compound unchanged.

e) Reliability test

Figure 6 reports the ΔV_{DS} values measured at the conditions of V_{DS} = 20 V, I_D = 2.5 A, at a power pulse duration of 100 ms and a delay time interval of 10 μs with the exception of glue b devices that have been measured at I_D = 1.5 A. The ΔV_{DS} values have been measured before and after 100 and 200 thermal cycles (-65 °C 30 minute, +150 °C 30 minute) respectively. The devices assembled with solder exhibit low variations of this parameter after thermal cycling indicating good stability, as known. In the case of glues, in particular the glue b), a worsening (increase) of the parameter is observed. For both glues the ΔV_{DS}

increase is accompanied by the values spreading and these behaviors are related to the increased electrical resistance of the die to leadframe joints induced by thermo-mechanical fatigue of the joints.

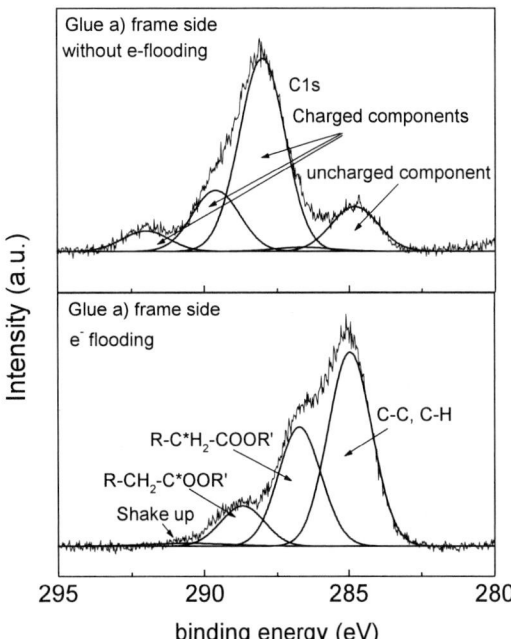

Fig. 8. C1s spectrum of glue a) from the frame side without and with e-flooding.

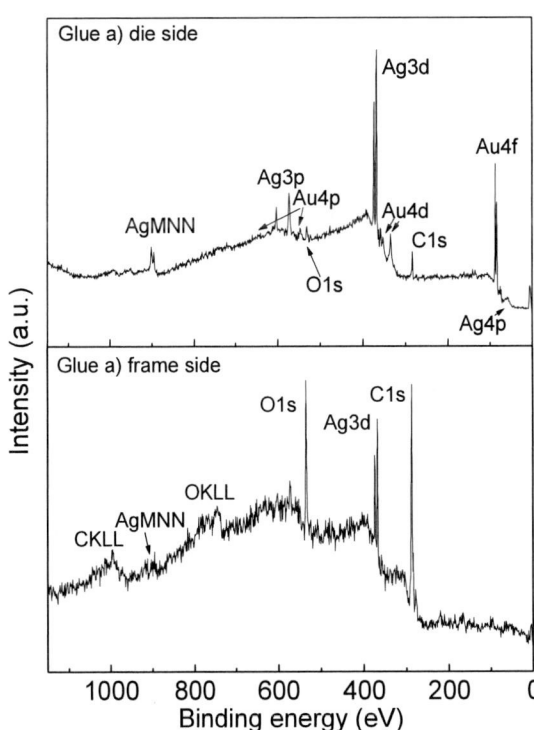

Fig. 7. XPS survey spectra of glue a) from die and frame side surfaces respectively.

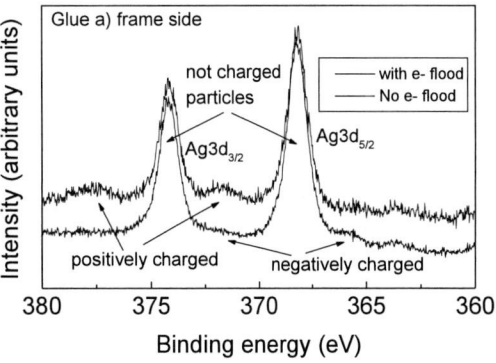

Fig. 9. Ag3d XPS spectral regions of glue a) frame side: without and with e-flooding.

2. Chemical characterization of interfaces of devices at t=0

The interfaces have been characterized after mechanical rupture of the die-leadframe joints (see Fig. 2).

Glue a)

Optical microscopy observations (not reported) revealed that the fracture is predominantly of adhesive type, according to the discussion of Table 2, with randomly distributed regions of cohesive type fractures, and occurs at the glue-die backmetal interface. Fig. 7 reports typical XPS survey spectrum of the die side (upper part) that shows Au in addition to C, O and Ag signals. The XPS detection of Au signals confirms that the fracture occurs very close to the interface with Au metal and that this layer is not dissolved by the glue, a conversely situation occurs with soft solder that is known to dissolve the capping layer by chemical reaction with tin. The high resolution C1s carbon signal acquired from this side surface (not reported) can be assigned, at least in part, to the segregation of additives and/or carbon based contaminants.

The XPS analysis of frame side (Figure 7 lower part), that retains the great amount of residual glue layer, shows the presence of Ag, C and O signals. Moreover the spectra with and without electron flooding (see reference [8] for the details of the method) indicate the presence of insulator material. In particular the charged components present in the C1s spectrum, reported in upper part of Figure 8, obtained without e-flooding, indicate the insulating nature of the XPS sampled outermost glue layer. The C1s spectrum reported also in Fig. 8 (lower part), obtained with e-flooding and after binding energy referencing, can be

fitted with Gaussian components centred at 285, 286.5, 289 eV, assigned to C-C/C-H, R-C*H$_2$COO-R' and R-CH$_2$C*OO-R' respectively, that are characteristics of nature of the polymeric matrix [9]. Fig. 9 reports the Ag3d spectral regions of the frame side both with and without charge balance. The presence of components at binding energy shifted, i.e. charged, indicate the presence of Ag particles floating and thus not forming metallic contacts to others Ag particles, but only with the insulating polymer matrix [8].

Glue b)

The fracture is mainly of adhesive type and occurs at the frame-glue interface. Again randomly distributed regions of cohesive type fractures can be observed. The XPS of die side (Fig. 10) that retains the glue layer shows the signals of carbon and oxygen. The Ag signals have not been detected (see the insert and the circle showing the Ag3d region). The absence of Ag at fracture surface leads to an insulator layer at interface between frame and glue which thickness is at least 3

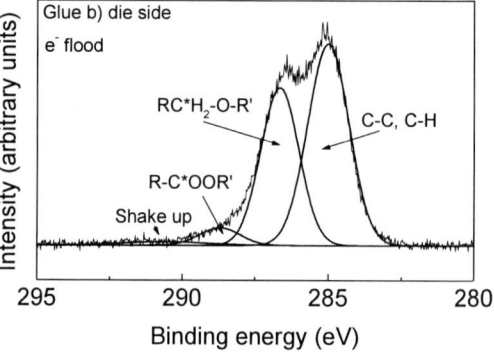

Fig. 11. C1s spectrum of glue b) die side.

times the inelastic mean free path of electrons at kinetic energy of about 885.3 eV (hv MgKα - Binding Energy Ag3d) in the polymer matrix of this glue estimable in about 9-10 nm [10, 11]. The high resolution C1s spectrum reported in Fig. 11, can be fitted with Gaussian components centred at 285, 286.7 eV indicating the nature of the polymeric matrix, and at 288.6 eV due to other glue additives and/or oxidation products.

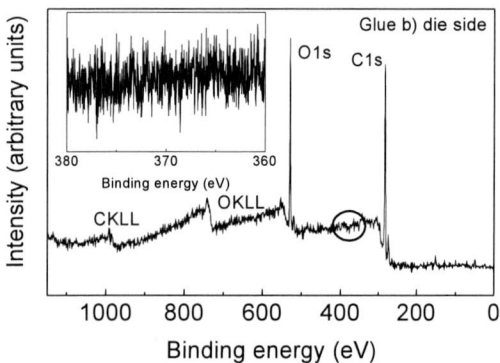

Fig. 10. XPS survey spectrum of glue b) from the die side surface. The circle and the insert indicate the spectral region of Ag3d doublets.

978-1-4244-4722-0/09 $25.00
© 2009 IMAPS-ITALY

Fracture surface quantitative chemical compositions of the two glues

Material	Surface compositions % wt.		
	C	O	Ag
Glue a) frame side	58.1	21.1	20.6
Glue b) die side	71.2	28.8	n.d.[1]

[1] Below the detection limit of the XPS technique.

Tab. 4. Compositions of fracture surfaces measured by XPS.

Table 4 shows typical fracture surface chemical composition (% weight), measured by XPS, of the corresponding survey spectra reported in Figures 7 and 10 respectively, that represent the fracture surface (interfaces before mechanical rupture) compositions of the glues layers. The compositions refers to average fracture surfaces including both adhesive and cohesive type regions (see experimental section). Particularly worthy of notice is the very low silver surface concentration of glue b) compared to the bulk compositions that contain more that 80 % weight of silver. Also in the case of glue a) the Ag content at interfaces is even below the threshold of conduction that is in the range 65-70 % of silver [12]. These interface effects are to be considered responsible of the reduced thermal and electrical performances particularly of glue b).

Conclusions

The results reported in this work show that the performances of the adhesive conductive materials in die-leadframe joints are strongly influenced by die-glue, glue-leadframe and Ag particle-Ag particles interfaces. These findings are valid also increasing the Ag content in the material formulation to obtain high electrical and thermal conductivities. The XPS analyses of die-glue-leadframe fracture surfaces revealed the presence of Ag depleted insulating layers at backmetal die – glue and leadframe-glue interfaces. These insulating layers in the analyzed cases have estimated thickness at least 3 times the IMFP i.e. of the order of 9-10 nm. The insulating layers have to be considered both at level of interfaces with glue-die and glue-leadframe as well as at level of Ag particles interfaces. The observed depleted regions influence both thermal and electrical resistance of the assembled devices and their variations under thermal cycling. The mechanism leading to the formation of these Ag depleted regions is still under study. The incompleteness formation of metallic bonds among the Ag particles is the most important cause of poor thermal and electrical conductivity of glues. New chemical formulations of conductive glues must be able to promote the formation of metallic bonds among Ag particles or, better, their sintering and this can help to improve greatly the glues performances.

Acknowledgements

The authors wish to thank Giovanni Corrente and Filippo Scrò of STMicroelectronics for the reliability measurements and Netzsch GmbH for the collaboration in the area of thermal conductivity measurements.

References

[1] Directive 2002/95/EC of the European Parliament and of the Council of 27 January 2003 on the restriction of the use of certain hazardous substances in electrical and electronic equipment, Official Journal of the European Union, 13.2.2003.

[2] C. Tschudin, O. Hutin, S. Arsalane, F. Bartels, P. Lambracht, M. Rettenmayr, "Lead Free Soft Solder Die Attach Process for Power Semiconductor" Proceedings of the Packaging, SEMICON China 2002 Technical Symposium, SEMI, Shanghai (2002) pp. AH1-AH5.

[3] G.Q. Lu, J. N. Calata, G. Lei, and X. Chen, "Low-temperature and Pressureless Sintering Technology for High-performance and High temperature Interconnection of Semiconductor Devices", Proceedings of the 8th Int. Conf. on Thermal, Mechanical and Multiphysics Simulation and Experiments in Micro-Electronics and Micro-Systems, EuroSime 2007, p. 609-613.

[4] Yi Li, C.P. Wong, "Recent advances of conductive adhesives as a lead-free alternative in electronic packaging: Materials, processing, reliability and applications", Materials Science and Engineering R 51 (2006) 1–35.

[5] L. Gobbato, A. Scandurra, T.Y.Tee, "Silver Glues and Lead-free Alloys in Die Attach of Power Devices", Proceedings of the 7th Electronic Packaging Technology Conference, Volume 1, Singapore, Dec. 7-9, (2005), pp. 363-366.

[6] V. Del Bo, F. Bartels, A. Scandurra, Ch: Luchinger and S. Radek, "More Insight into the Soft Solder Die Attach", Proceedings of the 11th European Microelectronics Conference, Venice, Italy, 308-319 (1997);

[7] A. Scandurra, A. Porto, L. Mameli, O. Viscuso, V. Del Bo and S. Pignataro, "Characterization and Reliability of Ti/Ni/Au, Ti/Ni/Ag and Ti/Ni back Side metallization in the Die Bonding of Power Electronic Devices", Surf. Interf. Anal. Vol. 22, p. 353, (1994);

[8] A. Scandurra, A. Cavallaro, S. Pignataro, R. Tiziani, L. Gobbato and C. Cognetti, "Curing and electrical conductivity of conductive glues for die attach in microelectronics", Surf. Interface Anal. (2006); 38: 429-432

[9] G. Beamson and D. Briggs (Eds.) The XPS of Polymers Database, Surface Spectra Publisher, 2000

[10] S. Tanuma, C.J. Powell, D.R. Penn, "Calculations of electron inelastic mean free paths for 31 materials", Surf. Interf. Anal. 11 (1988) 577

[11] S. Tanuma, C.J. Powell, D.R. Penn, "Calculations of electron inelastic mean free paths. V. Data for 14 organic compounds over the 50-2000 eV range", Surf. Interf. Anal. 21, (1993) 165.

[12] E. Allievi, "Thermally interface materials for microelectronic assembly", Proceedings of the Adhesive & Polymers in Microelectronics and Photonics, IMAPS Workshop, October 19, 2005, Milan, Italy.

Modeling of Flip Chip Bump Patterns to minimize Crosstalk on a BU-BGA package design.

Keith Sheach, Gordon Xiang, Senior Member, IEEE, Pierre Brunet
High Speed Package Design, Ottawa Design Center, STMicroelectronics, Inc.
16 Fitzgerald Road, Nepean, Ontario, K2H 8R6, Canada
Emal: keith.sheach@st.com, gordon.xiang@st.com, pierre.brunet@st.com,

Abstract

We provide a practical example of concurrent design practice employed to optimize system level design specifications. Through package level modeling it was possible to improve the performance of a die I/O assignment. 3D EM modeling of the package flip-chip interconnect revealed that a significant improvement in the differential cross talk levels could be achieved by simple re-ordering of the die pad assignments.

Key words: High-speed, X-talk, Interconnect, Flip chip package

I. Introduction

In the electronics industry commercial considerations are, as always, driving the performance requirements of electronics systems to increasingly rigorous levels. Die I/O counts continue to rise rapidly, faster clock rates are being implemented while cost demands continue to shrink die sizes. These are familiar challenges to the packaging design community, where the rapid advances in silicon technology must be harnessed and interfaced to the less volatile technologies of PCB design and assembly.

The nature of this challenge is not new, and indeed it has been a principle driving force behind the packaging design process with, for example, increasing convergence and integration of package design tools into the overall system design process. What lends new urgency to these trends are the expanding commercial markets for the highest performance silicon technology in relatively low cost consumer products. Performance specifications which relatively recently may have been required only in the highly specialized, and less cost sensitive, military and medical markets now appear in the datasheets of the latest gaming consol, or home PC video card.

The importance of an integrated design approach, and parallel development have been addressed widely, and are generally agreed. Packaging is the physical interface between the silicon technology and the PCB, and packaging designers are required to take greater responsibility for ensuring the performance achievable at the silicon is delivered to the board. This in turn requires that packaging design include aspects of the die level I/O and the PCB pads and interconnect structures connecting the package. In this article, we demonstrate a practical approach to these issues.

II. Methodology

Presented with design specification with particularly rigorous crosstalk requirements a built up package was designed allowing critical nets (differential pairs) to be routed on separate layers and hence minimizing cross talk between those pairs. Careful planning of the BGA ball assignments also separated those critical pairs and ball to ball cross talk was modeled and found to be negligible. The remaining cross-talk between the critical pairs was therefore entirely a resultant of the die I/O pattern, and the associated flip chip bumps and breakout via structures.

This package design and modeling process was carried out concurrently with the silicon development. Hence the package designers had input into the final I/O placement pattern and pitch, an obvious benefit to the parallel design approach, affording an opportunity to optimize the "macro" system level performance from consideration of "micro" design features.

Three different die bump patterns were modeled to compare the influence of the signal bump spacing and pattern on cross-talk between signals. The first pattern investigated modeled a total of 6 flip chip bumps, with a regular orthogonal spacing on a 225um pitch. The pattern consisted of two rows, the first row included one GND bump, and a TX_P and a TX_N bump. The second, adjacent, row contained a GND and an RX_P and

978-1-4244-4722-0/09 $25.00
© 2009 IMAPS-ITALY

RX_N bump. The TX pair was routed from the flip chip pad directly down through buried/blind vias to layer 4 of the package.

The RX pair was similarly routed to layer 2 of the package. Both pairs were then routed on their respective layers, separated by a continuous ground plane.

The 3D model of the flip chip bumps, vias, a short section of differential transmission lines, and the associated ground structures was a constructed in order to determine the crosstalk which could be attributed directly to the flip chip interconnect structures. The specification of interest was the differential cross talk between the RX pair and the TX pair. The model construction is shown in **Figure 1.**

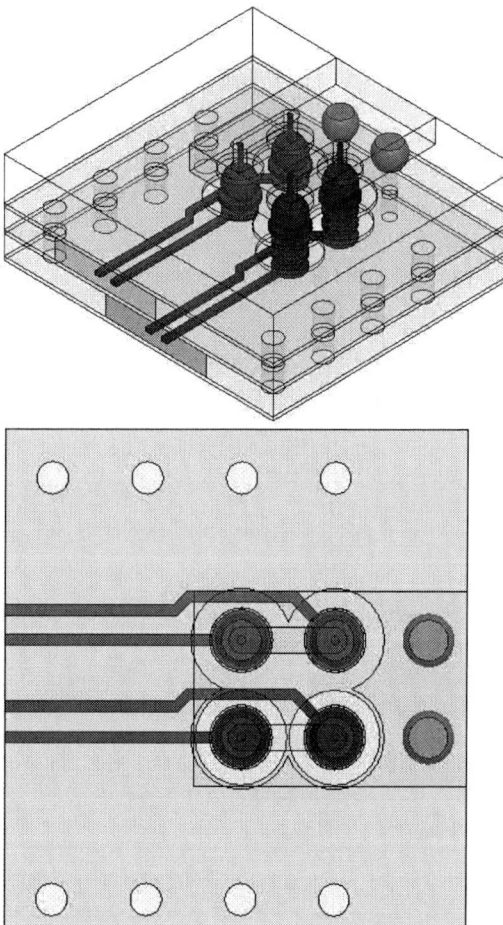

Figure 1:Standard orthogonal pattern I/O 225um pitch (RX shown in Red, TX in Blue, GND BGA Balls in Green)

A second 3D model was also created where the flip chip signal bumps used the same pitch but a staggered pattern, hence further separating the RX to TX pair distance without modifying the overall die

pad co-ordinates. The geometry of this model is shown in **Figure 2.**

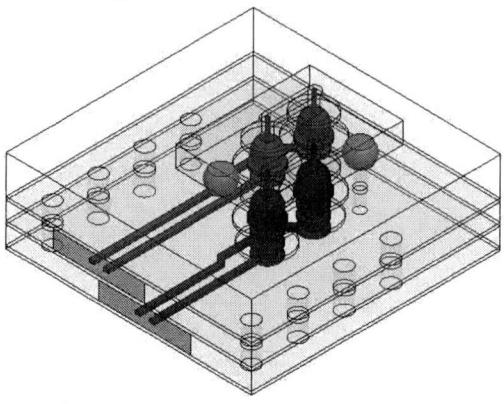

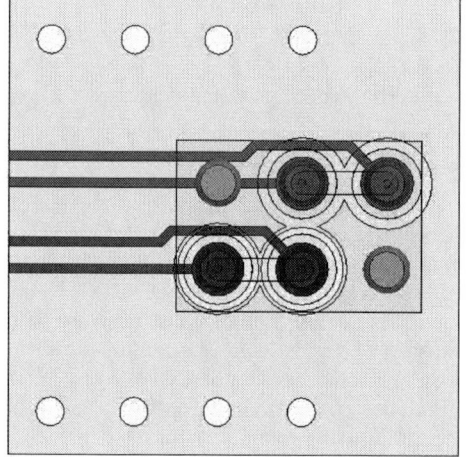

Figure 2: Staggered pattern I/O 225um pitch (RX shown in Red, TX in Blue, GND BGA Balls in Green)

The final model geometry which was constructed used the pattern of the standard (orthogonal) layout, but the pitch between RX and RX bumps was reduced from 225um to 200um, while the RX to TX pitch was retained at 225um. This geometry can be seen in **Figure 3**.

The simulation conditions, boundaries, and meshing parameters were identical in each of the three models. A meshing frequency of 16 GHz was used, in a 'driven terminal solution'. The output ports captured the stripline differential pairs on layer 2 (RX pair) and layer 4 (TX pair). The input ports were defined by constructing a quasi coaxial feedthu for each of the signal FC balls. These ports were also defined as differential ports for the RX and TX pairs respectively. The short length of 'quasi-coaxial' feed for the RX pair were isolated from the similar TX pair structures by grounded metal, therefore the port and launch structures made no contribution to the differential cross-talk.

978-1-4244-4722-0/09 $25.00
© 2009 IMAPS-ITALY

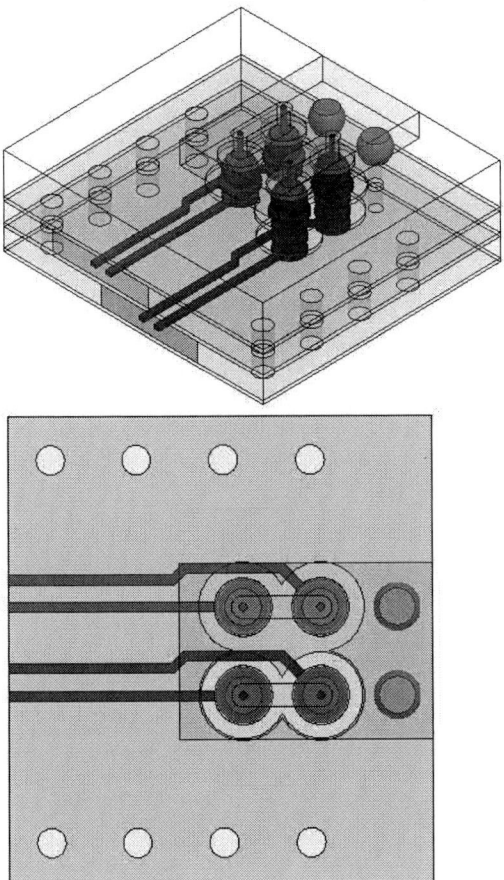

Figure 3: Orthogonal pattern I/O 200um pitch within row (RX-RX).
225um pitch between rows (RX-TX)
(RX shown in Red, TX in Blue, GND BGA Balls in Green)

Boundary conditions were applied identically in the three models with radiation boundaries defined on the edges of the geometries representing the gap between the die and the package substrate, and conductive boundaries representing infinite ground planes above and below the model geometries.

III. Results

Figure 4 shows the results for insertion loss, return loss and cross-talk for the RX pair of the standard (orthogonal bump pattern) model. Insertion losses (IN_RX to OUT_RX) on this small structure are as expected minimal, likewise return losses (IN_RX to IN_RX) are low and remain below -20dB up to 16GHz.

The most important result, the differential cross-talk, is shown as two curves, one shows the near-end cross-talk (IN_RX to IN_TX), and the other shows the far-end cross talk (IN_RX to OUT_TX). For both near and far end cross-talk the

S-parameters indicate minimal coupling is present between the RX and TX pair.

The maximum cross-talk apparent at the highest frequency of 16GHz remains in the region of -40dB, equivalent to a voltage variation of 1%. This level of cross-talk would normally be considered good in most applications, and in this case does meet the design specifications, but with little remaining margin which could easily be consumed by manufacturing tolerances and material variations.

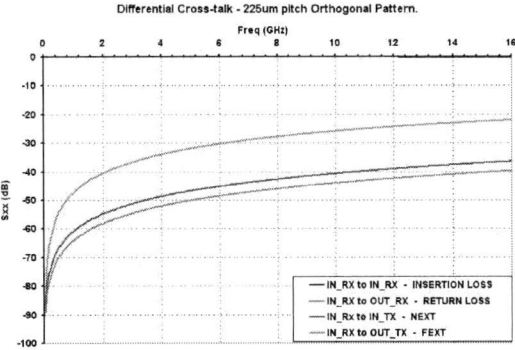

Figure 4 – Differential Crosstalk for 225um pitch Orthogonal model.

Figure 5 shows the same data sets for the 'staggered' bump pattern model. Insertion losses (IN_RX to OUT_RX) on this small structure are similar to those found in the standard model, and return losses are also nearly identical to the standard model as would be predicted. The crosstalk observed is again shown as two curves, one shows the near-end cross talk (IN_RX to IN_TX), and the other shows the far-end cross talk (IN_RX to OUT_TX).

In this case the cross talk indicated by the S-parameters data shows a marked reduction in comparison with the standard model. Overall the crosstalk is observed to be lower, and there is greater differentiation between near end and far end effects. Far-end cross-talk has been reduced by the greatest margin and does not rise above -70dB up to 16GHz, near-end cross-talk is also reduced and remains below -50dB to the maximum frequency.

These levels of cross talk are broadly equivalent to a voltage variation in the region of 0.1%, and represent a considerable reduction in differential cross talk from already low levels, allowing the design to meet the strict cross talk specifications with sufficient margin to eliminate any concern of material or manufacturing variations.

Finally, **Figure 6** shows the results for the mixed pitch orthogonal design. It can be observed that insertion losses remain largely unchanged for the RX pair. The crosstalk data however does

978-1-4244-4722-0/09 $25.00
© 2009 IMAPS-ITALY

indicate a small reduction in near end, and far end cross talk to the adjacent TX pair. The improvement however is minor; approximately 2dB at 10GHz, therefore modifying the bump pattern to provide an asymmetric bump pitch would only allow a significant performance improvement in the most marginal of design conditions.

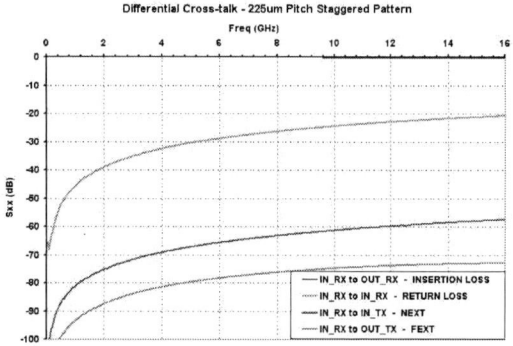

Figure 5 – Differential Crosstalk for 225um pitch Staggered model.

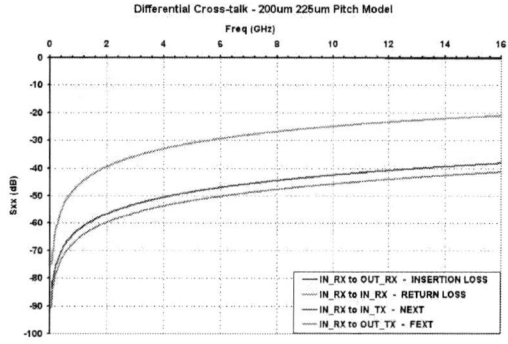

Figure 6 – Differential Crosstalk for 225um / 200um pitch Orthogonal model.

IV. Conclusions

HFSS models of three possible designs of SERDES flip chip bump pattern were constructed and the results compared. It was found that cross talk effects in all three patterns were low structures and would be acceptable in most cases. However the crosstalk associated with the staggered I/O pattern indicated a significant reduction in differential cross talk when compared with the standard orthogonal pattern. A much smaller reduction in the cross talk was observed in the I/O pattern which used a 225 inter-pair pitch and a 200um intra-pair pitch. However it should be noted that this result also indicates that a smaller I/O pitch could be implemented without compromising the existing cross-talk specification, thereby reducing the die size.

From consideration of the die I/O design requirements it was clear that the marginal improvement resulting from the mixed pitch design was insufficient to justify the new pattern, and this case the small reduction in die size that this pattern would allow was not a critical consideration. The staggered I/O pattern, using the regular 225um pitch did however offer the prospect of improved performance with minimal impact on the die I/O design.

Hence as a result of the early engagement of the package designers, and consideration of system level specifications, it has been shown that it was possible to significantly improve the overall performance of a design by providing recommendations on the layout of the die level I/O.

Joint Project for Mechanical Qualification of Next Generation High Density Package-on-Package (PoP) with Through Mold Via Technology

Moody Dreiza, Jin Seong Kim, and Lee Smith of Amkor Technology
Didier Campos and Eric Saugier of ST Microelectronics
Pauli Jarvinen of Nokia

Abstract

This paper will summarize joint work between ST Microelectronics, Amkor Technology and Nokia; to qualify Amkor's through mold via (TMV™) bottom package technology for next generation high density PoP applications. The 12 x 12mm daisy chain test vehicle reported in this joint work includes a thin flip chip die in a fully molded bottom package with 516 bottom BGAs at 0.4mm pitch and 168 top solderable through mold vias at 0.5mm pitch. This paper will report the package level (moisture resistance, temperature cycling) and board level (temperature cycle, drop) qualification data against IC and handheld application requirements.

Additional data for package warpage control and board level reliability for larger PoP applications using TMV technology will be included beyond what was reported at ECTC [1], SMTA International [2] during 2008 and IMAPS Device Packaging [3] in March 2009, based on a 14 x 14mm daisy chain test vehicle with 620 bottom BGAs at 0.4mm pitch and 200 top vias at 0.5mm pitch. Additional data on the TMV technology will be provided including: maximum die to package size design benefits for wirebond, stacked and flip chip die, coplanarity and package warpage measured by shadow moiré across lead free SMT reflow profiles. JEDEC standardization work for next generation PoP applications will be provided for mechanical and high density electrical interface requirements driven by low power double data rate 2 memory (LP DDR2), in single and dual channel architectures which require 0.5 and 0.4mm pitch interfaces respectively. [4]

Keywords: 3-D packaging, package-on-package (PoP), stacked package, high density interconnect

Background and First Generation PoP Technology

Nokia and Amkor played key roles in the development of the first generation of PoP technology, beginning with early work reported at the 2003 ECTC conference. [5] ST Microelectronics has been an early adopter of PoP for both memory [6] systems as well as mobile processors. The background or history of commercialization for this first generation of PoP technology was summarized in an article from Smith published by Semiconductor International in June of 2007. [7]

PoP has seen tremendous growth over the past four and a half years following the first adoption in a mobile phone. Recently, industry analysts estimate between 175 and 220 million bottom PoP units were shipped in 2008 with over 80% consumed by mobile phones, driven by the high silicon content required in smartphone applications. The first generation bottom PoP technology typically integrates the baseband or application processor device and uses either a center gate mold or an exposed flip chip die structure. The top single or combination memory package typically uses a perimeter 2 row solder ball array for the stacking or memory interface, using a ball diameter and pitch sufficient to provide stacking clearance over the center mold or FC die as shown in **Figure 1.** 0.65mm pitch stacked interfaces are typical with center mold bottom packages and 0.5mm pitch interfaces are common with use of thin exposed FC die bottom packages. These technologies have served the industry fairly well over the past 4 years but face challenges when new applications require higher integration such as stacked die in the bottom package and interconnect densities below 0.65mm pitch in the stacked interface. A high density PoP approach to support these requirements, explored the creation of a tall fine pitch solder column like structure with a ball on ball type stacked interface as reported by Dreiza et al [8]. This technology showed promise but has not seen wide commercial adoption due to the material changes required in established SMT stacking processes.

978-1-4244-4722-0/09 $25.00
© 2009 IMAPS-ITALY

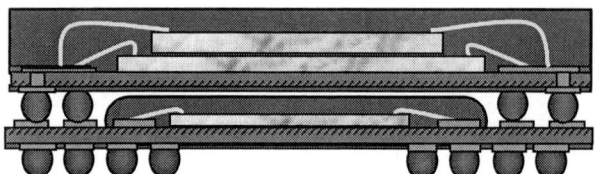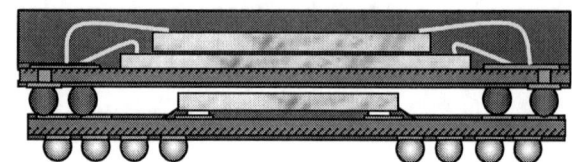

Figure 1: 1ˢᵗ generation of PoP stacked structures.

Requirements for Next Generation High Density PoP Applications

Next generation PoP technical requirements have been listed recently by Smith [9] along with Zwenger et al [3]. In summary, these publications state the market requires a next generation high density bottom PoP technology that provides – increased: integration, miniaturization and performance without requiring development of a new SMT stacking infrastructure or adding cost. These are challenging requirements to meet given the increased interconnect densities associated with new memory and signal processing architectures. Reports on the first generation of PoP technologies provides a baseline for improvements required in BGA pitch reduction with tighter warpage control, thinner overall stack ups, high stacking yields without impact or design restrictions for higher die to package ratio applications. The baseline data for stacking yields requirements by current PoP stacked interface pitches can be found in the publications by Ishibashi [10] of Nokia which explored the elevated warpage profile differences between the bottom (concave) and top convex warpage and the impacts on stacking yields as shown in **Figure 2**. Ishibashi concluded for high yield stacking, package reflow warpage levels in the PoP memory interface area should be controlled to 36um for 0.65mm and 33um for 0.5mm pitch stacked interfaces. The baseline data for the impact of bottom die size ratios can be found in the joint study by Amkor, Nokia, Panasonic, Senju and Sharp reported by Yoshida et al [11] expanded to include other design variables in the paper by Lin et all [12].

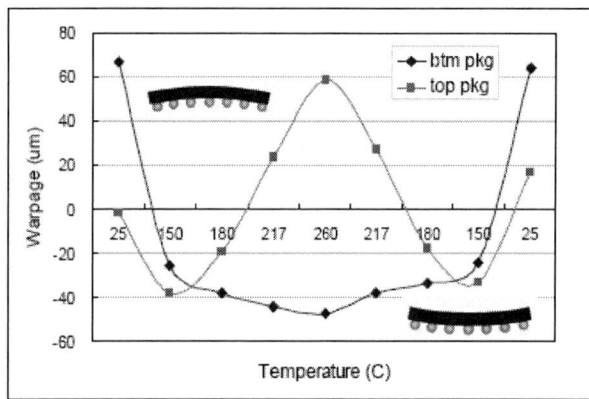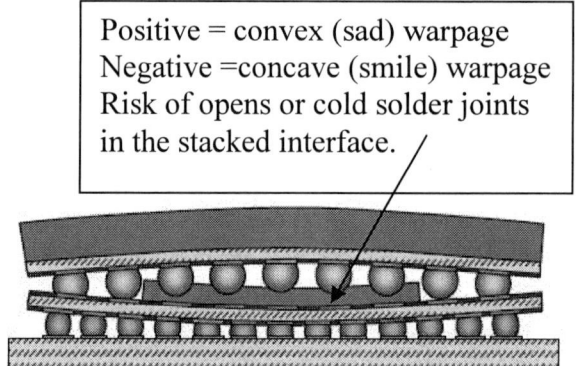

Figure 2 Shadow moiré warpage plot for bottom vs top PoP profiles along with cross section figure showing impact large liquidus to solidus warpage gap can have for opens or cold solder joints in the critical stacked memory interface.

Through Mold Via (TMV) Technology for Next Generation High Density PoP Requirements

Amkor has benefited from the strong growth in PoP applications [13] as a full service, high volume supplier of PoP technologies, which includes design and assembly of bottom, top packages and system in a package (SiP) modules with integrated PoP stacks assembled with a one pass reflow SMT stacking process flow. Due to this high level of business and broad participation as represented in the reported research, Amkor has been evaluating technologies which would address the challenges presented by next generation high density PoP applications. The application of solder vias through the bottom package mold cap was first reported by Kim et al [1] as a new bottom package structure and assembly method for fine pitch PoP requirements with improved warpage control. A joint board level reliability study based on this high density 14 x 14mm test vehicle with 620 bottom BGAs at 0.4mm

978-1-4244-4722-0/09 $25.00
© 2009 IMAPS-ITALY

pitch and 200 stacked solder joints at 0.5mm pitch reported at ECTC was reported at SMTA International [2]. The TMV PoP test vehicle is shown in **Figure 3** below.

14 x 14mm 6 net daisy chain next generation PoP test vehicle
- 0.4mm thick mold cap with molded underfill encasing
• 7 x 7mm Flip Chip daisy chain die at 220µm bump pitch
• 32 tiny 01005 size 0 ohm resistors (to represent decoupling caps)

Cross Section View

Top View
200 Solder Lands
@ 0.5mm pitch
memory interface

Amkor®
TMV™
14 mm 620 / 200

Bottom View
620 BGAs @
0.4mm pitch

TMV test vehicle reported at ECTC 2008
Joint tech paper at SMTAI August 2008

Figure 3, bottom TMV PoP test vehicle cross section top and bottom views as reported at ECTC and SMTAI in 2008.

Amkor's internal manufacturability and reliability qualification of the TMV PoP technology was reported earlier this year along with the official market introduction for the availability of this technology at the IMAPS Device Packaging Conference [3, 14].

Test Vehicle Description for This Joint Work on Mechanical Qualification of TMV PoP
For the purposes of this mechanical qualification study, a 12 x 12mm (144mm²) bottom TMV PoP was designed using a 64mm² ST Microelectronics flip chip daisy chain die, having an area array bump pattern at a 225um bump pitch. The bottom BGA pattern consists of 516 lead free 0.25mm diameter solder balls at 0.4mm pitch. The top TMV pattern consists of 168 lead free solder vias at 0.5mm pitch. The bottom TMV PoP test vehicle is shown in **Figure 4**.

12 x 12 mm multi net daisy chain next generation PoP test vehicle
- 0.25 mm thick mold cap with molded underfill encasing
8.25 x 7.75 mm Flip Chip daisy chain die at 225µm bump pitch

Cross Section View

Top View
168 Solder Lands
@ 0.5mm pitch
memory interface

Amkor®
TMV™
12 mm 516 / 168

Bottom View
516 BGAs @
0.4mm pitch

Figure 4, bottom 12mm TMV PoP test vehicle used in this joint work.

The use of a daisy chain FC die in the test vehicle allows for in-situ monitoring of the thin FC die and lead free bump to substrate interconnection through mechanical shock (drop) and temperature cycle (TC) stress testing. The design has separate bottom and top BGA nets for in-situ monitoring of the fine pitch bottom BGA to mother board and top package BGA to TMV connections during drop and TC testing.

The package substrate was designed to rules required for functional high density bottom PoP components with a 4 layer (1-2-1) blind/buried via thin core stack up resulting in a 300um overall substrate thickness. The flip chip wafer was thinned to be encapsulated by a thin 0.25mm mold cap which meets next generation PoP thickness reduction targets. The flip chip die was underfilled with a low stress material typical for advanced CMOS devices that use brittle low K dielectric inner-layers.

Material selection for This Joint Work for Mechanical Qualification of TMV PoP

Warpage control across the elevated temperature conditions required with lead free surface mount assembly, presents design and material selection challenges for extremely thin BGA packages using thin substrates and mold caps. These challenges increase for fine pitch bottom PoP technologies with high silicon die to package ratios - which is the case with this test vehicle. Due to the reduced epoxy mold compound (EMC) volume with this thin mold cap and large FC die to package ratio assembled on a thin (100um) core substrate, a design of experiments (DOE) was performed using several material sets as shown in **Table 1.** The DOE was used to select the optimum combination of EMC and substrate core material based on warpage control, with the coplanarity data reported as the averages for each leg. **Figure 5** shows the shadow moiré warpage measurement results of the 4 legs evaluated in a lead free reflow profile. It was decided to use Leg 2 material combination which provided the best coplanarity and warpage profile.

One of the principal goals of this study was to determine the stacking yield of dense (0.5mm pitch) top side TMV joints.

Leg	PCB core	Mold Compound	Average Room Temp Coplanarity
1	Core A	Compound A	94 um
2	Core B	Compound A	64 um
3	Core A	Compound B	120 um
4	Core B	Compound B	86 um

Table 1 DOE for bottom TMV PoP test vehicle material set with room temp coplanarity

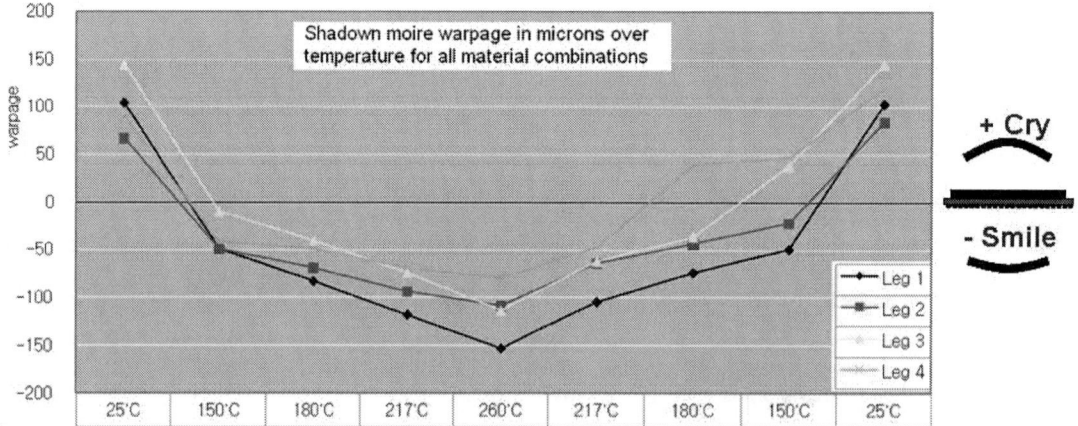

Figure 5: TMV PoP evaluation of different materials to determine optimal combination of mold compound and substrate core. Leg 2 material was targeted for further study.

Leg 2 material allowed for bottom package coplanarity to be well controlled at room temperature (64um typical). Further, the high temperature warpage shows that the top side TMV lands still remain within the guideline defined by Nokia's Ishibashi[10]. In his guideline Ishibashi states: *"Package warpage of PoP memory interface area should be smaller than 0.036mm (0.65mm top pitch) or 0.033mm (0.5mm top pitch) over solder liquidus temperature"*

In order to determine the warpage in the critical stacking TMV land interface zone on the top side of the bottom package we employed the method laid out by JEITA[15]. This method suggests adjusting the baseline to the maximum warpage in the measurement zone. This is detailed in **Figure 6**.

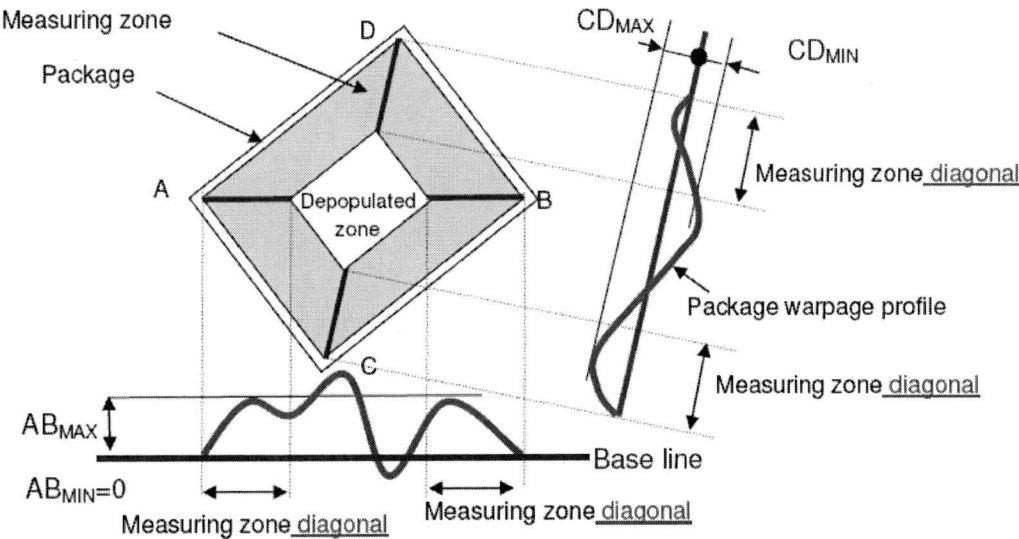

Figure 6: JEITA[15] method for determining warpage in the top side land area (excluding the depopulated regions.

Figure 7, provides the JEITA method analysis for the high temp shadow moiré warpage data from Leg 2 to evaluate the warpage in the TMV stacking zone across the diagonal of the bottom package. Through data interpretation we estimate the top side land area has a maximum warpage value of 30 microns, which is within Ishibashi's guideline for 0.5mm pitch stacked interfaces described above. Application of this JEITA analysis and Ishibashi's guideline is supported by the high 99.4% stacking yield (1 TMV open / 176 units stacked) to be reported in the board assembly summary.

978-1-4244-4722-0/09 $25.00
© 2009 IMAPS-ITALY

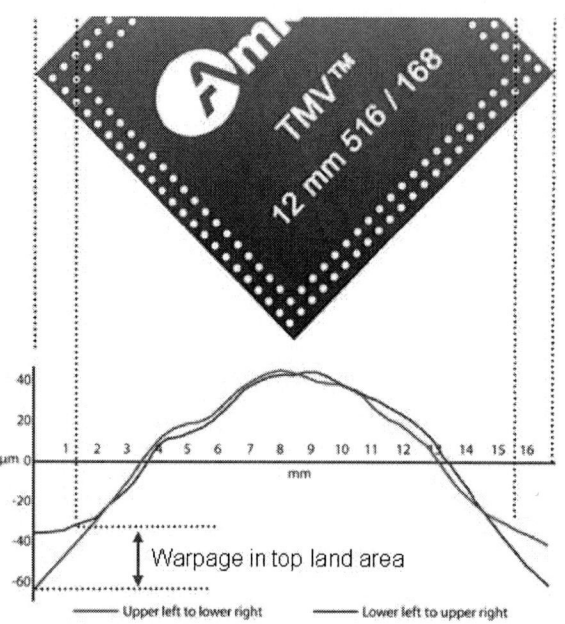

Figure 7: JEITA warpage analysis of Leg 2 elevated temperature shadow moiré data (each line represents one diagonal)

Package level reliability results

The packages were subjected to preconditioning (MSL L3) and reliability life stress testing TC "B" according to JEDEC specifications with sample sizes and results listed in **Table 2**.

Test	Condition	Number of packages	Result
MSL L3 (Moisture Reliability Test)	192hour soak 60% Relative Humidity 30°C	22	Pass
TC"B"500x (Temperature Cycling)	-55°C to -125°C	75	Pass
TC"B"1000x (Temperature Cycling)	-55°C to -125°C	75	Pass

Table 2: Package Level Reliability Tests and Results

After preconditioning and temperature cycle testing, the packages were monitored by CSCAN to check for delamination and continuity tested for opens and shorts.

Board assembly

Test configuration consisted of a TMV PoP bottom package (Figure 4) and a mating top DC package, resembling a typical memory component. The top component had 2 dies stacked on a 2 layer substrate. Board assembly was performed on Nokia's R&D line which uses the same conditions as mass production PoP stacking. The bottom TMV PoP test vehicle was assembled using paste printing and top component using paste dipping. One-pass reflow was done to solder both packages together and to the test board at the same time. Altogether 15 panels were assembled, each having three test boards and four sites/board (in total 176 sites assembled).

Board level assembly and reliability data was not available by the EMPC manuscript deadline. Thus results will be provided in the conference presentation material along with conclusions from this joint project.

978-1-4244-4722-0/09 $25.00
© 2009 IMAPS-ITALY

References

1. JinSeong Kim, et al, "Application of Through Mold Via (TMV) as PoP Base Package", The 58[Th] Electronic Components and Technology Conference, Lake Buena Vista, Florida, May 2008

2. Zwenger, C., et al, "Surface Mount Assembly and Board Level Reliability for High Density PoP Utilizing Through Mold Via Interconnect Technology", SMTA International Conference, Orlando, Florida, August 2008

3. Zwenger, C., et al, "Next Generation Package-on-Package (PoP) Platform with Through Mold Via (TMV™) Interconnection Technology", IMAPS Device Packaging Conference, Scottsdale, Arizona, March 2009

4. JEDEC JC-63 Item # 48.18 PoP 12x12mm body size, 0.5mm pitch for one channel LPDDR2 with 23 rows and columns. Item # 48.24 PoP 12x12mm body size, 0.4mm pitch for a two channel LPDDR2 with 29 rows and columns.

5. Yoshida, A., et al, "Design and Stacking of an Extremely Thin Chip-Scale Package," Electronic Components and Technology Conference, 2003

6. ST Microelectronics press release, Sept. 13, 2006, "ST Microelectronics Launches Package-on-Package Memory System Solutions for Mobile Applications"

7. Smith, Lee, "Package-on-Package: The Story Behind this Industry Hit," Semiconductor International, June 2007

8. Dreiza, M., et al, "High Density PoP (Package-on-Package) and Package Stacking Development," The 57[Th] Electronic Components and Technology Conference, Reno, Nevada, May 2007

9. Smith, Lee, "Driven by Smartphones, Package-on-Package Adoption and Technology Are Ready to Soar" by Lee Smith, Amkor Technology, Chip Scale Review Magazine, July 2008

10. Ishibashi, K., "PoP (Package-on-Package) Stacking Yield Loss Study," ECTC 2007

11. Yoshida, A., et al, "A Study on Package Stacking Process for Package-on-Package (PoP)," ECTC 2006

12. Lin, W., et al, "Control of the Warpage for Package-on-Package (PoP) Design," SMTAI 2006

13. Amkor Technology press release, April 18, 2007, "Amkor PoPs Cork for Fast Growing Package on Package Solution"

14. Amkor Technology press release, March 5, 2009, "Amkor to Introduce Next Generation Package on Package Technology at IMAPS Device Packaging Conference

15. JEITA document "Measurement methods of package warpage at elevated temperature and the maximum permissible warpage" section 3.4

Acknowledgements

The support of Amkor Technology Korea's TMV team was crucial in the package build stage including Yoon JuHoon, Park DongJoo and Kim KwangHo.

TMV is a trade mark of Amkor Technology, Inc.

Introduction of a unified equipment platform for UV initiated processes in conjunction with the application of electrostatic carriers as thin wafer handling solution

Dietrich Tönnies, Markus Gabriel, Barbara Neubert, Marc Hennemeyer, Margarete Zoberbier, and Ralph Zoberbier

SUSS MicroTec, Schleissheimer Strasse 90, D-85748 Garching, Germany

Phone: +49 (0)89 32007 -149, E-mail: dietrich.toennies@suss.com

Abstract

This paper introduces the new MA8 Gen3 Aligner generation designed specifically for the development of 3D and MEMS packaging technologies. Photolithography on the wafer backside, wafer bonding and replication of microstructures are specific 3D / MEMS processes. They have in common that all of them require precision alignment and often will have to run on ultra-thin wafers. The MA8 Gen3 allows to run all the processes above on a unified equipment platform and therefore offers the ideal equipment solution for 3D and MEMS technologies.

A second focus of this paper is at the applicability of electrostatic carriers as a thin wafer handling solution for the processes mentioned above. Electrostatic carriers are a straight forward and cost effective handling solution. They are not based on temporary adhesive wafer bonding and therefore do not require expensive adhesives or other consumables.

The electrostatic carriers and the multi-functional Aligner platform combined result in a very cost effective manufacturing solution for 3D and MEMS packaging. The advantages of this solution will be discussed in more detail for the wafer-level manufacturing and assembly of camera modules and for Plasma Dicing.

Key Words: Mask Aligner, Replication, UV Bonding, Image Sensors, Plasma Dicing, Thin Wafer Handling

Introduction

3D and MEMS packaging involve process steps that are not common to semiconductor manufacturing. Among these processes are photolithography on the wafer backside, wafer bonding and replication of microstructures. These processes have in common that they require precision alignment with a trend for the alignment accuracy towards the sub-micron range. As an additional challenge, these processes will have to run on ultra-thin wafers.

Presently, much of the research and development of 3D packages and ultra-thin wafer handling is geared towards a low cost and reliable manufacturing process for Through-Silicon-Vias (TSVs). There are, however, many more design features of 3D packages or MEMS devices that require innovative equipment solutions. Interestingly, wafer-level camera modules that are currently under development for use in next-generation mobile phones are early adopters of many of these new process techniques.

The wafer-level camera currently drives many equipment innovations that will become key for cost effective 3D and MEMS packaging later on. The reason why the wafer-level camera is a forerunner is that manufacturers don't have to solve the thin-wafer-handling problem: CMOS image sensor wafers are permanently bonded to glass wafers prior to back-grinding. These glass wafers act as mechanical support for the thinned CMOS wafers. Since the glass wafer becomes an integral part of the camera module the thin wafer does not have to be de-bonded, an important detail that significantly simplifies the manufacturing process.

Within the manufacturing process of wafer-level cameras one can identify two complementary technologies. One is the wafer-level packaging of image sensors [1] and the other is the wafer-level manufacturing of camera objectives [2]. Both the sensor device and the objective are then assembled on wafer level by a wafer bonding process. The wafer-level packaging of image sensors was pioneered by Shellcase/Tessera (Israel/USA) while

978-1-4244-4722-0/09 $25.00
© 2009 IMAPS-ITALY

the wafer-level manufacturing of micro objectives was pioneered by companies like Heptagon (Finland/Switzerland).

A Multi-functional Aligner Platform

As these technologies are under development for several years now, existing equipment platforms have been modified to address resulting new process requirements: For example, the manufacturing process of wafer-level camera modules includes several processes that require precision alignment:

- Through-Silicon-Vias
- Backside RDL (redistribution layer)
- Plasma dicing
- Wafer-level imprinting (micro lenses)
- Wafer-level bonding (lens stacks)

Unfortunately, compromises have often to be made when modifying existing equipment platforms leading to less-than-ideal solutions. Therefore, SUSS MicroTec has decided to completely redesign their manual Aligner product family in order to develop an equipment platform that integrates all innovations of recent years in a well thought-out tool design. The result is the MA8 Gen3, the third generation of our MA8 product line (Figure 1). The MA8 Gen3 manual Aligner supports Photolithography, UV replication, UV bonding, standard bond alignment and Nano-Imprint-Lithography processes. It has been designed to offer improved alignment capabilities in terms of accuracy, flexibility and user friendliness. In particular, many new features of the MA8 Gen3 lead to improved process control which in return makes this tool generation more suitable for production environments. In fact: Beyond its use in R&D environments the MA8 has always been utilized in specialized production environments. In particular, this is the case whenever non-standard semiconductor materials are used that cannot be handled easily by conventional automated equipment. As long as production volumes are not very high manual equipment can be more cost effective than using modified automated machines. The wafer-level camera is an example where manual equipment is very common because of the strong wafer warpage after wafer thinning. In consequence, the MA8 Gen3 has developed into a next-generation Aligner platform for many new research and production areas.

Wafer-Level Camera Modules

In the following the flexibility of the new MA8 Gen3 will be discussed with the example of manufacturing wafer-level cameras. Usually, the purpose of TSV technologies is to realize high-

density vertical interconnects between vertically stacked dice. TSVs are typically etched prior to wafer thinning starting from the front side of the wafer. On the other hand, with the image sensors discussed in this paper the TSV is part of the package design. Therefore, TSVs are etched after wafer thinning starting from the wafer backside. Consequently, the etch mask has to be patterned on the backside of the wafer. This requires an accurate alignment of the etch mask (photoresist) to the contact pads on the opposite side of the wafer. Backside lithography is possible by using Mask Aligners with Bottomside Alignment (BSA) systems. BSA systems have been available on Mask Aligners for many years and are commonly used in MEMS. They are, however, largely unknown in the Advanced Packaging industry.

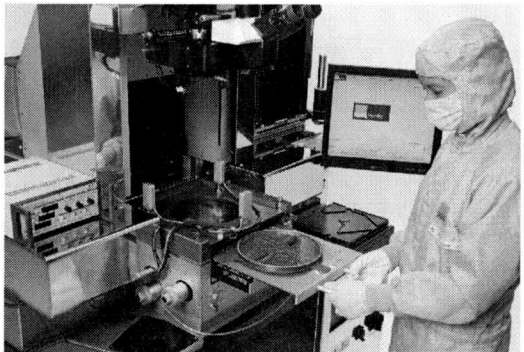

Figure 1: The MA8 Gen3 supports photo-lithography, UV bonding, UV replication bond alignment and NIL

Next to the photolithography function Mask Aligners are perfectly suited for replicating microstructures into UV curable polymers. This technique has been developed some years ago as part of the European Dondodem Project for replicating optical elements such as diffractive gratings, deflection mirrors or micro lenses [3]. In case of camera modules imprinting is used for replicating the lenses of the camera's objective. The process is shown in Figure 2. A UV transparent PDMS stamp is loaded onto a stamp holder. Next, a UV curable polymer with suitable properties (e.g. optical transmission, refractive index, thermal stability to survive a reflow process) is dispensed onto a glass wafer. Then wafer and stamp are brought into contact at a force strong enough to replicate the patterns of the PDMS stamp into the polymer. UV exposure cures / hardens the polymer and finally stamp and wafer can be separated again. The similarity to contact photolithography is obvious. The PDMS stamp corresponds to a photo mask while the UV curable polymer corresponds to the photo resist. In fact, this replication process can be done on a modified Mask Aligner with an increased contact force and a modified levelling of

978-1-4244-4722-0/09 $25.00
© 2009 IMAPS-ITALY

stamp and wafer. In case lenses have to be imprinted onto both sides of the wafer aligned imprinting is necessary. In order to guarantee a good optical imaging quality of the resulting lenses an excellent post replication alignment accuracy is a must. Because of the mechanical forces acting during the imprinting step shear forces that would deteriorate the alignment accuracy have to be minimized. This can be done by an accurate levelling of stamp and wafer in addition to a mechanically robust design of the alignment system. The alignment and levelling system of the MA8 Gen3 has been significantly improved compared to the previous tool generation in order to allow highly reliable imprinting processes.

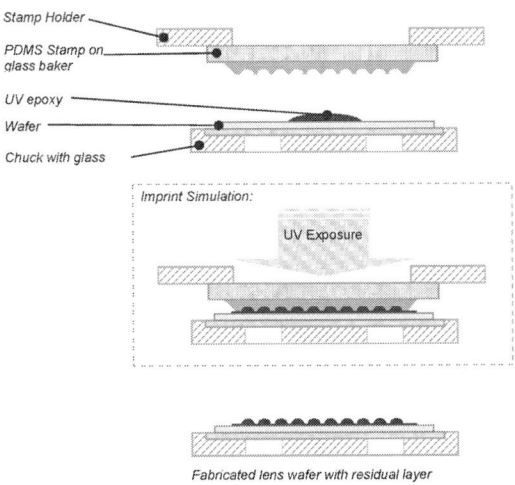

Figure 2: Micro lenses can be manufactured on wafer-level by imprinting a UV curable polymer. Hundreds of micro lenses can be replicated in only one process step.

Camera objectives usually are composed of a set of lenses. Assembling single lenses into a lens stack (objective) is time consuming and expensive especially if lenses become very small. With the lenses manufactured on wafer-level, however, it is straight forward to apply a wafer bonding process in order to assemble the lens stacks on wafer-level as well. Like in the previous example the alignment and contact printing function of Mask Aligners offer easy means to use this equipment for wafer bonding. In MEMS technology wafer alignment and wafer bonding are usually not implemented in an "in-situ" equipment setup. Instead, dedicated tools - the bond aligner and the wafer bonder - are used. The reason is that many bond processes are difficult to combine with delicate alignment mechanics especially when vacuum or high temperature processes are needed. While with many MEMS devices the wafer bond has to seal a vacuum cavity the stacking of micro-lens wafers has no such requirement. Therefore, adhesive bonding is a good option. Moreover, lens wafers are transparent which allows to cure the adhesive by UV

rather than curing the material at elevated temperatures. These two facts allow for a very simple and cost effective in-situ bonding process on a mask aligner type of equipment. The principle setup is shown in Figure 3. The Mask Aligner requires a modified mask holder that can hold the top lens wafer. Wafer alignment, wafer contact and UV exposure are the same operations as with standard contact printing. The main difference is in the wafer handling sequence as the wafer pair has to be removed from the machine after bonding. In a photolithography process the mask typically remains in the tool for subsequent exposures. Obviously, post bond alignment accuracy is critical for the overall optical performance of the objective. At the same time the accurate control of the distance of individual lenses in a lens stack is important. Therefore, the dispense volume and dispense pattern of the adhesive has to be controlled in such a way that an even and void free spreading of the material is achieved on the wafer. Finally, the Aligner has to ensure an excellent levelling of both wafers to avoid generating a wedge in the adhesive layer. And of course, the mechanical setup has to be robust enough to avoid shear movements which would deteriorate the post bond alignment accuracy.

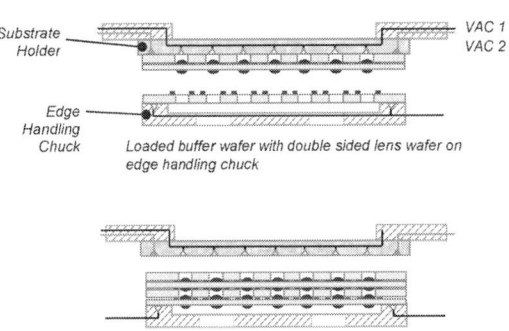

Figure 3: In-situ UV bonding on the MA8 Gen3 is very cost effective way to assemble complex micro objective at wafer-level.

The MA8 Gen3 is currently under evaluation by several manufacturers of wafer-level-cameras. At this point of time it appears that it will become a standard in this particular industry. As stated above the wafer-level camera can be considered cutting edge for other 3D packaging innovations except for the thin wafer handling issue. The question is whether replication technology and UV bonding can be instrumental to realize design features of other future 3D packages. It appears desirable as all these processes support wafer level packaging / assembly and in consequence a highly parallel and cost effective manufacturing.

978-1-4244-4722-0/09 $25.00
© 2009 IMAPS-ITALY

Thin Wafer Handling

As stated above many 3D packaging processes will require a solution for thin wafer handling. The processes discussed before will only be applicable to 3D packaging if they can be realized on very thin wafers with the possibility to de-bond the thin wafer from its supporting carrier. At this point of time it appears to be an industry consensus that very thin wafers will require such a supporting carrier. Several concepts exist, most of them based on temporary adhesive wafer bonding. Material suppliers that offer specialized adhesives include:

- Brewer Science [4]
- 3M [5]
- Thin Materials [6]

A key issue with all these temporary wafer bonding processes is how to separate the thin wafer at the end of the process chain without damaging the wafer. This problem does not come by surprise because on the one hand the adhesive has to ensure a mechanically, chemically and thermally strong adhesion during wafer processing but on the other hand has to allow for an easy separation at the end of the process flow. These somewhat contradicting requirements make it difficult to find a reliable and cost effective manufacturing process.

A very different approach is based on electrostatic carriers [7][8]. This technology has been pioneered by the Fraunhofer IZM and is currently commercialized by companies such as ProTec Carrier Systems. Electrostatic carriers hold the wafer by an electrostatic field generated by the carrier. Electrostatic forces are strong enough to hold the wafer over a long period of time and during standard semiconductor processes. Since no organic chemicals are involved electrostatic carriers can stand up to very high process temperatures (> 400°C). And since electrostatic carriers are typically manufactured from standard Silicon wafers using standard Silicon processes they can be considered compatible with semiconductor technology.

For SUSS MicroTec the reason for looking into electrostatic carrier technology was driven by a joint project with Panasonic Factory Solutions. As part of this project Panasonic and Suss develop a Plasma Dicing process involving photo lithography and plasma etching. Plasma Dicing allows extremely accurate and narrow dicing street and avoids edge chipping. As shown in Figure 4 the Plasma Dicing process requires a photo resist etch mask on the flip side of the thin wafer. It is a key characteristic of Panasonic's process that the plasma etching is starting from the wafer backside. Therefore, metal test structures in the dicing streets don't have to be plasma etched. In order to be compatible with standard back-end processes the back-grinding tape will remain on the wafer during the entire process. While the tape is stable enough to survive the photolithography process including resist bake the biggest challenge is in the plasma etching process as care has be taken that no gas is trapped in the tape/wafer interface that may burst during a high vacuum process. At the same time the thermal coupling of the thin wafer has to be good enough to avoid excessive heating of the wafer during etching. To have solved these issues is part of Panasonic's proprietary technology.

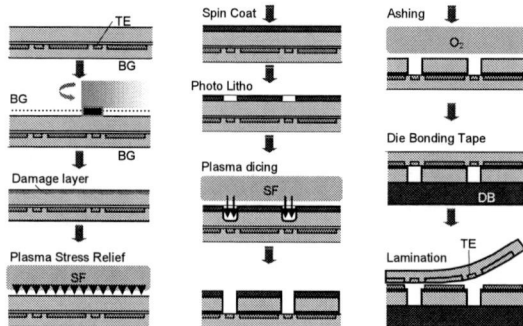

Figure 4: Plasma Dicing process flow (Panasonic)

While the tape lends some stability to the thin wafer it is still not rigid enough to be processed by standard lithography equipment. Therefore, we were looking into a carrier technology for the lithography process. If any of the temporary bonding techniques based on adhesives gains an industry wide acceptance such a solution might be very suitable for Plasma Dicing. However, at this point of time these technologies are not readily available. Therefore, we decided to evaluate the Transfer Electrostatic Carrier (T-ESC) technology.

There are uni-polar and bi-polar carrier designs. Figure 5 shows the principle design of a bi-polar carrier. A bi-polar carrier has two electrically isolated electrode structures on top. A voltage is applied between both electrodes and the carrier is charged similar to a capacitor. Charging the carrier generates an electrostatic field which has – due to the design of the carrier - the capability to attract a wafer located on top of the carrier. If sufficiently charged this attracting force is strong enough to hold the wafer during semiconductor processes. The de-bonding / de-chucking process is straight forward. It simply includes a de-charging of the carrier which eliminates the force pulling wafer and carrier together. The design of a uni-polar carrier is even more simple: The uni-polar carrier has only one electrode and the wafer acts as the second electrode. A voltage is applied between carrier and wafer and both are attracted by an electrostatic force similar to the electrodes of a capacitor.

978-1-4244-4722-0/09 $25.00
© 2009 IMAPS-ITALY

Figure 5: Principle design of a bi-polar Electrostatic Carrier

The T-ESC technology evaluated in this paper has been provided by ProTec Carrier Systems GmbH. A first run with standard High-Temperature-Carriers did not deliver good results during the photolithography processes. Problems occurred during any process involving wet chemicals of low viscosity. In case of photolithography such chemicals are used for resist developing and resist stripping. In both cases the wet chemical (solvent) is pulled efficiently between the carrier / wafer interface because of capillary forces. The effect could not even be prevented when spinning the wafer at high speed during the process. Figure 6 shows such a T-ESC carrier after de-chucking the wafer: Residues of the developer media are clearly visible at the edge of the carrier but easily can cover the entire carrier surface as well (and in consequence the backside of the thinned wafer). These residues typically will cause strong forces between carrier and wafer which makes the separation of carrier and wafer difficult.

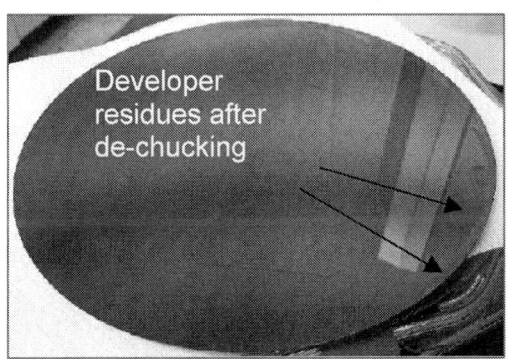

Figure 6: Standard T-ESC carriers should not be used in the lithography process because wet chemicals (e.g. developer) are pulled between carrier and wafer.

Therefore, in a joint project, ProTec and SUSS have developed a specialized carrier design for use in a photolithography process. In order to reduce the capillary forces between wafer and carrier a step was cut into the edge of the carrier and order locally increase the gap between wafer and carrier at the wafer edge. By this means the capillary forces are significantly reduced. If the carrier/wafer is rotated at an elevated speed during the process the centrifugal force acting on the liquid in addition to the reduced capillary forces keep the interface between wafer and carrier clean. This proved

essential for an easy separation of the thin wafer from the carrier.

Figure 7: Resist etch mask consisting of a 15μm thick AZ 9260 resist with 15μm and 45μm wide dicing street patterns.

The T-ESC carrier technology was then applied to the Plasma Dicing process. 100μm thin wafers mounted on back-grinding tape where chucked on the modified electrostatic "Litho"-carriers. Then the photolithography process was done on standard automated processing equipment (SUSS ACS200 Coat/Develop Cluster, SUSS MA200Compact Mask Aligner). The boundary condition of having to spin the wafer at high speed during develop requires the modification of the develop recipe and may potentially lead to increased consumption of developer. However, the process ran without problems and carrier and wafer could be separated after the process without any damages. In case of Plasma Dicing the resist is stripped in the plasma chamber by plasma ashing and therefore no wet chemical process is needed. Figure 7 shows the resulting resist edge mask. A 15μm thick AZ 9260 resist film was used in order to provide sufficient resist thickness to withstand the etching process. 15μm and 45μm wide dicing streets, respectively, have been exposed into the resist film demonstrating how narrow dicing street can be when applying Plasma dicing.

Conclusions

In this paper we have introduced a new Mask Aligner platform that supports a number of wafer-level processes than can be instrumental for 3D packaging. In conjunction with the utilization of Transfer-Electrostatic-Carrier technology this equipment solution has the potential to offer very cost effective means for realizing even very advanced 3D package designs. No matter how complex future packaging technologies may become, cost will always be a significant factor defining the commercial success. The MA8 Gen3 and T-ESC technology will make a good match.

978-1-4244-4722-0/09 $25.00
© 2009 IMAPS-ITALY

Acknowledgements

The authors would like to acknowledge the support of ProTec Carrier Systems for providing us with their T-ESC chucking / de-chucking technology.

References

[1] D. Shariff, N. Suthiwongsunthorn, F. Bieck, J. Leib, "Via Interconnections for Wafer Level Packaging: Impact of Tapered Via Shape and Via Geometry on Product Yield and Reliability", Proceedings of the 57th Electronic Components and Technology Conference (ECTC), Reno, Nevada, May 29 – June 1, pp. 858-863, 2007.

[2] R. Voelkel, R. Zoberbier; "Inside Wafer-Level Cameras", Semiconductor International, February 2009, pp. 28-32, 2009.

[3] M. T. Gale, S. Obi, N. de Rooij, "Replicated optical MEMS in sol-gel materials", International Conference on Optical MEMS, 2003 IEEE/LEOS Volume , Issue , 18-21, Aug. 2003, pp. 20-21, 2003.

[4] Dongshun Bai, Wenbin Hong, JoElle Dachsteiner, Amadine Jouve, Rama Puligadda, Chad Brubaker, and Tian Tang, "Temporary wafer bonding materials with adjustable debonding properties for use in high-temperature processing," IMAPS 2008: Proceedings of the International Microelectronics and Packaging Society 41st International Symposium on Microelectronics, pp. 222-227, 2008.

[5] C.R. Kessel, " 3M Wafer Support System – Premium Wafer Thinning Using Glass Support Carriers", Q1 2007 MEPTEC Report, Vol. 11, No. 1, pp. 23-25, 2007.

[6] J. Boudaden, M. Pieka, "Wafer Thinning Technology of Thin Materials AG", Forum 'be-flexible', Fraunhofer IZM, Munich, Germany, Dec. 2, 2008.

[7] C. Landesberger, S. Scherbaum, D. Bollman, and K. Bock, "Handling Ultra-thin Wafers", Advanced Packaging Magazine, May/June 2007, pp. 32-34, 2007.

[8] C. Landesberger, S. Scherbaum, K. Bock, "Carrier techniques for thin wafer processing", Proceedings of 2007 CS MANTECH, Austin, Texas, May 14-17, pp. 33-36, 2007.

Roll-to-Roll Manufacturing of Organic Photovoltaic Modules

M. Tuomikoski, P. Kopola, H. Jin, M. Ylikunnari, J. Hiitola-Keinänen, M. Välimäki, M. Aikio and J. Hast

VTT Technical Research Centre of Finland, Kaitoväylä 1, FI-90571 Oulu, Finland

Phone: +358 722 2195, Fax: +358 722 2320, E-mail: markus.tuomikoski@vtt.fi

Abstract

Organic photovoltaics are being explored for powering electronic devices by harvesting the sun's energy or indoor lighting. Such solar cells have shown promise as they can be deposited on thin flexible foils that enable flexibility of integration within products while keeping the weight light. Several R&D efforts are underway in creating improved organic solar cells in terms of their efficiency and other performance metrics. One such is the FACESS project ("Flexible Autonomous Cost Efficient Energy Source and Storage"), funded under the 7th Framework Programme of the European Commission. The main goal is to develop economically viable solar cells on large scale for mass production to achieve the power conversion efficiency (PCE) up to 2.5% thus producing 250 mW under AM1.5 illumination. This requires selection, modification and up-scaling of suitable solar cell materials, effective patterning technologies, development of the printing processes and improved encapsulation technologies. This development will be done keeping all the time the application, the potential and the required performance in mind. The development and experimental results of roll-to-roll techniques such as wet chemical etching of an anode electrode, gravure printing of PEDOT:PSS and the P3HT:PCBM photoactive layer and gas barrier lamination of the printed modules will be presented in detail. The performance status for lab-scale printed solar cells is currently 2.4% of PCE under AM1.5 illumination. In addition the major technological challenges and solutions for roll-to-roll manufacturing will be summarised.

Key words: roll-to-roll, printing, photovoltaic, organic

Introduction

Organic solar cells (OSCs) have a great potential to be future's renewable and environmentally friendly energy production technology. The operation of the organic solar cells is based on harvesting sun light and converting it to electrical energy. The intense development of organic photovoltaics during the recent years has improved significantly the power conversion efficiency. [1] Furthermore, one of the key issues on the route towards commercialisation is the cost-efficiency. Low manufacturing costs can be attained by processing photovoltaics with large-area manufacturing technologies such as high-throughout roll-to-roll printing techniques onto flexible substrates. [2]

The R2R manufacturing of OSCs is currently under investigation in the 7th Framework Programme EU-funded project entitled FACESS – "Flexible Autonomous Cost Efficient energy Source and Storage".[3] The general objectives of this project are following: manufacture efficient organic solar cells (OSC) and a thin film battery (TFB) on flexible substrate using commercially available materials and cost efficient roll-to-roll (R2R) mass production techniques like printing, as well as integrate a control transistor circuitry on a foil.

The ultimate goal is to integrate these three structures into a single assembly resulting in a flexible and fully autonomous energy source. In this assembly organic solar cells harvest the solar energy and charge the thin film batteries which provide the electricity for an external load. The Si-based transistor circuitry integrated on the foil controls effective charge operation. The target for organic solar cell module is to achieve the power conversion efficiency (PCE) of 2.5% thus producing 250 mW under AM1.5 illumination.

Materials and Methods

During the FACESS project, conventional roll-to-roll printing technologies such as gravure printing were used in the preparation of organic solar cells. Gravure printing can be considered as one of the fastest, the most simple-structured and most cost-effective printing techniques. The optimisation of printing parameters and formulation of inks for the gravure printed organic layers have been done in the lab-scale printing machine as well as in a larger scale pilot machine.

In this study, organic solar cells were prepared on PET substrates coated with indium-tin-oxide (ITO) layer. Poly(3,4-ethylenedioxythiophene) doped with poly(styrenesulfonate) (PEDOT:PSS) was processed on the top of ITO. The photoactive layer consisted of poly(3-hexylthiophene) (P3HT) as

978-1-4244-4722-0/09 $25.00
© 2009 IMAPS-ITALY

a donor material and [6,6]-phenyl-C$_{61}$-butyricacid methylester (PCBM) as an acceptor material. Then, Ca/Ag electrodes were vacuum evaporated through a shadow mask resulted the active area of 16 mm^2. All the cells were encapsulated under nitrogen atmosphere to prevent device degradation by oxygen and moisture. The schematic structure of the organic bulk heterojunction solar cell is presented in Figure 1.

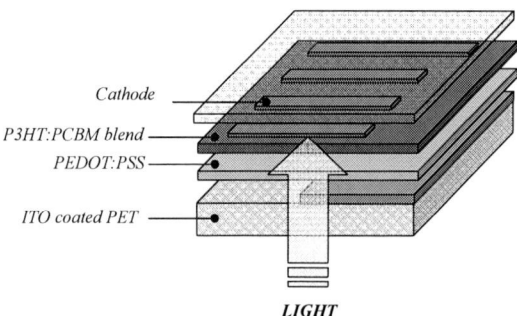

Figure 1: The architecture of an organic bulk heterojunction solar cell.

Characterisation

Photovoltaic devices are normally tested in a controlled environment, with illumination of 1 sun, which stands for solar illumination on a clear, sunny day. Hence, a solar simulator that provides stable and repeatable illumination of 1 sun is needed.

A measurement set-up for organic solar cells, which complies with the standards, was designed. The most important component of the measurement apparatus is the solar simulator that produces illumination defined in the standards as AM1.5 (Air Mass 1.5) illumination. A Cermax xenon short-arc lamp, which provides high irradiance and a broadband spectrum that resembles natural sunlight, was chosen as the light source for the simulator. A xenon lamp's colour temperature is about 6000 K, the same as that of natural sunlight. [4]

The optics for the solar simulator was designed to achieve uniform illumination of the test plane. The light from the short-arc xenon lamp is collected by a large elliptical mirror after it travels through an aperture in the casing of the lamp and AM1.5 global filters (AM0 and AM1.5), and then focused at the end of a light-mixing rod. The light-mixing rod mixes the light through multiple total internal reflections, thereby improving spatial and angular uniformity. The other end of the light-mixing rod is then imaged with another elliptical mirror and further focused on the test plane through a third mirror, which is used to turn the light beam 90 degrees. The irradiation of the test plane was fixed accurately to 1000 W/m^2 using the aperture therefore the specification in the IEC 904-3 standard was reached. Illumination uniformity over 90% over the area of 4 x 4 cm^2 was measured. Figure 2 shows the solar simulator in operation.

Figure 2: AM1.5 solar simulator.

R2R Etching of ITO

The architecture of solar cell is based on standard configuration starting with patterned anode on plastic substrate. The substrate was purchased as ITO coated PET in rolls having a roll width of 300 mm and a length of 100 m. The nominal sheet resistance was below 60 ohms/sq. The desired cell pattern for ITO electrode was prepared by R2R wet chemical etching process comprising flexography printing of a resist as a reverse pattern, acid solution based etching, and removing of the resist. Figure 3 shows the photograph of ITO patterning on PET foil.

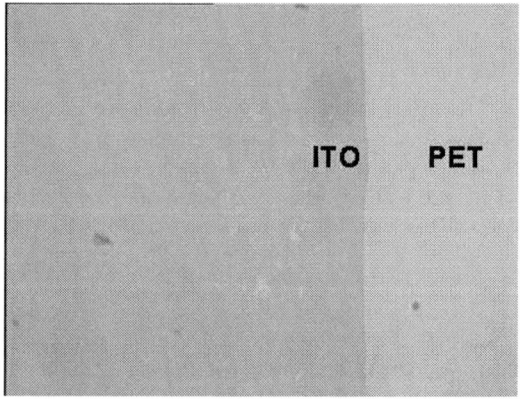

Figure 3: Wet chemical etched ITO patterning on PET foil.

R2R Printing of PEDOT:PSS and P3HT:PCBM

Experimental design was utilised for investigating the parameters which have the most significant impact on printability of the photoactive layer composed of P3HT:PCBM blend. The work was started by evaluating and optimizing the parameters affecting the printability of the photoactive material and choosing the inspected responses. The preliminary printability tests were carried out with a tabletop gravure printer. The examined responses were the film thickness, surface roughness and visual inspection of the film quality. The design of experiments held 29 randomized experiments and three factorial levels. As an example, a response surface for the film thickness of the photoactive layer is illustrated with the cell depth of the cylinder engravings and the solid concentration of the photoactive ink in Figure 4.

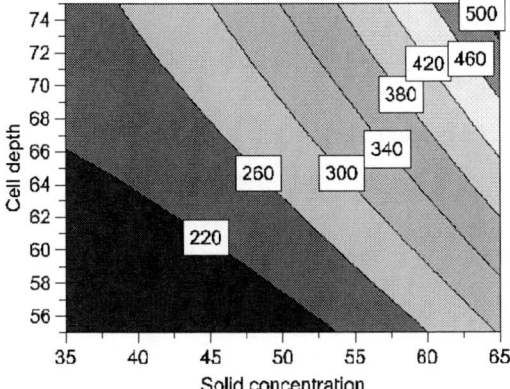

Figure 4: The response surface for the film thickness of the photoactive layer with the printing speed of 60 m/min (cell depth [μm], solid concentration [mg/ml], film thickness [nm]).

The results achieved with the experimental design were used for planning the printing parameters and ink formulations to the pilot printing runs. Two layers, PEDOT:PSS and P3HT:PCBM blend, of the organic solar structure were processed in a pilot line by gravure printing. The modification was done to the as-received PEDOT:PSS in order to improve the wetting properties of the ink on the ITO-PET foil and thus to prepare an uniform thin film. The modified PEDOT:PSS was printed by using a gravure printing cylinder which contained the line densities of 70L, 80L, 90L and 100L cell depths ranging from 36 μm to 42 μm. As a demonstration example, a photograph of the PEDOT:PSS roll printed in the pilot line is shown in Figure 5. On top of PEDOT:PSS, a photoactive layer of P3HT:PCBM was successfully prepared.

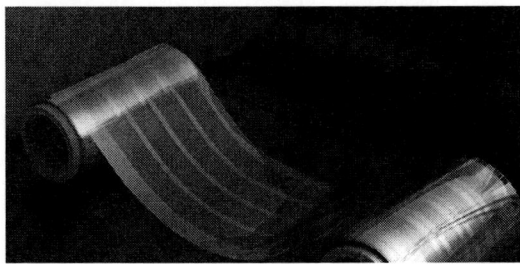

Figure 5: Modified PEDOT:PSS roll gravure printed in pilot printing machine.

Characteristics of Solar Cells

The reference OSCs were fabricated by spin-coating PEDOT:PSS and P3HT:PCBM on a patterned ITO glass substrates in air in order to compare spin coated and printed cells. The ITO-coated glass substrates were cleaned by ultrasonic treatment in deionized water, acetone and isopropyl alcohol sequentially. The Ca/Al cathode electrode was vacuum evaporated on the top of photoactive layer. The cells were encapsulated by the UV-cured epoxy in nitrogen-filled glove box.

Typical PCE values of 3.2-3.3 % were achieved for the five spin-coated reference cells. The average open-circuit voltage, short-circuit current and fill factor of 575 mV, 8.4 mA/cm^2, and 0.67 were obtained, respectively. I-V characteristics are shown in Figure 6.

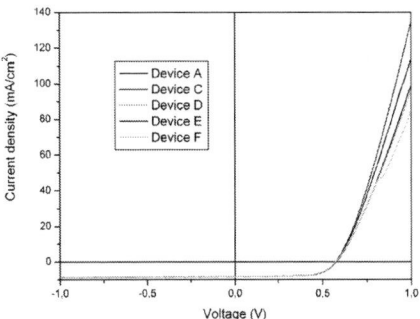

Figure 6: I-V characteristics of five spin-coated cells.

For the printed OSCs, the improvement of the print quality and the optimized film thickness of photoactive layer were observed as an enhancement of electrical performance, thus the performance of printed organic solar cells was improved from 1.4% to 2.4% under AM1.5 illumination. In addition, the film thickness was varied between the samples. The highest open-circuit voltage, short-circuit current and fill factor of 565 mV, 8.1 mA/cm^2, and 0.57 were obtained, respectively. The measurements proved that the performance of printed cells is very close with the spin coated reference cells. I-V characteristics are shown in Figure 7. Futhermore, the cell optimization is currently on-going in order to enhance the printed cell performance.

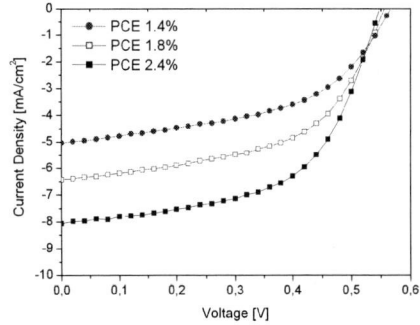

Figure 7: I-V characteristics of printed cells.

978-1-4244-4722-0/09 $25.00
© 2009 IMAPS-ITALY

Printed Solar Cell Module

The solar cell module for autonomous energy systems was fabricated by utilising aforementioned techniques and materials. The module requires a specific tailoring of cell interconnection in order to increase the output power. The photograph of the flexible solar module, which has three monolithic interconnections of single solar cells, is shown in Figure 8.

Figure 8: The photograph of 6 cm^2 sized printed solar cell module having power conversion efficiency of 1.6%.

The power conversion efficiency of 1.6% under AM1.5 illumination was reached on the active area of 6 cm^2. In addition, the open-circuit voltage, short-circuit current and fill factor of 1.8 V, 12.7 mA, and 0.41 were obtained, respectively. I-V characteristics of printed solar cell module are presented in Figure 9.

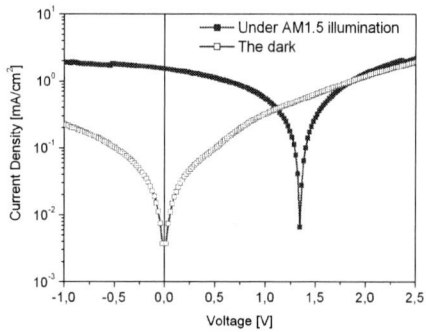

Figure 9: I-V characteristics of printed solar cell module (PCE 1.6%, Voc 1.8V, Isc 12.7 mA, FF 0.41).

Conclusions

In conclusion, we have used experimental design to do the first optimisation to the printing parameters and ink formulations in the lab-scale printing machine before moving to pilot scale printing machine. In addition, we have proved that the polymer photovoltaics including printed PEDOT:PSS layer and photoactive layer composed of P3HT:PCBM blend can be prepared with large scale roll-to-roll printing machines.

We have developed printed organic solar cell module consisting of three interconnected solar cells. The power conversion efficiency of 1.6% was obtained under AM1.5 solar illumination. The work will continue in order to reach the target of 100 cm^2 sized solar cell module having PCE of 2.5%.

Acknowledgements

The European Commission (FP7-2008-ICT-1-215271, FACESS project) is highly acknowledged for funding the work presented on this paper. We thank FACESS partners for the guidance in organic solar cell fabrication. Also, we like to thank VTT's Center for Printed Intelligance (CPI) for partial funding of this work.

References

[1] S. S. Sun, N. S. Sariciftci, "Organic photovoltaics Mechanisms, Materials and Devices", Taylor & Francis, USA, 2005.

[2] C. J. Brabec. "Organic photovoltaics: technology and market", Solar Energy Materials & Solar Cells, pp.273-292, vol. 83, 2004.

[3] http://www.vtt.fi/proj/facess

[4] Elion G. R. & Elion H. A. (1979) Electro-Optics Handbook. Marcel Dekker Inc. 376 p.

Encapsulation of the Next Generation advanced Mems& Sensor Microsystems

Arnold Bos, Lingen Wang, Ton van Weelden

Boschman Technologies B.V.

Stenograaf 3, 6921 EX DUIVEN, the Netherlands

Phone: +31 (0) 26 3194900, Fax: +31 (0) 26 3194999, Mail: TonvanWeelden@Boschman.nl

Abstract

The needs of ambient intelligence and miniaturization of electronics require extensive integration of semiconductor components such as MEMS and Sensor packages. A MEMS/sensor package needs access to environment, which is a challenge for MEMS or sensor encapsulation. Film Assisted Molding (FAM) technology can meet the demand of the functional area opening in the encapsulation. FAM technology can fulfill the requirement of encapsulation of the next generation advanced MEMS microsystem with low cost and excellent performance. A combination of FAM and wafer level molding enables MEMS/sensor encapsulation with low cost, good quality and high reliability. FE simulations are performed to investigate the film performance for the FAM cycle. The simulation results match the over-molding process quit well.

Key words: Film assisted technology, MEMS encapsulation, over-mold, wafer level molding and finite element simulation

Introduction

As the development of microelectronics is driving towards further miniaturization, integration is the way to more functionality in a device. Due to the needs of ambient intelligence, the micro system needs integration of standard chips, micro-electromechanical systems (MEMS) devices, sensors, actuators, min-cameras, etc.[1]. The usual approaches toward component packaging only focus on standard chip packages, but do not support the generation of more complex systems, integrating not only ICs, but also sensors, micro electromechanical systems (MEMS) devices, min-cameras, etc. [2] General microelectronic packaging has only been considered as an afterthought at the end of the design process. Electronic devices only need electronic contact with the external environment, while MEMS/sensors often need both protection from and access to environment. For example, a pressure sensor chip requires protection from mechanical stress, also needs access to the environment to make its measurement.

Packaging of MEMS devices has been proven to be costly and complex and it has been a significant barrier to the commercialization of MEMS and miniaturization of electronic system.

Ideally, package designers would like to use a semiconductor mass-production over-molding technology to encapsulate MEMS with free surfaces where the chip needs to sense the environment. Film assisted molding Technology of Boschman can address the MEMS/sensor encapsulation challenge with low cost and high reliability.

Wafer level package attracts more and more interest for better performance and low cost. The MEMS/sensor performance can be improved by dicing encapsulated wafers, especially for MEMS/sensor applications. Boschman wafer level over-molding technology combining with FAM provides the opportunity to realize the wafer encapsulation with/without MEMS/Sensors.

Finite element simulations were carried out to gain insight into the film performance during the Film Assisted Molding (FAM) process. The simulation results can be the basis of optimization the mold design.

Film Assisted Molding technology

The FAM technology is a microelectronic transfer molding technology with film lining the mold parts. The increasing use of the Film Assisted Molding (FAM) technology confirms the importance of this new encapsulation technology to the semiconductor industry. FAM deals with challenges of releasing components from the mold and keeping die or specific surfaces clear from molding compound. The applications can be general chips and MEMS/sensors with open windows. [3]

Figures 1a-e show the main process steps of film assisted molding cycle.

978-1-4244-4722-0/09 $25.00
© 2009 IMAPS-ITALY

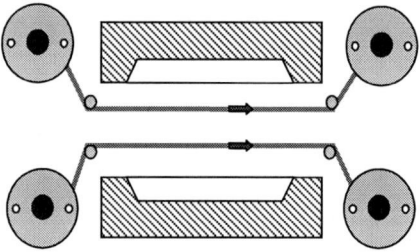

Fig. 1a: Intake the fresh film

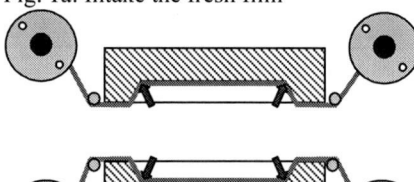

Fig. 1b: Move the mold to the film position and suck film to the mold inner surfaces by vacuum

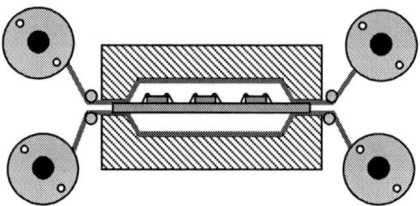

Fig. 1c: Load lead frames/substrates and close mold

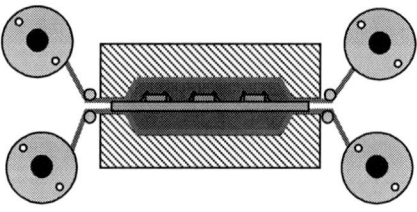

Fig. 1d: Fill mold cavity by transferring molding compound

Fig. 1e: Open mold and unload the molded components

From the above process schematics, the FAM cycle includes intaking fresh film, sucking the thin film to the mold inner surface, loading the substrates, closing the mold, transferring EMC to the mold cavity, mold opening and unloading the molded components. Before the next cycle is started, the vacuum on the film is removed. The film is then transported one length of the mold to renew the film. The Teflon-film has a high tearing resistance and can work at temperatures as high as 200°C. In the process, the film is sucked to the inner surface of the mold and follows the three-dimensional form of the cavities because of vacuum. After loading the lead frame or substrate into the mold, the molding compound encapsulates the devices. After curing the epoxy encapsulation material, the encapsulated components can be unloaded.

Releasing from the mold is tackled by a "seal film" made of a Teflon-based material. This overcomes many problems with other release techniques such as coating the mould with Teflon or using release agents. Teflon coated surfaces are easily damaged because mould compounds contain abrasive "filler" particles. Releasing agents are designed to prevent the encapsulant sticking to the mold metal, yet at the same time the molding compound is required to make a firm attachment to other parts of the component. In any case, these techniques further need mold cleaning after each production step. The problem is even more critical for those sensitive chips since micron-size particles can be sufficient to damage the active surface. With Boschman's seal film, the encapsulant never comes in contact with the mould metal. The protection for each production shot is provided by a fresh piece of film. Since the release problem has been essentially solved, one can consider different mould compounds with stickier properties or ultra low viscosity, even the extremely sticky compound without filler. The films also act as "gasket", providing a good seal and stopping encapsulant material flowing out of the molding area. Without film, the metal-on-metal contact areas of the two mould halves have to be pressed together at extremely high pressure, which creates more potential for damage of delicate components. The gasket effect allows these pressures to be dramatically reduced without EMC bleed and resin flash.

FAM applications

According to the applications, Film Assisted Molding (FAM) technology can be used as adhesive film technology, release film technology, seal film technology.

Adhesive Film Technology (AFT) is developed to protect large areas of LLP's, QFN and Array packages. A high temperature resistant, low cost PET-type polymer film carrier is developed as a hot seal adhesive film (see Fig. 2).The AFT process is a high volume, high quality, and low cost solution of lamination and delamination in the back end process flow. This adhesive film presents a reliable and low cost alternative to pre-tape on lead frames. The melting characteristics of this hot seal layer and the well defined layer thickness guarantee good

978-1-4244-4722-0/09 $25.00
© 2009 IMAPS-ITALY

adhesion to the lead frame by building up resistance against compound flow below the leads and therefore preventing EMC bleed and resin flash on the lead frames.

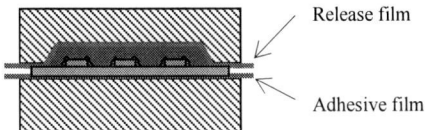

Fig. 2: The adhesive and release film

Film Technology ensures that the mold area where film is applied does not come into contact with the mold compound. Release Film Technology (see Fig. 2) allows mold without ejector pins at the film side and facilitates smooth ejection even with sticky molding compounds. The film works as a barrier between mold surface and encapsulant. Therefore, the release film technology keeps the mold free from the encapsulation materials, which can also avoid contamination of EMC and extend the mold life.

Film assisted molding technology can also encapsulate packages with exposed window (See Fig. 3). In this Seal film technology the film works as "gasket" to prevent the molding compound flowing to the open windows. The sealing film has enabled the creation of encapsulant-free surfaces. This technology meets the challenge of die encapsulation with open window without compound bleed and resin flash, which opens the opportunities for MEMS encapsulation with low-cost and reliable over-molding.

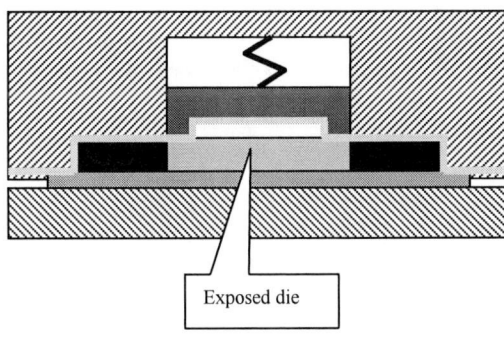

Fig. 3: Encapsulation of exposed die

As mentioned above, the microelectronic packaging trend requires extensive integration including die, MEMS/sensors, etc. However, the MEMS/sensor encapsulation is a challenge in the microelectronic industry. An open window is needed to access to the environment for encapsulated MEMS package like gas and liquid sensors. Generally, the exposed die, MEMS/sensor, is brittle and can not stand large stress. Seal film technology lends itself to encapsulate exposed die in such applications as MEMS/sensor or optical sensor,

pressure sensors or fingerprint sensors. If desired, this technique allows over molding a part of the chip. By clamping the film around the chip's functional area, the surface of the chip is protected from the compound and will be bleed and flash free. Comparing to steel clamp, the clamping force can decrease dramatically with soft film. Film assisted molding technology is a unique cost-effective MEMS/sensor encapsulation by transfer molding.

This technology can be used to encapsulate solder balls with or without MEMS open window. (See Fig. 4)

Fig. 4: Encapsulating the solder balls of die with exposed window

Boschman has developed film technique so that film can be used on both sides of the device with sealing film on the top and either sealing or adhesive film on the bottom, depending on applications. The technique with sealing film on both sides of the device is particularly useful for lead frame mounted chips requiring package structuring on both sides.

Simulations for film performance

In order to get more insight into the film performance during the molding cycle, FEM simulations for the molding process were performed including the steps of sucking film to the inner mold surface by vacuum, mold close and transferring EMC into mold cavities. As an example, here Fig. 5 shows the one side seal film performance during encapsulating the MEMS with open window.

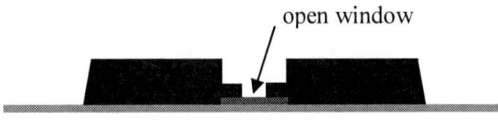

Fig. 5: The schematic of encapsulated MEMS with open window

The film is considered as elastic-plastic material. The elasto-plastic properties were derived from the measurement results of the film supplier. The actual stress-strain curve is not available. The film has good elongation properties with tearing elongation 650% at 175 $^{\circ}$C.

Boundary conditions are different at various molding steps. The vacuum sucking of the film in to the mold inner can be transferred into the pressure on the film surface and the EMC transferring pressure can be transferred into the pressure on the

978-1-4244-4722-0/09 $25.00
© 2009 IMAPS-ITALY

film surface. The pressure directions are perpendicular to the deformed film surface.

The x direction displacements are fixed for nodes of the middle and end of film because of the symmetries during the molding cycle.

Step 1: 0.7 bar vacuum pressure on the bottom side of film to push the film to the inner surface of the mold;

Step 2: bottom mold moving up, the die and substrate contact the film;

Step 3: compress the film thickness from 100 μm up to 50 μm to close the mold;

Step 4: 80 bar EMC transferring pressure on the bottom side of film because of transferring EMC pressure;

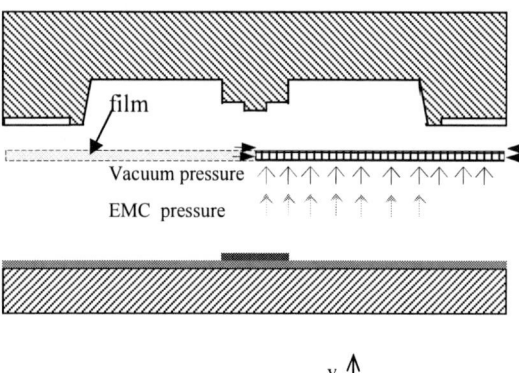

Fig. 6: The schematic of FE boundary conditions of film in the mold

The bottom curves of top mold, top curves of MEMS and substrate are modeled as rigid bodies. The seal film is considered as the deformable bodies. The sticking force between EMC and film can be transferred to the friction between film and top mold.

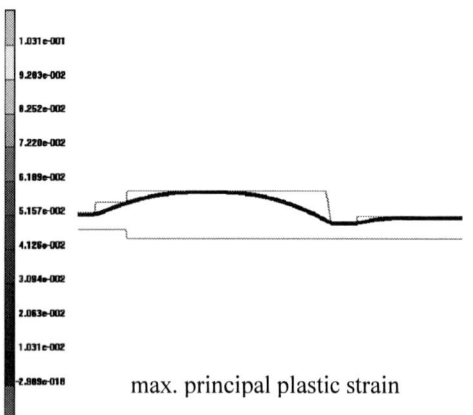

Fig. 7: The max. principal plastic strain of deformed film at vacuum sucking step

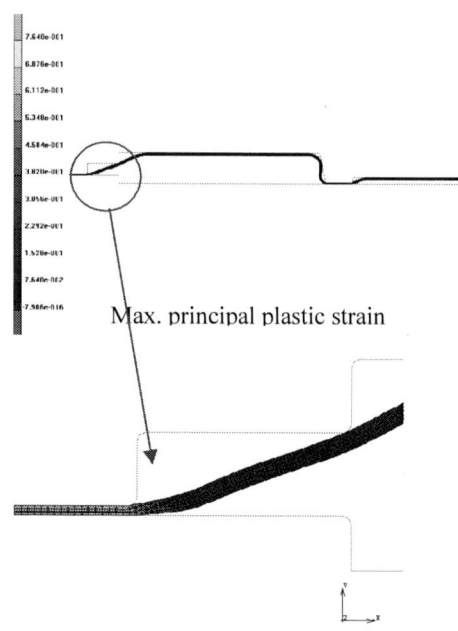

Fig. 8: The max. principal plastic strain of deformed film at mold close step

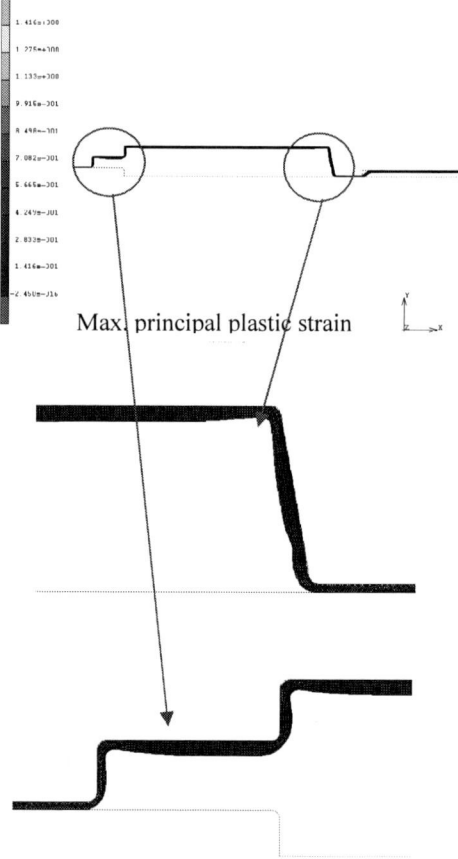

Fig. 9: The max. principal plastic strain of deformed film after EMC filling

To make the vacuum sucking film untouch the wire of the die or MEMS/Sensors during molding by proper design of mold cavities, selection of suitable film and application of vacuum pressure. (See Fig. 7) The required clamping force can be reached by suitable compressed thickness of the film. The compressed film between the top and bottom mold works as seal for the injected epoxy molding compound. (See Fig. 8) Proper clamping force can prevent the crack of brittle MEMS/Sensor by controlling the film deformed thickness. The EMC transferring pressure pushes the film in to the corners, matching the mold 3D surface, (See Fig. 9), The EMC will completely fill the mold cavities. Film Assisted Molding is effective and reliable encapsulation for microelectronics including MEMS/Sensor.

Wafer level transfer molding with FAM

M.Brunnbauuer introduced the wafer level package technology, which includes chip relocation, fixture at a new location and encapsulation [4]. One of the key process steps is encapsulating the relocated chips with molding compound. Although compression molding is a good candidate for wafer level molding, the reliable transfer molding has obvious advantages in better performance. Wafer can be encapsulated with MEMS/Sensors by combination of transfer molding and Boschman Assisted Film technology. The following figure 10 shows the schematic of molded wafer with open windows.

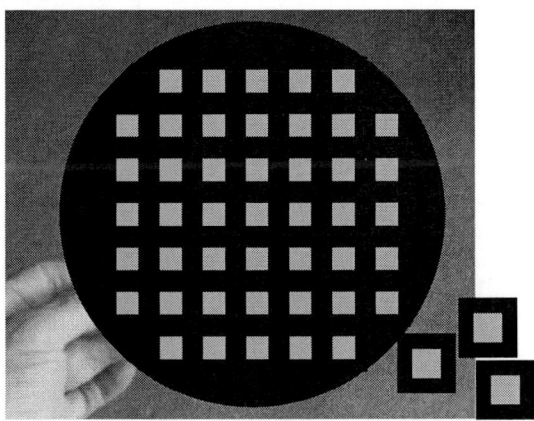

Fig. 10: Molded wafer with MEMS/Sensor

Conclusions

The Film Assisted Molding technology (FAM) for transfer molding is introduced for encapsulating microelectronic devices, FAM enables a bleed and flash free window for MEMS/sensors. The EMC bleed and resin flash can be prevented by the compressed film, not steel-steel direct contact. The mold life can be extended and mold cleaning is not needed with FAM because the EMC will not touch the mold. The vacuum pressure makes enough space to prevent the film touching the wires during the mold close. FE simulations are performed to understand the deformation of film during molding process. The simulations can be the basis of optimization of mold design. Wafer level molding with film assisted technology can be used to further reduce the cost and improve the performance. Film assisted technology can meet the MEMS/Sensors encapsulation challenge with low cost and good quality.

References

[1] G.Q. Zhang, F. van Roosmalen, M. Graef, "the Paradigm of "More than Moore" ", Proceedings of 6th International Conference on Electronics Packaging Technology (ICEPT2005), Shengzhen, China, August 30 – September 2, 2005.

[2] Karl-F. Becker, Erik Jung, Andreas Ostmann, Tanja Braun, Alexander Neumann, Rolf Aschenbrenner, Herbert Reichl, "Stackable System-On-Packages with Integrated Compounts", IEEE TRANSACTION ON ADVANCED PACKAGING, Vol. 27, No. 2, pp. 268-277, May, 2004.

[3] Boschman website: http://www.boschman.nl/

[4] Brunnbauer, M.; Furgut, E.; Beer, G.; Meyer, T., Embedded wafer level ball grid array (eWLB), Electronics Packaging Technology Conference, 2006. EPTC '06. 8th, 6-8 Dec. 2006.

Next Generation Leadless RF Packages Utilizing 1st Level Low Cost Flip Chip Interconnect Technology

S. Walczyk, P. Dijkstra, N. Kramer, J. Verspeek

NXP Semiconductors Netherlands B.V.
IS&O, CSC-Innovation
Gerstweg 2, 6534 AE Nijmegen, The Netherlands
Phone: +31-24-353 3080, Fax: +31-24-353 3642, Email: Sven.Walczyk@nxp.com

Abstract

There is an ongoing demand for low cost high performance RF components for high frequency applications. Current packages for this domain use plastic packages based on wirebonding. The RF performance is limited by parasitic effects due to the RLC network between the wirebond from the dies to the leadframe. One of the best ways to reduce these effects is the use of flip chip technology. Flip chip interconnections are essential as they provide minimal emitter inductance. Thus the device speed and signal integrity with the lower inductance interconnection makes flip chip most expedient for RF devices.

A low cost leadless packaging concept using flip chip interconnects will be presented. The package technology chosen is called UTLP (Ultra Thin Leadless Package). This technology is based on a leadframe, consisting of a three layer metal stack. The aluminum bondpads of the dies are plated with an Electroless Nickel Immersion Gold (ENIG) underbump metallization. The flip chip Pb-free solder bumps (SnAgCu) are created by means of a stencil printing process on the die rather than using expensive lithography as the 1st level interconnects.

The fine bump pitches of 200μm on the discrete dies offer not only cost advantages but also allow for further miniaturization. The challenges with the interaction of the resulting limited solder volume and the leadframe finish are illustrated. Solutions to circumvent the possible existence of brittle $AuSn_4$ intermetallic formations at this interface are described. The assembly flow and lifetesting results are presented.

Key words: Flip Chip; Intermetallic formation; Leadless; Reliability

Introduction

There is an increasing demand towards higher frequencies for applications in the RF domain while the cost has to be low for breakthrough into the commercial market. Possible applications can be found in microwave communication systems, satellite TV reception, consumer products and even in emerging automotive systems like car radar.

In the field of high frequency semiconductor packaging flip chip is the prime choice to improve the RF performance as it accomplishes smaller parasitics. Beside that flip chip is also perfectly suited for size reduction (e.g. in mobile phones) and improved thermal performance [1].

Flip Chip technology is industrialized since a long time, starting with the IBM C4 process in the early sixties. Many different choices of FC penetrated the market so far. In general there are 2 approaches: Wafer level package (1st level interconnect) and encapsulated Flip Chip

interconnect. The WLCSP approach requires a large die area to cope with PCB pitch standards whereas the encapsulated solution offers a low-cost bond pad pitch transformation from 1st to 2nd level interconnect.

Technology

For the domain of RF packages we are introducing and describing a discrete bipolar transistor as a first carrier. The coarse assembly process flow can be divided into 2 sections:
1. Waferbumping
2. Package assembly

Waferbumping

The aluminum bondpads of the die are plated with an Electroless Nickel Immersion Gold (ENIG) underbump metallization (Fig.1).

978-1-4244-4722-0/09 $25.00
© 2009 IMAPS-ITALY

After applying the plating leadfree solder (SAC) is attached to the prepared bumps by means of stencil printing (Fig.2).

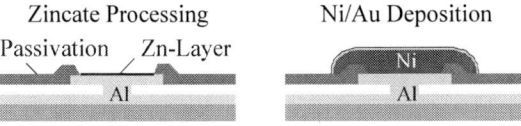

Figure 1: UBM process on top of Al pads

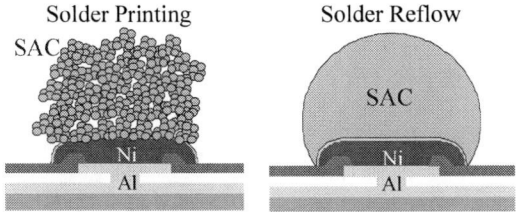

Figure 2: SAC solder process on wafer

SAC105 (Sn98.5Ag1Cu0.5) is chosen for maximum drop and vibration performance as repetitively reported in amongst others [2] [3]. The low amount of Ag (1%) in the interconnect leads to a more ductile solder joint (lower elastic modulus) acting as a mechanical buffer for these kind of external forces hence leading to improved brittle failure performance [4]. Type 6 (5-15µm) is used for the solderpaste to obtain a robust paste printing process, which leads to uniform solderbumps with pitches ranging from 150µm upwards.

The wideband transistor die has 4 pads with a bump pitch of 200µm as can be seen in Figure 3. On the bondpads solderbumps of roughly 80µm in diameter are present. Due to the approximation of a truncated sphere the height is somewhat lower and is a function of the soldervolume applied during the stencil process.

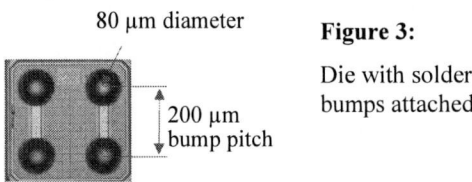

Figure 3:

Die with solder bumps attached

Package concept and assembly flow

The UTLP technology uses a leadframe with several characteristics. The basematerial consists of a 3-layer stack of Cu-etch barrier-Cu. The top side of the leadframe is etched to form the fine line/space pattern required for the flip chip die attach. The bottom copper layer is partially plated with NiPdAu. These area's are used as an etch mask in assembly when the bottom copper layer is etched to form the isolated solderlands. A schematic drawing of a leadframe is given in Figure 4.

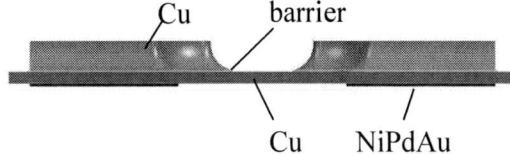

Figure 4: Schematic drawing of the etched leadframe

The assembly flow for the flip chip package is the same as for the wirebond version, however the conventional glue die attach and wirebonding is replaced by flip chip die attach. The package can be transfer molded and no underfill is required since the dies are small and a sufficient gap underneath the die is available. The process flow is presented in Fig 5.

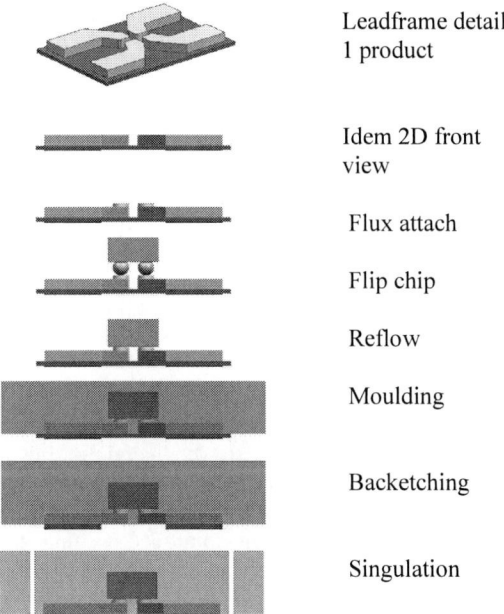

Figure 5: Package assembly flow

Comparable flip chip concepts have been published before, amongst others, in reference [5]. This design is based on more complex AuSn bumps requiring high temperature solder processing. Our approach is opting for mass volume, high speed hence low-cost in combination with optimum performance and reliability.

The complete package can be seen in Figure 6 while an internal sketch of the flip chip package of this wideband transistor is illustrated in Figure 7.

Figure 6: 4 pin package with a size of 1.2 x 0.8 x 0.4mm

Figure 7: Sketch of the Flip Chip Package

Experimental

The UTLP technology gives the opportunity to select different top finishes of the leadframe. In case of conventional glue die attach and Au wirebonding the top surface consists of NiPdAu [6]. In case of flip chip die attach a copper surface is preferred to avoid excessive solderwetting and the formation of brittle $AuSn_4$ intermetallic compounds[7][8].

The effect of the different top finishes on the properties of the bump after reflow are as expected. In case of a NiPdAu finish very good wetting is observed which leads to very low solder bumps and a brittle $AuSn_4$ intermetallic at the entire interface. A copper finish results in controlled wetting and significantly higher solder bumps after reflow. The different situations are shown in Figure 8.

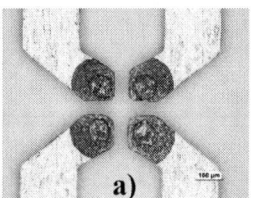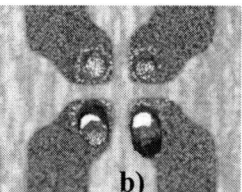

Figure 8: Wetting and bump shear behaviour

    a) Large area wetting on Au surface with low bump shear forces
    b) Restricted wetting on Cu surface with high bump shear forces

In case the bumps are mechanically tested by means of die shear large differences are observed. In the experiment with Cu the shear mode is within the solder and showing values in the range of the solder properties whereas for NiPdAu the shear mode is brittle fracture in the intermetallic at much lower values. In order to achieve a reliable interconnect the absence of brittle intermetallics is required and the bump itself should consist of ductile SAC and be as high as possible to survive mechanical stress.

The bumps on copper have been analyzed in order to verify the absence of the non-preferred intermetallics. The cross section together with a SEM analysis to identify the different metallic phases can be seen in Figure 9 and 10. The stand-off as well as the presence of ductile SAC solder results in a reliable interconnect. In more detail this part will be discussed in the paragraph qualification tests.

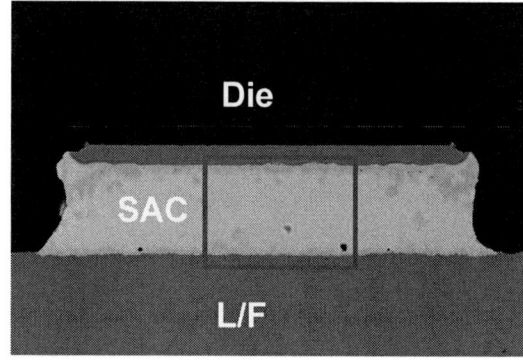

Figure 9: SEM cross section of the SAC interconnect on Cu-top leadframe. Free SAC is dominant within the bump with traces of Cu Sn intermetallics found near the leadframe

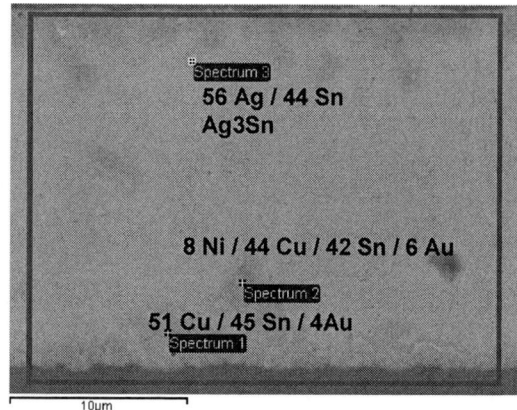

Figure 10: SEM cross section of the SAC interconnect on Cu-top leadframe. Free SAC is dominant within the bump with traces of $Cu_6Sn_5(Ni,Au)$ intermetallics found near the leadframe. Only small spots of $AuSn_4$ intermetallics are found which emerged from the Au-flash of ENIG

978-1-4244-4722-0/09 $25.00
© 2009 IMAPS-ITALY

Reflow conditions

In general three main parameters determine the amount of wetting on a copper surface during reflow: Reflow profile, gas environment and flux.

The reflow profile is given by the flux type and the solder composition (see Figure 11). The flux type has to be tacky to hold the die in place after flipping and prior reflowing.

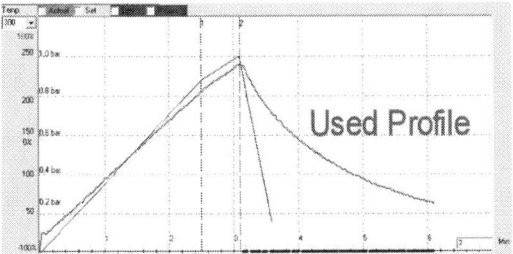

Figure 11: Reflow profile

To identify the influence of the gas environment and the amount of flux a DOE was carried out as can be seen in Table 1.

Table 1: Impact of reflow conditions on bump properties

Parameter settings		Test results		
Reflow gas	Flux diameter [μm]	Average height [μm]	Average die shear [g]	Shear interface
Air	150	39	182	solder
N2	150	21	277	solder
Air	100	54	125	leadframe
N2	100	27	219	solder

The result shows that reflowing in air leads to higher bumps however the situation of marginal wetting (shear interface at the leadframe) needs to be prevented by utilizing a certain minimum flux diameter (Figure 12). Higher die shear values can be obtained when reflowing in a nitrogen atmosphere with the advantage of lower sensitivity on flux diameter variations.

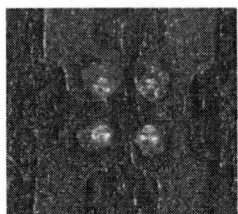 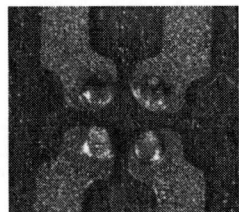

Figure 12: Shear mode of solder joint: Left at leadframe interface; Right within solder bump

Qualification Tests

Assembled parts have been subjected to several reliability tests. In table 2 an overview of these tests is given. The focus on these tests were based on the mechanical integrity of the interconnects rather than the die/package

performance (i.e. biased tests). For this reason, and to avoid RF oscillations, we used daisy chain structures as internal carrier to monitor possible resistance increases between terminals on board level. Since the devices are very small all the tests except thermal shock have been performed on test boards (see Figure 13). Although these boards are not according BLR standards the tests are more severe than if loose devices would have been used.

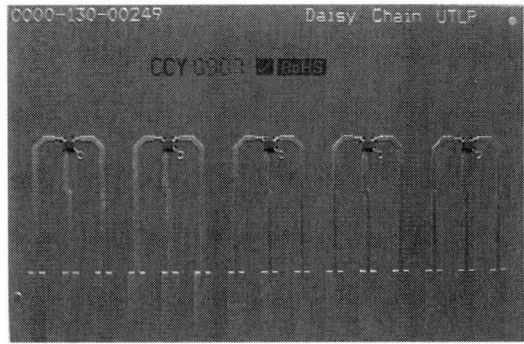

Figure 13: Test board for qualification of Flip Chip Daisy Chain products

Table 2: Summary of life tests applied for the daisy chain test structures to investigate the reliability of the flip chip interconnection

Daisy chain in FC UTLP (4 lead, 200 mm pitch)				
Lifetest	Condition		Read points	Result
MSL 1	Precon (85°C/85%RH) 10 x reflow (250°C)		0 hr, 3 x, 10 x	Pass
Thermal Shock liquid to liquid	-65°C to 150°C	1000 cyc	0 hr, 100cyc, 250cyc, 500cyc, 1000cyc	Pass
High Temperature Storage Life (HTSL)	150°C	1000hr	0 hr, 168 hr, 500 hr, 1000hr	Pass
Temperature Cycling (TC)	-65°C to 150°C	1000 cyc	0 hr, 200cyc, 500cyc, 1000cyc	Pass
Unbiased Highly Accelerated Steam Test (UHAST)	130°C/85% RH	192hr	0 hr, 96hr, 192hr	Pass
Drop test	1000 drops		0 hr, 1000 drops in situ	Pass

Accelerated testing was performed mainly for the following reasons:
Testing of the Moisture Sensitivity to the Level of 1 was accomplished to investigate possible remelt problems or delamination issues. The flip chip interconnects experience severe mechanical loads during temperature changes caused by the CTE mismatch between the leadframe and the die. Hence thermal cycling and thermal shock were used to accelerate possible creep damage or fatigue stress in the solder joint. In-situ drop testing on board level according international standard has been done to monitor the mechanical robustness of the solder

978-1-4244-4722-0/09 $25.00
© 2009 IMAPS-ITALY

bump, package construction and second level interconnect. Humidity testing was performed to identify potential corrosion effects and HTSL to get acquainted with possible IMC growth, and the effects thereof, at the interface between the bump and the leadframe or chip. To check the interfacial strength of the solder bump to leadframe and die, we have completed bump shear tests prior moulding.

MSL testing revealed no change of the shape of the solder joint or effects of recrystallization of the bulk material. The encapsulation with the "dark green" mould compound gives a very good adhesion to the leadframe which does not lead to deterioration of the solder joint even after 10 times of reflow at 250°C. As we also have not observed any voids within the bumps we can state that this concept flawlessly complies with MSL1.

We have not found any significant increase of the resistance or even any electrical discontinuity related to the first level (internal) flip chip interconnects during all accelerated testing performed.

Analysis

Visual inspection and thus verification of the quality of the solder joints after stress testing has been conducted with the use of microscopy, X-ray, SEM and EDX. Examples can be seen in Figure 14 and 15.

During some aging tests we observed the expected interfacial reactions between the SAC solder and the two adjacent surfaces (Cu and ENIG). As mentioned before, some minor portions of $AuSn_4$ intermetallic compounds were formed due to the Au finish on the Ni underbump in combination with enriched Sn from the solderbump. As the weight and atomic percentage thereof is very limited we do not have any reliability concern on the occurrence of this IMC.

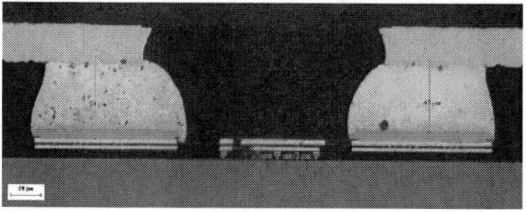

Figure 15: Typical cross section after MSL testing 10* reflow @ 250°C. The shape of the bump remains the same as prior testing.

Two intermetallic compounds were found close to the Cu leadframe due to copper diffusion into the SAC solder: Cu_6Sn_5 and Cu_3Sn (see Figure 16). As we have passed all former described lifetesting, we have not detected any negative side effects due to the formation of these 2 intermetallics.

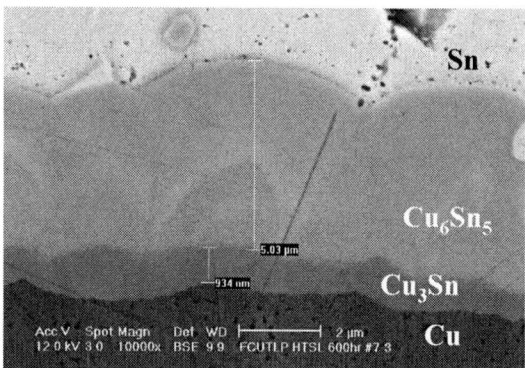

Figure 16: Found IMC's after 600hrs of HTSL. Cu_6Sn_5 and Cu_3Sn have been formed between the SAC solder bump and the Cu leadframe

Analysis of the performed bump shear tests uncovered ductile fractures within the bump as can be seen in Figure 8b. The inhomogenuity of the failure mode expressed in this figure is due to the combination of shear and tilt as the shear height was rather high with 50μm.

Conclusions

We have demonstrated a cost-efficient package concept well suited for low to medium pin count ranges. Cost adders like underfill and solder resist are not required. This concept is based on the in-house available leadframe technology UTLP and the mature and widely accepted wafer bumping process together with solder reflow. The package is dark green as it is RoHS compliant and halogen free, which serves todays' and next generations' needs. The Flip Chip interconnect and package construction is robust as shown in the reliability tests performed, which makes this package very suitable for the consumer market. This package can extend the RF application towards higher frequencies due to the

Figure 14: Typical cross section after 1000 cycles thermal shock (liquid to liquid –65°C – 150°C)

lower parasitics achieved with this flip chip technology.

Acknowledgements

The authors would like to thank Boud van Blokland, Erik Eltink and Victor Brouwer for their support with respect to Equipment and Product- and Process assembly, Roelf Groenhuis and Mark Luke Farrugia for their valuable discussions and package design contributions as well as Erik Janssen for his contribution in the field of reliability. Furthermore we greatly appreciate the work done by Ludo Krassenburg in the field of microstructure analysis. We are also grateful for the support delivered by the Fraunhofer Institute fuer Zuverlaessigkeit und Mikrointegration IZM.

References

[1] J. H. Lau, Ed., Flip Chip Technologies, New York: McGraw-Hill, 1995

[2] L.Garner et.al, "Finding Solutions to the Challenges in Package Interconnect Reliability", Intel Technology Journal, Electronic Package Technology Development, Volume 09, Issue 04, ISSN 1535-864X, DOI:10.1535/itj.0904

[3] T. Teutsch, "ENIG vs. ENEP(G) Under Bump Metallization for Lead-Free WL-CSP Solder Bumps – a Comparison of Intermetallic Properties Using High Speed Pull Test", IMAPS International Conference on Device Packaging, Scottsdale, Arizona, March 17-20, 2008.

[4] R. Dudek, "Fatigue Life Prediction and Analysis of Wafer Level Packages with SnAgCu Solder Balls", Proceedings of 2006 Electronics Systemintegration Technology Conference (ESTC), Dresden, Germany, pp. 903-911, 2006

[5] H.Theuss et al., "A Highly Reliable Flip Chip Solution based on Electroplated AuSn Bumps in a Leadless Package", Proceedings of 2005 Electronic Components and Technology Conference (ECTC), Lake Buena Vista, Florida, USA, May 31 – June 3, pp. 272-279, 2005

[6] P. Dijkstra, "UTLP: A New Platform for Leadless Semiconductor Packages", Semicon München, Germany, April 2006

[7] C.E. Ho et al., "Interactions Between Solder and Metallization During Long-Term Aging of Advanced Microelectronic Packages", Journal of Electronic Materials, Vol 30, No.4, 2001, pp.379-385

[8] N.Duan et al., "The influence of Sn-Cu-Ni(Au) and Sn-Au intermetallic compounds on the solder joint reliability of flip chips on low temperature co-fired ceramic substrates", Microelectronics Reliability, Volume 43, Issue 8, August 2003, pp. 1317-1327

BCB-Based Wafer-Level Packaging of Integrated CMOS/SOI Piezoresistive MEMS Sensors

Dominik Weiland*, Aboubacar Chaehoi*, Shona Ray**, Diarmuid O'Connell*, Mark Begbie*, Changhai Wang***,

*	Institute for System Level Integration, Livingston, EH54 7RG, U.K.
**	Semefab Ltd., Glenrothes, KY7 4NS, U.K.
***	School of Engoneering & Physical Sciences, Heriot Watt University, Edinburgh EH14 4AS, U.K

Abstract

We present a two-staged BCB- and anodic-bonding-based packaging approach used to package both a 3-axis piezoresistive accelerometer and an absolute pressure sensor both based on the same CMOS/SOI process with integrated on-chip amplification. A number of electrical connections run from the sensing element and the integrated amplification circuitry to the bond-pads, crossing the area used for the bond. Therefore, the bonding technique used for the top surface needs to provide good conformance over non-planar structures in order to create a sealed cavity. Zero-level packaging is achieved using anodic bonding of a pyrex wafer on the back-side and a BCB-based bonding approach to attach another pyrex wafer to the front-side. As the seismic mass of the accelerometer is formed by both the SOI-handle and -device layers, recesses to allow upwards and downwards movement of the mass are crucial for the performance of this device. Due to the heat-induced reflow process and the relative softness of the BCB material good conformance over non-planar connection tracks is achieved in the bonding process.

Key words: MEMS, Packaging, Accelerometer, Pressure sensor, Micromachining, BCB

Introduction

Silicon-on-Insulator (SOI) is a highly attractive material for both VLSI and MEMS sensor applications due to the reduction of leakage currents and parasitic capacitance for VLSI and the increased sensitivity of sensors due to the large forces achievable by bulk etching the material.

This work addresses the implementation of both a 3-axis accelerometer and an absolute pressure sensor fabricated by Semefab using its in-house MEMS and CMOS technologies on SOI under the SemeMEMS project [1]. The proof mass of the accelerometer is micromachined using Deep Reactive Ion Etching (DRIE) from both sides of the wafer and the membrane of the pressure sensor is micromachined using DRIE from one side only. This monolithic approach allows batch production and dramatic cost reductions over hybrid sensors which use separate MEMS and signal conditioning circuits. Furthermore, the embedded signal conditioning and in particular, the low noise level in polysilicon gauges enable acceptable performance, while the sensitivity can be improved by adding a dedicated on-chip amplification circuitry.

Possible applications are for instance in the automotive field [2] (airbag triggers, shock monitoring, tyre pressure monitoring systems etc.) where highly reliable and low-cost sensors are required.

BCB-based bonding approaches have been demonstrated for both chip-scale [3] and wafer scale

[4-6] packaging of MEMS devices. In the latter, thin layers of BCB (less than 3µm) were used and the outer perimeter of the Silicon-BCB-Silicon interface was covered with a Silicon-Dioxide layer to provide a hermetic seal, which cannot be achieved with the BCB layer alone.

In the work presented here, the aim was to provide a cheap, reliable means to attach a transparent pyrex wafer to the top-side of the wafer to create test-vehicles for characterisation of integrated CMOS/MEMS accelerometers while allowing the manufacture of absolute pressure sensors on the same wafer. A large stand-off distance of more than 10µm between the silicon- and the pyrex-wafer was vital to allow sufficient movement of the seismic mass of the accelerometer.

Three-axis piezoresistive accelerometer

The micromachined 3-axis accelerometer is composed of four suspended single-crystal silicon beams attached to a central proof mass. The acceleration is measured by means of embedded strain gauges which convert the stress at the anchor point of the beam into resistance variations. When an in-plane acceleration is applied (acceleration along the x or y axis), the seismic mass is twisted along the orthogonal axis; the aligned beams will therefore bend in opposite directions, as shown in Figure 1. In the case of an out-of-plane acceleration along the z axis, the mass undergoes a vertical

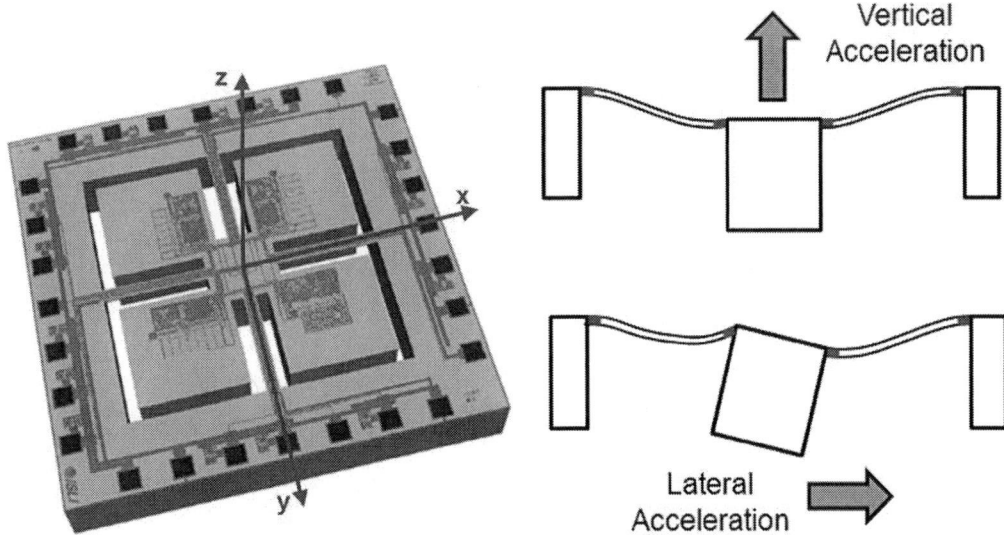

Figure 1: Schematic of the 3-axis accelerometer and working principle.

translation movement with all four beams bending in the same direction.

The dimensions of the beams are $750 \times 80 \times 20 \mu m$, while the overall chip is $3 \times 3 \times 0.4mm$ when unpackaged and $3 \times 3 \times 1.4mm$ when fully packaged.

The CMOS circuit has been characterized at operating temperatures between 20 and 120°C. The voltage reference was found to vary by around 2%, the reference current by around 20% as it is dominated by the negative temperature coefficient of the high-resistivity polysilicon used. This large variation in reference current is permissible however due to the amplifier design which allows for variations up to 200%. The amplification gain has been characterised, a gain of 65 and 40 has been measured respectively for the in-plane and the out-of-plain amplification chain. These characterization results match the simulation and design specifications. Mechanical characterizations (shock and vibration) are currently in progress.

Piezoresistive Absolute Pressure Sensor

The pressure sensor consists of a membrane formed by SOI device layer and four piezoresistive strain gauges placed at the centre of the edges of the membrane, as shown in Figure 2a. A pressure change deforms the membrane, thereby causing stress, which in turn is measured by the strain gauges by the means of a change in resistance. The signal from the change in resistance is amplified by on-chip amplification circuitry.

The dimensions of the membrane are $500 \times 500 \times 15 \mu m$, while the overall chip is $2 \times 2 \times 0.5mm$ when unpackaged and $1.5 \times 1.5 \times 0.9mm$ when fully packaged.

Figure 2b shows the sensitivity of the sensor to the pressure. The sensor exhibit a non-linearity – principally due to residual stress in the membrane - below 15PSI while a very good linearity is achieved up to 70PSI (maximum pressure allowed by our setup). A sensitivity of 60mV/PSI is measured in the linear range of the sensor.

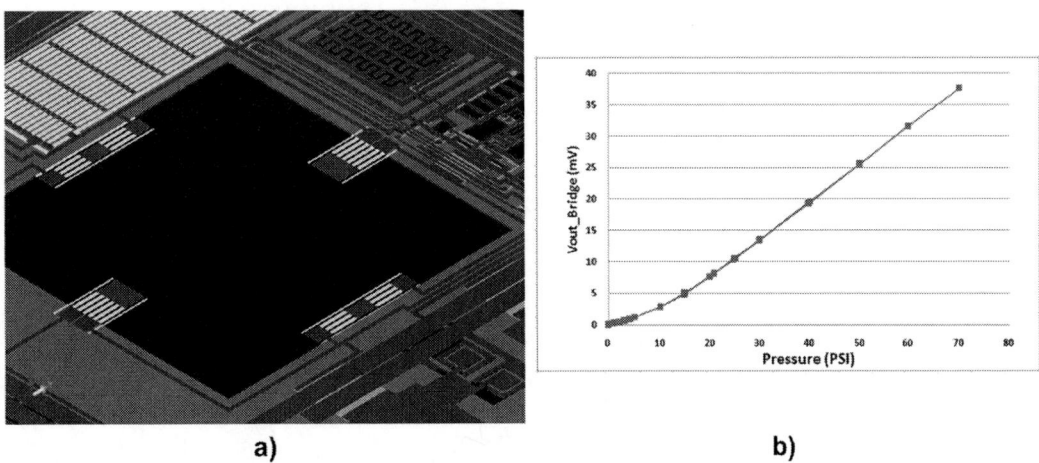

a)
b)

Figure 2: a) Pressure sensor membrane with four piezoresistive strain gauges and surrounding CMOS circuitry. b) Sensor output for absolute pressure sensor vs. ambient pressure.

978-1-4244-4722-0/09 $25.00
© 2009 IMAPS-ITALY

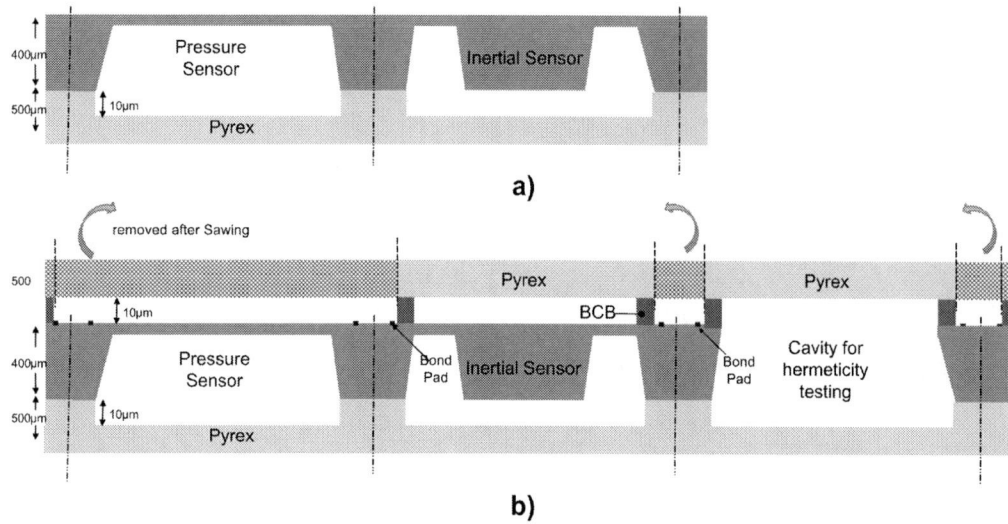

Figure 3: a) Diagram of one-staged packaging approach using anodic bonding b) Diagram of BCB-based packaging approach for the front-side and anodic-bonding-based approach for the back-side of wafer for fabrication of both intertial and pressure sensors.

One-stage packaging approach

After completion of the actual integrated CMOS/MEMS devices, a pyrex-wafer (Boroflat 33, 500μm thick) is wet-etched using hydrofluoric acid (HF) to form 10μm deep recesses in order to allow displacement of the silicon seismic mass of the accelerometer during device operation. This wafer is then bonded to the underside of the SOI-wafer using anodic bonding, thus providing protection for the accelereomter and a sealed cavity for the pressure sensor, as shown in Figure 3a. Due to the high-voltages and high temperatures required and the poor coverage of non-planar sufraces anodic bonding can only be used for the back-side but not the front-side of the devices.

This packaging approach allows for absolute pressure sensors and accelerometers to be packaged simultaneously. Protection against contamination of the sensors, however needs to be provided by a higher packaging level.

By etching holes into the pyrex forming the pressure sensor cavities, differential pressure sensors, absolute pressure sensors and acceleromters can be fabricated in parallel.

Two-Stage Packaging Approach

For the two-stage approach, a pyrex-wafer with 10μm deep recesses is also anodically bonded to the underside of the Silicon wafer. Then a 10μm thick BCB-layer (Cycoltene 4024-40 from Dow Corning) is deposited onto a bare pyrex-wafer and structured using photolithography leaving rings of BCB surrounding the cavity around the seismic mass on the accelerometer. A number of electrical connections run from the sensing element and the integrated amplification circuitry to the bond-pads, crossing this BCB ring. Therefore it is vital for the BCB to provide good conformance over non-planar structures in order to create a sealed cavity. Furthermore, the area covered by the BCB needs to be large enough to provide good adhesion between the two die to be joined and at the same time the distance between the BCB-ring and the bond pads needs to be large enough to facilitate wire-bonding.

The pyrex-wafer is then flipped over and bonded to the front-side of the SOI-wafer using thermal-compression bonding in an EVG501 bond tool. Both the bond temperature (300 °C) and the force applied in this step are essential parameters to ensure a homogenous high-quality bond across the wafer. For the front-side the BCB-ring creates enough stand-off distance between the pyrex and the silicon to allow free movement of the seismic mass in the specified acceleration range, as shown in Figure 3b.

To allow access to the bond pads on the device, the front-side pyrex cap needs to be smaller than the silicon-die, as shown in Figure 5. This is achieved in the wafer-dicing step by dicing the top-pyrex wafer along the dashed lines shown in Figure 3b with a 150 μm wide dicing-blade.

This cut proved difficult due to the small stand-off-distance of 10μm between the pyrex and the Silicon wafer. Therefore tight control of the depth of the cut was essential to prevent damage to underlying bond-pads and conncetion tracks. Furthermore, lateral control of the cut was also

978-1-4244-4722-0/09 $25.00
© 2009 IMAPS-ITALY

Figure 4: a) Micrograph of correctly diced, fully-packaged accelerometer. b) Micrograph of accelerometer damaged during dicing of the front-side pyrex.

essential in order to clear enough BCB to allow successive wire-bonding while at the same time maintaining enough BCB for a realiable bond.

A problem faced during dicing was the slight movement of the die which had already been diced along three of its sides. When cutting the fourth side of the die the rotation of the dicing blade caused an out-of-plane tilt of the die resulting in different cut-depths of the dice for different sides of the die. The effect of this is shown in Figure 4b, in which on the top side of the die all BCB has been removed by the dicing blasé and the underlying connection tracks were damaged. A correctly diced die is shown in Figure 4a.

In the second cut, the pyrex blocks in between accelerometer die and above pressure sensor die are washed away by the cooling liquid, leaving the bond-pads and the underlying silicon exposed. In a final dicing step, both the silicon and the back-side pyrex are then diced along the central dashed line using a 300 μm wide dicing blade.

The devices are then singulated and packaged into a Dual-Inline-Package (DIL), as shown in Figure 5. For devices packaged using the one-stage packaging approach, the lid of the DIL can provide the necesarry protection against contaminations.

Hermeticity Tests

To characterize the hermeticity of the BCB-bond, the seismic mass has been removed from a number of accelerometers in order to increase the volume of the cavity, as shown in Figure 3b. Using helium bombing they have then been tested according to MIL-STD-883, which they passed with an average He leak rate of 1.51×10^{-8} atm.cc.s-1 and a standard deviation of 4.05×10^{-9} atm.cc.s-1 after a bombing time of 4 hours at 4 Bars. As Helium can diffuse through pyrex at a rate of 10×10^{-8} atm.cc.s-1 the samples cannot pass MIL-STD-750, which has higher specifications than the MIL-STD-883 method [7].

Conclusions

We have demonstrated a two-staged BCB-based packaging approach allowing for the fabrication of piezoresistive accelerometers and both absolute and differential pressure sensors on a single-wafer.

Acknowledgements

The authors would like to acknowledge the support from Scottish Enterprise through the iDesign project and from Scottish Enterprise and the DTI through the SemeMEMS project. Furthermore, the authors would like to acknowledge the help of Suzanne Millar (HWU) for carrying out the hermeticity tests.

978-1-4244-4722-0/09 $25.00
© 2009 IMAPS-ITALY

Figure 5: Photograph of a packaged piezoresistive accelerometer.

References

[1] Semefab (Scotland) Ltd., Sensor & Silicon Solutions, www.semefab.co.uk.

[2] S. Renard, "Industrial MEMS on SOI", Journal of Micromechanics and Microengineering Vol.10, pp. 245-249, 2000.

[3] C.H. Wang, J. Zeng, K. Zhao, H.L. Chan, "Chip Scale Studies of BCB Based Polymer Bonding for MEMS Packaging", Proceedings of 2008 Electronic Components and Technology Conference, pp. 1869-1873.

[4] S. Seok, N. Rolland, P.-A. Rolland, "Mechanical and Electrical Characterization of Benzocyclobutene Membrane Packaging", Proceedings of 2007 Electronic Components and Technology Conference, pp. 1685-1689.

[5] J. Oberhammer, F. Niklaus, G. Stemme, " Sealing of adhesive bonded devices on wafer level", Sensors and Actuators A, 110, pp.407-412, 2004.

[6] F. Niklaus, H. Andersson, P. Enoksson, G. Stemme, "Low temperature full wafer adhesive bonding of structured wafers", Sensors and Actuators A, 92, pp. 235-241, 2001.

[7] V.O.Altemose, "Helium Diffusion through Glass", Journal of Applied Physics, Vol.32, No.7, pp. 1309-1316, 1961.

978-1-4244-4722-0/09 $25.00
© 2009 IMAPS-ITALY

New Flipchip Technology

Authors: Reinhard Windemuth (PFSE Europe)
and Takatoshi Ishikawa (PFSC Japan)

Panasonic Industrial Europe GmbH, PFSE,
Hans-Pinsel-Str. 2, 85540 Haar (Munich, Germany)
Phone +49 / 89 46159 – 365, Fax -260, Reinhard.Windemuth@eu.panasonic .com

Abstract

Flipchip Technology is getting more and more important for future packaging solutions. This presentation gives an overview of different Flipchip Technology Solutions provided by industry. Mainstream Technologies such as C4 and ACF / ACP processes are explained. In special focus will be recently developed new processes to improve: Thermosonic Gold to Gold Interconnect (GGI) Process and Encapsulant Solder Connect (ESC) Process. Those being very fast and highly reliable. This is why they are suitable for further future miniturization. They cover a wide range of industrie´s product applications. Typical characteristics and process parameters for ESC and GGI will be described and analysed. Reliability data will be shown and explained.
Both processes are suitable to be used in Chip on Board (COB), Waferlevel (COW) and Embedded Packaging Technology & Assembly. Some examples of how to use Flipchip processes for embedding active components to FR4 Printed Circuit boards (PCB) are shown and explained.

Key words:
P1. Flipchip, Embedded Technology Miniturization, METAL JOINT, CONTACTING,
P2. machine parameters, ESC process design, Process flow, Process & Quality criteria
P3. cross sections, GGI ultrasonic flipchip, Process conditions, Material parameters, machine parameters, process window
P4. Status of Industry, organic substrates (Flex and FR4) , heated bonding tool, substrate temperature
P5. Embedding components into Organic subtrates, high yield, shielding characteristics, SIMPACT, Laser Cavity , Conclusion

Introduction

Electronic industry faced extremely shrinking product size since the early 20th century. Especially the invention of transistors, logical switches and integration of those electronical circuits to semiconductor material chips enabled an extremely fast miniturization of products. This shrink of assembly dimensions required changing technologies. SMT (Surface mount technology) mounting of passive components, COB bare die assembly with die- and wirebonding. To further more reduce package size, Flipchip Technology is getting into focus of electronic modules and semiconductor devices for more than 20 years now. Target is to further more reduce package size to multichip assemblies, stacking die assemblies, 3D silicon stacking and 3D Assembly Embedded Technology.

Many of those assembly methods are already in Industrie´s High Volume Manufacturing (HVM) since years, others are still under development, optimization and prototype phase. Mainly Consumer-electronics markets such as Personal Computers and Mobile Phones are pushing those high density technologies to get more and more performance & yield in even less product size. Automotive, Aeronautic, Medical and other industrial applications push to highest longterm reliability.

Flipchip Processes

Flipchip dice require individual chip design. Metalic bumps on the connecting pads are typically assembled either by solder ball mounting, stud bumping technology or plated gold technology. We distinguish two kind of Flipchip interconnections:

1. METAL JOINT interconnections: metal interphase by either soldering or interconnecting with US bonding process or thermocompression. Underfill material is applied to protect the assembly from mechanical damage and to protect of the active chip surface underneath the flipchip die from contaminations.

2. CONTACTING interconnection: organic underfill material does the mechanical interconnection. Electrical contact is done by touching of chip electrodes and substrate landing pads.

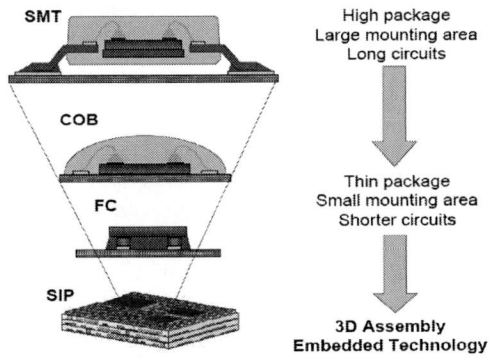

Figure 1: Miniturization trend

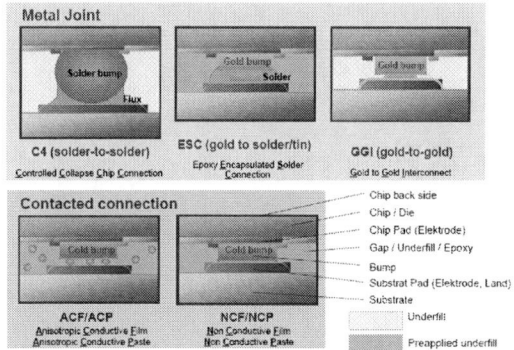

Figure 2: various Flipchip assembly Technologies, schematics

Each Flipchip process requires special substrate-pad and Chip-pad design (finish). Some pre-processing like Flux-dipping or Plasma cleaning. Bonding process typically requires machine parameters like pressure (bonding force), temperature, ultrasonic energy (US power) and their timing. Underfill material (applied before or after bonding) requires curing process to ensure good hardening of the material. High quality underfill interconnection needs to be ensured avoiding delaminations and voids.

Figure 3: Overview Flipchip processes: Process flow

Suitable design of all those process parameters will decide about reliability and performance of the electronical device later on.

Typically all products are different. Each of them requires special characteristics like mechanical reliability, electrical performance (contact resistance, etc.), process time, pitch dimensions, material price etc. Following those criteria an appropriate and suitable selection of materials and final Flipchip process is mandatory.

	Connection	Contact-Resistance	Typical Process time	Minimum Pitch
Conventional C4 [Mass Reflow] Controlled Collapse Chip Connection	Metalic	< 10 mΩ/bump	0.1 sec.	200 um
C4 Individual Bonding	Metalic	< 10 mΩ/bump	10~30 sec.	200 um
ACF / NCF Anisotropic Nonconductive Foil (Preapplied)	Contact	>100 mΩ/bump	5~ 10 sec.	< 50 um
ACF / NCP Anisotropic Nonconductive Paste (Preapplied)	Contact	>100 mΩ/bump	5 ~ 10 sec.	< 50 um
ESC (Au-Sn) Epoxy Encapsulated Solder Connection	Metalic	< 10 mΩ/bump	2 - 6 sec.	< 50 um
GGI Au-Au Ultrasonic Gold to Gold Interconnect	Metalic	< 10 mΩ/bump	0.3 sec.	< 50 um

Figure 4: Process & Quality criteria for various Flipchip processes.

ESC process design:

ESC stands for Encapsulant Solder Connection. This newly introduced process was designed to combine advantages of different processes like C4 soldering (Controlled Collapse Chip Connection) and preapplied ACP (Anisotropic Conductive Paste) interconnection.

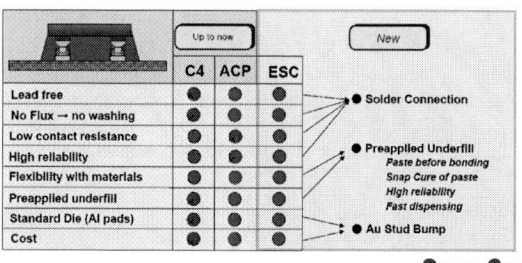

Figure 5: idea and concept of ESC process

Soldering process to ensure intermetallic interconnection. Preapplied underfill processing to avoid later underfill processing, as the gap underneath the chip is getting more and more narrow whereas the chips are getting larger and larger. When gap dimensions are getting less than 20μm height, underfill processing is getting even challenging after Flipchip assembly. Plasma Cleaning Process can help to improve underfill seap in factor. (see [1]: Plasma Cleaning, Nonomura)

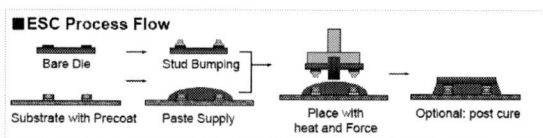

Figure 6: Process Flow of ESC process

Process flow is getting very simple with minimized process steps as stud-bumping, paste supply, Flipchip bonding and optional post cure process. High reliability of those assemblies could be proven for assemblies with less than 30μm pitch in High Volume Manufacturing.

978-1-4244-4722-0/09 $25.00
© 2009 IMAPS-ITALY

■ Cross-section examples

 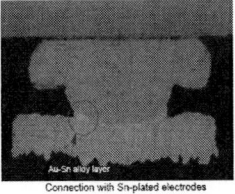

Connection with solder-precoated electrodes   Connection with Sn-plated electrodes

Figure 7: typical cross sections for ESC process with either SnAg solder pre-coated electrode or Sn-plated electrode. Both applications use Au stud bump on chip pad.

Design of ESC5 process

ESC process is using solder reservoir on the bondpad (Sn plating or SnAg solder precoating). This surface finish requires additional process step to add the solder or Sn material by screenprinting, chemical plating or galvanic plating. To simplify, solder material can be provided by special developed Epoxy Paste material. This material is designed with small solder material particles. Those are accumulating underneath the bumps while bonding process. Those accumulated particles are going to melt while the heated bonding tool reaches a temperature higher than solder melting temperature. This melting process requires special properties of the Epoxy material to resist the high temperatures and provide a proper soldering interconnection.

Prozess-steps ESC5

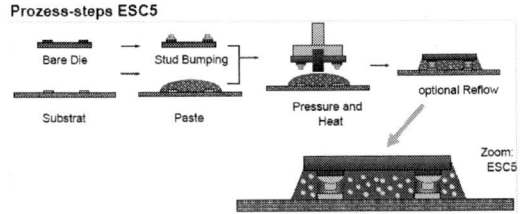

Figure 8: Process flow of ESC5 process.

Connection with ESC5 solder paste   Connection with ESC5 solder paste

Figure 9: Cross sections of ESC5 interconnections by using Namics -138 ESC5 Epoxy paste.

Similar as for ESC process assemblies, ESC5 assemblies could be proven to perform same high level reliability with pitches down to some 50µm.

GGI Flipchip (ultrasonic flipchip)

GGI Process (Gold to Gold Interconnection) is widely spread in Asian countries. Recently also in Europe and USA many activities with GGI Process started with good results. Following process overview (see figure 3) GGI process ist the fastest solution without solder reflow in case of intermetallic interconnection.

GGI interconnection US bonding process is well known from conventional Gold-Ball Wirebonding Technology. To get this bonding process reliable, both machine parameters and material characteristics need to be adjusted to each other in a proper way. After that very good and reliable interconnecting results can be achieved.

Process conditions

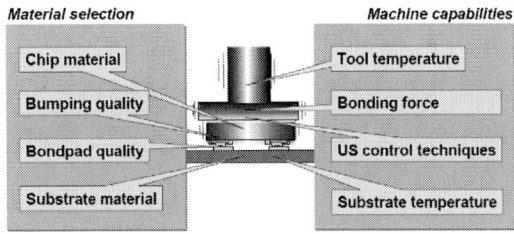

It is very important to optimize each process condition

Figure 10: Process conditions

Main Material parameters are:
- Chip material and Chip design
- Bumping quality
- Bondpad quality (substrate landing pad)
- Substrate material and substrate design

Main machine parameters are:
- Tool temperature
- Bonding force
- US control techniques (ultrasonic US power)
- Substrate temperature (stage)

It is crucial to get material parameters and machine parameters fine-tuned to each other to maximize the process window with optimum results. Basically it can be stated: As long as the substrate is well suitable to Au-Au wirebonding, good results in GGI flipchip bonding can be expected too. As there are several interconnections to be done same time a high demand on planarity and homogenity is mandatory. Typically substrate planarity should be in the range of less than +/- 3 µm, same with stud bump height consistency. It is very important to optimize each process condition.

In Asia a lot of HVM (high volume manufacturing) experience was collected over the last 10 years in the field of LED assembly, Saw-filter assembly, TCX0 assembly. Those are typically done on hard substrates materials such as ceramic, glass and silicon (COW). Chips are typically with pincounts from 2 to about 30 IO (low pin count).

978-1-4244-4722-0/09 $25.00
© 2009 IMAPS-ITALY

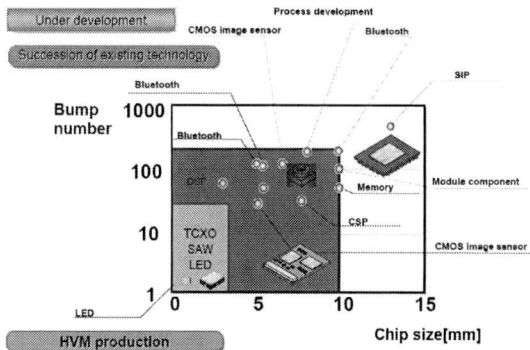

Figure 11: Applications for GGI process and status in industry

More challenging is GGI bonding with higher pincount and on organic substrates. Next chapters describe possibilities to approach bonding on organic substrates. Studies in using this GGI technology for higher Pincount such as 2.000 and more are ongoing.

Materials: Current Status in Industry

		Mass production	Test production	Development
FLAT & HARD	Silicon wafer	O		
	Glass	O		
	Ceramic	O		
	Flex (TAB) single layer one side	O		
	BT single layer one side	O		
	Flex (TAB) single layer both sides		O	
	BT single layer both sides		O	
	Flex (TAB) multi layer both sides			O
ROUGH & SOFT	BT multi layer both sides			O

Special machine features (such as heated bondtool) are needed, when Ultrasonic GGI-process is applied to organic PCB.

Figure 12: Status of Industry depending on substrate material. Especially on Flex and multilayer PCB manufacturing some special efforts are ongoing to prove wide process window and high reliability process with GGI Flipchip.

Compare different substrate materials for GGI process

By using organic substrates (Flex and FR4) much higher US power is needed to achieve comparable bonding results (die shear). On ceramic substrates even with low US power good results can be achieved. This is due to "energy absorption effects" of US power by oscillation of soft organic materials. Especially when substrate temperature is getting close to TG temperature of the material (glass transition point). High ultrasonic power should be avoided to widen the process window of the full system. The risc of pad damages or die cracking increases with high US power.

Dependency on PCB material.

⇒ Bonding strength of ceramics is larger than that of organic (FR4, Flex)
⇒ Bump deformation of organic is smaller than that of ceramics.

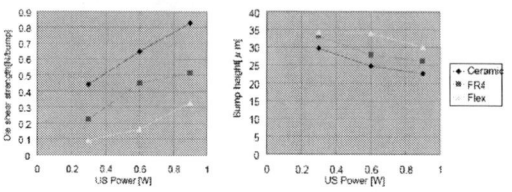

Chip : □2mm, Bump number : 12
Force : 0.98N/bump US time : 0.3s

Figure 13: Comparison of die shear strength depending on US power by using different substrate materials.

GGI quality in dependency on substrate temperature

When using organic FR4 substrate materials, bonding results decrease significantly with increase of temperature. This is due to materials are getting softer with higher temperature. This results in absorption of US energy. The absorbed US power cannot be used for interconnecting bonding process.

Dependency: Substrate Temperature

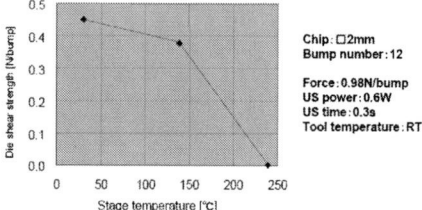

Depending on FR4 material quality (TG temperature) this charactristics may differ.

Figure 14: Bondability depending on substrate temperature

Use of heated bonding tool

Typically heat is entered to the bonding process by heating the substrate material before and while bonding. This results in higher substrate temperature. Some applications work much better if heat is entered from the top side via a heated bonding tool. Heated bonding tool is mechanically linked to the ultrasonic transducer / generator. Therefore special design of ridgit bonding tool is required (see [2]).

978-1-4244-4722-0/09 $25.00
© 2009 IMAPS-ITALY

Dependency: Heated Bondtool

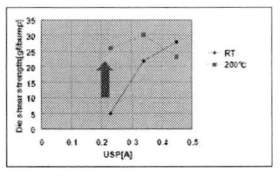

When using heated bondtool, good results can be achieved with less ultrasonic power

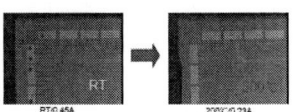

Pad damage can be reduced with combination of chip heating and lower power.

Left side: IR microscope image after IC bonding shows pad damages as black dots

Figure 15: Die shear depending on US power when using non heated and heated bonding tool. Pad damages can be observed and detected when using an IR microscope.

By using a suitable Flipchip bonding equipment with a heated bonding tool big improvements can be reached. Especially when using lower US power. Inspection of bond pad damage by using IR microscope helps in finding critical situations that may results in poor longterm reliability performance of the Flipchip assembly.

Use of Flipchip for Embedding components into Organic subtrates

Flipchip assemblies can be designed with a very small total package height. This is why Flipchip assemblies are very suitable for embedding active components into inner layers of either ceramic or organic multilayer substrates. Industry is investigating this embedding technology globally to come up with High Volume Manufacturing Solutions in near future. Different approaches are ongoing.
Main target is:
- By embedding conventional passive and active components total package size can be drastically reduced.
- When using individually checked components only, high yield can be ensured when using an appropriate Flipchip process (material selection, process parameter selection as mentioned above).
- Embedding components allows inner via structure network with very good shielding characteristics. This results in shorter wiring length and lower noise especially when using this technology for manufacturing of RF products or components.

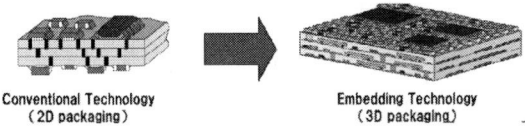

Conventional Technology (2D packaging)　　Embedding Technology (3D packaging)

Figure 16: Comparison of conventional 2D packaging technology with 3D packaging embedded technology.

As above figure 13 shows clearly, much higher density of electronic circuits can be reached by embedding active and passive components into inner layers of multilayer modules.

Examples for embedded technologies

Several technologies are under investigation and development all over the world to approach an HVM solution for embedding active and passive components into multilayer substrates. This chapter is just showing two examples.

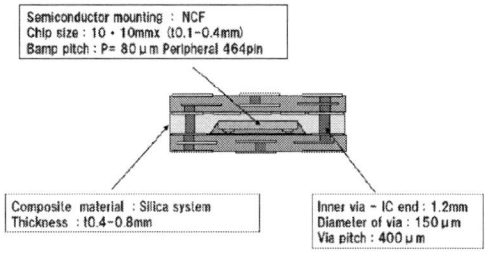

Figure 17: SIMPACT technology

SIMPACT technology introduced by Matsushita Electric Industrial (Panasonic Jisso Eng. Lab). This technology is using silica system composite inner layer material with thicknesses of 0,4 to 0,8 mm for embedding active components with Inner via interconnections with via diameter of about 150µm.

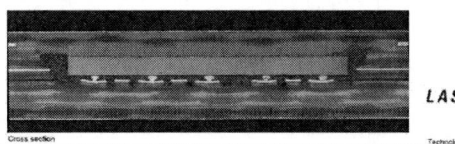

Figure 18: embedded active component by using Laser Cavity Technology introduced by Würth Elektronik (Germany). Flipchip interconnections is done by ESC5 process.

Laser Cavity Technology is being introduced by Würth Elektronik (see [3]). By special laser equipment a cavity inside the inner layer of a multilayer organic substrate is prepared. Flipchip assembly is performed inside those cavities by using ESC5 technology. After Flipchip process, further top layers are laminated on top of the inner layer to finalize the embedded assembly. This technology turned out to be very reliable and ready for high volume manufacturing in the industry.

Conclusion:

Flipchip Technology is getting more and more important in industry due to further miniaturization. Several Flipchip processes exist. Suitable process is to select for each application.
ESC / ESC5 process to be used for high reliability of high pin count & low pitch Flipchip applications
GGI process is one of the fastest and most simple fluxless flipchip process. Reliability on organic substrate materials can be improved by optimization of material properties and equipment capabilities.
Examples: Flipchip is used to mount active components (bare die) to Embedded Systems. High

978-1-4244-4722-0/09 $25.00
© 2009 IMAPS-ITALY

reliability could be proven. Processes are ready for HVM (high volume manufacturing).

Literature:
[1] Plasma Surface Modification Technology for flipchip packages, By Masaru Nonomura, Panasonic Factory Solutions Co, Tosu City, Saga Japan, EPP Europe issue 2007 5_6
[2] Development of high-quality flip chip bonding technology using a heating ultrasonic tool, by Makoto Okazaki and Masafumi Hizukuri, Panasonic Factory Solutions Co, Tosu City, Saga Japan, EPP Europe 1&2 / 2007
[3] Embedded Active Devices – Würth Elektronik, Dipl.-Ing. Roland Schönholz, Produktmanager Circuit Board Technology, Würth Elektronik Schopfheim GmbH, PLUS Nr.2 Feb09

New Plasma Cleaning Technology

Authors:
Reinhard Windemuth (PFSE Europe)
Masaru Nonomura (PFSC Japan)
Gene Dunn (PFSA USA)

Panasonic Industrial Europe GmbH, PFSE,
Hans-Pinsel-Str. 2, 85540 Haar (Munich, Germany)
Phone +49 / 89 46159 – 365, Fax -260, Reinhard.Windemuth@eu.panasonic .com

Abstract:
Plasma Cleaning Technology is used for different applications in backend packaging applications. Plasma Cleaning is used to improve capability and performance of
- *Wirebonding (etching)*
- *Thermosonic Flipchip (etching)*
- *Underfill (surface activation)*

This presentation sets focus on individual process solutions how to improve those backend processes. Different demands on process and equipment come from different applications. Both etching and surface activation and their special solutions are described. Some mathematialc simulations are described. Effect of special gas additive to improve eunderfill cleaning process results is shown. Challenges for underfill with large die and small gap and underneath the Die are shown and solutions explained. Challenges for bondpad cleaning by Argon etching for wirebonding and GGI Flipchip are shown and solutions explained. Methods of process and surface analysis like Auger Analysis and Contanct Angle measurement and others will be described.

Key words:
P1. Plasma cleaning in micro-packaging applications, vacuum, RF generator
P2. Wire bonding, physical etching effect, Argon cleaning, Auger Analysis, Parallel plate chamber
P3. Spectroscopy, depth profile, GGI Flipchip, surface modification, underfill, wettability, Oxygen
P4. Contact angle, viscosity, Seap in time, underfill capillary,
P5. Processing time, Epoxy fillet

Introduction

Plasma etching technology has been developed in semiconductor manufacturing process, and contributed a great deal to fine pattern etching, dry process and low temperature processes. Nowadays Micro-packaging backend applications such as wirebonding, molding, wirebonding, Flipchip and underfill getting more and more in focus as plasma cleaning applications.

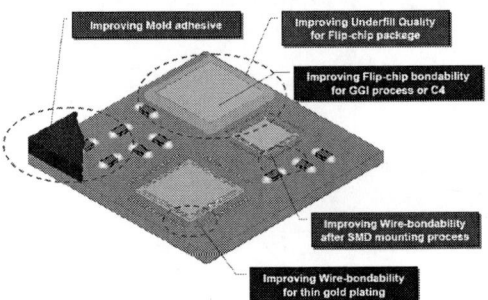

Figure 1: various micro packaging backend applications getting into focus for plasma cleaning.

This article describes four areas of interest: plasma etching to improve wire bonding quality, surface modification to improve surface wettability, Recently Plasma Dicing technology was developed to improve dicing results in comparison to conventional mechanical and laser dicing processes.

Vacuum Based Plasma Technology

A plasma etching system consists of an RF generator and a vacuum chamber with a parallel plate arrangement that keeps the lower electrode at a negative voltage bias with respect to the upper electrode (Figure 2). This design is most effective for directional ion flow from the grounded upper electrode to the negatively biased lower electrode where the object to be treated is located. Typical vacuum pressures are set at 8-12Pa (60-90mtorr). The electric field helps accelerate the free electrons to collide with the argon atoms. Ionization occurs as a free electron dislodges an electron from the Ar atom, resulting in a positively charged Ar+ ion plasma. These highly directional ions move in parallel fashion toward the negatively biased chamber electrode and thus provide a physical etching effect to the Au pads on the substrate. Argon is particularly effective since

978-1-4244-4722-0/09 $25.00
© 2009 IMAPS-ITALY

it is relatively heavy and has a high sputter yield*
with respect to gold and nickel.

■ Improvement of wire bondability

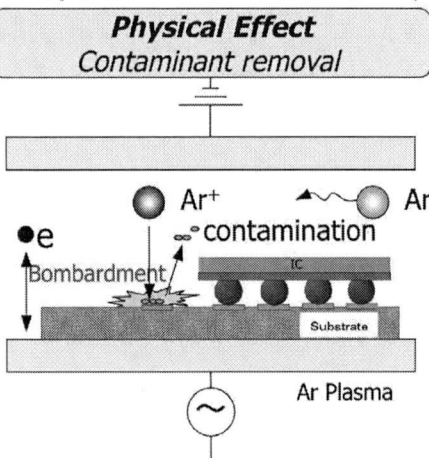

Figure 2: Argon plasma etching of Au bond pads

Plasma cleaning before wirebonding

A contaminant-free gold surface is critical in
achieving high quality wire bonds and stud bumps.
Introducing a plasma etch to the substrate prior to
wire bonding helps remove the contamination that
may cause defects. Typically, argon gas is employed
for ist ability to remove both organic and inorganic
compounds. The etching represents a combination of
thermal energy from electron bombardment and
physical energy from positive ion bombardment.
Figure 2 shows a parallel plate etching chamber in
which the substrate to be treated rests directly on the
negatively biased lower electrode.

Etching Effect for Ni cleaning

The positively charged Ar ions move toward the
lower electrode, creating an etching effect on the Au
bond pads. The Au etches at a rate of about
30nm/min. For organiccontaminants,the etch rate is
around 500nm/min. Typical etch times range from 8-
13 seconds, depending on chamber design. With this
recipe, 5-6nm of the Au surface is etched away. As
cost pressures force manufacturers to reduce Au
plating thicknesses, the contamination from the
underlying Ni becomes an issue. Die attach epoxy
cure temperatures can cause Ni compounds to diffuse
to the Au surface, thus contaminating the bond pads.
(See Figure 3.)

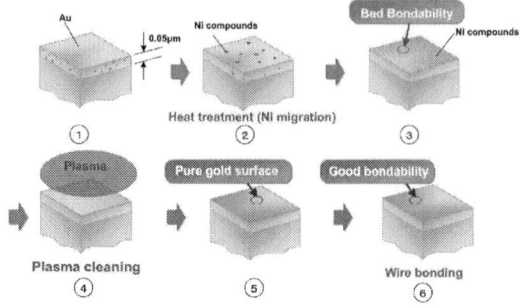

Figure 3. The effect of Plasma on Ni contaminants

It is important to point out that the parallel plate-type
vacuum chamber is particularly effective at removing
inorganic compounds such as nickel hydroxide
(NiOH).

Batch Vs. Parallel Plate

Unlike batch plasma systems, where full magazines of
substrates are loaded for treatment, parallel plate
machines allow the entire substrate surface to be
etched by the Ar+ ions, resulting in a more uniform
etch with subsequent improved wire bondability.

Auger Analysis

A common approach to detect contamination such as
Nickel or organics) — and the effectiveness of the
plasma etch — is to conduct an Auger analysis before
and after cleaning. The results can be later confirmed
by wire ball shear testing. Figure 4 shows the removal
of Ni compounds after plasma cleaning,while Figure 5
shows Machines for plasma etching can provide
benefits for a number of applications in assembly and
packaging of ICs. the result is improvement in shear
strengths after plasma etching.

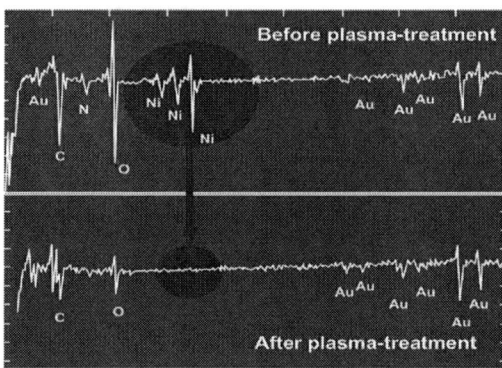

Figure 4. Auger spectroscopy before & after plasma
etch

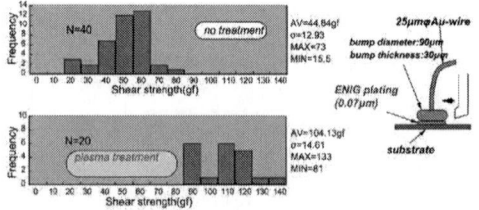

Figure 5. Ball shear improvement after plasma

To forecast optimum machine parameters the target etching depth needs to be evaluated. For this target a depth profile can be generated by several Auger analysis. Typically one Auger analysis after each less than 1 nm surface etching is used. Mainly elements like Au, Ni, O2 and C are important criteria. Thickness of contaminated surface can be identified and process paramteres (RF power, process time) to be adjusted to those findings.

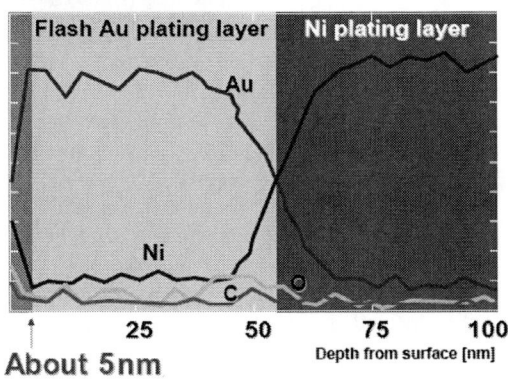

Figure 6. Depth profile of Ni Au plating showing Ni, Au, C and O2 curves. Top 5 nm layer shows significant contaminations.

Plasma before GGI Flipchip bonding
The use of plasma cleaning also extends to gold stud bump flip chip bonding. Thermosonic flip chip bonding uses gold stud bumps on die that are flipped and bonded to a variety of gold plated substrates ranging from ceramics to organics and even thin polyimides. Studies prove that Ar plasma treatment improves the die shear strength of these joints as well. Shear strength measured in grams/bump showed an increase from 30 to over 60 grams/bump by performing plasma treatment prior to flip chip bonding. Another indicator of joint quality is bump transfer after die shear. Evidence of gold bump material transferred to the substrate and a failure mode that shows fracture in the gold bump region is desirable. No transfer to the substrate or peeling of the bump entirely from the die is less indicative of good monometallic gold joint formation. Figure 7 shows bump transfer percentage change from a recipe of 15 seconds at 600W with Ar gas at 10Pa (75mtorr).

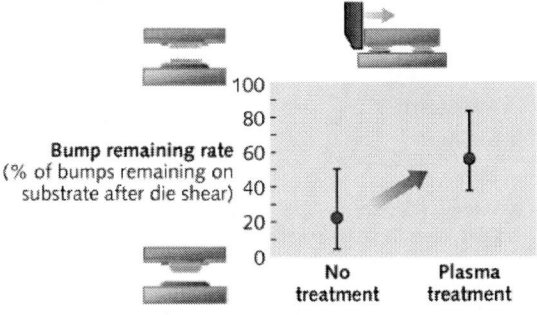

Figure7: Plasma improvement of Au bump flip chip bump transfer rate

Surface Modification for underfill
Plasma is also often used to improve the adhesion of underfills and mold encapsulants. Unlike the Ar ion etching approach, the plasma effect here is for surface modification or chemical action that occurs on the die/substrate interface. (See Figure 8.)

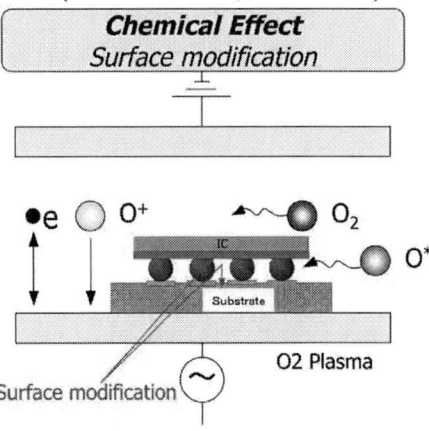

Figure 8. Various species of oxygen plasma

In this case, oxygen plasma is applied to alter the chemical bonds of the die and substrate surface with the goal of increasing the wettability of the surfaces. In this plasma, we are not so interested in the O+ ions nor the electrons. The operative species here is the neutral oxygen radical, a highly reactive particle that forms additional molecular bonds on the affected surfaces, effectively increasing the surface area. (See Figure 9.)

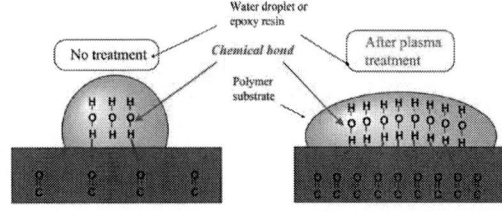

Figure 9. Increased molecular bonds after plasma etch

The extent to which the O radicals can penetrate the gap beneath the die determine the effectiveness of the surface modification, especially for improved underfill adhesion and the elimination of voids.

Contact angle measurement
Comparisons of untreated versus plasma treated surfaces are done using a contact angle goniometer. The contact angle is reduced dramatically from 80 degrees to less than 10 degrees. (See Figure 10.). This improvement typically decreases underneath the Flipchip die. The further away from the chip edge, the less plasma cleaning effect. In case of die size gets

978-1-4244-4722-0/09 $25.00
© 2009 IMAPS-ITALY

larger, and the gap between the die and the substrate gets smaller, this issues needs to be adressed.

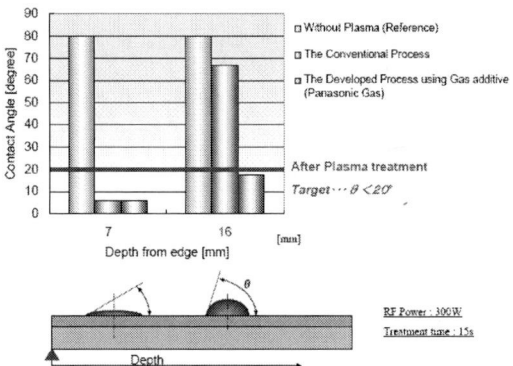

Figure 10. Reduction in contact angle after plasma treatment at different locations of a flipchip assembly.

The Flip-chip process asks for underfill technologies to improve mechanical strength and reliability of micro devices. Recently we started to focus on the underfill process to meet the high challenge of narrow gap and big die size: the future generation of Flip-chip devices.

The reliability of Flip-chip packaging depends on the underfill quality. Fig.11 shows the main difficulties of underfill process.

Peeling

Due to poor adhesion strength, the peeling occurs on the surface of die or substrate.

Voids

While underfill is penetrating, air bubbles occur and stay in the underfill.

Cracking

For stranded voids, cracking may occur due to thermal shock.

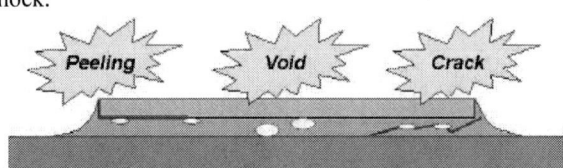

Fig. 11 main Problems of Underfill Process

These problems highly depend on bad surface conditions prior to the underfill process. Such as organic or inorganic pollutions or poor chemical activation status of both surfaces inside the gap. Thus it becomes more and more necessary to control the fluidity of the underfill paste.

Underfill flow characteristics

Fig.4 shows a capillary model to simulate an underfill process. Ignoring the influence of gravity for underfill paste, following Washburn's law the flow time T may be calculated as:[6]

$$T = \frac{k \cdot \eta \cdot z^2(t)}{h \cdot \gamma \cdot (\cos\theta_1 + \cos\theta_2)} \quad \text{--- (1)}$$

With

$z(t)$ penetrate distance

h height of the gap

γ surface tension of the underfill paste

η viscosity of the underfill paste

θ_1 static contact angle on the substrate

θ_2 static contact angle on the IC passivation.

Following this formula, four variables can be influenced. Those variables are γ, η, θ_1 and θ_2.

There are two possibilities of how to improve the flow time:

A) Changing the characteristics of the underfill paste itself by modifying the type of underfill material.

B) Changing the wettability inside the gap by performing plasma cleaning process.

Answer A) needs many long time reliability testings and requalification of the final product. Thus it might be difficult for the market to accept it. Answer B) refers to modifying the surface conditions. Underfill flow velocity can be improved without changing the underfill material by using our plasma technology. This is why we developed this new plasma surface modification technology for Flip-chip packaging.

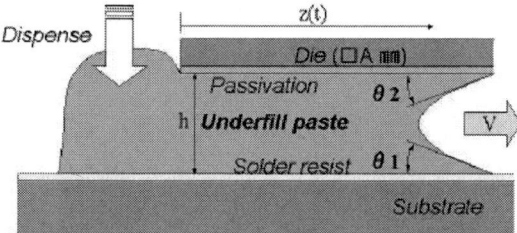

Fig. 12 Underfill Capillary Model

Diffusion factor

We looked for ways to improve the diffusion of the oxygen radicals into the small gap. Knowing that the rate of gas diffusion in vacuum chambers is proportional to temperature — but inversely proportional to gas density, molecular weight, and particle diameter— we succeed in mixing oxygen with an additional gas whose properties would increase the diffusion rate of the combination. This gas mixture helped to improve wetting angles, especially on larger die compared to single-gas oxygen plasma.

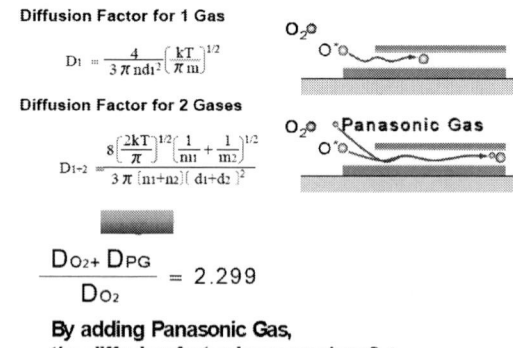

Figure 13: Diffusion factor to improve by using gas additiv3

Processsing Time

After successfully improving the flow characteristic by material and Plasma cleaning parameters significant improve of process time can be gained. Example (see figure 14.) shows improvement from 58 seconds to 39 seconds per flipchip die. By using gas additive the processing time could be even improved to 34 seconds. According to the higher fluidity a more continues shape of the Epoxy fillet could be observed.

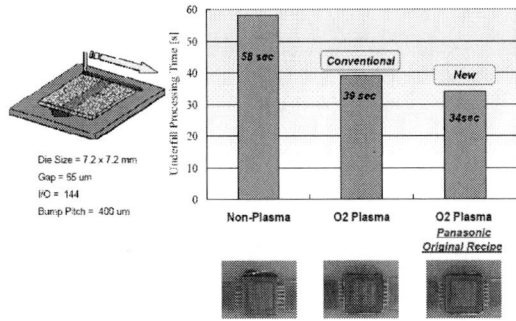

Figure 14: Decrease of underfill seap in time for test vehicle and shape of Epoxy fillet.

Conclusion

Plasma technology has become mainstream in today's microelectronics packaging. It is a proven solution for cleaning Au surfaces prior to wire bonding, and the same clean gold surfaces are necessary for Au/Au monometallic joints formed on flip chips that employ Au bumps. Plasma also offers benefits at the underfill and mold encapsulation steps. Oxygen radicals formed in the plasma modify surfaces to improve their wettability and help reduce voiding. Challenges exist in underfilling large die with small gaps. New gas mixtures help enhance the diffusion of the radicals to penetrate these small gaps.

References

• M. Nonomura, "Plasma Surface Modification Techniques for Underfill Process," *Proc. Known Good Die Workshop*, Napa, Calif., 2005.
• J.O'Hanlon, *A User's Guide to Vacuum Technology, Third Edition*,Wiley, 2003.

A Study on High-density High-speed SerDes Design in Buildup Flip Chip Ball Grid Array Packages

Gordon Xiang, Senior Member, IEEE, Keith Sheach, Pierre Brunet
High Speed Package Design, Ottawa Design Center, STMicroelectronics, Inc.
16 Fitzgerald Road, Nepean, Ontario, K2H 8R6, Canada
Email: gordon.xiang@st.com, keith.sheach@st.com, pierre.brunet@st.com

Abstract

A study on high-density high-speed SerDes (HHS) designs in buildup laminate flip chip ball grid array (fcBGA) packages is presented in this paper. Experiences have shown that three main capacitive discontinuities happen in flip chip bump area, core PTH via area and BGA transition area. Literature [1] have studied three PTH via configurations for differential pairs and suggested that a routing structure with PTH via on the top of BGA be the best design option, which implies that all Serdes routings have to be done in buildup layers above the core layer. However, with growing Serdes number, limited buildup layer number and specified X-talk number, one cannot route all Serdes differential pairs in the buildup layers above the core layer. In a result, routing some Serdes differential pairs in buildup layers below the core becomes a must. In our paper, the three PTH via structures in [1] are analyzed in detail. In order to obtain a more realistic electrical performance, we include die bump, package wiring and BGA transition with a small portion of PCB transmission line in our models. Ansoft full-wave HFSS tool is used to run for S-parameters. Measurement data from a test system were used to guarantee the correctness of our model setup including boundary condition, port and material definitions and so on. Our studies have shown that all three PTH via configurations can be optimized and provide a similar level electrical performances up to 15GHz, which means a smaller buildup layer number and a package cost reduction for a given application. Numerical results for a 4-4-4 buildup package optimization are presented to show the merit of our methodology

Key words: High-speed, Serdes, Ball grid array, Buildup substrate, Flip chip package

I. Introduction

Modern high-speed applications have been moving toward higher speeds and finer physical geometries. Those applications need hundreds of high-speed Serdes (HSS) data slices which are often packaged in a space-limited substrate to reduce the cost. Traditional negligible discontinuities and noise coupling become critical challenges in the high-speed and high-density-wiring packages.

In our study, we will present a methodology to design high-density high-speed SerDes (HHS) in buildup laminate flip chip ball grid array (fcBGA) packages. A typical multi-layer buildup laminate fcBGA package is shown in **Figure 1**.

Figure 1: A typical multi-layer buildup laminate fcBGA package.

Different applications may need 1~5 build up layers (organic material) in both front and back sides of the core [Bismaleimide Triazine (BT) material]. Mechanical drill is normally applied to the core layer for plated through holes (PTH), but laser drill is applied to buildup layers for advanced high density blind vias in multi-layer wiring substrates. Because of the different drill procedures, PTH has a diameter of 105um to 250um, but blind via diameter can be as small as 60um.

In our example, we used a 4-layer core and 4 buildup layers at each side of the core called 4-4-4 substrate. Flip chip bump and solder ball pads are on the top layer and bottom layer, respectively. Normally, the high-speed differential signals are routed on the buildup layers in the front side of the core (above the core) due to the limits from the high-density bump pitch which can be as small as 180um and from the core PTH pitch which is required to be at least 350um. In this study, we will present a wiring methodology to route differential signals on the buildup layers in both front and back sides of the

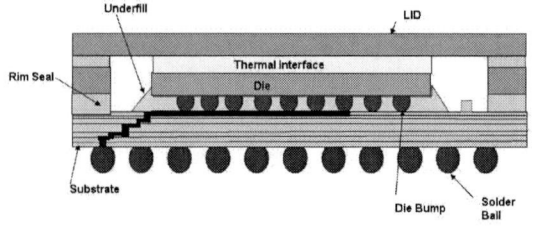

core meanwhile maintaining the same electrical performances.

This paper is organized as follows. In Section II, the transmission line discontinuity in the buildup substrate is described. Three main discontinuities are identified, and their electrical behaviors are analyzed. A wiring strategy for differential signals on the buildup layers in both front and back sides of the core is given. Electrical optimization methodology and numerical results are presented in Section III. Finally, some conclusions are drawn in Section IV.

II. Electrical Discontinuities and Wiring Strategy in Buildup Substrates

Experiences have shown that three main capacitive discontinuities happen in buildup fcBGA packages, i.e., flip chip bump area, core PTH area and BGA transition area as shown in Figures 3a~d, respectively [1].

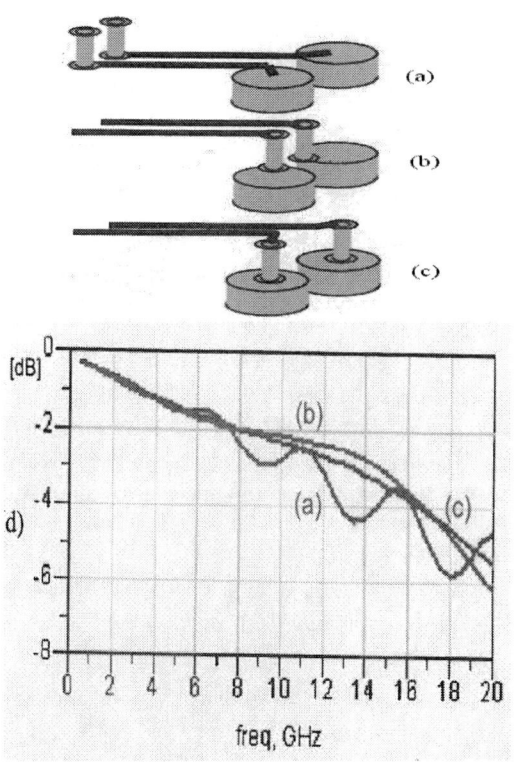

Figure 3: Three PTH configurations in buildup

Literature [1] have studied these three PTH via configurations for differential pairs and suggested that a routing structure with PTH via on the top of BGA be the best design practice, which implies that all Serdes routings have to be done in buildup layers above the core layer. However, with growing Serdes number, limited buildup layer number and specified X-talk number, one cannot route all Serdes differential pairs in the buildup layers above the core layer. Figure 4a shows a

situation where the wiring density of differential pairs is so high that results in a high X-talk between Serdes channels.

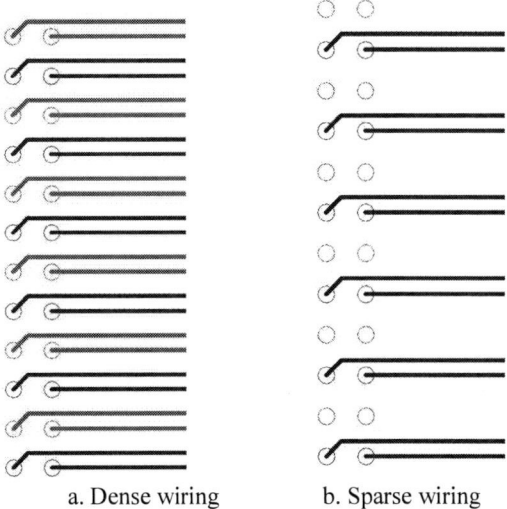

a. Dense wiring b. Sparse wiring

Figure 4: Serdes differential pair wirings in die escaping zones

In a result, routing some Serdes differential pairs in buildup layers below the core becomes a must. Figure 4b presents a routing strategy in which a sparse wiring with a lower X-talk can be obtained. In this example, half of the differential pairs are routed in lower buildup layers. Even if Serdes differential pairs can be routed in buildup layers above the core layer, manufacturing limit doesn't sometimes allow the PTH vias to be put on top of BGAs which are too close to the substrate edge. Figure 5 shows the limitation of manufacturing process. Because high-speed design needs a bigger anti-pad to control the impedance, the copper plane around the differential PTH lands is cut out and formed an open void close to the substrate edge. The unbalanced copper around PTH will result in an unbalanced PTH land etching. Those irregular PTH lands will possibly misalign to blind vias, thus producing open circuits. Therefore, the PTH's have to be moved away from the substrate edge. In these specific cases, the PTH's cannot be put on top of the ball pads that suggested in [1].

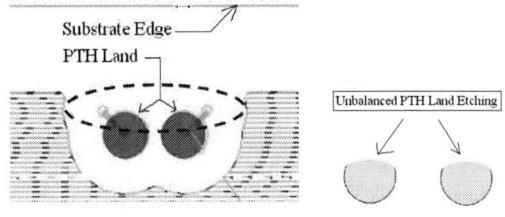

Figure 5: Unbalanced PTH Land Etching because of the unbalanced Cu around them.

978-1-4244-4722-0/09 $25.00
© 2009 IMAPS-ITALY

III. Electrical Optimizations for Buildup fcBGA Package

Studies have shown that three main capacitive discontinuities happen in buildup fcBGA packages. The biggest capacitive discontinuity occurs at BGA transition from the package to the PCB. The second biggest capacitive discontinuity appears at the core PTH transition from the front buildup layer to the back buildup layer. The third biggest capacitive discontinuity happens at the bump transition from the die to the package. As discussed above, there are three main PTH configurations for high-speed differential signals in buildup substrates as shown in **Figure 3**. In order to obtain a more realistic electrical performance, all our models included die bump, package wiring (about 20mm long) and BGA transition with a small portion of PCB transmission line (deembeded to be 2" long). Ansoft full-wave HFSS tool was used to run for S-parameters. Measurement data from a test system were used to guarantee the correctness of our model setup including boundary condition, port and material definitions and so on [2]. In order to speed up the modeling, the original models were split into two sub-models, called package segment and BGA transition segment as shown in **Figure 6**. The package segment started from the die bump and ended somewhere uniform tansmission lone appeared in the package. The BGA transition segment started from uniform transmission line split point and ended at uniform transmission line in PCB.

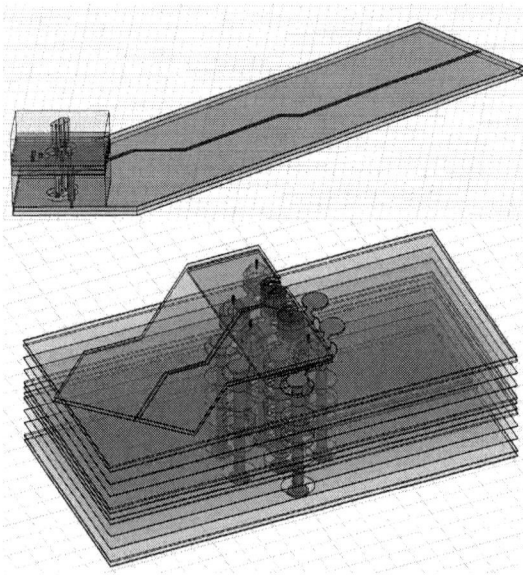

Figure 6: Two sub-models, i. e., package segment (upper) and BGA transition segment (lower) for case a with PTH under the bump.

In order to alleviate these discontinuities, one good practice is to enlarge the anti-pad size near those discontinuity zones and use the minimum bump/PTH/ball sizes to reduce the capacitances. Another strategy is to properly lower the transmission characteristic impedance to reduce the reflections at these discontinuities. We will show that all three PTH via configurations can be electrically optimized by using the above strategies and provide a similar level electrical performances up to 15GHz.

Figure 7~9 show the modeling results for the three PTH configurations illustrated in **Figure 3**. Please note that in three cases about 20mm and 2" transmission lines are included in the package and the PCB, respectively. Furthermore, all the three cases have the same channel path lengths. From **Figure 8** and **Figure 9** we can see that both PTH configurations provide almost the same electrical performance. The interesting observation happens when we look at the S-parameter curves in **Figure 7**. It is obvious that PTH under bump provides better return losses than other two PTH configurations. This can be further explained by their TDR plots.

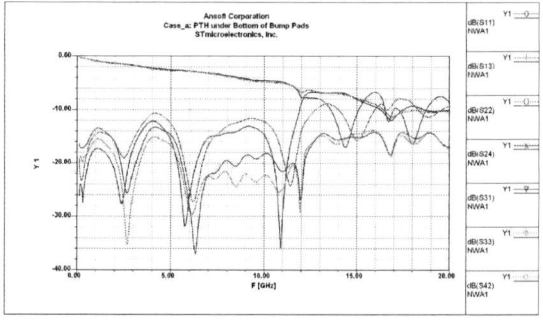

Figure 7: S-parameters for case a, i.e., PTH at bottom of the bump.

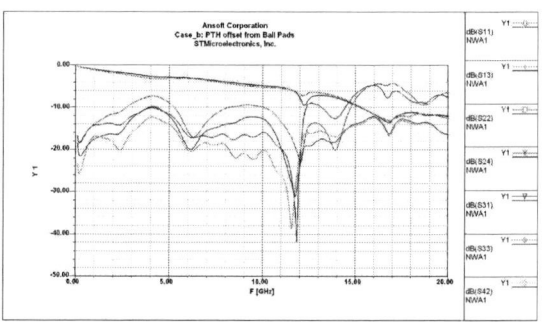

Figure 8: S-parameters for case b, i.e., PTH offset from the ballpad.

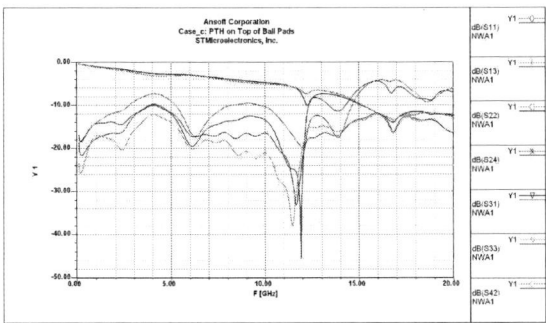

Figure 9: S-parameters for case c, i.e., PTH on top of the ballpad.

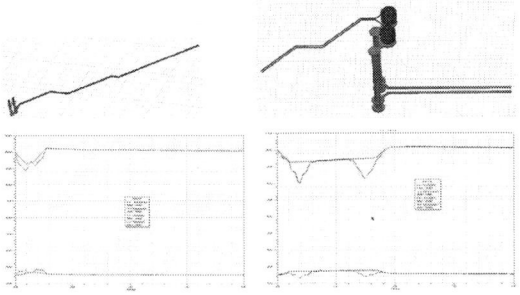

Figure 10: TDR for case a, i.e., PTH at bottom of the bump. Tr=100ps.

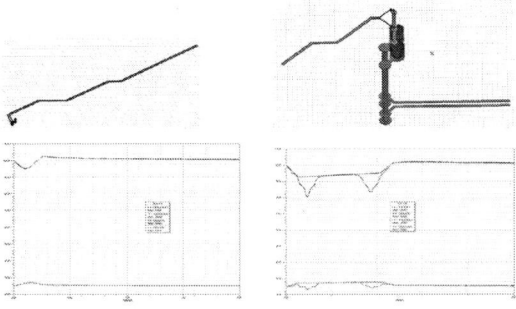

Figure 11: TDR for case b, i.e., PTH offset from the ballpad. Tr=100ps.

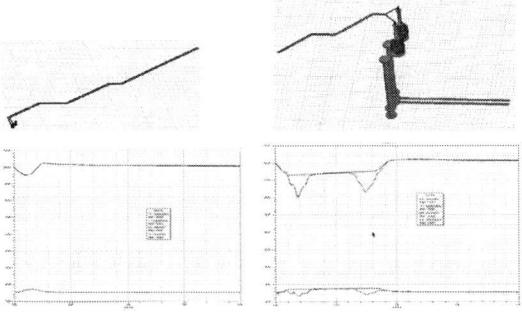

Figure 12: TDR for case c, i.e., PTH on top of the ballpad. Tr=100ps.

Figure 10-12 show the TDR plots for all the three PTH configurations. From those plots we can see that the BGA discontinuity dominates other discontinuities such as PTH, blind via discontinuities, etc. Furthermore, the bump discontinuity is much smaller than the BGA discontinuity. Case b and Case c provide almost the same impedance behaviors. This means that offsetting slightly the PTH from the ballpad will not impact the channel electrical performance up to 10Gb/s application. By observing Case a in **Figure 10** we can see that the discontinuity of combining both bump and PTH is smaller than BGA discontinuity, but a bit larger than the bump discontinuity. To best use these impedance features one can properly design the transmission line as an impedance transformer to reduce the reflections from those discontinuities.

IV. Conclusions

In our paper, the three PTH via structures have been analyzed in detail. Our models have included die bump, package wiring and BGA transition with a small portion of PCB transmission line to predict the electrical performance for a more realistic aplication. Our studies have shown that all three PTH via configurations can be optimized and provide a similar level electrical performances up to 15GHz. Interesting results have been observed that Case a in which the PTH is put under bump pad provides a better behavior if a proper transmission line design is applied. This conclusion provides us with a package design guideline that wiring differential pairs on buildup layers below the core is a good option in high-speed design while waching out the X-talk impact on the channel performance, which means a smaller buildup layer number and a package cost reduction for a given application.

References

[1]. N. Na, J. Audet, et al, "Design Optimization for Isolation in High Wiring Density Packages with High-speed Serdes Links", 2006 Electronic Components and Technology Conference.

[2]. Gordon Xiang, et. al, "High-speed Electrical Design of Flip Chip Buildup Package for STM Project W259", Internal Report, STMicroelectrics, Inc. Feburary, 2009.

A Novel Methodology for Analyzing Variation Risk Introduced by the Manufacturing Process in Microsystems

Yunfei Sun[1], Clifford R Fowkes[2] and Nabil Gindy[1]

[1]*School of Mechanical, Materials and Manufacturing Engineering, University of Nottingham, UK*

[2]*Integrated Products Manufacturing, Knowledge Transfer Network, UK*

Tel: +44(0)115 951 5998 Fax: +44(0)115 951 3800

e-mail: yunfei.sun@nottingham.ac.uk

Abstract

This paper describes a novel methodology for analyzing the variation risk introduced by manufacturing processes in microsystems using a Key Characteristic (KC) method and statistical analysis applied to microelectromechanical systems (MEMS) fabrication tolerances. The KCs of the microsystems are identified based on the product requirements and then propagated to the subsystem KCs and associated manufacturing processes. Statistical approaches are given for estimating the variation of the product KCs, the variation contribution of subsystem KCs and the associated manufacturing processes. The KC variation risk is predicted and the manufacturing processes that contribute most to the variation are identified.

This methodology is applied to an innovative design for a micro CMM (co-ordinate measuring machine) probe. One product KC and its subsystem KCs are identified and a systematic view of KC propagation structure is demonstrated. The variation in the product KC is predicted and the variation contribution of the MEMS probe manufacturing processes is presented. Variation risk analysis results can be used to implement a variation risk mitigation strategy.

Key words: Variation risk analysis, key characteristics, statistical variation analysis, KC propagation, MEMS fabrication tolerance.

1. Introduction

The functional performance of many microsystems is critically dependent upon the accuracy with which they can be manufactured. Thus it is increasingly important to evaluate the micro-components' geometrical accuracy and variation risk. [1-2].

Geometrical and size parameters of MEMS components are often limited by the manufacturing constraints of the associated processes. The principal challenge of the MEMS design is to meet the functional requirements while staying within the limits of micromachining technology [3]. It is important to know the critical MEMS characteristics and link them to the component parameters and the related manufacturing process in order to assess the process capability to deliver the required performance. The required assessment can be undertaken by the use of a systematic tool known as variation risk analysis.

This paper aims to address the MEMS fabrication variation issue by estimating the change in performance arising from variation in the structural dimensional parameters of an example device. The example used is a micro CMM probe [4] that is being developed at the UK National Physical Laboratory (NPL). This probe was chosen because its performance is affected by a number of features and tolerances that are dependant on a range of MEMS manufacturing processes.

The application of KCs method has increased in the last decade by industries that need to focus their efforts on critical product and processes characteristics due to cost and resource constraints. The success of variation risk analysis derives from it's ability to focus attention on the most critical KCs and their associated manufacturing processes [5].

Section 2 describes the KC concept, the identification of KCs and the systematic KC process. Section 3 deals with the application of the statistical variation analysis method. Section 4 presents application of the KC method and statistical variation analysis to a micro system to estimate the variation of product KC and the contribution from associated manufacturing processes. Finaly the results are discussed and conclusions are drawn.

2 Variation Risk Analysis

Manufacture of the MEMS uses a number of process steps, each of which can affect the dimensional accuracy and material characteristics and hence the performance of the MEMS. Variation risk analysis is applied to identify and assess the impact of variations introduced by the manufacturing processes in the MEMS from a holistic viewpoint. The KC method is used to enable a company to focus limited resources on the control of those manufacturing processes that most effect critical features and key characteristics of the product.

2.1. Key Characteristic (KC) Method

Although micro systems and components tend to have a simpler product structure, there is still a need to identify a reduced number of KCs due to the high cost required to achieve required accuracies. Sometimes it is impractical to achieve all of the design specification at once, therefore tradeoffs need to be established to achieve acceptable compromises between design and manufacture [3].

KCs can be defined at the product, subsystem or component levels of a product. The product KCs are those for which reasonably anticipated variation can significantly affect a product's safety, fit or function, or compliance with customer requirements. The subsystem/component KCs are those that affect the product KCs and thus contribute to the variation of product KCs. The manufacturing process KCs are those processes used in manufacture or assembly that affect subsystem KCs. Process key parameters are variables are those that can cause significant variation of the manufacturing KCs.

A systematic KCs process is needed to handle the product from a holistic point of view, in order to highlight critical manufacturing processes and understand the sources of variation. This systematic process can also help improve concurrent engineering activities by focusing decisions and attention on the critical product parameters and their interactions. Figure 1 illustrates the systematic KC propagation structure.

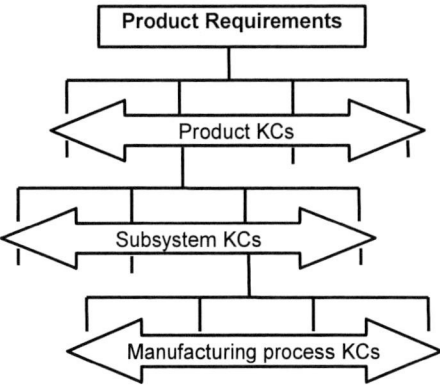

Figure 1: A systematic KC propagation structure.

The KCs method developed at MIT has established a base for this process [6], particularly by defining a propagation procedure for individual KCs. However, this approach may encounter difficulties due to the quantitative model used. A combination of tools and techniques for KC identification and propagation based on product decomposition level at the design or manufacturing phase is necessary for a successful application of the KCs process.

The first step of KC identification is to array customer-driven product goals against product design attributes that can achieve the goals. Customer goals are prioritised and key design characteristics required to achieve the critical goals are identified. Quality Function Deployment (QFD) is a very useful tool to convert customer requirements into product functional requirements, and accordingly identify which should be product KCs.

The second step of KC identification is to decompose the product KCs to the lowest level possible in order to highlight all required actions. Qualitative Risk analysis techniques based on using the Failure Mode and Effect Analysis method are recommended tools for the propagation process when it is impractical to use a quantitative approach.

Pande and Holpp [5] have shown that the KC identification methodology should be driven by data. Using tolerance chain analysis, modelling, simulation and design of experiments, product KCs can be flowed down to contributing parameters (subsystem/component KCs) and their associated manufacturing processes. The KCs propagation process can be used to identify KCs at any level in the product decomposition tree. That is, each level of subsystem can be flowed down to it's own subsystem/component KCs and thus there can be unlimited levels of KC propagation.

978-1-4244-4722-0/09 $25.00
© 2009 IMAPS-ITALY

A systematic KC propagation structure is one where statistical variation analysis can be applied to estimate the variation of the product KCs. It can also quantitatively identify the subsystem KCs and the related manufacturing processes' contribution to the variation of product KCs.

2.2. Statistical Variation Analysis

In the second phase of analysis, the assessment of KCs variation is performed using tools such as statistical variation analysis, Taguchi Loss Function, and Design of Experiments. The identified KCs are used as a base to measure and control the related process stability and capability in order to mitigate variation risks.

A variation propagation analysis can identify which parameters contribute most to variation in the product KCs. The variation of the MEMS structure's geometry and dimensions are applied with normal (Gaussian) distribution and different tolerances. This means that the tolerances correspond to the $\pm 3\sigma$ process capability, that is, $T_i = \pm 3\sigma$, where σ represents the standard deviation of the process [7].

For each product KC, the associated parameters (variables) must be identified by KC propagation. A statistical approach is applied to predict the variation in KCs by evaluating weighted variations of the parameters.

If we have f_i, a KC which is a function of variables x_j, the approximated variance of f_i can be estimated as follows:

$$Var[f_i(x)] \approx \sum_j \left(\frac{\partial f_i}{\partial x_j}\right)^2 \left(\frac{T_j}{3}\right)^2 \qquad (1)$$

Here the T_j represents the tolerance for the variable x_j. The partial derivative represents the sensitivity of a KC to its lower level KCs. The relative contribution of a KC to its upper level KC can be calculated as follows:

$$P_j = \frac{\left(\frac{\partial f_i}{\partial x_j}\right)^2 \left(\frac{T_j}{3}\right)^2}{\sum_j \left(\frac{\partial f_i}{\partial x_j}\right)^2 \left(\frac{T_j}{3}\right)^2} \qquad (2)$$

By applying Eq.1 and Eq.2 through all of KC propagation levels, the variation of product KCs can be estimated. Also, the contributions from subsystem KCs can be estimated thereby enabling identification of the processes that contributes most to the variation of product KCs.

Taguchi Loss Function is a statistical approach used to evaluate the losses incurred in product KCs due to variation; the losses increase as the characteristics' variation increases, and the loss value can be determined using cost or product performance measures. Characteristics with high loss value can be considered as high risk.

Design of Experiments (DOE) is a statistical method that can be used to identify key process parameters by indicating those that have a strong impact on the related manufacturing process KCs. These process parameters need to be controlled to achieve better process yield. It can also be used to study and find the variables and their values that most affect the performance of a product.

2.3. Variation Risk Analysis

By analyzing the overall variation of the identified KCs using the systematic KC process and statistical variation analysis, product KCs with significant variation compared to the designed tolerance and the related subsystem KCs are considered as having high risk and are identified for further variation mitigation actions. The associated manufacturing KCs that contribute most to the variation of the product KCs need to be controlled, improved or replaced. Thus the variation risk values can be used to prioritise the risk reduction actions.

3. Case Study

The micro CMM probe is designed to be used on a micro CMM platform [8]. The micro CMM probe contains components of sizes ranging from 2 μm to 2 mm. Certain of the designed dimensions are of critical importance in order for the micro CMM probe to function correctly. Manufacture of the micro CMM probe uses a number of process steps, each of which can affect the dimensional accuracy and material characteristics and hence it's performance. Variation risk analysis is applied to systematically identify and assess the impact of variation risk introduced by the micro CMM probe's manufacturing process.

The micro CMM probe consists of a mechanical structure combined with sensors that can be used to determine when the probe tip comes into contact with an object that is being measured [9]. Figure 2 shows the micro CMM probe conceptual design, which consists of a tip ball, stylus, suspension body and three suspension arms. Piezoelectric elements are deposited at each end of the suspension arms to act as sensors. Each of the three suspension arms is rigidly fixed at the end adjacent to the outer sensor.

978-1-4244-4722-0/09 $25.00
© 2009 IMAPS-ITALY

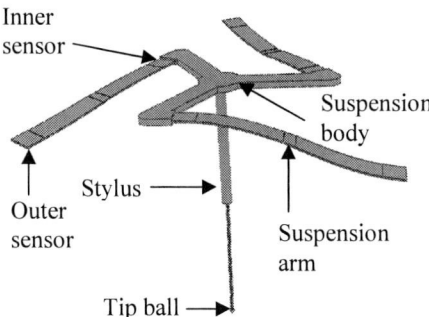

Figure 2: Micro CMM probe conceptual design.

Figure 3 and figure 4 show the geometry and structure of the probe and suspension arms. x_s, y_s correspond to the position of the free end of arm 1 with respect to the origin, l_{st} represents the length of the stylus. The micro CMM probe suspension arms are manufactured using a single material with a Young's modulus, E_s (with the exception of the piezoelectric layers). The piezoelectric material is lead zirconium titanate (PZT) and is characterised by its constitutive equations with the poling in the z axis.

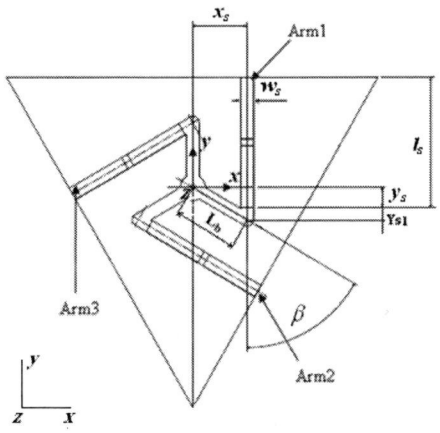

Figure 3: Geometry and structure of micro CMM probe.

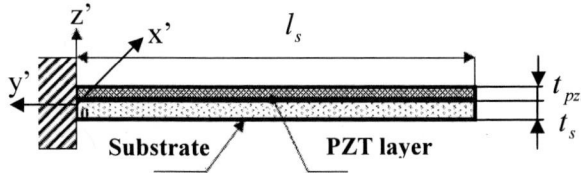

Figure 4: Geometry and structure of suspension arm.

In this case study, the aim is to demonstrate how to apply basic principle of KC methodology and variation risk analysis.

3.1. Product KC and Identification of Subsystem KCs

The detectable output of the micro CMM probe is the voltage generated by the sensors, thus the output voltage of sensors is chosen as a product KC for the probe and subsystem KCs need to be identified to build a systematic KC propagation structure.

During a measurement with the micro CMM the PZT sensors are subjected to a strain and produce an output voltage that is proportional to the movement of the probe tip. The normal stress vector on suspension arm 1 can be derived using analytical modelling. If we use a factor α indicating the position of suspension arm 1 along the y axis, $\alpha \in \left[-\frac{1}{2}, \frac{1}{2} \right]$ from the free end to the fixed end, the normal stress vector representing internal action for suspension arm 1 can be derived as follows:

$$
\sigma_{sa1} \equiv \begin{bmatrix} \sigma_{x1} \\ \sigma_{y1} \\ \sigma_{z1} \end{bmatrix}
$$

$$
= \frac{E_s t_s}{2 l_s^2 l_{st}} \begin{bmatrix} 0 \\ 12\alpha x_s X_{tx} + (6\alpha l_s - 12\alpha y_s - l_s)X_{ty} + 12\alpha l_{st}X_{tz} \\ 0 \end{bmatrix} \tag{3}
$$

The normal strain S_{sa} of PZT sensors can be calculated according to Hooke's law [10]:

$$
S_{sa} \equiv \begin{bmatrix} S_x \\ S_y \\ S_z \end{bmatrix} = \frac{\sigma_{sa}}{E_s} \tag{4}
$$

The PZT sensors detect the micro CMM probe movements by transferring the mechanical strain on the PZT sensor elements to an electric polarization proportional to that strain. The open output voltage of sensors on suspension arm 1 V_z can be derived using constitutive piezoelectric equation as follows [11]:

$$
V_z = g_{32} \frac{t_s t_{pzt}}{2 l_s^2 l_{st}} \times \\ \left(12\alpha x_s X_{tx} + (6\alpha l_s - 12\alpha y_s - l_s)X_{ty} + 12\alpha l_{st}X_{tz} \right) \tag{5}
$$

Where g_{32} is one of the piezoelectric coefficients in Vm^{-1}/Nm^{-2} and represent the voltage gradient in the PZT per unit pressure input.

978-1-4244-4722-0/09 $25.00
© 2009 IMAPS-ITALY

Using the above results, the subsystem KCs associated with the open circuit output voltage can be identified. In this paper, only the related dimensional parameters are chosen as subsystem KCs to demonstrate the variation risk analysis methodology. Other parameters such as material property can also be included in this analysis using the same approach. Figure 5 illustrates the results and the systematic KC propagation structure.

The sensitivities s_i of each parameter x_i to the product KC can be derived using the following equation:

$$s_i = \frac{\partial V_z}{\partial x_i} \qquad (6)$$

Where i denotes the ith parameter that influences the output voltage of sensors.

3.2. Fabrication Process Capability:

The micro CMM probe prototype is manufactured by a combination of composite sol-gel, photolithography, electroforming and one-pulse electro-discharge (OPED) processes. The output of each of these processes is affected by a number of process variables. The ease with which each of these process variables can be controlled and the sensitivity of the output to these is briefly considered in this analysis to indicate the likely sources of variance. A more detailed analysis would be required to fully assess the manufacturability of the micro CMM probe using these methods.

The suspension body and arm are manufactured using photolithography and electroplating process. A number of masks are used in this photolithography process. The accuracy of the masks used in photolithography has a direct affect on the dimensions of the features as does the accuracy of alignment (registration) of the masks to the substrate [12]. Several masks are used in manufacture of the micro CMM probe; therefore the repeatability of registration is important. As a result of its widespread use in semiconductor manufacture, photolithography at this scale is a mature process capable of a high degree of accuracy.

The primary output variable of electroplating is the thickness of the deposited material. Morphology, edge structure and dimensions may also be affected; however plating is a mature and well understood process, so these output variables can generally be well controlled.

A PZT thick film composite sol gel process [13] is utilised to fabricate the piezoelectric layer that will form the sensor. This has a number of process steps that will affect the piezoelectric characteristics of the sensor. Part of this process involves wet etching. Over or under etching can affect the edge form of the part and in extreme cases, the dimensions, for example the distance of sensors to the edge of the substrate of suspension arms. Although the piezoelectric constant is affected by the sol gel process only the variation of geometry is considered here, hence a single value for PZT constant is used throughout.

The tolerances of the manufacturing processes for the suspension body, arms and stylus considered in the variation risk analysis are shown in table 1.

Table 1: Tolerances of MEMS manufacturing processes.

Manufacturing process	Tolerance
Photolithography and wet etching	$\pm 5 \ \mu m$
Composite sol gel	$\pm 200 \ nm$
Photolithography and electroplating	Thickness: $\pm 100 \ nm$ Length: $\pm 50 \ \mu m$ (defined by mask) Angle: $\pm 1^\circ$ (defined by mask)

3.3. Variation Risk Analysis

The variation in the subsystem KCs resulting from the fabrication variation causes a change in the product KC which can significantly influence the performance of the micro CMM probe. Figure 5 shows the systematic KC propagation result for the product KC and variation contributions from its subsystem KCs and associated manufacturing process KCs. The first layer on the left shows the product KC, the middle layers show subsystem KCs and the layer on the right shows the associated manufacturing processes.

By applying Eq.1 and Eq.2 through all KC propagation layers, the variation of product KCs and the contributions from subsystem KCs can be estimated. An analysis of figure 5 shows that the variation of thickness of the PZT layer contributes most to the variation of the sensor output voltage (83.3%). Thus the composite sol gel process that is used to manufacture the PZT layer is identified as the most critical process for this assembly and needs to be carefully controlled.

A series of experiments have been implemented for both the inner sensor and outer sensor on suspension arm 1 for the displacement of tip ball within a cube of sides 2 μm. The results show that the ratio of the standard deviation is less than 5.1% of the product KC nominal value for both sensors.

The variation in subsystem KC PZT thickness always contributes most to the variation of the

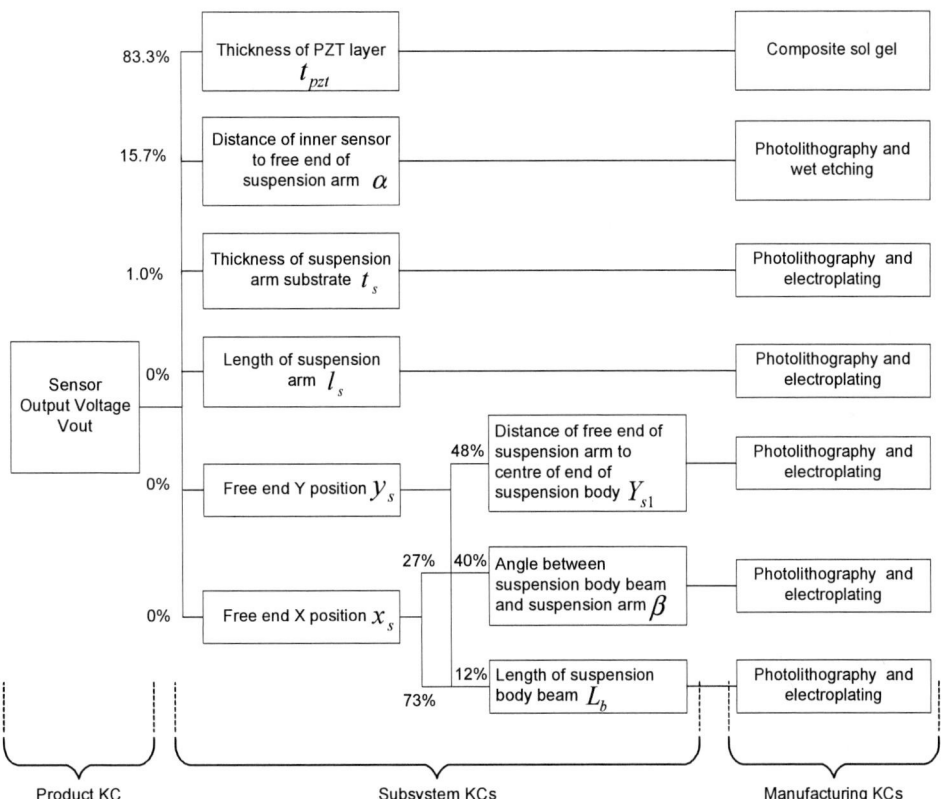

Figure 5: KC propagation for the open circuit output voltage of the inner sensor on the free end of suspension arm 1 and the variation contributions from its subsystem KCs.

product KC for both sensors. For the outer sensor, it is relatively constant. For the inner sensor, it is still the main contribution, but varies from 83.3% to 98.32%. Figure 6 shows the result of the percentage variation from contributing subsystem KCs. Similar results are obtained for sensors on other suspension arms.

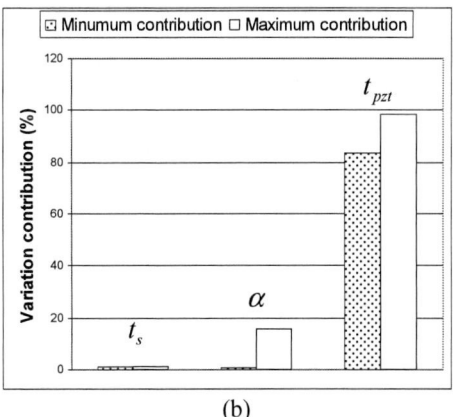

(b)

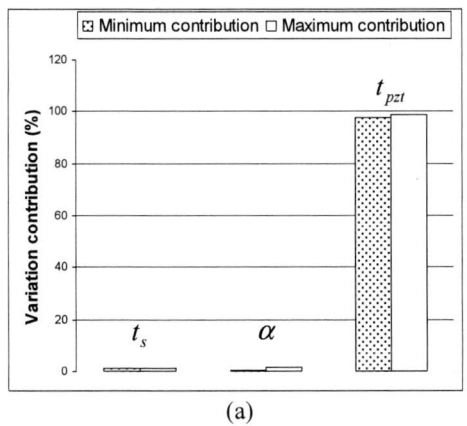

(a)

Figure 6: Variation contributions (contribution of thickness of substrate, t_s; contribution of distance of sensor to the edge of the substrate, α; percentage contribution of thickness of PZT layer, t_{pzt}), for (a): inner sensor (free end, $\alpha = -1/2$) (b): outer sensor (fixed end, $\alpha = 1/2$) on suspension arm 1.

978-1-4244-4722-0/09 $25.00
© 2009 IMAPS-ITALY

4. Conclusion

Variation risk analysis provides a basis for ensuring acceptable yields in manufacturing. This work can help to design products that are robust to variation and improve the product quality. It can also be used as part of a risk mitigation strategy. In order to provide a more complete Design for Manufacture (DFM) assessment, other factors must be considered, particularly the applicability and maturity of the various process steps. A full DFM analysis would need to include variation risk analysis, process capability knowledge and any other relevant factors.

In this paper, a KC identification process and a systematic KC propagation view was illustrated. It was demonstrated that variation risk analysis provides a valuable tool to identify which parts of the micro CMM probe require most care in manufacture. The results clearly illustrate that the variation in thickness of the PZT layer contributes most to the variation of sensor output voltage and hence the sensitivity of the micro CMM probe. Therefore the composite sol gel process is the most critical manufacturing stage in the current design. This variation risk analysis can also be used to predict the yield of the product KC based on the variations of subsystem KCs and its required design tolerance.

Acknowledgment

The authors gratefully acknowledge useful assistance and comments from Dr Richard Leach, Dr Stoyan Stoyanov and Dr Robert Dorey. The authors also thank to the 3D Mintegration project which provided part of the research funding.

References

1. Ha SK., Jeong H and KIM J, "Robust design of a decoupled vibratory microgyroscope considering over-etching as a fabrication tolerance factor", JSME International Journal 49:273-281, 2006.
2. Shavezipur M, Ponnambalam K, Khajepour A, Hashemi SM, "Fabrication uncertainties and yield optimization in MEMS tunable capacitors", Sensors and Actuators A 147: 613–622, 2008.
3. Epstein, A, "Millimetre-scale, MEMS gas turbine engines", ASM Turbo Expo, Gerorgia, USA, 2003.
4. Leach R, Chetwynd D, et al "Recent advances in traceable nanoscale dimension and force metrology in the UK", Measurement Science and Technology 17: 467-476,2006.
5. Pande, P. & Holpp, L. "What is Six Sigma", McGraw-Hill, USA, 2002.
6. Thornton AC, "Variation risk management focusing quality improvements in product development and production", John Wiley & Sons, New Jersey, 2004.

7. Chase KW, Drake P Jr, ed, "Dimensioning and tolerancing handbook" McGraw-Hill, New York, 1999.
8. Haitjema H, Pril WO, Schellekens PHJ, "Development of a Silicon-based Nanoprobe System for 3-D Measurements" CIRP Annals - Manufacturing Technology 50: 365-368, 2001.
9. Stoyanov S, Bailey C, Leach R, et al, "Modelling and prototyping the conceptual design of 3D CMM micro-probe", Proceedings of the 2nd IEEE International Electronics System-Integration Technology Conference 1: 193-198, 2008.
10. Coates RC, Coutie MG, Kong FK, "Structural Analysis", Chapman & Hall, London, 1988.
11. Neubert HKP, "Instrument transducers: An introduction to their performance and design", Clarendon Press, Oxford, 1975.
12. Lai JM, Chieng WH, Huang YC, "Precision alignment of mask etching with respect to crystal orientation", Journal of Micromechanics and Microengineering 8: 327–329, 1998.
13. Dauchy F, Dorey RA, "Patterned crack-free PZT thick films for micro-electromechanical system applications", Int J Adv Manuf Technol 33: 86-94, 2007.

Fracture Toughness Assessment of ACF Flip-chip Packages under High Moisture Condition with Moire Interferometry

Jin-Hyoung Park, Jae-Won Jang, Kyung-Woon Jang, Kyung-Wook Paik, and Soon-Bok Lee

KAIST 375 Gwahangno Yuseong-Gu, Daejeon 305-701, Korea

Tel: +82-42-350-3069, Fax: +82-42-350-5013, turbomb@kaist.ac.kr

Abstract

A primary factor of ACF package failure is delamination between the chip and the adhesive at the edge of the chip. This delamination is mainly affected by the normal strain at the edge of the chip. This normal strain was measured on various electronic ACF package specimens by micro Moiré interferometry with a phase shifting method. In order to find the effect of moisture, the reliability performance of an adhesive flip chip in the moisture environment was investigated. The failure modes were found to be interfacial delamination and bump/pad opening which may eventually lead to total loss of electrical contact. Different geometric size specimens in terms of interconnections were discussed in the context of the significance of mismatch in CME between the adhesive and other components in the package, which induces hygroscopic swelling stress. Interfacial fracture toughness between ACF and die chip were evaluated by peel test. The effect of moisture diffusion in the package and the CME mismatch were also evaluated by using the Moiré interferometry. From Moiré measurement results, we could also obtain the stress intensity factor K. Through an analysis of deformations induced by thermal and moisture environments, a damage model for an adhesive flip chip package is proposed.

Key words: Anisotropic conductive film (ACF), Coefficient of moisture expansion (CME), Moiré interferometry, Electronic packaging, Delamination, Reliability

Introduction

Flip-chip assemblies using an adhesive have been widely used to PDA devices, mobile phones, and LCD devices. Flip-chip technology offered several advantages of excellent electrical performance, very small chip-size packages, improved environmental compatibility, low process temperature, and lowered costs [1], [2]. In order to satisfy the high pin-count and performance requirements flip-chip will become predominant technology for chip-to-next level interconnect. However, various technological improvements will be required to support flip-chip on board (FCOB) implementation such as material, process development, metrology, and reliability. Especially, the reliability study based on hygro-thermal technologies, of a packaged microelectronic system through a well-designed, well-understood, and thoroughly applied accelerated tests are critical. Hygro-thermal reliability evaluation helps to determine which failure modes apply to a given part, how probable it is that these failure modes will occur while the part is in service, and how they might be prevented during the design and manufacturing stage.

Despite many advantages, ACF type flip-chip package bares serious mechanical reliability issues. High CTE (coefficient of thermal expansion) mismatch between the chip and the substrate causes high strain/stress at the interconnection as well as the chip [3], [4], [5], [6]. Another critical issue

among them is electrical performance deterioration upon moisture absorption of the polymeric resin. The latter is attributed to gradual delamination growth at the chip and adhesive film interface induced by CME mismatch driven peel stresses by the swelling deformations [7].

In this study, the hygroscopic swelling of ACF and PCB substrate induced by CME mismatch was evaluated to define the failure modes under high moisture condition. A fully moistured ACF specimen and a dried specimen are evaluated in order to calculate the stress intensity factor from the swelling deformations. A 90 degree peel strength test was performed to determine the loss of adhesion strength upon moisture absorption.

In order to quantitatively evaluate the amount of micro swelling deformations of ACF type packages, Moiré interferometric methods were readily adopted for its non-contact and whole field measuring capabilities, which are suitable for small specimens like electronic packages [8], [9], [10]. Relatively high sensitivity compared to other experimental approaches is appropriate for measuring localized small thermal deformation behavior, which provides useful data for extensive studies on determining the failure mode and mechanism [11], [12]. In order to clearly understand the thermal behavior of flip-chip packages under high temperature and moisture conditions, phase-shifting techniques was used for further sensitivity

978-1-4244-4722-0/09 $25.00
© 2009 IMAPS-ITALY

enhancement and accurate strain/deformation analysis.

Specimens

In this study, the hygro behavior of an ACF type package was evaluated (Fig. 1). A conventional ACF using di-functional epoxies is used for this study. This ACF has a general polymer matrix and a glass transition temperature of 108 °C. In order to observe the effect of different geometric shape, the following ACF specimens have been used (Table 1).

The bonding conditions for all specimens were equal at 45 MPa and 180 °C for 20 seconds. The geometric size of all specimens was not varied, except the thicknesses of the chip and PCB. The thickness of the ACF was 50μm, and the dimensions of the chip were 8 mm×8 mm. The dimensions of the gold bumps in the chip plate were 120 μm×120 μm×18 μm, and the pitches of the peripheral bumps were all 130 μm. The size of the PCB was 20 mm×20 mm. The properties of each material are listed in Table 2.

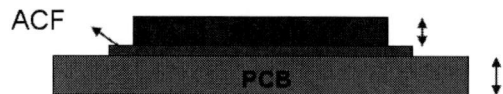

Figure 1: ACF package specimen

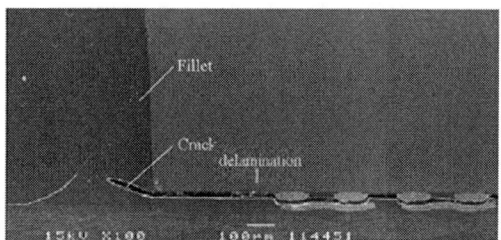

Figure 2: Failure induced by moisture

Table 1: ACF package specimens

specimen	Chip (μm)	PCB (μm)
S1	180	120
S2	180	550
S3	180	980
S4	480	120
S5	480	550
S6	480	980

Table 2: Material properties

	Modulus (GPa)	CTE (ppm/°C)	ν	Tg (°C)
Die	170	2.45	0.28	-
ACF	2.4 (under Tg)	a1 = 46 a2 = 158	0.34	108
Substrate	17.2	αx=αy=20 αz=50	0.15	180

Humidity tests

Hygroscopic swelling assisted by loss of adhesion strength upon moisture absorption is responsible for moisture-induced failure in the adhesive flip chip interconnects. The diffusion of the water molecules into the space in the polymer chain causes swelling deformations (Fig. 3). The polymer materials on the electronic package have different humidity properties from other materials in packages. The mismatch in CME between the adhesive and other components in the package induces hygroscopic swelling stress. Thus, understanding of the complicated hygroscopic swelling deformations is important to design adhesive packages [13], [14].

Humidity tests of the ACF type specimens were performed in a humidity chamber. The humidity test conditions were 85 °C/85 %. Fig. 4 shows the change in resistance of the ACF package specimens. S1 and S4 specimens have a high resistance change.

In order to understand the results of the humidity tests, it is important to determine humidity properties of the polymer materials. The polymer materials of ACF type package are ACF and PCB substrate. The silicon die chip does not absorb the moisture. The humidity properties are a saturated moisture content, diffusivity, and CME. The saturated moisture content and diffusivity of ACF were determined by TGA equipment to measure the weight gain. The CME of ACF was determined by TMA to measure the displacement change. In contrast, since PCB substrate are constructed with polymer resin and glass fiber, the CME of PCB substrate was determined by Moire interferometry to measure the micro average strain. Table 3 shows the humidity properties of ACF and PCB substrate at 85 °C/85 %.

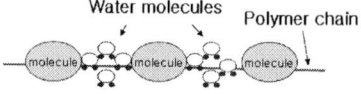

Figure 3: Swelling phenomena by moisture

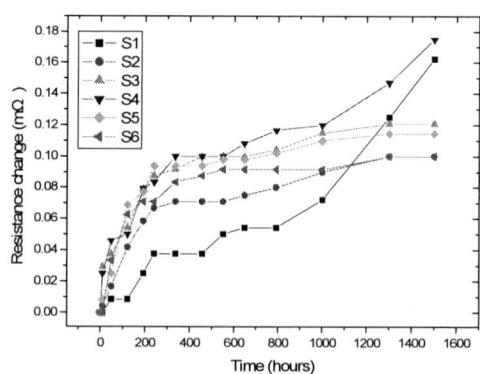

Figure 4: Change of resistance in 85 °C/85 % environment

978-1-4244-4722-0/09 $25.00
© 2009 IMAPS-ITALY

Swelling deformations of ACF type package

The bonding of ACF type specimens is conducted at high temperature. Therefore, the ACF type specimens at room temperature show a bent shape due to CTE mismatch of each material. The ACF and the PCB of the ACF package specimens were polymer materials. These polymer materials are swelled by absorbing water. At room temperature, the bending warpage decreased by the swelling deformation of the PCB, but peeling stress at the edge of the chip increased gradually. In order to shed light on these phenomena, experiments using Twyman-Green interferometry were performed. This optical system can measure the global warpage behavior in a die chip. The resolution of the Twyman-Green interferometry was the half of the wavelength of the laser light source, i.e., $\lambda/2=0.316\mu m/fringe$

After drying ACF package specimens at 100 °C for 24 hours, the warpage of the specimens in a 85 °C/85 % humidity chamber was measured by Twyman-Green interferometry. Fig. 5 shows the warpage results. The warpage of the chip side decreased considerably (Fig. 6). The total warpage of S2 specimen decrease was about 4 μm. This value is similar to the value obtained by increasing the temperature to 50 °C. At high temperature, the amount of diffusion increases dramatically. Generally, thin die chips showed large warpage, but the warpage value was not strongly related to the humidity life (Fig. 7). The very large warpage may lead to the delamination of the interconnection layer. The moderate warpage value does not affect the humidity reliability. Therefore, the warpage value is not a good moisture damage parameter.

The distribution of peeling stress along the interface of fully moistured specimen was determined by the humidity properties (eq. (1)), [15]. The different geometric shape affects the distribution of peeling stress. Fig. 8 shows the peeling stress along ACF interface. S1 and S4 package specimens fully saturated by moisture showed large peeling stress at the edge of the chip. Fig. 9 shows the relation between the Peeling stress and the reliability results of humidity tests. The peeling stress at the edge of the chip is an important damage parameter.

$$\sigma(x) = \frac{\mu \Delta \varepsilon}{\kappa} \frac{\cosh(kx)}{\cosh(kl)} \qquad (1)$$

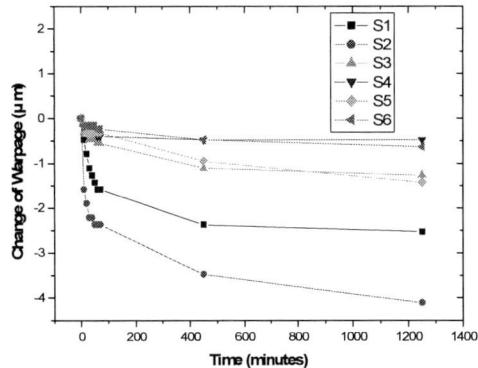

Figure 5: Change of warpage in 85 °C/85 % environment

Table 3: Humidity properties

85°C/85% condition	Msat	D (10^{-7}cm²/s)	CME (10^{-3}/%)
ACF	0.635	8.23	3.02
Substrate	1.403	3.94	0.35

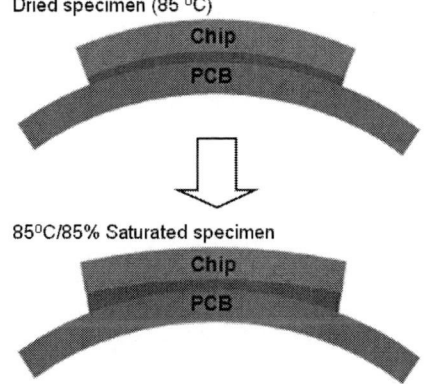

Figure 6: Warpage decrease by swelling

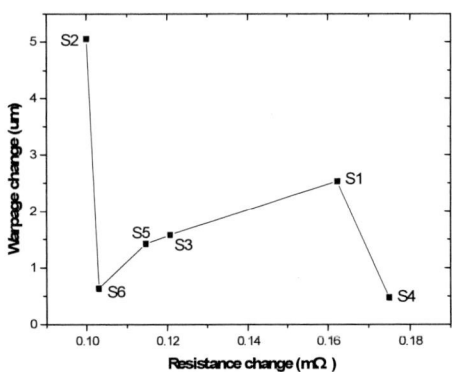

Figure 7: Relation between the warpage change and the resistance change

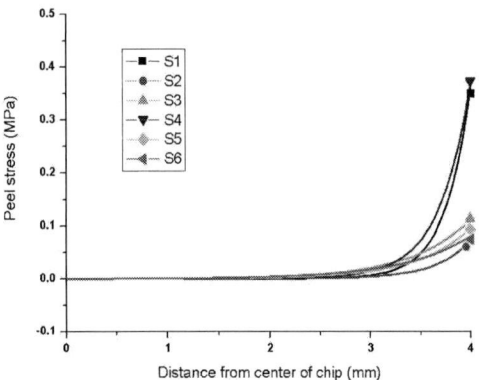

Figure 8: Peel stress along the ACF interface

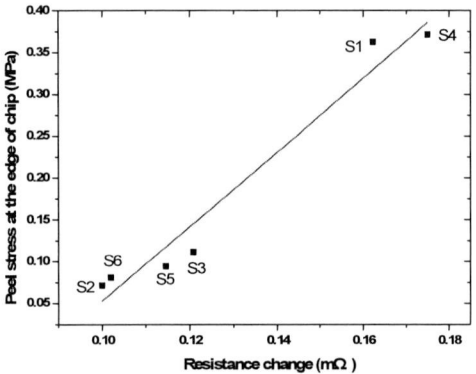

Figure 9: Relation between the peeling stress at the edge of the chip and the resistance change

$$\varepsilon_x = \frac{\partial U}{\partial x} = \frac{1}{f}\left(\frac{\partial N_x}{\partial x}\right) = \frac{1}{f}\left(\frac{\Delta N_x}{\Delta x}\right) \quad (2)$$

$$\varepsilon_y = \frac{\partial V}{\partial y} = \frac{1}{f}\left(\frac{\partial N_y}{\partial y}\right) = \frac{1}{f}\left(\frac{\Delta N_y}{\Delta y}\right) \quad (3)$$

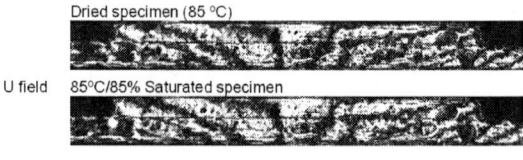

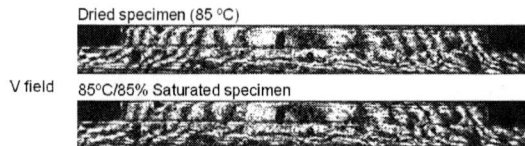

Figure 10: Moire fringe patterns of S5 specimen induced by moisture

The hygro-thermal behavior of the ACF type package is complicated. The ACF and the PCB of the ACF package specimens have the different humidity properties. The ACF layer showed larger hygro-swelling deformations than PCB in the ACF package specimens, and consequently a loss of adhesion strength occurred by moisture induced swelling. The ACF package specimens fully saturated by moisture showed large peeling stress at the edge of the chip. In order to elucidate these phenomena, experiments using Moiré interferometry were performed.

The sectioned specimens were attached to the specimen grating at 125 °C. The frequency of the specimen grating was 1200 lines/mm. The bi-thermal Moire tests of ACF specimens were performed at 85 °C temperature. The resolution of the Moiré fringe image was 0.417 µm/fringe displacement. The strain in ACF type package was calculated using Eq. (2), (3) from the Moiré fringe image.

After drying ACF package specimens at 100 °C for 24 hours, the strain distributions of the specimens in a 85 °C/85 % humidity chamber were measured by Moiré interferometry. Fig. 10 shows Moire images of moisture saturated ACF specimens and dried specimens. Each specimen had a complicated micro swelling deformation. Next chapter will deal with analysis of these Moire results.

Interfacial fracture toughness

From the Moiré fringe results, we can calculate the stress intensity factor K by eq. (4), [16], [17]. Since the specimens fully saturated by moisture showed large peeling stress at the edge of the chip, it was determined that these specimens had a larger K_1 than the dried specimens. Table 4 shows the K_1 result of each specimen. Energy release rate G value can be obtained by eq (6).

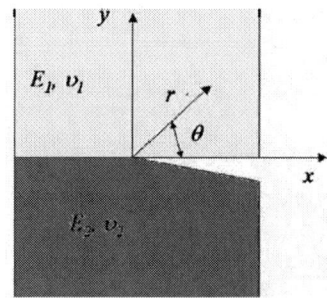

Figure 11: Interfacial delamination problem

978-1-4244-4722-0/09 $25.00
© 2009 IMAPS-ITALY

$$K'_1 = \frac{\left\{ A\cos\left[\varepsilon ln\left(\dfrac{r}{L}\right)\right] + B\sin\left[\varepsilon ln\left(\dfrac{r}{L}\right)\right]\right\}}{D}$$

$$K'_2 = \frac{\left\{ B\cos\left[\varepsilon ln\left(\dfrac{r}{L}\right)\right] - A\sin\left[\varepsilon ln\left(\dfrac{r}{L}\right)\right]\right\}}{D}$$

$$(4)$$

where

$$A = \delta_y - 2\varepsilon\delta_x$$
$$B = \delta_x + 2\varepsilon\delta_y$$
$$D = \frac{8}{E*\cosh(\pi\varepsilon)}\sqrt{\frac{r}{2\pi}}$$

$$(5)$$

$$G = \frac{(1-v^2)K_1^2}{E} + \frac{(1-v^2)K_2^2}{E}$$

$$(6)$$

$$K_1 = -0.434P_0 h^{-1/2} + 1.934M_0 h^{-3/2}$$
$$K_2 = -0.558P_0 h^{-1/2} - 1.503M_0 h^{-3/2} \qquad (7)$$

$$G_c = (1-v^2)(P_0^2 + 12M_0^2/h^2)/2Eh \qquad (8)$$

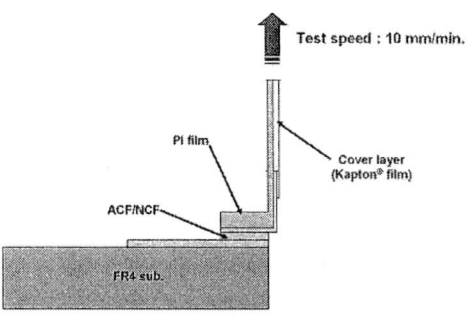

Figure 11: Schematic diagram of 90 degree peel test

Table 4: Stress intensity factor K

Specimen	Dried specimen K_1 (MPa m$^{1/2}$)	Saturated specimen K_1 (MPa m$^{1/2}$)
S1	0.12	0.37
S2	0.16	0.25
S3	0.16	0.21
S4	0.1	0.36
S5	0.1	0.15
S6	0.08	0.18

In order to determine the loss of adhesion strength upon moisture absorption, a 90 degree peel strength test was performed [18]. Fig. 12 shows the strength results of a moisture saturated ACF specimen and a dried specimen. From these results, the critical fracture toughness K_c was calculated by eq. (7), [19].

Critical energy release rate G_c value was obtained by eq (8). Fig. 13 shows the relations between the stress intensity factor K of ACF type specimens and the critical fracture toughness K_c. The dried ACF specimens had a lower K value than the Kc of the dried ACF interface. However, the moisture saturated ACF specimens had a similar K value with K_c of the saturated ACF interface. Since S1 and S4 package specimens fully saturated by moisture showed large peeling stress at the edge of the chip, S1 and S4 specimens had large K value. Fig. 14 shows the energy release rate G of ACF type specimens. The energy release rate G of S1, S4 specimens were close to the critical release rate G_c of fully saturated ACF. The G_c of dried ACF was 297 J/m^2.

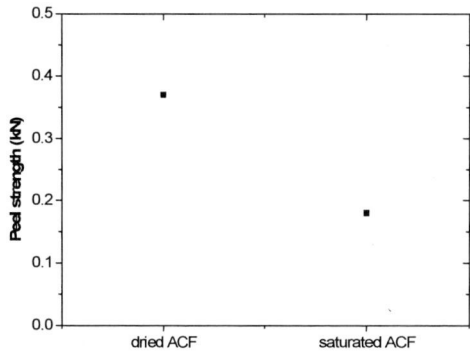

Figure 12: Change of Peel strength of ACF

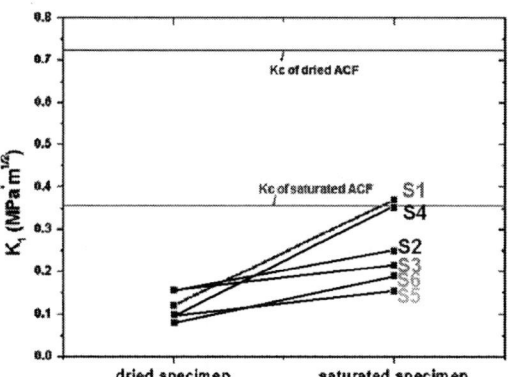

Figure 13: Relation between stress intensity factor K of ACF type specimens and the critical fracture toughness K_c

978-1-4244-4722-0/09 $25.00
© 2009 IMAPS-ITALY

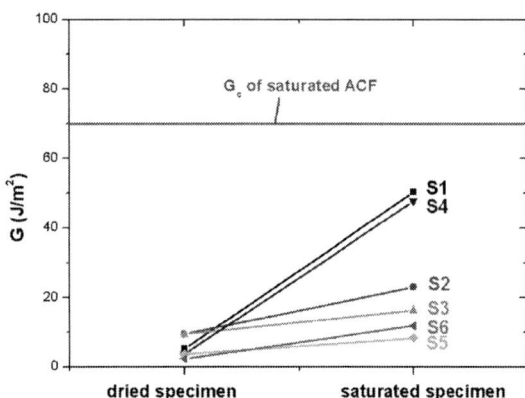

Figure 14: Energy release rate G of ACF type specimens

Conculutions

In this study, the hygro behavior of an ACF type package was evaluated. The polymer materials of ACF type packages are swelled by absorbing the humidity. Mismatch in the CME between the adhesive and other components in the package induces hygroscopic swelling stress. A primary factor of the ACF package failure is delamination between the chip and the adhesive at the edge of the chip. This delamination is mainly affected by the peeling stress at the edge of the chip.

The specimens fully saturated by moisture showed large peeling stress at the edge of the chip. Therefore, these specimens had a larger K1 than the dried specimens. Application of moderate geometric shape condition was shown to be a very powerful means for reducing the peeling stress. Consequently, low peeling stress was shown to improve the humidity reliability of the interconnection layer.

Acknowledgments

This research was supported by a grant from Center for Nanoscale Mechatronics & Manufacturing, one of the 21st Century Frontier Research Programs, which are supported by Ministry of Education, Science and Technology, KOREA

References

[1] A.M. Lyons et al., "A New Approach to Using Anisotropically Conductive Adhesives for Flip-Chip Assembly", IEEE Trans. on Comp. Pack. & Manuf. Tech.-Part A, Vol. 19, pp. 5-11, March, 1996.

[2] J.S. Rasul, "Chip on paper technology utilizing anisotropically conductive adhesive for smart label applications", Microelec. Reliab., Vol. 44, pp. 135-140, January, 2004.

[3] Han B. and Kunthong P., "Micro-Mechanical Deformation Analysis of Surface Laminar Circuit in Organic Flip-Chip Package: An Experimental Study", Journal of electronic packaging, Vol. 122, No. 4, pp. 294-300, December, 2000.

[4] Huimin, Xie, Boay, Chai Gin, Asundi, A., Yu, Jin, Yunguang, Lu, Ngoi, B.K.A. and Zhaowei, Zhong, "Thermal deformation measurement of electronic package using advanced moire methods", Proceedings of 3rd Electronics Packaging Technology Conference (EPTC 2000), Singapore, December 5-7, pp. 163-168, 2000.

[5] Miller, M.R., Mohammed, I., and Ho, P.S., "Quantitative strain analysis of flip-chip electronic packages using phase-shifting moire interferometry", Optics and lasers in engineering, Vol. 36, No. 2, pp. 127-139, August, 2001.

[6] Wang, Guotao, Zhao, Jie-Hua, Ding, Min, and Ho, P.S., "Thermal deformation analysis on flip-chip packages using high resolution moire interferometry", 2002 Thermal and Thermomechanical Phenomena in Electronic Systems (2002 ITHERM), San Diego, California, May 30- July 1, pp. 869-875, 2002.

[7] D. Post, B. Han, P. Ifju, "High Sensitivity Moiré", Springer-Verlag, New York, pp. 118-119, 1994.

[8] A.F. Bastawros, A.S. Voloshin, P. Rodogoveski, "Determination of Thermally Induced Deformations in Electronic Packages by Moiré Interferometry", In Proceeding of 39th Electronic Components & Technology Conference (ECTC '89), pp. 864-868, 1989.

[9] B. Han and Y. Guo, "Determination of an Effective Coefficient of Thermal Expansion of Electronic Packaging Components: A Whole-Field Approach", IEEE Components, Packaging, and Manufacturing Technology-Part A, Vol. 19, No. 2, pp. 240-247, June, 1996.

[10] S. Liu, et al., "Study of Delaminated Plastic Packages by High Temperature Moire and Finite Element Method", IEEE Trans. on Components, Packaging, and Manufacturing Technology -Part A, Vol. 20, No. 4, pp. 505-512, December, 1997.

[11] Y. Guo and S. Liu, "Development in Optical Methods for Reliability Analysis in Electronic Packaging Applications", ASME Journal of Electronic Packaging, Vol. 120, pp. 186-193, June, 1998.

[12] B. Han, Y. Guo, C.K. Lim, D. Caletka, "Verification of Numerical Models Used in Microelectronics Packaging Design by Interferometric Displacement Measurement Methods", ASME Journal of Electronic Packaging, Vol. 118, pp. 157-163, September, 1996.

[13] Seungbae, Park, "Predictive Model for Optimized Design Parameters in Flip-Chip Packages and Assemblies", IEEE Trans. on

Components and Packaging Technology, Vol. 30, No. 2, pp. 294-301, June, 2007.

[14] Lei L. Mercado, "Failure Mechanism Study of Anisotropic Conductive Film (ACF) Packages", IEEE Trans. on Components and Packaging Technology, Vol. 26, No. 3, pp. 509-516, September, 2003.

[15] M. Y. Tsai, C. H. Hsu, and C. N. Han, "A Note on Suhir's Solution of Thermal Stresses for a Die-Substrate Assembly", Journal of Electronic Packaging, Vol. 126, pp. 115-119, March, 2004.

[16] X. Q. Shi, "Effect of Hygrothermal Aging on Interfacial Reliability of Silicon/Underfill/FR-4 Assembly", IEEE Trans. on Components and Packaging Technology, Vol. 31, No. 0, pp. 94-103, 2008.

[17] X. Q. Shi, "Determination of Fracture Toughness of Underfill/Chip Interface With Digital Image Speckle Correlation Technique", IEEE Trans. on Components and Packaging Technology, Vol. 30, No. 1, pp. 101-109, March, 2007.

[18] M. J. Yim and K.W. Paik, "The Contact Resistance and Reliability of Anisotropically Conductive Film (ACF)", IEEE Trans. on Advanced Packaging, Vol. 22, No. 2, pp. 166-173, May, 1999.

[19] M. D. Thouless, "Fracture mechanics for thin-film adhesion" IBM Journal of Research and Development, Vol. 38, pp. 367 – 377, 1994.

978-1-4244-4722-0/09 $25.00
© 2009 IMAPS-ITALY

The Design and Improvement of LTCC-based Capacitive Pressure Sensors Employing Finite Element Analysis

Ciprian Ionescu[1], Paul Svasta[1], Cristina Marghescu[1],
Marina Santo Zarnik[2], and Darko Belavič[2]

[1] University "Politehnica" of Bucharest, Center for Technological Electronics and Interconnection Techniques, Spl. Independentei 313, 060042-Bucharest, Romania,

Phone: (+40)21-316 9633; Fax: (+40)21-316 9634; E-mail: ciprian.ionescu@cetti.ro

[2] HIPOT-RR, Šentjernej, Slovenia

Abstract

In recent years there was a raising interest in capacitive sensors. Their rate on today's sensor market is steadily rising due to the decreasing of the technology cost, their stability and the employment of simple conditioning circuits interfacing. These sensors can be used to measure a wide variety of physical quantities: flow, pressure, liquid level and others. The capacitive pressure sensors this paper focuses on are constructed on a LTCC base with a circular edge-clamped diaphragm (membrane) overlapped to a ring fixed on the substrate. Assembled this structures form a cylindrical chamber. One of the electrodes is deposited on the bottom side of the membrane; the other is placed opposite - directly on the substrate. Changes of the distance between the electrodes, which appear as a result of an applied pressure, imply a change in capacitance. A few samples of this sensor have been produced and characterized. In the construction of this sensor materials with different characteristics are used (LTCC Du Pont 951, AgPd, brass). A virtual model of the sensor was consequently developed using ANSYS. The results of the performed measurements were compared with the simulated ANSYS model. Changes in design and their influence on the sensitivity and reliability of the sensor have been implemented as well. The aim of this paper is to introduce the finite element model and some of the ways this model can be used to improve the performances of the sensor by optimizing its design.

Key words: Capacitive Pressure Sensor, Ceramic Pressure Sensor, Finite element modeling, LTCC

Introduction

The work presented in this paper has been conducted in frame of a bilateral project between University "Politehnica" Bucharest and "Jožef Stefan" Institute from Ljubljana. The aim of this project was the study and the improvement of the design of a capacitive pressure sensor realized using thick-film and LTCC and thick-film materials and technologies.

When developing a sensor a study regarding its sensitivity or it's susceptibility to changes of other physical dimensions than the measurand can lead to a design that proves to be more accurate. Finite Element Modeling is a viable prognosis solution.

Since sensors are being employed in increasingly more fields and applications and since they need to be increasingly reliable virtual prototyping becomes more commonplace.

Regarding pressure sensors, the market has been dominated for years by silicon based sensors, whose behavior is, as a result, well known.

However, there has been an increase in the use of ceramic pressure sensors. This can be explained by their low cost and stability. The basic structure of such a sensor is that of two parallel electrodes with one fixed electrode and one electrode bonded to a diaphragm [2]. The value of the capacitance is determined by the distance between electrodes, the area of the electrodes and the nature of the employed dielectric.

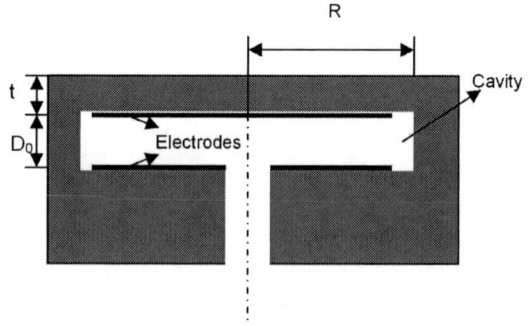

Figure 1 Sensor architecture

As previously stated the sensor that is the object of this paper is realized in thick-film technology and employs air as dielectric. Pressure applied on the diaphragm will have as effect a deflection which will lead in turn to a change in the value of the capacitance. The proposed sensor architecture can be observed in fig. 1. The sensor's characteristics depend on the construction, the dimension and the material properties of the sensor structure.

In order to realize the first prototypes the materials and technologies used have been established. Virtual prototyping allowed then the simulation of different dimensions of the geometrical parameters and temperature dependence (a problem for ceramic sensors) without the costs that producing several prototypes with several dimensions would imply. $ANSYS^{TM}$ provides a wide range of tools to perform this multi-field analysis.

Theoretical Basis

The first step in predicting the behavior of the sensor is building a mathematical model based on the related equations. The model shall be a simplified one, not taking into account all the factors and shall be regarded as a starting point.

Since it is one of the key factors in understanding the behavior of this kind of sensor, the characterization of the behavior of the diaphragm is analyzed. The sensor was designed with a circular, edge-clamped diaphragm. The deflection follows the following law:

$$y(r) = \frac{3P(1-v^2)(R_d^2 - r^2)^2}{16Et^3} = y_0(R_d^2 - r^2)^2 \quad (1)$$

Neglecting the fringing field the capacitance shall be:

$$C(P) = \varepsilon_0 \varepsilon_r \int_0^{2\pi} d\theta \int_0^{R_d} \frac{r\, dr}{D_0 - y(r)} \quad (2)$$

Where the following notations have been used: $y(r)$ the deflection at the position r from the center when a pressure P is applied, E the elasticity (Young) modulus, v Poisson's ratio, t the thickness and R_D the radius of the diaphragm. $C(P)$ is the value of the capacitance when a pressure P is applied. The following factors influence this value: the permittivity of vacuum - ε_0, the relative permittivity - ε_r, the radius of the electrode - R_e, the distance between electrodes at no deflection - D_0 and $y(r)$ - the deflection. It is possible to give a closed form to the integral (2) by replacing $y(r)$ from (1). In order to simplify the calculation we introduce a dimensionless parameter $\beta = y_0/D_0$ and we change the variable from r to x as follows:

$$r = \sqrt{\beta}\left(R_d^2 - r^2\right) \quad (3)$$

This leads to:

$$C(P) = \frac{\pi \varepsilon_0}{\sqrt{\beta} D_0} \int_{\sqrt{\beta}R_d^2}^{0} \frac{dx}{x^2 - 1} \quad (4)$$

Respectively:

$$C(P) = \frac{\pi \cdot \varepsilon_0}{2D_0\sqrt{\beta(p)}} \ln\left|\frac{\sqrt{\beta(p)}R_d^2 + 1}{\sqrt{\beta(p)}R_d^2 - 1}\right| \quad (5)$$

These equations are valid when pressure is applied from the top side.

The material constants used can be seen in the following tables:

Table 1 LTCC material properties

Characteristics	Value
Density (ρ)	3.1 [g/cm^3]
Thermal expansion coefficient (TEC)	5.8×10^{-6}[1/K]
Thermal conductivity (k)	3.3 [W/mK]
Elastic modulus (E)	110 [GPa]
Temperature coefficient of elastic modulus (TCE)	-240×10^{-6} [1/K]
Poisson's ratio (v)	0.17

Respectively:

Table 2 Ag/Pd material properties

Characteristics	Value
Density (ρ)	10.9 [g/cm^3]
Thermal expansion coefficient (TEC)	17.7×10^{-6}[1/K]
Thermal conductivity (k)	393 [W/mK]
Elastic modulus (E)	92.5 [GPa]
Poisson's ratio (v)	0.38

To simplify the data analysis a MathCAD Worksheet was used.

Optimization of the Basic Design using Finite Element Analysis

As previously stated the sensor's characteristics depend on the construction (form), the dimensions and the material properties of the sensor structure. Since the form was already established it remained to be seen which dimensions would enhance the performance of the sensor.

In the pre-processing phase the model was built and materials, loading, and boundary conditions defined. The model was defined using a top-down approach - the volumes that define the component were created first. For the optimization of this model the effects that different dimensions of the membrane or cavity had on the behavior of the sensor had to be analyzed. In order to allow easy changes of model data a parametric model was used. The dimensions were defined as parameters. A change of these parametric values does not require further modification of the model. The different material models were then defined. For this model three material models were used: one for the LTCC ceramic, the second for the air and the third for electrostatic LTCC.

The following types of analysis have been performed – structural analysis, electrostatic analysis and coupled analysis: structural (load pressure surface load – a distributed load applied over a surface) and thermal strain (load temperature – body load a volumetric load). All loads are applied on the solid model volumes and areas.

Applying pressure will have as a direct effect a deflection of the diaphragm and as indirect effect a change of the value of the capacitance. It is important to see how the capacitance changes with pressure. The output variable (capacitance) is a function of the input variable (pressure) and the model should use the same simulation environment. To do this "mesh morphing" was used.

The model was 3D meshed with the following element types:

- SOLID 92 3-D 10-Node tetrahedral structural solid, defined by 10 nodes having three degrees of freedom at each node; other capabilities include plasticity, strain, stress stiffening, and

- SOLID 123, 3-D electrostatic tetrahedral solid.

The number of mesh elements used depends on the level of accuracy desired, but must also take into account time management.

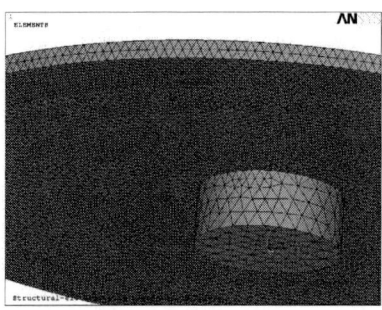

Figure 2 Finite Element Model

The LTCC ceramic block was modeled as structural element and the air cavity was modeled as electrostatic air (no influence for the structural domain). Structural analysis is performed first and the displacement resulted at the membrane interface

is transferred to the volume that models the cavity using the morphing feature.

The coordinates of the nodes change but the nodes remain associated with the surfaces. The deflection ratio is relatively small.

The deflection of the membrane was read by using the so called path operations in which a variable from the model is plotted against a path defined in the model. The path in this case was the radius of the diagram. The deflection was then plotted and compared to the values calculated using formulae or if available with the measured data.

For the electrostatic solution the capacitance was computed (electrostatic analysis) using the *CMATRIX* macro-command.

Analysis results

For the membrane radius dimensions between 4 mm and 6 mm were attempted. For small radiuses small deflections were observed which in turn leads to the sensor being very sensitive to parasitic influences. For bigger radius the sensitivity improves but one has to keep in mind the economic costs. Since this was not a concern we did not attempt to draw a conclusion.

Knowing that ceramic sensors are influenced by temperature, an analysis regarding the sensor's behavior with temperature changes was considered.

The change of the thickness implies a change of the characteristics of the LTCC tape, so that the data corresponds to an existing available tape with that thickness. From that point of view the results regarding the optimal thickness could not be considered conclusive (material data was not the same).

Both the realized samples and the simulated sensor showed a pretty good linearity.

Some results can be observed in the following table and graphics. In the table the sensitivity for three different radiuses for two different temperatures can be observed.

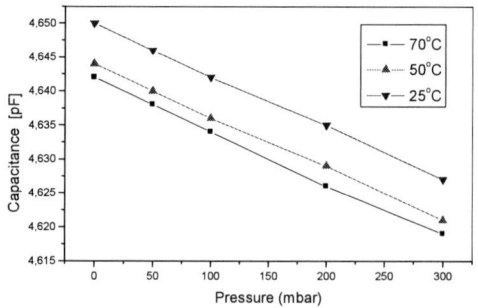

Figure 3 Variance of the capacitance for three values of the temperature measured with AD7746 evaluation board from *Analog Devices.*

978-1-4244-4722-0/09 $25.00
© 2009 IMAPS-ITALY

Table 3 Synthesis of data from simulation results

Diaphragm Radius [mm]	3				4				5			
Thickness [mm]	0.2				0.2				0.2			
Gap [mm]	0.2				0.2				0.2			
$\Delta C/C_0$ [%]	-0.3		-0.15		-0.17		-0.18		-0.71		-0.76	
$(\Delta C/C_0)/\Delta p$ [ppm/Pa]	-0.06		-0.03		-0.034		-0.036		-0.142		-0.15	
Temperature [°C]	0		100		0		100		0		100	
Applied Pressure [kPa]	20	70	20	70	20	70	20	70	20	70	20	70
Capacitance [pF]	2.462	2.2388	2.2493	2.2459	3.5216	3.5153	3.5358	3.5294	5.0943	5.0577	5.1218	5,0833

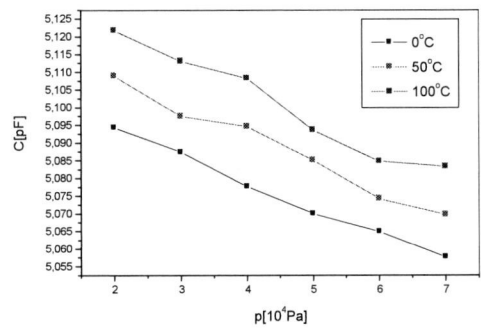

Figure 4 Variance of the capacitance for a radius of 5 mm, for three temperatures; results as obtained with ANSYS.

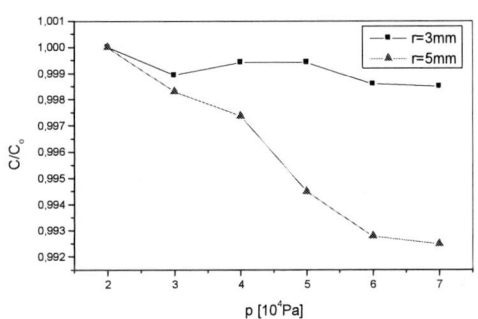

Figure 5 Capacitance change in function of pressure at 100o C represented for two radii values.

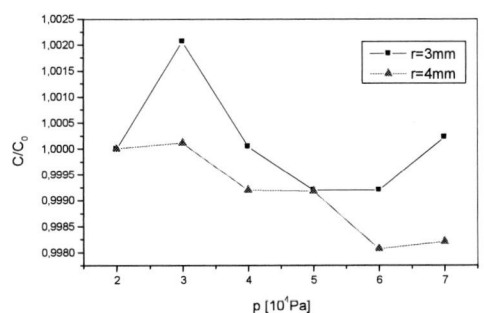

Figure 6 Capacitance change in function of pressure at 0o C represented for two radii values

Acknowledgements

This paper is a result of activities that took place in the frame of the project DAPST funded by The Romanian National Authority for Scientific Research (ANCS).

References

[1] ANSYS 6.1 Theory, Reference Manual documentation.

[2] Stephen Beeby, Graham Ensell, Michael Kraft, Neil White, "MEMS Mechanical Sensors", Artech House, 2004, Cap.6.

[3] J.I. Pavlič, Marina Santo Zarnik, Darko Belavič, "Feasibility study of a ceramic capacitive pressure sensor", Proceedings. Ljubljana: MIDEM – Society for Microelectronics, Electronic Components and Materials, pp.95-100, 2006.

[4] Darko Belavič, Marina Santo Zarnik, Mitja Jerlah, Marko Pavlin, Marko Hrovat, Srečko Maček, "Capacitive thick-film pressure sensor : material and construction investigation", Proceedings of the XXXI International Conference of IMAPS Poland 2007, Rzeszóv, Krasiczyn, Poland, September 23-26, pp.249-253, 2007.

[5] Darko Belavič, Marina Santo Zarnik, Srečko Maček, Mitja Jerlah, Marko Hrovat, and Marko Pavlin, "Capacitive Pressure Sensors Realized With LTCC Technology", ISSE 2008 Proceedings, May 7-11, pp. 271-274, 2008.

[6] L.K.Baxter, "Capacitive Sensors", *www.capsense.com*

Investigation of Solder Joints by Thermographical Analysis

P. Svasta[a], C. Ionescu[a], N.D. Codreanu[a], D. Bonfert[b]

[a] University "Politehnica" of Bucharest, Center for Technological Electronics and Interconnection Techniques, Spl. Independentei 313, 060042-Bucharest, Romania,

Phone: (+40)21-316 9633; Fax.: (+40)21-316 9634; E-mail: ciprian.ionescu@cetti.ro

[b] Fraunhofer Institute Reliability and Microintegration / IZM-M, Munich, Germany,

Abstract

The present stage of development of electronics technology there is an increased interest in the study of new materials used as solder that replaces the already banned tin-lead alloy. Many investigations were done in direction of metallurgic compatibility between the printed circuit board finishing, component terminal finishing and solder itself. Other studies are taken in direction of mechanical characterization of lead free alloys. We propose a study from electrical point of view. So we will analyze the current capabilities of solder joints by thermographical investigations. The finite conductivity of the solder alloy will act as a dissipating media and will heat the adjacent region to the joint. In order to make this effect more pregnant it is necessary to dispose of very precise low-ohm resistors and to pass through them relatively high currents. Based on a high resolution infrared camera the temperature gradient will be better observed. The measured results are compared to the results derived from finite element modeling and simulation. In the simulations, the whole 3D structure of solder joint is modeled. It has practically a very complex shape and is difficult to be analyzed using other methods. Our tests will be realized on organic rigid substrates using a configuration that was intended to characterize the solder joint dissipation. The data derived from this analysis will be very useful in the design process of high power circuits that could be used in advanced electronic modules.

Key words: solder joints, thermography, finite element analysis, coupled field modeling and simulation

Introduction

In actual stage of electronics the major attachment technique of components to printed circuit boards (PCB's) is represented by the soldering technology. The various method for soldering electronic components include hand soldering, wave soldering, reflow soldering using infrared radiation or convection ovens. Due to the European RoHS regulations the vapor phase soldering process, although known for many years come in foreground for use in lead-free applications. Modern soldering techniques as laser soldering are also investigated. Regardless the soldering process a solder joint of a component has mechanical, thermal and electrical functions that must ensure the reliability of the electronic assembly.

For our investigation we have decided to use SMD chip type components. The solder joint of rectangular type components is presented in figure 1.

The significance of parameters from figure 1 is: Maximum Side Overhang -A , Maximum End Overhang- B, Minimum End Joint Width -C, Minimum Side Joint Length- D, Maximum Fillet Height- E , Minimum Fillet Height F, Solder Fillet Thickness G, Height of Termination- H, Minimum End Overlap J, Width of Land-P, Width of Termination –W. In our model we have tried to model accordingly to these drawings the solder parameters.

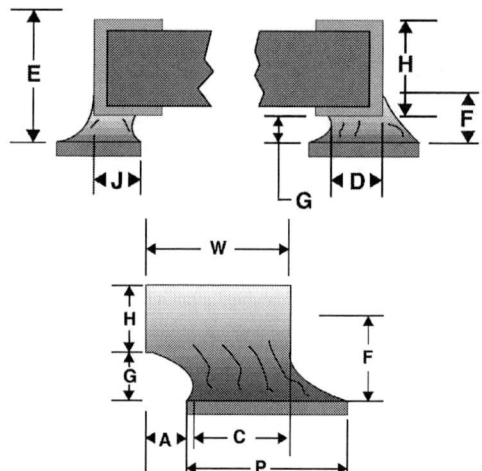

Figure 1: Solder joint geometry for rectangular components according to IPC ref [3]

The IPC standards give intervals for these parameters not strictly values. Some parameters are noted to be correlated to the technology or the design. We have chosen a 2512 chip type geometry for the solder joints to be investigated.

978-1-4244-4722-0/09 $25.00
© 2009 IMAPS-ITALY

Modeling for electric-thermal analysis

We have the intention to do a comparison between measured and simulated values in this solder joints study. To obtain the thermal solution, i.e. the temperature map, a coupled-field analysis is required. For this type of analysis the interaction (coupling) between two or more types of physical phenomena (fields) is considered. Such analyses may involve direct or indirect coupling of fields.

When performing a directly coupled analysis, the variables from both fields (e.g., heat generation rate and temperatures) are computed simultaneously. This method is necessary when the individual field responses of the model are strongly dependent upon each other. Directly coupled analyses are usually nonlinear since equilibrium must be satisfied based on multiple criteria. The finite element model requires more computational resources in this case.

An indirectly coupled analysis involves the solution of single-field models in a particular sequence. The results of one analysis are used as loads for the following analysis. This is also known as the sequential method of coupled analysis. This method of analysis is applicable when there is one-way interaction between fields.

In our case, for example, if we consider that the resistivity of conductive materials is not temperature dependent we could also apply this method. This method is usually more efficient than the direct method, and it does not require use of special coupled finite elements and no multiple iterations are required. We have used ANSYS[TM] software which supports both type of simulations.

Electric-Thermal coupling is presented in figure 2:

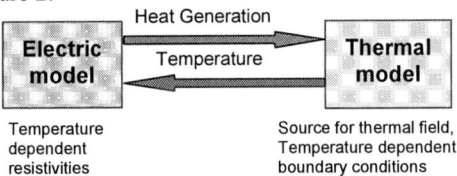

Figure 2: Coupled field electric-thermal simulation

Only for one simulation scenario we have developed a different model using a different solving approach. This is called "Multi-field solver". In this modeling technique there are created two overlapped solid models with different finite element types which may have different meshes. Each model and the associated parameters and boundary conditions is saved as a "field". There is the need to define the interaction surface between the fields, in our case we have a transfer which occurs from the electrical domain to thermal domain as volumetric transfer and the transfer from thermal field to electrical field is done also inside the metallic conductive elements (change in resistivity). The solver converges more rapidly for this solver as in direct coupling, but there were no differences found in results up to the third

decimal position. For the ease of model creation and the ease of parameters change we have realized the models using the first presented method, i.e. direct coupling using thermal-electric element called SOLID69. This permits us to combine mapped meshing where this was possible to be done with the free meshing using pyramids (tetrahedral elements)

Modeling considerations

A little simplification was necessary for the purpose of the analysis. In this direction, resistors constructive details were ignored as not important. The resistors were modeled as simple bricks.

The software that was used is ANSYS, a finite element analysis (FEA) software for which the modeling and simulation flow includes: building the solid model, defining and assigning material properties and proper finite elements, meshing the model, applying the loads and boundary conditions, and finally solving and postprocessing the results. A characteristic of the model is that the full 3D structure was modeled. In all cases parametric type model was built which permit us to realize a series of runs without to re-create the solid model.

A major problem in modeling planar structures, as the copper traces, is the large number of elements that can be generated by the very thin layers that model the conductive, dielectric or resistive depositions used in electronics. We have used a special modeling technique, which implies the building of the solid model by extrusion of areas along "z" direction. In this way, hexahedral elements and not tetrahedral are built, and the number of finite elements can be dramatically reduced.

For our models presented here which include the large FR4 substrate, there were up to 1300000 elements, with 294000 nodes, a large number absolutely sufficient for the electrical field which requires a finer mesh than the thermal field. The running time for one data was about 2 hours. A model, for one structure, as seen in FEA program is presented in figure 3.

Figure 3: Solid model for one tested structure

The boundary conditions involve the applying of heat transfer coefficients on the exterior surfaces. For the convection coefficients we have chosen to take some result from literature and our previous papers. The board was hold suspended and there was also convection from the bottom side of the board. We have used temperature dependent film coefficients. The values were derived from values at room temperature with the assumption of variation according to $\sim(\Delta T)^{0.25}$ relation.

Temperature dependent resistivities were used for copper and for solder alloy too. The parameters that were used in simulations are presented in table 1:

Table 1: Material properties used in analysis

Mat. nr.	Material	Thermal constant (W/mK)	Resistivity (Ω·m) at 25 C
1	Copper	390	1.72e-8
2	FR4	0.3	∞
3	Solder Sn-Pb (63-37)	57.9	14.9 e-8
3*	Solder Sn-Ag (96,5-3,5)	55.3	12.3e-8

The issues for determining the heat convection coefficients are presented in [5]. The source of heat is the electrical power dissipated in the volume of electrical components, copper traces, solder joints, resistors.

The loads are applied to the model as volume (body) loads, this means a heat generation rate (HGEN) or other named power density. These Joule heat generation has a specific distribution for a certain geometry and is difficult to be predicted without using software simulation tools.

Experimental vehicle

The test board was supposed to emphasize the heating of solder joints. Regarding this it is very difficult to emphasize only the heat produced by the current inside the joints because in a real assembly the current is supplied through relatively thin PCB traces. There is a trade-off in this study, one point is to apply large currents and to dissipate the energy in the joint and not in the resistor or in the electronic component. On the other side a good electrical contact means also a good thermal contact so the heat will be conducted away from the solder joint region.

We have used a structure with the intention to emphasize different the joint properties compared to copper track heating.

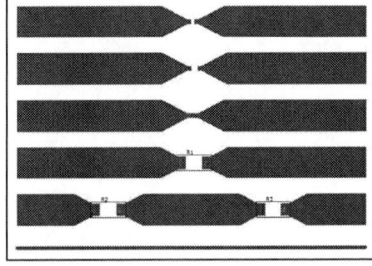

Figure 4: Test board for investigations

The structure presented in figure 4 has being used for comparative investigation of copper tracks and solder joints heating. In order from upper to lower side we have a) tracks with gap 0.75 mm, b) with gap 1.5 mm, c) copper necked track with 1 mm minimal width, d) One copper part like resistor type 2512, e) Two resistor similar to previous, f) one copper track with constant width of 1mm.

Three different substrate materials were tried, generally known as: ½ oz (17 µm), 1 oz (35 µm), 2 oz (70 µm) with laminate thickness of 0.5 mm, 0.8 mm and 1.5 mm respectively. For each of the six structures we have built parametric models. In figure 5 is a part of the model realized for structure 1.

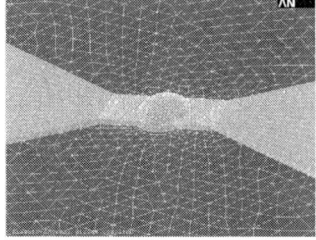

Figure 5: Finite element model for structure 1

It presents a hemispherical solder joint, realized with the intention to read the temperature distribution in it. In figure 6 a mesh detail for the solder fillet modeled for structures 4 and 5, is presented.

Figure 6: Detail of meshed model for structure 4

For the measurement under load we have applied constant current through our probes.

Electrical characterization of solder joints

It is difficult to estimate by hand calculus the solder joint resistance especially because of the complex contact geometry between the component and the solder.

From the variety of SMD components we have chosen rectangular type components. This is because these have applications in power electronics, as power resistors. Our component is a bulk copper part with the dimensions corresponding to SMD 2512 resistor. Regarding the model of solder joint we have built a model based on IPC reference [3]. In order to approximate the real situation, the input current is applied through a very short PCB stub, as in figure 7.

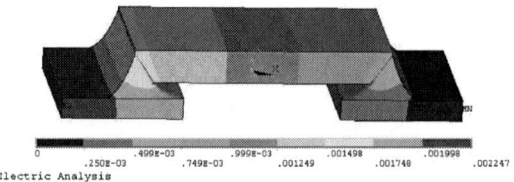

Figure 7: Model used for electrical analysis

The electrical circuit for this assembly, which is very simple, is presented in Figure 8

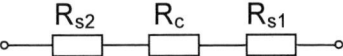

Figure 8: Equivalent electric circuit of SMD chip resistor assembly

In the picture above Rs1 and Rs2 are the solder joint resistances and Rc is the resistance of the component. If we apply constant current to the structure and read the obtained potential we can read twice the solder joint resistance.

We have run the simulations for different material resistivities and for different type of solder joints geometries. The parameters that were changed are F and G from figure 1. Parameter F which determines the percentage of solder fillet that covers the height of resistor was changed to realize a cover percentage from 20% to 90 % of component height H. For parameter G three values were tried 0,1mm, 0.2 mm and 0.3 mm. For G=0.3 mm the obtained resistance of the structure for different fillet height (F parameter) was between 113.4$\mu\Omega$ and 115.9$\mu\Omega$. It was observed a relatively stronger dependence on solder fillet thickness, as expected small values are more convenient.

In this case the analysis is pure electrical, no temperatures or other coupling was involved. From the potential values we can extract the resistance of the solder joint. We can examine also the Joule heat generation in the solder, but the relevance of this may be reduced in a real circuit. In this situation the temperature is the most important factor, not the heat generation rate.

Other possible results from the electrical simulation are the plots where the current densities in the circuit can be examined. In figure 9 two qualitative situations of current densities vector plot are presented. The size of the solder fillet for the two cases are 0.4 and 0.9 from component height.

Figure 9: Current distributions for different solder fillet height (not at the same scale)

Experimental setup

All the measurements were done using the boards presented in figure 3. A high current DC voltage source was used to supply the probes. The source was operating in constant current mode (current limiting). A low resistance shunt resistor made from parallel connected wirewound resistors was also used for precise reading of the current, the instrument front panel meters presenting a low accuracy. The shunt is necessary also to permit the operation of the power supply in a point with convenient voltage level, slightly higher than 0 V.

We have stopped the measurements when the obtained temperature on the board becomes unusual hot.

Because the boards were not provided with coating material there were problems detected in measurement of bare copper tracks and solder joints. It is a well known issue of the infrared measurement that the shiny surfaces are difficult to measure. The observed phenomenon was that the temperature of the tracks seams to be higher than the rest of the board, although their emissivity is much lower. This is due to reflected heat from ambient. A solution possible to be tried in latter experiments will be to do the measurements in a closed (dark) box. We have decided to coat the boards with a mate dye, sprayed from a tube. This has given a picture with uniform temperatures at room temperature which means that the effects of ambient reflections were eliminated. The very thin painting is expected to not produce any change in the heat transfer distribution. The thermovision camera was placed at 30 cm above the board and the current was incremented in 1 ampere step. The measurements were taken according to IPC standards [2] at three minutes after a current supply change.

Sample results

We present the result of simulations and of measurements made on the structure 4 from figure 4. In figure 10, the FEA result in form of a thermal map is presented.

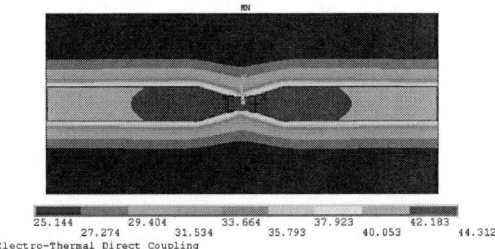

Figure 10: Results from FEA, structure 4, at 10 A

A thermogram picture captured in the measurement of the same structure is presented in figure 11.

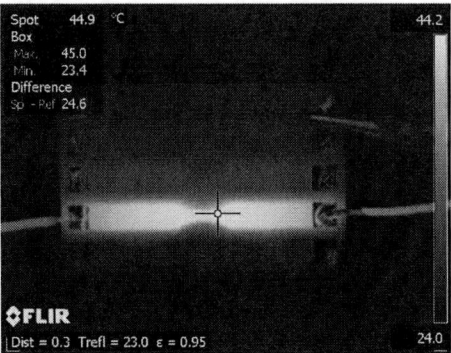

Figure 11: Thermogram picture at 10 A, same structure as in fig.10

The result from measurement and from simulation are very closed one to each other, only for very high temperatures differences of about 2 degrees being observed, FEA showing a higher temperature that the real one. This can be due to the fact that radiation effects were included in the convection film coefficient.

From a deeper analysis it can be seen the heat generation takes place in the copper traces and in the solder joints. There is a uniformity effect of the PCB laminate and of the conducting copper traces, so the expected hot spot in the solder joint is not at all so obvious. Although the resistivity of the solder alloy can be up to ten times higher than the resistivity of copper, the relatively large dimensions of the solder joints section, compared to the copper traces leads to a low Joule heat generation in the joint itself.

In this direction we will be re-orienting toward the heat dissipated by the copper traces. Results from simulation, measurements and reference literature are presented in table 2.

Table 2: Temperature in degree Celsius of a 1mm (40 mil) PCB copper track

Current (A)	Data source	1/2 oz (18µm)	1 oz (35µm)
2	Measurement	45.6	33
	FEA	47.1	35
	Ref [2]	50	30
3	Measurement	76.1	46.7
	FEA	77.2	47.4
	Ref [2]	85	37
4	Measurement	122	67
	FEA	124.3	67.8
	Ref [2]	125	49
5	Measurement	178.5	92.6
	FEA	180.2	94.2
	Ref [2]	N/A	65
6	Measurement	N/A	131.4
	FEA	N/A	133.3
	Ref [2]	N/A	85

For the 35 µm copper thickness board, the results are plotted against the current, in figure 12.

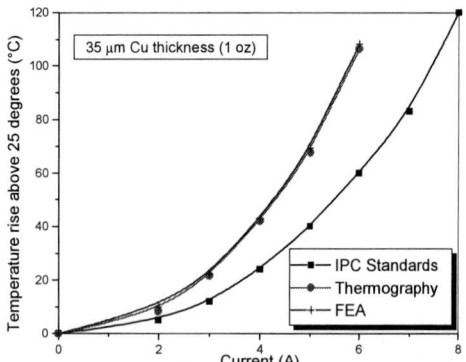

Figure 12: Temperature rise above 25 C from IPC standard compared to measurements and simulation on our test board

From the graph we see the very good agreement between simulation and measurements results and a large difference from the values presented by IPC. It is possible to obtain such different results because in the IPC standard the clad laminate thickness is not clearly shown. According to different sources [1] the current carrying capabilities of the copper traces on PCB depend on the copper thickness and material type and thickness (epoxy or phenolic) laminate.

Anyhow, the IPC graphs are intended to be used backwards, this means to determine the current for a given track cross-section. Also the graphs have already applied some de-rating derived from practical experience.

Conclusions

The investigation started in order to characterize the solder joints have been re-oriented towards investigations on PCB copper traces. In order to characterize from thermal point of view only the heating of solder joints, a new test structure must be developed using a spatial structure not a PCB. From the simulations, but also from thermographical measurements it has resulted that the limiting factor for high power applications is not the solder joint itself, but the PCB copper track.

The measurement of self-heating PCB traces has shown large differences between the IPC standard values and the measurements/simulations based on the test structure from figure 4.

Regarding the electrical behavior, the resistance of the solder joint depends not only on the material and geometry of the solder alloy itself, but depends also on the configuration of the PCB assembly including PCB traces and geometry of component. The reason for this is the different field distribution on solder edges, or briefly saying the entry and the exit surfaces of the current. The current distribution on the surface of the solder is in this sense, determined by the "exterior" circuit.

References

[1] G.W.A Dummer, "Materials for Conductive and resistive functions", Hayden Book, New York, Chapter 6, pp. 91-106, 1970.

[2] ****, IPC-2221, "Generic Standard on Printed Board Design", pp. 38.

[3] ***, IPC/EIA J-STD-001C, "Joint Industry Standard".

[4] ANSYS 6.1, "Theory Reference Manual", 2002.

[5] W. M. Rohsenow, J. P. Hartnett, Young I. Cho (eds.), "Handbook of heat transfer", 3rd edition, McGraw-Hill, 1998.

[6] K. Puttlitz, K. Stalter (editors), "Handbook of Lead-Free Solder Technology for Microelectronic Assemblies", Marcel Dekker, New York, 2004.

978-1-4244-4722-0/09 $25.00
© 2009 IMAPS-ITALY

Fully embedded optical and electrical interconnections in flexible foils

E. Bosman, G. Van Steenberge, P. Geerinck, J. Vanfleteren and P. Van Daele

Ghent University - IMEC

Phone: +32 92645560, Fax: +32 92645374, E-mail Address: Erwin.Bosman@Ugent.be

Abstract

This paper presents the development of a technology platform for the full integration of opto-electronic and electronic components, as well as optical interconnections in a flexible foil. A technology is developed to embed ultra thin (20 µm) VCSEL's and Photodiodes in layers of optical transparent material. These layers are sandwiched in between two Polyimide layers to get a flexible foil with a final stack thickness of 150 µm. Optical waveguides are structured by photolithography in the optical layers and pluggable mirror components couple the light from the embedded opto-electronics in and out of the waveguides. Besides optical links and optoelectronic components, electrical circuitry is also embedded by means of embedded copper tracks and thinned down Integrated Circuits (20 µm). Optical connection towards the outer world is realized by U-groove passive alignment coupling of optical fibers with the embedded waveguides.

Key words: Opto-electronic packaging, flexible, thin chip, VCSEL, optical interconnect, embedding

Introduction

Embedding of optical and electrical circuitry in flexible foils is not just an extension to rigid carrier systems, but it enables a whole range of new dedicated applications and adds cost, speed, weight and volume improvements to existing rigid applications. The complete stand-alone flexible package presented in this paper adds an increase in compactness and weight beyond the possibilities of rigid substrates.

The work presented can also be considered as the development of a generic research platform for flexible optical communication. Two clear trends in the world of electronics have been witnessed in the past 10 years. On one hand, optical communication is no longer serving only long distances but has also been introduced in the on-board communication for chip to chip data transfer [1-2]. On the other hand, flexible electronics have proven their worthiness in portable applications and their profit in cost reduction because of roll to roll production feasibility [3-4]. One can expect that both trends will meet each other in the near future.

Several research groups have developed a hybrid rigid / flexible optical link, consisting of a flexible waveguide connection between two rigid boards containing the electronics and the opto-electronics. What follows describes a fully flexible package of multimode waveguides and active opto-electronic devices.

Materials

The development of opto-electrical PCB's on rigid substrates is growing towards maturity, however their use in applications is still limited to some high-end purposes. The main target for optical PCB's are optical backplanes, demanding very low light propagation losses of the optical path. Bulk materials for these optical layers are chosen to have low losses at a typical wavelength of 850 nm for data communication and 1.3 or 1.55 µm for telecommunication applications. They must have the right processibility properties in view of UV-crosslinking ability, spin-coating, temperature- and chemical resistance and mechanical brittleness. Special care must be taken to ensure the compatibility of the production process with the standard PCB production processes. This means the materials should be inert to production solvents and temperatures used during the production steps of the electronic assembly afterwards. The substrate should be physically and chemically stable in temperatures of about 280 degrees during solder reflows and in temperature cycling's from -40 to +85 degrees.

Truemode Backplane™ Polymer [5], Ormocer® [6], LightLink and Epocore [6] are materials which meet these requirements and have shown good results when applied on rigid substrates in the past. The existing optical materials are however not flexible and strong enough to be bended without cracking or damaging. Therefore these layers are sandwiched between two spin-coated Polyimide (PI) [8] layers, one at the top and one at the bottom, which absorb all stress and pressure during bending. This way the flexibility and durability of the substrate is significantly increased. The final mechanical reliability and robustness can however not be extended to the high demands of the flex market. Further material development is ongoing within the consortium of the FAOS Project (IWT funded Flemish project: Flexible Artificial

978-1-4244-4722-0/09 $25.00
© 2009 IMAPS-ITALY

Optical Skin) and the Phosfos Project (EU FP7 Funded project: Photonic Skins for Optical Sensing) to create novel cross-linkable polymers with improved properties for the flexible applications. During the past 4 years a lot of flexible optical materials have been developed by global material providers also, but these materials are often commercially unavailable and under development.

Development of a flexible optical foil

A flexible optical foil is considered to be a flexible, bendable foil containing optical transparent layers with light confining tracks to route the light signals over the entire foil size. Here fore we fabricate a stack of a cladding-layer, a core-layer and another cladding-layer. Isolating tracks in the core-layer and consequently surrounding the track completely with cladding material, results in the creation of optical waveguides. This is a well proven principle for optical interconnections on rigid boards. Experiments have shown that the flexibility, strength and reliability of the completed layer-structure is significantly improved when the optical layers are sandwiched between two Polyimide layers as described in the "materials"-section. This approach of consecutive layer stacking, results in a symmetrical build-up. The symmetry of the final stack cannot be underestimated since it has a big impact on the warpage of the foil.

The processing of the optical foil is performed on a rigid temporary glass carrier to ensure the compatibility of the fabrication scheme with the standard PCB fabrication processes.

The proposed process layout results in a very light, thin (160 μm total thickness) foil with a high tensile strength due to that of Polyimide. High flexibility is achieved and the minimum mechanical bending radius before damaging the structure is set to 2.5 mm. Figure 1 shows a photograph and a cross-section of the optical foil.

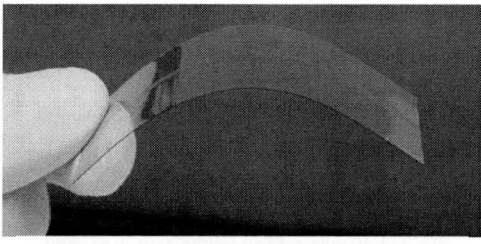

Figure 1: Photograph and cross-section of a flexible optical foil in Epocore material.

The foil has been realized and the adhesion matters have been optimized for the 4 optical materials discussed in the material section of this paper.

Thinning of opto-electronic components

Embedding rigid components like VCSEL's and Photodiodes inside flexible substrate asks for special measures to ensure the flexibility. Standard commercially available optoelectronic components have a typical thickness of 150 μm. This means that the total substrate thickness will be larger than 150 μm. Since we have to apply Polyimide and waveguide layers on top of the embedded die, the total thickness will become way too large to ensure good flexibility of the substrate. To counter this problem, we developed a thinning process for the dies, to reach thicknesses smaller down to 20 μm. This way the dies have even proved to be bendable, which increases the reliability of the package.

The naked dies are mounted face down onto a temporary rigid glass carrier with a dedicated wax, together with large dummy GaAs chips to spread the applied pressure on the carrier (the VCSEL and Photodiode array chip are very small : 1000 x 250 μm) the applied load on the die is too high.

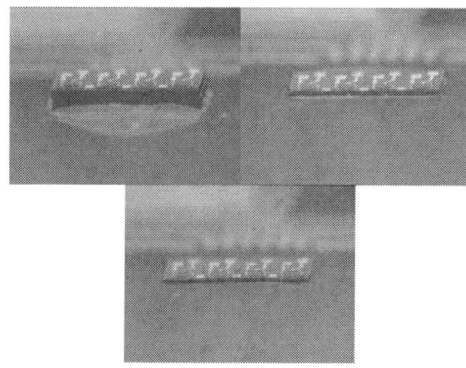

Figure 2: Tilted side view of a 150 μm thick photodiode 1*4 array (*top left*), a lapped PD array (35 μm) (*top right*) and a lapped + polished PD array (20 μm) (*bottom*).

Due to the thinning process, damage is done to the backside of the thinned die. Any initial microcracks can cause failure of the die when it is bended over a small bending radius. Therefore we need to achieve low roughness of the back side which imply that the initial cracks are very shallow and then negligible. This is the reason why the last 15 μm of the total thickness is removed by very slow, low damage polishing step. Figure 2 shows a photograph of a non-thinned, a lapped and a lapped + polished photodiode array. Figure 3 shows the roughness of the complete backside of a thinned VCSEL after the lapping step (top left) and after the polishing step (top right), measured with a non contact profilometer (WYKO). For a 350*1000 μm2 VCSEL, we see a difference in height of 3 μm

978-1-4244-4722-0/09 $25.00
© 2009 IMAPS-ITALY

between the lowest and highest point, which is a negligible height difference.

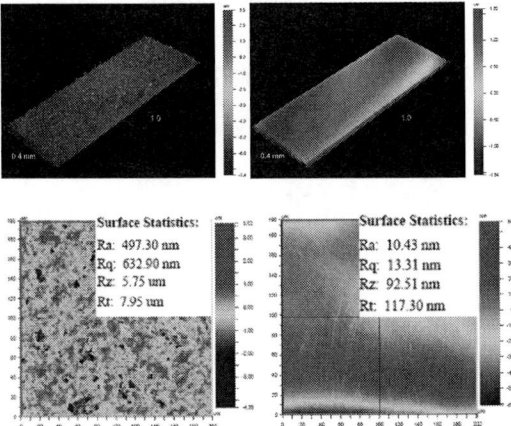

Figure 3: Profile of the backside of a lapped (*top left*) and a lapped + polished (*top right*) VCSEL array. Roughness measurement on a 50*50 μm x μm area for a lapped (*bottom left*) and a lapped + polished (*bottom right*) VCSEL array.

A roughness measurement for a rectangular area of 200*200 μm2 of a thinned VCSEL after the lapping step (left) and after the polishing step (right), measured with a non contact profilometer (WYKO) shows that the roughness of the lapping step is clearly reduced from 500 to 10 nm Ra by the polishing step (Figure 3 bottom).

The thinning process for opto-electronic components does not allow damage to be done to active layer of the die. However the thinning could induce stress inside the die and the active layers. To test the influence of this stress on the optical characteristics of the VCSEL, the VCSEL was attached to an FR-4 substrate and then wire bonding to metal tracks on the substrate. This was done for a non thinned VCSEL array (150 μm), a lapped VCSEL array (50 μm) and a lapped + polished VCSEL array (30 μm). Each array consists of 4 VCSEL's. The LI curves of the VCSEL's were measured and compared. Also the current threshold of each mode produced by the VCSEL are measured and compared. Averaging the results showed us that there is no noticeable change in these parameters due to the thinning of the components, so we can say that the thinning of opto-electrical components induces no change in optical behavior of the VCSEL.

Embedding of thin opto-electronics in the flexible foil

The flexible optical foil presented so far is a passive substrate which can only be used for guiding, splitting, multiplexing, etc. Many reports have been published in the past 2 years, demonstrating foils like presented here to link two rigid boards or connectors with a flexible optical

link. This paper however shows "the next step to flexibility" by integrating the actives inside this foil. By thinning the actives, they are physically bendable. The connectorisation electrically and optically of these actives with each other, with the passive waveguides and with the outer world is discussed in this research.

The embedding of opto-electronic components in the optical foil has been realized so far for the Truemode Backplane Polymer™ only. Figure 4 shows the schematic overview of the process flow used to embed thinned VCSEL and PD arrays in the cladding layer of the flexible optical foil.

For planarization reasons, the VCSEL's and Photodiodes are placed in a cavity inside the under cladding layer.

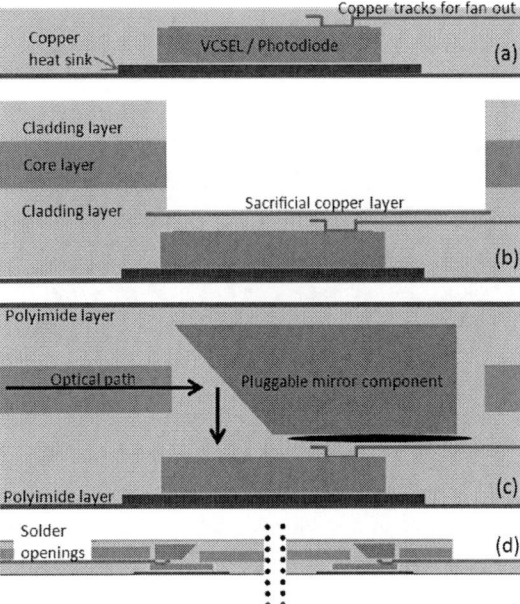

Figure 4: Schematic overview of the process flow for the production of an active flexible opto-electrical foil.

The cavities are laser ablated (KrF Excimer laser; 248 nm wavelength) because of the high-accuracy-needs for the dimensions of the cavity. The cavity must be obviously larger than the die, but cannot be too large because the cavity must perform passive alignment of the die. This way a 10μm positioning accuracy of the die can be obtained. Fine pitch die placers can even lower this accuracy to several microns. In between the under cladding layer and the PI layer, a metal island is deposited. This acts as a heat sink for the active components and also as a laser stop. This way we achieve good depth control of the cavity and fast processing.

The active components are mounted inside the cavity with a low temperature curable adhesive. The adhesive needs to be heat conductive but not electrical conductive and needs to show very low viscosity to fill the whole cavity with a thin adhesive

978-1-4244-4722-0/09 $25.00
© 2009 IMAPS-ITALY

film. To reach the high alignment requirements and coupling efficiencies, it is necessary that the die is perfectly leveled with the substrate surface and not tilted. In the next step the die is covered with a cladding layer of 10 μm to finish the embedding of the die. This layer has proven to be well planarised, meaning that the cavity principle works well. Laser ablated micro-via's are then made to the embedded contact pads and metalized. Metallization is needed to fan out the small pitch contact pads of the embedded VCSEL and Photodiode array's (250 μm pitch) towards larger pitch contact pads (2 mm pitch) on the substrate surface.

The production of waveguides on top of the embedded dies is done with standard lithography and an alignment error in relation to the active area of the optoelectrical components is smaller than 5 μm. 45 degrees coupling structures are fabricated externally in flexible Polyimide material and embedded in the flexible optical foil on top of the embedded opto-electronics.

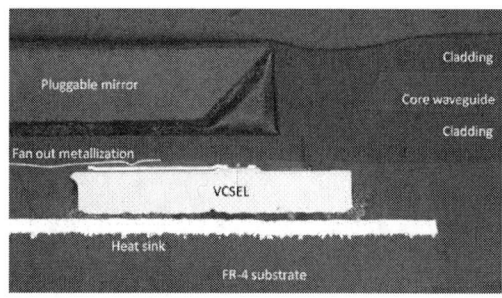

Figure 5: Cross-section of an embedded VCSEL array in the flexible optical foil.

The whole structure is then again covered with a cladding layer to cover the embedded component and to obtain final planarization. Figure 5 shows a cross-section of an embedded VCSEL array in Truemode Backplane Polymer™ together with the coupling component and the galvanic interconnection. As a demonstrator, a 2 cm long optical link between a thinned VCSEL and Photodiode was embedded inside the optical foil, which shows a total optical loss of about 5 dB for the complete link.

Embedding of IC's in the flexible optical foil

To enhance the functionality and intelligence of the flexible optical foil, Integrated Circuits (IC's) can be embedded in the layer build-up in the same way as the opto-electronic components. The embedding of IC's causes no demand for extra processing steps and the mounting of electrical and opto-electrical active devices can be done in one process step.

Figure 6: Photograph of an embedded thin (25 μm) IC in the flexible optical foil stack on an FR-4 substrate.

The thinning of IC's is done in a comparable way as the opto-electronic chips, but some inherent differences demand for some changes in process parameters and set-up. IC's are much larger in size and the base material is Si instead of GaAs for the opto-electronic devices, used in the previous described work. Because of the large size of the IC's, the dispensing of the glue underneath the chips must be well controlled. Figure 6 shows an embedded IC in the same layer build-up as the flexible optical foil, however the processing was done on an FR-4 substrate instead of a temporary glass carrier. 4 points measurements show a micro-via resistance lower than 20 mOhm.

Conclusions

The development of a series of new technologies is presented: Technologies needed to make the step from rigid opto-electronic boards towards completely flexible opto-electronic boards. First a flexible foil optical foil was realized with embedded optical waveguides, fabricated in commercially available optical transparent material, enhanced in durability by two mechanical supporting spin-on Polyimide layers. In a second phase, thinned (20 μm) VCSEL's and Photodiodes are embedded inside the under cladding layer of the waveguides. The embedded opto-electronics are electrically connected with the outer world using micro-via's, they are cooled down by the provision of heat sink features and they are optically coupled with the waveguide with embedded externally fabricated flexible coupling components. As a third phase, the functionality of the flexible opto-electronic board is enhanced by the embedded of drivers, amplifiers and intelligence IC's inside the same layer and in the same process as the opto-electronic components.

Acknowledgements

This work was supported in part by the Institute for the Promotion of Innovation by Science and Technology (IWT), Flanders, Belgium, in the framework of the project "Flexible Artificial Optical

978-1-4244-4722-0/09 $25.00
© 2009 IMAPS-ITALY

Skin" (FAOS), and in part by the EU FP7 project "Photonic Skins for Optical Sensing" (Phosfos).

References

[1] G. Van Steenberge, N. Hendrickx, E. Bosman, J. Van Erps, H. Thienpont, P. Van Daele, "Laser Ablation of Parallel Interconnect Waveguides", IEEE Photonics Technology Letters, Vol. 18, No. 9, May 2006.

[2] A.L. Glebov, M.G. Lee, K. Yokouchi, "Integration technologies for board-level optical interconnects", Proceedings of SPIE Photonics Europe, April 2006.

[3] Joseph Fjelstad, "Flexible Circuit Technology", Third edition, 2007

[4] www.bpaconsulting.com

[5] www.exxelis.com Truemode Backplane™ Polymer data sheets

[6] http://www.microresist.de Ormocer® and Epocore material datasheets

[7] Rohm & Haas: www.rohmhaas.com

[8] www.hdmicrosystems.com/tech/polyimid.html

Metal Trace Impact Life Prediction Model for Stress-Buffer-Enhanced Package

C. Y. Chou, C. J. Huang, M. Sano, and K. N. Chiang

Advanced Microsystem Packaging and Nano-Mechanics Research Lab. PME

Advanced Packaging Research Center

National Tsing Hua University, HsinChu, Taiwan 300, R.O.C.

Phone: 886-3-5742925   Fax: 886-3-5745377

E-mail: knchiang@pme.nthu.edu.tw

Abstract

In this study, the objective is to develop a proper impact life prediction model for stress-buffer-enhanced package which has a specific failure mode, the broken metal trace, during board level drop test. The so called stress-buffer-enhanced package applies a thick and soft buffer layer to protect solder joints; however, metal traces embedded inside the buffer layer are instead broken. The broken metal trace is not common for board level drop test, and there are few life prediction models to predict the performance of the package. Unlike thermal cycle test, the dynamic response of drop impact is irregular and not cyclic; therefore, the concept of cumulative damage is considered in the life prediction model. The results showed that the cumulative plastic strain of metal trace could accurately predict impact life.

Key words: drop test, life prediction, fatigue, trace failure

1. Introduction

The handheld and portable devices are widely carried out with more capabilities and functions; on the other hand, they are more prone to be dropped during their useful service life. It therefore becomes imperative to study the dynamic response of packages together with the board on which the packages are mounted. The board-level drop test is then developed to evaluate the reliability of electronic components in an accelerated test environment. In line with this, the board level drop-test regulations formed by JEDEC [1], [2] were published to assess drop performance for surface mounted IC package.

About the study of board level drop test, several researches have been published. Jenq et al. [3] investigated the high-G impact under JEDEC board level drop test regulations experimentally and numerically. The results showed that the first failure of most test boards occurred within 20 drops, and the failed packages were located near the center of test board. Tee et al. [4] proposed "Input-G method" to reduce the huge computation amount of drop test simulation. The input-G method ignores the drop table, fixture and contact surface, and instead, the impact pulse is taken as input acceleration loading at 4 screws of test board. However, few studies mentioned about the life prediction model for board level drop test. Because of the highly uncertainty of

the drop test and the lack of precise simulation model, it is difficult to develop a stable life prediction model with good experimental correlation. Tee et al. [5] proposed impact life prediction model using the maximum peeling stress of critical solder as failure criteria to determine the drop life of packages. Results showed that the characteristic impact life is a more consistent parameter compared with first-failure life, as the process and testing variation is minimized through averaging test data.

This paper discloses the development of impact life prediction model under board level drop test. The stress-buffer-enhanced package [6]-[8], which has high reliability for both drop test and thermal cycle test, is chosen as the test vehicle of the prediction model. For a better understanding, Fig.1 illustrates the sketch of the stress-buffer-enhanced package. The IC chip is attached on the chip carrier. Filler polymer is applied around the chip, disclosing the package's capability of fanning out connections. Afterward, the redistributed interconnects are laminated on a soft and thick dielectric layer. The whole package is finally mounted on a PCB board by solder joints. The thick dielectric layer provides a stress buffer to protect solder joints from mechanical deformations. The board level drop test for this stress-buffer-enhanced package was accomplished both experimentally and numerically. After correlating the experimental data, the impact life

978-1-4244-4722-0/09 $25.00
© 2009 IMAPS-ITALY

prediction model was constructed as described in the following sections.

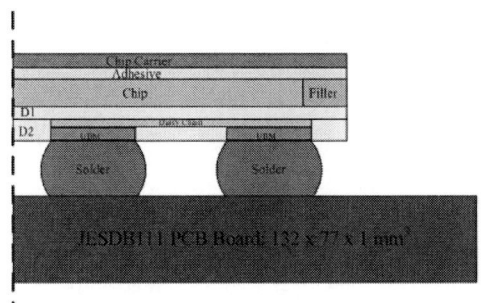

Fig.1: Illustration of stress-buffer-enhanced package

2. Board Level Drop Test Experiments and Results

The drop test experimental setup is illustrated in Fig.2. 15 stress-buffer-enhanced packages (5mm x 4.5mm) were mounted on standard JEDEC drop test board (132mm x 77mm x 1mm). In this study, JESD22-B111 drop test condition B was adopted. The peak acceleration, pulse duration and drop height are 1500 G, 0.5 ms and 112 cm respectively.

The drop test result is shown in Fig.3; from this, several phenomena different from foregoing board level drop test studies were observed. The first failure occurred after 129 drops at the corner chip of PCB (U1), while most chips survived 240 drops. The proposed stress-buffer-enhanced package reveals a much better reliability of drop test compared with conventional wafer level package (WLP). In addition, the position of first failure varied from that of previous studies [3] which was located near the center of PCB (U3, U8, U13). The drop test analysis for conventional WLP disclosed the solder joint cracks along intermetallic compound (IMC) layer as the main cause of failure; in this study, however, broken metal lines in packages instead of solder joint fractures were perceived to be the reason of failure. It was thus presumed that the soft stress buffer layer absorbed the impact, thereby ensuring that the solder joints are protected. Nevertheless, the metal lines within stress buffer layer suffered relatively larger deformation.

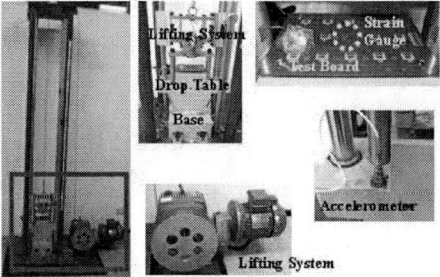

Fig.2: Drop test experimental setup

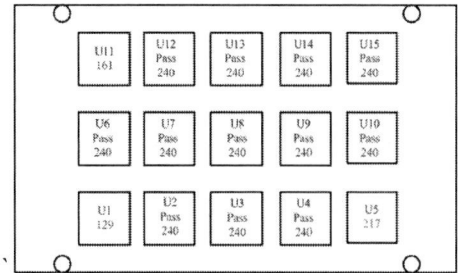

Fig.3: Drop test result

Because the failure mode of stress-buffer-enhanced package focused on the broken metal lines, the configuration of trace layout becomes a significant factor affecting the impact reliability of this package. Therefore, two kinds of trace layout, the straight trace and the curved trace, were then experimented. The broken metal line failures were recorded as shown in Fig.4. And the experimental results were listed in Table1. In Table1, P1 represented packages pass 90 drops, while P2 represented packages pass 240 drops. Interestingly, the failure occurred in the order of U1, U11, U5, and U15. It is resulted from the asymmetric configuration of solder joints and trace layout, shown in Fig.5. Figure5 shows the straight trace layout which has sparse solder joints at bottom-left corner. The U1 package has relatively the shortest distance between the sparse solder joint area and the fixed screw. As a result, the impact wave from the screw attacks the weak sparse area directly, causing the first failure of U1 package. Therefore, in this research, we assume that every position of the packages, i.e. U1, U2, ..., U15, should be treated as an individual experiment during board level drop test, even though they are all mounted on the same test board.

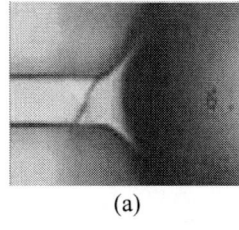

(a)

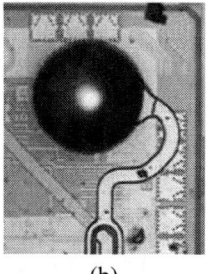

(b)

Fig.4: Failure analysis of broken metal line (a) failure of straight trace (b) failure of curved trace

Table1: Experimental results of board level drop test

Straight Layout	U1	U2	U3	U4	U5	U6	U7	U8	U9	U10	U11	U12	U13	U14	U15
Board#1	58	P1	47	P1	31	P1	P1	P1	P1	P1	25	P1	26	66	26
Board#2	67	P1	72	P1	39	P1	P1	P1	P1	P1	29	85	36	75	25
Board#3	48	72	53	P1	46	P1	P1	P1	P1	P1	20	P1	26	46	21
Board#4	71	P1	53	P1	42	P1	P1	P1	P1	P1	18	P1	29	41	25
Curved Layout	U1	U2	U3	U4	U5	U6	U7	U8	U9	U10	U11	U12	U13	U14	U15
Board#1	116	P2	138	P2	P2	P2	P2	P2	P2	P2	151	P2	P2	P2	196
Board#2	129	P2	P2	P2	217	P2	P2	P2	P2	P2	161	P2	P2	P2	239
Board#3	102	P2	166	P2	127	P2	P2	P2	P2	P2	102	P2	P2	P2	221

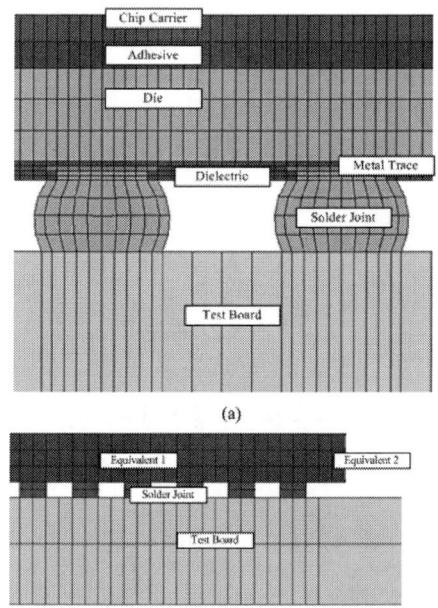

Fig.5: Relations of trace layout and board position

3. Drop Test Simulation and Experimental Validation

From the aforementioned results, the experimental data revealed that every package on the test board should be treated as an individual experiment, because the dynamic responses of the packages are all different. In the finite element modelling, the quarter model of the test board with 15 packages mounted on it was constructed using FEA software LS-DYNA, shown in Fig.6. The input-G method [4], which applied experimental acceleration on the fixed screw as input loading, was adopted in this research. Several simplifications were applied to reduce the extremely large calculation of the finite element model. For the correlation package, i.e. package U15, the detailed finite element model was built, while simplified finite element model was applied for the rest of the packages, i.e. packages U8, U9, U10, U13, and U14. Figure7(a) shows the detailed model include the chip carrier, adhesive layer, die, filler polymer, dielectric layer, metal trace, solder joint, and test board. Figure7(b) shows the simplified finite element model; the solder joint was considered as a rectangular column. Besides, the complicated structure of the stress-buffer-enhanced package was replaced by equivalent material obtained by rules of mixture [9]. Equivalent1 layer represents the replacement material for the die region, and equivalent2 layer represents for the filler polymer region.

The confirmation of the equivalent model is shown in Fig.8, representing the strain propagation of test board. Results show that the strain variation between complete model and equivalent model is within acceptable range; however, the computational time can greatly reduce from 438 hours to 57 hours. Similar results were also proposed by Yeh et al. [10].

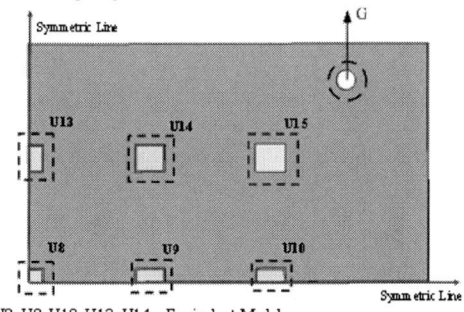

U8, U9, U10, U13, U14 – Equivalent Model
U15 – Trace Model

Fig.6: Finite Element Modeling of Test Board

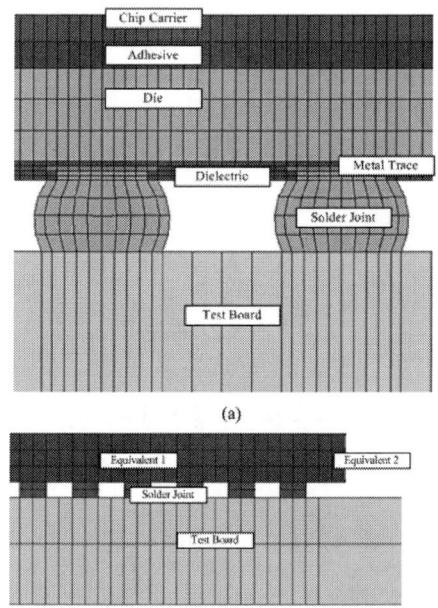

Fig.7: (a) Illustration of trace model, (b) Illustration of simplified model

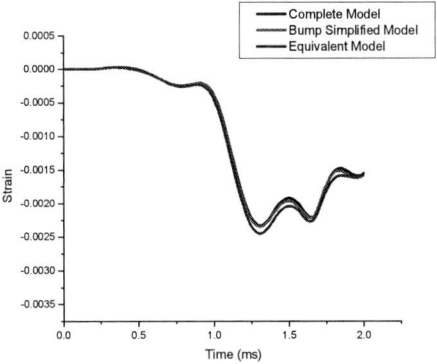

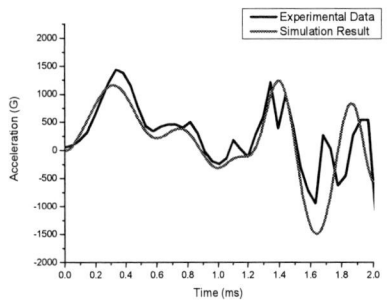

Fig.10: Validation of acceleration at package U15

Fig.8: Confirmation of equivalent model

For ensuring the accuracy of dynamic simulations, some experimental validations were performed. The response of longitudinal strain at the center of PCB where the strain gauge was attached at the same position was used to validate the simulation results, shown in Fig.9. The validation of experimental and simulation results are within reasonable range. Subsequently, the acceleration of package U15 (package side) was further validated because package U15, the detailed model, was chosen as the correlation of life prediction model. The more accurate of the simulation result of package U15, the more reliable of the impact life prediction model. Figure10 shows the acceleration validation of package U15. The validation shows good agreement between experimental data and simulation result. Interestingly, the first peak of the acceleration at package U15 is almost the same with the impulse of the input loading, i.e. 1500G. Based on the simulation results, the development of life prediction model was made as described in the following section.

4. Metal Trace Impact Life Prediction Model

In order to quantitatively estimate the impact performance of packages, the impact life prediction model is developed in this section. Five drop experiments, 2 straight trace layout and 3 curved trace layout, were selected to construct the related finite element model as shown in Fig.11. According to the aforementioned experimental results, every position of the packages, i.e. U1, U2, ..., U15, should be treated as an individual experiment during board level drop test due to their asymmetric trace layout and solder joint distribution. Case1 and case2 represented models of straight trace layout at different position, U1 and U15. Case3, case4 and case5 represented models of curved trace layout at different position; especially, case5 has a thinner dielectric layer. Besides, in order to decrease the computation amount, all the five cases only have one critical trace where the failure was observed in drop experiments, shown in Fig.11.

All the simulations were conducted from 0ms to 6ms, which represented the period of the fundamental natural vibration of test board. The simulation results are shown in Fig.12. Results show that larger strain concentrated at the neck area of metal trace and pad. Besides, the effective plastic strain propagations on traces are shown in Fig.13. The strain propagations would be the foundations of the impact life prediction models.

Fatigue fractures are the most commonly identified kinds of failure of structural metals. It is defined as the phenomenon of failure of a material under cyclic loading. However, the dynamic responses of board level drop test, i.e. the strain propagation of metal trace, are irregular and not repeated. The conventional Coffin-Manson relation [11] may not be suitable for the impact life prediction. According to the cumulative damage theories [12], [13], the fatigue process is considered to be a process of damage accumulation until a certain maximum tolerable damage. In other words, fatigue is an exhaustion process of a material's inherent life. Therefore, this paper developed impact

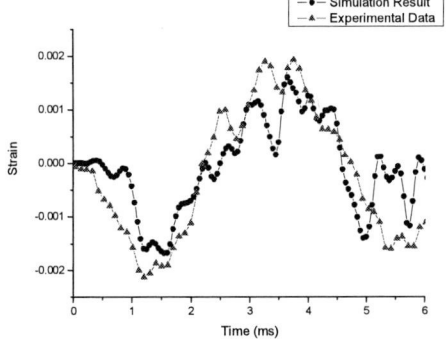

Fig.9: Validation of strain at center of test board

978-1-4244-4722-0/09 $25.00
© 2009 IMAPS-ITALY

life prediction model, based on the cumulative damage theories.

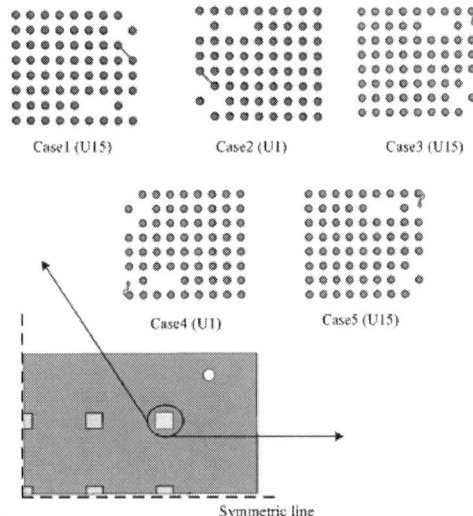

Fig.11: Illustration of correlation models

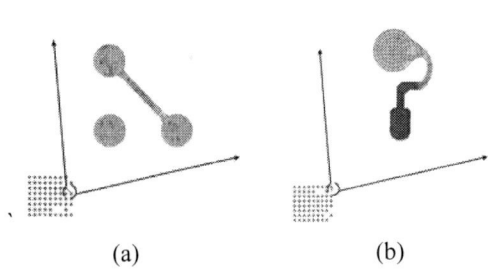

(a)   (b)

Fig.12: Strain distribution of metal traces (a) straight trace (b) curved trace

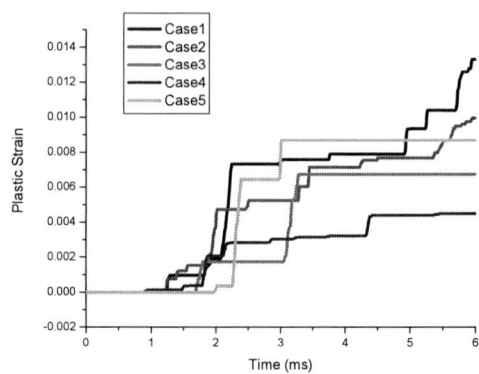

Fig.13: Maximum effective plastic strain propagation of metal traces

The impact life prediction models proposed in this study is related to the cumulative plastic strain, shown in Eq.1. In Eq.1, $\Delta\varepsilon_{pl}$ represents the incremental plastic strain in every time step; C_1 and C_2 are correlation parameters.

$$N_f = C_1 \left(\sum_{t=0ms}^{t=6ms} \Delta\varepsilon_{pl} \right)^{C_2} \quad (1)$$

Following, case1, case2, and case3, were chosen to estimate the correlation parameters, C_1, and C_2. After correlations, C1 and C2 are determined to be 57.64 and -2.049 respectively. Case4 and case5 were chosen to be the predicted models using the anterior correlation parameters. Table2 list the prediction results. The characteristic life in Table2 was obtained using Weibull distribution. Results show that there is a good correlation for impact life prediction model. Accordingly, the cumulative plastic strain is suitable for the impact life prediction model, because the failure of drop test belongs to low cycle fatigue which is dominated by plastic strain.

Table2: Metal trace impact life prediction results

	Trace Layout	Position of Package	Dielectric Thickness (μm)	Experimental Characteristic Life (Drops)	Predicted Characteristic Life (Drops)
Case1	Straight Trace	U15	40	24.82	------
Case2	Straight Trace	U1	40	64.77	------
Case3	Curved Trace	U15	40	128.46	------
Case4	Curved Trace	U1	40	234.21	298.7
Case5	Curved Trace	U15	30	72	77

Through the proposed impact life prediction model, one can determine the drop impact performance of a specified package quantitatively. This approach provides both accurate simulation comparison and absolute impact life prediction. Moreover, designers may execute computational experiments to greatly reduce expansive experiments, and accelerate time-to-market procedures. However, similar to most finite element models, the value of plastic strain obtained from dynamic simulations depends on the element size and other modelling assumptions. The uncertainties of board level drop test are also affecting the reliability of this impact life prediction model. There are no universal correlation parameters; similar situations occurred in thermal cycling predictions. For more accurate impact life predictions, it is suggested that both dynamic simulations and drop experiments need to be detailed considered.

5.   Conclusions

This study intended to develop a reliable impact life prediction model for metal traces which becomes a must to estimate the performance of packages subjected to drop impact. Several LS-DYNA simulations were conducted to elucidate the mechanical behavior of test board and packages. After validating the simulation results with experimental data, the development of impact life prediction model were then accomplished. The impact life prediction model was proposed using the cumulative plastic strain as failure criterion related to impact life. Results showed that the cumulative plastic strain is suitable for the impact life prediction model, because the failure of drop test belongs to low cycle fatigue which is dominated by plastic strain. There is good correlation between impact life predicted by simulations and measured by experiments. This approach provides both accurate simulation comparison and absolute impact life prediction. Based on the impact life prediction model, designers may execute more computational experiments to accelerate time-to-market procedures.

Acknowledgements

The authors would like to thank the National Science Council NSC97-2222-E-007-007-MY3 and Advanced Chip Engineering Technology Inc. for supporting this research.

References

[1] JEDEC Solid State Technology Association, JESD22-B110: Subassembly Mechanical Shock, 2001.

[2] JEDEC Solid State Technology Association, JESD22-B111: Board Level Drop Test Method

of Component for Handheld Electronics Products, 2003.

[3] S. T. Jenq, H. S. Sheu, C. L. Yeh, Y. S. Lin, and J. D. Wu, "High-G drop impact response and failure analysis of a chip packaged printed circuit board," International Journal of Impact Enginnering, Vol. 34, pp. 1655-1667, 2007.

[4] J. E. Luan, T. Y. Tee, E. Pek, C. T. Lim, Z. Zhong, and J. Zhou, "Advanced numerical and experimental techniques for analysis of dynamic responses and solder joint reliability during drop impact," IEEE Transactions on Component and Packaging Technologies, Vol. 29, pp. 449-456, 2006.

[5] T. Y. Tee, H. S. Ng, C. T. Lim E. Pek, and Z. Zhong, "Impact life prediction modeling of TFBGA packages under board level drop test," Microelectronics and Reliability, Vol. 44, pp. 1131-1142, 2004.

[6] C. Y. Chou, T. Y. Hung, S. Y. Yang, M. C. Yew, W. K. Yang, and K. N. Chiang, "Solder joint and trace line failure simulation and experimental validation of fan-out type wafer level packaging subjected to drop impact," Microelectronics and Reliability, Vol. 48, pp. 1149-1154, 2008.

[7] M. C. Yew, C. Y. Chou, and K. N. Chiang, "Reliability assessment for solders with a stress buffer layer using ball shear strength test and board-level finite element analysis," Microelectronics and Reliability, Vol. 47, pp. 1658-1662, 2007.

[8] M. C. Yew, C. Y. Chou, C. S. Huang, W. K. Yang, and K. N. Chiang, "The Solder on Rubber (SOR) Interconnection design and its reliability assessment based on shear strength test and finite element analysis," Microelectronics and Reliability, Vol. 46, pp. 1874-1879, 2006.

[9] R. F. Gibson, "Principles of Composite Material Mechanics", second edition, Florida, USA, CRC press, 2007

[10] T. Y. Tsai, C. L. Yeh, Y. S. Lai, and R. S. Chen, "Transient submodeling analysis for board-level drop test of electronic packages," IEEE Transactions on Electronics Packaging Manufacturing, Vol. 30, pp. 54-62, 2007.

[11] L. F. Coffin, Jr., "A Study of the Effects of Cyclic Thermal Stress on a Ductile Metal," ASME Transactions, Vol. 76, pp. 931-950, 1954.

[12] A. Palmgren, "Endurance of ball bearings," Z. Ver. Deuts. Ing., Vol. 68, pp. 339-341, 1924.

[13] M. A. Miner, "Cumulative damage in fatigue," ASME Transactions, Vol. 67, pp. 159-164, 1945.

978-1-4244-4722-0/09 $25.00
© 2009 IMAPS-ITALY

3D Packaging and Supply Chain Management

Paul Collander

Poltronic Ltd., Espoo, Finland & HDPUG, Scottsdale, AZ, USA

Phone +358 400 608074, paul@poltronic.fi

Abstract

As integration density on chip level is slowing down the importance and added value of advanced packaging is rising. Even semiconductor manufacturers admit that the way forward is to work on packaging and interconnect.

One of the promising solutions is to go to 3D packaging but most of these technologies have been applied slower than anticipated and pioneering companies may run into financial trouble. One additional reason for these problems are the growing supply chain complexities, isolating material suppliers, package developers and producers from end product manufacturers like OEMs. Globalization and outsourcing add to these problems.

This paper will look at some 3D packaging technologies, analyzing some solutions and implementations including some viewpoints on how they fulfil the requirements of system houses. We will also stress the importance of technology organizations like IMAPS and HDPUG and how they will bridge gaps in the supply chain knowledge.

Indeed package developers need to assure a customer base for their developing work, but that is not enough! Competing suppliers need to coordinate efforts to assure they fit in the global community and OEMs need to assure they have multiple sources to rely on and that the suppliers' development expenses are shared among multiple OEMs! Networking is needed both up and down the supply chains but also laterally between competitors.

Key words: 3D packaging, embedded passives, TSV, supply chain management

Introduction

Integration in semiconductors has been the main road to miniaturisation up to recently. However, physical limits start to play an increasing role for slowing down IC technology development and simultaneously increasing investment needs for each IC technology generation.

As investments in Packaging and Interconnects have always been orders of magnitude lower than in IC processing there is still a lot of room available for improvements on this level. A die package contains still more fan out area than die area to make second level assembly easier.

At the same time the field of packaging is occupied by many small players, none of whom have much room for individual creativity and development without considering the rest of the players up and down the supply chain.

Reducing dimensions tend to have a negative impact on yield and cost per area. Therefore was the step to Quad packages from Dual in lines as big as the step from Quad to Area Array. Miniaturisation was achieved by moving to a new concept. Within each concept, incremental miniaturisation is possible if board level technology and manufacturing sites are updated.

Why Go 3D?

All electronics manufacturing has for a long time been 2-dimensional. When the 2 dimensions are well enough utilised a natural but radical step is to utilise the 3rd dimension. The concept of moving from 2D to 3D is only starting to see volume production as multiple players feel the impact on tools, materials and process development. The change is more radical than the incremental improvements and it creates new pressure on both design logic, product partitioning into modules and on all handling and inspection methods.

Naturally the impact on miniaturisation can be radical as components and chips are moved atop each other instead of aside each other. In best case also the interconnections are greatly simplified by going 3D, as in a case of stacked memory chips.

The impact on the size of small hand held gadgets is however minimal as the volume is often to be miniaturised, not the area. Here the secondary benefit may be driving; close 3D spaced chips with simple and short interconnects can improve electrical performance due to smaller resistance, capacitance and inductance in interconnects.

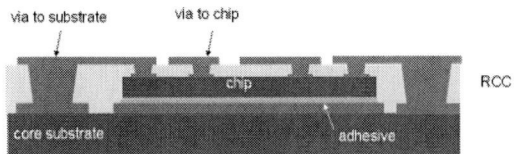

Figure 1. Embedding a die according to Hiding Dies project concept. Board area can be used for other components. [1]

3D Packaging Consortia

As so many technologies have been suggested for 3D integration and so many players are dependent on each other, a lot of 3D consortia have been established around institutes or applications or solutions. These consortia are essential in taking ideas all the way to sustainable production. But this implementation process is difficult also for consortia as so many parameters need to be fit together and consortia members typically have some competing values to defend. Some members are competing against each other for business; others differ in case of application demands or other issues.

It has already been found that 3D may be a solution for very high volume and low cost applications on one hand, but simultaneously also for low volume high demanding military or extreme high speed performance applications. Although 3D constructions for these applications may look similar, the solutions may differ radically.

Among most demanding tasks for such consortia is to estimate manufacturing costs of non existent processes, or even worse, estimate OEM's cost of ownership of final products based on a variance of technologies and number of players. Manufacturing cost in such a network is impossible to forecast accurately as expenses are radically dependent on total volume and utilization level of concerned subcontractors. Therefore, any outstanding solutions are necessarily expensive and keeping in the mainstream is of high priority. This strategy is not pushing for development, except in perfect orchestration with suppliers, customers and competitors.

The best consortia are those giving full value to all players from materials to OEM specifications and being able to communicate the impacts of each player on the others.

Fraunhofer IZM

Fraunhofer Geshellshaft is the biggest research organisation in Europe but their research is spread on many very different topic areas and electronics is only one. Electronics is mainly taken care of in IZM, Institute Zuverlässigkeit und Microintegration. IZM is divided in 9 departments,

one of which is High Density Interconnect & Wafer Level Packaging. This is the group for 3D packaging. The most frequent speaker there is Jürgen Wolf. He is deeply involved in nanotechnologies for interconnects, narrow and high Cu pillars for stacking chips, re-distribution metal layers on wafers, different kinds and sizes of TSVs and system interconnect design taking electrical and thermal performance into consideration. The 3D activity is more or less an extension of the long term work on "Hetero System Integration".

IZM is, however, a complex organisation and packaging and interconnect programmes are cross department activities. On IZM level the R&D areas are listed as:

> 3D-System Integration
> Large Area Electronics
> MEMS Packaging
> Micro Reliability and Lifetime Estimation
> Photonic Packaging
> RF & Wireless
> Thermal Management
> Wafer Level Packaging
> Sustainable Technical Development

The active role of industrial players in their 3D development is less apparent, except for providing application specifications and subcontracted material development and services.

IMZ has been IC centric but the 3D packaging is reviewed as a solution away from too expensive to design SoCs to improved electrical performance in stacked dies, packages or substrates.

IMEC 3D programme

IMEC in Belgium is probably the biggest European research institute on electronics. Their three biggest areas of research are Circuit Design (including all semiconductor materials and organic electronics), Packaging and Photovoltaic. The Packaging and Interconnect R&D is headed by Eric Beyne and is composed of three main areas:

> * 3D integration
> * Multilayer thin-film packages
> * Flexible and stretchable electronics

Of these is the 3D programme the largest. IMEC has been pushing vocabulary and classification of 3D interconnects based on JISSO levels, thus aiming at taking a grip on the myriad of alternatives offered for solutions.

IMEC says clearly that 3D packaging is very different from 2D. Both supply chain, responsibilities, priorities, functionality and design architectures are different. The aim is to reduce chip cost by limiting SoC complexity using in stead commodity chips adding performance and customisation on packaging level only! Chips are specialised on different operations (memory, processor, DSP, RF) but combined in one package. Short 3D interconnects may improve functionality and save energy. Modules are fully tested before

978-1-4244-4722-0/09 $25.00
© 2009 IMAPS-ITALY

board assembly. The main drawback of modules may be yield loss as function of number of chips.

3D packaging demands a new design architecture and policy. The supply chain becomes more complex including fuzzy boundaries between players. Management of chains puts higher pressure on OEMs. But savings in SoC cost and design time should overcome management problems when the 3D concept has been introduced and experience is building up.

IMEC is pushing for TSV to enable direct stacking of either identical or different kinds of chips but are indicating 3D modules can be built without interfering with chip- and wafer design (to enable TSV) but using traditional interconnect methods like FlipChip and Wirebonding.

IMEC is active on Technology Roadmaps for 3D as their vision is that nomenclature, classified solutions and players' interfaces all need some standardisation to enable utilisation and benefiting of the new concepts. Through these discussions players should find and refine their niches and boundaries.

High Density Packaging User Group

The HDPUG group is a consortium of some 30 industrial companies globally spread on all continents and representing OEMs, their suppliers, PWB and component manufacturers and material suppliers. The companies are playing with equal rules although the top of the supply chains are setting the final goals. Activities are organized in projects, run by member company personnel and focused on assessing newly introduced technologies, materials and packages. The activities are strictly industrial and new development is limited to a minimum in order to limit IP issues. In fact in most cases the aim is that any member company could introduce or benefit from the results immediately after the project.

The environmental issues like Pb-free and Halogen free have been the basis of most recent projects. Both reliability testing methods and materials tested have provided project participants hands-on experience to be utilised in products and processes.

To this end 3D packaging has not been a hot issue in this telecom and computer dominated consortium. However, with a conscious move back to the original high density packaging, the group started 2008 to get involved in a 3D System in Package project divided in three branches:
1) A 2D 4 chip RF module is re-designed and realised as a much smaller 3D module having all 4 chips stacked on top of each other, using simultaneously PWB embedded chip, FlipChip and piggy backed wirebonded chip. This demonstrator tries to show that these new technologies are viable and industrial.

2) A joint technology roadmap analysis reviewing technologies and solutions from all different supply chain levels, materials to OEMs. This analysis is used as an internal guide for the demonstrators and will lead to a guideline for members, including suggestions for where to use which technologies and what roadblocks to avoid.
3) A second demonstrator proofing if and how Integrated Passive Devices can be used for improving electrical performance when passives are located close to active chip pads.

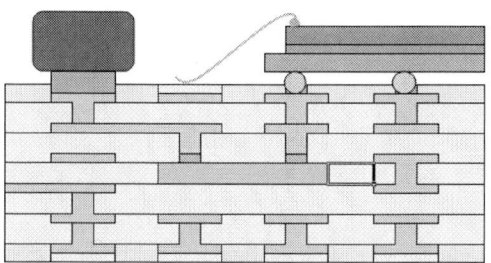

Figure 2. HDPUG planned SiP test vehicle. One chip embedded, one FlipChipped and one back to back wirebonded.

The Embedded Die SiP project is scheduled to provide results in summer 2009 and finished by the end of the year. This swift schedule is typical for our projects as we are not really creating new technology but we implement jointly existing bits and pieces. Having material developers, process owners, board manufacturers and EMS providers working together erases many question marks each player would have working alone. Our typical phrase is that not only working up and down the supply chain is a must but additionally we need collaboration between competitors on each level. We have taken years of building up confidence between member companies and their employees to achieve competitors to collaborate.

Stacked Die Constructions

For minimal board space, placing chips on top of each other is probably the best solution. Indeed there are tens of widely diverse constructions, from Package on Package, PoP, to directly stacked chips interconnected through Through Silicon Vias, TSV.

Not only the technologies are very different but also the industrial work load is spread very differently. In recent years the industrial community has been optimizing workload sharing in the supply chain for minimal cost of ownership for the OEMs and their first level suppliers. Outsourcing has had an impact on all players. In 2009 a new, or at least previously forgotten mechanism has changed all roles and shares of duty: The economic crisis has pushed OEMs and their suppliers to take home surprisingly large amount of outlaid work in order to protect their employees from being laid off. The trend in last decade of dividing the supply chain into numerous thin layers of highly specialised independent companies have suddenly seen slices merged upwards saving the employees of the upper end. Question is, will these in the end manage to produce what specialists have provided them so far?

Module Manufacturing Infrastructure

For decades have the module manufacturers been a middle man between distant IC fabs and system integrators, being both geographically and volume vice closer to the system houses. However, especially in Europe this trend has come to an abrupt end already years ago. The number of (multichip /hybrid) module manufacturers in Europe has probably dropped by 80 % during last few years. This trend has taken place due to the fierce financial pressure applied by OEMs on all their suppliers, taking module manufacturers to low labour rate countries in the far east. They have simultaneously benefited from the fact that more and more material and component manufacturers are based in Asia. The collaboration in this end of the supply chain has been improved leading to similar loss of interaction in the upper end of the chains.

The newest trend of OEMs taking home manufacturing from subcontractors may thus be more difficult than only few years ago.

Product Responsibility Issues

Subcontracting has always comprised a certain amount of delegating responsibility and in many cases this delegation has been a difficult job, especially when non-mature technologies of unknown cost structure and yield have been employed. The problem is largely to estimate new processes and their cost and yield while still only demonstrated on laboratory level. The problem is multiplied by separate estimates at each level in the supply chain adding uncertainty to each level and even more on the sum. In principle each supplier carries responsibility for his own process in a stable environment, but the request is to also carry responsibility in a changing world.

Intellectual Property, IP, plays a double role here. On one hand manufacturers are not able to fully reveal his own process parameters to his suppliers and customers due to sensitive IP. On the other hand each manufacturer receives processed materials or components from his supplier upstream and may ruin part of that material as yield loss. The situation may be worst in case of expensive chips being included in still developing 3D processes. Who covers the cost of miss-processed chips?

Supply Chain Management

The essential here is to realise that each player in a developing supply chain is to some degree biased, believing (too much) in his proposed process. To overcome the risk of blindness a consortium based on multiple players, representing both different steps in the supply chain and alternative, parallel solutions, may be very well worth the effort of collaboration.

Conclusions

In spite of the great opportunities of 3D integration the complex manufacturing hinders a rapid development and implementation of such solutions. The paper tries to stress the fact that here we are not only looking at a bunch of technology challenges and problems but even more at a complex industrial network that need concerted actions forward. Only by working together can we get to the promising goals of really high density packaging in an affordable and predictable way. Here is the challenge for EMPC and similar events to bring the pieces together in spite of travel bans and suppliers going bankrupt.

Information sources

3D packaging is developing fast and there are so many consortia involved that at least to find a starting point Internet is recommended for searching.

Some outstanding sources with some neutrality are:
Phil Garrou's blog at
http://www.semiconductor.net/blogger/3068.html
Advanced Packaging at http://ap.pennnet.com/
Françoise von Trapp blog at
http://francoisevontrapp.blogspot.com/

Other sources:
www.imaps.org
www.imapseurope.org
http://www.imaps.org/programs/gbc09spring.htm
www.hdpug.org
http://www.imecevents.be/program12.aspx
http://www.emc3d.org/
http://www.prc.gatech.edu/events/3dassm/
http://www.jiep.or.jp/icep/

References

[1] A.. Ostmann, Fraunhofer IZM, "Industrial and Technical Aspects of Chip Embedding Technology", IMAPS & IEEE ESTC2008 conference, September 2008, London, UK

[2] P. Collander, "3D packaging technologies and applications", IEEE CPMT Finland. Seminar 20.11.08 Imbera, Espoo, Finland

[3] P. Collander, "3D packaging technologies and applications", IMAPS India EMIT08 conference, Bangalore, India 16.12.08

[4] International Conference on Electronics Packaging, ICEP, Kyoto, Japan, April 14 – 16, 2009. Sponsored by Japan Institute of Electronics Packaging (JIEP) and IEEE CPMT Society Japan Chapter

978-1-4244-4722-0/09 $25.00
© 2009 IMAPS-ITALY

The necessity of corrosion protection for solderable pure tin deposits on IC outer leads

Jürgen Barthelmes, Paolo Crema*, Peter Kühlkamp, Olaf Kurtz

Atotech Deutschland GmbH, Erasmusstrasse 20, 10553 Berlin, Germany

*ST Microelectronics, Via C. Olivetti 2, 20041 Agrate Brianza, Italy

+49-30-34985746, juergen.barthelmes@atotech.com

Abstract

Due to its ease of application pure tin has been selected by the industry as the main alternative to solderable tin/lead deposits following RoHS legislature. After several years of industry experience several issues have developed which are due to the corrosion of the top layer during storage. The most serious of effects has been the growth of whiskers during heat/humidity (55°C/85% r.h.) storage. In the vicinity of exposed Copper, at the dam bar or toe cut, whiskers appear after an induction period, whose lengths may exceed the tolerable limit after 4000 hour storage. Another deficiency of equal concern is the deterioration of solderability after long term storage or, as an equivalent, after steam ageing. An improvement of these defects will be achieved by applying a corrosion resistant post-treatment, which has to combine corrosion preventive properties with being able to withstand the reflow procedure, while still being present on the surface afterwards to protect the deposit from oxidation. Whisker and solderability studies show the effectiveness of these new coatings. Solderability after heat/humidity storage can be considerably improved and even at low temperatures excellent wetting is obtained. TOF-SIMS (time-of-flight secondary ion mass spectroscopy) is being used to clearly demonstrate that these coatings will withstand multiple reflow operations. In the search for effective corrosion protection, the hydrophobicity of the final tin layer is being investigated. Furthermore this water-repellent property has to be maintained throughout prolonged contact with water. Thus a clear indicator for a compound to be suitable as a protective layer is its property to avoid discoloration during steam ageing.

Key words: IC, corrosion, whisker, solderability, discoloration, TOF SIMS

Introduction

Since the start of its introduction as a lead-free solderable finish the whisker propensity and the solderability of matt pure tin layers have been the major concerns for its implementation in IC assembly. Whilst the other main alternative – preplated leadframes with a thin Nickel-Palladium-Gold coating – suffers from the disadvantages of unpredictable, fluctuating cost and reduced reliability, considerable additional effort is being given to further improve the pure tin surface solderability, while reducing its tendency to grow whiskers.

On IC outer leads the heat/humidity storage (55°C at 85% rel. humidity) is now being rated the most critical for whisker growth [1]. The formation of these protrusions from the surface finish during temperature cycling or room temperature storage is either not significant or it can be mitigated by the common practice of a postbake or temperature annealing [2].

Comparing laboratory samples to singulated IC units after trim and form clearly shows the detrimental effect of corrosion for whisker growth in the vicinity of exposed copper during hot and humid conditions.

Figure 1: Corroded IC lead tip

Figure 2: Whisker growth near dam bar cut

As a model for corrosion induced whiskering serves the excessive formation of tin oxide at elevated temperatures and humidity. This new surface layer of lower density will exert additional pressure onto the underlying tin deposit. The more noble Copper will accelerate this oxidation, explaining why in IC assembly these whiskers are found near open Copper areas.

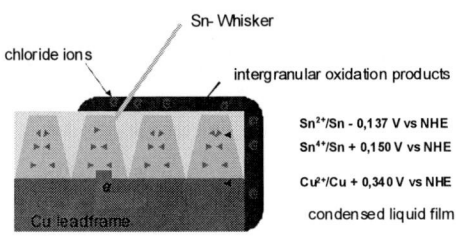

Figure 3: Corrosion whisker model

The corrosion theory depicted in figure 3 has been supported by investigations with different levels of ionic contamination present on the tin surface prior to storage. It was clearly shown that ICs artificially contaminated by sulphate and chloride grow substantially more and longer whiskers during in hot and humid environment [3].

The corrosion of tin surfaces also has an unfavourable influence on the solderability of these systems. The negative effect of long term storage is often associated with humidity, the reason why it is simulated by applying the steam ageing or pressure cooker test. Under these conditions matt tin layers have a particular weakness due to its larger surface area and thus the more pronounced formation of non-solderable oxides. This disadvantage is easily detected after heat/humidity storage through the appearance of a darkened, tarnished or, in other words, discolored surface. Any tin surface which retains its clean and white appearance will exhibit good solderability after storage.

Improvement of Corrosion Resistance

Following the alternatives known for other metals (i.e. Zn), the corrosion resistance can be enhanced by an alloying element, which may or may not be metallic, or by applying a post-treatment, which is then called passivation or sealer. The former option has been excluded, since the known alloys of tin contain either hazardous elements (Pb, Bi) or are difficult to handle and expensive (Ag, Cu, Bi). The main prerequisites for any solution were:

retained solderability
applicability in existing equipment
economical cost

The novel solution recently developed replaces the conventional neutralization step, which follows directly after the electroplating of the tin layer, by an immersion process, during which a thin protective and water repellent organic coating is applied onto the tin surface. The process sequence is summarized in Figure 4.

Temp.°C	Process Step
25-50	**Tin Plating**
RT	Rinse
RT-50	**Organic Coating** (Neutralization)
RT	Rinse
	Dry

Figure 4: Plating sequence for corrosion resistant tin surfaces

Solderability

This final finish will cover the metal surface within a matter of a few seconds. The solderability of the freshly plated units will not be negatively affected, as can be seen in Figure 5.

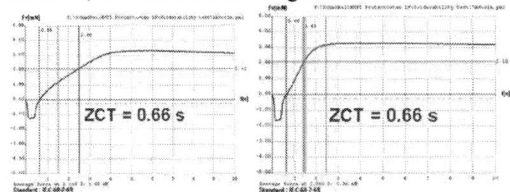

Figure 5: Solderability comparison of as plated samples by wetting balance measurement; left: conventional neutralization, right: organic coating; non activated flux, T=245°C

However and as expected by its intended function, the solderability after steam ageing of an organically coated sample will be strongly improved. Less tin-oxide is formed during the the hot and humid storage, leading to faster wetting and higher wetting forces. Figure 6 quantifies the described phenomenon.

978-1-4244-4722-0/09 $25.00
© 2009 IMAPS-ITALY

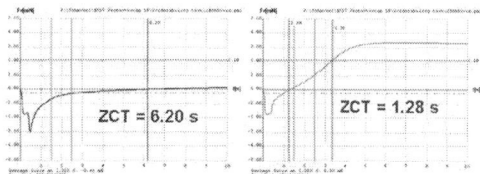

Figure 6: Solderability comparison of pressure cooker (8 hours) aged samples by wetting balance measurement; left: conventional neutralization, right: organic coating; non activated flux, T=245°C

Whisker Mitigation

During the development of the protective post-treatment suitable agents were selected according to their rating in the so-called Kesternich test (DIN 50018). While the conventional alkaline neutralization showed s2 or 10-25% corroded tin after 72 hours, the best organic top-coat exhibited a result one order of magnitude better (s5, 1-2.5% corrosion).

The actual whisker mitigation by this novel post-treatment was firstly tested in a production line using three different packages FW25, PDIP32 and LQFP of different base alloys FPG, C194 and C7025. Following the outline of previous examinations [3] one set of each package was artificially contaminated by dipping the units into a chloride ion containing solution. After 4000 hours storage at 55°C and 85% relative humidity the plated samples showed little signs of whisker growth. As can be witnessed in Figure 7, only the ionically contaminated samples showed critical whisker lengths of up to 40 µm.

Figure 7: Toe cut of FW25 (top row), PDIP32 (center row) and LQFP (bottom row) packages after 4000 hours heat/humidity storage (55°C/85% RH) without (left) and with (right) chloride contamination

Stability of the coating

The corrosion resistant organic coating was proven to be an effective way to improve solderability and mitigate whisker growth. For this to occur it must be able to withstand several excessive heat treatments while being adsorbed on the tin surface, namely an annealing step of one hour at 150°C and at least one lead-free reflow operation during assembly onto the PCB. This sequence incl. the whisker testing is shown in Figure 8.

Temp. [°C]	Process Steps
25-50	Tin Plating
RT-50	**Organic Coating**
	Dry
150	Annealing 1 hour
260	Reflow
55°C/85% RH	Storage of 4000 hours

Figure 8: Plating, storage and testing sequence for plated tin surfaces in electronic components

Strictly following this sequence an assembled IC package was measured using time-of-flight secondary ion mass spectroscopy (TOF SIMS) to identify the organic coating after annealing and refow. The unit was assembled on a PCB and investigated directly in the vicinity of the dam bar cut, where the corrosion whisker risk is at its maximum. Figure 9 displays the examined area.

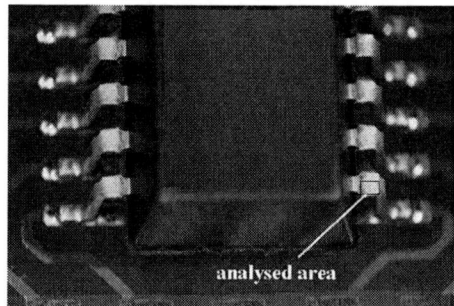

Figure 9: Area on the assembled IC package for the TOF SIMS examination, near dam bar cut

The TOF SIMS revealed the extraordinary stability of the corrosion resistant organic coating. Even after reflow at 260°C strong signals from the IC surface were obtained. The results and relative intensities are summarized in table 1.

978-1-4244-4722-0/09 $25.00
© 2009 IMAPS-ITALY

After reflow	Without coating	With coating
Main compound	-	++
Main compound as dimer	-	+

Table 1: Relative TOF SIMS intensities for the characteristic masses of the main organic compound and its dimer; measured after lead-free reflow and assembly onto a PCB; with and without coating

The search for even more effective coatings

Following the development and the implementation of the first corrosion resistant organic top-coat it is evident that a milestone improvement with respect to solderability and whisker mitigation has been achieved. However in spite of the obvious progress signs of corrosion can still be found after whisker storage. On top of that factors favouring the oxidation of the packages such as the subsequent trim&form procedure and handling errors call for even more effective solutions.

Current R&D work thus centers around strengthening the anti-oxidizing effect of the organic protection to reduce or even iliminate the visible corrosion and ease the detrimental effect of scatches by the forming tools.

As sensors for the effectiveness of a hydrophobic agent, readily measureable properties of the coatings, such as the prevention of discoloration and a high contact angle with water, are assessed.

To witness the power of a given test electrolyte to prevent the oxidation the tin plated specimen are only partially immersed into these solutions prior to pressure cooking. The tin surface with the organically coated areas should after pressure cooking ideally remain as white as the original surface.

Figures 10 and 11 compare this effect for the currently applied organic coating Protectostan®LF and a new and promisable test formulation.

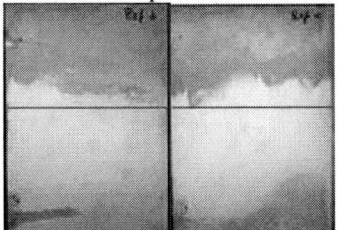

Figure 10: Discoloration prevention by the currently applied organic coating Protectostan LF, coated at room temperature (left) and at 60°C (right); immersion below blue line

Figure 11: Discoloration prevention by a newly formulated organic coating, coated at room temperature (left) and at 60°C (right), immersion below blue line

The anti-discoloration property of the test formulation resembles the one of Protectostan®LF. Quite conspicuous is the fact that the tin coating remains white and clean even above the immersion level. This illustrates that the surface active ingredient may have creeped up the tin surface to protect even remote areas which have not been in contact with the electrolyte. This special 'self healing' property may potentially be of value, when protected tin layers are being scratched during trim&form.

Summary

A novel type of post-treatment has been developed to protect plated tin layers on IC packages from corrosion. This type of organic coating will improve the packages' solderability and mitigate the whisker growth in hot and humid environment. To be effective within the finished electronic component the coating will withstand the reflow process. The search for even stronger corrosion protection continues as detrimental handling errors and scratches during trim&form may not be avoided.

References

[1] P.Oberndorff, M.Dittes, P.Crema, P.Su, E.Yu, Electronics Packaging Manufacturing, **29**, 4, 239-245 (2006)

[2] J. Barthelmes, F. Lagorce-Broc, P. Kühlkamp, S.W. Kok and D.G. Neoh, Proceedings of the IEMT 2006, Malaysia, B4-4

[3] J. Barthelmes, P. Crema, J. Gauci, A.Borg, R. Caruana and P. Kühlkamp, IPC International Conference of Leadfree Electronics, P10, Austin, TX, December 2007

Pot Life Improvement of Low Temperature and High-speed Curable Anisotropic Conductive Adhesive (ACA)

Jong-Hyun Lee*, Ju-Hyung Kim, and Chang-Yong Hyun

Department of Materials Science & Engineering, Seoul National University of Technology,

Seoul 139-743, Korea

82-2-970-6612, 82-2-970-6565, pljh@snut.ac.kr

Abstract

In this study, the imidazole-based curing accelerator powders were coated with a chosen agent to increase the pot life of ACA resin formulation. To accomplish this simple procedure, the coating processes were tried with a vapor or molten state of the coating agent. The coating processes with the vapor state could not be considered as an effective conformal coating method. However, a coating process using the molten state was effective at increasing the pot life, and showed minor effect with regard to the curing rate and processability of final formulation.

Key words: RFID assembly, chip bonding, ACA (anisotropic conductive adhesive), high-speed curing, pot life

Introduction

Rapidly enlarged radio frequency indentification (RFID) industry and interest in low temperature bonding materials in the electronic interconnection technology has been inducing technical improvement in the polymer-based conductive adhesives [1-5]. Although the conductive adhesives reveal relatively poor reliability in the comparison with the solder joints, it is estimated that the adhesive materials will be adopted more widely in low cost interconnection processes due to their low bonding temperature, fast bonding speed and so on. Furthermore, the formulation of conductive adhesive materials is one of the core technologies in the RFID assembly industry in which high-speed chip bonding is very beneficial in reduction of assembly cost.

The author has already demonstrated the research results about the low cost formulations having high-speed (snap) curing properties as a chip bonding material for the assembly of RFID inlays [6,7]. The formulations were optimized to demonstrate high-speed curing properties that the bonding process is completed within several seconds using usual RFID chips. The suggested formulations were consisted of bisphenol-F resin, anhydride-based curing agent (~80 % of resin in molecular weight), imidazole-based curing accelerator (~15 % of resin in molecular weight), spherical Ag fillers, etc.

However, pot life of the suggested formulations was relatively short, about 90 min. This means that the formulations had a considerable problem for the industrial applications. Several ideas have been considered to enlarge pot life of the

formulations with regard to the curing accelerator. One is using imidazole derivatives as the curing accelerator and the other is to coat the accelerator with non-reactive material [8]. The cause of pot life increase in the latter case could be considered as the barrier effect by coating material. A curing accelerator serves to promote cross-linking reaction between base resin and curing agent or between resins. Therefore, curing reaction could be restrained at room temperature in the case that curing accelerator powders were coated with the specific agent having nothing to do with the cross-linking reaction. Meanwhile, the coating agent is melted and dissolved into the formulation and uniformly through the convection during the ramping when the bondline temperature is increased to 150 ~ 170 °C in the bonding process for a RFID chip. Finally, the curing accelerator was exposed and reacted with the mixture of resin and curing agent. Consequently, the high-speed curing properties of the final formulation are not greatly affected by coating of the curing

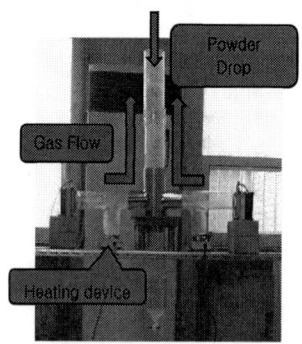

Figure 1: Vapor state coating apparatus using the free falling of curing accelerator powders.

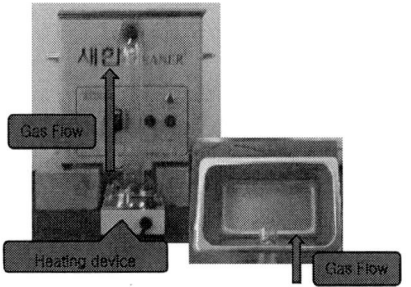

Figure 2: Vapor state coating apparatus incorporating an ultrasonic vibration chamber.

accelerator powders. Therefore, this research was tried to improve pot life of ACA formulations by applying simple coating methods on curing accelerator powders.

Experimental Procedure

The imidazole-based curing accelerator powders were coated with a chosen agent having a melting temperature of about 45 °C and a boiling temperature of about 130 °C. To accomplish the physical coating procedure simply, the coating agent was applied in the molten or vapor state, respectively.

For the vapor state coating process, the accelerator powders traveled across the hollow pillar from the top to bottle through the free falling. The vaporized coating agent was continuously generated from both sides of the pillar structure and transported into the center of the hollow pillar during the falling.

As another vapor state coating process, the accelerator powders were maintained within the ultrasonic vibration chamber to face with the coating material of vapor state that was injected to the chamber. This process guarantees longer reaction time. Furthermore, the ultrasonic vibration motion prevents the agglomeration behavior of the accelerator powders during coating process.

For molten state coating processes, manual stirring and mechanical alloying methods were attempted. After the coating agent was melted at about 75 °C, we sprinkled imidazole-based curing accelerator powders into the melted coating agent while stirring. The weight ratio of accelerator to coating agent was 1: 0.15. The mixture was cooled down to ambient temperature, and pulverized to make a new powder shape. Mechanical alloying was performed to proceed the coating process more uniformly and efficiently. The curing accelerator and coating agent powders were put into a jar containing steel balls, ground, and mechanically combined repeatedly during several minutes using a planetary mill. After a certain milling time, the coating agent powders were melted, and covered the ground fine curing accelerator powders. The weight ratio of

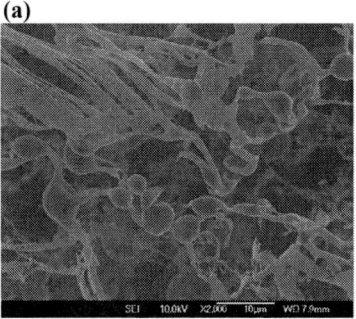

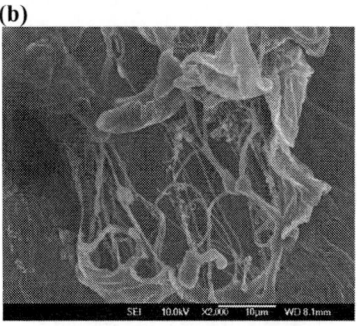

Figure 3: Surface of imidazole-based curing accelerator powders coated by vapor state coating methods using an apparatus (a) for the free falling of curing accelerator powders and (b) incorporating an ultrasonic vibration chamber.

accelerator to coating agent was varied from 9: 1 to 7: 3.

From above various attempt, the best coating process was selected through the measuring of apparent viscosity of formulations incorporating imidazole-based curing accelerator powders as a function of storage time at 30 °C. The final formulation obtained from the best effective coating method was converted to anisotropic conductive adhesive (ACA) by adding of Ag fillers. Finally, RFID chip bonding properties were evaluated using the ACA.

Results and Discussion

Figure 3 is surface images of an imidazole-based curing accelerator powder observed by using a FE (field emission)-SEM as a function of vapor state coating process. Figure 3(a) is the surface image of a raw curing accelerator powder. And Figure 3(b) and 3(c) are surface images of a coated curing accelerator powder by free falling or ultrasonic vibration method, respectively. The surface image of a raw curing accelerator powder was observed as twine structure. Therefore, it was observed that the vaporized coating agent can not penetrate easily into the inside of curing accelerator powders. Free falling coating method can be characterized as relatively high vapor temperature and short reaction time. From the characterization, intensive morphological

978-1-4244-4722-0/09 $25.00
© 2009 IMAPS-ITALY

(a)

(b)

Figure 4: SEM images of used imidazole powders: (a) before coating and (b) after coating using specific agent.

change and relatively small coating material volume were identified (Fig 3(b)). It was analyzed that the morphological change occurs when reaction between the vaporized coating agent and the surface of a curing accelerator powder do not happen uniformly. Meanwhile, the coating process through ultrasonic vibration presents relatively low vapor temperature and adequate reaction time. Fig 3(c) indicates relatively portly coating material volume and lessened morphological change. Nevertheless, both

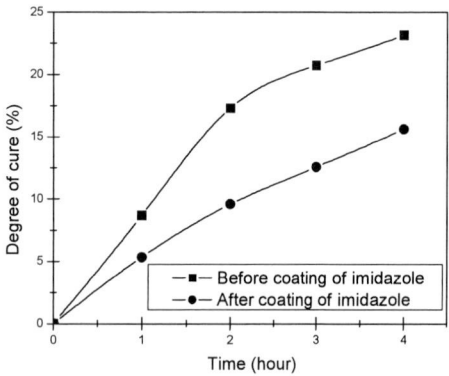

Figure 5: Variation of degree of cure as a function of aging time at 30 °C: (a) formulation incorporating normal imidazole powders and (b) formulation incorporating specially coated imidazole powders.

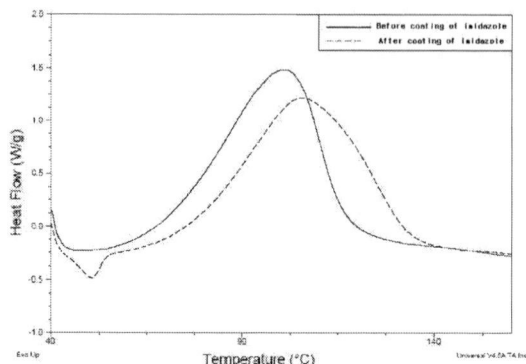

Figure 6: DSC result of the formulation incorporating coated imidazole powders.

processes could not be considered as an effective conformal coating method.

Authentic coating was observed through the molten state process. Figure 4 shows the SEM images of initial pulverized curing accelerator powders and coated powders fabricated by re-pulverizing after the manual stirring, respectively. Both samples were pulverized into the size of 200 ~ 300 um. It was observed that the coated powders showed smooth surface, indicating the covering of coating agent in comparison with initial pulverized powders.

Figure 5 shows variation of degree of cure in the ACA binder formulation as a function of aging time at 30 °C. The degree of cure was determined as the ratio of the exothermic area was observed after specific aging time to the area of a peak observed during dynamic scan. Degree of cure after 1 hr and 2 hr was 8.69 and 17.29 % with the formulation incorporating non-coated curing accelerator powders. Degree of cure in the formulation incorporating non-coated powders was increased relatively rapidly with increase of storage time. Meanwhile, degree of cure in the formulation incorporating coated curing accelerator powders was 5.34, 9.59, 12.57, and 15.62 %, respectively, according to increase with interval of 1 hr. Consequently, the molten state

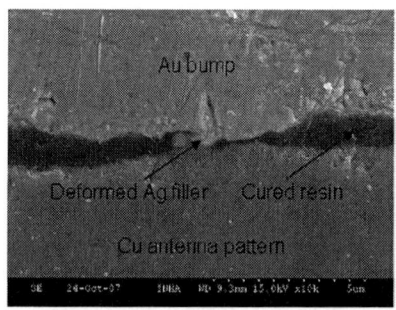

Figure 7: SEM image showing the electrical path between metal conductors formed by the plastic

978-1-4244-4722-0/09 $25.00
© 2009 IMAPS-ITALY

deformation of Ag powder in ACP after the RFID chip bonding.

(a)

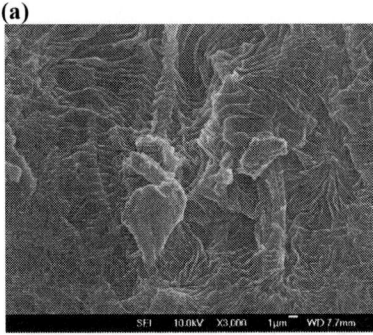

(b)

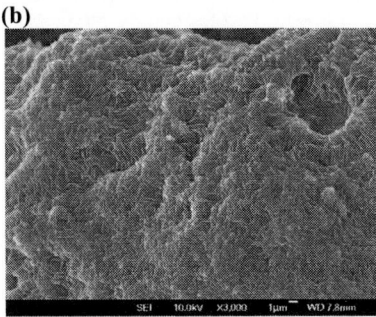

Figure 8: SEM images of a coating agent/ imidazole-based curing accelerator mixture fabricated using mechanical alloying method; the weight ratio of the accelerator to coating agent is (a) 9: 1 and (b) 8: 2.

coating process using manual stirring method was observed to be effective in the increase of pot life of ACA binder formulation. Apparent pot life at 30 °C was measured to be increased form just 120 min to about 510 min.

The reason that pot life in the formulation incorporating coated curing accelerator powders did not last over 1 day was analyzed due to the random pulverization process. Our pulverization process using a rotary mill could reveal the inside of imidazole-based curing accelerator powders. If the pulverization step could be progressed only in the coated agent, pot life of our formulation would be much longer.

Figure 7 shows a dynamic DSC (differential scanning calorimeter) scan's result measured in the formulation incorporating coated curing accelerator powders. The endothermic peak observed at 50 °C indicates melting of the coated agent. The exothermic peak was entirely shifted to a higher temperature region, thus a little reduction in curing rate was estimated. However, the increase of curing time during a real bonding process was observed to be under 1 sec.

Figure 8 is a SEM image showing a physically and electrically interconnected part of RFID chip formed using the ACA incorporating coated curing accelerator powders. During the bonding procedure, temperature of hot bar was increased from 170 to 180 °C, and that of bottom plate was also increased from 120 to 150 °C. Bonding pressure was 10 N and bonding time was 5 sec. The complete operation of the fabricated RFID inlay assembly was confirmed by testing using reader.

Meanwhile, the mechanical alloying method introduced as a more effective process did not show similar or enhanced latency in ACA binder formulation. The reason was judged due to the transformation into extremely fine curing accelerator powders and pulverization step. Mechanical alloying process basically brings about formation of fine particle or platelet shape. Excess increase in the surface area of curing accelerator powders can trigger the curing reaction rapidly even at room temperature. Figure 8 is a SEM image showing the mixtures processed as a function of the ratio of accelerator to coating agent for 6 min. The distance between wrinkles measured in the case of 8: 2 was shorter in comparison with the case of 9: 1. The result implied that finer particles or layers were formed in the case of 8: 2 during mechanical alloying. The Case of 9: 1 was observed to have a little more latency than that of 8: 2. Consequently, formation of finer curing accelerator particles was judged to lead decrease of pot life.

Conclusion

The imidazole-based curing accelerator powders were coated with a chosen agent to increase the pot life of ACA resin formulation. To accomplish the simple procedure, the coating processes were tried with a vapor or molten state of the coating agent. Coating processes in vapor state could not be considered as an effective conformal coating method. However, a coating process in molten state was effective for the increase of pot life, and showed minor effect with regard to the curing rate and processability of the final formulation. Nevertheless, a more effective molten state and coating process along with better pulverization processes are still required.

Acknowledgements

The authors acknowledge the financial support from the Ministry of Commerce, Industry and Energy (MOCIE) of Korea.

References

[1] D. Wojciechowski, J. Vanfleteren, E. Reese and H.-W. Hagedorn, Microelectronics Reliability, Vol. 40, pp. 1215, 2000.

[2] J.S. Rasul, Microelectronics Reliability, Vol. 44 pp. 135, 2004.

[3] C.-M. Cheng, V. Buffa, W. O'Hara, B. Xia and J. Shah, Global SMT & Packaging, Vol. 5, pp.17, 2005.

[4] M.J. Yim and K.W. Paik, Int. J. Adhesion & Adhesives, Vol. 26, pp. 304, 2006.

[5] M. Fairley, "RFID Smart Labels", Tarsus Exhibition and Publishing, London, pp.19, 2005.

[6] J.-S. Lee, J.-K. Kim and J.-H. Lee, "Formulation and Evaluation of Isotropic Conductive Paste for the Bonding of RFID Chip ", Collection of the Awarded Research Papers in RFID/USN Korea, Korea Association of RFID/USN, Seoul, pp. 43, 2007.

[7] J.-S. Lee, J.-K. Kim, C.-W. Lee, J.-H. Lee and M.-S. Kim, "Study on the Manufacturing Process of Latent Curing Agent for Improved Pot-life of ACP in RFID Application", Proceedings of the 2008 Autumn Annual Meeting of Korean Welding and Joining Society, Incheon, Korea, November 20-21, pp. 20, 2008.

[8] L. Wang and C.P. Wong, "Studies on Latent Catalyst Systems for Pot-life Enhancement of Underfills", Proceedings of the 1999 International Symposium on Advanced Packaging Materials, pp. 67, 1999.

Local Hardening Behavior of Free Air Balls and Heat Affected Zones of Thermosonic Wire Bond Interconnections

C. Dresbach*, G. Lorenz*, M. Mittag*, M. Petzold*, E. Milke**, T. Müller**

* Fraunhofer Institute for Mechanics of Materials, Walter-Hülse-Str.1, 06120 Halle, Germany

** W. C. Heraeus GmbH, Heraeusstr. 12-14, 63450 Hanau, Germany

+49(0)345-5589-174, +49(0)345-5589-101, Christian.dresbach@iwmh.fraunhofer.de

Abstract

The local deformation behavior of the free air ball (FAB) and the heat affected zone (HAZ) of thermosonic wire bond interconnections is of great interest in reliability considerations for current highly integrated microelectronic devices. The mechanical properties of the HAZ have significant influence on the loop stability, which is very critical in fine pitch and long loop applications. On the other hand, knowledge of the hardening behavior of the FAB is essential to avoid chip damage risks, particularly if bonding is applied to state-of-the-art ICs containing mechanically sensitive low K dielectric materials. The significance of this reliability risk is even more increased if gold wires were replaced by copper wires. In this study, we characterized the hardening behavior of FABs from three typical gold bonding wires using a modified micro compression test. The stress/strain behavior was calculated via inverse finite element simulations from the experimental force/displacement curves. The mechanical properties of the FABs were compared to these of the related HAZ and the unaffected wire, and were correlated to the wire microstructure that was investigated by electron backscatter diffraction (EBSD). Since the EBSD results for the FAB grain structure indicate a possible anisotropic behavior, a capillary compression test setup was developed allowing to mechanically characterize the FAB in a loading situation comparable to the bond process itself and the results were compared to the experiments with a loading in perpendicular direction. Both these approaches allow an extended characterization of bonding wires considering also the FAB properties, and support therefore material selection for application and wire material development.

Key words: gold, wire bonding, EBSD, capillary compression test

Introduction

In current highly integrated microelectronic devices including System in package and stacked die solutions, the system reliability is strongly influenced by the reliability of the wire bond interconnections. Loop stability becomes an important factor especially in fine pitch and long loop applications. The loop stability is affected by loop geometry, length and mechanical properties of the heat affected zone (HAZ) just above the ball where grains were recrystallized during the flame-off process. Additionally, controlling the mechanical properties of the free air ball (FAB) is of utmost significance due to chip damage risks if bonding is applied to state-of-the-art ICs containing mechanically sensitive low-K dielectric materials. Unfortunately, there are no proven methodologies for characterizing the mechanical properties of the heat affected zone and free air ball available. In traditional approaches, micro- or nanoindentation experiments are often used for determination of these local mechanical properties. Indentation testing is well established for hardness measurements, which is e.g. very useful in advanced

failure analysis. On the other hand, calculations of meaningful material parameters required in advanced material laws for component simulations are up to now still questionable [1] and under development; especially if to be applied to small sample dimensions such as the heat affected zones and free air balls of wire bonding interconnects. An alternative to nanoindentation is the micro compression test of small wire sections as described in [2]. In this paper, results of a modified micro compression test applied for a systematic characterization of the mechanical properties of free air balls from different Au wire materials and different free air ball sizes are presented. The mechanical properties were characterized in terms of stress/strain behavior from inverse finite element simulations. They are compared to those of both the unaffected wire and the heat affected zone. In addition, the microstructure is investigated by electron backscatter diffraction (EBSD) method and correlated with the mechanical properties. To consider a potential anisotropic material behavior of the free air ball during the actual loading situation in ball bonding, a new test setup was developed which is called capillary compression test. In this test a

978-1-4244-4722-0/09 $25.00
© 2009 IMAPS-ITALY

bond capillary tool is used for support of the free air ball, which allows loading the ball in normal direction along the wire axis instead of loading in a perpendicular direction. For this reason, the capillary compression test is more comparable to the bond process itself. Even though the experiment is realized at room temperature and without ultrasonic power application yet, the derived stress/strain behavior, and consequently, the extended understanding of wire material properties and micro structure can serve as a basis for component simulations during the design process, for material selection and failure analysis, and also for further wire material development.

In this investigation three selected gold bonding wire materials with a diameter of 25μm were used. Wire A and B are doped gold wires (4N) and wire C is an alloyed gold wire (2N). Free air balls with four different sizes ranging from 45μm to 75μm in diameter were produced using the semi automatic gold ball bonding equipment Delvotek 5610. The actual diameter of each free air ball was measured by light optical microscopy with a Leica DMRXE-650H microscope and considered in the investigations. Figure 1 shows a typical produced free air ball.

Figure 1: Light optical image of a gold free air ball

Mechanical Characterization of Free Air Balls Using a Micro Compression Test

For the mechanical characterization of free air balls (FAB), we used an adapted micro compression test in a micro indenter which was originally developed for mechanical characterization of small wire cylinders [2]. For this investigation, the FAB is placed on a flat diamond support in the micro indenter Shimadzu DUH 202 and is compressed with a flat diamond punch. A principle of the micro compression test (MCT) on a FAB is shown in Figure 2. The mechanical properties of the free air balls were determined via inverse finite element simulations from the experimental force/displacement curves as described below. The applied load and corresponding displacement are recorded during the load controlled experiment. The initial contact in the force/displacement plots of each experiment is corrected through analyzing the maximum slope of the first part in the loading curve and shifting the whole curve by the displacement of the intersection from the linear slope with the x-axis.

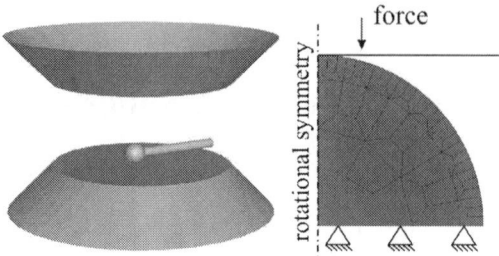

Figure 2: Principle of micro compression test (left), simplified rotational symmetric finite element model for inverse calculation of the stress/strain behavior (right)

In addition, the displacement is normalized to the nominal FAB diameter $d_{nominal}$ and the force is normalized to the square of the nominal FAB radius $r_{nominal}$ to consider small differences in free air ball size.

$$u_c = u_m \cdot \frac{d_{nominal}}{d_{actual}} \quad F_c = F_m \cdot \left(\frac{r_{nomial}}{r_{actual}} \right)^2$$

A mean curve with reduced data points is calculated from the corrected force/displacement curves corresponding to each FAB diameter. The small scatter of experimental results and thus, high reproducibility of the test, is exemplarily shown in Figure 3.

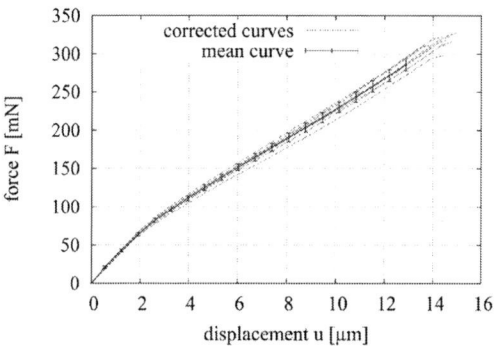

Figure 3: Reduced mean curve of the force/displacement behavior of FAB A, diameter 55 μm, in micro compression test

A finite element model with rotational symmetry was used for inverse calculation of the stress/strain behavior from the mean force/displacement plots using Abaqus6.8. In the model the wire end is neglected for decreasing calculation time of the optimization process. In the inverse finite element simulation, the mechanical properties are systematically changed until the minimum error square between experiment and simulation is found. For calculation of the minimum error square, the displacement data points of the reduced mean curves are used as load steps and the simulated forces are compared with the experimental forces. An adaptive surface response method is applied using the software optiSLang3.0 for

978-1-4244-4722-0/09 $25.00
© 2009 IMAPS-ITALY

optimization. To achieve uniqueness in the optimization process, only the plastic material properties were varied and the elastic properties were determined from the experiment and kept constant in the simulation. Therefore, we calculated an effective modulus E_{eff} from the stiffness S of the unloading curve at maximum load in analogy to the indentation modulus in nanoindentation experiments [3]. Instead of the projected contact area calculated from contact depth and indenter geometry in nanoindentation experiments, we used the contact area A_p of the FAB after unloading measured by light optical inspection.

$$E_r = \frac{S\sqrt{\pi}}{2\sqrt{A_p}} \quad S = \frac{dF}{du}$$

$$E_{eff} = \frac{1-v_s^2}{\dfrac{1}{E_r} - \dfrac{1-v_i^2}{E_i}}$$

For the Young's modulus of the indenter E_i=1140GPa was used. The Poisson's ratio was chosen as v_i =0.07 for the indenter and v_s=0.42 for the FAB. The mean effective Young's modulus was used as a constant in the optimization process, and a friction coefficient of 0.2 was assumed. The Ramberg-Osgood equation was used as hardening law [4], so that only three parameters K_0, K_y and M_y have to be varied during the optimization process.

$$\sigma_y = K_0 + K_y \cdot \varepsilon_p^{\frac{1}{M_y}}$$

At first, free air balls with a diameter of 55μm of the three typical gold bonding wires A, B and C were analyzed in a micro compression test to investigate the influence of the wire material on the mechanical properties of the FAB. A comparison of the experimental and simulated force/displacement curves is shown in Figure 4. A good accordance was achieved for all optimizations of the experiments.

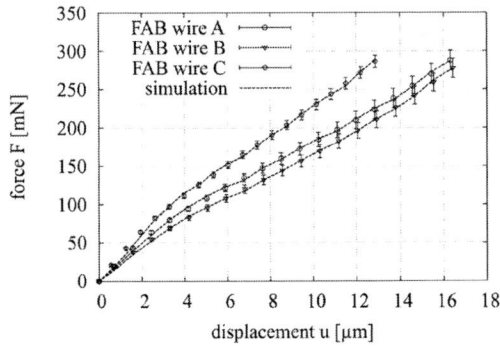

Figure 4: Comparison of measured and simulated force/displacement behaviors of 55μm FABs of the wires A, B and C

The FAB of wire material C shows a much harder deformation behavior compared to the FABs from the wire materials A and B, which behave

nearly similar, except a slightly harder deformation behavior of the material A compared to B. From the simulated stress/strain behavior presented in Figure 5 it can be seen that the free air ball of the alloyed wire C shows higher initial yield stress and also higher amount of hardening compared to the FAB of the doped wires A and B. This is reasonable since wires A and B are 4N materials while C is a 2N material.

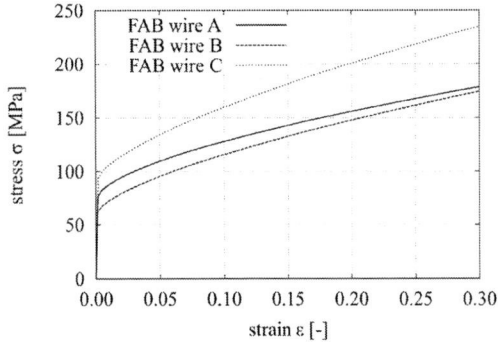

Figure 5: Calculated stress/strain behavior of 55μm FABs of wire A, B and C

Using FABs of the wire material A with additionally varied diameters (45μm, 65μm, 75μm), the influence of the flame-off process on the deformation properties was investigated. Comparing the results found for the FABs of the wire A, there is no clear and distinct effect of FAB size on the mechanical properties found. However, it could be noticed that the two smaller FABs show slightly higher yield stresses than the larger FABs.

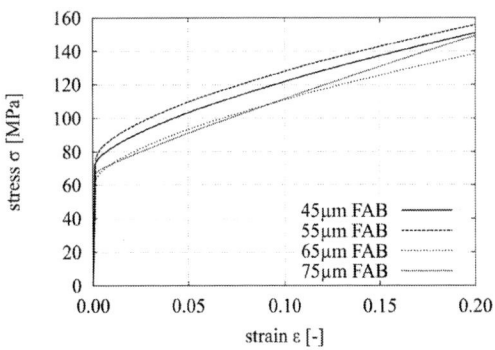

Figure 6: Calculated stress/strain behavior of wire A with different FAB sizes

From these results it can be concluded that there is a significant influence of the wire material on the mechanical properties of the free air ball, even after re-solidification, and rather a comparable small influence of the flame-off process parameters.

Comparison of Mechanical Properties of Free Air Ball, Heat Affected Zone and Unaffected Wire Region

In order to validate the testing approach, and for a comparative characterization of the mechanical properties of the free air balls with those of either the initial wire material before bonding and the heat affected zone above the ball, additional tensile tests and micro compression test of unprocessed wires and samples prepared from the heat affected zones were performed. For micro compression testing small cylinders with a length between 50μm and 70μm were metallographically prepared (Figure 7) and tested like the FABs described before.

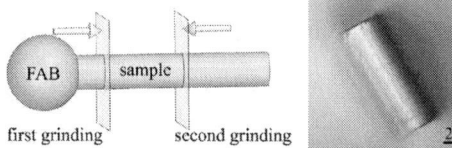

Figure 7: Principle of sample preparation for micro compression tests on wire cylinders (left) and light optical microscope image of a wire cylinder (right)

The stress/strain behavior was calculated via optimization using a 3D finite element model (Figure 8). As hardening law a modified Chaboche model was used for representing also the higher strains which could be realized with micro compression tests on small cylinders. In the following diagrams only the first 20% of the stress/strain behavior are shown to improve clearness, even though results were determined up to 60% compression.

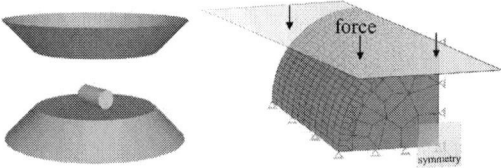

Figure 8: Principle of micro compression test on wire cylinders (left) and 3D finite element model for inverse determination of stress/strain behavior (right)

The results of these investigations are summarized by presenting the stress/strain curves calculated from the experimental results for wire A, B and C in Figure 9 to Figure 11. It becomes evident, that the results derived from inverse finite element simulation of the micro compression test are in good agreement with the results of conventional tensile experiments, thus validating the approach used. In addition, it is found for all tested wire materials that the initial yield stress determined in a micro compression test of the unaffected wire is systematically slightly lower compared to these of the tensile test. It is not clear yet whether these differences are due to anisotropic plasticity caused by the elongated grain structure of the drawing texture. However, such effects have to be taken into consideration.

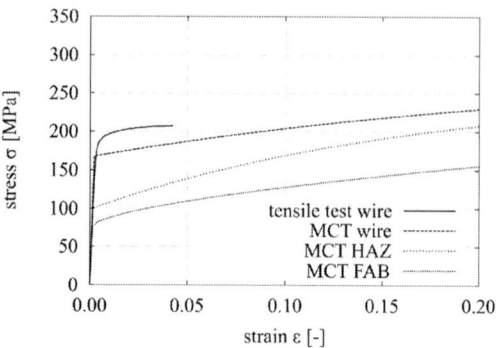

Figure 9: Stress/strain behavior of unaffected wire, HAZ and FAB of wire A

Comparing free air ball, heat affected zone and initial wire, it can be shown that the initial yield stress of the free air ball is only approximately 46% of the initial yield stress of the unaffected wire. The mechanical properties of the heat affected zones are as expected in between these of the free air ball and the unaffected wire having an initial yield stress in the order of 64% of the initial wire.

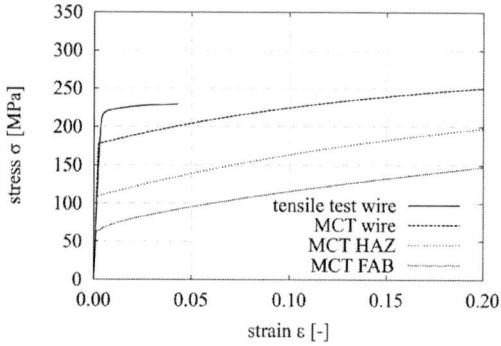

Figure 10: Stress/strain behavior of unaffected wire, HAZ and FAB of wire B

The amount of hardening is similar for all compression experiments, except for the HAZ of material A. The hardening determined in tensile experiments is rather small.

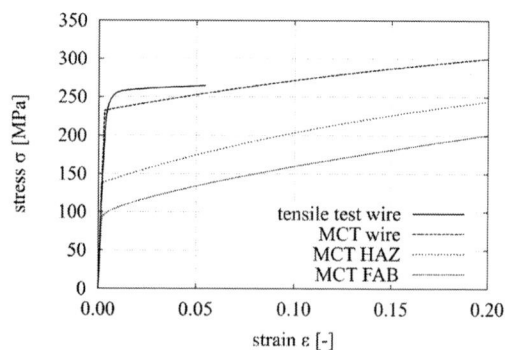

Figure 11: Stress/strain behavior of unaffected wire, HAZ and FAB of wire C

978-1-4244-4722-0/09 $25.00
© 2009 IMAPS-ITALY

Microstructure of Free Air Ball, Heat Affected Zone and Unaffected Wire Region

The microstructure of the free air balls, heat affected zones and unaffected wires were characterized by electron backscatter diffraction method (EBSD). The unaffected wire shows typically very small elongated grains. The mean grain diameter in a cross section perpendicular to the wire axis is approximately 0.6µm-1.1µm, whereas the lengths of the grains are several 10µm as could be seen from the EBSD results on lengthwise cross sections in Figure 12. The grains in the heat affected zone are more globular and not elongated; so that a recrystallization and not only a grain growth must have been occurred during the flame-off process. The mean grain diameter of the HAZ (HAZ A: 6.2µm; HAZ C: 2.6µm) is approximately 4 to 5 times the grain diameter of an unaffected wire.

Figure 12: Microstructure of free air ball, heat affected zone and unaffected wire region determined with EBSD (top: wire A, bottom wire C; color code: inverse pole figure in wire direction)

The microstructure of the free air ball is characterized by having only few grains which are orientated with a preferred ⟨001⟩ orientation in wire direction. The mean grain diameter of the FABs (FAB A: 20µm; FAB B: 12µm) is 18 times the mean grain diameter of the unprocessed wire. Especially the grains from FAB C seem to be grown from the wire end, so that it could be concluded that the non-melted wire end acts as a seed crystal during the solidification. Because of these elongated grains also a possible anisotropy in the plastic deformation behavior has to be considered. In particular, a test is required that allows to load the ball in wire axis direction, as it is the case during the actual ball bonding process itself.

Mechanical Characterization Using a Capillary Compression Test

A new test method for the determination of possible anisotropic material behavior was developed using a bond capillary tool. In the following, the test is called capillary compression test. For the capillary compression test, a free air ball with absolutely straight wire end has to be produced. After separating the sample from the wire spool it has to be re-positioned into the (now wire-less) test capillary without damaging or hardening the FAB material (Figure 13).

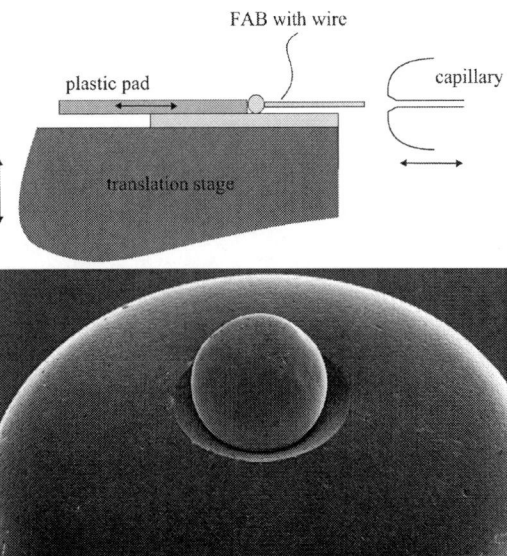

Figure 13: Principle of re-positioning of free air ball in a bond capillary and SEM image of free air ball before capillary compression test

The test capillary with the well-placed free air ball can then to be positioned upright on the sample holder of the microindenter Shimadzu DUH 202. A parallel arrangement of the capillary to the diamond punch, and a very stiff fixing of the capillary are of utmost significance. The FAB can be loaded in wire direction using a flat diamond punch, see Figure 14. The stress/strain behavior is calculated like for the micro compression test but with consideration of the capillary geometry and the presence of the wire end in a rotational symmetric finite element model. The effective Young's modulus can be determined from the unloading curves and optical measured contact areas.

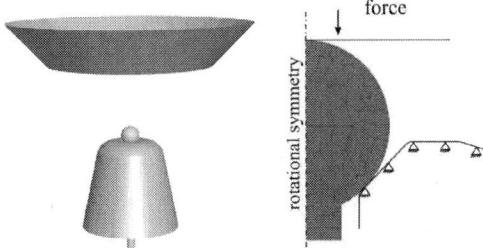

Figure 14: Principle of capillary compression test on free air balls in a capillary (left), 2D rotational symmetric FE model for inverse determination of stress/strain behavior (right)

Capillary compression tests were exemplarily performed on free air balls with a diameter of 55µm. The good agreement of the experimental and optimized force/displacement behavior of all tested wire materials is shown in Figure 15.

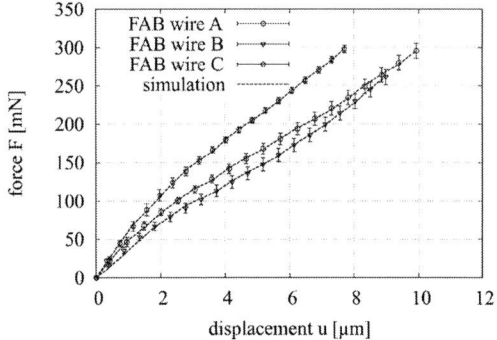

Figure 15: Mean force/displacement curves of capillary compression tests on 55μm FABs of wire material A, B and C

The resulting stress/strain curves determined from the inverse finite element calculations are presented in Figure 16. In addition, results of micro compression tests in perpendicular direction as discussed in the previous section are also displayed. Comparing the calculated stress/strain behavior determined from a capillary compression test in wire axis with a micro compression test of the FABs perpendicular to the wire axis it can be seen that the initial yield stresses and the first amount of hardening are in excellent agreement for all the tested three wire materials. For more than 10% plastic strain, the FABs show a more pronounced hardening when testing in a capillary compression test.

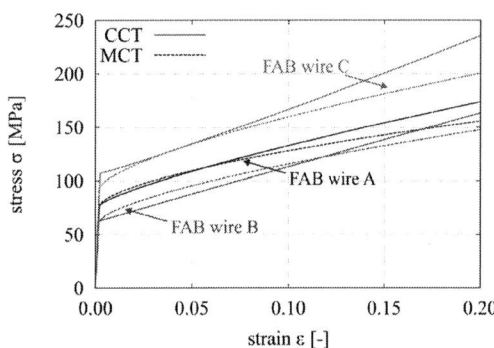

Figure 16: Comparison of stress/strain behavior calculated via inverse FEM simulations from MCT and CCT on FABs

It is up to now not known if this effect is a real microstructure based phenomenon, or only an artifact resulting from the approach in these experiments. However, since the capillary compression test is more similar to the conditions of the bonding process itself, the resulting material properties should be used when simulating the deformation behavior of free air balls. A possible explanation for determining a similar initial yield stress and primary hardening, but finding differences in yield stresses at higher strains for the same material could be the uncertainty in assuming the

friction coefficient. This effect should, however, have nearly the same influence on the results independently on the wire material. This is not the case for these experiments where FAB C shows larger differences in the hardening behaviors than FABs A and B. To discuss possible microstructure effects, the grain structure of FABs A and C were analyzed in cross sections in a lengthwise direction and perpendicular to the wire axis. A principle of both sections is shown in Figure 17.

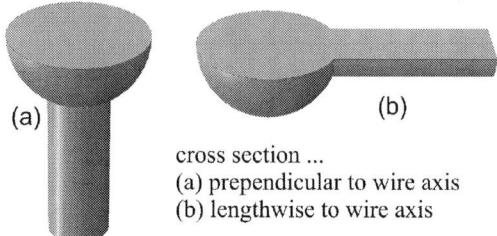

cross section ...
(a) prependicular to wire axis
(b) lengthwise to wire axis

Figure 17: Principle of cross sections for EBSD analysis (left: perpendicular section; right: lengthwise section)

The grain structures of the FABs from wire material A and C are shown in Figure 18. The area weighted mean grain diameter for wire A is approximately 12μm in perpendicular section and 20μm in lengthwise section. The mean grain diameter for wire C is 7μm in perpendicular section and 12μm in lengthwise section.

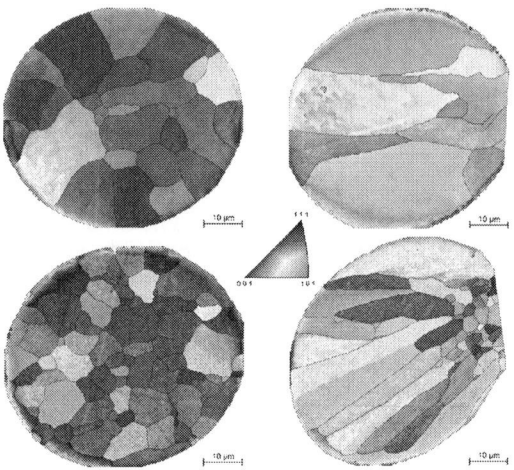

Figure 18: Microstructure of the FABs A (top) and C (bottom) in a cross section lengthwise to the wire axis (left) and parallel to the wire axis (right); color code: inverse pole figure in view direction

Although, due to lacking statistical significance, these values may only be used as a first estimation, and not for exact quantification, it is indicated that the free air ball material behavior cannot be explained by simple grain size correlations. This becomes clear when comparing

978-1-4244-4722-0/09 $25.00
© 2009 IMAPS-ITALY

the results of wire A in a capillary compression test with these of wire C in a micro compression test. Because of similar mean grain diameters of 12μm, a similar deformation behavior would be expected, but this is not the case. In contrast, the determined mechanical properties of wire A in capillary compression and wire C in micro compression are significantly different, whereas the mechanical properties determined in capillary compression and micro compression for each FAB material are rather similar, although they have different mean grain diameters. To discuss possible reasons, the differences in grain shapes have to be considered. For wire A in capillary compression test the effective maximum dislocation storage length should be approximately half the grain diameter. In contrast, this should not be the case for the elongated grains when testing FAB C in micro compression test. From this comparison it can be concluded that - for each FAB tested here - the effective mean dislocation storage length should be nearly the same independently from the loading direction because of the elongated grains, so that the initial yield stress and primary hardening are similar for both directions. The higher amount of hardening could be explained by the non-uniform deformation of the FAB, whereby interacting grain boundaries in the effective volume could rise faster when compressing the FAB in the capillary compression test than in a micro compression test. All these first hypothetical considerations have to be proven in further investigations by additional compression tests and EBSD analyses. Furthermore the influence of friction between the tool and the FAB has to be investigated in more detail, too.

Summary

The mechanical properties of free air ball, heat affected zone and unaffected wire of three typical 25μm gold bonding wires (two 4N materials and one 2N material) were analyzed in micro compression tests. The stress/strain behaviors were calculated from the corresponding force/displacement curves using inverse finite element simulations. This approach allows to determine the local mechanical properties and thus, to quantify the deformation behavior of different process-affected regions, such as the free air ball or the heat affected zone in terms of parametric elasto-plastic material laws for further simulation of reliability properties. Comparing selected bonding wire materials it was shown that, for the same wire material the flame-off process led to a reduction in initial yield stress in the FAB to about 46% of the initial value. This behavior corresponds to a distinct increase in grain size through solidification by a factor of 18 compared to the initial wire. In spite of the drastic increase of grain size the differences between wire materials are not leveled out during flame-off process. In this study it was furthermore shown that FABs of the investigated 2N wire

material showed both a higher yield stress compared to the investigated FABs of 4N materials as well as a different grain structure with reduced grain size. A reason for the smaller grains of the 2N FAB could be the higher amount of alloying elements or the smaller initial grain size acting as seed crystals. In addition, the parameter set of the flame off-process, determining the ball size for a given bonding wire material, is another influencing factor, although its effect on the free air ball deformation properties was found to be less compared to the influence of wire material.

Not only the initial wire that is affected by the drawing process, but also the free air ball shows a distinct anisotropy in grain structure with elongated grains in wire axis. Thus, also mechanical anisotropy must be considered during a determination of the plastic deformation behavior. A new test setup for characterizing the mechanical properties in direction of the wire axis was developed based on a modified micro compression test of a free air ball supported by a bonding capillary. By comparing the results for the capillary compression test loading the ball in wire axis with results of the micro compression test loading the ball in perpendicular direction, a very close agreement in the stress/strain curves was found for a range up to about 10% strain. For higher strains the amount of hardening was found to increase more in the CCT. Although not verified yet, these results indicate a possible anisotropic material behavior of FABs that has to be considered in further more detailed studies. For such investigations the newly developed capillary compression test is the first methodology for a quantitative characterization of the mechanical properties of the free air ball itself in the required loading orientation. All other experimental procedures, e.g. nanoidentation are susceptible to effects from sample preparation (which is not necessary for these micro compression tests), and are less sensitive to small changes in the mechanical properties. The parametric stress/strain results shown here provide an advanced basis to understand bonding wire properties in more detail and can thus support wire developments, material selection and component design. To improve the significance of the quantitative deformation parameters derived for simulations of the bonding process the influence of temperature and ultrasonic power on the mechanical properties has to be considered in addition. First ideas and experimental results have been demonstrated elsewhere [5] and will be analyzed in further investigations.

Acknowledgements

The authors would like to thank German Ministry for Education and Research (BMBF) for funding part of this work in the PIDEA project "Failure Detection and Analysis for Complex Micro- and Nano-System Applications".

978-1-4244-4722-0/09 $25.00
© 2009 IMAPS-ITALY

References

[1] L. Liu, "Can Indentation Technique Measure Unique Elastoplastic Properties?", Journal of Materials Research, Vol. 24, No. 3, March 2009.

[2] C. Dresbach et al., "Test Methods for characterizing the Local Plastic Deformability of Bonding Wires", Proceedings of the 1st Electronics Systemintegration Technology Conference (ESTC), Dresden, Germany, September 5-7, pp. 732-740, 2006.

[3] W. C. Oliver and G.M. Pharr, "An improved technique for determining hardness and elastic modulus using load and displacement sensing indentation experiments", Journal of Materials Research, Vol. 7, No. 6, pp. 1564-1583, June 1992.

[4] J. Lemaitre and J. Chaboche, "Mechanics of Solid Materials", Cambridge University Press, Cambridge, Chapter 5, pp. 163-176, 2002.

[5] C. Dresbach, "Local Mechanical Deformation Properties of Gold Bonding Wires and Free Air Balls", Talk given at 1st conference on Microreliability and Nanoreliability in Key Technology Applications (MNR), Berlin, Germany, September 5-7, 2007.

Mechanical Properties and Microstructure of Heavy Aluminum Bonding Wires for Power Applications

C. Dresbach*, M. Mittag*, M. Petzold*, E. Milke**, T. Müller**

* Fraunhofer Institute for Mechanics of Materials, Walter-Hülse-Str.1, 06120 Halle, Germany

** W. C. Heraeus GmbH, Heraeusstr. 12-14, 63450 Hanau, Germany

+49(0)345-5589-174, +49(0)345-5589-101, Christian.dresbach@iwmh.fraunhofer.de

Abstract

Heavy aluminum wire bonding is of great interest as a first level packaging technology for automotive and power electronic devices. The related mechanical and thermo-mechanical cyclic loading conditions require an adequate consideration of reliability aspects and lifetime estimations based on in-depth knowledge of material properties, involving an understanding of the correlation between mechanical deformation behavior and microstructure. In this paper, we present results of mechanical and microstructure investigations using advanced data evaluation approaches. For characterization of static mechanical properties, tensile tests were performed to determine the stress/strain behavior of selected wire materials. The results derived showed a pronounced hardening effect which could be adequately described by classical hardening laws. The grain structure of the wires was analyzed using the electron backscattered diffraction method (EBSD). It is shown, that data statistics and data evaluations have considerable effects on the results represented in terms of grain size. Furthermore, the correlation between deformation properties and microstructure could be described by the Hall-Petch relation predicting a linear dependence of the initial yield stress on the square root of grain size. Since in many cases a knowledge of local changes in the mechanical properties due to either process-induced effects or due to application conditions is required, we also compared results of microindentation and micro compression experiments on small wire cylinders with those of the tensile tests. A very good agreement of the flow curves determined from the micro compression test with the tensile test data was found while the correlation with the indentation results was rather limited. Thus, the newly developed micro compression test is a very powerful tool for characterization of local material properties of bonding wires used in real components. It may also serve as a basis for more sophisticated cyclic hardening investigations as it is necessary for establishing advanced lifetime predictions in future.

Key words: aluminum, wire bonding, Hall-Petch relation, EBSD, indentation, micro compression test

Introduction

The reliability of heavy aluminum wire bond interconnections is of fundamental interest in power electronics and automotive applications. In the last few years there have been multiple publications considering empirical reliability aspects of aluminum wire bonds, see e.g. [1]. However, a more generalized understanding that also allows developing reliable lifetime models requires considering the thermo-mechanical cyclic material behavior and damage evolution as well as local changes in material properties during the bond process and during the application conditions. In addition, a deeper understanding of the microstructure-based deformation mechanisms could help in future wire developments, and also in design process and failure analysis of real components. In most cases, only the maximum load in gram or cN and the maximum elongation at break in percentage are given as the mechanical properties of aluminum bonding wires. In some cases an ideal plastic or bilinear material behavior is assumed for finite element simulations of real components [2] [3], but these models do not adequately represent the elasto-plastic material behavior.

In this paper, we present the results of mechanical and microstructure characterization of different aluminum wire materials. Using extended evaluation approaches we focus on the determination of quantitatively meaningful material parameters from tensile and compression tests which can be used for advanced component simulations. For microstructure characterization, the electron backscatter diffraction method was used allowing to characterize the grain structure and grain size effects. This approach allows us to discuss the initial yield stress and hardening behavior of aluminum bonding wires in correlation to the microstructure. We used four different bonding wire materials (A, B, C, D) from different suppliers while the range of wire diameters varied from 150μm to 500μm. In order to consider local changes in material properties, both microindentation experiments and

978-1-4244-4722-0/09 $25.00
© 2009 IMAPS-ITALY

micro compression tests were performed. A comparison of the results indicates that the latter are more powerful and reliable for determination of local mechanical properties. In addition, the micro compression test could also be used for characterizing wire sections of bonded interconnections in failure analysis, and shows large potential for further cyclic tension/compression tests for determining the necessary cyclic hardening behavior.

Macroscopic Mechanical Characterization of Aluminum Bonding Wires

The macroscopic mechanical properties of the used bonding wires were first analyzed by tensile tests. The gauge length was chosen to be 30mm and the wires were directly fixed by the clamping jaws of the universal testing machine so that no sliding of the wire could occur, and it was proven that the failure takes place nearly in the center of the gauge length. The straightening of the bonding wires in the initial part of the engineering stress/strain curves - resulting from the slightly curved wire geometry at beginning of the test - was corrected by fitting of the highest slope in the first part of the curve, and shifting of the whole curve by the intersection of the linear slope with the x-axis. Mean curves of at least ten experiments were evaluated and will be shown in the following diagrams. From the engineering point of view, the strength of the material (maximum load before rupture) is not the essential parameter to characterize the deformation of ductile materials. Instead, particularly the initial yield stress, but also the hardening behavior and the elongation at break are more important, and will be considered in further discussions of the results. The results of initial yield stress $\sigma_{Y0.2}$ (at 0.2% offset) determined in the tensile tests are summarized in Figure 1. It can be seen that there are substantial differences between the investigated wire materials. In general, there is a tendency within each wire material for an increase of the initial yield stress with decreasing diameter.

The wire A, C and D show only a small amount of increase in initial yield stress with decreasing diameters, while for wire B there is a considerable shift from a relative low value in $\sigma_{Y0.2}$ at 450µm diameter to distinctly higher values at diameters of 300µm and below. Wire materials are optimized by balancing a required ductile behavior (high plastic strain at break) and a high strength (yield stress). In general, increasing the elongation at break causes the yield stress to decrease. When plotting the elongation at break ε_B against the initial yield stress $\sigma_{Y0.2}$ for the wires considered during these investigations (Figure 2), not a single unique relationship but a grouping of the materials was found. The tendency of decreasing elongation with increasing initial yield stress becomes obvious within each group. This grouping indicates strong differences in microstructure and therefore different deformation mechanisms. These differences are not solely dominated by the wire material itself but by annealing procedures as could be seen from the wire B, which is part of both wire groups.

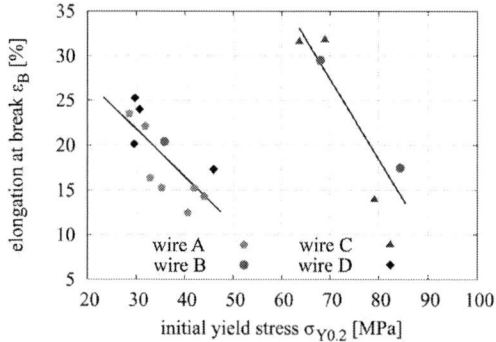

Figure 2: Correlation between elongation at break ε_B and initial yield stress $\sigma_{Y0.2}$

In Figure 3, the mean stress/strain curves for different wire diameters are shown exemplarily for wire A and C.

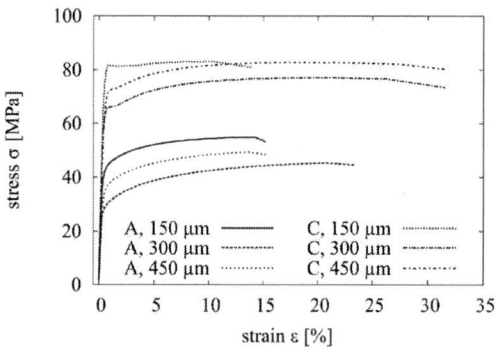

Figure 3: Stress/strain behavior of wire A and C for different diameters

It can be seen that for most bonding wires (in that case all except wire C, 150µm diameter) the assumption of an ideal plastic or bilinear material

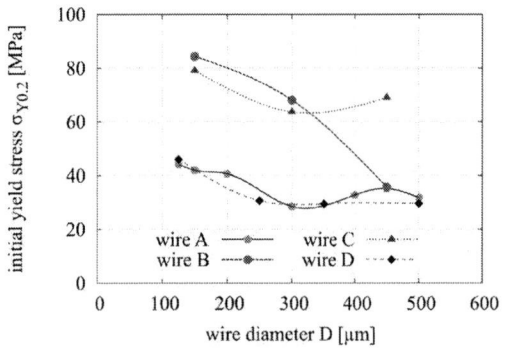

Figure 1: Initial yield stress $\sigma_{Y0.2}$ for different heavy Al bonding wires in dependence on the wire diameter D

978-1-4244-4722-0/09 $25.00
© 2009 IMAPS-ITALY

behavior is only a rough estimation because the aluminum bonding wires show nonlinear hardening behavior. For finite element simulations of the deformation behavior of real components it is useful to define the plastic material behavior in form of parameterized material laws. For this purpose, we evaluated classical hardening laws to represent the hardening behavior of the measured flow curves. First, the engineering strain ε_{eng} was corrected with respect to a theoretical Young's modulus of E_{theo}=64GPa to minimize possible testing machine compliances effects.

$$\varepsilon_c = \varepsilon_{eng} - \frac{\sigma_{eng}}{E_{eng}} + \frac{\sigma_{eng}}{E_{theo}}$$

The true stress σ_{true} and logarithmic strain ε_{ln} were then calculated from the assumption of a constant volume and the consideration of the actual length [4].

$$\sigma_{true} = \sigma_{eng}(1 + \varepsilon_c) \quad \varepsilon_{ln} = \ln(1 + \varepsilon_c)$$

Now, the plastic strain ε_p was determined by subtraction of the actual elastic strain.

$$\varepsilon_p = \varepsilon_{ln} - \frac{\sigma_{true}}{E}$$

The resulted mean flow curves where fitted by the well-known hardening laws of Ramberg-Osgood and Chaboche. The dislocation theory states that the yield stress is proportional to the square root of the density of dislocations ρ_d which leads in analogy for macroscopic strains to the Ramberg-Osgood equation [5].

$$\sigma = K_0 + K_y \cdot \varepsilon_p^{\frac{1}{M_Y}}$$

Lemaitre and Chaboche presented in [6] an additive viscosity-hardening law which consists of the initial yield stress, a hardening function and a viscous stress part. In the following we only used the part containing the initial yield stress and the hardening function named here as the elasto-plastic Chaboche model.

$$\sigma = K_0 + K_\infty \cdot \left(e^{-b \cdot \varepsilon_p}\right) + H \cdot \varepsilon_p$$

The hardening behavior of all tested aluminum wires can be well represented by both hardening models, as is shown exemplarily in Figure 4, even though no model could rebuild the pronounced yield point of the wire material C.

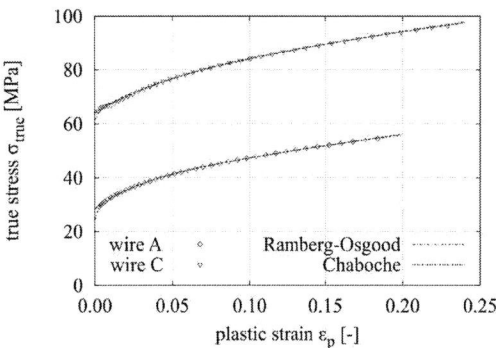

Figure 4: Fit of Ramberg-Osgood and Chaboche models of flow curves for the wire A and C with diameters of 300µm

The initial yield stress $\sigma_{Y0.2}$ is more accurately represented by K_0 of the Chaboche model for nearly all tested bonding wires, as could be seen in Figure 5. Instead, the Ramberg-Osgood equation underestimates the initial yield stress for nearly all experiments, but represents the remaining flow curve in a similar very satisfying way.

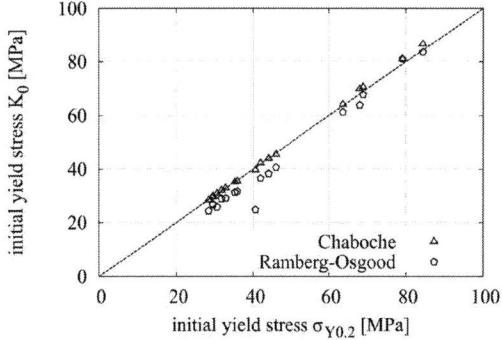

Figure 5: Comparison of fitted initial yield stress K_0 of Ramberg-Osgood model and Chaboche model to the experimental value $\sigma_{Y0.2}$

Micro Structure of Aluminum Bonding Wires

The electron backscatter diffraction method (EBSD) has become the most powerful tool for microstructure characterization of crystalline materials during the last years. However, due to its sensitivity to surface properties one has carefully to assure that the sample preparation has no effect on the microstructure of the material of interest. This becomes of utmost significance if EBSD is applied to quantitative characterization of ductile materials with less stable grain configurations like aluminum bonding wires. Hence, we developed a specific metallographic preparation technique with iterative etching steps after polishing, and final ion polishing. In addition to mechanical effects, care has also to be taken in assuring that temperature effects from ion polishing do not alter the grain microstructure. The correlation between mechanical properties and microstructure is typically given in terms of the

Hall-Petch relation, where it is assumed that the yield stress increases linearly with the inverse square root of the mean grain diameter. When determining the mean grain diameter one has to assure that preparation artifacts are avoided and a statistically adequate quantity of EBSD measurements has been performed. Furthermore, also the evaluation procedure of calculating the mean grain diameter from multiple measurements has to be chosen very carefully, in particular if the grain size distribution is very broad and the mean grain diameter is in the same dimensions as the scan field size (e.g. wire diameter). Thus, we developed a statistical tool for mean grain diameter calculation from the overall population of multiple measurements. In this work we will display the results of three different mean grain diameter calculations. The arithmetic average of the grain diameter, which is the standard in EBSD analysis software, is not used, because it effectively weights smaller grains stronger than larger grains, which is not in agreement with the physical interpretation the Hall-Petch relation is referring to. Instead, different area fraction-weighted values are used. The area-weighted grain diameter D_{dA}, which some EBSD analysis softwares also supply, is the sum of the grain diameters D_i multiplied by the area fraction A_i/A_{ges} of the grain:

$$D_{dA} = \sum_{i=1}^{k} D_i \cdot \frac{A_i}{A_{ges}}$$

A weak point of that procedure could be the involved weighting of a one dimensional argument (diameter) by a two dimensional fraction (area). Because of that the mean grain diameter D_{Aw} calculated from area-weighted grain area A_{dA}, which is not integrated in standard EBSD software, is also used.

$$D_{Aw} = 2 \cdot \sqrt{\frac{A_{dA}}{\pi}} \quad A_{dA} = \sum_{i=1}^{k} A_i \cdot \frac{A_i}{A_{ges}}$$

Provided that the sample preparation is of sufficient quality, and an adequate quantity of EBSD measurements is performed it was found during this investigation, that the grain size distribution could be represented by a log-normal distribution if the area fraction is plotted against the grain diameter. Consequently, we also calculated the grain diameter D_{ln} from fitting the lognormal distribution to a histogram of the class width normalized area fraction with 30 logarithmic equidistant classes of the diameter.

$$f(x) = \frac{A}{\sqrt{2\pi} \cdot \sigma \cdot x} \cdot e^{\frac{-(\ln x - \mu)^2}{2\sigma^2}}$$

$$D_{ln} = e^{\mu + \frac{\sigma^2}{2}}$$

In general, the microstructure of heavy aluminum bonding wires can be divided into two different types. The first type of aluminum wires

consists of very large grains relative to the wire diameter (Figure 6). The ratio of mean grain size and wire diameter $R = D_{grain}/D_{wire}$ is approximately 0.15. These wires typically show a preferred $\langle 001 \rangle$ orientation. At least five EBSD measurements should be used for quantitative characterization of such an aluminum wire.

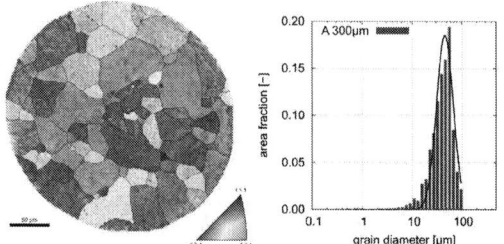

Figure 6: Results of EBSD analysis of wire A with a diameter 300µm (left: typical grain structure image using an inverse polfigure color code; right: overall grain size distribution of EBSD measurements from seven wire sections)

The second type of aluminum wires consist of a very fine grained microstructure with a preferred $\langle 111 \rangle$ orientation with an additional core of preferred $\langle 001 \rangle$ orientation, as shown in Figure 7. For such an aluminum wire only three EBSD analyses are necessary to get a representative log normal distribution.

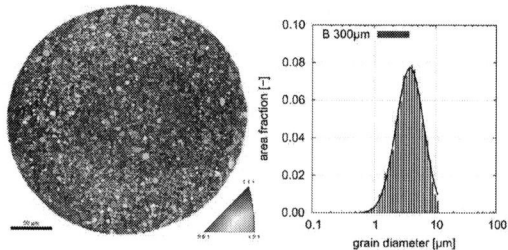

Figure 7: Results of EBSD analysis of wire B with a diameter 300µm (left: typical grain structure image using an inverse polfigure color code; right: overall grain size distribution of EBSD measurements from three wire sections)

In Figure 8 it can be seen that the mean grain diameters D_{Aw} and D_{ln} are always higher than the diameter D_{dA}. The difference between D_{Aw} and D_{dA} is with 13±4% nearly constant. In contrast, the difference between D_{ln} and D_{dA} is negligible for the wires with small grain sizes but immense for wires with large grain sizes. Here, the according grain size distributions seem to be cut off at larger grain diameters, so that a boundary constraint for the grain growth during thermal treatment could be assumed when the grain size is in the range of the wire diameter. From that it is concluded that D_{dA} and D_{Aw} are more realistic values to characterize the grain diameters. However, because of the comparability to other experimental results, the mean value D_{dA} will

978-1-4244-4722-0/09 $25.00
© 2009 IMAPS-ITALY

be used in the following discussions; even though the - in our opinion - more realistic value D_{Aw} should be used when quantifying the results.

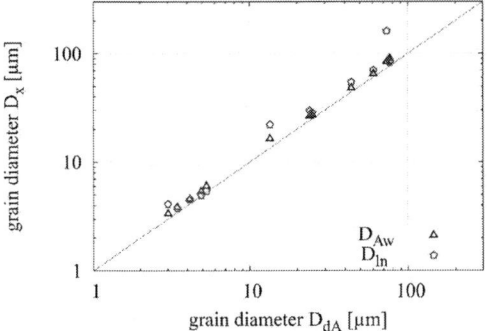

Figure 8: Comparison of mean grain diameters D_{Aw} and D_{ln} to the mean grain diameter D_{dA}

The results of the EBSD analyses of all aluminum wires are summarized in Figure 9. There is no global correlation between the mean grain diameter D_{dA} and the wire diameter, but a slightly decrease of the grain diameter with decreasing wire diameter is found within each wire material. Furthermore the grouping into fine-grained and coarse-grained wire materials is evident.

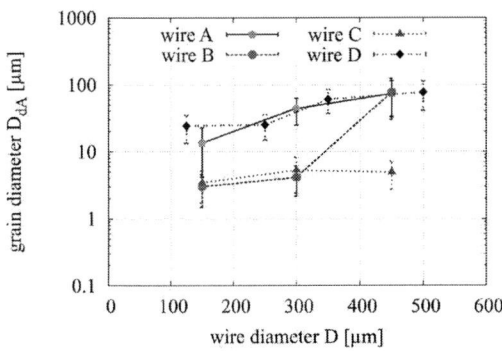

Figure 9: Correlation of mean grain size and wire diameter for different Al wire materials

Correlation of Microstructure and Macroscopic Mechanical Properties

Comparing the macroscopic mechanical properties and microstructure shown before it is conspicuous that the found groupings in mechanical properties and microstructure correlate with each other. The wire group with the lower initial yield stress corresponds to the large-grained microstructure and the group with the higher initial yield stress accords to the fine-grained microstructure. As mentioned before the classical approach when correlating the microstructure and the mechanical properties is the Hall-Petch relation. Hall and Petch showed independently from each other that the lower yield point, or the cleavage strength of polycrystalline steel and iron lies on a straight line when plotted against the inverse of the

square root of the grain size [7] [8]. The intersection of the linear fit with the y-axis corresponds to the yield stress of a single crystal. They concluded that these finding correlate to the dislocation theory of Eshelby, Frank and Nabarro [9]. Here, it is assumed that the dislocations are stored before a grain boundary until the critical shear stress to cross the boundary is achieved. From each additional dislocation, the force $K=\tau b$ is acting on the already stored dislocation so that for a maximum storage length of $D/2$ the relationship between the yield stress and the inverse square root of the grain diameter follows [4]. Considering additionally the single crystal yield stress for relatively large grains, the Hall-Petch relation can be derived.

$$\sigma_Y = \sigma_0 + k \cdot d^{-\frac{1}{2}}$$

When accordingly plotting our results, the predicted linear relation between initial yield stress and square root of grain size is found (Figure 10).

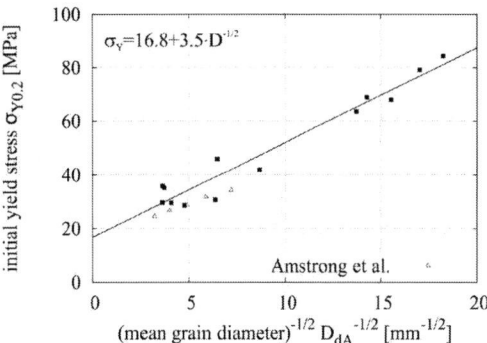

Figure 10: Hall-Petch relation of aluminum bonding wires compared to macroscopic aluminum specimen taken from [11]

The Hall-Petch parameters for the aluminum bonding wires were determined to be $\sigma_0=16.8$ und $k=3.5$. The parameter σ_0 is in the range of elsewhere published data for aluminum, whereas the constant k is slightly higher, see [10] [11] [12]. The difference is rather small when taking into account that the data presented in literature only observe a upper range in grain size from 20µm to 600µm ($7mm^{-1/2}$-$1.3mm^{-1/2}$), whereas in this study the Hall-Petch relation is investigated also at significantly smaller grain sizes down to 3µm ($18\ mm^{-1/2}$).

Local Mechanical Characterization using Microindentation Experiments

In many application cases, such as in failure analysis for real components, tensile testing of a relatively long sample section is not possible. A classical approach for characterizing rather local mechanical properties of real components is micro- or nanoindentation testing. For comparing different real components in failure analyses, indentation methods have proven to be indispensable. There have been a lot attempts for additionally converting the complex local material response of an indentation experiment to a macroscopic uniaxial material behavior since Tabor showed the comparability of indentation results to tensile tests [13]. Especially with the potential of finite element methods and optimization routines, calculations of the uniaxial stress/strain behavior from spherical indentation experiments have often been used. Nevertheless, the transferability of local deformation behavior to uniaxial material properties and the sensitivity of indentation experiments to artifacts are still in discussion and subject of scientific investigations [14]. Instrumented indentation tests on metallographically prepared cross sections of the aluminum bonding wires were performed using a Shimadzu DUH 202 tester equipped with a Vickers indenter. In analogy to [15], we calculated the indentation hardness H_{IT} considering the real indenter tip geometry. At a maximum load of 20mN indentation depths of 1.2μm to 1.6μm were achieved. Force/displacement curves are exemplarily shown for wires B and D in Figure 11.

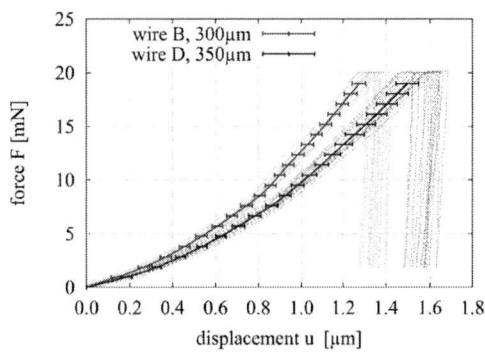

Figure 11: Force/displacement curves of wire B and D (measured curves and mean curve with standard deviation)

The differences in maximum indentation depths are rather small compared to the standard deviations, even though these wires have large differences in yield stress (wire B: 68.0±0.7MPa; wire D: 29.5±0.3MPa). The resulting hardness shows also significant differences (wire B: 417±21MPa; wire D: 302±16MPa), but lower than the difference in yield stress. From Figure 12 it can be seen that an increase of hardness corresponds to an increase of yield stress and ultimate tensile stress, as expected. Unfortunately, there is a great scatter of the data points in that small range of hardness so that no precise correlation can be derived from indentation experiments. This becomes even more obvious when the correlation between initial yield stress and hardness is compared with classical empirical approaches like the one of Tabor [13]. Here, the hardness is assumed to be approximately three times the yield stress for ideal plastic material, or three times the yield stress at 8% plastic strain for materials which work harden. From Figure 12 it can

also be seen that the results do not correspond to Tabor's correlation.

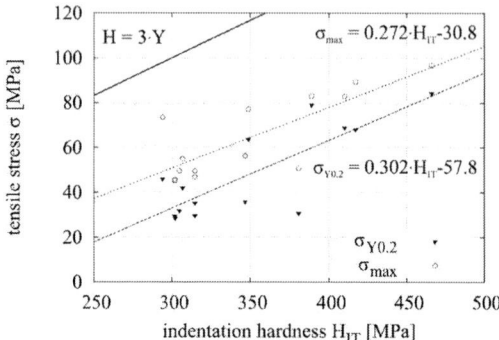

Figure 12: Correlation of initial yield stress $\sigma_{Y0.2}$ and ultimate tensile stress σ_{max} to indentation hardness H_{IT} and comparison to Tabor's empirical approach H=3·Y (plotted as σ=1/3·H)

It can be concluded that indentation experiments could be used qualitatively for the comparison of local material behavior, but a quantitatively transferability to macroscopic homogeneous material behavior for heavy aluminum wires is questionable. A possible reason might be that the indentation size is in the range of the grain size for these materials.

Local Mechanical Characterization using Micro Compression Test

As an alternative to microindentation experiments, the micro compression test can be used for determining the local mechanical properties of real components, see Figure 13. In this experiment a small cylinder is metallographically prepared from the wire region of interest. When the wire is at least 300μm thick and the length of the cylinder is approximately twice the diameter, these wire segments can be tested like in a conventional compression test in upright position; whereas cylinders of smaller wire diameters can be tested by loading them perpendicular to the wire axis.

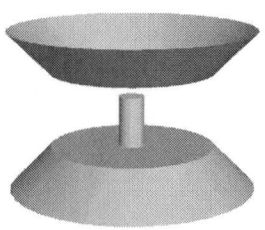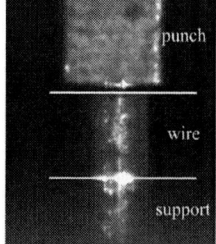

Figure 13: Principle of micro compression test (left) and optical image of sample before compression test

Because of the small dimensions, experimental boundary conditions such as plan parallelism and friction effects become even more important than for macroscopic specimen. Even

978-1-4244-4722-0/09 $25.00
© 2009 IMAPS-ITALY

though the feasibility of this test has been shown before [16], the proof of the correctness was missing before this investigation. Small cylinders with a length of approximately twice of the diameter were prepared from each wire with a diameter equal or larger than 300µm. For support and loading, two polished steel cylinders were used in a plan parallel arrangement in a universal testing machine. After placing the wire cylinder on the support, the sample was loaded under displacement control up to 60% compression. The force/displacement curves were corrected in a similar way as the tensile experiments described before, but without the Young's modulus correction. For estimating the possible effect of friction the Siebel correction as described in [17] was used. The micro compression test is very reproducible as it could be seen from the mean engineering stress/strain curves and the standard deviation of seven experiments of each wire material shown in Figure 14.

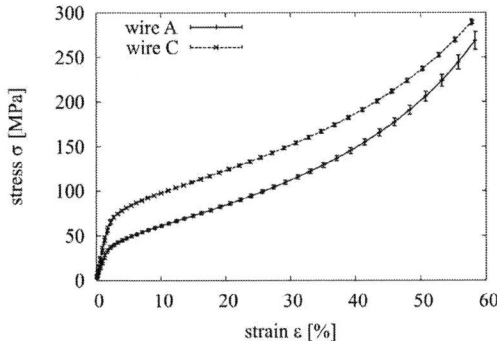

Figure 14: Mean engineering stress/strain curves of wire A and C with diameters of 450µm

The calculated flow curves for these wires are exemplarily shown in Figure 15. The difference in true stress considering friction becomes even more important the higher the strain, but the influence is rather small for the assumed friction coefficient of µ= 0.2 in the range up to 15% strain.

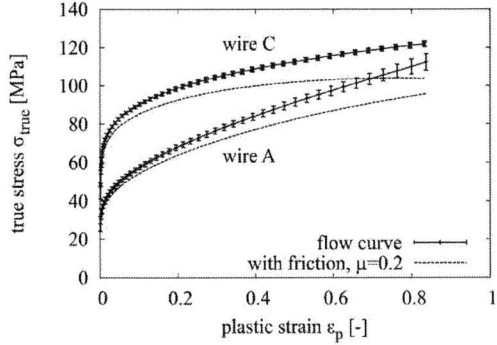

Figure 15: Mean flow curves of wire A and C with diameters of 450µm (dotted lines: flow curves considering a friction coefficient of µ=0.2)

Figure 16 summarizes the results of the micro compression tests for all investigated wire materials and wire diameters. Here, the initial yield stress determined in the micro compression test is presented as a function of the corresponding yield stress determined in the conventional "macroscopic" tensile test. It can be seen that the micro compression results are in a very good agreement with the tensile yield stress results for all tested aluminum bonding wires, even though they are slightly lower. The standard deviation of the yield stress is in the same range as for the tensile tests.

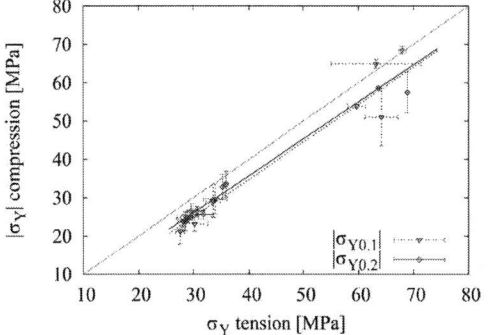

Figure 16: Comparison of initial yield stresses determined in tensile tests and compression tests

In addition, also the complete flow curves derived from the micro compression tests are in a very good agreement with the tensile flow curves (Figure 17). A pronounced yield point such as in the tensile test results for wire material C was not found in any micro compression experiment.

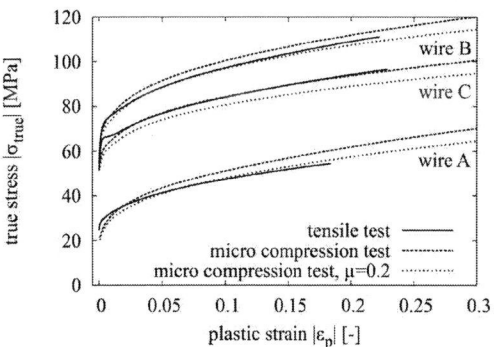

Figure 17: Comparison of flow curves determined by tensile tests and compression tests

It is currently not understood if this effect of a pronounced initial yield point depending on the loading situation could be explained by the typical explanation for pronounced yield points - where dopants pin on dislocations - or another deformation mechanism resulting from the wire drawing is responsible for this effect.

Summary

Using an advanced analysis of the macroscopic mechanical properties determined in tensile tests, we have shown that the plastic material behavior is well reproduced by hardening laws of both Ramberg-Osgood and Chaboche. In finite element simulations of real components, these hardening laws should be used replacing the simplifications of ideal plastic or bilinear material laws. Analyzing the microstructure determined by the electron backscatter diffraction method it could be seen that there is a qualitative variation in the mechanical behavior of the aluminum wire materials. One group of wires with large grains compared to the wire diameter resulted in significant lower yield stresses while a second group of wires with relatively small grains resulted in higher yield stresses. However, a quantification of electron backscatter diffraction results requires an adapted statistical analysis of the grain size distributions. During these investigations, specific procedures were elaborated which could be based on a log normal distribution of grain sizes. The number of required EBSD scans to characterize a sufficient number of grains depends on the grain diameter. Based on these results, the relationship between the initial yield stress and the grain diameter was shown to follow a Hall-Petch type relationship with a linear dependence of initial yield stress on the square root of grain diameter. Since in many cases also the local mechanical properties have to be determined, results of indentation and micro compression tests were compared to tensile testing. It was shown that microindentation methods failed to represent the mechanical properties with the required accuracy and sensitivity, probably due to the large grain size compared to indentation size. In contrast, the micro compression test using small wire sections tested in compression showed very high potential for quantitative characterization of the initial yield stress and also the hardening behavior of small wire samples. Also, the flow curves derived from the micro compression tests are in very good agreement with those of tensile tests. Consequently, the micro compression test can be used as an alternative for characterizing the hardening behavior when it is not possible to perform adequate tensile tests, such as in failure analysis of specific bond loops. In addition, the micro compression test forms the basis for further development of cyclic tension/compression tests for determining the cyclic hardening behavior, so that more realistic device simulations and lifetime predictions are possible.

References

[1] L. Merkle et al., "Mechanical Fatigue Properties of Heavy Aluminum Wire Bonds for Power Applications", Proceedings of the 2nd Electronics Systemintegration Technology Conference (ESTC), Greenwich, UK, September 1-4, pp. 1363-1368, 2008.

[2] S. Ramminger et al. "Reliability Model for Al Wire Bonds subjected to Heel Crack Failures", Microelectronics Reliability, Vol. 40, pp. 1521-1525, 2000.

[3] J. Wilde, "Lebensdauerprognose von Drahtbond-Verbindungen für die Mechatronik mittels FEM", Proceedings of DVS/GMM-Conference, Fellbach, Germany, February 6-7, 2002.

[4] G. Gottstein, "Physikalische Grundlagen der Materialkunde", Springer, second edition, Berlin, Chapter 6, pp. 190-197, 2001.

[5] J. Lemaitre and J. Chaboche, "Mechanics of Solid Materials", Cambridge University Press, Cambridge, Chapter 5, pp. 163-176, 2002.

[6] J. Lemaitre and J. Chaboche, "Mechanics of Solid Materials", Cambridge University Press, Cambridge, Chapter 6, pp. 288-300, 2002.

[7] E.O. Hall, "The Deformation and Aging of Mild Steel: III Discussion of Results", Proceedings of the Physical Society, Vol. 64, No. 9, pp. 747-753, March, 1951.

[8] N.J. Petch, "Cleavage Strength of Polycrystals", Journal of the Iron and Steel Institute, Vol. 174, pp. 25-28, Mai 1953.

[9] Eshelby et al., "The Equilibrium of Linear Arrays of Dislocations", Philosophical Magazine, Vol. 42, No. 327, pp. 351-364, 1951.

[10] K. K. Ray and K. Chakraborthy, "Grain Size Dependence of Flow Stress in Aluminum", Journal of Material Science Letters, Vol. 13, pp. 919 – 921, 1994.

[11] R. Armstrong et al., "The Plastic Deformation of Polycrystalline Aggregates", Philosophical Magazine, Vol. 7, No. 73, pp. 45-58, 1962.

[12] I. Kovács, "Grain Size Dependence of the Work Hardening Process in Al99.99", physica status solidi (a), Vol. 194, No. 1, pp. 3-18, 2002.

[13] D. Tabor, "The Hardness of Solids", Review of Physics in Technology, Vol. 1, pp. 145-178, 1970.

[14] L. Liu, "Can indentation technique measure unique elastoplastic properties?", Journal of Materials Research, Vol. 24, No. 3, March 2009.

[15] EN ISO 14577, "Instrumented indentation test for hardness and materials parameters", German version, 2002.

[16] C. Dresbach et al., "Test Methods for Characterizing the Local Plastic Deformability of Bonding Wires", Proceedings of the 1st Electronics Systemintegration Technology Conference (ESTC), Dresden, Germany, September 5-7, pp. 732-740, 2006.

[17] T. Schenk et al., "A simple Analogous Model for the Determination of Cyclic Plasticity Parameters of Thin Wires to Model Wire Drawing", Journal of Engineering Materials and Technology, Vol. 129, pp. 488-495, July 2007.

Effect of Microstructure Design on Reliability of FBGA Lead-Free Solder Joints

F.X. Che*, and J.E. Luan

STMicroelectronics, 629 Lorong 4/6 Toa Payoh, Singapore 319521

*Email: faxing.che@st.com

Abstract

In this paper, the fine pitch ball grid array (FBGA) assembly with lead free solder was tested under thermal cycling condition. Finite element modeling and simulation was conducted. The microstructure designs of via in pad of substrate and through hole (via) in PCB under the solder joint were simulated. The solder mask definition (SMD) and non-solder mask definition (NSMD) on PCB board side was also simulated. The global-local modeling technique was implemented in FE simulation for FBGA assembly. Based on global model result, the critical solder joint location was detected and then the fine meshed local model was created for critical solder joint including the detailed microstructure design. Fatigue life prediction of solder joint was carried out based on FE simulation results. Some summaries can be made from FE simulation results. Through hole in PCB changes the failure site from package side solder joint interface to board side solder joint interface. For the case of solder joint with SMD on PCB side, through hole in PCB reduces the solder joint fatigue life significantly. Via in pad of substrate slightly reduces the solder joint fatigue life.

Key words: Thermal cycling reliability test, Finite element modeling and simulation, Lead-free solder joints, Fatigue life prediction, Microstructure effect.

Introduction

Miniaturization is the trend for microelectronic product to cater for the portable design requirement. Fine pitch of solder joint, via in pad of substrate design, and through via in PCB board is widely used in microelectronic packaging design. These designs affect the solder joint reliability significantly. Therefore, design for reliability is the big challenge for such microelectronic products. Thermal cycling test is commonly used for accelerated testing of electronic assemblies to assess the solder joint reliability. Cyclic stress-strain will deform the solder material when subjected to thermal cycling due to CTE (coefficient of thermal expansion) mismatch among different electronic materials, thus inducing thermal fatigue failure in solder joints. Another fast way to evaluate the solder joint reliability is using finite element (FE) modeling and simulation to simulate the solder joint stress strain behavior.

It was known that combination of experiment and finite element modeling and simulation is useful and powerful way to assess the reliability of electronic assemblies. Finite element modeling and simulation is a powerful tool and take an important role in solder fatigue life prediction. Further miniaturizations of component size, microstructure design in packaging and higher I/O counts are expected trends in electronic packaging applications. In order to reduce the element size in the FE simulation, some reduced models were used by researchers, including slice or strip model [1], one-eighth or octant model [2-4], 2D model [4-6]. Some trade-off in accuracy is expected in these simple models. Submodeling technique and global-local modeling were widely used in FE simulation for electronic assemblies to reduce requirement of computer resource and time-saving without lose of accuracy [7-9]. FE simulation results showed that 2D and slice models underestimate the solder joint fatigue life compared to 3D quarter model [10]. In this study, 3D quarter model was used for FE modeling and simulation.

In this paper, the fine pitch ball grid array (FBGA) assembly with Sn-Ag-Cu lead free solder was tested under thermal cycling condition. Finite element modeling and simulation was conducted for thermal cycling load. The microstructure designs of via in pad of substrate and through hole (via) in PCB under the solder joint were simulated to investigate the effect of microstructure on solder fatigue life and failure distribution. The solder mask definition (SMD) and non-solder mask definition (NSMD) on PCB board side was also investigated through FE simulation. Fatigue life prediction of solder joints was carried out based on FE simulation results.

978-1-4244-4722-0/09 $25.00
© 2009 IMAPS-ITALY

Reliability Test and Analysis

Fine pitch ball grid array (FBGA) samples was used for thermal cycling test. FBGA package has the geometry size of 7mm×7mm×0.6mm. Sn-Ag-Cu solder ball with diameter of 0.3mm was used in FBGA assembly with solder joint pitch of 0.5mm. The geometry size of die is 4.8mm×4.8mm×0.17mm, which is mounted on 0.2mm substrate through die adhesive and then encapsulated by epoxy mold compound. Solder mask define (SMD) is designed on substrate side with SMD opening of 0.28mm. There are 157 solder joints used in FBGA assembly as shown in figure 1. Via-in-PCB designed (also called hole in PCB in this paper) was used for many solder joints as indicated with blue dots in figure 1. Figure 2 shows the detailed cross-section view for via-in-PCB design. Cross-section view in figure 2 is from A-A cutting line shown in figure 1. Via-in-pad design is also used for some substrate pads as shown in figure 3. Therefore, different via types are designed for solder joints: via-in-pad design for both substrate and PCB sides, via-in-pad on substrate side and no via on PCB side, no via-in-pad on substrate side and via-in-pad on PCB side, no via-in-pad design for both substrate and PCB sides. The effect of different via designs on solder joint reliability will be investigated by thermal cycling test and FE modeling and simulation. Thermal cycling test with temperature from -40 °C to 125 °C and 2 cycles per hour was conducted for FBGA assembly. Figure 4 shows the typical solder joint fatigue failure with failure site close to intermetallic compound (IMC) layer when subjected to thermal cycling. Failure analysis also shows that the failure rate is higher for solder joint with hole under joint, which indicates that hole in PCB affects the solder joint reliability.

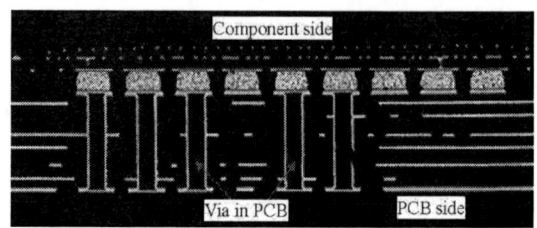

Figure 2: Via-in-PCB design for FBGA assembly.

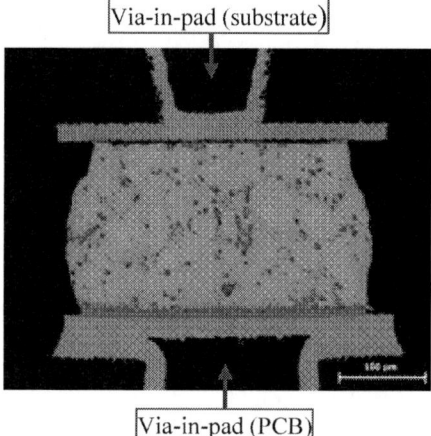

Figure 3: Via-in-pad for both substrate and PCB sides.

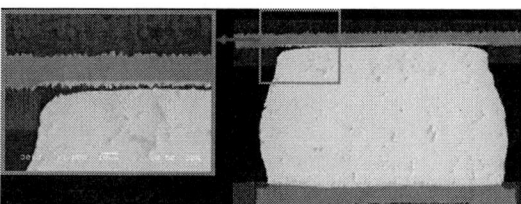

Figure 4: Typical solder joint fatigue failure.

Finite Element Modeling and Simulation

Comprehensive finite element modeling and simulation was conducted to investigate the effect of microstructure on solder joint fatigue life. Total 8 cases were simulated as shown in Table 1. SMD or NSMD is for PCB side, hole means through hole (via) in PCB board, via means via-in-pad of substrate. So the effect of hole, via, SMD on solder joint reliability will be investigated for FBGA assembly. Consider NSMD-no hole-no via case as control case in this study. Figure 5 shows the quarter FE model for FBGA assembly. Symmetry boundary condition was applied onto cutting planes along centerline.

Global-local modeling technique was used in this study to reduce the requirement of computing resources and save the solving time without loss of accuracy. The coarse global model was simulated

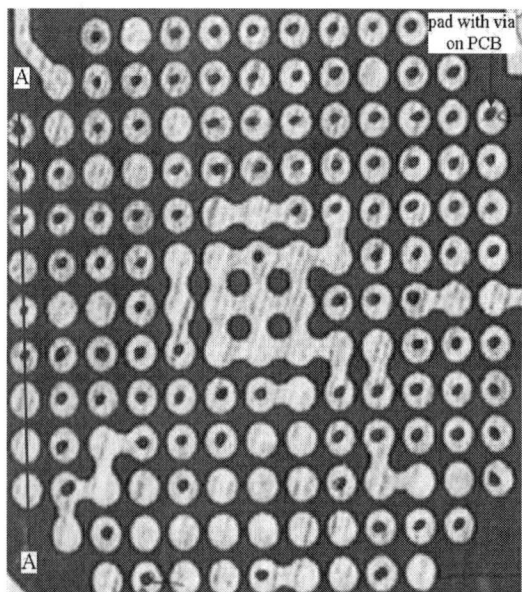

Figure 1: Layout of solder joint and via in PCB.

978-1-4244-4722-0/09 $25.00
© 2009 IMAPS-ITALY

for entire structure to find the critical solder joint location. Then, the detailed fine meshed local model for critical solder joint was created to extract the accurate FE simulation result. Two different modeling methods were compared in this study. The first one is called submodeling method, that is, the individual submodel was created for critical solder joint. The DOFs from global model result were transferred to the corresponding cut boundary of submodel as applied boundary condition. The simulation result for solder joint can be obtained from the submodel. The second one is called global-local modeling method. In this method, the global model was solved firstly, and then the critical solder joint was found from global model result. The fine meshed local model for critical solder joint was created and then was connected with global model through constrain equation along the interfaces between global and local models. It was found from simulation results that both modeling methods lead to almost the same results. So the global-local with constrain equation was chosen for this study.

Different local models listed in Table 1 were created to investigate the effect of microvia and hole on solder joint failure site and life prediction. Figures 6 and 7 show the local models with both via-in-pad on substrate and hole in PCB for both PCB pad designs of NSMD and SMD, respectively. For other local models of without via and/or hole, the element meshing is the same with just considering change of via and/or hole materials to corresponding substrate or PCB materials. In FE simulation, material properties, especially for solder material, are important issue and they will affect solder fatigue life prediction significantly. Table 2 list the material properties used in FE modeling and simulation. The temperature dependent modulus and viscoplastic Anand constitutive model was simulated for Sn-Ag-Cu solder [11]. Volume averaging technique is frequently introduced for calculating strain energy density to minimize the effect of mesh sensitivity and stress concentration on fatigue life prediction. The whole interface layer elements based averaging method was chosen for fatigue life control parameter calculation. Figure 8 shows the interface layer elements for both top solder/component interface and bottom solder/PCB interface. The solder joint fatigue life prediction will be conducted for both component and PCB side interface layers to observe critical failure location.

Table 1: FE simulation cases.

Case 1-Control		no hole, no via
Case 2	NSMD-PCB side	hole, via
Case 3		hole, no via
Case 4		no hole, via
Case 5	SMD-PCB side	no hole, no via
Case 6		hole, via
Case 7		hole, no via
Case 8		no hole, via

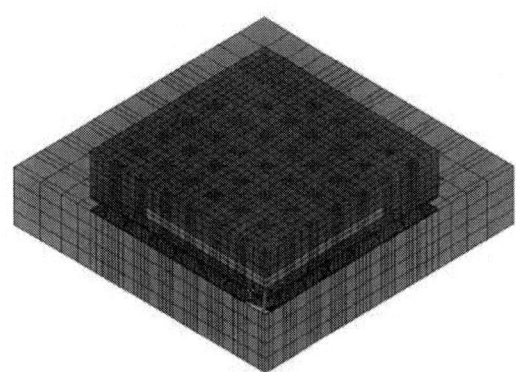

Figure 5: Quarter FE global model for FBGA.

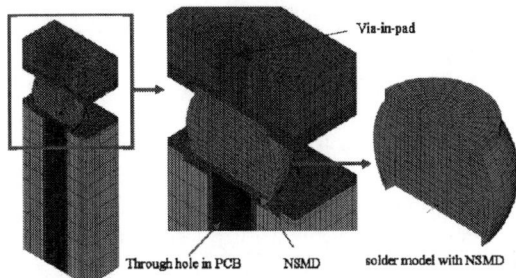

Figure 6: Detailed FE local model with via and hole for NSMD on PCB side (case 2).

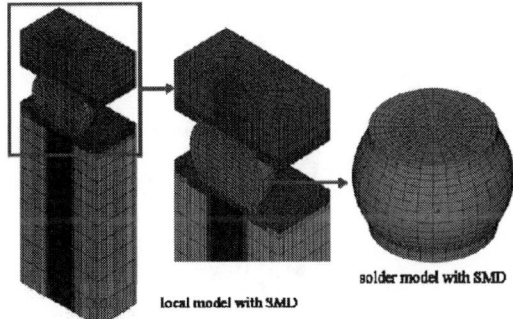

Figure 7: Detailed FE local model with via and hole for SMD on PCB side (case 6).

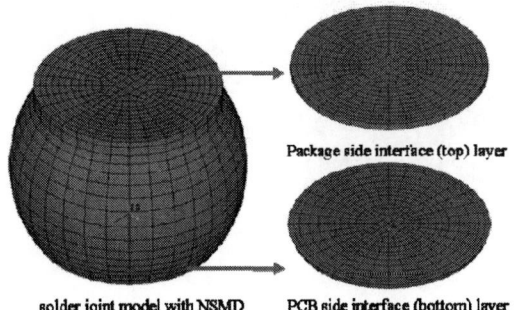

Figure 8: FE solder joint model and top and bottom layer elements for volume-averaging.

Table 2: Materials and material properties used in finite element modeling and simulation.

Materials	Modulus (GPa)	Poisson ratio	Tg (C)	CTE1 (ppm/C)	CTE2 (ppm/C)
Si die	131	0.3	-	2.8	-
Mask AUS380 (via fill)	2.4	0.3	100	60	161
BT Core (Mitsubishi CCL-HL832NX-A)	x,y:26 z:11	xy:0.19 xz,yz:0.39	170	x,y:15 z: 28.4	---
Die attach (Loctite QMI 536)	0.3	0.3	-31	93	174
Mold GE100LF1-2	24/0.45	0.3	140	8	34
FR4 PCB	x,y:25 z:11	xy:0.11 xz,yz:0.39	--	x,y:17 z: 60	---
Solder	45.8@-35C, 39.7@25C 27.9@75C, 23.3@125C	0.4		22	
Hole fill (THP-100DX1)	4.5@25C, 2.9@100C 0.56@150C, 0.25@200C	0.31	160	32	115

Result and Discussion

In FE simulation, three thermal cycles were simulated because it is enough to obtain converged strain or strain energy density results [11]. The critical solder joint can be obtained based on maximum strain energy density accumulation during thermal cycling simulation. Figure 9 shows the accumulated plastic strain energy density for all solder joints and the critical solder joint locates under silicon die corner, which is selected for further local model analysis. Figure 10 shows the critical solder joint result for control case (NSMD on PCB, no via in pad and no hole in PCB). It can be seen from figure 10 that strain energy density of solder/component top interface is higher than that of solder/PCB bottom interface, which indicates that component/solder interface layer is more prone to failure than PCB/solder interface layer for control case. The volume-averaged method was used for plastic strain energy density calculation:

$$\Delta W_{ave} = \frac{\sum \Delta W \bullet V}{\sum V} \quad (1)$$

where the volume is usually from the interface layer as shown in figure 10.

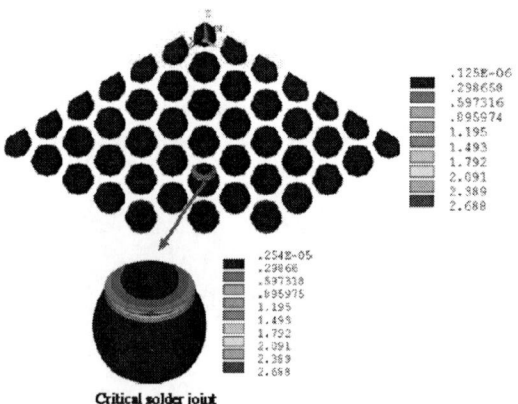

Figure 9: Plastic strain energy density contour of solder joints from global model result.

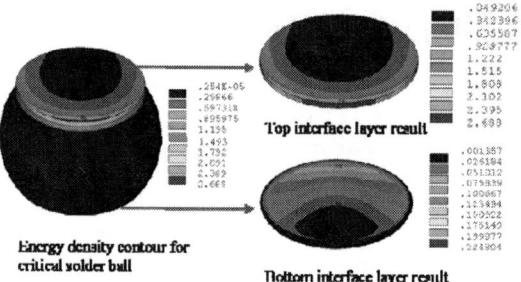

Figure 10: Plastic strain energy density contour of critical solder joint from local model for control case (without via and hole).

Figures 11 and 12 show the critical solder joint result for case 2 (NSMD on PCB, via in pad and hole in PCB) and case 6 (SMD on PCB, via in pad and hole in PCB), respectively. It can be seen that strain energy density of solder/PCB bottom interface is higher than that of solder/component top interface, which indicates that PCB /solder interface layer is more prone to failure than component /solder interface layer when the designed hole in PCB is under the critical solder joint. Therefore, the existing hole under solder joint changes failure location from component/solder interface to PCB/solder interface due to large CTE mismatch between hole filling material and solder material. Compared case 2 (NSMD) with case 6 (SMD), it can be seen that SMD design on PCB reduces solder joint reliability significantly. Solder joint fatigue life prediction was conducted for each case using energy-based fatigue life model for Sn-Ag-Cu solder [12]:

$$N_f = 327.115 \times (W_p)^{-1.288} \quad (2)$$

where W_p is the volume-averaged plastic strain energy density from simulation result. Plastic strain energy density results based on both component and PCB side interface layers were calculated for solder joint life prediction. FE simulation result and solder joint life prediction for all cases are summarized in Table 3. Some conclusions can be made from Table 3. For solder joint without PCB hole, the failure site locates the solder/component interface layer. When PCB hole is designed underneath the critical solder joint, failure site will be changed from solder/component interface to solder/PCB interface. For NSMD cases, PCB hole design leads to similar solder joint fatigue life (above 3000 cycles) with different failure sites: solder joint failure at component site for no-PCB hole case and solder joint failure at PCB site for PCB hole case. For SMD cases, PCB hole design reduces solder joint fatigue life significantly (from about 3300 cycles to about 1700 cycles). For solder joint with SMD and without PCB hole design, solder joint fatigue life increases more than 25% compared to the corresponding cases without NSMD. The effect of via-in-pad of substrate on solder joint fatigue life is not significant. For solder joint without PCB hole, via-in-pad reduces solder joint fatigue life by less than 8%. For solder joint with PCB hole, the effect of via-in-pad on solder joint fatigue life can be ignored because solder joint failure site is not in via side but in PCB hole side.

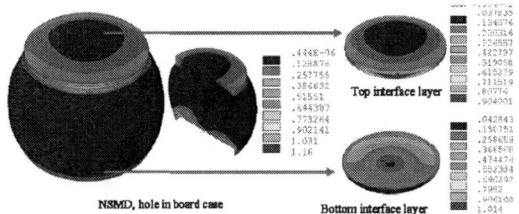

Figure 11: Plastic strain energy density contour of critical solder joint from local model for case 2 (NSMD, via and hole).

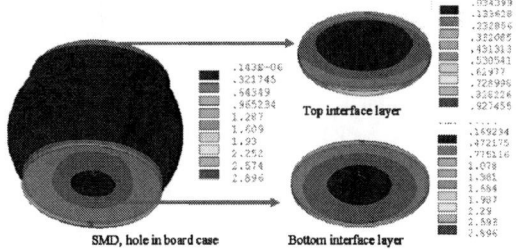

Figure 12: Plastic strain energy density contour of critical solder from local model for case 6 (SMD, via and hole).

Table 3: Finite element simulation result and fatigue life prediction summary.

Different cases		Energy density (Mpa)		Life prediction	
		pkg side	PCB side	pkg side	PCB side
NSMD	no hole, no via	0.1658	0.0199	**3309**	50910
	hole, via	0.0793	0.1633	8559	**3377**
	hole, no via	0.0758	0.1665	9065	**3291**
	no hole, via	0.1748	0.0186	**3092**	55275
SMD	no hole, no via	0.1376	0.1240	**4208**	4815
	hole, via	0.0662	0.2667	10808	**1794**
	hole, no via	0.0639	0.2710	11306	**1758**
	no hole, via	0.1464	0.1198	**3886**	5030

Summary and Conclusion

In this paper, the fine pitch ball grid array (FBGA) assembly with lead free solder was tested under thermal cycling. Finite element modeling and simulation was conducted to investigate the effect of microstructure design, such as via in pad of substrate and through hole (via) in PCB, on solder joint thermal fatigue reliability. The solder mask definition (SMD) and non-solder mask definition (NSMD) on PCB board side was also simulated. Two modeling methods, global-local modeling with constrain equation on cut boundaries and submodeling technique, lead to almost the same results. The critical solder joint location was detected from FE simulation and consistent with failure analysis from test. Through hole in PCB changes the failure site from package side solder joint interface to PCB board side solder joint interface. For NSMD cases, PCB hole design leads to similar solder joint fatigue life with different failure sites. For SMD cases, PCB hole design reduces solder joint fatigue life significantly. Via in pad of substrate design slightly reduces the solder joint fatigue life. The effect of SMD on solder joint fatigue life has different trend. The worst case is solder joint with PCB hole design and SMD design on PCB side.

Acknowledgements

Authors would like to thank Xavier Baraton from STMicroelectronics, Corporate Packaging, Engineering & Automation (CPA) for management support and acknowledge colleagues, Luc Petit and Daniel Yap, for their support on reliability test and failure analysis.

978-1-4244-4722-0/09 $25.00
© 2009 IMAPS-ITALY

References

[1] Pang, J.H.J., Chong, D.Y.R., and Low, T.H., "Thermal Cycling Analysis of Flip-Chip Solder Joint Reliability", IEEE Transaction on Components and Packaging Technologies, Vol. 24, No. 4, pp. 705-712, 2001.

[2] Pang, H.L.J., Seetoh, C.W., and Wang, Z.P., "CBGA Solder Joint Reliability Evaluation Based on Elastic-Plastic-Creep Analysis", Journal of Electronic Packaging, Vol. 122, pp. 255-261, September, 2000.

[3] Pang, H.L.J., and Chong, Y.R., "Flip Chip on Board Solder Joint Reliability Analysis Using 2-D and 3-D FEA Models", IEEE Transaction on Advanced Packaging, Vol.24, No. 4, pp. 499-506, November, 2001.

[4] Yao, Q., and Qu, J., "Three-Dimensional Versus Two-Dimensional Finite Element Modeling of Flip-Chip Packages", Journal of Electronic Packaging, Vol. 121, pp. 196-201, 1999.

[5] Che, F.X., Low, T.H., Pang, H.L.J., Lin, W. C.C., Chiang, C. L.S., and Yang T. K. A., "Modeling Thermo-Mechanical Reliability of Bumpless Flip Chip Package", Proceedings of 54th Electronic Components and Technology Conference (ECTC), Las Vegas, Nevada, June 1-4, pp.421-426, 2004.

[6] Che, F.X., Zhang, X., Zhu, W.H., and Chai, T.C., "Reliability Evaluation for Copper/Low-k Structures Based on Experimental and Numerical Methods", IEEE Transactions on Device and Material Reliability, Vol. 8, NO. 3, pp. 455-463, September, 2008.

[7] Zhu, J., "Three-Dimensional Effects of Solder Joints in Micro-Scale BGA Assembly", Journal of Electronic Packaging, Vol. 121, pp. 297-302, 1999.

[8] Zhu, J., Quander, S., and Reinikainen, T., "Global/Local Modeling for PWB Mechanical Loading", Proceedings of 51st Electronic Components and Technology Conference, Orlando, FL, USA, May 29-June 1, pp. 1164-1169, 2001.

[9] Che, F.X., Pang, H.L.J., Zhu, W.H., Sun, W., Sun, Y.S., Wang, C.K., and Tan, H.B., "Development and Assessment of Global-Local Modeling Technique Used in Advanced Microelectronic Packaging", Proceedings of *EuroSimE 2007*, London, UK. Apr. 15-18, pp. 375 -381, 2007.

[10] Che, F.X., and Pang, H.L.J., "Thermal Fatigue Reliability Analysis for PBGA with Sn-3.8Ag-0.7Cu Solder Joints". Proceedings of 2004 Electronics Packaging Technology Conference (EPTC), Singapore, December, pp. 787-792, 2004.

[11] Che, F.X., Luan, J.E., and Baraton, X., "Effect of silver content and nickel dopant on mechanical properties of Sn-Ag-based solders", Proceedings of 58th Electronic Components and Technology Conference, May 27-30, pp.485-490, 2008.

[12] Che, F.X., Luan, J.E., Yap, D., Goh, K.Y., and Baraton, X., "Thermal cycling fatigue model development for FBGA assembly with Sn-Ag-based lead-free solder", Proceedings of 33rd International Electronics Manufacturing Technology Conference (IEMT), Penang, Malaysia, November 5-6, pp. 1-6, 2008.

Thermoelastic Properties of Printed Circuit Boards: Effect of Copper Trace

Hu Guojun, Goh Kim Yong, Luan Jing-en, Lim Wee Chin and Xavier Baraton

STMicroelectronics, 629 Lorong 4/6 Toa Payoh, Singapore 319521, Singapore

Tel: +65 63897056. E-mail address: guojun.hu@st.com

Abstract

After encapsulation, thermo-mechanical deformation builds up within the electronic packages due to coefficient of thermal expansion (CTE) mismatch between the respective materials within the package as it cools to room temperature. Printed circuit boards (PCB) are designed and manufactured with a variety of polyamide materials such as solder mask, metallic material such as copper trace, composite materials such as prepreg and core material. Polyamide materials such as solder mask and composite materials such as prepreg play important factor on the total deformation of laminate package due to the large CTE. On the other hand, the patterning of the copper layers also exerts important influence to the thermal mechanical behavior of the substrate due to the consistent large Young's modulus of copper at both room temperature and reflow temperature compared with the small Young's modulus of polyamide materials. Some approximate methods based on rule of mixtures have been used for estimating material properties in layers of copper mixed with interlayer dielectric material. However, few techniques include the effect of copper trace pattern and copper percentage. The detailed comparison of different approximate methods has been done in this paper and a new methodology has been developed to include the effect of Poisson's ratio, copper trace pattern and copper percentage. The equivalent properties of copper trace layer using the new methodology have been compared with the results using the detailed finite element simulation of the actual laminate substrates. The results show that the contribution of the copper trace on in-substrate-plane modulus becomes more and more important with the increase of copper percentage.

Key words: PCB, substrate, copper trace, FEA, Young's modulus, CTE

1. Introduction

Warpage of IC packages after encapsulation is a major concern in package development because of the risk of die cracking if a tensile stress is built up in the die particularly in thin packages [1]. Also excess warpage can cause problems with mounting the package onto a PCB or another package such as package-on-package (PoP). Realistic modeling and analysis of the mechanical performance and reliability of electronic packages requires sophisticated constitutive models for the many complex, non-traditional engineering materials that compose these intricate devices. The proper modeling of the behavior of such plastic materials is becoming important in the reliability studies of these IC packages.

One challenge is the proper simulation of the thermal mechanical properties of the copper trace layer which consist of metallic material and dielectric material. The dielectric material such as solder mask and prepreg play important factor on the total deformation of laminate package due to the large CTE (see Table 1). On the other hand, the metallic material such as the copper trace exerts important influence to the thermal mechanical behavior of the substrate due to its consistent large

Young's modulus from room temperature to solder reflow temperature. There are a few studies on the approximation and computation methods of equivalent laminate properties, but few techniques include the effect of Poisson's ratio, copper trace pattern and copper percentage. The effective thermo-mechanical properties of the trace layer are dependent upon copper trace pattern, copper percentage and properties of the constituent materials. Some study uses one direction orthogonal copper layout to equivalence the nearly symmetrical copper trace pattern [2] which may overestimate the orientation effect of copper trace pattern. On the other hand, some methodology is developed to include both the copper trace pattern density and orientation in actual substrates based on Matlab [3].

With the real circuit drawing file, the copper trace pattern (circuit) will be discretized and a finite element mesh will be generated for analysis in this study. At the same time, equivalent material properties based on theoretical predictions have been calculated and compared with the detailed finite element analysis (FEA). The detailed comparison of different approximate methods has been done in this paper and a new methodology has been proved efficiently with the consideration of copper trace

978-1-4244-4722-0/09 $25.00
© 2009 IMAPS-ITALY

pattern and copper percentage. It is shown that CTE and modulus of elasticity (Young's modulus) are influenced significantly by the copper trace percentage in the copper trace layers of the substrate, while the effect of copper trace orientation on the anisotropic material property is not so obvious due to the nearly symmetrical copper trace pattern. At the same time, it has been found that the contribution of the copper on in-substrate-plane and out-substrate-plane modulus becomes more and more important with the increase of copper percentage. Furthermore, it is found that the unbalance of copper trace percentage has important effect on the package warpage.

2. Micromechanical Analysis of Mixed Materials

From a unidirectional trace pattern, take a representative volume element that consists of equal spaced horizontal traces interspersed with dielectric material (Figure 1).

Based on the rule of mixtures, the equivalent thermo-mechanical properties of the trace layer can be approximated with the following equations [4, 5]:

$$E_1 = E_f V_f + E_m V_m \qquad (1)$$

$$\frac{1}{E_2} = \frac{V_f}{E_f} + \frac{V_m}{E_m} \qquad (2)$$

$$v_{12} = v_f V_f + v_m V_m \qquad (3)$$

$$\alpha_1 = \frac{\alpha_f E_f V_f + \alpha_m E_m V_m}{E_f V_f + E_m V_m} \qquad (4)$$

$$\alpha_2 = \alpha_f V_f + \alpha_m V_m \qquad (5)$$

where E_1 is the longitudinal Young's modulus, E_2 is the transverse Young's modulus, E_f and E_m are the elastic modulus of copper and dielectric material respectively; V_f and V_m are the volume fraction of copper trace and dielectric material respectively; v_{12} is the equivalent Poisson's ratio, v_f is the Poisson's ratio of the copper and v_m is the Poisson's ratio of the dielectric material; α_1 is the longitudinal CTE, α_2 is the transverse CTE, α_f and α_m are the CTE of copper and dielectric material respectively.

For the transverse CTE α_2 and the transverse modulus E_2, there is another description which includes the effect of Poisson's ratio [5]:

$$\alpha_2 = \left(1 + v_f\right)\alpha_f V_f + \left(1 + v_m\right)\alpha_m V_m - \alpha_1 v_{12} \qquad (6)$$

$$\frac{1}{E_2} = \frac{V_f}{E_f} + \frac{V_m}{E_m} \\ -V_f V_m \frac{v_f^2 E_m / E_f + v_m^2 E_f / E_m - 2v_f v_m}{V_f E_f + V_m E_m} \qquad (7)$$

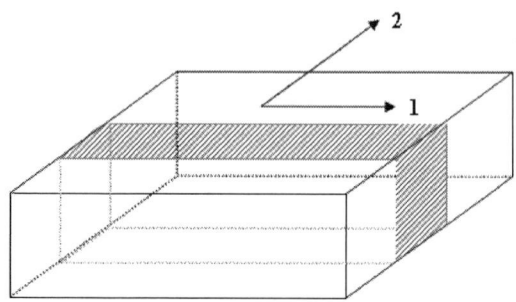

Figure 1 Representative volume element of unidirectional trace pattern.

3. Simulation Based on Finite Element Analysis

The theoretical predictions using equations (1-5) are limited to 'straight line" or other regular simple distribution pattern. It is hard to predict the material property of the complicated pattern type (see Figure 2(a)). On the other hand, finite element analysis is a more general technique to get the equivalent material property of the real circuit layer which is quite complicated. The simulation tool is based on the scripts using ANSYS APDL language. Figure 2(a) is a typical type of copper trace pattern. To generate the FEA mesh for this copper trace pattern, the first step is to discretize the copper trace pattern represented by key points. The material of each element is determined by the majority of the key points in that element. All the analysis mentioned above are using the real circuit drawing file. Due to this approximation process, the copper contents in the FEA model may vary slightly. Figure 2(a) is the original copper trace pattern drawing and Figure 2(b) shows the corresponding FEA mesh. From Figure 2(a) and Figure 2(b), it can be seen that the FEA mesh in this study still matches well with the original circuit drawing.

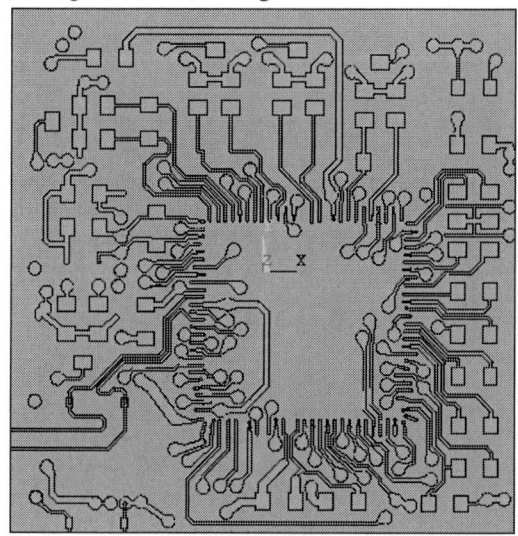

Figure 2(a) Copper trace pattern before FEA meshing.

978-1-4244-4722-0/09 $25.00
© 2009 IMAPS-ITALY

Figure 2(b) FEA mesh for the copper trace pattern.

The equivalent material properties are calculated using the detailed simulation based on the FEA mesh of the copper trace pattern. The equivalent Young's moduli are got from the tensile test simulations and CTE in three axial directions are calculated using the thermal expansion simulation. This study uses one copper trace layer which consists of metallic material such as copper and dielectric material such as solder mask. The material properties solder mask and copper are listed in table 1. It is worth mentioning that the copper trace pattern in the real substrate is relatively orthotropic rather than one direction in unidirectional trace pattern. Thus the in-substrate-plane Young's modulus in X direction E_X will be the same order of magnitude with the other in-substrate-plane Young's modulus in Y direction E_Y.

Table 1 Material Property of Solder Mask, Prepreg and Copper Trace.

Material	CTE (ppm/°C)		Young's Modulus (MPa)	
	<Tg	>Tg	<Tg	>Tg
Solder Mask	60	130	2400	110
Prepreg	25	16	16000	7300
Copper	17.3	-	117000	-

A few different copper trace patterns have been selected for this study. Figure 3 shows the relationship between the in-substrate-plane Young's modulus in X direction E_X and copper trace volume fraction (copper percentage) V_f based on the detailed FEA for the actual laminate substrates. On the other hand, the in-substrate-plane Young's modulus is also calculated using equation (1), equation (2) and equation (7). From Figure 3, it can be seen that also shows that the results using equation (2) and equation (7) are close to the FEA

data points in the range of low copper trace volume fraction, but the results using equation (2) and equation (7) are more and more away from the FEA data points with the increase of copper trace volume fraction. Furthermore, the Young's modulus using equation (1) is not the same order of magnitude with FEA data points.

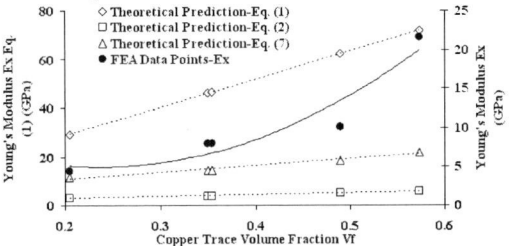

Figure 3 Values of Young's modulus E_X as function of copper percentage.

In Figure 4, the in-substrate-plane CTE in X direction CTE_X is plotted as a function of copper trace volume fraction (copper percentage) V_f based on the detailed FEA for the actual laminate substrates. At the same time, the in-substrate-plane CTE is also calculated using equation (4), equation (5) and equation (6). From Figure 4, it can be seen that also shows that the results using equation (5) and equation (6) are parallel to the FEA data points in the range of low copper trace volume fraction, while the CTE using equation (4) only matches well with the FEA data points for the high copper trace volume fraction case.

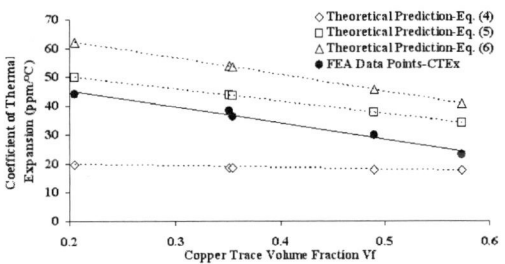

Figure 4 Values of CTE CTE_X as function of copper percentage.

From Figure 3, it can be seen that the relationship between the in-substrate-plane Young's modulus and copper trace volume fraction is not linear anymore. The contribution of the copper on in-substrate-plane modulus becomes more and more important with the increase of copper percentage. At the same time, there is still some distance for in-substrate-plane CTE between the FEA data points and theoretical predictions using equation (5) and equation (6) though the trend based on theoretical prediction matches well with the FEA data points. Based on equation (5) and equation (7), a new methodology has been developed to strengthen the effect of copper trace volume fraction V_f:

$$\alpha^i_{i=1,2} = \frac{\left(\alpha_f V_f + \alpha_m V_m\right)}{\left[1 + V_f\left(v_f V_f + v_m V_m\right)\right]} \qquad (8)$$

$$\frac{1}{E^i_{i=1,2,3}} = \frac{V_f^2}{E_f} + \frac{V_m^2}{E_m}$$

$$\qquad (9)$$

$$-V_f V_m \frac{v_f^2 E_m/E_f + v_m^2 E_f/E_m - 2v_f v_m}{V_f E_f + V_m E_m}$$

where E_1 is equal to E_X, E_2 is equal to E_Y and E_3 is equal to E_Z; α_1 is equal to CTE_X and α_2 is equal to CTE_Y.

Figure 5 shows the variation of out-substrate-plane modulus E_Z, the in-substrate-plane Young's modulus E_X and E_Y with copper trace volume fraction V_f based on the detailed FEA for the actual laminate substrates. On the other hand, the in-substrate-plane and out-substrate-plane Young's moduli are also calculated using equation (9). With the comparison of the results in Figure 3 and Figure 5, it can be seen that also shows that the results using equation (9) are more close to the FEA data points for the whole range of copper trace volume fraction compared with the other methods based on equation (2) and equation (7). This may be mainly due to the reason that the new method can include the quadratic polynomial relationship of the in-substrate-plane Young's modulus and copper trace volume fraction.

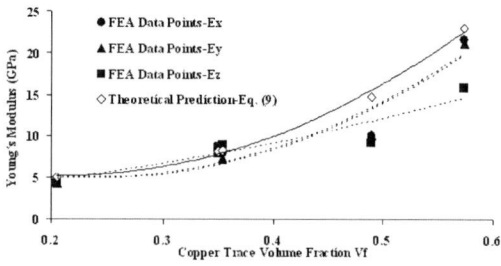

Figure 5 Values of Young's modulus using FEA and theoretical predictions.

In Figure 6, the detailed simulation have been done for the equivalent properties of out-substrate-plane CTE CTE_Z, the in-substrate-plane CTE CTE_X and CTE_Y for a range of copper trace volume fraction V_f. On the other hand, the out-substrate-plane CTE is calculated using equation (5) and in-substrate-plane CTE is calculated using equation (8). With the comparison of the results in Figure 4 and Figure 6, it can be seen that also shows that the values of in-substrate-plane CTE using equation (8) are more close to the FEA data points for the whole range of copper trace volume fraction compared with the other methods based on equations (4-6). However, the values of out-substrate-plane CTE using equation (5) match well with the FEA data points which mean that equation (5) is still

suitable for the calculation of out-substrate-plane CTE. At the same time, it can be seen from Figure 6 that the relationship of CTE and copper trace volume fraction is still linear. Furthermore, it can be seen from Figure 6 that the values of out-substrate-plane CTE are much higher than the values of in-substrate-plane CTE which is the same conclusion gotten by some other studies [6, 7]. This difference is necessary to distinguish in-substrate-plane and out-substrate-plane thermal expansion properties in the thermal mechanical analysis. From Figure 5 and Figure 6, it can be seen that the in-substrate-plane Young's modulus and the in-substrate-plane CTE in the X and Y direction are quite close to each other. The results show that the copper trace pattern in the real substrate is relatively orthotropic and the effect of copper trace orientation on material property is not so obvious due to the nearly symmetrical copper trace pattern.

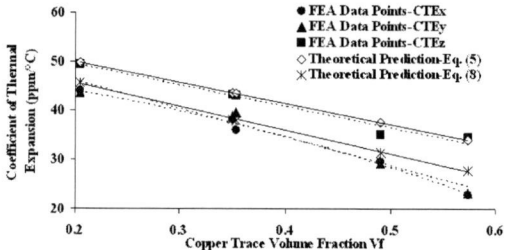

Figure 6 Values of CTE using FEA and theoretical predictions.

4. Application of Detailed FEA and Theoretical Prediction for PoP Substrate

Fast growing PoP technology provides cost effective format for mobile phones, typically consisting of the logic processor in the bottom package and memory device stack in the top package (Figure 7). Controlling package reflow warpage is essential to get good SMT yield in PoP stacking since the solder joints have uneven deformation due to package warpage (see Figure 8). The proper simulation of the thermal mechanical properties of the copper trace layer will be quite important for the understanding of the material property of laminate substrate and warpage control of PoP bottom package. Figure 9 is a cross-sectional photograph of a 4-layer substrate, comprising of four copper trace layers each of which consists of metallic material and dielectric material such as solder mask or prepreg. Cross-sections of traces remaining after etching of the four copper trace layers are visible.

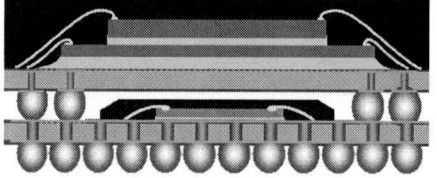

Figure 7 Schematic of PoP.

978-1-4244-4722-0/09 $25.00
© 2009 IMAPS-ITALY

Figure 8 Solder joints uneven deformation due to package warpage.

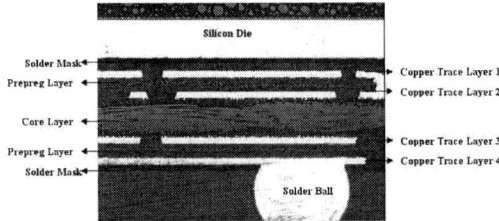

Figure 9 Cross-section of 4 layer substrate of PoP bottom package.

Figure 10 is a typical type of copper trace pattern of PoP bottom package substrate. From Figure 10, it can be seen that the laminate substrate are usually designed to have nearly symmetrical copper patterns. Three substrate designs are simulated and the copper percentages of each copper trace layer are listed in Table 2. Based on the new equations (8-9), the copper percentage in conjunction with the dielectric properties is then used to calculate the equivalent thermoelastic properties for each copper trace layer. The equivalent material properties are used for the warpage calculation of PoP bottom package using FEA.

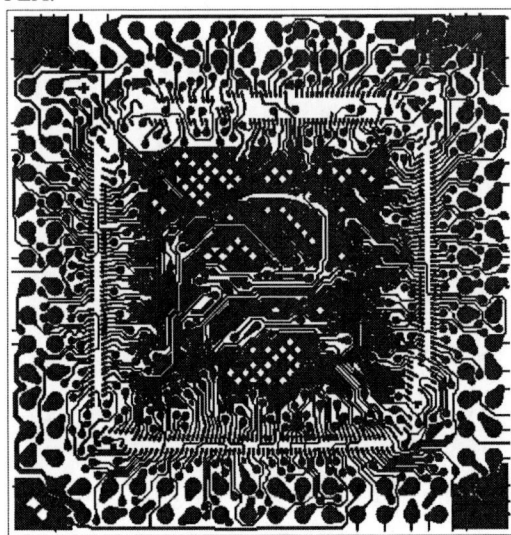

Figure 10 Top view of a typical copper trace pattern.

Table 2 Comparison of Copper Percentage in Substrate.

Substrate	Copper Trace Layer 1 (%)	Copper Trace Layer 2 (%)	Copper Trace Layer 3 (%)	Copper Trace Layer 4 (%)
S1	48.67	49.02	66.46	65.08
S2	59.19	66.46	64.86	65.08
S3	65.08	66.46	49.02	48.67

One kind of PoP bottom package (Size: 14 X 14 mm) was used for this study and the warpage were obtained from measurements on a serial samples using different build of materials (BOM). Warpage direction of the bottom package was crying (convex) which is typical for cavity-molded bottom package. Comparisons of warpage at room temperature based on FEA and experiment measurement using shadow moiré are shown in Fig. 11. The result shows that the warpage based on simulation matches well with the experiment measurement with the consideration of the thermal mechanical properties of the copper trace layer based on equations (8) and (9). Comparisons of warpage using the same BOM and three different substrate designs are shown in Figure 12. From Figure 12, it can be seen that the unbalance of copper trace percentage has important effect on the package warpage. The warpage of PoP bottom package using substrate design S1 is around 20%-30% lower compared with that using substrate design S3. As well as the thermal mechanical properties of the copper trace layer, for the molding compound the chemical cure shrinkage value of 0.05% is used in the warpage calculation and the chemical cure shrinkage has been proved important for warpage study [8].

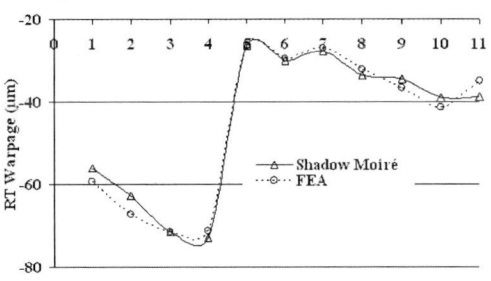

Figure 11 Comparison of warpage based on shadow moiré and FEA at 25°C.

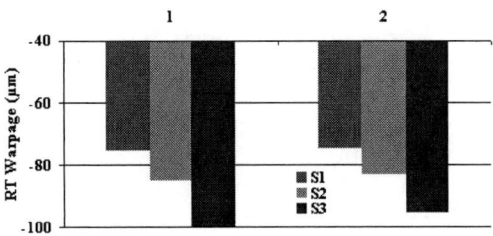

Figure 12 Effect of copper trace on warpage at 25°C.

5. Conclusions

In this paper one new theoretical methodology has been developed for the estimating the equivalent properties of copper trace layer and the results show the results using the new methodology are more close to the FEA data points for the whole range of copper trace volume fraction compared with the other methods. The results show that CTE and modulus of elasticity (Young's modulus) are influenced significantly by the copper trace percentage in the copper trace layers of the substrate, while the effect of copper trace orientation on the anisotropic material property is not so obvious due to the nearly symmetrical copper trace pattern. At the same time, it has been found that the contribution of the copper on in-substrate-plane and out-substrate-plane modulus becomes more and more important with the increase of copper percentage. Furthermore, it is found that the unbalance of copper trace percentage has important effect on the package warpage.

Acknowledgments

The authors would like to thank Dr. Carlo Cognetti from STMicroelectronics' Corporate Packaging, Engineering & Automation (CPA) for management support and acknowledge colleagues from STMicroelectronics (Grenoble)-Nathalie Maurice, Patrick Laurent, Jerome Lopez and his team members (Alexandre Coullomb, Pierino Calascibetta and et al.) on the help of experiment and substrate design.

References

[1] G.J. Hu, J.E. Luan and Xavier B, "Characterization of Silicon Die Strength with Application to Die Crack Analysis", Proceedings of 33rd International Electronics Manufacturing Conference, Malaysia, pp. 1–7, 2008.

[2] D.M. Thomas, L.J. John, "The Effects of In-plane Orthotropic Properties in a Multi-chip Ball Grid Array Assembly", Microelectronics Reliability, Vol. 42, No. 6, pp. 943-949, 2002.

[3] McCaslin L, S.K. Sitaraman, "Methodology for Modeling Substrate Warpage using Copper Trace Pattern Implementation", Proceedings of 58th Electronic Components and Technology Conference (ECTC), Florida, USA, pp. 1582 – 1586, 2008.

[4] A.K. Kaw, "Mechanics of Composite Materials". CRC Inc, second edition, New York, 2006.

[5] S.W. Tsai, H.T. Hahn, "Introduction to Composite Materials", Technomic Publishing, Pennsylvania, 1980.

[6] B.Z. Hong, L.S. Su, "On Thermal Stresses and Reliability of a PBGA Chip Scale Package", Proceedings of 48th Electronic Components and

Technology Conference (ECTC), Seattle, WA, USA, pp. 503–510, 1998.

[7] T.D. Moore, J.L. Jarvis, "Failure Analysis and Stress Simulation in Small Multichip BGAs", IEEE Transactions on Advanced Packaging, Vol. 24, No. 2, pp. 216–223, 2001.

[8] G.J. Hu, J.E. Luan, S. Chew, "Characterization of Chemical Cure Shrinkage of Epoxy Molding Compound with Application to Warpage Analysis", ASME Journal of Electronic Packaging, Vol. 131, No. 1, 011010 (6 pages), 2009.

New developments in high performance solder products for power die assemblies

M. Fenner, A. Mackie, G. Wilson

Indium Corporation

Phone: +441 908 580 400, Fax: +441 908 580 411, E-mail: Europe@indium.com

Abstract

The use of special purpose solder pastes in power die attach is well-established offering low voiding and reliable bonding in volume manufacturing. However these materials are designed around high lead alloys and applied by dispensing. IGBT circuits are made by printing high tin alloys to multiple die sites, placing die and reflowing in a process more similar to conventional PCB or hybrid thick film assembly. This paper describes how the opportunity was taken to make use of the latest developments in Pb-free SMT flux technology and re-optimize them to the different requirements of IGBT die attach.

We rehearse the attributes and requirements of IGBT circuitry and then go on to show how a high performance Pb-Free solder paste has been developed to meet the requirements of large power die attachment (LDA) in IGBT module manufacturing processes. The paste has excellent print and handling characteristics and routinely returns less than 0.5% voiding under large die over a wide range of vacuum reflow conditions. The flux vehicle chemistry offers ease of cleaning to be compatible with the next stage processes of wire bonding & circuit encapsulation.

Key words: solder paste, IGBT, die attach, void free, wire bonding, cleaning

Introduction

IGBT (Insulated gate bipolar transistor) modules are fast becoming one of the most important electronic power switching tools in a variety of industries from trains to military vehicles to trucks. They are being adopted because they allow control of large currents from small actuator currents, and also exhibit increased reliability over analogue switching units such as relays and knife-switches.

In the ideal IGBT die, the high conductivity at high current density seen with bipolar gate junction transistors (BJT's) is combined with the high speed switching capability of metal oxide semiconductor FET's (MOSFET's).

Figure 1 shows a typical cross section through the functional areas of an IGBT module, excluding the thermal interface material (TIM) and its associated heat spreader, and the dielectric potting gel/compound.

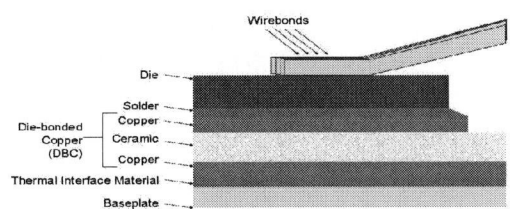

Figure 1: IGBT cross section

Die attach requirements

Significant amounts of I^2R heat are generated during operation of the module. Any materials used in the assembly of the module must therefore have high bulk (material) thermal conductivity and high electrical conductivity. Processes used for assembly must also ensure that the measured thermal and electrical conductivities are maintained as close as possible to the bulk properties. This means that voiding (bubbles in the finished solder joint, usually found using x-ray inspection) must be reduced to as low a level as possible. Standard solder-based power die-attach processes may generate up to 5% or even 10% total voiding, however IGBT die have a much higher current density (2 to 2.5 times that of standard power MOSFETs [1]), so voiding reduction is crucial. Typically, less than 1.0% or even 0.5% voiding may be desired.

The primary factors [2] for reducing voiding are given in figure 2. To ensure low voiding using a solder paste die attach process it is necessary to optimise the paste chemistry, metal loading; reflow profile and preferably vacuum processing.

Bond line thickness has to be controlled to between typically $250 - 375$ μm (0.010-0.015"). The need for a higher thermal conductance using a thinner solder layer is balanced by the requirement for a sufficiently high stand-off to reduce strain on the solder joints induced by the CTE mismatch

978-1-4244-4722-0/09 $25.00
© 2009 IMAPS-ITALY

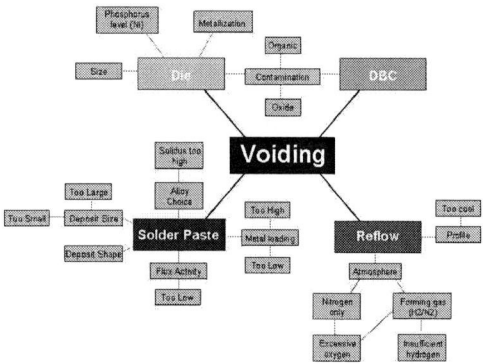

Figure 2, contributors to voiding

between the substrate ceramic (typically alumina Al_2O_3, or, more commonly, aluminium nitride, AlN) and its copper pad.

Wire bonding

Wire bond strength is also central to reliable performance. Cracks in the bond, bond lift-offs, wire corrosion and coarsening of the aluminium microstructure have been identified as major causes of wire bond failure [3.]

Aluminium bonding wire is used instead of the higher-conductivity gold, to keep costs to a minimum. Typically there are three or four bonding wires diameter 500 µm (0.020") or ribbon of equivalent area, for each die. As with all components and interfaces in the entire module, the resistivity must be kept to a minimum, so in some assemblies each wire or ribbon is stitch bonded (multiple bond sights) to the die surface

It is well known that contaminants such as organic materials or ionic compounds on the surface of the die can cause a decrease in the pull-strength of the final wire bond [4]. A bond that may pass the pull-strength test, yet entrap a small amount of oxide or similar material that acts as a dielectric, creating a small capacitance that, together with the known inductance problems of the wire bonding wire itself, may increase the duration of a transient (LC) voltage spike.

In many assembly processes using fluxes and solder paste, the trend for the last 15 years has been towards the use of no-clean materials. However, with the notable recent exception of clip-bonding assembly, most Power Semiconductor assembly processes, particularly those using wire bonding, still require cleaning. It is crucial that all contaminants are removed not just to ensure good wire bonding but also to ensure proper adhesion and curing of encapsulants.

The flux cleaning process is a complex function of a variety of factors to ensure safe removal of all residue components without damage to materials of construction. It is usually assessed and controlled by monitoring ionic residues, as they are the principal cause of flux related in-field failures. They are also the easiest to detect and monitor in (near) real time on a production line using ROSE (Resistance of Solvent Extract) testing. In power die assembly this is insufficient as absence of ionics is not necessarily absence of all residues. Non-ionic residues may not be harmful in the service life of the product (the usual concern), but could interfere with subsequent process such as wire bonding, or damage integrity of encapsulants or their adhesion. More comprehensive evaluation is required.

Solder Paste Selection and optimisation

Table 1 summarises the assembly functions outlined above and relates them to die attach process requirements.

Table 1: Assembly function vs. soldering requirements

Function	Soldering requirement
Electrical Conductivity	Void Free
Thermal conductivity	Void Free
Wire bonding	Clean
Encapsulation	Clean
Reliable	Void Free and Clean

We surveyed users throughout Europe; this showed that most users screen print solder paste, a smaller number stencil print. The most widely used solder alloy is 96.5Sn/3.5Ag and post solder cleaning is virtually entirely semi-aqueous. Thus our objectives were to develop a paste capable of being screen or stencil printed through 200µm emulsion 80 mesh, and then batched for vacuum reflow and cleaning.

Project Criteria

- Product in combination with process will produce less than 1% voiding
- A minimum 8 hours print/open life
- Good response to pause (RtP)
- Post solder cleanliness levels able to routinely meet or exceed established industry criteria for wire bonding
- Capable of being used in existing standard production equipment.

The project emphasis was on reflow and cleaning as these areas have the greatest impact on yield and in-service performance.

Candidate formulations

The recent imposition of Pb-free soldering in electronics assemblies has not been without its issues and these have been thoroughly aired at

978-1-4244-4722-0/09 $25.00
© 2009 IMAPS-ITALY

virtually every industry conference for at least the last 5 years. On a positive note this lead to a very large amount of research into flux and paste technology using high tin content solders in SMT. The increased knowledge from this work illuminated the development of the differently specialised LDA (Large Die Attach) pastes used in IGBT soldering.

After review two material types were identified and went forward for development

Paste A,   New technology Rosin/resin.
Paste B   Print variant of high Pb die attach type

Paste A is formulated to be minimum process change material. As an established die attach material, Paste B acted as a control and as a low residue material also offered possibilities of simplifying the cleaning process.

Printing trials

Modifications made to the pastes were not foreseen to have impact on print characteristics, nevertheless, candidate types were tested at Dek's UK facility using the range of metal percents to be investigated in the reflow, voiding and cleaning trials. These initial print trials were done to establish that materials would print and that response to pause was unaffected. All performed with no measurable variation from their existing properties.

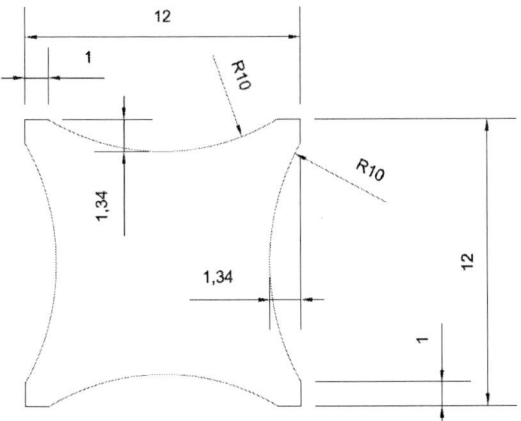

Figure 3, Screen aperture used in trials

The print pattern used was designed specially for this project. It is a generic pincushion representative of that used by many companies.
The screen material was SD 245/65 mesh at 22 deg, plus 200 um emulsion. Print method was off contact with 2mm gap /snap off.

Reflow and voiding Study

Reflow and voiding studies were carried out in cooperation with Pink, using a Pink VADU100 oven. Materials used were plain, un-patterned copper DBC tiles. Die were either Infineon 12 x 12 mm or ABB 14 x 14 mm.

Paste was printed using the screen and print parameters previously described. Candidate materials were reflowed in air, nitrogen or vacuum with metal loadings ranging from 84.5 to 90 % and evaluated visually and by X-ray.

The principal purpose of this study was to determine effects of metal loading on voiding using a variety of profiles. 'Short and long' profiles are illustrated below, Figures 4A and 4B. The secondary purpose was to determine the effect of different atmospheres on void levels for possible later work.

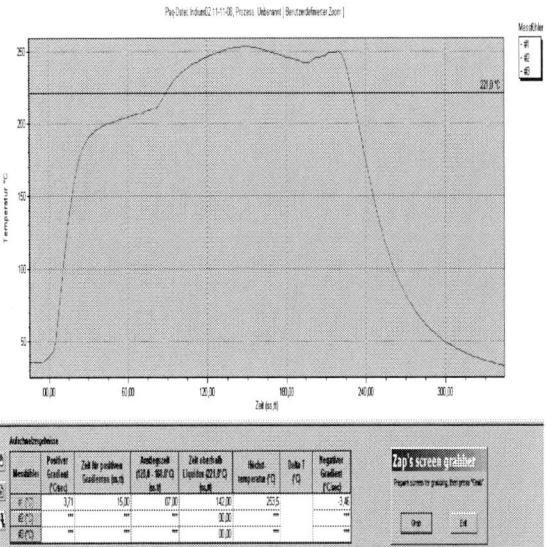

Figure 4A: long profile

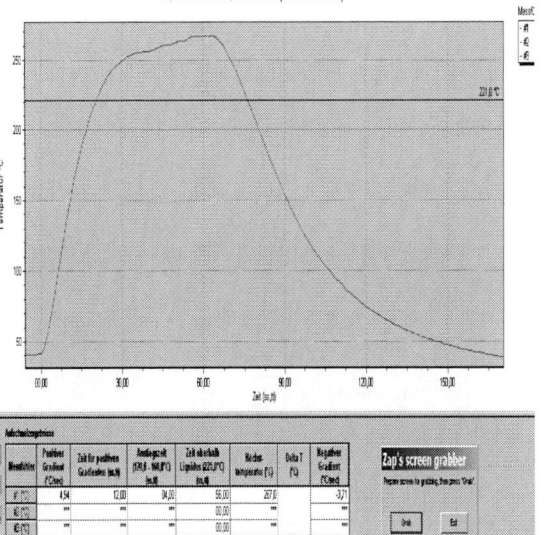

Figure 4B: Short Profile

Table 2: Summary of Voiding % all profiles in Vacuum; measured by Dage XIDAT XD7500VR, grey scale calculation

Metal %	Paste A	Paste B
84.5	0.725	-
85.74	-	-
86	0.45	-
87	-	-
88.5	0.25	0.675
89	0.6	0.575
89.5	0.5	0.225
90	0.55	0.45

Table 3: Paste A, metals loading and voiding

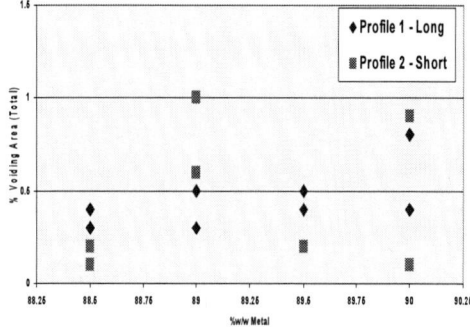

Examples of voiding results are given in Tables 2 and 3. Table 2 shows the aggregate of voiding performance for all atmospheres for various metals loadings. Table 3 shows the in-vacuum performance of Paste A across a range of metal loadings according to profile. Figures 5a, 5b and 5c show these results pictorially. Figure 5a is Paste type A, and 5b is paste B. These show voids <0.25%, compared to the die in 5C at a little over 3% voiding.

Visual Appearance

All Work samples were assessed visually for flux spatter and discolouration/staining around die area, and ranked on a scale of 1-15. (15 being lowest spatter)

15 – 11　Lowest
10 – 6　　Medium
 5 – 1　　Highest.

Paste A　Lowest Spatter score 12 at 89.5% M/L.
Paste B　Lowest Spatter score 14 at 88-90% M/L.

It was found that different atmospheres did not significantly affect flux spatter. There was some correlation of spatter with metal loading (higher loading marginally lower spatter) on flux types A, but this was not strong enough to be able to draw any conclusions or utilise. Flux B showed no spatter variation with metal loading but gave unacceptable discolouration in surrounding area as shown in. Figure 6. This is attributed to the lower process temperatures of Sn/Ag soldering compared to high Pb alloys for which the material was originally formulated.

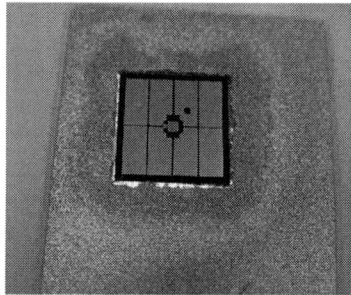

Figure 6: Unacceptable discolouration

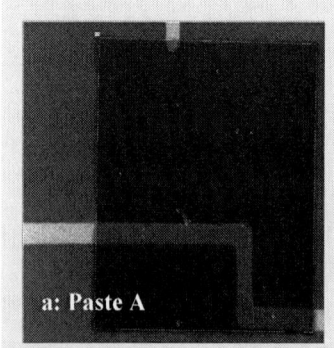
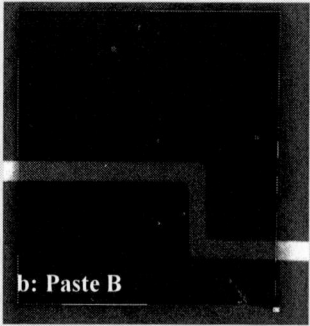
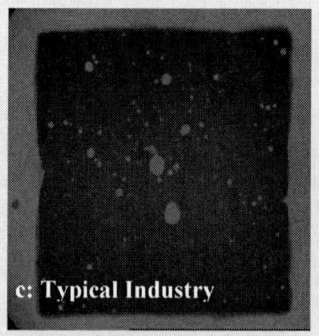

Figure 5: X–rays of voiding under die

978-1-4244-4722-0/09 $25.00
© 2009 IMAPS-ITALY

Cleaning Study

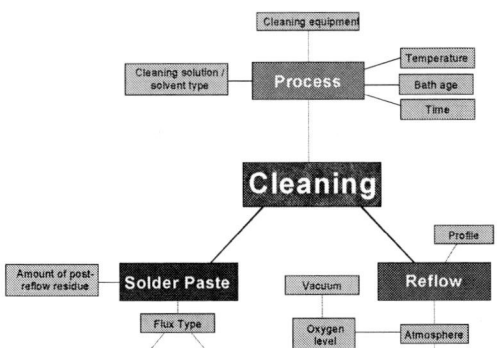

Figure 7: Major Control Parameters for Post-Reflow Flux-Cleaning Process

Post solder flux removal or "cleaning" is a complex function of a number of inter-related factors, Figure 7. As discussed in the Introduction post cleaning of die circuitry has to be carried out to a higher standard than is normal in electronics assembly and different criteria are needed to monitor its success. Most cleaning control procedures are centred on measuring residual ionics, or monitoring conductivity of final rinse water. Non ionics are of equal concern in die attach assemblies; non ionic residues can impact on subsequent processes such as wire bonding or encapsulation. A series of tests has been developed which monitor total resides [5] and these were employed in this project.

The cleaning study was carried out at Zestron's Ingolstadt facility.

Pastes were printed using the projects screen described above using further die and substrates as used in the voiding study. Test assemblies were then reflowed under nitrogen with ~500ppm oxygen. Profile was a straight ramp to a peak temperature of 255C, time above solidus 70 seconds. Nitrogen reflow is "similar to worse" in terms of its effects on flux pryolosis/cleanability compared to vacuum,

Two cleaning processes were used for the trials to represent standard industrial processes in common use. Cleaning cycles 1 and 2 were used with pastes A and B. Cleaning cycle 1, with cleaning chemical at 0% was used for water soluble paste C

Cleaning cycle 1: VA201
Miele industrial dishwasher type IR6002
1. 10 mins 50°C 20% concentration of VA201
2. 3 mins 21°C (room °C) descaled water
3. Repeat 3 mins 21°C (room °C) descaled water
4. 3 mins 50°C DI water
5. Repeat 3 mins 50°C DI water
6. 45 mins 85°C air dry

Cleaning cycle 2:
FA Ultrasonic Process
1. 10 mins 50°C ultrasonic 40kHz
2. 5 mins 21°C (room °C) DI water
3. repeat 5 mins 21°C (room °C) DI water
4. 10 mins 85°C dry

Results of the study are listed in Table 3 and show the pastes meet established criteria for successful implementation industrially.

Conclusion

Paste type A meets the requirements of large power die attachment (LDA) in IGBT module manufacturing processes. The paste has excellent print and handling characteristics and returns less than 0.5% voiding under large die over a wide range of vacuum reflow conditions. The flux vehicle chemistry offers ease of cleaning to be compatible with the next stage processes of wire bonding & circuit encapsulation.

Acknowledgements

The authors gratefully acknowledge the provision of facilities, assistance and advice by Dage, Dek, Pink and Zestron, and the supply of working materials by ABB and Semelab.

Table 3: results of cleaning tests

Cleaning	Paste type	TEST				
		Surface tension	Resin test	Organic layer Test	Interference Contrasting	Flux residue testing
Uncleaned	Paste A	10	5	5	10	10
VA201	Paste A	15	15	15	15	15
FA+	Paste A	15	15	15	15	15
Uncleaned	Paste B	10	10	5	10	5
VA201	Paste B	15	15	15	15	15
FA+	Paste B	15	15	15	15	10
	Criteria	>40N/m - review of the liquid contact angle	No indication of residual die under magnification	<15 seconds to liquid black	visual assessment of Cu metallization under magnification	No indication of residual die under magnification

978-1-4244-4722-0/09 $25.00
© 2009 IMAPS-ITALY

References

[1] Covi, "IGBTs Challenge MOSFET's in Switching Power Supplies", pp. 28-29, *Switching Power Magazine*, Winter 2002

[2] Dr Ning-Cheng Lee, "Reflow soldering processing and troubleshooting SMT, BGA, CSP, and Flip Chip Technologies", Newnes, pp.288, 2001.

[3] Schuetze et al., "The new 6.5kV IGBT module: a reliable device for medium voltage applications" PCIM, August 2001

[4] http://ap.pennnet.com/display_article/293844/36/ARTCL/none/none/1/Clip-Bonding-on-High-power-Modules/

[5] Strixner, "Optimization due to proper cleaning wire bonding processes", EPP Europe May/June 2007

Robust LTCC/PZT Sensor-Actuator-Module for Aluminium Die Casting

M. Flössel[1], U. Scheithauer[1], S. Gebhardt[2], A. Schönecker[2], A. Michaelis[1]

[1] Technische Universität Dresden, Institut für Werkstoffwissenschaft, 01062 Dresden, Germany
Phone: +49 (0) 351 2553-611, Fax: +49 (0) 351 2554-309, e-mail: Markus.Floessel@ikts.fraunhofer.de
[2] Fraunhofer Institut für Keramische Technologien und Systeme, Winterbergstraße 28, 01277 Dresden, Germany

Abstract

The present paper reports on a new module design based on LTCC-PZT laminates. The assembly of modules is achieved by packaging of ceramic PZT plates between LTCC green layers and subsequent sintering of the evolving multilayer. The challenge exists in avoiding tension cracks at shrinking of LTCC layers on the already sintered piezoceramic during the firing process. Thermal characteristics of several LTCC green ceramic products were investigated, systematically, to obtain crack free ceramic modules. The advantages of a module with full integrated PZT ceramic tiles are the mechanical stabilisation of the piezoceramic, the electrical insulation and the shielding of external environmental influences. More general, our approach combines LTCC microsystems technology and piezo technology and allows for a tremendous improve of functional integration, e.g. sensing, actuation, buried electronic circuits, and strain-stress transformation. After preparation, the ceramic modules were introduced in the manufacturing chain of aluminium die casting. Thus aluminium components with integrated sensor-actuator modules could be prepared by die casting for the first time (in cooperation with the University of Erlangen-Nuremberg). The piezoelectric modules survived this manufacturing step without deterioration and fortify the concept of adaptive metal structures in automotive and machine building industry. The functions of the moulded LTCC/PZT modules are estimated by electro-mechanical characterisation methods (e.g. measuring and determination of dielectric coefficient, loss angle tan δ, remanent polarisation and deflection).

Key words: LTCC/PZT, Sensor-Actuator-Module, Multilayer-Packaging, Aluminum Die Casting

Introduction

Usually, active structures are produced in separated steps, covering production of functional modules with sensing and actuation capability, production of the mechanical load carrying structure and final assembly by the application of the functional module on the structure. This process has disadvantages, as for example limited productivity, limited functionality due to restriction of allowed module positions, and reliability risks due exposure to environmental influences.

An alternative approach with improved achievement potential is seen in the integration of functional modules directly during the fabrication of the load carrying structure. Aluminium die casting can serve as example. Here, modules are cast-in during structure fabrication. But as clear requirement, the module has to be robust enough to survive this production step. During metal die casting high thermal and mechanical loads occur: aluminium melt temperature is higher than 660 °C, filling time is in less than 50 ms and causes high temperature gradients (thermo shock conditions) and pressures is up to 1000 bar.

Polymer based packaged piezoelectric modules were already die casted [2, 3 and 6], but they have a limited performance because of there low temperature stability and low couplimg rigidity of the polymer. Also you have a property mismatch between metal and polymer.

Here we report on our new fully ceramic multilayer-packaged piezoelectric-module with the focus on design and fabrication such as charaterization.

Materials and Methods

Design and Fabrication

For fabrication of fully piezoelectric modules made by multilayer-packaging technology a 3-ply LTCC (= Low Temperature Cofired Ceramic) was used which is the carrier material for mechanical stabilization and electrical termination of the PZT in one. The module size was choosen as (L x W x T: 45 mm x 20 mm x <1 mm).

The design is shown in fig. 1 and 2.

Figure 1: Explosion image of LTCC/PZT-module - without electrical termination; small plate: PZT (schematic)

Figure 2: LTCC/PZT-module with electrical termination and soldering pads

Two kinds of LTCC were used: DuPont 951 Green Tape™ [4] and Heraeus HeraLock® Tape-HL2000 green sheets [5]. Before laminating, LTCC green sheets have to be preconditioned in a convection box oven by 10 minutes at 80 °C. Lamination of the LTCC/PZT-modules with an already sintered PZT-plate from "CeramTec" (Sonox® P53) of (L x W x T: 26 mm x 11 mm x 0,2 mm) was done in an isostatic lamination system (IL 4008) with recommended parameters of approximately 210 bar at 70 °C for 10 minutes for LTCC DuPont 951 respectively approximately 170 bar at 75 °C for 10 minutes for LTCC HeraLock® Tape-HL2000.

Afterwards the modules were sintered in a box oven with a special burnout and firing profile. Firing peak temperature was 850 °C for DuPont 951 Green Tape™ and 865 °C for HeraLock® Tape-HL2000, respectively.

Aluminium die casting was done at the Institute of Science and Technology of Metals, University of Erlangen-Nuremburg.

Microstructure evaluation and measurements

"Procon X-Ray CT-Compact" device (acceleration voltage of 130 kV, current of 100 μA and analysis software "XRay-Office") was used for making cross-section images of 3-ply LTCC/PZT-modules. As well conventional polished microsection preparation was done.

For deflection measurements the sintered LTCC/PZT-modules were glued with a 2-component resin (EP 20A-natur + Hardener 158) on steel benders (1.4301 – V2A) as shown in fig. 3. The steel benders were fixed in a measuring set-up (fig. 4) at one side and deflection measurement took place on the free end of the bender using laser triangulation (Micro-Epsilon). Measurements were performed in static mode with electrical field strength of E_S = 2 kV/mm and dynamic mode with E_D = 0,05 kV/mm.

Figure 3: LTCC/PZT-module glued on steel bender

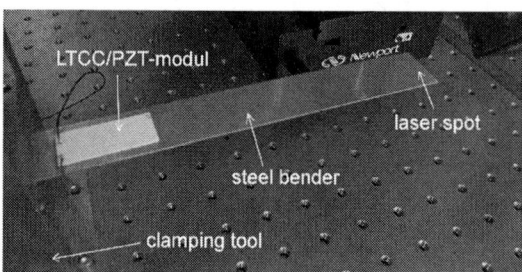

Figure 4: Steel bender with LTCC/PZT-module in measuring set-up

The dielectric constants were determined at 1 kHz, using a Hewlett Packard 4194A Impedance Analyzer before and after polarisation at air and room temperature at E_{pol} = 2 kV/mm and t = 5 min. Measurement of the parameters were done at least 24 h after poling.

Results and Discussion

From a multiplicity of tested carrier materials in previous examinations [7] two kinds of Low Temperature Cofired Ceramics LTCC DuPont 951 Green Tape™ and LTCC Heraeus HeraLock® Tape-HL2000 were selected which are containing alumosilicates [8].

LTCC/PZT-modules with LTCC DuPont 951 as carrier structure showed crack formation and bowing after sintering caused by approximately 13 % shrinkage in x/y-plane of the LTCC (fig. 5).

978-1-4244-4722-0/09 $25.00
© 2009 IMAPS-ITALY

Figure 5: LTCC DuPont 951 with embedded PZT-plate after sintering

To suppress these defects LTCC HL2000 could only be used (fig. 6). It has only a shrinkage in z-direction effected by the special fabrication of the self-constrained tape [1] combined with a near zero shrinkage in x/y-plane. As shown in fig. 6 a carrier material for production of LTCC/PZT-modules was found. With this kind of tape fitting steps which have to be used in general in multilayer-packaging technology could be avoided.

Figure 6: LTCC HL2000 with embedded PZT-plate after sintering

Sonox® P53 PZT-material in polarised situation reveals significant lower capacity values when it´s embedded and sintered in the LTCC carrier material. For this purpose clamping of the PZT-plate through the LTCC carrier material is responsible. Standard polarisation field strength is too small to exhaust full dielectric properties.

For functionality tests the produced piezo-modules were glued on steel bender and deflection measurements were done. The deflection in static mode is around 130 µm and in dynamic mode with a resonance frequency f_R = 19,70 Hz is around 550 µm as shown in fig. 7 and 8.

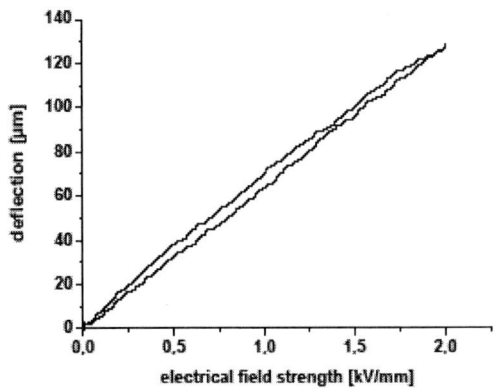

Figure 7: Deflection measurement in static mode

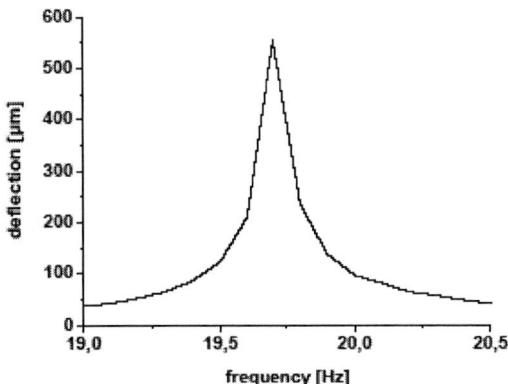

Figure 8: Deflection measurement in dynamic mode

3-ply LTCC/PZT-modules with embedded PZT-plates were used for aluminium die casting tests to evaluate crack formation during aluminium die casting. Also the bonding of LTCC on aluminium metal matrix was investigated.

Fig. 9 and 10 show X-ray analysis and polished microsection with the aluminium die casted plates with integrated LTCC-modules. As observed, there is almost no crack formation in LTCC as well as in PZT-plates that can be attributed to the die casting process.

978-1-4244-4722-0/09 $25.00
© 2009 IMAPS-ITALY

Figure 9: Aluminium die casting plate (white box: position of LTCC/PZT-module)

Figure 10: X-ray image of aluminium die casting plate with integrated LTCC/PZT-module

The multilayer-packaged technology is considered as a successful approach for producing ceramic LTCC/PZT-modules which is shown by the smooth interface between LTCC/PZT-plate and LTCC/aluminium. No air locks or cavities were detected (fig. 11) and so you have a perfect bonding to aluminium.

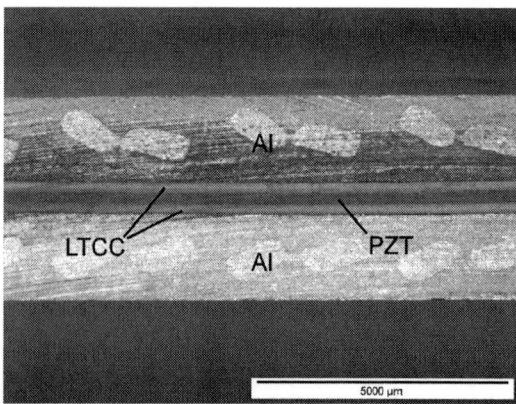

Figure 11: Cross-section of an integrated LTCC/PZT-module

Conclusion and Outlook

Multilayer-packaging technology was successfully used to obtain a robust LTCC/PZT-sensor-actuator-module with the advantages of mechanical stabilization and electrical termination of the piezo-ceramic material, frictional connection to metal matrix, electrical insulation and withstanding mechanical and thermal loads during metal die casting.

It was made out of an already sintered PZT ceramic plate with LTCC Heraeus HeraLock® Tape-HL2000 green layers with a near 0 % shrinkage in x/y-plane and subsequent sintering. Moreover LTCC Heraeus HeraLock® Tape-HL2000 is an outstanding material for aluminium die casting.

Functional demonstration of the installation was done by dielectric measurements which showed less performance than an unembedded PZT-plate. The explanation is the clamping of the PZT in the substrate material. Nevertheless deflection measurements in static and dynamic mode were successfully done, so actuator function was demonstrated.

Further examinations will be focused on clamped PZT and it´s dielectric properties and aluminium die casted piezo-modules with full electrical termination.

Long-time performance tests of LTCC/PZT-modules are in progress.

Acknowledgements

The authors wish to thank Mr. Matthias Rübner from University of Erlangen-Nuremberg (Institute of Science and Technology of Metals) for the aluminium die casting of LTCC/PZT-modules and Fraunhofer IKTS Dresden.

The financial support of the German Research Foundation is gratefully acknowledged.

References

[1] T. Rabe, W. Schiller, T. Hochheimer, C. Modes, and A. Kipka, "Zero Shrinkage of LTCC by Self-Constrained Sintering," Int. J. Appl. Ceram. Technol., 2 [5] 374–382: 2005.

[2] http://www.smart-material.com

[3] http://www.invent-gmbh.de/s08x_duraact.htm

[4] Datasheet DuPont 951 Green Tape™ (http://www2.dupont.com/MCM/en_US/ techinfo/datasheets.html)

[5] Datasheet Heraeus HeraLock® Tape-HL2000 (http://www.wc-heraeus.de)

[6] Matthias Rübner, Carolin Körner, Robert F. Singer, "Integration of Piezoceramic Modules into Die Castings – Procedure and Functionalities", Advances in Science and Technology Vol. 56 (2008) pp 170-175

[7] Markus Flössel, diploma thesis "Untersuchungen zum Einfluss der chemischen Zusammensetzung des Substratmaterials auf die Eigenschaften einer Bleizirkonattitanat (PZT) Dickschicht ", TU Dresden, 2007

[8] W. Kinzy Jones, Yanqing Liu, Brooks Larsen, Peng Wang, Marc Zampino, "Chemical, structural and mechanical properties of the LTCC tapes", The International Journal of Microcircuits and Electronic Packaging, Volume 23, Number 4, Fourth Quarter, 2000 (ISSN 1063-1674)

SMD Pressure and Flow Sensors
for Industrial Compressed Air in LTCC Technology

Y. Fournier, A. Barras, G. Boutinard Rouelle, T. Maeder, P. Ryser

Ecole Polytechnique Fédérale de Lausanne (EPFL), Laboratoire de Production Microtechnique (LPM)
BM 2.137, Station 17, CH-1015 Lausanne, Switzerland - http://lpm.epfl.ch/tf

Tel: +41 21 693 7846, Fax: +41 21 693 3891, yannick.fournier@a3.epfl.ch

Abstract

In this work, we propose an SMD (surface mount device) pressure sensor in LTCC technology specially designed for standard industrial compressed air. It combines the measurement of pressure and temperature, with its integrated signal conditioning electronics for linearization, adjustment and temperature compensation. For the first time, such a sensor can be mounted on an integrated electro-fluidic platform like a standard component using surface mount technology, obviating the need for both wires and tubes. Manufacturing of the SMD sensor is described, as well as its performance and limitations.

Furthermore, a proposal for an anemometric flow sensor is described. The heating / sensing elements of the sensor consist of PTC co-fired resistors suspended on bridges across the fluidic channel; different geometric and processing variants have been manufactured for tests. In the future, we intend to integrate this flow measurement function and corresponding electronics to the pressure sensor, resulting in an all-in-one integrated industrial compressed air flow –pressure – temperature sensor.

Key words: LTCC, sensor, SMD, pressure, flow

1. Introduction

Over the past years, the fields of sensors and microfluidics in LTCC (low-temperature co-fired ceramic) technology have been developed considerably, adding new possibilities to this material initially developed for high-density electronics and packaging. Research has lead to the emergence of micro-heaters [1], flow sensors [2], pressure sensors, micro-reactors, fluidic mixing channels and bioreactors. However, these devices were mainly developed as stand-alone products without signal amplification, and not suited for industrial applications with surface mounting technology (SMT). It would then be an improvement to have a sensor mountable on a fluidic platform as if it were a standard component, obviating the need for both wires and tubes

Our laboratory had previously developed different kinds of sensors in standard thick-film technology and in LTCC [3], aimed for low-cost, high-volume applications. For instance, an SMT micro-flow sensor for liquids [1] was integrated in a disposable microreactor driven under *LabView*.

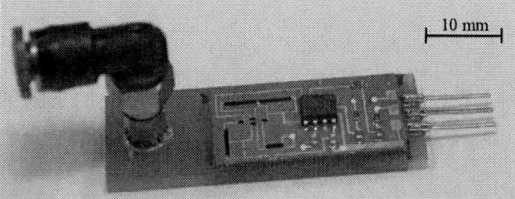

Figure 1: LTCC pressure sensor soldered on test fluidic PCB with M3 inlet and 2.54mm connector.

In this work, we propose for the first time an SMD sensor in LTCC for measuring compressed air pressure and accessorily temperature, which integrates signal conditioning electronics for linearization, adjustment and temperature compensation (Figure 1). The sensor is not intended for precise measurements, but rather as a safety device in a complex fluidic circuit involving "industrial air" (pressure up to 6 bars nominally). The relative pressure measurement is based on thick-film piezoresistors mounted in Wheatstone bridge on an LTCC membrane (Figure 7); the nominal range is 0...6 bars, for a repeatability of 0.1%. The next chapters cover in detail the design, manufacturing and testing of the device.

Toward the Next Step

In addition to the pressure sensor, a demonstrator of airflow sensor was developed, with the view to merge the two sensors in a future step.

978-1-4244-4722-0/09 $25.00
© 2009 IMAPS-ITALY

Here the measurement is anemometric and based on the calorimetric principle, with a central heater and two external thermistors (Figure 2); this disposition allows measuring the sense of the flow as well, as we presented in [2]; the intended range is between 0 and 100 [Nl/min] when using a bypass (only a fraction of the total flow is measured). An extra thermistor upstream in the bulk senses the airflow temperature. The study of this sensor is the object of chapter 3.

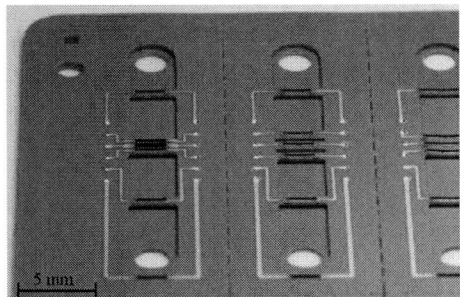

Figure 2: LTCC flow sensors in batch, fired here without lid to show the thermistors. Channel width is 3 mm, length 24 mm, and inlet 2 mm.

2. Pressure Sensor: Design Considerations

The device was designed with the following guidelines:

a. Piezoresistors in full Wheatstone bridge on a membrane; LTCC able to sustain an air pressure of at least 10 bars (nominally 6).
b. Mounting-induced stress should not be an issue.
c. Device must be compatible with surface mount technology (flip chip). No external wires and no tube for connections; all connections must be at the bottom.
d. Integrated electronics for processing the bridge signal, and maximum three electrical connections: power, signal, and ground.
e. Laser trimming should be avoided, or limited to rough pre-trimming operations.

The tape system chosen is the *DuPont (DP) 951 GreenTape™ P2* and *PX* (165 and 254 μm unfired thickness respectively). The bulk of the device is made out of 254 μm tape only; the membrane with 254 μm, and 165 μm for testing purposes.

Piezoresistors

For point a), it was decided to place resistors $R_1...R_4$ on top of the membrane to insure media separation, and to limit the stress in the membrane to ~100 MPa (admitting flexural strength ≈ 320 MPa). For a thick circular membrane, the dominant stress is the radial one at the edge, which can be written:

$$\sigma_{r,edge} = -\frac{3r^2}{4h^2} \cdot \Delta P ,$$ where r is the radius, ΔP the

differential pressure, and h the thickness. This latter is determined by the chosen tape thickness; for a vertical shrinkage of 20%, the fired thicknesses will be 132 μm and 200 μm. The pressure retained is 10 bars (security coefficient 1.6), i.e. 1 MPa. Hence, we can only play with the membrane diameter, keeping in mind that the placement and size of resistors strongly affects the bridge response.

Thus, the membrane diameter was chosen at 3.6 mm, giving a nominal tensile stresses at the edge of $\sigma_{254} = $ -61 MPa and $\sigma_{165} = $ -139 MPa. Figure 3 displays schematically the placement of the resistors (rectangles) on the circular membrane. The dimensions are: width $x = 0.4$ mm, length $y = 0.83$ mm, central spacing of inner (tensile) resistors $z = 0.4$ mm, outer (compressive) resistors placed 1/3 outside. The expected value is around 10 kΩ for each resistor.

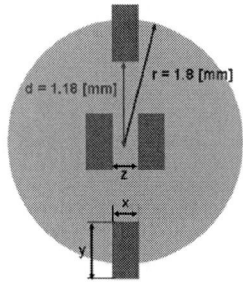

Figure 3: Schematic placement of piezoresistors (pink) on the 3.6 mm-diameter LTCC membrane.

Fluidic Channel Variants

To meet point b), three fluidic channel variants were considered: simple, direct (Figure 4) and zigzag (Figure 5).

Figure 4: Semi-transparent 3D schematic views of fluidic channel between orifice (mauve) and membrane (green). <u>Left</u>: simple variant. <u>Right</u>: direct variant. Notice at the top the resistors (red) and tracks (blue), displayed in an early version.

Although prototypes of each variant were produced for destructive tests, the zigzag version was retained for the development of the sensor, because it seemed the most conservative. The goal being to minimize the stress induced by soldering on the membrane and to avoid delamination at the bottom of the membrane cavity, the first two variants were rejected for the following reasons:

978-1-4244-4722-0/09 $25.00
© 2009 IMAPS-ITALY

- The simple channel leads to the membrane cavity on an edge, possible source of delamination. Furthermore, a long, thin channel is more prone to squeezing upon lamination than two short, fragmented ones.
- The direct variant is the shortest, but can conduct too much assembly stress to the membrane. It also forces the orifice to be directly beneath the membrane, which can be problematic when layouting.

 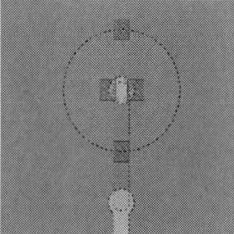

Figure 5: Semi-transparent schematic views of zigzag variant. Left: in 3D. Right: Top view.

Five tapes are necessary for the zigzag channel variant; from top to bottom:
1. Top contacts and membrane
2. Membrane cavity
3. Channel upper layer
4. Channel lower layer
5. Orifice and bottom contacts

Compatibility with Flip-chip Assembly

To satisfy point c) (no tubes, no wires), wire bonding was directly excluded for obvious reasons; in our case, only vias can bring the signals at the bottom. However, components must be placed on surface. The presence of both discrete SMD components soldered on top of the sensor, and soldering pads at the bottom for the fluidic and electrical connections (Figure 7) is made possible by the use of two solder pastes with different melting points (lead free SnCuAg for the former (221°C), and SnBi (138°C) or SnPb (179°C) for the latter). The use of the same solder paste is also possible, provided the sensor is not moved and it is not handled during reflow.

Integrated Electronics: *ZMD* Signal Conditioner

The guideline d), concerning integrated electronics, is fulfilled thanks to the *RBicLite™ ZMD31010*. As its datasheet [5] says, it is a low-cost sensor signal conditioner in a small SOP8 package, which enables easy and precise calibration of resistive bridge sensors via EEPROM. It will digitally correct their offset and gain with the option to correct offset and gain coefficients and linearity over temperature. A second-order compensation can be enabled for temperature coefficients of gain or offset or bridge linearity. It communicates via a one-wire serial interface to the host computer and is easily mass calibrated. Once calibrated, the output

pin *Sig™* can provide selectable 0 to 1 V, rail-to-ail ratiometric analog output, or digital serial output of bridge data with optional temperature data.

The chip requires only one external capacitor of 0.1 μF to filter the power line, but we also added passive filtering elements to the bridge and to the output *Sig* pin for electromagnetic interference (EMI) considerations. This latter capacitor turned out to be an issue, as it can drain too much current and prevent communication between the sensor and the computer during trimming, hence the red mark on Figure 7.

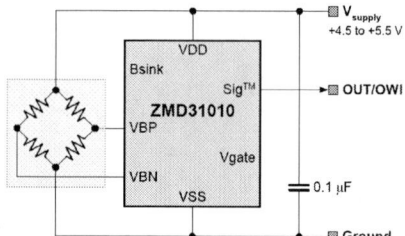

Figure 6: *ZMD31010* application circuit diagram.

Coarse Offset Adjustment by Laser Trimming

The last issue, laser trimming (e), cannot be totally avoided yet at this stage of development. For a first iteration, it is impossible to control the piezoresistor values precisely and to make the bridge signal to fall in the offset tolerance of the ZMD chip, which is inversely proportional to the input span and dependent on the pre-amp gain (selectable between 6, 12, 24 and 48). Therefore, it was decided to use extra resistors; but instead of trimming them directly, Ag:Pd tracks leading to resistors are cut ("digital trimming"), which allows better stability and sufficient precision for the required coarse pre-trimming.

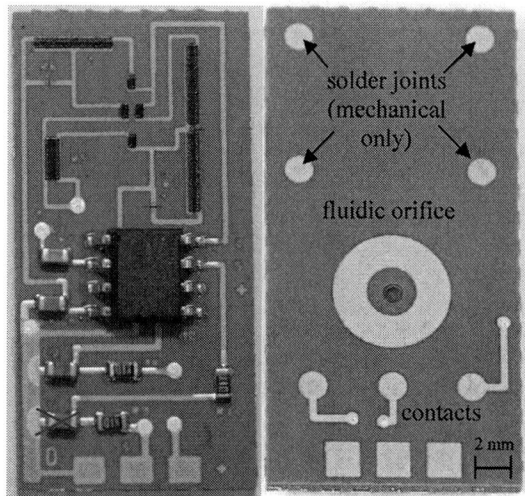

Figure 7: LTCC pressure sensor ready to be mounted, after trimming and initial soldering. Left: top view; note the signal conditioner in the center (*ZMD31010*), the four resistors forming

the full bridge at the top, and the tracks laser cut for offset trimming. The capacitor marked red must be strongly limited. **Right**: bottom view; note inlet and contact solder pads, as well as mechanical joints.

Accordingly, three additional resistors of decreasing value were added ($R_5...R_7$, see Figure 8), and initially short-circuited with conductive tracks that allow each resistor to be activated either way, to increase or decrease the offset. This "ternary" (offset decrease, no change or offset increase) digital trimming scheme can theoretically reduce the offset range by a factor of $3^3 = 27$.

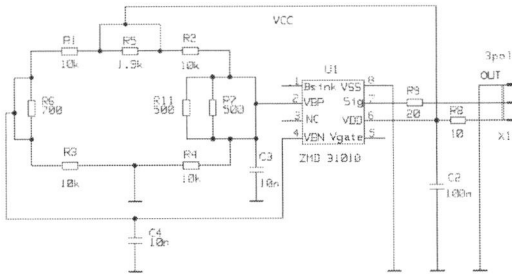

Figure 8: Schematic of electrical circuit. R_7 and R_{11} form a single 250-Ω resistor, but were physically separated for layout reasons.

Manufacturing Steps

LTCC was cut with a 1064 µm Nd:YAG laser, screen-printed on unfired tapes, and processed with rather standard parameters, except for the lamination (160 bars and 10', between metal plates at room temperature). The screen-printing pastes used were the *DP 6141* for vias, the *DP 6146* for tracks and pads, and the *DP 2041* for the resistors. The firing profile was standard, with a 5 K/min sintering ramp and 30' dwell time at 875°C, and occurred in a lamp air furnace.

After firing, lead-free 96.5Sn-3Ag-0.5Cu solder was screen-printed, SMD components placed with a semi-automatic placer, and the sensor was fired in a reflow oven. Finally, the raw bridge offset was measured and the offset adjusted by cutting the necessary tracks. The *ZMD* accepts a remaining offset of ~60 mV. A great disparity was observed between circuits of a same substrate, but it was possible to trim the majority of them. It is also worth noting that despite the fact that the resistor values turned out to be about 70% lower than nominal, it did not affect the bridge much because of the common variation.

Proof Testing by Mounting the Sensor on PCB

Once trimmed, eight sensors were soldered onto test fluidic PCBs, consisting of three layers each, as depicted on Figure 9. The result is shown on Figure 1. Variations were made by alternating the purely mechanical joints in use (to evaluate the assembly stress, i.e. the front and rear soldering pads), while the contacts and the fluidic inlet pad were always soldered. This time an SnPb solder was employed for simplicity.

All sensors and PCB survived the application of air pressure (up to 7 bars, limited by our supply). The chips proved to be easily programmable, except two sensors for unknown reasons.

Influence of assembly and external perturbations

Rapid tests were carried out by pushing at various positions on the LTCC sensor, to determine the influence on the sensor response. The signal was acquired through the *ZMD SSC Evaluation Kit* by using the digital output; the pressure was furnished by a *Druck DPI520* pressure regulator (max 5 bars gauge, precision of 0.1 mbar).

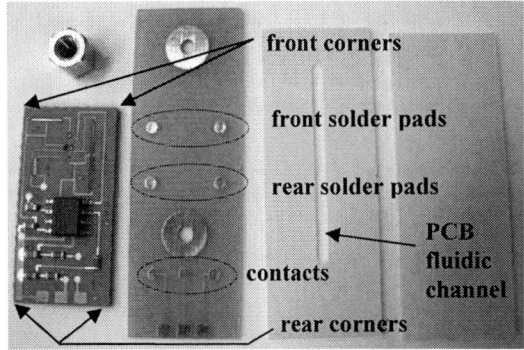

Figure 9: Top view of the sensor and of the not-yet assembled test PCB, made of three epoxy layers with contacts and diverse solder pads.

The differences are not surprising between the 165 and 254 µm versions: the 165 µm have a bigger raw bridge signal (7% of the 40 mV/V chip limit) than the 254 µm (3.7%), but are more sensitive to external perturbations. At 5 bars with a calibration at 90% of the full span, a 1 N vertical force on the front corners of the sensor induces an error of 0.30% for the 165 µm, but only 0.18% for the 254 µm variant.

For a given tape thickness, the best soldering variant against front mechanical perturbations is the one with the two front points soldered (max 0.14% of deviation). The absence of front or rear soldered points protects better from pressing at rear corners (0.06% against 0.23%), but less from pressing at the front -0.22% instead of 0.14%). Testing the variant with two rear points soldered could not be performed, due to a defective sensor. The variants with all points soldered were too variable to conclude properly; more testing is required. Finally, the worst case encountered was the only sensor not mounted on a PCB but directly soldered to a M3 nut for testing purposes: strong perturbations induced by pressing (up to 2.3%) were measured.

978-1-4244-4722-0/09 $25.00
© 2009 IMAPS-ITALY

Hysteresis was also tested around 2.5 bars: the error is between 0.02% and 0.10% for all sensors; indeed, this is rather the hysteresis of the *Druck* regulator that is measured than the error of the LTCC sensor.

Results of Tests with a *Druck* Pressure Regulator

The sensor #05-3, with 165-µm membrane and no front or rear point soldered, was connected to the *Druck DP1520* pressure regulator, and the sensor ratiometric output voltage was measured by a *Keithley 2000* multimeter; both being steered by GPIB in *LabView*. The sensor had been calibrated beforehand with the *ZMD SSC Evaluation Kit* between 10 and 90% of V_{DD} = 5 V for 0…5 bars, i.e. the output voltage should be between 0.5 and 4.5 V. Afterwards, it was supplied by a 5 V precision voltage reference.

Three full ramps (of pattern 0-5-0 [bar] each) were run to measure the hysteresis and repeatability of the sensor. The ramps were not continuous, but were indeed a succession of 50 steps per ascent / descent. The measures were done as follow: every 500 ms, the *Druck* pressure and the sensor output voltage were scrutinized. The standard deviation and the variance were calculated from their ten last values. If the standard deviation of the *Druck* pressure was inferior to 0.2 mbar (relative stability), all data were recorded and the next pressure was assigned. In practice, this stability was reached quite rapidly, in 6-7 seconds.

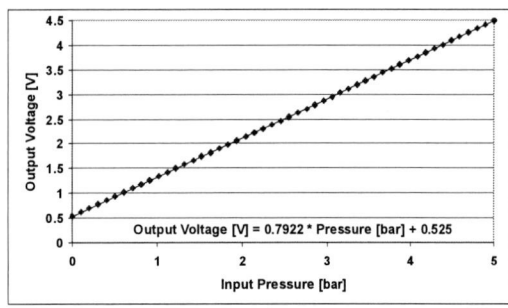

Figure 10: Graph of pressure sensor ratiometric output voltage in function of input pressure for three ramps of pattern 0-5-0 [bar] each. The trend line is perfectly linear, and no noise visible.

Figure 10 shows results beyond expectations: the sensor output is perfectly linear, and there is no visible noise. The normalized standard deviation plots revealed that most of the voltage measures lied below 0.05%, and those of pressure below 0.005%. It is then not abusive to state the repeatability of the sensor as better than 0.1%.

The calibration seemed to be not too bad too; the offset of 0.525 instead of 0.5 is probably due to the difference of voltage reference between the *ZMD SSC* and the external reference. We can thus for now only talk about the repeatability of the sensor, and not about its absolute precision.

3. Flow Sensor: Design Considerations

As explained in chapter 1, the sensor is based on thermistors suspended on bridges tightly cut amid the fluidic channel. This is possible by means of laser cutting the tape after the screen-printing operations, the goal being to minimize the heat conduction into LTCC to get the highest possible signal on the sensing elements.

Seven geometric variants were designed from three channel widths (2, 3 and 4mm), three resistor lengths (1.6, 2.1 and 2.6mm), and three central resistors spacing (0.4, 0.8 and 1.2mm). Three LTCC tapes were used: *DP 951* and zero-shrinkage *Heraeus HeraLock HL800* and *HL2000* (Figure 11). The pastes used were the Ag *DP 6141* for vias, Ag:Pd *DP 6146* for tracks and pads, and *DP 5092D* for the thermistors.

Stacking was also varied between no lid (3 layers), closed chamber with naked resistors (5 layers) and "sandwich" version with 50µm that insulate the resistors (6 layers, Figure 14). In addition, metallized test bridges were co-fired on each batch to verify the absence or presence of deformations, as depicted on Figure 13.

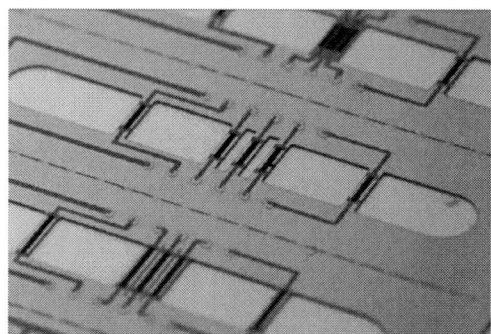

Figure 11: Thermistor bridges *HL2000* tape screen-printed and laser cut. The vias, tracks and resistors are clearly visible.

On Figure 12 are the schematic views showing the thermistors and the connectors. The vias are placed exactly above the footprint of the bottom connector.

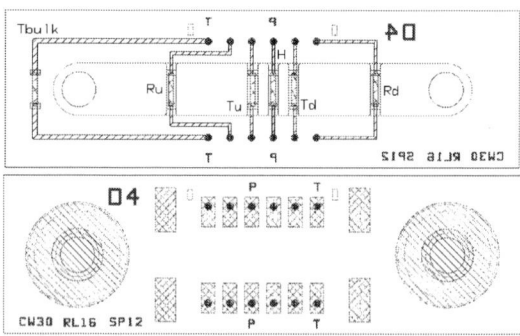

Figure 12: Schematic top and bottom views of the flow sensor. Note the thermistors placement, and the connector solder pads with the vias beneath.

978-1-4244-4722-0/09 $25.00
© 2009 IMAPS-ITALY

Processing and Outcome

The circuits were laser cut for vias, screen printed, and tailor cut again to make the sleekest bridges possible. The lamination occurred at a temperature of 40 or 55°C, between two metal plates between 115 and 160 bars, for duration of 10'. The firing profile was adapted to the *HeraLock* tapes with a sintering ramp of 3 [K/min] and a peak at 880°C. Unsurprisingly, the *DP 951* output was very good, profiting from the "drum skin" effect when lamination was a bit excessive. It was the contrary for the *HL800*: despite great care and laminations with bare tapes or increased tape count, it suffered very strong deformations, rendering it useless. The older *HL2000* was better, except for one lamination test with a high tape layer count, where excessive deformation was observed (too high temperature and too much pressure, cf. Figure 13). Otherwise, the *Heraeus* tapes tolerated the *DP* pastes rather well.

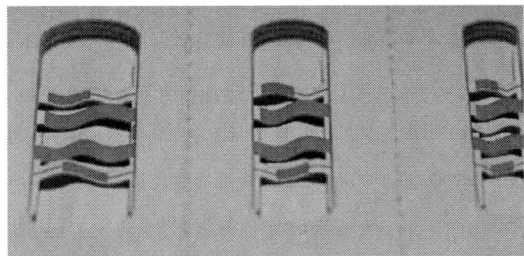

Figure 13: *HL2000* 11-layer circuit with moralized test bridges showing strong deformation after lamination @ 55°C, 160 bars and during 10'.

The sandwich version gave very good results (Figure 14), which is promising for use with aggressive fluids.

Figure 14: Fired "sandwich" 6-layer *DP 951* circuit showing perfect test bridges; electrodes are visible through the 40μm protective layer.

The final step before testing the flow sensor was the soldering of the fluidic connectors (M3 nuts) and of the connector, depicted in oven in Figure 15:

Figure 15: Flow sensors undergoing final processing step: soldering M3 fittings and *Erni* low-profile 12-pin male connector.

Results of Initial Tests

A flow sensor of the 6-layer *DP951* variant D7 (channel width 4 mm, resistor length 1.6 mm, spacing 0.8 mm) underwent a quick test with external measuring equipment linked by GPIB and controlled under *LabView*. A homemade PID drove the heater through a DC power supply (*HP6625a*), and a *NI USB-9219* universal 4-channel analog input module monitored the bulk as well as the sensing thermistors. The heater thermistor was maintained throughout the test at constant temperature difference at around 100°C. Practically, it was achieved by setting its resistance 30% greater than at ambient, because of a TCR of 3000 ppm/K. The airflow was manually set by coupling a manual pressure reducer with a float flow meter, and was varied between 8 and 61 Nl/min.

The results are presented on Figure 16: the left scale is for the half-bridge normalized output voltages in [mV/V], and the right scale is for the heating power in [W]. Not surprisingly, the upstream and downstream half-bridge behaviors are non-linear with the flow: the signals show the greatest variations up to 30 Nl/min, then somewhat peak until 50 Nl/min, and finally sink beyond that. The sum of the half-bridge signals peaks at 10 mV/V, which is reasonably good without amplification.

The heating power presents a much more continuous characteristic, almost linear up to flows at 30 Nl/min. For this measure, it varied from 0.33 to 0.77 W. In itself, measuring the heating power could be sufficient for calculating the flow. For high flows and flows higher than measured, it could even be simpler to computer. Nevertheless, let us do not forget that the final sensor should use a bypass to measure only a fraction of the total flow. Thus, the maximum flow that the sensor would undergo could be limited to the linear portion of the graph.

978-1-4244-4722-0/09 $25.00
© 2009 IMAPS-ITALY

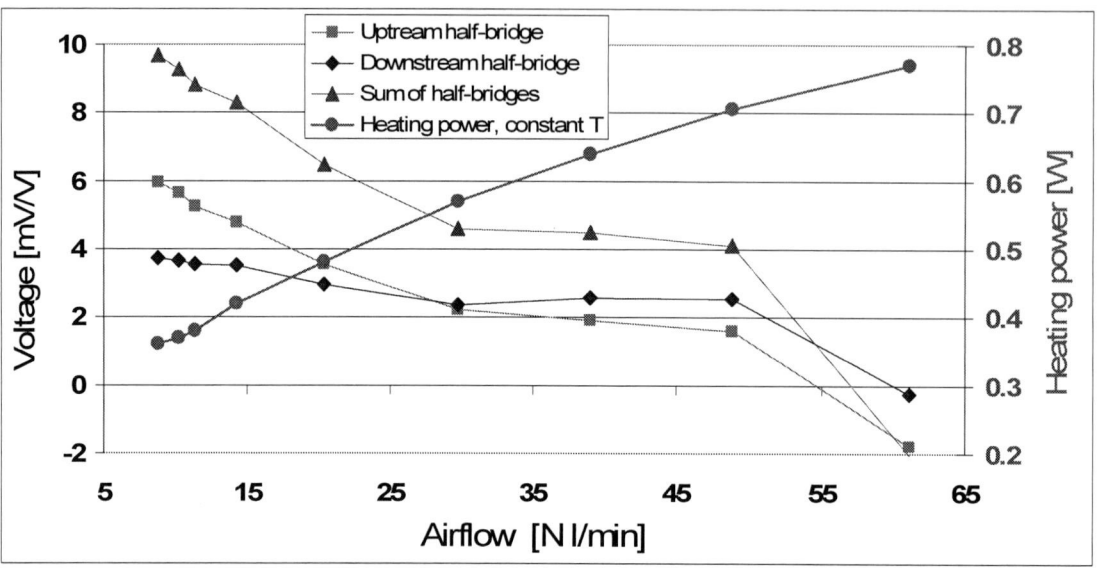

Figure 16: Graph of flow sensor upstream and downstream half-bridges output voltages, sum of half-bridges, as well as heating power in function of airflow, showing non-linear behaviours.

Thoughts on Suitable Integrated Electronics

The *ZMD31010* signal conditioner would also be suitable for this flow sensor: the sensing thermistors would be mounted in full Wheatstone bridge, while the heater would be controlled separately. When merging the pressure and flow sensor, this allows to optimize cost and space by exchanging the two *ZMD31010s* for one *ZMD31012*, containing two independent *31010* dies in a SSOP14 package for just a slightly bigger footprint.

4. Conclusions

The feasibility of an integrated SMD pressure sensor for industrial compressed air in LTCC technology was demonstrated. Eight sensors using four piezoresistors placed on a membrane where successfully produced and tested in diverse variants, showing a remarkable repeatability <0.1% for a low-cost sensor fitted with a low-cost signal conditioner from *ZMD* (~4€). The sensors, developed for 6 bars of nominal pressure, sustained at least 7 bars. The *ZMD* chip allows an easy programming of gain, offset and temperature correction of the bridge raw signal, provided the raw offset falls into chip tolerances.

In a second step, a calorimetric flow sensor for air in LTCC was developed and quickly tested, but still lacks integrated electronics and optimized dimensions. While it was designed to measure a fraction of the total flow (up to 100 Nl/min) through a bypass, it was possible to measure a direct flow up to 61 Nl/min at 6 bars.

It is intended to merge the two sensors in a close future to build a truly integrated, multi-function low-cost industrial sensor able to measure

air pressure, flow and temperature with enough precision to make it suitable for a safety device in a complex fluidic platform.

Acknowledgements

The following people from the LPM are warmly thanked: Mr. M. Garcin for his help with screen-printing operations, Mr. N. Craquelin for the extensive development of the flow sensor test bench, and Mr. N. Dumontier for hardware support.

References

[1] J. Kita, F. Rettig, R. Moos, K. Drue, H. Thust, "Hot-Plate Gas Sensors – are Ceramics Better?", Proceedings of CICMT 2005.

[2] Y. Fournier, R. Willigens, T. Maeder, P. Ryser, "Integrated LTCC micro-fluidic modules - an SMT flow sensor", 15th European Microelectronics and Packaging Conference – IMAPS, pages P2.06, 577-581, 2005.

[3] Y. Fournier, O. Triverio, T. Maeder, P. Ryser, "LTCC free-standing structures with mineral sacrificial paste", Proceedings of Ceramic Interconnect and Ceramic Microsystems Technologies, Munich (DE), pages 11-18 (TA12), 2008.

[4] H. Birol, T. Maeder, P. Ryser, "Application of graphite-based sacrificial layers for fabrication of LTCC (low temperature co-fired ceramic) membranes and micro-channels", Journal of Micromechanics and Microengineering 17, 50-60, 2007.

[5] ZMD31010_RBic_Lite_DataSheet_1.97_24-Sep-08.pdf, from www.zmd.biz

ALX Permanent Polymer Dielectrics
For Microelectronic Packaging Applications

Philip Garrou*, Alan Huffman** and Jeffery Piascik**
*Microelectronic Consultants of NC and **RTI International
* Research Triangle Park, NC 27709
*philgarrou@att.net, 919-248-9261

Abstract

In this paper we present the results of a study done to compare the new Asahi Glass ALX polymers with other microelectronic polymers in typical WLP structures. We present property data on ALX-211, including spin speed curves and resolution plots, planarization, thermal stability and stress data and adhesion as evaluated using polymer bump shear testing. We have examined the processing of ALX-211 and developed a standard process flow for these materials. We have used ALX-211 in a typical bump-on-polymer process flow, with eutectic Sn/Pb solder bumps, and compared its performance to that of BCB through solder bump shear testing.

Key Words: ALX, photo polymer, WLP, bump-on-polymer

Introduction

The materials used as permanent polymer dielectrics in microelectronic packaging and interconnect today are: (1) polyimides (PI); (II) benzocyclobutene (BCB); (III) epoxies and to some extent (IV) polybenzoxazole (PBO). There are several texts on these materials as they apply to microelectronics [1-2] and several reviews on the use of such materials in microelectronic applications [3-4].

Spin-on polymers are widely used in microelectronic packaging applications. For bumping and wafer level packaging (WLP), BCB and PI are the predominant spin-on dielectric materials used for repassivation and redistribution [5, 6]. Each material has it's respective strengths and weaknesses. BCB, manufactured by Dow Chemical, has low moisture absorption, low shrinkage on cure, low curing temperature (210 – 250 °C), a low dielectric constant and develops an easily metallized sloped via side wall when exposed in proximity, but has low elongation to break and tensile strength and undergoes brittle fracture which limits its application in bump-on-polymer applications as chips become larger than 6 mm. In fact, Amkor has recently commented that "First-generation WLCSP technologies were not capable of supporting larger die sizes due to use of materials such as BCB which have low tensile strength and low elongation to break and thus proved incapable of surviving BLR tests such as drop and temp cycle testing..." [7].

Polyimides are supplied commercially by Asahi Chemical, Toray, HD Micro and Fuji Film among others. They are superior mechanically to BCB but suffer from much higher moisture absorption (which can lead to blistering if not carefully processed), higher shrinkage on cure, higher dielectric constants, and have cure temperatures of 320 - > 400 °C [1-4].

Asahi Glass Co.'s new ALX polymers are negative tone, photoimagable, fluorinated aromatic polymers [8]. ALX polymers show dielectric constant, shrinkage, and moisture absorption superior to PI and comparable to that of BCB. They also show a significantly lower cure temperature (190°C) than BCB or PI and 2X+ higher elongation than BCB. Mechanical, thermal, and electrical properties of ALX-211 are shown in Table 1, and compared to published information on BCB [9-11] and PI [12].

Low moisture is a significant property for those involved in fabrication of high frequency devices and integrated passive devices where dielectric constant must remains stable in non-hermetic, varying humidity environments. Although the water absorption for ALX is reported as 2X the BCB number, this has little consequence on processing since the processes (similar to BCB) vs PI materials which are well known to need bakeouts after every exposure to water (such as an SRD [spin/rinse/dry] step) [2].

978-1-4244-4722-0/09 $25.00
© 2009 IMAPS-ITALY

Material Property	BCB	PI	ALX-211
Modulus (GPa)	2.9	2.9-3.5	1.3
Elongation (%)	8	20-75	20
Tensile Strength (MPa)	87	120-200	90
CTE (ppm/°C @ 25°C)	42	35-55	60
Stress (MPa)	28	28-30	32
Dielectric Const. (1KHz-1GHz)	2.6	2.9-3.4	2.6
Resistivity (Ω·cm)	10^{19}	$>10^{16}$	$>10^{16}$
Water Absorption (% @ 85°C/85%RH)	0.2	~ 1-3%	0.4

Table 1: Physical Properties ALX-211 vs BCB and PI

Film Stress

Thin film stress consists of two components: (1) Thermal Stress (σ_1) - stress resulting from temperature change and miss match in CTE between the cured polymer film and the substrate and (2) Intrinsic stress - stress built into the film during deposition due to the shrinkage and contraction that occurs after the film has adhered to the interface surface. Film stress is usually determined by measuring substrate "bow" via the change in the radius of curvature. Figure 1 shows the film stress measured for ALX-211 cured at 190 °C (standard process cure) and 250 °C.

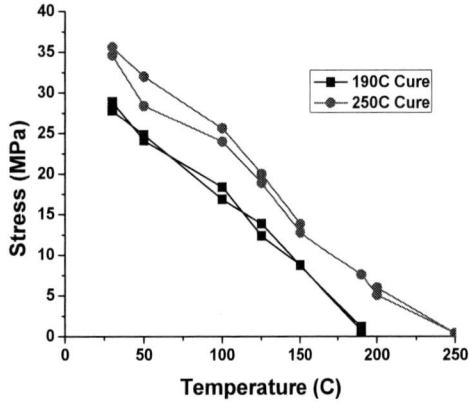

Figure 1. Stress in Cured ALX-211 Films

Thermal Stability

The thermal stability of a polymer is the temperature above which the material chemically degrades and/or its physical or mechanical properties degrade. The thermal stability required of a polymer is dependent on the application use temperature and/or the time/temperature experienced by the polymer during subsequent processing steps.

Thermal stability is typically determined by thermogravametric analysis (TGA), data which tracks weight loss vs temperature as the temperature is ramped. Such data is very dependent on the ramp rate and the thickness of the sample. The faster you ramp and the thicker the sample, the more thermally stable the material appears to be. "Ramp TGA" data is therefore misleading since it can make polymers appear to be several hundred degrees more thermally stable then they actually are. TGA data should be reported Isothermally, i.e weight loss vs time at the temperature of interest. The thermal stability of ALX is shown in Figure 2 and Table 2 . The sample in this case was AL-X215 (~25um thick); cure at 190 °C for 2 hrs.

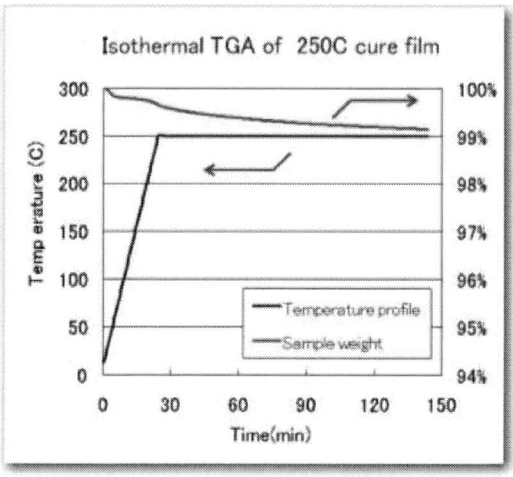

Figure 2 Thermal Stability of ALX

	Wt loss / hr (%)	
Isothermal Temp (°C)	190 °C cure	250 °C cure
250	< 0.5	< 0.5
300	< 0.5	< 0.5

Table 2. Isothermal Weight Loss of ALX

Planarization

We examined the planarization capability of ALX-211 on a test vehicle containing isolated lines of plated Cu. Cu lines were plated to 1, 3.5, and 5 μm thicknesses, coated with 7.5 μm of ALX-211 (90 second 60°C softbake, 50 mJ/cm^2 exposure dose, then cured at 190°C for 2 hrs) Figure 3 summarizes the results of the planarization study. Degree of planarization (DOP) was calculated as shown in Figure 4.

978-1-4244-4722-0/09 $25.00
© 2009 IMAPS-ITALY

Profilometry was used to measure the thickness of the polymer over the lines.

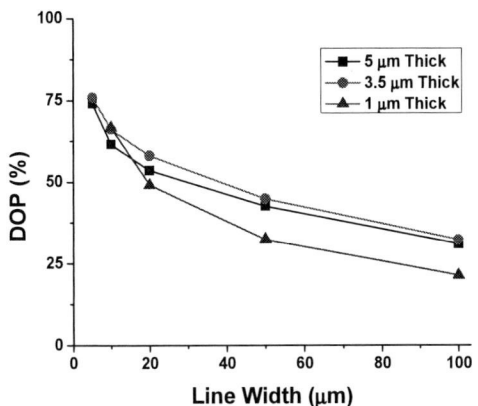

Figure 3: ALX-211 DOP over isolated Cu lines

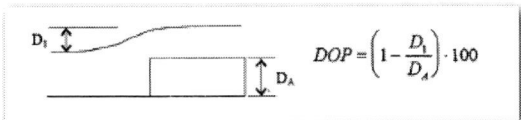

Figure 4 Measurement of DOP [13]

Spin Curves

Figure 5 shows the spin speed curve generated for the ALX-211 formulation after softbake (60°C for 90 seconds) and after cure (2 hrs at 190 °C). The spin coating quality of ALX-211 was excellent, free from particles or other defects with good wafer thickness uniformity.

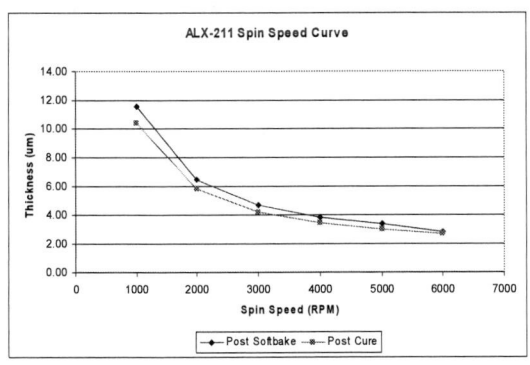

Figure 5. ALX-211 Spin Speed Curve

Soft Bake Temp vs Via Resolution and Film Retention

We evaluated the soft bake process to determine the effect of bake temperature on via resolution and film retention following develop (Figs 6 and 7). The soft bake temperature was varied from 40°C to 80°C with the bake time

held constant at 90 seconds. For this experiment, 7.9 μm of ALX-211 was spun on wafers with a blanket aluminum layer following the baseline coating process. The wafers received an exposure dose of 200 mJ/cm^2 (measured @ 365 nm). The exposed polymer films were developed with AGC's PS-201 develop solvent, which is a mixture of PGMEA and ethyl lactate. A puddle develop process was used, consisting of two 10 second static puddles with a 5 second spin rinse in between, followed by a 20 second rinse at 500 RPM (P10x2R20x1) and a 1500 RPM drying spin. A 100°C stabilization bake (post develop bake) for 90 seconds followed develop.The wafers were then cured under nitrogen at 190°C for 2 hours. An oxygen descum process followed, designed to remove 2500-3000Å of polymer.

The results of the via resolution study vs. soft bake are shown in Figure 6. Diameter measurements were made at the bottom of the vias imaged from 3 different mask feature sizes (50, 30, and 20 μm). While via resolution improved with increasing soft bake temperature, cracking was noted in the films following soft bake at temperatures higher than 70°C. 20 μm features were not opened at all, but it was later found that this was due to the exposure dose used in the experiment. Nonetheless, 60°C appeared to be the optimal soft bake temperature.

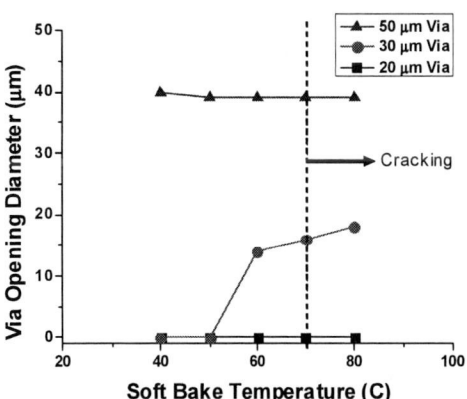

Figure 6: Via Resolution vs. Soft Bake Temperature

The effect of soft bake on film retention after develop is shown in Figure 7, with the film retention at 94% of the as-spun thickness when the 60°C soft bake temperature is used.

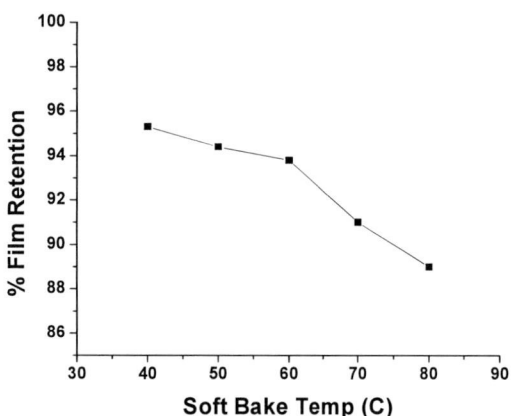

Figure 7: Post-Develop Film Retention vs.
Soft Bake

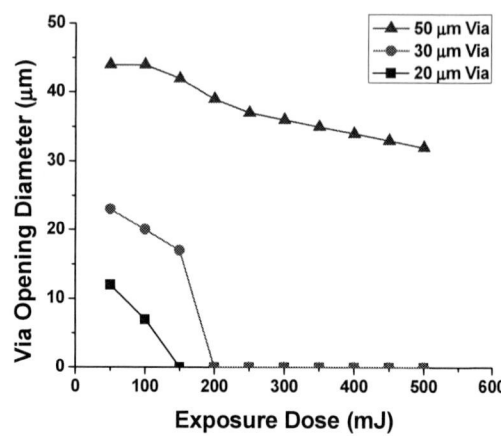

Figure 8: Via Resolution vs. Exposure Dose

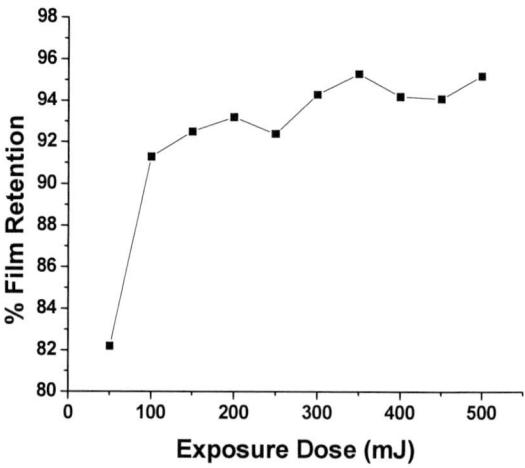

Figure 9: Post-Develop Film Retention vs.
Exposure Dose

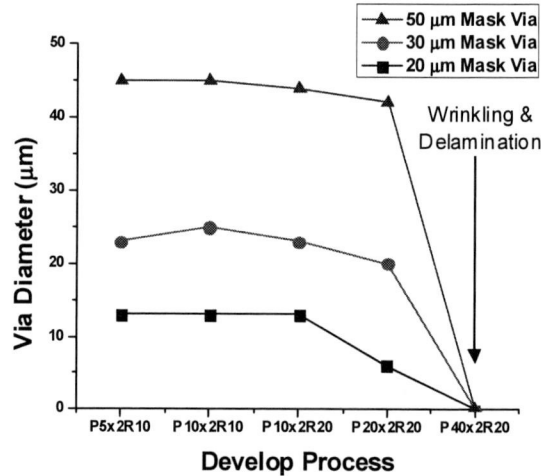

Figure 10: Develop Time vs. Via Resolution

Exposure Dose vs Via Resolution and Film Retention

Exposure dose was examined for its effect on via resolution and post-develop film retention. The as-spun film thickness was 7.9 µm and the soft bake process was held constant at 60°C for 90 seconds for these experiments. The develop (P10x2R20x1), cure and descum processes were the same as outlined previously. The results are shown in Figures 8 and 9. To achieve maximum via resolution, the optimal exposure dose (measured at 365 nm) was found to be 50 mJ/cm2 which retained ~82% of the original film thickness after develop. 15 and 20 µm mask features were easily resolved and well-opened, indicating a 2:1 aspect ratio (AR) was obtainable.

Develop Time

Five different develop processes using PS-201 were evaluated, as shown in Figure 10. Via resolution was relatively constant for the three shortest develop processes of 20 seconds (P5x2R10x1), 30 seconds (P10x2R10x1), and 40 seconds (P10x2R20x1). Wrinkling and delamination of the film occurred when the develop time reached 100 seconds. Time to clear on blanket development monitors is less than 10 seconds.

Via Resolution vs Exposure Gap

Exposure gap was varied from 10 to 100 µm as the diameter of the bottom of the vias were measured and SEM cross sections were taken to observe the via sidewall profile with change in exposure gap .

978-1-4244-4722-0/09 $25.00
© 2009 IMAPS-ITALY

Figure 11 shows the effect of exposure gap on via resolution. The test wafers were metallized with blanket aluminum and then coated with 7.5 µm of ALX-211 polymer using a 60°C, 90 second soft bake and were exposed with a 50 mJ/cm2 dose. Wafers were then developed with PS-201 developer using the P10x2R20x1 puddle develop process. A reduction in via resolution was seen as the exposure gap was increased, which is typical with negative acting photoimagable materials.

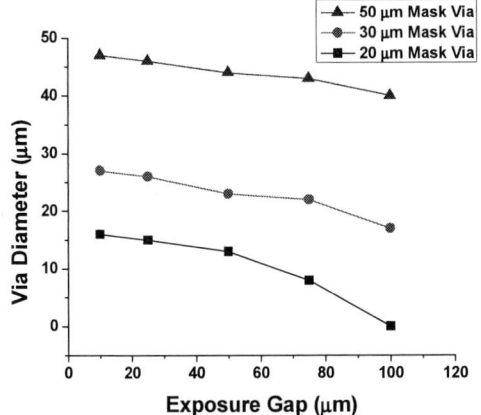

Figure 11: Via Resolution vs. Exposure Gap

Figure 12 shows SEM images of cross sectioned vias. As the exposure gap increases the via sidewall angle becomes more shallow. The sidewall slope is approx. 83° at a 10 µm exposure gap. When the exposure gap reaches 75 µm, the sidewall angle is ~60°. The foot at the via base is approximately 2-3 µm wide.

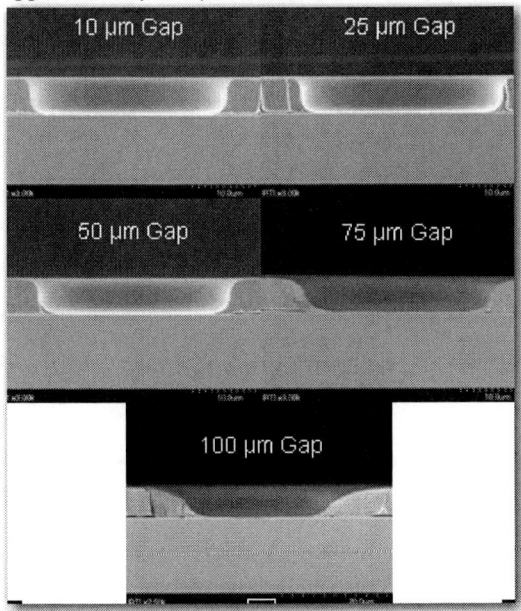

Figure 12: SEM Cross Sections of ALX-211 Via Sidewalls

Based on our evaluations, a standard set of process conditions were established that were subsequently used for the fabrication of the test structures discussed in the following sections.

ALX-211 Recommended Standard Process Flow

- Spin on adhesion promoter followed by 100°C hot plate bake for 90 seconds
- Spin coat ALX-211 followed by 60°C hot plate bake for 90 seconds
- Exposure at 50-100 mJ/cm^2 (measured at 365 nm)
- Puddle develop with PS-201 using P10x2R20x1 process
- Cure in nitrogen atmosphere at 190°C for 2 hrs
- O$_2$ plasma descum, for target 2.5-3.0K Å polymer removal

Measurement of % Cure

Cure measurements were made using transmission mode FTIR comparing the relative heights of a peak at 721cm^{-1} (stable reference peak) and a product-specific peak at 912cm^{-1} as shown in Figure 13.

For most polymer systems, the use of a cure temperature lower than the maximum processing temperature of the device typically results in change of properties of the polymer, as it's further cured during the reflow process, such as decreased chemical resistance or changes in tensile strength, modulus or elongation which can cause reliability problems.

A unique characteristic of ALX-211 in solder bumped applications is that the recommended cure temperature of the polymer (190 °C) is lower than the reflow temperature used in the fabrication of the Pb/Sn or Pb free bumps. To insure that further curing was not happening to ALX-211 during subsequent processing, we monitored the cure percentage after multiple solder reflow cycles. A fully processed ALX-211 film was cured at 190°C for 2 hours. The cure percentage was measured prior to reflow and then again after 1, 6, and 11 reflow cycles. Figure 14 indicates, the cure % remains relatively constant even after 11 reflow cycles.

978-1-4244-4722-0/09 $25.00
© 2009 IMAPS-ITALY

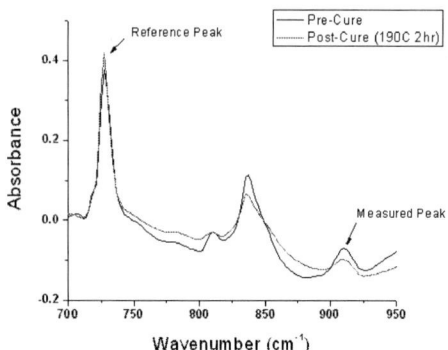

Figure 13. %Cure by FTIR

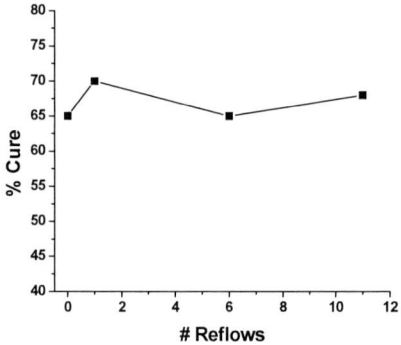

Figure 14: Cure % Stability After Multiple Solder Reflow Cycles

Adhesion from Polymer Bump Shear Measurements

It is normally accepted that shear testing better mimics failure modes (like the drop test) than tensile testing (i.e stud pulls). The adhesion of ALX polymers to typical materials encountered in WLP processes were therefore measured in the shear mode by photo-patterning 100 μm square polymer bumps of ALX-215 films on Si wafers with blanket base layers of Si_3N_4, Cu, or Al. (ALX-215, a higher viscosity formulation of ALX, was used to achieve the required thickness of the polymer bumps in a single spin coating). The same 190°C/2 hour cure process was used for the ALX-215.

Table 3 shows the results of the polymer bump shear tests. The bump shears were done using a 4 μm tool lift and a 100 μm/sec shear speed. Failure occurred at the polymer-base layer interface for all tests meaning we were measuring the adhesion strength (shear mode) for the interfaces in question.

Material	Ave. Shear Strength (g)	Std. Dev. (g)
SiN	45.1	.93
Cu	48.4	1.7
Al	47.1	0.9

Table 3: Polymer Bump Shear Strength

The adhesion between two layers of ALX polymer was also tested in a similar manner by fabricating polymer bumps of ALX-215 on a blanket layer of ALX-211. The blanket layer (7.8 μm) was applied to Si wafers with ~3000Å of thermal SiO_2 and cured at 190°C for 2 hours. AP-903 was applied to the first layer of ALX-211 before applying the second polymer layer of ALX-215. Figure 15 shows a polymer bump-on-polymer structure before and after shear. The failure mode of these structures was at polymer-SiO_2 interface, with a large portion of the blanket polymer layer pulling off of the SiO_2 surface around the sheared bump. No failures were seen at the interface between the polymer layers. We interpret this as excellent polymer-to-polymer adhesion, which is critical for multilayer polymer applications. Table 4 gives the shear strength data for the polymer bump-on-polymer shear experiment. The polymer- polymer bump shear adhesion is > 68 grams since the failure in this case was in the weakest interface, i.e. the SiO_2/ALX-211 interface.

Figure 15: Polymer Bump-on-Polymer Structure Before (left) and After Shear (right)

	Ave. Shear Strength (g)	Std. Dev. (g)
ALX-215 to ALX-211	> 68.3	1.8

Table 4: Polymer Bump-on-Polymer Shear Strength

WLP Test Structure Evaluations

WLP test structures were fabricated with AL-X211 and BCB and solder bump shear was used to compare the performance of the two materials.

Fig 16 shows a repassivation/solder bump structure with a single layer of polymer applied and patterned over a surface that replicated a typical device wafer. The substrates were Si wafers coated with an evaporated stack of ~500Å of Ti followed by 1000-3000Å of aluminum. Approximately 1.5 μm of PECVD Si_3N_4 was then deposited and 75 μm openings were made down to the underlying Al. For ALX-211, the previously outlined processes were used. The BCB polymer was processed using a standard process developed at RTI [14]. The target final thickness of the polymer layers was 5 microns. Exposure gap for the AL polymer wafers was set at 75 μm, in order to achieve a via sidewall slope that was close to that achieved with BCB at a 50 μm exposure gap. The vias formed in the ALX and BCB layers were opened over the Al pad in the Si_3N_4 opening.

After polymer processing, all wafers were metallized with a sputtered UBM consisting of 1000A of Ti and 1500A of Cu. A thick photoresist was applied and patterned on the UBM to define the bump pattern. A diffusion barrier of Ni (approximately 1.5 um thick) was

plated over the UBM, followed by eutectic Sn/Pb solder. After plating, the field UBM was chemically etched away and the bumps were reflowed.

Two different bump structures were evaluated, both having a 100 μm bump base diameter and a bump height. The first solder bump structure had no via, (called 0/100) and sat on a blanket polymer layer. This structure was evaluated to determine the adhesion of the polymer to the underlying layers of the structure. The second structure had a via under the solder bump imaged from a 75 μm mask feature (called 75/100). This 3:4 via:bump base diameter ratio is representative of the design of bumped structures utilized commercially throughout the industry.

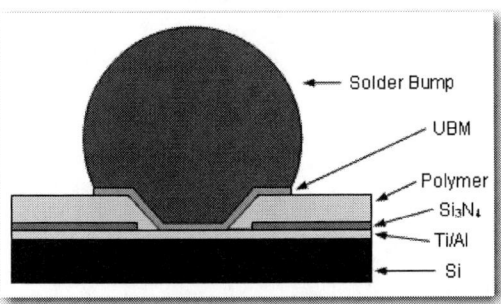

Figure 16: Repassivation Bumped Test Structure

Solder bump shear testing was done in accordance with JEDEC specification JESD22-B117A Condition A [10], with a shear tool lift of 10 μm, a shear speed of 100 μm/second, and a test end point at 70% of the maximum shear force. Table 5 shows the results of the bump shear tests performed.

The shear strengths of these structures with BCB and ALX-211 were very similar. The pad lift failure mode seen for the 0/100 bump structures was a separation between the UBM pad of the bump and the polymer surface. For the

	Via Bump Diameter	
	0/100	75/100
Avg. BCB shear strength (g)	38.4	35.7
Std dev (g)	2.6	1.0
Failure mode	Pad lift	Pad lift
Avg ALX shear strength (g)	35.2	34.0
Std dev (g)	1.1	0.9
Failure mode	Pad lift	Ductile shear

Table 5: Repassivation Test Structure Bump Shear

75/100 bump structures there was a significant difference in the failure modes, with the ALX-211 polymer samples exhibiting solder shear while the BCB polymer samples exhibiting a pad lift failure mode similar to that observed in the 0/100 bump structures.

As a further evaluation of the performance of the ALX-211 polymer, the repassivation test structures were subjected to additional reflow cycles beyond those required for the bump fabrication process (2 reflow cycles). Bump shear tests were done after an additional 5 and 10 reflows and the results are shown in Tables 6 and 7 below.

Once again, the failure strengths of both bump structures with both polymers are similar. The failure mechanism on the 75/100 remains ductile solder shear on ALX-211 samples compared with a pad lift failure mode on the BCB samples.

	Via Bump Diameter	
	0/100	75/100
Avg. BCB shear strength (g)	39.0	46.8
Std dev (g)	2.1	1.5
Failure mode	Pad lift	Pad lift
Avg ALX shear strength (g)	46.4	44.7
Std dev (g)	2.1	1.4
Failure mode	Pad lift	Ductile shear

Table 6: Repassivation Test Structure Bump Shear After 5 Additional Reflow Cycles

	Via Bump Diameter	
	0/100	75/100
Avg. BCB shear strength (g)	40.3	45.0
Std dev (g)	2.4	0.9
Failure mode	Pad lift	Pad lift
Avg ALX shear strength (g)	42.3	41.7
Std dev (g)	1.7	0.6
Failure mode	Pad lift	Ductile shear

Table 7: Repassivation Test Structure Bump Shear After 10 Additional Reflow Cycles

The second WLP structure fabricated to compare ALX-211 and BCB was the bump-on-polymer structure shown in Figure 17. The bump-on-polymer structure is commonly used to improve the reliability performance of solder bumped devices by increasing the thickness of the stress buffer polymer dielectric material and moving the bump location away from the device I/O pad. The first polymer layer was an unpatterned 5 μm thick ALX film applied to Si wafers over ~3000Å thermal SiO_2. The thickness of the electroplated Cu RDL line is approximately 2.5 μm. The second polymer layer is 5 μm and has photo-defined via structures over the plated Cu RDL lines.

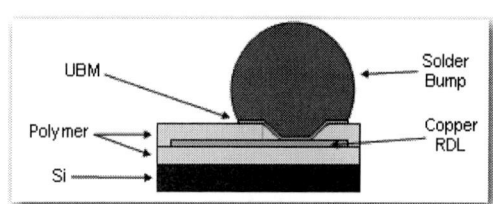

Figure 17 Bump on Polymer Bumped Test Structure

The processing of the two ALX-211 polymer layers follows the previously described optimal process conditions except that the exposure dose for the second layer was increased to 200 mJ/cm^2 in order to achieve sufficient exposure to prevent delamination during develop (due to absorption of light energy by the underlying polymer layer). As for the polymer bump-to-polymer experiment test structures, a 190°C/2 hour cure process was used for both layers.

The BCB polymer was processed using a standard process developed at RTI which results in vias approximately 15 μm smaller than the mask feature [14]. Similar to the ALX-211 processing, the exposure dose of the second BCB layer was increased by approximately 50% in order to achieve a full exposure dose over the

first polymer layer. In accordance with the BCB processing guidelines from Dow Chemical, the cure of the first BCB layer was done at 210°C/40 minutes in order to achieve good adhesion between the first and second polymer layers and the second layer cure was done at 250°C/1 hour [15].

Following the polymer processing, standard RTI sputtered thin-film Ti/Cu UBM was deposited and patterned to define the areas where solder would be electroplated. Following electroplating, the resist template was stripped, the UBM layers were chemically etched away and the solder bumps were reflowed. As in the previous experiment with repassivated solder bump structures, 0/100 and 75/100 bump structures were evaluated.

Solder bump shear testing was done in accordance with JEDEC specification JESD22-B117A Condition A, with a shear tool lift of 10 μm, a shear speed of 100 μm/second, and a test end point at 70% of the maximum shear force [16]. Table 8 summarizes the results of the bump shear tests conducted on the ALX-211 and BCB bump-on-polymer test samples.

While the failure modes of the two structures are the same regardless of the polymer used, the shear strengths of the ALX-211 samples are consistently higher than the BCB samples.

	Via Bump Diameter	
	0/100	75/100
Avg. BCB shear strength (g)	45.5	58.5
Std dev (g)	2.2	1.9
Failure mode	Pad lift	Ductile shear
Avg ALX shear strength (g)	63.5	69.8
Std dev (g)	2.9	3.2
Failure mode	Pad lift	Ductile shear

Table 8: Bump-on-Polymer Solder Bump Shear Results

Conclusions:

We have evaluated Asahi Glass ALX-211 spin-on permanent dielectric polymer. We have developed a standard process flow for this material.

Adhesive shear testing was conducted on several ALX-211 interfaces, although there is not a lot of other thin film shear test data with other materials to compare to, no issues were encountered. ALX-211 exhibited significantly

978-1-4244-4722-0/09 $25.00
© 2009 IMAPS-ITALY

higher bump shear strengths, presumably due to its higher elongation and lower modulus compared to BCB.

Measurements of ALX-211 cure percentage after multiple eutectic solder reflow cycles (approximately 220°C) indicates that the level of cure is not affected which should ensure consistent mechanical properties during multiple solder reflow processes.

Our studies indicate that ALX-211 polymer has desirable physical properties, an easily implemented process flow , good planarization properties, and good mechanical performance when subjected to shear stresses and solder bump shear testing.

References:

[1] M. Goosey, <u>Plastics for Electronics, 2nd Ed.</u> Springer, 1999.

[2]. G.Czornyj, ; M. Asano; R.L. Beliveau; P. Garrou; H. Hiramoto; A. Ikeda; J.A. Kruez; O. Rohde, "Polymers in Packaging " In *Microelectronics Packaging Handbook,* 2nd ed.; R.R. Tummala et. al. Eds.; Chapman & Hall: New York, 1997; pp 509–623.

[3] P. Garrou, "Thin Film Packaging and Interconnect", Chapter 9 in <u>Thin Film Technology Handbook</u>, A. Elshabini-Riad and F. Barlow Eds., McGraw Hill, 1998.

[4]. P. Garrou, " Thin Film Polymeric Materials in Microelectronic Packaging and Interconnect", *Proc.Int. Symp on Advanced Packaging Materials, 1998, p. 53*

[5] P.Garrou,"Wafer Level Chip Scale Packaging (WL-CSP): An Overview", IEEE Trans CPMT, vol. 23, 2000, p. 198.

[6] P. Garrou, "Wafer Level Packaging Has Arrived" Semiconductor International, October 2000, p. 119

[7] D. Hayes, "The Changing Landscape of WLCSPs", Int..Wafer Level Pack.Conf, SMTA, San Jose, 2007.

[8] T. Eriguchi, et. al., "Low Temperature Curable
Photosensitive Polymer Dielectric for Wafer Level Packaging Applications", *Proceedings of 2007 IMAPS Device Packaging Conf., Scottsdale*, AZ, 2007.

[9] J. Im et. al." On the Mechanical Reliability of photo- BCB- Based Thin Film Dielectric Polymer for Electronic Components", Trans. Amer Soc of Mechanical Eng, 2000, p. 28.

[10] J. Im et. al., "Physical and Mechanical Properties Determination of Photo-BCB-Based Thin Films" Proceed. ISHM, 1996, p. 168.

[11] A Strandjord et. al., "Process Optimization and Systems Integration of a Copper/Photosensitive Benzocyclobutene MCM-D", Int. Journal of Microcircuits and Electronic Components, 1996, p. 260

[12] PI data taken for : Asahi 8100S, Toray/Dow Corning PW1000; HD Micro 4000; Fuji Film 7000.

[13] L.B. Rothman, "Process for Forming Passivated Metal Interconnection System with a Planar Surface", J. Electrochem. Soc., Volume 130, 1983, p. 1131.

[14] D. Mis, G. Rinne, P. Deane, G. Adema, "Flip Chip Production Experience: Design, Process, Reliability and Cost Considerations", Proceed. ISHM, 1996, p. 291

[15]"Process Proceedures for Cyclotene 4000 Series Photo-BCB Resins", Dow Chemical, Feb 2005.

[16] JEDEC Solid State Technology Association, JEDEC Standard JESD22-B117A, "Solder Bump Shear," © JEDEC Solid State Technology Association, 2006, available at www.jedec.org

978-1-4244-4722-0/09 $25.00
© 2009 IMAPS-ITALY

Experimental analysis on the mechanism of moisture induced interface weakening in ACF package

Gi-Dong Sim, Chang-Kyu Chung, Kyung-Wook Paik and Soon-Bok Lee

KAIST, 373-1 Guseong-dong, Yuseong-gu,

Daejeon 305-701, Republic of Korea

82-42-350-3069, 82-42-350-5013, gggaami@kaist.ac.kr

Abstract

In this paper, we have determined the influence of moisture absorption to the adhesion property of ACF for flip chip interconnection. An attempt was made to quantitatively measure the interfacial fracture toughness of both silicon/ACF and FR4/ACF sandwiched specimens exposed to different humidity conditions using 4-point bending test. Evidence of crack propagating along the interface was checked by cross section images of the specimens. For moisture absorption, specimens were stored inside the humidity chamber(85 ℃,85% RH) until they were fully saturated. Pressure Cooker Test (PCT) was performed to observe vaporization effect on the adhesion strength under real service condition. By comparing the 4-point bending test results between different humidity conditioned specimens, interface weakening by moisture absorption has been observed quantitatively and the mechanism was considered. Scanning Acoustic Microscope (SAM) images were taken to observe whether delamination exists at the specimen interface.

Key words: anisotropic conductive film (ACF), interface fracture, 4-point bending, delamination, moisture induced weakening

1. Introduction

Integrated circuit (IC) packages are requiring smaller, lighter with higher I/O and better electrical performance at lower cost. To satisfy these requirements, flip chip interconnection using ACF (Anisotropic conductive film), has become a trend in electronic packaging due to its low bonding temperature, low cost capability (no flux and underfilling), fine pitch capability and green process technology [1]. Consequently, reliability of ACF is emerging as an important issue and the degradation mechanism is becoming an important research topic.

One of the most important failure mode observed is delamination along the chip/ACF and substrate/ACF interface, as shown in Fig.1(a). Once delamination occurs, crack propagates along the interface and may lead to abrupt failure of the microelectronic package. Especially, shear failure in thermal cycling test and delamination failure in moisture absorption test are the two major reliability problems in Chip-on-Board packages using ACF flip chip interconnection. Both failure condition can be analyzed as bending mode owing to CTE (coefficient of thermal expansion) or CME (coefficient of moisture expansion) mismatch between IC chip and the substrate (Fig.1(b)). Therefore, quantitative evaluation of interface delamination for the electronic structure which has been exposed to high temperature or absorbed moisture is a critical issue. Between these two

issues, we focused on the influence of moisture absorption to the adhesion property of ACF.

(a)

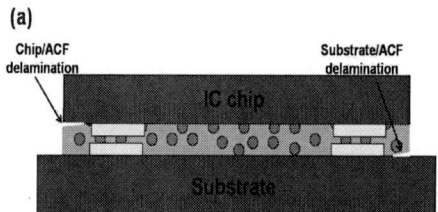

(b)

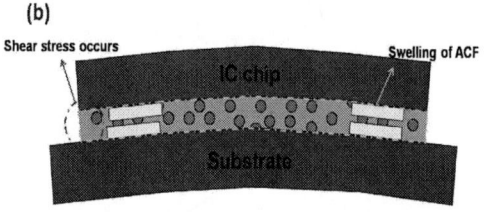

Figure.1 (a) Delamination along the bimaterial interface, (b) bending mode at low temperature

Moisture penetration into the flip chip package is known to induce crucial effects on the package reliability. Moisture induced swelling, popcorn crack are well known phenomena that may cause degradation of the package performance and interface delamination could occur during operation [2]. Therefore, precise information about moisture induced interface weakening acquires greater importance. As polymeric materials and inorganic materials have different values of CME, former researcher analyzed CME mismatch induced stress, strain problems as same as CTE mismatch problems [3]. These results are excellent to evaluate moisture induced failure of plastic integrated circuit (IC) package. But, in this package type structure, interface weakening effect caused by ACF property itself is hard to be abstracted.

In our experiment, we employed sandwiched specimen consisting of equivalent material as the top and bottom substrate and ACF as the adhesive. Consequently, CME mismatch were not expected and our specimens were appropriate for the evaluation of ACF property degradation due to moisture.

Between various adhesion test methods which can be utilized to measure adhesion force or energy, we adopted four point bending test using notched specimens. Upon all other tests, delamination sandwiched specimen is noticed to be quantitative and reproducible while others are qualitative and dependent to residual stresses [4]. In the present work, interface toughness of ACF were evaluated precisely and quantitatively for both silicon/ACF and FR4 (Flame Retardant 4)/ACF specimens in different humidity conditions. As we have obtained the intrinsic adhesion property of ACF at room temperature and room humidity condition from our previous research [5], we can analyze out the moisture effect by comparing the test results.

In this paper, we'll further discuss the mechanism of moisture induced interface weakening by comparing the results between saturated specimens and PCT (Pressure Cooker Test) specimens. Through analysis, we can also decide whether adhesion strength weakening by hygroswelling or vaporization effect during service condition would be the dominant effect to the failure of ACF.

2. Experimental Methods
2.1 Materials

Anisotropic conductive film (ACF) is an adhesive between electronic components which consist of epoxy based polymer matrix and conductive fillers for electrical conduction and mechanical interconnection. Schematic figure of flip chip interconnection using ACF is shown in Fig.2.

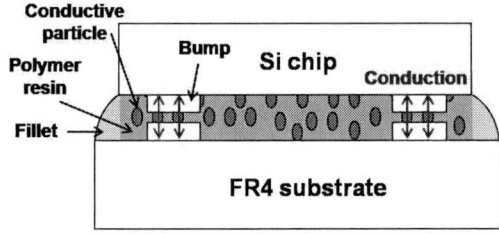

Figure.2 ACF flip chip interconnection

The thickness of the ACF used in our experiment was 35 μm and the conductive particles were five micron-diameter metal-coated polymer balls.

Materials chosen as the substrate for four-point bending test were Si and FR4. From Chip-on-Board (COB) packages, Si represents integrated circuits while FR4 represents printed circuit board. Size of the upper and lower beam were both 30 mm * 3mm * 1mm for Si and FR4.

For the specimen fabrication, thermo-compression bonding procedure was imported [6]. Schematic illustration of the bonding process is shown in Fig.3. Briefly, first the lower part was prebonded with ACF at 80℃ for 3 seconds. Next, the upper part and the lower part were aligned over the ACF bonding plate. Finally, the whole structure was bonded at 190℃ and 50 N for 20 seconds. Unlike the whole flip chip package in Fig.2, we excluded the metal bump and fillet to remove the effect to the adhesion property.

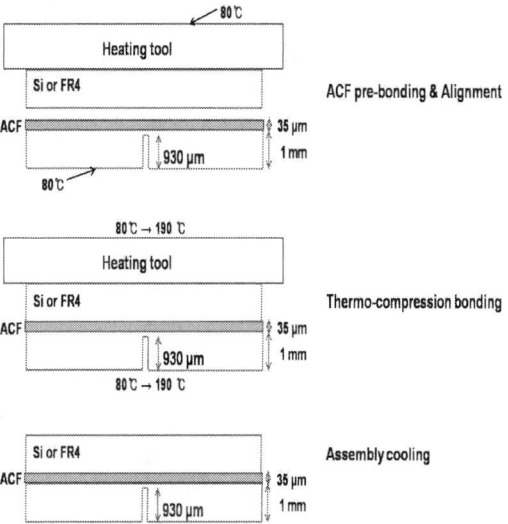

Figure.3 Schematic illustration of the ACF bonding process

2.2 Experimental
2.2.1 Moisture absorption

Figure.4 Humidity chamber for moisture absorption

Cured sandwiched specimens were stored in the humidity chamber (Fig.4) of 85℃, 85% relative humidity (RH) condition until they are fully saturated. Saturated state was checked by measuring the weight of the specimen from time to time as shown in Fig.5. Unlike the moisture diffusion of FR4 specimen, which is Fickian, moisture diffusion of Si specimen has step looking weight increment. This is due to the fact that Si itself does not absorb moisture. That is, weight increase inside Si specimen totally occurs by ACF. As the weight increment was smaller than the resolution of measurement, clear Fickian curve was unavailable in this case. However, as our object was to check the threshold time for saturated specimen, this test result is adequate. Si and FR4 specimens stored longer than 300 hours were applied to four-point bending test.

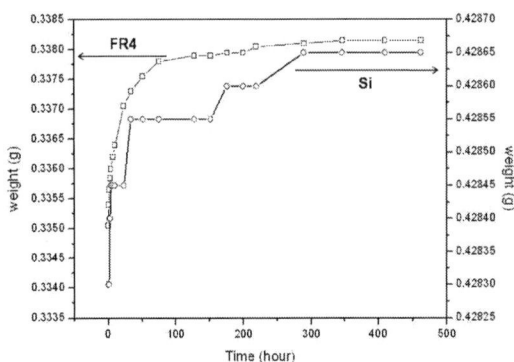

Figure.5 Saturation estimate using weight measurement

2.2.2 Pressure cooker test (PCT)

PCT was performed to investigate moisture-related reliability of ACF flip chip assemblies. The PCT condition was 121℃, 100% relative humidity (RH), and 2 atm. Specimens were exposed to harsh environmental conditions: 99 hours. Through comparison between moisture absorbed specimen and PCT specimen, we expect to separate the effect of moisture induced swelling and popcorn to the interfacial strength of ACF.

2.2.2 Four-point bending test

Four-point bending adhesion test was carried out using a standard sandwiched structure, as shown in Fig.6. For crack initiation, notch was produced in 930 μm depth. Notch depth is decided to minimize the effect of glass fiber existing inside the FR4 substrate. The experiment was performed in a displacement control mode with optimized displacement rate for each specimen. Further explanation will be described at the next section.

By applying elastic beam theory, the strain energy release rate can be calculated as [7] :

$$G_c = \frac{21\left(1-\nu^2\right)M_c^2}{4Eb^2h^3} = \frac{21\left(1-\nu^2\right)P_c^2l^2}{16Eb^2h^3} \quad (1)$$

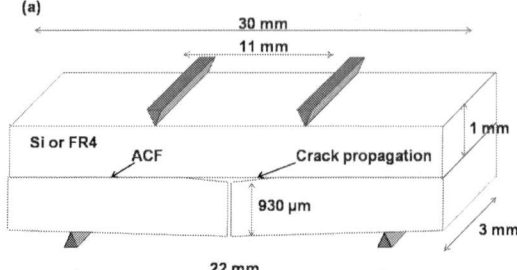

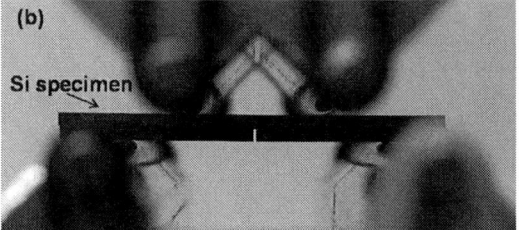

Figure.6 Four-point bending setup of the sandwiched specimen: (a) illustration and (b) test image

where $M_c = P_c l/2$ represents the critical bending moment, P_c is the plateau load, l is the spacing between the inner and outer loading lines, b is the sample width, h is the specimen height, E and ν are the Young's modulus and Poisson's ratio of the substrate. Properties for the materials analyzed in our experiment and the test conditions are shown in Table I and Table II, respectively. However, property of ACF is never applied actually in our experiment because the equation is based on homogeneous solution.

Table I. Properties of materials used

	Thickness (mm)	Young's modulus (GPa)	Poisson's ratio
ACF	0.035	1.2	0.35
Si	1	130	0.27
FR4	1	17.2	0.15

Table II. Test conditions for experiment

	Dimension (mm)
Length	30
Width	3
Pin spacing	5.5

2.2.3 Displacement rate optimization

In our previous experiment, displacement rate conditions were 0.1 μm/s for both Si and FR4 sandwiched specimen. However, former researches

978-1-4244-4722-0/09 $25.00
© 2009 IMAPS-ITALY

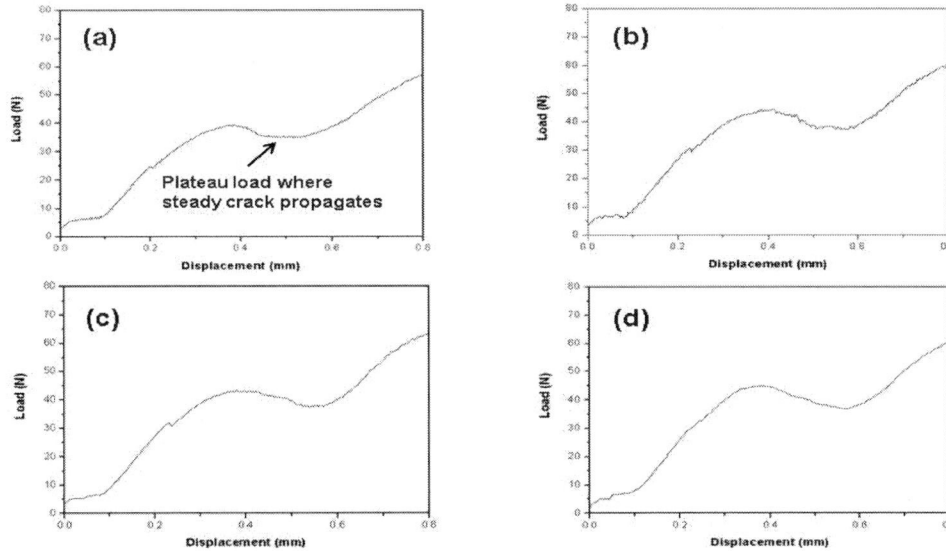

Figure.7 Load-displacement curve for various displacement rate conditions :
(a) 0.2 μm/s, (b) 0.3 μm/s (c) 0.4 μm/s, (d) 0.5 μm/s

suggest experiments should be done before moisture spreads out in laboratory humidity condition [8]. As an example, a 30 minute standard has been introduced. In our case, Si specimen copes well with the time limit with displacement rate 0.1 μm/s. However, as the elastic modulus is 10 times lesser than Si, experiment time exceeds the standard in case of FR4. Therefore, we decided to increase the displacement rate for FR4 specimen to match up with experiment time limit.

To optimize the displacement rate, less experiment time and reliable, reproducible test results were both considered. Load-displacement curve for various rate conditions are shown in Fig.7. Especially at 0.5 μm/s, few test results were hard to observe load plateau region.

Suitability of the rate for moisture absorbed specimens is summarized in Table III. Between 0.3 μm/s and 0.4 μm/s, we chose 0.3 μm/s as the optimized displacement rate for FR4 specimen.

Table III. Optimization of the displacement rate

	0.1 μm/s	0.2 μm/s	0.3 μm/s	0.4 μm/s	0.5 μm/s
Experiment Time	X	X	O	O	O
Reliable Test results	O	O	O	O	X

3. Results and Discussions

3.1 Interfacial fracture toughness

As the bending moment increases, a crack initiates from the notched tensile side surface and propagates vertically to the interface. At the instance crack meets the interface, two options may be expected to occur which are penetration or debonding. As our interest is the debonding situation, we had to check the evidence of crack propagation through the interface region during experiment. For FR4 specimen, there is no problem because crack propagation can be seen directly during the experiment. However, as Si is a brittle material, bending and crack propagation are hard to been checked during experiment for the Si specimen. Therefore we stopped the test when steady crack growth was expected and checked the evidence of crack propagation. Especially, two points were concerned. First, whether the notch is broken and crack started to propagate were checked. And afterwards, the presence of actual crack at the interface was confirmed. Results are shown in Fig.8. Through these evidence, we were confident about interface debonding during the experiment.

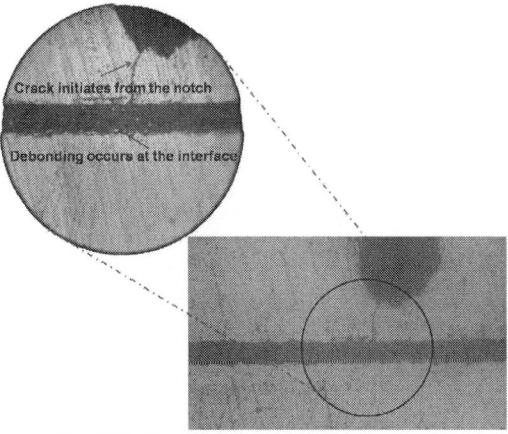

Fig.8 Evidence of crack initiation and propagation at the interface

By observing the load-displacement curve, plateau load is decided and from equation (1), energy release rate is calculated. This energy release rate is the superposition of pure mode I and mode II. Energy release rate for each case is represented as [9]:

$$G_{\text{I}} = \frac{3\left(1-v^2\right)P_c^2 l^2}{4Eb^2 h^3} \qquad (2)$$

$$G_{\text{II}} = \frac{9\left(1-v^2\right)P_c^2 l^2}{16Eb^2 h^3} \qquad (3)$$

Stress intensity factor K at the interface for mode I and mode II in plane strain condition can be obtained from Irwin's universal relation,

$$G = \frac{\left(1-v^2\right)K^2}{E} \qquad (4)$$

the total stress intensity factor Ki can be calculated taking account of the effect of mode I and mode II

$$K_i = \sqrt{K_{\text{I}}^2 + K_{\text{II}}^2} \qquad (5)$$

Critical energy release rate calculated from equation (1) can be applied to represent adhesion strength of ACF at different interfaces. Energy release rate and corresponding stress intensity factors are shown in Table IV. Detail explanation of the interface fracture energy decrease due to moisture absorption and PCT will be further discussed.

In four-point bending test, remote phase angle ψ is:

$$\psi_\infty = \tan^{-1}\left(\frac{K_{\text{II}}}{K_{\text{I}}}\right) \approx 41° \qquad (6)$$

and the local phase angle is known as

$$\psi = \psi_\infty + \omega + \varepsilon \ln(l/h) \qquad (7)$$

where ω is an angle that depends on the elastic properties of the films and the substrates and ε is the bimaterial constant defined as [10]

$$\varepsilon = \frac{1}{2\pi} \ln\left[\frac{(3-4v_1)/\mu_1 + 1/\mu_2}{(3-4v_2)/\mu_2 + 1/\mu_1}\right] \qquad (8)$$

However, as the elastic mismatch is not too large for most cases, it is common to specify the phase angle as the remote phase angle. In real electronic products, CME mismatch between silicon die and the FR4 substrate may induce bending of the

package during moisture absorption. Therefore, mixed mode condition is acceptable compared to real service reliability problem.

3.2 Moisture absorption effect

Two major reliability problems should be discussed for the reliability of moisturized specimens. First is the adhesion strength weakening by hygroswelling of ACF. And secondly, vaporization effect and delamination owing to vapor pressure should be considered. The object of our current research is to focus on the effect of moisture to the intrinsic adhesion property of ACF.

Most of the currently produced ACFs generally consist of a thermosetting epoxy with minor thermoplastic polymers. Thermoplastic polymers are known to have porous chains that moisture can work its way between the polymer chains [11]. Therefore, swelling may occur inside the polymer resin. As shown in Table IV, the adhesion property of ACF seemed to weaken by hygroswelling during moisture absorption. Especially, this effect was gross for the Si specimen. Compared to Si specimen, ACF/FR4 interface adhesion strength was observed to be much more endurable.

In our previous research, we mentioned that the interface fracture toughness of ACF/FR4 interface is much larger than ACF/Si interface because both ACF and FR4 materials are epoxy based. Similarly in moisture absorbed specimen, hygroswelling stress in the FR4 package was not strong enough to intercept the inter- and intra-hydrogen bonding between ACF and FR4. On the other hand, ACF/Si interface was observed to be prone to hygroswelling. From our experiment result, moisture uptake can be thought to attribute to the degradation of the intrinsic adhesion property of the polymer resin inside ACF. However, in order to check that hygroswelling is not the main reason for delamination, Scanning Acoustic Microscope (SAM) images were taken for both specimens before (Fig.9) and after moisture absorption (Fig.10). Reflection mode was applied to silicon specimens which shows delmination in bright images. On the other hand, as FR4 consist of glass fiber, diffraction occurs for the FR4 sandwiched specimen. Therefore, transmission mode was applied to the specimen where black area represents delamination occurrence. From our result, delamination did not appear after moisture absorption for both Si and FR4 specimens.

Consequently, SAM figures convinced that swelling is not the main factor for delamination. Therefore, we may think that moisture absorption induced hygroswelling, without delamination occurrence, attributes to the degradation of the adhesion strength of ACF.

978-1-4244-4722-0/09 $25.00
© 2009 IMAPS-ITALY

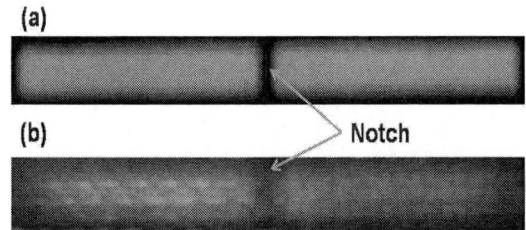
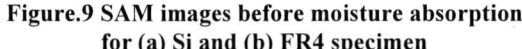
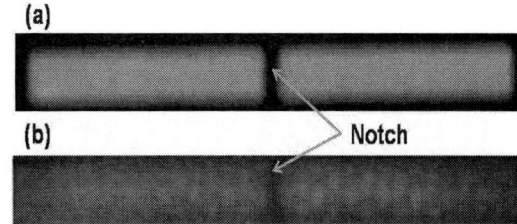

Figure.9 SAM images before moisture absorption for (a) Si and (b) FR4 specimen

Fig.10 SAM images after moisture absorption for (a) Si and (b) FR4 specimen

Table Ⅳ. Interface fracture energy and stress intensity factors for various environment conditions

	Si/ACF interface			FR4/ACF interface		
	room	85℃/85%	PCT 99hr.	room	85℃/85%	PCT 99hr.
Interface fracture energy G (J/m^2)	2.16	1.26	1.25	414.7	350.68	314.71
K_I (MPa·m1/2)	0.42	0.319	0.318	2.04	1.88	1.78
K_{II} (MPa·m1/2)	0.36	0.276	0.275	1.77	1.63	1.54
Relative strength (%)	100	58.33	57.87	100	84.56	75.89

3.3 Vaporization effect

PCT specimens were compared with saturated specimens to abstract the vaporization effect. Compared to saturated specimens, PCT specimens are additionally exposed to high temperature (121℃) which may cause vaporization at the interface of the substrate/ACF. And the experiment results for the PCT specimens are shown in Table IV. From our result, we observed that Si/ACF interface fracture energy were almost constant for the specimen saturated at 85℃/85% condition and the specimen exposed to PCT condition. However, for the FR4/ACF interface, significant decrease of the adhesion strength was observed.

We think two factors may be the main reason for this phenomenon. The first reason is that moisture fraction is very small in case of Si/ACF specimen owing to the fact that Si itself does not absorb moisture. Secondly, temperature would be different between the free surface and the inner part of the package. Which indicates vaporization may not have occurred directly at the interface. Brought together, in case of FR4 specimen, vaporization could occur near the free surface of the specimen due to moisture absorption of FR4 itself. And this vaporization process would deliver more moisture up to the interface through diffusion. As a result, moisture penetration could occur at the interface or

vaporization of larger quantity of water molecules could occur at the interface and cause delamination. For the Si specimen, on the other hand, these conditions may be restricted by the absence of moisture inside Si itself.

Further discussions may be required to understand the precise mechanism. However, based on our experiment results, we conclude that moisture absorption during service load of an ACF package may cause severe adhesion strength degradation even when delamination is not coevolved.

4. Conclusion

Recently, ACF materials have been widely used in microelectronic packages for flip chip interconnection and already are major packaging methods for flat panel displays. However, epoxy based ACF materials may introduce several reliability problems by mechanical, environmental (e.g., thermal, moisture). Therefore, fracture mechanics based reliability analysis is frequently applied to solve these issues in microelectronic packages.

Fracture mechanics based four-point bending test was introduced to quantitatively calculate the interface weakening due to moisture absorption. Displacement rates were optimized to obtain reliable and reproducible results for both Si and FR4 specimens. Cross sectional image was taken to check the evidence of crack propagation through the

978-1-4244-4722-0/09 $25.00
© 2009 IMAPS-ITALY

interface. Additionally, we took SAM images to observe whether delamination occurred inside the sandwiched specimen. Although interface fracture energy decreased significantly for the Si specimen, delamination occurrence was not observed. Therefore, we can make a conclusion that moisture uptake may attribute to degradation of intrinsic adhesion property of the polymer resin inside ACF by hygroswelling.

PCT experiments also revealed the possibility of vaporization and moisture diffusion inside the package for the FR4 specimen.

However, through our experiment result, the dominant effect to the failure of ACF package during service condition may be thought as the hygroswelling of ACF owing to moisture absorption.

Acknowledgement

This work was financially supported by the Fundamental R&D Program for Core Technology of Materials funded by the Ministry of Knowledge Economy, Republic of Korea.

References

[1] Yim, M. J., Jeon, Y. D., Paik, K. W., "Reduced thermal strain in flip chip assembly on organic substrate using low CTE anisotropic conductive film," IEEE Trans Electron Pa M, Vol. 23, No. 3, pp. 171-176, July, 2000.

[2] Shi, X. Q., Zhang, Y. L., Zhou, W., Fan, X. J., "Effect of hygrothermal aging on interfacial reliability of silicon/ underfill/FR-4 assembly," IEEE Trans CPMT, Vol. 31, No. 1, pp. 94-103, 2008.

[3] Zhang, G.Q., van Driel, W.D., Fan, X.J., "Mechanics of Microelectronics", Springer, first edition, New York, 2006

[4] Evans, A. G., Rühle, M., Dalgleish B. J., and Charalambides, P. G., "The fracture energy of bimaterial interfaces", Materials Science and Engineering A, Vol. 126, No. 1-2, pp. 53-64, June, 1990.

[5] Gi-Dong Sim, Chang-Kyu Chung, Kyung-Wook Paik and Soon-Bok Lee, 2009, "Interfacial fracture toughness of anisotropic conductive films (ACFs) flip chip interconnection," Jpn. J. Appl. Phys., submitted.

[6] Kwon, W.-S., Paik, K.-W., "Fundamental understanding of ACF conduction establishment with emphasis on the thermal and mechanical analysis," Int. J. Adhes. Adhes., Vol. 24, No. 2, pp. 135-142, April, 2004.

[7] Charalambides P. G., Lund J., Evans A. G., and McMeeking R. M., "Test specimen for determining the fracture resistance of bimaterial interfaces," J. Appl. Mech., Vol. 56, No. 1, pp. 77-82, March, 1989.

[8] Tanaka, N., Kitano, M., Kumazawa, T., Nishimura, A., "Evaluating IC-package interface delamination by considering moisture-induced molding-compound swelling," IEEE Trans CPMT, Vol. 22, No. 3, pp. 426-432, September, 1999.

[9] Hutchinson, J.W., Suo, Z., "Mixed Mode Cracking in Layered Materials," Advances in Applied Mechanics, Vol. 29, pp. 63-191, 1992.

[10] Dunders, J., "Edge-bonded dissimilar orthogonal elastic wedges," J. Appl. Mech., Vol. 36, pp. 650-652, 1969.

[11] Ferguson, T., Qu, J., "Effect of moisture on the interfacial adhesion of the underfill/solder mask interface," Journal of Electronic Packaging, Vol. 124, No. 2, pp. 106-110, June, 2002.

3D IC products using TSV for mobile phone applications:
An industrial perpective

Yann Guillou, Anne-Marie Dutron

ST-ERICSSON, 12 rue Jules Horowitz, 38019 Grenoble Cedex, France

Phone: +33 476 58 58 77, Fax: +33 476 58 60 50, E-mail address: yann.guillou@stericsson.com

Abstract

3D integration and configurations based on More than Moore approaches are becoming very popular in recent years. The first applications that could use this breakthrough technology are making more and more sense for implementation in the next generations of cellular phones. This paper gives explanations on what is 3D Integration and divides this denomination into 2 main categories: 3D at the packaging level and 3D at the IC level. Presentations of R&D efforts in both categories are given in order to illustrate the main differences. Insights about the complementarities of 3D ICs with some latest advanced packaging solutions such as Fan-Out Wafer Level Packaging are discussed. Finally, a global roadmap of applications that may use these advanced solutions is shown with respective challenges and timeframes.

Key words: Cellular phones, Through Silicon Via (TSV), 3D integration, Fan-Out Wafer Level Packaging (FOWLP)

Introduction

The mobile phone and wireless industries have been growing at a very fast pace in the last 10 years. With more than 1.1 billion phones sold in 2008, the wireless market has been one of the main contributors in driving the development of the most advanced semiconductor technologies. In effect, after the military era in the 70's, the PC era in the 80's and early 90's, the wireless era is now considered as the driving application to push the limits of Moore's law. The convergence of mobile smartphones, PC and Mobile Internet Devices is now obvious. The computing power for portable devices is becoming mandatory to enable mobile internet browsing, mobile TV, and all increasing multimedia mobile applications. In that context, miniaturization and computing power are the driving forces for silicon technology development. Following the ITRS roadmap with CMOS downscaling is the traditional way to proceed. But the R&D expenses to sustain such a roadmap are extremely selective. Only a few companies or consortia can afford it. That's why advanced CMOS technology could start to be a kind of commodity and integration with new solutions appears as a much stronger differentiator element [1]. Packaging contributions in the final product gained in importance and begins not to be only considered as a vulgar piece of plastic. Major improvements and progresses have been done and a new solution is gaining more and more importance: 3D Integration. This paper will give some explanations and guidelines to argue why using 3D Integration with

TSV could make sense for the next generation of wireless products.

3D Packaging

Despite the recent buzz in the industry for a couple of months, 3D configurations are not new at all. In effect, 3D configurations at the packaging level have been around for years.

Stacked dies with wire bonds in BGA are an example of this. Based on marketing reports, actual mid- and high-end phones contain several 3D packages.

A stack of memory dice connected to the same substrate with wire bond is something very common. For instance, in 2003, STMicroelectronics announced a stack of 10 dice, a world record at that time (fig 1). Since then, some packaging houses and IDMs have prototyped stacks with up to 40 dice. No one would argue that 40 dice on the vertical axis is not 3D.

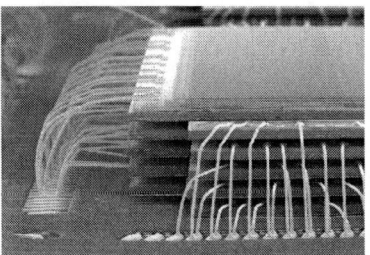

Fig 1: 3D packaging, 10 dice stacked

978-1-4244-4722-0/09 $25.00
© 2009 IMAPS-ITALY

Other 3D packaging configurations are called package-on-package (PoP), or package-in-package (PiP) in nomad devices. PoP (fig 2) or PiP is very popular to combine digital base band or application processor with the memory ICs [2]. Some packaging evolutions of PoP and PiP still using the third dimension have been lately announced by top SATS providers (Fan-In PoP [3], MaPPoP, Through-Mold-Via [4]…). The idea behind all these new 3D packages is to enable better integration with a smaller footprint, smaller BGA ball pitches and at the lowest cost.

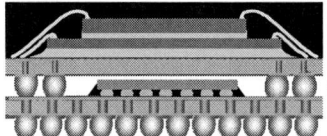

Fig 2: Flip Chip PoP

Emerging technologies using embedding concepts enabling 3D configurations, such as embedded die in laminate or re-built wafers, have been presented by substrate players or IDMs, for instance. Some of them are very promising and could be *the* packaging platform beyond BGA. Fan-Out Wafer Level Packaging (FOWLP) is one of them. FOWLP goal is to construct a package as small as possible to enable all the BGA balls to fit on it. The packaging is built around the individual known good die and provides a way to make the laminate substrate vanish (fig 3). FOWLP improves signal integrity, shortens interconnects, reduces line/space for rerouting, and as a consequence, allows a reduction in the package footprint. Infineon, or Freescale, released major announcements in the past few months, respectively with eWLB and RCP technologies [5] [6]. Since then, STMicroelectronics partnered with Infineon and STATS Chip Pac to develop the 2nd generation of FOWLP, the 3D-eWLB, and enable 2-sided and consequently, 3D configurations [7].

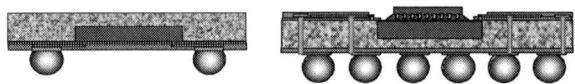

Fig 3: FOWLP 1 side (left) , 2 sides (called 3D-eWLB, right)

3D at IC level using TSV

However, all the configurations mentioned above are 3D at the packaging level. None of them uses TSV, which could allow "real" 3D at the IC level in some configurations.

TSV is another matter; although definitively interesting, it is not compulsorily enabling 3D in all cases. Imaging products are an illustration of this (fig 4).

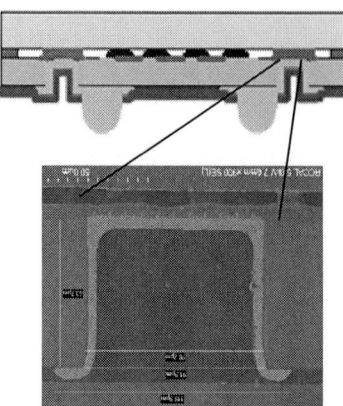

Fig 4: Schematics and cross section of STMicroelectronics CIS with TSV

The first mass products using TSV came about a couple of months ago when CMOS Imaging Sensor (CIS) companies saw an opportunity to use this new type of interconnection. The possibility to substitute wires by connections coming from the backside of the die would gain them a reduction in the camera volume and its cost [8]. However, in this example, we can't speak about TSV enabling 3D configuration because, in fact, no chips are connected along the vertical axis. The top layer is only a glass carrier, not an active die; this can be called 2.5D but definitively not 3D. Soon maybe new CIS using TSV and true 3D at the IC level may happen with the integration of the processor.

TSVs are nothing more than a techno brick that allows dice to interconnect on the vertical axis at the IC level. The impact of TSV is that these short interconnections could enable a new split of functions and new product partitioning. However, it is important to notice that TSV interconnections can be done at different levels and the impact on final applications is not the same. Either the TSV is done at the bond pad level or at the global interconnect level. We will not consider TSV at the local interconnect level in this paper due to its immaturity for implementation in cell phones in the near future.

At the bond pad level, TSV is done in a via last process in a wafer level packaging fab environment. The pitch of TSV is in the range of a bond pad pitch. Typical values from 50 to 150μ in term of pitch can be considered. An aspect ratio can reach 5:1 and filling is conformal or full. RLC electrical performances are better than wire bonds, except for the capacitance that may remain high. This type of TSV can be based on an improvement of technologies developed for CIS.

978-1-4244-4722-0/09 $25.00
© 2009 IMAPS-ITALY

At the global interconnect level, we consider TSV with higher aspect ratio done in foundy environment, preferably via middle (after FEOL but before BEOL) or first (before FEOL). TSV pitch is in the 10-20µ range with the aspect ratio ranging from 5:1 to 10:1 or even a bit more. For this category of TSV, dice are not only connected at bond pad level but also at the IP block or memory bank levels for instance. This type of TSV enables the achievement of structures with very short interconnections and good electrical performances. True heterogeneous integration is made possible and SoC–like features can be realized but also using the third dimension and not only the x-y directions. Dice interconnected with this type of TSV are closer to SoC than SiP. For these reasons, 3D SoC is potentially a good term for this configuration [9].

An understanding of these differences is key to really be able take into consideration the potentialities of this disruptive TSV technology. Futhermore, it helps in discussions between players and in definition of roadmaps and standards. This is a first and compulsory step: a shared nomenclature needs to be defined and adopted.

As of today, only a few applications using 3D IC TSV can be found in products. However, 3D configuration at IC level without TSV can be seen in recent cell phone or other nomad devices now on the market. Structures such as Face-to-Face (F2F) bring new solutions to some specific products (see fig 5). The main driver for using F2F is performance. For instance, it enables a logic die, such as an application processor or digital baseband, to be connected with a very short connection and high density with its memoy ICs. The idea behind this concept is to have a fast and direct access between dice and thus very high bandwidth. In effect, the requirement of the applications for high bandiwth is exponentially increasing and mainly driven by video features: 1080p30 playback, 1080p30, 60, 120 Camcorder, 3D gaming… All these requirements will be mandatory for all high end cell phones in the future.

Fig 5: Face-to Face configuration

F2F is a configuration that may need to be considered for applications other than logic/memory. However, in all cases the wires will remain and limit some performance.

Complementarity of 3D packaging and 3D IC

Except for some wafer-level chip-scale packages (WLCSP) or silicon interposer with TSV directly mounted on Printed Wiring Board (PWB) with a BGA balls, TSV is not a packaging solution by itself. TSV uses only "back-end world" skills such as bonding, fine pitch bumping, back grinding and thin wafer handling, for instance. In effect, final packaging is required to connect the device to the PWB. And due to assembly constraints, the choice of the final package solution could impact the entire TSV process flow. This final packaging could be a BGA package, a Fan-Out WLP type, an embedded die in laminate, or other. Here, it is interesting to notice how complementary 3D IC configurations with TSV and 3D packaging can be. In effect, FOWLP can enable designs to fully benefit from the 3D IC integration, and for instance, reduce the package footprint with more aggressive design rules than BGA packages. TSV is a new technology and FOWLP a new packaging technology as well. Only considering BGA-type packages for structures with TSV might be a wrong reasoning.

Some constraints can even be relaxed by coupling 3D IC TSV with 3D advanced packaging and then the full benefits of the TSV can emerge. Issues such as thin wafer handling can, for instance, be simplified. In the following schematic, figure 6 illustrates a 2 die stack with TSV in a FOWLP package.

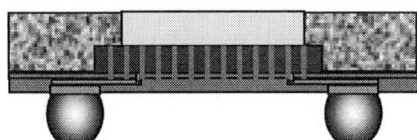

Fig 6 : 3D IC TSV stack in FO WLP

3D IC Design

In order to be able to manufacture 3D IC products, the capability to design them is mandatory. It seems obvious, but that does not mean it is ready today. In effect, as of today the focus in 3D is on process rather than design tools. CAD solutions are considered as a weak area and clearly delay 3D IC products from coming to market.

3D design tools can be separated into 2 main categories: 3D partitioning tools and 3D implementation tools. A 3D partitioning tool helps to optimize the full design with the position of blocks, the number of tiers, and globally all the new integration possibilities due to the 3rd dimension. Furthermore, it enables the designer to get a rough picture of the final cost and the performance of the system in a short time.

978-1-4244-4722-0/09 $25.00
© 2009 IMAPS-ITALY

3D implementation is mainly floorplanning, placement and routing, and timing. Electrical and thermal models need to be developed for this goal as well as all verification analysis (fig 7).

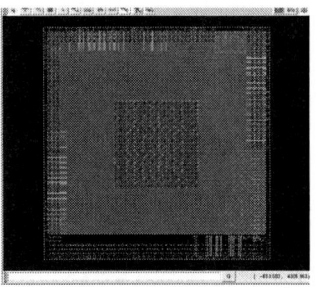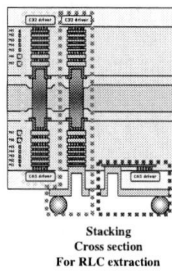

**Stacking
Cross section
For RLC extraction**

Fig 7: 3D TSV place & route and RLC extraction

3D IC TSV is the convergence of silicon and packaging with the design. New architectures can in effect be considered and achieved. In fact, new architectures have to be considered if cost effectiveness is to be reached.

In order to fully benefit from 3D TSV and make this technology cost effective, 3D architecture needs to be considered at a very early stage. However, designers are facing a gap between TSV technology process and TSV system design. This gap is due to the fact that there is no clear TSV technology roadmap in the industry. With a scaling-based approach and a classical follow-up of Moore's law, it was easier to focus the R&D efforts and predict the size and performance of a new techno node. With 3D TSV, the industry is facing a new paradigm. Designers' mentalities have to be modified and the former constraints of 2D have to be partially forgotten.

Roadmaps exist, but are not necessarily very relevant. For this reason, many options can be considered, and process technologists do not know by themselves where to go and on what to focus. The only solution is to reinforce collaborations and discussions between the design community and the hardware technology community. A holistic approach is necessary to find the technology / design sweet spot and the application needs to drive developments.

Application roadmap

Applications that could use TSV and 3D remain a hot question. Only a few people have clear views on products that could use such innovations. Many niche markets would exist, for sure. However, when we think about mass market products particularly in wireless, the scope of applications tends to reduce. The following roadmap (figure 11) is based on ST-Ericsson's and STMicroelectronics' view and portfolio for wireless products. Memory

stacks with TSV such as DRAM are not mentioned as they are not part of the company business area anymore. Combinations of 3D IC TSV with advanced packaging such as FOWLP are not detailed on the schematics.

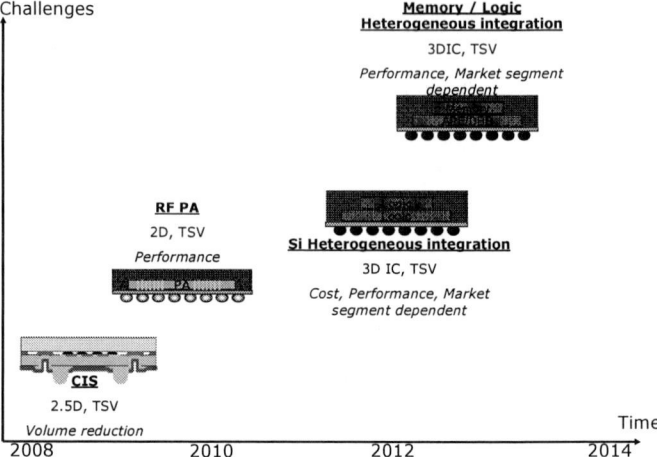

Fig 11: Application roadmap for cellular phones

The first CMOS Imaging Sensors are now on the market and ramping up. Details have already been discussed in the 3D IC section.

Power Amplifiers (PA) built in SOI are likely to use TSV technology in the future in order to improve performance and reduce die size. For PA, TSV is only used for parasitics and each is connected to a common ground on the metallized backside of the chip. Wires remain for I/O. In that case, a very low cost TSV technology is compulsory. Small thermal dissipation improvement is foreseen as well.

The first true 3D ICs using TSV are forecast for 2012. New partitioning of chips with IP in the best techno node will appear. A smart split of functions will be done in order to achieve the right cost/performance trade-off with TSV as the new enabler. An intermediate step based on a silicon interposer for the bottom die, containing only routing and few functions, is likely to happen. It will help in bridging the gap toward 3D SoC and a full readiness of design tools. With 3D IC and TSV, new topics will need to be considered. Numbers of options in silicon wafers, such as the type of ESD protection and test strategy are a few. A key advantage of 3D IC for this scheme of integration is clearly the time reduction of critical IP development in an advanced techno node. With a smart partitioning, complexity will be reduced and no longer on the critical path.

A memory / logic stack using TSV is a type of application the industry often refers to. The main reason for this is the increased bandwidth required

978-1-4244-4722-0/09 $25.00
© 2009 IMAPS-ITALY

by the final applications. With a new memory / logic interface architecture, based for instance on a wide I/O approach, this bandwidth challenge might be overcome. Furhtermore, this new wide I/O interface with parallel access to the memory will enable lower power consumption in the memory bus [10]. For cellular phones, this bandwidth bottleneck is foreseen after the LP DDR2 memory generation. However, many challenges are rising with this application. Thermal management is definitively a critical point in this approach. In effect, the power dissipation of the logic die, typically an application processor or a digital baseband, can heat the memory directly stacked on top. As memories have a lower Tj than a logic die (85°C or 105°C), the memory will receive too much heat and won't work correctly. Power dissipation of the bottom die will be range from 1 to 3 W in low to high end 3G platforms.

Another main issue with the wide I/O memory is the standardization and supply chain. Most of the time, the memory and logic die will come from different companies. Standardization will be required to enable the final OEM integrator to source different memory types or any double sourcing. Discussions are on-going between major players at this time for a definition of standards. For the first products, a non-standardized wide I/O memory might be available for high-end products.

Applications for mobile phones using 3D IC and TSV listed in this paragraph are not exhaustive but illustrate different on-going activities as of today.

Illustration with a Silicon demonstrator

Full 3D structures from design to packaging are illustrated in this paragraph. A stack of 45 and 130nm die has been achieved by STMicroelectronics (fig 8). Active structures of both die have been tested and are functional.

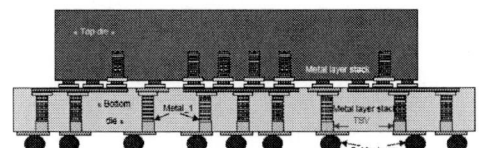

Fig 8: Schematic of 3D TSV stack

The top die is in 45nm technology and includes SRAM cuts, ROM, and standard cells. The bottom dic is in 130nm technology and includes the voltage regulator, buffers, thermal and mechanical sensors, daisy and delay chains as well as all the routing from the top die to the package.

A layout of both die can be seen in the following figure (fig 9).

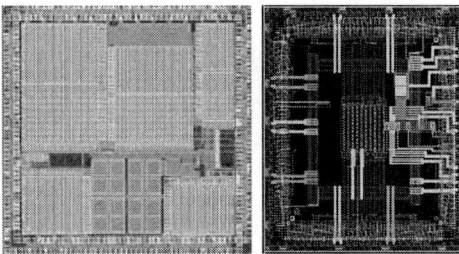

Fig 9: Top (left) and bottom (right) die layout

The stack of dice has been packaged on a BGA substrate and tested (fig 10).

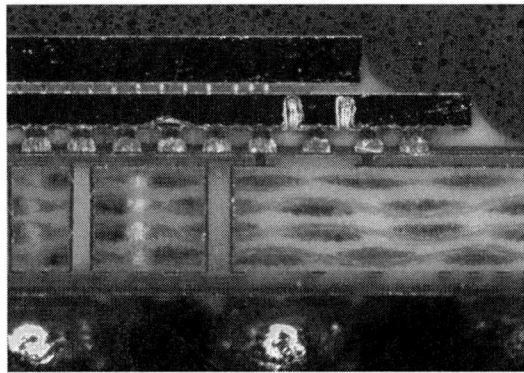

Fig 10: Die stack with TSV on BGA substrate

Conclusions

Roadmaps will mature during the next few years as people begin to understand all the capabilities of 3D Integration. As of today, only the emerging part of the iceberg is visible. 3D thinking is only at its early beginning and much more will be discovered and understood in up-coming years. The only thing to avoid with 3D Integration is to continue thinking based on the past experiences of downscaling. Applying Moore's law based on scaling to a More than Moore approach such as 3D Integration TSV is not necessary at all. The application should help in the roadmap definition.

It is crucial to differientiate 3D packaging and 3D at the IC level using TSV. TSV by itself is not a packaging technology apart from some few exceptions. Consequently, 3D TSV and FOWLP packaging platforms do not have to be considered as competitiors but more as complementary areas. In recent years, the semiconductor industry has expressed some growing interest in these ideas and put some significant efforts in allowing the emergence of these new breakthrough technologies. Still, some challenges remain ahead for a wide adoption. Most of them are cost, a shift in the design method paradigm, system co-design (heterogenous

978-1-4244-4722-0/09 $25.00
© 2009 IMAPS-ITALY

functions, packaging), new CAD tools, new architectures, and more new challenges. The 3D IC TSV combined with the FOWLP packaging platform appears as the next wave for future integration and should initiate some new integration schemes. We expect to expand the innovation landscape through lower cost and better electrical and thermal performances enabled by new partitioning and architectures, higher flexibility, better integration with easier software implementation and a shorter time to market.

Acknowledgements

The authors would like to acknowledge all the different teams involved in 3D and TSV at ST-Ericsson and STMicroelectronics. The authors acknowledge the partner R&D centers that participate in the different technological efforts.

References

[1] J. Walker, "Front End / Back End: Morphing the Manufacturing Model for Tough Times", Proceeding of IMAPS GBC, Scottsdale, Arizona, March 2009.

[2] K. Kujala, "3D Integration and Packaging: Requirement and Expectations", Proceedings of RTI TechVenture 2007, San Francisco, California, October 2007.

[3] F. Carson, "The development of the Fan-In Package in Package", Proceedings of ECTC, Lake Buena Vista, Florida, USA, May 2008.

[4] C. Zwenger, "Next Generation Package on Package Platform with Trhough Mold Via Interconnection Technology", Proceedings of IMAPS DPC, Scottsdale, Arizona, USA, March 2009.

[5] T. Meyer "Recent development in WLB and eWLB Technology", Proceedings of SEMICON Europa, Stuttgart, Germany, October 2008.

[6] B. Keither, "Redistributed Chip Packaging", Semiconductor International, April 2007.

[7] STMicroelectronics Press release, "STMicroelectronics, STATS ChipPAC and Infineon to Set New Milestone in Establishing Wafer-Level-Packaging Industry Standard", August 2008, www.st.com

[8] J. Michailos, "Through Silicon Via Technology for Wafer Level Camera: From R&D to Manufacturing", Proceedings of SEMICON Europa, Stuttgart, Germany, October 2008.

[9] E. Beyne, "3D Integration Technology and the microelectronics supply chain: Barriers and Solutions", Proceedings of IMAPS Global Business Council, Scottsdale, Arizona, USA, March 2009.

[10] M. Facchini et al., "System-level Power/performance Evaluation of 3D stacked DRAMs for Mobile Applications", accepted for publication in DATE, Nice, France, April 2009

978-1-4244-4722-0/09 $25.00
© 2009 IMAPS-ITALY

PROCESS CHARACTERISATION OF DUPONT MXA140 DRY FILM FOR HIGH RESOLUTION MICROBUMP APPLICATION

Werner Liebsch, Hao Yun, Chester E. Balut, DuPont Electronic Technologies
Andrew Ahr, Anna Phung, DuPont EKC Technologies
Alan Huffman, RTI Technologies,

DuPont de Nemours s.á.r.l.
Rue General Patton
Contern, L 2984 Luxembourg
Tel.:+49 6026 1328 Fax: +49 6026 1346
e-mail: werner.liebsch@deu.dupont.com

Abstract

The dry film has shown improved thickness uniformity and cost competitiveness versus liquid photoresists. However, traditional dry film photoresists for high pillar bump plating show limited resolution capabilities. DuPont has developed the MXAdvance100 series dry film photoresist for high resolution applications for high density wafer level packages. It offers unique characteristics like the smooth side wall, high aspect ratio and high resolution. It is an ideal photoresist material for sharply defined metal plating on wafer. The paper describes how to achieve the high resolution and how process conditions impact on the resolution and material performance including the micro pillar plating. The detailed process evaluation includes the entire lithographic process including the remove process. Cross section SEM analysis, process window and process linearity were studied for copper micro pillar fabrication. The key factors were identified for achieving a good high density micro pillar structure.

Introduction

The electronics market is faced with constant pressure from designers who want to develop "leading edge" products and from customers who want more functionality in smaller packages for lower prices. This drives the need for miniaturized 3D packages which requires an increase in I/O density and thus further shrinkage of the bump dimensions and pitches. The uniform precoated thickness of dry-film with minimum waste in lamination and usage capability to the edge of the wafer because of its no-edge-bead characteristic has been accepted by the industry for standard wafer bumping applications for bumps in the 80μm – 120μm height range. Based on customer experience and learnings by manufacturing with the thick dry-film resists of the WBR2000 series the higher resolution MXAdvance100 series has been developed to meet the needs for a micro bump pitch of 50μm and below.

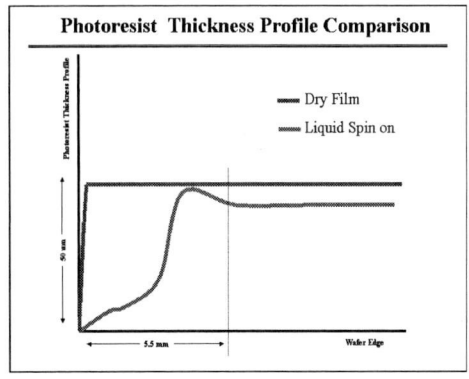

978-1-4244-4722-0/09 $25.00
© 2009 IMAPS-ITALY

Material

Dry-film photosensitive material is coated on a polyester foil with uniform constant thickness and after drying protected with a polyethylene foil.

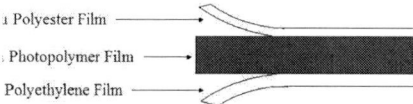

In exposure where the polyester foil is still on the film protecting the emulsion, the particles in the polyester are imaged when using an aligner. This after plating produces bumps with striations at the sides of the pillar.

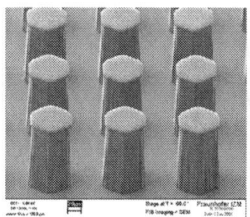

The imaging of these particles was eliminated by using steppers and exposing with appropriate focus adjustment. For quality improvement of the bump shape also when using aligners as an exposure tool, the high resolution dry-films are coated on a specially produced high quality polyester foil with a high clarity.

The current dry-film for wafer bumping application, WBR2000 series is formulated as a thick dry-film for plating pillars up to 120μm height. Formulated also for chemical and thermal resistance it is also used for stencil printing of solder paste and subsequent reflow. However, the resolution capabilities of the thinner versions of WBR2000 are limited and an aspect ratio of 2:1 could not be achieved. Only formulated for plating and etching but no longer for reflow, the MXAdvance100 series offers the required resolution aspect ratio again in any thickness. With the 40μm thick MXA140 a resolution of 17μm diameter could be achieved.

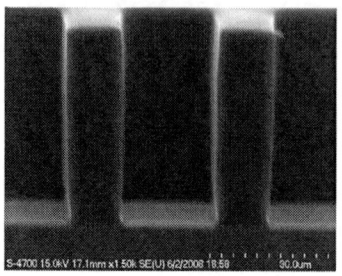

Lithographic Process

Lamination
Lamination has been conducted at different temperatures and speeds on a manual laminator as well as using different chuck temperatures and speeds on an automatic lamination tool. In all cases pressure was used to achieve a good adhesion. The variables in lamination, as long as a good adhesion to the substrate is achieved, have only a very minor or no influence on the final result.

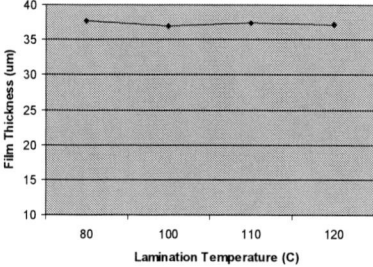

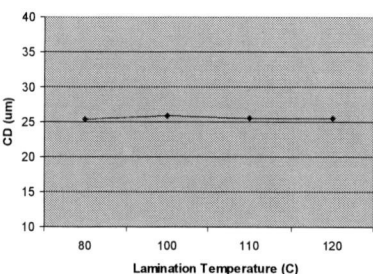

Exposure and Development
Exposure and development is a "system". Both variables influence each other and count for the final result of the shape and quality of the plated pillar. The exposure is the main factor for the final size of the image. For determining the correct exposure and its processing window, a value for development had to be fixed. For MXA140 the optimum exposure level is at 175 mJ/cm^2 with a working range between 150 mJ/cm^2 and 250 mJ/cm^2.

978-1-4244-4722-0/09 $25.00
© 2009 IMAPS-ITALY

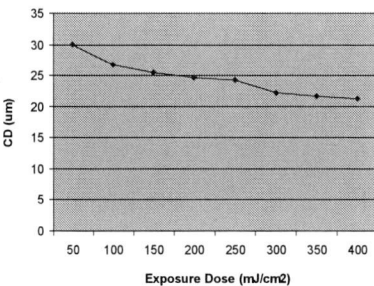

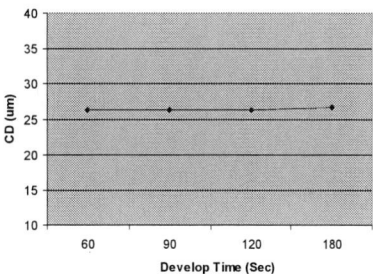

The angle of the sidewall is also determined by the "system" as well as the shape. For plating applications the angle should be from rectangular to slightly trapezoid to assure that the foot of the plated pillar has maximum adhesion to the substrate.

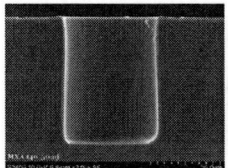 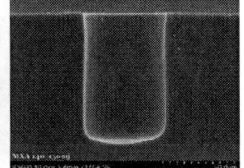

These SEMs demonstrate, that the critical dimension is achieved with an energy of 150 mJ/cm² but the hole is not developed cleanly down to the bottom. To achieve a clean development and this is mandatory for aspect ratios > 1:1, the development time had to be increased. A good result with a minimum resist-foot has been achieved using three times of the initially set development time.

Measurement of the critical dimension under these development conditions indicate, that the exposure dose is sufficient for a strong polymerisation of the resist, because the critical dimension - in this study 25µm diameter has been used - is basically not influenced by the development conditions

Bakes

The influence of a post exposure bake has also been studied. As the temperature is far below thermal polymerisation there was no measurable influence on the critical dimension or on the sidewall shape. The theoretically expected improvement in definition and sharpness of the structure could not be demonstrated.

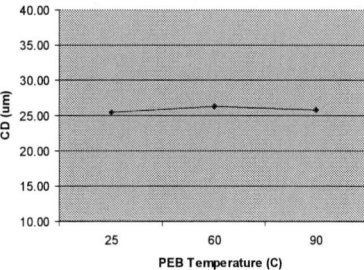

A post development bake has been studied as well by baking for 20min at 80°C. No difference vs. no baking or improvement of the plating result could be observed. By electro-plating other metals than copper and tin the PDB might be necessary to achieve the required chemical resistance.

In this study all processed wafers through plating had neither a PEB nor a PDB

Descum and Electroplating

Like for any photoresist the descum processing step has been performed. O₂-plasma was applied for 2min at 250W in order to remove any possible organic residues but more important to smoothen the sidewall of the resist and make it hydrophilic for optimum wetability in copper-plating. Standard electroplating has been performed including chemical pre-plate-cleaning of the sputtered copper of the substrate. Cu-pillars and Cu-pillars with a Sn-cap have been plated.

978-1-4244-4722-0/09 $25.00
© 2009 IMAPS-ITALY

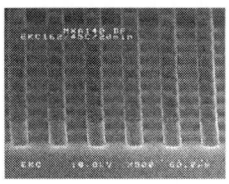 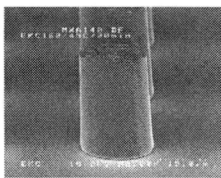

Removing

Using a wet bench-like application, both types of wafers were successfully cleaned using EKC162™ remover. The removal proofed to be very easy and the wafers were clean at all processing times and temperatures. The conditions were between 45°C and 55°C and between 10min and 20min. The plated height of the pillar was slightly below film-thickness, which is a requirement for this clean and easy removal.

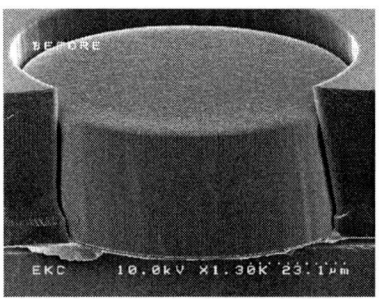

The short dwell times and the low temperatures minimize any potential etch loss issues. There has not been observed any attack on the Cu or the Sn in this study. The MXA140 dry-film does not dissolve, it is removed in a lift off process, thus the recommended remover tool would require a skin filtration system to remove the dry-film particles.

Conclusion

The results of this study show that the MXAdvance140 dry-film is capable for producing 20µm diameter microbumps. This can be achieved within a wide processing window which is ideal for production because minor process variations will not influence the final result of the plated bump. With its excellent resolution capabilities and easy removal performance it can be used in thinner versions for producing even smaller diameter and lower height microbumps.

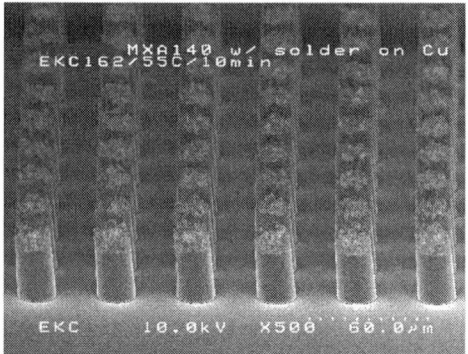

978-1-4244-4722-0/09 $25.00
© 2009 IMAPS-ITALY

RELIABILITY STUDIES ON HIGH CURRENT POWER MODULES WITH PARALLEL MOSFETS

Sarma G H , Nitin G, Ramanan, Manivannan,
Kaushik Mehta* and Arya Bhattacharjee*
sarma@si2micro.com, Si2 Microsystems, EPIP, whitefield, Bangalore, India
(* Si2 Microsystems SanJose USA)

ABSTRACT

In the current range upto 300 amperes, MOSFETs have proved tobe the most acceptable and cost effective devices in power modules for power and auto sectors; beyond 600 amps, IGBT are the preferred devices. To extend the usable range, MOSFETs in parallel are becoming more and more acceptable, provided the design, manufacturability and long term reliability have been fully evaluated for built in current stresses, thermal stresses under steady and transient current loads and ruggedness under the harsh load environments.

A half H bridge module with two mosfets in parallel (per arm of the bridge) with a combined current rating of 220 amps have been evaluated for various manufacturing approaches like heavy Al wirebonding, TAB contacts and Au bump approach. The studies have been done under current and thermal stressing with a view to identify possible failure modes.

Studies under SEM / SAM indicate sensitivity to damages, if there is improper current distribution along the die as well as substrate; these local variation in current distributions are improper locations of the die contacts (in wire bonding approach), inherent potential variations arising out of current crowding, variations in the effective conductance / inductance of the source and drain circuits. The studies have resulted in the optimization of a power module design addressing critical second order reliability considerations.

This work is significant, as the industry is looking for extension of MOSFET usage in higher power needs and at the same time would like to identify and eliminate all long term failure modes.

Introduction

Both in power conversion and inversion areas, there is continued focus on increasing the efficiency levels without making any compromise on the long term reliability of the modules. With further emphasis on energy saving and cost reduction, the technology is going through many innovative ideas involving new materials, processes and architectures [1,2,3].

Power Module structure

For generic power modules, the H bridge is a widely accepted architecture especially for UPS and motor drive applications. Power MOSFETs,

both NMOS and PMOS devices fit this approach for reasons sited widely in literature.
MOSFETs with current ratings around 100 – 150 amps are commercially available with the desired ruggedness and reliability. For applications for beyond these ratings, devices in parallel is an accepted approach [4,5]; however, this brings in new reliability issues and failure modes.
One such commercially usable module has been chosen for reliability studies under this paper
The devices in parallel are stressed both thermally and electrically and the failure modes thereof have been studied. Passive temperature aging test [6] refers to heating up and cooling down over several cycles and arrive at a lifetime estimation. This addresses the externally triggered modes of failure and covers expansion

978-1-4244-4722-0/09 $25.00
© 2009 IMAPS-ITALY

mismatch, defects aggrevated due to aging, thermally generated defects etc.

Devices in Parallel

Extensive reporting has been done on the issues related to current rating enhancement through devices in parallel [5]. The single factor that comes out predominantly on the studies is the emphasis on device matching and layout matching; the former contributes to threashold voltage and current balancing while the latter contributes more to parasitic conductive / inductive balancing thus making the module more resistant to transients.

Module under study

This paper is the result of the continued reliability related studies on a commercially used power module marketed by Si2 microsystems. The power module targets UPS and Automotive applications and hence has to meet stringent quality requirements and ruggedness. It is a half H bridge module (see Fig. 1,2) and a 'set of two' would meet the application needs of auto and

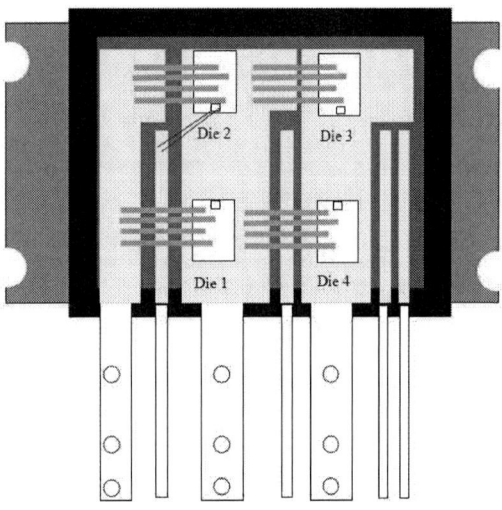

Fig: 1. Schematic of the power module

power sectors. Each arm of the bridge has two NMOS devices in parallel each of about 100 amperes catalogued rating at room temperature. With a rating of 60 V, the module has a custom designed IMS (Insulated Metal Substrate), suitable for cost effective industrial applications.

The thermal resistance is in the range of 0.4 C / watt and has been achieved with optimization in the processes involved and the materials used.

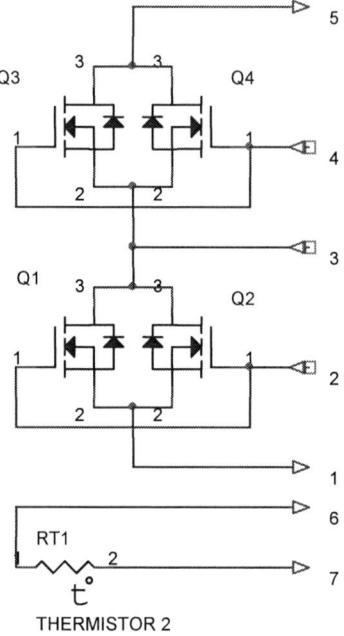

Fig 2: Power module schematic and pinout

Well proven production technologies of Solder mounted die with appropriate heatspreader layout and multiple heavy aluminium wire bonding on a custom designed substrate ensure process repeatability and good long term reliability. The module has successfully gone through the industry standard qualification procedures. Under continued evaluation of the layout and processes, certain design of experiments are being carried out. One such exercise is the failure modes and remedies thereof.

Design of Experiment

The module described above has been stressed both thermally and electrically till the point of breakdown and the resulting failure modes have been observed and studied in detail. This work will address only steady state related stressing. The experimental setup for current stressing is shown in Fig 4. The module is water cooled (Fig .3) to ensure a constant temperature at case level. Electrical stressing refers to current excursions

978-1-4244-4722-0/09 $25.00
© 2009 IMAPS-ITALY

Fig 3: Water cooled module to ensure constant case temperature during stressing

till the device fails; the change in substrate temperature during the current excursions is used to track the change in R_{DSON} (due to change in device junction temperature). It may be noted that there is change in R_{DSON} both due to I_{ds} and junction temperature [5]. As could be seen later, these changes are instrumental in early failures of parallel devices.

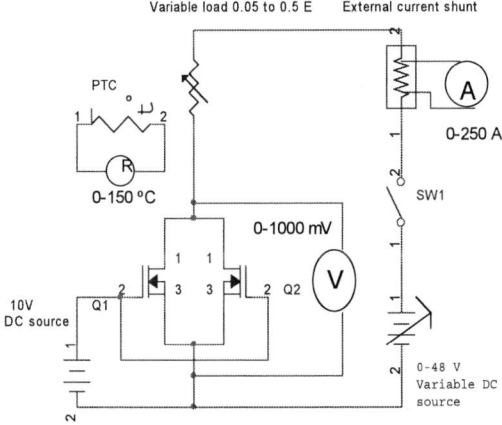

Fig 4: Experimental setup for current stressing

Some causes leading to earlier failures

Based on some earlier reliability studies done by the authors on the said modules, it has been found that variation in the local current distribution (during device conduction mode) may cause early failures. This is mainly due to the fact that the imbalance in the current sharing of parallel devices is further aggravated (say, two devices in parallel) by

- Increase in R_{DSON} spread during stressing
- Increased effect of conductance path mismatch during current stressing
- Increased effect of thermal path difference during thermal stressing

The spread in R_{DSON} is generally contained within acceptable limits while choosing the devices for parallel operation. However, during current stressing, the effect of Ids as well as the device heating (and the resulting increase in junction temperature) on R_{DSON} values can be seen in Fig 5. As against a single device value varying from 6.1 to 7.5 milli ohms, two devices in parallel has demonstrated a change from 3.2 to 6.0 milliohms with a spread of near zero at low currents and as high as 1 milliohm at higher currents (>150 amps). What is detrimental to long term reliability is the possibility that the 1 milliohm spread may be due to the enhancement of device mismatch due to stressing (rather than a balanced equal increase between devices), which in turn will stress one of the devices much more than the other, leading to earlier failure of the said device. This has been confirmed by earlier failures of device sets with higher R_{DSON} mismatch during our stress tests.

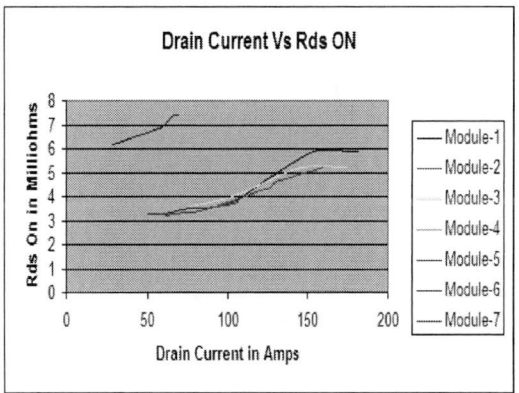

Fig 5 : The spread in R_{DSON} values among different modules of the same lot

Effective conductance path for each device includes
- Parallel conductance paths due to 4 wirebonds
- Path from wirebond shoes (through Al metal film) to the drain metal film

The wirebond positions on the large die are generally chosen after careful analysis towards equal and efficient current paths from source to drain; that is the reason why the 4 bondsites are staggered on the die surface.

However, this staggering introduces differential resistance paths in the current flow. Further, the substrate layout, optimized towards device positioning (thermal requirements) and pin configurations, introduces certain current flow path differences. Current stress failure modes clearly indicates (see Fig. 8,9,10) the early onset of failures in the longer cumulative current paths.

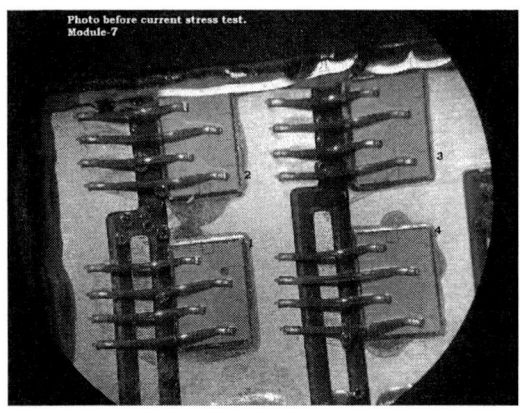

Fig 6: A typical module before test

Fig 7: A good die with wirebond contours

Fig 8: Uniform damage under current stress – balanced die conductance

Fig 9: Non uniform damage due to high current stress (imbalance in die conductance) See also Fig 10 below.

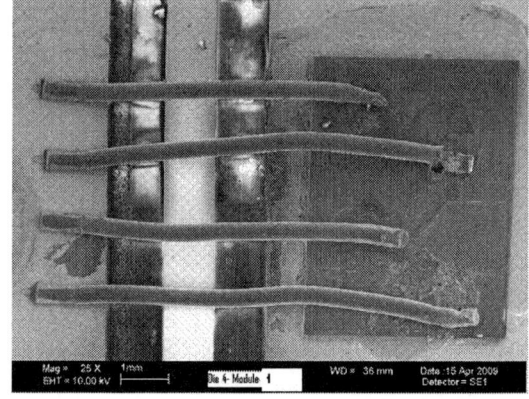

Fig 10: SEM view of damage due to current stress

In Fig.9 & 10, the top wire in the top first device has longer current path compared to bottom device in parallel (not seen in the picture). It may be noted that the bottom wirebonds (of the first device) itself is not damaged at all, thereby confirming that the current sharing has been aggravated due unaccepatable local high imbalance. If only these local imbalances are reduced withing limits, these parallel devices would have withstood longer stressed life.

For normal functioning, these layout and wirebond variations are acceptable. But under current stressing experiments, these trigger earlier failures.

New Sample structures for detailed studies

Some of the variations attempted are towards achieving a balanced current entry into the source :

- bumped die (for source) to ensure distributed contacts (Fig 11)
- TAB contact for bumped source face (Fig 12 & 15)
- bumped and flipchipped device (Fig 14 & 15)

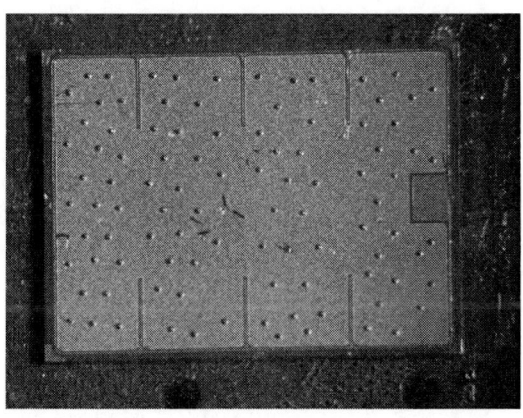

Fig 11: Bumped die for uniform current entry

In order to achieve a pervasive current entry into the die, a gold bumped die has been used. The bumping generated by an ASM auto bond machine, has an area array of 100 bumps, configured into pre chosen matrix, thus providing a desired curren inflow selectivity. One such die (~ 6mm X 4 mm) bump pattern is shown in Fig 11 .

Fig 12 : Bumped die with wirebonded OR TAB source contacts

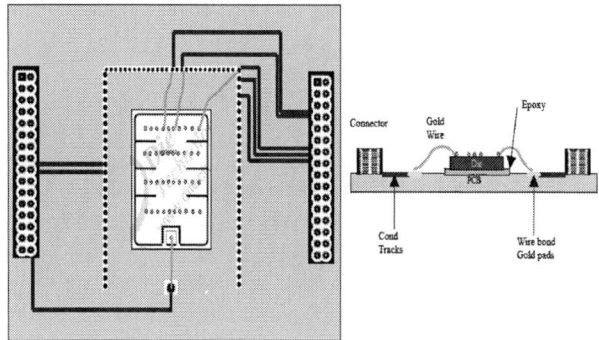

Fig 13 : Schematic of wirebonded array contact test board

In one of the variations for study purpose, instead of bumping the die, the die is mounted on a patterned PWB, multiple wirebonds done from die to PWB pads (an array of 100 wirebonds to cover the entire die surface) such that through two connectors mounted on the PWB, the current can be driven into the die at a preselected pattern. (Fig 12 & 15).

978-1-4244-4722-0/09 $25.00
© 2009 IMAPS-ITALY

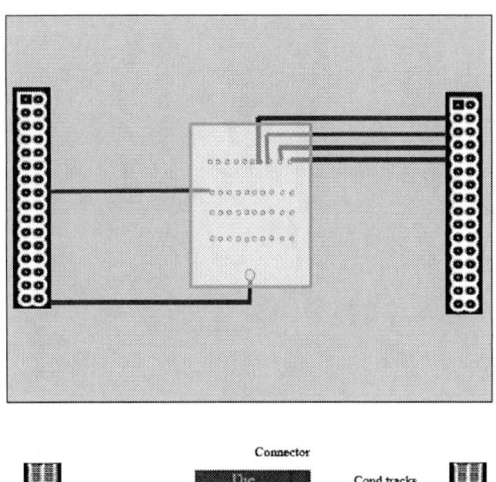

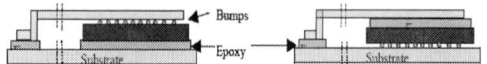

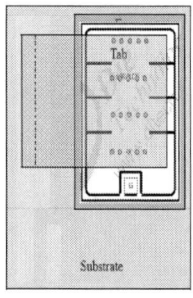 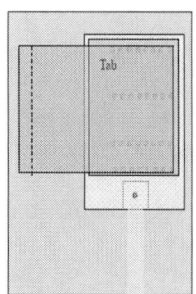

Fig 14: Schematic of flipchipped array contact test board

Fig 15 : Schematic of Bumped die with TAB (left) and Bumped die flipchipped (right)

The aim of the experiment is to locate failure points in case of current stress or thermal stress on the die. (The wire bonded assembly is targeted only for steady stage current and cannot be used for transient studies).
Both bumped die with top TAB contact and bumped die with flipchip mounting are potential assembly approaches in area array power device mounting.

Thermal Stressing

Results with new structures

The initial experimental results are very encouraging and has confirmed the earlier failure mode setting due to conductive path imbalances, especially due wirebond locations, wire bond lengths and copper paths. These experiments are expected to provide rich data for future layout and wirebond optimizations. Also the wirebonded array die can give a quick experimental data for practical optismization of layouts; this is being attempted and the results will be reported.

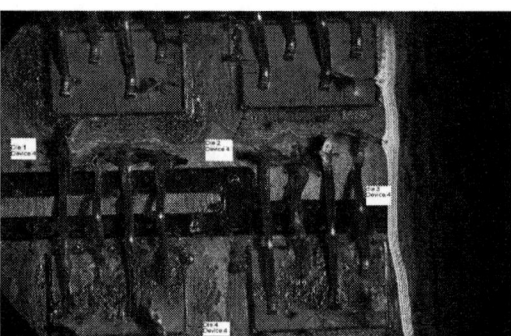

Fig 16 : A half bridge module showing failure of one arm due to thermal stressing

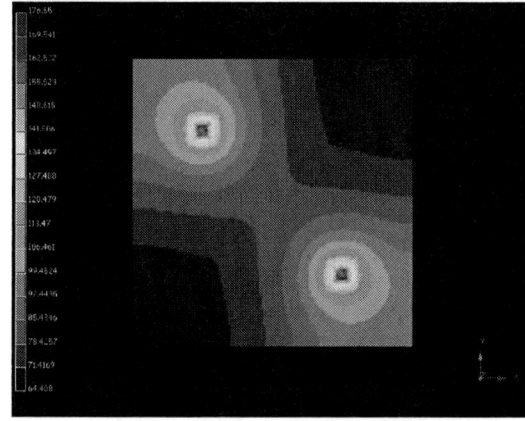

Fig 17 : Thermal analysis of half bridge module

978-1-4244-4722-0/09 $25.00
© 2009 IMAPS-ITALY

Thermal analysis of the parallel devices, as they function in a half bridge module is shown in Fig 16. The junction temperature under worst case analysis does not exceed 171 C, well within acceptable device specifications.

The temperature cycling (standard for industrial applications – minus 40 to + 125 C) has been accepted as the best method for thermal stressing. After each cycling, the module is subjected to current stressing (130 amperes) till failure occurs. Thermal stress damages are generally indicative of least damage to wirebonds and predominantly manifested by die cracks, source metal damages etc. (Fig 18, 19)

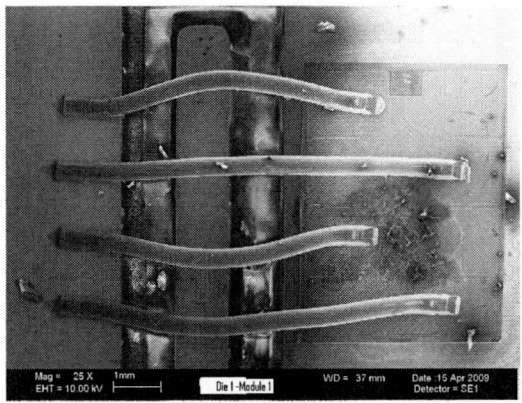

Fig 18: SEM view of a thermally stressed device (notice wirebonds are intact)

Fig 19: SEM view of Thermally stressed damage (originated from voids below die)

Conclusions

This work, as reported here, confirms earlier onset of failure modes in power modules with parallel devices,due to aggravated imbalances in current share between parallel devices. These are mainly due to spread in R_{DSON} and due to conductance path variations. Design of experiments have been done through innovative assembly approach (bumped array wirebond contacts, bumped array TAB contact and bumped flipchipped contacts), which will provide data for further insights into early module failures involving parallel devices. This data will form the basis for optimization of layouts and wirebond sites, towards minimizing long term failures.

Acknowledgement

The authors would like to thank Mr Gangadhar for providing enough samples from production lots for the study and SITAR management for providing access to SEM facility. The work is supported by Si2 microsytems.

References :

[1] John Mookken, " Next Generation Power Electroncics " , Power Electronics Europe 2005

[2] Alberto Castellazi etal, IEEE Transaction on Power Electronics, May 2006

[3] Valentin von Tils, Robert Bosch, 18th Sym. Power SCs and ICs, June 2006

[4] Toni Lopez etal., "Static Paralleling of Power MOSFETs in Thermal Equilibrium", Proc. IEEE 2006

[5] James B. Forsythe, "Paralleling of Power MOSFETs for Higher Power Output", Application Note, www.irf.com

[6] R Schacht etal., " Accelerated Active High Temperature Cycle Test for Power MOSFETs", Proc. IEEE 2006

[7] D C Katsis etal., "A Thermal, Mechanical, and Electrical Study of voiding in the Solder Die attach of Power MOSFETs", Trans. IEEE CPT, vol 29, Mar 2006

[8] Alfio Consoli, "Thermal Instability of Low voltage Power MOSFETs", Trans. IEEE Power Electronics, vol 15, May 2000.

978-1-4244-4722-0/09 $25.00
© 2009 IMAPS-ITALY

Solder Process Optimization: Influence of Heating and Cooling Rate on the Thermo-Mechanical Stress Generated in Components

Michael Hertl, Diane Weidmann, and Jean-Claude Lecomte

INSIDIX – 24 rue du Drac – F-38180 Grenoble/Seyssins - France

Phone: +33-4.38.12.42.80 – Fax: +33-4.38.12.03.22 - E-mail: michael.hertl@insidix.com

Abstract

Optimisation of the reflow solder cycle is a critical task, in the sense that the consequences of poor process definition may affect the reliablitiy of the assembly on the one side, as well as the industrial profitability on the other side. The present paper analysis the thermo-mechanical stress behavior of a typical electronic component under thermal profiles with heating and cooling rates varying from 0.15 to 3.0°C/s. The thermo-mechanical stress in the components is assessed by real-time measurements of the components surface topography (warpage) and deformation during temperature increase and decrease cycles typical for JEDEC type reflow profiles. The consequences for the solder process are discussed.

Key words: temperature stress component warpage reflow

Introduction

Reflow solder profiles are classified by internationally recognized standards, as for example the ICP/JEDEC J-STD-020D standard. These standards define parameters such as maximum and minimum solder temperature, total cycle time, and heating and cooling rates.

However the process manager keeps a large freedom to choose the finally used values for each parameter within the given ranges. For instance in the J-STD-020D standard, only maximum values are specified for temperature ramp up (+3°C/s) and ramp down (-6°C/s), but no minimum values are given.

From a standpoint of pure process efficiency, the optimum choice would be to use the maximum ramp rates allowed, in order to minimize the total cycle time. On the other hand, the larger the temperature ramp rates are, the larger will be the spatial thermal gradients induced in the components during the solder cycle, thus maximizing the thermo-mechanical stress generated in the components. Consequences might range from high process failure rates (delamination and/or opens at the end of the process) to poor reliability expectations by premature aging. Therefore for full process optimization chosing adapted heating and cooling rate values is at least as important as the total cycle time.

In the present paper we will apply different heating and cooling gradiants to a given BGA component, in order to analyse the influence of these gradients on the warpage behavior of the components. This information will enable the process design engineer to chose the best heating and cooling rate, for optimizing the process speed

without compromising the assembly reliability expectations.

Experimental Set-Up

The experimental set-up is shown in Figure 1. The electronics package to be studied is illuminated by structured light (stripe pattern) on the sample surface. The stripe pattern is more or less deformed by the sample's surface structure. The resulting image is captured by a CCD camera.

Thermal stress generation is available by top and bottom heating and cooling elements. The current sample temperature is monitored. User defined temperature profiles with heating gradients up to +3°C/s and cooling gradients down to -6°C/s may be imposed to the component, within a temperature range from -60°C up to +300°C.

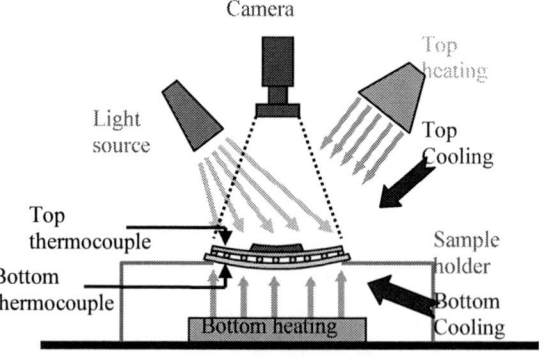

Figure 1: The TDM (Topography and Deformation Measurement) experimental set-up

978-1-4244-4722-0/09 $25.00
© 2009 IMAPS-ITALY

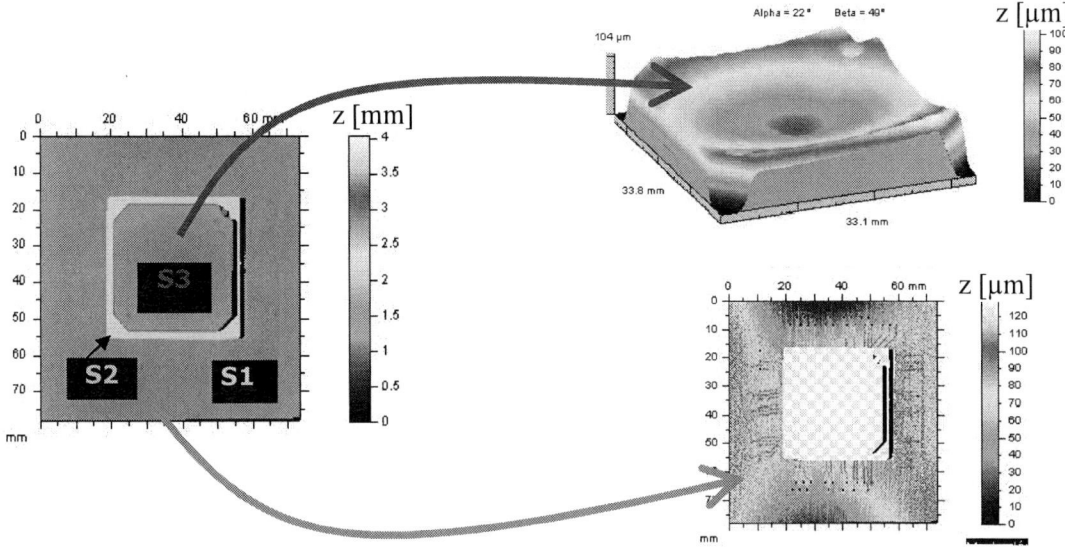

Figure 2: Topography acquisition by TDM: Left: Full field absolute BGA on PCB topography acquired within 5 seconds. Right: Software zoom on either BGA top level (S3) or PCB (S1).

The primary obtained information consists of high resolution 3D topography images of the entire assembly. The high depth of view (up to 32 mm) of the optical set-up allows to acquire all relevant levels of virtually every component or assembly simultaneously, as for example PCB, BGA substrate and BGA top surface, see Figure 2. For each pixel, the (x,y,z) coordinates are absolutely known. After acquisition, software zoom on each level is possible and will reveal detailed surface information on a μm scale.

Applied Thermal Profiles

The TDM system has been designed for being able to reproduce all relevant reflow solder temperature profiles, including those defined by the ICP/JEDEC J-STD-020D standard. In particular, heating up components with maximum grandients of +3°C/s and cooling them down with up to -6°C/s is possible, as shown in Figure 3.

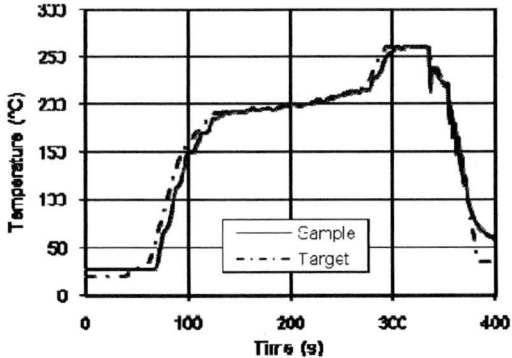

Figure 3: Application of a typical JEDEC type reflow profile with + 3°C/s heating and -6°C/s cooling to a component.

The temperature progession is fully regulated, both during heating and during cooling. Therefore the real sample temperature profile follows the target profile very closely.

While JEDEC profiles in general are characterized by varying heating and cooling rates (fast initial ramp up, plateau, fast ramp up to maximum, cooling down), we use simplified profiles in the present study. The idea is to focus on the pure influence of different ramp rates, which are variable for different experiments, but constant during each experiment. Therefore, the four profiles shown in Figure 4 have been used.

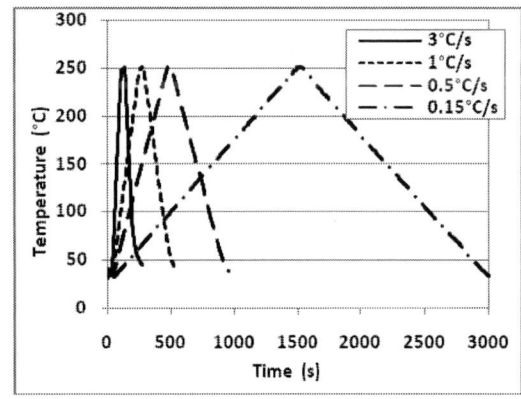

Figure 4: The four temperature profiles used in this study, with temperature gradients ranging from 0.15°C/s up to 3°C/s.

The plotted profiles are those measured on the samples by thermocouple. All profiles have a constant heat up rate, from room temperature to a maximum temperature of 250°C. At this temperature, a 30 second flat plateau follows. Then, the component is cooled down with the same ramp rate as used for heating up.

978-1-4244-4722-0/09 $25.00
© 2009 IMAPS-ITALY

5 identical samples have been used, each sample only for one measurement. The slowest profile (with a temperature gradient of +/-0.15°C/s) is done twice, on two different samples, in order to check for reproducibility. The other profiles are done on one sample each.

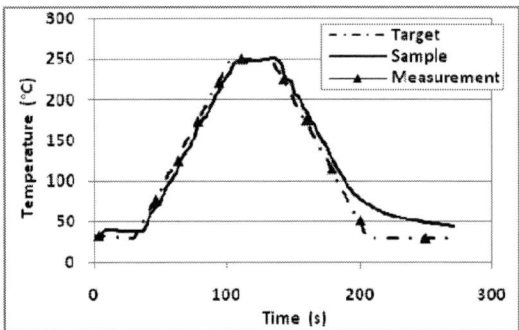

Figure 5: Comparison of target and sample temperature in the case of the fastest used profile. The instances where topography data are acquired are indicated.

It should be noted that the fastest profile done in the present study represents a severer thermo-mechanical load to the sample as the severest standard JEDEC profile, due to the absence of the plateau phase which usually serves for elimination of residual humidity, and for temperature homogeneisation in the component. The TDM system has therefore preliminarily been tested for its capability to reproduce this kind of severe profile. The result is shown in Figure 5. During the entire heating cycle, and during the cooling cycle down to below 100°C, the sample temperature follows the target temperature within some few degrees. We can therefore conclude that the generation of all the selected profiles will be possible with high reproducibility.

Components

One of the BGA components under test is shown in Figure 6, from bottom side. All measurements in this study have been done bottom side, because this is the side where the components will be soldered. In fact, previous studies have revealed different expansion behaviours for top and bottom sides on BGA components [1]. Therefore, relevant results for practical applications should be obtained on the bootom side of the BGA.

Figure 7 shows the BGA topography as acquired by TDM. The solder ball structure is fully resolved. This is a prerequisite for reliable measurements, because removing the balls prior to doing the topography analysis would alter the stress distribution in the component, and therefore might generate results which do not represent the real component warpage situation during the solder process, where the balls are in place.

Figure 6: Photograph of the BGA component, bottom side.

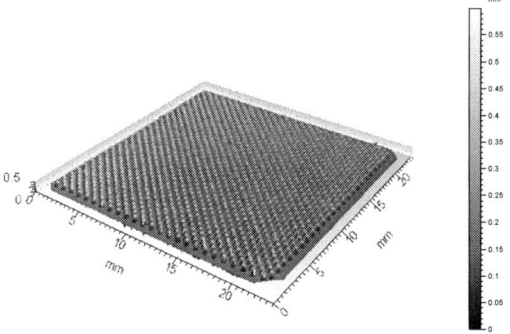

Figure 7: Topography of the component, acquired by TDM.

Warpage Measurement Results

For result assessment, the data acquisition and analysis procedure is as follows:

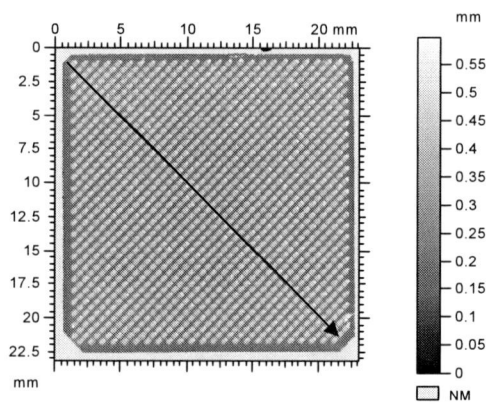

Figure 8: Primary topography data. A 2D diagonal profile along the arrow is extracted for data analysis.

Component topographies have been measured (as indicated in Figure 5) at 30°C, 75C, 125°C, 175°C, 225°C, 250°C, 225°C, 175°C, 125°C, 75°C, and 30°C for all four profiles indicated in Figure 4. The primary obtained data is a topography

978-1-4244-4722-0/09 $25.00
© 2009 IMAPS-ITALY

image as shown in Figure 8. Through this topography, a 2D diagonal profile is extracted following the arrow indicated in Figure 8. On the obtained profile the amplitude min->max is calculated automatically by the software. The amplitude is positively indexed (e.g. +22μm) in case of convex shape of the 2D profile, and negatively indexed (e.g. -6μm) in case of concave shape of the profile.

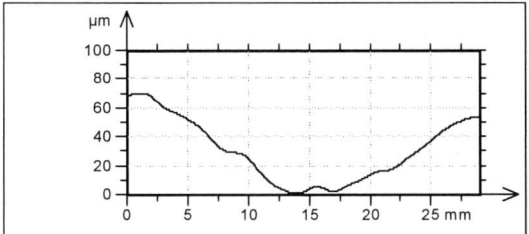

Figure 9: 2D diagonal profile extracted along the arrow indicated in Figure 8.

The extracted profile follows a diagonal line inbetween the solder balls, therefore no bump is visible in the profile. Both the profile is extracted and the amplitude is calculated automatically, allowing fast data assessment. This amplitude is the value which is referred to as the component's warpage in the following.

Figure 10 shows the warpage of two components A and B, when following a temperature profile with a 0.15°C/s gradient. For each temperature and each profile there are two markers in the graph, one corresponding to the respective temperature during the heating period, the other one to the same temperature during the cooling down periode.

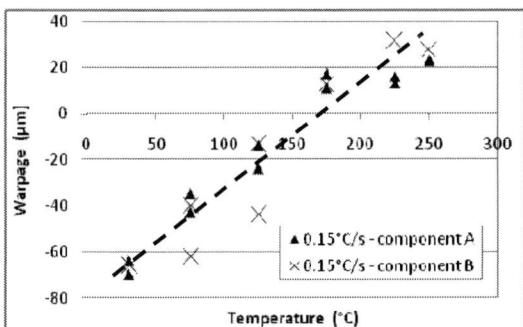

Figure 10: Component warpage vs. temperature for two components measured individually with a heating/cooling rate of 0.15°C/s. The dashed line is added to guide the eye.

There is no significant warpage difference indicated for these two components, at all temperatures. For each component there is no significant warpage difference between heating and cooling cycle, neiter. Initially, when coming out of production, both components are concave with a warpage of about -70μm. When heating up to

250°C, the warpage concavity decreases, and at about 170°C the components become flat (Tg). At higher temperatures, the components become convexe, with an absolute warpage of about +25μm at 250°C. We can conclude the following:

- Components A and B have identical stress distribution when coming out of production.This is insofar normal that they are supposed to be identical.
- For identical components, and identical temperature progression, the measured warpage vs. temperature is identical. In other words, the measurement is reproducible (this should be self-evident, but will be an important notice in the final part of this work).
- For the given components, and the given temperature progression, all deformations that the components undergo during the temperature profile are elastic. No residual plastic deformation remains.

Figure 11 shows the warpage vs. temperature behavior of four individual components, with temperature gradients from 0.15°C/s to 3°C/s. Only one of the two curves presented in Figure 10 is reproduced in Figure 11, to keep it simple.

All components are concave just after production (i.e. the initial room temperature warpage is negative).

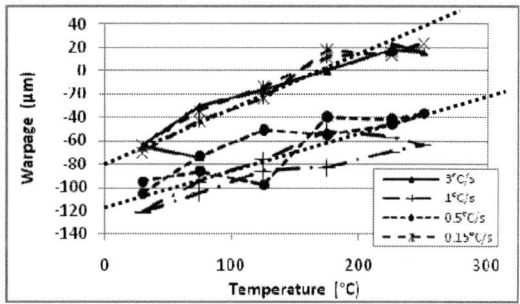

Figure 11: Component warpage vs. temperature for four different heating/cooling rates, measured on four different components. Those components were supposed to be identicall. The two dotted lines are added to guide the eye.

However, two of the components have an initial warpage of about -65μm (those used with the temperature gradients 0.15°C/s and 3°C/s), wheras the two others have an initial warpage of about -110 μm.

Within each of these two groups, the warpage vs. temperature behaviour is similar for both components, though both components within each group arc heated and cooled with different gradients. Moreover, the components of the first group (0.15°C/s and 3°C/s) change from concave to convexe shape (at about 170°C), whereas the components of the second group (0.5°C/s and 1°C/s) stay concave at all temperatures.

978-1-4244-4722-0/09 $25.00
© 2009 IMAPS-ITALY

Conclusions

The initial aim of the present study was to identify variations in component warpage behavior as a function of component heating and cooling rates.

The ability of the TDM equipment for those types of measurements, with both very high and very low temperature gradients, has been proven, and reproducible warpage vs. temperature measurements have been obtained on individual but identical components.

However, during the measurement, significantly different initial warpage values (-65μm for some of the components, vs. -110μm for the others) have been found for different components which were supposed to be identical.

Therefore, the tentative interpretation of the results as shown in Figure 11 is the following:

- The components supplied for the present study are not identical. Either they are assembled by using e.g. two different types of mould compound, or, if the used materials are identical, they have been assembled in two different lots with different process parameters. This is deduced from the two different warpage values obtained at initial room temperature.
- For the components under test, the heating and cooling temperature gradient is of no influence on the warpage vs. temperature characteristics (within the gradient limits used in this study). This is deduced from the fact that the components submitted to the maximum (3°C/s) and minimum (0.15°C/s) gradients applied in this study show the same warpage vs. temperature characteristics (Figure 11).
- All deformations observed in the present study are elastic deformations, and thus fully reversible.

With reference to the initial aim (optimization of solder speed without compromising the assembly reliability), the result is:

- The present components can undergo the maximum heating rate allowed by the ICP/JEDEC J-STD-020D standard without compromising the solder quality due to excessive stress accumulation.
- However, independently of the solder behavior, an additional study should be undertaken to understand the differences of initial warpage measured on supposingly identical components, in order to avoid fluctuating reliability expectations of the components due to poorly defined component manufacturing conditions.

Acknowledgements

This work has been partially supported by the Europeen Union under the PIDEA+ research program entitled "Failure Detection and Analysis for Complex Micro- and Nano-System Applications – FULL CONTROL"

References

[1] M. Hertl, D. Weidmann, J.-C. Lecomte, "Advanced Assessment of failure modes of high pin count BGAs due to thermal stress during the reflow solder cycle", Proceedings of the 2009 Interconex Conference, Paris, France, April 7-8

Wafer Post-Processing for a Reconfigurable Wafer-Scale Circuit Board

Moufid Radji[1], Ahmed Lakhssassi[2], Mohammed Bougataya[2], Anas Hamoui[1], Ricardo Izquierdo[3]

(1) Electrical and Computer Engineering Department, McGill University
(2) Département d'Informatique et d'Ingenierie, Université du Quebec en Outaouais
(3) Département d'Informatique, Université du Quebec à Montréal

Abstract

The WaferBoard^{TM} rapid prototyping platform for electronic systems is proposed as a tool to help meet today's tight delivery time, performance and reliability constraints. At the core of WaferBoard^{TM} is the WaferIC^{TM}, a wafer-scale reconfigurable CMOS circuit. At the surface of this complex circuit is a sea of identical contacts, any pair of which can be interconnected through a mesh grid network called WaferNet^{TM}. The user can simply deposit packaged integrated circuits on the smart active surface, and then a complex interconnect pattern between these ICs can be established in a matter of minutes. As is the case with development of any novel technology, design of the platform poses several technical challenges. Postprocessing tasks to be accomplished on the CMOS wafer are laid out hereafter. A .18µm CMOS TestChip fabricated to validate the WaferIC^{TM} concept on a 1/100th scale is outlined. Furthermore, sample microfabrication results, such as TSV etching, are presented along with thermo-mechanical investigation outcomes.

Key words: Electronic Prototyping, Wafer-Scale Integration, Through Silicon Vias, Thermo-Mechanical Analysis

I. INTRODUCTION

In the recent years, electronic devices have been required to provide an ever-growing set of functionality. As the field of circuit design is relatively mature, pressure to deliver these full-featured electronic systems is increasingly born by the systems engineer. At the prototyping stage several iterations of a complex interconnection network often need to be implemented, each requiring the tedious and time consuming redesign of a Printed Circuit Board (PCB). To meet the tight constraints on size, power efficiency and time-to-market, new techniques for prototyping of electronic systems need to be put in place. WaferBoard^{TM1}, a rapid prototyping technology based on a wafer scale circuit, WaferIC^{TM}[1], is proposed to mitigate this development and prototyping bottleneck. The WaferIC sits on a bed of power-supply circuitry inside a robust mechanical housing. The WaferIC is an active substrate that can transmit digital information between any package balls of a set of conventional chips placed by the user (user ICs, or uICs) anywhere on its surface. The surface is alignment-insensitive, so hand-placement is sufficient. The WaferIC detects uIC contacts, and

based on the package footprint and minimal user input, the WaferConnect^{TM} software identifies the components present on the smart surface and provides power and ground to the uICs through the proper contact nodes. During a configuration step, the user specifies inter-chip connections which are downloaded thereafter in a manner similar to programming a Field Programmable Gate Array (FPGA). Signal paths contain repeaters to ensure signal integrity. The WaferIC provides connectivity to uICs on the top side, and to the power supply circuitry on the back side. Topside contacts to uICs are made through metal plated bumps on top of WaferIC^{TM}. The bumps are required to provide sufficient ductility and deformability to attain a good electrical contact. They must also offer good thermal transfer to maintain temperature homogeneity on WaferIC^{TM} and uICs. Through-Wafer Vias (TWV) are used to feed the WaferIC from underneath directly into an internal WaferIC power distribution grid that starts on metal 1 (M1). Through silicon vias (TSVs) technology is currently under development [2], and figure 1 shows a sample etch result. However, unlike these samples, the TSV's required for the WaferBoard^{TM} platform must be etched through the silicon from the backside and stop on the lowest metal layer of the standard CMOS process.

Characterization of key features in the fabrication of the WaferIC is being done on full size wafers; in addition a 2x1.9mm^2 .18µm CMOS chip, TestChip1.0, was fabricated through TSMC to

---

[1]WaferBoard™ is a brand name for a Rapid Prototyping Platform for Electronic Systems from Gestion TechnoCap Inc., DreamWafer Division.

978-1-4244-4722-0/09 $25.00
© 2009 IMAPS-ITALY

validate functionality on a 1/100th scale. In this paper, the physical structure of WaferBoard™ is first laid out, along with a description of some of the postprocessing challenges. Then focus is placed on TSVs, a key postprocessing step. Next is an overview of the Testchip1.0 along with preliminary investigations made on it. Lastly thermo-mechanical investigations undertaken so far are shown before concluding remarks.

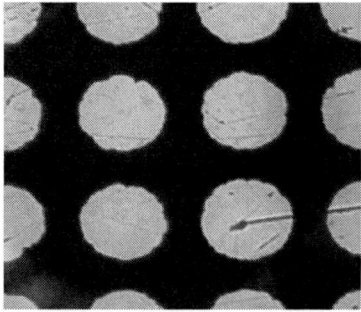

Figure 1: Optical image of backside illuminated TSVs: 200µm diameter-600µm thick wafer

II. PHYSICAL STRUCTURE OF WAFERBOARD™

A. Overview

The WaferBoard™ platform is a reconfigurable circuit board whose main component is a wafer scale CMOS circuit, WaferIC. The physical housing is designed with a strong emphasis on heat and stress mitigation. Atop WaferIC is a pouch filled with thermally conductive fluid attached to the lid. The fluid not only dissipates heat through the top of the structure, but it also enables the application of uniform pressure onto the uICs which can have various heights. The applied pressure is controlled by a plunger which can generate up to 10 atmospheres, as needed to ensure proper detection and connection to the BGA-bumped uICs. The surface of the WaferIC consists of a very dense sea of contacts (NanoPad™) each capable of providing power or routing a signal from any pin of a uIC. Since the nanopads' size is much smaller, and their density is much higher than that of uIC BGA solder balls, each component contacts several nanopads at once. An internal wafer-scale interconnect network, WaferNet™, is used to link any two given nanopads, by carrying signals between them. Power for the WaferIC and the uICs is provided though a network of connections on the backside. By means of TSVs and solder bumps, the WaferIC is connected to the power supply assembly. The entire structure sits a on a backside heat sink which dissipates some of the heat through the bottom. Figure 2 depicts a schematic cross-section of the device.

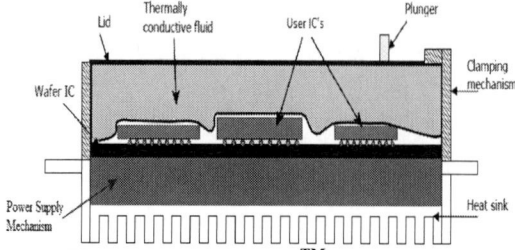

Figure 2: WaferBoard™ cross-section

B. A few postprocessing challenges

The first and probably least trivial step is the design and fabrication of the wafer scale CMOS circuit. Besides all RTL and transistor level issues, one bottleneck comes from the limited reticule size. A characteristic of WaferIC that greatly helps with this is its repetitive cell-based architecture. The single building block is as cell containing an array of nanopads. WaferNet ensures the complex interconnections between these cells which are tiled on the reticule scale; reticule stitching techniques [1] are then used to obtain the wafer scale interconnect network.

After the circuit is manufactured by the CMOS foundry, several postprocessing steps still remain to be performed to attain a functional device.

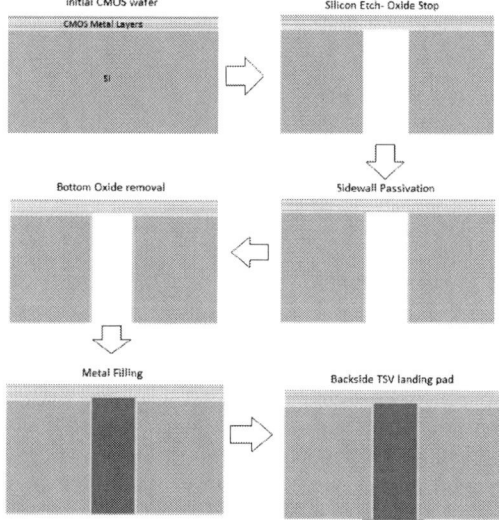

Figure 3: TSV process flow

On the top surface, gold columns up to 50µm in height have to be plated above each 80x80µm nanopad. A barrier layer such as nickel needs to first be deposited between the top layer aluminum and the gold columns. The columns' height is chosen to provide enough compliance to accommodate the small variability in BGA sizes underneath uICs by deforming the soft gold. On the bottom surface TSVs of aspect ratio 5:1 need to be etched and metal filled. In doing this, the oxide underneath metal 1 must be used as etch stop, similarly to the Buried Oxide layer (BOX) used on Silicon-on-Insulator (SOI) wafers for manufacturing of MEMS devices

978-1-4244-4722-0/09 $25.00
© 2009 IMAPS-ITALY
384

such as accelerometers[3]. Figure 3 describes the process flow to be followed to attain these TSVs. Next, solder bumps are placed to provide connection to power supply assembly. Given the large number of contacts and the coarse nature of alignment done with bumping tools, this step will prove quite challenging[4].

III. TSVs, A KEY POSTPROCESSING STEP

A. TSV etching

Etching is to be done using an Inductively Coupled Plasma (ICP) tool capable of achieving Deep Reactive Ion Etching (DRIE). The standard Robert Bosch GmbH process is chosen to obtain deep trenches using the ICP-DRIE tool [3]. As the process results from cycling through an etch step followed by a deposition step, the Bosch process suffers from the scalloping effect; this effect along with the roughness of the sidewalls is clearly visible on figure 4. This rough surface may prove detrimental during the via filling step.

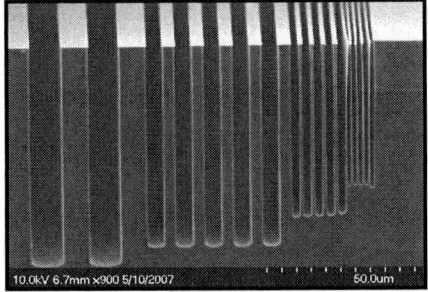

Figure 4: SEM image showing side wall roughness

Another characteristic of the process which can prove troublesome in some applications is its Aspect Ratio Dependant Etching (ARDE); figure 5 clearly illustrates this. Using the Bosch process, attempting to etch different via widths at the same depth using a single mask can be cumbersome.

Figure 5: ARDE: 10um, 5um, 2um and 1um Openings
Courtesy of: **Philippe Vasseur**-*LMF-Ecole Polytechnique de Montreal*

In order to design a reliable recipe for etching through the 50μm-dia vias present on the testchip, several iterations had to be made on centimeter-size samples. Each such sample was covered with patterns for various via sizes and shapes. Figure 6 below describes the process flow used.

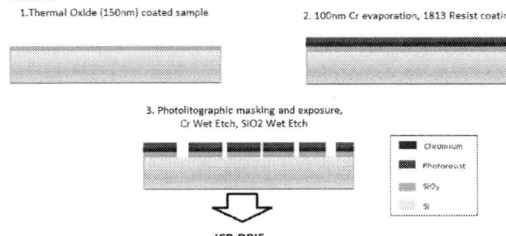

Figure 6: Bosch Process Characterization sample preparation

Initially, the recurring issue was the formation of "black silicon", at the bottom of the vias, which eventually halts the etching. Figure 7 shows these grass roots, which are due to overpassivation. The effective solution was found to be the significant increase of the etch step duration with respect to the passivation step. Figure 8 shows square vias with rounded corner, and figure 9 is a set of sample etch profiles obtained with the optimized recipe. As can be understood from table 1, this recipe is optimized for a given via width, and depth. In very narrow vias, the sidewall forms an acute angle with respect to the horizontal, whereas this angle is obtuse in very large openings. One can expect sidewall angle below 1 degree for the target 50um vias. The aspect ratio is not a linear function of width, rather the aspect ratio increases rapidly as the width decreases.

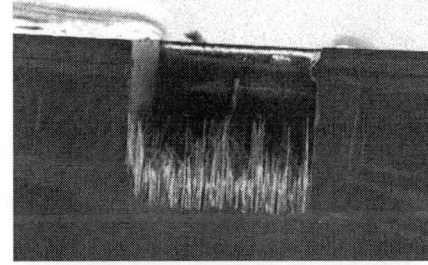

Figure 7: SEM Image showing black silicon issue

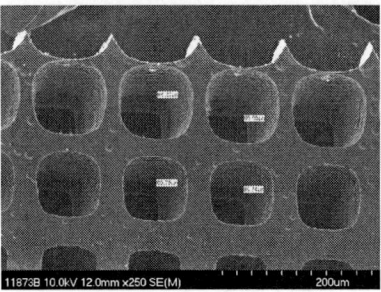

Figure 8: Sample Etching output, 45 degree angle view

C. Via filling

Once the vias have been successfully etched, the next task is to insulate their walls and metalize them to ensure connection of M1 to the backside landing pads.

978-1-4244-4722-0/09 $25.00
© 2009 IMAPS-ITALY

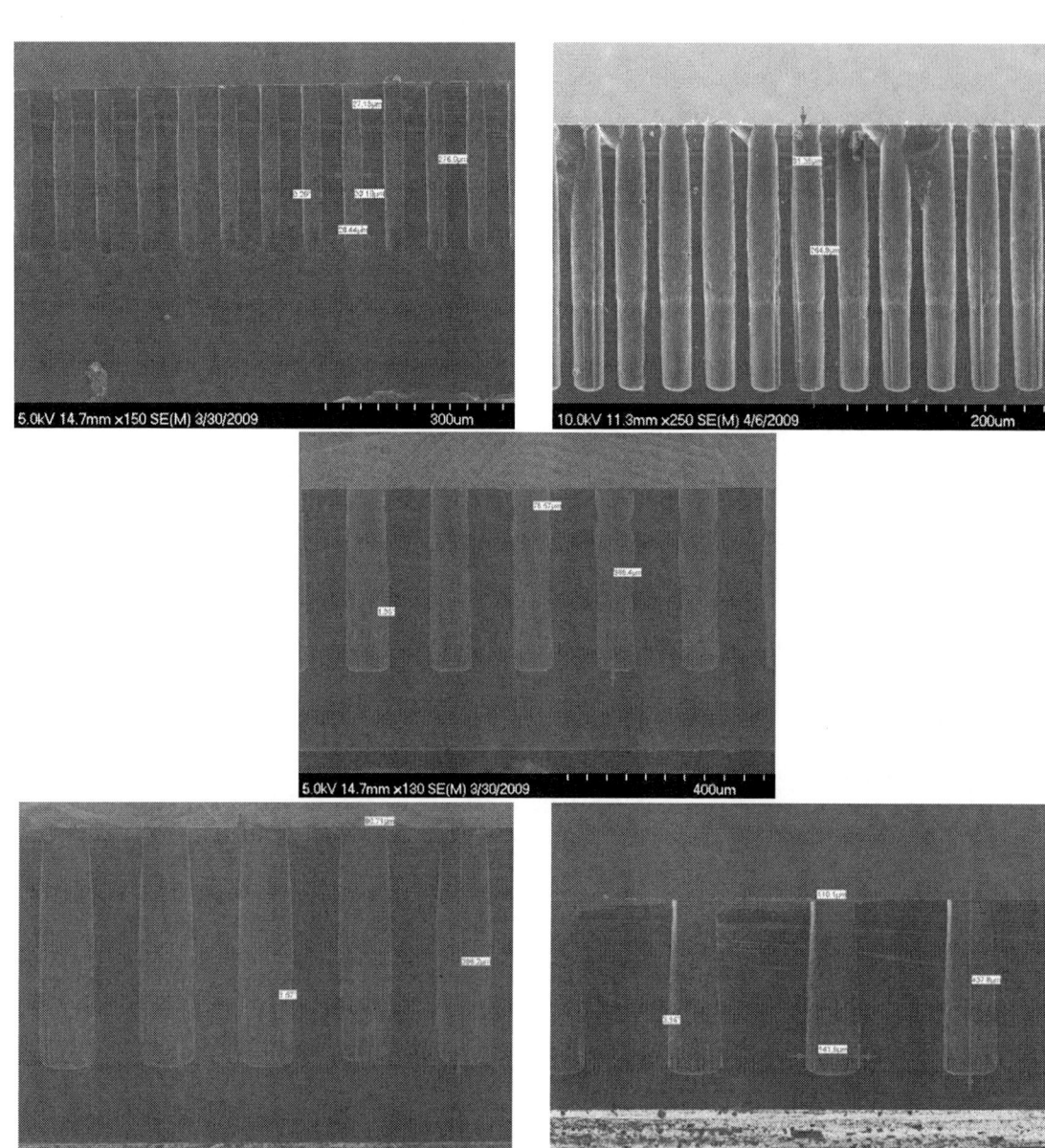

Figure 9: SEM profile of sample etch results

Width(μm)	Depth(μm)	Aspect ratio	Sidewall Angle(deg)
10	153	15.3	-1.26
30	269	8.97	0.29
40	316	7.9	0.88
80	381	4.8	1.51
110	437	4.0	3.14
200	496	2.5	4.24

Table 1: Summary of Etch results

978-1-4244-4722-0/09 $25.00
© 2009 IMAPS-ITALY

Coating of a sidewall insulation dielectric is done to limit leakage current from any via into the silicon substrate. Achieving a conformal coating of SiO_2 while remaining below the CMOS temperature limit of 400ºC poses a few difficulties. In a conventional Low Pressure Chemical Vapor deposition (LPCVD) system, the thermal energy required for deposition dictates temperatures of 700ºC or higher. Other techniques of direct deposition suffer from line-of-sight limitations given the required depth of the vias. The envisaged solution to overcome the high temperature requirement of LPCVD is the use of plasma to enhance the chemical reaction (PECVD) at temperatures of 400ºC or lower. There is effectively a tradeoff between coating quality and deposition temperature. The source gas can be either Oxygen or Tetra-Ethyl-Ortho-Silicate (TEOS). Inductively coupling the plasma (ICPECVD)[7] has also proven to result in deposition of a quality film.

To remove the oxide underneath M1, dry etching has to be performed. The ICP-DRIE tool will be used with a proper combination of gases for selective etching of SiO_2. Given the highly uneven topologies, dry film resist will need to be used as masking layer for this step. Metal filling will then be done using standard electroplating techniques or simply a conductive paste as best suited for low contact resistance. Lastly the backside pad layer will be electroplated or patterned from a copper foil.

IV. TESTCHIP1.0

The Test-Chip1.0 on figure 10 is an array of 3x3 WaferIC unit cells each made of 4x4 nanopads. To the right of the active area is a column used for characterization of two postprocessing tasks: reticule stitching and TSVs. All around the chip are wire bonding pads used for packaging of the device. The entire circuit cover covers an area of $2x1.9mm^2$.

For electrical testing and validation, the chip is packaged using the wire bonding pads. For postprocessing, on the other hand, bare dies are utilized. The rightmost column consists of passive patterns, most of which serve the purpose of characterizing reticule stitching. Figure 11 below outlines the simple pattern used to emulate the overlap of metal lines for inter-reticule stitching[1].

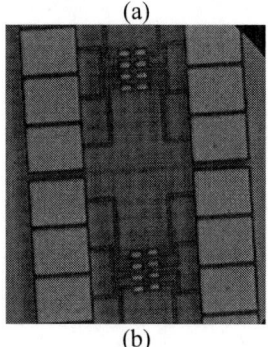

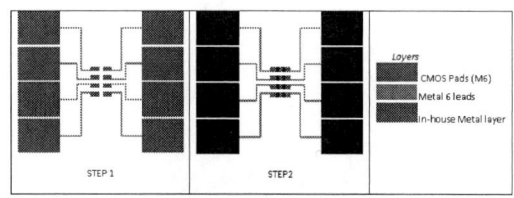

(a)

(b)

Figure 11: Inter-reticule stitching patterns (a) layout of process, (b) optical image of Step 1 pattern

Within the same column, a structure is designed to verify the attainment of M1 and the continuity of metal in the TSV. Besides this TSV characterization pattern, TestChip1.0 features 16 TSV connections pads on M1 of size 50x50μm in the active circuitry. For this test circuit, the power supply is routed to predetermined bonding pads, not just to TSVs as will be the case on the actual WaferIC. For probing purposes, the M1 TSV contact is raised to M6 through a series of vias. The FIB image on figure 12 shows a side view of the layer stack above a TSV. The oxide layer underneath M1 is apparent. Work is currently being done to apply the aforementioned etch recipe to TestChip1.0 using this oxide layer as etch-stop. A slight overetch will be performed to ensure silicon removal with minimal damage to the oxide layer due to the 150:1 selectivity to SiO_2[3]. Another useful feature of the designed recipe is the presence of a small amount of passivation gas present during the etch step, this help to reduce any notching effects [3].

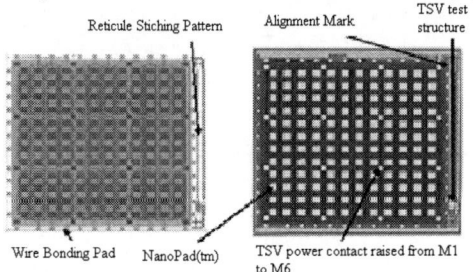

Figure 10: TestChip1.0 Layout and Micrograph

978-1-4244-4722-0/09 $25.00
© 2009 IMAPS-ITALY

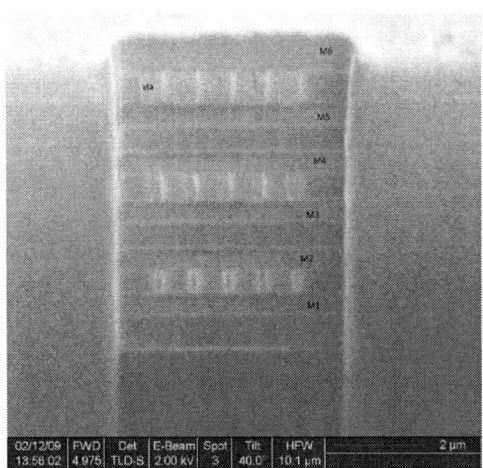

Figure 12: FIB slice of TSV top pad

V. THERMO-MECHANICAL ANALYSIS

The reduced feature sizes, increased power requirements and package contact densities all contribute to make thermal issues extremely critical in designing WaferBoard[TM] platform. Thermal analysis is a crucial vehicle for predicting the change in the electrical characteristics or possible stress-induced failure of the platform. The complexity of the device dictates the need for a detailed analysis and optimization of coupling between both the heat-transfer fluid pouch used to apply even pressure on uICs, and to the backside heat sink through WaferIC and its supporting structures; i.e. the complete thermal coupling from components to ambient. The accurate and fast evaluation of heat flow patterns becomes an essential step in the overall design verification. This consists in studying a wide variety of possible thermal scenarios of operation of the WaferBoard[TM] to obtain a preliminary assessment of its thermo-mechanical behavior.

A. Wafer Scale Thermal Analysis

In this section, the estimation of thermal peaks and the stress induced on the WaferIC[TM] are presented. These have become a major issue with the ever increasing power density of uICs deposited on its surface. The investigation is done using a thermal heat sources placement approach to estimate and predict working temperature of WaferBoard[TM] structures. In a second step, estimated temperature gradients are used to calculate stress profiles. Finite element analysis is used for peak temperature prediction during WaferBoard[TM] operation. NISA finite element software (Numerical Integrated elements for System Analysis) [8] is used to predict thermal behavior. The power dissipated in heat sources (components) placed on the WaferIC[TM] is modeled by heat flux produced inside the components. The final results are dependent on several parameters and thermo-mechanical components of the system, such as the uIC package and heat sinks (geometry and materials), to be

designed at a later stage of this research. The main heat source is the uICs, and based on on-going detailed electrical design [9] the bulk of the power consumed by the WaferIC[TM] is dissipated directly underneath these uICs. The specific benchmark for which results are presented in this paper is a 60W IC (a Pentium is used for reference) for which the package has a 40*40mm^2 area (0.0016 m^2). The problem formulation is presented graphically in figure 13.

Figure 14 shows the evolution of the temperature at Y1-Top solders balls- WaferIC[TM]. Hence, a non-uniform temperature profile and thermal peaks of 42.2 °C exists through the device thickness at the Top solders balls- WaferIC[TM] interface. Under these conditions, the maximum thermal variation ΔT_{max} is 17.2 °C in the structure.

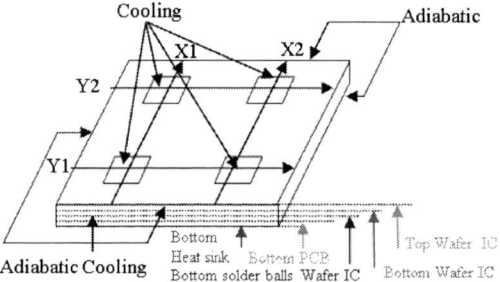

Figure 13: WaferBoard[TM] inside thermal boundary condition (BC)

However, at the Top solders balls- WaferIC[TM] interface the maximum local thermal variation $\Delta T_{max(loc)}$ is 3 °C induced locally during WaferIC powering. Between the components, there is subsequent relaxation on temperature gradients through the structure leading to an essentially uniform temperature variation. The cooling of the device is effectively controlled by conduction through to the WaferIC and through to the component.

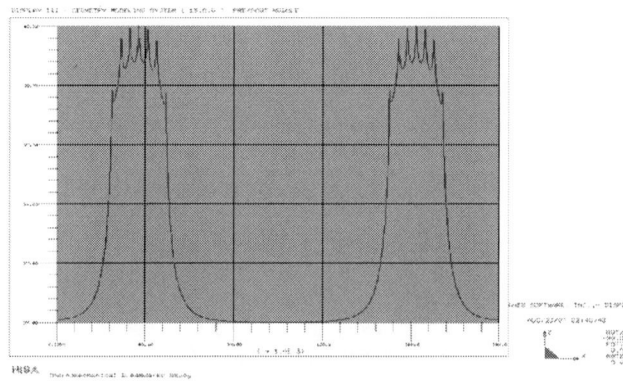

Figure 14: Sandwich forced convection through PCB surfaces Temperature profile Y1-Top solders balls- WaferIC[TM]

978-1-4244-4722-0/09 $25.00
© 2009 IMAPS-ITALY

B. Thermo-mechanical stress analysis of WaferIC in the presence of TSV's in 70 micron model

For stress analysis, the 70 micron thick model is used to obtain critical thermal and mechanical analysis data. Furthermore, a quarter model is designed to reduce the total number of finite elements and allow for a manageable sized model.

Multilevel structures with TSVs are becoming commonplace in the electronic industry. Because these structures are made of materials that have different properties, specifically a different coefficient of thermal expansion (CTE), thermal stresses, distortion, and warpage are a source of concern. In the case of thin multilayered devices with TSV such WaferIC, the surface may be assumed to be in a state of plane stress [10]. Hence, an extensive finite element analysis is needed to investigate the WaferIC surface, especially at solder balls interfaces. In this section, the approach proposed captures the thermo-mechanical stress singularity at the WaferIC level to predict the critical tensile stress. Thus, this approach is only valid for steady-state heat conduction and is limited by simplifying assumptions. Hence, predicted hot spot temperature and realistic associated stress distribution can be used to guide thermo-mechanical design. Figure 15 shows shear stress results in the Y1-Bottom solders balls-WaferIC interface in the case of sandwich forced convection through PCB surfaces and simply supported in the bottom boundary conditions.

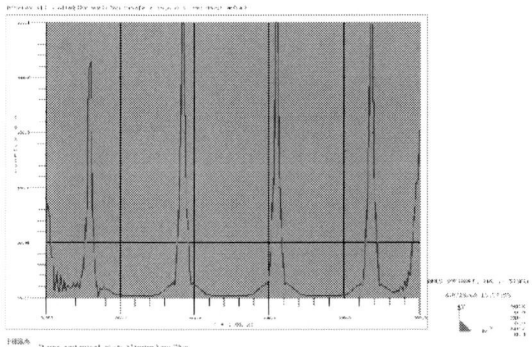

Figure 15: Sandwich forced convection through PCB surfaces simply supported in the bottom SHEAR STRESS through Y1-Bottom solders balls-WaferIC (70µm)

The maximum shear stress is of the order of **+16.1** MPa, acting on the bottom solders balls-WaferIC interface and the extreme variation is located on the solder ball interface. This is justified by the fact that the heat is accumulated in the waferIC substrate, therefore the maximum thermal variation is found at the solder ball interface. On the other hand, the induced thermal stress is combined with the intrinsic one due to the fabrication processes as well as to the stress due to the TSV holes. Therefore, cumulative damage is likely to occur at the critical interface regions. As such layers are very thin; any imperfection in structure may lead to cracking and subsequent shear-initiated delamination of the WaferIC structure. The nature of such interface materials must therefore be considered very carefully in WaferBoard™ design for intense planar activity applications.

VI. CONCLUSION

The WaferBoard™ rapid prototyping platform for electronic systems, based on WaferIC, a wafer scale circuit, shows great promise. Its use in an electronics manufacturing process flow will result in significant time and cost savings. Several challenges in the design and fabrication of the platform have been outlined.

Postprocessing from a microfabrication standpoint requires careful planning and characterization, as several tasks have to be accomplished sequentially. Amongst other achieved results, the developed DRIE process achieving a 15:1 aspect ratio will prove very instrumental to the fabrication of TSVs.

From a thermo-mechanical standpoint, the effects of heat and stress on the WaferBoard™ structure's behavior are of great significance to the development of the technology. Large values of stress, distortion and warpage can be induced by various steps during operation and during post processing the TSVs. The highest effective stress appears in the corners if the device is clamped and the maximum variation is found at solder ball interface.

Work is currently being done to make the postprocessing requirements less stringent. Larger via sizes in a thicker wafer are being considered for easier filling and greater strength, and a wider pitch is being envisaged to limit overall stress.

The integrated multidisciplinary design approach utilized so far will certainly help overcome most of these challenges.

ACKNOWLEDGEMENTS

The authors would like express their gratitude to Gestion Technocap Inc., DreamWafer division for their financial support and Intellectual Property. Indirect contributions made by the entire DreamWafer™ research team have been greatly appreciated. Special thanks go to the very resourceful Laboratoire de Microfabrication (*LMF-Ecole Polytechnique de Montreal*) staff for provision of the facilities and tools necessary to this research. The financial support of NSERC and Precarn is acknowledged. The authors would also like to thank CMC Microsystems for providing design tools and access to technologies.

978-1-4244-4722-0/09 $25.00
© 2009 IMAPS-ITALY

REFERENCES

[1] Norman, R. U.S. Patent Application Number 11/611,263.

[2] Takahashi, K., Sekiguchi, M., "Through Silicon Via and 3-D Wafer/Chip Stacking Technology" Symposium on VLSI Circuits, Digest of Technical Papers, pp. 89-92, 2006.

[3] Walker, M., "Comparison of Bosch and cryogenic processes for patterning high-aspect-ratio features in silicon", Proc. SPIE, Vol. 4407, 89, 2001.

[4] Hutter, M., Oppermann, H., Engelmann, G. Reichl, G. "High Precision Passive Alignment Flip Chip Assembly Using Self-alignment and Micromechanical stops", Electronics Package Technology Conference,2004

[5] **McAuley, S. A., Ashraf, H., Atabo, L., Chambers, A., Hall, S., Hopkins J., Nicholls G. "Silicon micromachining using a high-density plasma source", Journal of Applied Physics, pp. 2769–2774, 2001**

[6] Z. Cui " Micro-nanofabrication: Technologies and Applications", Birkhäuser, pp. 216-224, 2005

[7] A. Boogaard, A.Y. Kovalgin, I. Brunets, A.A.I. Aarnink, J. Holleman, R.A.M. Wolters, J. Schmitz, "Characterization of SiO2 films deposited at low temperature by means of remote ICPECVD", Surface and Coatings Technology, Volume 201, Issues 22-23, Euro CVD 16, 16th European Conference on Chemical Vapor Deposition, pp. 8976-8980, 25 September 2007.

[8] NISA II "user's manual" EMRC, Troy, Michigan

[9] R. Norman, E. Lepercq, Y. Blaquière, O. Valorge, Y. Basile-Bellavance, R. Prytula, Y. Savaria, "An Interconnection Network For A Novel Reconfigurable Circuit Board", IEEE/NEWCAS08 (accepted).

[10] C. Yeh, C. Ume, R. E. Fulton, K. Wyatt, and J. W. Stafford, "Correlation of Analytical and Experimental Approaches to Determine Thermally Induced PWB Warpage", IEEE TRANSACTIONS ON COMPONENTS, HYBRIDS, AND MANUFACI'URING TECHNOLOGY, VOL. 16, NO. 8, pp. 986-995, December 1993.

[11] Bougataya, M., Ahmed Lakhsasi, Y. Savaria, D. Massicotte, "Mixed Fluid-Heat Transfer Approach for VLSI Steady State Thermal Analysis" IEEE CCECE02 Proceedings, Winnipeg, Manitoba, 403-407. (2002).

[12] Corinne P *et al*. "Analytic Modeling, Optimization, and Realization of Cooling Devices in Silicon Technology", IEEE Trans. on components and packaging Technologies, Vol 23, No 4, June 2000.

[13] Cheristopher.J *et al.* "A Simulation Study of IC Layout Effects on Thermal Management of Die Attached GaAs ICs", IEEE Trans, on components and packaging Technologies, Vol 23, No 2, June 2000.

[14]] A. Lakhsasi, A. Skorek," Dynamic Finite Element Approach for Analyzing Stress and Distortion in Multilevel Devices ", SOLID-STATE ELECTRONICS, PERGAMON, Elsevier Science Ltd., Volume 46/6 pp. 925-932, May 2002.

[15] Bakir, M.S. *et al*, "Sea of Leads Compliant I/O Interconnect Process Integration for the Ultimate Enabling of Chips With Low-*k* Interlayer Dielectrics," Advanced Packaging, IEEE Transactions on, vol.28, no.3, pp. 488-494, Aug. 2005.

Evaluation of Printed Electronics Manufacturing Line with Sensor Platform Application

Eerik Halonen[1], Kimmo Kaija[1], Matti Mäntysalo[1], Antti Kemppainen[2], Ronald Österbacka[3], and Niklas Björklund[3]

[1]Tampere University of Technology, Institute of Electronics, P.O. Box 692, FIN-33101 Tampere, Finland

Phone: +358 40 849 0102 Fax: +358 3 3115 3394 E-mail: eerik.halonen@tut.fi

[2]VTT Technical Research Centre of Finland, P.O. Box 1100, FIN-90650 Oulu, Finland

[3]CoE for Functional Materials and Department of Physics, Åbo Akademi University, Porthaninkatu 3, 20500, Turku

Abstract

The increasing demands in electronics manufacturing are driving the development of new manufacturing processes. Printing processes, such as gravure, flexography or inkjet printing, have emerged as novel manufacturing methods of electronics that also enable new type of applications to be designed. A clear benefit of printing is that the process is additive, i.e. material is deposited only on the areas where it is needed, which reduces material consumption. Utilization of several printing techniques allows patterning of large area circuitry with high throughput and down to 20µm line widths. The variety of functional inks for printed electronics is continuously increasing and includes e.g. metallic and organic conductive, dielectric, piezoelectric, and semiconductive inks. This paper studies the integration of suitable printing methods on a hybrid production line where the evaluation is based on a sensor platform application that consists of an antenna, organic transistors, printed conductors, printed keyboard, and a land pattern for an ASIC component. The proposed production line is a combination of different printing methods due to the requirements for throughput, accuracy, print thicknesses, surface roughness and suitability of available inks for different printing methods. The sensor application contains several different electrical interconnections that are formed on the hybrid production line. The electrical performance of the interconnections needs to be studied separately in order to find the optimal process parameters at the technology interfaces.

Key words: Printed electronics, Electronics manufacturing, Technology integration, Electrical interconnection

Introduction

Printed electronics is a very fast growing industry. During the past several years, the research in the field of printable electronics has been vast. The fabrication of passive components [1] and transistors [2; 3] by using inkjet technology has been represented. In printed electronics it is possible to utilize in practice any of the available printing methods. In addition to inkjet printing, faster R2R methods have also been studied. Those techniques are ideal for manufacturing large area applications, like photovoltaic or lighting solutions.

When printed electronics, in the future, is intended to bring in a cost effective direction, it has to be analyzed which are the most sensible means to fabricate different kinds of applications. There are several variables that have an influence on the final productivity and the cost savings. Unfortunately some of those factors are unsolved at the present. The first organic electronics products reached the markets in 2005. [4] Many visions of possible solutions and applications where the printed electronics can be implemented are developed. In the next few years, the printed electronics is expected to be more widely transported from a laboratory and research environment to electronics manufacturing. It is predicted that within 2-5 years printing methods are widely used in mass production of e.g. flexible displays, RFID tags, organic photovoltaic cells, and organic memories. [4] This development requires lots of study and work on the issues related to the integration of printing methods as a part of a manufacturing line.

The right manufacturing method should be found for every part of every application. This naturally means the cheapest and the fastest method that still can guarantee an adequate level of performance. This is not however quite enough. The whole manufacturing process has to be taken into account as totality. The integration of different manufacturing techniques into a functional production line is a huge challenge. Very little information and experience considering the subject is available worldwide so far. At this point, the most

978-1-4244-4722-0/09 $25.00
© 2009 IMAPS-ITALY

significant values are the production volumes of the technologies and the synchronizing of the whole production process. This relates to products fabricated totally by printing methods, and products where the printing techniques and the conventional electronics manufacturing are integrated together.

Printing Methods

Flexography printing is one of the most common printing methods nowadays. The basic idea of flexography is to supply an ink from an embossed cylinder to a substrate. Patterns are on separate plates and those are attached on a plate cylinder. The materials of the plates are usually rubber or photopolymers. The ink supplied on the plate cylinder with an anilox roll. Anilox roll is an engraved cylinder with full of little cups. Ink fulfills the cups and the anilox roll transfers ink on the flexo cylinder. Typically the cups are engraved with a laser and the size of the cups varies depending on the printed and ink. The role of a doctor blade is to wipe out extra ink from the surface of the anilox roll. Impression roll sets the desired pressure between substrate and printing plate. [5]

There are several things that have an impact on the print quality e.g. surface roughness of a substrate, the material of a printing plate and its mounting tape, the type of an anilox roll, printing speed, and printing pressure. With flexography printing, it is very hard to print patterns that consist of large areas and fine shapes at the same time. The parameters have to be adjusted for one fature at a time. E.g. smaller anilox roll cups are required for accurate patterns or thin layers, whereas larger cups are needed for thick layers. [5]

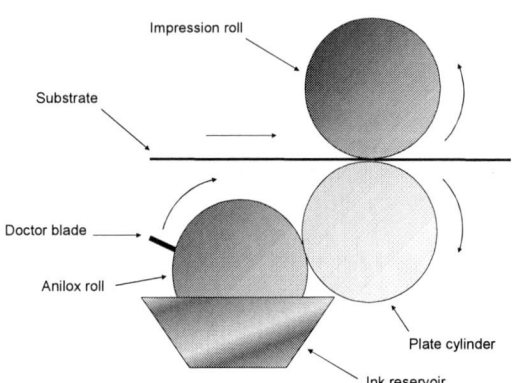

Figure 1: Structure of the flexography printer

Gravure printing differs from the flexography printing so that there are only two rolls in the process. In gravure printing "anilox" roll and a printing roll is the one and the same cylinder. Printied pattern is engraved straight on the gravure cylinder that transports ink from it to a substrate. A doctor blade levels the ink the same way as in flexography printing as well as substrate pressure is set with an impression roll. Generally three different methods are used to pattern a gravure cylinder: chemical etching, electromechanical engraving or laser engraving. Electrommechanical engraving is the most common at the present but laser engraving is a fairly new technique and is going to develop into a very usable option in the future. [5]

The main parameters that affect print quality of gravure are the substrate, the gravure cylinder, the quality of the engraved pattern, doctoring, and the viscosity of the ink. There are strict requirements for the surface smoothness of the substrate and the gravure cylinder. The right choice of the doctor blade and the shape of the blade are maybe the most important factors for the quality. Ink viscosity should be also low enough to achieve uniform structures but, on the other hand, high enough that it doesn't flow off the cells of the cylinder. [5]

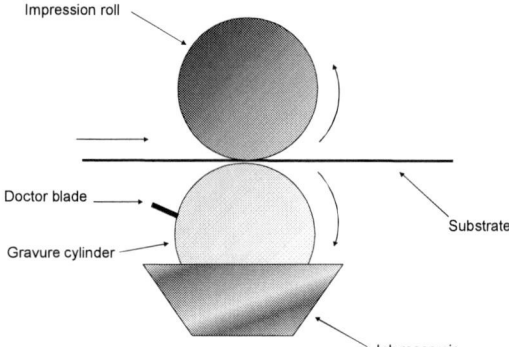

Figure 2: Structure of the gravure printer

Inkjet printing, is a digital printing method, which has been found out very practical and flexible way to make printed electronics. Inkjet doesn't request mechanical contact with a substrate in contrast to the aforementioned printing methods. Inkjet printers can be roughly divided into two categories: 1) continuous inkjet (CIJ) and 2) drop-on-demand (DoD). DoD printheads apply pressure pulse to ink with e.g. piezo- or thermoelement that ejects a drop from a nozzle orifice when required. Inkjet printers that are suitable for electronics manufacturing operate mainly at the piezo pulse principle. [5]

Inkjetted print quality depends on the ink (e.g. viscosity, surface energy, particle size), the substrate (its surface features), the temperature of a plate, and the parameters of the printhead (e.g. temperature, voltage). Different inks behave differently when they come out from the printhead or when they land on the substrate. Substrate materials also have different surface energies that affect ink spreading and size and shape of a drop on it. Printhead parameters affect the velocity, volume, formation time, shape, and flight angle of ejected drops.

978-1-4244-4722-0/09 $25.00
© 2009 IMAPS-ITALY

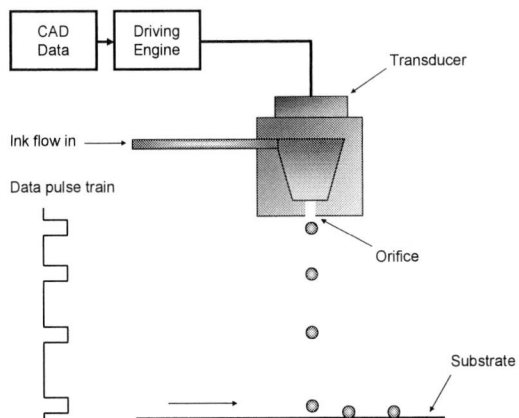

Figure 3: Drop-on-Demand inkjet printer

Screen printing is an old and pretty simple printing method that has been used in electronics for printing solder pastes, conductor insulators, passive components, adhesives, and markings. The foundation of the technique is a weaved screen that can have different thickness and thread density. The pattern is usually made to an independent stencil. To make a print, the squeegee blade is brought down forcing the screen into contact with the substrate. This forces the ink through the open areas of the screen on the substrate and the desired pattern is formed. The most common screen materials are polyester and stainless steel. Squeegees are normally made of polyurethane compound. There are mainly three different printing methods in screen printing: flat bed, cylinder, and rotary. Flat bed is the most common screen printing method. Cylinder screen printing is otherwise similar to the flat bed but the substrate moves as attached to roll. In rotary screen, ink and a squeegee assembly are inside a rolled screen. Impression cylinder produces pressure to substrate. [5] Rotary screen enables way higher throughput capacity than flat bed screen and the integration into a R2R line.

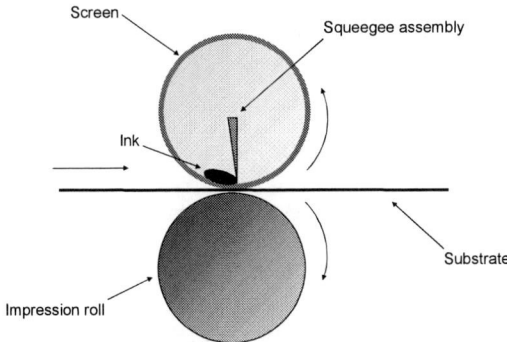

Figure 4: Structure of the rotary screen printer

Table 1 summarizes some crucial attributes of the mentioned printing methods.

Table 1: Some characteristics of the printing techniques. [5; 6; 7; 8; 9]

	Flexo	Gravure	Inkjet	Screen / Rotary Screen
Lateral Resolution [μm]	>40	>15	20-50	50-100
Ink thickness [μm]	0,8-8	0,8-8	0,3-2	<100 / 5-80
Ink viscosity [mPas]	50-500	50-200	< 20	500-50000
Throughput [m²/sec.]	10	60	0,01	<10
Printing speed [m/min]	100-500	100-1000	15-100	10-15 / 15-100
Layer Roughness [nm]		5-20	<100	Rough, ink depended
Registration [μm]	<200	>10	<5	>25
Ease/€ of making plates	Easy	€	N/A	Moderate
Amount of material	Medium-High	Medium-High	Small	Small / Medium
Substrate requirements		Smooth (paper, board, plastics)	3D possible	

Sensor Platform Application

The principled layout of the product is represented in Figure 5. The application has been divided into five sub-parts. There are basically three different antenna types for different frequency ranges represented. The organic transistor type in this study is a bottom-gate. The circuitry is split in two. There are a basic single layer rough circuitry and a more challenging multilayer circuitry. The two other parts that are under investigation in this paper are a keypad that is made with piezo electric structure and ASIC connections that requires dense circuitry. In each part there are some different requirements for the techniques and at the same time every feature has its own effects on the whole manufacturing of the product.

Antenna is an essential part of portable communication applications. Nowadays e.g. most of the RFID antennas are manufactured by etching method but printing could be more cost effective, faster, and a more environmentally friendly option in a mass production. [10] There are some experiences of antenna printing and results have been promising. The manufacturing requirements, such as conductivity and physical dimensions (thickness), of antennas vary between different frequency ranges. Also the antennas in different frequency ranges have different operational principles. In this paper are studied three frequency ranges of antennas. These three are High frequency (HF), Ultra High Frequency (UHF), and Super High Frequency (SHF) ranges. These have been chosen because they are the most typical frequencies that are used with RFID applications.

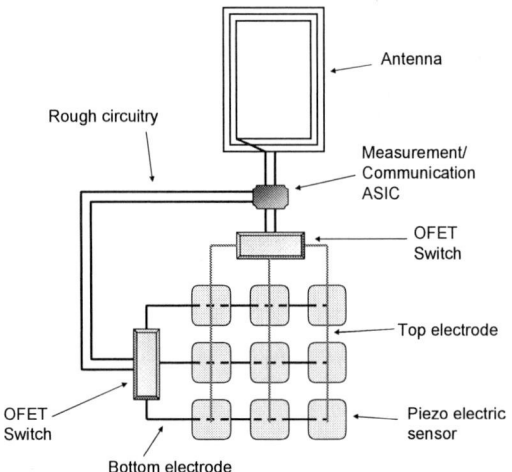

Figure 5: A principled layout of the sensor platform application

The first antenna type under study is a HF antenna. It is inductive coupled, which means that it uses the energy of a magnetic field to transfer data. In practice the antennas are coil shaped and there are normally way less than ten conductor loops. The size of an antenna is somewhere around credit card or even larger. This antenna type might be the hardest to manufacture of all these three options. High conductivity is desirable which usually requires thick layers. Post processing, such as sintering, also has a significant role in the high conductive antenna fabrication. Secondly, a spiral patterned shape requires a certain resolution from the printing technique. However there is seldom a need for smaller gaps than 100 µm between conductors. Typically, the gaps are wider, sometimes even millimeters.

The second represented antenna type is an Ultra High Frequency (UHF) antenna. A coupling method of this type of antenna is an electromagnetic backscatter. There are several different shaped antenna patterns. One of the most general shapes is a bow-tie antenna. It is a typical dipole antenna and its dimensions are the size of a few centimeters. Normally there is no need for very high accuracy manufacturing technique when producing UHF antennas. Another advantage compared to HF antenna is that UHF antenna can be thinner to achieve adequate conductivity. [11] In fact the thickness needs to be just about a fourth of HF antenna thickness, which brings direct saving to material costs and makes printing technologies an appealing manufacturing method.

A frequency range of 2.45 GHz is designated in some literature still as UHF range but some others are calling it as a Super High Frequency (SHF) range. In fact the frequency of this antenna type is already in the area of microwaves. The coupling method of the antenna is also electromagnetic backscatter. In this group there are many different antenna designs too, and the most common type is a dipole antenna. As the operation frequency increases the physical size of an antenna decreases. The antennas of SHF area have many similarities with UHF antennas. The only remarkable difference is a size which is little smaller in SHF area on average. This may lead to a need for somewhat better lateral resolutions in some antenna designs.

Organic transistors are still very much at a development stage and their performance cannot compete with traditional semi-conductor transistors. It is however assumed that organic transistors will develop into the level that they are feasible and economical to manufacture for certain types of applications. With the studied application transistors are the most complex part in many ways. First problem is several overlapping layers that consist of different materials. An interaction between different materials is a major challenge in this kind of multilayer structures. Each layer also requires one production stage with sintering or curing processes. Secondly, the accuracy requirements for some layers are very high. Precision matters are going more and more significant when layer count increases. In upper layers not only resolution but also registration plays an important role.

The bottom-gate structure of the transistor is shown in Figure 6. The first layer of the OFET is a gate electrode. The most promising transistor structure in the project has been obtained when aluminum gates are evaporated onto a substrate, and the aluminum oxide that forms onto the surface of the aluminum operates as a part of the following dielectric layer. The Al_2O_3 layer can be grown further with for example anodization or plasma treatment.

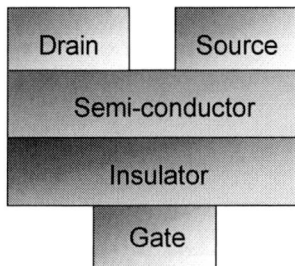

Figure 6: Bottom gate OFET structure

Besides aluminum oxide there is a need for other dielectric layer also. The most significant requirements are that it is thin and defect-free. A thin dielectric layer (only tens of nanometers) is required to achieve an operation voltage of a few volts. [3]

The third layer of the bottom-gate transistor is a semi-conductor layer that should be as smooth as possible without defects. [12] Polymer based semiconductors can be formulated as inks and used with various printing techniques. Small molecule based semiconductors can also be used, but the deposition method (e.g. evaporation) is much harder to implement in R2R processes.

The uppermost layer consists of drain-source electrodes. This is a very critical phase in the bottom-gate transistor fabrication. The channel length should be small and offset between the gate and the drain-source electrodes should be minimized. This requires very good resolution and alignment ability for the processing technique.

The circuitry section is divided into two separate parts. Firstly there is discussed a simple, rough, single layer circuitry that is manufactured straight on the substrate. In the application these kinds of examples represent all main conductors between the antenna, the ASIC and the switches. The pattern is quite rough and lateral resolution requirements are something around a few hundred of microns.

Secondly is a multilayer circuitry that is not necessary fine but it requires little different or more features for the manufacturing technique. This kind of example in the application could be the upper electrodes of the keypad. It is highly application specific matter but for example good registration and capacity to print on different surfaces are two important requirements. Sometimes only one of those could be enough but in most cases they are both crucial. This study concentrates especially on registration capacity. Always in multilayer cases layout design and material interfaces need to be considered carefully in addition to techniques. The most complex and the most difficult layer should be designed at the lowermost because the lowest layer is normally easiest to manufacture.

One of the subparts of the application is a keypad. The piezo layer is a part of the product that doesn't have very high requirements for the accuracy or registration among others. The most essential demand for piezo layer would be that it is uniform and has a smooth surface. These two features affect positively both piezo electric behavior and success of the conductors that are printed on top of it.

It would be desirable that the amount of ASIC I/O pins is quite small and the pitch is at least 200 µm when connecting on a flexible substrate. As a result of electronics miniaturization the requirements of connecting density are increasing. Therefore the resolution limitation in this work is set below 100 µm. This emphasizes the registration requirements also.

Selection of the Methods

This part of the paper extends the previous study and suggests manufacturing techniques for each sub-part of the application. The whole picture of the manufacturing process is tried to keep in mind in the selections. Besides these deposition stages, it is important to notice that after each phase some kind of treatments, like sintering or curing, are usually needed. These process steps actually may take the major part of the total manufacturing time.

The first stage is the evaporated aluminum gate electrodes as explained earlier. The idea is that these gates have already been done before the roll comes to the "printing unit". The gate aluminum oxide can be further grown as explained earlier.

In the second phase an antenna, the bottom circuitry of the keypad, and the other rough circuitries are printed by gravure, flexo or (rotary) screen printing. They all are fast methods and the accuracy capabilities are sufficient to these structures. When manufacturing an HF antenna the thickness of gravure or flexography printing might not be enough. So the suggestion to this stage is rotary screen printing. Gravure and flexography printing are at least equal in the case of higher frequency antennas however.

The third step is to supplement the piezo electric material for the keypad. It highly depends on the ink that is chosen for this application but at least for the meantime every method mentioned in the previous stage could be usable also into this one. The samples during this project have been screen printed, so the rotary screen printing is selected to this stage too.

The fourth step is the dielectric layers of the transistors. The main challenge is the thickness of the layer of the deposited material. There might be some material technical issues between aluminum oxide and dielectric ink that are not known yet. Those can affect to the selection of the manufacturing method. The technology for the dielectric layer printing is inkjet because it is the only one of these methods that can theoretically provide such thin layers.

The fifth task in this production is the semiconductors for OFETs. The thickness of the semiconductive layer is not as crucial as in the dielectric layer. Inkjet surely is the most potential choice of the printing methods. Also gravure or flexography cannot completely be forgotten. In the laboratory dielectric and semiconductor layers have been done with spin coating, but it is not possible to integrate into a R2R line.

The final stage is the making of the drain-source electrodes onto the organic transistors, the upper keypad circuitry, and the pads for the ASIC connection. If these parts are wanted to manufacture in one phase, which is reasonable, pretty far the only suitable method is inkjet printing. Multilayers, which in this application mean upper keypad electrodes, can likely be printed with R2R techniques, like gravure, flexography or rotary screen, but they don't suit for the manufacturing of the ASIC connections and particularly not for the drain-source electrodes. After these selections the production line for the application could be as shown in Figure 7.

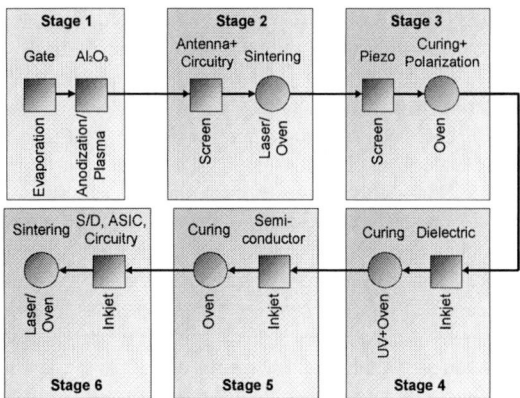

Figure 7: Production stages based on the selections

Electrical Interconnections

As there can be seen, the manufacturing process is quite complicated and there are not only many different kinds of technological interfaces, but also several material interfaces. The essential matter for the functionality of the application, are the electrical contacts between connected materials. One of them is the contacts between the aluminum gates and the silver connectors. Second interesting interface at the moment is the upper and the lower silver electrodes of the keyboard and the piezo material between them.

The Al-Ag interconnection was studied with two different kinds of ink-substrate combinations. The combinations were chosen so that there is possible to test sintering temperatures on a large scale. The first combination was ink 1 on a Polyethylene terephthalate (PET) substrate and the second one was ink 2 on a Polyimide (Kapton®) substrate. Ink 2 requires higher temperature for sintering than ink 1, and Polyimide has a lot better temperature tolerance than PET.

The test structure consisted of twelve evaporated aluminum traces. The width of the traces was one millimeter. Also one millimeter wide silver connectors were inkjet printed across the aluminum lines. The attribute that was measured, was the contact resistance of the Al-Ag interconnection, which in this structure was about a size of one square millimeter. The structure is shown in Figure 8.

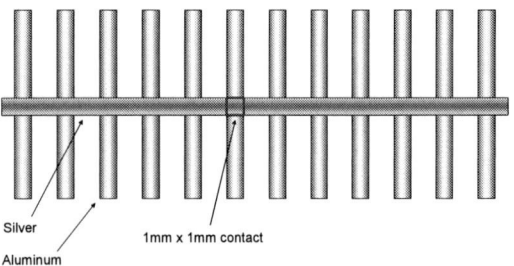

Figure 8: Al-Ag interconnection test structure

The first aim was to find the best sintering parameters for the both combinations. The sintering method was a convection oven sintering. Four sintering times and three temperatures were chosen for the samples. The samples were cleaned with isopropanol before printing. However, the ink 1-PET combination also needed some surface activation to improve wetting of the ink. The results of these sintering tests are summarized in Figure 9 and Figure 10.

Contact resistances were determined with four-point measurements. Relatively big variations in some cases were caused from the print quality. Especially the ink 1 has a couple of jettability problems with the used printheads. Most clearly those can be seen in Figure 9 with printing parameters 160 °C and 60 minutes.

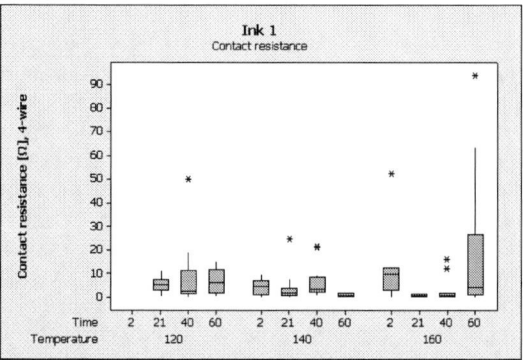

Figure 9: Contact resistance of the Ink 1

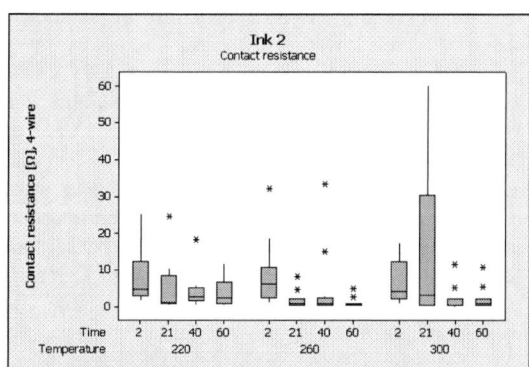

Figure 10: Contact resistance of the Ink 2

Connections between silver electrodes and the piezo material were also shortly studied. Before any electrical tests with the structure are worthwhile, it needs to be tested are the upper electrodes possible to make with inkjet printing. The first doubt is how the inkjet can cross the edges of the piezo material layer.

The piezo layers are screen printed and a single layer is tens of microns thick. Tests were made for a single layer and two layers samples. First the single layer samples were tested with the ink 2. 600dpi and 1270dpi resolution print files were used.

978-1-4244-4722-0/09 $25.00
© 2009 IMAPS-ITALY

1270dpi seemed to be a little better but the crossing of the edge managed with the both resolutions.

The results were quite promising and the tests were also tried onto the two layer samples. In Figure 11 can be seen that there are no problems with the edges of the piezo material. One observed problem was the porosity of the piezo material that can cause nanoparticles to sink inside it.

Figure 11: Inkjet printed Ag line over the edge of the screen printed piezo material layer.

Conclusions

Printing technologies has a great potential for electronics manufacturing in the future. This paper discusses about fully printable application, but of course there also exist plenty of alternative production methods for many subparts. The more likely scenario in the near future is that printed electronics will be integrated as a part of the production lines. However, it is about the time to consider all different kinds of threats and challenges that could appear in the printed electronics integration. This study speaks out this matter and offers sort of a selection procedure for choosing the appropriate printing technique.

The Al-Ag interconnection tests are going to continue with the optimal sintering parameters represented in this paper. These initial tests have shown that the ordinary aluminum oxide layer onto the aluminum doesn't prevent its electrical connection with silver. The further trials will study, how the various printing parameters, as well as different kinds of pre-treatments affect the results. Also the whole process capability, including treatments and printing, is analyzed and process improvement and control plans are made. Currently the process causes too much variation to the contact resistance values.

In addition to the Al-Ag- and silver-piezo-silver interconnections, other related material and technological interconnections are developed as well. One of those is the attachment of the ASIC onto the silver pads. The two most probable connecting methods at the moment are conductive adhesive and silver pad coating followed by soldering.

Acknowledgements

This work was supported by the Finnish Funding Agency for Technology and Innovation (TEKES), VTT Technical Research Centre of Finland, Ciba Finland, UPM, StoraEnso, Savcor, Pulse Finland and Modines.

References

[1] Redinger, D., Molesa, S., Yin, S., Farschi, R. & Subramanian, V. An Ink-Jet-Deposited Passive Component Process for RFID. IEEE transactions on electron devices, Vol. 51, No. 12, December 2004. IEEE. pp. 1978-1983.

[2] Subramanian, V., Chang, P.C., Lee, J.B., Molesa, S.E. & Volkman, S.K. Printed Organic Transistors for Ultra-Low-Cost RFID Applications. IEEE transactions on components and packaging technologies, Vol. 28, No. 4, December 2005. IEEE. pp.742-747.

[3] Molesa, S.E. Ultra-Low-Cost Printed Electronics. 15.03.2006. Technical Report No. UCB/EECS-2006-55. University of California at Berkeley, Electrical Engineering and Computer Sciences. 206 p.

[4] Hecker, K. Organic Electronics Association. OE-A Roadmap for Organic and Printed Electronics. VDMA. [WWW]. May 2008. [Cited 29.10.2008]. Available at: http://www.electronicsincanada.com/images/stories/whitepapers/PDFs/oe-a_roadmap_whitepaper2008_maypublic.pdf.

[5] Gamota, D.R., Brazis, P., Kalyanasundaram, K. & Zhang, J. Printed organic and molecular electronics, USA 2004, Kluwer Academic Publishers. 695 p.

[6] Kahn, B. E. Technical Overview of the Emergence of Printing Electronics and Displays, 4th Printable Electronics and Displays Conference, IMI, Las Vegas, NV, October 26, 2005. Presentation.

[7] Schottland, P. Printing Conductive Layers, Printable Electronics and Displays Conference, 10/26/2005. Presentation.

[8] Bisges, M. Conductive Nanoimprint Process for TFT backplane production, 4th Printable Electronics and Displays Conference, Las Vegas, NV, October 26, 2005. Presentation.

[9] VTT, Flexidis training workshop 2006

[10] American Printer. The truth about RFID printing. [WWW]. 01.06.2007. [Cited 10.06.2008]. Available at: http://americanprinter.com/binding-finishing/printing_truth_rfid_printing/.

[11] Merilampi, S., Ukkonen, L., Sydänheimo, L. Ruuskanen, P. Kivikoski, M. Analysis of Silver Ink Bow-Tie RFID Tag Antennas Printed on Paper Substrates. International Journal of Antennas and Propagation. Volume 2007, Article ID 90762. Hindawi Publishing Corporation. 9 p.

[12] Hecker, K. Organic Electronics 1st Edition. VDMA Verlag GmbH [WWW]. 2006. [Cited 02.05.2008]. Available at: http://www.vdma.org/wps/wcm/resources/file/eb6cc14189f8e31/organic_electronics_6MB.pdf.

Development of Matrix Clip Assembly for Power MOSFET Packages

Martien Kengen[1], Wil Peels[1], David Heyes[2]

[1] IS&O CSC-Innovation, Nijmegen, the Netherlands
[2] BL Power Management, Hazel Grove, England

NXP Semiconductors, Gerstweg 2, NL-6534AE Nijmegen, the Netherlands
Phone +31-243532214, Fax +31-243533642, martien.kengen@nxp.com

Abstract

New developments in trench technology for power MOSFET's drives intrinsic electrical silicon resistance to a minimum. This implicates that the contribution of the package resistance becomes more significant in the total electrical resistance (Rdson) of the product. Low Rdson for power packages is an important characteristic. Within the package the electrical interconnect between die top and leadframe is the main contributor to the Rdson value. In order to achieve a low Rdson the industry uses therefore (thick) wire bonding, ribbon bonding and clip bonding. The latter leads to the lowest Rdson, mainly because of the size of the clip and the low spreading resistance at the bondpad surface. In general, clip bonding is done by soldering a Cu clip to one or both bondpads and their corresponding leads. This requires a good control of the soldering process especially the positioning of the die and clips during the molten phase of the solder reflow process.

At NXP a new technology has been developed that makes it possible to place clips for both, the gate and source, for a matrix of products simultaneously. The matrix-wise processing is suited for leadless packages and is applicable for a large range of power products. This paper reveals the method of Matrix Clip Assembly applied to a power MOSFET package. A statistically significant large amount of packages has been assembled, electrically evaluated and subjected to life testing. The verification run showed a very low Rdson, high yield, wide process window and good lifetest performance. The electrical performance has been simulated and a direct comparison between real wire bonded, ribbon bonded and clip bonded samples has been made.

Key words: packaging, matrix clip assembly, Rdson, solder process optimization

Introduction

Power devices are highly driven by a low on-state resistance (Rdson). The Rdson is determined by the die –and package resistance. The die resistance has for long been a major contributor to the total Rdson of the product. However, new developments in trench technology for power MOSFET's has led to a significant reduction of the intrinsic electrical silicon resistance. This draws the attention towards package resistance as becoming the main contributor to the total product resistance. The package resistance is mainly determined by the top interconnect between die bondpad and lead, and the surface resistance of the top metallization of the die bondpad. Several solutions are developed to reduce the top interconnect resistance. Thick and multiple wires, ribbon bonding and soldered clips are seen in the market. Thick wires have a limited cross section and do still add the surface resistance of the die top metallization to the total package resistance. Ribbons have a larger cross section and reduce the surface resistance losses to some extend,

and a soldered clip will enable largest cross section and makes use of the total top die metallization, leading to the lowest package resistance. However, soldering clips on a die that itself is soldered to the diepad is a challenging process. In addition, today's developments in die technology leads to better performing dies, and the lower intrinsic resistance enable even smaller die designs. As a consequence, die bondpads become smaller, and so do clip sizes, leading to an even more challenging process of soldering a clip onto a die. One challenge is how to deposit the right amount of solder paste and how to handle the small clips. A bigger challenge is how to guarantee that, during solder reflow, when the whole system of die and clip is floating in molten liquid solder paste, die and clips stay in place. In the molten phase there is only solder surface tension and wetting dynamic behaviour that can lead to equilibrium. The outgassing of the flux system furthermore hampers this process.

At NXP a new technology has been developed that solves the challenges of handling and placing the clips. The new technology also solves

978-1-4244-4722-0/09 $25.00
© 2009 IMAPS-ITALY

the challenge of keeping clips and dies at the right position. This technology is given the name Matrix Clip Assembly (MCA). In MCA a matrix of products are clip bonded simultaneously, using clips that are positioned in a matrix held together by a polyimide foil.

This paper describes the MCA concept, using a power MOSFET as the carrier package. A production run has been made to validate this technology. Samples were electrically evaluated and subjected to life testing. The electrical performance has been compared with a simulation model and a direct comparison with measured wire bonded and ribbon bonded samples was made.

MCA Concept

The package that is demonstrated is a leadless NanoPAK 3333-8 package, figure 1.

Figure 1: NanoPAK 3333-8 Package

The package contains a MOSFET die where both gate –and source contact are established with a Cu-clip. Figure 2 shows the package design:

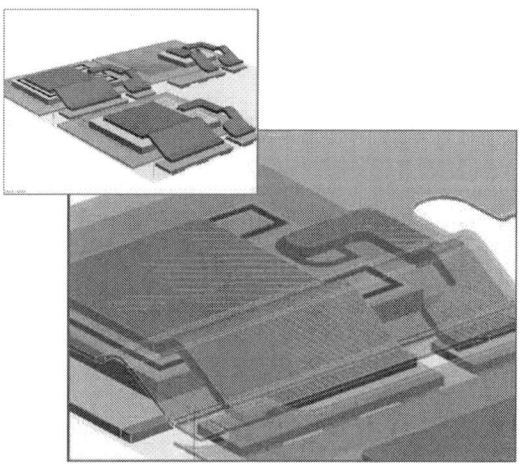

Figure 2: Package and Clip Designs

The clip frame is a Cu foil laminated on a carrier foil. By using photo-etching and a cut –and bending tool, clips can be made in any dimension or shape that is required, enabling maximum design flexibility. One single foil consists of a matrix of gate –and source clips, corresponding to the matrix of products on the lead frame, see figure 3.

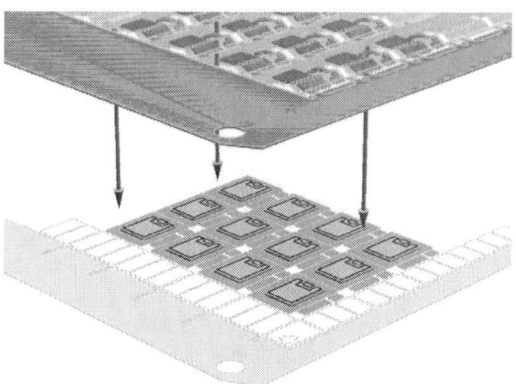

Figure 3: Positioning of Matrix Clip Frame

Process Flow

The process steps are visualised in figure 6. Each process step handles a complete leadframe, so a batch wise processing at each process step is achieved. It starts with stencil printing solder paste onto the diepads and leads of the lead frame. The lead frame has a Au finish.

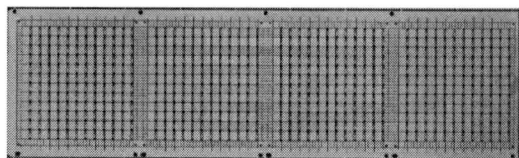

Figure 4: Test-Leadframe (576 Products)

Then dies, with Ag top –and backside metallization, are placed using a high-speed diebonder. Diebonding is the only serial process step, all other process steps remain batch wise. After diebonding, stencil printing is used to deposit solder paste on top of the die, for the source -and gate contact. Next, the clip frame is placed on top of the leadframe, using a vacuum jig with centering pins to guarantee proper alignment.

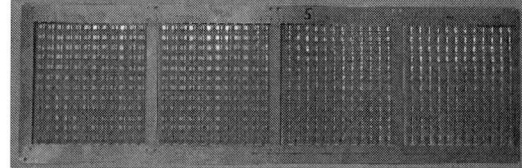

Figure 5: Clip frame

A solder jig is used during the solder reflow process. The solder jig will press down the clips and ensures contact between solder paste and clip during the reflow process. A double hot plate reflow oven was used, and reflow was done in a Nitrogen atmosphere. After flux cleaning, the leadframe is molded. Back etching is applied to remove the Cu at the backside of the molded leadframe, clearing off the Cu in the sawing lane. After singulation (sawing only through molding compound), individual, finished products are left on the sawing foil.

978-1-4244-4722-0/09 $25.00
© 2009 IMAPS-ITALY

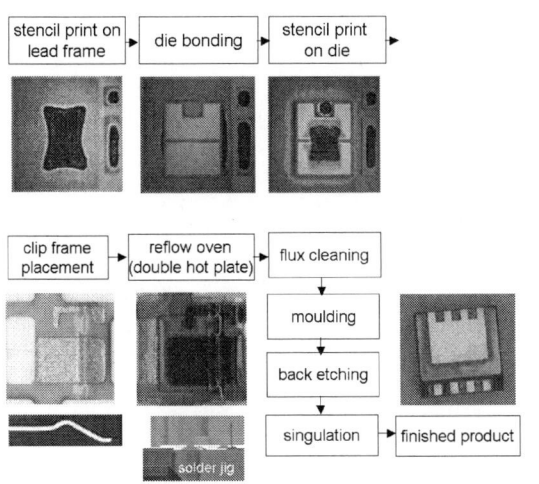

not optimized   optimized

Figure 6: Process Flow MCA Assembly

Process optimization

The solder process is the most critical part in the process flow, because at this point the position of die and clip are determined. Several types of solder paste and flux systems were tested in combination with various solder reflow profiles. As a solder paste, a high lead Pb95Sn5 solder paste was used. High lead solder paste is nowadays the only reliable die attach material for power applications and is therefore exempted from the RoHS list. A special formulation was developed, in order to limit the wetting activity of the solder paste, thereby giving a better control of the positioning of the die. Using a matrix clip frame with all the clips held together by the carrier foil, kept control of the position of the clips, and only the die could move in x –and y direction. By making use of symmetrical wettable areas above and below the die, and applying an optimized temperature reflow profile, the die could be kept at the right position. A solder paste printing - and reflow process was determined giving an optimal and reproducable result with respect to die position, bond line thickness and void rate.

Figure 7 shows the optimization result. The corresponding X-ray photos are also shown. The upper left photo shows solder paste spreading out over the diepad, further outside the area where the die is positioned. This leads to some die shift and die rotation. The die rotation can lead to a short when the edge of the die touches the clip. The photo to the right shows the result after optimization of the solder paste formulation and reflow process. In the optimized situation, the solder paste remains at the die position and dies and clips stay at the right position. The X-ray photos show that void rate could be significantly reduced by choosing the right reflow process.

Figure 7: Optimizing Solder Reflow Process for Die Position Control and Void Rate Reduction

One other concern was the filling behaviour of the molding compound, especially underneath the narrow areas between clip and die. Various cross sections were made, but no significant voids were observed. Figure 8 shows a typical cross section.

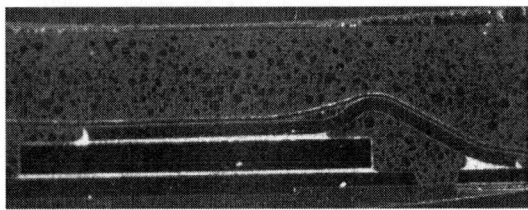

Figure 8: Filling Behaviour of Molding Compound

Verification run

A test batch of 36 leadframes, equivalent to ~ 20k products, was assembled. After singulation, all products remained on the dicing tape for electrical testing with a wafer probe. The electrical yield was > 98%. Some failures were die related or caused by the wafer probing method. Observed assembly failures were due to handling of the leadframe during the soldering process. The reflow oven was operated by moving the solder jig manually from a hot plate back onto a cold plate. Improper handling of the leadframe led to solder paste extending over the edge of the diepad, occasionally causing a short between drain and source lead. In a fully automated production line these failures can be limited. It emphasizes the importance of the solder reflow process. The amount of solder paste deposit was verified to be consistent within 2% variation.

Rdson

Electrical resistance was measured for all MCA packages made in the verification run. In parallel, the MCA package was simulated using the ANSOFT simulation package. The measured Rdson was in agreement with the electrical simulation.

A direct comparison could be made with thick Au wirebonded (6 x 50 μm) and Al ribbon bonded (1.0 x 0.1 mm) samples, based on the same die type and leadframe. The distributions of Rdson values for the three different top-interconnect technologies are shown in figure 9.

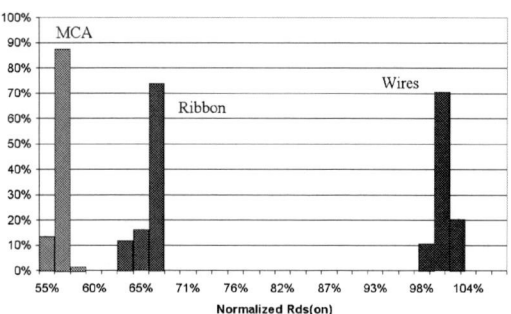

Figure 9: Normalized Rdson Distribution for Clip –Ribbon –and Wire Bonded Top Interconnects

The clip bonded package gives the lowest Rdson, followed by the ribbon bonded product. The clip bonded package has the advantage of making use of almost the complete source area, thereby reducing the surface resistance to a minimum. It is clear that thick Au wires cannot compete because of the limited number of Au wires that can be applied.

Life testing

The MCA samples were subjected to life testing. A sample size of 30 products was pre-conditioned according to MSL-1. The most critical tests for this application are temperature cycling test (TMCL) and Autoclave (PPOT). TMCL, at −55/ +150 °C, showed 0/30 fails and the Rdson shift was < 2% after 1000 cycles. Autoclave (PPOT) gave 0/25 fails at 96 hrs.
So, no failures were found, giving a sound foundation to the MCA package concept.

Alternative designs

The MCA concept can also be applied to QFN based leadless packages. Instead of a carrier foil, clips are mutually connected by tiebars. Figure 11 shows an example.

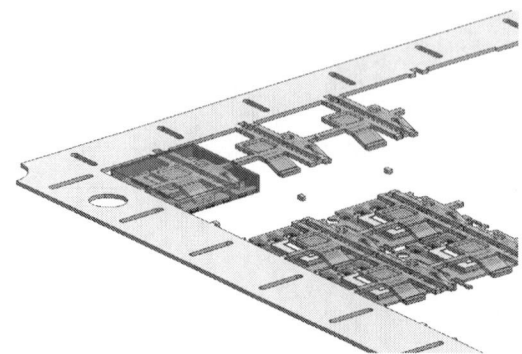

Figure 11: Alternative Clip Frame solution

Conclusions

Matrix Clip Assembly is a reliable and feasible packaging technology for clip bonding in leadless power applications. The clip frame has maximum design flexibility making this technology suitable for other clip bonded leadless packages.

The MCA technology enables good position control of dies and clips during the solder reflow process and further improvement is achieved by designing the right solder paste formulation combined with an optimization solder reflow process.

Acknowledgements

The authors would like to thank Orthodyne Electronics Corporation for their contribution in making ribbon bonded samples for our evaluation. Thanks also to Umicore for their cooperation in formulating a solder paste optimized for our application. InnoteQ Technical Projects BV is thanked for their development of the cut –and bending tool.

978-1-4244-4722-0/09 $25.00
© 2009 IMAPS-ITALY

Electrostatic wafer handling for thin wafer processing

C. Landesberger, R. Wieland, A. Klumpp, P. Ramm, A. Drost, U. Schaber,
D. Bonfert, K. Bock

Fraunhofer-Institute for Reliability and Microintegration IZM, Hansastrasse 27d, 80686 Munich, Germany

e-mail: christof.landesberger@izm-m.fraunhofer.de, phone: +49 89 54759 295, fax -100

Abstract

Mobile electrostatic carriers, so called e-carriers, offer a new and promising technical solution for simple and reversible attachment of thin wafers onto support substrates. The paper reports on latest developments in manufacture of e-carriers based on silicon wafers with through substrate vias and backside contact pads. Thermal and electrical characterizations of e-carries have proven very low leakage currents at temperatures up to 300 °C. These new types of e-carriers enable long term electrostatic holding capabilities. It is also shown that electrical properties of transistor devices are not changed when CMOS wafers are attached onto e-carriers, neither in face up nor face down attachment configuration. Furthermore, the paper proposes a technical concept how electrostatic support technique can be used for thin wafer processing in wet-chemical environments. E-carriers can also be applied for reversibly bonding of single chip devices onto a carrier substrate. The paper explains how this feature can be used in chip to wafer stacking for 3d integrated systems.

Key words: thin wafer technology, carrier substrates, e-carriers, ultra-thin wafer processing

Introduction: Thin wafer technology

Manufacture technologies for thin wafers of a thickness of 50 to 150 μm have become a basic need for a wide variety of new microelectronic products, like for instance power devices, stacked die applications, opto-electronic components, 3d integrated systems and wafer stacks with embedded micro-mechanical structures.

An outstanding task in the field of thin wafer technology relates to manufacture concepts which enable additional process steps at the backside of an ultra-thin wafer. Appropriate technical solutions are given by wafer support systems like for instance temporary bonding of a rigid carrier substrate onto the front side of a device wafer. Known techniques use polymeric bonding materials like wax, solvable glues or thermally releasable adhesives [1]. Other concepts are based on bonding materials that can be released after UV laser irradiation through a transparent glass carrier [2]. However, application of reworkable polymers is generally limited to the temperature range below 250 °C. Further increased temperature stability would be required to allow process steps like sintering of backside metal or plasma etching of dielectric layers

In order to circumvent the low temperature stability of polymeric adhesives an alternative carrier technique is currently being developed which uses electrostatic forces for reversible attachment of fragile wafers to a rigid support substrate: so-called mobile electrostatic carriers [3]. It is based on carrier substrates which are fully compatible to standard wafer handling equipment and it allows for very short cycle times for attach and de-attach of a thin device substrate. It was already shown that electrostatic carriers (e-carriers) can be used for thin wafer processing during lithography and plasma etching and also for bumping and solder reflow processes performed at wafers of a thickness of some 50 μm [4].

Technical concept of mobile electrostatic carriers

The basic concept for handling thin wafers by means of electrostatic carrier plates (e-carrier) is shown in fig. 1.

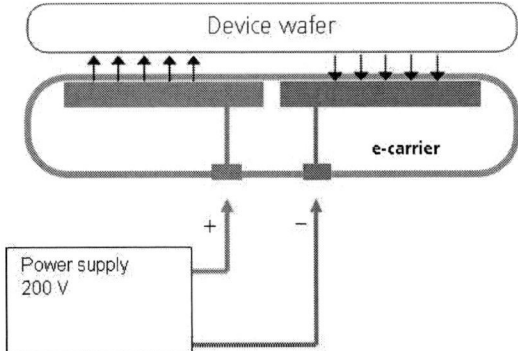

Figure 1: Working principle of mobile electrostatic wafer carrier, so-called e-carrier.

At the front side of the e-carrier identical pairs of large electrode areas are formed. On top of the electrodes a multi layer of dielectric material is deposited which allows for both storage of electrical charges as well as electrical insulation at the surface of the e-carrier. A thin semiconductor wafer can be placed on top of the e-carrier substrate and then the electrodes are charged by an external power supply. The resulting electrostatic fields provoke a separation of charge carriers (electrons and holes) at the backside of the semiconductor wafer and thereby cause an attractive force between e-carrier and thin wafer. After initial charging the external power supply is disconnected. The electrostatic forces remain active for a longer period of time.

The configuration of the stacked wafer pair is similar to a plate capacitor. The attractive force between wafer and carrier can be calculated to:

$$F = \varepsilon \, A \, U^2 / 8 \cdot d^2$$

The holding force F depends on the dielectric constant $\varepsilon = \varepsilon_r \cdot \varepsilon_0$ of the dielectric layers, the electrode area A (approximately half the wafer surface), the distance d between electrodes and wafer and the applied voltage U.

In principle the base carrier plate can be made of different materials and also by different manufacturing technologies like for instance thin film technology on silicon or glass wafers or thick film technology on ceramic plates.

Manufacture of e-carriers with backside contact pads

Choosing silicon as base material for the carrier plate offers several advantages: high thermal conductivity, same coefficient of thermal expansion when thin silicon wafers are to be processed, full compatibility with common fabrication technology and availability of a large variety of high quality thin film layers. Last point is of strong relevance with respect to the functional performance of e-carriers for two reasons: First, long duration times of electrostatic attraction require perfect electrical insulation between the electrodes and the thin wafer and also between the electrodes and the carrier substrate. Secondly, high attractive forces can be achieved when the insulating layers are very thin. Thin film technology on silicon wafer substrates with thermally grown oxide layers and plasma or CVD deposited dielectric layers fulfills these two requirements.

Most applications of e-carriers require electrical contact pads at the backside of the carrier plate. Recent research work was carried in order to realize through substrate vias (TSV) in the silicon base plate. Formation of holes was done by laser drilling and subsequent etching steps which are necessary to ensure a smooth side wall of the vias. Most important task concerns the electrical

insulation of via holes. During operation of e-carrier a voltage of some 200 V will be applied and any microscopic defect in the insulation layer will result in an electrical breakdown of the e-carrier.

Fig. 2 shows a first demonstrator of an e-carrier with TSV and corresponding electrical contact pads at the backside of the silicon substrate. The visible rewiring at the backside of e-carrier is completely insulated except for two contact pads which are supposed to be connected to an external power supply during initial charging of e-carrier. Location of contact pads can be chosen according to customer requirements and the contact geometry of a charging station.

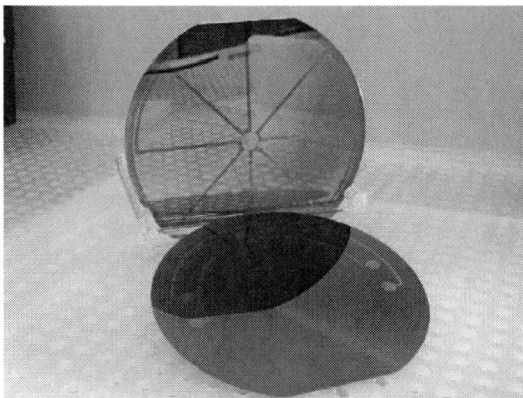

Figure 2: Electrostatic carriers with electrodes at the front side and rewiring and contact pads at their backside.

Figure 3: Electrostatic carrier with electrode areas on both sides of the carrier substrate and with backside contact pads.

A further possible configuration can be achieved when electrode areas are realized at the backside of e-carriers as well. Thereby, charging capacity of the e-carrier substrate is doubled and this helps to ensure long duration times of electrostatic attraction later on. Contact pads at the backside are defined by a lithographic patterning and etching process which opens the insulation layers at desired positions. The photograph in fig. 3 shows such e-carrier with electrode areas on both sides. Backside pattern is visible in the reflection of an underlying mirror substrate.

Electrical and thermal characterization of e-carriers with backside contact pads

Most important application examples for e-carriers relate to high temperature processes which need to be performed at the backside of a very thin device wafer. In such cases any alternative reversible bonding technique based on glue layers would suffer from unstable polymeric adhesives.

In order to determine the time duration of electrostatic fields and their dependence on temperature a thin silicon wafer was attached onto an e-carrier and the wafer stack was placed onto the heated chuck of a wafer probing equipment. After charging of e-carriers at a voltage of 100 V the external power supply was disconnected and the wafer stack was heated to the final test temperature of 150 °C and 200 °C. The remaining voltage at the backside contact pads was measured by an electrometer (Keithley 617) after certain periods of time. Results are shown in fig. 4. After 1 hour at room temperature the voltage is reduced by less than 1 %, after 1 hour at 200 °C the voltage is reduced by 6,5 %. Such low decays in voltage practically do not influence the attractive force between e-carrier and thin wafer attached on it.

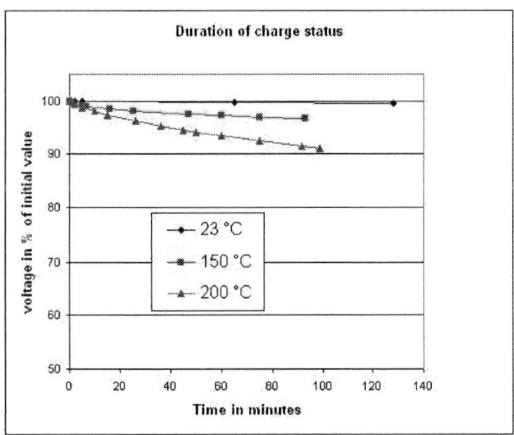

Figure 4: Time depending decay of clamping voltage of e-carrier at 23 °C, 150 °C and 200 °C.

As a second test method we measured the leakage current at 100 V DC while heating the wafer stack up to 300 °C which is the highest temperature allowed at the wafer probe equipment. Measurement unit was a precision semiconductor parameter analyzer, Agilent 4156 A. Fig. 5 shows the measured increase of leakage currents with increasing temperature.

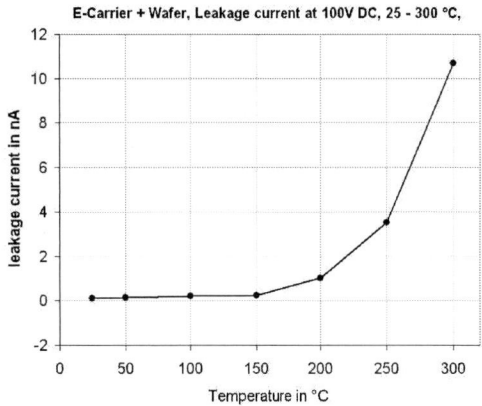

Figure 5: Temperature dependence of leakage currents of a wafer stack consisting of an e-carrier substrate and a thin wafer attached onto it.

In principle, the measurement of leakage currents at 300 °C can be used to estimate the duration time of charge storage. The electric capacity of e-carrier was measured to be 140 nF. A leakage current of 11 nA would discharge this capacity within 20 min approximately. However, the holding capability of e-carriers also depends on the polarizability of the dielectric layers. Actually, even after discharge of the electrodes strong holding forces can be detected. Possible explanations of this behavior would be the existence of a constant polarization of dielectrics or the presence of trapped charge carriers. This would result in oppositely charged surfaces which are separated by the few nanometer thin air gap between the surface of e-carrier and the thin wafer attached onto it. This effect would explain why strong attractive forces may be present although the measured voltage at the electrodes is in the range of just a few volts. So, the observed increase of leakage current at 300 °C does not necessarily prevent the holding capabilities at even much higher temperatures. Further research work will be carried out to experimentally verify the high temperature behavior of these new e-carriers.

Investigation of possible influences of electrostatic clamping on the performance of CMOS devices

A frequently asked question concerns the possible influence of electrostatic fields on the electrical properties of CMOS devices. To clarify this point we measured CMOS devices before and after chucking on mobile e-carriers. Test wafers were 200 µm thin CMOS wafers. First, we measured the threshold voltage V_t when placing the wafer active side face-up onto the e-carrier. Maximum change of V_t was in the range of the repeatability of the measurement itself (see Table 1). We repeated

978-1-4244-4722-0/09 $25.00
© 2009 IMAPS-ITALY

the measurements after placing the CMOS wafer face down onto e-carrier. In this configuration the distance between IC devices and the electrodes of e-carrier substrate is just some 5 μm. In order to simulate a long term attachment the wafer was electrostatically fixed for more than 16 hours. The resulting wafer map of percental change of threshold voltage of transistors is shown in fig. 6. No significant change in V_t was found. So it can be concluded that the applied electrostatic fields generally won't affect the characteristics of CMOS transistors. The reason for this independence can be found in the moderate strength of the electric fields provoked by the e-carrier when a charging voltage up to 200 V is selected. The strength of the resulting field is lower by a factor of 5 to 10 compared to the maximum allowed electric field strength at the gate dielectric of a standard CMOS transistor.

			0,01	0,03	-0,02				
		0,01	0,04	0,04	-0,04	-0,03	0,01	0,00	
	0,00	0,01	0,05	0,01	0,02	0,02	0,03	0,03	0,09
	0,05	0,04	0,03	0,03	0,07	0,03	0,06	0,03	0,04
0,02	-0,07	-0,10	0,00	0,01	0,02	0,08	0,06	0,07	-0,09
0,03		0,05	-0,10	0,03	-0,04	0,05	0,05		-0,02
	0,03	0,04	0,05	0,02	0,05	0,02	0,02	0,08	-0,03
	0,05	0,07	0,03	-0,01	-0,09	0,04	0,01	0,07	-0,01
		0,03	-0,10	0,05	-0,01	-0,06	0,01	0,01	
			0,05	0,05	-0,03	0,03	0,01		

Figure 6: Wafer map of percental change in threshold voltage of CMOS transistors after chucking onto a mobile e-carrier at 200 V for 16 hours.

Table 1: Change of transistor characteristics

status of CMOS wafer	% change of threshold voltage V_t
After e-chucking, active side face up, 200 V, 10 min	< 0,08 %
After e-chucking, active side face down, 200 V, 16 hours	< 0,10 %
Repeatability of measurement (no e-carrier involved)	< 0,08 %

Concept and possibilities for application of e-carriers in fluid environments

Required process steps for very thin wafers may comprise wet-chemical treatments in etching bathes or deposition of metal layers like for instance an under bump metallization. Of course, electrostatic attraction is also active in fluids. However, fluids of low electrical resistivity would discharge the electrodes if the contact pads are not protected. Furthermore, there is another physical principle that must be taken into account: the tendency of dielectric materials to move towards the regions of highest electrical field strength. In the case of e-carrier wafer stacks the strongest fields exist between thin wafer and e-carrier. Therefore, fluids will try to penetrate along the interface between wafer and carrier substrate. It is assumed that the potential wetting of the wafer surfaces might be reduced or inhibited by choosing an appropriate surface passivation for the e-carriers. Further possibility would be the application of a sealing ring along the edge of the thin wafer attached. Fig. 6 schematically shows the basic idea of such concept. For this application it would be preferable to use e-carriers of a diameter slightly larger than the thin wafer itself.

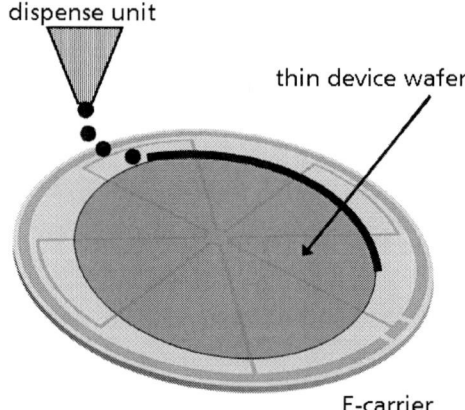

Figure 6: Edge sealing concept for applications of e-carriers in wet processing of thin wafers.

The polymeric sealing might be of a material that can be either dissolved chemically or removed after thermal treatment, laser irradiation or exposure to UV-light or plasma gas. The described combination of electrostatic carrier technique with polymeric sealing rings would allow for running a very thin wafer through a process sequence which comprises high temperature steps as well as wet-chemical treatments.

978-1-4244-4722-0/09 $25.00
© 2009 IMAPS-ITALY

Electrostatic handling of thin dies for 3d integration

A growing field of interest is the temporary handling of thin dies for 3D integration applications, especially for the so-called chip-to-wafer or chip-to-chip stacking approach. The use of an electrostatic chip carrier wafer allows for accurate die placement with conventional flip chip tools and for a quick and easy die release after the 3D process, which, for example, can be a soldering process under vacuum at elevated temperatures.

In principle electrostatic clamping of single dies works the same way like the attachment of full wafers. Sole restriction is that each die need to feel the electric fields from both polarities. Therefore, electrode design of an e-carrier for single die attach looks like a regular matrix of electrode pairs over the entire substrate surface. It was already shown that also strongly bent chips of a thickness of some 25 μm can be held electrostatically [5]. One additional application example for such wafer level chip configuration would be electrical testing of single thin dies.

For accurate thin chip placement, optically detectable adjustment marks should be created during the manufacturing of such an electrostatic carrier wafer. Fig. 7 shows several thinned dies held by electrostatic force on a mobile e-carrier wafer.

Figure 7: Single dies electrostatically clamped onto a mobile e-carrier substrate of a diameter of 200 mm.

Conclusions

Electrostatic attraction of thin wafers or dies on support substrates enables a simple and reliable handling technique for fragile semiconductor devices. Outstanding advantages of the e-carrier concept are the lack of any polymeric bonding material and the applicability for processing of very thin wafers at temperatures even above 300 °C. In this paper it was shown for the first time that mobile e-carriers based on silicon wafers with through substrate vias, backside contact pads, good electrical insulation and long duration times of the electrostatic status even at elevated temperatures can be realized.

It should also be noted that the functionality of e-carriers depends on both the electric properties of the carrier substrate as well as the mechanical behavior of the thin device wafers that should be attached onto it. Therefore, joint development activities are proposed where semiconductor manufacturers, equipment companies and research groups may work together in order to prove the usability of e-carriers in fabrication processes for very thin device wafers.

Acknowledgements

Part of this research work was funded by Bayerische Forschungsstiftung under contract number AZ-691-06.

References

[1] A. Smith, R. Puligadda, W. Hong, T. Matthias, C. Brubaker, M. Wimplinger, S. Pargfrieder, "High Temperature-Resistant Spin-On Adhesive for Temporary Wafer mounting Using an Automated High-Throughput Tooling Solution", CS Mantech Conference, May 14-17, pp 29-32, 2007, Austin, TX, USA.

[2] C. Kessel, K. Saito, F. Weimar, "Wafer Thinning by using 3M Wafer Support System"; Workshop "Thin Semiconductor Devices", November 2006, Munich, Germany; (www.be-flexible.de).

[3] US patent US 7,027,283 B2; European Patent: EP 1 305 821 B1.

[4] C. Landesberger, S. Scherbaum, K. Bock, "Carrier techniques for thin wafer processing", Conference CS Mantech, Austin, TX, 2007.

[5] R. Wieland, E. Hacker, C. Landesberger, P. Ramm, K. Bock, "Thin Substrate Handling by Electrostatic Force"; Conference Smart Systems Integration, Barcelona, 2008.

AUTHOR INDEX

Aasmundtveit, K. 92
Aasmundtveit, K. E. 723
Ahr, A. 367
Aikio, J. 1
Aikio, M. 217
Alajoki, T. 1
Alderman, J. 447
Ali, W. 559
Ansorge, F. 497, 502
Argento, C. 646
Arnold, J. 782
Astier, A. 676
Auchere, D. 78
Avenas, Y. 425
Axisa, F. 697
Azzopardi, C. 167
Azzopardi, M. 58
Bailini, A. 180
Baldo, L. 58
Balogh, B. 729
Balut, C. E. 367
Baraton, X. 321
Barras, A. 338
Barthelmes, J. 290
Batut, N. 87
Bauer, J. 688
Baumgartner, T. 7
Bechtold, F. 798
Becker, K. 688
Beelen-Hendrikx, C. 12
Begbie, M. 232
Belavic, D. 266, 735
Belharet, D. 20
Belmonte, M. 35
Bembnowicz, P. 27
Bennemann, S. 682
Berg, R. V. D. 767
Bhattacharjee, A. 371
Bhatti, N. S. 31
Bjorklund, N. 391
Bock, K. 403
Boehme, B. 713
Boettcher, L. 451
Bonazzoli, M. 35
Bonfert, D. 270, 403
Bonino, S. 35
Bonnot, L. 676
Bos, A. 221
Bosman, E. 275
Bossuyt, F. 697
Bougataya, M. 383
Brannen, C. 69
Brun, J. 40
Brunet, P. 199, 248
Brunet-Manquat, C. 676

Brunke, O. 163
Brunner, S. 159
Burkard, H. 776
Bursik, M. 513
Byun, K. Y. 413
Caccioli, D. 128
Campaniello, M. 180
Campos, D. 203
Canegallo, R. 487
Cardu, R. 487
Caswell, G. 591
Cavallaro, A. 191
Chaehoi, A. 232
Chan, W. L. 581
Chang, L. 555
Chang, Y. 555
Charbonnier, J. 676
Chau, H. 563
Chausse, P. 676
Che, F. X. 315
Chen, W. 555
Cheramy, S. 676
Chiang, K. 280, 528
Chin, K. 555
Chin, L. W. 321
Chou, C. 280, 528
Christiaens, W. 671
Chung, C. 354, 703
Chung, Q. H. 413
Chvatal, M. 763
Cobussen, H. 151
Codreanu, N. D. 270
Colin, D. 20
Collander, P. 286
Connors, M. 569
Conway, P. P. 548
Copeland, D. 569
Corradi, U. 587
Couderc, P. 628
Coyle, R. 787
Crema, P. 290
Cristaldi, G. 45
Curran, B. 624, 688
Debono, J. 78
Dekker, J. 7
Delaney, K. 109
Demosthenous, A. 447
Dietzel, A. 534
Dijkstra, P. 226
Ding, J. P. 581
Dohle, R. 98
Donaldson, N. 447
Doriol, P. J. 52
Dreiza, M. 203
Dresbach, C. 299, 307

AUTHOR INDEX

Drost, A. 403
Dubreuil, P. 20
Dunn, G. 243
Dunne, T. 98
Dutron, A. 361
Eidner, I. 688
Elfving, A. 723
Endrinal, L. 151
Farcy, A. 651
Fenner, M. 327
Fiori, F. 437
Fischer, T. 688
Fledderus, H. 534
Flossel, M. 333
Fontana, F. 58
Fornes, T. D. 74
Forzan, C. 52
Fournier, Y. 122, 338, 666
Fowkes, C. R. 252
Frassati, F. 40
Frisk, L. 466, 661
Fu, S. L 703
Gabriel, M. 211
Gacusan, R. 139
Gagnard, X. 676
Galan, J. V. 575
Galeotti, R. 35
Garrou, P. 345
Gatt, S. 58
Gebhardt, S. 333
Gee, H. 741
Gerrinck, P. 275
Gindy, N. 252
Giry, J. P. 191
Gobbi, L. 35
Golonka, L. 27
Gonthier, L. 87
Gordon, P. 729
Goßler, J. 98
Graff, J. M. 723
Granier, H. 20
Graziosi, G. 52, 78, 139
Griol, A. 575
Gromala, P. 474
Grosse, C. 682
Grubl, W. 491
Gualandris, D. 65
Guedon, S. 425
Guillou, Y. 361
Guttowski, S. 624, 688
Gyenge, O. 617
Hakansson, A. 575
Halonen, E. 391
Hamoui, A. 383
Han, C. 613, 637

Hansen, U. 617
Hao, J. Y. 581
Harjunpaa, H. 103
Hast, J. 217
Hauck, T. 646
Healey, R. 787
Hegde, P. 171
Heikkinen, M. 1
Hein, M. 601
Heino, P. 103
Hejatkova, E. 513
Helfenstein, M. 457
Hennemeyer, M. 211
Henry, D. 676
Henshall, G. 787
Hertl, M. 378
Heyes, D. 399
Hiitola-Keinanen, J. 217
Hildebrandt, S. 692
Hirte, M. 688
Ho, L. N. 443
Ho, S. C. 581
Hoffmann, C. 159
Hoivik, N. 723
Holc, J. 735
Holland, P. 741
Hong, W. 613, 637
Hough, P. 69, 74
Howell, K. 787
Hraiz, W. 122
Hrovat, M. 735
Hsu, H. C. 703
Hu, G. 321
Hua, F. 787
Huang, C. J. 280
Huang, H. M. 581
Huffman, A. 345, 367
Humbla, S. 601
Hung, T. 528
Hurtado, J. 575
Hurtony, T. 729
Hutt, D. A. 548
Huwel, W. 671
Hyun, C. 294
Imbs, Y. 78
Imran, M. 31
Indelli, G. F. 191
Innocenti, M. 487
Ionescu, C. 266, 270
Ishikawa, T. 237
Izquierdo, R. 383
Jacq, C. 116, 122, 666
Jacques, S. 87
Jang, J. 259
Jang, K. 259

AUTHOR INDEX

Jansen, K. M. B. 713
Jarvinen, P. 203
Jaud, M. 651
Jerlah, M. 735
Jesudoss, P. 759
Jiang, Y. J. 581
Jin, H. 217
Johannessen, R. 92
Juntunen, E. 1
Kaija, K. 391
Kaiser, S. 425
Kang, S. 569
Kapischke, W. 776
Kapitanova, P. 601
Karaszkiewicz, S. 451
Kashiwagi, Y. 443
Kattelus, H. 7
Kellomaki, M. 103
Kemethmuller, S. 98
Kemppainen, A. 391
Kengen, M. 399
Kholodnyak, D. 601
Kiilunen, J. 661
Kim, J. 203, 294, 613
Kim, S. C. 413
Kim, Y. 413
Kimura, K. 518
Kittel, H. 497
Klein, M. 7
Klengel, R. 682
Klug, G. 814
Klumpp, A. 403
Knauf, B. J. 548
Knodler, D. 7
Kocjan, S. 735
Kokko, K. 103
Kopola, P. 217
Koponen, M. 1
Kosonen, T. 1
Kramer, N. 226
Kristiansen, H. 92
Ku, C. W. 703
Kuah, E. 581
Kuhlkamp, P. 290
Kuisma, H. 7
Kurtz, O. 290
Kusters, R. 534
Kuusiluoma, S. 466
Lakhssassi, A. 383
Landesberger, C. 403
Lang, F. 408
Lartigues, P. 628
Lecomte, J. 378
Leduc, P. 651
Lee, C. 509

Lee, J. 294, 413
Lee, S. 259, 354, 509, 563
Leib, J. 617
Lepine, B. 40
Leroy, R. 87
Li, Q. F. 581
Liebsch, W. 367
Lim, L. A. 607
Link, J. 776
Lishchynska, M. 109
Liu, C. 548, 555
Lorenz, G. 299
Luan, J. 315, 321
Luebbehuesen, J. 163
Macek, S. 735
Mach, M. 159, 418
Mackie, A. 327
Maeder, T. 116, 122, 338, 666
Maggi, L. 128, 133
Magni, P. 139
Majzner, J. 763
Makinen, J. T. 1
Malgioglio, G. 45
Manessis, D. 145, 451
Manivannan 371
Mantysalo, M. 391
Marechal, L. 78
Marechal, Y. 425
Marghescu, C. 266
Marsala, J. 569
Marti, J. 575
Martins, O. 425
Marty, A. 651
Masto, A. 569
Mathewson, A. 759
Matsutani, H. 810
Maus, S. 617
Mavinkurve, A. 151
Mazenq, L. 20
McCaffrey, C. 759
Mehta, K. 371
Memis, I. 746
Merlin, M. 437
Metasch, R. 706
Michaelis, A. 333
Milke, E. 299, 307
Min, C. 559
Mittag, M. 299, 307
Mourey, B. 40
Mozek, M. 735
Mueller, W. 587
Muller, J. 159, 418, 601
Muller, T. 299, 307
Muller, W. H. 646
Nakamoto, M. 443

AUTHOR INDEX

Ndip, I. 624, 688
Neubert, B. 211
Neubrand, T. 163
Neyret, M. 676
Nguyen, H. 92
Niehoff, K. 497
Nishikawa, H. 443
Nitin, G. 371
Nonomura, M. 243
Noren, M. 159
Nowakowska, D. 27
Nurmi, S. 7
O'Connell, D. 232
Ogurtsov, V. 759
Oh, C. 613, 637
Ohashi, H. 408
Oldervoll, F. 92
Ollila, J. 1
O'Malley, G. 782, 787
Ong, S. 767
Osterbacka, R. 391
Ostmann, A. 145, 451
Owzar, A. 457
Paik, K. 259, 354
Pandher, R. S. 787
Pandini, D. 52
Park, J. 259, 461
Park, N. 613, 637
Parviainen, A. 466
Passagrilli, C. 167
Patzelt, R. 145
Peels, W. 399
Peltier, N. 425
Perala, J. 466
Perrone, R. 601
Perugini, L. 487
Petaja, J. 1
Petersen, W. 457
Petzold, M. 299, 307, 682
Pfahl, B. 782
Phung, A. 367
Piascik, J. 345
Pieters, P. 656
Pignataro, S. 191
Podprocky, T. 534
Pohlner, J. 98
Pot, A. 767
Preve, G. B. 575
Radji, M. 383
Ramanan 371
Ramkumar, M. 607
Ramm, P. 403
Ray, S. 232
Rebholz, C. 502
Reichelt, J. 474

Reichl, H. 145, 451, 624, 688
Reznicek, M. 482
Reznicek, Z. 482
Reznicek Jr., Z. 482
Roehm, H. 767
Roellig, M. 706
Rosser, S. G. 746
Roth, H. 163
Rotigni, M. 52
Roubion, J. 87
Rouelle, G. B. 338
Rousseau, M. 651
Rubingh, E. 534
Russo, S. 191
Ruythooren, W. 597
Ryser, P. 116, 122, 338, 666
Rzepka, S. 474, 713
Sack, T. 787
Saeed, U. 587
Saeidi, N. 447
Sala, S. A. 180
Sanchis, P: 575
Sandera, J. 513
Sano, M. 280
Santospirito, G. 187
Sarma, G. H. 371
Saugier, E. 203
Scandiuzzo, M. 487
Scandurra, A. 191
Schaber, U. 403
Schachler, R. 7
Scheel, W. 688
Scheithauer, U. 333
Schindler-Saefkow, F. 497
Schischka, J. 682
Schmadlak, I. 646
Schmid, B. 7
Schmitz, S. 491
Schneider-Ramelow, M. 491
Schonecker, A. 333
Schreier-Alt, T. 497, 502
Schuch, B. 491
Scrofani, E. 45, 191
Sedlakova, V. 754, 763
Serra, N. 666
Shanmugam, T. 767
Sharma, U. 741
Sheach, K. 199, 248
Shigetou, A. 641
Shin, Y. 509
Sidiki, T. P. 767
Sikula, J. 754
Silberschmidt, V. V. 171
Sillon, N. 676
Sim, G. 354

AUTHOR INDEX

Sitomaniemi, A. 1
Smetana, W. 543
Smith, L. 203
Snugovsky, P. 787
Song, B. 613, 637
Stam, F. 759
Stapleton, R. 69
Starkey, D. 563
Stary, J. 513
Stephan, R. 457
Suga, T. 641
Suh, M. S. 413
Sun, Y. 252
Svasta, P. 266, 270
Svetly, A. 587
Sweatman, K. 787
Szendiuch, I. 482, 513, 543
Takemoto, T. 443
Tanimoto, S. 408
Terzoli, A. 187
Ticozzi, G. 133
Tisdale, S. 787
Tiziani, R. 139, 167
Tofel, P. 754, 763
Tonnies, D. 211
Topper, M. 7, 617, 688
Torfs, T. 671
Tsai, C. 555
Tuomikoski, M. 217
Twomey, K. 759
Tyldum, H. 92
Uno, T. 518
Upadhya, G. 569
Val, C. 628
Valimaki, M. 217
Van Daele, P. 275
Van Den Brand, J. 534
Van Der Burgt, F. 767
Van Dort, M. 151
Van Driel, W. D. 151
Van Hoof, C. 671
Van Steenberge, G. 275
Van Weelden, T. 221
Vanek, J. 543
Vanfleteren, J. 275, 671, 697
Vath III, C. J. 607
Vendik, I. 601
Verrun, S. 676
Verspeek, J. 226
Vervust, T. 697
Vicard, D. 40
Villa, C. 65, 139
Villain, J. 587
Villavicencio, Y. 52
Vitali, B. 167

Vittu, J. 676
Vogel, M. 569
Von Hofen, H. 746
Walczyk, S. 226
Wang, C. 232
Wang, K. 723
Wang, L. 221
Webb, D. P. 548
Weidmann, D. 378
Weiland, D. 232
Weilguni, M. 543
Weippert, C. 587
Whalley, D. C. 171
Whitney, B. 569
Wieland, R. 403
Wilson, G. 327
Windemuth, R. 237, 243
Wolter, K. 692, 713
Wolter, K. J. 706
Wright, W. 759
Xiang, G. 199, 248
Yamada, T. 518
Yamaguchi, H. 408
Yamamoto, M. 443
Yew, M. 528
Ylikunnari, M. 217
Yong, G. K. 321
Yoo, S. 509
Yun, H. 367, 563
Zafarana, R. 191
Zarnik, M. S. 266, 735
Zhang, J. 563
Zhang, M. 563
Zhang, W. 597
Zoba, D. 69
Zoberbier, M. 211
Zoberbier, R. 211

2009 European Microelectronics and Packaging Conference

(EMPC 2009)

Rimini, Italy
16 – 18 June 2009

Pages 408-817

IEEE Catalog Number: CFP0954H-PRT
ISBN: 978-1-4244-4722-0

Copyright © 2009, International Microelectronics and Packaging Society-ITALY
All Rights Reserved

***This publication is a representation of what appears in the IEEE Digital Libraries. Some format issues inherent in the e-media version may also appear in this print version.*

IEEE Catalog Number: CFP0954H-PRT
ISBN 13:   978-1-4244-4722-0

Additional Copies of This Publication Are Available From:

Curran Associates, Inc
57 Morehouse Lane
Red Hook, NY 12571 USA
Phone: (845) 758-0400
Fax: (845) 758-2633
E-mail: curran@proceedings.com

TABLE OF CONTENTS

In-mould Integration of Electronics into Mechanics and Reliability of Overmoulded Electronic and Optoelectronic Components...1
T. Alajoki, M. Koponen, E. Juntunen, J. Petaja, M. Heikkinen, J. Ollila, A. Sitomaniemi, T. Kosonen, J. Aikio, J.T. Makinen

A 3-D Packaging Concept for Cost Effective Packaging of MEMS and ASIC on Wafer Level...............7
T. Baumgartner, M. Topper, M. Klein, B. Schmid, D. Knodler, H. Kuisma, S. Nurmi, H. Kattelus, J. Dekker, R. Schachler

Trends in IC Packaging...12
C. Beelen-Hendrikx

Temporary Adhesives for Wafer Bonding: Deep Reactive Ion Etching Application.............................20
D. Belharet, P. Dubreuil, D. Colin, L. Mazenq, H. Granier

Integrated LTCC-Glass Microreactor and µTAS with Thermal Stabilization for Biological Application...27
P. Bembnowicz, D. Nowakowska, L. Golonka

Fabrication & Characterization of S-Band Power Amplifier using GaAs Die......................................31
N.S. Bhatti, M. Imran

High Speed Packaging Solutions for LiNbO₃ Electro-Optical Modulator...35
S. Bonino, R. Galeotti, L. Gobbi, M. Belmonte, M. Bonazzoli

Packaging and Wired Interconnections for Insertion of Miniaturized Chips in Smart Fabrics..........40
J. Brun, D. Vicard, B. Mourey, B. Lepine, F. Frassati

Aluminum Ribbon on a Power Device...45
G. Cristaldi, G. Malgioglio, E. Scrofani

Advanced Modeling Techniques for System-level Power Integrity and EMC Analysis........................52
G. Graziosi, P.J. Doriol, Y. Villavicencio, C. Forzan, M. Rotigni, D. Pandini

MEMS Pressure Sensors – New LGA Packagings..58
F. Fontana, L. Baldo, M. Azzopardi, S. Gatt

Wafer Level Packaging Fan Out Thermal Management: Is Smaller Always Hotter?...........................65
D. Gualandris, C.M. Villa

Multifunctional Coatings for Wafer-Level Chip Scale Packaging...69
R. Stapleton, D. Zoba, C. Brannen, P. Hough

Highly Conductive Adhesives via Novel Heterogeneous Structures..74
T.D. Fornes, P.W. Hough

Application of 3D Modeling Tools for Advanced Packaging on a Broad Range of Industrial Applications..78
Y. Imbs, L. Marechal, D. Auchere, G. Graziosi, J. Debono

Experimental Characterization of Thermo-Mechanical Properties of Lead-based Solders for Power Electronics Packaging Reliability Applications...87
S. Jacques, J. Roubion, N. Batut, R. Leroy, L. Gonthier

Investigation of Compliant Interconnect for Ball Grid Array (BGA)..92
R. Johannessen, F. Oldervoll, H. Kristiansen, H. Tyldum, H. Nguyen, K. Aasmundtveit

Reliability of 100 µm Bi- and In- Solder Balls...98
S. Kemethmuller, R. Dohle, J. Pohlner, T. Dunne, J. Goßler

Hydrolysis Testing of ACF Joined Flip Chip Components with Conformal Coating...........................103
K. Kokko, H. Harjunpaa, P. Heino, M. Kellomaki

Package Design for Alleviating Stress in Materials Embedded with Electronic Systems....................109
M. Lishchynska, K. Delaney

Development of Low-firing Lead-free Thick-film Materials on Steel Alloys for Piezoresistive Sensor Applications..116
C. Jacq, T. Maeder, P. Ryser

Structuration of Zero-shrinkage LTCC Using Mineral Sacrificial Materials.....................................122
T. Maeder, C. Jacq, Y. Fournier, W. Hraiz, P. Ryser

Process Development for a Very Precise Placement of a Lens for Micro-Optics Based Components......128
D. Caccioli, L. Maggi

Electromagnetic Simulations for the Packaging Design of Telecommunication Component................133
L. Maggi, G. Ticozzi

WPLGA: New Package Family for Medium Pin Count with Design Flexibility....................................139
P. Magni, G. Graziosi, C. Villa, R. Tiziani, R. Gacusan

Advancements in Bumping Technologies for Flip Chip and WLCSP Packaging 145
D. Manessis, R. Patzelt, A. Ostmann, H. Reichl

Assembly - Chip Interactions Leading to PPM-level Failures in Microelectronic Packages 151
A. Mavinkurve, H. Cobussen, W.D. Van Driel, L. Endrinal M. Van Dort

Small Size LTCC FlipChip-Package for RF-Power Applications 159
J. Muller, M. Noren, M. Mach, S. Brunner, C. Hoffmann

X-ray nanoCT of Interconnections in IC Packages: Visualizing of Internal 3D-Structures with Submicrometer Resolution 163
J. Luebbehuesen, H. Roth, T. Neubrand, O. Brunke

Cu Wire Bonding: Reliability Improvement for High Temperature in Plastic Packages 167
C. Passagrilli, B. Vitali, R. Tiziani, C. Azzopardi

Size and Microstructure Effects on the Stress-Strain Behaviour of Lead-Free Solder Joints 171
P. Hegde, D.C. Whalley, V.V. Silberschmidt

Experimental Study of Polymers as Encapsulating Materials for Photovoltaic Modules 180
S.A. Sala, M. Campaniello, A. Bailini

Fine Die-Attach Delamination Analysis by Scanning Acoustic Microscope 187
G. Santospirito, A. Terzoli

Materials Science Challenges in Green High Power Density Devices 1. The Ag Loaded Glues for Die Bonding 191
A. Scandurra, G.F. Indelli, R. Zafarana, A. Cavallaro, E. Scrofani, S. Russo, J.P. Giry, S. Pignataro

Modeling of Flip Chip Bump Patterns to Minimize Crosstalk on a BU-BGA Package Design 199
K. Sheach, G. Xiang, P. Brunet

Joint Project for Mechanical Qualification of Next Generation High Density Package-on-Package (PoP) with Through Mold Via Technology 203
M. Dreiza, J.S. Kim, L. Smith, D. Campos, E. Saugier, P. Jarvinen

Introduction of a Unified Equipment Platform for UV Initiated Processes in Conjunction with the Application of Electrostatic Carriers as Thin Wafer Handling Solution 211
D. Tonnies, M. Gabriel, B. Neubert, M. Hennemeyer, M. Zoberbier, R. Zoberbier

Roll-to-Roll Manufacturing of Organic Photovoltaic Modules 217
M. Tuomikoski, P. Kopola, H. Jin, M. Ylikunnari, J. Hiitola-Keinanen, M. Valimaki, M. Aikio, J. Hast

Encapsulation of the Next Generation Advanced Mems& Sensor Microsystems 221
A. Bos, L. Wang, T. Van Weelden

Next Generation Leadless RF Packages Utilizing 1st Level Low Cost Flip Chip Interconnect Technology 226
S. Walczyk, P. Dijkstra, N. Kramer, J. Verspeek

BCB-Based Wafer-Level Packaging of Integrated CMOS/SOI Piezoresistive MEMS Sensors 232
D. Weiland, A. Chaehoi, S. Ray, D. O'Connell, M. Begbie, C. Wang

New Flipchip Technology 237
R. Windemuth, T. Ishikawa

New Plasma Cleaning Technology 243
R. Windemuth, M. Nonomura, G. Dunn

A Study on High-density High-speed SerDes Design in Buildup Flip Chip Ball Grid Array Packages 248
G. Xiang, K. Sheach, P. Brunet

A Novel Methodology for Analyzing Variation Risk Introduced by the Manufacturing Process in Microsystems 252
Y. Sun, C.R. Fowkes, N. Gindy

Fracture Toughness Assessment of ACF Flip-chip Packages under High Moisture Condition with Moire Interferometry 259
J. Park, J. Jang, K. Jang, K. Paik, S. Lee

The Design and Improvement of LTCC-based Capacitive Pressure Sensors Employing Finite Element Analysis 266
C. Ionescu, P. Svasta, C. Marghescu, M.S. Zarnik, D. Belavic

Investigation of Solder Joints by Thermographical Analysis 270
P. Svasta, C. Ionescu, N.D. Codreanu, D. Bonfert

Fully Embedded Optical and Electrical Interconnections in Flexible Foils 275
E. Bosman, G. Van Steenberge, P. Gerrinck, J. Vanfleteren, P. Van Daele

Metal Trace Impact Life Prediction Model for Stress-Buffer- Enhanced Package 280
C.Y. Chou, C.J. Huang, M. Sano, K.N. Chiang

3D Packaging and Supply Chain Management 286
P. Collander

The Necessity of Corrosion Protection for Solderable Pure Tin Deposits on IC Outer Leads 290
J. Barthelmes, P. Crema, P. Kuhlkamp, O. Kurtz

Pot Life Improvement of Low Temperature and High-speed Curable Anisotrpic Conductive Adhesive (ACA)..294
  J. Lee, J. Kim, C. Hyun

Local Hardening Behavior of Free Air Balls and Heat Affected Zones of Thermosonic Wire Bond Interconnections..299
  C. Dresbach, G. Lorenz, M. Mittag, M. Petzold, E. Milke, T. Muller

Mechanical Properties and Microstructure of Heavy Aluminum Bonding Wires for Power Applications..307
  C. Dresbach, M. Mittag, M. Petzold, E. Milke, T. Muller

Effect of Microstructure Design on Reliability of FBGA Lead-Free Solder Joints................315
  F.X. Che, J.E. Luan

Thermoelastic Properties of Printed Circuit Boards: Effect of Copper Trace.......................321
  G. Hu, G.K. Yong, J. Luan, L.W. Chin, X. Baraton

New Developments in High Performance Solder Products for Power Die Assemblies...............327
  M. Fenner, A. Mackie, G. Wilson

Robust LTCC/PZT Sensor-Actuator-Module for Aluminium Die Casting..............................333
  M. Flossel, U. Scheithauer, S. Gebhardt, A. Schonecker, A. Michaelis

SMD Pressure and Flow Sensors for Industrial Compressed Air in LTCC Technology............338
  Y. Fournier, A. Barras, G. Boutinard Rouelle, T. Maeder, P. Ryser

ALX Permanent Polymer Dielectrics For Microelectronic Packaging Applications.................345
  P. Garrou, A. Huffman, J. Piascik

Experimental Analysis on the Mechanism of Moisture Induced Interface Weakening in ACF Package...354
  G. Sim, C. Chung, K. Paik, S. Lee

3D IC Products Using TSV for Mobile Phone Applications: An Industrial Perspective............361
  Y. Guillou, A. Dutron

Process Characterisation of Dupont MXA140 Dry Film for High Resolution Microbump Application...367
  W. Liebsch, H. Yun, C.E. Balut, A. Ahr, A. Phung, A. Huffman

Reliability Studies on High Current Power Modules with Parallel MOSFETs..........................371
  G.H. Sarma, G. Nitin, Ramanan, Manivannan, K. Mehta, A. Bhattacharjee

Solder Process Optimization: Influence of Heating and Cooling Rate on the Thermo-Mechanical Stress Generated in Components..378
  M. Hertl, D. Weidmann, J. Lecomte

Wafer Post-Processing for a Reconfigurable Wafer-Scale Circuit Board..............................383
  M. Radji, A. Lakhssassi, M. Bougataya, A. Hamoui, R. Izquierdo

Evaluation of Printed Electronics Manufacturing Line with Sensor Platform Application.........391
  E. Halonen, K. Kaija, M. Mantysalo, A. Kemppainen, R. Osterbacka, N. Bjorklund

Development of Matrix Clip Assembly for Power MOSFET Packages....................................399
  M. Kengen, W. Peels, D. Heyes

Electrostatic Wafer Handling for Thin Wafer Processing...403
  C. Landesberger, R. Wieland, A. Klumpp, P. Ramm, A. Drost, U. Schaber, D. Bonfert, K. Bock

Long-Term Joint Reliability of SiC Power Devices at 330°C...408
  F. Lang, S. Tanimoto, H. Ohashi, H. Yamaguchi

Chip to Chip Bonding using micro-Cu Bumps with Sn Capping Layers...............................413
  J.S. Lee, K.Y. Byun, Q.H. Chung, M.S. Suh, S.C. Kim, Y. Kim

Thermal Design Considerations on Wire-Bond Packages..418
  M. Mach, J. Muller

A New Methodology for Multi-Level Thermal Characterization of Complex Electronic Systems : From Die to Board Level...425
  O. Martins, N. Peltier, S. Guedon, S. Kaiser, Y. Marechal, Y. Avenas

Impact of Package Parasitics on the EMC Performance of Smart Power SoCs......................437
  M. Merlin, F. Fiori

Characteristics of Electrically Conductive Adhesives Filled with Copper Nanoparticles with Organic Layer..443
  L.N. Ho, H. Nishikawa, T. Takemoto, Y. Kashiwagi, M. Yamamoto, M. Nakamoto

Design and Fabrication of Corrosion and Humidity Sensors for Performance Evaluation of Chip Scale Hermetic Packages for Biomedical Implantable Devices...447
  N. Saeidi, A. Demosthenous, N. Donaldson, J. Alderman

Realisation of Embedded-Chip QFN Packages - Technological Challenges and Achievements...451
  A. Ostmann, D. Manessis, L. Boettcher, S. Karaszkiewicz, H. Reichl

Far-End Maximum Crosstalk for Coupled Lines as Function of Load....................................457
  A. Owzar, R. Stephan, W. Petersen, M. Helfenstein

Effects of Test Conditions on Bending Impact of Lead Free Solder..461
   J. Park

Connector Reliability Testing Using Salt Spray..466
   A. Parviainen, J. Perala, L. Frisk, S. Kuusiluoma

Accelerating the Temperature Cycling Tests of FBGA Memory Components with Lead-free Solder Joints without Changing the Damage Mechanism..474
   J. Reichelt, P. Gromala, S. Rzepka

Bisected Thermodynamic Sensor as the Power AC/DC Transmitter..482
   M. Reznicek, I. Szendiuch, Z. Reznicek Jr., Z. Reznicek

3D Integration with AC Coupling for Wafer-Level Assembly..487
   M. Scandiuzzo, L. Perugini, R. Cardu, M. Innocenti, R. Canegallo

Kirkendall Voiding in Au Ball Bond Interconnects on Al Chip Metallization in the Temperature Range from 100-200°C After Optimized Intermetallic Coverage..491
   M. Schneider-Ramelow, S. Schmitz, B. Schuch, W. Grubl

Thermo-Mechanical Stress Analysis..497
   K. Niehoff, T. Schreier-Alt, F. Schindler-Saefkow, F. Ansorge, H. Kittel

Simulation and Experimental Analysis of Substrate Overmolding..502
   T. Schreier-Alt, C. Rebholz, F. Ansorge

Mechanical and Microstructural Properties of SiC-Mixed Sn-Bi Composite Solder Bumps by Electroplating..509
   Y. Shin, S. Lee, S. Yoo, C. Lee

Some Facts from Lead-free Solders Reliability Investigation..513
   I. Szendiuch, J. Stary, J. Sandera, M. Bursik, E. Hejatkova

Surface-Enhanced Copper Bonding Wire for LSI and Its Bond Reliability under Humid Environment..518
   T. Uno, K. Kimura, T. Yamada

A Study of Thermal Performance for Chip-in-Substrate Package on Package..528
   T. Hung, M. Yew, C. Chou, K. Chiang

Novel Interconnection Processes for Low Cost PEN/PET Substrates..534
   J. Van Den Brand, R. Kusters, H. Fledderus, E. Rubingh, T. Podprocky, A. Dietzel

Characterization of PTC Resistor Pastes Applied in LTCC Technology..543
   J. Vanek, W. Smetana, M. Weilguni, I. Szendiuch

Processes for Integration of Microfluidic Based Devices..548
   D.P. Webb, P.P. Conway, D.A. Hutt, B.J. Knauf, C. Liu

High Linearity and Broadband WiMAX Power Amplifier Design Using Board Level Integration Technology..555
   W. Chen, K. Chin, C. Tsai, L. Chang, Y. Chang, C. Liu

Influence of the Fabrication Errors on Multilayer Thick Film Circuits..559
   W. Ali, C. Min

Correlation between Material Selection and Moisture Sensitivity Levels of Quad Flat No-lead (QFN) Packages..563
   M. Zhang, S.W.R. Lee, J. Zhang, H. Yun, D. Starkey, H. Chau

Passive Phase Change Tower Heat Sink & Pumped Coolant Technologies for Next Generation CPU Module Thermal Design..569
   M. Vogel, D. Copeland, A. Masto, S. Kang, B. Whitney, G. Upadhya, M. Connors, J. Marsala

Packaging of Silicon Photonic Devices: Grating Structures for High Efficiency Coupling and a Solution for Standard Integration..575
   J.V. Galan, A. Griol, J. Hurtado, P:. Sanchis, G.B. Preve, A. Hakansson, J. Marti

Encapsulation Challenges for Wafer Level Packaging..581
   E. Kuah, J.P. Ding, Q.F. Li, J.Y. Hao, W.L. Chan, S.C. Ho, H.M. Huang, Y.J. Jiang

Mechanical Behaviour of SAC-Lead Free Solder Alloys with Regard to the Size Effect and the Crystal Orientation..587
   J. Villain, W. Mueller, U. Saeed, C. Weippert, U. Corradi, A. Svetly

NanoBond® Assembly – A Rapid, Room Temperature Soldering Process..591
   G. Caswell

Characterization of Oxidation of Electroplated Sn for Advanced Flip-chip Bonding..597
   W. Zhang, W. Ruythooren

Miniaturisation of a LTCC High-Frequency Rat-Race-Ring by Using 3-Dimensional Integrated Passives and Embedded High-K Capacitors..601
   R. Perrone, P. Kapitanova, D. Kholodnyak, I. Vendik, S. Humbla, M. Hein, J. Muller

DreamPAK – Small Form Factor Package..607
   L.A. Lim, M. Ramkumar, C.J. Vath III

Fatigue Life Prediction of Plated Through Holes (PTH) Under Thermal Cycling 613
N. Park, J. Kim, C. Oh, C. Han, B. Song, W. Hong

Thin Hermetic Borosilicate Glass Layers for Highly Reliable Chip-Passivations in Wafer-Level-Packaging 617
U. Hansen, J. Leib, S. Maus, O. Gyenge, M. Topper

Modeling and Quantification of Conventional and Coax-TSVs for RF Applications 624
I. Ndip, B. Curran, S. Guttowski, H. Reichl

Stacking of Full Rebuilt Wafers For SiP and Abandoned Sensors/Applications 628
C. Val, P. Couderc, P. Lartigues

Creep Mechanism Fractography Analysis on SnPb Eutectic Solder Joint Failure 637
C. Oh, C. Han, N. Park, B. Song, W. Hong

Direct Interconnection of Chemical Mechanical Polishing (CMP)-Cu Thin Films at 150ºC in Ambient Air 641
A. Shigetou, T. Suga

Damage Risk Assessment of Under-Pad Structures in Vertical Wafer Probe Technology 646
T. Hauck, I. Schmadlak, C. Argento, W.H. Muller

Impact of Substrate Coupling Induced by 3D-IC Architecture on Advanced CMOS Technology 651
M. Rousseau, M. Jaud, P. Leduc, A. Farcy, A. Marty

Versatile MEMS and MEMS Integration Technology Platforms for Cost Effective MEMS Development 656
P. Pieters

Reliability Testing of Frequency Converters with Salt Spray and Temperature Humidity Tests 661
J. Kiilunen, L. Frisk

Screen-Printed Polymer-Based Microfluidic and Micromechanical Devices Based on Evaporable Compounds 666
N. Serra, T. Maeder, C. Jacq, Y. Fournier, P. Ryser

3D Integration of Ultra-thin Functional Devices Inside Standard Multilayer Flex Laminates 671
W. Christiaens, T. Torfs, W. Huwel, C. Van Hoof, J. Vanfleteren

3D Integration Process Flow for Set-top Box Application: Description of Technology and Electrical Results 676
S. Cheramy, J. Charbonnier, D. Henry, A. Astier, P. Chausse, M. Neyret, C. Brunet-Manquat, S. Verrun, N. Sillon, L. Bonnot, X. Gagnard, J. Vittu

Advanced Failure Analysis Methods and Microstructural Investigations of Wire Bond Contacts for Current Microelectronic System Integration 682
R. Klengel, S. Bennemann, J. Schischka, C. Grosse, M. Petzold

Electrical Modeling and Analysis of the Impact of Slits on Microstrip Lines in Thin Film Polymer Layers 688
I. Ndip, M. Topper, K. Becker, M. Hirte, I. Eidner, T. Fischer, B. Curran, J. Bauer, W. Scheel, S. Guttowski, H. Reichl

3D Integration Technologies for Ceramic Substrates in a SHM Application 692
S. Hildebrandt, K. Wolter

A New Low Cost, Elastic and Conformable Electronics Technology for Soft and Stretchable Electronic Devices by use of a Stretchable Substrate 697
F. Bossuyt, T. Vervust, F. Axisa, J. Vanfleteren

Comparison between Die Attach Film (DAF) and Film over Wire (FOW) on Stack-die CSP Application 703
C.L. Chung, C.W. Ku, H.C. Hsu, S.L. Fu

A Novel Thermo-Mechanical Test Method of Fatigue Characterization of Real Solder Joints 706
R. Metasch, M. Roellig, K.J. Wolter

Thermo Mechanical Characterization of Packaging Polymers 713
B. Boehme, K.M.B. Jansen, S. Rzepka, K. Wolter

Au–Sn SLID Bonding: Fluxless Bonding with High Temperature Stability, to Above 350ºC 723
K.E. Aasmundveit, K. Wang, N. Hoivik, J.M. Graff, A. Elfving

Optimization of Flip-chip Laser Soldering for Low Temperature Stability Substrate 729
T. Hurtony, B. Balogh, P. Gordon

Low Energy Consumption Thick-film Pressure Sensors 735
D. Belavic, M.S. Zarnik, M. Mozek, S. Kocjan, M. Hrovat, J. Holc, M. Jerlah, S. Macek

Reliability Comparison of Aluminum Redistribution based WLCSP Designs 741
U. Sharma, H. Gee, P. Holland

Miniaturization of Printed Wiring Board Assemblies into System in a Package (SiP) 746
S.G. Rosser, I. Memis, H. Von Hofen

Long Term Stability of Polymer Based Resistors Tested by Noise, Non-Linearity and Electro-Ultrasonic Spectroscopy 754
V. Sedlakova, P. Tofel, J. Sikula

System Packaging & Integration for a Swallowable Capsule Using a Direct Access Sensor 759
P. Jesudoss, A. Mathewson, W. Wright, C. McCaffrey, V. Ogurtsov, K. Twomey, F. Stam

Interface Resistance between Polymer Based Conducting and Resistive Layers 763
P. Tofel, V. Sedlakova, M. Chvatal, J. Majzner

New Packaging Technology Enabling Integration of Magnetics and Semiconductors in One Component 767
A. Pot, H. Roehm, R.V.D. Berg, T. Shanmugam, S. Ong, F. Van Der Burgt, T.P. Sidiki

Large Panel, Highly Flexible Multilayer Thin Film Boards 776
H. Burkard, W. Kapischke, J. Link

Closing Technology Knowledge Gaps: Projects Arising from the iNEMI Technology Roadmap 782
B. Pfahl, J. Arnold, G. O'Malley

Addressing Opportunities and Risks of Pb-Free Solder Alloy Alternatives 787
G. Henshall, R. Healey, R.S. Pandher, K. Sweatman, K. Howell, R. Coyle, T. Sack, P. Snugovsky, S. Tisdale, F. Hua, G. O'Malley

A Comprehensive Overview on Today's Ceramic Substrate Technologies 798
F. Bechtold

Compression Molding Solutions for Various High End Package and Cost Savings for Standard Package Applications 810
H. Matsutani

Advanced Solutions for Ultra-Thin Wafers and Packaging 814
G. Klug

Author Index

Long-Term Joint Reliability of SiC Power Devices at 330°C

Fengqun Lang[1], Satoshi Tanimoto[2,3], Hiromichi Ohashi[1], Hiroshi Yamaguchi[1]

[1] Energy Semiconductor Electronics Research Laboratory (ESERL), National Institute of Advanced Industrial Science

and Technology (AIST),

Central 2, 1-1-1,Umezono,Tsukuba, Ibaraki,305-8568,Japan

Tel: +81-29-861-3140; Fax:+81-29-861-3397;   E-mail: fengqun-lang@aist.go.jp

[2] Advanced Inverter Laboratory, Research and Development Association for Future Electron Devices (FED),

in ESERL of AIST

[3] Technology Research Laboratory No. 1, Nissan Research Center, Nissan Motor Co., Ltd,

Natsujima 1,Yokosuka, 237-8523, Japan

Abstract

SiC power devices were die bonded to a AlN/Cu/Ni(Au) direct bonded copper (DBC) substrate with a Au-Ge eutectic solder in a vacuum reflow system. The long term joint reliability of the bonded chips was evaluated at 330°C in air for up to 1600 hours. The bonded samples were inspected with a micro focus X-ray TV system. The microstructure of the samples was observed and analyzed by the scanning electron microscope (SEM) equipped with an energy dispersed X-ray analyzer (EDX). The mechanical and electrical properties of the bonded samples were evaluated before and after high temperature aging. After reflow, the die shear strength of the bonded samples reached up to 72MPa. In the initial aging stage, the die shear strength sharply decreased with aging time. This was due to the formation of NiGe intermetallic compound (IMC) at the solder/DBC substrate. The NiGe IMC was resulted from the reaction of Ge in the solder and and Ni in the DBC substrate. After 400 hrs-aging, little change was observed. After aging for 1600 hrs, the die shear strength decreased to 22 MPa, which was about 3 times higher than the standard value 6 MPa (IEC 749, Japan). Little change was observed in the electrical resistance between the cathode of the chip and the DBC substrate. Oxidation of the DBC substrate was also observed.

Key words: High temperature electronics, SiC power devices, Reliability, Die bonding, Diffusion.

1. Introduction

The needs of power electronics with much lower power loss and smaller volume push the development of new power devices and new power electronics packaging technologies [1]. The advent of silicon carbide (SiC) devices presents a possibility to realize the higher efficiency and miniaturization of the power electronics. SiC devices posses a potential which can be operated under extreme conditions, such as high power, high temperature and high frequency operations due to its wide band gap, high electric breakdown field, high thermal conductivity, and high carrier saturation velocity [2]. SiC materials are the best candidates for high power applications. High power devices have been demonstrated on SiC with better performance than similarly rated Si devices [3]. Compared with Si device, about 50% power loss can be avoid. Benefit from the excellent high temperature reliability, the application of SiC devices in power modules could reduce the cooling system size or even avoid the use of cooling systems. As a result, the output power density can be significantly increased. To realize the high temperature functions of SiC power devices, the development of high temperature packaging technologies becomes more and more important. In this paper, we report the result of the long- term joint

978-1-4244-4722-0/09 $25.00
© 2009 IMAPS-ITALY

reliability of SiC devices at 330℃.

2. Experimental Procedures

The SiC-SBD devices with 2.2 m m □ and 0.4 mm-thick were used in this experiment.. The die side (cathode) of the devices was metalized with Ti/Ni/Ag. The SiC-SBD devices were die bonded to a widely used AlN/Cu/Ni(Au) DBC substrate with Au-3.15wt.%Ge eutectic solder （melting point 356 ℃, hereafter termed as AuGe）. A vacuum reflow system was used to bond the devices. After reflow, the samples were inspected with a micro focus X-ray TV system. The samples with void percent below 5 % were used in this experiment.

High temperature aging experiment was carried out at 330 ℃ in air. After certain aging time, the samples were inspected with a micro focus X-ray TV system and some samples were die-sheared to evaluate the die bond strength. The microstructure of the samples was also observed and analyzed with the scanning electron microscope (SEM) equipped with an energy dispersed X-ray analyzer (EDX). The forward I-V curves of the samples after some time aging were measured using a four point probe method. The slope of the I-V curve at I=3A was adopted as the electrical resistance.

3. Results and Discussion

3.1 Cross-section of the as-bonded samples

Figure 1 shows the cross-section of the as-bonded chip to a DBC substrate with AuGe solder. The SiC chip was well bonded to the DBC without any flaws. An with about 1 μm-thick NiGe layer formed at the solder/DBC interface during reflow, under which was a Ni layer and Cu circuit. The AuGe solder exhibited the typical microstructure: the Ge-rich phase (black) dispersed within the Au-rich phase martrix (white) evenly.

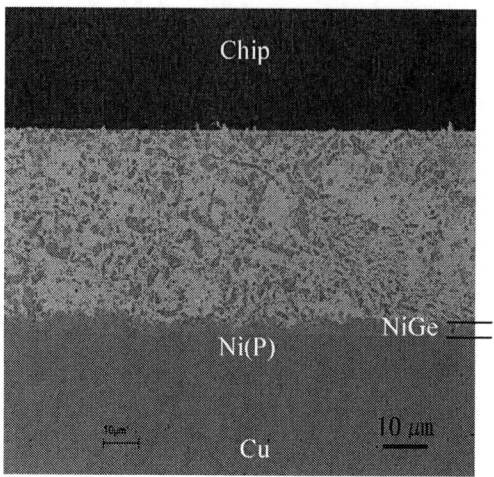

Figure 1. SEM image of the cross-section of the as- bonded SiC chip to a DBC substrate.

3.2 High Temperature Aging at 330 ℃

Figure 2 illustrates the relationship between the die shear strength and aging time. The as-bonded chip exhibited a high die shear strength of up to 70 MPa, which was about 12 times higher than the limit value of die bond strength of 6.2 MPa (IEC749, JEITA) [4]. This was benefit from the good

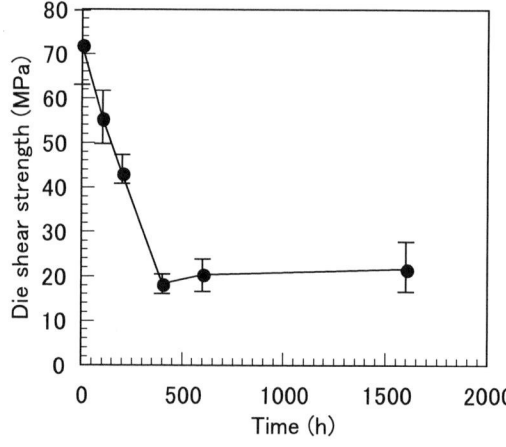

Figure 2. Influence of aging time in the die shear strength the bonded SiC chip.

microstructure of the as-bonded samples. The die shear strength sharply decreased the initial aging stage and after 400 hrs, little change was observed in the die shear strength. After aging for 1600 hrs, the

978-1-4244-4722-0/09 $25.00
© 2009 IMAPS-ITALY

die shear strength was 22 MPa, which was still 3 times higher than the standard value.

Figure 3 (a),(b) and (c) give the SEM images of the bonded samples aged for 200 hrs, 400 hrs and 1600 hrs, respectively. After aging for 200 hrs, the

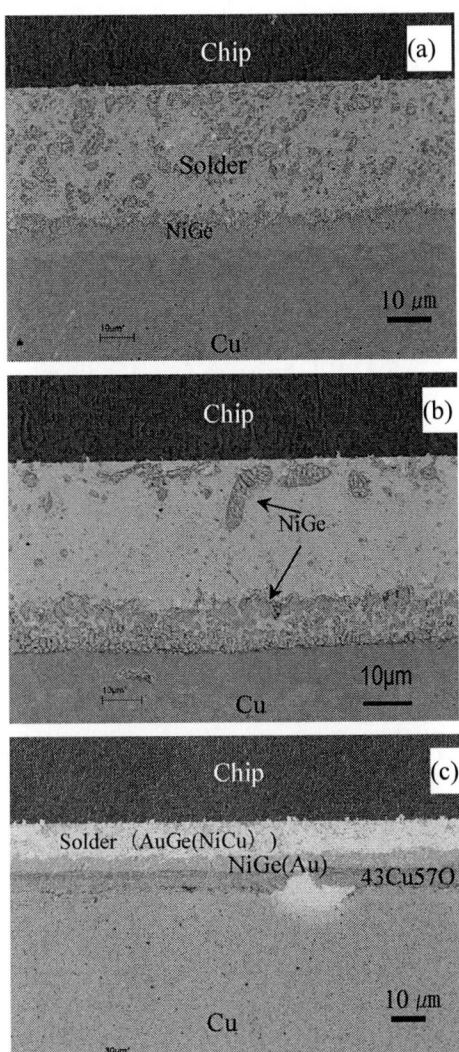

Figure 3. SEM images of the bonded SiC chips after aging at 330°C for (a) 200 h, (b) 400 h and (c) 1600 hrs, respectively.

NiGe IMC at the solder/DBC interface increased to a thickness of about 10 μm, the decomposition of Ni layer was also observed. The Ge-rich phase in the solder also aggregated. Further increasing aging time to 400 hrs, the NiGe IMC layer grew to a thickness of 15 μm, and was composed of NiGe particles. This layer degraded the

bond strength of the SiC chip. In fact, examination of the fractured samples revealed that fracture occurred in the NiGe IMC layer. As a result, as shown in Fig. 2, the bond strength reached the lowest value after aging for 400 hrs. On the other hand, NiGe phase was also observed near the chip in the solder, as shown in Figure 3 (b).

After aging for 1600 hrs, the NiGe IMC layer was not observed. A 43Cu57O (all in at.%, unless indicated) layer formed on the Cu circuit, on which was a NiGe layer doped with about 3 at.% Au. Cu and Ni diffused into the AuGe solder to form a AuGeNiCu alloy. Au was also observed to diffused into the Cu circuit.

Figure 4 shows the influence of aging time in the electrical resistance of the bonded SiC chips. Little change was observed in the electrical resistance during aging, the reasons are under investigation.

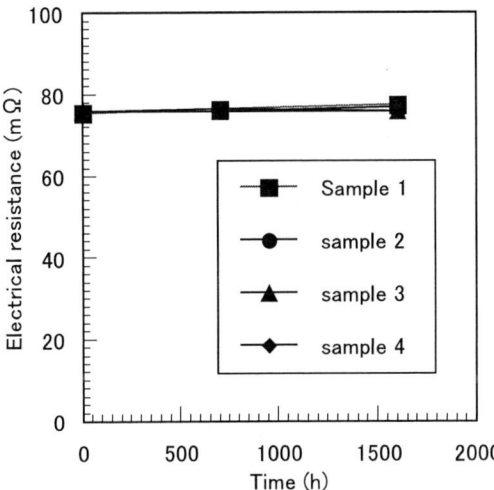

Figure 4. Influence of aging time in the electrical resistivity of the bonded SiC chips.

3.3 Fractured Morphologies

Figure 5 presents the surface morphology and elemental mappings of Cu, Ni and O of the substrate after die shear test. The sample was aged at 330 °Cfor 600 hrs. Copper along with oxygen was mainly detected under the chip after shear test, it is concluded that fracture happened at the

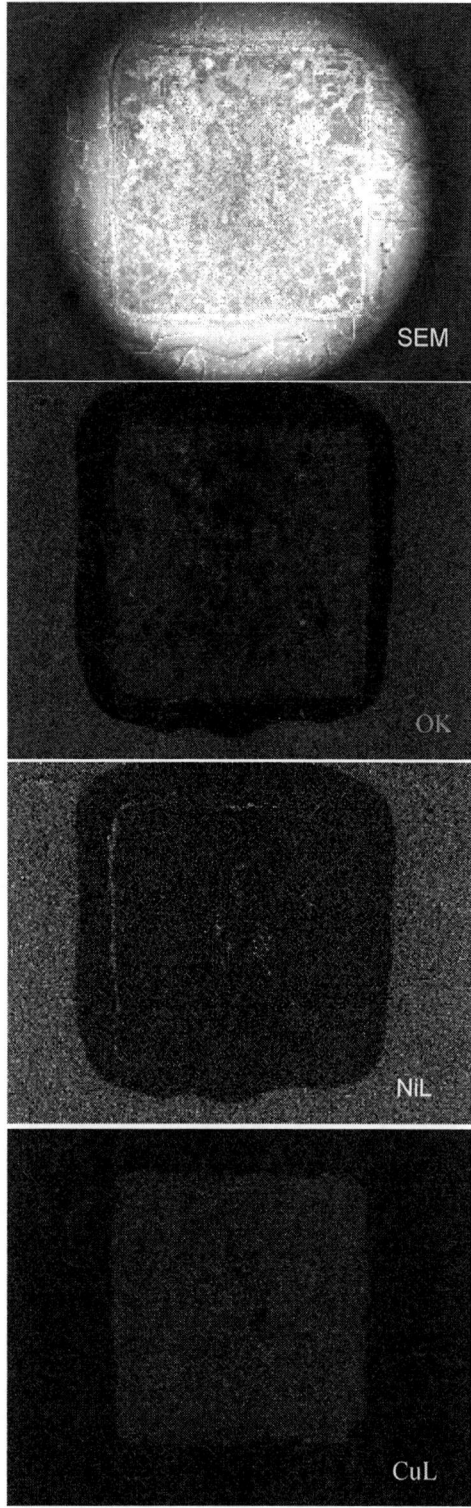

Figure 5. Surface morphology and elemental mappings of the substrate after die shear test. The sample was aged at 330°C for 600 hrs.

Cu/copper oxide interface. Ni was also observed to

be oxidized.

4. Conclusions

SiC power devices can be bonded to a AlN/Cu/Ni(Au) DBC substrate with a Au-Ge eutectic solder in a vacuum reflow system. After reflow, the die shear strength of the bonded samples reached up to 72MPa. In the initial high temperature aging stage at 330 °C, the die shear strength sharply decreased with aging time due to the formation and rapid growth of a NiGe intermetallic compound (IMC) at the solder/DBC substrate. The NiGe IMC was resulted from the reaction of Ge in the solder and and Ni in the DBC substrate. After 400 hrs-aging, little change was observed. After 1600 hrs, the die shear strength decreased to 22 MPa, which was about 3 times higher than the standard value 6 MPa of die shear strength. Very little change was observed in the electrical resistance between the cathode of the chip and the DBC substrate.

Acknowledgements

This work was supported by the NEDO (New Energy and Industrial Technology Development Organization) project, Development of Inverter Systems for Power Electronics.

References

[1] H. Ohashi, "Recent Power devices trend", IEEJ Trans. IA, vol. 122, No.3, pp.168-171, 2002 (in Japanese)

[2] S. Abedinpour, K. Shenai, "Power electronics technologies for the new millennium", IEEE Third International Caracas Conference on Devices, Circuits, and Systems (ICCDCS2000), IEEE Catalog Number: 00TH8474C, pp. P111-1–P111-9.

[3] K.Shenai, "High-power robust semiconductor electronics technologies in the new millennium" Microelectronics Journal, Vol32, No.5-6, 2001, pp.397-408.

[4] Standard Comparison Table of Quality and Reliability Test Methods for Semiconductor Devices, EIAJ EDR-4702. Ed. Technical Standardization Committee on Semiconductor Devices. Issued: Electronic Industrial Association of Japan, Mar.. 1996, pp.86.

Chip to Chip Bonding using micro-Cu Bumps with Sn Capping Layers

Jin Soo Lee, Kwang Yoo Byun[1], Qwan Ho Chung[1], Min Suk Suh[1], Seong Cheol Kim[1],

and Young-Ho Kim*

[1]Hynix Semiconductor Inc., San 136-1 Ami-ri Bubal-eub Icheon-si Kyoungki-do 467-701 S. Korea

Division of Materials Science & Engineering, Hanyang University, Seoul, 133-791, Korea

Phone: 82-2-2220-0405, Fax: 82-2-2293-7445

*E-mail: kimyh@hanyang.ac.kr

Abstract

The chip to chip bonding technique using a Cu bump capped with thin Sn layers has been frequently applied to 3D chip stacking technology. We studied the effect of the microstructure on the joints. The joints were fabricated by joining micro-Cu bumps capped with Sn-Ag solder with sizes of 10 um X 10 um, 20 um X 20 um, and 30 um X 30 um to Cu pads capped with Sn-Ag solder at 245 °C-330 °C using a thermo compression bonder. Three different types of microstructure were formed in the joints depending on the bonding condition: an Sn-rich phase with Cu_6Sn_5 in the Cu interfaces, Cu_6Sn_5 in the interior with Cu_3Sn in the Cu interfaces, and one single Cu_3Sn phase. The joint with only the Cu_3Sn phase had the highest shear strength. Specimens were aged at 150 °C for up to 1000 h. During aging, the microstructures of all the joints became Cu_3Sn phase only. The shear strength of the joints was very sensitive to the formation of Cu_3Sn and microvoids.

Key words: Sn-Ag, Lead-free, TSV, Micro-bump, Shear test

Introduction

There is increasing demand for the production of low power, low weight and compact packaging technologies for aerospace and military applications. Three-dimensional (3D) packaging technologies are now emerging to shrink the size of electronics [1]. These 3D packaging technologies have several advantages over conventional planar 2D packaging technology, such as significant size and weight reduction, lower noise, lower power consumption, smaller delays, and higher operating frequencies [2-5].

The wire bonding method has been widely used for the interconnection of stacking chips. However, the long wiring length of the chip to chip interconnection may limit the size reduction and cause a deterioration of the high frequency performance [2, 4]. As a new interconnection technology for 3-D stack packaging, chip to chip vertical interconnection technology using through-silicon via (TSV) has been extensively investigated in recent years [6-9]. This technology can minimize the interconnection length and enables ultrafine-pitch interconnections. A chip to chip bonding method using Cu bumps with thin Sn capping layers has been developed [6-8]. The microstructure in the solder joints is very sensitive to the bonding conditions. Based on these bonding conditions, three different types of microstructure can be formed in the joints [10]. A previous study was performed with Cu bumps of 100 *um* X 100 *um*. A micro-bump under 60 *um* in diameter was used for TSV technology [11]. The Sn-Ag capping layer was used for preventing whiskers, improving mechanical properties, and lowering the melting point.

In this study, we characterized the microstructure change of the solder joints formed using micro bumps capped with a Sn-Ag solder of 10 *um* X 10 *um*, 20 *um* X 20 *um*, and 30 *um* X 30 *um*. Sn-Ag solder thickness was fixed at 2 *um* in all specimens. We also investigated the effect of the microstructure on the shear strength of the solder joints.

Experimental procedure

Ti (50 *nm*), Cu (1 *um*), Au (50 *nm*), and Ti (50 *nm*) thin films were sequentially deposited on an oxidized Si wafer for under bump metallization (UBM) using a DC magnetron sputtering system. Via holes of 10 *um* X 10 *um*, 20 *um* X 20 *um*, or 30 *um* X 30 *um* were patterned using thick PR. Cu bumps with a thickness of 10 *um* were formed using an electroplating process after the top Ti layer was removed. The thin Sn-2.5% Ag capping layers were then electroplated on Cu bumps. The Cu pads

978-1-4244-4722-0/09 $25.00
© 2009 IMAPS-ITALY

covered with Sn-2.5% Ag (2 *um*)/Cu (10 *um*) were also fabricated. The basic process flow is summarized in Fig. 1. The solder joints were formed by joining the Sn-2.5% Ag/Cu pad and the Sn-2.5% Ag/Cu bump at 245℃ for 60 s, 270℃ for 90 s, and 330℃ for 150 s under 60 MPa using a thermal compression bonder. The test specimens were aged in an oven at 150℃ up to 2000 h. The shear strength of the solder joints was measured by shearing the test chips, as shown in Fig. 2. The microstructure of the solder joints before and after aging and the failure loci were analyzed using scanning electron microscopy (SEM) with energy dispersive spectrometry (EDS).

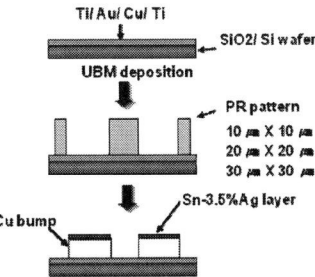

Figure 1: Schematic diagram showing Sn-2.5%Ag/Cu bump formation

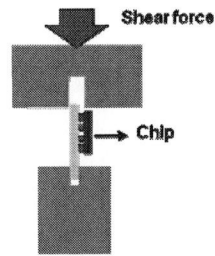

Figure 2: Schematic of the chip shear test

Results and Discussion

Formation of different microstructure

After joining the chips and pads, three different microstructures were obtained in the solder joints by controlling the bonding conditions. Figs. 3-5 show the microstructure in three specimens of different sizes.

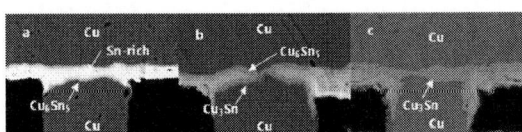

Figure 3: SEM images showing the cross sections of the solder joints of the 10 *μ*m specimen (a) Type □ (b) Type □ (c) Type □

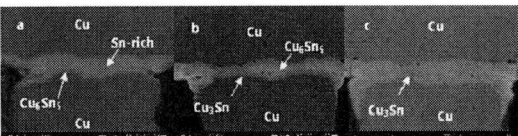

Figure 4: SEM images showing the cross sections of the solder joints of the 20 *μ*m specimen (a) Type Ⅰ (b) Type Ⅱ (c) Type Ⅲ

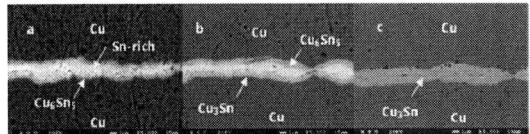

Figure 5: SEM images showing the cross sections of the solder joints of the 30 *μ*m specimen (a) Type Ⅰ (b) Type Ⅱ (c) Type Ⅲ

The type Ⅰ microstructure, which consisted of an Sn rich phase in the center region and a Cu_6Sn_5 intermetallic in the Cu interface was obtained by bonding at 245℃ and 60 s under 60 MPa in all sizes of specimens. In Type Ⅰ, the Cu_6Sn_5 phase was the first phase to form at the liquid Sn/Cu interface [12]. This occurred because the diffusion rate of the Cu was faster than that of the Sn in a Cu-Sn system, which results in more Sn contained within IMCs [13]. The type Ⅱ microstructure consisted of a Cu_6Sn_5 phase in the center region and a Cu_3Sn intermetallic in the Cu interface. It can be obtained by bonding at 270℃ and 90 s under 60 MPa in all sizes of specimens. The type Ⅲ microstructure, which consisted of only Cu_3Sn, was observed by bonding at 330℃ for 150 s under 60 MPa in all sizes of specimens. In Type Ⅲ, only the Cu_3Sn phase of the Sn/Cu solder joints was observed. Three different types of microstructure were formed at the interface between Cu and Sn through the interdiffusion of Cu and Sn atoms. In general, the Cu_6Sn_5 IMC appears first, since the activation energy of the Cu_6Sn_5 phase was lower than that of the Cu_3Sn IMC [13, 14].

Effect of aging treatment

The microstructure change of the joints with three different microstructures in three specimens of different sizes during aging is shown in Figs. 6-8.

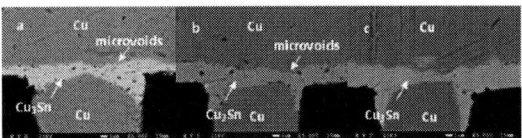

Figure 6: SEM images of the solder joints of the 10 *μ*m specimen after 150℃ aging treatment of (a) Type □ (b) Type □ (c) Type □

978-1-4244-4722-0/09 $25.00
© 2009 IMAPS-ITALY

After the 1000 h aging treatment, the microstructures of Type I and Type II were transformed into the Cu_3Sn phase.

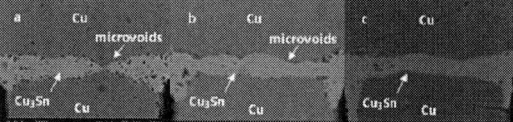

Figure 7: SEM images of the solder joints of the 20 μm specimen after 150°C aging treatment of (a) Type □ (b) Type □ (c) Type □

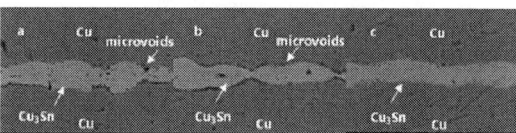

Figure 8: SEM images of the solder joints of the 30 μm specimen after 150°C aging treatment of (a) Type □ (b) Type □ (c) Type □

The Type III microstructure of the solder joints, which consisted of only the Cu_3Sn phase, did not change with aging treatment.

Rapid diffusion of one material into another could cause crystal vacancies to form in the bulk material. These vacancies attracted each other resulting in the formation of voids [15]. The microvoids were caused by the migration of Cu atoms into the solder since each of the Cu and Sn atomic forms moved at a different velocity [16]. The voids grew up during aging and aggregated into both the interface between Cu and Sn and the inside of Cu_3Sn in all sizes of specimen. Microvoids were observed at the interface of the Cu_3Sn phase and the Cu layer in the joints of Type I and Type II after a 1000 h aging treatment, as shown clearly in Figs. 6-8. For the specimen with only the Cu_3Sn phase, regardless of size, microvoids were not visible in the solder joints. The microstructure, which consisted of Sn and Cu_6Sn_5 or Cu_6Sn_5 and Cu_3Sn, tended to transform to the Cu_3Sn and Cu_3Sn phase and became a dominant phase with aging in all specimens. The IMC growth rate was fast as the bump size was decreased.

Shear strength of the solder joints

Shear test were performed for the solder joints and the relation between the solder joints and the shear strength was investigated. Fig. 9 shows the shear strength results of the solder joints with aging treatment.

Before aging treatment, the solder joint, which consisted of only Cu_3Sn (Type III) had the highest shear strength, and the strength of the specimen that consisted of Cu_6Sn_5/Cu_3Sn (Type II) was next.

If the microstructure is the same type, bump size does not affect shear strength. Regardless of

bump size, the shear strength of the Cu_3Sn phase (Type III) was higher than the other two types of phases.

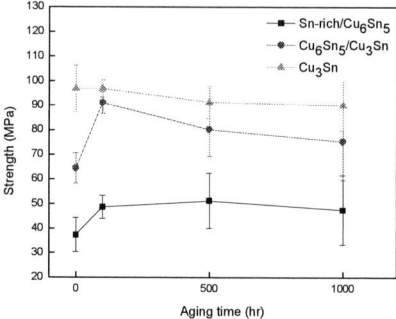

(a) 10 μm specimen

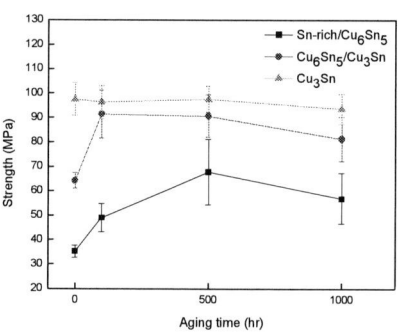

(b) 20 μm specimen

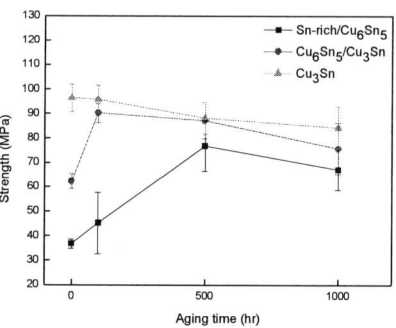

(c) 30 μm specimen

Figure 9: Shear strength of the solder joints as a function of aging time

The shear strength of the joints, which consisted of Sn/Cu_6Sn_5 and Cu_6Sn_5/Cu_3Sn phases, increased rapidly, then decreased. The shear strength of the solder joints was at the maximum when all solder joints were totally transformed into the Cu_3Sn phase. The shear strength of the specimen with only the Cu_3Sn phase slowly decreased during aging. The microvoids, which are formed because of phase transformation, have a detrimental effect on the shear strength of the solder joints. The specimen, which consisted of only the Cu_3Sn phase, had the highest resulting shear strength in all sizes of specimen.

Before the aging treatment, two fracture modes occurred in the Type I specimens. One was the Sn cohesive failure, the other was the mixture of Sn cohesive failure and interfacial failure between the Sn and Cu_6Sn_5. The specimens revealed that the cohesive failure of the Sn layer had low shear strength since Sn is soft. The specimens with the failure mode as the cohesive failure in the Cu_3Sn layer had the largest strength values. However, specimens contained many microvoids in the Cu_3Sn/Cu interface after aging for a long time. The failure occurred in the Cu_3Sn/Cu interface in three specimens. Consequently, the shear strength decreased because of the phase transformation and the microvoids. There were also two fracture modes in the Type II specimens before the aging treatment. One was interfacial failure between Cu_3Sn and Cu_6Sn_5, while the other was a mixture of interfacial failure between Cu_3Sn and Cu_6Sn_5 and cohesive failure of Cu_6Sn_5. After 1000 h of aging treatment, these fracture modes also changed into Cu_3Sn and Cu interfacial failure for the same reason as Type I. In Type III specimens, the failure mode was the cohesive failure of Cu_3Sn before aging. After aging, the interfacial failure between Cu and Cu_3Sn was sometimes observed, but the cohesion failure of Cu_3Sn was still the predominant fracture mode. The silent decrease of shear strength in the Type III specimens was related to the Cu/Cu_3Sn interface failure.

Conclusions

Chip-to-chip bonding was performed using micro-Cu bumps capped with an Sn-Ag thin layer and the effect of the microstructure on the shear strength of the solder joints was investigated. Three different types of microstructures were obtained in the solder joints by controlling the bonding conditions. The main results are summarized as follows:

(1) Before aging, the Sn rich phase in the center and the Cu_6Sn_5 phase in the Cu interfaces were formed in the solder joints by joining at 245℃ for 60 s. The Cu_6Sn_5 phase in the center and the Cu_3Sn phase in the Cu interfaces were formed in the solder joints by joining at 270℃ for 90 s. Only the Cu_3Sn phase was formed in the solder joints by joining at 330℃ for 150 s.

(2) The solder joint consisting of the Cu_3Sn phase had the highest shear strength before aging.

(3) The Sn/Cu_6Sn_5 or Cu_6Sn_5/Cu_3Sn phase in the solder joints transformed to the Cu_3Sn phase during aging, and the shear strength of the

specimens that contained only the Cu_3Sn phase had the highest shear strength.

(4) Microvoids were observed in the Cu_3Sn layer and the interface between Cu_3Sn and Cu with aging, and the shear strength decreased when the specimens were aged for a long time.

(5) The effect of bump size was not significant in all specimens.

Acknowledgement

This research was financially supported by Hynix semiconductor, Korea.

References

[1] Y. Yano, T. Sugiyama, S. Ishihara, Y. Fukui, H Juso, K. Miyama, Y. Sota, and K. Fujita, "Three-dimensional Very Thin Stacked Packaging Technology for SiP", Proceedings of the 2002 Electronic Components and Technology Conference (ECTC), San Diego, California, May 26-30, pp.1329-1334, 2002.

[2] S. F. Al-sarawi, D. Abbott, and P. D. Franzon, "A review of 3-D packaging technology", IEEE Transactions on Components Packaging, and Manufacturing Technology B, Vol. 21, No. 1, pp. 2-14, 1998.

[3] J. Miettinen, M. Mantysalo, K. Kaija, and E. O. Ristolainen, "System Design Issues for 3D System-in-Package (Sip)", Proceedings of the 2004 Electronic Components and Technology Conference (ECTC), Las. Vegas, Nevada, June 1-4, pp. 610-615, 2004

[4] S. Sheng, A. Chandrakasan, and R. W. Brodersen, "A portable multimedia terminal", IEEE communications magazine, Vol. 30, No. 12, pp.64-75, 1992.

[5] R. E. Terrill, "Aladdin: Packaging lessons learned", Multichip Modules, pp.7-11, 1995.

[6] K. Tanida, M. Umemoto, Y. Tomita, M. Tago, Y. Nemoto, T. Ando, and K. Takahashi, "Ultra-high-density 3D Chip Stacking Technology", Proceedings of the 2003 Electronic Components and Technology Conference (ECTC), New Orleans, Louisiana, May 27-30, pp.1084-1089, 2003

[7] M. Umemoto, K. Tanida, Y. Nemoto, M. Hoshino, K. Kojima, Y. Shirai, and K. Takahashi, "High-performance vertical interconnection for high-density 3D chip stacking package", Proceedings of the 2004 Electronic Components and Technology Conference (ECTC), Las. Vegas, Nevada, June 1-4, pp.616-623, 2004

[8] K. Hara, N. Hashimoto, K. Yamaguchi, T. Kobayashi, Y. Yokoyama, and M. Fukazawa, "The Optimization of Terminal Structure for Interconnections in 3D packaging Technology with Through silicon Vias", Proceedings of the 2004 Electronic Components and Technology Conference (ECTC), Las. Vegas, Nevada, June 1-4, pp.161-166, 2004.

978-1-4244-4722-0/09 $25.00
© 2009 IMAPS-ITALY

[9] N. Tanaka, Y. Yoshimura, and T. Naito "Low-Cost Through-hole Electrode Interconnection for 3D-SiP Using Room-temperature Bonding", Proceedings of the 2006 Electronic Components and Technology Conference (ECTC), San Diego, California, May 30-June 2, pp.814-818, 2006.

[10] Y. H. Kim, H. Y. Cho, J. S. Lee, and Y. -H. Kim, "Chip to Chip Bonding using Cu Bumps Capped with Thin Sn Layers", Proceedings of the 2009 The 16th Korean Conference on Semiconductors, Daejeon, pp.40-41, 2009.

[11] K. Kumagai, Y. Yoneda, H. Izumino, H. Shimojo, M. Sunohara, T. Kurihara, M. Higashi, and Y. Mabuchi, "A Silicon Interposer BGA Package with Cu-Filled TSV and Multi-Layer Cu-Plating Interconnect" Proceedings of the 2008 Electronic Components and Technology Conference (ECTC), Orlando, Florida, May 27-30, pp.571-576, 2008.

[12] T. Laurila, V. Vuorinen, and J. K. Kivilahti, "Interfacial reactions between lead-free solders and common base materials", Journal of Materials science & engineering, Vol. 49, No. 1/2, pp.1-60, 2005.

[13] H. T. Lee and M. H. Chen, "Influence of intermetallic compounds on the adhesive strength of solder joints", Journal of Materials science & engineering. properties, microstructure and processing. A, Structural materials,Vol. 333, No. 1/2 , pp. 24-34 , 2002.

[14] A. C. K. So, Y. C. Chan, and J. K. L. Lai, "Aging Studies of Cu-Sn Intermetallic Compounds in Annealed Surface Mount Solder Joints", IEEE Transactions on Components, Packaging, and Manufacturing Technology-Part B, Vol. 20, No 2, pp. 161-166, 1997.

[15] P. Sun, C. Andersson, X. Wei, Z. Cheng, D. Shangguan, and J. Liu, "High temperature aging study of intermetallic compound formation of Sn-3.5Ag and Sn-4.0Ag-0.5Cu solders on electroless Ni(P) metallization", Journal of alloys and compounds , Vol. 425, No. 1/2, pp. 191-199, 2006.

[16] S. -H. Lee, H. R. Roh, Z. G. Chen, and Y. -H. Kim, "Contact Resistance and Shear Strength of the Solder Joints Formed Using Cu Bumps Capped with Sn or Ag/Sn Layer", Journal of Electronic Materials, Vol. 34, No. 11, pp.1446-1454, 2005

Thermal Design Considerations on Wire-Bond Packages

M. Mach and J. Müller

TU Ilmenau, ZIK MacroNano, Institute for Micro- and Nanotechnologies, D-98693 Ilmenau, Germany, Gustav-Kirchhoff-Str. 7

Tel. ++493677 69 3385 e-mail: matthias.mach@tu-ilmenau.de

Abstract

Due to their excellent microwave behaviour (low or moderate dielectric loss, highly conductive inks) and their hermeticity, LTCC materials are being widely used for MMIC packaging. The latter is often required to protect bare dies against the impact of harsh working conditions on the field. Several applications, such as front end modules for mobile communication or transmit modules for point-to-point systems, require packages with very low thermal resistance, since up to 70% of the supplied power is converted into heat.

Although glass ceramic materials have a better thermal conductivity compared to organic boards, it is still insufficient. Thermal vias are one way to locally improve the thermal conductivity. However, their dimensions, bulk material, arrangement and number of layers have a strong impact on the resulting performance. Heat sources (hot spots) on the MMIC, the MMIC thickness and optional heat spreaders need to be included in the evaluation as well. Previous simulations showed that as many thermal vias as possible should be placed below the heat source. In addition, thermal vias should be directly located under the hot spot.

The whole system with the environment from the bare die to the heat sink has to be taken into account, as well the heat path in the package. In this investigation various constructions are considered. The principal build-up consists of an Al-plate as heat sink (thermal ground) with a glued PCB on the top, the mounted package and the bare die as the heat source in the package. A simulation model of a power amplifier TGA9083-SCC from TriQuint was used as a bare die to realize a realistic scenario.

The paper considers different thermal concepts by means of simulation. Variables in the studies are LTCC metallization materials, PCB designs and mounting materials for wire-bond packages.

The optimal system for lowest thermal resistance is a LTCC-housing with an inserted MoCu-plate which is directly mounted onto the board heat sink.

Key words: LTCC Wire Bond Package, Microwave Package, Thermal Simulation, Thermal Design, Thermal Via

Introduction

Typical power dissipation in RF-applications achieves 2 to 5 Watts and in high power cases it may reach more than 20 Watts. LTCC materials are frequently used for constructing RF-structures. These materials offers thermal conductivities in the range from 2 to 5 W/m*K, which is insufficient for an efficient thermal management in such applications.

In order to prevent the microwave device from overheating, the carrier substrate should provide the highest possible thermal conductivity. Thermal vias in LTCC can be used to improve the integral value of the thermal conductivity. Depending on the array formation (diameter of vias and pitch), comparable values to or even better than alumina have been practically achieved [1-3]. Vias filled with silver paste achieved the best thermal performance. However, due to reliability aspects gold conductors and vias are usually required for a variety of applications.

[4] and [5] recommend the use of thermal vias with constant, large via diameters directly underneath the hot spots. However, this recommendation is in contrast to potential hermeticity demands. Therefore, to guarantee hermeticity, moderate design rules for thermal vias and layer counts for thermal package models should be used. RF-chips have often several heat sources and hot spots on the die. The position of thermal vias is very critical due to their interference. GaAs based MMICs suffer additionally from the relatively poor thermal conductivity and therefore reduced heat spreading. The aim of this paper is to characterise different packages, mounting methods and materials for RF-power applications by thermal simulation.

RF-Packages on PCBs

General build-up model

The general build-up scenario comprises a LTCC package with a MMIC and a printed circuit board which is glued to an aluminium baseplate (Figure 1). The die is attached to the LTCC package with 60 μm thick AuSn-solder labelled as thermal

978-1-4244-4722-0/09 $25.00
© 2009 IMAPS-ITALY

interface material 1 (TIM 1). The package itself is mounted on the PCB with either solder or adhesive (TIM 2). The printed circuit board is glued to the aluminium plate with Ag-epoxy or non-conductive epoxy (TIM 3).

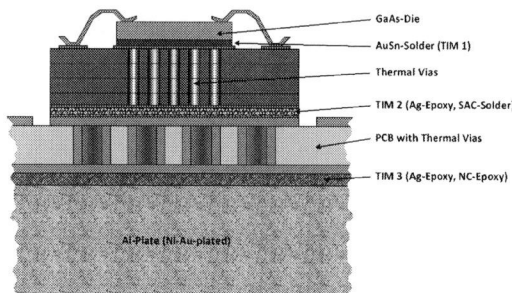

Figure 1: Principle build-up model

Model of heat source

The heat source used for all package simulations is a model of a 2-stage GaAs power amplifier (PA). It was derived from the PA TGA9083-SCC from TriQuint. The size of this PA is 4.52x3.05x0.1 mm³. The power added efficiency (PAE) at 4.8W RF-output power is 40 %. Therefore, a total (thermal) power loss of 7.2 W on both stages with a split-ratio of 18:82 is used for the thermal model. The dimension of the first stage is 0.69x0.23 mm² and the second stage measures 2.6x0.23 mm². The stages are situated on the top side of the die (Figure 2). The supplier recommends a maximum junction temperature of 150°C to achieve the maximal lifetime. Junction temperatures up to 200°C are also possible, but with restrictions on lifetime.

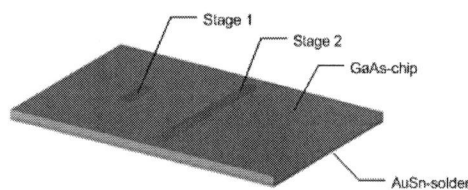

Figure 2: Chip-model of the power amplifier

LTCC Package Options

Three LTCC package types were included in the investigation (Figure 3). All designs use 210 μm thick LTCC tapes and a metallization thickness of 15 μm. Thermal vias have a diameter of 170μm and a pitch of 348 μm. The LTCC package size is 10 mm x 10 mm.

Three via/conductor material configurations were also considered for the simulation:
    M1) Ag-via/Ag-conductor (requires plating)
    M2) Mixed metal (AgPd-Via/AgPt-conductor)
    M3) Au-via/Au-conductor.

The first LTCC package (package A, Fig. 3a) has stacked thermal vias underneath the die. Package B (Fig. 3b) contains an additional printed heat spreader in the middle of the package and an extended thermal via array from the heat spreader to the bottom of the package. In package C (Fig. 3c), the die is directly mounted to a 500 μm thick copper-molybdenum plate which is attached to the cavity frame of the LTCC.

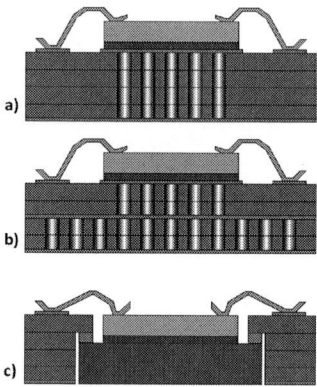

Figure 3: LTCC packages for RF-power applications: a) with thermal vias, b) with thermal vias and additional heatspreader, c) with MoCu heatsink.

PCB Assembly Options

The PCB consists of 250 μm thick FR-4 with 18 μm copper cladding. For thermal reasons it is glued to an aluminium plate. Four different PCB-designs are implemented in order to extract the impact of the PCB on the overall thermal behaviour:

    I)   PCB with thermal vias (unfilled, Fig. 1)
    II)   PCB with plugged thermal vias (Fig. 4a)
    III) PCB without vias (Fig. 4b)
    IV)   PCB with cut-out for the package (Fig. 4c).

The thermal vias have a diameter of 400 μm and a pitch of 1mm. The PCB is fixed on aluminium plate using either non-conductive (standard) or thermally enhanced (conductive) epoxy (TIM 3). The standard thickness of the bond line is 100 μm but variations down to 40 μm are considered as well.

In case of the direct package-to-base plate attach (design IV) SAC-soldering is considered as the third alternative.

Due to the large variety of design options the simulation experiments will be described based on case studies.

Simulation

All package versions were prepared as 3D-models and simulated with FloTherm® from Flomerics. Boundary conditions are still air with the possibility of free convection but no radiation. The

978-1-4244-4722-0/09 $25.00
© 2009 IMAPS-ITALY

air and the lower part (1 mm) of the 5 mm thick aluminium heat sink are set to the ambient temperature of 25°C.

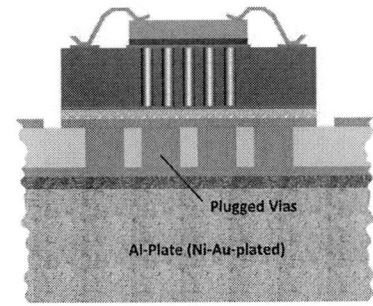

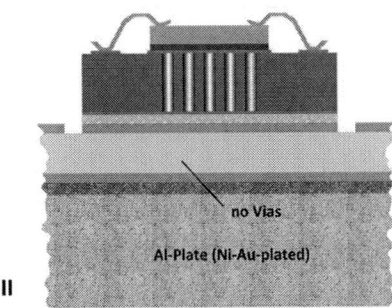

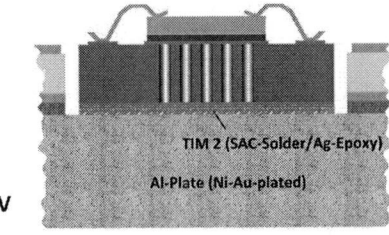

Figure 4: LTCC packages on different PCB-Designs (with plugged thermal vias (design II), without thermal vias (design III), with cut-out in PCB (design IV)).

In the simulation models the round and cylinder shaped vias were replaced by bottom and top area equal cuboids. For example, the vias with a diameter of 170 μm and 400 μm were replaced by cuboids with a ground area of 154x154 μm² and 354x354 μm² respectively (Fig. 5). This simulation trick helps to save mesh nodes and therefore computational time.

Table 1 shows the thermal conductivity parameters for all models. The thermal conductivity of the silver conductor seems to be unexpectedly low compared to the silver-palladium via. However, these parameters are based on laser flash measurements of printed films. GaAs exhibits a very strong temperature dependency of its thermal conductivity. This behaviour was approximated by the equation in table 1 (Θ = temperature [°C]).

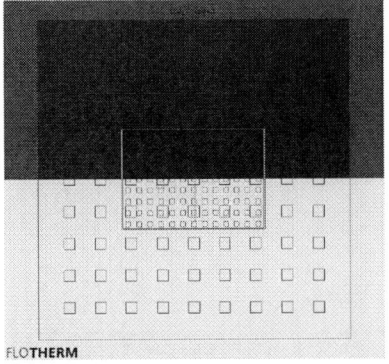

Figure 5: Thermal via distribution underneath the PA in the LTCC-package and underneath the package in the PCB.

Table 1: Simulation parameters

Material	Thermal conductivity [W/mK]
AgPt-conductor	35
Ag-conductor	63
Ag-via	289
AgPd-via	85
Au-conductor	53
Au-via	78
LTCC	3
Al 6061	180
Copper	385
FR 4	0,3
unfilled epoxy	0,2
silver filled epoxy	2
Au80/Sn20-solder	59
SAC-solder	57
GaAs	$\lambda = -0.1476 \dfrac{W}{m \cdot K^2} \cdot \Theta + 48.385 \dfrac{W}{m \cdot K}$
CuMo	170

The simulations were carried out considering a power loss of 1.3 W for stage 1 and 5.9 W for stage 2. The obtained junction temperatures and the calculated thermal resistance for the different build-ups are presented and discussed in the following sections.

Results

Standard Construction

To compare the different build-up versions the construction shown in Fig. 1 is defined as "standard" . The material set is given in table 2.

The first stage of the PA heats up to about 97°C. The junction temperature of the second stage achieves almost 127°C due to its higher power dissipation. The hot spot-to-heatsink thermal resistance (R_{th} related to the size of the amplifier stage) is about 17 K/W. Since the hottest region on the die is of main interest only the second stage is evaluated in the further scenarios.

978-1-4244-4722-0/09 $25.00
© 2009 IMAPS-ITALY

Table 2: Material option for the "standard" structure

Option	Material
LTCC-metal	Ag-via + Ag conductor (M1)
TIM 2	SAC solder
TIM 3	Silver filled epoxy
PCB	with thermal vias (type I)
LTCC design	Package A

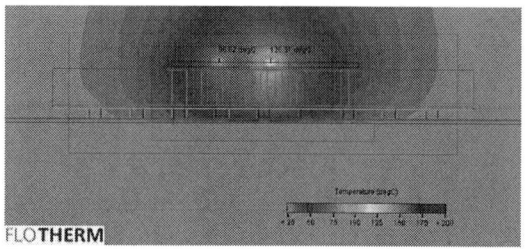

Figure 3: Thermal simulation model of package A in the standard configuration according to table 2.

Influence of the internal heat spreader – LTCC Package B

This scenario is similar to the build-up scheme from the previous section (materials etc.). The only modification is related to the internal LTCC-package design.

The printed heat spreader together with the added vias in the lower layers of LTCC package B (Fig. 3b) has almost no influence on the heat flow. The junction temperature of stage two is only reduced by approximately two Kelvin (Θ_{jct}=125°C) related to the package A design, which means that the complex package B can be replaced with the less complex package A. Further evaluations are therefore carried out without package B.

Influence of LTCC Metallization

In addition to the full silver LTCC package the two other material options (M2 and M3) are included in this case study, which is also based on the standard construction.

As expected the gold and the mixed metal package have a higher thermal resistance ($R_{th,Au}$=21.9 K/W, $R_{th,MM}$=21.2 K/W). Due to the considerably low thermal conductivity of the AgPd transition via, the mixed metal design is almost similar to the Au-version. It needs to be mentioned that the entire via stack was changed to AgPd (parameterized thermal model). The general behaviour shows the important impact of thermal via materials. All remaining investigations are based on the silver material set for optimum performance.

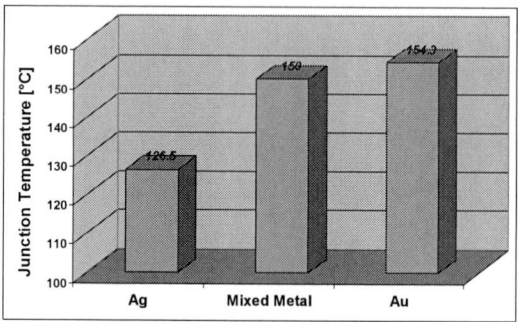

Figure 7: Junction temperature of the second stage for LTCC packages made of silver, mixed metal and gold mounted on PCB.

Influence of PCB-Design

The simulation parameters are given in table 3. In addition to the different PCB-designs also the impact of the PCB attach material is evaluated in this case study.

Table 3: Simulation test matrix for PCB-designs

TIM3	PCB-Design according Fig. 1 and 4			
	I	II	III	IV
Ag-Epoxy	✔	✔	✔	✔
unfilled	✔	✔	-	✔

The simulation results for the silver epoxy glued PCBs are consistent. The PCB version without thermal vias exhibits the highest (Fig. 8) and the plugged via PCB has the lowest thermal resistance in this group (Fig. 9). Without thermal vias in the PCB the junction temperature of stage two increases to approximately 220°C. This build-up scenario cannot survive under ambient temperatures higher than 25°C. The variant with poorly thermally conductive adhesive was not considered in this study.

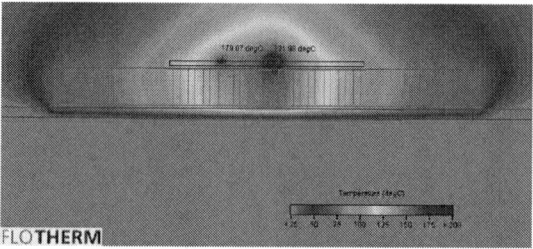

Figure 8: Thermal simulation result of PCB design III (PCB without thermal vias) in combination with Ag filled epoxy and package A.

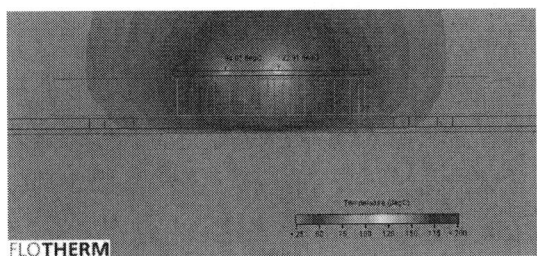

Figure 9: Thermal simulation result of PCB design II (PCB with plugged thermal vias) in combination with Ag filled epoxy and package A.

The package directly glued to the aluminium core (Fig. 10) was expected to have the lowest thermal resistance. However, this build-up version becomes worst for the mounting case with unfilled epoxy. One possible explanation is that there is no heat spreading achieved on the path from the die to thermal ground (aluminium core). The unfilled epoxy between the package and the Al-plate is a huge thermal barrier in such a case. On test structures which are soldered to the PCB, the solder (TIM 2) provides a low thermal interface resistance and the copper conductors and vias optimum heat spreading. The interface resistance of the path PCB/unfilled adhesive/Al-core is then reduced (in comparison to case IV) due to the increased interface area contributing to the heat flow. This effect might be overestimated by the simulation conditions and needs to be confirmed by real measurements.

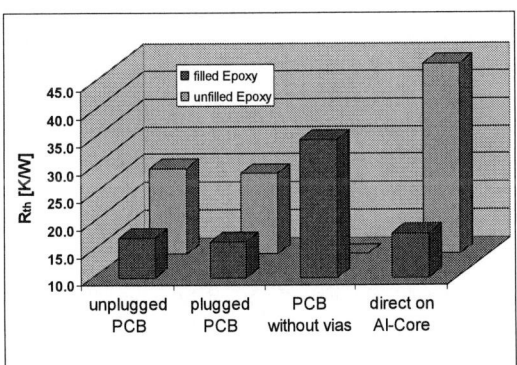

Figure 10: Thermal resistance (second stage of PA) for package A, various PCB designs and two TIM 3 materials.

Standard LTCC Package vs. Heat Sink Package mounted directly on the aluminium plate

The copper-molybdenum base plate package (LTCC package design C, Fig. 3c) provides optimum conditions for spreading the heat underneath the GaAs die. Main thermal barriers are the die itself and the AuSn solder between die and heat sink (TIM 2). The best thermal performance is expected for the direct attachment on the aluminium

plate (PCB design IV). This version is compared to the "standard" LTCC package A (Fig. 11).

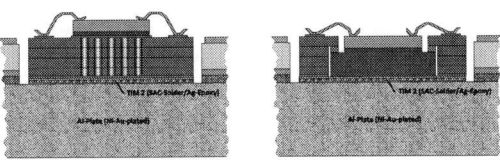

Figure 11: Comparison of standard package A and heat spreader package C with TIM 2 as variable.

Results of the simulation are given in Fig. 12. In the best case (SAC soldering to the aluminium plate) the junction temperature is not exceeding 64°C ($R_{th,HS}$=6.5 K/W) which is approximately half the value of the standard package directly soldered on the Al-core. However, due to reliability issues (difference in TCE between LTCC and Al) direct soldering is not recommended. If conductive adhesive is applied for package C attachment (Fig. 13) the junction temperature of stage two will be slightly higher (Θ_{jct}=73°C). The application of unfilled epoxy increases the thermal resistance by more than 100 % and thus the junction temperature (Θ_{jct}=111°C). However, the values for package C are not comparable to the impact observed on package A (same case as in previous section). This behaviour supports the theory about the spreading influence.

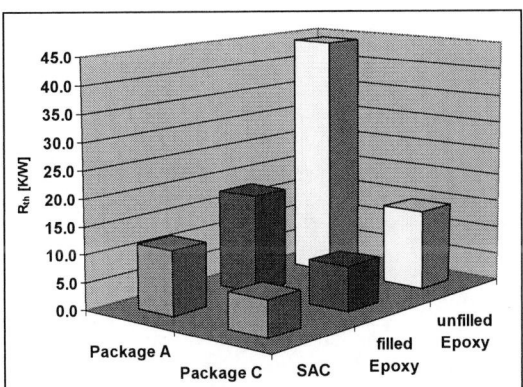

Figure 12: Simulated thermal resistance of package A and C mounted with various thermal interface materials on the aluminium core.

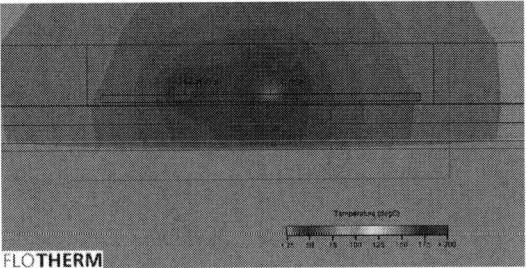

Figure 13: Temperature distribution of package C mounted with filled epoxy.

Influence of bond line thickness for direct package assembly

All previous simulations were carried out with a 100 µm bond line thickness for the interface to the aluminium plate. The relatively high thickness is required to compensate surface imperfections (e.g. roughness, scratches and waviness). If better surfaces would be available the thickness could be reduced. This scenario is considered for direct assembly of the heat sink package with epoxy (Fig. 12, package C). The reduction of the epoxy thicknesses from 100 µm to 40 µm lowers the junction temperature of the critical stage by about 21 K for unfilled and about 5 K for filled epoxy. The linear behaviour in Fig. 14 reveals that the epoxy has no contribution on heat spreading. In reality however, a reduced epoxy layer thickness will not result in linear improvement. Investigations showed that the epoxy thermal resistance is the sum of the two boundary resistances (between the adhesive and the two mounting partners) and the epoxy bulk thermal resistance [6]. The boundary effects will not change with reduced layer thickness.

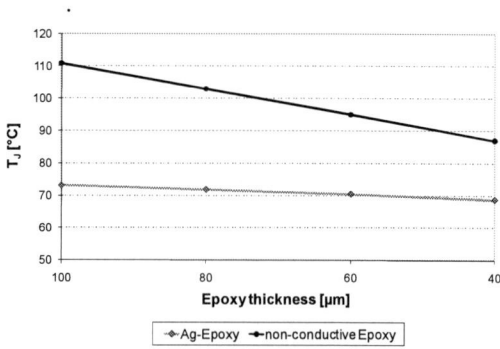

Figure 14: Influence of epoxy thickness on T_J of stage 2 of package C

Summary and Outlook

The simulation results of various build-up schemes for GaAs microwave power-packages based on LTCC have shown the broad range of possible measures to thermally optimise the assemblies. GaAs integrated circuits with localised heat sources (hot spots) suffer from poor internal thermal spreading due to their low thermal conductivity. Therefore, these dies are usually rather thin compared to silicon devices. The investigation in this paper was conducted to demonstrate the general potential of materials and technologies. It also revealed factors with high and low impact on the thermal management. Among the factors with low impact are printed internal LTCC heat spreaders and plugged vias in the printed circuit board. The use of thermal vias itself is very important for both the LTCC package and the PCB. Other factors with high thermal impact are the LTCC metallization system and the thermal interface material used to attach the package or the PCB to the heat sink. A pure silver system (vias and conductors) is recommended for LTCC packages for optimum heat transfer. Since silver should not be used without protection on the outer layers, the silver process needs to be combined with electroless NiAu or NiPdAu plating.

The best performance was achieved with the package C directly mounted on the Al-core. The packages A and B soldered on the Al-core with SAC solder could also be used with restrictions on reliability.

The chip attach material was not varied in this study (AuSn-solder). It is expected to have an important impact especially in combination with GaAs devices.

The best thermal package design is based on a housing with an integrated MoCu-metal plate. It offers very good heat spreading which provides a larger thermal interface area to the following assembly levels (e.g. system heat sink or PCB). In reality, the designers have limited degrees of freedom for the specific applications. The package is usually determined by its electrical interfaces (e.g. BGA or LGA type). In such cases it is recommended to perform a multi-parameter simulation based on statistical experimental planning. In the presented studies only static application cases were considered. In real applications the PAs are used in combination with transient loads, such as in radar applications. Upcoming studies will also consider transient loads.

Future investigations will focus on the use of alternative materials for chip attach and heat sinks. Nano-particle silver has a great potential for either highly conductive adhesives or low temperature sinter-interfaces. Novel heat sink materials based on silver- or copper-diamond offer much higher thermal conductivities than traditional used metals such as tungsten, molybdenum or AlSiC.

Acknowledgements

The authors would like to thank the German Government for the financial support. The work carried out in the project MultiSysTeM (Centre for Innovation Competence MacroNano) is funded by the BMBF. Additional financial support was provided by the Thuringian Ministry of Culture.

References

[1] M. Ray Fairchild et al., "LTCC Technologies for Harsh-environment Automotive Applications Utilizing Soldered Interconnections", Proc. of the 37th International Symposium on Microelectronics, Nov. 14-18, 2004, Long Beach, CA.

[2] Dan Amey, Mimi Keating, "Thermal Performance of 943 Low Loss Green Tape LTCC with Thermal Via and Internal Metallization Enhancements", Proceedings of the 2004 Ceramic Interconnect Technology

978-1-4244-4722-0/09 $25.00
© 2009 IMAPS-ITALY

Conference, April 26-28, 2004, Denver/CO, USA.

[3] Marc A. Zampino, Ravindra Kandukuri, W. Kinzy Jones, "High Performance Thermal Vias in LTCC Substrates", 2002 Inter Society Conference on Thermal Phenomena, pp. 179-185.

[4] Victor A. Chiriac, Tien-Yu Tom Lee, "Thermal Assessment of RF-Integrated LTCC Front End Modules", IEEE Transactions on Advanced Packaging, Vol. 27, No.3, August 2004.

[5] Jens Müller, Matthias Mach, Heiko Thust, Christoph Kluge, Dieter Schwanke, "Thermal Design Considerations for LTCC Microwave Packages", Proceedings of the 2006 EMPS, May 21-24, 2006, Slovenia

[6] R. Schacht, D. May, B. Wunderle, O. Wittler, A. Gollhardt, B. Michl, H. Reichl: Characterization of thermal interface materials for thermal simulation, Proc. Deutsche IMAPS-Konferenz 2005, München, 5.-6. Oktober 2005.

A new Methodology for Multi-Level Thermal Characterization of Complex Electronic Systems : From Die to Board Level

O. Martins [1], N. Peltier [2], S. Guédon [2], S. Kaiser [2], Y. Marechal [1], and Y. Avenas [1]

[1] G2ELAB, UMR 5269 INPG-UJF-CNRS, BP 46, 38402 Saint Martin d'Hères Cedex, France
[2] DOCEA Power, 166B, rue du Rocher du Lorzier, 38430 Moirans, France

Phone: +33 4 76 82 64 38 - Email: Olivier.Martins@g2elab.grenoble-inp.fr

Abstract

Thermal management is becoming a major concern in microelectronics because of transistor technology reduction and power density increases within complex packages. Temperature rise due to power dissipation worsens harmful clock skew, jeopardizes reliability and leads to over-consumption because of leakage current dependence on temperature. To limit these risks, electronics engineers have to perform thermal simulations at an early stage of the design flow and for several granularity levels (die, package, PCB, ...). To speed up and ease the thermal characterization process, the engineers need small, accurate and easy-to-generate thermal models, which can be reused at every integration step.

Several macro-modeling techniques exist (DELPHI, HotSpot, ...), but they cannot satisfy all the points mentioned above. This paper presents a new methodology called Flex-CTM for Flexible Compact Thermal Modeling to build and to interface compact thermal models at different granularity levels. Each part of an electronic system is prepared to be plugged into any other environment and reduced to save memory and time, resulting a thermal micro-model. Therefore, a fast-to-simulate macro-model of a full system can be obtained by assembling the micro-models.

The Flex-CTM is found to have number of advantages over both current resistive models (junction-to-case and junction-to-board) and Dynamic Compact Thermal Models. The first advantage of coupling models together allows multi-source and dynamic simulations at any design level. The second is the control on the accuracy. The third advantage is the Boundary Condition Independence property to allow architecture exploration. Finally and the most important, micro and macro-models are shared by teams to be reused and completed.

Key words: Compact Thermal Modelling, model coupling, package thermal characterization, Boundary Condition Independence, multi-level modelling

Introduction

In microelectronics, device designers are increasingly miniaturizing the electronic components, to design smaller products and to add new functionalities. This miniaturization is at the origin of a strong rise of the power consumption in electronic systems. Therefore, the transistor technology reduction and the rise of the operating frequency, have caused a dramatic increase of power density in Integrated Circuits (IC). The electro-thermal phenomenon leads to a high temperature elevation with the power consumption increase. Now, the thermal problematic associated with the power consumption becomes a major issue for the microelectronic design, and particularly within the framework of complex circuit design like the System on Chip (SoC) and the System in Package (SiP) circuits.

High temperatures of the components cause thermal and mechanical stresses which affect circuit reliability. The manufacturing cost of the product increases owing to the need of cooling systems. Furthermore, thermal gradients within the die, due to local hot spots, bring on delay errors in logical gates and thereby limit expected performances. As discussed before, the temperature rise leads to an overconsumption that decreases the autonomy of nomad systems. Moreover, it can lead to a component destruction caused by thermal runaway.

In order to avoid these risks, electronic engineers have to perform transient thermal simulations at an early stage of the design flow, and at several granularity levels (die, package, board). Many teams in a single company are in charge of building their own thermal model (package model, die model, board model...). Consequently, it is difficult to put together the whole set of models to simulate the thermal behavior of a full circuit. The Flex-CTM methodology described here points out the share of these models. The models can be seen as bricks that are assembled to build more complex models.

978-1-4244-4722-0/09 $25.00
© 2009 IMAPS-ITALY

To speedup the thermal characterization process, the model must be compact and accurate to run fast simulation allowing a bias up to 10%. On the other hand, to cover the entire phenomenon analysis, the models must be dynamic and multi-sources. To summarize the need, a model must meet the following four criteria. First, the models have to be dynamic to cover a large set of phenomena. Second, multiple power sources can be applied to fit with real cases exploration. Third, to save modeling and simulation time, the models must have a reduced number of unknowns. Fourth, to be pluggable and reusable in different use cases, the models have to be boundary condition independent (BCI).

The paper deals with 5 items. First, a short background of existing thermal models is presented. Second, the Flex-CTM methodology is described. Third, the build of elementary pluggable compact thermal models is detailed. Fourth, the modeling and the simulation of the whole system is explained. Finally, the speed and accuracy performances of the methodology are evaluated for a typical integration of systems.

Background

Several models already exist to simulate the thermal behavior of an electronic system. First, numerical methods (Finite Element Method) partition a volume into elementary units. The method transforms the continuous domain into a discreet one in order to compute the temperature at each node of a mesh. According to the JEDEC convention, the numerical thermal model is also called detailed model. These ones may be difficult to build if the geometries are complex. Moreover, these numerical models are huge, then slow to simulate. For example, figure 1 shows the geometry of a complex package and its associated FEM mesh.

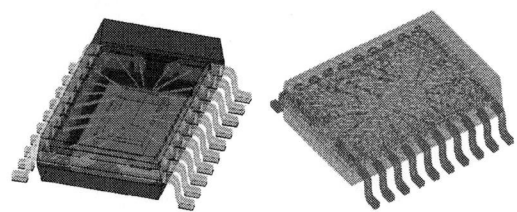

**Figure 1: Finite Element Model of a TSSOP
Package**

Pioneering DELPHI methodology has been introduced to generate smaller models, in terms of number of unknowns **Error! Reference source not found.**. This is a fitting method that creates Compact Thermal Models (CTM). A CTM is made of a network of resistors between key points of a package (a junction and outers). A DELPHI model is illustrated figure 2. Other fitting methodologies have been introduced to add capacitive terms in the DELPHI compact models **Error! Reference source not found.**, like the European project PROFIT **Error! Reference source not found.** and **Error! Reference source not found.**. The building of these models takes a lot of time because many simulations of the detailed model must be performed in transient domain combining heat transfer coefficients applied on exchange surfaces. In addition, the junction modeling in DELPHI-like models is quite rough because it is considered as a uniform heat dissipation source.

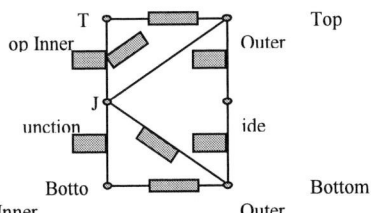

**Figure 2: A Typical DELPHI Compact Thermal
Model**

Recently, HotSpot analytical models **Error! Reference source not found.**, have been introduced to enhance the thermal model capabilities. The static and dynamic behavior are addressed by a resistive and capacitive network between blocks at different levels. These models are mostly specific for the die. The package and board models are quite coarse and valid for specific heat flux repartitions. Figure 3 shows a resistive HotSpot model of a die divided into 4 blocks on a substrate.

978-1-4244-4722-0/09 $25.00
© 2009 IMAPS-ITALY

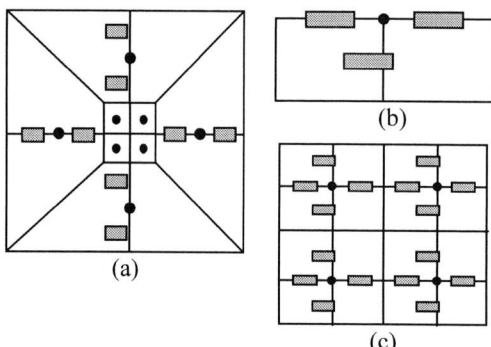

**Figure 3: Resistive HotSpot Model of a Die
divided into 4 Blocks on a Substrate**

**(a) Connections in the Substrate
(b) Connections in a block thickness
(c) Connections on the Junction Surface**

None of the existing methods meet all the specifications quoted previously (Introduction section §4). In the following section, a new methodology called Flex-CTM is introduced.

The Flex-CTM Methodology

Usually, thermal models are built by different actors at different integration levels and the current methodology does not allow to easily reuse and couple these models. The CTM is "Flexible" in that way it allows to combine models together and to adapt them to any environment. Therefore, the Flex-CTM methodology relies on the split of the system and processes 6 steps, which are illustrated figure 4.

978-1-4244-4722-0/09 $25.00
© 2009 IMAPS-ITALY

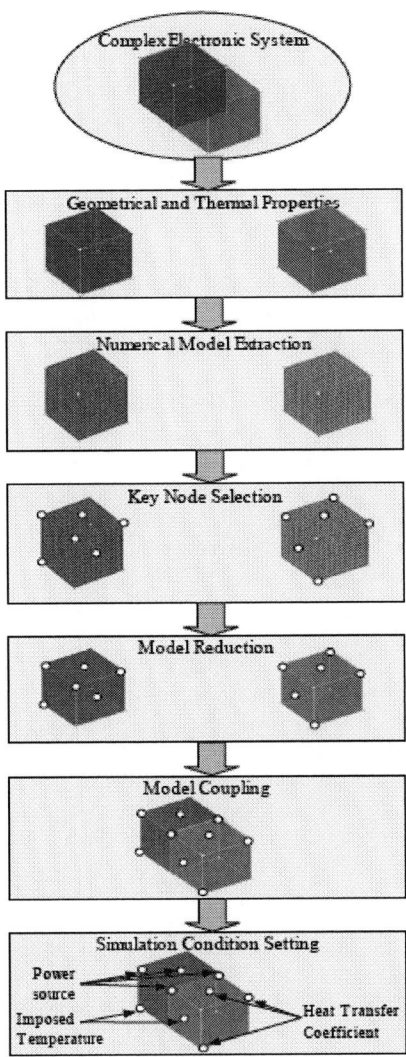

Figure 4: The Flex-CTM Methodology

Firstly, given the geometry and the thermal properties of a global system, a new geometrical and thermal description is created for each homogeneous material part. For instance, a BGA package (figure 5) is split into four descriptions (die, substrate, encapsulant and bondwires). The description contains the interfaces defining the surfaces which allow power dissipation (interface P) and/or heat exchange (interface E). The next 3 stages are applied on each description.

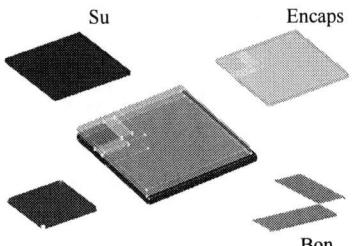

**Figure 5: Geometrical and Physical Description
of each part of a BGA package**

978-1-4244-4722-0/09 $25.00
© 2009 IMAPS-ITALY

Secondly, given the description of a component part, the numerical model is extracted. The numerical model is finely meshed (N nodes) and built without boundary condition. The results of the extraction are the conduction matrix G, the diffusion matrix C, the geometrical properties of all meshing nodes (identifier, matrix index and coordinates), and the nodes identifiers belonging to the interfaces (P, E) of the model.

Thirdly, the huge extracted model has to be prepared before being reduced. The reduction will decrease the size of the numerical system and so, the interface nodes to keep are selected. The whole nodes (ne) of interfaces (P, E) cannot be kept, otherwise the reduction will not operate. To be efficient, the reduction needs the model to verify the rule ne²<<N. So, sub-sampled interfaces (P' and E') are defined to substitute for the original interfaces (P and E) in order to reduce the number of external nodes. A trade-off scale between very compact model to very accurate model is available and must be set by the user. That leads to define the subset of external nodes ne'. Then, the substituted interface nodes are a function of the original interface nodes following a linear interpolation. The resulting model is given by the numerical model in which selected interfaces (P' and E') are substituted for external interfaces (P and E).

Fourthly, the matrices G and C are ordered to place the nodes of respectively J' and E' at the ne' first columns. The matrices G and C are reduced to decrease the dimension of the system, preserving the first moments orders. The method is called a Moment Order Reduction (MOR) **Error! Reference source not found.**. However, the MOR is enhanced by adding a post-processing that controls the accuracy of the reduction. The reduced model has ne' external nodes and N' unknowns where N' is approximately 10% of N.

Fifthly, the micro-models built through the previous four steps are coupled to simulate the thermal behavior of the original system. These micro-models are coupled together through their respective interface nodes. It allows teams to build only their micro-model of interest and to reuse existing ones to model a whole system. The resulting model is called Flex-CTM.

The final step is the setting of the simulation conditions in order to simulate the thermal behavior of the original system. In consists in applying power sources on P' interface, heat transfer coefficients and imposed temperatures on E' interfaces of the Flex-CTM.

Such models meet the specifications because Flex-CTM are small, reused and shared between the different trades. For instance, both package manufacturer and die designer can build their own Flex-CTM and share them to simulate their models in a more realistic environment; against to apply approximative heat transfer coefficients or basic bad-known model. Finally, the Flex-CTM have the "Flexible" property. Meaning that each micro-model of the Flex-CTM can be modified independently.

Build of Pluggable Compact Thermal Models

The Flex-CTM methodology begins with the creation of micro-models for each part of the whole system. A micro-model is a BCI compact thermal model with external connections for power sources, imposed temperatures, convection models, and other conductive models. This section explains how to build a micro-model.

Model splitting is the first stage of the methodology which breaks up the global model into elementary homogeneous material parts. At this stage, the geometries and the physical properties of each part is described. The interfaces of the elementary parts are classified in two groups: power sources (interface P) and exchange interface with the environment (interface E).

Then, the **extraction process** begins with the build of a numerical model of each component of the whole system, without any boundary condition. The meshing of a component is well adapted due to the model splitting step. The finite element method (FEM) is chosen to retrieve the physical transfer functions at any meshing node. The transfer functions between nodes are written as a thermal admittance system. This system conjugates the thermal conductance sub-system G and the thermal capacitance sub-system C. The matrices G and C are square, symmetric and strictly definite positive. Equation (1) shows the computation of each element of the matrices G and C where k is the thermal conductivity of the material (in $W.m^{-1}.K^{-1}$), ρ is its density (in $kg.m^{-3}$) and C_p is its specific heat capacity (in $J.kg^{-1}.K^{-1}$). Ω is the volume of the FEM meshing element. αi is the form function, which value is equal to 1 on the node i and 0 on the other nodes.

$$G[i, j] = \int k.div(\alpha_i).div(\alpha_j) d\Omega$$
$$C[i, j] = \int \rho C_p . \alpha_i \alpha_j d\Omega \qquad (1)$$

Moreover, the extraction process gives the geometrical properties of all the meshing nodes. Each node identifier, index in the FEM matrices and 3D coordinates of the node are extracted. Finally, all the nodes belonging to the interfaces P and E of the model are identified.

The **node selection** step consists in preparing the numerical model before its reduction. To perform it, a subset of external nodes which will be kept after reduction, has to be defined. The external nodes represent the inputs-outputs of the model after reduction. These nodes are used to apply environmental conditions as power sources, imposed temperatures, convection models, and conduction models. The other nodes of the model are called internal nodes. Intuitively, we could define all the surface nodes as external nodes, but it would lead to a very bad reduction. So, the user has to define sub-sampled interfaces which are used in place of the original ones. The number of external nodes on this new interface is ne'. The user controls the trade-off between accuracy and the size of the reduced model for his study case, modifying the number of external nodes ne'. The replacement of the original interface by the sub-sampled one is performed through a coupling. This coupling is detailed in the following section ("Whole System Modeling and Simulation").

The **reduction** of a thermal detailed model must be conducted by a robust method because the size of the detailed model is large (few hundred thousand equations). The Model-order reduction (MOR) techniques have been used intensively to reduce drastically the time of the simulations in the thermal analysis. The pioneering method AWE **Error! Reference source not found.** matches the explicitly calculated moments of the system but suffers from numerical instability with higher orders (eg >10). Therefore, other methods based on projections on the Krylov subspace have been proposed to reduce the dimension. The most popular methods are the Padé approximation via Lanczos PVL **Error! Reference source not found.**, and the well known Arnoldi process of multipoint moment matching approximants of the general discretized thermal system **Error! Reference source not found.**, **Error! Reference source not found.**, **Error! Reference source not found.**. The Arnoldi algorithm is numerically stable, contrary to the Padé approximation, and the implementation of the Implicitly Restarted Arnoldi Method (IRAM) is available in ARPACK **Error! Reference source not found.**.

The discreet expression of the detailed model in time domain has a static member and a dynamic member (2).

$$G\,x \square\, C\,\dot{x} = B\,p \qquad (2)$$

Where G is the thermal conductance subsystem, x the vector of the unknown temperatures, C the subsystem of thermal capacitances, B the input-output driver matrix, and p the vector of the discreet inputs-outputs.

G and C are sparse and symmetric definite positive of size m x n. The vector p is a dense vector of size m x ne as well as x. Where ne is the number of external nodes and m the number of equations.

To reduce the dimension of the system (2) the moment matching procedure will be applied in the frequency domain using the Laplace transformation.

$$Y(\omega) = B^T (G + j\omega C) B$$
$$\text{if} \quad R = G^{-1} B \text{ and } A = G^{-1} C$$
$$\text{then } Y(\omega) = B^T (I + j\omega A) R \qquad (3)$$
$$\text{and } T(\omega) = Y^{-1}(\omega) p$$

Where Y is the thermal admittance of the system, and T the vector of the unknown temperatures in the frequency domain.

The matrix B must be chosen to perform the average of temperatures and the average of power sources according to a set of nodes for each surface, elsewhere it is the identity. Therefore, Arnoldi algorithm is computed on A and R to extract the orthogonal bases U that spans the Krylov subspace.

$$K_{n'} \square A, R \mathrel{\square} col-span\left\{ R, A^{-1} R, \dots, A^{\pm n' \square 1 \square} \right\} \qquad (4)$$

Once U is calculated for a first number of moments (n' = ne+ni'), ni' starting at 10, the reduction is done by projecting G, C and B onto the base.

$$G'_{n'} = U^T G U$$
$$C'_{n'} = U^T C U \qquad (5)$$
$$B'_{n'} = U^T B$$

To control the accuracy, the matrix Y' of the transfer functions of admittances is calculated (6) in the frequency domain. Note that Y' is very fast to compute.

$$Y'(\omega) = B'^{T}(I + j\omega A')R'$$

$$\text{where} \quad R' = G'^{-1}B' \text{ and } A' = G'^{-1}C' \tag{6}$$

While the mean of the Y' (6) values are changing, (3), (4) then (5) are calculated increasing ni'. This loop allows to ensure to extract the optimum first order moments that match the Krylov subspace. The reduction rate, comparing the number of values in the detailed system with the reduced system is approximately 90%.

Whole System Modeling and Simulation

This section deals with the coupling of micro-models to build a model of a full system and then its simulation. The interfaces of the models to connect are not supposed to have the same geometrical configuration. Therefore, the models cannot be coupled by just linking the interface nodes one by one. In the followings, the coupling method to connect two micro-models with different interfaces is described.

Two micro-models (G_1, C_1) and (G_2, C_2) are assembled in a non-coupled system equation (7). Where G_1 (resp 2) is the conduction matrix and C_1 (resp 2) the diffusion matrix of the micro-model 1 (resp 2). T_1 (resp 2) is the vector of temperature at each node of the micro-model 1 (resp 2). F_1 (resp 2) is the load vector at each node of the micro-model 1 (resp 2).

$$\begin{bmatrix} C_1 & 0 \\ 0 & C_2 \end{bmatrix}\begin{bmatrix} \dot{T}_1 \\ \dot{T}_2 \end{bmatrix} + \begin{bmatrix} G_1 & 0 \\ 0 & G_2 \end{bmatrix}\begin{bmatrix} T_1 \\ T_2 \end{bmatrix} = \begin{bmatrix} F_1 \\ F_2 \end{bmatrix} \tag{7}$$

The repartition of the nodes of the two uncoupled micro-models is described figure 6. The "e" index is for the external node block, and the "i" index is for the internal node block. The structure of the nodes of the two models is modified in order to isolate the external nodes of the model 2, see (8).

$$T_{1i} \qquad T_{1e} \qquad T_{2e} \qquad T_{2i}$$

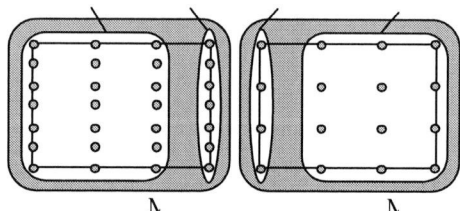

Figure 6: Nodes Repartition of two Uncoupled Models

$$T'_1 = T_{1e} \, \square \, T_{1i} \, \square \, T_{2i}$$
$$T'_2 = T_{2e} \tag{8}$$

The goal of the coupling process is to create links (G_{12}, G_{21}, C_{12}, C_{21}) between two models as shown in equation (9).

$$\begin{bmatrix} C_{11} & C_{12} \\ C_{21} & C_{22} \end{bmatrix}\begin{bmatrix} \dot{T}'_1 \\ \dot{T}'_2 \end{bmatrix} + \begin{bmatrix} G_{11} & G_{12} \\ G_{21} & G_{22} \end{bmatrix}\begin{bmatrix} T'_1 \\ T'_2 \end{bmatrix} = \begin{bmatrix} F'_1 \\ F'_2 \end{bmatrix} \tag{9}$$

with

$$G_{11} = G_1 \square G_{2i} \quad G_{22} = G_{2e}$$
$$C_{11} = C_1 \square C_{2i} \quad C_{22} = C_{2e} \quad \quad (10)$$
$$F'_1 = F_1 \square F_{2i} \quad F'_2 = F_{2e}$$

Where G_{2i} (resp C_{2i}, F_{2i}) is the internal node sub-block of G_2 (resp C_2, F_2) and G_{2e} (resp C_{2e}, F_{2e}) is the external node sub-block of G_2 (resp C_2, F_2).

This link is realized by the method of Lagrange multipliers **Error! Reference source not found.**, a well known mathematical method. It transforms a minimization problem under constraints in a minimization problem without constraint. The Lagrange multipliers enable to solve the heat equation (7), ensuring the continuity of temperature at the interfaces of the two models (11).

$$\dot{T}'_2 = X \dot{T}'_1$$
$$T'_2 = X T'_1 \quad \quad (11)$$

Where X is the cross coupling between the nodes of the two model interfaces.

This mathematical method applied to the problematic of thermal model coupling is described in **Error! Reference source not found.**. The looking for the singular points of the Lagrangian gives the following matrix system:

$$\begin{bmatrix} C_{11} & C_{12} & -X^T & 0 \\ C_{21} & C_{22} & I & 0 \\ -X & I & 0 & 0 \\ 0 & 0 & 0 & 0 \end{bmatrix} \begin{bmatrix} \dot{T}'_1 \\ \dot{T}'_2 \\ \square_1 \\ \square_2 \end{bmatrix} \square \begin{bmatrix} G_{11} & G_{12} & 0 & -X^T \\ G_{21} & G_{22} & 0 & I \\ 0 & 0 & 0 & 0 \\ -X & I & 0 & 0 \end{bmatrix} \begin{bmatrix} T'_1 \\ T'_2 \\ \square_1 \\ \square_2 \end{bmatrix} = \begin{bmatrix} F'_1 \\ F'_2 \\ 0 \\ 0 \end{bmatrix} \quad (12)$$

Eliminating $\square_1$, $\square_2$ and T'_2 in (12), it leads to a new equation system (13) which describes the thermal behavior in transient state of the two coupled models.

$$\bar{C}\dot{T}'_1 \square \bar{G}T'_1 = \bar{F} \quad \quad (13)$$

where

$$\bar{C} = C_{11} \square X^T C_{21} \square C_{12} X \square X^T C22 X$$
$$\bar{G} = G_{11} \square X^T G_{21} \square G_{12} X \square X^T G22 X \quad (14)$$
$$\bar{F} = F'_1 \square X^T F'_2$$

In this methodology, the coupling matrix X is an application of the nodes of interface 2 to interface 1. For each node Node2$_j$ of the eliminated interface 2, surrounding interface-1-nodes are identified (see figure 7).

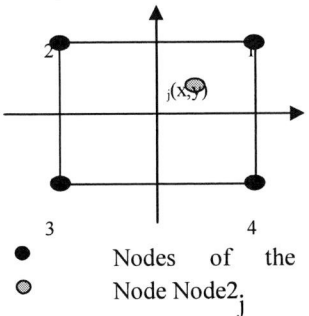

Figure 7: Neighbor Nodes Configuration for a Hexahedral Mesh of the Interface 1

Thus, the neighbor nodes of Node2$_j$ are weighted, computing the form factors at the coordinates of the node Node2$_j$. Equation (15) shows the general form factors for a hexahedral mesh of first order.

$$\phi_1(x,y)=\frac{1+x}{2}\frac{1+y}{2} \quad \phi_2(x,y)=\frac{1-x}{2}\frac{1+y}{2}$$
$$\phi_3(x,y)=\frac{1-x}{2}\frac{1-y}{2} \quad \phi_4(x,y)=\frac{1+x}{2}\frac{1-y}{2} \quad (15)$$

So, the temperature T_j of the node Node2$_j$ is computed through equation (16).

$$T_j(x,y)=\phi_1(x,y)T1+\phi_2(x,y)T2$$
$$+\phi_3(x,y)T3+\phi_4(x,y)T4 \quad (16)$$

The equation (16) expresses the temperature T_j as the linear combination Λ_j of the second interface temperatures.

$$T_j=\Lambda_j T'_1 \quad\quad (17)$$

Finally, the coupling matrix X between the two models is built by all the line matrices Λ_j.

$$X=\begin{bmatrix} \Lambda_1 \\ \Lambda_2 \\ \vdots \\ \Lambda_{ne'_2} \end{bmatrix} \quad\quad (18)$$

Where ne'$_2$ is the number of external nodes in the model 2. X is a rectangular matrix of size ne'$_2$ x (n1+ni$_2$), where n1 is the number of nodes of the model 1 and ni$_2$ is the number of internal nodes of the model 2.

The Flex-CTM resulting from the coupling process is a RC network which can be solved with a basic Spice like transient simulator. Dirichlet conditions are applied on the model, to impose temperatures Ti on several external nodes of the Flex-CTM. Neumann conditions are applied on the model to impose heat transfer coefficients on surfaces of the system. That leads to plug convection resistors R$_{conv}$ (19) on the convection surface nodes.

$$R_{conv}=\frac{1}{hS} \quad\quad (19)$$

Where h is the heat transfer coefficient applied on the surface S of the model. Then, power sources Q are applied on junction nodes and the Flex-CTM can be simulated in a transient scenario (see figure 8).

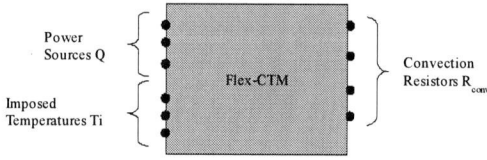

Figure 8: Simulation Condition Application

Evaluation of the performances of the Flex-CTM Methodology

The methodology is now evaluated with a simple test case. The studied system is a die in a Ceramic Pin Grid Array (CPGA) package. The geometry of the IC package is described figure 9.

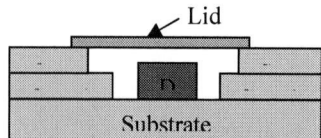

Figure 9: CPGA Description

978-1-4244-4722-0/09 $25.00
© 2009 IMAPS-ITALY

The set of material properties used to model the CPGA package are listed in table 1.

Table 1: Material Properties of the Different CPGA Parts

Part	Material	Thermal Conductivity $[W.m^{-1}.K^{-1}]$	Volumetric Heat Capacity $[J.m^{-3}.K^{-1}]$
Substrate		18	3247200
Step 1	Typical Alumina	18	
		Step 2	
		8	
	Lid	Kovar 17,5	3704400
Die	Silicon	159.5	1631000

The air layer surrounding the die is assumed to be insulating due to its very weak thermal conductivity. So, in this case a unique die-to-package heat flow path through the bottom face of the die is considered. The CPGA is simulated applying a uniform power source of 1W on the junction surface. The heat transfer coefficients applied on the external package surfaces are computed with a CFD tool. A reference measure of the average temperature on the Die junction is obtained by simulating a FEM model of the full circuit.

The first step of the Flex-CTM methodology splits the CPGA geometry according to the physical properties (see figure 10).

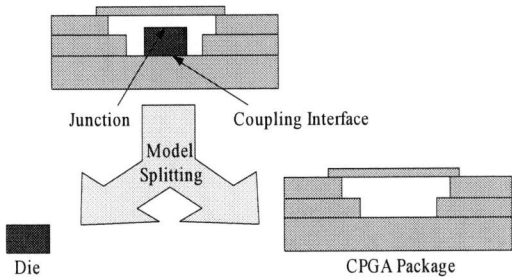

Figure 10: CPGA Description

The Junction interface is on the top of the Die and the coupling interface links the Die with the Substrate.

FEM models of both the Die and the CPGA package are extracted to retrieve the numerical systems G and C of both and their FEM node properties. Concerning the heat transfer coefficients, they are simulated with a commercial CFD tool to embed them in the CPGA Package. The nodes belonging to the Junction surface in the Die model, and those belonging to the coupling interface of both models are kept during the reduction process. The FEM Die model has 10 K nodes and the FEM package model has 150 K nodes. The Junction and the coupling interfaces have both 300 nodes. The simulation time of the reference measure takes 4 hours.

The node selection is performed to keep 64 nodes per interface. This is a medium trade-off between accuracy and complexity.

The reduction method is applied on the two numerical models. After the reduction process, the reduced model of the Die has 168 nodes and the reduced model of the package has 278 nodes. The two reductions take 15 minutes.

Once the two micro-models are available, they are coupled together through the coupling interface and the size of the resulting Flex-CTM is 382 nodes.

Finally, a power step source of 1 W is uniformly distributed on the Junction interface P'. The transient simulation of the Flex-CTM takes 2 seconds. This simulation and the reference are compared figure 11. The maximum of error reaches 0,1°C at t = 1,5e-4 s because certain frequencies are smoothed.

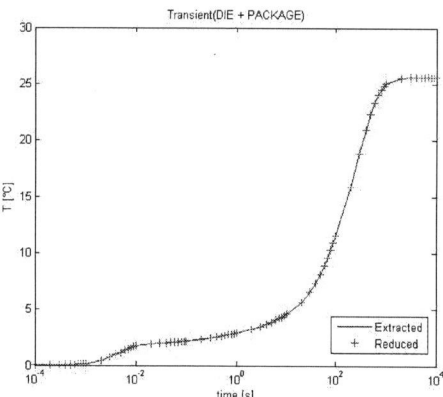

Figure 11: Original Model and Flex-CTM Temperatures

The comparisons in terms of accuracy, model size and simulation time between the original CPGA and the Flex-CTM are summarized in table 2. The time to build this Flex-CTM is about 25 minutes.

Table 2: Material Properties of the Different CPGA Parts

Model	Node Number	Simulation Time
Numerical CPGA model	150000	4 hours
Flex-CTM	382	2 seconds

Conclusions and Future Work

The Flex-CTM methodology meets the needs of electronic engineers to perform a fast temperature analysis of a complex electronic system at different integration levels. Flexible Compact Thermal Models are BCI, so they can be reused whatever the environment is. Many power sources can be applied on junction nodes allowing hotspot detection on a die. Moreover, Flex-CTM have a few node number, which allows multiple exploration or electro-thermal simulation in a constant window of time. Finally, the methodology enables system designers to share their work at different integration levels.

The methodology has been evaluated with a simple test-case at the package modeling level. The results show that Flex-CTM meet the specifications required specifically in terms of accuracy, and simulation time saving.

The next step is now to enhance the methodology with an automated selection of the number of nodes at the interfaces.

Besides, several test cases covering multi-level (die, package, board) design aspects are to be run.

References

Articles from conference proceedings

[1] C. Lasance, "The European Project PROFIT: Prediction of Temperature Gradients Influencing the Quality of Electronic Products", Proceedings of the SEMITHERM XVII, pp. 120 – 125, 2001

[2] W. Huang, K. Sankaranarayanan R.J. Ribando, M.R. Stan and K. Skadron, "An Improved Block-Based Thermal Model in HotSpot 4.0 with Granularity Considerations", Proceedings of the Workshop on Duplicating, Deconstructing, and Debunking (WDDD), in conjonction with the 34th International Symposium on Computer Architecture (ISCA), 2007.

[3] Hang Li, Pu Liu, Zhenyu Qi, Lingling Jin, Wei Wu, Sheldon X.D.Tan, and Jun Yang, "Efficient Thermal Simulation for RunTime Temperature Tracking and Management", Proceedings of the 2005 International Conference on Computer Design (ICCD'05).

Journal articles

978-1-4244-4722-0/09 $25.00
© 2009 IMAPS-ITALY

[4] H. Vinke and C. Lasance, "Compact Models for Accurate Thermal Characterization of Electronic Parts", IEEE Transactions on Components, Packaging and Manufacturing Technology – Part A, Vol. 20, NO. 4, December 1997

[5] F. Chrisitiaens, B. Vandevelde, E. Beyne, R. Mertens and J. Berghmans, "A Generic Methodology for Deriving Compact Dynamic Thermal Models, Applied to the PSGA Package", IEEE Transactions on Components, Packaging and Manufacturing Technology – Part A, Vol. 21, NO. 4, December 1998

[6] C. Lasance, "Highlights from the European Thermal Project PROFIT", Journal of Electronic Packaging, Vol 126, pp 565 – 570, December 2004

[7] L. T. Pillage and R.A. Rohrer, "Asymptotic waveform evaluation for timing analysis," IEEE Transactions CAD, Vol. 9, pp. 352-366, Apr. 1990.

[8] P. Feldman and R.W. Freund, "Efficient linear circuit analysis by Padé approximation via the Lanczos process," IEEE Transactions. CAD, vol. 17, pp. 645-654, Aug. 1998.

[9] L. Codecasa, D. D'Amore and P. Maffezzoni, "An Arnoldi Based Thermal Network Reduction Method for Electro-Thermal Analysis", IEEE Transactions on Components and Packaging Technologies, Vol. 26, No. 1, March 2003.

[10] D. Celo, X. Guo, P. K. Gunupudi, R. Khazaka, D.J. Walkey, T. Smy and M.S. Nakhla, "The Creation of Compact Thermal Models of Electronic Components Using Model Reduction," IEEE Transactions on Advanced Packaging, Vol. 28, No. 2 May 2005.

[11] T. Bechtold, E. B. Rudnyi, M. Graf, A. Hierlemann, J.G. Korvink, "Connecting Heat Transfer Macromodels for MEMS Array Structures", Journal of Mechanics and Microengineering, 15(6), pp 1205 – 1214, 2005

Book

[12] G. Strang, "Introduction to Applied Mathematics", Wellesley Cambridge. Press USA, 1986.

Web sites

[13] Francisco M. Gomes and Danny C. Sorensen, http://www.caam.rice.edu/software/ARPACK/

978-1-4244-4722-0/09 $25.00
© 2009 IMAPS-ITALY

Impact of Package Parasitics on the EMC Performance of Smart Power SoCs

Marco Merlin†, Franco Fiori‡

†Istituto Superiore Mario Boella, via P.C. Boggio 61, 10138 Torino, Italy

‡Politecnico di Torino, Corso Duca degli Abruzzi 24, 10129 Torino, Italy

†{+39 011 2276711, merlin@ismb.it}, ‡{+39 011 5644141, franco.fiori@polito.it}

Abstract

The paper deals with the propagation of EMI in Smart Power SoCs through high frequency parasitic paths, i.e. silicon substrate, and it focuses on the interaction between on-chip parasitic capacitors and package parasitic elements, that negatively affects IC electromagnetic emission. The paper highlights such unwanted parasitic effects through computer simulations and experimental test results. Finally, a new grounding scheme to improve the EMC performance of such integrated circuits is presented.

Key words: electromagnetic emission, package parasitics, switching noise, smart power, substrate grounding

Introduction

In the last decades, the strong development of semiconductor technologies, has led to smaller and more performing electronic circuits, which nowadays often include a microcontroller that senses and controls the system which is intend for, through analog, digital and power front-end devices. Aiming at the reduction of costs, while improving system performance and reliability, such front-end circuits are often integrated in the same silicon die, usually called "smart power" System-on-Chip (SoC) or front-end ASICs (see Fig. 1).

Such Integrated Circuits (ICs) are often directly connected to sensors and actuators through printed circuit board (PCB) traces and cables, so that disturbances originated by on-chip switching circuits (charge pumps, power transistors and core logic gates) drive such interconnects that behave as unintentional antennas, delivering unwanted electromagnetic emission (EME).

To this purpose, it should be noted that interference driven by switching power transistors directly connected to cables can be easily figured out, while those related to on-chip switching circuits could be difficult to explain.

To accomplish this task, the parasitic coupling of analog and digital building blocks of such SoCs with the power section through the silicon substrate and through package parasitic elements should be taken into account. To this purpose, the paper analyzes the parasitic coupling among components integrated in smart power SoCs through the silicon substrate, and based on that a new grounding scheme to reduce electromagnetic emissions is presented. The paper is organized as follows: Section II summarizes some important issues of smart power technology

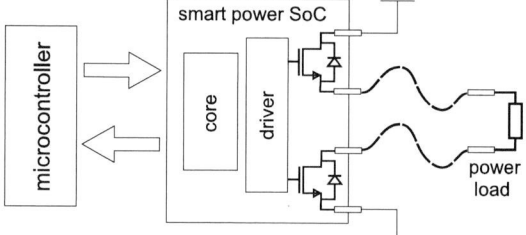

Figure 1: Electronic unit including a smart power SoC.

processes, hence Section II.A shows an equivalent circuit describing the parasitic coupling of components through the silicon substrate. Section II.B describes a new grounding scheme that significantly reduces electromagnetic emissions of smart power ICs. Section III shows the procedure we followed to build up an electric EME equivalent model of a real smart-power SoC. In section IV, experimental results that prove the effectiveness of the proposed solution are shown. Finally, Section V draws some concluding remarks.

Electromagnetic Emission in Smart Power SoCs

As mentioned in the introduction, improvements in mixed-signal technologies, have boosted the integration into a single chip of power, analog and digital sections, leading to lower costs and improved performance. In this context, the technology process to be employed for the design of application specific ICs should be carefully selected, in order to fulfill both electrical and thermal specifications, as well as intra-compatibility issues, while keeping the overall chip area at a minimum. To this purpose, smart power technology processes, can be qualified through several basic features among which the list

978-1-4244-4722-0/09 $25.00
© 2009 IMAPS-ITALY

of available components (design kit), voltage classes, lithography and substrate isolation type [2]. Currently, junction-isolated processes are largely used in mass production basically for their low cost, even though they are critical in terms of substrate parasitic coupling (latchup, parasitic bipolar transistors), high temperature and leakage currents. Such problems can be overcome with silicon-on-insulator (SOI) technology processes, in which oxide isolation is exploited. As a drawback, such processes show greater thermal resistance and higher production costs. However, both technologies show a parasitic coupling of surface components through the silicon substrate, which could be effective at high frequency affecting the EMC performance of smart power SoCs.

A. Substrate Parasitic Coupling

With reference to junction-isolated process, it should be noted that parasitic coupling among SoC subcircuits can take place through substrate contacts (resistive) as well as through p-n reverse-biased isolation junctions (capacitive) [3][4]. Based on that, and considering those smart power SoCs that include at least a digital core block and a power transistor, the parasitic coupling through the silicon substrate can be modeled by the equivalent circuit of Fig. 2. Here, the digital core is represented by a time-variant current source (I_{core}), in parallel with the series of the core capacitance C_{core} and the resistance R_{core}. In this circuit, the capacitance C_{sc} describes the parasitic coupling of core reverse-biased isolation junctions with the substrate, while R_{sc} models the substrate contacts. Similarly, parasitic coupling of a power transistor with the substrate is modeled by the capacitance C_{pwr}. In the same figure, the package model is recalled by means of inductors, that connect pads on silicon (crossed square) to printed circuit board components. Furthermore, this circuit includes PCB level components like the logic gates power supply bypass capacitor C_b, an EMI connector filter (C_f) and the common mode inductance L_{cm}. This last element is strictly related to the PCB layout.

With reference to such an equivalent circuit, it can be noticed that a part of the switching current (I_{core}) flow through the common-mode inductance (L_{cm}) so that common-mode EME delivered by cables is experienced. To this purpose, it is worth mentioning that EME spectra are strictly related to PCB parasitic elements (like L_{cm} in Fig. 2), which cannot be controlled or taken into account by IC designers. Nonetheless, it can be observed that most of the switching current sunk by the core block is taken from the internal core capacitance C_{core} and the

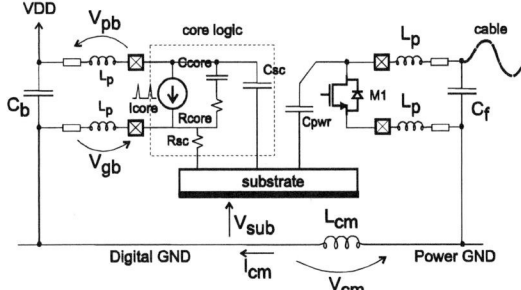

Figure 2: Simplified schematic view of the parasitic substrate coupling in common smart-power system-on-chips.

external bypass capacitance C_b, while only a small part of it is taken from off-chip components and parasitic paths. Switching currents that flow through package interconnections are mostly responsible for electromagnetic emissions, particularly those flowing in the IC power supply interconnects [5], because the ground and power bounce they drive, could reach all the IC pins through the substrate parasitic couplings.

Depending on the overall resistance of substrate contacts of the digital block (R_{sc} in Fig. 2), the current loop that involves the common-mode impedance can be closed (predominantly) either through the V_{DD} pin or through the *GND* pin. In case of low-resistance substrate contact (i.e. $R_{sc} < 1\Omega$), a part of the core supply current flows through the external bypass capacitor C_b and substrate voltage fluctuations are mostly driven by the *ground-bounce* (V_{gb}). On the contrary, for high-resistance substrate contacts, the substrate voltage fluctuations are mostly driven by the power bounce (V_{pb}). In both cases, the substrate is involved by unintentional current loops that drive electromagnetic emission.

B. EME Mitigation

In order to reduce the magnitude of the common mode current I_{cm}, which is mostly responsible of EME, different strategies have

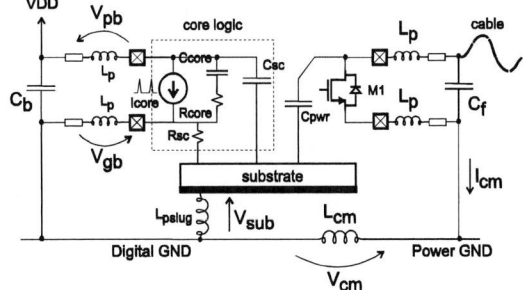

Figure 3: Substrate to ground connection.

978-1-4244-4722-0/09 $25.00
© 2009 IMAPS-ITALY

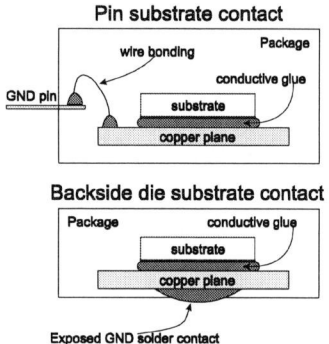

Figure 4: Proposed grounding solutions.

already been explored, among which:

- use of package with reduced size and/or increased number of parallel power supply and ground pins to reduce the equivalent inductance of power supply loops [7],
- increment of on-chip bypass capacitor to reduce the magnitude of the current flowing through package interconnects [8],
- use of circuit topologies and proper architecture of digital building blocks for power supply current shaping [9],
- use of proper grounding topologies which reduce substrate voltage bounce [6].

This paper points out through the analysis of experimental test and computer simulation results the key role played by the silicon substrate on the propagation of switching noise in smart-power SoC.

To this purpose the effect of connecting the die backside contact to the PCB ground has been explored. This additional connection, which is described in Fig. 3 by the parasitic inductance L_{pslug} acts as a current shunt, preventing the switching noise current to flow through the common mode inductance L_{cm}, thus reducing substrate voltage bounce as well as EMEs. In this circumstance the digital ground pin should not be connected to the PCB ground avoiding the ground bounce to drive common mode currents. Unluckily, the mutual coupling of package parasitic inductances reduces such a decoupling effectiveness, especially at high frequency. For this reason, the way the package realizes the substrate contact significantly affects the emission levels.

EME Parasitic Model

With the aim to analyze the effect of different substrate grounding solutions on the EME of smart power SoCs, two identical integrated circuit encapsulated into the same package type, with and without an exposed backside pad have been

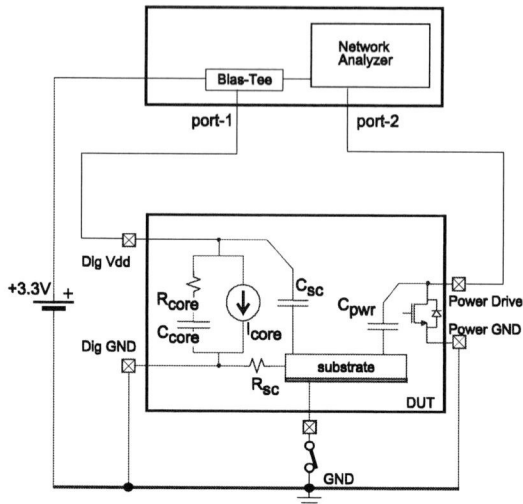

Figure 5: Set up for scattering measurements.

considered. In particular, the substrate of the first device under test (DUT) has been grounded through one of the package pin, which has been connected to the die backside with a down-bond, while in the second DUT, the die backside is directly accessible for soldering on the package backside, as it is sketched in Fig. 4. Based on that, three different solutions for substrate grounding have been considered:

(a) substrate connected to ground by mean of metal-substrate contacts through the GND pin of core logic gates,
(b) die backside contact connected to ground through a dedicated pin,
(c) die backside contact connected to ground through an exposed package backside pad.

These two DUTs have been mounted in two identical test boards, making their pins accessible through RF probe contacts. The digital *GND* pin has been tied to the PCB ground. Then, the scattering parameter matrix between the digital core power supply port and the drain contact of one of the SoC power transistors has been measured, using the Network Analyzer HP8753E.

The above mentioned substrate grounding connections have been considered for experimental tests, which have been carried out providing to the DUT input and output ports proper DC bias voltages (see Fig. 5). Furthermore, to avoid the measurement being affected by time-variant current absorption of switching logic gates, the input clock signal has not been provided to the DUT and the clock input pin has been connected to ground.

978-1-4244-4722-0/09 $25.00
© 2009 IMAPS-ITALY

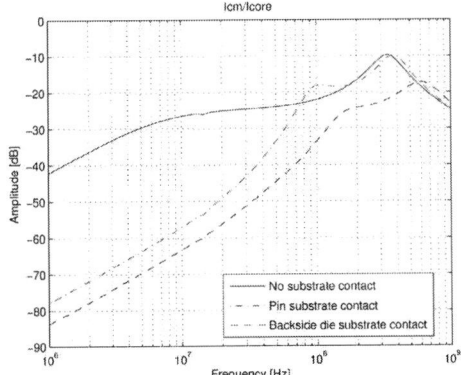

Figure 6: Comparison of scattering measurements in different grounding configurations.

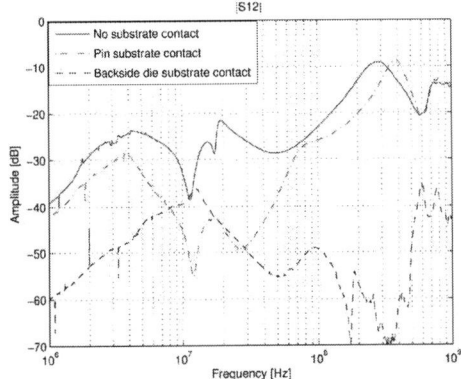

Figure 7: Transfer function $\dfrac{I_{cm}}{I_{core}}$ **based on the equivalent passive model, for different grounding solutions.**

S-parameter measurements have been performed for each one of the above mentioned substrate grounding scheme, and it has been observed that the transmission of RF power from the logic core to one of the power transistor is significantly reduced if the package backside ground contact is connected to the PCB ground. This result is highlighted in Fig. 6 where the transition scattering parameter of the different grounding schemes are reported.

This result confirms that the more the substrate is well-connected to ground, the less disturbances are allowed to spread around the integrated circuit.

The remainder of this section shows the procedure we followed to evaluate the parameters of the equivalent circuit (like that shown in Fig. 3) that describes the parasitic coupling between signal and power sections. The same DUTs of S-parameters measurements have been considered. To this purpose, package parasitics have been obtained by

means of the Ansoft Q3D Extractor® [10], the resistance R_{sc} has been obtained by DC measurements, the capacitances C_{sc}, C_{pwr} and substrate network have been extracted on the basis the SoC layout view, technology process parameter using the method described in [3], while the core equivalent parameter C_{core} and R_{core} have been extracted to get a good fitting with S-par measurements. Using this model (like that in Fig. 3) the transfer functions $\dfrac{I_{cm}}{I_{core}}$ reported in Fig. 7 have been obtained. These simulation results highlight the effectiveness of the substrate grounding through the package backside exposed pad.

In order to complete the above described model, the current source I_{core} has been evaluated in the frequency domain under the assumption that logic functions were implemented into the logic core by means of a synchronous architecture. Based on that, such a current has been assumed to be a triangular shaped periodic signal (Fig. 8), with period $T = \dfrac{1}{f_c}$, (f_c is the clock frequency), peak value I_p and pulse width t_p. This last parameter is usually ranges from $\dfrac{T}{10}$ to $\dfrac{T}{4}$. This waveform can be written as the convolution of a triangular function with a Dirac train, hence

$$i_c(t) = I_p \cdot tri\left(\frac{t}{t_p}\right) * \left[\sum_n \delta(t - nT)\right].$$

Alternatively, this current can be expressed in the frequency domain as

$$I_c(f) = \frac{2I_p}{t_p} \Im\left[rect\left(\frac{t}{t_p}\right) * rect\left(\frac{t}{t_p}\right) * \sum_n \delta(t - nT)\right]$$

$$= \frac{2I_p}{t_p} \left\{ \Im\left[rect\left(\frac{t}{t_p}\right)\right]\right\}^2 \cdot \left[\sum_n \delta(t - nT)\right]$$

$$= \frac{2I_p}{t_p} sinc^2\left(t_p f\right) \cdot \left[\sum_n \delta\left(f - \frac{n}{T}\right)\right]$$

which can be approximated by the low-pass filter asymptotic function having cutoff frequency at $f_p = \dfrac{1}{t_p}$ and magnitude rolling off -40 dB/dec.

Finally, the magnitude of the current spectrum can be related to the DC current sunk by switching logic gates through the following expression

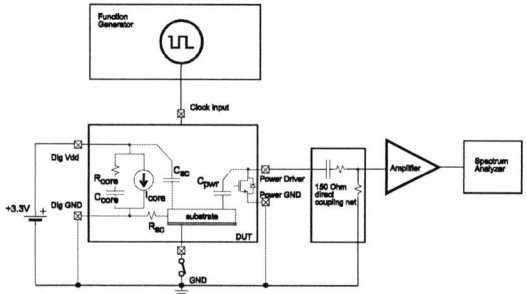

Figure 9: Set up for conducted emission measurements compliant to IEC61967-4, 150 Ω direct coupling method.

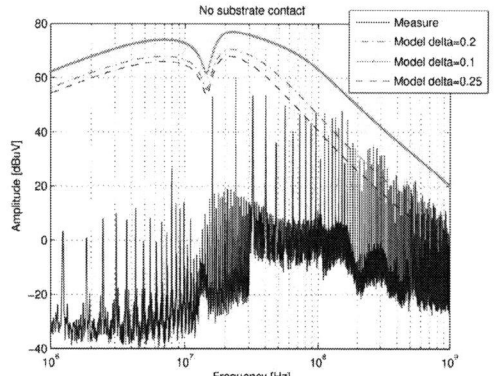

Figure 10: Conducted emissions measurements with no substrate contact.

$$I_{DC} = \frac{1}{T}\int_0^T i_c(t)\cdot dt = \frac{I_p}{2}\cdot \frac{t_p}{T},$$

which can be rewritten as

$$I_p = \frac{2}{\Delta}\cdot I_{DC},$$

in which $\Delta = \dfrac{t_p}{T}$.

Next section shows the comparison of model prediction with experimental results for three different substrate grounding solution and a few values of the Δ parameter.

Experimental Validation

The conducted EME of the DUTs, which have been previously characterized in terms of scattering parameters, have been evaluated with of a test setup sketched in Fig. 9, which is compliant to the standard IEC-61967-4 (150 Ω method) [1]. In this setup, The power supply voltage ($V_{DD}=3.3V$) and the clock signal ($f_c=8MHz$) are provided to the DUT, while conducted emissions has been measured at the drain terminal of one of the no switching DUT power transistors using a 32dB low-noise amplifier and a spectrum analyzer.

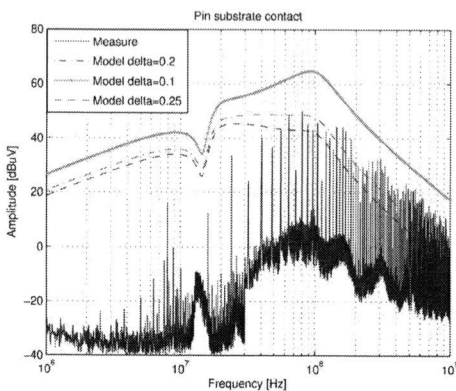

Figure 11: Conducted emissions measurements with substrate connected to GND through a dedicated pin.

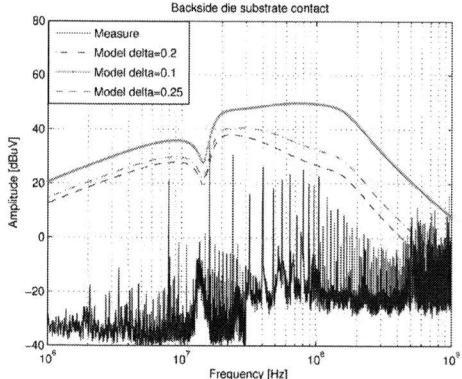

Figure 12: Conducted emission measurements in case of the package backside pad connected to GND.

Fig. 10, 11, 12 show the results of emission measurements, which refer to three different substrate grounding solutions.

Such experimental results confirm what has been obtained with scattering parameter measurements: a low impedance connection of the die backside to the PCB signal ground strongly attenuate conducted EME. Furthermore, these last figures show the comparison of experimental results and model predictions, which have been obtained for different values of the Δ parameter.

Conclusions

In this paper, the propagation of switching noise in Smart Power SoC through unwanted parasitic paths has been investigated and the interaction of package parasitics with on-chip parasitic capacitors has been analyzed.

An equivalent model for EME analysis has been proposed and the important role of the substrate grounding to control electromagnetic emission has been highlighted. To this purpose, three different

978-1-4244-4722-0/09 $25.00
© 2009 IMAPS-ITALY

substrate grounding solutions have been analyzed through measurements and computer simulations, and it has been shown that the substrate grounding through a package backside exposed pad provides significant EME reduction.

References

[1] IEC 61967 "Integrated circuits – Measurement of electromagnetic emissions – 150kHz to 1GHz", 2005.

[2] B. Murari et al., "Smart Power ICs: technology and applications", 2nd ed., Springer-Verlag, Berlin, 2002.

[3] P. Crovetti, F.Fiori, "Efficient BEM-based Substrate Network Extraction in Silicon SoCs", Microelectronics Journ., Vol.39 pp.1774-1784, 2008, ISSN: 0959-8324.

[4] M. Van Heijningen, M. Badaroglu, S. Donnay, G. Gielen, H. De Man, "Substrate noise generation in complex digital systems: efficient modeling and simulation methodology and experimental verification", Journ. of Solid-State Circ., Vol.37, no.8, pp. 1065-1072, Aug 2002.

[5] P. Larsson, "Resonance and Damping in CMOS Circuits with On-Chip Decoupling Capacitance", IEEE Transactions on Circ. and Systems-I: Fundamental Theory and Applications, Vol. 45, No. 8, Aug. 1998, pp. 849-858.

[6] F. Fiori, "Reducing the Electromagnetic Emissions of Smart Power System-on-Chips by Design", 20th Int. Zurich Symposium on EMC, Zurich 2009.

[7] E. Sicard et al. , "EMC of Ics: Techniques for low emission and susceptibility", Springer, 2006, ISBN: 0-387-26600-3.

[8] I. Blunno, F. Gregoretti, C. Passerone, D. Peretto, L. M. Reyneri, "Designing low electromagnetic emissions circuits through clock skew optimization", 9th IEEE International Conference on Electronics, Circuits and Systems, Dubrovnik, Croatia, September 15-18, pp. 417-420, 2002.

[9] F. Gregoretti, N. Andrikos, F. Musolino, "Comparative measurements and analysis of electromagnetic emissions of synchronous and asynchronous processors", in MIPRO 2008 31st International Convention on Information and Communication Technology and Electronics, Opatija, Croatia, May 26-30, pp. 202-207, 2008, ISBN/ISSN: 978-953-233-036-6.

[10] "Q3D Extractor v8 – 3D/2D Parasitic Extraction for High-Performance Electronic Designer (datasheet)", available at www.ansoft.com/products/si/q3d_extractor/.

Characteristics of Electrically Conductive Adhesives filled with Copper nanoparticles with Organic Layer

L-N. Ho[*], H. Nishikawa[*], T. Takemoto[*], Y. Kashiwagi[**], M. Yamamoto[**], M. Nakamoto[**]

[*]Joining and Welding Research Institute, Osaka University, Osaka 567-0047, Japan
[**]Osaka Municipal Technical Research Institute, Osaka 536-8553, Japan

+81-6-6879-8691, nisikawa@jwri.osaka-u.ac.jp

Abstract

Electrically conductive adhesives (ECAs) have been investigated for the use in microelectronics packaging as a lead-free solder substrate. As metal filler in conductive adhesives, silver is the most commonly used due to its high conductivity and stability. However, the cost of conductive adhesives with silver fillers is much higher than conventional lead-free solders and silver has poor electro-migration performance. Copper is a promising candidate for conductive filler metal due to its low resistivity and low cost, but oxidation causes this metal to lose conductivity. In this study, ECAs using surface-modified copper fillers were developed. In particular, to overcome the problem associated with the oxidation of copper, copper nanoparticles were coated with organic substance from when nanoparticles were made. The organic-coated copper was tested as a filler metal. Especially the effect of organic layer on the electrical resistance just after curing and after high-temperature exposure was investigated. As a result, it was found that the electrical resistance of ICA with organic-coated copper filler was more stable than that of ICA with pure copper filler.

Key words: Electrically conductive adhesives, copper nanoparticle, organic layer, oxidation,

1. Introduction

Electrically conductive adhesives (ECAs) have been investigated for use in microelectronics packaging and interconnections as a lead-free solder substitute due to such advantages as low curing temperature. They are usually composed of conductive metal fillers and polymer resin. The metal fillers and polymer resins provide electrical and mechanical interconnections between surface mount device components and a substrate. For polymer resins, thermosetting resins such as epoxy, polyimide and polyurethane are commonly used. For the metal fillers, gold (Au), silver (Ag), copper (Cu) and nickel (Ni) in various sizes and shapes are used. Among them, silver is the most commonly used due to its high conductivity and stability. However, the cost of conductive adhesives with silver filler is much higher than the conventional lead-free solders. Therefore, currents ECAs still have some limitations in terms of cost as well as electrical and mechanical properties as replacements for lead-free solder.

Many research efforts have focused on the improvement of electrical conductivity and the reliability of ECAs joints [1-5]. Additionally, the replacement of expensive Ag fillers by other metal fillers or new materials has been examined for wider applications of ECAs [6-8]. Copper is a promising candidate for conductive filler metal due to its low resistivity, low cost and good electro-migration performance. However, a problem for copper fillers

is oxidation of the copper surface and deterioration of the ECA electrical properties. There are basically two approaches for the surface treatment of copper fillers to prevent oxidation of the metal surface. On is an inorganic material coating and the other is an organic coating. For example, silver-coated copper powders have been investigated as metal fillers in ECAs [9,10]. However, the thermal stability of these coatings is a concern because their effectiveness is easily lost when exposed to curing conditions and various environmental conditions.

In this study, ECAs using surface-modified copper fillers have been developed. In particular, to overcome the problem of high electrical resistance associated with the oxidation of copper, copper nanoparticles were coated with organic substance from when nanoparticles were made. The organic-coated copper was tested as a filler metal. The effect of organic layer on the electrical resistance of ECAs just after curing and after reliability tests was investigated.

2. Experimental

2.1 Materials

In order to overcome the degradation of electrical resistance associated with the oxidation of copper, the organic-coated copper nanoparticles were prepared through a controlled-thermolysis method [11, 12]. Copper nanoparticles were coated with organic substance from when nanoparticles were made. Besides, commercial copper filler

978-1-4244-4722-0/09 $25.00
© 2009 IMAPS-ITALY

provided by Mitsui Mining & Smelting Corp. was used as a control sample to compare with the organic-coated copper filler. Phenolic resin was applied as a polymer matrix. Metallic filler was incorporated into the thermoset resin as the adhesive matrix by mixing in an agate mortar. It was then mixed and defoamed in a hybrid defoaming mixer (Model: ARE250) from Thinky Corp. The contents of ECAs prepared in this study are shown in Table 1. ECA Cu-1 was mainly composed of commercial copper filler and phenolic resin. ECA Cu-2 was mainly composed of organic-coated copper nanoparticles and phenolic resin.

2.2 Characterization of metallic fillers

Morphology of metallic fillers is observed by using field emission scanning electron microscopy (Model: JSM-6500F) from JEOL. Powder X-ray diffraction of metallic fillers and as-cured ECAs was performed on a standard instrument JDX-3530M from JEOL. The samples were measured at 2θ from 10° to 100°. Besides, thermal gravimetric analysis of metallic fillers was done on a standard instrument TG/DTS 6200 of Exstar 6000. The sample was heated at a heating rate of 5 °C/min under an air purge of 50 ml/min from ambient temperature to 200 °C and hold at 200 °C for 1 h.

2.3 Preparation of sample for electrical resistivity measurement

FR-4 board with copper pads at both ends was used to measure the electrical resistivity of ECAs. Two parallel strips of cellophane tape were placed apart along the length of a standard 50 × 95 mm. Then, another two strips of the tape were placed perpendicular to the parallel strips in order to create a test specimen with 25 mm length and 5 mm width as shown in Fig. 1. Using a clean stainless steel spatula, the adhesive was spread on to the space between the tape strips. A total of 5 specimens

Table 1   Content of electrically conductive adhesives in this study.

Conductive adhesives	Metallic fillers	Filler content (mass %)	Curing temperature (°C)
ECA Cu-1	Commercial Cu filler	80	175
ECA Cu-2	Cu nanoparticles	80	175

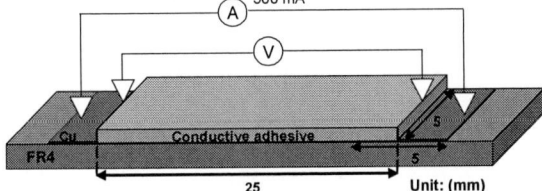

Fig. 1   Schematic diagram of sampe for measurement of electrical resistance of ECA.

were prepared on a single FR-4 board for each sample. After that, all the tapes were removed and the FR-4 board with ECA was place into a convection oven for curing at 175 °C for 1 h. Then, some samples were subjected to high-temperature exposure at 125 °C for 1000 h in air to investigate the thermal reliability and stability of the ECAs. The four-point probe method, shown in Fig. 1, was applied to measure the electrical resistance of the ECAs using a nanovoltmeter (Model: 2182A) and a precision current source (Model: 6220) from Keithley. Besides, the size of the cured ECA samples on the FR-4 board were measured by using a CCD laser displacement sensor (Model: LK-G series) from Keyence together with software MAP-3D from COMS.

3. Results and discussion

3.1 Morphology of metallic filler

The morphology of the metallic fillers applied in ECAs in this study was observed by using SEM. Figure 2 (a) and (b) show the SEM images of commercial copper filler and organic-coated copper filler respectively. For the commercial copper filler,

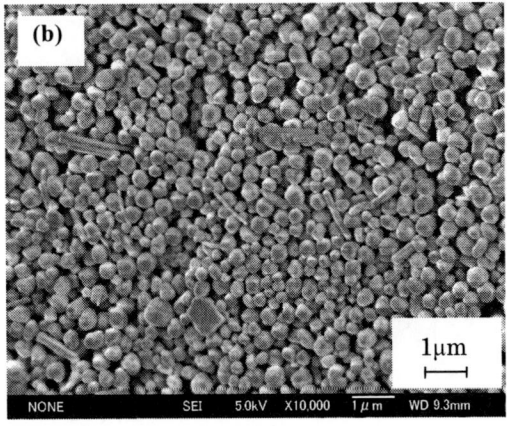

Fig. 2   SEM images of commercial copper filler (a) and organic-coated copper filler (b).

978-1-4244-4722-0/09 $25.00
© 2009 IMAPS-ITALY

agglomerates formed by nanocrystalline (46 – 110 nm) could be observed and the size of the agglomerates was about 0.14 to 0.43 µm. For organic-coated copper filler, some particles in the form of rod could be observed among the uniform spherical particles. Most of the spherical particles are within the size range of 0.14 – 0.5 µm, and a very small amount of particles with size up to 0.9 µm could be observed.

3.2 X-ray diffraction (XRD)

Figure 3 shows the XRD pattern of the metallic fillers used in this study. The diffraction peaks at 2θ = 43.30º, 50.43º and 74.13º correspond to the *fcc* structure of Cu (h k l) (1 1 1), (2 0 0) and (2 2 0) planes with the space group of *Fm3m* (JCPDS No. 04-0836) could be observed in both commercial copper filler and organic-coated copper filler.

3.3 Thermal gravimetric analysis (TGA)

The metallic fillers used in this study were investigated by using TGA method. The metallic fillers were heated in the atmosphere of air up to 200 ºC and held at 200 ºC for 1 h. Figure 4 depicts the TGA profiles of the metallic fillers. It could be observed that both metallic fillers show weight gain

during holding at 200 ºC for 1 h due to the oxidation of the particles in the atmosphere of air. The commercial copper filler showed slightly small weight gain compared to the organic-coated copper filler. This may be due to the commercial copper fillers are more susceptible to oxidation compared to the organic-coated copper filler. This means that copper fillers with organic substance at the surface of the particles are more effective in preventing rapid oxidation of the particles at high temperature compared to that of the commercial copper filler.

3.4 Resistivity of as-cured ECAs

Figure 5 shows the electrical resistivity of ECAs immediately after curing at 175 ºC for 1 h. By comparing the resistivity of the as-cured ECAs, the ECA filled with the commercial copper filler (ECA Cu-1) possessed relatively high electrical resistivity, which is about 30 times of the ECA filled with the organic-coated copper filler (ECA Cu-2). This could be attributed to the presence of oxides in the as-cured ECA Cu-1 as shown in Fig. 6.

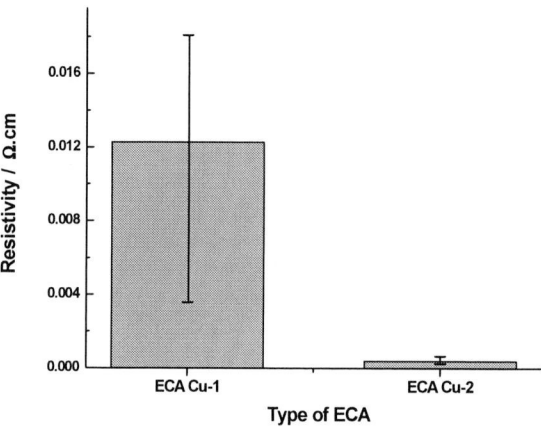

Fig. 5 Resistivity of as-cured ECAs in this study.

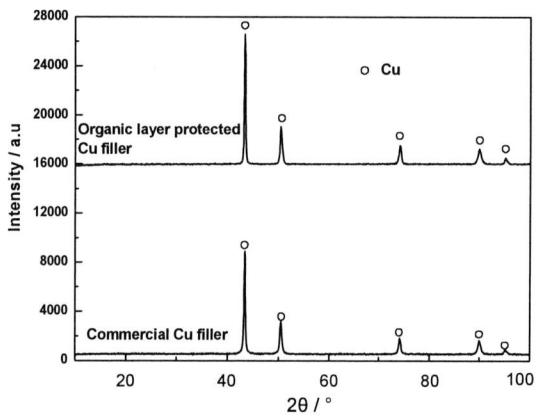

Fig. 3 XRD patterns of commercial Cu filler and organic –coated Cu filler, respectively.

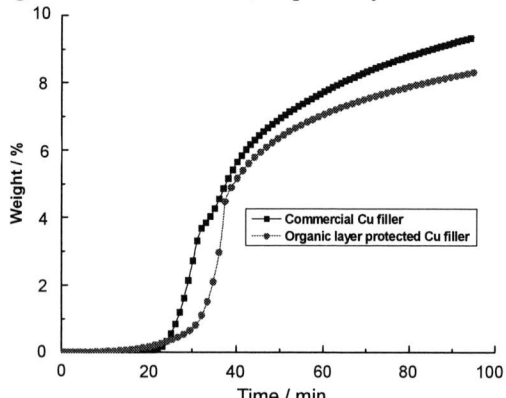

Fig. 4 TGA profiles of commercial Cu filler and organic-coated Cu filler in an atmosphere of air

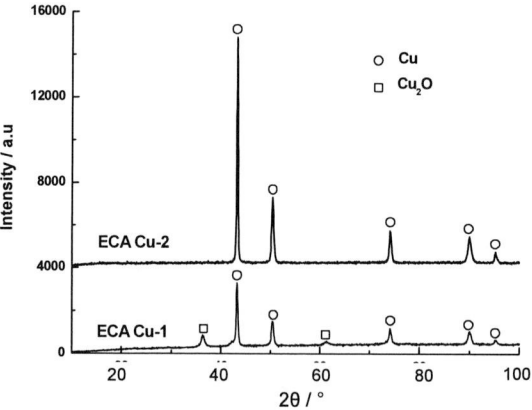

Fig. 6 XRD patterns of as-cured ECA Cu-1 and as-cured ECA Cu-2, respectively.

978-1-4244-4722-0/09 $25.00
© 2009 IMAPS-ITALY

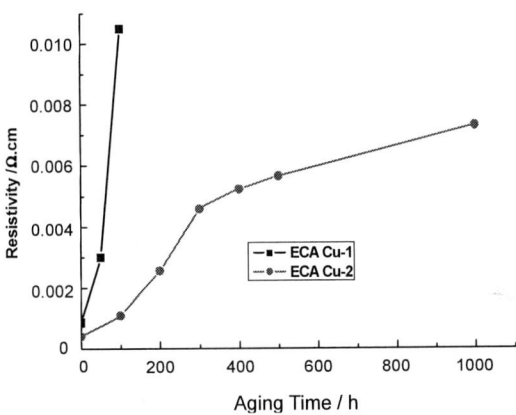

Fig. 7 Resistivity of ECAs during high temperature exposure at 125 °C for 1000 h.

3.5 High-temperature exposure

ECAs prepared in this study were subjected to high-temperature exposure at 125 °C for 1000 h in order to investigate the long-term thermal stability and reliability of the ECAs. The change of the electrical resistivity of the ECAs as a function of the high-temperature exposure test is shown in Fig. 7. The resistivity of the ECA filled with the commercial copper filler (ECA Cu-1) increased rapidly in the first 100 h of high-temperature exposure, indication the lost of electrical conductivity. On the other hand, the electrical resistivity of the ECA filled with the organic-coated copper filler (ECA- Cu-2) remained stable throughout 1000 h with a final resistivity of 7.2×10^{-3} $\Omega \cdot$ cm. This means that the ECA Cu-2 is thermally stable and resistant to oxidation at temperature as high as 125 °C up to 1000 h. Since Cu is well known for being easily oxidized in air as could be observed in the ECA filled with the commercial copper filler, the results of the ECA Cu-2 could be appreciated. This could be due to the existence of an organic substance on the surface of the copper nanoparticles, which prevents rapid oxidation of the ECAs under high-temperature exposure and hence lead to consistent resistivity over 1000 h.

4. Conclusion

In summary, it can be concluded that the as-cured ECA filled with the commercial copper particles (ECA Cu-1) has a relatively high resistivity compared to the ECAs filled with the organic-coated copper nanoparticles (ECA Cu-2) due to the presence of copper oxides in the as-cured ECA Cu-1. Compared to the ECA Cu-1, the ECA Cu-2 was more thermally stable with consistent resistivity at 125 °C over 1000 h. The low and consistent resistivity observed in the ECA Cu-2 during high-temperature exposure may be due to the existence of the organic substance on the surface of the copper

particles, which could effectively prevent oxidation of copper inside the ECA.

Acknowledgements

This work is a part of Osaka Central Area Industry-Government-Academia Collaboration Project on "City Area Program" sponsored by MEXT, 2007-2009. Special thanks are given to N. Terada from Harima Chemical Inc. for kindly providing us the phenolic resin in this study.

References

[1] H.K. Kim and F.G. Shi, "Electrical reliability of electrically conductive adhesive joints: dependence on curing condition and current density", Microelectronics Journal, Vol. 32, pp. 315-321, 2001.

[2] H. Jiang, K-S. Moon, J. Lu and C.P. Wong, "Conductivity enhancement of nano silver-filled conductive adhesives by particle surface functionalization", Journal of Electronic Materials, Vol. 34, No. 11, pp. 1432-1439, 2005.

[3] W.J. Jeong, H. Nishikawa, H. Gotoh and T. Tadashi, "Effect of solvent evaporation and shrink on conductivity of conductive adhesive", Materials Transactions, Vol. 46, No. 3, pp. 704-708, 2005.

[4] W.J. Jeong, H. Nishikawa, D. Itou and T. Takemoto, "Electrical characteristics of a new class of conductive adhesive", Materials Transactions, Vol. 46, No. 10 , pp. 2276-2281, 2005.

[5] F.M. Coughlan and H.J. Lewis, "A study of electrically conductive adhesives as a manufacturing solder alternative", Journal of Electronic Materials, Vol. 35, No. 5, pp. 912-921, 2006.

[6] S. K. Kang, S. Buchwalter and C. Tsang, "Characterization of electroplated bismuth-tin alloys for electrically conducting materials" Journal of Electronic Materials, Vol. 29, No. 10, pp. 1278-1283, 2000.

[7] H-H. Lee, K-S. Chou and Z-W Shih, "Effect of nano-sized silver particles on the resistivity of polymeric conductive adhesives", International Journal of Adhesion & Adhesives, Vol. 25, pp. 437-441, 2005.

[8] H.P. Wu, J.F. Liu, X.J. Wu, M.Y Ge, Y.W. Wang, G.Q. Zhang, J. Z. Jiang, "High conductivity of isotropic conductive adhesives filled with silver nanowires", International Journal of Adhesion & Adhesives, Vol. 26, pp. 617-621, 2006.

[9] M. Yamamoto and M. Nakamoto, "Novel preparation of monodispersed silver nanoparticles via amine adducts derived from insoluble silver myristate in tertiary alkylamine", Journal of Materials Chemistry, Vol. 13, pp. 2064-2065, 2003.

[10] Y. Kashiwagi, M. Yamamoto and M. Nakamoto, "Facile size-regulated synthesis of silver nanoparticles by controlled thermolysis of silver alkylcarboxylates in the presence of alkylamines with different chain lengths", Journal of Colloid and Interface Science, Vol. 300, pp. 169-175, 2006.

978-1-4244-4722-0/09 $25.00
© 2009 IMAPS-ITALY

Design and Fabrication of Corrosion and Humidity Sensors for Performance Evaluation of Chip Scale Hermetic Packages for Biomedical Implantable Devices

Nooshin Saeidi[1, 2], Andreas Demosthenous[1], Nick Donaldson[1], John Alderman[2]

[1] University College London - Gower Street - London - WC1E 6BT -UK- ☎+44 (0)20 7679 2000
[2] Tyndall National Institute - Lee Maltings - Prospect Row – Cork – Ireland - ☎+353 21 4904177

nsaeidi@ee.ucl.ac.uk

Abstract

In this paper we report on the development of a set of test chips that can be used to verify the sealing techniques with biomedical implants as their target application. These test chips have been designed and fabricated based on the Tyndall's in-house CMOS 1.5 micron technology. Each chip includes humidity, temperature and/or corrosion sensors. Different designs and sizes have been considered for these sensors to compare the sensors sensitivity and to select the optimum ones for the final chip. The sensors are being tested and the results obtained for two of the humidity sensors are presented.

Key words: Biomedical Packaging, Hermeticity Testing, Test Chip, Capacitive Humidity Sensors,

Introduction

Packaging of biomedical devices is very crucial due to strict requirements posed by their target applications and differs significantly from those of conventional ICs. The reliability of a biomedical device is highly depended on the performance of its package. Also, more challenges emerge when the size of the device decreases drastically or the device is intended to work in human body for decades [1, 2].

To verify a packaging technology which has been chosen or developed for biomedical devices, a certain mechanism is required to validate the performance of the technology. In this paper, we report design and fabrication of a series of test chips, consisting of humidity sensors and corrosion detectors which will be used to evaluate the packaging technology for miniaturized biomedical implantable devices. These sensors are able to monitor moisture and corrosion levels inside the package Thus they can be used as a measure for the reliability of the packaging technology.

Three test chips have been designed based on CMOS 1.5 micron technology, consisting of a number of humidity, corrosion and temperature sensors. Corrosion detectors and temperature sensors are single metal layer structures while humidity sensors are single metal layer capacitive structures covered by a layer of plain or porous polyimide. Also, structures with different line widths and line gaps have been designed to evaluate the effect of

these factors on the accuracy and stability of the sensors.

Apart from discussing design and process steps for these test chips, we also demonstrate experimental results using these structures.

Fabrication

The test chip was designed based on three individual layouts each of which consists of various sensors to measure humidity, temperature and corrosion.

Overall size of each layout is 6 mm × 4.4 mm and three wafers were used for fabrication of the test chips. All the sensors in the test chips were designed based on single metal layer structures. To simplify the process and reduce the steps, for the first wafer only metal deposition and patterning steps were performed, while for the other two wafers, a layer of polyimide was also deposited which acts as a dielectric layer for the humidity sensors.

The fabrication process for the test chips is straight forward. As depicted in Figure 1 the sensors are fabricated on top of a 4-inch p-type silicon wafer used as the substrate. A 1 μm thick layer of oxide is grown on top of the substrate and then 0.5 μm-thick layer of Aluminum (1% Silicon) is deposited and patterned to form the corrosion and temperature sensors, and also interdigitated electrodes of the humidity sensors. The final step is deposition of polyimide (HD Microsystems PI2545) which acts as the dielectric material for the humidity sensors. The final thickness of the polyimide layer deposited on the metal structures is 2.65 µm. The polyimide layer is also patterned to form openings with different geometries in the dielectric layer of the humidity sensors.

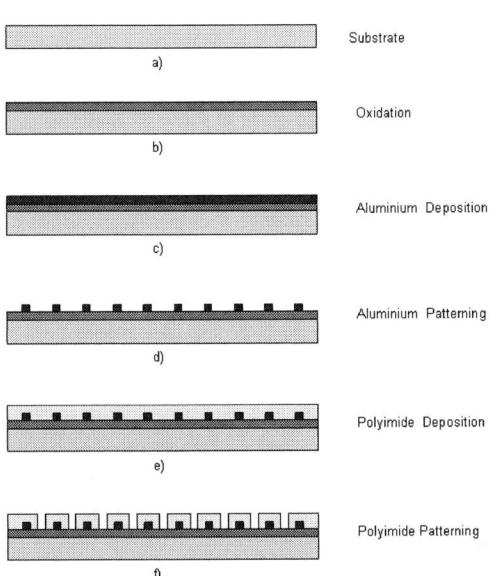

Figure 1: Fabrication Flow

As mentioned before, the temperature and corrosion sensors are metal only structures, so polyimide deposition and etching steps are not performed for the first wafer, to avoid unnecessary steps for these sensors and make the process easier.

All the steps shown in Figure 1 are carried out for all the humidity sensors in this test chip. Table 1 summarizes the list of fabricated sensors.

Table 1: List of Fabricated Sensors

Name	Description
HS-1u	Humidity Sensor (without openings)
TS1	Temperature Sensor
CR-1u	Interdigitated Corrosion Sensor
LD-1u	Ladder Corrosion Sensor
TT-1u	Triple-Track Corrosion Sensor
SP1	Spiral Corrosion Sensor
HS-3u-p	Humidity Sensor (4µm openings)
HS-3u-cover	Humidity Sensor (without openings)
HS-3u-holes	Humidity Sensor (2µm openings)
Edge-SP1	Spiral Corrosion Sensor
Edge-SP2	Spiral Corrosion Sensor
Edge-SP3	Spiral Corrosion Sensor
TripleTrack-5u	Triple-Track Corrosion Sensor
TripleTrack1	Triple-Track Corrosion Sensor
Ladder1	Ladder Corrosion Sensor
Ladder3	Ladder Corrosion Sensor
Corrosion1	Interdigitated Corrosion Sensor
Comb2	Comb Like Corrosion Sensor

Corrosion Sensors

Corrosion of the wire bonds, pads and metal tracks due to moisture penetration is a common problem with devices that need to operate in corrosive and humid environments. In fact, corrosion is a major cause of failure in implantable devices when their packages are not truly hermetic.

As illustrated in Figure 2, several corrosion detectors have been designed and included in the test chips. The main idea for designing the corrosion sensors was to observe the increase in the resistance of the metal tracks which occurs due to corrosion. However, as corrosion is a slow process, a long term accelerated test is used to characterize these sensors.

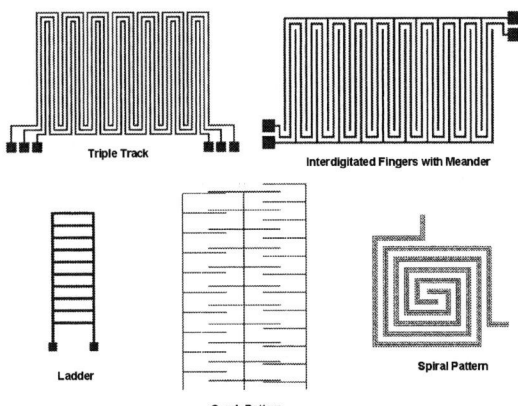

Figure 2: Corrosion Sensors

Humidity Sensors

The sensitivity and response time of a capacitive humidity sensor can be improved by changing the shape, dimensions and type of the moisture sensing layer or the electrodes [3-6]. The humidity sensors designed for these test chips are all single layer interdigitated fingers covered by a thin layer of polyimide. The humidity sensors of the test chips only differ in electrode widths, gaps and the openings in the polyimide layer.

Experimental

In order to facilitate characterization of the fabricated sensors, some of the dies were wire bonded into 24-pin ceramic packages.

The test setup which is being used to characterize the structures consists of a climatic test chamber (Heraeus Votsch HC7033), a multimeter (Agilent 34410A) and a LCR meter (Agilent E4980A). The test chips were put into the climatic test chamber which provides a controlled environment for the experiments.

HS-3u-p: Humidity Sensor Characterization

HS-3u-p is a humidity sensor in this test chip with the overall size of 2 mm × 2 mm. The electrode width is 3 µm and the gap between two adjacent

978-1-4244-4722-0/09 $25.00
© 2009 IMAPS-ITALY

electrodes is 8 μm. To increase the surface which is in contact with the outside environment, a 4 μm width rectangular opening was made in the polyimide layer between each pair of adjacent electrodes. Figure 3 shows a three dimensional model of this sensor.

Figure 3: HS-3u-p Humidity Sensor

To investigate the sensor response, a series of experiments were carried out. Figure 4 shows the result obtained from one of these experiments, with the temperature fixed at 37 ºC and the humidity ramped up and down between 55 to 85 %RH.

As shown in Figure 4, when the humidity is increased by 30%, the capacitance of the sensor changes from 22.5 pF to 42.5 pF.

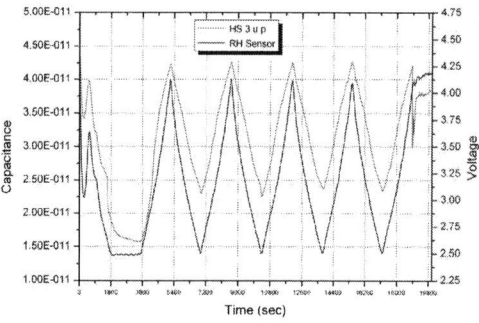

Figure 4: HS-3u-p Response

The sensitivity HS-3u-p is calculated by Equation (1)

$$S = \frac{\Delta C}{\Delta RH} \qquad (1)$$

where ΔC is the change in the capacitance of the sensor and ΔRH is the change of the relative humidity; therefore sensitivity of this sensor is 0.67 pF per % RH.

The sensor size is a key factor which affects the absolute capacitance change; while we are interested to investigate the effect of the other design factors (e.g. change in the structure of the dielectric layer) on the sensor sensitivity..

To exclude the influence of the sensor size, the normalized capacitance change, which shows the sensitivity independent of the sensor size is calculated by Equation (2)[7].

$$NPCC = \frac{C_f - C_i}{C_i} \times \frac{1}{\Delta RH} \times 100 \qquad (2)$$

where $NPCC$ is the normalized capacitance change, C_i is the capacitance at an initial relative humidity, C_f is the capacitance at a final relative humidity. The normalized capacitance change for HS-3u-p is 2.96/%RH.

Figure 5 shows the sensor response versus change in the humidity. As previously shown in Figure 4, at the beginning of the experiment when the sensor is dry a delay is observed during the first cycle until the sensor reaches the proper capacitance level. But for the following cycles the sensor settles in and changes in the capacitance level when humidity ramps up and down remains almost the same for each cycle.

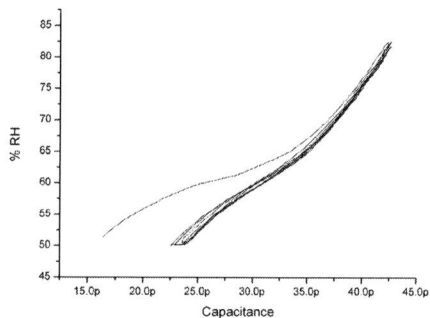

Figure 5 HS-3u-p Hysteresis

HS-3u-Cover: Humidity Sensor Characterization

Figure 6 shows a three dimensional model of HS-3u-cover humidity sensor. Like HS-3u-p, the electrode width is 3 μm and the gap between two adjacent electrodes is 8 μm. Unlike to HS-3u-p, there is no opening in the polyimide layer in this sensor. The main purpose for including this design in the test chips was to investigate the effect of openings in the dielectric layer on the sensitivity and the response speed of the sensors.

As expected, the less surface in contact with the air, the lower the moisture absorption and desorption rates. Figure 7 shows the response of HS-3u-cover when exposed to changing humidity. In this test the temperature was fixed at 37 ≻C and the humidity was ramped up and down between 55 to 85 % RH.

Figure 6: HS-3u-cover Humidity Sensor

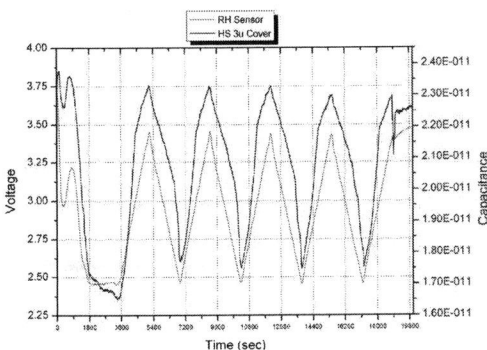

Figure 7: HS-3u-cover Response

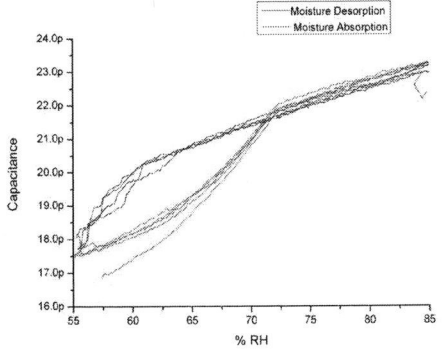

Figure 8: HS-3u-cover Hysteresis

When the humidity is increased by 30%, the capacitance of the sensor changes from 17.5 pF to 23.5 pF and the resulting sensitivity of this sensor is around 0.2 pF per % RH. The normalized capacitance change for HS-3u-cover is 1.14/%RH.

Figure 7 also shows that the desorption rate is smaller than absorption rate; therefore the sensor response in dehumidifying cycles is slow. Figure 8 shows the sensor response versus change in the humidity. As mentioned before the difference in absorption and desorption rates for this sensor causes a hysteresis in the sensors response.

Conclusion

A set of test chip consisting of a number of humidity, temperature and corrosion sensors has been designed and fabricated. The results obtained from experiments show a high sensitive humidity sensor that can measure the level of moisture inside the package of biomedical implants. It was also shown that the change in the dielectric layer of capacitive humidity sensors would considerably affect the sensors sensitivity.

Acknowledgements

The present work is a part of a research project funded by EPSRC and the European Commission to develop new packaging methods for chip scale implanted devices.

References

[1] R. A. M. Receveur, F. W. Lindemans, and N. F. de Rooij, "Microsystem technologies for implantable applications," *Journal of Micromechanics and Microengineering*, vol. 17, pp. R50-R80, May 2007.

[2] K. Najafi, "Packaging of implantable microsystems," *2007 IEEE Sensors, Vols 1-3*, pp. 58-63, 2007.

[3] J. Laconte, V. Wilmart, J. P. Raskin, and D. Flandre, "Capacitive humidity sensor using a polyimide sensing film," in *Symposium on Design, Test, Integratives and Packaging of MEMS/MOEMS*, Cannes, FRANCE, 2003, pp. 223-228.

[4] H. H. Zeng, Z. Zhao, H. F. Dong, Z. Fang, and P. Guo, "Fabrication and test of MEMS/NEMS based polyimide integrated humidity, temperature and pressure," *2006 1st IEEE International Conference on Nano/Micro Engineered and Molecular Systems, Vols 1-3*, pp. 788-791, 2006.

[5] U. Kang and K. D. Wise, "A high-speed capacitive humidity sensor with on-chip thermal reset," *IEEE Transactions on Electron Devices*, vol. 47, pp. 702 - 710, 2000.

[6] C. Y. Lee and G. B. Lee, "Humidity sensors: A review," *Sensor Letters*, vol. 3, pp. 1-15, Mar 2005.

[7] Y. H. Kim, J. Y. Lee, Y. J. Kim, and J. H. Kim, "A Highly Sensitive Humidity Sensor Using a Modified Polyimide Film," *JOURNAL OF SEMICONDUCTOR TECHNOLOGY AND SCIENCE*, vol. 4, pp. 128-132 2004.

978-1-4244-4722-0/09 $25.00
© 2009 IMAPS-ITALY

Realisation of Embedded-Chip QFN Packages - Technological Challenges and Achievements

A.Ostmann[1], D. Manessis[2], L. Boettcher[1], S. Karaszkiewicz[1], and H. Reichl[2]

[1]Fraunhofer Institute for Reliability and Microintegration (IZM)

[2]Microperipheric Research Center, Technical University of Berlin (TUB)

Gustav-Meyer-Allee 25, 13355 Berlin, Germany

E-Mail: ostmann@izm.fhg.de, Tel: +49-30-46403187

Abstract

The chip embedding technology achieved significant progress the last years. After various research activities the main focus is today on industrialisation and implementation of new business models. In the project HERMES European partners from industry and research aim to bring embedding technology based on low-cost PCB /Printed Circuit Board) processes to a market-ready product flow, demonstrated by automotive, power electronics and telecommunication applications. The research part of the project aims to overcome current limitations and to achieve even higher levels of miniaturisation. This paper will describe the embedding process flow and will discuss the process steps in detail. Three devices realised, a 2-chip module, a 3D System-in-Package and a QFN (Quad Flat No-Lead) package will be described and discussed. Especially the realisation of the QFN is a strong challenge for today's machine capabilities, since it contains a chip with a pitch of 100 μm.

Key words: *chip embedding, packaging PCB technology*

Introduction

In most electronic systems, there is a continuing pursuit of further miniaturisation and increased functionality. Current technologies provide organic substrates with high-density build-up layers and microvias, equipped on both sides with surface mount passive components and active chips in packages. The technological front has rapidly advanced from 2D systems integration to 3D SiPs (System-in-Packages) with stacked dies to keep up with miniaturisation trends. Towards this technology direction, chip and component embedding turns out to be a very promising technology route for even higher 3-dimensional integration of components. System requirements for signal frequencies in the order of several GHz can not be met by long bond wires and extensive interconnect paths on a board. In order to maintain signal integrity, much shorter and impedance-matched interconnects are required. Embedding allows to have conductors not only under but also over a component leading to a 3-dimensional packaging also on top of the embedded components. The component can be electrically connected to the top or bottom conductive layer or to both of them, e.g. in case of power ICs with contacts on both sides.

There are different ways of embedding an active chip. The die can be placed with its active side down to the substrate (face-down) Then the dies are embedded in a dielectric layer by lamination. Such a technology is described in [1]. An other way is to place a chip with its backside on a substrate /face-up), followed by embedding in a laminate layer. This technology, also called "Chip in Polymer", has been initially shown by Fraunhofer IZM [2].

In the EU-funded project HIDING DIES partners from industry and research organisations have shown the suitability of face-up embedding for different application demonstrators and could demonstrate the realisation of modules in a high-volume manufacturing line [3]. In the project chips with contact pad pitches down to 150 μm were embedded.

As a successor, a new EU project HERMES has inaugurated with wide participation of European industries and research institutes with a broader scope of furthering the embedding technology borders at R&D level and more importantly of bringing embedding technology in real manufacturing PCB production. By the industry partners application demonstrators from the areas automotive, power electronics and telecommunication will be will be realised. For the demonstrators face-up as well as face-down embedding will be considered, always having in mind industrial high-volume manufacturing.

The research aim of HERMES is to further extend the technological limitations of chip

978-1-4244-4722-0/09 $25.00
© 2009 IMAPS-ITALY

embedding. A major limitation is the contact pitch. Today embedding of chips with small peripheral pads requires the application of an additional redistribution layer (RDL). The ambition of HERMES is to achieve embedding of chips with 60 μm pitch only. As a first step in this direction test chips with a pitch of 100 μm is under investigation, as described more in detail below.

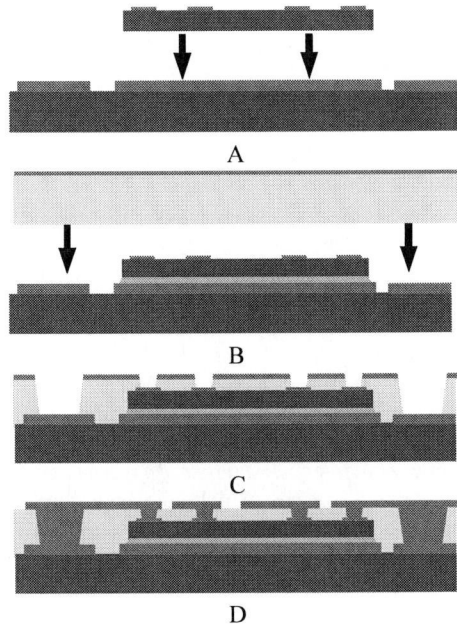

Figure 1: Process steps of chip embedding: (a) face-up die bonding, (b) embedding in a polymer layer by vacuum lamination, (c) laser drilling of vias to chip and substrate, (d) metallisation of vias and Cu structuring

Embedded chip packages

A future vision is the realisation of complex, highly miniaturised systems by a 3D stack with various active and passive components, embedded side by side and over each other. Before this vision can become real many technical questions and open issues have to be solved, e. g. a very high yield level is required. Today more realistic is the manufacturing of packages with one or two embedded chips. Their major advantages compared to traditional packages are:

- Packages can be very thin. An embedded chip of 50 μm thickness and Cu wiring layers on bottom and top together have around 100 μm thickness only.
- Interconnects to the chip are planar and geometrically well defined. In contrast to bond wires this allows a very precise impedance control.
- Each layer has a planar surface. This allows either a sequential stacking of embedded dies in one SiP or the easy stacking of

- electrically tested single chip packages or multi-chip SiPs.
- Production panel size is not limited, compared to package production based on chip&wire and molding. The full PCB panel size (18"x24" or larger) can be utilised, which leads to cost reduction.

The basic structure of a package realised by face-up chip embedding is delineated in Figure 2. The illustration shows embedding of two chips for the creation of an embedded dual chip package. The embedding technology focuses on the use of standard PCB (Printed Circuit Board) processes. The main advantage of such processes is the capability of using large substrate sizes for the package manufacturing and by that a significant cost reduction. By using thin silicon dies and a direct interconnects with microvias very small and thin packages can be realised, enabling a further miniaturisation of the final application.

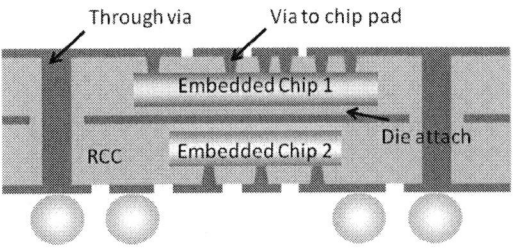

Figure 2: Schematic of package with two embedded chips.

Technology aspects & challenges

Chip embedding technologies are about to be brought to a more advanced level of interconnection at even finer chip pitches without use of additional RDL layers for pitch enlargement by trying either face-up or face-down chip interconnection. It seems apparent that the face-up chip interconnection through vias may reach different "chip pad pitch" limits at prototype and industrial level. Then the face-down chip embedding approach may offer further flexibility to lower the pitch limits. However, the common pursuit in all technologies is even more precise die placement and bonding for better microvia alignment in combination with high speed die placement at high yield. The following section elaborates on the technology steps and challenges encountered for the realisation of embedded single or multi chip packages.

Wafer preparation

Laser drilling of microvias and the PCB metallisation process is not compatible with standard contact pads of semiconductor chips. Therefore an additional metal layer (bump) has to be applied on wafer level. Typically 5 μm electroplated Cu is used, but also electroless NiPd can be used [4, 5]. Chips

978-1-4244-4722-0/09 $25.00
© 2009 IMAPS-ITALY

with very small pitches, originally designed for wire bonding, can also be compatible with embedding after an application of an additional RDL provided by subcontractor services. For embedding of the silicon chip with the active side facing-up, the wafers should be thinned to 50 µm, using subcontractor services if needed. Thinning of the dies is necessary since most RCC (Resin Coated Copper) layers used for embedding have a 70 µm to 90 µm resin thickness.

Chip placement and bonding

Placement accuracy is extremely crucial for chip embedding. The process tolerances for sequential die bonding, via drilling and Cu structuring have to be very low in order to achieve an acceptable yield. One of the requirements is that the machines for these three process steps should use the same alignment fiducials on the core substrate. Prior to chip placement, die ejection from a dicing tape takes place and should be optimised according to the adhesiveness of the tape, size and ultimate thickness of the chip. Adhesiveness of the tape is regulated by UV exposure and is to be reduced for component pick-up and placement without endangering the chip integrity.

Chip bonding has been developed by using printable pastes and die attach films (DAF). Screen printing allows a precise control of volume and location of the adhesive paste, which is rather a problem for dispensing. Electrically conductive Ag-filled pastes or B-stage pastes can be used. Another method is the use of a dicing die attach film (DDAF). It is a UV dicing tape which has two layers, a conventional UV dicing foil and an adhesive layer on top. Wafers are mounted on the adhesive layer. The dicing blade has to cut the silicon and the top layer of the tape. In the picking process this layer remains at the chip and serves as adhesive. Die attach films have shown superior adhesive co-planarity compared to printed pastes, which is extremely important for precise epoxy thickness over the chip after RCC lamination. By using a Datacon evo machine a placement accuracy of ±10 µm at 3σ was achieved on 18"x12" panels.

Embedding by RCC lamination

The core substrate with the die bonded chips is covered from the top side with a RCC layer as shown in Figure 1B. Temperature and pressure profiles should be adjusted carefully to promote epoxy adhesion at all interfaces and avoid chip breakage. A thickness of 15-20 µm over the chip surface is desirable for the subsequent microvia opening and filling. The overlaying Cu layer serves as the base for package routing. Curing of the epoxies take places at about 185 °C for 60 min. Detailed description of the lamination process is given in [6]. New developments in RCC laminates can improve significantly the adhesion on the chip surface and the reliability of the embedded packages by new epoxy formulations with adjusted thermo-mechanical properties. New classes of glass-reinforced RCCs have emerged in the market for lowering the CTE of the resin and bringing it closer to the copper values.

Via opening and filling

In the Chip-in-Polymer technology (face-up embedding), interconnections are achieved via microvia laser drilling to chip pads and subsequent metallisation, similarly to the established formation of microvias on PCBs. For microvia drilling a pulsed 355 nm UV laser has been used. It can ablate Cu as well as the RCC dielectric. Accurate alignment of the microvia drilling with respect to the underlying Cu pattern remains challenging for the yield of the interconnection. With an I/O pad pitch of 150 µm and enlarged Cu pads of 100 µm size, alignment based on fiducials on 100x100 mm² (sub-)panels is sufficient. For smaller pitches a higher number of local fiducials, closer to each chip, will be necessary. After drilling, a via cleaning (desmear) process is needed to remove remaining epoxy resin and sufficiently roughen the via wall to promote good adhesion of the subsequent electroplated copper. Then a conductive palladium seed layer is deposited on the epoxy surface which allows a direct via metallisation without need for an electroless copper layer prior to copper electroplating. A minimum thickness of 10 µm Cu is required in the microvias. By the use of special Cu plating chemistry a nearly complete filling of the microvias can be achieved, as shown in Figure 5.

Subtractive Cu line patterning

The plated Cu on top of a RCC layer is subtractive structured by acidic etching. A dry film resist of 15 µm thickness is applied by a roll laminator and UV exposed by an Orbotech Paragon 9000 LDI (Laser Direct Imaging) system. LDI has several advantages compared to film exposure. It allows the use local fiducials, i. e. smaller sub-panels can be exposed separately each with individual adaptation of x and y shift and rotation. Furthermore stretch factor compensation can be done for each panel or sub-panel individually.

Subtractive Cu etching in a spry tool is the PCB standard. However the under-etch always result geometry tolerances, limiting the smallest line and space widths. Due to the lab type spray etch tool currently used at IZM, minimum line width which can be achieved reliably is limited to 40 µm.

Semi-additive Cu line patterning

In order to reduce the feature size of Cu wiring a semi-additive, pattern plating technology is currently under development. The goal of the ultra-fine line patterning is to produce structures down to 15 µm lines and spaces, which is also the maximum resolution of the LDI system. The basic principle of a semi-additive process is to use a RCC layer with extremely thin Cu. After lamination and via drilling

978-1-4244-4722-0/09 $25.00
© 2009 IMAPS-ITALY

first a resist is applied and structured. Then Cu is electroplated, starting from the thin Cu layer, filling the trenches formed by the resist mask. Finally the thin base Cu is removed by differential etching.

The RCC used in the investigation had a Cu thickness down to 1.2 µm, They were supplied by Circuit Foil. At structure sizes of 15 µm, the adhesion of the resist layer to the under laying copper becomes critical. A suitable surface treatment of the copper layer is applied, that generates a very fine and homogeneously distributed surface roughness. A dry-film resist of 15 µm thickness is used. After LDI exposure the resist pattern is developed in a spray development tool. For the subsequent semi-additive copper plating and the needed resist removal after plating, it is important to control the profile of the resist structure. The preferred profile for the resist opening is an almost 90° angle or a slightly negative undercut opening to enable a reliable plating process and resist removal. In Figure 3 structures with 16 µm line and space widths and a Cu thickness of 12 µm are shown after plating.

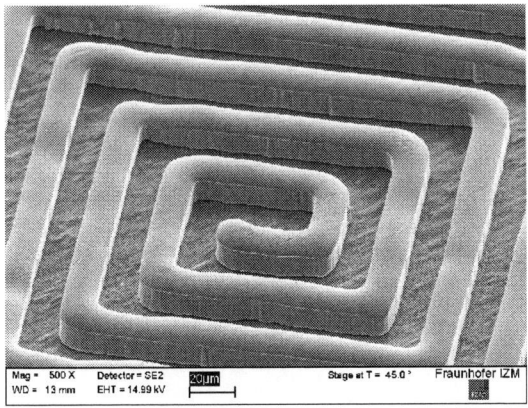

Figure 3:. Electroplated Cu structure with 16 µm lines and spaces.

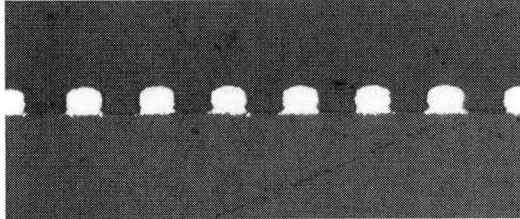

Figure 4: Cross-section of Cu lines with 12 µm thickness and 16 µm lines and spaces.

The fine-line structures require a well controlled differential etching of the base copper. Figure 4 shows a cross-section of the same pattern after etching of the thin base Cu. A micro-etch solution with low etch rate has been used. Nevertheless a slight attack of the Cu line edges is visible. The initial evaluations have shown encouraging result and will be optimised regarding Cu thickness uniformity.

Realised embedded packages

In this section the current state of embedding technology at Fraunhofer IZM will be illustrated by three examples of packages realised.

Dual chip device

In the frame of the German R&D project MST-SmartSense a device with two embedded chips in one layer was realised. It is used as a pre-study for more complex modules with active chips. The minimum pitch is 150 µm at a bondpad diameter of 50 µm.

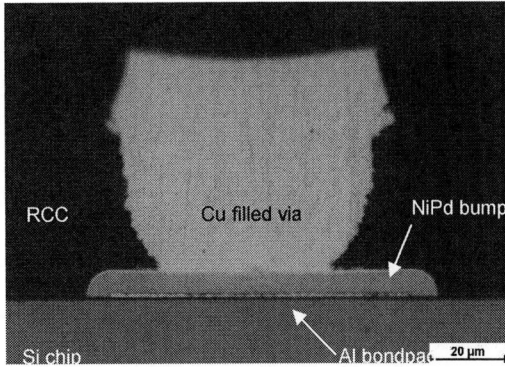

Figure 5: Close-up of a microvia on an electroless NiPd pad metallisation.

As shown in a cross-section of a single microvia to an embedded chip in Figure 5 the Al contact pads were metallised with electroless NiPd bumps prior to embedding. The vias formed by UV laser drilling have a diameter of 65 µm at the top and 45 µm at the bottom. The electroplated Cu forms good contacts to the NiPd bumps without interfacial voids. The typical alignment tolerance via to pad is ± 10 µm. Figure 6 shows a row of microvias with 150 µm pitch. Figure 7 shows a x-ray image of a part of the device.

Figure 6: Cross-section of contacts to embedded chip (150 µm pitch).

978-1-4244-4722-0/09 $25.00
© 2009 IMAPS-ITALY

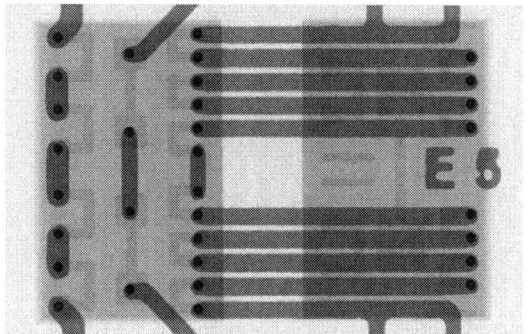

Figure 7: X-ray image of two embedded test chips (smallest pitch 150 μm).

3D System-in-Package

<to the drawing in Figure 2. At first one chip with 5 μm Cu bumps in an area array layout of 150 μm pitch was die bonded on a Cu foil, followed by embedding in a RCC layer. Then in the Cu foil openings were structured in order to give space for through vias drilled later. Next the second chip was die bonded to the bottom side of the initial Cu foil, followed by embedding in RCC, Microvias were drilled to the chip pads and vias outside the chips were drilled through both RCC layers. In the following plating step, the through vias were metallised and microvias to the chips were filled with Cu. After Cu structuring, on both sides further RCC layer were laminated, drilled, Cu plated and structured. The resulting sequence of five Cu wiring layers and two embedded chips is shown in Figure 8. These dual chip SiPs were realised on a panel format of 350x250 mm².

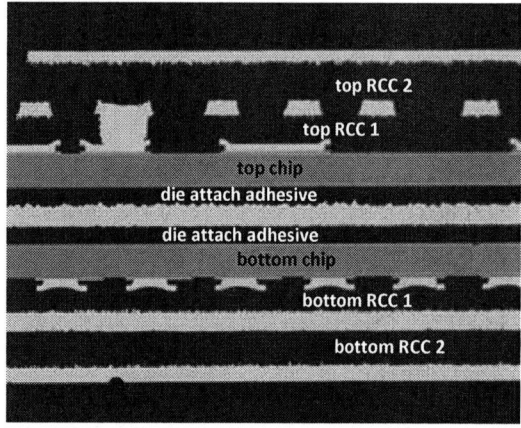

Figure 8:: Cross-section of a SiP module with two embedded chips in-between five Cu wiring layers.

QFN package

In the research part of the HERMES project a QFN package was selected as evaluation vehicle. The QFN size is 10x10 mm². The package has 84 contacts at 400 μm pitch. Inside the package a test chip is embedded. The chip has a size of 5x5 mm², a thickness of 50 μm and peripheral pads of 100 μm pitch. The Al pads have a size of 40 μm. By 5 μm Cu bumps they were enlarged to 85 μm. Figure 9 illustrates the two different variations of the QFN package "A" and "B". The variation A realises a redistribution of the chip bondpads to the peripheral QFN contacts and the connections to the chip pads are done by laser microvias.

Figure 9. QFN package variations "A" and "B"

Variation B also incorporates routing of the contacts to the backside of the substrate using through via contacts, realising a standard QFN footprint including the thermal die contact in the middle. Both variations can yield in a low thickness of the final package.

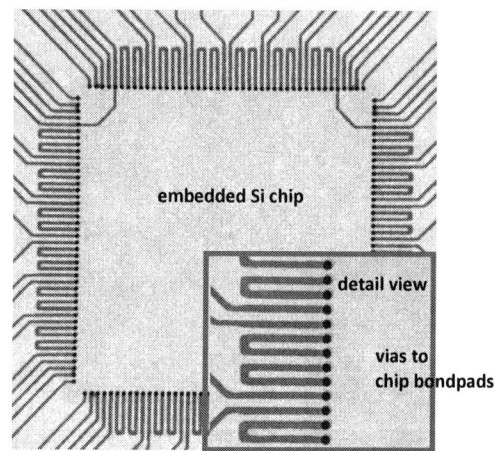

Figure 10. X-ray image of embedded chip in a QFN package.

The processing of the QFN packages is done in a large panel format of 350x250 mm². A total of 256 packages are realised on this panel size. The thickness of the used core is 50 μm, realising an ultra thin package but also providing good mechanical stability. The 50 μm thin dies are die bonded face-up on the core substrate using a dicing die attach film (DDAF) with 20 μm thickness. With the given pad pitch of 100 μm the positioning of the die bond must be extremely precise in a range of ± 10 μm. The accurate position of the embedded die is essential for the precise laser via position. The given pad size of 85 μm and the current via diameter of 50 μm determines the maximum acceptable tolerance.

978-1-4244-4722-0/09 $25.00
© 2009 IMAPS-ITALY

Figure 10 shows the x-ray inspection of the area with the embedded chip in a ready processed package. The picture shows, that the alignment of the microvia to the bondpad is precise, only a slight offset is visible. The ready processed panel, that contains the packages, is then separated by laser cutting.

The realised package, shown in Figure 11, (without solder mask) has an overall thickness of 160 µm and a standard QFN footprint of 400 µm. In order to study the yield and reliability of the embedded QFN packages currently larger numbers of packages are under manufacturing. A test setup will be used for a quick functional testing of the packages. This setup is based on a QFN test socket and automated, parallel daisy chain and four point testing, using a LabVIEW environment.

Figure 11: QFN package with embedded chip (total thickness: 160 µm).

The next step in development is to reduce the microvia diameter to 30 µm in order to achieve a better yield in alignment of Cu patterning to via drilling. A further important yield improvement is expected by the introduction of semi-additive Cu patterning instead of subtractive etching.

Conclusions

Three different modules with embedded chips recently realised were described: a two chip evaluation device, a dual chip stacked SiP and a QFN package. The QFN is part of the research activities in the EU project HERMES. It has a size of 10x10 mm² and 84 I/Os at a pitch of 400 µm. The embedded chip in the QFN is 5x5 mm² in size and has a peripheral pad configuration at 100 µm pitch. The total thickness of the package is 160 µm. The packages have been realised in 350x250 mm² panels at prototype level. Embedding of chips with 100 µm pitch is getting close to the limitations of current process technology. Therefore a semi-additive Cu patterning process is under development. Using RCC with ultra-thin 1.2 µm Cu the realisation of

16 µm lines/spaces structures at a Cu thickness of 12 µm was demonstrated.

Acknowledgements

The authors would like to thank the European Commission for the financial support of the HERMES project (FP7-ICT-224611), Björn Grallert for the evaluation work on semi-additive plating process and the German BMBF for the financial support of the MST-SmartSense project.

References

[1] T. Karila, "Thermal Performance of the Embedded IC Structure", Proc. EMPC 2007- the 16th European Microelectronics and Packaging Conference & Exhibition, June 2007, Oulu, Finland.

[2] A. Ostmann, A.Neumann, E. Jung, R. Aschenbrenner, H. Reichl, „Stackable Packages with Integrated Components", Proc. 5th Electronics Packaging Technology Conference EPTC 2003, Singapore, October 7-8, 2003.

[3] A. Kriechbaum, W. Bauer, "Embedding of active components into multilayer printed circuit boards", Proc. 4th European Microelectronics and Packaging Symposium ESTS, 22 – 24 May 2006, Terme Čatež, Slovenia.

[4] L. Boettcher, D. Manessis, A. Ostmann, S. Karaszkiewicz, H. Reichl, "Embedded Chip Packages – Technology and Applications", Proc. International Surface Mount Technology Association Conference-SMTA 2008, Orlando, FL, August 2008, pp 43 – 48.

[5] A. Ostmann, D. Manessis, H. Stahr, M. Beesley, J. De Baets, M. Cauwe, "Industrial and technical Aspects of chip embedding", Proc. 2nd Electronics System Integration Technology Conference (ESTC) 2008, Greenwich, September 1-5, 2008, , pp. 315-320.

[6] D. Manessis, S-F.Yen, A. Ostmann, R. Aschenbrenner and H. Reichl, "Technical Understanding of Resin Coated Copper (RCC) lamination processes for realisation of reliable chip embedding technologies", Proc. Electronics Components & Technology Conference (ECTC) 2007, Reno, NV, May 29- June 1, 2007, pp. 278-285.

Far-End Maximum Crosstalk for Coupled Lines as Function of Load

Amir Owzar, Ralph Stephan, Wesley Petersen , Markus Helfenstein
STERICSSON, Switzerland
Phone : ++41-44-465 1 273, Fax : ++41-44-465 1 806

E-mail Address : amir.owzar@stericsson.com

Abstract

The move to 45nm Technology has seen a further reduction in the noise margin for CMOS devices. This has resulted in severe constraints on crosstalk between adjacent lines on the SIP (System-In-Package) substrate. The known formula in the literature for calculation of crosstalk between two coupled lines includes only the impact of the driver impedance but not the impact of the load capacitance. In this paper a formula has been developed to calculate the amplitude of coupled noise between two adjacent signal lines as a function of load capacitance.

Key words: crosstalk, coupled lines, load capacitance, driver impedance, SIP.

Introduction

Package design for SIP applications requires an analytical expression for the estimation of the maximum crosstalk between two adjacent lines. The exsiting formula from literature, for the calculation of the crosstalk includes only the impact of the driver impedance and delivers a conservative estimation in terms of the crosstalk amplitude. This does not include the impact of the load capacitance [1]. In this paper a formula for the calculation of the coupled amplitude between two adjacent signal lines with geometrical dimension below 50um, has been developed, which determines the crosstalk amplitude as a function of termination at both ends.

1 Coupled lines

In this study the investigated structures are embedded in a homogeneous dielectric with a shielding plate on both sides (triplate structure). Further as shown in Figure 1 it is assumed, that both lines have identical termination ot both ends. The far-end of both lines are terminated with the capacitive loads, which model the input strucucre of a cmos driver input. The maximum far-end crosstalk in a homogeneous dielectric can be calculated by using

$$U_{max} = K_C \frac{2r_D}{\left(1+r_D\right)^2} \cdot U_D \qquad (1)$$

where U_D is the input voltage, r_D is the normalized driver impedance[1] and the coupling coefficient[2]

---

[1] $r_D = \dfrac{R_D}{Z_0}$

between two adjacent signal lines [1]. As mentioned before (1) doesn't show the dependency of the crosstalk amplitude with respect to the load impedance.

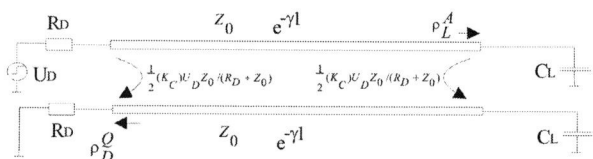

Figure 1:Coupled lines with identical termination [3].

The importance of adjusting the right value for the driver impedance is due to the fact, that by this means the dissipated energy in the transmission line could be minimized [4].

2 Calculation of the line parameters

For the determination of the crosstalk amplitude, we used the SPICE model of the coupled lines, which includes the termination element on both sides. The SPICE model includes the frequency dependency of the line parameters R'(f), L'(f) and C'. R' is line resistance per unit length, L' is the line inductance per unit length and C' is the line capacitance per unit length. By this mean the attenuation of the signal on the active lines has been taken into account.

---

[2] $K_C = \dfrac{C_{12}}{C_{10}}$

978-1-4244-4722-0/09 $25.00
© 2009 IMAPS-ITALY

The line parameters have been determined by using a field simulation tool. The applied tool required the geometrical dimentions of the lines and the surrounded material, as well as the material parameters. It delivers the parameters for coupled lines: R11,R12,R22,L11,L12,L22,C11,C12,C22. Here we used the G30 as dielectric material and the line structure as shown in Figure 2 has been assumed as being a triplate structure. The line width and distance between the two lines have been varied from 20-200um. The distance to the ground level has been constant for all line geometries. Table 1 includes the calculated line capacitance delivered by the field simulator and the calculated coupling factor based on the calculated capacitance between the two signal lines for different dimentions

Table 1: Coupling factor k_C (W=S: 20-200um).

W=S	C_{12} [PF/m]	C_{11} [PF/m]	k_C
20	45	95	0.47
40	32	100	0.32
50	28	104	0.27
75	20	116	0.17
100	15	129	0.12
150	8	158	0.05
175	6	172	0.03
200	4	187	0.02

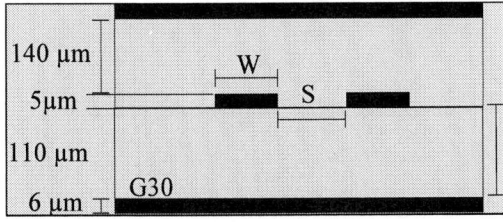

Figure 2 : Cross section of investigated structure

Based on the calculated line parameters SPICE model for lines with different geometries are generated and simulated.

3 Spice model for the coupled lines

For the generation of a time domain SPICE model for the investigated structures the LineGen[5] has been used. This tool delivers a SPICE netlist based on frequency dependent line parameters and line length. The accuracy of the generated line model depends heavily on the number of frequency points for all line parameters. To increase the accuracy of the model for the investigated structure, 30 frequency points of coupled line have been calculated and used for the model generation. The frequency dependend line parameters for two coupled lines with W=S=20um is shown in Table 2. columns 2-3 includes the resistance per line length, columns 4-6 include the inductance per unit length and the last 3 columns the capacitance per unit

length. The capacitance shows no frequency dependency. The frequency points cover the whole range between 1MHz – 1THz. To investigate the accuracy of the line models the coupled lines have been measured.

4 Measurement of the coupling voltage

To investigate the accuracy of the calculated line parameters and generated SPICE model the coupled far- and near-end voltage for two coupled lines with W=S=75um have been measured. The measurement has been done by using a vector network analyzer in the frequency domain. By using the IFFT(Inverse Fourier Transformation) built-in function of the equipment, the measured voltage at both ends of two coupled lines has been transferred to the time domain. The results for the crosstalk of two coupled line structures are shown in Figure 3. As shown in Figure 3 for two lines with the length of 5cm, and normalized driver impedance and normalized load impedance of 1 the simulated and the measured far-end and near-end correspond very well.

Table 2: Line parameters for W=S = 20um.

f[MHz]	R11[Ω/cm]	R22[Ω/cm]	L11[nH/cm]	L12[nH/cm]	L22[nH/cm]	C11[pF/cm]	C12[pF/cm]	C22[pF/cm]
1.00E+00	1.73	1.73	8.37	4.45	8.37	0.95	-0.45	0.95
1.00E+01	1.75	1.75	6.95	3.06	6.95	0.95	-0.45	0.95
2.00E+01	1.76	1.76	6.82	2.94	6.82	0.95	-0.45	0.95
3.00E+01	1.76	1.76	6.79	2.91	6.79	0.95	-0.45	0.95
4.00E+01	1.77	1.77	6.77	2.90	6.77	0.95	-0.45	0.95
5.00E+01	1.77	1.77	6.77	2.89	6.77	0.95	-0.45	0.95
6.00E+01	1.78	1.78	6.76	2.89	6.76	0.95	-0.45	0.95
7.00E+01	1.78	1.78	6.76	2.89	6.76	0.95	-0.45	0.95
8.00E+01	1.79	1.79	6.75	2.89	6.75	0.95	-0.45	0.95
9.00E+01	1.80	1.80	6.75	2.88	6.75	0.95	-0.45	0.95
1.00E+02	1.81	1.81	6.75	2.88	6.75	0.95	-0.45	0.95
2.00E+02	1.91	1.91	6.71	2.88	6.71	0.95	-0.45	0.95
3.00E+02	2.04	2.04	6.66	2.88	6.66	0.95	-0.45	0.95
4.00E+02	2.16	2.16	6.63	2.89	6.63	0.95	-0.45	0.95
5.00E+02	2.29	2.29	6.60	2.89	6.60	0.95	-0.45	0.95
6.00E+02	2.40	2.40	6.57	2.89	6.57	0.95	-0.45	0.95
7.00E+02	2.51	2.51	6.55	2.89	6.55	0.95	-0.45	0.95
8.00E+02	2.62	2.62	6.54	2.89	6.54	0.95	-0.45	0.95
9.00E+02	2.72	2.72	6.52	2.89	6.52	0.95	-0.45	0.95
1.00E+03	2.82	2.82	6.51	2.89	6.51	0.95	-0.45	0.95
1.20E+03	3.01	3.01	6.49	2.89	6.49	0.95	-0.45	0.95
1.50E+03	3.29	3.29	6.46	2.89	6.46	0.95	-0.45	0.95
1.80E+03	3.56	3.56	6.44	2.89	6.44	0.95	-0.45	0.95
2.00E+03	3.73	3.73	6.43	2.89	6.43	0.95	-0.45	0.95
2.20E+03	3.90	3.90	6.42	2.89	6.42	0.95	-0.45	0.95
2.80E+03	4.39	4.39	6.39	2.89	6.39	0.95	-0.45	0.95
3.00E+03	4.55	4.55	6.39	2.89	6.39	0.95	-0.45	0.95
3.50E+03	4.92	4.92	6.37	2.89	6.37	0.95	-0.45	0.95
4.00E+03	5.28	5.28	6.36	2.89	6.36	0.95	-0.45	0.95
5.00E+03	5.93	5.93	6.34	2.89	6.34	0.95	-0.45	0.95
6.00E+03	6.51	6.51	6.32	2.89	6.32	0.95	-0.45	0.95
7.00E+03	7.05	7.05	6.31	2.89	6.31	0.95	-0.45	0.95
1.00E+04	8.46	8.46	6.28	2.89	6.28	0.95	-0.45	0.95
1.50E+04	10.34	10.34	6.26	2.89	6.26	0.95	-0.45	0.95
2.00E+04	11.88	11.88	6.24	2.89	6.24	0.95	-0.45	0.95
3.00E+04	14.49	14.49	6.22	2.89	6.22	0.95	-0.45	0.95
4.00E+04	16.71	16.71	6.21	2.89	6.21	0.95	-0.45	0.95
5.00E+04	18.79	18.79	6.21	2.89	6.21	0.95	-0.45	0.95
6.00E+04	20.68	20.68	6.20	2.90	6.20	0.95	-0.45	0.95
7.00E+04	22.49	22.49	6.20	2.89	6.20	0.95	-0.45	0.95
8.00E+04	24.24	24.24	6.20	2.90	6.20	0.95	-0.45	0.95
9.00E+04	25.84	25.84	6.18	2.90	6.18	0.95	-0.45	0.95
1.00E+05	27.64	27.64	6.20	2.89	6.20	0.95	-0.45	0.95
1.00E+06	97.80	97.80	6.16	2.89	6.16	0.95	-0.45	0.95
1.00E+06	97.80	97.80	6.16	2.89	6.16	0.95	-0.45	0.95

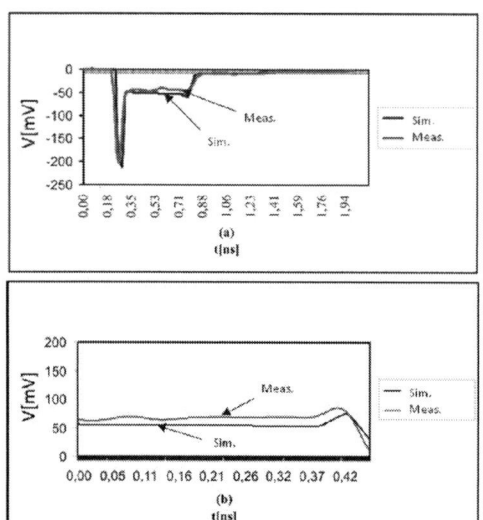

Figure 3: (a) coupled voltage far-end. (b) coupled voltage near-end. S=W=75um.

The good matching between the simulated and corresponding measured results proves the accuracy of the generated SPICE model for two coupled lines.

5 Coupled lines with different termination

For the determination of an expression for coupled voltage between two coupled lines with identical termination the SPICE line model has been built. Due to the fact that in most of the designs the matching between the line impedance and the driver impedance is assumed, for all simulated structures the matching at the driver end has been assumed. So we considered only the case of $r_D = 1.0$. The investigation has been done for lines with 3 different line geometries: W=S=75, 40 and 20um.

Figure 4 shows the simulation results of lines with normalized load impedance[3] varying between 0.1-2.0 and with line width and distance between the lines of 75um. tpd is the delay time given by :

$$t_{pd} = \sqrt{L'C'}\,\ell \;[2].$$

By using the equation (1) which has been developed in [3] we get for the maximum crosstalk of this line at an amplitude of 9%. This value, as shown in Figure 4, matches very well with the maximum crosstalk amplitude determined by the simulation. It is shown in Figure 4 that by increasing the load capacitance the peak of the crosstalk voltage experiences a delay. The reason for the delay is that the charge time is increased by the increased load capacitance. Furthermore the amplitude of the

---

[3] $c_L = \dfrac{C_L}{C'\ell}$

crosstalk is decreased. For cost reduction normally the line dimension has to shrink, so the practically interesting geometrical structures, are lines below 50um. For that reason we focus our investigation on crosstalk between lines with W=S=40 and 20um.

As shown in the Figure 5 and Figure 6 the simulated maximum crosstalk for coupled lines with S=W=20 and 40um, are equal to 24% and 16% respectively. These values correspond well to equation (1). But the equation (1) is a conservative estimation. In reality the coupled crosstalk as shown in Fig4-6 is a function of the load capacitance. For developing a new expression we have to take into account the load impedance.

The equation (2) has been developed for the calculation of the coupled voltage as a function of coupling factor and load impedance. In Figure 7 the simulation results are compared with the results by using (2). For the whole range of the normalized load capacitacnce 0.1-2.0 the coupling amplitude is predicted very accurately by using equation (2).

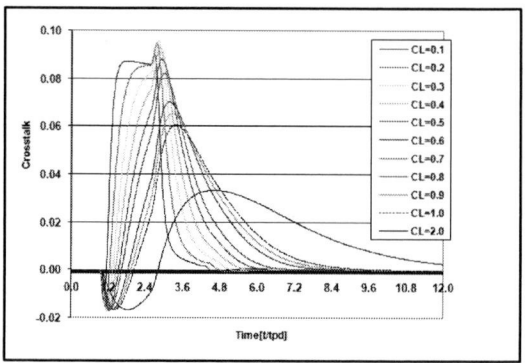

Figure 4: Simulated crosstalk vs. load capacitance $c_L = 0.1 - 2.0, r_D = 1.0, \ell = 5cm$. W=S=75um.

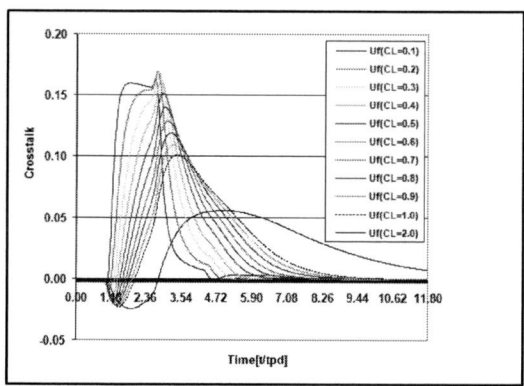

Figure 5: Simulated crosstalk vs. load capacitance $c_L = 0.1 - 2.0, r_D = 1.0, \ell = 5cm$. W=S=40um

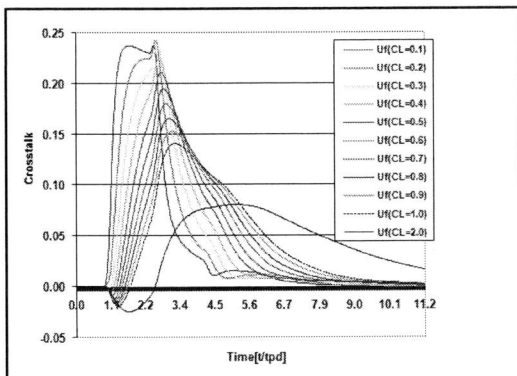

Figure 6: Simulated crosstalk vs. load capacitance $c_L = 0.1 - 2.0, r_D = 1.0, \ell = 5cm$. W=S=20um

According to the results shown in Figure 4-6 the equation (1) is more accurate as the peaks of the crosstalk occur closer to the normalized delay of one.

$$U_{max} = (-0.25 * K_C^2 * c_L - 0.5 * K_C * c_L + K_C)U_D \quad \textbf{(2)}$$

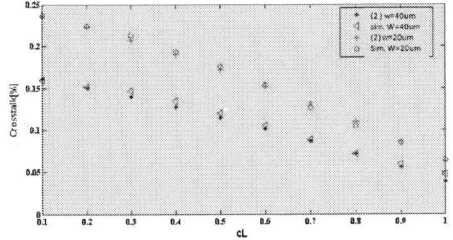

Figure 7: Simulated and calculated crosstalk vs. load capacitance $\ell = 5$cm $r_D = 1.0$.

Conclusion

In this paper we developed based on simulation results a new analytical expression for the calculation of the crosstalk between two coupled lines for small geometrical dimension. The comparision with the simulated and calculated results show the high accuracy of the proposed expression.

Acknowledgements

The authors would like to thank Mr. John Mallon for his valuable suggestions.

References

[1] Y. Yang: Crosstalk Estimate for CMOS-Terminated *RLC* Interconnections, IEEE Trans. Circuits and systems -: Fundamental theory and applications, vol 44, pp 82-84, Feb 97.

[2] J. R. Brews: Overshoot-Controlled RLC Interconnections, IEEE Transaction on Electron Devices: 32(1) 77, 1991.J. John, "A Terrific Paper", Journal of Materials, Vol. 2, No. 3, pp. 22-33, November, 1992.

[3] Y. Yang: Design guidelines for high-performance electric interconnections- A simple approach, IEEE Trans. Comp. Packg. Manufact. Technol. Part B, vol 19, pp 230-237, Feb 96

[4] A. Owzar, E. Baykal, M. Helfenstein, " Power Optimization by Using Automatic Adjustment of the Driver Output Resistance for SIP Applications ", *8-th VLSI Packaging Workshop*, vol. 8,no.1 pp. 35-38.2006

[5] S. Cyrusian: Time domain modeling of lossy transmission lines, Dissertation TU Berlin: 1996

978-1-4244-4722-0/09 $25.00
© 2009 IMAPS-ITALY

Effects of test conditions on bending impact of lead free solder

J. Park.

Research Institute of Industrial Science & Technology, Pohang 790-330, Korea

pjhyun@rist.re.kr

Abstract

Drop reliability of portable IT products such as PDA's, notebooks and mobile phone is issue with increase of using for lead free solder by environment regulation. Intermetallic compound formed at interface between lead free solder and pad is main cause of weakness on drop impact. To evaluate drop reliability, drop impact test methods is widely used by JEDEC standard, which is time consuming and expensive. In this study, we performed bending impact tests of a soldered joint as one of indirect test methods and studied the effects of variables, such as components size and types, aging treatments, bending variables, and number of tests by statistical analysis. Minitab software were used to conduct statistical ananysis. From the bending impact results at different component size or type, p-values were smaller than 0.05, and we knew that this test method quantitatively could be represent bending impact reliability of lead free solder joints under given conditions. We could statistically distinguish the effects of thermal shock or aging treatment of solder joints by bending impact tests under given conditions. we know that proper plus amplitude range and minus amplitude are between 5mm and 10mm, and minus 3 mm, respectively. Suitable frequency ranges were 2.5-10Hz. Bending impact life of 10 tests was statistically equivalent that of 5 tests and we could select 5 tests instead of 10 tests.

Keywords: Bending impact test, Lead free solder, Reliability

Introduction

It is necessary to replace lead solder, which has been used in the past, with lead-free solder because environment regulations governing the use of lead have intensified in many countries. Increasing of use of portable IT products such as mobile phones, navigation systems, PDA's, laptop's and portable gaming devices the drop reliability of microelectronics has become an important issue.[1] PCB of These products have many solder joints which connected by lead free solder paste. The evaluation methods of the drop reliability of the board level has published in 2003 by JEDEC standard.[2] But this method is time and cost consuming. Besides, this methods itself is not a clear test methods of lead free solder joints of cpmponents. It is necessary to standardize the easy and low cost method to measure the drop reliability of a soldered joint , especially, connected by lead free solder paste. In this study, we performed bending impact tests of a soldered joint as one of indirect test methods and studied the effects of variables, such as components size and types, aging treatments, bending variables, and number of tests by statistical analysis.

Test specimen and methods

Test specimen

Table 1. Specifications of the materials used for the tests

| | PCB | BGA and QFP |

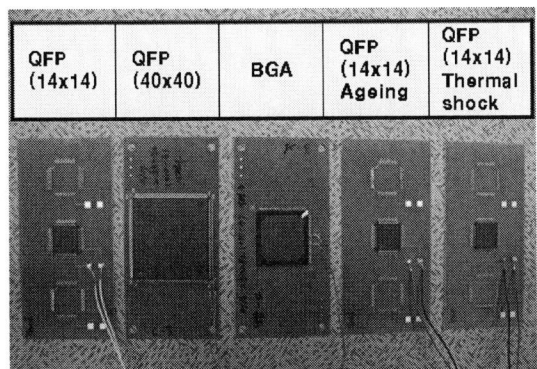

Figure 1 Used various component on PCB with daisy chain

The properties of the materials used are listed in Table 1. Figure 1 shows used various component on PCB with daisy chain. The lead-free solder paste is Sn–3Ag–0.5Cu alloy system with RMA type flux. The reflow profile consisted of three stages, including preheating at 150 to 180 °C for 110 to 120 s, soldering at 217°C for 40 s, and heating above a peak temperature of 245°C for 4s. Test equipment is micro electromagnetic testem(Model : MMT-101NB made by Shimadzu). Figure 2 shows apparatus of test equipment. To detect of resistance change caused by crack initiations or propagations, Keithley 2700 multi-meter is used, and bending jig is 3 points type

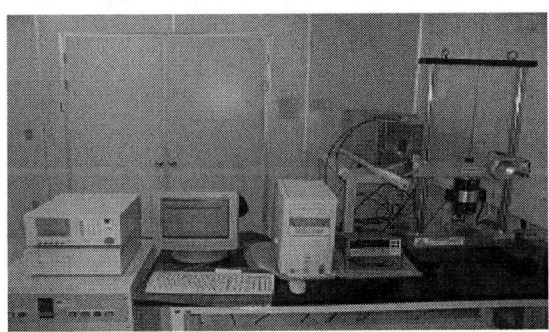

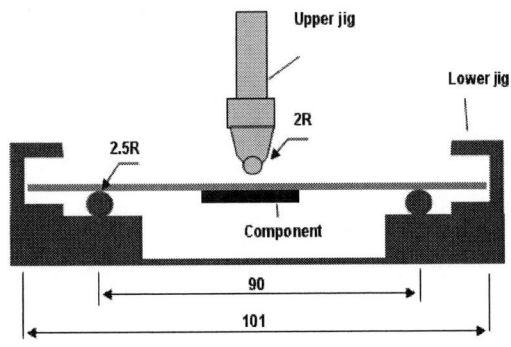

Figure 2 Bending impact equipment and Jig dimension

Test methods

Effect of components size and type

Effect of QFP components size was studied and size of tested components were 14X14 mm and 40X40 mm. Comparison of bending impact life for BGA and QFP component also conducted to analysis of effect on type of components under same test conditions. Failure criterion is 1000 ohms.

Effect of thermal shock and aging treatment

Intermetallic layer between solder and PCB pad mainly produced under high temperature environment such as repeated thermal cycle and long time exposure at high temperature conditions cause a decrease of drop life. So, to simulate high temperature conditions, specimen with thermal shock tested between -40°C and 125°C for 800 cycles and aging tested at 140°C for 1000 hours also investigated.

Effect of bending parameters

Amplitude and frequency of bending impactloading are important variables and these were also studied. Plus amplitude range was between 1mm and 10 mm, and minus amplitude range was between 0mm and -3 mm. Figure 3 shows schematic drawing of bending impact and definition of plus or minus amplitude. Frequency range is between 2.53Hz and 10 Hz.

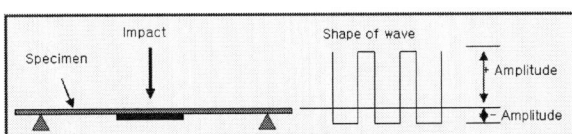

Figure 3 Schematic drawing of bending impact and definition of plus or minus amplitude.

Effect of specimen Numbers

Effects of test numbers for component on PCB were studied to obtain optimum number of tests. The tests were repeated 5 times or 10 times under the same conditions to determine the standard deviation of the bending impact life as a measure of the test repeatability. For data analysis statistical softare Minitab was used and estimation was statistically decided by p-value of 0.05.

Results and discussions

Effect of components size and type

Effect of component size on bending impact life to failure is shown in Figure 4. According to statistical results of 2 sample t-Test of Minitab software, p-value is 0.05 or smaller than 0.05 , generally say that number of cycles to failure of 14x14 QFP component is statistically different from that of 40x40 QFP component(assumption of 95% confidence interval). Different component sizes probably have different bending impact life. So if any given test method would be adopted as a standard test methods, number of cycles to failure of 14x14 QFP component should be statistically different from that of 40x40 QFP component. Based on this fact, we know that this test method quantitatively can represent bending impact reliability of lead free solder joints with different QFP size. Under given test condition, bending impact life of 40x40 QFP component has too lower value, so, in that case even though there is some difference of solder material property between two solder joints, we can not easily distinguish difference of property for two solder joints by bending impact test. We conclude that 14x14 QFP component size is better than 40x40 QFP component size to evaluate general impact reliability of solder joint. Figure 5 shows effects of component types on bending impact life. According to this result, p-value is 0.001 and we also know that this test method quantitatively can represent bending impact reliability of lead free solder joints with different component type.

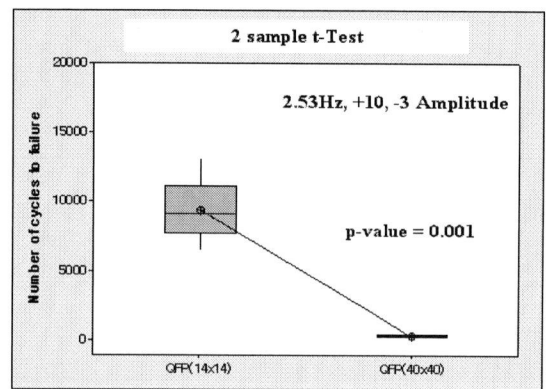

Figure 4 Effect of component size on bending impact life

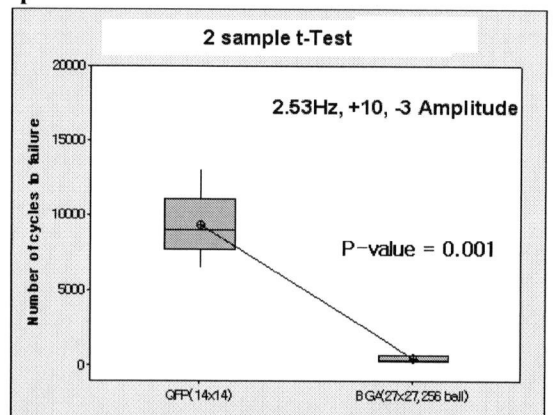

Figure 5 Effect of component types on bending impact life

Effect of thermal shock and aging treatment

Effect of thermal shock on bending impact life to failure is shown in Figure 6. In most case , mechanical properties of solder joints decrease after thermal shock. Figure 6 shows that bending impact life of solder joint without thermal shock is statistically different from that of with thermal shock. Mechanical properties of solder joints decrease after aging treatment. Figure 7 shows that bending impact life of solder joint without aging is statistically different from that of with aging. we know that this test method quantitatively can represent bending impact reliability of lead free solder joints with or without thermal shock or aging

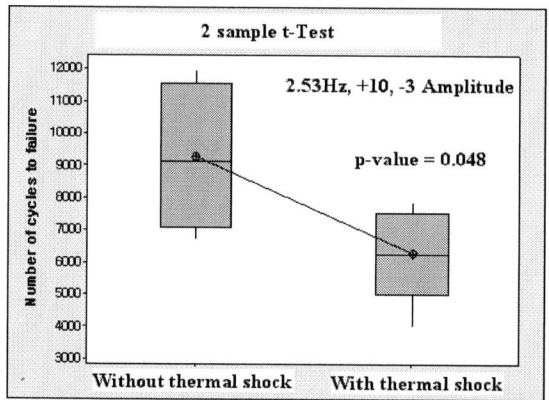

Figure 6 Effect of thermal shock on bending impact life

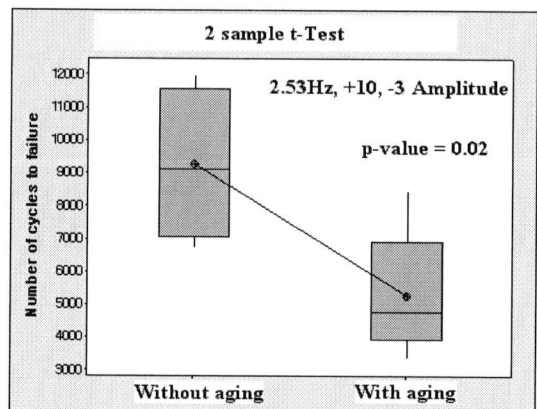

Figure 7 Effect of thermal shock on bending impact life

Effect of bending parameters

Effect of plus(+) amplitude on bending impact life to failure is shown in Figure 8. According to results, under the condition of plus 1 mm amplitude with 2.53 Hz and minus 3 mm amplitude, there is no failure and we know that proper plus amplitude range is between 5mm and

10mm. Effect of minus(-) amplitude on bending impact life to failure is shown in Figure 9. According to results, under the conditions at minus 0 mm and 1.5 mm amplitude with 2.53 Hz and plus 10 mm amplitude, there are no failures and we know that proper minus amplitude is 3 mm.

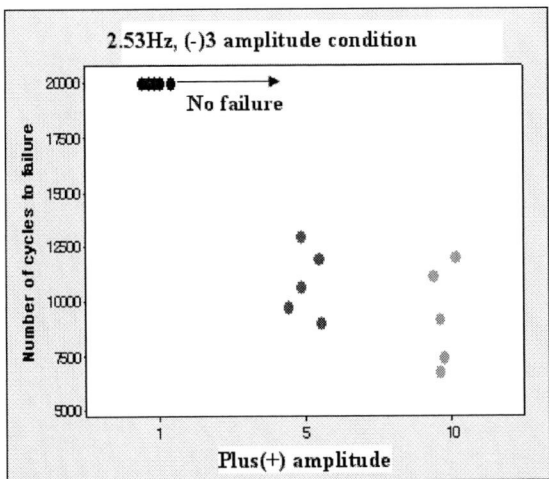

Figure 8 Effect of plus amplitude on bending impact life

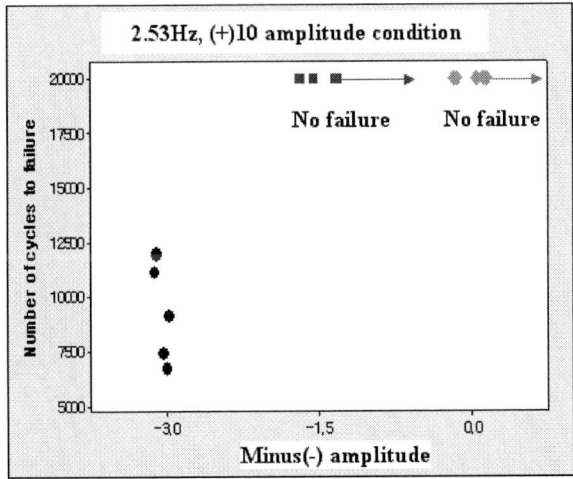

Figure 9 Effect of minus amplitude on bending impact life

The frequency effect has been studied to ensure the reduction of test cost and time. Effect of frequency on bending impact life to failure is shown in Figure 10. we know that bending impact life of solder joint with 2.53 Hz is lower than that of with 10Hz. It means that as frequency is increase, impact damage also increase.

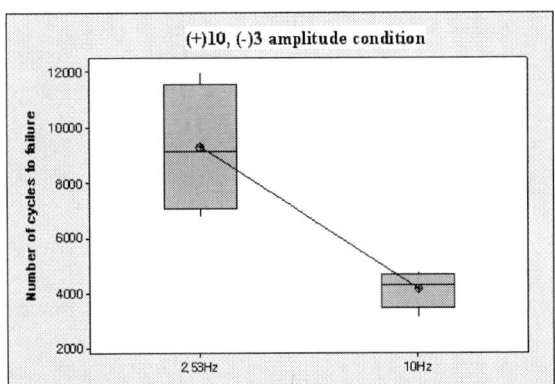

Figure 10 Effect of frequency on bending impact life

Effect of specimen Number

Figure 11 shows comparison of bending impact life between 10 tests and 5 tests. The p-value is 0.651. As mentioned above, p-value is greater than 0.05, number of cycles to failure are statistically equivalent. Therefore, bending impact life of 10 tests is statistically equivalent that of 5 tests and from view point of cost and time, we could select 5 tests instead of 10 tests.

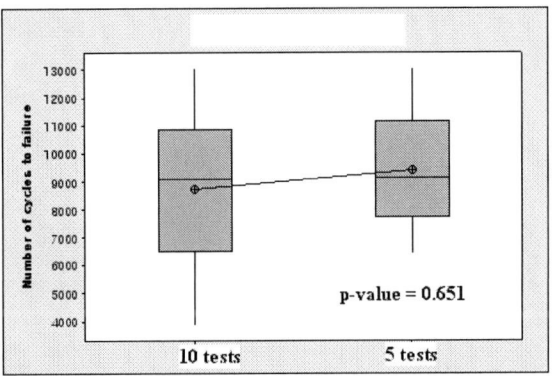

Figure 11 Effect of test number on bending impact life

Conclusions

Cyclic bending impact tests have been performed to standardize as a substitution of drop impact test method of solder joints connected by lead free solder pastes.

From the bending impact results at different component size or type, p-values are smaller than 0.05, and we conclude that this test method quantitatively could be represent bending impact reliability of lead free solder joints under given conditions. We can statistically distinguish the effects of thermal shock or aging treatment of solder joints by bending impact tests under given conditions. we know that proper plus amplitude range and minus amplitude are between

978-1-4244-4722-0/09 $25.00
© 2009 IMAPS-ITALY

5mm and 10mm, and minus 3 mm, respectively. Suitable frequency ranges were 2.5-10Hz. Bending impact life of 10 tests is statistically equivalent that of 5 tests and we could select 5 tests instead of 10 tests.

References

[1] J.J.M. Zaal, H.P. Hochstenbach, W.D. van Driel and G.Q. Zhang, An Alternative Solder Interconnect Reliability Test to Evaluate Drop Impact Performance, *ECTC 2008*, p 1181-1186
[2] Board Level Drop Test Method ofComponents for Handheld ElectronicProducts, JESD22-B111, 2003

Connector Reliability Testing Using Salt Spray

A. Parviainen, J. Perälä, L. Frisk and S. Kuusiluoma

Department of Electronics, Tampere Univeristy of Technology, P.O. Box 692, FIN-33101 Tampere, Finland

Phone +358 40 849 0620, Fax +358 3 3115 3394, E-mail address: anniina.parviainen@tut.fi

Abstract

In this study a salt spray test was used to compare the reliability of electrical connectors from two different manufacturers. The difference between connectors was in the coating materials of the connector pins. Manufacturer A used tin for the outermost coating layer, nickel for the second layer, and brass for the pin material. Manufacturer B also used tin for the outermost coating and brass for the pin material, but copper for the second coating layer.

The test was executed according to SFS-ISO standard. The connectors were placed in a salt spray test chamber for 2000 hours. The test consisted salt spray operation modes of 300 and 1100 hours and a period of 600 hours when the chamber was halted between the salt spray operation modes. During testing the connectors were measured using continuous real-time measurements. A 10 % increase in the measured voltage was used as a failure criterion. Significant differences between the manufacturers could be found related to materials during testing.

Cross sections were analyzed by means of a microscope and a scanning electron microscope (SEM). An elementary analysis was performed to determine the material changes. The pin itself is made of brass, which contains copper and zinc. In the elementary analysis of the untested connector it was observed that the brass contained about 65 wt% of copper and 35 wt% of zinc. The elementary analysis of the damaged connector revealed that the damaged area contained 95 wt% of copper and 5 wt% of zinc. The disappearance of zinc in the damaged connectors can be explained by dezincification.

Key words: Reliability, Salt spray, Connector, Dezincification

Introduction

During the design of a product the reliablity of components should be carefully considered. Every component has a function and its operation usually has an effect on several other components or even on the entire device. To make matters worse, even a single mistake in selecting components can undermine the reliability of the device and adversely affect both the sales of the product and the future of the company producing it.

Connectors in general have a significant role in the operation of electrical devices. Connectors are used in connecting together a variety of components, modules, subsystems, or devices. If even one connector fails to operate correctly, the entire device may malfunction. A connector may have one or several contacts and this means that the failure of just one contact can lead to the failure of the whole connector.

The object of this paper was to examine and compare the reliability of electrical connectors. A common failure mechanism of connectors is degradation caused by corrosion. The salt spray test induces corrosion and also has a long history of use in testing. The salt spray test also enables a comparison to be made between different types of

connector and it makes possible to study the effect of a corrosive environment on operating behavior.

Experimental

In this study reliability of connetors were tested using a a salt spray chamber. The tested connectors consisted of header, plug, crimp, and conductor. The crimp was first connected to the conductor. Each connector required six crimped contacts. When all the conductors were crimped, the crimp was then connected to the plug housing. When all the crimps were in the plug housing, it was attached to the header.

Two types of connectors from two different manufacturers A and B were tested. Connectors correspond to each other as follows: A1 and B1 (Spox), and A2 and B2 (Minifit). Same types of connectors from different manufacturers were compared to each other. A total of 8 pieces of both type of Spox connectors and 25 and 30 pieces of Minifit connectors were tested in a salt spray chamber and the performance results of the different connectors were compared.

978-1-4244-4722-0/09 $25.00
© 2009 IMAPS-ITALY

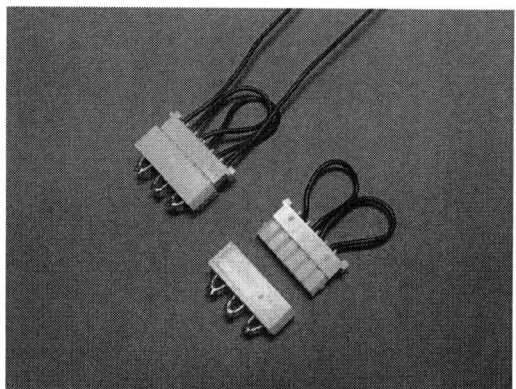

Figure 1: The structure of A1 and B1 Spox connectors.

Figure 2: The structure of A2 and B2 Minifit connectors.

Table 1: Materials and number of the connectors.

	A1 Spox	B1 Spox	A2 Minifit	B2 Minifit
Header housing	Polyamide	Polyamide	Polyamide	Polyamide
Plug housing	Polyamide	Polyamide	Polyamide	Polyamide
Plug contact (crimp)	Tin plated Phosphor Bronze	Tin plated Phosphor Bronze	Brass	Tin plated Phosphor Bronze
Header contact	Brass	Brass	Brass	Brass
Header contact coating	Tin over Nickel	Tin over Copper	Tin over Nickel	Tin over Copper
Amount	8	8	25	30

The materials of the connectors are presented in Table 1. The housings in every connector were made of polyamide. Also, the contact pin of header parts was made of brass in every type of connector. Brass is an alloy of copper (over 50 %) and zinc. The difference in materials was in the contact plating. The manufacturer 'A'

uses tin over nickel coating and the manufacturer 'B' uses tin over copper coating.

The salt spray test was performed with automatic Ascott salt spray chamber S450xp. The correct parameters were entered into the chamber's program and all the required materials were placed inside. The salt spray chamber is shown in Figure 3.

Figure 3: Ascott salt spray chamber S450xp.

The testing procedure was mainly determined by a testing standard for neutral salt spray, NSS. A suitable standard for salt spray testing and Ascott's salt spray chamber was SFS-ISO 9227 standard. According the standard, the temperature in the test chamber must be kept in the range of 35 ºC ± 2 ºC during the test. Recommended testing periods were 2, 6, 24, 48, 96, 168, 240, 480, 720, or 1000 hours. The standard also stated that the duration must be suitable for the test item. If there is no determined test time for the test items, the duration can be decided among the testers. Concentration of sodium chloride (NaCl) in solution must be 50 g/l ± 5 g/l, equivalent to 5 % concentration of NaCl in deionized water. The value of pH of the solution must be between 6.5 and 7.2. [1]

At first the connectors were tested for 300 hours. After this test was halted for 600 hours while the connectors remained in place. After the halt testing was further continued for 1100 hours. Thereby, the overall duration of salt fog testing was 1400 hours.

Only visual inspections were done during the test, and maintenance if required. The measurement of the contact resistance was done continuously in real time. The measurement system for one contact is shown in Figure 4.

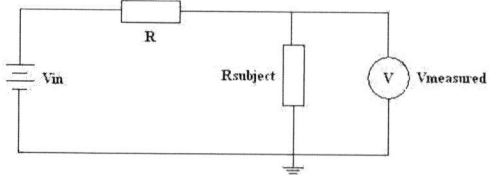

Figure 4: Measurement system for one contact resistance.

In Figure 4, $R_{subject}$ is the contact resistance of the connector. $V_{measured}$ is the measured voltage over the contact of the connector. R is the bias resistor and V_{in} is the supply voltage. The resistance of the connector, $R_{subject}$ can be calculated as

$$R_{subject} = \frac{V_{measured}R}{V_{in} - V_{measured}}$$

Real time measurement was achievable because of the sealable entry port of the salt spray chamber. The entry port makes it possible for the wires to go in the chamber during the test. The resistance or voltage was measured every second during testing. The data of the voltage are saved continuously and analyzed afterwards. From the data, it is possible to see when the resistance of the connector is increasing and operation is no more acceptable

To measure the resistance from a contact of the connector certain arrangements were necessary. The capacity of the chamber and the test setup does not make it possible to measure each contact separately. Consequently, the resistance from each connector was measured by making a daisy chain structure to the connector.

The connections in the plug were made with crimps and conductors. The connections in the headers were made with solder in every type of connector. Two long conductors were also connected to the plug with crimps and the other ends to the measurement system. The total measurement system consisted of a circuit board system and data logger. The measurement wires of the connectors were connected to a circuit board, which routes the information to the data logger system. Data was stored with PC-based PXI controller from National Instruments.

Results after Salt Spray Testing

The object of the test was to compare connectors from two different manufacturers: A and B. The salt spray test was performed according to SFS-ISO 9227 standard. With the selection of this test type, the experimental focus was on the accelerated occurrence of corrosion in the contact pins or crimps. A 10 % increase in the measured voltage served as a failure criterion. The logged data was examined and failures were identified from the data. Failed connectors were molded in acryl, ground and polished, and analyzed by means of a microscope and a scanning electron microscope (SEM).

A total of 19 Minifit connectors from manufacturer 'A' failed and also 25 Minifit connectors from manufacturer 'B'. All of the A1 Spox and B1 Spox connectors failed. Amounts of the failed connectors are represented in Table 2.

Table 2. Amount of the failed connectors.

	A1	**B1**	**A2**	**B2**
Failed connectors	8	8	19	25
Connectors in the test	8	8	25	30
Failure percentage	100 %	100 %	76 %	83 %

Figure 5 and 6 illustrate the failure percentage of the connectors as a function of time.

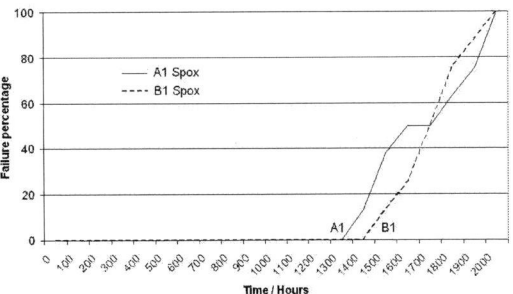

Figure 5: The failure percentage of A1 and B1 connectors as a function of time.

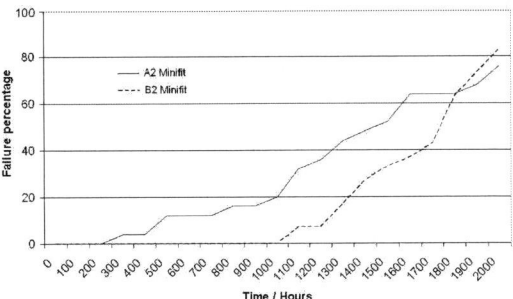

Figure 6: The failure percentage of A2 and B2 connectors as a function of time.

All of the A1 and B1 Spox connectors failed and failures started to occur after 1300 hours of testing. A2 connectors begin to fail much earlier than the B2 connectors. The test time in the figure includes the 600 hours during which the chamber was halted and salt spray mode was discontinued.

Since the voltage of the connector is measured and data stored at every second, the moment of failure can be determined directly from the data file. The data contains much interference, which was caused the laboratory environment. In order to gain an image of the logged data free of background interference, the data is filtered with Matrix Laboratory (MATLAB). Median filtering was used to filter the noise of the data and this can be seen in Figure 7. From the lower image it can be seen that the level of the measured voltage is about 0,1 volts. The y-axis represents the voltage and the x-axis, the time in hours.

978-1-4244-4722-0/09 $25.00
© 2009 IMAPS-ITALY

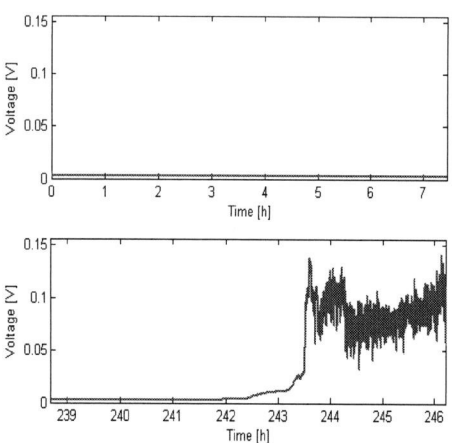

Figure 7: Filtered data. The upper graph shows voltage of an A2 Minifit connector at the start of the test. The lower graph shows voltage of the same connector at the time of failure.

The upper graph presents the data at the start of the test and the lower one shows the data at the moment of failure. It can be seen that the failure occurred after about 242,5 hours of testing. This also indicates that the failure is permanent and irreversible as was the case with every failed connector. When this connector remained in the test despite its failure, it was observed that the voltage value increased over time. This indicates a failure mechanism in which the degradation progresses and eventually destroys an entire component or a part of it.

Failure Analysis

Failure analysis began with an examination of the measured data. After identifying the failed connectors, visual inspections were carried out. Measurement wires and solder connections were examined to ascertain that they were not the cause of failure. Variation in the color of the connectors was also detected. After the data analysis and visual inspection, a number of the failed connector samples were prepared to gain a cross-section and magnified image. The failed connector was molded in acryl and cured under pressure. After curing, the sample was ground and polished, and examined under an optical microscope.

In order to obtain a reference image, a connector which had not been used in the test was also prepared for microscopic examination. Figure 7 and 8 shows untested connector header pins of Spox and Minifit connectors. In Figure 8 is the Spox connector pin and in Figure 9 is the Minifit connector pin.

Figure 8: Cross-sectional image of untested connector header pin of A1 Spox.

Figure 9: Cross-sectional image of untested connector header pin of A2 Minifit.

The pins are made of brass and have a tin over nickel coating; tin forming the outermost layer. The different materials can be seen more clearly in the figures below. Data on the materials were provided by the manufacturer and verified by elementary SEM analysis.

The cross-sectional image of an untested A1 Spox connector is shown in Figure 10. The different material layers can be seen clearly. Figure 11 shows the cross-sectional image of an untested A2 Minifit connector. The outermost layers are tin, other thin layers are nickel, and the body materials are brass.

978-1-4244-4722-0/09 $25.00
© 2009 IMAPS-ITALY

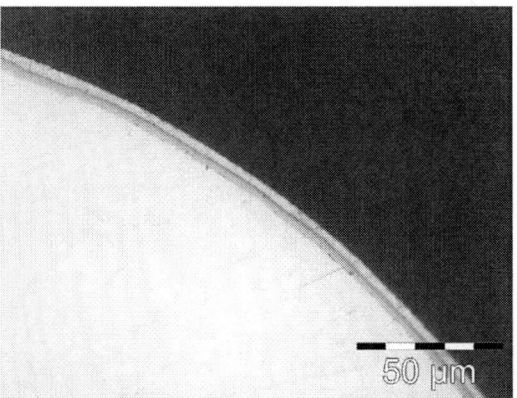

Figure 10: Magnified cross-sectional image of an untested A1 Spox connector header pin.

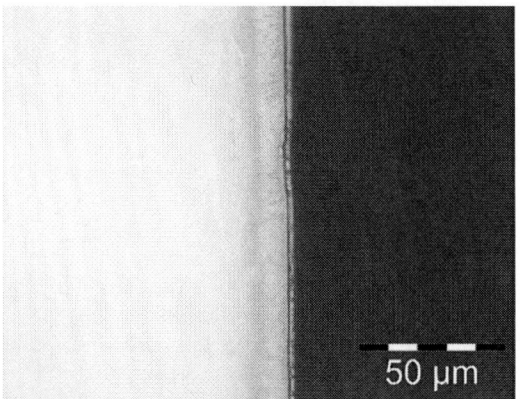

Figure 11: Magnified cross-sectional image of an untested A2 Minifit connector header pin.

Figure 12 shows a cross-sectional image of a tested Spox connector pin with degradation of the coating layers. The pin becomes vulnerable to corrosion when coating materials degrade. The degradation can be caused from galvanic corrosion when materials with different nobilities are in contact.

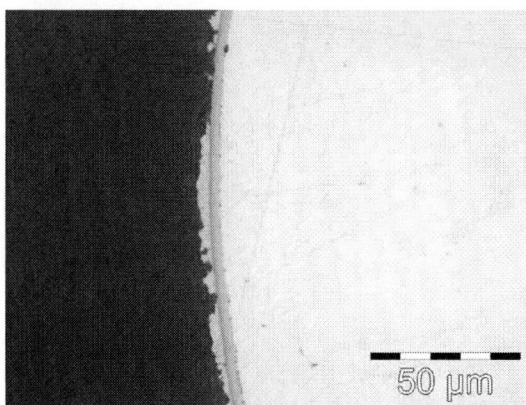

Figure 12: Cross-sectional image of A1 Spox connector, which has some degradation in the coating materials.

Figure 13 shows corrosion product in header pin of A1 Spox connector. Corrosion is able to progress from the surface area to the body material of the pin when the coating layers are degraded.

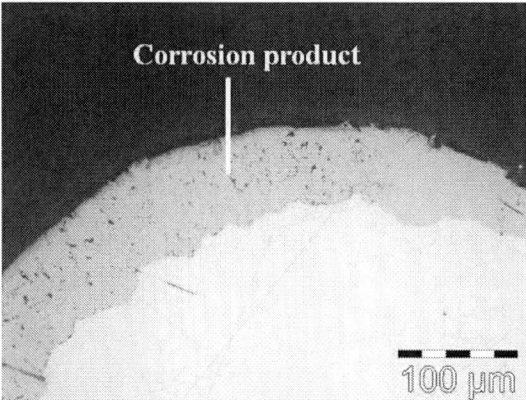

Figure 13: Cross-sectional images of a damaged A1 Spox connector after spending 2000 hours in the salt spray test chamber.

Examination of the cross-sectional area of the failed connectors showed that a new material had formed on the pin surface and in some cases this had spread into the pin.

Figure 14 presents an image of another sample, an A2 Minifit, which showed more corrosion products. The data on this particular connector shown below was analyzed earlier. It failed to operate after about at 242.5 hours of testing, but was kept in the chamber for another 300 hours during the interruption period and for a further 300 hours of actual testing. Figure 15 shows a badly damaged header pin with greater magnification.

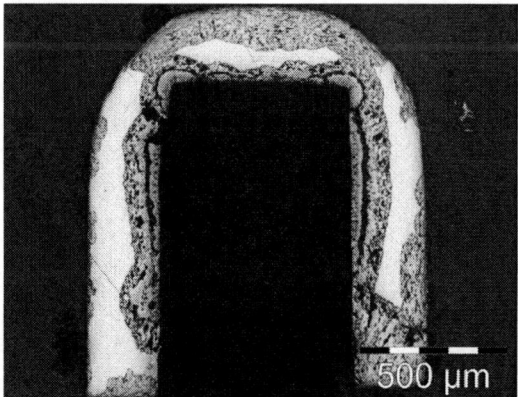

Figure 14: Cross-sectional images of a badly damaged A2 Minifit connector after spending over 800 hours in the salt spray test chamber.

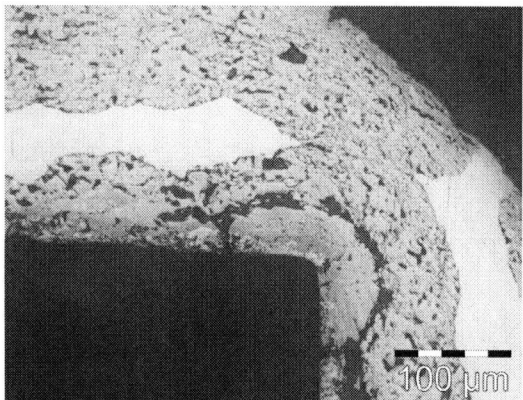

Figure 15: Cross-sectional images of a badly damaged A2 Minifit connector after spending over 800 hours in the salt spray test chamber with greater magnification.

All four different types of connector were tested for the same duration. The connectors were all placed in the chamber in the same way and in the same orientation. All were subjected to uniform test conditions. The single difference between the connectors, and capable of affecting the test results, is the different coating materials of the header pins. The coatings are shown in the figure below.

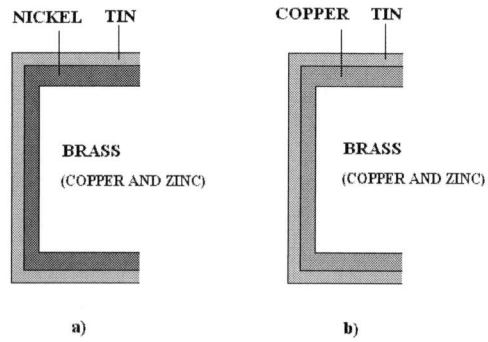

Figure 16: Coatings of different manufacturer's connectors. a) Manufacturer 'A' uses tin over nickel coating and b) manufacturer 'B' uses tin over copper coating.

Galvanic corrosion occurs where two or more dissimilar metals form an anode and a cathode, and an electrolyte forms the contact between them. The less noble anode oxidizes or releases electrons and the nobler cathode reduces or receives electrons. The greater the difference is between the metals' nobilities, the greater the risk is of galvanic corrosion. According to ASM Handbooks [2], galvanic corrosion is usually the result of poor design and choice of materials. [2] [3] [4]

As with the corrosion products of the badly damaged connector in Figure 14 and 15, the coatings of the connectors received careful study. After examination under a microscope, they were subjected to scanning electron microscopy (SEM).

Figure 17 presents an SEM-image of the coatings of an untested A2 header pin connector. The uppermost layer is tin, the middle layer is nickel, and the lowest body material is brass. The materials were identified by elementary analysis.

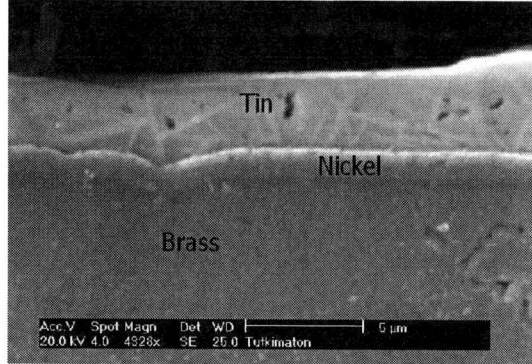

Figure 17: Coating layers of an untested A2 connector's header pin.

Inspection of an untested connector with SEM revealed fractures in the coatings at the inner corners of the connectors pin. Both coating layers of this connector were fractured, as can be seen in Figure 18. This kind of fracture can be caused by the manufacturing process; for example, if the pin is first coated and then bent to the desired shape. In this case, both protective coating layers were fractured, rendering the damaged area vulnerable to corrosion. Corrosion can start from such damaged areas and progress into the body material. The smaller amount of failures among Spox connectors can be explained with this phenomenon. Spox connector pins are different shapes and no fractures in the coatings were detected before testing.

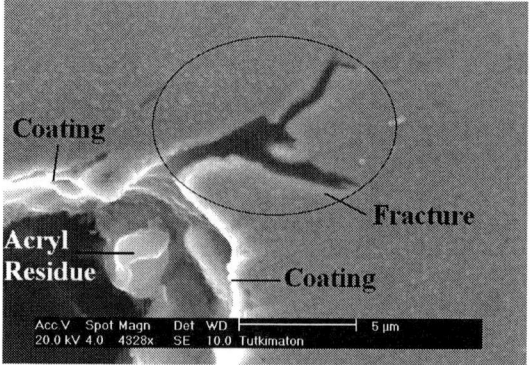

Figure 18: Fracture of the coatings at the corner of the pin.

The badly damaged A2 Minifit connector presented in the microscope images above was also examined with SEM. An elementary analysis was performed to determine the material changes to the damaged parts of the connector. The pin itself is made of brass, which contains copper and zinc. In the elementary analysis of the untested connector it was observed that the brass contained about 65 wt% of copper and 35 wt% of zinc. The elementary analysis of the damaged connector revealed that the damaged area contained 95 wt% of copper and 5 wt% of zinc.

According to the definition of brass, the brass content of the untested connector was high in zinc. The disappearance of zinc phenomenon with the damaged connector can be explained by dezincification. This is a common mode of galvanic corrosion in brasses containing less than 85 % copper [5]. Zinc corrodes preferentially and leaves porous remains of copper and corrosion products. In the dezincification process the brass dissolves, the zinc ions stay in solution, and the copper plates back on. According to ASM Handbooks [5], the areas suffering from dezincification usually contain 90-95 % copper. This was also seen in the case of the badly damaged connector. The badly damaged pin and a magnified mirror-image with SEM of the marked area are shown in Figure 19 and 20. [5]

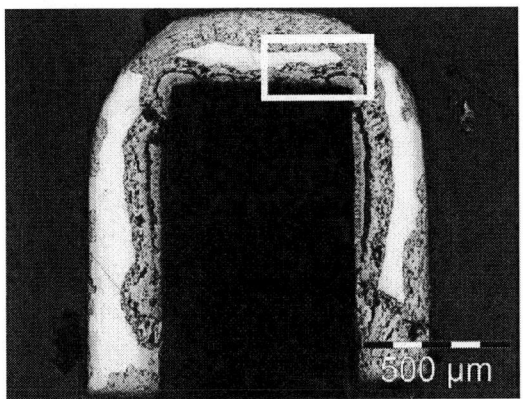

Figure 19: Magnified image made with optical microscope.

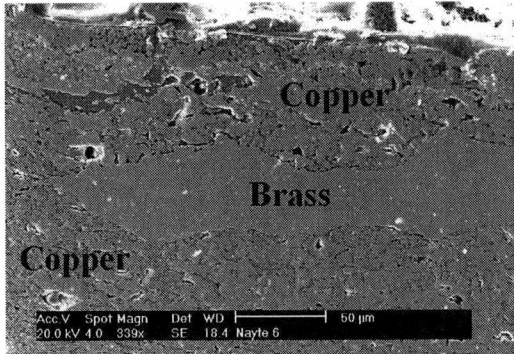

Figure 20: Magnified image made with SEM.

In the case of manufacturer 'A' connector's coatings, the layers were tin, nickel, and brass. Tin and brass have almost the same nobility whereas nickel is nobler than either tin or brass. In this case, if there is a fracture where the salt solution can penetrate into the material surfaces, the nickel layer corrodes the tin layer as the tin is the anodic and nickel is the cathodic material. As the coating

corrodes and vanishes, the brass is exposed to salt solution and corrodes rapidly. Galvanic corrosion causes dezincification in the brass material. [6]

In the case of manufacturer 'B' connector's coatings, the layers are tin, copper, and brass. The materials are of almost equal nobility. It is difficult to distinguish the anodic from the cathodic material since tin and copper form a fairly stabile compound. If a fracture occurs on the surface area this can cause salt solution to penetrate to the brass and induce a loss of zinc. Fractures can result from wear or other mechanisms. [6]

Conclusion

The aim of this paper was to use salt spray testing to determine the reliability of connectors and to compare the same type of connectors from different manufacturers. The salt spray test incurs degradation, mainly corrosion, in the test items. This test is a suitable reliability test method for connectors, whose main degradation or failure mechanism is corrosion in addition to wear. The connector housing does not fully protect the contacts from pollutants or impurities. When the impurities enter the connection, the corrosion process accelerates. The resistance of the electrical connection increases and operation becomes less predictable as the connector degrades.

In the test two different types of connector from two different manufacturers were placed in a salt spray chamber for a period of 1400 hours. After 300 hours the test was halted for 600 hours before the test was continued for 1100 hours. Once started, corrosion is an inexorable process so failures could also occur during the interruption.

The failures were detected from the logged data of measured resistance. A 10 % increase in the measured resistance was deemed to be an instance of failed operation. Usually, the resistance increased rapidly as a failure occurred. In no case was there recovery from failure – when operation failed the state was permanent. The difference between these connectors was in the materials of the header pins. The connectors with a tin over copper coating endured better than the connectors with a tin over nickel coating.

The pins of Spox connectors from both manufacturers had contact pins shaped of a solid circle. Probably, the pin coating has remained undamaged during the manufacture process. The coating degrades because of galvanic corrosion. The homogeneity of the coating explains late failures.

The pins of Minifit connectors from both manufacturers had contact pins shaped of a horseshoe. As revealed in the SEM analysis of an untested pin, the coating material of manufacturer A connectors has fractures in the inner corners of the connectors pin. This kind of fracture can be caused by the manufacturing process; for example, if the pin is coated before bending it to the desired shape. Failures occur earlier as the coating is damaged in

978-1-4244-4722-0/09 $25.00
© 2009 IMAPS-ITALY

advance. Both fractures of coatings and galvanic corrosion increase the occurrence of the failures towards the end of the test.

The Minifit connector from manufacturer B did not have early failures and the reason might be caused by the manufacturing process. The coating is remained undamaged if the pin is coated after bending. This may explain why there were no early failures with the manufacturer B connectors.

There are no remarkable differences between the results of the Spox connectors. However, connectors from manufacturer A started degrading earlier than connectors from manufacturer B.

A greater difference between manufacturers can be seen in the case of Minifit connectors. Connectors from manufacturer A started degrading substantially earlier than the connectors from manufacturer B. This is a result from the fractures of the pin coatings with the connectors from the manufacturer A.

Acknowledgements

We would like to thank TEKES (Finnish Funding Agency for Technology and Innovation) and the following companies which supported this work: Vacon, Kone, and Konecranes. Additionally, we would like to thank the staff at the Institute of Materials Science at Tampere University of Technology.

References

[1] Finnish Standards Association SFS, SFS-ISO 9227. 2001. Corrosion tests in artificial atmospheres. Salt spray tests. Federation of the Finnish Metal and Engineering Industries, Standards Department. 21 p.

[2] ASM Handbooks Online. [Cited 31st January in 2008]. Document: Corrosion of Copper and Copper Alloys. Available:
http://products.asminternational.org/hbk/index.jsp

[3] Dasgupta, A. & Pecht, M. "Material Failure Mechanisms and Damage Models." IEEE Transactions on Reliability, Vol. 40, NO. 5. pp. 531-536. 1991

[4] Song Z. G., Neo S. P., Oh C. K., Redkar S. and Lee Y. P., "Copper Corrosion Issue and Analysis on Copper Damascene Process", Proceedings of 11th IPFA, Taiwan, 2004.

[5] ASM Handbooks Online. [Cited 31st January in 2008]. Document: Failures from Various Mechanisms and Related Environmental Factors Available:
http://products.asminternational.org/hbk/index.jsp

[6] Hack, H.P. Metals Handbook, Vol.13, Corrosion 9th edition, ASM International. 1415 p. 1987

Accelerating the Temperature Cycling Tests of FBGA Memory Components with Lead-free Solder Joints without Changing the Damage Mechanism

Jan Reichelt, Przemyslaw Gromala, Sven Rzepka

Qimonda Dresden GmbH & Co. OHG, QD BET CMI[†]
Postfach 10 09 64; D-01079 Dresden; Germany

Abstract

The paper reports a comprehensive study on accelerated temperature cycling tests. Based on a DoE test plan, multiple accelerated temperature cycling tests were performed on different SMT soldered 512Mb FBGA daisy chain packages. The thermo-mechanical behavior of two SnAgCu solder alloys was studied. A sensitivity analysis assessed the effects of the TC stress tests factors on the characteristic lifetime. No change in failure mode was seen. The solder interconnect fracture due to solder creep is most sensitive to the magnitude of extreme temperatures and the temperature difference only. Dwell time and ramp rate may be chosen to minimize test time to failure. Applying a log-linear response surface analysis, a quantitative correlation between all factors has been established. In addition, three acceleration factor models have been assessed. They are all applicable to transform the TC test result among the parameter space of different thermal profiles but show slightly different accuracy.

Key words: Temperature Cycling, Acceleration Factor Modeling, Response Surface Analysis, Solder Bulk Fatigue

Introduction

Under service conditions, memory components and modules are exposed to thermo-mechanical stresses like all other micro and nano electronic systems as well. These stresses are caused by the mismatch in thermal expansion between the materials involved and the change in temperature during service. The consequence is fatigue and ultimately a mechanical failure of the electrical joint. In order to assure adequate thermo-mechanical endurance, each new products needs to pass the temperature cycling (TC) test [1]. A temperature cycle swing is defined by minimum and maximum temperature, the dwell periods at both temperatures, and the ramp rate during heating and cooling as shown schematically in figure 1. Usually, the test is designed to accelerate the desired damage mechanism without changing it. The acceleration needed for allowing the test to fit into the schedule of the component development, whose cycle time is continuously shrinking. Hence, increasing the acceleration factors (AF) would save time to market and cost in the component development phase. The natural challenge is to preserve the failure mode, i.e., really just to accelerate the damage but not to change its mechanism.

[†] The authors can now be reached by email:
Jan Reichelt <Jan.Reichelt@BakerHughes.com>
Przemyslaw Gromala
<PrzemyslawJakub.Gromala@de.Bosch.com>
Sven Rzepka <Sven.Rzepka@ENAS.Fraunhofer.de>

Procedure

Devices Under Test (DUT)

This study focused on the memory FBGA packages designs. Three 512Mb products (DDR, DDR-2, and DDR-3) of different technology levels have been selected to represent the current product portfolio in terms of package and die sizes as well as of solder ball patterns. Figure 2 depicts the DDR package from the bottom side. It can be seen that the adhesive fully covers not only the square die but also the ball area. Figure 3 represents the solder ball array of the DDR-3 component. The perfect symmetrical distribution of the solder ball can be seen. This time, the die is quite slander and the adhesive is not larger than it. At each side of the array, one row of solder joints is left uncovered.

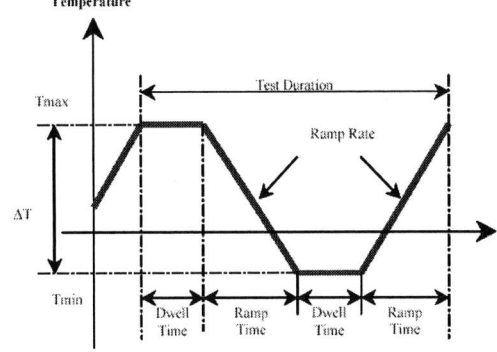

Figure 1: Schematic of a temperature swing

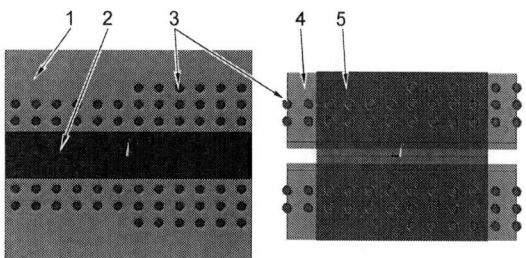

Figure 2: 512Mb DDR package
1 – Substrate solder mask, 2 – bond channel,
3 –solder ball, 4 - adhesive coverage, 5 – die

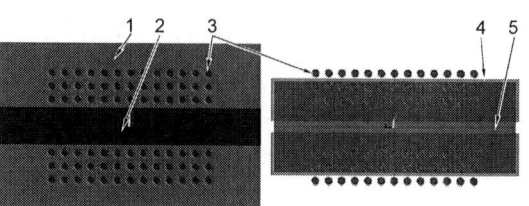

Figure 3: 512Mb DDR-3 package
1 – Substrate solder mask, 2 – bond channel,
3 –solder ball, 4 - adhesive coverage, 5 – die

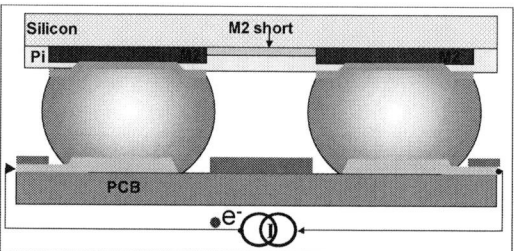

Figure 4: Qimonda daisy chain module concept

In addition, the number of solder balls has changed from 60 (DDR, DDR-2) to 78 (DDR-3).

In order to assess the 2nd level reliability, meaning the robustness against fatigue degradation on board level, the packages were mounted onto test boards using a halogen-free PCB material (ITEQ140G) and two different lead-free solder alloys, SnAg1.0Cu0.5 (SAC105) without and with Nickel doping (SAC105ND). As test board, a daisy chain structure has been used. It allows in-situ measurements of the resistance during the TC tests, figure 4. The failure is captured by an event detector with a resistant threshold. After end-of-life testing, the distribution of the failure events over time is analyzed statistically.

Design of Experiment (DoE)

One goal of this study has been the determination of acceleration factors that allow translating the TC results in terms of time and number of cycles to failure between the different thermal profiles. This would allow testing the samples by the TC profile that leads to the failure most rapidly and transferring the results to any standard profile afterwards in order to minimize the test time during the development of new packages. Of course, this approach needs to be validated before it can be applied. Hence, the study focuses on showing the accuracy and the robustness of the factorial relationships, and the acceleration models involved.

A DoE plan was set-up considering the five factors characterizing a TC profile: minimum and maximum temperature, dwell time at both temperatures and ramp rate (table 1). Two values were chosen for each parameter complying with the JEDEC standard [1] and the constraints resulting from the single chamber test equipment usually used in industry. Initially, an N-1 fractional factorial scheme with center point was compiled to capture the experimental space by nine legs (TC_9…17). In addition, four typical Qimonda accelerated TC test conditions were included (TC_18…21). Experiments were conducted for all these TC profiles. Later-on, the legs TC_1…8 (table 1) were added to expand the scheme into a full factorial test plan, which provides an even higher level of confidence when validating the modeling solutions.

Life Data Analysis

The Weibull distribution is most commonly used in reliability engineering. It covers the situation of failures according to a dominant wear-out mechanism being stochastically distributed because of numerous small effects contributing to it. Therefore, it can model the situation of TC failures very well – across all profiles included in this study. The Weibull distribution involves two parameters as given by probability density function (PDF), eq. (1):

$$f(T) = \frac{\beta}{\alpha} \cdot \left(\frac{T}{\alpha}\right)^{\beta-1} \cdot \exp\left(-\left(\frac{T}{\alpha}\right)^{\beta}\right) \qquad (1)$$

Table 1: DoE matrix concerning the T-t profiles

DoE	T_max [C]	T_min [C]	Dwell_max [min]	Dwell_min [min]	Ramp rate [K/min]
TC_1	100	-40	20	20	10
TC_2	100	-40	5	5	20
TC_3	125	0	20	20	20
TC_4	125	-40	20	5	10
TC_5	125	0	5	5	10
TC_6	125	-40	5	20	20
TC_7	100	0	20	5	20
TC_8	100	0	5	20	10
TC_9	113	-20	13	13	15
TC_10	125	-40	20	5	20
TC_11	100	-40	20	20	20
TC_12	100	0	20	5	10
TC_13	100	-40	5	5	10
TC_14	100	0	5	20	20
TC_15	125	0	20	20	10
TC_16	125	0	5	5	20
TC_17	125	-40	5	20	10
TC_18	100	0	10	10	10
TC_19	125	0	10	10	10
TC_20	125	-40	10	10	10
TC_21	100	0	10	10	20

978-1-4244-4722-0/09 $25.00
© 2009 IMAPS-ITALY

Here, f(T) is the probability over time, α is the characteristic lifetime and β is the shape parameter of the distribution. The estimates of the parameters of the Weibull distribution can be found graphically by plotting the data on probability paper or analytically by either least squares or maximum likelihood methods, respectively. Thereby, the PDF function will usually be transformed into the cumulative probability density function (CDF), eq. (2). The Weibull parameters are determined for each DoE leg by linear regression after logarithmic scaling was applied to eq. (2).

$$F(T) = 1 - e^{-\left(\frac{T}{\alpha}\right)^{\beta}} \quad (2)$$

The characteristic lifetime α is also known as N_{63} indicating the failure of 63% or $(1-e^{-1})$ of the specimens by this time. The abbreviation MTTF is also used synonymously although not being correct. It actually stands for mean-time-to-failure, which implies 50% rather than 63%. If the shape parameter β is larger than unity it quantifies the width of the distribution, i.e., the difference in time between the first and last failing specimen, somehow inversely to the standard deviation. It is interpreted as estimator for process quality and material creep / relaxation behavior because the differences in time to individual failure can only be caused by scatter and deviations in materials, manufacturing, or testing. Thus, constancy of this parameter across different Weibull distributions may also indicate that one single failure mode occurs and that it is the same in all tests.

Response Surface Analysis

When a test like temperature cycling involves multiple accelerating stresses or requires the inclusion of engineering variables, a general multivariable relationship is needed. Such a relationship is the general logarithmic / linear approach as described by eq. (3).

$$L(X) = \exp\left(a_0 + \sum_{j=1}^{n} a_j X_j\right) \quad (3)$$

It describes the life characteristics L as a function of the stress vector with n contributors, $X = (X_1, X_2 \ldots X_n)$, each being scaled by a model coefficient a_j. For the sake of simplicity, eq. (3) can be converted into eq. (4):

$$\ln(L(X)) = a_0 + a_1 X_1 + a_2 X_2 + a_3 X_3 + a_4 X_4 + a_5 X_5 \ldots (4)$$

This relationship can now be customized by substituting the actual effects to the generic stress factors X_j as listed in table 2. On the left side, the characteristic life α is taken as response parameter. By means of the model coefficients $a_{1\ldots5}$, an empirical factorial relation is constituted. It allows estimating the lifetime at any point of the parameter space defined by the ranges of the five stress factors. This concept is also known as response surface analysis. Maximizing the correlation coefficient R^2 and mini-

Table 2: Life-Stress Relation transformation

Factors	Stress / Life	Unit	Transformation	Exponential LSR
X_1	T_{MAX}	[K]	ln(T)	Power
X_2	T_{MIN}	[K]	ln(T)	Power
X_3	$Dwell_{MAX}$	[min]	1/t	Arrhenius
X_4	$Dwell_{MIN}$	[min]	1/t	Arrhenius
X_5	Ramp Rate	[K/min]	ln(R)	Power
L(X)	char. Life	[Cycles]	ln(α)	Power

mizing the root mean square (RMS) error, the values of the model coefficients $a_{1\ldots5}$ are computed.

Acceleration Factor Modeling

The concept of quantitative accelerated life testing analysis consists of two parts, the test determining the lifetime parameters of the product in the minimum amount of time and a model that transforms the test results to any condition, for which the reliability estimate is needed. In solder joint fatigue, a first empirical equation for this transformation was established by Coffin and Manson (CM) [2], eq. (5). It computes the acceleration factor AF_{CM} based on the ratio of temperature differences occurring under laboratory test ΔT_L and field ΔT_F conditions, respectively. The exponent A depends on the material used.

$$AF_{CM} = \left[\frac{\Delta T_L}{\Delta T_F}\right]^A \quad (5)$$

An extended approximation has been developed by Norris and Landzberg (NL) [3]. Besides the temperature differences ΔT_F, ΔT_L, the computation of the acceleration factor AF_{NL} is based on cycle frequencies f_F, f_L and maximum temperatures T_{MAX_F}, T_{MAX_L}, eq. (6). The model also involves an Arrhenius term with the activation energy E_a and Boltzmann constant k_B that refers to the diffusion based nature of the failure mechanism. Originally, the NL model parameters $B_1 \approx 1.9$, $B_2 \approx 0.333$ and the activation energy were calibrated for SnPb solder alloys. Re-calibrating these values, the model can also be applied to other solder materials.

$$AF_{NL} = \left[\frac{\Delta T_L}{\Delta T_F}\right]^{B_1} \cdot \left(\frac{f_F}{f_L}\right)^{B_2} \exp\left(\frac{E_a}{k_B}\left\{\frac{1}{T_{MAX\_F}} - \frac{1}{T_{MAX\_L}}\right\}\right) (6)$$

An alternative modification of the CM approach has been introduced by Pan HP (HP) [4], eq. (7). It accounts for the dwell times t_F, t_L instead of cycle frequency as it is most effective to the creep damage. Originally, the HP model has been applied to a SnAgCu solder alloy with the model coefficients of $C_1 \approx 2.65$ and $C_2 \approx 0.136$ being most appropriate.

$$AF_{HP} = \left[\frac{\Delta T_L}{\Delta T_F}\right]^{C_1} \cdot \left(\frac{t_L}{t_F}\right)^{C_2} \exp\left(\frac{E_a}{k_B}\left\{\frac{1}{T_{MAX\_F}} - \frac{1}{T_{MAX\_L}}\right\}\right) (7)$$

As seen from equations (5)-(7), all approaches come with specific parameters. They need to be adjusted before trustworthy transformations of the test results to other conditions will also be possible in situations other than those at model set-up. More generally, the validity needs to be shown before any AF model can be applied to a new situation [5]. The goal of this study has been to find a suitable AF model for transforming the lifetime results of BGA memory modules across the different temperature profiles of the TC tests. The evaluation of the three AF model concepts has been based on the differences between the result for TC profile i obtained by the AF model, x_{Mod_i}, and by the experiment, x_{Exp_i}, condensed in a root-mean-square (RMS) error value capturing all n TC profiles as seen in eq. (8).

$$RMS_{error} = \sqrt{\frac{\sum_{i=1}^{n-1}\left(x_{Mod\_i} - x_{Exp\_i}\right)^2}{n-2}} \qquad (8)$$

Results

Failure Mode Assessment

When assessing the life data statistically, the value of the shape parameter is dependent on the failure mode, the process stability, and the material quality. Hence, large jumps and big differences in the shape parameter would indicate changes in the failure mode among the test series.

Figure 5 shows the β distribution of four DUT investigated in form of box plots. The mean values, β_{MEAN}, are 6.5±1.0, which marks a rather narrow range. The standard deviations σ_β are 1.0±0.1, which are also quite uniform across the test series. That means, no significant differences exist between the four test series. As material quality and process conditions had been kept constant for all specimens, the failure mode must have been very similar in all four series as well. Independent of the package type, solder fatigue occurred, eventually causing electrical failures due to cracks expanding to complete fractures.

Since β_{MEAN} and σ_β are quite similar in all test series, an estimator can be deduced for predicting early fails based on the characteristic lifetime. It is valid for all the packages types investigated, eq. (9).

$$\beta_C = \beta_{MEAN} - \sigma_\beta \qquad (9)$$

The value of the estimator is kept on the low side in order to avoid overestimations of the real lifetimes. Across this study, a value of 5.5 has been determined for β_C. Such conservative approaches can reduce the measurement effort quite safely. They even allow estimating the probability of early fails based on N_{63} results predicted by calibrated FEM simulations as shown in [6]. This really speeds up the package development phase as it replaces some of the time consuming tests completely.

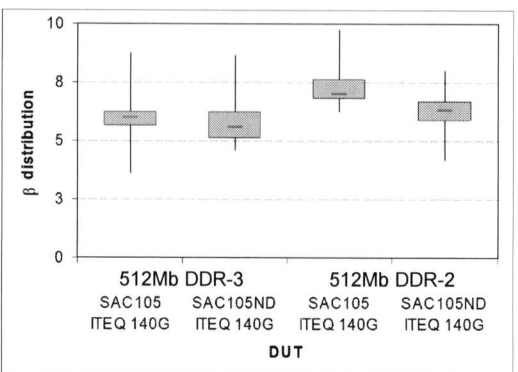

Figure 5: Box plot of shape parameter β

Besides the statistical evaluation, destructive analyses like 'dye and pry' and cross-sectioning have been performed. This allows finding physical justifications for the remaining scatter of the shape parameter still seen in the box plots (fig. 5). As typical example, the cross-section of a Nickel doped SnAg1.0Cu0.5 joint is shown in figure 6. The crack had been propagating through the bulk of the solder and along the interface between solder and the intermetallic phase next to the pad. In creep fatigue, the cracks usually follow the grain and phase boundaries as diffusion is fastest and flux divergences are most likely there. At the interface to the CuSn intermetallics at the pads, however, blocking crystals may redirect the cracks to grow across the crystal.

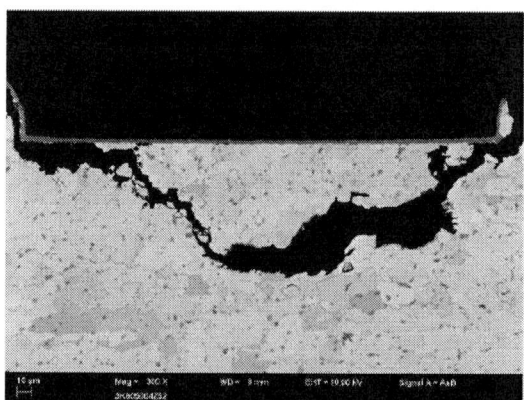

Figure 6: Ion-polished cross-section of a nickel doped SnAg1.0Cu0.5 sample

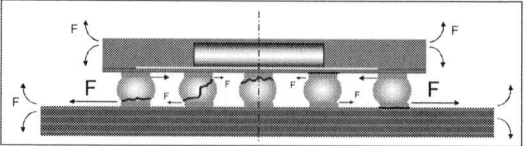

Figure 7: Schematic of fatigue degradation on BGA solder interconnection

In Sn1.0Ag0.5Cu solders, the Ag_3Sn precipitates promote the intra-crystalline crack growth in addition. The Nickel dopants have not been visible in the microstructure. Still, they tend to increase the 2nd level TC lifetime compared to the plain solder alloy. Most likely, Nickel strengthens the interface adhesion at the intermetallic phase, which delays the crack initiation, and further promotes the redirection of the crack propagation inside the bulk, which lowers its rate.

All these effect, size and shape of the grains, orientation of the grain boundaries, location of the large intermetallic crystals, and distribution of the precipitates are stochastic phenomena. Therefore, the amount of scatter the shape parameters show in figure 5 really seems a reasonable magnitude.

While cross-sections reveal details at one particular plane cut through a limited number of joints, the 'dye & pry' analysis provides survey information on the fracture mode across the entire component site. This way, the dominant fracture mode can be determined. Figure 7 depicts the possible fatigue degradation sites. In this study, the joints always failed in the solder close to the pads, i.e., at the PCB side or at the component side. Neither diagonal cracks through the joints nor pad lifts were observed causing initial failure. Failures at PCB side were preferentially found at corner joints of the DDR-3 component, i.e., outside the die area. Here, tensile stress components are assumed to contribute to this fracture mode in addition to the shear load. The tensile stress is generated by cyclic warpage of the component during the test, which is caused by the internal mismatches between mold compound, substrate, and die. Packages whose dies cover the entire ball matrix (e.g., the 512Mb DDR component, fig. 2) do not show this warpage so much due to the stiffness effect of the silicon die. Then, the global shear determines the failure almost exclusively based on the thermal mismatch between the PCB and the entire component. In this case, the cracks preferentially occur at the component side because it is stiffer than the PCB side, where the pad may be able to tilt a bit more.

Response Surface Analysis

The response surface analysis has been done applying Cornerstone™. Fitting the model coefficients, an acceptably high correlation ($R^2 >0.9$) and low RMS error of the number of cycles to failure (less than 0.1 orders of magnitude, i.e. about 20%) was achieved. Examples of the predicted response graphs are shown in figure 8 for both 512Mb DDR-3 series. The corresponding model coefficients are listed in table 3. Both illustrations demonstrate the effects of each stress factor on the TC life. In the graphical representation (fig. 8), the absolute magnitude of the slope indicates the size of the effect. In the table, the value of the coefficient quantifies it precisely. The sign stands for the direction. Positive

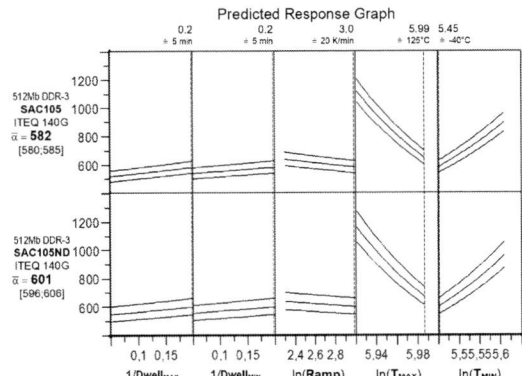

Figure 8: Response Surface Analyze

Table 3: Model coefficients of the log-lin. response surface equation, eq. (4)

Coefficient · Factor	512Mb DDR-3	
	SAC105 ITEQ 140G	SAC105ND ITEQ 140G
a_0 = Constant	41.5±4.5	40.5±5.9
$a_1 \cdot \ln(T_{MAX})$	-8.3±0.7	-8.3±0.9
$a_2 \cdot \ln(T_{MIN})$	2.7±0.3	2.9±0.4
$a_3 \cdot 1/Dwell_{MAX}$	0.8±0.3	0.6±0.4
$a_4 \cdot 1/Dwell_{MIN}$	0.5±0.3	0.6±0.4
$a_5 \cdot \ln(Ramp)$	-0.1±0.1	-0.1±0.1
Correlation R^2	98%	97%
RMS error = $f_{\ln(\alpha)}$	0.07	0.09

slopes and coefficients mean a direct dependency of the response parameter on that particular effect, i.e., the characteristic lifetime increases with an increase in the magnitude of the factor. Inverse relations have negative slopes. However, it is worth noting that this sign convention applies to the factors as used eventually, i.e., after performing the transformation according to tab. 2.

According to the results of the response surface analysis, the number of cycles-to-failure is mainly determined by T_{MAX} and T_{MIN}. The dwell time is also influential but to a much smaller extend. Long dwell times somewhat decrease the characteristic lifetime as it leads to more creep and, thus, to more creep damage, especially at the maximum temperature as creep diffusion follows Arrhenius law. Still, the effect of differences in dwell time is rather small. The effect of the ramp rate is almost negligible even. Only in cases of thermal shock, the ramp rate may develop a significant impact on the number of cycles-to-failure. Such dependencies have also been reported by other studies like [7, 8].

Still, the factors dwell time and ramp rate are important as they determine the total duration of each thermal cycle and, therefore, the overall TC test time. Hence, optimization should not just focus on TC life (in cycles-to-failure) but also on TC speed (in cycles per hour).

Consequently, short dwell times and high ramp rates are recommended for speeding up the TC test process. Maximizing the temperature interval in addition, the fastest TC test is designed yielding the test results in minimum time. The optimization done based on the response surface model is very reliable as R² is higher than 95% and RMS error is very low (tab. 3).

Acceleration Factor Modeling

Probability plots and life vs. stress plots are the most important methods of data representation in the quantitative accelerated life testing analysis.

Slightly different to the life data analysis, the probability plots used in accelerated life testing arrange the failure data according to the magnitude of a stress factor. This shows the acceleration associated with this particular stress factor clearly. Figure 9 depicts such a probability plot for all 512Mb DDR-3 DUT with nickel doped SnAg1Cu0.5 solder. It focuses on the temperature difference as this is the main stress factor in the TC tests. A single equivalent value is taken for all the slopes β. This follows the idea that accelerating the failure without changing its mechanism shifts the distributions along the ordinate (scaled in cycles-to-failure) but shall not affect their shape. In the example shown in fig. 9, the equivalent shape parameter has a magnitude of 5.3. It fits well to β_C of 5.5 estimated by eq. (9). In addition, the data points are well captured within narrow bands around the distribution lines. Both results, the good match of equivalent and predicted shapes and the low scatter along the distributions, are quantitative indications, which further strengthen the assumption of solder bulk fatigue near the pads being the mechanism that determines the failure in all DoE legs. Qualitatively, the dominance of this mechanism has already been observed in the 'dye & pry' analysis discussed above.

Applying a 3-cell matrix, life vs. stress plots can be compiled. For the 512Mb DDR-3 DUT with Nickel doped SnAg1Cu0.5 solder, the life vs. stress relationship (also known as S-N curve) is shown in figure 10. Such plots are widely used for estimating the parameters of the life-stress relationships, which usually are inverse power laws. In figure 10, the value of the power n is 2.0. It matches the exponent A of 1.9 fitted by the Coffin-Manson model, eq. (5), perfectly (tab. 4). This is yet another indication for the validity and applicability of the hypothesis, the different thermal profiles studied by the DoE are accelerating or decelerating the failure propagation, respectively, but do not change the failure mode.

Within the range of TC parameters covered by this study, the failure has always been caused by solder bulk fatigue near the pads on each side, PCB or substrate. No change in failure mechanism was seen. Therefore, the AF model concepts, eq. (5)-(7), should be able to reliably transform the TC results, i.e., the life data, across all thermal profiles within

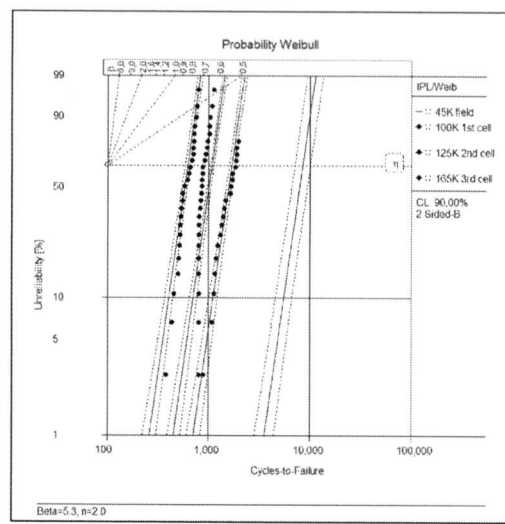

Figure 9: Summarized probability plot for 512Mb DDR-3 DUT with solder SAC105ND

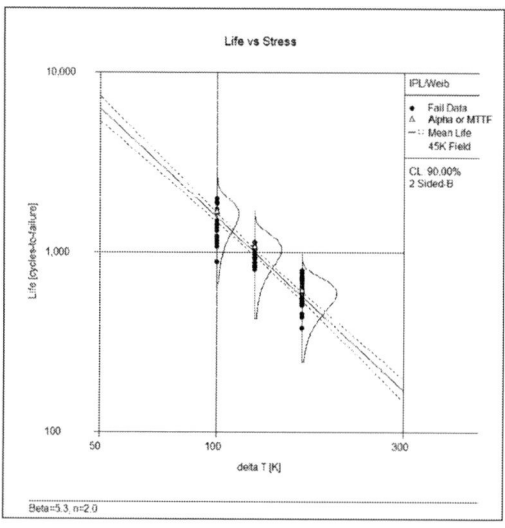

Figure 10: Life vs. stress (S-N) plot for 512Mb DDR-3 DUT with solder SAC105ND

the parameter space of this study including the 'maximum testing speed' profile optimized by response surface analysis. Hence, it is now worth assessing accuracy of each of the three AF model concepts.

Table 4: AF model coefficients

Solder alloy	Coffin-Manson		Norris-Landzberg			
	A	RMS	B_1	B_2	E_a	RMS
SAC105	1.60	0.166	1.51	0.05	0.66	0.157
SAC105ND	1.94	0.254	1.47	0.14	0.83	0.153

Solder alloy	Pan-HP			
	C_1	C_2	E_a	RMS
SAC105	1.51	0.10	0.71	0.146
SAC105ND	1.50	0.14	0.89	0.141

The original CM model already allows transforming the TC results. The differences between the acceleration factors estimated according to the CM model (AF_{CM}) and acceleration found in the experiment AF_{EXP} is very low. For the two solder alloys involved in this study, the RMS error based on these differences is just 17% or 25%, respectively (tab. 4).

Increasing the sophistication, the AF models are even more precise. The approach of Pan HP provides the acceleration factor AF_{HP} whose RMS is as low as 14...15%. This is in the order of the experimental noise, i.e., the characteristic lifetime N_{63} scatters in about this range when the tests are repeated.

A further aspect is the behavior of solder alloy. The results for the coefficients A, B_1, C_1 and E_a seem very reasonable as they are similar but not identical to the values of the original studies [3, 4] due to the change in solder material. The values obtained for E_a [eV] are larger that the typical activation energy of grain boundary diffusion (0.5...0.6 eV) but smaller than that of dislocation motion (>1.0 eV). This matches the practical observation, in which the cracks have been found propagating along grain and phase boundaries but also across the bulk of the grains. Consequently, the effective activation energy should indeed be at an intermediate level. Interestingly, more activation is needed in case of Nickel doping. This supports the hypothesis on the effect of the dopants discussed before.

Figures 11, 12, and 13 directly compare the acceleration factors observed by experiments and those predicted by modeling for 512Mb DDR-2 and DDR-3 DUT. In case of perfect match, all data points would fall onto the solid 45° line. The distance of the actual location quantifies the remaining error. According to the comparison, the CM model is recommended for assessments in the early stages of the development. This simple model requires the least number of coefficients to be determined. Yet, it already provides an adequate accuracy. No data point exceeded the ±40% error band (fig. 11). Later on in the development process, the advanced approaches (NL, Pan) are more favorable. Having more degrees-of-freedom, these models transform the results at higher accuracy. Usually, the HP model would be applied as it gives the most precise results. In this study, no data point has been more than ±20% away from the target line (fig. 13). In cases where the cycle frequencies are known only but not the dwell times, the NL approach would still be available – whose maximum transformation error has also not been larger than ±25%. In practical use, the accuracy will even be better than indicated by these extreme error levels. On average, none of the three AF models exceeded ±15% error. The predictions are confident with a stability index of $R^2 > 0.9$ at 95% confidence level. On the other hand, [9, 10, 11] point out that the applicability of AF models is

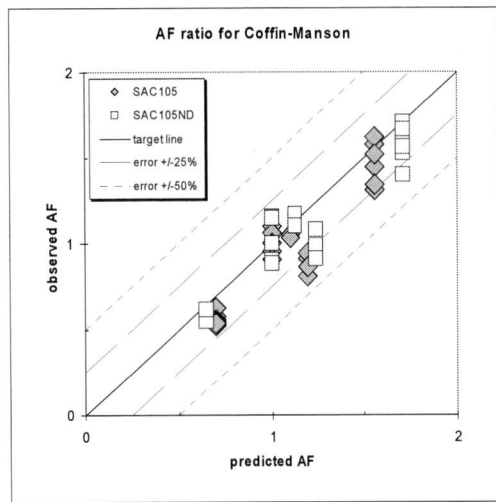

Figure 11: AFs observed by the experiments and predicted by the Coffin-Manson model

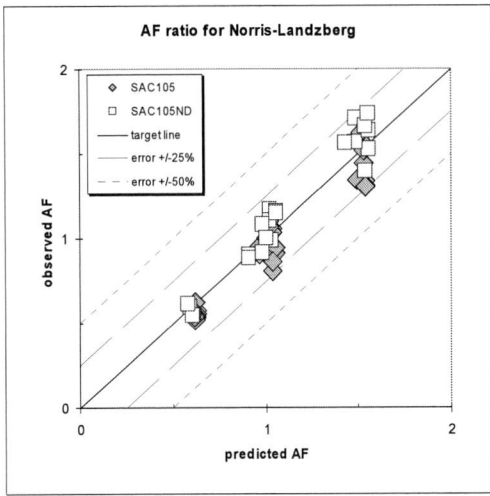

Figure 12: AFs observed by the experiments and predicted by the Norris-Landzberg model

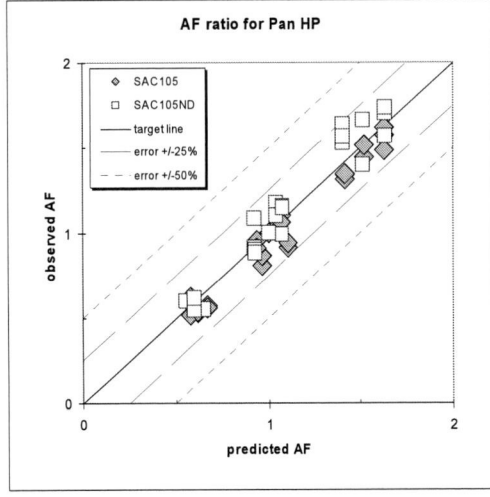

Figure 13: AFs observed by the experimens and predicted by the Pan HP model

restricted to specific product designs, i.e. FBGA, CSP, PBGA etc., for which the coefficients had been determined. Thus, the accuracy of the AF models should be checked carefully before applying to other package types. In some cases, adjustments or even recalibration may be needed.

Conclusion

Based on a DoE approach, a statistical test plan was derived for assessing a quantitative relationship between the lifetime of FBGA memory modules and the stress factors of the TC test. Across all thermal profiles involved, the dominant failure mode has been 'solder bulk fatigue' leading to cracks and fractures close to either interface of the solder and the pad (component or PCB sides). In addition, the shape parameter β slightly varied with solder alloy only.

The response surface model, which has been used to evaluate the test results, was able to capture the DoE data points with an R^2 as high as 95%. Hence, the deductions drawn from that model are trustworthy and reliable. The maximum test acceleration can be achieved by choosing maximum temperature difference, minimum dwell periods (at both temperatures, minimum and maximum) and maximum ramp rate. All three acceleration factor model approaches are applicable to transform the results of this highly efficient test to any other test condition with good precision. The practical inaccuracy has not exceeded 15%, which already comes close to the typical TC test reproducibility. Hence, the simple Coffin-Manson model is already applicable in fast assessments. The more advanced models (Norris Landzberg, Pan HP) provide twice the transformation accuracy but require some parameters more to be calibrated.

Acknowledgement

The authors would like to thank the colleagues of late Qimonda Backend Development organization in Dresden for their brave assistance in sample preparation and testing as well as for the fruitful technical discussions in all the past years.

References

[1] JESD22-A104C, "Industrial Standard for Temperature Cycling Tests"

[2] R.T Tummala, E.J. Rymaszewksi, A.G. Klopfenstein, "Microelectronics Packaging Handbook", Kluwer Academic Publishers, 2nd edition, 1997

[3] K.C. Norris and A.H. Landzberg, "Reliability of Controlled Collapse Interconnections", IBM J. Res. Develop. 13, 1969, pp. 266-271.

[4] N. Pan et al., "An Acceleration Model for Sn-Ag-Cu Solder Joint Reliability under various Thermal Cycle Conditions", Proc. SMTAI 2006, pp. 876-883

[5] R. Zhang, J.P. Clech, "Applicability of various Pb-free Solder Joint Acceleration Factor Models", Proc. SMTAI 2006

[6] P. Gromala, J. Reichelt, S. Rzepka, "Accurate Thermal Cycle Lifetime Estimation for BGA Memory Components with Lead-free Solder Joints", Proc. 10th EuroSimE; April 27-29, 2009; Delft, Netherlands

[7] C.J. Zhai, S. & R. Blish, "Board Level Solder Reliability vs. Ramp Rates & Dwell Time during Temperature Cycling", IEEE 2003

[8] J. Bartelo, S.R. Cain, D. Catelka, K. Darbha, "Thermo-mechanical Fatigue Behavior of selected lead-free solders", IPC SMEMA Council APEX[SM] 2001

[9] O. Samela, "Acceleration Factors for Lead-free Solder Materials", IEEE 2003

[10] J.P. Clech", Acceleration Factors and Thermal Cycling Test Efficiency for Lead-free Sn-Ag-Cu Assemblies", SMTAI 2005

[11] J.P. Clech" Accelerated Thermal Cycling Conditions and Acceleration Factors for Lead-free Surface Mount Assemblies", Proc. SMTAI 2006

Bisected Thermodynamic Sensor as the Power AC/DC Transmitter

M. Reznicek, I. Szendiuch, Z. Reznicek, Jr., and Z. Reznicek

Department of Microelectronics, FEEC, Brno University of Technology, Czech Republic, Udolni 53, 60200 Brno

xrezni14@stud.feec.vutbr.cz

Abstract

Bisected thermodynamic sensor in temperature shift balance circuit as the power AC/DC transmitter is described. This paper deals with new thick film sensor used for thermodynamic direct power AC/DC transmission in industrial systems. In previous papers the basic principle of thermodynamic sensor was introduced and its construction on alumina ceramic substrate with commercial thick film pastes was demonstrated. Generally this sensor has plenty of application possibilities in various areas of industry. Some final users especially from RF area are asking for galvanic isolated sensors independent to contiguous environment effects. Such applications can be realized by the bisected thermodynamic sensor probe. The basic principle of this solution is bisected thermodynamic sensor with the thermodynamic environment temperature shift between both active sensor sections to influence of ambient temperature compensation. The simulation model of bisected thermodynamic sensor as particular segment elements of thermodynamic system was simulated. Values obtained from simulation are compared and discussed with experimental results. There is a good conformity between calculated and measured values. Results obtained from simulations enables construction of new family of segmented thermodynamic sensor probes in balance circuitry models. These could be integrated into many industrial thermodynamic metering systems designed in standardized design environments. They can be used as technical solution in galvanic isolated power measuring systems in integral industrial designs of any complex thermodynamics systems. Namely power and heat distribution, overpower protection systems, thermal circuit breaking systems, HIRF detection systems, etc.

Key words: thermodynamic sensor, AC/DC transmitter, HIRF, simulation, balance circuit

Thermodynamic sensors in balance circuits

The Thermodynamic sensors (TDS) are active thermodynamic sensing elements that are used to define and monitor thermal processes or conditions in thermodynamic systems. They represent new family of sensors used for thermodynamic monitoring and controlling of process in industrial systems.

The basic model and theory of ideal thermodynamic sensor integration as an ideal element in large models of thermodynamic system was presented in [1]. The original theory of ideal thermodynamic sensor as a process and media energy activity monitoring device was presented in [2, 3, 4].

Each separate thermodynamic sensor is monitore all activities in system along with the system environment effect.

The double thermodynamic sensor probes and theory of thermodynamic shield to environment effect suppresse was described in [5 and 6]. Two (one external and second one internal) thermodynamic sensors with different itself working

temperatures are used in the independent balance circuits and operate as reciprocal thermodynamic shield.

Bisected thermodynamic sensor

The ideal thermodynamic sensor in balance circuit according to fig.1 is continuously balanced on itself constant temperature by defined equivalence in [1].

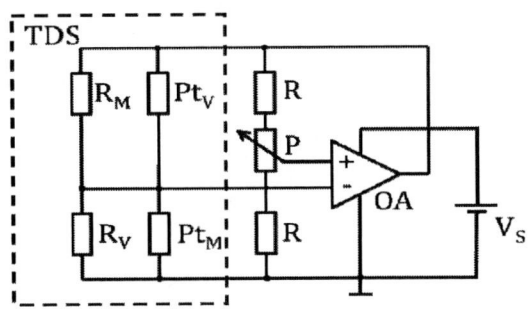

Figure 1: Ideal TDS in balance circuit

Original TDS sectioned to two equal temperature dependent parts TDS$_A$ and TDS$_B$ ensembled of resistor heaters R$_{HA}$ and R$_{HB}$ according to fig.2 is projected as bisected thermodynamic sensor.

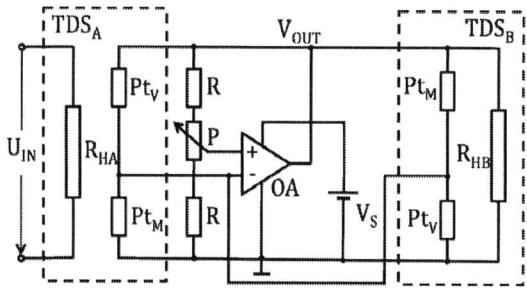

Figure 2: Bisected TDS in balance circuit

In this configuration in balance circuit is bisected TDS operated so, that its part TDS is observed actual thermal situation on part TDS$_A$. Initial temperature conditions are adjustable by potentiometer P brash position. They are after it balanced by actual OA output voltage V$_{OUT}$ so, that R$_{HB}$ heating resistor thermal activity is following actual the actual thermal activity of independent resistor heater R$_{HA}$. Actual DC output voltage U$_{out}$ in this case is equal to effective value of AC input voltage non-dependently of its actual shape form. It can be periodical sinus, triangle, rectangle or quite irregular or destroyed.

The actual temperature dependent Pt$_M$ and Pt$_V$ resistors activities contributory values on both TDS sections (TDS$_A$ and TDS$_B$) are each case identical without relevant effect to circuit transmition characteristics.

AC/DC Effective sonde realization

The principal half TDS sections of effective transmitter sonde can be realized.

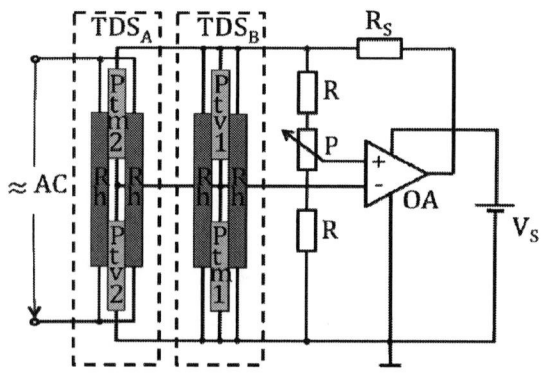

Figure 3: Assembled half TDS principle version

One of the practicable assembled versions is in principle visibled on the fig.3.

It was realized as assembled bonded thermodynamic body from integral miniature multi resistor structures on thick film alumina substrates.

In real applications is the balance circuit extended of some components to expand functionality, increase reliability and better calibration. Both bi-sected TDS parts are fabricated on one Alumina substrate as identical thick film multi-motive unified components. They are used to sonde thermosystem body assembling. The physical size is seeing as the illustration photos on fig.4. and fig.5.

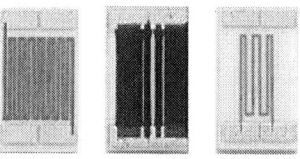

Figure 4: Half TDS sample photo demonstration

Figure 5: Sonde sample photo demonstration

AC/DC transferability

The Bi-sectioned TDS was used to the experimental AC/DC sonde fabrication.

Its self-balancing output signal time record as the response to step by step increasing/degreasing input AC voltage amplitude was used as global data source to basic transfer characteristics determination. The example of it is given on fig.6.

978-1-4244-4722-0/09 $25.00
© 2009 IMAPS-ITALY

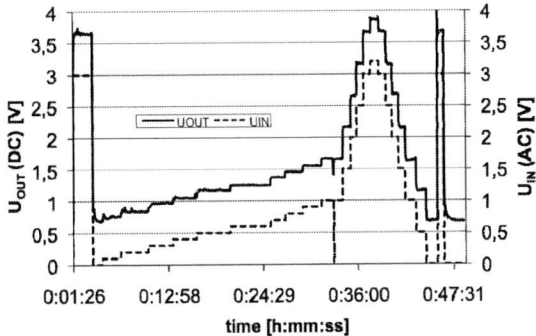

Figure 6: AC/DC output/input time record

The input signal intenzity switch point is amplificantly margined by the long vertical line added to record for better visualization.

AC/DC transfer characteristics

It is very easy to convert previous AC/DC output/input voltage signals to actualpower interpretation using the real heaters Rh known resistivities. Their power (thermal) subactivities are on both part TDS sections practically identical, Output voltage transfer sensitivity can be intensified (calibrated) by the right balance circuit upgrade. The example of sonde response power transfer interpretation is given on fig.7.

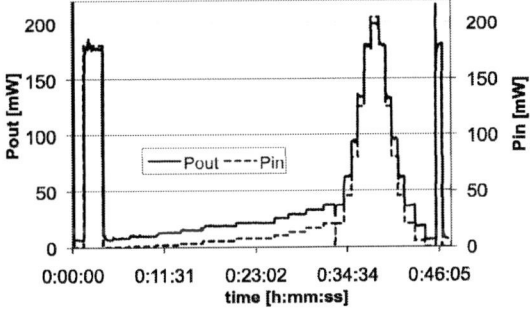

Figure 7: AC/DC transfer record in power interpretation

The speed of sonde response

The short time zooming to final input voltage pulse time duration analyse indicate quick output signal reaction to high input signal individual steps. One time owershot is very well-marked, but for all that the otput is comming to stay very quickly right after.

To precision differentiation was the digital 10x response speed degreasing firmware filter to the micro-computed balance circuit installed.

The zoom of both defined versions of response in power activity graphic representation is on the fig.8.

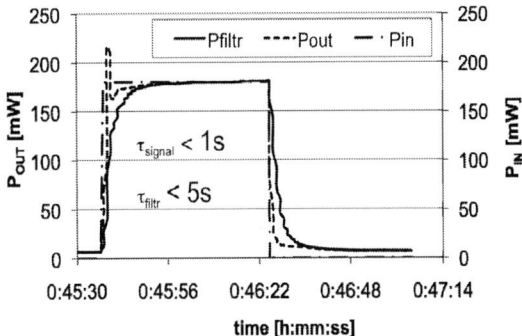

Figure 8: Sonde response speed voucher

The analog output signal response delay constant is shorter 1 and the filtered digital data response delay constant is shorter 5 second.

The output transfer characteristics

To easy find and output signal value assignment to input activity was siple software routine of last otput signal value before input activity switch point margined to previous of it assignment. It was applied to step-up respectively step-down of input activity changes separately. The graphic representation of sonde power transfer characteristic with hysteresis is on the fig.9.

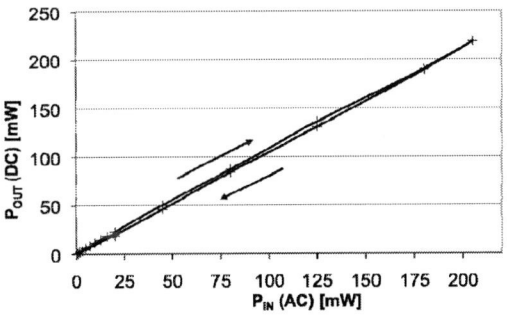

Figure 9: Sonde transfer characteristic

The specified firmware or software correction routine to increase transfer characteristic linearity can be applied as needed.

The short time low signal stability

The short time zooming to low input voltage signal intensity part of AC/DC response record time duration analyse indicate good short time low signal stability. The zoom of defined response part in power activity graphic representation is on the fig.10.

978-1-4244-4722-0/09 $25.00
© 2009 IMAPS-ITALY

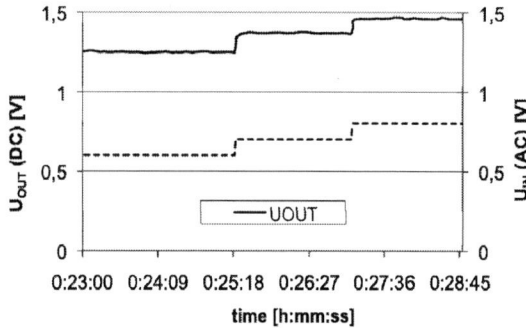

Figure 10: Sonde short time low signal stability

Analogical zoom of response to high input voltage signal intenzity analyse indicate a little worse, but acceptable short time high signal stability. Its graphic representation is on the fig.11.

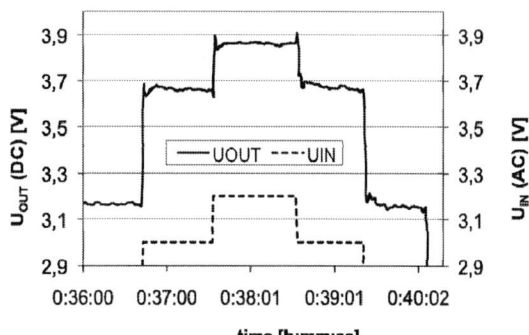

Figure 11: Sonde short time high signal stability

AC signals large frequency scale matching

Large frequency scale matching is dependent of steric arrangement and impedance match. The best one is outcome of intuitive design in combination with rightly used simulation environmet and objective testing procedures during the sonde development procedure.

The test to power (impedance) match on the spectral analyzer result as the PSV representation is on the fig. 12.

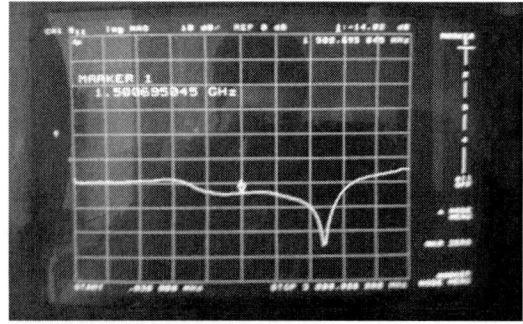

Figure 12: Sonde PSV response

The real sonde response frequency dependence

The real output signal to constant amplitude input frequency changes was tested with resutts on the fig 13.

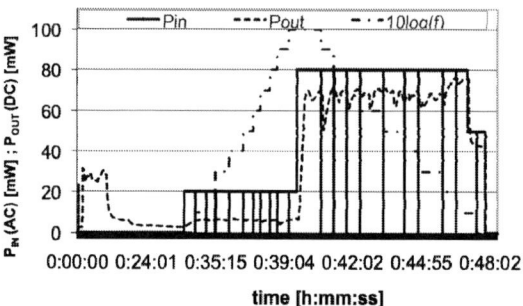

Figure 13: Sonde frequency non dependence test

Conclusions

The new design of bi-sected thermodynamic sensor in balance circuit was developed in version for AC/DC transmitters based on one section TDS AC thermo activity compenzated by by DC thermoactivity in second one identical TDS section. The two identical TDS separation warrants the transfer characteristic environment temperature non dependency, right transfer linearity and good sensitivity in principle used. All transfer characteristic parameters are simply guaranteeable the both sections design identity and symertricity rule enforced.

The transfer sensitivity is increasing with graded stage of miniaturization and the simple design make the large frequency scale impedance match easier.

These real advantages predetermined presented design to many actual applications and heads for trasmiters based on thermo effect development reactivation.

It is important to integrate it into many galvanic isolated industrial power metering systems, heat distribution metrology, owerprotection systems, hazardous electro-magnetic field intensity monitoring and detection systems etc.

Acknowledgements

This research has been supported by the Ministry of Industry and Trade under the program TANDEM, contract FT—TA4/115 Basemo and taking advantage of patent pending rights [1]

References

[1] Patent pending Nr. CZ-297066, „Technique of referential temperature and temperature difference measurement, asymetric

978-1-4244-4722-0/09 $25.00
© 2009 IMAPS-ITALY

temperature sensor and asymetric referential unit to technique application", Bulletin Nr. 3/2001

[2] Řezníček M., Szendiuch I.:, " Process energy balance monitoring ", Proceedings of the 11th conference Student EEICT, vol. 1, ISBN-80-214-2888-0 no. 1, pp. 33-44, 2005.

[3] Řezníček Z., Tvarožek V., Řezníček M., Szendiuch I., „Temperature balanced process media energy activity monitoring", International Conference EDS-IMAPS CS 2005 Brno, Czech Republic, September 15-16 20

[4] Řezníček Z., Tvarožek V., Řezníček M., Szendiuch I.: Hybrid Constant Temperature Regulator, International Conference EuroSimE It 2006 COMO, Italy, April 15-16 2006

[5] Tvarožek V., Vavrinský E., Řezníček Z.: Novel approach in ratiometric technique of sensing, J. of Electrical Engineering, 58 (2007), 2, 98-103

[6] Řezníček Zdeněk, Sr., Szendiuch Ivan, Řezníček Michal, Tvarožek Vladimír, Řezníček Zdeněk, Jr., Thick Film Sensor for Temperature Balanced Process Monitoring, International Conference EDS-IMAPS CS 2005 San Chose, USA, September 15-16 20

[7] Řezníček Zdeněk, Sr., Szendiuch Ivan, Řezníček Michal, Řezníček Zdeněk, Jr., Thick Film Double thermodynamic sensor, International Conference EDS-IMAPS CS 2005 Greenvich, UK, September 1-5 2008

3D Integration with AC coupling for Wafer-Level Assembly

M. Scandiuzzo, L. Perugini, R. Cardu[*], M. Innocenti, R. Canegallo

STMicroelectronics, Agrate Brianza, Italy

[*]ARCES, University of Bologna, Italy

Mauro.Scandiuzzo@st.com

Abstract

This paper presents a solution of stacked chips using a capacitive communication from electrodes at the last metal layer with a wafer level assembly process. The wafer level approach instead of the die level allows high throughput and enables further optimization of the capacitive structures. To reach a good AC coupling an additional passivation layer was deposited then planarized and at the same time the dielectric thickness was monitored. An inter-electrode oxide around 400nm was proved and then bonded by molecular direct bonding. The alignment accuracy of $\pm 1\mu m$ and the bonding quality were checked by infrared microscopy. The upper silicon wafer was thinned around $50\mu m$. The buried I/O pads of both chips were opened by dry plasma etching through the back of the top wafer. The stacked chips were diced and packaged in a standard ceramic cavity and bonded with gold wires. Good performance in term of low power and a large communication bandwidth of 1.23Gbps/pin with $8x8\mu m^2$ electrodes has been measured.

Key words:

Capacitive coupling, Wafer Level, 3D Face to Face

Introduction

The steady progress in integrated electronics up to the present has been driven by technology scaling of devices and wiring. This trend has enabled an ever increasing level of functional integration on-chip and it has aimed to the development of new and more complex systems. As semiconductor technology continues to scale, the ever increasing density, performance and reliability requirements in IC realization create significant process integration challenges for future products. For most large logic applications, the interconnect delays dominate the system performance and the I/O limitation contributes to increase the performance gap in communication bandwidth between internal and external devices. In order to allow the integration in the same system of mixed technologies such as logic with memory, mixed signal or RF transceivers with sensing arrays several hurdles have to be overcome. A wide range of solutions have been proposed to address some of these challenges. Three-dimensional (3D) integration is growing in interest as key technology for the implementation of complex systems: heterogeneous process integration, simplified routing, volume and footprint reduction are some of the features that take advantage of stacking and vertically connecting dies. Many 3D fabrications concepts have been investigated in recent years such as System-in-Package (SiP), through substrate vias or contactless interconnections. These options differ in their impact on integration parameters, such as number of layers (tiers), method of assembly, interconnection size and impact on manufacturing technology [1].

While 3D die packaging or SiP is used in portable electronic applications, such as cell phones, to reduce the footprint required for ICs or to improve the memory density, the high inductance from the wire bonding at the edge of the die to the PCB , limits performance and I/O density and therefore this approach is not suitable to build an high performance computing system that require large bandwidth, data rates and interconnect density, along with very low power consumption. Through Silicon vias offer the greatest interconnection density and the best performance but also the greatest cost due to manufacturing process steps. This work focuses on the implementation of a wireless interconnection scheme based on capacitive coupling. Die or wafer are stacked and aligned face-to-face; communication electrodes are realized in the upper metal layer of each die and they are connected to dedicated communication circuits.

3D Capacitive Interface

In Fig. 1 it is sketched the basic scheme of a 3D capacitive interconnect network. It based on placing two ICs in a 3D stacked Face to Face configuration in order to create capacitors between

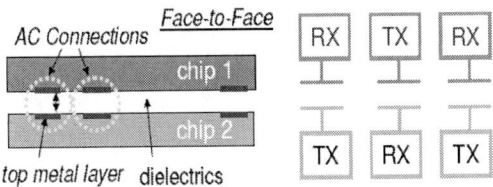

Figure 1: AC capacitive coupling principle

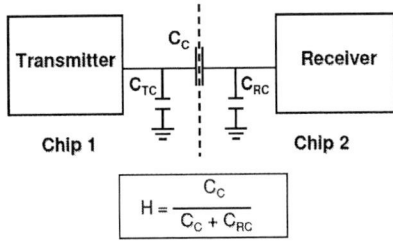

Figure 2: 3D capacitive channel model

the two opposite chip surfaces. The capacitor electrodes are made using the upper metal layer with the top silicon oxide passivation acting as dielectric. A transmitter and a receiver circuits are connected to the bottom and top electrodes to provide AC coupling communication. The interface design is aimed to minimize the interconnection area by implementing a very sensitive receiver and a transmission approach that maximizes communication efficiency. The importance of a sensitive receiver can easily be grasped from the expression of interconnection attenuation H determined using the inter-chip model sketched in Fig. 2: C_C is the inter-electrode coupling, C_{RC} is the parasitic coupling to ground of the receiver electrode and C_{TC} is the parasitic coupling to ground of the transmitter load (it is on the low impedance node so it does not affect the signal transmission). The inter-electrode coupling (and hence H) is reduced when the interconnection size reduces; thus, a receiver that can manage a significant attenuation is mandatory for the reduction of interconnection pitch. In addition to provide the maximum coupling a high accuracy stacking have to be implemented which means high accuracy in alignment and small inter-electrode gap.

Synchronous communication circuits

This section presents a communication approach that exploits synchronism in order to reduce size and guarantee communication reliability. A pipelined synchronous communication interface has been proposed previously [2]. Fig. 3 and Fig. 4 show a different design solution for transmitter (TX) and receiver (RX) circuits: a dynamic topology provides high speed functionality, while an appropriate mix of high threshold (*VTH*) transistors (depicted with thicker gate) and low threshold (*VTL*) transistors enable to manage leakage current and to preserve the correct functionality in low speed operating condition.

Transmitter:

The transmitter is based on a 2-stage dynamic circuit: when the clock *CKt* is low, the output node (*Qt*) is precharged to *VDD*. When *CKt* rises, the transmitter input stage, formed by *P1, P2, P3, N1, N2* mos, samples the *Dt* value on *CKt* rising edge: the input data is evaluated and *Qt* voltage falls

according to the sampled data while the state is latched at the dynamic node *int TX*. If the sampled data is '1', *Qt* does not fall during the evaluation phase and its value is dynamically kept; *N4* is a *VTH* device in order to prevent undesired voltage drop on the electrode in high impedance state because of the pull-up leakage due to *P5* (*VTL*) is orders of magnitude larger than the pull-down leakage due to *N4*. Moreover, *P4* and *N2* are in a feedback that preserve the correct behavior of *int TX*: when its voltage is high, there is an active feedback (*P4* on) that makes leakage negligible; on the contrary when it is low, the pull down network has a much larger leakage because it is implemented with low threshold devices. *P1* and *N4* enable a synchronous preset of transmitter, controlled by the active-high signal *SDt*, by forcing the first stage of the circuit to drive *int TX* low.

Receiver:

The receiver is implemented with a precharge and evaluation scheme that matches the transmitter topology. When the clock (*CKr*) is low, the input node (*Dr*, connected to the 3D electrode) is precharged to *VDD*; during this phase, the first stage of the receiver (*P6, N6* and *N7*) is disabled while the subsequent stages keep the value received during the previous clock cycle. In particular the stage formed by *P8,P9, N9, N10, N11* is a latch scheme which keeps the *QNr* low value in low impedance mode while the high value is kept in high impedance mode by the leakage current. The evaluation of received data takes place when *CKr* is high: during this phase *Dr* is kept in high impedance state, so that the electrode voltage can change according to the signal coming from transmitter. If the transmitted value is '1', then the transmitter does not generate any transition along the vertical channel thus *Dr* is left high; this high-level is guaranteed despite leakage on *Dr* since pull-down leakage (*N6*) is lower than the pullup leakage (*P6*). If the transmitted value is '0', a falling edge is generated by the transmitter and is propagated to *Dr* with an attenuation dependent on channel characteristics; then its value is amplified by the subsequent stage and finally latched by the output stage. *N6* provides a feedback that drives low the electrode voltage if a low level is correctly received (*Dr* falls over the logic threshold of the receiver). This circuit doesn't need a dedicated reset signal to set its state; if the transmitter is in reset

978-1-4244-4722-0/09 $25.00
© 2009 IMAPS-ITALY

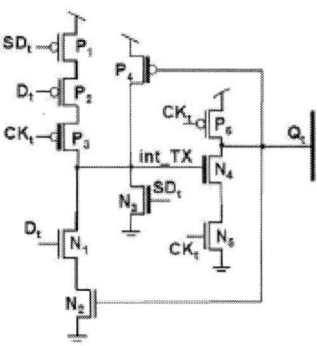

Figure 3: Synchronous Transmitter Circuit

mode, then the receiver output *QNr* keeps a logic '0' after one *CKr* cycle. In order to guarantee functionality and performance, TX and RX need to be synchronized. A clock signal is propagated from the receiver chip to the transmitter chip and two symmetrical clock trees for receivers and transmitters have been implemented so that the clock skew between RX and TX is directly related to the vertical propagation delay of clock. With this approach the synchronization is respected since transmitters are always delayed with respect to receivers by a fixed amount of time. Thus, a programmable delay paths (digitally trimmed) can be used to tune the inter-chip clock skew and achieve the best operating conditions. Moreover, the clock transmission can be accomplished with non critical transceivers: they do not require strong optimization in power and area since they can be shared between a number of clocked connections.

3D Wafer Level Assembly

In order to provide a 3D Capacitive Interconnection network with performance comparable to more common 3D solutions (for instance through-silicon vias) the interconnection size must scale as well as the power consumption. This means that the stacking procedure must provide high inter-electrode coupling. Previously works that demonstrate die level assembly package have been proposed [3-5]. This paper is focused on wafer level assembly that consists in post-processing standard CMOS wafers.

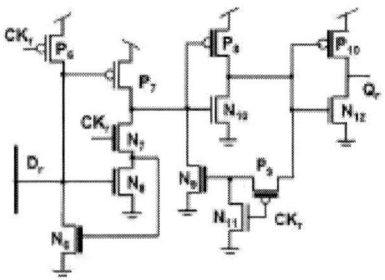

Figure 4: Synchronous Receiver Circuit

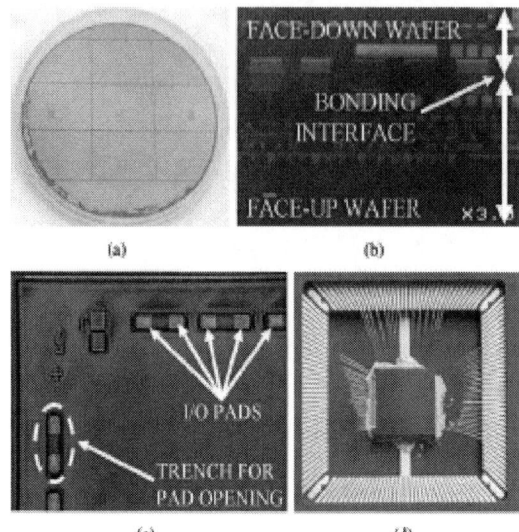

Figure 5: Wafer-to-wafer assembly: (a) wafers are aligned, stacked face-to-face and connected via direct molecular bonding; (b) SEM highlights the bonding interface; (c) via opening for I/O connections and (d) wire-bonding to standard ceramic package.

The basic concept of this strategy can be summarized as follows: two wafers with symmetrical layout are conditioned for molecular direct bonding; then, they have a precise alignment and direct bonding of top surfaces. The top substrate of the stacked wafers is then thinned down and the buried I/O pads of both chips are opened by creating trenches (Fig. 5(a) shows the processed wafers). For this approach to be implemented a sequential process flow is required. The passivation oxide of each wafer is planarized and thinned2; then, an additional passivation layer is deposited on top of the devices and the resulting insulation layer is adjusted to the value defined for AC interconnections. In order to create the stack, the wafers are aligned face-to-face and bonded by molecular direct-bonding; the SEM photograph in Fig. 5(b) highlights the bonding interface with its thin inter-electrode oxide (the target of this technology is 400 nm). At that point, the back of the top wafer is thinned down and alignment marks are lithographically defined in order to prepare the structure for I/O via opening. The vias are finally opened through the bulk silicon (from the back of the top wafer), making the buried I/O of both wafers bondable; the result of this process step is shown by the microscope image in Fig. 5(c). Finally, the stacked chips are diced and packaged in standard ceramic cavities and bonded with gold wires (Fig. 5(d)). This approach is more complex than the die-level one and is strongly affected by KGD issues, since it is not possible to sort the chips before the assembly step. Moreover, if the stack includes devices of different area, the silicon is wasted on the

978-1-4244-4722-0/09 $25.00
© 2009 IMAPS-ITALY

Figure 6: Test chip conceived for wafer-level-assembly in 0.13 μm CMOS technology

wafer that includes the smaller die For these reasons, a die to wafer approach is currently considered. Nevertheless, this approach shows a high throughput and enables further optimization of the capacitive structures thanks to direct-bonding: the larger dielectric permittivity and the smaller inter-electrode gap (down to 400 nm) provide better communication performance with respect to the die-to-die approach.

Experimental results

A silicon prototype Fig. 6 has been designed in a $0.13 \mu m$ CMOS technology ($1.2V$ power supply, six metal layers and dual *Vth*) for the validation of the proposed circuits. This silicon implementation demonstrate a throughput per pin of $1.23Gbps$ with electrodes area down to $8x8 \mu m^2$ and the power consumption is $0.11 \mu W/MHz$ plus $16.13 \mu W$ of leakage. The measured performances are fully summarized in Table I with different electrode sizes.

Conclusion

The work presented so far describes a complete communication scheme based on capacitive coupling. Communication circuits have been designed and implemented in 0.13 μm CMOS technology and wafer-level assembly have been investigated. A peak bandwidth of 1.23Gb/s/pin has been measured with an electrode area down to 8x8μm; this provides state-of-the-art bandwidth and interconnection pitch with respect to other contactless interconnection options. The power consumption proves to be 0.14 mW/Gb/s.

Size $[\mu m^2]$	Throughput/pin [Gbps/pin]	Dynamic Power $[\mu W/MHz]$	Static Power $[\mu W]$
8x8	1.23	0.11	16.13
15x15	1.2	0.13	14.7
25x25	1	0.16	14.7

Table 1: Measured performance for different electrode sizes

Acknowledgment

Authors thanks CEA-LETI for supporting 3D wafer level package

References

[1] W.R. Davis, et al. "Demystifying 3D ICs: The Pros and Cons of Going Vertical". In *Design & Test of Computers, IEEE*, Volume 22, Issue 6,pp.496-510, June 2005.

[2] A. Fazzi et all, "A 0.14 mW/Gbps high-density capacitive interface for 3-D system integration," in *Proc. IEEE Custom Integrated Circuits Conf. (CICC)*, 2005, pp. 101–104.

[3] A.Fazzi, et al, "3-D Capacitive Interconnections With Mono-and Bi-Directional Capabilities". *IEEE Journal of Solid-State Circuits*, Vol.43, NO.1, January 2008

[4] R. Canegallo et all, "3D Contactless communication for IC design", IEEE Proc. ;ICICDT 2008 Conference, pp. 241-244, June 2008

[5] A. Fazzi et al., "3D Capacitive Interconnections with Mono- and Bi-Directional Capabilities". In *ISSCC2007*, February 2006, pp.356-357.

Kirkendall voiding in Au ball bond interconnects on Al chip metallization in the temperature range from 100 – 200°C after optimized intermetallic coverage

M. Schneider-Ramelow*, S. Schmitz*, B. Schuch**, W. Grübl**

* Fraunhofer Institute for Reliability and Microintegration (IZM), Gustav-Meyer-Allee 25, 13355 Berlin, D

** Conti Temic microelectronic GmbH, Nürnberg, D

++49 03 46403-172, Fax -271 and martin.schneider-ramelow@izm.fraunhofer.de

Abstract

The presentation addresses the reliability of Au ball bonds of different Au wire qualities on Al chip metallizations of different thicknesses and compositions at temperature storage from 100 to 200°C up to 4000 h. In this context the interfacial reactions and intermetallic phase coverage directly after the bonding process was optimized to get the best starting condition for phase growth at elevated temperatures and to avoid critical Kirkendall void growth.

This failure mechanism is influenced by numerous factors, such as aging temperature and time, Au wire and Al metallization composition and ratio of mixture as well as the percental area of interconnection formation under the ball. These influences are mainly responsible for ball lift offs under operating conditions. In many cases lift offs already occur at Al metallization thicknesses > 1 μm and temperatures in the range of 175°C, while temperatures up to 150°C or at 200°C are less critical.

Investigations include mechanical tests of Au loops and ball contacts as well as microstructure observations of the contacts in correlation to material composition, aging temperature and Al metallization thickness. Au/Al intermetallic phase thicknesses below the Au contacts on Al metallization are typically a few hundred nanometers thick directly after the bonding process, depending on bonding conditions like process parameters and material combination. These phases grow under temperature influence and Kirkendall voiding can occur. A most significant result in this context is that pull and shear lift offs occur if the chip metallization is clearly thicker than 1 μm and intermetallic phase coverage (after bonding) is less than 2/3 of the bottom side ball area.

These results will considerably contribute to a better understanding of Kirkendall voiding failure mechanisms.

Key words: Au wire bonding, Al chip metallization thickness, interface formation, Au-Al intermetallic phases

Introduction

Wire bonding continues to be popular and dominant in the field of bonding technologies in the industry. Thermosonic bonding is used for Au and Cu wires and currently comprises about 90% of all wire bonding interconnections. Interdiffusion, intermetallic phase (IP) growth and Kirkendall voiding in Au ball bond on Al chipmetallization systems at temperature storage is in discussion since ball/wedge wire bonding in microelectronic packing was developed. Today this is of very special interest in various microelectronic application fields (e.g. automotive) because more and more maximum operating temperatures exceed 150°C.

The failure mechanism Kirkendall voiding often results in ball lifting from chip metallization. The most influential factors are the bonding parameters [1-3] including tool geometry (capillary) [4-5] and the transducer frequency [6]. The ball bond intermetallic coverage is one of the most important boundary conditions for interconnection reliability at high temperature storage (≥ 150°C) where massive IP growing occurs [2, 3]. In addition to temperature and storage duration, thickness and chemical composition of Al metallizations have to be considered. Alloying state of Au wires and contaminations are other typical influencing factors relating to growth of IPs and - more critical - Kirkendall voids [7-9].

The paper presents results of tempering Au ball bonds of different Au ball qualities on various Al chip metallizations after optimizing the interconnection formation and IP coverage. Main finding is that Au ball lift offs from Al metallizations can be eliminated even in the critical temperature range of 170 – 180°C if metallization is AlSiCu and the thickness is smaller then 1 μm. Another important result is the recommendation to optimize the IP coverage up to 90% and don´t use pure Al metallizations if possible.

Bonding of different Au wires on different chips

TS Au wire bonding (25 μm) was carried out on a standard bonding machine (ESEC 3088 Ball/Wedge-Bonder) with a transducer frequency of 120 kHz at 125°C work holder temperature and capillary UTS-38-C-1/16-XL (SPT). Chips with 4 different Al metallizations were glued on ceramic substrates with Au thick film metallizations and interconnected with 4 different Au wires (Table 1).

Table 1: Materials and properties

material	notation	parameter
Au bond wire 25 μm	K&S Radix Plus (3N pure)	break. load 10.8 cN elongation 4.6%
	K&S AW99 (4N pure, Be free)	break. load 12.6 cN elongation 4.0%
	Heraeus HA3 (2N plus 1% Pd)	break. load 13.3 cN elongation 4.3%
	Heraeus HA9 (4N pure)	break. load 11.8 cN elongation 4.6%
chip metal-lization	chip A	2 μm Al
	chip B	4 μm AlSiCu
	chip C	4.4 μm AlSiCu
	chip D	0.8 μm AlSiCu

The evaluation of bonding series was performed by shear testing (ball bond) on XYZTec test equipment and KOH etching (10 min) of Al metallization to determinate the IP coverage beneath the balls. Pull testing was done on Dage 4000 test equipment.

The failure codes of pull and shear test before and after temperature storage up to 200°C are summarized in Table 2 and Table 3. Views of characteristic pull and shear codes are shown in Figure 1 and Figure 2. The initial pull and shear test results after parameter optimization are summarized in Table 4 (20 tests each combination).

Table 2: Pull test codes

pull code	failure mode
2	neck crack (over ball)
6	ball lift off including metallization / IP
7	ball lift off (interface IP)

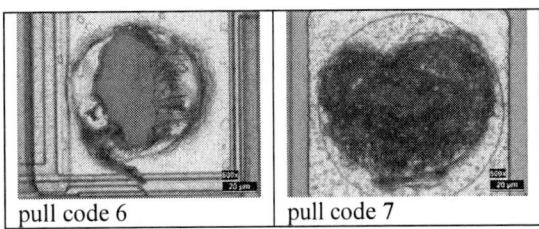

Figure 1: Typical appearance of critical pull codes

Table 3: Shear test codes

shear code	failure mode
3	ball shear (shearing through the ball)
4	metallization lift off
5	cratering
6	mixed mode* at high shear forces
7	mixed mode* at lower shear forces

* **Partial crack in chip metallization, in the ball and at the interface**

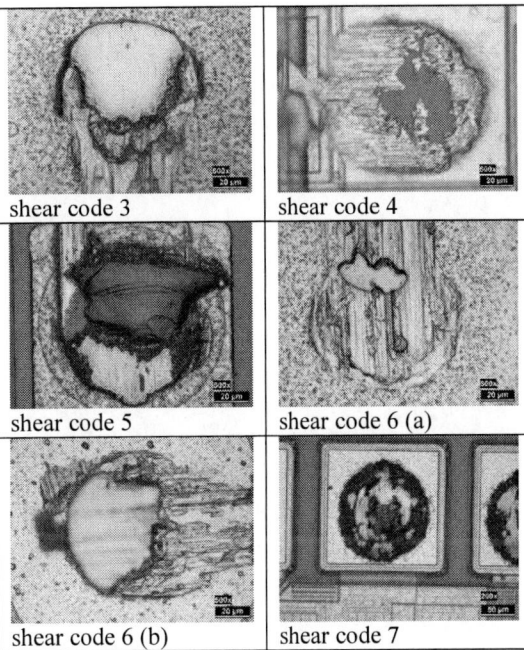

Figure 2: Typical appearance of shear codes

After bonding parameter optimization the initial state of bonding quality was characterized by the absence of pull or shear lift offs which indicates a good interconnection. These good initial bonding results were confirmed by a high degree of IP formation at the interface Au ball / Al chip metallization as to be seen at the bottom side of the Au balls after KOH etching (see **Figure 3**, chip D). For all 12 chip / wire combinations the IP coverage directly after the bonding process was proofed to be greater than 2/3 of the bottom side ball area.

At last cross sections were made to investigate the IP growth and void formation after temperature storage and to compare it with the mechanical test results.

Table 4: Initial pull and shear test results (n = 20)

wire/chip combination	pull test *		shear test			
	X	σ	X	σ	s3	s6
HA3 chip A	12.7	0.4	62.5	5.8	0	20
HA3 chip B	13.5	0.4	56.4	4.9	0	20
HA3 chip C	12.8	0.4	51.9	3.5	0	20
HA3 chip D	13.4	0.7	62.2	2.8	0	20
HA9 chip A	11.2	0.3	76.7	2.9	0	20
HA9 chip B	11.6	0.4	57.3	4.9	15	5
HA9 chip C	11.1	0.3	62.4	4.5	14	6
HA9 chip D	11.8	0.3	65.5	8.7	6	14
AW99 chip A	10.8	0.2	58.1	3.2	2	18
AW99 chip B	11.4	0.3	58.2	7.5	6	14
AW99 chip C	10.9	0.3	52.4	4.4	0	20
AW99 chip D	11.6	0.3	58.2	6.1	0	20
Radix chip A	11.2	0.3	48.9	3.0	0	20
Radix chip B	12.0	0.4	51.1	4.5	0	20
Radix chip C	11.1	0.6	56.4	5.5	0	20
Radix chip D	11.9	0.3	52.4	4.3	0	20

* only neck breaks; X: Mean value in cN; σ: standard deviation in cN; s3/s6: number of shear code 3 resp. 6

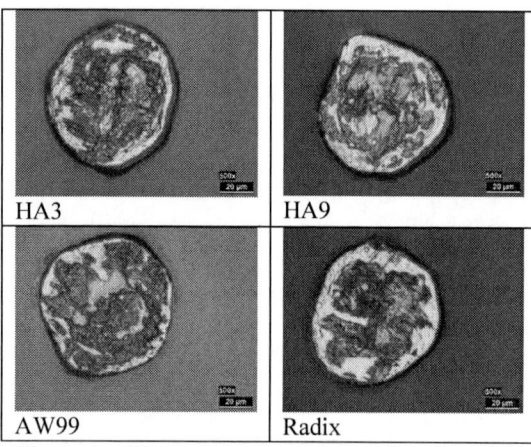

Figure 3: IP coverage underneath the balls directly after bonding and KOH etching (chip D)

Temperature storage test results and discussion

All bonding samples for reliability investigations were bonded with optimized bonding parameters and annealed in laboratory furnaces. Test temperatures and time intervals for mechanical testing were ascertained as follows:

- 100°C – 150°C: 24, 96, 250, 500, 1000, 1500, 2000 and 4000 h (140°C up to 5000 h)

- 160°C – 200°C: 4, 8, 24, 48, 96, 250, 500, 1000, 2000 and 4000 h

Best results could be achieved with the combination of chip D with the thin AlSiCu metallization and the 3N pure Radix Plus Au wire as to be seen in Figure 4 and Figure 5 where at all temperatures and test intervalls pull and shear forces were more or less at the same level and no pull and shear lift offs occurred. At all other 2.0 – 4.4 μm

thick Al metallizations pull lift offs appeared. Worst case occurred with the thickest AlSiCu metallization (4.4 μm) of chip C with poorest pull test results in combination with the 4N pure AW99 Au wire (Figure 6 and Figure 7) and poorest shear test results in combination with the 2N pure HA3 Au wire (Figure 8 and Figure 9).

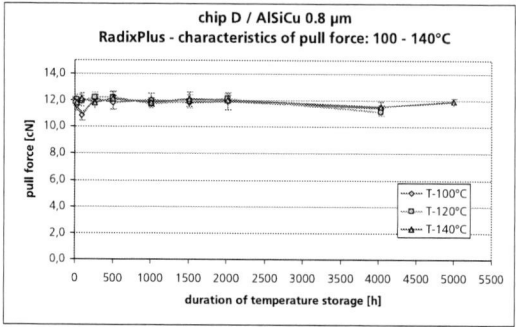

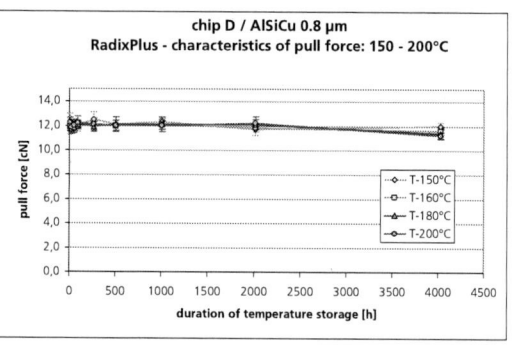

Figure 4: Pull test results Radix / chip D

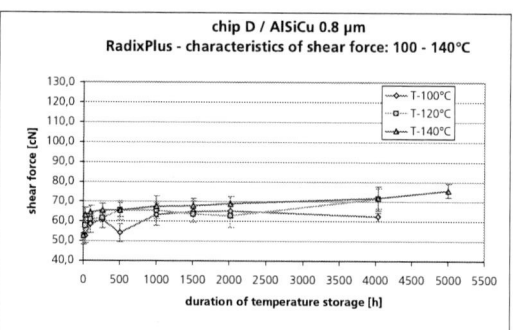

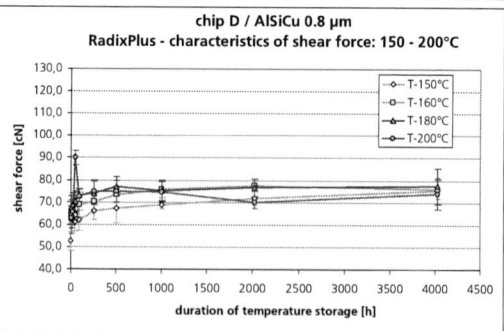

Figure 5: Shear test results Radix / chip D

978-1-4244-4722-0/09 $25.00
© 2009 IMAPS-ITALY

Pull lift offs with pull force values below 4 cN are very critical. This was especially the case at 200°C above 1000 h storage duration where 7 of 11 pull lift offs occurred with pull force values between 2.1 and 3.7 cN (s. the red bars and high standard deviations in Figure 7).

The red bars in Figure 8 and Figure 9 indicate the shear code 7 with a fracture running partially in the chip metallization, in the ball and at the interface (Figure 2) at shear force values less than 60 cN. In these cases starting interface degradations caused by Kirkendall voiding can be assumed.

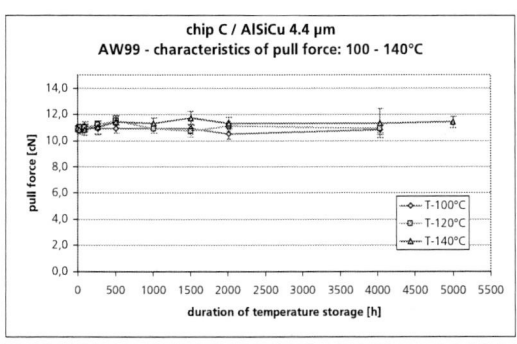

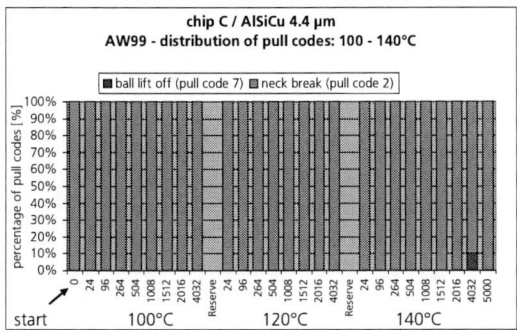

Figure 6: Pull test results AW99 / chip C (100 - 140°C)

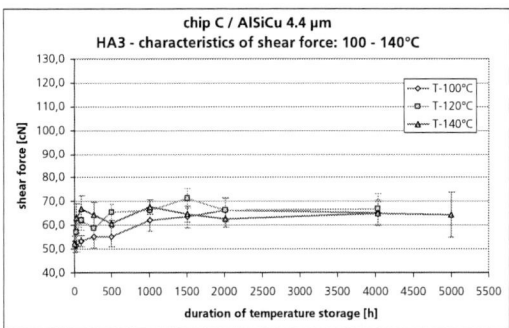

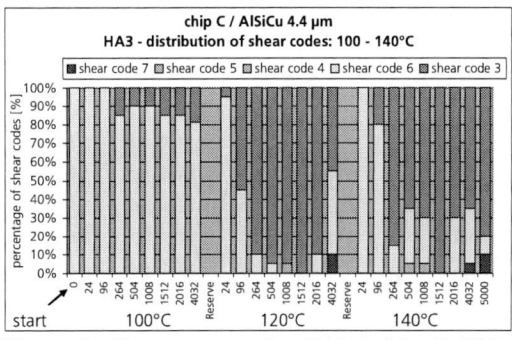

Figure 8: Shear test results HA3 / chip C (100 - 140°C)

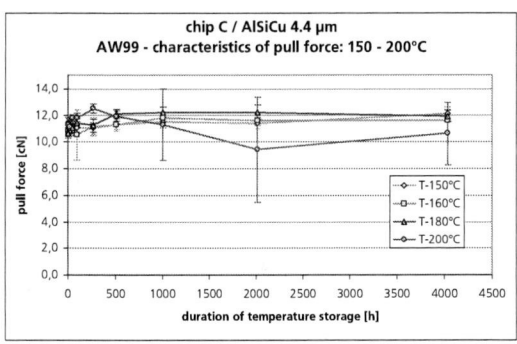

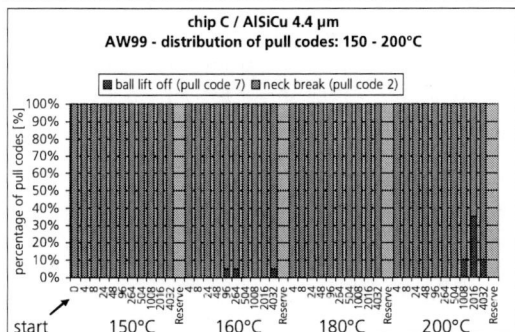

Figure 7: Pull test results AW99 / chip C (150 - 200°C)

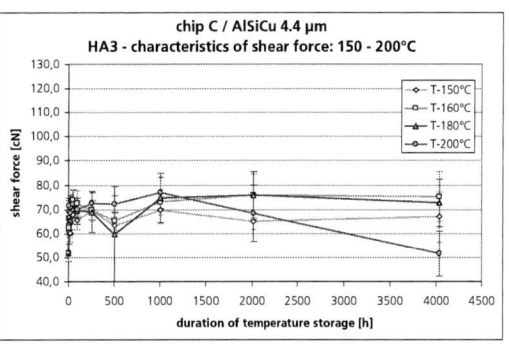

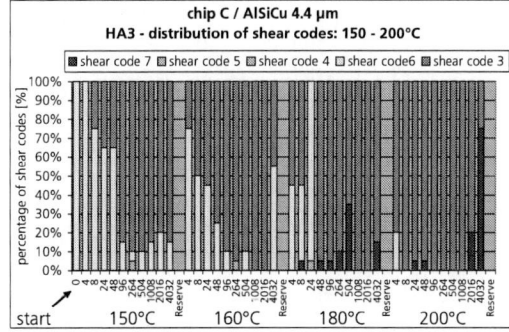

Figure 9: Shear test results HA3 / chip C (150 - 200°C)

Cross sections of Au HA3 ball bonds on chip C after 4000 h of temperature storage at 200°C give an example for Kirkendall voiding in Figure 10. Different intermetallic phases are visible between ball and chip. Despite the very long duration at high storage temperature the interdiffusion processes were not finished. This is caused by the chip pad size (165 µm x 165 µm) and Al thickness of 4.4 µm and therefore availability of Al for interdiffusion and phase formation as well as the lower diffusion coefficients of Au wires which are alloyed with Pd. But none of all measured shear force values of this shear code 7 (on all chips) was smaller than 25 cN. If these pore constellations underneath the balls are critical under application and realibility aspects needs to be discussed.

More critical to evaluate are pull lift offs with single pull force value below 4 cN for all investigated Au wires with initial breaking loads above 10 cN. Therefore all quantified pull lift offs are summarized in Table 5. Values are red-marked if pull force undercut 4 cN.

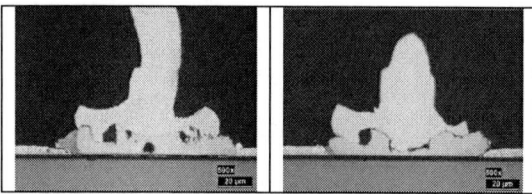

Figure 10:   Cross sections of HA3 Au balls on chip C after 4000 h at 200°C

Table 5: Compilation of all pull lift offs

chip	wire	t [h]	T [°C]	pull force [cN]
A	HA9	250	200	4.5
B	HA3	1000	140	11.9
B	HA3	1000	180	7.0
B	HA3	24	150	7.1
B	HA3	96	150	8.6
B	HA3	500	150	10.9
B	HA3	2000	150	12.4
B	HA9	500	200	10.3
B	Radix	4000	150	7.2 and 12.1
C	HA3	500	200	3.0
C	HA3	500	200	7.0
C	HA3	2000	200	9.4
C	HA9	4000	200	7.7 and 8.0
C	AW99	4000	140	7.5 and 9.0
C	AW99	96	160	2.5
C	AW99	250	160	9.0
C	AW99	4000	160	7.8
C	AW99	1000	200	3.0 and 3.7
C	AW99	2000	200	2.1 and 3.2
C	AW99	2000	200	3.2 and 3.3
C	AW99	2000	200	4.5 and 7.2
C	AW99	2000	200	8.0
C	AW99	4000	200	3.0
C	AW99	4000	200	5.0

As to be seen very clearly in Table 5 critical Au ball pull lift offs occurred only on the thickest Al pad metallization of chip C with the biggest pad dimension of approx. 165 µm x 165 µm additionally. The pad sizes of the other chips A, B and D were smaller with edge lengths in the range of 100 µm – 120 µm. Some other pull lift offs at pull force values above 4 cN appeared on chip B with Al thickness of 4 µm and one on chip A with 2 µm Al. On chip D with only 0.8 µm Al thickness no lift offs occurred. After longer storage duration at higher temperature all Al of the chip metallization was transformed to intermetallic phases respectively in Au_4Al after 4000 h at 200°C (Figure 11).

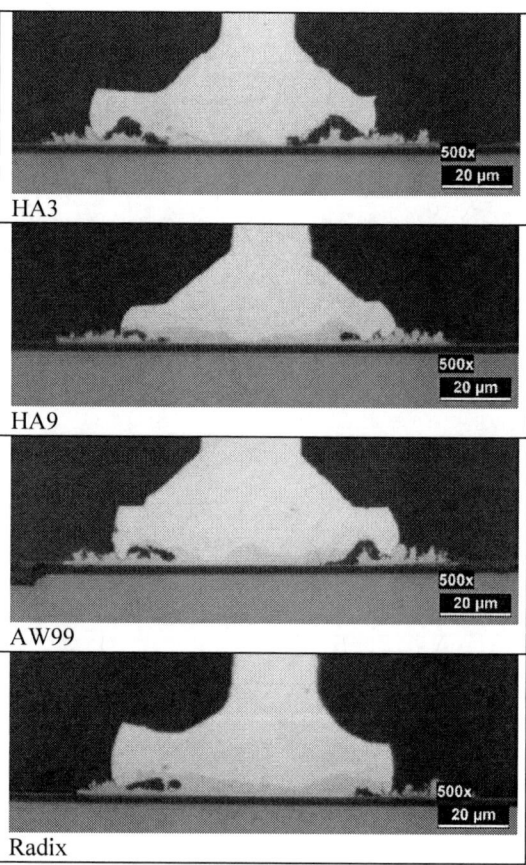

Figure 11:   Intermetallic phase formation on chip D after 4000 h at 200°C

The strength of the ball bond interconnections was high enough despite a lot of gaps at the interface (s. Figure 11) as a result of Kirkendall voiding as well as volume expansion through interdiffusion resulting in intermetallic phase formation and growth. This is a result of higher strength of intermetallic Au/Al phases compared to metallic Au or Al. No pull lift offs appeared on chip D with all wire types and all pull force values were above 10 cN in all cases.

Compared to the results in [9] where a hard degradation with lift offs at very low pull force values was determined there is only a marginal count of Au ball lift offs in the investigation presented in this paper on chips with thicker Al tempered in the critical temperature range from 160 – 180°C. Presumed reason for this divergence is that in [9] the IP coverage underneath the balls wasn´t optimized and controlled with KOH etching. Ball bond optimization in [9] was checked through mechanical tests only.

The conclusion can be drawn that there must be a direct correlation between the reliability of Au ball bonds on Al chip metallization at temperature storage and the Al thickness on the one hand and the intermetallic coverage directly after bonding on the other.

Conclusions

Reliability test results of different Au wires which were ball/wedge bonded with optimized parameters on different Al metallizations and tempered in the range from 100 - 200°C have been presented in this paper. The initial IP coverage underneath the balls was inspected after KOH etching and found to be in the range from 75 - 90% of the bottom side ball area. The interface strengths were tested before and after temperature storage by pull and shear tests and correlated to intermetallic phase growing and Kirkendall voiding by microstructural investigations on cross sections.

Main result is a correlation between the thickness and the alloying state of Al chip metallization and the appearance of ball lift offs. The failure mechanism is Kirkendall voiding during intermetallic phase formation and growth until gap formation underneath the ball is complete and ball lift off occurs. This failure mechanism is influenced by numerous factors, such as aging temperature and time, Au wire type and Al metallization composition and thickness as well as the percental area of interconnection formation (IP coverage) under the ball directly after bonding. Most lift offs occur with pure Al metallizations and thicknesses $\gg 1$ µm. But it can be confirmed that it is possible to temper Au ball bonds on Al Chip metallization up to 200°C up to 4000 h without the occurrence of lift offs at mechanical tests if AlSiCu chip metallizations with thicknesses smaller than 1 µm are used and the IP coverage directly after the bonding process is nearly 90% of the bottom side ball area, even at the well known very critical storage temperature in the range of 175°C.

For practical work it can be recommended to use alloyed Al finish metallizations on chips (e.g. AlSi1Cu0.5) with reduced Al thickness (max. 1 µm) and to optimize the initial IP coverage to minimize the risk of ball lifting while tempering at elevated temperatures.

References

[1] Dittmer, K.; Kumar, S.; Wulff, F.: Influence of bonding conditions of small ball bonds due to intermetallic phase (IP) growth. Int. Conf. on High Density Packaging and MCMs 1999. P.403-408.

[2] Beleran, J.; Wulff, F.; Breach, C. D.: Gold ball-bond mechanical reliability at 40µm pitch: squash height and bake temperature effects. In technical papers on K&S web pages www.kns.com (July 2007).

[3] Breach, C. D.; Wulff, F.; Dittmer, K.; Calpito, D.R.M.; Garnier, M.; Boillot, V.; ToK Chee Wei: Reliability and failure analysis of gold ball bonds in fine and ultra-fine pitch applications. SEMICON 2004 Singapore and in technical papers on K&S web pages www.kns.com (July 2007).

[4] Singh, I.; Levine, L.; Brunner, J.: Reliablility Ground Rules Change at < 50µm Pitch. IEEE Int. Electronics Manufacturing Technology (2003) P.1-5.

[5] Tamala, K.; Wee, L.S.; Kwon, O.D.; Cheung, K.K.; Levine, L.: Resolution of a fine pitch wire bonding reliability problem. SEMICON 2006 Singapore und in technical papers on K&S web pages www.kns.com (July 2007).

[6] Victor P. Jeacklin, ESEC SA and members of the ESEC wirebond process research and development in CH: Room Temperature Bonding Using Higher Frequencies. http://www.smallprecisiontools.com/frames-publications.htm.

[7] Schneider-Ramelow, M.; Lang, K.-D.; Geißler, U.; Scheel, W.; H. Reichl: Interface Reactions during Au-Ball/Wedge and AlSi1-Wedge/Wedge Bonding at room temperature. Lecture and paper on the 16th EMPC 2007 in Finland. Proceedings EMPC 2007, June 17-20. 2007 Oulu, Finland. P.128-133.

[8] Forschungsbericht: Thermosonic Drahtbonden bei Verfahrenstemperaturen unter 100°C. BMWi/AiF-Abschlussbericht AiF 13.309 B (Laufzeit des Vorhabens: 05.2002 - 04.2004).

[9] Schneider-Ramelow, M.; Schuch, B.; Lang, K.-D.; Reichl, H.: High temperature and element alloying influences on Kirkendall voiding in Au ball bond interconnects on Al chip metallization. Proceedings of the 40th IMAPS International Symposium on Microelectronics. November 11-15, 2007 in San Jose, CA, USA. P.642-647.

Thermo-Mechanical Stress Analysis

Katrin Niehoff[1], Thomas Schreier-Alt[1], Florian Schindler-Saefkow[2], Frank Ansorge[1] und Hartmut Kittel[3]

[1]Fraunhofer IZM, Micro Mechatronic Systems, Oberpfaffenhofen, Germany
phone:+49 (0)8153-9097-840; fax:+49 (0)8153-9097-511 katrin.niehoff@mmz.izm.fraunhofer.de
[2]Fraunhofer IZM, Micro Materials Center, Berlin, Germany
[3]Robert Bosch GmbH, Chassis Systems Control, Abstatt, Germany

Abstract

This paper reports about the development and application of a new test chip for stress analysis in microelectronic packaging processes and reliability tests. Special focus will be on transfer molding with epoxy molding compound (EMC). The CMOS based stress sensor is able to measure in-plane normal stress and shear stress on the chip surfaces with an absolute stress resolution of ± 3MPa. By the use of a multiplexer and an array of measuring cells on the chip surface we are able to sense temperatures and mechanical stresses space- and time resolved. We analyzed 1st and 2nd level encapsulation methods by transfer and injection molding and performed an in-situ relaxation and stress monitoring of the packaged sensor during temperature load and moisture uptake. The experimental results will be compared with numerical simulations. In substitutive application for electronic parts, the test chip implicates a major potential for investigating production processes and lifetime impacts on MEMS. Adjusting material combinations and optimizing packaging processes to achieve reliable products will be the applications on focus in future.

Key words: Stress-Measurement; CMOS test chip; Transfer Molding; Reliability

Introduction

Today´s microelectronic fabrication has to cope with demanding trends, like miniaturization, high level integration and high density functionality. One effect of these trends is an extent in different material combinations used in microelectronic systems, which become harder to adjust in uniform thermal expansion. Therefore, the increasing complexibility in microelectronic assembly – and packaging processes are causing new challenges in reliability and quality assurance of microsystems. Temperature loads, harsh operational conditions, but also single process steps are inducing certain amounts of mechanical or thermo-mechanical stresses that become essential to analyze in order to avoid major reliability problems like:

- Package cracking and delaminating
- Wire damage
- Packaging warpage
- Moisturized packaging swelling etc.

Mechanical stresses can significantly influence the sensor sensibility and can lead to a signal offset e.g. at inertial and magnetic sensors. Therefore, monitoring forces during MEMS assembly and encapsulation processes is an essential method to prevent an in built sensor drift due to inapplicable packaging parameter or time-depending, viscoelastic material properties.

Especially for the reliable development of future device generations characterization methods that are capable of sensing mechanical and thermo-mechanical stresses have to be found. A promising approach in this sense will be a combination of experimental measuring methods with FEM simulations [1,2].

Stress measurement system

In the BMBF funded project "iForceSens" a calibrated stress measurement system was developed consisting of test chip and ASIC control unit. The MST chip was fabricated using complementary metal-oxide semiconductor (CMOS) technology and contains orthogonal stress sensitive current mirrors. The change in drain current within these orthogonal Si-MOSFETs due to applied mechanical stress depends linearly on the piezoresistivity of silicon [3,4].

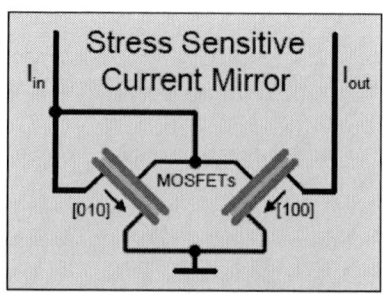

Figure 1: Stress sensing principle

978-1-4244-4722-0/09 $25.00
© 2009 IMAPS-ITALY

Two different current mirror orientations determine the stresses: nMOS transistor channels along [010] and [100] silicon crystal direction are used for calculating the in-plane shear stress σ_{xy}; pMOS transistor channels in [-110] and [110] direction are used for calculating the difference in in-plane normal stress $\sigma_{xx} - \sigma_{yy}$

$$\sigma_{xy} \approx \frac{-1}{\pi_{11}^{(n)} - \pi_{12}^{(n)}} \frac{I_{out} - I_{in}}{I_{out} + I_{in}}$$

$$\sigma_{xx} - \sigma_{yy} \approx \frac{2}{\pi_{44}^{(p)}} \frac{I_{out} - I_{in}}{I_{out} + I_{in}}$$

with I as current and π_{ii} as piezoresistive coefficients. The later are characterized through four point bending tests of silicon strips bended along defined orientations.

By determining the sum of the in-plane normal stresses $\sigma_{xx} + \sigma_{yy}$

$$\sigma_{xx} + \sigma_{yy} \approx \frac{2}{\pi_{11}^{(n)} + \pi_{12}^{(n)}} \left(1 - \frac{I_{in} + I_{out}}{2I_0} - \pi_{12}^{(n)} \sigma_{zz} \right)$$

it is possible to calculate the single normal in-plane stress components σ_{xx} and σ_{yy}. These components can be calculated with an accuracy of 13% while shear or normal stress difference even have stress resolutions of 4.5% and 1.2%. These accuracies apply as long as the stress component σ_{zz} normal to the chip surface is negligible (< 100 MPa) or known.

Four different sensor layouts were established with different arrays of measuring cells: Model 1 consists of 6 x 10 cells with dimensions of 2.83 x 1.71 mm^2, Model 2 is having 6 x 6 cells and sizes of 1.67 x 1.71 mm^2, Model 3 with 7 x 9 measuring cells has dimensions of 1.90 x 3.52 mm^2 and Model 4 consisting of 4 x 4 measuring cells has dimensions of 1.13 x 1.11 mm^2. Additionally each cell contains a npn-temperature sensing element.

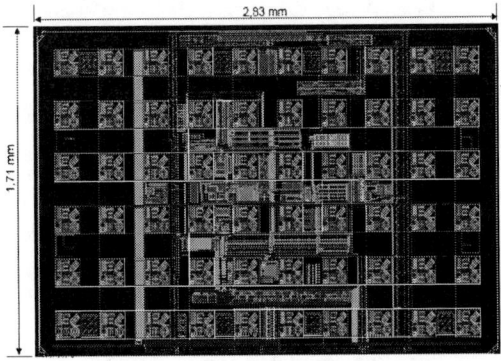

Figure 2: One out of four sensor layouts

The working temperature of the chip is between -40 and +180°C. It can be assembled either in flip-chip

technology or via wire bonding on existing AlSiCu metallization pads. The measuring cells have a grid of 260 μm while a lateral space resolution of 50 μm can be reached with interpolation.

The ASIC control unit provides high speed data acquisition leading to a time resolution for single cell readout of 50 milliseconds.

Figure 3: Compact ASIC control unit for mobile application

Parameter Identification and Finite Element Simulations

The main component of this work was to analyze the stress distribution on MEMS chips as accurate as possible while going through backend processes. For verification of the stress measurement system, a Finite Element Model was established that described the material behavior in a realistic manner. Molding compounds show time and temperature depending properties. Mechanical material properties, like Young-Modulus, vary strongly in high temperature ranges. Especially at the glass transition temperature strong change in material strengths and an incline of coefficient of thermal expansion will be observed.

In order to model viscoelastic materials with Finite Element Methods, the approach of a Prony Series development is used. Hence, unknown parameter of the material model are identified mathematically by adjusting the model response with experimental data. To define the relaxation behavior experimentally, the material response of a stress- or strain step needs to be found. To shorten the typical time period for experimental data acquisition in time domain (1-10^6 s), a harmonic material response as a function of frequency is found using DMA (dynamic mechanical analysis). [5]

The relaxation modulus is defined as sum of parallel arranged Maxwell elements with linear elastic spring and viscose damper.

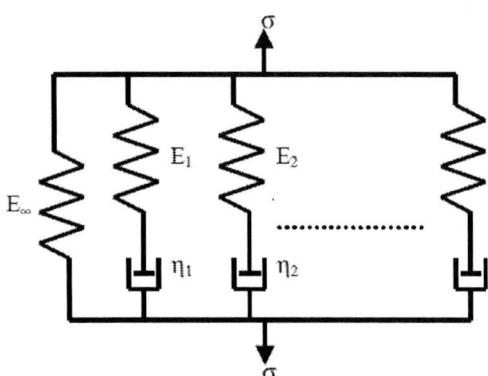

Figure 4: Generalized Maxwell model

The shear modulus can be defined as

$$G(t) = G_\infty + G_0 \sum_{i=1}^{n} \alpha_i e^{\left(\frac{-t}{\tau_i}\right)}$$

in time domain and

$$G'(\omega) = G_\infty + G_0 \sum_{i=1}^{n} \alpha_i \frac{\tau_i^2 \omega^2}{1 + \tau_i^2 \omega^2}$$

in frequency domain with

$$\alpha_i = \frac{G_i}{G_0} \text{ as Prony coefficient}$$

which can be found with curve fitting methods (e.g. least square fit) of the DMA data.

The Finite Element simulations were performed in Ansys Multiphyiscs with viscoelastic stress definition of

$$\sigma = \int_0^t 2G(t-\tau)\frac{d\gamma}{d\tau}d\tau + I \int_0^t K(t-\tau)\frac{d\varepsilon}{d\tau}$$

where

σ = stress

γ = shear strain

ε = volumetric strain

G(t) = shear modulus

K(t) = bulk modulus

t = current time

τ = relaxation time

I = unit tensor

We additionally reduced data at various temperatures by using time-temperature equivalence to obtain one general master curve for a reference temperature. The WFL (William-Landel Ferry) shift function applied as a factor to the master-curve is

mapping single temperature depending stress relaxation curves.

$$\log a(T) = \frac{c_1\left(T - T_{ref}\right)}{c_2 + T - T_{ref}}$$

with c_1, c_2 as material dependent constants and T_{ref} as reference temperature.

Moldflow encapsulation simulations were performed in order to analyze and optimize the transfer molding process parameter like encapsulation time and pressure to achieve minimal warpage.

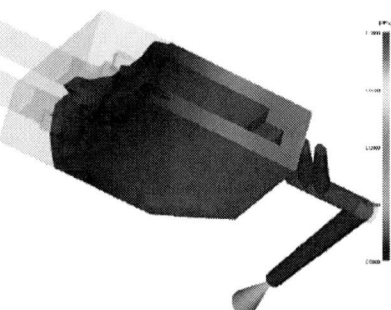

Figure 6: Due to numerical flow simulations the transfer molding encapsulation process was predicted as uncritical

The temperature change for polymer cure after the transfer molding process is inducing thermo-mechanical stress on the die due to CTE mismatch of the different system materials.

In the structural numeric simulation model, the copper lead frame was implemented as elastic plastic material model, the silicon die as elastic- and the mold and die attach as viscoelastic material model. In order to investigate the package induced stress, the system was defined as stress-free at tool form temperature of 180°C. The induced stress due cooling down from process (180°C) to room temperature (23°C) was evaluated.

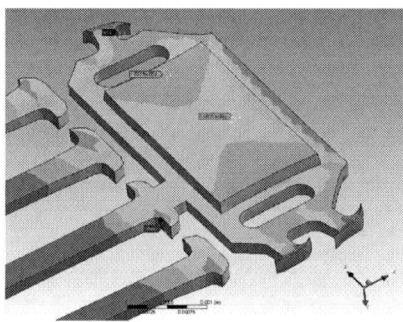

Figure 5: Induced stress during thermosetting encapsulation process (molding cap not shown)

Assembly and packaging of the sensor ASIC

The test chip was assembled on copper lead-frame via die attach and aluminum wire bonding.

Figure 7: (a) Die attached to lead frame and aluminium wire bonded; (b) transfer molded system with epoxy resin after laser beam cutting

A tool form was designed to package the chip as PSSO4-housing. After transfer molding, the chip was separated from the lead frame via laser beam cutting. The single chips were post cured at molding temperature for stress relaxation. As 2nd level encapsulation method the test chip was injection molded with fiber reinforced thermoplastic material. This assembly method is typical for rotational wheel speed sensors in vehicles.

Results from simulation and experimental measurement

After wafer cutting, the single bare chips were measured at room temperature in order to identify the initial stress state. The final stress state was measured at room temperature of the packaged transfer molded chip after post curing. The stress difference therefore comprises the cure of die attach, the crosslinking of the molding compound as well as temperature cooling from process to room temperature during post curing.

The result of the experimentally measured and numerically calculated lateral distribution of the in-plane shear stress – and the difference of in-plane normal stress components on the chip surface is shown in Figure 8.

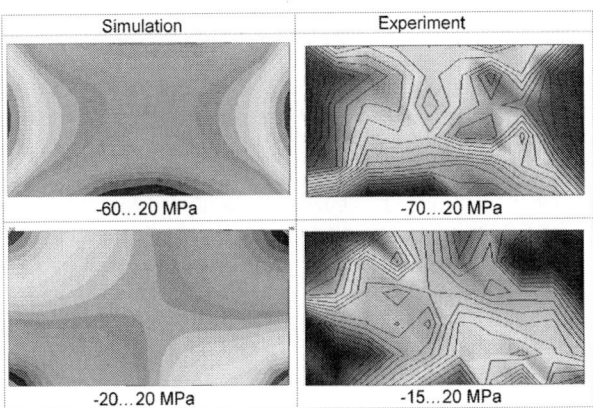

Figure 8: Difference in in-plane normal stress (σ_{xx} - σ_{yy}) (above) and in-plane shear stress σ_{xy} (below) by simulation and experiment

Experiment and simulation show a good agreement in stress values and stress distribution after transfer molding and post curing. Maximal values of normal stress difference can be detected along the middle axis of the chip while shear stress values are highest at the chip corner. In Figure 8, red color corresponds to maximal and blue color to minimal stress values.

Furthermore, the induced stress through moisture uptake was analyzed. When the package material absorbs moisture, it will diffuse through compound material and die attach and change material properties to reduced tensile strength and elasticity. Moreover it can lead to corrosion of the chip component. If moisture reaches void or cavity areas and will be heated above boiling temperature, the so called popcorn – effect causes material swelling with strong mechanical expansion.

The packaged chip was exposed to moisture ambiance in order to investigate the induced stress through moisture swelling. Additional numerical simulations have been performed that resulted in -8 to 18 MPa induced stress at saturated state in 85% relative moisture at 85°C. Comparable results were achieved in simulation and measurement (saturation was achieved after 18 hours).

2nd level encapsulation of the molded chip via injection molding and cooling from process to room temperature was performed resulting in maximal values of normal stresses difference of 190 MPa. The lateral stress distributions showed a less symmetric manner compared to the 1st level encapsulation process. Details can be found in [6].

Conclusion and Outlook

The project results verify the high accuracy and performance of the measurement system. The substitutive application of the test chip instead of electronic parts provides new insides in packaging processes in order to adjust and optimize the

978-1-4244-4722-0/09 $25.00
© 2009 IMAPS-ITALY

associated parameters and material systems. The test chip is capable in giving a good experimental validation to numerical simulations and can run through the main qualification and reliability tests in order to give an exact lifetime prognoses of the microelectronic system.

Due to its high flexibility in assembly (wire bonding, flip chip or stud bump prototyping technology), chip on board as well as PCB chip integration can be evaluated. Additionally backside chip thinning can be performed which enables the CMOS test chip integration in innovative System in Package applications.

Acknowledgements

The authors would like to thank the German BMBF for financial support and VDI/VDE Innovation + Technology for project monitoring.

References

[1] J. Schwizer, W. H. Song, M. Mayer, O. Brand H. Baltes; *Packaging test chip for flip-chip and wire bonding process characterization*, 12th International Conference on Solid State Sensors, Actuators and Microsystems, Bosten, USA,2003

[2] S. Majcherk, T. Lenke, L. Hirsch; *A silicon test chip fort he thermomechanical analysis of MEMS packagings*; Micorsyst Technol 2009, 15, pp 191-200

[3] R. C. Jäger, J. C. Suhling, R. Ramani, A. T. Bradley, J. Xu; *CMOS Stress sensors on (100) Silicon*, IEEE J. Solid State Circuits 2000, 35 (1), pp. 85-95

[4] Y. Chen, R. C. Jäger, J. C. Suhling; *Multiplexed CMOS Sensor Arrays for Die Stress Mapping*; in Proceedings of the 32nd European Solid-State Circuits Conference, ESSCIRC. 19.-21. Sept. 2006, pp. 424-427

[5] K. M. B. Jansen, *Thermomechanical modelling and characterisation of polymers*, lecture notes Mechanical Engineering, TU Delft, Netherlands, 2007

[6] H. Kittel; *Novel stress measurement system for evaluation of package induced stress*, SMART SYSTEMS INTEGRATION, Barcelona, Spain, 9.-10. April 2008

978-1-4244-4722-0/09 $25.00
© 2009 IMAPS-ITALY

Simulation and experimental analysis of substrate overmolding

Thomas Schreier-Alt, Christian Rebholz, Frank Ansorge

Fraunhofer IZM, Micromechatronic Systems, Argelsrieder Feld 6, 82234 Oberpfaffenhofen, Germany

Phone: +49-(0)8153-90975-52, Fax: -11, thomas.schreier-alt@mmz.izm.fraunhofer.de

Abstract

This paper presents new experimental and numerical methods to characterize the transfer overmolding of substrates with epoxy polymer. We investigated Multi Chip Modules on ceramic panels as well as on printed circuit boards encapsulated by a MAP-type molding process. Experiments show that the polymer flow during the overmolding process depends significantly on the mold height: While standard MAP-type mold cavities are filled homogeneously and symmetrically by the polymer, low cavity heights (< 500 µm) can cause the flow front to concentrate on a few flow paths. We developed a numerical method to describe this inhomogeneous polymer flow. The reason for this unusual behavior seems to be the shear thinning of epoxy molding compounds: Within preferred flow paths the polymer shear rate is increased, resulting in reduced viscosity which enforces a necking on distinct flow paths. Inhomogeneous flow behavior can cause the formation of air traps and excessive wire sweep. We also developed a new experimental method to measure the pressure distribution within cavities: our sensor is based on commercially available, passive pressure sensitive films and is operational at temperatures up to 200°.

Key words: Electronic encapsulation, MAP-type molding, Polymer flow, Pressure distribution

Introduction to MAP type molding

Transfer molding with epoxy molding compound (EMC) is the standard process for reliable chip encapsulation since decades. Currently the trends of the market clearly drive towards Systems in Package (SIP). They not only allow much higher component integration rates compared to classical single chip technologies but have significant advantages for the manufacturers as they can offer complete functional modules to the OEM instead of single components, increasing the added value and uniqueness of their product.

One upcoming encapsulation standard for SIP can be identified to be the substrate overmolding process. The substrate to be overmolded typically is a Printed Circuit Board (PCB), a ceramic panel or a leadframe. The substrate consists of several electronic modules arranged in a map-type configuration. Each of them is composed of numerous passive or active components. The assembled surface of the substrate is encapsulated completely by EMC and singulated by dicing. A simple construction of such a tool for substrate overmolding and an overmolded PCB are pictured in Figure 1. Details of the molding process are given at the end of this chapter.

The advantages of these Mold Array Package (MAP-type) geometries for production seem to be obvious:

- one mold tool can be used for all kinds of module dimensions as long as their height and substrate size remains the same.
- the length of polymer flow can be minimized because runners for material supply can be omitted. Only one gate is necessary per substrate.
- reduced package size
- increased integration density of components
- reduced packaging costs

Typically the singulated modules are QFN (Quad Flat No Leads) or ball less package types. A very promising technology derived from ball less MAP-type molding is the embedded Wafer Level BGA (eWLB) approach [1]. eWLB starts with redistribution of a diced, ball less silicon wafer on a carrier tape. After overmolding the tape is removed and the exposed pads can be electrically connected.

Both encapsulation technologies, MAP-type molding as well as eWLB combine the overmolding of large areas ($\varnothing$ = 25 - 100 mm) with a relatively thin mold cover (height < 1mm).

978-1-4244-4722-0/09 $25.00
© 2009 IMAPS-ITALY

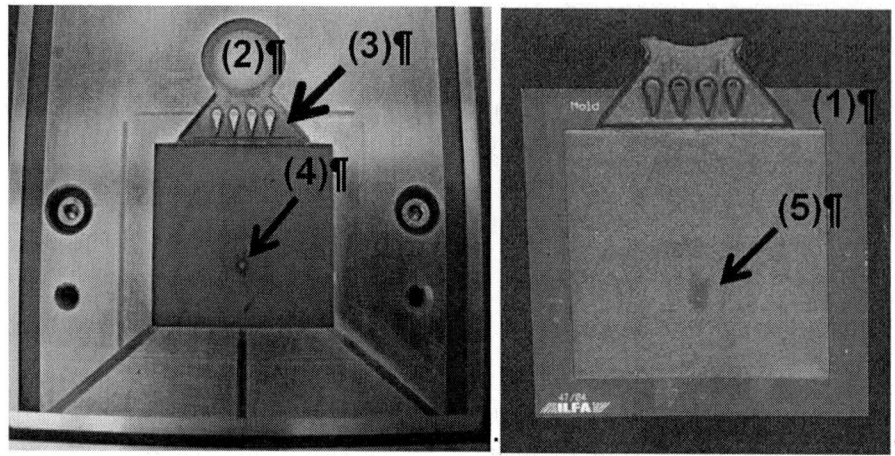

Figure 1: Left: Molding steel tool. Right: Overmolded FR-4 substrate with film gate
(1) substrate, (2) pellet position, (3) flow shaping structures, (4) PZT pressure sensor, (5) mark of sensor

The MAP type overmolding process of PCB substrates consists of several production steps: A substrate (labeled "1" in Figure 1) and an epoxy pellet (labeled "2") are inserted into the mold tool. Both can be preheated, if a satisfactory warming within the mold tool can not be guaranteed. The pellet is compressed by a plunger therefore forced to melt and fill the cavity on top of the PCB through the film gate. Specially designed shaping structures ("3") can be used to achieve sufficient shear heating within the EMC and a homogenous flow front at the gate. After filling of the cavity the EMC is pre-cured for about a minute to guarantee sufficient gelation and therefore adequate contour accuracy after ejection of the substrate.

The integration of sensors and actuators within MAP type modules narrows the molding process window significantly, because their sensitivity to stress is much higher than monolithic silicon chips or passive components. The aspect of thermo-mechanical stress during reactive molding is often neglected, as the encapsulation process itself is considered stress free. Even for epoxies with low viscosity during processing, this assumption is often disproved in practice. The need to monitor the pressure acting on the electronic parts during the polymer encapsulation process is essential considering e.g. the spreading usage of MEMS sensors within automotive industry and their increased demand for signal stability, production tolerances and cost [1].

The pressure within the cavity is typically measured by PZT pressure sensors mounted flushing with the mold cavity wall, chronologically recording pressure values at the measurement spot (labeled "4"

in Figure 1). Disadvantages of this sensor are the missing ability to record an areal pressure distribution and the occurrence of unattractive marks on the part's surface ("5"). As their size can be even larger than one single singulated module, their usage within MAP-type transfer molding is rarely accepted. Another disadvantage of active sensors – electrical as well as optical - is that the signal transmission requires holes or grooves within the metal tool. This is undesirable, especially if the pressure only should be measured during the developmental stage.

Multilayered Pressure sensitive Films

To overcome these disadvantages of conventional pressure sensors, a Prescale MW pressure sensitive film (sensitivity 100-500 bar) has been modified to enable its usage within the transfer molding process. We appended two additional polymer layers onto the commercially available film and mounted this multilayer on top of the PCB: an isolating polyimide coating and an adhesive layer. This multilayer has to prevent the Prescale film from degassing or swelling, fix it mechanically and increase its stiffness. As the softening temperature of the PET resin within the Prescale film is 80°C, the modification has also to guarantee that the pressure within the cavity is almost entirely transferred to the pressure sensitive particles within the Prescale film. They show only a predictable color change if a certain pressure level is exceeded that must not be relieved by the soft PET matrix. With our multilayer film we can enable test durations at 180°C of several minutes without detraction of the sensor function.

978-1-4244-4722-0/09 $25.00
© 2009 IMAPS-ITALY

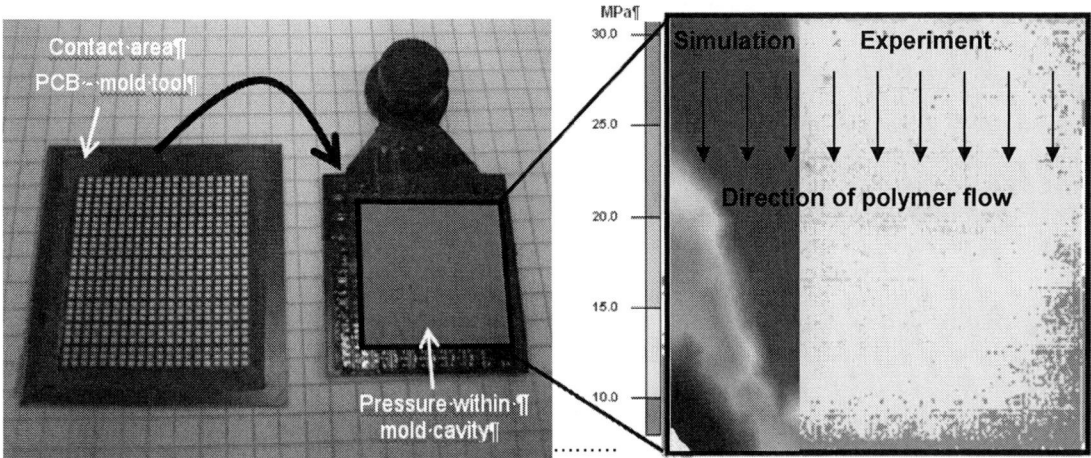

Figure 2: Apart from few elevated pressure values at substrate notches the pressure value of 125 bar recorded by the Kistler sensor could be replicated by averaged Prescale results (data analysis by "Fujifilm Pressure Distribution Mapping System FPD-8010E").

The first process we investigated was an FR-4 board overmolded with EMC through a film gate as shown in Figure 1. The polymer melt has been injected directly onto a Prescale-IZM-multilayer mounted onto the PCB within the cavity. The color distribution on the FR-4 board (Fig. 2) is very homogenous with a slight pressure drop with increasing distance from the gate. The pressure distribution is in accordance with numerical simulations. The pale red colored structure on the sensitive film does not necessarily represent the maximum pressure distribution within the cavity: the pressure sensitivity of the Prescale film depends on temperature which can chance within wide limits. Therefore the thermal behavior on the substrate's surface should be regarded, e.g. by numerical simulations. In most cases, with a proper pre-heating of the substrate and the polymer, the temperature remains constant and the color pattern will represent the maximum pressure distribution.

The temperature regime of operation within transfer molding is far away from the recommended temperatures for pressure sensitive films. Additionally the multilayer structure is influencing the pressure sensitivity of the sensor.

The accuracy of this measurement technique has been analyzed by comparing its pressure values with conventional sensors based on piezoelectric measurement systems. Apart from FR-4 substrates we tested the technology with ceramic substrates and on different transfer molding machines with different positions of the PZT sensor within the cavity. We derived a semi-empirical formula that describes the sensitivity of the pressure sensitive film at various temperatures and pressures. For transfer molding with T≈175°C and p≈100 bar the following formula was used for pressure derivation.

$$S = \frac{p_{Fuji}}{p_{korr}} = \left(1 - \frac{T - 115°C}{35°C} \cdot \frac{25\ bar}{p_{Fuji}} \right)^{-1}$$

The maximum pressure value, recorded by several PZT sensors within the cavity, has been varied between 55 bar and 125 bar. The measurement values of the Kistler PZT pressure sensors are displayed in Table 1 and compared with the multilayer film values. The percentage values refer to these multilayer pressure values in relation to the Kistler sensor. We found both values to agree within ±10%.

Table 1: Accuracy of cavity pressures values of multilayer film recorded during transfer molding and correlated with PZT sensors

Sensor Position / Pressure	57 bar	75 bar	85 bar	90 bar	125bar
Near gate	100%	97%	105%	90%	98%
Far away from gate	102%	104%	102%	98%	98%

MAP type molding of FR-4 substrates

The flow behavior of EMC on substrates, containing a heterogeneous variety of devices is far more complicated than the single chip encapsulation. Even on small FR-4 substrates (edge length: 25 mm) without any components a lot of surprising results can be observed.

Figure 3 shows the time dependent evolution of the flow front during substrate overmolding. The short shots have been produced by increasing the amount of EMC material step by step. The flow front (picture A) splits into two separate flow paths shortly after injection into the cavity. First, the material flows uniquely along the edges of the cavity. Then an additional flow path starts in the centre of the substrate (B, C), but keeps sharply separated from the outer paths. They reach the PCB corner opposite of the gate slightly before the central path and continue their way along the cavity border (D, E). If the cavity is not evacuated the enclosure of air traps is unavoidable.

The reason for this behavior is not yet fully understood but is definitely rooted in the reduced stiffness of the substrate above its glass transition temperature. As the coefficient of thermal expansion (CTE) is rising sharply above Tg even small temperature differences between the mold tool (175°C) and the PCB cause a significant elongation of the substrate after the mold tool is closed. Differences in the distribution of copper on top and bottom side of the substrate can also cause a bending of the substrate due to the CTE mismatch. The substrate is mechanically fixed between the halves of the mold tool and the elongation of the substrate must lead to a convex, ∩-shaped bending and forces the polymer to flow around this central elevation. With increasing pressure at the gate and further increasing temperature of the substrate, buckling of the PCB starts in the middle leading to a ∩∩-shaped substrate with three distinct flow paths. The curvature of the substrate can still be measured after ejection of the part, because it is frozen by the cured epoxy polymer.

If the experiment is conducted with a substrate assembled with electronic components, this behavior is normally suppressed by the components. They stiffen the PCB mechanically, suppress the convex bending of the substrate due to their smaller CTE and finally work as spacers between PCB and mold tool that are preventing the soft substrate to buckle by mechanically touching the upper mold tool – definitely another source for reliability problems.

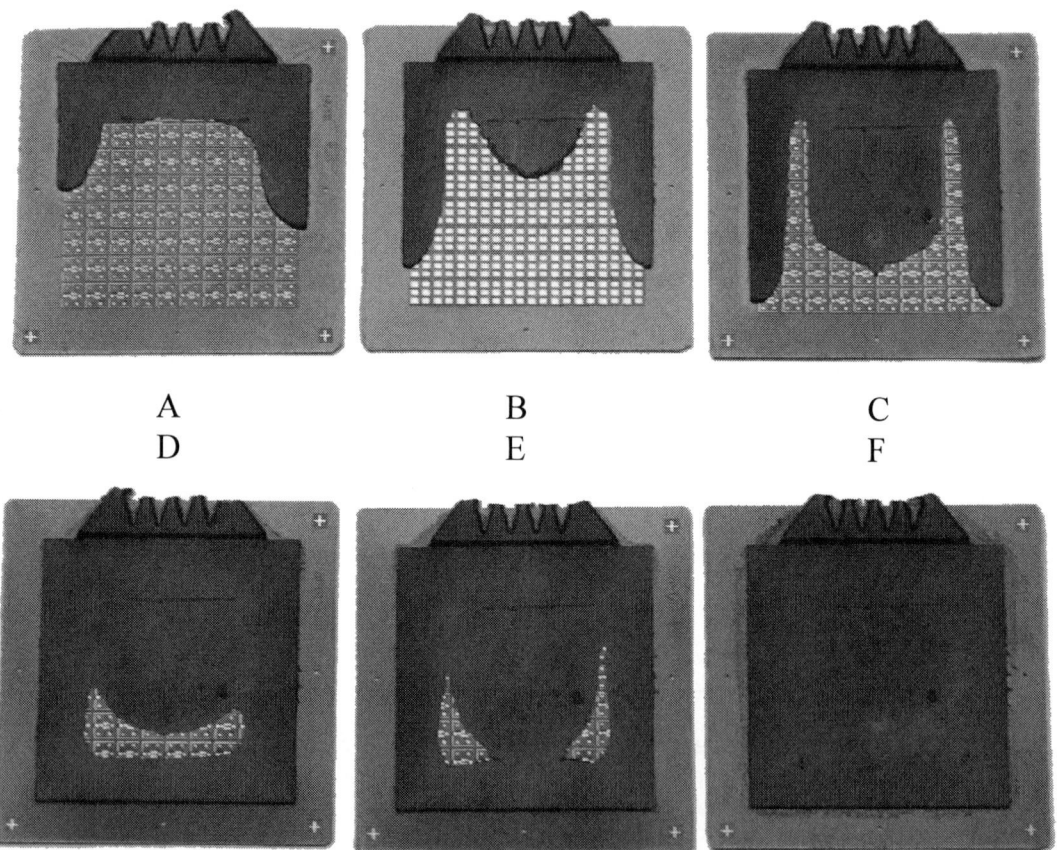

Figure 3: The filling behavior of EMC on FR-4 substrates can be surprisingly complicated which can be illustrated by short shots with a reduced amount of polymer.

Simulation of Polymer Flow during MAP type overmolding

Numerical simulation of the polymer flow within MAP-type cavities is until now a challenging task for a process engineer. First the large amount of small gaps between electronic parts, mold wall and PCB has to be meshed precisely, heavily increasing the FEM computation time. Secondly, the amount of filler particles within the polymer (typically up to 90 wt-%) can increase the flow resistance of the epoxy within small gaps. Thirdly, the flow front can get very complicated, even if the EMC is molded on a stiff substrate without any components assembled on it.

Commercially available tools for simulation of transfer molding are – as far as we know – not able to predict the filling behavior of polymers within small, thin gaps precisely. In case of a point gate they mostly show a more or less symmetric circular pattern around the gate which does not change significantly if one of the following parameters changes: filler degree, viscosity or temperature of the melted epoxy, cavity height. Nevertheless one important fact is predicted by the FEM simulation: the polymer velocity is significantly increased near the gate. Regarding the governing laws of polymer flow shows that the definition of shear rate dependent viscosity is basically leading to a flow described analytically by Hagen-Poiseuille's law. In the specific form of Navier-Stokes it leads to a hyperbolic velocity distribution within the cavity of gap size h and Viscosity η. The FEM-values of the velocity within gaps of 250 µm and 1000 µm thickness do not differ significantly from analytical values derived by Poiseuille's theory.

$$v(y) = \left(y^2 - hy\right)\frac{\Delta p}{2\eta} + v_{wall}$$

The pressure drops linearly along the flow path x and shows a quadratic dependency of the gap size.

$$\Delta p = 8\eta\frac{v_{max}}{h^2}x = 12\eta\frac{\bar{v}}{h^2}x$$

Consequently, the polymer penetration within two simultaneously filled gaps of height h1 and h2 can simply be described by

$$\frac{x_1}{x_2} = \frac{h_1}{h_2}\sqrt{\frac{\eta_2}{\eta_1}}$$

It is obvious that his correlation can not be used if gap sizes are of the same dimension as the filler particles.

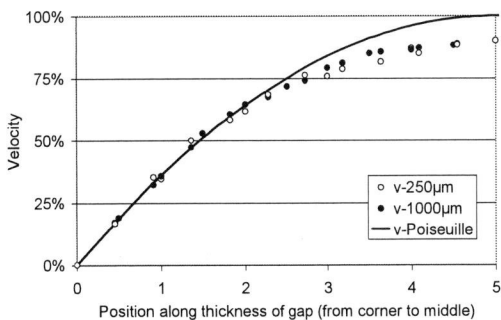

Figure 4: Numerical and analytical calculation of polymer velocity within gaps of 250 µm and 1000 µm.

Our experimental observations suggest another important effect during MAP-type overmolding that is not predicted correctly by standard numerical simulation software. If a film gate is used, a straight flow front is moving over the panel. If flow paths are not too long (< ~50 mm), no disturbance of this behavior could be observed. If a pin gate is used, the velocity of the melt decreases monotonously with increasing free surface area. Consequently the shear rate at the flow front decreases steadily, increasing viscosity. Only within narrow gaps the polymer experiences significant shear which enhances flowability at these regions – an effect that is generally used when the gate diameter is designed.

This locally increased shear rate increases flowability even more and leads to a necking of the flow to few distinct paths. Consequently the circular flow front splits up into several flow paths, an effect that could also be observed even on plane substrates without any assembled components, because little random variations of polymer velocity are enough to break the symmetry of the filling behavior.

A lot of experiments and simulations with different mold materials and mold geometries have been analyzed. It could be demonstrated, that the following parameters significantly influence the occurrence of flow path concentration:

- thickness of overmolding
- distance between electronic parts and between parts and mold walls
- viscosity of the polymer
- size of the filler particles

978-1-4244-4722-0/09 $25.00
© 2009 IMAPS-ITALY

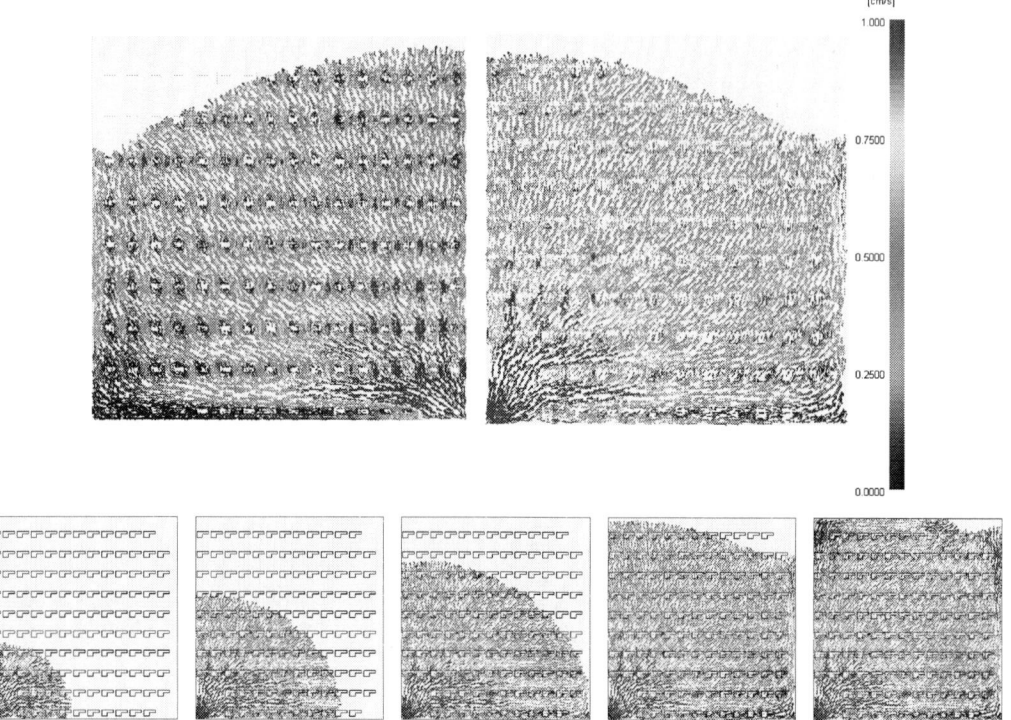

Figure 5: Simulation of filling behavior (polymer velocity) on substrates with a mold cavity height of 750 μm (top, left) and 450 μm (top, right). No significant difference between both flow fronts can be observed. The row below shows polymer velocities after 4s, 7s, 10s, 14s and 15s.

This necking behavior during transfer molding can not easily be implemented into standard transfer molding simulation tools. We numerically adjusted the material behavior by a thixotropic shear thinning parameter that is highest at the gate location and monotonically drops with proceeding time.

Not only could the detailed form of the flow front be predicted (Fig. 6, left), but also the locally reduced viscosity within a few preferred flow paths

between the components (Fig. 6, right). At the beginning of overmolding the regions with slow material movements are additionally filled by the preferred flow paths, but the occurrence of these shear thinned regions is already the forerunner of an asymmetrical filling behavior where regions far away form the flow paths remain unfilled.

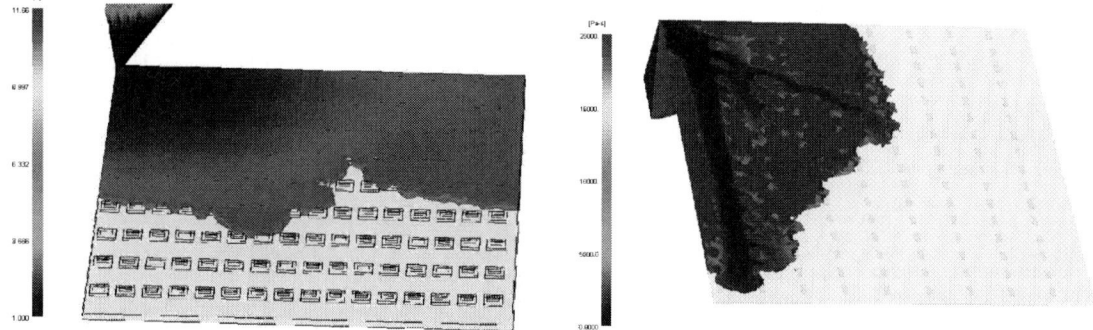

Figure 6: Simulation of filling behavior on a substrate with a mold cavity height of 750 μm (left: time dependent flow front, right: viscosity) and point gate. The flow front gets highly inhomogeneous during the filling process.

Summary

Homogenous material flow and pressure distribution within the cavity is of particular importance for encapsulation of mechanically sensitive parts, e.g. MEMS. With the use of pressure sensitive films a fast and inexpensive estimation of the pressure distribution can be achieved. Our multilayer film design is based on commercially available color changing papers (e.g. Prescale), an adhesive layer and an isolating film. This assembly enables tests for several minutes at 180°C without detraction of the sensor function. The accuracy of this pressure measurement method is ~10%, checked by conventional pressure sensors located at different positions within the mold walls. The pressure results are also fitting well with numerical simulations.

By using a standard molding process, the polymer melt has been injected directly on this multilayer, generating red color changes on the Prescale film depending on the local maximum pressure value. Also the contact pressure between solder pads and bottom mold wall could be measured with the pads emerging clearly from the flat PCB surface.

We could clearly distinguish two different kinds of flow behavior during MAP type overmolding: 1. Homogenous filling of the cavity with a closed flow front and a continuous pressure drop along the flow path. 2. Contraction of the polymer flow within few, narrow paths with increased shear rate, reduced viscosity and excessive velocity.

Additionally we could show that modifications of the computational fluid simulation tool increase the accuracy of the flow behavior prediction, especially of the filling patterns with concentrated flow paths. The models have been validated with short shots within the transfer molding machine.

Acknowledgements

We would like to thank Ralf Andussies from FUJIFILM Recording Media GmbH for valuable support concerning measurement strategies at elevated temperatures. We appreciate the assistance of Dr. Frank Rehme from EPCOS AG concerning implementation of our developments.

References

[1] R. Müller-Fiedler, V. Knobloch. "Reliability aspects of microsensors and micro-mechatronic actuators for automotive applications". Microelectronics Reliability, Vol. 43, pp. 1085–1091, 2003

[2] T. Meyer, G. Ofner, S. Bradl, M. Brunnbauer, R. Hagen, "Embedded Wafer Level Ball Grid Array (eWLB)", Proceedings of the 2008 Electronic Packaging Technology Conference (EPTC), Singapore, December 9-12, pp. 994-998, 2008.

[3] S. C. Chen et al., "Preleminary study of polymer melt rheological behavior flowing through micor-channels", Int. Comm. Heat and Mass Transfer, Vol. 32, pp. 501-510, 2005

Mechanical and Microstructural Properties of SiC-Mixed Sn-Bi Composite Solder Bumps by Electroplating

Yue-Seon Shin, Sehyung Lee, Sehoon Yoo*, Chang-Woo Lee

Micro-Joining Center, Korea Institute of Industrial Technology, Incheon 406-840, Korea

*Phone: 82-32-850-0268, Fax: 82-32-850-0210, E-mail address: yoos@kitech.re.kr

Abstract

SiC-nanoparticle-mixed Sn58Bi solders were successfully produced by electroplating and the mechanical and microstructural properties of Sn58Bi+SiC solders were observed and analyzed in this study. Dispersed SiC nanoparticles by ultrasonic horn were added into Sn-Bi plating solution. With the SiC added Sn-Bi plating solution, solder bumps were electroplated. After electroplating, the solder bump specimens were aged up to 400 hrs at 100 ℃. The melting temperature of SiC-mixed Sn-58Bi solder was same as non-mixed Sn-58Bi. The shear strength of Sn58Bi+SiC solder bumps increased by 6% as compared with Sn58Bi solder bumps. The thicknesses of intermetallic compound were almost same for both Sn58Bi and Sn58Bi+SiC samples. The Sn58Bi+SiC solder bumps had finer eutectic structures and grain sizes than Sn-Bi without SiC nanoparticles. From the fracture surface analysis, fracture occurred at solder side not at the joint interface. Therefore, the shear strength values represent the solder materials itself and the finder microstructure of Sn58Bi+SiC improved the shear strength.

Key words: Sn-58Bi, SiC nanoparticles, Pb-free solders, Solder bump, Electroplating

Introduction

Recent small, light, and multi-functional trends of electric devices require fine pitch solder bumps for flip chip joining [1-3]. The mechanical reliabilities of solder interconnection become more important for such fine pitch flip chip packages. In addition, various Pb-free solder materials are being used in solder interconnection due to global environmental regulations[4]. Most of the lead-free solders have higher melting temperature than eutectic Sn37Pb, therefore, the high soldering temperature frequently damaged the device performance and joint reliablilities. The low temperature soldering is beneficial for reducing thermal damages in electronic packages. Especially, flexible substrates need low temperature bonding process because of warpage problems at high temperature. Among the various solder materials, the eutectic Sn58Bi solder has lower melting temperature (138 ℃) than the eutectic Sn37Pb solder. However, the brittleness and low fatigue resistance of Sn58Bi should be overcome for obtaining high reliability[5-7]. To improve the mechanical properties of the Pb-free solder, many researchers have studied on composite solders with Fe[4], carbon nanotubes[8], Al_2O_3[9], ZrO_2[10]. The previous studies utilized mechanically mixing of solder particles and reinforced materials and the application of the mixture was mostly solder paste or solder ball. Since solder paste and solder ball was not applicable for fine pitch flip chip application under 100μm, an alternate method has to be applied to the process of composite solders for the fine pitch applications.

In this study, SiC mixed Sn-58Bi solder bumps were fabricated with electroplating, which is widely used for the fine pitch solder bump process. During the electroplating, the dispersed SiC nanoparticles in the plating solutions are entrapped within the metal deposit on the cathode[11]. The mechanical and microstructural effects of SiC nanoparticles mixing with Sn-58Bi solders were investigated with SEM, TEM and EDS in this study.

Fabrication of SiC-mixed Sn-58Bi solder bump

Substrate for the solder bump formation was Si wafer. Ti/Cu layers were sputtered onto the Si wafer. After sputtering, dielectric layers were deposited and electrical lines were formed by photolithography and etching process. For the seed layer of electroplating, 0.3 μm Cu was deposited on patterned wafer. The wafer was screened by photoresist except for pad areas.

After preparing patterned wafer, Sn58Bi+SiC solder bump was formed by electroplating. The SiC nanoparticles were β-SiC obtained by Nand Materials Inc. As shown in Fig. 1, the diameters of as-received SiC nanoparticles were 45-55nm. The SiC nanoparticles were added into the SnBi plating solution. The amount of added SiC was 0.5g per 1L of SnBi plating solution. The solution was then ultrasonicated by ultrasonic homogenizer (VCX-750, Sonics, Japan) to disperse SiC particles in the plating solution.

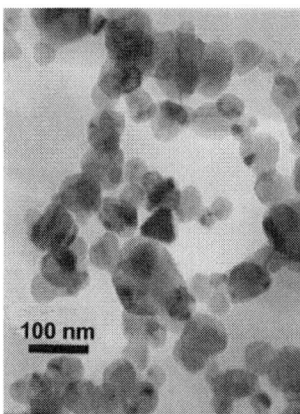

Fig. 1. TEM of as-received SiC nanoparticles

Before plating, the Si substrate was plasma-treated to remove oxide on Cu pads and to activate the surface. After plasma cleaning, the Si wafer was dipped into acid liquid for surface microetching and then the Sn58Bi+SiC was plated by the condition shown in Table 1. After electroplating, the solder bumps were reflowed and the reflow profile is shown in Fig. 2. The peak temperature was 170°C. For the aging tests, the reflowed solder bumps were aged for 100 and 400 hrs at 100℃. Microstructure of the solder bump was characterized by scanning electron microscopy(SEM) and transmission electron microscopy(TEM). The TEM samples were prepared by a focued ion beam (FIB) instrument. The shear strength was measured by shear tester (DAGE-4000).

Table 1. Bath composition and plating condition

Bath composition		Plating condition	
Sn^{+2}	55 g/L	Current density	10 A/dm^2
Bi^{+3}	123.6 g/L	Temperature	40 ℃
pH	-0.13	Time	15 min

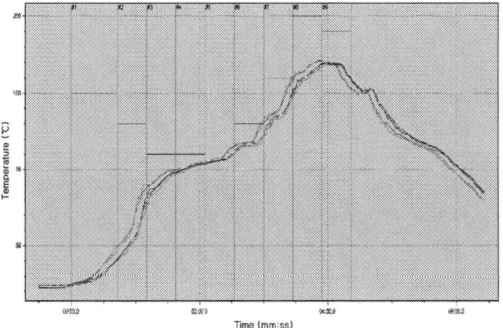

Fig. 2. Reflow profile of Sn58Bi solder bumps

Dispersion of SiC particles in Sn58Bi

SiC-mixed Sn58Bi solder was observed with TEM and shown in Fig. 2. It was observed that nanoparticles are well-dispersed in a grain. The size of the nanoparticle in the grain was about 50 nm which was same as as-received SiC particles. Selective area diffraction and EDS was carried out on the grains and it was determined that the nanoparticles were SiC.

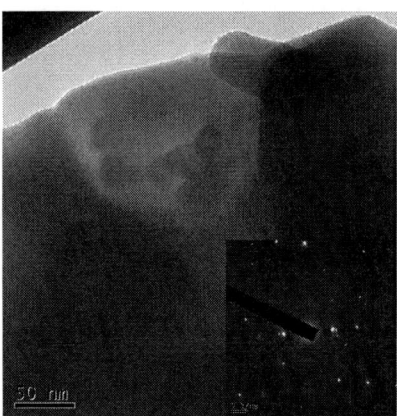

Fig. 3. Cross-sectional TEM of SiC mixed Sn58Bi solder

Microstructure refinement by SiC addition

The comparative microstructures of aged solder bumps of Sn58Bi solder with or without SiC nanoparticles were observed with SEM and shown in Fig. 3. The Sn58Bi solder without SiC nanoparticles exhibited extensive coarsening of microstructure. On the other hand, the eutectic microstructure of Sn58Bi+SiC solder bumps was much finer than that of no SiC added Sn58Bi solder. Table 2 shows average sizes of the eutectic lamellar structure of Sn58Bi and Sn58Bi+SiC. The average size of the eutectic structure decreased up to 48.4% by adding SiC nanoparticles. Such refinement of eutectic microstructure might have been due to the increased heterogeneous nucleation sites and impeded phase or grain boundary movement by SiC nanoparticles.

The average thicknesses of intermetallic compound(IMC) were determined from the cross-sectional SEM and shown in Fig. 5. The thicknesses of both Sn58Bi and Sn58Bi+SiC bumps increased with increasing aging time. However, IMC thicknesses were almost same for both Sn58Bi and Sn58Bi+SiC solders.

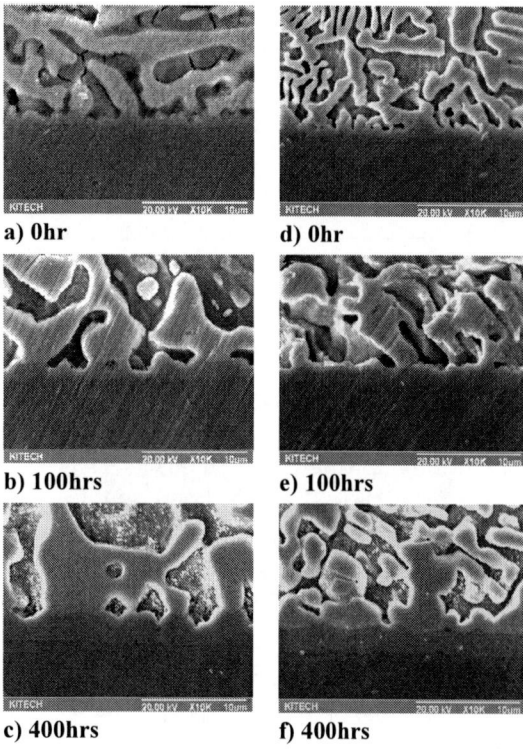

a) 0hr   d) 0hr

b) 100hrs   e) 100hrs

c) 400hrs   f) 400hrs

Fig. 4. SEM of Sn58Bi and Sn58Bi+SiC solders with varying aging time. a, b, and c: Sn-58Bi solder and d, e, and f: Sn58Bi+SiC.

Table 2. Average size of lamellar structures

	Sn-58Bi	Sn58Bi+SiC	Decreased %
0hr	1.97μm	1.32μm	32.9%
100hr	3.09μm	2.05μm	34.6%
400hr	4.59μm	2.37μm	48.4%

Shear strength increase by SiC addition

Ball shear test was carried out to understand the dependence of shear strength of solder bumps on SiC nanoparticle addition and shown in Fig. 6. After reflow, the shear strengths of Sn58Bi and Sn58Bi+SiC were 89.56 gf and 94.33 gf, respectively. Approximately 5% of shear strength was increased by adding SiC nanoparticles. To understand the effects of intermetallic compounds, the solder bumps was aged for 100hrs and 400hrs at 100℃. The shear strength of solder bumps decreased with aging time. However, the shear strength of 100hr-aged Sn58Bi+SiC solder bump was 9% higher than that of Sn58Bi solder bump. After 400hr aging, the shear strength of Sn58Bi+SiC solder bump was 10% higher than Sn-58Bi solder bump.

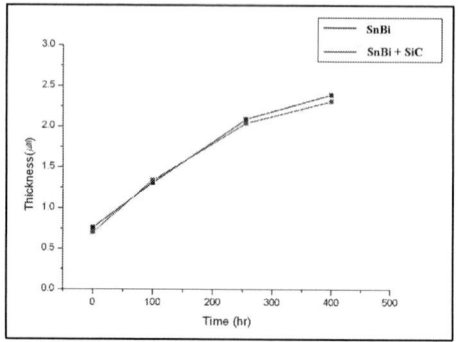

Fig. 5. Comparative thickness of intermetallic compounds

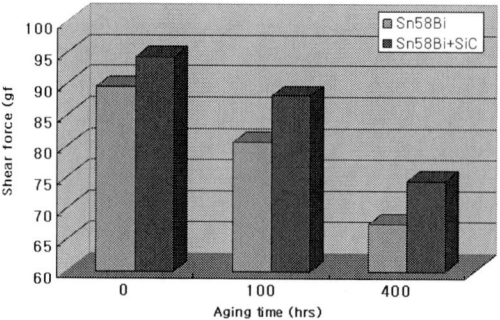

Fig. 6. Shear strength of solder bump

Fig. 7 shows that fracture surface of solder bumps. Both surfaces indicate ductile failure mode. Although SiC nanoparticles were added in Sn058Bi solder bumps, there was not change of failure mode. Fig. 8 shows a result of energy dispersive spectroscopy (EDS) analysis. Sn and Bi were detected on fracture surface but Cu was not detected. Such EDS results indicated that the fracture occurred at solder side not joint interface. Therefore, the value of shear strength represented the solder properties itself, not the joint interface properties.

There are possible mechanisms of increased shear strength of Sn58Bi+SiC solder bump. First, fine lamellar structures impede dislocation movement by grain boundary because higher energy is required for dislocation movement across the phase boundaries. Accordingly, Sn58Bi+SiC solder bump showed the higher shear strength than Sn58Bi without SiC. Another reason may be the effect of dispersion strengthens by SiC nano particles. If SiC nano particles are locally agglomerated, those could become crack sources. But the uniformly dispersed SiC nano particles make the solder bumps hardened.

Conclusions

SiC nanoparticles were successfully incorporated into Sn58Bi solder by electroplating. SiC nanoparticles were well-dispersed in the grains of Sn58Bi. Microstructural coarsening was

978-1-4244-4722-0/09 $25.00
© 2009 IMAPS-ITALY

suppressed by SiC nano particle addition. The shear strength of Sn58Bi+SiC solder bumps was higher than Sn-58Bi solder bumps. The increased shear strength was attributed to grain refinement and dispersion hardening by SiC particle dispersion.

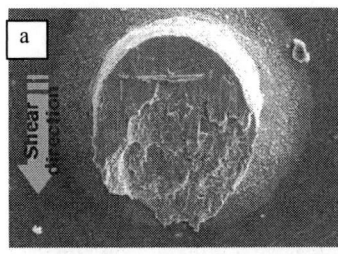
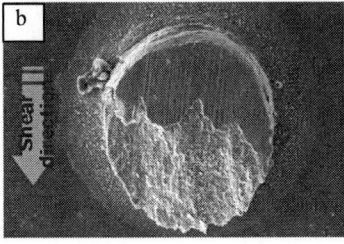

Fig. 7 Fracture surface of shear tested solder bumps. a) Sn-58Bi solder bump and b) Sn58Bi+SiC solder bump

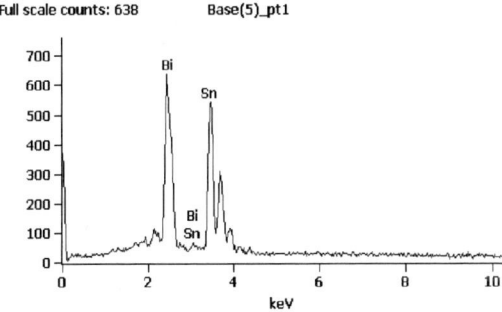

Fig. 8. EDS result of fracture surface

Acknowledgements

This work is supported by a grant from cooperative research program of Ministry of Knowledge Economy, Korea.

References

[1] European Union WEEE. Directive, 3er Draft, May, 2000

[2] R. Smerons and R. Strauss, Elect. Commun, 57, pp. 148 (1982)

[3] Z. MEI, "Characterization of Eutectic SnBi Solder Joints", Journal of Electronic Materials, Vol. 21, No. 6, 1992, pp. 599-607

[4] S. Jin and M. McCoramack, "Dispersoid additions to a Pb-free solder for suppression of microstructural coarsening", Journal of

Electronic Materials, Vol. 23, No. 8, 1994, pp735-739

[5] The Measurement of Zeta potential Using an Autotitrator : Effect of pH, Application Note by Malvern Instruments, 2005

[6] H. K. Lee, "Codepositon of mocro- and nano-sized SiC particles in the nickel matrix composite coatings obtained by electroplating", Surface & Coating Technology, 201, 2007, pp. 4711-4717

[7] D. Frear, "A Microstructural Study of the Thermal Fatigue Failures of 60Sn40Pb Solder Joints", Journal of Electronic Materials, Vol. 17, No. 2, 1988, pp. 171-180

[8] K. Mohan Kumar, V. Kiropesh, Andrew A. O. Toy, "Single-wall carbon nanotube(SWCNT) functionalized Sn-Ag-Cu lead-free composite solders", Journal of alloys and compounds, 450, 2008, pp229-237

[9] H. Mavoori, S. Jin, "New, creep-resisant, low melting point solders with ultrafine oxide dispersions", Journal of Electronic Materials, Vol. 27, No. 11, 1998, pp1216-1222

[10] Jun SHEN, Yougchang LIU, Dongiang WANG, Houxiu GAO, "Nano ZrO2 particulate-reinforced lead-free solder composite", J. Mater. Sci. Technol., Vol. 22, No. 4, 2006

[11] C. T. J. Low, R. G. A. Wills, F. C. Wlash, "Electrodeposition of compositie coatings containing nanoparticles in a metal deposit", Surface & Coatings Technology, 201, 2006, pp371-383

[12] The Measurement of Zeta potential Using an Autotitrator : Effect of pH, Application Note by Malvern Instruments, 2005

[13] H. K. Lee, "Codepositon of mocro- and nano-sized SiC particles in the nickel matrix composite coatings obtained by electroplating", Surface & Coating Technology, 201, 2007, pp. 4711-4717

[14] D. R. Frear, "The Mechnics of Solder Alloy Interconnects", Van Nostrand Reihold, 1994, pp.21-38

Some Facts from Lead-free Solders Reliability Investigation

Ivan Szendiuch, Jiri Stary, Josef Sandera, Martin Bursik, Edita Hejatkova

Brno University of Technology, Faculty of Electrical Engineering and Communication, Department of Microelectronics, CZ-602 00 Brno, Udolni 53, E-mail: szend@feec.vutbr.cz, stary@feec.vutbr.cz

Abstract

There are done many research works following investigation of lead-free solders, and differently results are achieved for identical solder materials. The reason is simple, the same solder compound can show diverse results when other material and process parameters are different. That is reason why we have addicted on investigation of some critical areas which has significant impact on solder joint reliability and lifetime. At length the soldering area should be divided in three principal parts that are materials and components, process and equipments, and design and testing. Our actual research work at Brno University of Technology is focused on soldering process parameters impact on solder joint structure, on wetting angle changes in dependence on material/process parameters and on thermo mechanical strain of soldered joints

Key words: lead-free solder, solders joint structure formation, wettability, solder joint reliability testing.

Introduction

The soldering is process used in widespread throughout the world many years. In electronics, solder are used for component attachment, electrical as well mechanical. From 1958, when was introduced the first IC in the market made electronic devices strong progress, and the same applies to solder joints. Main changes in solder joint technology during last decades can be defined by following facts:
- solder joints are all the time smaller,
- solder joints must have defined shape (defined solder volume),
- solder joints must have high reliability (there are often hidden under components or substrates),
- solder joints are produced by various techniques,
- new lead-free solder materials are in use.

In electronics there were long time used tin-lead alloys showing very good long-time chemical stability and reliability. One of main reasons for it was good known structure of SnPb solders with intermetalic layer based on SnCu compounds. This true Pb is highly resistant to corrosion and is a very stable part of solder compound containing in its volume 30 – 40%. Long time stability was reason to use it many other applications as for example in paints, batteries, bullets etc. But on the other hand lead was identified as a toxic environmental material, which use in products must be step by step discarding. In electrical and electronics products is lead still a key component, not only for tin-lead solders, but also for thick film pastes, etc., that are used in production of equipment and components. Actually directives RoHS define restriction of use of Pb, which means that SnPb solder compounds must

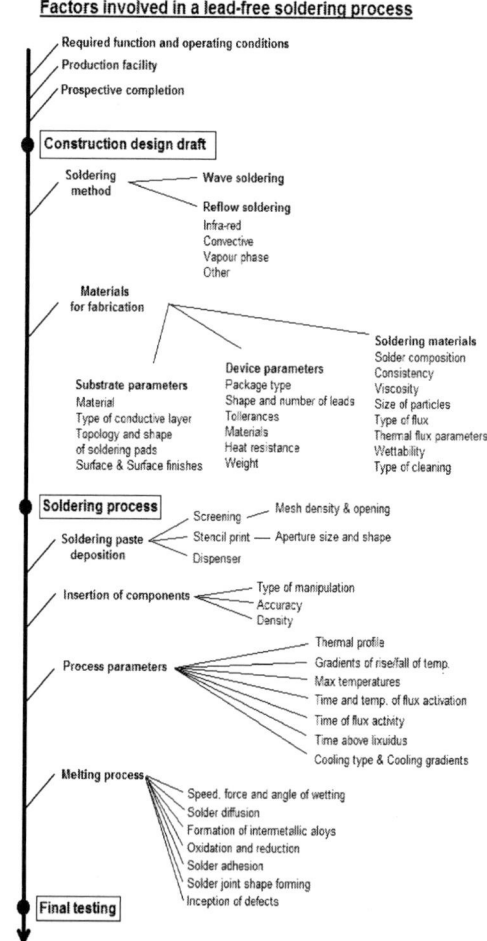

Figure 1: Lead-free soldering Roadmap

978-1-4244-4722-0/09 $25.00
© 2009 IMAPS-ITALY

be supplied by lead-free solder compounds. This step is bringing lot off unknown factors in electronic system production and in more some doubt in long term stability and life cycle duration.

There are many studies and works about lead-free solders and their applications over the entire world, but there are not defined unique final instructions and technological rules. Reason is simple, there are many various factors acting the final solder joint quality. In addition, different combination of these factors that occur during technological process brings different possibility and different results too. To achieve acceptable results asks to control necessary technological parameters during the whole process.

There are three basic parts deciding final solders joint quality:

- Materials and Components
- Process and Equipments
- Design and Testing

Formation of solder joint structure

Previous research work shows that the structure of solder joint has significant impact on solder joint parameters, especially on its reliability and life time. That's reason why we have addicted on this problematic. Additionally in the last time we have during our experimental work learn that one of the significant factors is temperature decrease in cooling zone of reflow profile. That was reason to observe the microstructure of various solder compounds under different cooling regimes (Fig. 2).

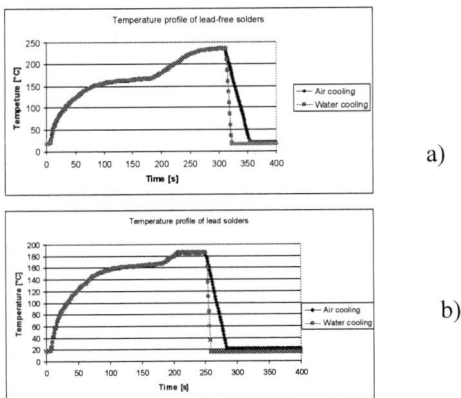

Figure 2: Temperature profiles with different cooling zone a) lead-free b) SnPb

There are together with structure homogeneity two significant characters to observe, the solder voiding and intermetalic layer formation. Even if voids can not be direct reason of the failure, they can evoke nucleus for defects that effect catastrophic failures in working system. Existence of intermetalic layer is necessary to ensure solder joint creation, but its formation reduces the bond strength of solder joint and accelerates formation of cracks.

Experimental procedure was done with solder joint specimens that were prepared as solder balls on PCB without surface treatment. We have studied five solder alloys to have comparison of various solder alloy systems (Sn60Pb40, Sn95.5Ag3.8Cu0.7 , Sn96Ag4, Sn99Ag+, Sn100C). SEM micrographs of various solder alloys are shown in Figure 3, 4, 5, 6 and 7.

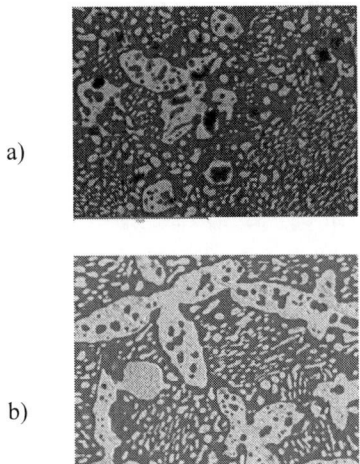

Figure 3: SEM micrograph of SnPb structure (2000x) a) air cooling A b) water cooling

In Fig. 3, where the determinative compound is Pb, we can observe significant influence of cooling profile B. The growth of Pb islets is much more significant and the complete structure is more compact. The same situation is for SAC solder compound (Fig. 4), although the structure is different, more rugged (see Fig.5). But determinative material in this structure is Sn instead Pb and there is significant intermetalic layer Ag_3Sn, which creates some like chains. In every case the voids occurrence seams to be blanked.

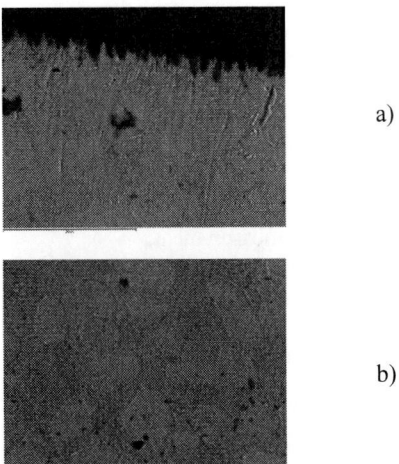

Figure 4: SEM micrograph of Sn95.5Ag3.8Cu0.7 structure (2000x) a) air cooling b) water cooling

The identical situation is by SA solder compound, where is not significant difference regarding to SAC alloy (Fig. 5a). Other situation is for SN100C solder alloy, which structure is very compact and regular (Fig. 5b). Intermetalic layers are a little significant, mostly in solder – Cu boundary district.

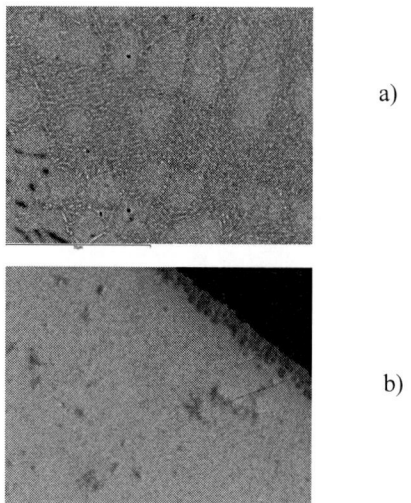

a)

b)

Figura 5: SEM micrograph of SAC structure a) and SN100C structure b) under cooling profile B (2000x)

The first results have shown significant influence of cooling zone on solder joint structure. Actually are prepared samples for temperature cycling where will be observed impact of this parameter on solder joint life cycle time.

Wettability significancy

Wettability of different surfaces, with different solders and chemistry using different process parameters is the key factor to reach a good solder joint reliability. Wetting balance method/globule test method is one of the most important methods for wettability measuring of solder alloys (see Fig. 6). Wettability needs to be quantitatively and qualitatively measured and evaluated. There are two types or configurations of wetting balance measurement:

- Bath method (solder bath is suitable for large components, THT components, base metals, substrates etc., disadvantage is larger amount of used solder alloy).
- Globule method (suitable for SMDs, individual leads of SMDs or THT components, base metals, substrates etc.).

The specimen is subjected to time variant vertical forces, which consist of the surface tension force and buoyancy force. The equilibrium wetting force of wetting balance F_w, is given by following equation:

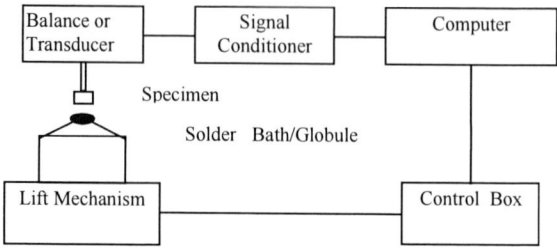

Figure 6: Wetting balance test using solder bath/globule method

$$F_w = c \cdot \gamma_A \cdot \cos\theta - g \cdot \rho \cdot v \qquad (1)$$

where: F_w is equilibrium wetting force (not moving specimen) [N]
c is specimen perimeter [m]
γ_A is surface tension between molten solder and flux [Nm^{-1}]
θ is contact angle [rad]
g is gravitational acceleration [9.81 ms^{-2}]
ρ is density of molten solder $\tau_M + 25\ ^\circ C$ [kg.m^{-3}]
τ_M is melting temperature [C]
v is specimen immersed volume [m^3]

Chemical (reaction) wetting through the chemical reaction between solder and base metal (substrate) results in changing composition of base metal and the solder mostly in the interface area. Solubility of base metal in strong reactive Sn solders through diffusion mechanism leads to formation intermetalic area in the interface. Lead free solders are accomplished with different wetting characteristic and changed with ageing. Diffusion mechanism and solubility of base metal changes wetting morphology and leads to different wetting characteristic curves. Wetting force, wetting speed, shape of wetting curve and possible type of diffusion mechanism influencing wetting is investigated and evaluated.

We have appointed on measuring and studying of wetting characteristics for two the most popular solders in electronics industry which are SAC 305 and SN100C with VOC free flux Litton Kester 979T (code 9) and ROL0 flux type 950E (code 0). Substrate material was copper 99,5% and dimensions of tested coupons (8 x 8) mm. Process parameters were arranged on controlled temperature of liquid solder: $\tau_M + 25^\circ C$, preheating: 0, speed 10 mm/s, depth 3 mm, delay of test: 10 s. Test were controlled in the following conditions:

1n – flux application and wetting characteristic measuring ,

978-1-4244-4722-0/09 $25.00
© 2009 IMAPS-ITALY

2n – sample 1n + flux application and wetting characteristic measuring,

2n1a – sample 2n and ageing 150 C/1hr + flux application and wetting characteristic measuring.

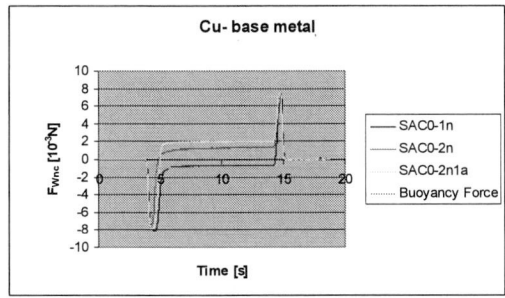

Fig. 7: SAC305, ROL0 flux

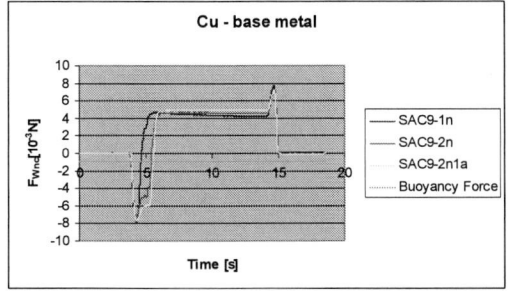

Fig. 8: SAC305, ROL0 VOC free

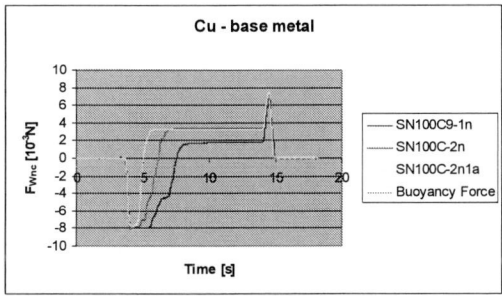

Fig. 9: SN100C, ROL0 VOC free

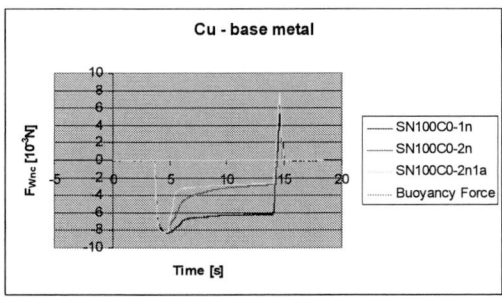

Fig. 10: SN100C, ROL0

The main aspect in the first step was to compare wetting characteristic of Cu base metal with SAC305 and SN100C, ROL0 and ROL0 VOC free fluxes in repeating wetting balance tests and with influence of ageing on wetting characteristics.

The results have shown the following facts (Fig. 7, 8, 9 and 10):

- ROL0 flux used for tin lead soldering is not strong enough for lead-free process especially with SN100C solder
- SN100C has lower wetting speed and lower wetting force comparing to SAC 305 with both types of tested fluxes
- thickness of solder after wetting balance test minimizes influence of intermetalic layer in the interface between solder and substrate, diffusion mechanism is studied
- ageing was positive on all of tested samples
- intermetalic layer and oxides are less important than flux application and flux residues retained on samples.

Tests continue and will be finished in few months.

Testing of solder joint reliability

The aim of this work is to establish unambiguous measurement method for testing of solder joint reliability by temperature cycling. To verify reproducible test method we have targeted on SAC 305 solder alloy. There was realized test pattern, which dimension is 70 x 70 mm. On this were mounted 40 pieces of SMD 2512 chip resistors by wave and vapor phase soldering under standard temperature profile (Figure 1). The material of PCB is FR 4 (Cu thickness 35 μm) with different surface treatment (HAL, Au, Sn immers).

Fig. 11: Test pattern for verification of solder joint reliability testing

Samples are tested in temperature chamber CTS T-40/25 with temperature range arranged from 0 to 100°C, eventually from -20 to 100°C. There is incorporated system for spatial and local heating of PCB. During the testing the interruption of single solder joint conjunction is automatically evaluated through electronic system connected with tested sample as is shown in Figure 12. There is connected powered LED diode parallel to examining solder joint failure (Fig.12a). When we connect LED with bistable trigger (BT), light of LED indicates during long time testing the first failure in connection field (Fig. 12b).

978-1-4244-4722-0/09 $25.00
© 2009 IMAPS-ITALY

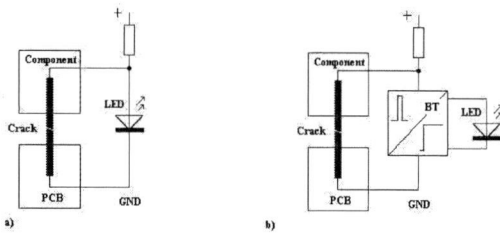

Fig. 12: Schematic diagram showing identification of solder joint crack

There are some results obtained by experimental work through this method:

- surface finish HAL has higher reliability in comparison to Sn immersion and Galvanic Au surface,
- the higher reliability was detected for PCB with higher thickness of Cu,
- the local heating is for failure exciting much more critical as spatial, for example by spatial heating the number of cycles to failure was 12000 compared to 8500 cycles by local heating.

Conclusion

Soldering is an important part of today's microelectronics, and it is used almost everywhere. However, creating of solder joint is very complicated process; therefore, it is good to know how to do it. Solder joint reliability and assembly integrity depend not only on materials and components, but also on the process arrangement that puts all them together. In more there are other factors influencing results of solder process, especially these creating in the design phase period. This paper shows some parts of this soldering area which can assist to get soldering process under control.

Acknowledgment

This research is supported by the Czech Ministry of Education in the frame of Research Plan MSM 0021630503 MIKROSYN „New Trends in Microelectronic Systems and Nanotechnologies" and Grant project GACR 102/09/1701 "Investigation of new principles for improvement of lead-free solder joint reliability".

References

[1] ELFNET, European Lead Free Soldering NETwork , http://www.europeanleadfree.net/

[2] iNEMI, The International Electronics Manufacturing Initiative, http://www.inemi.org/

[3] SZENDIUCH, I., STARÝ, J. Some New Results from Investigation of Lead-free Solders Application In Proceedings of IMAPS Nordic 2004 Annual Conference. Helsingor, Denmark , 2004, s. 49 - 55,

[4] BULVA, J., NOVOTNÝ, M., SZENDIUCH,I: Investigation of Lead-Free Solder Joints Reliability by Thermal Modelling In 15th European Microelectronics and Packaging Conference & Exhibition. Brugge, Belgium: IMAPS , BENELUX, 2005, s. 546 – 552

[5] DUSEK,M., SZENDIUCH, I. Investigation of Creep and Stress Relaxation of Lead-free Solder Joints 37th IMAPS Symposium on Microelectronics. Long Beach: USA, 2004

[6] SMT, Surface Mount Technology Magazine published by PennWell Corporation, www.smtmag.com

[7] GLOBAL SMT&Packaging, The global assembly journal for SMY and Advanced Packaging Professional

Surface-Enhanced Copper Bonding Wire for LSI and Its Bond Reliability under Humid Environment

Tomohiro Uno [a], Keiichi Kimura [a], Takashi Yamada [b]

[a] Advanced Materials & Technical Research Laboratories, Nippon Steel Corporation, 293-8511, JAPAN

[b] Nippon Micrometal Corporation, 158-1, Sayamagahara, Iruma-city, Saitama, 358-0032, JAPAN

E-mail: uno.tomohiro@nsc.co.jp, Phone:+81-439-80-2932, FAX:+81-439-80-2746

Abstract

There is growing interest in Cu wire bonding for LSI interconnection due to cost savings and better electrical and mechanical properties. Cu bonding wires, in general, are severely limited in their use compared to Au wires; such as wire oxidation, lower bondability, forming gas of $N_2+5\%H_2$, and lower reliability. It is difficult for conventional bare Cu wires to achieve the target of LSI application. A surface-enhanced Cu wire (EX1) has been developed. It is a Pd-coated Cu wire and has many advantages compared to bare Cu wires. Stitch strength was much better under fresh conditions and maintained without any deterioration after being stored in air for a prolonged period of time. EX1 had a lifetime of over 90 days in air, although it was 7days for the bare Cu wire. Spherical balls were formed with pure N_2 (hydrogen-free), whereas the bare Cu produced off-center balls. Cost-effective and secure gas, pure N_2 was only available for EX1. The reliability for Cu wire bonding under humid environment was investigated in pressure cooker test (PCT). The lifetime for EX1 and the bare Cu was over 800h and 250h, respectively. Humidity reliability was significantly greater for EX1. Continuous cracking was formed at the bond interface for the bare Cu wire. Corrosion-induced deterioration would be the root cause of failure for bare Cu wires in PCT. EX1 improves the bond reliability by controlling diffusion at the bond interface. The excellent performance of Pd-coated Cu wire, EX1 is comparable with Au wires and suitable for LSI packaging.

Key words: Copper bonding wire, stitch bond, ball formation, humidity reliability, PCT

1. Introduction

In semiconductor packaging, wire bonding is the main technology for making electrical connections between chips and substrate. Gold is the most popular interconnection material in wire bonding. Gold prices have risen significantly over the last few years so that demand for lower material cost has increased. There is growing interest in Cu wire bonding for LSI interconnection due to cost savings and better electrical and mechanical properties [1-3].

Cu wires with large diameters (>38μm) have been used for discrete and low-I/O power devices for many years. However, there are difficulties when implementing Cu wires in advanced LSI packaging such as fine pitch bonding [1,2]. A lower yield than the gold wire bonding process and long-term reliability are most crucial. A decrease in yield would eventually offset the cost saving for Cu wires. The yield to use Cu wire bonding must be as good as or better than that for the current gold wire bonding process. In addition, the qualification of high volume production requires a variety of long-term reliability tests.

Conventional Cu bonding wires, in general, are severely limited in their use compared to Au wires, for example: (1) oxidation on the wire surface, (2) storage lifetime before bonding, (3) stitch bondability, (4) the running cost issue due to gas formation, and (5) long-term reliability in a humid environment. These inherent problems limit the usage of thin Cu wires for LSI packaging.

(1) The major reason for the hindrance in the usage of Cu wire is that Cu is readily oxidized when exposed to air. Cu oxide deteriorates bondability at first bond (ball bond) and second bond (stitch bond). This lowers the manufacturability for Cu wires.

(2) Oxidation progresses further during the wire bonding process, leading to non-stick-on-lead (NSOL) failure. In order to maintain a fresh surface, the winding length of the Cu wire is limited to less than 500m, which is much shorter than that for Au wires. As the life expectancy of Cu wire is limited before and after unpacking, the issue of scrap occurs when the wire is unexpectedly stored for longer periods of time. Prolonging the life expectancy of Cu wires is strongly demanded from the manufacturing process.

978-1-4244-4722-0/09 $25.00
© 2009 IMAPS-ITALY

(3) Stitch bonding is a great challenge for Cu wire bonding. Advanced machine conditions and capillary design should be specifically optimized for Cu wires [2]. The optimum can sometimes bring about good results at the development stage or in small-volume pre-production. There are hurdles to overcome when shifting to high-volume production. These are mainly due to narrower process parameter windows at stitch bonding. It is necessary to develop a high performance Cu wire to improve the stitch bonding.

(4) Formation of spherical and high quality free-air-balls (FABs) is a critical requirement for Cu wires. A forming gas is necessary to provide an oxygen-free environment in ball formation. $N_2+5\%H_2$ mixture is a standard forming gas for Cu wire. H_2 has the advantage that the FAB shape is readily stabilized [1,4]. $N_2+5\%H_2$ also has several disadvantages: higher running cost, initial investment for setting up specified pipes, and safety issues concerning flammable gas. Pure N_2 gas is preferable since such a hydrogen-free gas is cost-saving and secure in manufacturing. Some papers have reported that pure N_2 is not allowed for Cu wires [1,4,5]. If pure N_2 is available, it would be significantly beneficial and provide a good manufacturing process control for packaging companies.

(5) The bond reliability for Cu wire bonding has been reported in many papers [3,6-8]. They are mostly focused on thermal reliability under a dry atmosphere in high temperature storage (HTS) testing. It has been pointed out that Cu/Al bonding produced better results in HTS testing because of the slower intermetallic (IMC) growth rate compared to Au/Al bonds. Recently greater focus has been placed on humidity reliability for Cu wire bonding in pressure cooker test (PCT) or highly accelerated temperature and humidity stress test (HAST). Cu wire's performance in PCT and HAST testing and its failure mechanism have not yet been disclosed.

Cu wire material has been developed to improve the bonding properties. It has been reported that soft Cu wire improves the bondability at first and second bonds [2]. Softness mainly depends on the purity of the Cu material. As Cu purity increases from 4N to 6N, its cost increases dramatically. The softening effect would be limited even with higher purity [9]. Current commercial Cu wires are mostly 4N in purity to create a balance for whole bonding performances.

The conventional products of the Cu bonding wire, so-called "bare Cu wire" were covered by copper itself at the wire surface. Preventing surface oxidation and improving bondability are trade-offs for Cu wire bonding. It is difficult for bare Cu wires to achieve the higher targets of LSI application.

The industry demands an advanced Cu bonding wire for LSI packaging. Metal-coated bonding wires, for example, are expected to enhance the surface performance [10]. But they have not been utilized so far because it is difficult for them to have the appropriate high quality and stable performance to meet all the requirements.

We developed a surface-enhanced copper bonding wire, namely EX1 wire, which is a Pd-coated Cu wire. Desirable characteristics of Pd-coating are oxidation-free, good adhesion to Cu wire, and good bondability. The high quality of the Pd-coated Cu wire allows for consistent productivity and high yield.

In this study, the basic bonding performance of the Pd-coated Cu wire and the bare Cu wire is compared. Stitch bondability, FAB formation with pure N_2 and humidity reliability at Cu/Al bonds in PCT aging are investigated.

2. Experimental

Two Cu bonding wires were employed: a commercial bare Cu wire and a surface-enhanced Cu wire. The bare Cu wire had no metal-coating and its purity was 4N Cu. The surface-enhanced Cu wire was a Pd-coated Cu wire, which is referred to as EX1 wire in this paper. The core part of EX1 was the same as the bare Cu wire. The wire diameter was 25 μm. Table 1 shows the basic properties of the two wires used. The mechanical and electrical properties of EX1 wire were similar to those of the bare Cu wire. Copper is easily oxidized in air. The wires were stored in air up to 90days before bonding to evaluate the lifetime of the wires.

An automated bonding machine Eagle 60AP (ASM Co., Ltd.) was employed and equipped with a "copper kit" to supply forming gas so as to shield the entire area of the wire tip and torch electrode during FAB formation. Two types of forming gases were used, pure N_2 and $N_2+5\%H_2$. $N_2+5\%H_2$ is popular as a standard forming gas for Cu wire bonding [1]. The N_2 gas was a commercially standard nitrogen gas with 4N purity. The gas flow rate was 0.4 to 0.8 liter/min.

The first bonding of FAB was conducted on pure aluminum pads 1μm thick. The second bonding, stitch bonding, was conducted on silver-plated lead frames. The testing package was 208-pin QFP and the bonding temperature was 200°C. The size of FAB was around 50μm. Bonded ball size was around 65μm on average. Bonding conditions such as applied pressure, ultrasonic vibration and

Table 1 Properties of copper bonding wires used.

Cu Wire	EX1	bare Cu
	Pd-coated Cu	(no-coted)
Breaking load /mN	112	107
Elongation /%	11.8	12.1
Vickers hardness of wire	55	54
Electrical resistivity /$10^{-8}\Omega$m	1.9	1.9

978-1-4244-4722-0/09 $25.00
© 2009 IMAPS-ITALY

touching speed were optimized. The capillary was of a commercial standard type. The bonded samples were encapsulated with two kinds of molding resin such as conventional resin including Br and Green resin without Br for reliability testing.

PCT aging was performed for reliability testing. Test conditions were 121°C and 100% relative humidity (RH). The decapping process to remove the molding resin after PCT aging was conducted with a mixture of fuming nitric acid and fuming sulfide acid followed by acetone washes. The bond strength was measured by wire pull testing and ball shear testing with using Daze4000.

Cross-sections of ball bonds after PCT aging were polished and observed by optical microscopy, SEM, and TEM. Field emission TEM (FE-TEM) used was HF2000 (Hitachi corporation). IMC phases and some products at the bond interface were analyzed by Auger electron spectroscopy (AES) and electron diffraction and electron dispersive spectroscopy (EDS) with FE-TEM.

3. Results

3.1 Stitch bondability

To utilize a copper wire in high volume production, the issue of improving wire stitch bondability must be resolved. Cu wire, in general, has disadvantages in lower stitch bond strength and lifetime before bonding.

Fig.1 (a) shows the stitch pull strength after various storage durations at normal temperature (25°C). Wires were stored without packing in air before bonding in a clean room. When comparing initial strength at time zero, EX1 produced greater strength than the bare Cu. This indicates that the bare Cu wire had poor bond strength even under fresh conditions. Surprisingly, EX1 maintained high pull strength up to 90 days in air without any deterioration. On the contrary, the pull strength of the bare Cu declined significantly after several days before bonding. In addition bonding problems occurred with the bare Cu like non-stick-on-lead (NSOL), which ruins the product yield. The NSOL failure ratio for the bare Cu exceeded 0.1% after 7days, not allowing for successful bonding. The decline in stitch bond strength and NSOL failure were all due to the oxidation progressing at the Cu surface.

Fig.1 (b) shows accelerated testing results for stitch pull strength after stored without packing at 50°C in an oven. With bare Cu, the pull strength declined drastically after 5 days and NSOL failure

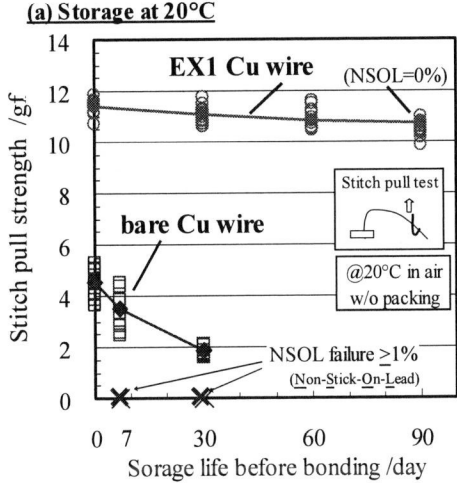

(a) Storage at 20°C

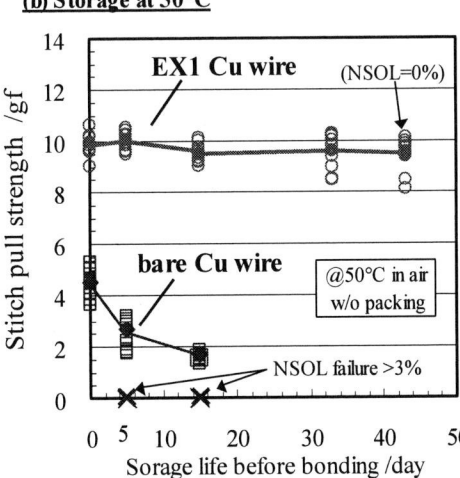

(b) Storage at 50°C

Fig.1 Stitch pull strength in duration of storage in air before bonding: stored at (a)25°C, (b)50°C.

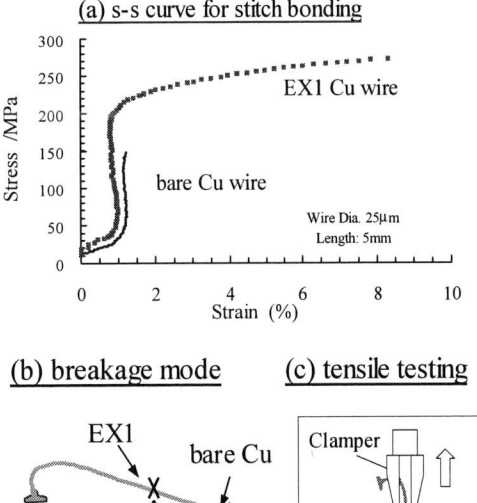

Fig.2 Tensile test for stitch bonding of Cu wires.

978-1-4244-4722-0/09 $25.00
© 2009 IMAPS-ITALY

occurred too much. On the other hand EX1 maintained high stitch strength even after 40 days. EX1 can prevent oxidation even at elevated temperature.

Tensile properties for stitch bonding are crucial to predict the reliability in TCT (temperature cycle test) and solder reflow testing. Neither pull testing nor wire peel testing can evaluate the strain at stitch bonds. Tensile test for stitch bonding was performed to evaluate the stress-strain (s-s) curve of Cu wires. The measurement was shown in Fig.2 (c). A bonded wire was clamped and pulled along in the wire direction. Constant angle between a stitched wire and inner lead should be maintained to evaluate quantitatively the strain. Fig.2 (a) shows quite different s-s curves of Cu wires. EX1 had extremely higher strain by 8%, although less than 1% for bare Cu. Breakage mode was also different. Breakage was occurred in the wire itself for EX1, while at stitch bond for bare Cu. EX1 had apparently great strain. These results indicate that EX1 would have good reliability in TCT and reflow testing.

It was confirmed that EX1 had significantly greater stitch bond performance in terms of bond strength, tensile strain, and prolonged lifetime, which were comparable to those of Au wires. The Pd coating layer on the Cu wire is effective in protecting the wire surface from oxidation and improving bond strength, which would lead to robust and wider process windows. Fig.3 shows a TEM image of the stitch bond interface between EX1 wire and Ag plating. The Pd layer remained in the entire interface as an extended thin layer. No void or cracking were observed. This Pd thin layer plays an important role in enhancing the adhesion of Cu and Ag plating.

3.2 FAB formation and ball bonding

Fig.4 shows FAB formation by comparing pure N_2 and $N_2+5\%H_2$. When using $N_2+5\%H_2$, both EX1 (c) and the bare Cu (d) produced normal FABs whose shape was spherical. With pure N_2, EX1 (a)

produced a spherical and symmetrical FAB, whereas the bare Cu (b) produced a tilted and off-center FAB. The tilted FAB (b) results in a center-deviated ball bond as shown in Fig.5, which is not acceptable for LSI packaging. An examination of the direction of the center-deviated deformation for the bare Cu bonds with $N_2+5\%H_2$ revealed that bonded balls

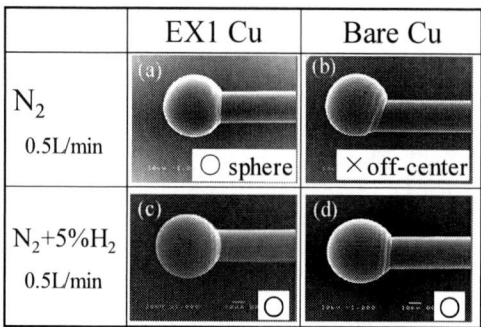

Fig.4 FAB formation with two types of forming gases: N_2, $N_2+5\% H_2$.

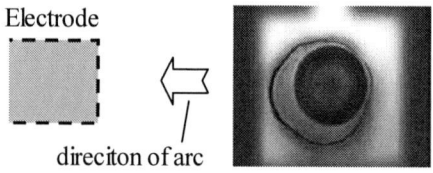

Fig.5 Bond shape failure (bare Cu with pure N_2).

Table 2 Failure ratio of bonded ball shape (200pin).

Gas Flow	N_2		$N_2+5\%H_2$	
(L/min)	EX1	bare Cu	EX1	bare Cu
0.8	0%	0%	0%	0%
0.6	0%	0.5%	0%	0%
0.5	0%	31%	0%	0%
0.4	0%	98%	0%	0%

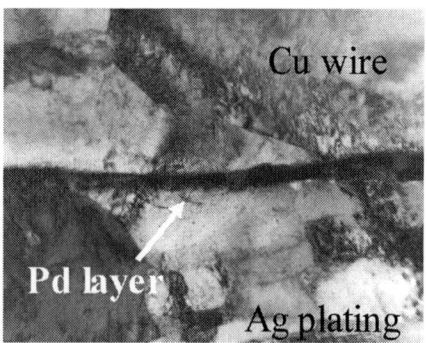

Fig.3 TEM image at stitch bond interface of EX1 wire / Ag plating.

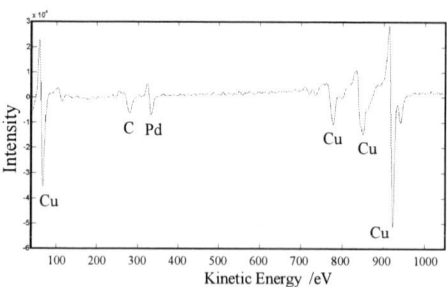

Fig.6 AES analysis at surface of FAB (EX1 wire).

978-1-4244-4722-0/09 $25.00
© 2009 IMAPS-ITALY

often tend to deform toward the electrode. Table 2 shows failure ratio of the center-deviated bonds with various gas flow rates, ranging from 0.4 to 0.8 l/min. Only a combination of the bare Cu and pure N_2 caused several center-deviated bonds and its failure ratio increased as the flow rate decreased. EX1 produced round bonded balls even as the flow rate was decreasing to 0.4 l/min with both N_2 and $N_2+5\%H_2$. The bond shape for EX1 was sufficiently stable even with minimum usage of pure N_2.

Pure N_2 is available only for EX1, whereas the bare Cu wires require $N_2+5\%H_2$. A combination of EX1 and pure N_2 is beneficial in total running cost and is readily controllable. H_2 in the forming gas is essential to the conventional bare Cu wires and the effect of H_2 is discussed later.

The FAB surface should be clean to prevent the oxidation during FAB formation. Fig.6 shows AES analysis at the surface of FAB for EX1 with pure N_2. It revealed that the FAB was prevented from oxidation even with hydrogen-free gas. It has been expected that hydrogen will serve as a reducing agent in gas formation [1]. The inclusion of hydrogen is not essential for preventing the oxidation of copper. How to create a perfect shield surrounding is more crucial to eliminate oxygen in atmosphere. Forming gas should cover the entire arc-spreading area from the wire tip to the electrode by adjusting the copper-kit.

Table 3 shows FAB sizes. It was found that wire materials and forming gas affected the FAB size. In terms of the forming gas, the FAB size for pure N_2 was slightly smaller than that for $N_2+5\%H_2$ by about 1μm. In terms of wire material with $N_2+5\%H_2$, EX1 produced slightly larger FABs than the bare Cu. The same size of FABs can be obtained by adjusting electronic flame-off (EFO) parameters

Table 3 FAB size of Cu wires (μm).

Forming gas	EX1	bare Cu
N_2	48.7	48.4
$N_2+5\%H_2$	51.5	50.9

Table 4 Ball bond size and shear strength.

Cu wire	EX1	bare Cu
Forming gas	N_2	$N_2+5\%H_2$
Ball bond size /μm	64.7	64.2
Shera strength /mN	382	375
Shera strength per area /MPa	116	115

Table 5 Vickers hardness of FAB (load:3gf).

Forming gas	EX1 Cu		bare Cu
	N_2	$N_2+5\%H_2$	$N_2+5\%II_2$
Vickers, Hv	65 +/- 3	68 +/- 3	67 +/- 3

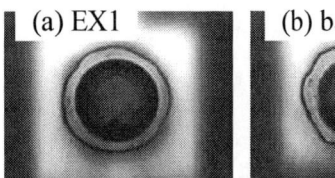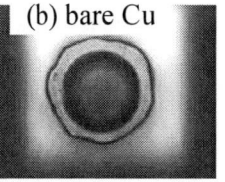

Fig.7 Bonded ball shape: (a) EX1 with N_2, (b) bare Cu with $N_2+5\%$ H_2.

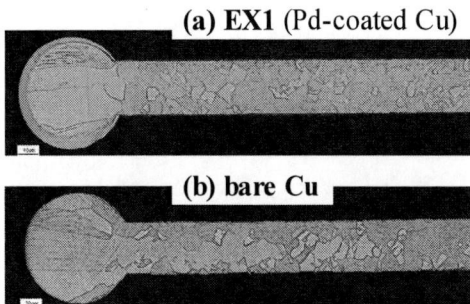

Fig.8 Cross sections of FAB and HAZ: (a) EX1 with N_2, (b) bare Cu with $N_2+5\%$ H_2.

such as EFO current and discharge time. The difference in FAB size is attributable to the heat energy provided during arc discharge, which is discussed later.

Table 4 shows the bonded ball size and shear strength. Bonded ball size and shear strength of EX1 were the same as those of bare Cu. Ball deformability was comparable between EX1 and bare Cu. N_2 gas as well as Pd coating on the wire would not exert any influence on the bond strength.

Bonded ball shape is shown in Fig.7. The bond shape of EX1 was rounder and more stable than the bare Cu, which was sufficient for fine pitch bonding in LSI packaging. Ball deformation is usually dominated by the grain structure in FABs. Table 4 shows the Vickers hardness in the cross-sections of FABs. There was no difference between EX1 with N_2 and the bare Cu with $N_2+5\%H_2$, The influence of forming gas and Pd-coating on FAB hardness was negligible. Fig.8 shows optical views of cross-sections of two wires. Both wires exhibit similar grain size and microstructure in FAB and the heat-affected-zone (HAZ). This implies that the Cu core part of EX1 wire dominates the grain structure of FAB and HAZ rather than the Pd outer layer does.

The Pd outer layer at the wire surface would melt and mix with the Cu core during arc discharge. Cu solid solution including Pd was formed but not distributed uniformly inside FAB. Fig.8 (a) shows that FAB for EX1 had a layer of the Cu-Pd solid solution at the periphery of FAB, where the color differs. Pd concentration in this part was higher compared to the core area within FAB. This Cu-Pd

978-1-4244-4722-0/09 $25.00
© 2009 IMAPS-ITALY

solid solution area would contribute to a rounder and more stable bond shape of EX1, as shown in Fig.7.

3.3 Bond reliability in PCT

Long-term reliability for Cu wire bonding is a major concern in replacing Au wires. Cu wire bonds are expected to have better thermal reliability since the growth rate of Cu-Al IMC is slower [3,6,7]. The humidity reliability for Cu wire bonding has not yet been disclosed. PCT aging was performed in this study to evaluate humidity reliability for EX1 wire

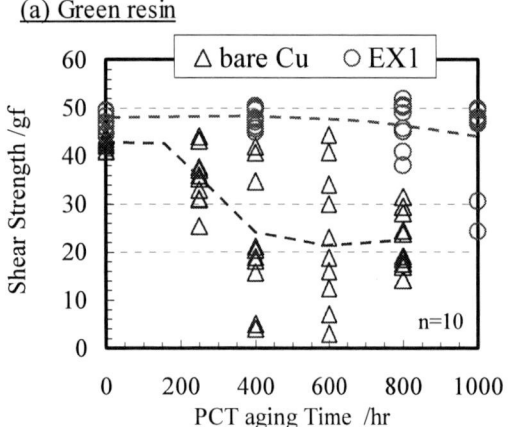

(a) Green resin

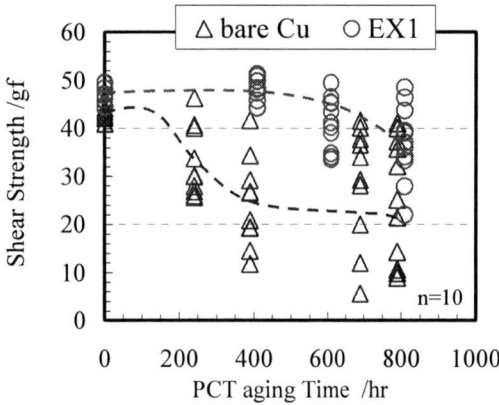

(b) Conventional resin

Fig.9 Shear strength after PCT (121°C, 100% RH): (a)Green resin, (b)conventional resin.

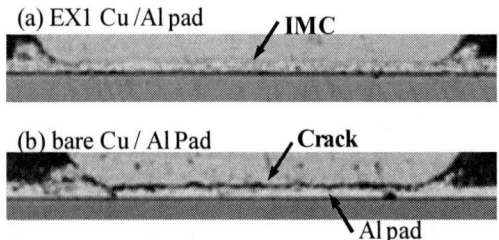

(a) EX1 Cu /Al pad   IMC

(b) bare Cu / Al Pad   Crack
Al pad

Fig.10 Cross sections of ball bonds after PCT for 400h with Green resin: (a) EX1, (b) bare Cu.

bonding. Two kinds of molding resins were employed such as conventional resin including Br and Green resin without Br.

Fig.9 shows the shear strength in the bonds of Cu wires and Al pads after PCT aging. It was surprising to see that wire materials dominated the bond reliability under the humid environment. With Green resin in Fig. 9 (a), shear strength for the bare Cu started to decrease from 40gf to about 35gf at 250h and reached less than 10gf at 400h. For EX1, the shear strength maintained steadily over 600h, starting to decline slightly from 800h to 1000h. If the acceptable criterion is around 25gf (half of initial value), the lifetime for EX1 and the bare Cu was over 800h and 250h, respectively. With conventional resin in Fig. 9(b), the result was the same as that of the Green one. The lifetime for EX1 and the bare Cu was 800h and 250h, respectively. PCT aging revealed that the humidity reliability for EX1 bonds was significantly better than that of the bare Cu, irrespective of the molding resin.

Fig.10 shows the optical views of cross-sections at the bond interface after PCT aging for 400h with Green resin. Failure analysis was focused on Green resin in this paper. Thin IMC was formed at the bond interface. No crack or void were observed for EX1. For the bare Cu, a crack was formed and propagated across the entire bond interface. This cracking appears to be the cause of failure for the bare Cu in PCT aging. The cracking was apt to initiate from the ball periphery and progress inwards to the ball center. Shear data for the bare Cu were broadened in Fig.9 (a), which seems to depend on the degree of crack propagation.

Cross sections of the bonds after PCT aging were analyzed to examine IMCs and the products present at the interface. Fig.11(a) shows the TEM image at the bond of the bare Cu after PCT for 400h, when the bond strength considerably declined. It was revealed that the interfacial structure after failure in PCT was " Cu / crack / Al oxide / IMC / Al." Continuous cracking was interposed between Cu and Al oxide. There was no IMC present adjacent to the crack. The Al oxide layer was 50-100nm in width and mainly composed of nano-crystal structure of Al_2O_3. It was characteristic that the Al oxide layer also included Cl by around 1mol% and small Cu particles (fcc crystal structure). Such a complex of Al_2O_3, Cu, and Cl could not be formed by mere interdiffusion of Cu and Al. Electron diffraction revealed that the IMC above the Al pad was identified as the CuAl phase. This IMC was interposed between the Al oxide and Al pad. Hence the CuAl phase obtained after PCT would not be an original phase at the early stage of aging, but transformed from another IMC phase after cracking occurred.

Feature products present were Al oxide, chlorine and cracking, neither of which existed in the original bond interface of Cu/Al. These analyses

978-1-4244-4722-0/09 $25.00
© 2009 IMAPS-ITALY

(a) bare Cu wire

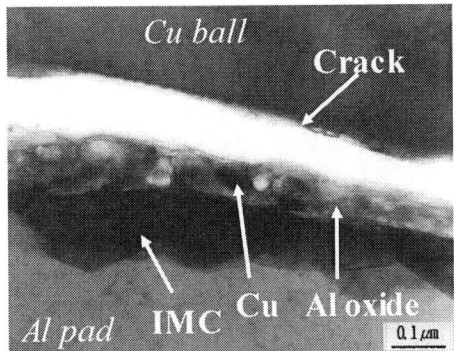

(b) EX1 Cu wire

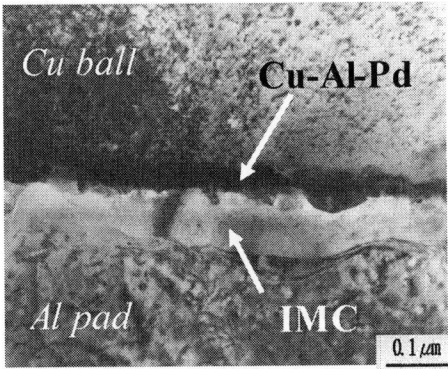

Fig.11 TEM images at bond interface after PCT aging for 400h with Green resin: (a) bare Cu , (b) EX1.

imply the assumption that corrosion would occur at Cu/Al bond during PCT aging. The failure mechanism is discussed later.

Fig.12 shows AES line analysis across the bond interface of EX1 after PCT. Pd concentration was higher at the interface. The Pd-enriched layers were divided into two parts. One was the Cu-Pd solid solution (Pd 3-5mol%) inside Cu ball close to the bond interface. The other was a Cu-Al-Pd compound (Pd: 2-3mol%).

Fig.11 (b) shows the TEM image at the bond interface of EX1 after PCT for 400h. A white and a dark contrast layer were present at the interface. The white layer was Cu-Al IMC, which was mostly identified as CuAl phase. The dark layer was the Cu-Al-Pd compound. Its concentrations of Pd and Cu were 1.5 to 2.5mol% and 58 to 75mol%, respectively. The dark layer was partially estimated to be Cu_3Al_2 by electrical diffraction. In addition, EDS analysis revealed that Cu-Pd solid solution including around 4mol% Pd was formed in the Cu ball closed to the bond interface. These Pd distribution and concentration agree well with the AES data shown in Fig. 12.

The result was that the structure at the interface of EX1 bonding would be "Cu / Cu-Pd solid solution / Cu-Al-Pd compound / CuAl / Al."

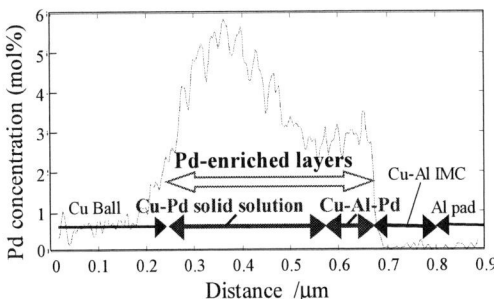

Fig.12 AES line analysis at bond of EX1/Al after PCT aging for 400h.

There was neither Al_2O_3 nor Cl found, both of which are characteristic of the corrosion products for the bare Cu wire bonds. Furthermore, no crack or void were observed for EX1, which results in high bond reliability.

4. Discussion

4.1 Cu ball formation mechanism

It is of great interest to know how hydrogen affects FAB formation. H_2 gas present in $N_2+5\%H_2$ can produce larger FAB size than pure N_2, as shown in Table 3. In addition, H_2 gas in $N_2+5\%H_2$ tends to produce spherical FABs, although off-center FABs with N_2 for bare Cu wire, as shown in Fig. 4 (b). In order to understand the effect of H_2 gas on these phenomena, two characteristics of arc discharge should be considered such as heating power and arc constriction. The larger FAB size would be attributed to higher heating power provided by gas formation around the wire tip. The constricting arc discharge helps to produce spherical FABs.

The greater heating power of the H_2-shielded arc would be caused by the increased arc voltage and heat flux density. Fig.13 shows a comparison of the

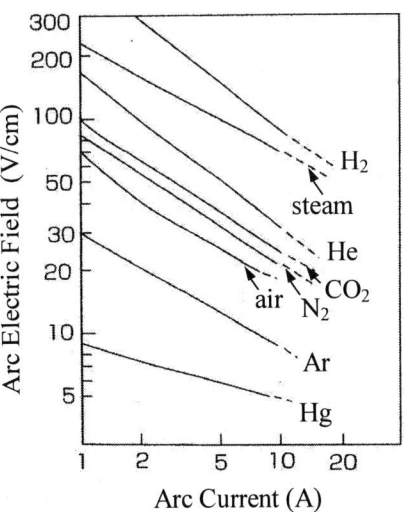

Fig.13 Effect of gases on arc electric field [13].

978-1-4244-4722-0/09 $25.00
© 2009 IMAPS-ITALY

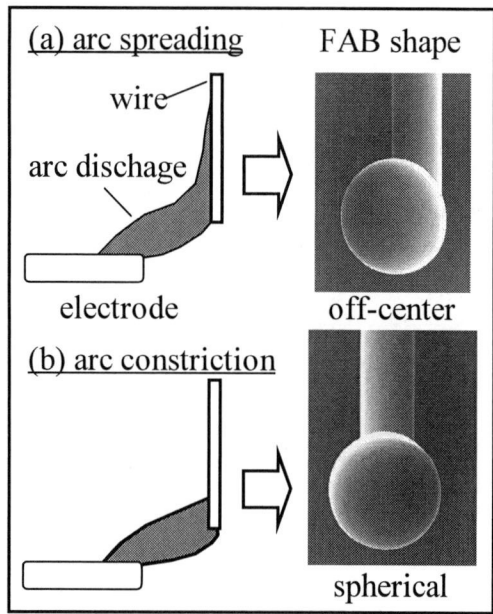

**Fig.14 Arc discharge behavior on the wire tip
(a) arc spreading, (b) arc constriction.**

arc electric field for various gases, corresponding to arc voltage [11]. It appears that arc voltage increases as its atomic weight of gas species becomes lighter. H_2 has the highest arc voltage among the gas species. As a consequence, H_2 in the forming gas is expected to provide increased heating energy to the Cu wire rather than pure N_2. This results in the larger FAB size for $N_2+5\%H_2$.

Another influence of H_2 during FAB formation is the arc constriction. The basic features of H_2 are lighter atomic weight and lower binding energy of H-H. H_2 gas tends to expand outward in the arc discharge compared to N_2 gas, which leads to greater cooling of H_2. Hence the temperature on the arc periphery decreases. Arc charge tends to spontaneously localize the arc path to the arc axis area so as to minimize the heat loss. The net effect of the cooling of H_2 is in fact the consequent arc constriction. This is known as the "thermal pinch effect" in welding technology [12]. The thermal pinch effect is considered to be significant with H_2 in the forming gas. With pure N_2, the pinch effect would not be so high that the arc tends to spread over for bare Cu wires.

Fig.14 illustrates the assumption of arc discharge close to the tip of the bonding wire during FAB formation: (a) arc spreading type representing the case of pure N_2, (b) constriction type representing $N_2+5\%H_2$. H_2 gas in $N_2+5\%H_2$ can constrict the arc discharge around the wire tip. The result is the spherical FAB formation. On the other hand, the arc constriction with pure N_2 is not so great that the arc tends to spread over the wire surface upward, which results in the center-deviated ball deformation.

The arc spreading would be directed at the electrode side of the wire tip. This agrees well with the fact that bonded balls tend to often deform toward the electrode with pure N_2, as shown in Fig.5. Arc distribution could be roughly estimated by electrostatic field and arc discharge parameters. The electric field around the wire tip prior to arc discharge was simulated in Fig. 15. Higher electric field was widespread at the electrode side of the wire. This deviation of electric field would enhance the arc spreading in Fig. 14 (a). It verifies that the FAB formation is dominated by the arc spreading.

Pd coating is expected to depress the arc spreading over the wire, which would be a similar performance to H_2 in the forming gases. Arc discharge usually depends on the surface of wires (anode material) rather than the core part of wires. It was also confirmed in other evaluations that Pd-coating was more effective in suppressing the failed FAB shape for pure N_2 than other metal-coating. Pd-coating would help the arc constricting instead of the thermal pinch effect for H_2. It would mainly be due to the Pd properties such as higher electrical resistivity, higher melting temperature, and higher ionization potential. Furthermore, Pd-coated wire produced larger FAB than the bare Cu, as shown in Table 3. This is because of higher heating energy or higher heat flux density by arc constriction in the procedure of melting the Pd coat. In addition, wire surface conditions such as oxide-free, microstructure, and roughness were also of great importance for stable arc constriction. These surface properties of Pd-coating should be optimized for consistent productivity and high quality of FAB formation.

4.2 Failure mechanism for Cu-Al bonds in PCT

Cu wires are believed to produce good reliability in regular isothermal aging of HTS up to 200°C [3,6]. There have been few reports on the bond failure for Cu wires in reliability testing.

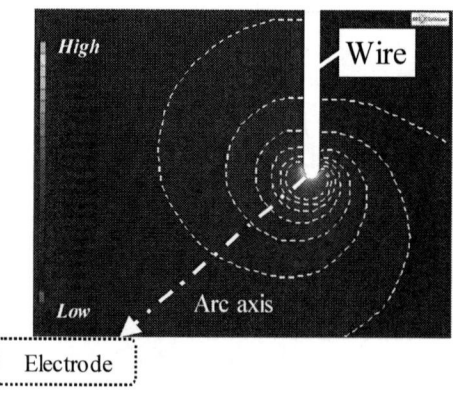

**Fig.15 Simulation of electric field around
wire in arc discharge (bare Cu wire).**

978-1-4244-4722-0/09 $25.00
© 2009 IMAPS-ITALY

Cracking occurred in HTS aging at 250°C [8]. With the unencapsulated test chip, bond strength at Cu/Al bonding declined in PCT aging and CuO is detected at the outer bond interface [8]. The failure mechanism of encapsulated Cu-Al bonds in a humid environment such as PCT or HAST testing has not been clarified yet. Fig.16 shows schematic diagrams of Cu/Al bonds after PCT aging for 400h in a comparison between bare Cu wire and EX1 wire, which is based on the findings in this study.

TEM and AES analysis mentioned above revealed that feature products after PCT aging were Al oxides including Cl and metal Cu. They were not present at the original bond interface before aging. This suggests that corrosion of Cu-Al IMCs occurred during PCT aging. The corrosion would be a kind of chemical reaction of IMCs and Cl. A small amount of Cl element is a typical impurity in epoxy molding resin. High moisture could facilitate the transfer of the Cl ion toward the bond inside the molding resin. The reaction between Al in IMCs and Cl results in producing Al oxide and Cu precipitation from Cu-Al IMCs. There is a possibility that Al chloride like $AlCl_3$ is formed as an intermediate in the reaction and $AlCl_3$ will be immediately transformed into Al oxide. As Al halide is considered chemically unstable, it can be readily hydrolyzed by moisture and consequently changed into Al oxide.

Similar corrosion behavior is involved in the reliability issue for Au-Al bond in HTS testing. Br included as flame retardant in molding resin attacks Al atoms in Au-Al IMC [14]. The major reaction product for Au/Al bond is a mixture of Al oxide and metal Au (fcc crystal) [15]. In addition, crack formation at Au/Al bonds causes bond deterioration. These products and cracks for Au/Al bonds agree well with those for Cu/Al bonds in this study. This suggests that the corrosion would be a cause of cracking at Cu-Al bonds after PCT aging. Further investigation is necessary to clarify the mechanisms for corrosion procedures and crack formation.

Improved PCT reliability of EX1 is due to the Pd-enriched layers at the bond interface. The Pd-enriched layers were divided into two parts such as Cu-Al-Pd compound and Cu-Pd solid solution inside the Cu ball. This double structure of Pd-enriched layers is crucial to the bond reliability. FAB itself has part of the Cu-Pd solid solution at the surface of FAB, as shown in Fig.8 (a), before bonding. This Cu-Pd solid solution formed during FAB formation may remain at the Cu/Al interface after bonding and provide Pd atoms for the Cu-Al-Pd compound. This specific diffusion system could help the growth of the Cu-Al-Pd compound even at a low temperature of 121°C in PCT aging.

There are several possibilities for the Pd-enriched layers suppressing the corrosion for Cu wire bonding. One is to protect the Cu/Al bond interface from corrosive ions. Another is to control

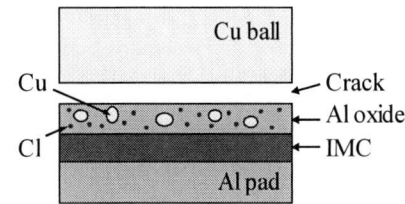

(a) bare Cu / Al pad

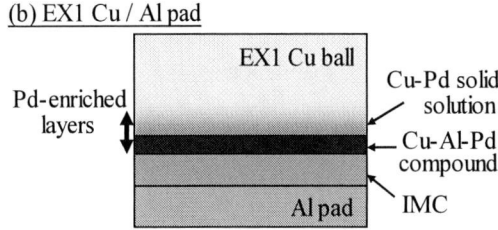

(b) EX1 Cu / Al pad

Fig.16 Schematic diagrams of Cu/Al bonds (a) bare Cu wire, (b) EX1 Cu wire (after PCT aging for 400h with Green resin).

the interdiffusion of Cu and Al atoms, which leads to a change in the growth of IMC phase. In the latter case, the Pd-enriched layers can serve as a diffusion barrier and retard corrosive IMC.

A predominant IMC at the bonds of EX1 after PCT aging was identified as CuAl phase, which was different from the phase for conventional Cu wires reported in previous papers. It is reported that $CuAl_2$ is the phase formed after aging at 175°C for 2h-200h [15]. Other papers report that Cu_9Al_4 and $CuAl_2$ are identified after HTS aging at 250°C [8]. On the other hand, the IMC phase after PCT has not yet been reported. Such phases as $CuAl_2$ and Al_2O_3 were not present at the bonds of EX1 after PCT. This comparison in terms of IMC formation indicates that Pd-enriched layers could dominate the diffusion behaviors and promote the IMC growth different from that for bare Cu bonds. The specific IMC growth for EX1 would be helpful to improve the PCT reliability.

5. Conclusions

Surface-enhanced Cu bonding wire, namely EX1 wire, was developed. EX1 is a Pd-coated Cu wire and has several advantages compared to conventional bare Cu wires. The basic bonding performance obtained in this study is listed in Table 6. Excellent bonding performance and reliability of EX1 is suitable for LSI application and sufficient to replace Au wires. The results obtained are summarized as follows:
1. Stitch strength was significantly better than that of bare Cu wire even under fresh conditions. Tensile strain for stitch bonding of EX1 wire was much greater than that of the bare Cu wire.

978-1-4244-4722-0/09 $25.00
© 2009 IMAPS-ITALY

2. Stitch bondability was maintained significantly better when stored in air before bonding. Lifetime for EX1 at normal temperature was over 90 days in air before bonding, although 7days for bare Cu wires. Pd coating is effective in suppressing oxidation at the wire surface and prolonging lifetime.

3. Spherical FABs were formed with pure N_2 (hydrogen-free), whereas the bare Cu wire produced an off-center ball shape. Cost-effective and secure gas, pure N_2 is only available for EX1.

4. Ball bond shape was better than that of the bare Cu wire. Bond strength was sufficiently high.

5. High humidity reliability was significantly greater in PCT aging. The lifetime for EX1 and bare Cu was over 800h and 250h, respectively.

6. Continuous cracking was formed at the bond interface for bare Cu wire, although there was no cracking for EX1. The cracking was a cause of failure for Cu/Al bonds in PCT aging.

7. There were mixture products of Al oxide, Cl, and metal Cu formed at the bond interface for bare Cu wires in PCT aging. Corrosion-induced deterioration would be the root cause of failure for bare Cu wires.

8. Pd-enriched layers were formed at the ball bonds for EX1. The Pd-enriched layers improve the bond reliability under humid environment by controlling diffusion and IMC formation at the bond interface.

References

[1] Singh, J. On, L. Levine, "Enhancing fine pitch, high I/O devices with copper ball bonding", Proc 55th Electronic Components and Technology Conf, (2005), pp. 843-847.

[2] J. Kam, H. Meng, D. Stephan, D. Calipito, K. Dittmer, L. Jamin,, G. Mui, "Materials characteristics of soft copper wires designed for advance application", Proc SEMICON Singapore, (2007), pp. 49-54.

[3] Saraswati, E. Tehint, D. Stephan, H. Goh, E. Pasamanero, D. Calpito, F. Wulff, C. Breach, "High temperature Storage(HTS) performance of copper ball bonding wires", Electronic Packaging Technology Conf, Singapore, (2005), pp. 602-607.

[4] C. Hang, C. Wang, M. Shi, S. Wu, H. Wang, "Study of copper free air ball in thermosonic copper ball bonding", 6th International Conf on Electronic Packaging Technology, (2005), pp. 414-418.

[5] H. Ho, J. Tan, Y. Tan; B. Toh; P. Xavier, "Modeling energy transfer to copper wire for bonding in an inert environment", Proc 7th Electronics Packaging Technology Conf EPTC, (2005), pp. 292-297.

[6] F. Wulff, C. Breach, D. Stephan, Saraswati, K. Dittmer, M. Garnier, "Further characterization

Table 6 Bonding performances of Cu wires.

Cu Wire	EX1 wire	Ref. Cu wire
	Pd-coated Cu	bare Cu
Stitch strength (arb unit)	1.3 - 2	1
Lifetime in air	$\geq$ 90days	$\leq$ 7days
Winding length	$\geq$ 3000m	100 - 500m
PCT reliability	> 800h	250h
Forming gas	N_2	--
	$N_2+5\%H_2$	$N_2+5\%H_2$

of intermetallic growth in copper and gold ball bonds on aluminum metallization", Proc SEMICON Singapore, (2005), pp. 1-9.

[7] C. Hang, C. Wang, M. Mayer, Y. Tian, Y. Zhou, H. Wang, "Growth behavior of Cu/Al intermetallic compounds and cracks in copper ball bonds during isothermal aging", Microelectronics Reliability, 48(2008), pp. 416-424.

[8] C. Tan, A. Daud, M. Yarmo, "Corrosion study at Cu-Al interface in microelectronic packaging". Applied Surface Science, 191(2002), pp. 62-73.

[9] N. Srikanth, J. Premkumar, M. Sivakumar, Y. Wong, C. Vath, "Effect of wire purity on copper wire bonding". Proc 9th Electronics Packaging Technology Conf, (2007), pp. 755-759.

[10] S. Kaimori, T. Nonaka, A. Mizoguchi, "The development of Cu bonding wire with oxidation-resistant metal coating", IEEE Transactions on Advanced Packaging, 29(2) (2006), pp. 227-231.

[11] C. Suits, "High pressure arcs in common gases in free convection", Physical Review, 55(1931), pp. 561.

[12] P. Poritskii, "Thermal contraction of arc discharge in mixtures of inert gases: special features", High Temperature, 44(2006), pp. 326-335.

[13] M. Drozdov, G. Gur, Z. Atzmon, W. Kaplan, "Detailed investigation of ultrasonic Al-Cu wire-bond: Microstructural evolution during annealing", J Material Science, 43(2008), pp. 6038-6048.

[14] T. Uno, K. Tatsumi, "Thermal reliability in gold-aluminum bonds encapsulated in bi-phenyl epoxy resin", Microelectronics Reliability, 40(2000), pp. 145-153.

[15] T. Uno, K. Kimura, K. Tatsumi, "Thermal bond reliability of high reliability gold alloy wires for automotive ICs", Proc 2005 International Symposium on Microelectronics, (2005), pp. 557-565.

A Study of Thermal Performance for Chip-in-Substrate Package on Package

Tuan-Yu Hung, Ming-Chih Yew, Chan-Yen Chou., and Kuo-Ning Chiang

Advanced Microsystem Packaging and Nano-Mechanics Research Lab., Department of Power Mechanical Engineering, National Tsing Hua University, HsinChu, Taiwan 30013, R.O.C.

Phone: 886-3-5742925, Fax: 886-3-5745377 and E-mail: knchiang@pme.nthu.edu.tw

Abstract

The three-dimensional package (3D package) is one of the popular designs for high-density packages. The chip-in-substrate (CiS)-type structure is one of the popular manners in 3D package because of the resulting improvement in package-stacking ability. To study its thermal performance, the finite element (FE) analysis is applied in this study. The designed dummy solder bumps and relatively better power arrangement conditions are proposed to improve the thermal performance of the package. The dummy solder bumps under the chip can improve the efficiency of heat dissipation from the chip to the printed circuit board (PCB). Moreover, the highest power dissipation is suggested to be placed at the lower chip. Thus, the thermal performance of the CiS packaging technology can be further enhanced, and it is suitable for applications on high-power IC devices.

Key words: 3D package, thermal performance, CiS packaging technology, FE analysis

Introduction

Among the important factors for packaging technology are the intergral circuit (IC) package cost, the drop performance of the paclage, the reliability of the paclage, and the thermal performance of the package. Nowadays, the subject of thermal management for the package is consistently becoming critical because the density of the package and the I/O on each chip increase progressively. There has been a proliferation of studies on the subject. In 2000 an article was published by Kim et al. [1] that provides extensive discussions on the thermal performance of thin quad flat j-forming (TQFJ). Kim's study utilizes the FLOTHERM simulation and thermal measurement. The results show that the maximum junction temperature rises noticeably as the stacking number of packages increases. In 2003, Chen et al. [2] proposed an effective methodology that integrates an infrared (IR) thermpgraphy measurement and finite element (FE) model for thermal characterization of packages. The thin qual flat package (TQFP) is the test vehicle in Chen's study, and the melodology is benchmarked by a thermal test die measurement. Chen et al. [3] and Chang et al. [4] used the FE method to discuss the thermal performance of the flip chip-plastic ball gird array (FC-PBGA) and the quad flat non-leaded (QFN) package, respectively. Nowadays, computer-aided engineering (CAE) methodology is being utilized to improve the efficiency of packaging technology.

Because of the improvement of IC manufacturing process and the demand of system integration, 3D package is one of the popular designs for the high density packages. Moreover, the thermal management for high density packages has also become critical. In this study, a packaging technology wich retans the advantage of 3D package, chip-in-substrate (CiS) type structure, is purposed to develop the packaging capability of signal fan-out for the fin-pitched IC and the improvement of stacking ability for packaging. Besides, it also looks for the better thermal dissipation performance by designed dummy ball under the chip. Figure 1 shows the schematic structure of CiS packaging technology. In the CiS packaging technologe, the chp is placed in the substrate's designed opening after wafer dicing. The filler material is selected to fill the trench between chip and substrate and to provide the smooth surface for the redistribution line. The interconnecting through hole inside the substrate is defined toconect the signal between two sides of substrate. The solder bumps can be located on both the chip and the substrate surface, and the pitch of the chip is faned out. Furthermore, there are singnal pads connected by the interconnecting through hole in both sides of the substrate. Therefore, the packaging technology of package-on-package (PoP) can be achieved.

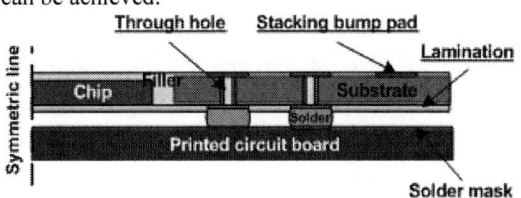

Figure 1: The schematic structure of CiS Packaging technology

978-1-4244-4722-0/09 $25.00
© 2009 IMAPS-ITALY

A number of studies [5] have investigated the thermo-mechanical behavior of designed packaging which uses the solft material around the chip. The studies indicate theat the thermo-mechanical behavior of the packaging is different from that of conventional wafer level package (WLP) because of the designed packaging structure. The soft filler and lamination material can offer a stress buffer layer for solder joints. Therefore, the solder joint reliability of the designed structure is prominent. Moreover, the accumulated stress/strain from the coefficient of thermal expansion (CTE) mismatch at rhe metal trance can be efficiently released through a proper layout technology. The purpose of this study is to investigate the thermal performance of CiS packaging, which also applies solft material around the chip. The route of heat dissipation is observed by the three-dimensional FE analysis. On the other hand, the thermal management of stacking chip-in-substrate package on package (CiSPOP) is also dicussed. Likewise, the effects of dominate design parameters in CiSPOP technology are extracted. Based on the numerical analysis, the thermal characteristic of designed 3D packaging techmique is discussed herein.

FE modeling and Thermal Properties Determination

Figure 2 shows the established FE model which refers to the CiS packaging technology. Meanwhile, the quarter model contains the silicon chip, the filler material, the lamination material, the solder bumps, copper pad, through hole, the solder mask, and the standard thermal test board [6]. Besides, the model includes 7 solder bumps. The chip is embedded by the CiS technology, and the package is mounted on the test board. The chip size of CiS package is 3.75mm × 3.6mm × 0.15mm. The external diamension after packaging is 9.6mm × 7.2mm × 0.252mm, and the pitch of solder bumps is 0.8mm. Moreover, the diamension of 95.5Sn/3.8Ag/0.7Cu lead-free solder bumps is 0.34mm.

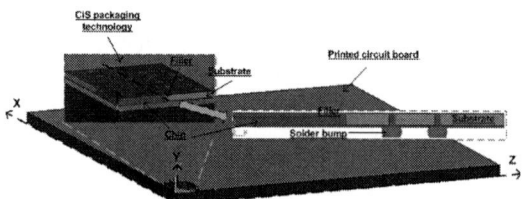

Figure 2: The established three-dimensional finite element model for thermal characterization of CiS packaging technology

In the FE analysis, the boundary consisting of heat transfer coefficient are applied to the top and bottom of the package, the top and bottom of the test board, the side wall of the package and the test board, and the side wall of the solder bumps. The heat

transfer coefficient is a nonlinear function of exterior surface temperature applied to natural and forced convection regime and includes the contribution of radiation. The total heat transfer coefficient, h_T, is calculated from the following equation [3, 7].

$$h_T = \left(h_{NC}^3 + h_{FC}^3 \right)^{1/3} + h_{rad} \tag{1}$$

$$h_{NC} = a \left(\frac{T_s - T_{amb}}{L_{ch}} \right)^n \tag{2}$$

$$h_{FC} = 3.79 \sqrt{V/L_{FC}} \tag{3}$$

$$h_{RAD} = \varepsilon \sigma \left[\left(T_s + T_{amb} \right) \left(T_s^2 + T_{amb}^2 \right) \right] \tag{4}$$

where T_s and T_{amb} are the external surface temperature of the package and the ambient temperature, respectively, ε is emissivity, σ is the Stefan-Boltzman constant, and the L_{ch} is characteristic length. The emissivity of the package device is set at 0.9. For horizontal plates, $L_{ch} = 0.5WL/(W + L)$, where W and L are the width and length of the plate, respectively. For vertical plates, $L_{ch} = H$, where H is vertical hight of the plate. Furthermore, the constant a and n are given as $a = 0.83$ and $n = 0.33$ for a horizontal plate facing upward, $a = 0.415$ and $n = 0.33$ for a horizontal plate facing downward, and $a = 1.09$ and $n = 0.35$ for a vertical plate. In the forced convection regime (Eq. 3), V describes the free-flow air velocity. L_{FC} is the characteristic length (given as equal to the package length).

According to the JEDEC standard [6], the test board is a four-layer printed circuit board (PCB) in which copper-patterns are printed in both of the surfaces. The effective thermal conductivity of the test board can be calculated by the following equation [8]:

$$k_{in-plane} = \frac{\sum_{i=0}^{N} k_i \times t_i}{\sum_{i=0}^{N} t_i}, \ k_{cross-plane} = \frac{\sum_{i=0}^{N} t_i}{\sum_{i=0}^{N} \frac{t_i}{k_i}} \tag{5, 6}$$

where t_i and k_i are the thickness and thermal conductivity of each layer, respectively. Besides, the applied thermal material properties in FE analysis are listed in Table 1.

Table 1: Material properties applied in FE model

Material	Thermal Conductivity K (W/m℃)
Filler	0.2
Cu trace & pad	380
Chip	150
Substrate	0.34
Lamination	0.2
Solder mask	0.2
SAC387 solder	57
PCB (x, y/ z)	(x, y/ z)=(22.8/ 0.34)

FE analytic results and discussion of CiS package

Through the established FE model of CiS package, the numerical analysis is initially executed as the power dissipation is 0.5W. The ambient temperature is set at 25℃, and the predicted stable temperature distribution is shown in Figure 3. The results show that the junction temperature under natural convection is 59.9℃.

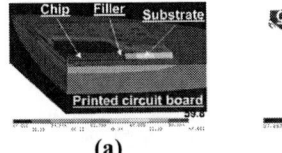 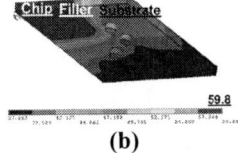

(a)   (b)

Figure 3: Thermal performance analysis of the CiS packaging technology. (a) predicted temperature distribution; (b) temperature distribution on package (natural convection, power dissipation = 0.5W)

As the substrate and filler with worse thermal conductivity are around the chip, the generated power can not be dissipated by these materials. For this reason, the dummy solder bumps are suggested to be placed under the chip as shown in figure 4. Because the dummy solder bumps in the chip region provide a better route of heat dissipation in the CiS package (as shown in figure 5(b)), the junction temperature could be reduced effectively (59.9℃ to 47℃).

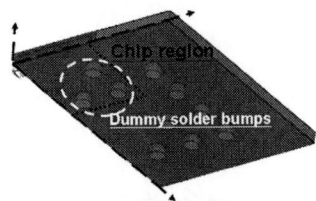

Figure 4: The established finite element model of CiS Packaging technology with dummy solder bumps

Traditionally, the thermal performance of a given package is described by the junction-to-ambient thermal resistance, θ_{ja}, defined bellow:

$$\theta_{ja} = \frac{T_j - T_{amb}}{P} \qquad (7)$$

where T_j and T_{amb} are the junction and ambient temperature, respectively, and P is the power dissipation at the given chip. In this study, the calculated thermal resistance of the CiS package structure and CiS package structure with dummy solder bumps are 69.8℃/W and 44℃/W respectively.

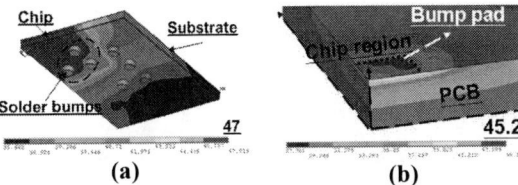

(a)   (b)

Figure 5: Thermal performance analysis of the CiS packaging technology with dummy solder bumps. (a) predicted temperature distribution on package; (b) temperature distribution on printed circuit board (natural convection, power dissipation = 0.5W)

In CiS packaging technology, the filler material is selected to fill the trench between chip and substrate. Moreover, the lamination provides the electrical insulation for the redistribution line. These substances have relative poor thermal conductivity (K), and their effects are studied by FE analysis. Figure 6(a) shows the presicted packaging thermal resistance as the thermal conductivity of the filler is adjusted. The results show that the thermal resistance reduces by 32% as the K of filler modies from 0.2 to 100 (W/m℃). Besides, the effective thermal conductivity of lamination layer consisting of dielectric and metal lines is calculated by Eq. (5, 6). The ratio of metal/laminatin can affect the effective thermal conductivity of lamination layer directly. Figure 6(b) shows the 35% improvement of thermal performance as the metal ratio increases from 20% to 80%. When the lamination has larger thermal conductivity, the heat dissipated through solder bumps is much easier.

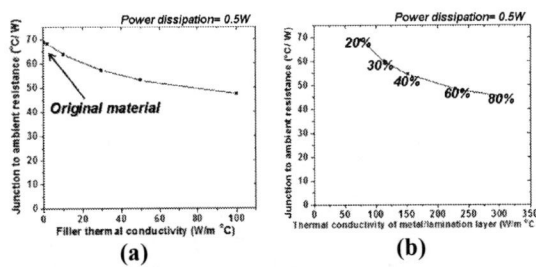

(a)   (b)

Figure 6: Thermal conductivity effect of materials in the CiS packaging technology. (a) filler material; (b) metal/lamination layer

CiS packaging technology may contain different sizes of the chips. As the size of the embedded chip decreases, the number of the solder bump under it reduces. In this study, the effect of chip/package ratio in the CiS packaging technology is also discussed. Different sizes of chips with the same powe dissipation, i.e. 0.5W, are analyzed. Figure 7 shows that the thermal resistance reduces by 46% as the ratio of chip/package size changes from 20% to 52%. The larger chip size which contains 4 solder bumps could provide effective thermal channeals from the package to the PCB and have relatively smaller power dissipation.

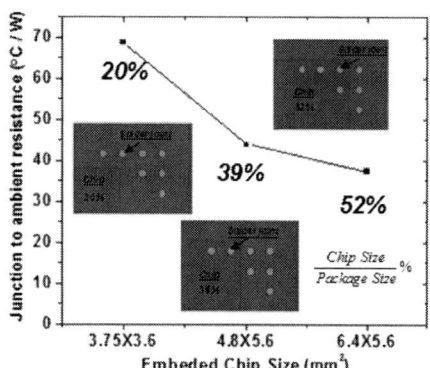

Figure 7: Thermal resistance value under differet (chip size/ package size) ratios

FE analytic results and discussion of CiSPOP

After the numerical analysis of CiS packaging technology, the stacking CiS packaging technology, CiSPOP, is also investigated. Figure 8 shows the established FE model which refers to the CiS packaging technology. In this model, the designed dummy solder bumps are applied under the chips as shown in figure 9. The quarter model contains 10 solder bumps under chip No.1; 14 solder bumps and one-half solder bump under chip No.2, and 14 solder bumps and one-half solder bump under chip No.3. The size of the chip No.1, No.2, and No.3 of CiSPOP are 3.75mm × 3.6mm × 0.15mm, 7.2mm × 4.8mm × 0.15mm, and 7.2mm × 4.8mm × 0.15mm respectively. The external diamension after packaging is 9.6mm × 7.2mm × 1.062mm, and the pitch of solder bumps is 0.8mm.

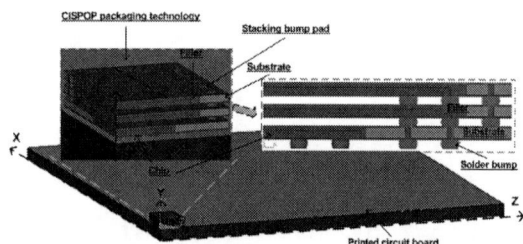

Figure 8: The established three-dimensional finite element model for thermal characterization of CiSPOP packaging technology

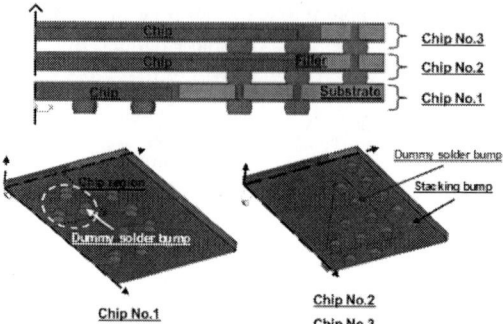

Figure 9: The established finite element model of CiSPOP with dummy solder bumps

Through the established FE model of CiSPOP, the numerical analysis is executed as the power dissipation is 0.5W at every chip. The ambient temperature is set at 25℃, and the predicted stable temperature distribution is shown in Figure 10. The results show that the junction temperature under natural convection is 76.7℃. The junction temperature occurs at the chip No.3. As the filler with worse thermal conductivity is placed over the CiS package, the generated power can not be dissipated from the upper CiS package to the lower one efficiently. In another words, the upper CiS package has worse ability of heat dissipation. For this reason, the underfill is suggested to be placed between the CiS packages as shown in figure 11. Because the underfill between the CiS packages provide a better route of heat dissipation from the upper CiS package to the lower one (as shown in figure 12), the junction temperature could be reduced effectively (76.7℃ to 69℃). Moreover, the thermal resistance could also be improved from 103.4℃/W to 88℃/W.

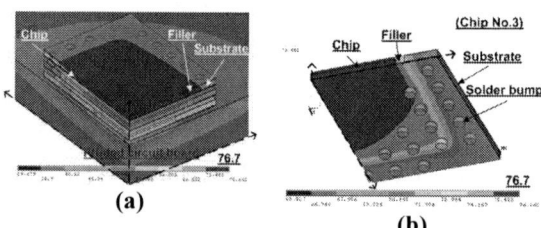

Figure 10: Thermal performance analysis of the CiSPOP packaging technology. (a) predicted temperature distribution; (b) temperature distribution on package (natural convection, total power dissipation = 0.5 × 3 = 1.5W)

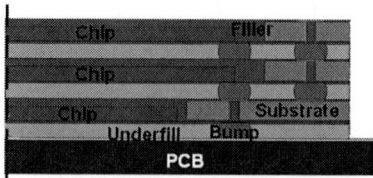

Figure 11: The established finite element model of CiSPOP with Underfill (thermal conductivity of underfill = 2 W/m℃)

978-1-4244-4722-0/09 $25.00
© 2009 IMAPS-ITALY

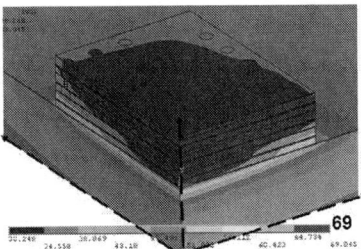

Figure 12: Predicted temperature distribution of the CiSPOP packaging technology with underfill. (natural convection, total power dissipation = 0.5 × 3 = 1.5W)

CiSPOP packaging technology may contain different power dissipations at the given chips. In this study, the effect of power dissipation is also discussed. Different power dissipations of chips with the same total powe dissipation, i.e. 1.5W, are analyzed. There are three arrangement conditions of power dissipations: 0.75W/ 0.375W/ 0.375W, 1W/ 0.25 W/ 0.25 W, and 1.25W/ 0.125 W/ 0.125 W. Figure 13 shows the thermal resistance value under different power dissipations. The X-axis indicates the position of the highest power dissipation. As the lower CiS package has better ability of heat dissipation, the highest power dissipation is suggested to be placed at the chip No.1 (black and red lines). However, the chip No.1 with a small size has a larger power dissipation density. When the excessively high powe dissipation is placed at the chip No.1, the extremely high junction temperature may be occurred (green line). On the other hand, thermal performance under forced convection is also compared in figure 14. The further improvement of thermal performance can be achieved through the air flow effect.

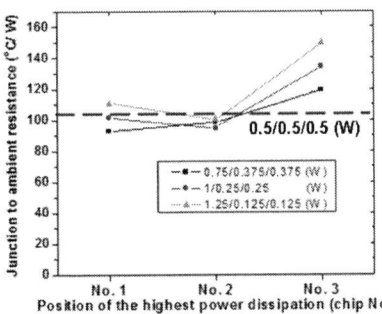

Figure 13: Thermal resistance value under differet arrangements of power dissipations

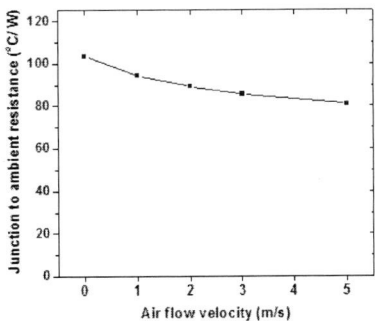

Figure 14: Air flow velocity effect in the CiSPOP packaging technology

Conclusion

The design concepts of the POP using CiS manner can be applied to the high-density IC devices. In the CiS packaging technology, I/Os with small pitch can be expanded through the lamination layers on the designed filler material and substrate material. In this study, the designed dummy solder bumps and the relatively better condition of the power arrangement are proposed to improve the thermal performance of the package. Based on the simulation results, the dummy solder bumps under the chip can improve the heat dissipation from the chip to the PCB efficiently. As the thermal comductivity of the filler and the lamination increase, the thermal performance can be improved. In the investigation of CiSPOP packaging technology, the highest power dissipation is suggested to be placed at the lower chip. Besides, the underfill between CiS package can effectively conduct the accumulated heat from the upper package to the PCB. Therefore, the thermal performance of the CiSPOP can be further enhanced, and it is suitable for applications on high power IC devices.

Acknowledgements

References

[1] T. H. Kim, J. H. Baek, S. H. Seol, J. J. Kim, Y. S. Kim, Y. B. Sun and S. Y. Oh, "Thermal Characterization of 3-Dimensional Memory Module," Twelfth IEEE SEMI-THERM™ Symposium, pp. 42-49, Austin, TX, USA, March 1996.

[2] W. H. Chen, H. C. Cheng and H. A. Shen, "An Effective Methodology for Thermal Characterization of Electronic Packaging," IEEE Trans on Components and Packaging Technologies, Vol. 26, No. 1 (2003), pp. 222-232.

[3] .K. M. Chen, K. H. Houng and K. N. Chiang, "Thermal resistance analysis and validation of flip chip PBGA packages," Microelectronics Reliability, Vol. 46 (2006), pp. 440-448.

[4] C. L. Chang and Y. Y. Hsieh, "Thermal Analysis of QFN Packages Using Finite

Element Method," Proc 5th EuroSimE, Brussels, Belgium, May 2004, pp. 499-503.

[5] M. C. Yew, H. P. Wei, C. S. Huang, D. C. Hu, W. K. Yang and K. N. Chiang, "A Study of Failure Mechanism and Reliability Assessment for the Panel Level Package (PLP) Technology," Proc 8th EuroSimE, London, England, April 2007, pp. 475-482.

[6] EIA/JEDEC Standard, "Test Boards for Area Array Surface Mount Package Thermal Measurements," Tech. Rep., EIA/JESD51-9, Arlington, VA, 2000.

[7] G. N. Ellison, Thermal Computations for Electronic Equipment, R. E. Krieger Publishing Company (Malabar, FL, 1989)

[8] S. W. Park, J. M. Kim, H. G. Baik, S. H. Kim, J. K. Hong, and H. S. Chun, "Thermal and Electrical Performance for Wafer Level Package," Proc 50th Electronic Components and Technology Conf, Las Vegas, USA, May 2000, pp. 301-310.

Novel interconnection processes
for low cost PEN/PET substrates

Jeroen van den Brand[1], Roel Kusters [1,2], Henri Fledderus[1], Eric Rubingh[1] , Tomas Podprocky[3], Andreas Dietzel[1]

[1] Holst Centre/TNO - Netherlands Organisation for Applied Scientific Research, HTC31, Postbus 8550, 5605 KN Eindhoven, the Netherlands.
[2] TNO Science and Industry - Netherlands Organisation for Applied Scientific Research, De Rondom 1, Eindhoven, the Netherlands
[3] Imec Gent, Technologiepark Zwijnaarde, Grote Steenweg Noord, B-9052 Zwijnaarde, Belgium

Abstract

Recently a new class of flexible electronics is starting to emerge which is most effectively termed 'printed electronics'. This term often refers to all-printed, cost effective, smart electronic products that will find a wide range of applications in large quantities in our society. The substrate material for these applications will be low cost materials like PEN or PET. Because of their lower thermal stability, novel interconnection technologies need to be developed. The current paper describes research into two of these interconnection technologies. The first one is 'embedded circuitry'. It will be shown that with this low cost technology it is possible to make conducting lines with widths down to 10 μm in foil and specific resistances down to $3 \cdot 10^{-7}$ Ω·m based on conductive pastes. The second topic is about filling of through-vias for the purpose of front-to-backside electrical contacting. Both inkjet printing and a combination of stencil and screen printing were investigated. The best results were obtained using a combination of stencil and screen printing. Through-vias with a diameter of 100 and 200 μm could be filled with a high yield, a good resistance and good mechanical and thermal stability. Key words: PEN, interconnection, conductive pastes, circuitry, low cost

Introduction

Recently a new class of flexible electronics is starting to emerge which is most effectively termed 'printed electronics'. This term often refers to all-printed, cost effective, smart electronic products that will find a wide range of applications in large quantities in our society. Examples include cheap sensor packages attached to food packaging to measure the ripeness of food, smart bandages that monitor the healing of wounds or smart active or passive RFID tags. These cost effective packages include all-printed functionality like conductive circuitry, OLED pixels, photodiodes and logics based on organic materials. There is a huge market for these products. According to a market report of Frost and Sullivan [1], the market for organic and printable electronics is expected to be a $35 billion US dollar industry by 2015 and reach over $300 billion US dollar in 2025. This is twice the size of the silicon industry today. For Europe, this is a new opportunity to be in the leading role in the development of this new class of products. Several major research institutes (Holst Centre, VTT, Fraunhofer IPMS, Fraunhofer IZM) in Europe have recognized this and are actively performing research in this field [2-4].

To make these printed electronic devices cost effective, the substrate material will likely no longer be the poly(imide) that is most commonly used in flexible electronics. By using polyesters like PEN or PET, the substrate costs can be reduced by a factor of 5-10 [5]. An additional advantage of these materials over poly(imide) is that they are optically clear, making them suitable for transmitting light which for example originates from OLEDs which can be directly printed onto the foil.

Disadvantages of these materials are however that they are considerably less thermally stable (PEN has a Tg of ~120 °C and PET of ~80 °C while poly(imide) has a Tg of ~350 °C) [5]. This limitation in thermal stability excludes many well established processes for making electronic products and/or renders the use of existing processes much more challenging. It is for this reason that it will be necessary to develop novel interconnection technologies for such low cost foils. The current paper describes research into two of these interconnection technologies. The first topic discussed is 'embedded circuitry'. This technology uses parts of the technology used for screen/stencil printing but allows going to much lower conductive line widths. In the technology the electronic circuitry is embedded inside the volume of the substrate. It can be used to make the full electronic circuitry in a substrate or to make parts of the circuitry where fine lines are required. The current work describes research to investigate the linewidth / depth possibilities of this technology. Secondly, research will be discussed on filling 50-200 μm diameter through-vias in PEN substrates for the purpose of front- to backside electrical contacting.

978-1-4244-4722-0/09 $25.00
© 2009 IMAPS-ITALY

supplier / type	Emerson&Cuming CE3104WLV	Dupont 5025	Cabot Ag-IJ-G-100-S1
material type	isotropic conductive adhesive	screen printing paste	inkjet printing ink
in current work used for	- embedded circuitry - filling of 100 and 200 µm vias	- embedded circuitry - printing of conducting lines - filling of 50 µm vias	filling of 50, 100 and 200 µm vias
conductor type	microsized Ag-particles	microsized Ag-particles	nanosized Ag-particles
shrinkage	~4 volume%	~60 volume%	~98 volume%
bulk resistivity	~$5 \cdot 10^{-6}$ Ω·m	~$3.8 \cdot 10^{-7}$ Ω·m	~$1.2 \cdot 10^{-7}$ Ω·m
viscosity	~65 Pa·s	~20-30 Pa·s	~0.144 Pa·s
recommended curing schedule	15 minutes, 125 °C	5 minutes, 120 °C	30 minutes, 130 °C

Table 1. Overview of relevant material properties for the conductor materials used in the current work

Figure 1. Schematic overview of the manufacturing process for embedded circuitry.

Two technologies have been investigated to achieve this: inkjet printing using Ag nanofilled inks and stencil printing using isotropic conductive adhesives and screen printing pastes.

These technologies combined allow to make both intrafoil (through-via filling) and interfoil interconnects (embedded circuitery).

Experimental

Used materials

The base substrate that was used throughout the current work is flexible, transparent, 125 µm thick, heat-stabilized, biaxially oriented PEN (polyethylene naphtalate) from Dupont Teijin Films (Teonex Q65). For the first part of the work *embedded circuitry*, two different conductive filling materials were evaluated. The first one was an epoxy-based Ag-filled isotropic conductive adhesive (ICA) CE3104WLV from Emerson and Cuming. The second material was an Ag-filled screen printing paste 5025 from Dupont. Both ICA's and screen printing pastes are readily commercially available and might both be suitable candidates for usage in this process. These filling materials were also used for the second part of the work: *through-via filling*. Additionally, for this second part, a Ag nanoparticle filled Cabot Ag-IJ-G-100-S1 was used to fill through-vias using inkjet printing. A summary of the used materials with the associated material properties is given in Table 1 together with the relevant material properties.

Embedded circuitry

Process flow

The process flow that is followed for the embedded circuitry is shown schematically in Figure 1. In a first step a focused laser beam is used to photoablate the circuitry pattern into the foil. In a second step a squeegee is used to squeeze-in a conductive paste into the grooves made in the foil. In a final step, the circuitry is dried and cured after which it is finished.

Methods

An eximer laser (λ = 248 nm) was used to photoablate grooves into the foils. An energy of 10 mJ/pulse, a repetition rate of 300 Hz and a demagnification of 10x was used. This setting gave an etch rate of 0.5 µm/pulse. Test structures were made which consisted of lines with varying widths (between 10 and 100 µm) and depths (between 10 and 50 µm) and lengths of 6 mm. More details on the laser work on PEN is discussed elsewhere [6].

Measurement pads were present at the ends of the lines to be able to measure the conductivity of the lines. Squeegee-filling of the samples was performed using a custom build tabletop sized squeegee tool which allowed accurate control of the speed and squeegee angle. The filling was performed using a metal blade at an angle of 45 degrees with respect to the surface area and also 45 degrees with respect to the structures on the foil. After filling, the substrates were cleaned from residual paste contamination by moving a clean metal blade across the foil. The pastes were then dried and cured by heating in a conventional oven for the required period, see Table 1. Finally, the foils were given a final cleaning treatment through gentle wiping with an acetone-soaked tissue. This cleaning was found to not remove material from the grooves.

Chip bonding

To evaluate the usability of the embedded circuitry, a fanout circuitry was made onto which a

978-1-4244-4722-0/09 $25.00
© 2009 IMAPS-ITALY

Si die can be flipchip bonded. A thinned (50 μm) 5 x 5 mm Si daisychain chip with 176 IO's was bonded using a Finetech flipchip bonder. A DELO AC163 anisotropic conductive adhesive was used as the adhesive. The adhesive was cured according to the supplier recommended schedule.

Through-via filling

Process flow

Filling of through-vias in PEN foils was evaluated using two different approaches: inkjet printing and screen/stencil printing. The same test design, see Figure 2, was used for both. The design (4 x 4 cm) consists of conductive structures printed on both sides of the foil which are interconnected by through-vias and which together form several daisy chains. All daisy chains combined, the design contains a total of 1100 vias. For the inkjet filling methodology, both the conductive structures and the via filling was performed using inkjet printing. For the screen/stencil printing methodology, first the vias were filled using a stencil aligned over the vias and after this the conductive tracks were printed using conventional screen printing.

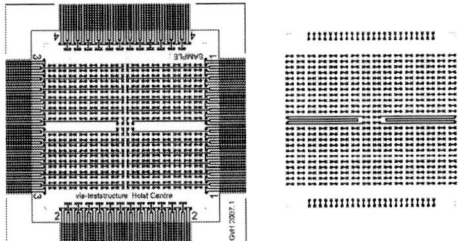

Figure 2. Through-via filling test design (size 4 x 4 cm). Left image shows design printed on top side of foil. Right image shows design printed on bottom side.

Methods

A DPSS Nd:YAG, 3rd harmonic laser operating at a wavelength of 355 nm was used to drill through-vias in the foil. The laser was operated in trepanning mode, at an energy of 3μJ/pulse, a mark speed of 100 mm/s (100 scans per via) and a pulse repetition rate of 30 kHz. Through-via diameters of 50, 100 and 200 μm were laser-drilled in the foils according to the test design shown in Figure 2. More details on the laser work on PEN is discussed elsewhere [6].

For the inkjet printing approach, a Dimatix DMP-2800 inkjet printer (Fujifilm Dimatix Inc., Santa Clara, USA) research inkjet printer was used. A print head with a nozzle diameter of 21 μm was employed which for this material should provide 10 pl droplets with an estimated diameter of 25 μm.

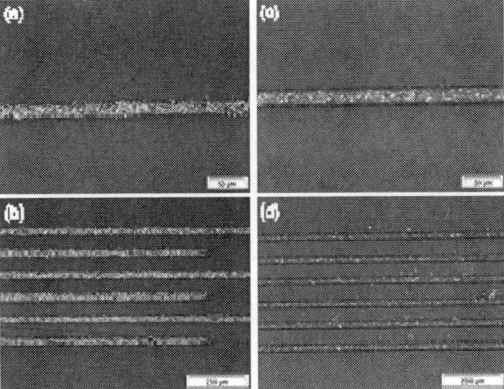

Figure 3. Microscopical photographs of filled grooves. (a) 10 μm wide filled with 3104WLV, (b) 25 μm wide filled with 3104WLV (c) 10 μm wide filled with Dupont 5025 (two filling steps), (d) 25 μm wide filled with Dupont 5025 (two filling steps). All grooves had a depth of 25 μm.

After printing the top part of the design, the substrate was mildy dried in an oven for 4 minutes at 110°C. Next the substrate was again placed on the inkjet printer and the bottom part of the design was printed. Finally, the ink was sintered by putting the foil in an oven for 30 min at 135°C.

For the screen/stencil printing approach, an EKRA E4-STS was employed. For the filling of the 100 and 200 μm diameter vias, a stencil with a thickness of 100 μm and respectively 150 and 250 μm openings was used. For the 50 μm diameter vias, a stencil with a thickness of 20 μm and 75 μm openings was used. As filling materials both the Emerson and Cuming CE3104WLV and the Dupont 5025 were evaluated, see also Table 1. Backside vacuum assistance was used to suck the paste into the vias. After filling, the conducting lines were printed using a 400 mesh screen and using a Dupont 5025 as the printing paste. Finally, the pastes were dried and cured by putting the foil in an oven for the required period, see Table 1.

Results and discussion

Embedded circuitry

Filling results

A matrix of line widths between 10 and 100 μm and depths between 10 and 50 μm were lasered into the foil and subsequently filled. For both of the filling materials that were used it was found to be possible to fill the full range of groove widths, as observed by visual inspection. As an illustration, Figure 3 shows microscopical top view photographs of 10 and 25 μm width for both filling materials. It can be seen that the grooves are well-resolved and without residual contamination.

Subsequently, an analysis was performed of the filling degree of the grooves.

978-1-4244-4722-0/09 $25.00
© 2009 IMAPS-ITALY

Emerson&Cuming, CE3104 WLV

		line width (µm)			
		10	25	50	100
depth (µm)	10	4	5	5	4
	25	4	5	5	4
	50	4	5	5	4

Dupont 5025, first filling

		line width (µm)			
		10	25	50	100
depth (µm)	10	6	5	7	5
	25	10	12	12	11
	50	13	23	27	24

Dupont 5025, second filling

		line width (µm)			
		10	25	50	100
depth (µm)	10	4	5	5	5
	25	6	8	8	7
	50	5	6	14	16

Table 2. Filling results depths as a function of line width and depth. Results in the table are the largest measured distance (in µm) from the foil surface to the top of the filling in the line. A small value indicates a better filled line.

Table 2 gives the measured distance (in µm) from the foil surface to the top of the filling in the groove. A small value indicates a better filled groove.

The results show that the ICA fills the grooves up to around 4-6 µm below the surface of the foil in a single filling step. Additional filling steps were found to not increase the filling degree. The screen printing paste did not give a well-filled groove after a single filling step. Prior to drying and curing, the grooves were filled but the high shrinkage of the material, see Table 1, causes an incompletely filled groove after drying and curing. After the second filling step, most of the groove depths are acceptably filled. Figure 4 shows some microscopical cross section images of the filled grooves. It can be seen that the ICA fills homogeneously across the groove width. The screen printing paste on the other hand has a more concave shape which follows the groove edges. This shape is the result of the shrinkage of the material. The bright parts in the photographs are the Ag-particles. It can be seen that the ICA has regions where no Ag-particles are present – see also furtheron.

Four point measurements were performed to determine the conductivities. Table 3 summarizes the results that were obtained. To make the results intercomparable, the measurement data is recalculated to specific resistances. These were calculated using the measured resistances, groove widths, depths and filling heights. The results for the ICA are not very good. The obtained specific resistances are on average a factor of 300 worse than the bulk value of the material.

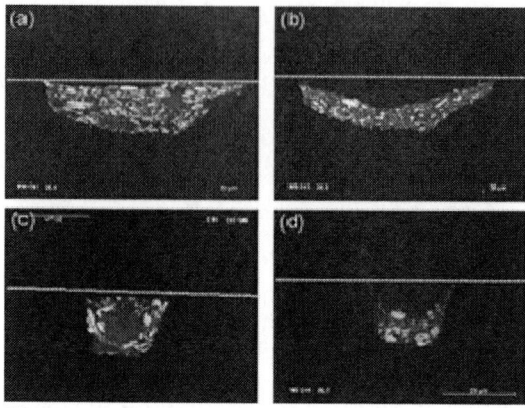

Figure 4. Cross section microscope images of filled grooves. (a) + (b) 100 µm wide, 50 µm deep grooves filled with CE3104 (a) and 5025 (b), (c)+(d) 25 µm wide, 25 µm deep groove filled with CE3104 (c) and 5025 (d). For clarity, the foil interface has been indicated by a dashed line.

The best result is obtained for the lines with a width higher than 50 µm and a depth higher than 50 µm.

Emerson&Cuming, CE3104 WLV

		line width (µm)			
		10	25	50	100
depth (µm)	10	open	4.1E-03	8.0E-04	2.4E-04
	25	2.2E-03	9.7E-04	4.7E-04	1.3E-03
	50	5.6E-03	8.7E-05	2.5E-05	3.1E-05

Dupont 5025, second filling

		line width (µm)			
		10	25	50	100
depth (µm)	10	1.4E-06	6.5E-07	4.0E-07	1.2E-06
	25	1.3E-06	5.2E-07	3.2E-07	6.0E-07
	50	3.8E-06	1.3E-06	4.1E-07	3.7E-07

Table 3. Specific resistivities ($\Omega \cdot m$) of the filled grooves. The CE3104WLV has a specified specific bulk resistance of $5 \cdot 10^{-6}$ $\Omega \cdot m$. The 5025 has a specified specific bulk resistance of $3.8 \cdot 10^{-7}$ $\Omega \cdot m$, see Table 1.

Even at these widths and depths, the specific resistance is still a factor 5 worse than the specific bulk resistivity of the material. The 10 µm wide and 10 µm deep lines did not show any conductivity. Some trends are visible in the results: in general, the specific resistance decreases with increasing groove width and depth.

The results for the screen printing paste are much better. On average, the specific resistances are a factor of 3 worse than the bulk value of the material. The worst results are obtained for the highest aspect ratio lines (10 µm wide, 25 and 50 µm deep) and for the widest and least deep line (100 µm wide, 10 µm deep).

An explanation for the bad performance of the ICA as compared to the screen printing paste could be found through a closer microscopical investigation of the filled grooves. Two types of problems could be identified with the ICA. The first problem was that the ICA showed the presence of

978-1-4244-4722-0/09 $25.00
© 2009 IMAPS-ITALY

'conglomerates' with a size of a few μm in the filled grooves. Figure 5a shows a photograph of such a conglomerate. These conglomerates became visible only after curing. The conglomerates were also visible in the cross section images of Figure 4 as the regions were no Ag-particles are present. Such conglomerates are likely not a problem for normal application of this material in which it is used to conduct over smaller distances and larger cross sectional areas. For the narrow grooves studied here, these conglomerates however almost completely block the conducting groove thereby limiting the conductivity.

The second problem identified was that directly after filling, before cleaning and curing, part of the material was immediately pushed out of the groove. Figure 5b shows an example of this. This is likely caused by entrapped air at the bottom of the groove which becomes compressed by the filling process and escapes through the filling material.

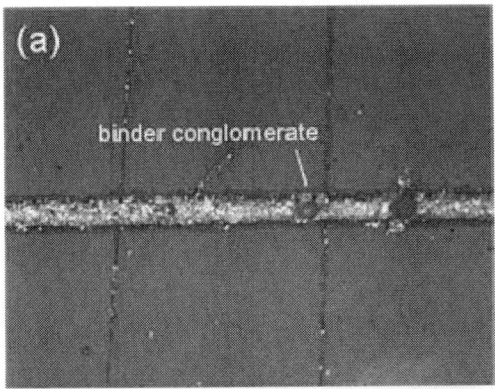

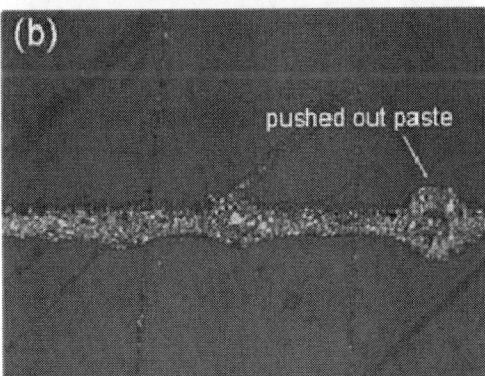

Figure 5. Microscopical photograph of 10 μm wide, 25 μm deep groove filled with CE3104. (a) Showing binder conglomerate particle, (b) Showed site with popped out material

This causes the formation of small interruptions of the conducting lines, thereby significantly increasing the line resistance. The phenomenon was found to be worse for higher aspect ratios. This phenomenon was not observed for the screen printing paste. Probably, as a result of the lower viscosity, air can easier escape from the bottom of the groove through the filling material. Moreover, the material shrinks around 60%, making it possible to have some macroscopic reorganization of the filling material in the groove.

Bonding of the chip

To investigate the usability of the circuitry, a Si die was flipchip bonded onto a fanout circuitry test circuitry. The test circuitry was made with 50 μm wide and 25 μm deep grooves. The grooves were filled using the screen printing paste in two sequential filling/curing steps. Figure 6 shows several photographs of this test circuitry with the Si die attached to it. All of the chip to groove interconnection worked and had a low interconnect resistance of less than 0.2 Ω.

Discussion and conclusion

The limitations and possibilities of the embedded circuitry technology have been investigated using two different filling materials.

The best results were obtained using a quite common Ag-filled screen printing paste, the Dupont 5025. It is shown to be possible to make lines down to a width of 10 μm which show a good conductivity. The conductivities that could be obtained approached that of the bulk material. Moreover, it was found to be possible to bond and interconnect a Si die onto the circuitry. These results shows that with this technology it is possible to make fine conducting lines in foil without the need for difficult manufacturing processes or expensive equipment. The circuitry that is made can be directly used for further processing like bonding a Si die onto it. As far as known, there is no other technology that has all these advantages.

A disadvantage of the usage of the screen printing paste as filling material is that it has a high shrinkage. For applications which require a completely filled groove, it will be necessary to perform multiple filling steps. This disadvantage could be prevented by using a filling material that does not show such a high shrinkage. The second tested filling material, an ICA has this characteristic. Indeed, it was possible to completely fill the groove in a single filling step. However, this material was found to be not really suitable for filling of grooves below a width of 50 μm because of problems related to the formation of conglomerates and material popping out due to entrapped air.

One could think of a combination of these materials where where needed the ICA is used to completely fill the groove while at the remainder of the circuitry the screen printing paste is used to conduct current over larger distances.

978-1-4244-4722-0/09 $25.00
© 2009 IMAPS-ITALY

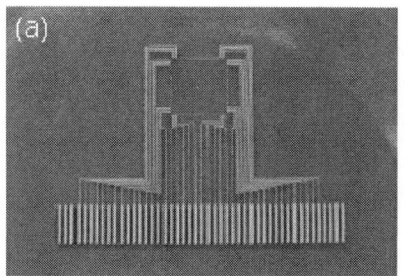 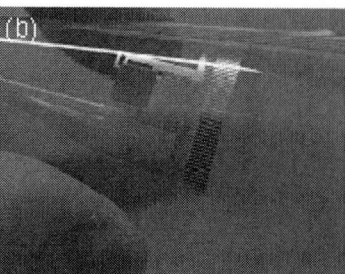 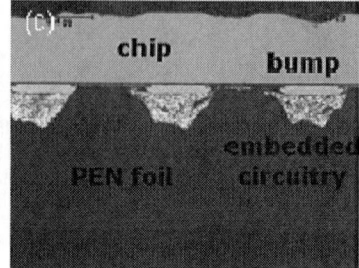

Figure 6. Fanout demonstrator (3 x 3 cm) of embedded circuitry. (a) overview image, (b) bended view with chip attached, (c) cross section of chip bonded onto circuitry

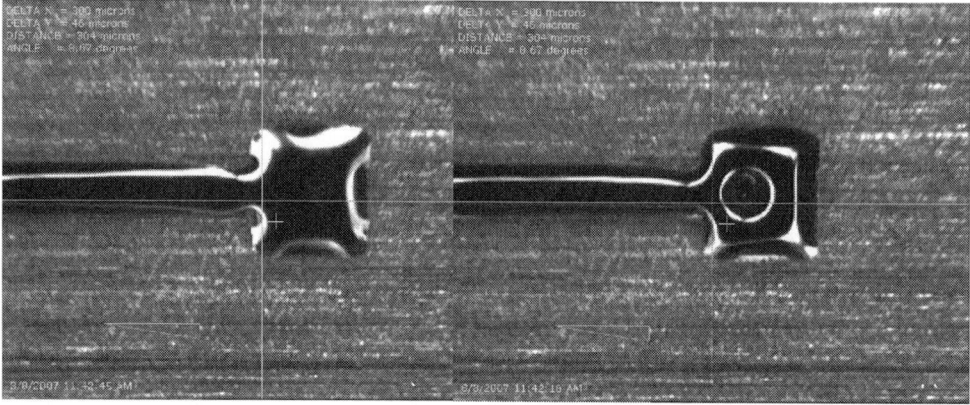

Figure 7. Inkjet filling of through-vias (left) without and (right) with backside vacuum assistance

Filling of through-vias in PEN substrates

Inkjet printing

Inkjet filling of through-vias was executed on samples with via diameters of 50, 100 and 200 μm made into the test structure shown in Figure 2. The experiments showed that it is necessary to use backside vacuum assistance as otherwise the ink would not penetrate into the vias. A microscopical photograph of filling with and without vacuum assistance is shown in Figure 7.

Filling of the 100 and 200 μm diameter vias was not very successful. The ink was sucked completely through the via and landed onto the support substrate. This is the consequence of the very low viscosity of the material, see Table 1. Figure 8 shows microscopical images of filled and dried/cured through-vias with the different tested diameters. The left side set shows top view images with front side illumination while right side set of images show the same with backside illumination. For the 100 and 200 μm diameter it can be seen that one can look through the vias, implying not much material was left behind in the via. Filling of the 50 μm was more successful. The microscopical images in Figure 8 show that the vias appear to be filled.

Stencil printing/screen printing

The same test structures with 50, 100 and 200 μm vias in PEN were also filled using a combination of stencil printing to fill the vias and screen printing to make the conductive lines. The best via filling results for the 100 and 200 μm were obtained with the ICA. The screen printing paste was found to have a too low viscosity and was sucked largely through the via onto the support. Filling of the 50 μm vias on the other hand was found to be not possible using the ICA. The ICA would not enter the via as a result of its high viscosity. Filling was found to be possible using the screen printing paste. As an illustration, Figure 9 shows a photograph of a test structure sample as made using stencil/screen printing. Figure 10 shows a microscopical cross section photograph of a 100 μm diameter via. The bump that is visible at the top is the result of releasing of the stencil. For the 50 μm diameter vias, some alignment issues of the stencil with respect to the vias on the foil were encountered. This caused part of the vias not to be filled. The used EKRA E4-STS has an estimated accuracy of 20 μm which turned out to be quite critical for the smallest via diameters.

Electrical measurements and process yield

Subsequently, the test samples were electrically characterized using four point testing methodology. In the design, the vias are measured in chains of multiple vias, see Figure 2.

978-1-4244-4722-0/09 $25.00
© 2009 IMAPS-ITALY

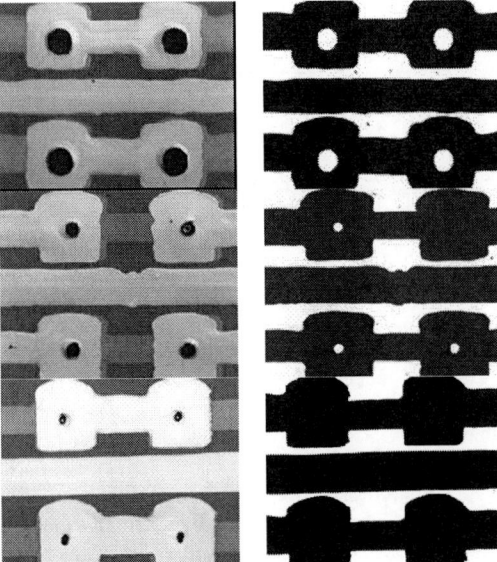

Figure 8. Inkjet filled through-vias in PEN. From top to bottom 200, 100 and 50 μm via diameter. Right side shows same image with backside illumination.

Figure 9. Photograph of via filling test structure as made using stencil/screen printing.

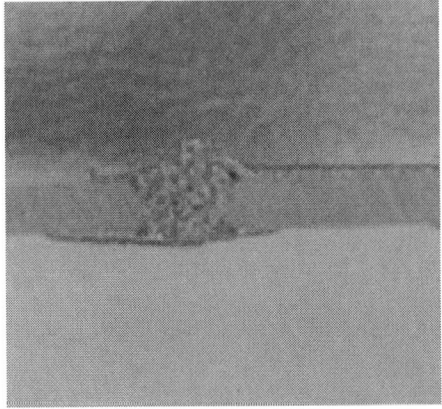

Figure 10. Cross section of foil with filled 100 um via. Bump at the top is the result of the release of the stencil.

Multiple samples were made for both inkjet printing and stencil/screen printing for the different via diameters. Overall, measurements were conducted on more than 2400 chains, containing more than 20000 filled vias. The averaged results are summarized in Table 4.

The daisy chains in the test structure also contain conducting line parts which interconnect the vias. The results in the Table are not corrected for the line resistances. The estimated contribution per via is around 0.2 Ω for the screen printed circuitry and 0.4 Ω for the inkjet printed circuitry.

For the calculation of the average values, via resistances higher than 5 times the average were not taken into account. These vias were also substracted from the process yield. The given process yields thus reflects the total of filled vias minus the combined amount of the vias that could not be measured (open circuit) and the vias that had a resistance higher than 5 times the average.

The inkjet printing results show a very low yield for the 100 and 200 μm vias. As discussed, this is due to the fact that the ink was sucked completely through the via and landed onto the support substrate. See also Figure 8.

The microscopical investigation of the 50 μm vias appeared to indicate that they are filled, see Figure 8. Despite this, the yield for this via diameter is low and the measured resistances quite high. An explanation for this is likely as follows. Inkjetting requires a low viscosity of the material. As a consequence, the solid loading is very low and the solvent content is very high. Upon curing, this results in a very high shrinkage, see Table 1. The likelihood of forming cracks around the irregularly shaped transition from the foil surface into the via is therefore high. Indeed, such cracks were observed around the circumference of the via. These cracks can both cause a completely malfunctioning via or a via with a high resistance.

The 100 and 200 μm diameter stencil/screen print filled vias show a very good yield. It is believed that with minor process optimizations, this can quite readily be improved to 100%. The yield for the 50 μm diameter is significantly lower. As discussed, this can be attributed to alignment issues of the stencil onto the foil. The via resistances for all diameters are quite good and more than acceptable for the envisaged low cost applications.

Lifetime and mechanical stability tests

The stencil/screen print filled vias of 100 and 200 μm were further tested to evaluate their stability in harsh environments and their mechanical stability. Because of their low yield, the inkjet filled and 50 μm via diameter stencil/screen print filled samples were not evaluated. Samples with 100 μm and 200 μm via diameters were subjected to 60 °C/90% RH for a period of 250 hours.

978-1-4244-4722-0/09 $25.00
© 2009 IMAPS-ITALY

filling process	via diameter [µm]	via resistance [Ω]	standard deviation [Ω]	process yield [%]
inkjet printing	50	2,20	0,80	49
	100	45,00	75,40	8
	200	2,87	1,87	8
screen/stencil printing	50	1,00	0,80	34
	100	0,38	0,12	95
	200	0,22	0,15	94

Table 4. Summary of electrical measurements performed on through-via filling using inkjet and screen/stencil printing. Via resistances contain a contribution from the connecting line. Estimated resistance of the line is 0.2 Ω for the screen printed circuitry and 0.4 Ω for the inkjet printed circuitry

To evaluate the thermal cycling stability, samples were also subjected to thermal cycling between -40 °C and +85 °C for 2 cycles per hour for a total of 250 cycles. To evaluate the mechanical stability, samples were bended for 250 cycles in both directions around rods with diameters of 60 and 20 mm. Before and after each test, electrical characterization was performed.

The obtained results are summarized in Table 5. It can be seen that the thermal stability and the stability in humid environments is very good for the studied conditions. The samples even showed improvement of the resistances. This is believed to be caused by further curing of the pastes due to the prolonged period at higher temperatures. The mechanical stability of the interconnects was also found to be quite good. The samples easily survived 250 cycles of bending around a cylinder with a diameter of 60 mm without significant changes in via resistances. Bending around a cylinder with a diameter of 20 mm did result in damage to the interconnects. The via resistances were found to increase by on average 100%.

Discussion and conclusion

Both inkjet and a combination of screen and stencil printing were investigated for the purpose of filling through-vias with diameters of 50, 100 and 200 µm for back to frontside contacting.

The combination of screen and stencil printing was found to be a successful route. It is possible to fill 100 and 200 µm diameter with a yield close to 100%. The filled vias have a resistance which is very well acceptable for the considered low cost applications. Moreover, the vias showed a good performance at elevated temperatures and increased humidity and also performed well in thermal cycling. Finally, the foils in which the vias were made could be bended down down to a radius of 30 mm without damage to the interconnects.

test	condition		results
temperature/ humidity	60°C/90%	250 hrs	improvement of resistance (16%)
thermal cycling	-40°C/+85°C, 2 cycles/hour	250 cycles	improvement of resistance (8%)
mechanical bending	folding around 60 mm diameter cylinder	250 cycles	no significant change in resistance
mechanical bending	folding around 20 mm diameter cylinder	250 cycles	increase of resistance by about 100%

Table 5. Summary of lifetime and bending test results for screenprint/stencil filled vias

The filling of vias with a diameter of 50 µm was less successful. This could be attributed to alignment issues of the used equipment. The vias that were successfully also showed a good via resistance.

Inkjet printing was found to be not very successful. This could be mainly ascribed to the ink that was used. This ink has a low viscosity which made it impossible to fill the 100 and 200 µm diameter vias. The ink was largely sucked through the via and landed on the substrate support. The 50 µm was found to be successfully filled but the yield was found to be low. This is likely caused by the low solid content of the ink with the associated high shrinkage of the material. This results in the formation of various microcracks around the vias and as a result of this, a high resistance or a completely failing via.

Conclusion

In the current work, research on two different interconnection technologies for low cost applications has been discussed. These interconnect technologies cover both intra and interfoil interconnects. The first one, embedded circuitry, involves making low cost electronic circuitry for the purpose of intrafoil (in-plane) interconnects. The results show that with this technology it is possible to make conducting lines with widths down to 10 µm in foil. Two filling materials were evaluated: a commercially available ICA and a commercially available screen printing paste. The screen printing paste was found to give the best results. It was also demonstrated that the circuitry that is made can be directly used for further processing like bonding a Si die onto it.

The second part of the work involved research on filling of through-vias using both inkjet printing and a combination of stencil and screen printing for the purpose of making interfoil interconnects. The best results were obtained using screen/stencil printing. Through-vias with a diameter of 100 and 200 µm could be filled with a high yield, a good resistance and good mechanical and thermal stability. The filling of through-vias using inkjet printing was not successful as a result of the low viscosity and the high shrinkage of the ink that was used.

978-1-4244-4722-0/09 $25.00
© 2009 IMAPS-ITALY

Both the embedded circuitry and the through-via filling combined allows to make low cost inter- and intrafoil interconnects. Future research will involve a further optimization of the conductivities that can be achieved.

References

[1] "Printed Electronics - Technologies and Applications," Frost and Sullivan2007.

[2] J. van den Brand, J. de Baets, T. van Mol, and A. Dietzel, "Systems-in-foil - Devices, fabrication processes and reliability issues," *Microelectronics Reliability,* vol. 48, pp. 1123-1128, 2008.

[3] G. Klink, E. Hammerl, A. Drost, D. Hemmetzberger, and K. Bock, "Reel-to-Reel Fabrication of Integrated Circuits Based on Soluble Polymer Semiconductor," in *5th International Conference on Polymers and Adhesives in Microelectronics and Photonics, Polytronic* 2005, pp. 1- 6.

[4] K. Bock, "Polymer Electronics Systems— Polytronics," *Proceedings of the IEEE,* vol. 93, p. 6, 2005.

[5] J. Fjelstad, *Flexible Circuit Technology,* 3rd ed.: BR Publishing Inc, 2007.

[6] R. Mandamparambil, H. Fledderus, J. v. d. Brand, M. Saalmink, R. Kusters, T. Podprocky, G. V. Steenberge, J. D. Baets, and A. Dietzel, "A Comparative Study of Via Drilling and Scribing on PEN and PET Substrates for Flexible Electronic Applications Using Excimer and Nd:YAG Laser Sources," in *8th Annual Flexible Electronics & Displays Conference and Exhibits* Phoenix, Arizona, USA, 2008.

978-1-4244-4722-0/09 $25.00
© 2009 IMAPS-ITALY

Characterization of PTC resistor pastes applied in LTCC technology.

J. Vanek, W. Smetana[1], M. Weilguni[1], I. Szendiuch

Department of Microelectronics, Brno University of Technology, Udolni 53, 602 00 Brno, Czech Republic

Phone: +420541146159, Fax +420541146298, jan.vanek@phd.feec.vutbr.cz

[1] Institute of Sensor and Actuator Systems, Vienna University of Technology, Gusshausstrasse 27 - 29 Vienna, Austria

Abstract

This study deals with analyses of PTC (Positive Temperature Coefficient) thick film resistor elements applied in LTCC (Low Temperature Co-fired Ceramics) technology. This study concerns the characterization of the electrical performance of PTC-elements placed in a different arrangement on a LTCC-multilayer substrate. Resistors are built-up on LTCC ceramic substrates and sintered in different configurations - on the surface of LTCC tape, in structure with embedded air cavity and co-fired in between two tapes. Three different cermet (ceramic –metallic) based thermistor compositions are used for this study. Electrical measurements were carried out in order to determine the temperature profile of the electrical resistance. TCR (Temperature Coefficient of Electrical Resistance) and the length / width ratio dependence on sheet resistance are evaluated. The influence of sintered configuration on the sheet resistance and TCR is discussed.

Key words: LTCC, PTC paste, resistor, TCR

Motivation

PTC resistor (thermistor) pastes are usually matched for conventional thick film technology using alumina substrate. Pastes show typical advantages which can be utilized for various applications. These materials are designed for applications for which thermistors are required to be intimately bonded to the substrate. Resistor pastes are often used in applications where temperature compensation of hybrid circuits and fast response sensors are required. Their properties are also key aspect of the proposal of various types of sensors based on calorimetric principle. In recent years, thick film resistor pastes are also used in LTCC technology to build up microsystems where sensors and actuators are included.

Resistor pastes have not identical performance on different substrates. This reflected in their electrical (e. g. sheet resistance and its temperature profile, TCR) as well as mechanical performance. They are usually adjusted on alumina substrates and their performance varies if they are applied on different LTCC-substrates especially in the case of co-firing. For non standard applications (e. g. LTCC technology), they are exposed to a specific processing, which often deviate from the standard processing recommended by the supplier. Consequently their performance must be determined by the user.

PTC thermistors may be implemented in a LTCC stack in different configurations. With regard to their interaction (such as diffusion) with the LTCC substrate, this study is based on following

assumption: Sheet resistance and temperature dependence of resistance may vary in dependence on the arrangement of the resistors in a multilayer LTCC-stack. They are applied on the top of an LTCC-stack, embedded between two tapes or placed in cavities inside of a LTCC-multilayer block.

Introduction

Resistors with a large temperature dependence of resistivity are required for temperature-sensing applications. Thick-film thermistors with positive TCR have usually a positive and linear temperature dependence of resistivity in contrast to thermistors with negative TCR (NTC – Negative Temperature Coefficient) which have a large and strongly non-linear dependence.

The dependence of the resistance vs. temperature for PTC resistors is described by the following linear equation:

$$R(T) = R_0(1 + \alpha T), \quad (1)$$

where R_0 [Ω] is the resistance at reference temperature, T is the temperature [K] and a is the TCR [K^{-1}]. PTC thick-film resistors have a metallic-like dependence of resistivity vs. temperature characteristic. The conductive phase in thick film PTC thermistors has a relatively low specific resistivity and a positive, linear metallic-like dependence of resistivity vs. temperature behavior, with a TCR of 7000 ppm. K^{-1} for single crystals and a few 1000 ppm K^{-1} for sintered micro-crystalline samples [1, 2].

978-1-4244-4722-0/09 $25.00
© 2009 IMAPS-ITALY

Experiments

Three different thermistor thick film pastes have been selected for sensor sample preparation. All tested pastes are supplied by ESL and they are based on cermet compositions. All the considered pastes show the same sheet resistance (Table 1).

Table 1: Properties of used pastes.

Producer - type	ESL-PTC	ESL-PCT	ESL-PTC
Type	2611-SP	2611 -I	2611 – 1A
Sheet resistance [$\Omega/\square$]	10	10	10
TCR [ppm.K^{-1}]	4000±100	3000±400	3000±400

Three different configurations were considered and proposed for testing. Each structure consists of three tapes laminated together in order to provide the same conditions for each configuration.

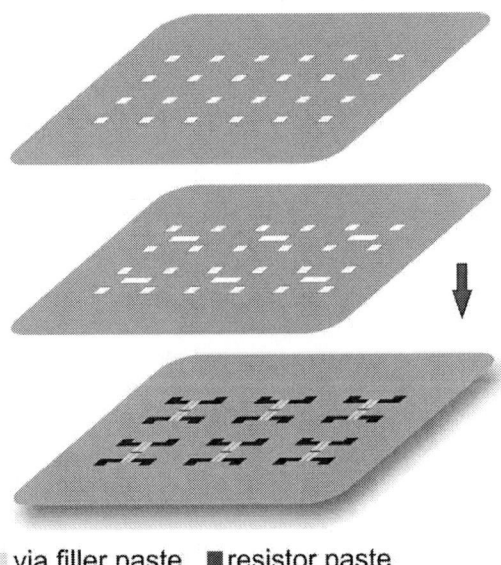

via filler paste ■resistor paste
tape ■ conductor paste ☐ air gap

Figure 1: The stacking principle of LTCC ceramic tapes in order to provide the embedment of resistors structures on the inside of cavities.

Principle of the structure stacking is shown in Figure 1 which illustrates the configuration with integrated cavity. The testing pattern which is applied on bottom tape (base) has the same dimensions for each configuration. The perforations on the middle tape form the embedded cavities and the top tape layer constitutes the cover. All s applied samples are built up of the same number of tapes. For the second test configuration the resistor array is embedded between two tape layers and an additional tape layer is applied on the bottom layer in the order

to provide a sufficient mechanical strength of test samples. Similar configuration is used for the testing structure without cavities but the middle tape is not present and the sample is complemented with one "blank" tape at the bottom. The third configuration consists of three layers of tape acting as a substrate for the resistor array. It corresponds to a conventional substrate arrangement. The base complemented with two blank tapes in order to each configuration consists of the same number of layers and to provide the same condition (e. g. after firing shrinkage) for each case.

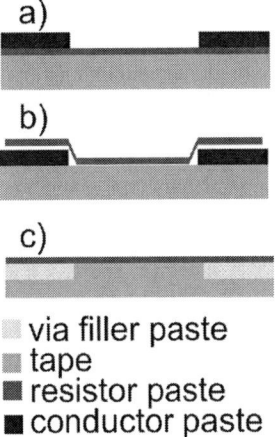

via filler paste
tape
resistor paste
conductor paste

Figure 2: Principle of formatting of the resistor bridge. a) Printed resistor is overlapped by conductive paste, b) conductive paste overlapped by resistor (wrong definition of the resistor length), c) approach chosen in this study – filled grooves overprinted by resistor.

The set-up of resistor array enables a resistor measurement by means of the Kelvin-method. Each sample is carrying a resistor array comprising 6 elements with the same aspect ratio or dimensions, respectively. Resistor arrays with four sets of different aspect ratios (N=1, 2, 3, 5) are provided for measurement. A substrate test base is designed like an array of six patterns formatted in Kelvin clips bridged by resistor (see Figure 1, 2). Each pattern on single base has the same dimension. That allows carrying out six measurements for each sample and setting arithmetic average of each design and detecting and eliminating possible wasters. Four different dimensions of resistors for each configuration were built-up: one, two, three and five squares.

In order to avoid geometrical distortion of resistor elements by overprinting the bars of contact pads which may affect the geometrical resolution especially of resistor elements of small dimensions an innovative approach has been applied. The bar of the conductor pad has been formed by laser machining a groove in the LTCC-tape which has been filled up with a conductor paste. It has been proved advantageously to use for groove filling a via

978-1-4244-4722-0/09 $25.00
© 2009 IMAPS-ITALY

filler paste. The embedded contact bar forms a uniform surface with that of the tape. This enables the overlapping of conductor bars without any deformation of the resistor path in the contact area which is usual for conventional applications. These procedures enable an exact definition of resistor geometry on the LTCC-substrate. This method is chosen in order to define the accurate resistor dimensions and to avoid warping of the structure (Figure 1, 2). Two conductor lines are bridged by resistors. The resistor is normally printed onto a tape and overlapped by conductor paste to make the interconnection with other circuitry (see Figure 2b). However, the risk of after-burnout warping due to different thermal expansion of materials and stacking of metallization is increased in this case. Therefore, the conductor lines were designed as grooves filled by LTCC via filler paste.

The CT700 LTCC ceramic tape material supplied by Heraeus company is used in this study. The thickness of this material in "green" state is 240 µm. A "ROFIN SINAR RSM 100D" computer controlled diode pumped NdYAG-laser equipped with a galvo-beam deflection-system has been used micro machining of the LTCC-tapes, whereas the groove machining is carried out by a ablation process as a serpentine laser cut. Each cut is performed several times with low power laser settings in order to attain smooth grooves.

Grooves are filled up with via filler paste by commonly used stencil printing technique. Via filler is dried for 10 minutes at a temperature 75 °C. This process is performed two times in order to eliminate the volume shrinkage of via filler paste caused by drying. Resistor pastes are screen printed over the conductor lines and the rest of circuitry is constituted by conductive contact pads carried out by screen printing technique as well. Applied conductive and via filler pastes are silver based compositions.

Each sample is prepared by stacking and laminating with the same parameters: at a temperature of 70°C, for 2 minutes and at a pressure of 60 bars. Samples are fired in the belt furnace in accordance with the temperature profile recommended by supplier (for 10 minutes at a temperature 850°C). Measured profile is shown in Figure 3.

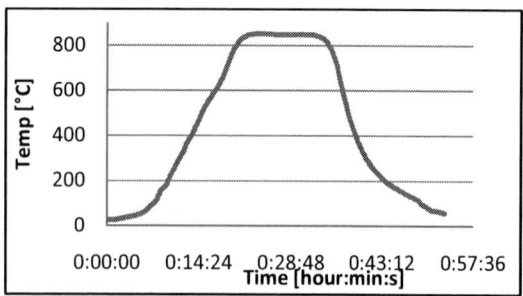

Figure 3: Temperature profile used for firing of samples.

Measurement

A fully automatic test set-up has been established to record the temperature dependence of PTC-elements. The samples to be tested are placed on a precision heating chuck. To provide a uniform temperature distribution across the sample area an additional heater is placed above the sample to avoid any heat dissipation by convection or heat conduction by the attached prober needles for carrying resistance measurement.

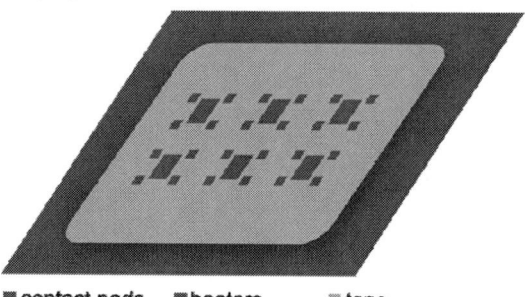

■contact pads   ■heaters   ▓tape

Figure 4: The principle set-up for measurement.

The so-called Kelvin (4-wire) method is applied for the resistance profile measurement whereas a computer controlled multi-channel electronic measurement unit (HBM Spider 8) is used. A sampling period 2s is chosen for temperature reading.

Results

As mentioned above, resistors values were designed with aspect ratios: 1, 2, 3 and 5. The dependence of the average sheet resistance in temperature was evaluated. Results of measurement of the resistance temperature profile for tested paste compositions are shown in graphs in Figures 5, 6 and 7 and their corresponding evaluated TCRs are summarized in Table 2. The influence of length / width ratio is shown at Figures 8, 9 and 10.

Table 2: Evaluated TCR for each tested pastes and configurations.

	2611-SP	2611 -I	2611 – 1A
Nominal [ppm.K⁻¹] Adjusted on alumina substrates:	4000±100	3000±400	3000±400
On surface [ppm.K-1]	4156	3256	2894
place din cavity [ppm.K-1]	3799	2979	2781
embedded [ppm.K-1]	3245	2379	2680

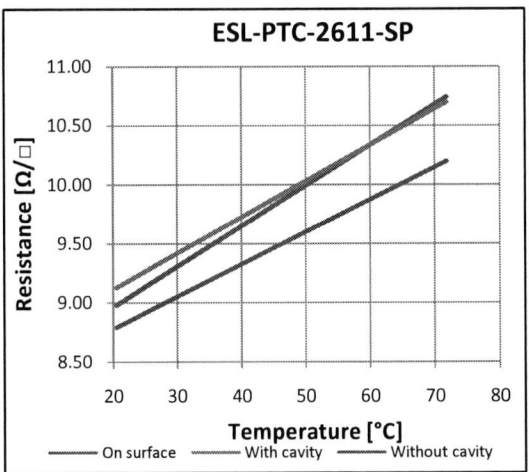

Figure 5: Sheet resistance vs. temperature characteristics of 2611-SP paste.

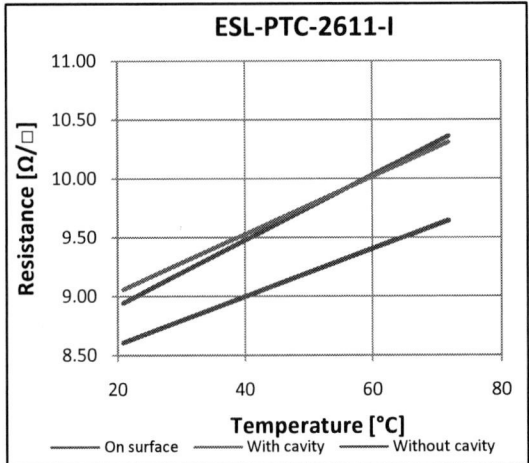

Figure 6: Sheet resistance vs. temperature characteristics of 2611-I paste.

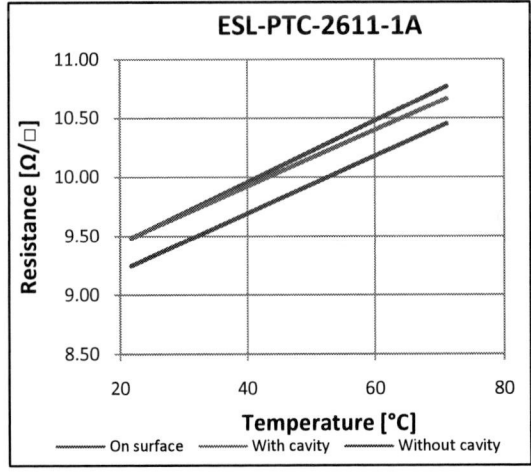

Figure 7: Sheet resistance vs. temperature characteristics of 2611-1A paste.

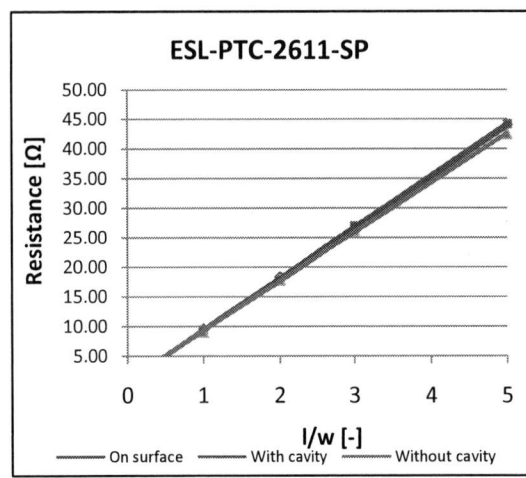

Figure 8: Resistance characteristic in dependence on aspect ratio (samples built-up with PTC-2611-SP paste) at a temperature 20°C.

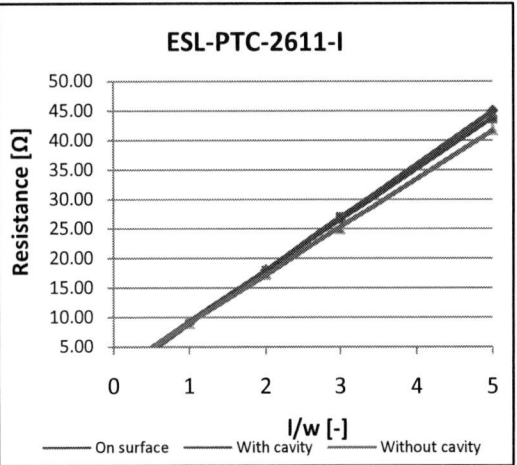

Figure 9: Resistance characteristic in dependence on aspect ratio (samples built-up with PTC-2611-I paste) at a temperature 20°C.

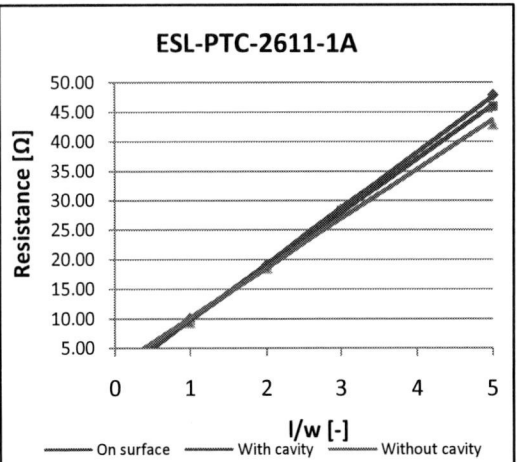

Figure 10: Resistance characteristic in dependence on aspect ratio (samples built-up with PTC-2611-SP paste) at a temperature 20°C.

978-1-4244-4722-0/09 $25.00
© 2009 IMAPS-ITALY

Summary

All the tested PTC resistor compositions are good candidates for acting as thermal sensing elements or compensation in integrated LTCC devices. The TCR values attained for PTC-elements placed on the surface of the LTCC-tape or within a cavity correspond to the values attained on alumina substrate within limits of tolerance. Only samples embedded in tape layers show a decrease in TCR-values. The integration of PTC elements in on LTCC-tape layers does not affect the linearity of the resistance vs. temperature characteristic of the resistive temperature sensor elements. In comparison with PTC-elements applied on alumina substrates samples applied on LTCC-substrates show a reduced sheet resistance. This drop in sheet resistance is pronounced for resistor elements embedded in LTCC-tapes.

Acknowledgements

The authors are very grateful to Dr. A. Kipka and Ch. Modes (Heraeus) for the supply of the ceramic tapes.

This work has been integrated within EU 4M Project (Contract Number NMP2-CT-2004-500274).

References

[1] P. R. van Loan, "Conductive ternary oxides of ruthenium, and their use in thick film resistor glazes",Ceram. Bull., Vol. 51, No. 3, pp. 23 1-233, 1972

[2] W. Pierce, D. W. Kuty, J. R.Larry, "The chemistry and stability of ruthenium based resistors", Solid State Technol., Vol. 25, No. 10, pp. 85-93, 1982

Processes for Integration of Microfluidic Based Devices

D.P.Webb[*], P.P.Conway, D.A.Hutt, B.J.Knauf and C.Liu

School of Mechanical andManufacturing Engineering, Loughborough University,
Loughborough, UK, LE11 2RN

*Contact details: D.P.Webb@lboro.ac.uk, +44 1509 227678

Abstract

Lab-on-a-chip and microfluidic device technology is in the early stages of commercialization. A major market is medical point-of-care (POC) and other kinds of portable diagnostics. Such systems require assembly and fluidic interconnection among the microfluidic elements and other components. However, microfluidic chips have until now been produced in relatively small numbers. While the structuring methods to make the fluidic channels are based on well established micro-manufacturing techniques such as lithography and embossing and are in general suited to mass manufacture, packaging and assembly methods require development for larger manufacturing volumes. One problem is with reliable sealing of a capping layer to the microfluidic layer for large area polymer wafers. In conventional thermocompression bonding the high pressures and need to heat the whole thickness of the polymer stack leads to potential distortion of the microfluidic structures. We present results on trials on the use of low frequency induction heating to deliver heat directly to the bonding interface thus permitting the use of lower bond pressures. The selection and structuring of susceptor materials is reported, together with analysis of the dimensions of the heat affected zone. Acrylic plates have been joined using a thin (<10 μm) nickel susceptor providing a fluid seal that withstood a fluid pressure of 590kPa. Another problem is how to provide support structures for connection of capillary tubing or for directly connecting microfluidic chips together to form systems. We propose the use of polymer overmoulding to create a meso-scale fluidic manifold, with anticipated advantages for manufacturing of ease of assembly and low part count. Results on the reliability of the fluid seal achieved by direct adhesion between overmould materials and glass are presented. A demonstrator overmoulded structure to connect a glass microfluidic chip to capillary tubing, using a compliant grommet for sealing, is also reported.

Key words: Microfluidic device, integration, interconnection, sealing, overmoulding, induction heating

Introduction

Microfluidics (also known as lab-on-a-chip or LOC) is the technology of handling micro-litre to pico-litre volumes of fluids. Fluids flow through micron-sized channels formed in the surface of silicon, glass or polymer substrates. Fabrication is by lithography for silicon or glass, and by moulding or forming techniques for polymers. The small size of the channels means the fluid dynamics are dominated by wall interactions, so that the surface properties of the channel walls have a strong influence on device functioning. Devices can rely entirely on capillary forces to drive flow, or external fluid pressurisation and control can be used. In general microfluidic networks are passive, with integrated pumps and valves being still relatively rare [1].

When used for chemical and biochemical analysis the advantages of microfluidics include highly controllable reaction conditions, consumption of reduced amounts of reagents and samples, improved sensitivity over macro-scale assays and small physical size.

The process advantages have lead to microfluidics being used extensively for life-science research, along with the closely related area of microarrays (biochips) [2]. Microfluidics supports multiple sequential chemical or biochemical operations on a single substrate, whereas micro-arrays support multiple, parallel identical or similar reactions. Microfluidic chips used for research are typically made in small batches, often in university and research facilities rather than by commercial manufacturers.

The miniaturisation capability makes microfluidics attractive as a technology to support point-of-care (POC) diagnostics, in which analysis is carried out as close as possible to the point of sampling, i.e. the patient. The advantages of POC analysis include reduction of costs, streamlining of healthcare, improvement of clinical outcomes and enabling rapid intervention in critical situations [3]. Exploitation of microfluidics for POC and consumer health requires scale-up of processes used for manufacturing, packaging and integration.

978-1-4244-4722-0/09 $25.00
© 2009 IMAPS-ITALY

Integration and Packaging Requirements

The ideal POC device would require no sample pre-treatment, could be operated by a lay user, and give unambiguous results. The most successful current POC and consumer health technology, which fulfils these requirements, is the lateral flow test [4], used for example in consumer pregnancy testing products. These work using a wicking substrate that causes the analyte to flow over test regions functionalised with an antibody or antigen. However the range of assay types serviceable with this principle and the capacity for quantitative analysis is limited.

Some commercial microfluidics based POC diagnostic systems already exist, such as the i-STAT system (www.abbottpointofcare.com). Disposable i-STAT cartridges also rely on capillary forces, to drive a blood sample through a fluidic channel and past a set of sensors. The cartridge plugs into a reader which outputs the clinical data. Packaging requirements are thus limited to assembly of the cartridge and electrical connections to the reader. Future POC microfluidic applications will involve assays of increasing complexity [3] requiring interconnections that support pressure driven flow [5], and methods of integrating multiple miniaturised fluidic elements together.

A typical microfluidic chip might be as depicted in Figure 1 (although there are variations, for example both the substrate and lid may be structured). Techniques for creating the microfluidic channel network in the fluidic substrate are mostly borrowed from the silicon processing (photolithography and micromachining)[6] or plastics processing (embossing, injection-moulding) [7] sectors and are suited to large production volumes. Integration processes are more problematic to productionise. These include bonding the lid to the fluidic substrate, and making of fluid connections to the fluidic input/output parts. Fluidic connections can be to tubing, typically polymer or glass capillaries, or to a fluidic backplane (see the Current Microfluidic Connectors section below).

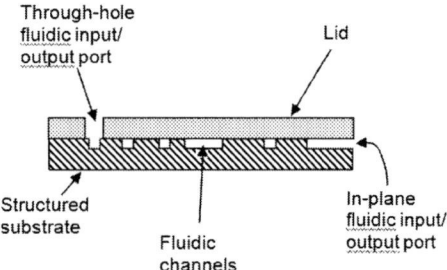

Figure 1. A typical microfluidic chip.

Lid Bonding Processes

Lid bonding of microfluidic systems can be regarded as first level packaging. Hence, many bonding methods for glass and silicon microfluidics are similar to those used with MEMS, i.e. thermocompression and adhesive bonding. The sealing process must make the microfluidic system leak-proof at working pressures, while not causing blockages or distortion of the microfluidic channels or affecting the wall properties. If an additional wetted material is introduced to make the bond, it must be benign to the working fluid.

Lid bonding for thermoplastic polymer microfluidics is of particular interest, because the much lower material costs make thermoplastics the materials of choice for large manufacturing volume applications such as POC. However, uniform bonding of thermoplastic plates over large areas is more difficult than with glass or silicon because of they are less flat. Hence polymer devices tend to be bonded individually, increasing cost. However the economies of scale from making multiple devices on the one substrate are less significant with microfluidic devices compared to microelectronics, because microfluidic chips are relatively large, typically 25 mm × 75 mm.

There are multiple methods of bonding thermoplastics including thermocompression, adhesive, ultrasonic, solvent, and microwave bonding, and laser welding. The use of these techniques for microfluidic lid bonding was reviewed by Tsao et al [8]. Tape bonding has also been reported [9]. Thermocompression bonding is the most direct method but requires accurate control of time, pressure and temperature to produce good bonds without distorting microstructures.

Induction Heating for Lid Bonding

The authors have proposed the use of low frequency induction heating to lid thermoplastic microfluidic devices [10]. The anticipated advantages of the method are rapid joining without the requirement for a chamber and low distortion of channels, potentially allowing large area or high throughput bonding with moderate pressures. LFIH is well established for bottle cap sealing, which is a very high throughput application. It is also extensively used in the steel industry for hardening, melting, soldering, welding, and annealing, and increasingly is finding application in other areas like heating fillings in dental medicine. Additionally the process is energy efficient. To join plastics a susceptor (a metal component that absorbs energy from the induction field) placed at the joining interface can be used. Hence heat is delivered directly to the joining interface, reducing the potential for heat distortion in the part.

In initial feasibility trials [10] PMMA plates 2 mm thick and 25 mm×25 mm in area were joined by placing a thin film susceptor between the plates, clamping and heating in an induction field. Visible melting of the polymer and strong joints were formed in seconds with use of a 7.5 µm nickel foil as susceptor, see Figure 2. The joints were pressure tested with air to 5.9 bar without failing.

978-1-4244-4722-0/09 $25.00
© 2009 IMAPS-ITALY

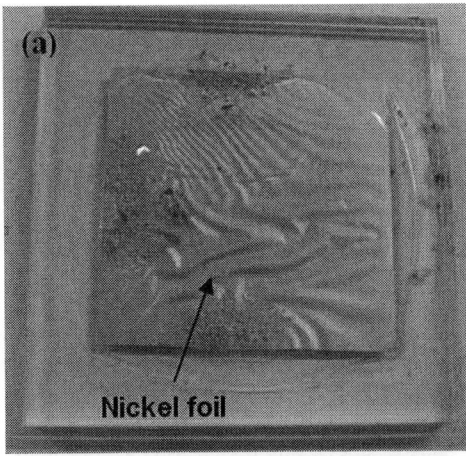

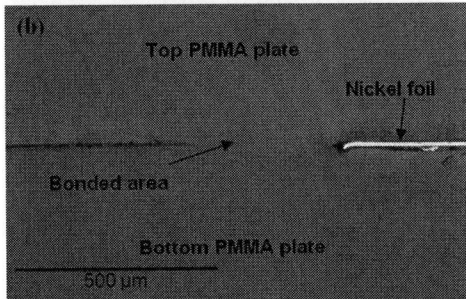

Figure 2. (a) 2mm thick PMMA plates joined using a 7.5 µm nickel foil susceptor, (b) cross section of joint.

A 5µm plated nickel coating was also found to heat sufficiently to melt the PMMA. However, because the layer covered the whole of the surface of one of the plates no plastic-plastic bond was formed, and the plastic-metal bond was found to be weak. Because it is desirable that the susceptor be as thin as possible, a 50-100nm evaporated nickel coating was also tested and was found to heat but not sufficiently to melt the PMMA. While ferromagnetic materials such as nickel have the strongest coupling to the induction field, a 5 µm thick coating of sputtered aluminium which is paramagnetic was also found to heat rapidly. Current work concentrates on optimising the LFIH process parameters to achieve bonding with minimum heat affected zone (HAZ) width.

Current Microfluidic Connectors

Fluid interconnection has been recognised for some time as a key area for increased research effort [1,11]. A fluidic connector system should withstand pressures from a few kPa to up to several MPa depending on the application, and have as little dead volume as possible. Again depending on application the ability to break and remake connection may be required.

While microfluidics has been a defined research area [1] for around 20 years, since the development of micro-total analysis systems (µTAS) [12], the field is only now in the early stages of commercialisation. Because of this standardized

methods and formats for fluidic connections, analogous to back end processes for microelectronics, have not yet emerged. Microfluidics fabrication is also more diverse than electronics, with devices made from a variety of materials including silicon, metal and ceramics, but the majority being made from polymers and glass.

The fragmented state of the industry makes information of what connector schemes are currently in use commercially hard to come by. One of the few general purpose connectors for microfluidic chips on the market are Nanoport couplers (www.upchurch.com). These are PEEK compression fittings to connect capillary tubing to typically glass microfluidics. The Nanoport (female part) is bonded to the chip using an adhesive preform. A compression nut screws in and compresses a compliant ferrule to hold the capillary in place and produce a fluid seal, as shown in Figure 3.

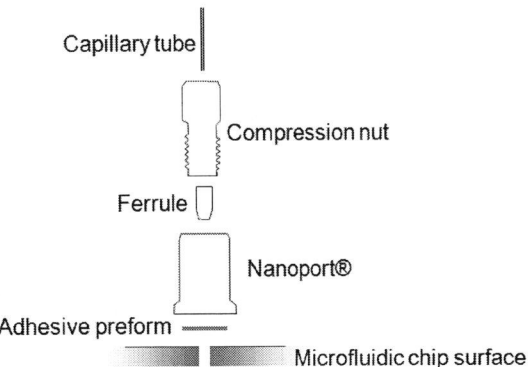

Figure 3. Nanoport® commercially available microfluidic coupler

There are many reports of proposed fluidic connection schemes in the academic literature. These are frequently referred to as the "world-to-chip interface" [13,14], and may include other types of connection e.g. electrical. While most of the concepts are intended only for research and laboratory users of microfluidic chips, some have been conceived as solutions for manufacturing on a large scale and/or are patented. The solutions can be categorised as modular systems, clamps and frames and couplers.

Modules

Modular concepts provide for the interconnection of multiple modular microfluidic elements, e.g.; chips, sensors, pumps and valves, using a fluidic backplane. An example was reported by Lammerink et al [15], who built a mixed fluid/electrical circuit board (MCB) to which the modules could be permananently connected by various bonding methods depending on the material of the module (plastic, silicon or glass). Flexible connectors for modules have also been proposed, for example by Man et al. [16], who manufactured flexible plastic connectors with large access holes on both ends which contained one or more capillaries

978-1-4244-4722-0/09 $25.00
© 2009 IMAPS-ITALY

by a multilayer process. Liu et al. [17] described a flexible connector made of three plastic layers that would be feasible for mass production.

A concept similar to levels in electronic packaging was described by P.C. Galambos and G.L. Benavides [18], for a system using two scales of packaging. The primary layer is a microfluidic chip put into an Electro-Microfluidic Dual Inline Package (EMDIP), which could be manufactured as a moulded plastic part. An example of 3D stacking approaches was described by Hasegawa and Ikuta [19]. They contained the modules in a holder unit using silicone rubber films to seal the joints.

Clamps and Frames

Clamps and frames are frequently offered by individual service microfluidic manufacturers as a rapid and reliable way of making connection to chips produced with a standardised fluidic i/o layout. As such they are useful for relatively low manufacturing volumes. An example in the academic literature was reported by Yang and Maeda [13] who used a socket with 28 pins integrated for the connection of electric signals and power. The fluid connectors in the hinged lid were equipped with built-in valves to control the flow rates.

Couplers

Couplers are a widely used method to connect microfluidic systems to external tubing. They can either be designed directly on the top layer of the microfluidic chip or be manufactured separately and bonded onto the chip afterwards. The possibilities of fixing microfluidic tubes / capillaries in these ports are manifold – some attachments are permanent, others are easy to remove. The Nanoport product described above is an example of a commercial coupler. In the academic literature Li and Chen [20] presented a PDMS based interconnector which could be bonded onto glass and PDMS microfluidic devices. The bond was made with oxygen reactive ion etching (RIE) and UV curable epoxy - the latter could be broken to detach the connector. The machinability of silicon is exploited for various designs e.g. by Meng et al. [21]. In their paper Gray et al. [22] came up with two different approaches: in the first the coupler was manufactured directly in the top layer of the microfluidic system using deep reactive ion etching (DRIE). In essence it was a hole with a stop position containing the tubing which could be fixed with glue. The second idea was moulded plastic coupler which was fixed to the microfluidic chip with heat-staked pegs and requiring a silicone gasket for the fluid seal. Moulded couplers arranged in a standardised layout integrated with the lid of a polymer microfluidic device are available from Microfluidic ChipShop in Germany [23]. The custom microfluidic network which accesses the ports is formed into the mating layer (see following section for a diagram of the parts of a microfluidic device).

Overmoulding Concept for Integration

A generic packaging method to convert a microfluidic chip from a stand-alone device, into a module that can be integrated into a larger system, has been proposed [24]. The concept is illustrated in Figure 4. The microfluidic device is embedded in a polymer overmould by injection moulding or other plastics moulding technique. During overmoulding, pins within the mould tool are used to create fluidic access channels to the microfluidic device input/output ports. The access channels provide the basis for engineering of robust connections to tubes, or other means of fluidic interconnect, by addition of suitable structures to the moulding.

The concept thus supports both couplers, and modular schemes as described in the Current Microfluidic Connectors section above. The method is intended to be compatible with mass manufacture, featuring use of high throughput processes (injection moulding), low part count and ease of assembly.

A fluid seal is required between the moulding and the microfluidic chip. This can be provided either by a flexible element such as a gasket, or by adhesion between the overmould and the material of the microfluidic device. The latter mechanism requires that the moulding material forms an adhesive bond to the material of the insert on solidification. While products incorporating insert moulded parts usually rely on mechanical interlocking for product integrity, exploitation of the phenomenon of direct adhesion of moulding resins to metals has been reported in the literature, for example by Grujicic et al [25] for automotive parts.

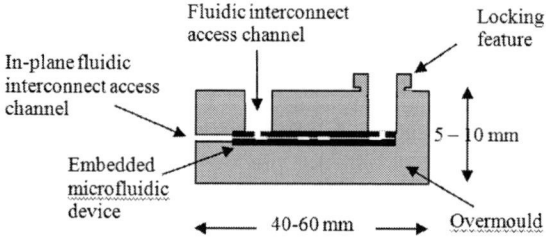

Figure 4. Concept of moulded interconnect to an embedded microfluidic device.

Overmoulding Trials

Overmouldings of glass microfluidic chips were manufactured in polypropylene, according to the design shown in Figure 5, to connect a microfluidic device with in-plane fluidic i/o ports to capillary tubing. The design features a thread and interior dimensions fitting an Upchurch 10/32 Nanotight compression fitting. The fitting consists of a PEEK compression nut and a perfluoroelastomer ferrule as shown in Figure 5. In designing the moulding there was some concern that the much lower stiffness of polypropylene compared to PEEK

978-1-4244-4722-0/09 $25.00
© 2009 IMAPS-ITALY

would mean that the ferrule would not be compressed sufficiently in use to ensure a fluid good seal. The overall thickness of the overmould was therefore set at 7mm, allowing a high wall thickness of 1.5mm in the region of the screw thread.

A finished moulding with a glass microfluidic chip insert is shown in Figure 6. The capillary was 1/16 inch outer diameter Teflon tubing. Water was observed to flow through the chip and emerge at the far end when the assembly was pressurised. Three mouldings using glass blanks instead of a microfluidic chip were pressure tested with water at 6 bar for 5 hours. A red dye was added to the water and the assemblies visually inspected at intervals for leaks. No leak was observed during the test.

In another characterisation test the force required to pull the capillary tubing out of the compression assembly was measured. The test was performed using a polypropylene moulding without a glass insert to simplify the test setup. The results were compared to pull out measurements using a Upchurch Nanoport fitting instead of the moulding as the male part of the compression assembly. For each test the capillary was inserted and the compression nut screwed in 3½ turns. This was enough to make the nut fingertight for both the Nanoport and the moulding. The tube was then pulled out and the same parts reassembled for the next test. The force measurements are shown in Figure 6. The maximum load seen in five tests with the moulding was 8.8N and with the Nanoport 7.5N. The similarity of the result demonstrates that the wall thickness of the overmould is sufficient to allow the compression fitting to work as designed.

Studies of the adhesion of several injection moulding grade commercial resins (PA12, LCP, PPS, COC, ABS) to glass have been published elsewhere [24]. The polyamide was found to be the best performer. In the most recent work the fluid seal between polyamide 12 and glass was demonstrated to survive pressurisation with water at 1.4 bar (20 psi) at room temperature for 116 hours. Current work focuses on reducing the variability in the performance observed and selection of adhesion promotion coatings.

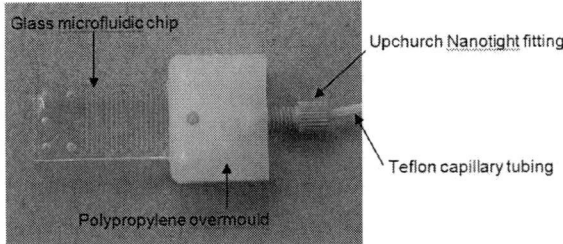

Figure 6. Glass microfluidic chip connected to Teflon capillary tubing via a polypropylene overmould

Extension of Integration Concepts

An extension of both the overmoulding and LFIH methods would be to combine them to create a process for integration of microfluidics as illustrated in Figure 7. Separate microfluidic elements, for example a channel network and a flow-through sensor, could be embedded in an overmould. The surface of the overmould would itself be structured with an interconnecting fluidic channel network. Induction heating would then be used to seal the interconnection network. An integrated module would thus be created in only two manufacturing steps. Simultaneous attachment and sealing of capillary tubing with lid sealing could also be achieved by, for example, using capillary tubing tipped with a susceptor.

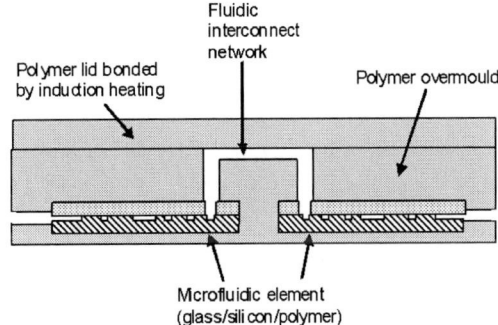

Figure 7. A concept for manufacture-friendly integration of microfluidic elements

Conclusions

Lab-on-a-chip and microfluidic device technology is in the early stages of commercialisation. A major projected market is medical point-of-care (POC) and other kinds of portable diagnostics. Such systems require assembly and fluidic interconnection among the microfluidic elements and other components. Because of the early stage of development of the industry standardized methods and formats for fluidic connections, analagous to back end processes for microelectronics have not yet emerged.

We have reviewed feasibility trials on low frequency induction heating for lidding of polymer microfluidic substrates. The anticipated advantages of the method are rapid joining without the

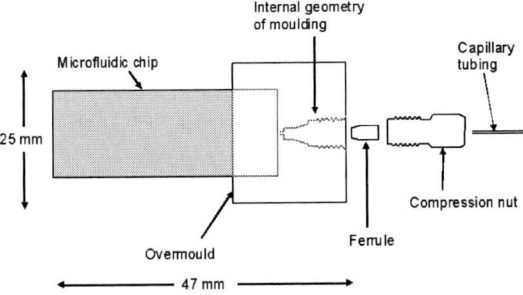

Figure 5. Design of moulding for microfluidic chip connection to a capillary tube

requirement for a chamber and low distortion of channels, potentially allowing large area or high throughput bonding with moderate pressures. The trials have shown that polymer-polymer bonds created by induction heating are stronger than metal-polymer bonds.

We have also reviewed proposed schemes in the academic literature on fluid interconnection, categorizing them into modules, clamps and frames and couplers. We have presented an overmoulding concept for integration of microfluidics, capable of supporting both modular and coupler interconnection. Connection of a glass microfluidic chip to a capillary tube has been demonstrated by overmoulding with polypropylene. The connection withstood 6 bar of fluid pressure for 5 hours without leaking.

Acknowledgements

The authors would like to thank Chandrahas Parmar for the preparation and testing of the polypropylene overmoulds. This work forms part of the 3D Mintegration Grand Challenge, funded by the United Kingdom EPSRC, grant reference: EP/C534212/1.

References

[1] Verpoorte, E. and De Rooij, N. F., Microfluidics meets MEMS *Proceedings of the IEEE*, vol. 91, pp. 930-953, Jun 6, 2003.

[2] Schena, M. *Microarray Biochip Technology*, Natick, MA, USA: Eaton publishing, 2000.

[3] Sia SK and Kricka LJ, Microfluidics and point-of-care testing. *Lab On A Chip*, vol. 8, no. 12, pp. 1982-1983, Dec 4, 2008

[4] Haeberle, S. and Zengerle, R., Microfluidic Platforms for Lab-on-a-Chip Applications *Lab on a Chip*, vol. 7, pp. 1094-1110, 2007.

[5] Ellis, M., Plastic diagnostic lab-on-a-chip *Materials World*, vol. 17, Jan, 2009.

[6] PRIME Faraday Technology Watch. *An introduction to MEMS*, Loughborough: PRIME Faraday Partnership, 2002.

[7] Becker, H. and Gaertner, C., Polymer microfabrication technologies for microfluidic systems *Analytical and Bioanalytical Chemistry*, vol. 390, no. 1, pp. 89-111, Jan 1, 2008.

[8] Tsao, C. W. and Devoe, D. L., Bonding of Thermoplastic Polymer Microfluidics *Microfluidics and Nanofluidics*, vol. 6, pp. 1-16, Jan, 2009.

[9] Nestler, J., Morschhauser, A., Hiller, K., Otto, T., Bigot, S., Auerswald, J., Knapp, H., Gavillet, J., and Gessner, T., Polymer lab-on-chip systems with integrated electrochemical pumps suitable for large-scale fabrication *The International Journal of Advanced Manufacturing Technology*, vol. 2009.

[10] Knauf, B. J., Webb, D. P., Changqing Liu, and Conway, P. P., "Packaging of polymer based microfluidic systems using low frequency induction heating (LFIH)," *International Conference on Electronic Packaging Technology & High Density Packaging, 2008. ICEPT-HDP 2008.,* Shanghai, PRC, pp. 1-6, 2008.

[11] Fredrickson, C. K. and Fan, Z. H., Macro-to-Micro Interfaces for Microfluidic Devices *Lab on a Chip*, vol. 4, pp. 526-533, 2004.

[12] Manz, A., Graber, N., and Widmer, H. M., Miniaturized total chemical analysis systems: A novel concept for chemical sensing *Sensors and Actuators B: Chemical*, vol. 1, no. 1-6, pp. 244-248, Jan, 1990

[13] Yang, Z. and Maeda, R., A world to chip socket for microfluidic development *Electrophoresis*, vol. 23, pp. 3474-3478, 2002.

[14] Yang, Z. and Maeda, R., Packaging for microfluidic devices and systems *Journal of the Japan institute of electronics packaging*, vol. 5, pp. 116-121, 2002.

[15] Lammerink, T. S. J., Spiering, V. L., Elwenspoek, M., Fluitman, J. H. J., and Vandenberg, A., "Modular concept for fluid handling systems - A demonstrator micro analysis system," *IEEE 9th Annual International Workshop on Micro Electro Mechanical Systems,* San Diego, USA, pp. 389-394, 1996.

[16] Man, P. F., Jones, D. K., and Mastrangelo, C. H., "Microfluidic plastic interconnects for multi-bioanalysis chip modules," *3rd Conference on Micromachined Devices and Components,* Austin, USA, pp. 196-200, 1997.

[17] Liu, T., Masood, S., Iovenetti, P., and Harvey, E., Development of Packaging Systems and Interconnects for Microfluidics Applications *Packaging India*, vol. 40, pp. 9-14, 2007.

[18] Galambos, P. and Benavides, G., "Electrical and Fluidic Packaging of Surface Micromachined Electro-Microfluidic Devices," *Microfluidic Devices and Systems Iii: Conference on Microfluidic Devices and Systems Iii,* Santa Clara, Ca, pp. 200-207, 2000.

[19] Hasegawa, T. and Ikuta, K., " Novel Interconnection for Micro Fluidic Devices," *Proceedings of the 2002 International Symposium on Micromechatronics and Human Science,* Nagoya, Japan, pp. 169-174, 2002.

[20] Li, S. F. and Chen, S. C., Polydimethylsioxane fluidic interconnects for microfluidic systems *IEEE Transactions on Advanced Packaging*, vol. 26, pp. 242-247, Aug, 2003.

[21] Meng, E., Wu, S. Y., and Tai, Y. C., Silicon Couplers for Microfluidic Applications *Fresenius Journal of Analytical Chemistry*, vol. 371, pp. 270-275, Sep, 2001.

[22] Gray, B. L. , Jaeggi, D., Mourlas, N. J., Van Drieenhuizen, B. P., Williams, K. R., Maluf, N. I., and Kovacs, G. T. A., Novel Interconnection Technologies for Integrated Microfluidic Systems *Sensors and Actuators a-Physical*, vol. 77, pp. 57-65, Sep 28, 1999.

[23] Gaertner, C., Klemm, R., and Becker, H., PCR-on-a-chip - a flexible microfluidic platform as the fast way to the product *MST News*, vol. 06, no. 1, pp. 40-42, Feb, 2006.

[24] Webb D.P., Hutt, D. A., Hopkinson, N., Conway, P. P., and Palmer, P. J., Packaging of microfluidic devices for fluid interconnection using thermoplastics *Journal of Microelectromechanical Systems*, vol. 18, pp. 354-362, 2009.

[25] Grujicic, M., Sellappan, V., Omar, M. A., Seyr, N., Obieglo, A., Erdmann, M., and Holzleitner, J., An overview of the polymer-to-metal direct-adhesion hybrid technologies for load-bearing automotive components *Journal of Materials Processing Technology*, vol. 197, no. 1-3, pp. 363-373, 2008.

High Linearity and Broadband WiMAX Power Amplifier Design Using Board Level Integration Technology

Wei-Ting Chen, Kuo-Chiang Chin, Cheng-Hua Tsai, Li-Chi Chang,
Yung-Chung Chang, and Chang-Chih Liu

Industrial Technology Research Institute (ITRI), Rm. 168, Bldg. 14, 195, Sec. 4, Chung
Hsing Rd., Chutung, Hsinchu, Taiwan 31040, R.O.C.

chenwt@itri.org.tw

Abstract

WiMAX is an innovative communication technology which is generally applied in latest commercial communication products. It uses of Orthogonal Frequency Division Multiple Access (OFDMA) modulation to suit multi-path environments that gives network operators higher throughput and capacity. Although this technology is of much benefit to users and applications, it is a real challenge of hardware circuit design, especially on power amplifier (PA) circuits. One reason is that OFDMA modulation scheme in WiMAX system requires very high linearity of PAs. Thus, the design methodology of board level integration which is one solution of system in package (SiP) is proposed to solve non-linear effect and thermal problem of PAs. Multilayer organic substrate process and high dielectric constant material (DK 40) is used to realize the design. Eventually, 50% of the entire passive components, such as the output matching circuit, RF chokes, and some of the decouple capacitors of power amplifier have been integrated into the substrate. The measured data of small-signal response for the designed PA is with 29dB gain, -12dB input reflection, -22dB output reflection, and the reverse isolation is more than 45dB between 2.5 and 2.7GHz. And with WiMAX OFDMA/64-QAM modulation signal testing, under error vector magnitude (EVM) conditions at -30dB, the designed PA can output 24.5dBm at 2.499GHz; 24dBm at 2.599GHz; 24dBm at 2.68725GHz, fully satisfying the standard requirements. These studies show that board level integration technology can be utilized in linearity demands and high power circuit designs, and can substantially shrink the sizes of the circuis and the cost. Furthermore, it also provides a new solution for next generation communication systems.

Key words: WiMAX, System in Package (SiP), Embedded Passives (EP), Power Amplifier (PA).

Introduction

Worldwide Interoperability for Microwave Access (WiMAX) is an emerging communication technology based on the IEEE 802.16 standard. It can coexist and inter-work with existing and innovative technologies, both wired and wireless. For example, WiMAX and Wi-Fi are complementary and are expected to be incorporated in dual-mode chipset in mobile devices, as WiMAX provides wider coverage, while Wi-Fi is better suited for high-throughput and indoor LAN application. WiMAX also addresses the requirements of those subscribers that want to be able to use their broadband connection regardless of location, functionality that DSL and cable modem services do not support [1].

The adoption of Orthogonal Frequency Division Multiple Access (OFDMA) modulation makes WiMAX provide high speed data rate and blanket large area. This complex WiMAX signal presents a very tough challenge to the circuit designers. One reason is that OFDMA modulation requires very high linearity, especially on the power

amplifier circuits. Moreover, since the primary WiMAX application is mobile internet access, the power amplifiers have to meet other basic requirements, such as large band width, high efficiency and low operation voltage [2].

This paper presents a board level integration technology to design a WiMAX power amplifier. Originally, SMD components are adopted in the power amplifier design, and then the plenty of the components are finally realized by the multi-layer substrate structure with Hi-DK material in this research. The final power amplifier circuit meets the WiMAX system requirements in the frequency band of 2.5 - 2.7GHz.

Sustrate Structure and Materials

To reduce the size and cost of the power amplifier, the circuit is designed by using Hi-DK material, and laminated in organic substrate. The Hi-DK material can efficiently help designers to shrink the electrical length of the circuits and easily create the capacitive components. The characteristic of the Hi-DK material is measured by RF cavity fixture and the result is shown in Fig. 1. The DK value is 30

978-1-4244-4722-0/09 $25.00
© 2009 IMAPS-ITALY

and the DF value is 0.06 at 2.6GHz. This material is compatible with standard printed circuit board process. Fig. 2 displays the substrate structure that consists of six copper layers and five dielectric layers and is symmetrical. The dielectric layer 1-2 and 5-6 are built up with the lower dielectric material (ITEG 1078) which mainly for the RF signal processing; and dielectric layer 2-3 and 4-5 are also build up with the Hi-DK material to reduce the size of the embedded components. The dielectric layer 3-4 is a material carrier (EM320), supporting the mechanical strength for other materials [3, 4, 5, 6].

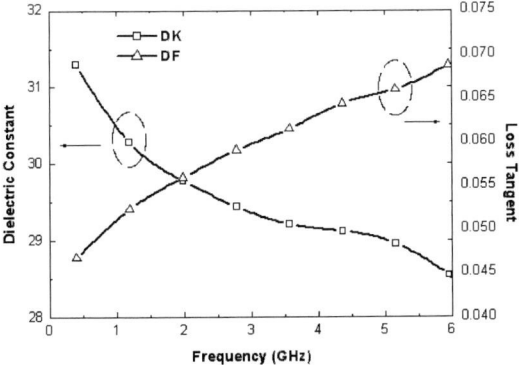

Figure 1: Characteristic of the Hi-DK material.

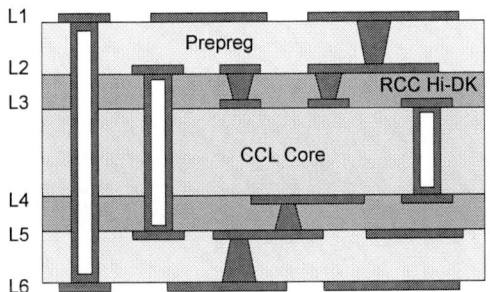

Figure 2: Substrate structure applied to the power amplifier circuit.

Power Amplifier Design

The AP3011 power amplifier (produced by RFIC Technology Corporation) is served as a test vehicle in this deign, and it is a discrete component in 16 pin QFN package. Fig. 3 is the scheme of the power amplifier offered by the manufacturer. In this research, according to the conditions of the material and substrate structure, there are several evaluated components can be embedded into the substrate, including 2 inductors (The values are 1.2nH and 2.2nH) and 6 capacitors (The valsues are 1.5pF, 3.3pF, 10pF, 47pF and 10nF). The arrangement of layer definition is listed as follows:

1. L1 & L6: Signal layer that provides RF signal processing and pads for SMD components.
2. L2 & L5: Ground layer for DC power and RF signal reference, containing one electrode for

MIM embedded capacitors.

3. L3 & L4: DC power layer for power amplifier and provides another electrode for MIM embedded capacitors.

Through the proposed structure, the inductive or capacitive components can be easily accomplished. Fig. 4 shows the Layer1 layout of the power amplifier circuit, the 1.2nH meander inductor is designed to be inductor L2, and MIM capacitor structure is used to realize the capacitor C6 and C9 with capacitance 47pF. The power amplifier output matching circuit consists of L1, C1, C2 and C3 which are also made from the multi-layer substrate structure and Hi-DK material, is displayed in Fig. 6. One of the decoupling capacitor C8 with value is 10nF can be achieved between power plane and ground plane. Other decoupling capacitors C4, C5 and C7 still use the SMD components, because their values are larger than 1uF. Through this board level integration technology, 6 capacitors and 2 inductors are embedded into the substrate, occupying 50% of the total passive components, and the size of the power amplifier circuit is $330 \times 300 \, \text{mil}^2$ [7, 8, 9].

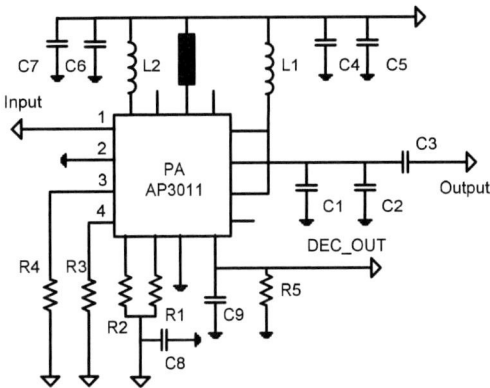

Figure 3: Scheme of the power amplifier.

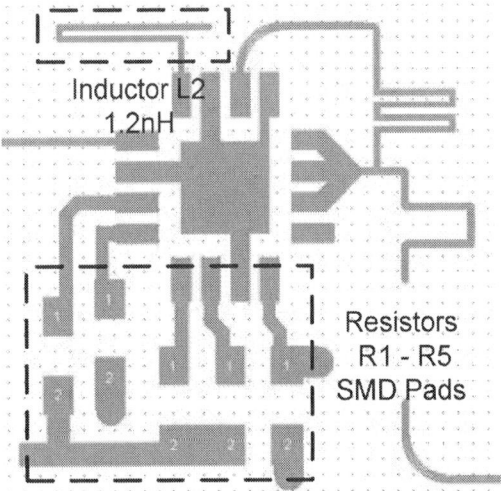

Figure 4: Layer1 layout of the power amplifier circuit.

978-1-4244-4722-0/09 $25.00
© 2009 IMAPS-ITALY

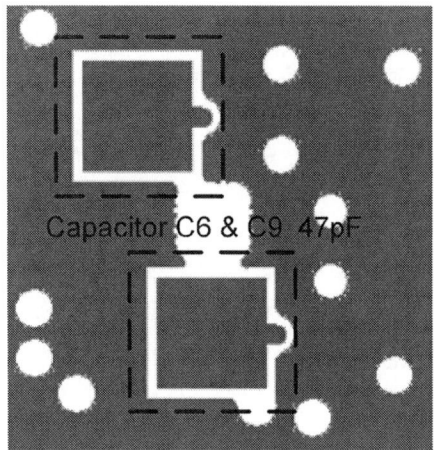

Figure 5: Layer4 layout of the power amplifier circuit.

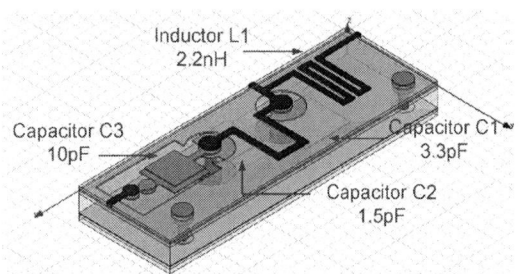

Figure 6: 3-D EM structure of the power amplifier output match.

Measurement Results

Fig. 7 shows the photograph of the power amplifier matching circuit. The comparison with 3-D EM simulation and measurement result is illustrated in Fig. 8. Because of the process variation, the matching point slightly moves to higher frequency. The final power amplifier circuit under test is composed by SMD resistors for circuit bias and SMD capacitors for DC decoupling. The small-signal response of the designed power amplifier is tested by the vector network analyzer from 45MHz to 6GHz. The measured data is shown in Fig. 10; 29dB gain is obtained with -12dB input reflection, and at least -22dB output reflection from 2.5GHz to 2.7GHz. The reverse isolation S_{12} is more than 45dB for the entire testing bandwidth. Fig. 11 shows the power amplifier performance of the modulation distortion quantities which is measured by the vector signal analyzer with 802.16e OFDMA/64-QAM signal [10] between 2.5 and 2.7GHz. Under error vector magnitude (EVM) conditions at -30dB, the power amplifier can output 24.5dBm at 2.499GHz; 24dBm at 2.599GHz; 24dBm at 2.68725GHz, fully satisfying the WiMAX standard requirements. Table 1 summarizes the circuit performances of the embedded and SMD power amplifier. One can find that the characteristic of embedded version power amplifier is similar to the SMD version power amplifier.

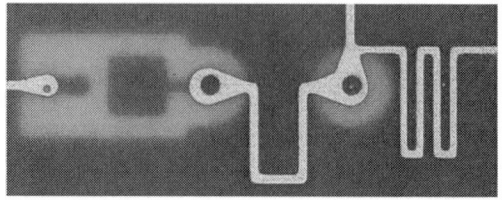

Figure 7: Photograph of the power amplifier output match circuit.

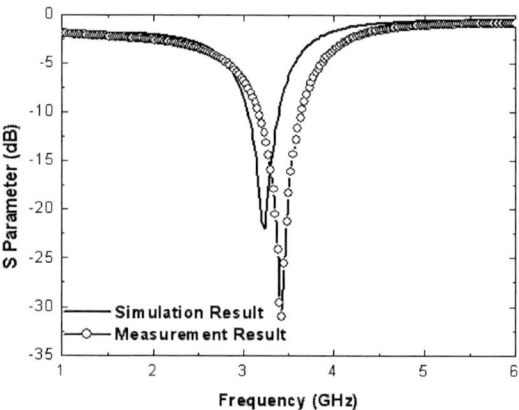

Figure 8: Measured results of the power amplifier output match circuit.

Figure 9: Photograph of the power amplifier circuit.

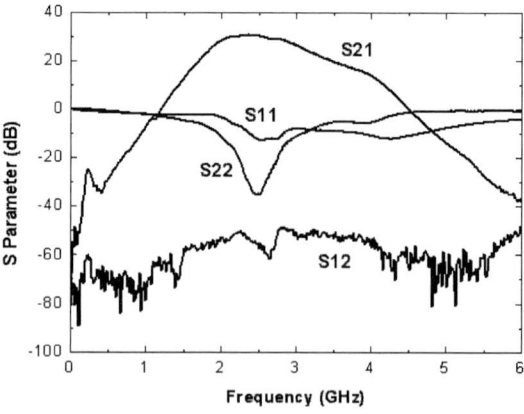

Figure 10: Measured small-signal response of the power amplifier.

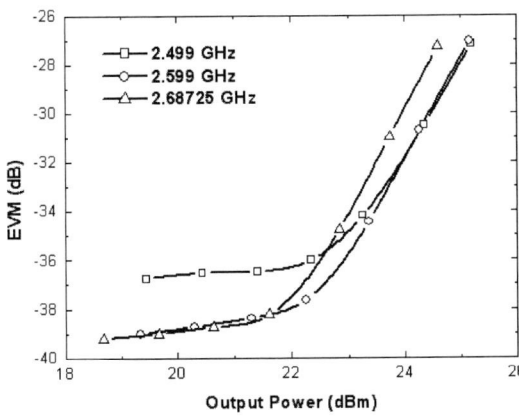

Figure 11: Measured results for 802.16e modulation signal.

Table 1 Summary of power amplifiers performance.

PA Version	SMD	EP
Frequency	2.5 - 2.7GHz	2.5 - 2.7GHz
S11	>22dB	>12dB
S22	>12dB	>22dB
S21	>29dB	>29dB
EVM (-30dB) 2.499GHz	25dBm	24.5dBm
EVM (-30dB) 2.599GHz	24.5dBm	24dBm
EVM (-30dB) 2.68725GHz	23.5dBm	24dBm

Conclusion

In this work, a high linearity and broadband WiMAX power amplifier has been presented and the 50% passive components have been integrated into the substrate to shrink the circuit size and the cost. This proposed board level integration technology by the multi-layer structure with Hi-DK material is verified via the final measurement results. The power amplifier reveals good EVM performance and can be used in WiMAX communication system. Further integration on other embedded sub-circuit components, such as filters, Balun and diplexer in the same process will still be developed.

Acknowlededment

The authors gratefully acknowledge ITRI/MCL, CCP and HannStar Board Corporation for their technical support with the materials and circuit fabrications during this work.

Reference

[1] WiMAX Forum, "Mobile WiMAX: The Best Personal Broadband Experience", WiMAX Forum, June 2006.

[2] Ping Li, and Paul DiCarlo, "Overview of WiMAX System and Related Power Amplifier Design", *Proc. 9th International Solid-State and Integrated-Circuit Technology Conf.*, Beijing, China, Oct. 2008, pp1361-1364.

[3] C. L. Weng, P. S. Wei, etc, "Embedded Passives Technology for Bluetooth Application in Multi-layer Printed Wiring Board (PWB)," *Proc. 54th Electronic Components and Technology Conf.*, Las Vegas, Nevada, May 2004, pp. 1124-1128.

[4] U. M. Jow, Y. J. Lai, C. L. Weng, C. S. Cheng and C. S. Shyu, "Functional Embedded RF Circuits on Multi-layer Printed Wiring Board(PWB) Process," *Proc. 55th Electronic Components and Technology Conf.*, Lake Buena Vista, Florida, May 2005, pp. 1634-1641.

[5] U. M. Jow, C. L. Weng, Y. J. Lai, C. S. Cheng and C. S. Shyu, "Embedded Passives on Multi-Layer Printed Wiring Board (PCB) for 5 GHz Front-end Module," *Proc. 56th Electronic Components and Technology Conf.*, San Diago, California, May 2006, pp. 1331-1337.

[6] C. H. Tsai, C. S. Chen, C. L. Wei, K. C. Chin, W. T. Chen and C. S. Shyu, "Embedded Passives in Multi-dielectric Layer Printed Wiring Board for IEEE 802.11a/b/g Tri-mode Dual-band Wireless Carbus Adapter," *Proc. 16th European Microelectronics and Package Conf.*, Oulu, Finland, June 2007.

[7] D. Pozar, Microwave Engineering, 2nd Ed., John Wiley, 1998.

[8] B. Razavi, RF Microelectronics, Prentice Hall, 1998.

[9] S. C. Cripps, RF Power Amplifiers for Wireless Communications, Artech House, 1999.

[10] Agilent Technologies, "IEEE 802.16e WiMAX OFDMA Signal Measurements and Troubleshooting, AN_1578", Agilent Technologies, June 2006.

Influence of the Fabrication Errors on Multilayer Thick Film Circuits

Wesam Ali[1], and Chunwei Min

[1]Public Authority of Applied Education & Training, College of Technological Studies, State of Kuwait

wesamma@hotmail.com

Abstract

The effects of fabrication errors in multilayer millimetre-wave components have been investigated, both through simulations and measurements on practical circuits. This paper provides useful guidelines to the circuit designers on the magnitude of fabrication errors that are acceptable, and presents data not previously reported in the literature. A particularly significant error that was quantified was that of skew between conductors on different layers, where it was found that a skew angle of only 0.1˙ resulted in very significant changes in bandwidth and insertion loss. The work was supported by a detailed investigation on a 35GHz, multilayer edge-coupled band-pass filter, which was fabricated on alumina substrates using photoimageable thick film process.

Key words: Multilayer, Photoimageable Thick Film Technology, Fabrication Error, Millimetre-Wave

Introduction

New data are presented on the microwave effects of fabrication errors in multilayer passive components. This has particular significance for modern hybrid integrated circuits, where developments in materials technology allow the designer to simultaneously improve both electrical performance and circuit packing density through the use of multilayer structures. Whilst these multilayer techniques provide the microwave and millimetre-wave circuit designer with the opportunity to design components in three-dimensions, and often at low-cost, these advantages can be lost if there are significant errors in the fabrication process. The present work has provided a critical analysis of fabrication errors, both through simulation and measurements on practical components at millimetre-wave frequencies.

Multilayer, edge-coupled band-pass filters were chosen as the principal type of component for the investigation because they are especially sensitive to errors in the fabrication process. For cost effectiveness in modern circuits, the filters need to be compact with high electrical performance [1]. In single layer filter designs, considerable adverse performance effects arise from fabrication errors, particularly in the gap size of the end-coupled sections. Multilayer structures offer more flexibility in the design of circuits, and the strong coupling is frequently needed in the design of filters. Multilayer structures can be obtained easily by overlapping multilayer coupled lines, without the need for small gaps [2]. Because of the limitations of fabricating small gaps, multilayer circuits require precise layer alignment with high resolution between the conductor layers. Photoimageable thick-film technology was chosen for current investigation because it enables both single layer and multilayer filters to be produced conveniently so that a direct comparison could be made. The work had shown that the photoimageable thick film technology is capable of realizing the circuit quality necessary for high performance microwave and millimeter-wave components. It had been shown that the multilayer filters offer benefits in overcoming the limitation of the conventional single layer components.

The multilayer edge-coupled band-pass filters (ECBPFs) reported in this paper were fabricated using photoimageable thick film process. Printed layers were deposited on 96% alumina substrates that had a thickness (h_1) of 635µm and relative permittivity (ε_{r1}) of 9.5. The conductors forming the coupled sections of the filter were separated by a printed dielectric layers (h_2= 10µm and ε_{r2}= 3.9), as shown in Figure1.

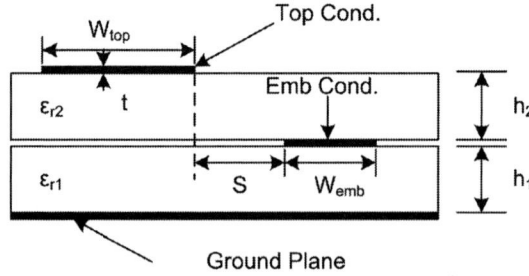

Figure 1: Configuration of a Multilayer Coupled-Line Structure

Filter Design and Fabrication Process

The design procedure for the multilayer ECBPFs has been well documented in the literature [2, 3]. A full-wave based analysis in conjunction with the specified modal parameters enabled a procedure for the filter design to be developed. There are three main electrical parameters used for the design of coupled-line structures, namely, modal impedance, effective dielectric constant, and mutual coupling coefficient. Modal impedances ($Z_{c,\pi}$) decide the corresponding widths for the top and embedded lines, which can be obtained using equation (1).

$$Z_0 = \sqrt{Z_{ct}Z_{\pi t}} = \sqrt{Z_{ce}Z_{\pi e}} \qquad (1)$$

The electrical length (θ) of the coupled lines is theoretically a quarter guide-wavelength ($\lambda_g/4$, $\pi/2$) long. Open-end effects of this structure can be represented by an additional length (ΔL) of the line that needs to be subtracted from the nominal $\lambda_g/2$ length. The physical lengths (L) of the lines used in the filter design are thus obtained by calculating the effective dielectric constant ($\varepsilon_{eff}(f)$) at the frequency of interest through the following equation

$$L = \left(\frac{c}{2\pi f}\right)\left(\frac{\theta}{\sqrt{\varepsilon_{eff}(f)}}\right) - \Delta L \qquad (2)$$

where c is the speed of light.

The coupling factor (k_c) may be obtained according to the modal impedances appropriately chosen for the two lines using

$$k_{c(dB)} = 20 \log\left(\frac{Z_{ct,e} - Z_{\pi,e}}{Z_{ct,e} + Z_{\pi,e}}\right) \qquad (3)$$

The multilayer ECBPFs have been designed with a centre frequency of 35GHz, a fractional bandwidth of 20%, and a roll-off specified by an insertion loss greater than 30dB at 25GHz. The physical dimensions for each coupled-line section of the design were obtained with the aid of a simulation package *ADS*® *Momentum*® (Advanced Design System from Agilent), and the results are shown in Table 1.

Photoimageable thick film technology provides a viable fabrication process for producing planar components with the high quality required for high frequency applications. The photoimageable process is an extension of conventional thick film technology, in which the standard thick film pastes are replaced by photosensitive materials. The ability to directly photo-image the printed layers means that the technology can provide the high line and space

resolutions required for planar components operating at millimetre-wave frequencies, easily down to 10μm [4]. Figure 2 shows the sequential stages of a photoimageable thick film process. Figure 3 is a photograph of the fabricated circuit of the proposed multilayer ECBPF investigated in the current work.

Table 1: Dimensions of the Filter

Sec., n	1	2	3	4
W_{top}	247	309	309	247
L_{top}	814	793	793	814
W_{emb}	210	260	260	210
L_{emb}	761	755	755	761
S	25	165	165	25

Unit:μm

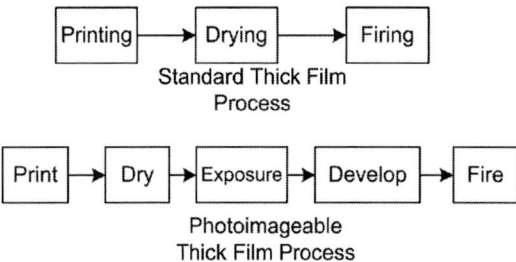

Figure 2: The photoimageable thick-film process

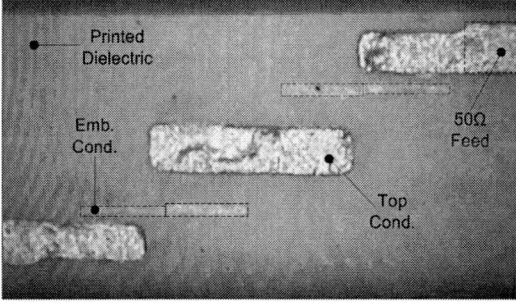

Figure 3: Photograph of the Circuit

Figure 4 shows a comparison between the simulated and measured responses of the filter. The measured filter had a centre frequency of 34.85GHz with a 3dB bandwidth about 6.75GHz (19.3%) and the roll-off of $|S_{21}|^2 < -30$dB at 25GHz. The maximum insertion loss in the passband is around 0.631dB. The return loss is below 10dB which represents reasonably good matching of the circuit around the operating frequency, and meets the required specification.

978-1-4244-4722-0/09 $25.00
© 2009 IMAPS-ITALY

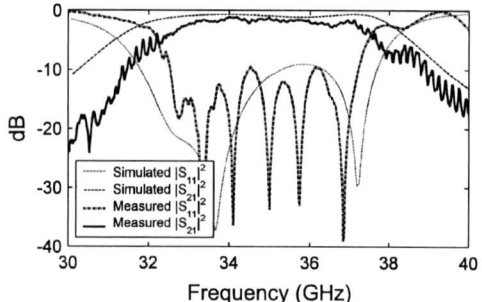

Figure 4: *S*-Parameters of the Filter

Misalignment Analysis

One of the most important aspects in fabricating multi-layer circuits is to achieve precise alignment between the conductors on different layers of the structure. This issue is very crucial as misalignment can cause a significant loss in the circuit performance, thereby loosing the advantages of multi-layer over the single-layer filter circuits. The effects of the registration error (misalignment) on the RF performance were investigated. Both horizontal (L_{err}) and vertical (S_{err}) errors between the top and embedded conductors were considered, as shown in Figure 5, for the filters designed at 35GHz.

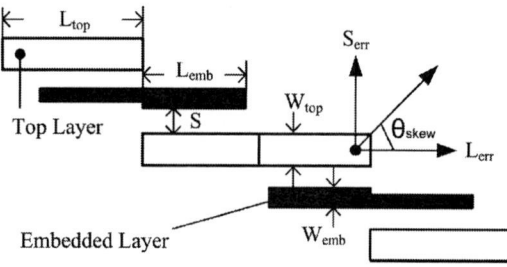

Figure 5: Layout of Multilayer ECBPF

This investigation had been demonstrated by using the simulation package (*ADS*). It can be seen from Figure 6 and Figure 7 that increase of registration errors in the horizontal direction shrank the working bandwidth and resulted in greater insertion loss, respectively. It was found from Figure 6 that a misalignment of 0.1mm in the L_{err}-direction caused a decrease of 0.8% in the bandwidth of the filter. It can be seen from Figure 7 that a misalignment of 0.1mm in the L_{err}-direction caused an increase of 0.5dB in the insertion loss.

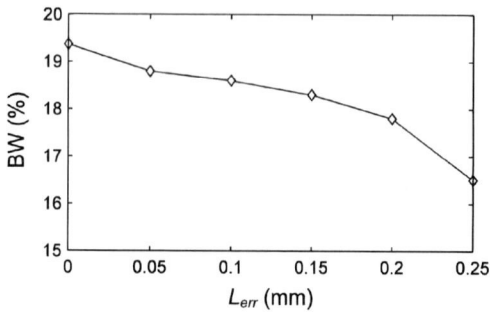

Figure 6: Effect of Horizontal Misalignment (L_{err}) on Bandwidth

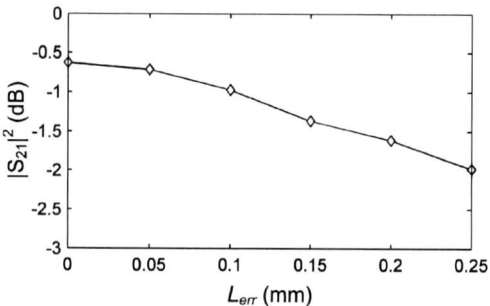

Figure 7: Effect of Horizontal Misalignment (L_{err}) on Insertion Loss

Figure 8 shows that a misalignment of 0.01mm in the S_{err}-direction causes an increase of 0.31dB in the insertion loss. The analysis showed that the misalignment in both (L_{err}) and (S_{err}) can cause significant loss in the filter performance due to the significant change in the coupling between the filter sections. Clearly, the results indicate that registration between the layers is the key issue when fabricating such coupling structure in multilayer configuration, and the performance of multilayer filters are more sensitive to error when the operating frequency goes higher.

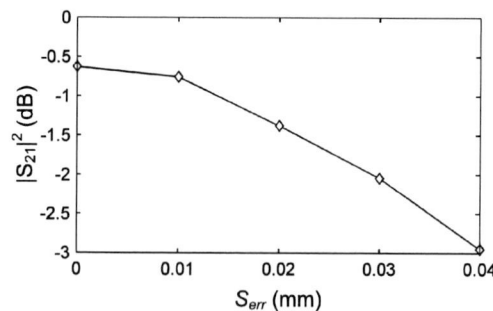

Figure 8: Effect of Vertical Misalignment (S_{err}) on Insertion Loss

978-1-4244-4722-0/09 $25.00
© 2009 IMAPS-ITALY

Another major misalignment error, which is often ignored, is the skewing of the top conductors, relative to those on the embedded layer. Figure 5 shows the definition of skew angle (θ_{skew}), and this is illustrated in Figure 9 for clarity. It was found that the skew angle altered the coupling behavior of the filter sections.

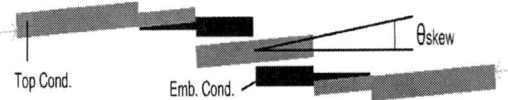

Figure 9: Illustration of Skew between the Conductor Layers

Figure 10 and Figure 11 show the effects of skew of conductor layers on bandwidth and insertion loss of the filter at 35GHz, respectively. It can be seen from Figure 10 when the skew angle increases to ($\theta_{skew} = 0.1°$), the bandwidth of the filter decreases by 5.6%. It was found from Figure 11 when the skew angle increases to ($\theta_{skew} = 0.1°$), the insertion loss increases by 0.133dB. It is clearly seen that any registration error due to skewing between the conductors may result shrinkage of the working bandwidth and an increase of the insertion loss. This analysis will be particularly useful for millimetre-wave designs, where fabrication errors become critical issues, and can provide useful practical guidelines for the circuit designers.

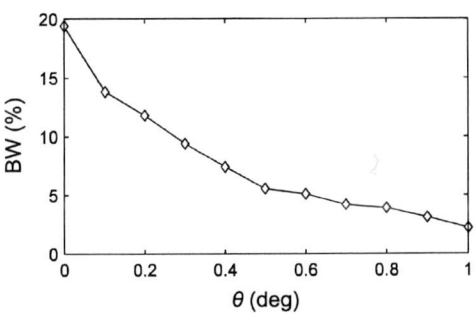

Figure 10: Effect of Skew (θ_{skew}) on Bandwidth

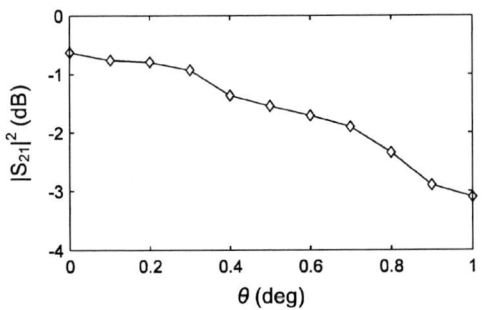

Figure 11: Effect of Skew (θ_{skew}) on Insertion Loss

Conclusion

The work has identified some critical aspects of the fabrication process that have significant influence on the performance of edge-coupled band-pass filters, and has not previously been reported in the literature. In particular, the work has shown the significance of skewing between the conductor layers of the structure, and has provided the circuit designer with typical guidelines for the degree of misalignment that would be acceptable in practical circuits.

Acknowledgements

The authors would like to acknowledge the financial support from Kuwait Foundation for the Advancement of Sciences.

References

[1] T. C. Edwards, and M. B. Steer, "Foundation of Interconnect and Microstrip Design", J. Wiley & Sons, 2000.

[2] C.-M. Tsai, and K. C. Gupta, "A Generalized Model for Coupled Lines and its Applications to Two-Layer Planar Circuits", *IEEE* Transaction on Microwave Theory & Techniques, Vol. 40, No. 12, 1992.

[3] C. Cho, and K. C. Gupta, "Design Methodology for Multilayer Coupled Line Filters", *IEEE MTT-S Digest*, Vol. 2, No. 8, 1993.

[4] D. Stephens, P. R. Young, and I. D. Roberston, "Design and characterization of 180GHz filters in photoimageable thick-film technology", *IEEE MTT-S Int. Microwave Symp. Dig.*, pp. 451-454, 2005.

[5] D. M. Pozar, "Microwave Engineering", J. Wiley & Sons, 3rd ed., 2005.

[6] C. Ng, M. Chongcheawchaman, M. Aftanasar, I. Robertson, and J. Minalgiene, "X-band Microstrip Bandpass Filter using Photoimageable Thick-Film materials", *2002 IEEE MTT-S Int. Microwave Symp. Dig.*, vol. 3, pp.2209-2212, 2-7 June 2002.

[7] V. Bengin, and D. Budimir, "Integrated Waveguide Bandpass Filters Using Thick-Film Technology", *IEEE, Microwave Review*, June 2004.

[8] P. Barnwell, and J. Wood, "Fabrication of Low Cost Microwave Circuits and Structures using An Advanced Thick Film Technology", *1998 IEMT/MIC Symp.*, pp. 327-332, 15-17 April 1998.

Correlation between Material Selection and Moisture Sensitivity Levels of Quad Flat No-lead (QFN) Packages

Minshu Zhang[1], S. W. Ricky Lee[1], Jack Zhang[2], Howard Yun[2], Dale Starkey[2] and Hung Chau[2]

[1]Department of Mechanical Engineering

Hong Kong University of Science and Technology

Clear Water Bay, Kowloon, Hong Kong

Phone: +852-23588444, Fax: +852-23588357

Email: Zhangms@ust.hk

[2]Henkel Corporation

15350 Barranca Parkway

Irvine, CA 92618, USA

Phone: +1-949-789-2500, Fax: +1-949-789-2595

Email: Jack.Zhang@us.henkel.com

Abstract

Moisture Sensitivity Level (MSL) test is the well-known industrial standard to classify the level of moisture sensitivity of plastic packages. However, except for the classification, MSL test provides no suggestion for improving package MSL performance from the package design point of view. In order to achieve an expected MSL performance, further investigation of the correlation between package MSL and material selection is required. Based on this major objective, four kinds of molding compounds and two types of die attach materials were studied in this paper. Commercial 4x4 Quad Flat No-lead (QFN) packages were fabricated as the test vehicle. The stress ratio criterion was employed to evaluate the failure of 4x4 QFNs under MSL-1 tests. In addition, MSL-1 tests and related failure analysis were implemented to provide the experimental results. From the comparison between the stress ratio analysis and the experimental results, it is concluded that the molding compound with lower adhesion strength and higher Young's modulus may lead to higher risk of delamination.

Key words: QFN, button shear test, stress ratio criterion, MSL test, C-SAM inspection

Introduction

Interfacial delamination is a long existing problem in plastic packages subjected to moisture preconditioning and reflow heating. In general, the cause of delamination or popcorn cracking is the competition between interfacial strength and hygrothermal stress. Therefore, research in this field is usually divided into two groups. One group is focused on the study of adhesion at molding compound/lead-frame and die attach/die pad interfaces [1] [2]. The other group is interested in the numerical stress calculation including thermal stress, swelling stress and vapor pressure [3] [4].

Since the moisture related failure was first reported in 1985 [5], many papers from both groups have been published to investigate the failure mechanism. It seems that the package would easily achieve the expected MSL performance. However, the challenges are also obvious. Usually material suppliers will suggest using materials with better adhesion but the stress level during reflow cannot be guaranteed. The simulation will show some advantages at stress calculation considering the package geometric effects. Nevertheless, the difficulty in obtaining material properties of mini-size packages and the lack of experimental investigation of interfacial adhesion will always bring many challenges to simulations. As a result, a

comprehensive system including both simulation and experiments is required to bridge the gap between package MSL performance and material selection. Zhang and Lee have tried to build such a system and employed the stress ratio analysis into the evaluation of the failure of 10x10 dummy QFNs in 2008 [6] [7]. The study on dummy QFN has showed the confidence to apply the stress ratio approach into the risk estimation of delamination. Therefore, a similar research methodology is adopted in this study with more molding compounds and die attach materials involved on a new test vehicle.

The resarech methodology is illustrated in Figure 1, which runs with two routes. One follows

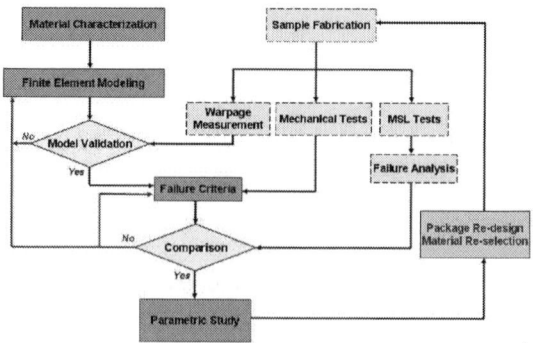

Figure 1. Research Methodology

the simulation route (purple background), starting from material characterization to finite element modeling. The interfacial stress can then be calculated from FEA. Combining with the interfacial strength obtained from mechanical tests, the failure criterion is established. Once the failure criterion is applied successfully to match with the MSL tests from the other experimental route (yellow background), then this validated FEA model will be used to conduct the parametric study and provide the guidelines to industry for package design.

Sample Preparation

A commercial 4x4 QFN package selected as the test vehicle in this paper includes four major parts, molding compound, die, die attach and copper lead-frame with silver ring on the die pad surface. The configuration and dimensions are shown in Figure 2 and the material properties are listed in Table 1. Since the major objective of this paper is to identify the correlation between the MSL performance and material selection of plastic packages, four kinds of molding compounds and two kinds of die attach were selected. Therefore, in total eight types of QFN were fabricated and investigated in this study.

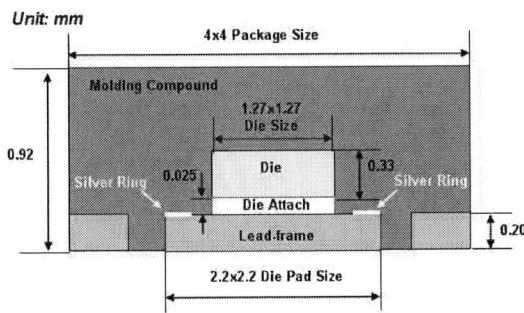

Figure 2. Schematic of 4x4 QFN Package

Table 1. Material Properties

Material	Modulus at 260°C (MPa)	v	α (ppm/°C)
Lead-frame	117000	0.34	18
Die (Si)	160000	0.28	4
DA_A	42	0.30	216
DA_B	136	0.30	153
MC_A	572	0.30	31
MC_B	530	0.30	31
MC_C	1434	0.30	35
MC_D	426	0.30	31

Moisture Sensitivity Level Tests

MSL-1 tests of 4x4 QFNs were implemented following the JEDEC standard [8]. Eleven samples of 4x4 QFNs went through moisture preconditioning under 85°C & 85%RH for 168 hours followed by three times of reflow heating with a peak temperature at 260°C. Both C-Scan and T-Scan

images of each type of QFN samples are presented in Figure 3. Delamination was found only in 4x4 QFNs with MC_C; i.e., Package Types 3 and 7.

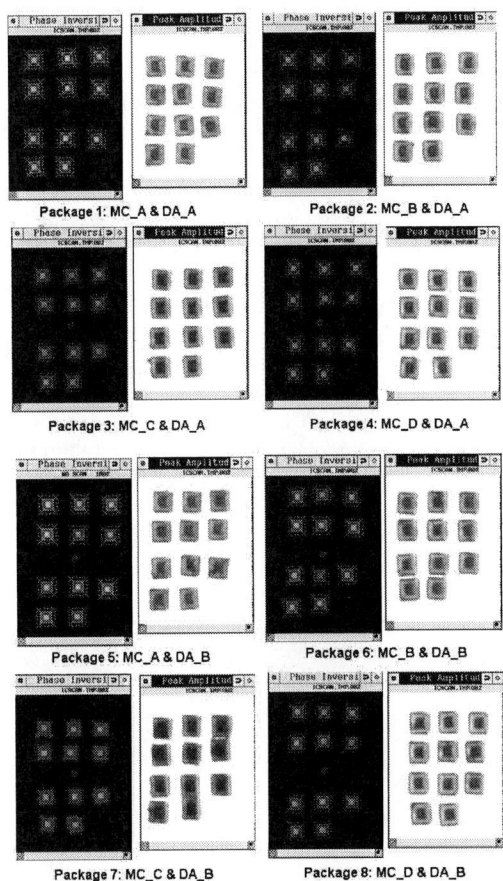

Figure 3. C-SAM Inspections of 4x4 QFNs after MSL-1 Tests

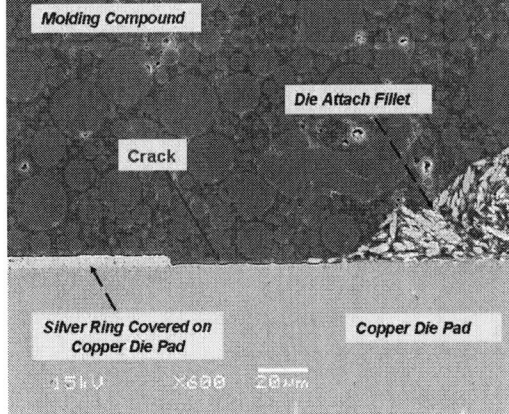

Figure 4. SEM Inspection of Failure Modes

A cross-section analysis of failed sample was conducted and the failure modes were photographed under Scanning Electron Microscope (SEM). An obvious crack was found at the molding compound/die pad interface around the junction of die attach fillet as shown in Figure 4, which confirmed the observation in C-SAM inspection. In addition, no delamination was found at molding compound/die top surface. Therefore, the interested failure area for subsequent study was focused at the molding compound/die pad interface.

Finite Element Analysis

A 3-D finite element model of a quarter of a 4x4 QFN package was established using ANSYS code including 17010 brick 8-node elements as shown in Figure 5. The reference temperature and the loading temperature were set at 175°C (the curing temperature of molding compound) and 260°C (the peak temperature in Pb-free reflow), repsectively. Then the elastic thermo-mechanical stress can be calculated and the stress contours of τ_{xz} are plotted in Figure 6. Subsequently, interfacial shear stress distribution along the path AB is obtained and shown in Figure 7. Stress concentration was found at the molding compound/die pad interface around the die attach fillet and die pad edge. It should be noted that, in reality, there was a silver ring on the copper die pad surface. However, the thickness of that silver ring was about 2 μm, which was too thin to consider in the finite element model. Therefore, the stress distribution appears continuously across the junction between the MC/Cu and the MC/Ag regions.

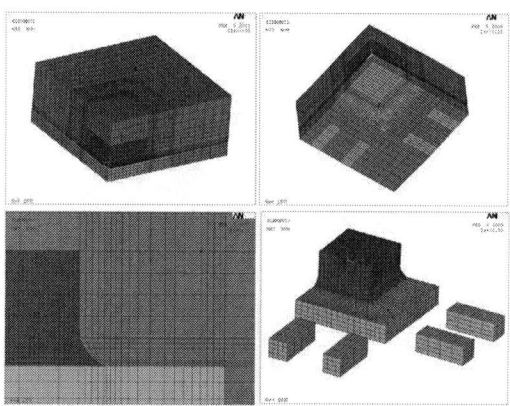

Figure 5. 3-D Finite Element Model of 4x4 QFN

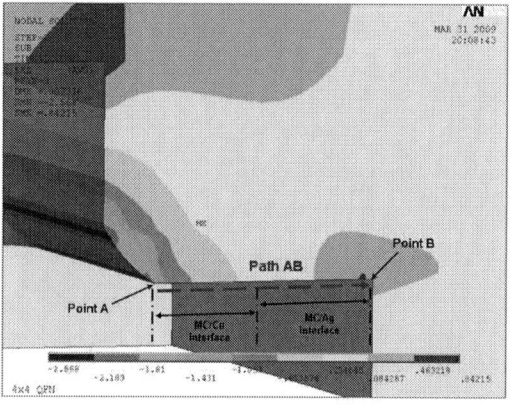

Figure 6. Typical Stress Contours of τ_{xz} in Molding Compound

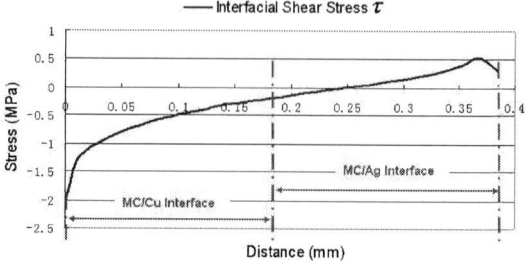

Figure 7. Typical Interfaical Shear Stress Distribution along Path AB

Besides the thermal stress, package with MSL preconditioning should be also subject to moisture related stress during the reflow heating; i.e., swelling stress and vapor pressure. Swelling stress is induced by the polymer material expansion after moisture absorption and the vapor pressure is induced by the moisture vaporizing. In reality the thermal and moisture induced stresses are mixed together during the reflow, making the stress analysis very difficult. Some relevant studies with finite element modeling, such as T. Y. Tee [3] and Zhang and Lee [6], can be found in the literature. Both of investigations showed that the thermal stress should be much higher than the swelling stress in the QFN package during reflow. Therefore, in the present study only the thermal stress was considered for the finite element modeling. In addition, the present study used the stress ratio criterion to determine the onset of delamination. There was no intention to investigate the interfacial crack propagation behavior. Therefore, the vapor pressure was not considered in this paper either.

Button Shear Test

The interfacial shear strength of molding compound/die pad interface was measured using the button shear test [9] [10]. Since the lead-frame used in 4x4 QFN had a silver ring on the die pad surface, there were two different strengths to be determined, molding compound/Cu (MC/Cu) and molding compound/Ag (MC/Ag). The first group of button samples was mounted on the Cu lead-frame whose surface treatment was identical to the Cu surface used in 4x4 QFN as seen in Figure 8(b). The shear test results represented the interfacial shear strength at the MC/Cu interface. The second group of button samples was mounted on the Ag lead-frames as seen in Figure 8(c) whose surface treatment was identical to the silver ring on the die pad. The shear test results represented the interfacial shear strength at the MC/Ag interface. It should be noted that the bottom side of those button specimens was glued by adhesive on a flat foundation in order to avoid the lead-frame buckling problem during the button shear test shown in Figure 8(a).

The diameter of the button was 3.57 mm and the button base area was about 10.00 mm². Button shear tests were conducted on Dage 4000 following two test flowcharts shown in Figure 9. Following

978-1-4244-4722-0/09 $25.00
© 2009 IMAPS-ITALY

route A, buttons were sheared under room temperature without any treatment. In route B, buttons went through reflow three times with MSL-1 preconditioning and then were sheared under room temperature. It should be noted that, from C-SAM inspections, all button shear test samples exhibited perfect bonding conditions. There was no delamination found before the shear tests. Finally, the interfacial shear strength was obtained from dividing the ultimateshear force during the test by the button base area.

(b) Button Shear on Cu

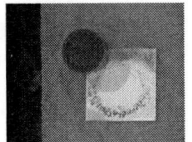

(a) Shear Test on Dage 4000   (c) Button Shear on Ag

Figure 8. Button Shear Test Setup and Typical Failure Modes

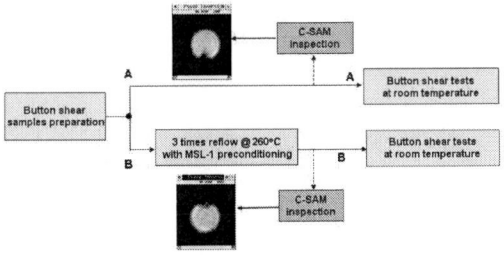

Figure 9. Flow Chart of Button Shear Test

The interfacial shear strengths are listed in Table 2 where Z_{Cu} and Z_{Ag} represent the interfacial shear strengths of molding compound on Cu and Ag, respectively, without any treatment. In addition, Z_{Cu}' and Z_{Ag}' represent the interfacial shear strengths of molding compound on Cu and Ag, respectively, with MSL-1 preconditioning followed by three times of reflow heating. The comparison of shear strengths is shown in Figure 10. From the comparison, two conclusions can be made: a) the interfacial strength is usually higher at MC/Cu than at MC/Ag interfaces; b) the MSL-1 test condition has significant effect on the interfacial strength. Without the moisture preconditioning and reflow treatment, MC_A and MC_B showed the best adhesion on Cu while MC_C gave the worst performance. In addition, MC_A showed the best adhesion on Ag while MC_C still gave the worst performance. After the button samples went through MSL-1 preconditioning followed by three times of reflow, MC_B and MC_D showed the best adhesion on Cu while MC_C still gave the worst performance. In addition, MC_A showed the best adhesion on Ag while MC_C gave the worst performance. It should be noted that, the shear strengths, Z_{Cu}' and Z_{Ag}', are to be used in the subsequent stress ratio analysis.

Table 2. Interfacial Shear Strength Measured by Button Shear Tests (MPa)

	Z_{Cu}	Z_{Cu}'	Z_{Ag}	Z_{Ag}'
MC_A	12.14	6.84	11.95	5.49
MC_B	12.44	8.74	7.14	2.58
MC_C	5.71	1.83	5.96	1.61
MC_D	10.79	8.57	6.89	4.69

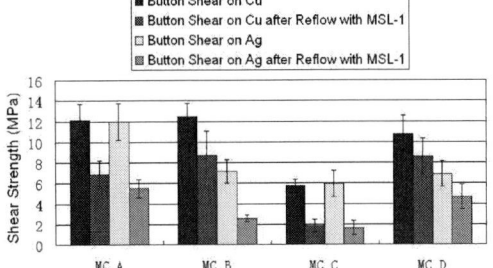

Figure 10. Interfacial Shear Strength Comparison

Stress Ratio Analysis

After obtaining the interfacial shear stress and shear strength, the stress ratio can be defined as follows:

$$Y_A = \left| \frac{\tau}{Z'_{Cu}} \right|; \quad Y_B = \left| \frac{\tau}{Z'_{Ag}} \right| \quad (1)$$

where the numerator was the interfacial shear stress calculated from FEA and the denominator was the interfacial shear strength obtained from button shear tests. The stress ratios are plotted in Figures 11-18. The higher value of stress ratio represents the higher risk of delamination. The maximum stress ratio values are selected from the stress ratio distributions and listed in Table 3.

From the comparison of maximum stress ratios, it is obvious to see that the package types (3 and 7) with MC_C appear to be much worse than the others. Therefore, these QFN samples are expected to have much higher risk for delamination than the other types of QFN samples under the MSL tests and this indication matches well with the previous MSL test results.

The higher Young's modulus in MC_C could generate higher stress in FEA, which increases the value of numerator in the stress ratio analysis. In addition, the poorest performance in MC_C in button shear tests decreases the value of denominator in the stress ratio analysis. As a result, higher stress ratio value was found in the QFNs with MC_C, which lead to the higher risk for delamination under the MSL tests than the QFNs with other tested molding compounds.

978-1-4244-4722-0/09 $25.00
© 2009 IMAPS-ITALY

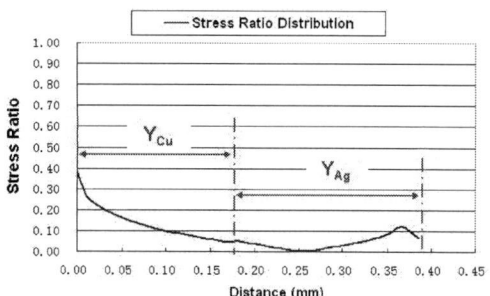

Figure 11. Stress Ratio Distribution along Path AB of QFN with MC_A and DA_A (Package Type 1)

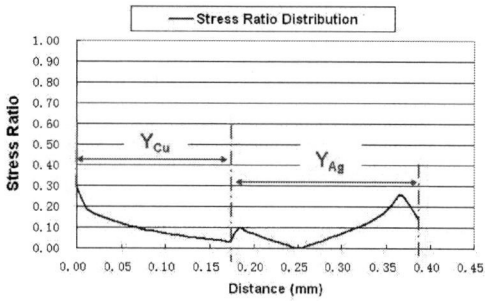

Figure 12. Stress Ratio Distribution along Path AB of QFN with MC_B and DA_A (Package Type 2)

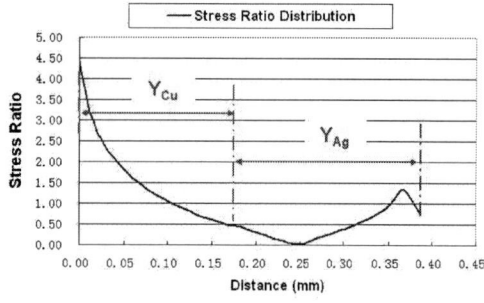

Figure 13. Stress Ratio Distribution along Path AB of QFN with MC_C and DA_A (Package Type 3)

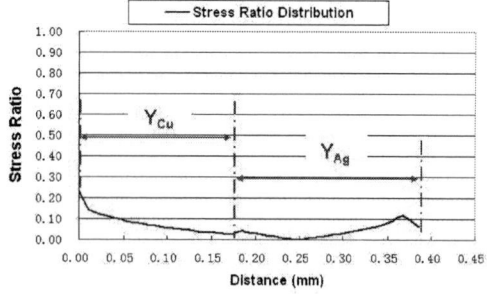

Figure 14. Stress Ratio Distribution along Path AB of QFN with MC_D and DA_A (Package Type 4)

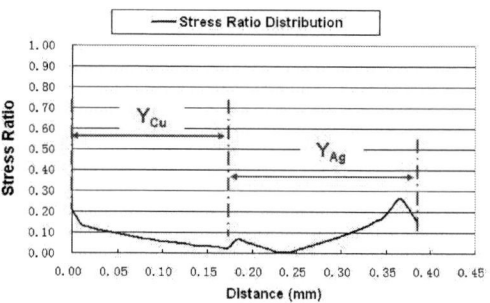

Figure 15. Stress Ratio Distribution along Path AB of QFN with MC_A and DA_B (Package Type 5)

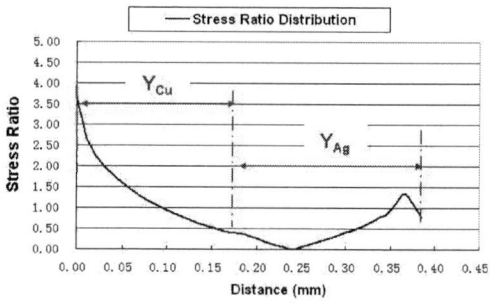

Figure 16. Stress Ratio Distribution along Path AB of QFN with MC_B and DA_B (Package Type 6)

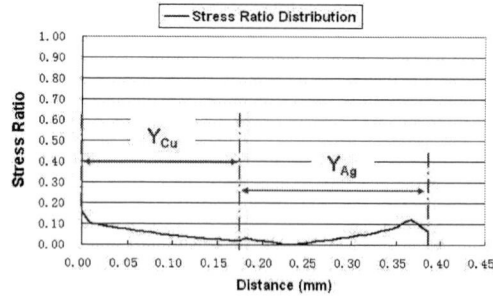

Figure 17. Stress Ratio Distribution along Path AB of QFN with MC_C and DA_B (Package Type 7)

Figure 18. Stress Ratio Distribution along Path AB of QFN with MC_D and DA_B (Package Type 8)

978-1-4244-4722-0/09 $25.00
© 2009 IMAPS-ITALY

Table 3. Maximum Stress Ratio of 4x4 QFNs

Package Type	Maximum Stress Ratio	Package Type	Maximum Stress Ratio
1	0.39	5	0.27
2	0.30	6	0.21
3	4.50	7	3.76
4	0.24	8	0.16

Conclusions and Discussion

In this study, a research methodology was established to investigate the correlation between the 4x4 QFN MSL performance and material selection. From the comparison between three stress analysis and experiments, several conclusions can be drawn:

1. Eight types of 4x4 QFNs were fabricated as the test vechiles. After MSL-1 tests, delamination was found in 4x4 QFNs with MC_C while others passed the MSL-1 tests. The delamination was detected with C-SAM and confirmed by the cross-section inspection.
2. The 3-D finite element model was established and the interfacial shear stress distribution was obtained. Stress concentration was found at the molding compound/die pad interface around the die attach fillet and die pad edge.
3. Interfacial shear strength at the molding compound/die pad interface was measured under two different conditions. The shear strength at the MC/Cu interface was usually higher than that at the MC/Ag interface. The MSL-1 test condition had significant effects on the interfacial shear strength.
4. The stress ratio criterion was employed to evaluate the risk of delamination. The results showed that QFNs with MC_C would give the worst MSL performance under the MSL-1 condition. This result matched well with the experimental observation.
5. MC_C has a Young's modulus much higher than other molding compounds. In the meanwhile, MC_C also showed the poorest interfacial strength in button shear tests among all tested molding compounds. Both of the above effects lead to the high risk of delamination in QFN with MC_C in MSL-1 test.

The study of commercial 4x4 QFNs in this paper would be a good supplementary application of the stress ratio analysis. However, the current stress ratio analysis used the stress calculated at 260°C while the interfacial strength was measured at room temperature. In the future, the high temperature button shear tests should be performed to make up the drawback in the current analysis. Nevertheless, even though the current prediction of moisture related failure by stress ratio is considered a qualitative approach, from the comparison of stress ratio values, package design engineers can still adopt the present methodology and select the most suitable packaging materials to reduce the risk of delamination under the MSL tests. This exhibits the major contribution of the current study.

Acknowledgments

This study was sponsored by Ablestik and Henkel through the grant ICI001N to HKUST. The authors would like to acknowledge this support.

References

[1] J. K. Kim et al., "Interface Adhesion between Copper Lead Frame and Epxy Molding Compound: Effect of Surface Finish, Oxidation and Dimples," Proc. 50th, ECTC., 2000, pp. 601-608.

[2] J. T. Huneke, "Die Attach Adhesion on Leadframes Treated with Antioxidants," Proc. 47th ECTC, 1997, pp. 208-214.

[3] T. Y. Tee and Z. W. Zhong, "Integrated Vapor Pressure, Hygroswelling, and Thermo-mechanical Stress Modeling of QFN Package during Reflow with Interfacial Fracture Mechanics Analysis," Microelectronics Reliability, Vol. 44, 2004, pp. 105-114.

[4] X. J. Fan et al., "A Micromechanics-Based Vapor Pressure in Electronic Packages," Journal of Electronic Packaging, Vol. 127, 2005, pp 262-267.

[5] I. Fukuzawa, S. Ishiguro and S. Nanbu, "Moisture Resistance Degradation of Plastic LSI's by Reflow Soldering," Proc. International Reliability Physics Symposium, 1985, pp. 192-197.

[6] M. S. Zhang and S. W. R. Lee, "Stress Analysis of Hygrothermal Delamination of Quad Flat No-lead (QFN) Packages," Proc. ASME IMECE, Boston, Massachusetts, 1-6 November, 2008. (IMECE2008-68110)

[7] M. S. Zhang and S. W. R. Lee, "Investigation of Moisture Sensitivity Related Failure Mechanism of Quad Flat No-lead (QFN) Packages," Proc. ASME IMECE, Boston, Massachusetts, 1-6 November, 2008. (IMECE2008-68120)

[8] JEDEC Solid State Tech. Assoc., IPC/JEDEC J-STD-020D, Moisture/Reflow Sensitivity Classification for Nonhermetic Solid State Surface Mount Devices, 2007.

[9] H. B. Fan, M. M. F. Yuen and E. Suhir, "Prediction of Delamination in Bi-material System Based on Free-Edge Energy Evaluation," Proc. 53rd ECTC, New Orleans, 2003, pp. 1160-1164.

[10] W. K. Szeto, M. Y. Xie and J. K. Kim, et al., "Interface Failure Criterion of Button Shear Test as a Means of Interface Adhesion Measurement in Plastic Packages," Int'l Symp. on Electronic Materials & Packaging, 2000, pp. 263-268.

978-1-4244-4722-0/09 $25.00
© 2009 IMAPS-ITALY

Passive Phase Change Tower Heat Sink
&
Pumped Coolant Technologies
for
Next Generation CPU Module Thermal Design

M. Vogel, D. Copeland, A. Masto (Sun Microsystems),
S. Kang, B. Whitney (Aavid Thermalloy), G. Upadhya (Cooligy)
M. Connors (Thermacore), J. Marsala (Thermal Form and Function)

Sun Microsystems, Santa Clara, CA, USA

01-650-352-6483, 01-408-276-4550, marlin.vogel@sun.com

Abstract

Increasing thermal demands of high-end server CPUs require increased performance of air-cooling systems to meet industry needs. Improving the air-cooled heat sink thermal performance is one of the critical areas for increasing the overall air-cooling limit. One of the challenging aspects for improving the heat sink performance is the effective utilization of relatively large air-cooled fin surface areas when heat is being transferred from a relatively small heat source (CPU) with high heat flux. Increased electrical performance for the computer industry has created thermal design challenges due to increased power dissipation from the CPU and due to spatial envelope limitations. Local hot spot heat fluxes within the CPU are exceeding 100 W/cm2, while the maximum junction temperature requirement is 105 C, or less.

Key words: heat sink, heat pipe, vapor chamber, coolant, pump, cold plate, heat exchanger, embedded heat pipe

Background

In order to meet the next generation CPU thermal requirements with a phase change heat sink, two heat sink technologies and their associated prototypes will be described. Each of the heat sink technologies use internal liquid-to-vapor phase change to efficiently spread the local CPU power to the air-cooled fin structure. The two passive phase change heat sink technologies are: multiple embedded tower heat pipes; and a hybrid vapor chamber / muliple tower heat pipe design.

Next generation CPU thermal designs will incur additional increases in overall power, or increases in local power density, or reduction in junction temperature requirements, or a combination of the above. Maintaining the same CPU module spatial envelope and air flow requirements for follow-on CPU designs in air-cooled servers will require on board, self contained, pumped coolant solutions that incorporate micro-channel cold plates with relatively low coolant pressure loss due to the current practical performance limitations of the passive phase change heat sink evaporator and condenser. Three pumped coolant technologies and their associated prototypes that met the below thermal performance requirements will be described.

Thermal Design Requirements

sink-to-air thermal resistance:
0.065 C/W (passive phase change heat sink)
0.045 C/W (pumped coolant)

heat source size:
25mm x 25mm (passive phase change heat sink)
22mm x 22mm (pumped coolant)

module air pressure loss: 140 Pa

module air flow rate: 120 cfm

multi-flow direction:
front-to-back (perpendicular to gravity),
bottom to top (parallel / same direction as gravity)
top to bottom (parallel / opposite direction to gravity)

module spatial envelope:
200mm height x 100mm width x 220mm flow length

mass: 1800 grams

altitude: sea level

978-1-4244-4722-0/09 $25.00
© 2009 IMAPS-ITALY

multi-orientation: bottom heating, gravity assisted fluid return,

side heating, non-gravity assisted fluid return

Description of Prototypes

Embedded Tower Heat Pipe Heat Sink design:

The embedded heat pipe [1], [2] heat sink prototype is shown in Figure 2. The design was optimized by the supplier through the use of internally developed design tools as well as a commercial CFD software tool. The prototype supplier indicated that the performance advantage comes from design methods that balance the internal and external heat pipe geometry to minimize the intrinsic temperature drop in the heat pipes to distribute the heat over the base of the heat sink and the joining processes to minimize the interfacial temperature drops to get the heat into and out of the heat pipes. Other embedded designs have a lower performance because they do not successfully achieve this balance.

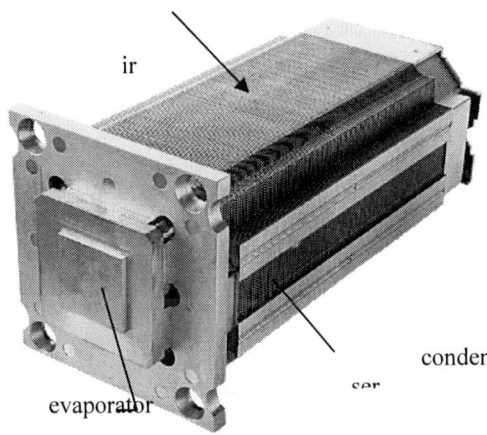

Figure 1: Aavid Thermalloy Embedded Tower Heat Pipe Heat Sink

Vapor Chamber Tower Heat Pipe Heat Sink design:

The prototype vapor chamber [3], [4] tower heat pipe heat sink is shown in Figure 2. The vapor chamber is a 3-dimensional heat pipe located in the heat sink base and is a relatively new technology that became commercially available during the mid-1990s, as compared to traditional unidirectional heat pipe technology that has been available for over 25 years. An aggressive development effort was carried out by the prototype supplier which allowed the wick thermal resistance to decrease by 50%. This provided a competitive edge over other heat sinks that incorporated vapor chamber

technology. Vapor chamber allows consistent extremely flat (no gaps) interface to heat sink. Pedestals designed to be included in the vapor space of the heat sink. This prevents having to conduct through large amounts of copper before being dissipated. Power is scalable to higher levels. With additional power input, a vapor tower will increase in resistance less than a conventional heat sink. Using four condenser tower heat pipe tubes allows for increased fin efficiency. Integrating the coolant flow path between the vapor chamber with the tower heat pipes allows for fluid return and vapor flow in three dimensions between the condenser and the evaporator, as shown in Figure 3.

Figure 2: Thermacore Vapor Chamber Tower Heat Pipe Heat Sink

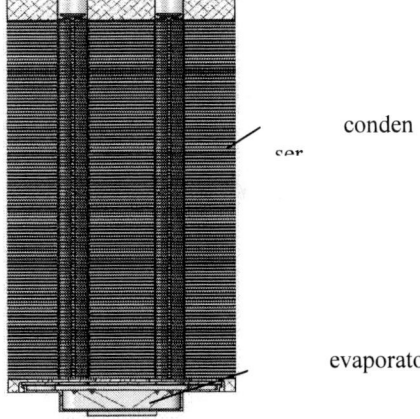

Figure 3: Illustration of the Thermacore Vapor Chamber Tower Heat Pipe Heat Sink

Pumped Coolant System designs:

The Aavid Thermalloy pumped coolant prototype is shown in Figure 4. The design [5] consists of a microchannel liquid cold plate, two liquid to air heat exchangers, a long life DC pump and associated tubing to connect all the components into a common liquid flow circuit. This heat sink uses single-phase liquid cooling with a water based coolant for high performance. The cold plate and liquid to air heat exchangers were optimized based on the DC pump performance curve using internally developed design tools as well as IcePak and Fluent CFD software.

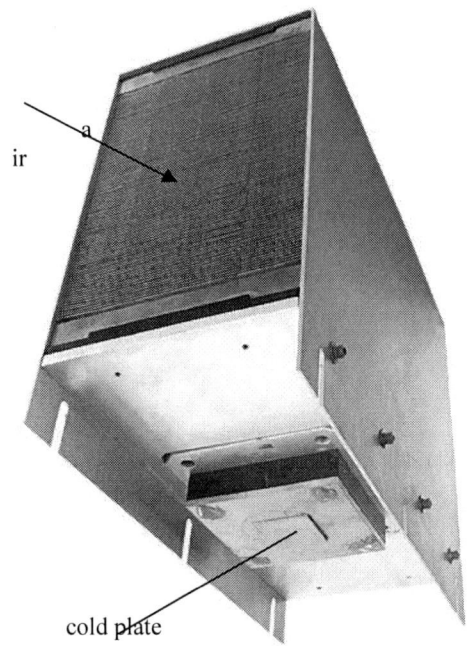

Figure 4: Aavid Thermalloy's pumped coolant system prototype.

The cold plate that contacts the heat source, seen at the bottom of the picture, is mounted on a plate that floats relative to the frame of the heat sink so that only the mass of the cold plate is seen by the heat source and circuit board. Four holes are provided on the floating plate to attach the cold plate to the heat source. The liquid to air heat exchangers are situated at the front and rear faces of the heat sink where the air flow enters and exits. An aluminum sheet metal duct encloses the heat sink. The pump us mounted on the inside surface of the aluminum duct in a location where the pump creates the least air flow blockage.

Figure 2 shows a schematic of the liquid and air flow paths through the heat sink. The working fluid absorbs heat in the cold plate (LCP) and then circulates through the two heat exchangers (HX-1

and HX-2) where the heat is transferred to the air. The liquid flow path is arranged so that the liquid enters in the heat exchanger at the downstream end of the air flow path and then flows to the heat exchanger at the upstream end of the flow path thus creating a counter-flow configuration between the two fluid streams. Within each heat exchanger the liquid and air flow streams are in a cross-flow configuration.

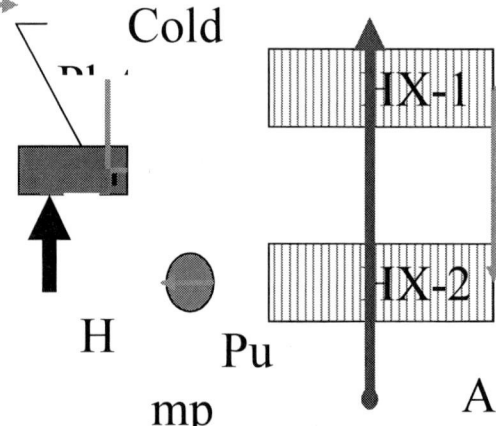

Figure 5: Schematic of Aavid Thermalloy's pumped coolant system prototype.

The Cooligy pumped coolant prototype [6], [7], [8], [9], and [10], is shown in Figure 6, and a side view illustration is shown in Figure 7. The CPU power is dissipated into an attached liquid cooled cold plate. Coolant is a water-based, single phase liquid. Cold plate is a manifolded, low pressure loss micro-channel fin designed for high volume manufacturing while maintaining a low thermal performance STD. A gimballed cold plate allows the thermal system mass to be detached from the CPU. Ambient air enters both heat exchangers by being routed in a separated, parallel flow path.

Figure 6: Cooligy's pumped coolant system prototype.

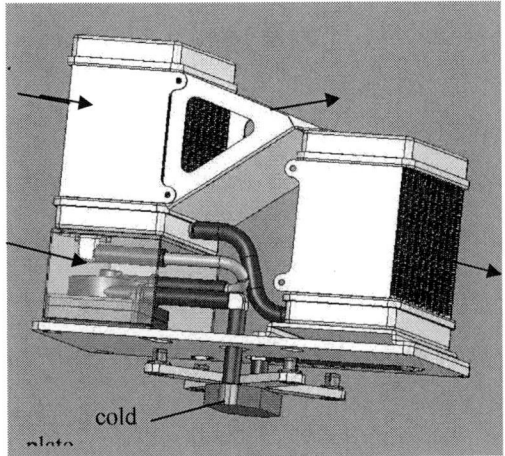

Figure 7: Illustration of Cooligy's pumped coolant system prototype.

Figure 8: Thermal Form & Function's pumped coolant system prototype.

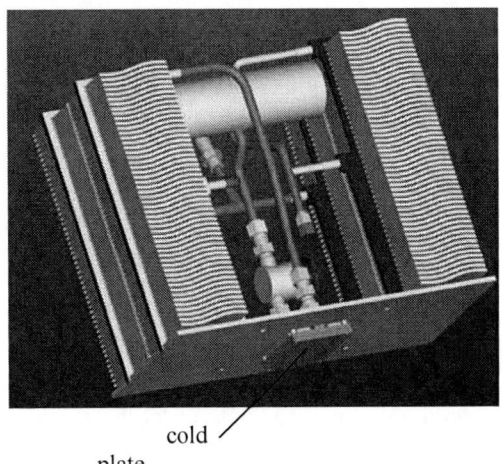

Figure 9: Illustration of Thermal Form & Function's pumped coolant system prototype.

The Thermal Form & Function pumped coolant prototype [11], [12], and [13] is shown in Figure 8, and an illustration is shown in Figure 9. Similar to the above pumped liquid system designs, the CPU power is dissipated into an attached liquid cooled cold plate. The pumped coolant is a dielectric refrigerant (R134a or equivalent), two phase (liquid-vapor) fluid. The coolant heat transfer mode in the cold plate is a forced convection boiling. The coolant vapor exiting the cold plate is condensed within the two coolant to air heat exchangers. Air and coolant flows in series through the heat exchangers. Pump is hermetically sealed. System is constructed through utilization of a mixture of materials (aluminum, copper, and refrigerant compatible plastics).

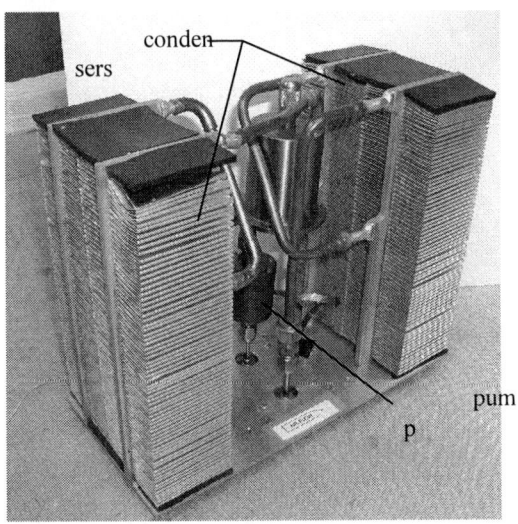

Results:

A hybrid, all-metal heat sink design (not shown) was optimized to yield the minimum sink-to-air thermal resistance while not exceeding the pressure loss and mass requirements. A commercial CFD software tool was used to optimize the heat sink design. The optimized design had a 15mm thick copper base, with 44 6063 Aluminum fins. The fin thickness is 0.5mm. The analysis results showed that the sink-to-air thermal resistance (0.11 C/W) of the all-metal heat sink design is 70% greater than the measured thermal resistance of the passive phase change prototype heat sinks. The measured thermal sink to air resistance for the passive phase change heat sinks met the requirement, 0.065 C/W, for multiple samples of each prototype design. Likewise, the measured thermal sink to air resistance for the passive phase change heat sinks met the requirement, 0.045 C/W, for multiple samples of each prototype design.

Reducing the air flow rate from the design requirement, 120 cfm, to 60 cfm yielded a 0.01 C/W to 0.015 C/W increase in the sink to air thermal resistance for all of the prototype designs. The pump power required for the pumped coolant designs was less then 15W. Reducing the pump power by 40% yielded a 0.005 C/W, or less, increase in the sink to air thermal resistance.

Conclusion:

The passive phase change heat sink prototypes yielded a 40W increase in the CPU coolable power as compared to the all-metal heat sink design.

Likewise, utilizing the same CPU design, the three pumped coolant designs increased the CPU coolable power by 40W as compared to passive phase change heat sinks. The pump coolant designs are a self contained design which doers not require coolant to be transported into and out of the electronic module. The electronic module spatial envelope is the same for both the passive phase heat sink and the pumped coolant design, allowing the CPU to be upgraded with increased performance without revising the system design. The increase in supplied power to the pump is less than 5% of the total input power to the CPU module is considered acceptable. One of the primary risks for incorporating a module level self contained pump coolant system is the reliability of the pump. Discussing the pumped coolant system reliability performance is beyond the scope of this paper and is addressed in [14].

Acknowledgements:

The Sun Microsystem co-authors acknowledge the contributions in technical innovation and the effort of support provided by the non-Sun authors and their respected, associated companies, for designing and manufacturing the prototypes that attained the aggressive thermal performance specification requirements for this product development project.

References

[1] Wu, R., " Heat-pipe type radiator and method for producing the same", United States Patent 6435266, August 20, 2002.

[2] Cheung, C., Moore, M., Prosperi, R., "Channel Connectionfor Pipe to Block Joint", United States Patent Application No. 20010050165, December 13, 2001.

[3] Grubb, K., "CFD Modeling of a Therma-Base Heat Sink",
http://www.thermacore.com/papers.htm.

[4] North, M., "Advances in Heat Sinks, Cold Plates, and Heat Spreaders", ASME Interpack Panel Session, 2005.

[5] Kang, S., Miller, D. and Cennamo, J., "Closed Liquid Cooling for High Performance Computer Systems", Proc. ASME Interpack, 2007

[6] Upadhya, G., "Active Micro-channel Cooling System for High Heat Flux Processor Cooling Applications", Cooligy White paper, 2006.

[7] Upadhya, G., Zhou, P., Hom, J., Goodson, K., Munch, M., "Electro-Kinetic Micro Channel

Cooling System for Servers", Proceedings of ITHERM 2004, Las Vegas, pp. 367-371, June, 2004.

[8] Upadhya, G., Zhou, P., Hom, J., Goodson, K., Munch, M., "Electro-Kinetic Micro Channel Cooling System for Desktop Computers", Proceedings of SEMI-ITHERM, San Jose, pp. 26-29, March, 2004.

[9] Upadhya, G.; Munch, M.; Peng Zhou; Horn, J.; Werner, D.; McMaster, M., "Micro-scale liquid cooling system for high heat flux processor cooling applications", Proc. Semiconductor Thermal Measurement and Management Symposium, IEEE Twenty-Second Annual IEEE, pp. 116 – 119, March, 2006.

[10] Upadhya, G.; Zhou, P.; Goodson, K.; Munch, M.; Kenny, "Closed-loop cooling technologies for microprocessors", Electronic Devices Meeting, Proc. IEDM '03 Technical Digest, IEEE International, pp. 32.4.1 – 32.4.4, Dec., 2003.

[11] Marsala, J., "Pumped liquid cooling system using a phase change refrigerant", US Patent 6,679,081, January 20, 2004.

[12] Howes, J.C.; Levett, D.B.; Wilson, S.T.; Marsala, J.; Saums, D.L, "Cooling of an IGBT Drive System with Vaporizable Dielectric Fluid (VDF)", Proc. Semi-Therm XXIV, pp. 9 – 15, Mart., 2008.

[13] Kelkar, K.M.; Patankar, S.V.; Kang, S.S., "Computational method for system-level analysis of two-phase pumped loops for cooling of electronics", Proc. ITHERM, pp. 95 – 104, May, 2008.

[14] Stern, M., Copeland, D., Vogel, M., Dunn, J., Kearns, D., Lindquist, S., "Reliability Specs for Closed Loop Liquid Cooling to the Board", IMAPs ATW for Advance Thermal Technologies, 2008.

978-1-4244-4722-0/09 $25.00
© 2009 IMAPS-ITALY

978-1-4244-4722-0/09 $25.00
© 2009 IMAPS-ITALY

Packaging of silicon photonic devices: grating structures for high efficiency coupling and a solution for standard integration

J. V. Galan,[1] A. Griol,[1] J. Hurtado,[1] P. Sanchis,[1] G. B. Preve,[1] A. Håkansson,[2] J. Marti[1]

[1]Nanophotonics Technology Center, Universidad Politécnica de Valencia, Camino de Vera s/n 46022 Valencia (Spain)

T: +34 96 387 97 68, F: +34 96 387 78 27, E: jogaco@ntc.upv.es

[2]DAS Photonics S.L., Camino de Vera s/n 46022 Valencia (Spain)

Abstract

Efficient packaging in silicon photonics requires a previous development of high performance fiber coupling structures. One of the most suitable fiber coupling structures in silicon is the grating coupler. Main advantages of using such a vertical coupling technique with respect to horizontal techniques are fiber alignment tolerances and wafer scale testing. We report design, fabrication and experimental measurements of conventional SOI grating couplers. Around 40% coupling efficiency is obtained when coupling a standard singlemode fiber to a singlemode SOI waveguide with 250nmx500nm section dimensions, with a fiber tilt angle of 10°. With respect to alignment tolerances, the variation effect of different fiber positions in the grating plane was evaluated as well as different tilt angles. It was obtained that coupling efficiency is almost constant for angle errors of ±2°. In addition, a tolerance of fiber position deviations of about ±2μm was reached, making possible multifiber alignments with the use of standard fiber-arrays. As the obtained vertical orientation is not always easy to adapt to standard layouts, we also introduce here a solution for allowing the use of standard DIL or butterfly packages with horizontal orientation, illustrating the design concepts and briefly describing the technologies involved.

Key words: silicon photonics, silicon-on-insulator technology, fiber coupling, grating couplers, packaging

Introduction

Silicon photonic integration technologies are mainly focused on adapting microelectronic industry tools to develop very large scale integration (VLSI) photonic components and circuits with low cost. The commitment of many companies all over the world towards developing low cost silicon photonics has increased in the last years, and even first silicon based laser approaches have been recently proposed in literature [1].

One of the most standardized silicon photonic technologies is Silicon on Insulator (SOI). When trying to inject light from an optical fiber to an SOI chip, ultra high coupling losses are obtained, mainly due to high differences between size and geometry of both fiber and SOI chip access waveguides. Standard singlemode fiber core diameter is around 10μm, whilst singlemode SOI waveguide core is about 500nm wide and 250nm height.

One of the most suitable solutions for efficiently light coupling between optical fiber and SOI waveguide is the grating coupler. The most important benefits of using vertical coupling techniques, such as grating couplers, compared to other in-plane coupling techniques is the compactness of the structures (about 10μmx10μm area), and the possibility of wafer testing [2]. Furthermore, due to its relatively high alignment

tolerances, multiple fiber attachment is also achievable [3]. However, coupling efficiency of conventional design configurations is commonly below 30%. Although examples of silicon photonic packaged components attached to grating couplers have recently been presented in literature [3], obtained vertical orientation is not always easy to adapt to standard layout for device packaging and assembly.

Here, we present experimental realization of conventional SOI grating couplers. We also introduce a solution for allowing the use of standard DIL or butterfly packages with horizontal orientation, illustrating the design concepts and briefly describing the technologies involved.

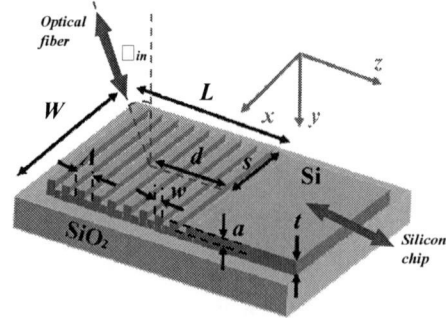

Figure 1: Schematic of a conventional SOI grating coupler and its main design parameters.

Conventional SOI grating coupler design

Schematic of a conventional unidimensional SOI grating coupler is depicted in Fig. 1. Main design parameters, such as grating period (Λ), etching depth (a), fiber tilt angle (θ_{in}), and corrugation width (w) are also shown in Fig. 1. Thickness of the SOI wafer is t. The grating is considered to be covered by air. Grating size (W, L) is mainly related to the size of the fiber optical mode. As we want to use standard singlemode fibers (10µm mode field diameter, MFD), we chose $W=L=12$µm.

For coupling to singlemode fibers, and taking into account we want to couple by means of 1st order of diffraction and neglect 2nd diffraction order, fiber is slightly tilted with respect to the normal direction to the grating ($\theta_{in}=10°$). The fiber is centered into the grating along x-axis direction, so $s=W/2$. Optimum fiber position along z-axis direction (d) is related to the coupling length of the grating ($d=L_c$). Coupling length is also related to the fiber beam size according to [4]. As we want to use singlemode fibers, optimum coupling length should be close to 4µm por optimum coupling to fiber [4]. A 1D grating just work for one polarization, so, our design is optimized for working with Transverse Electric (TE) polarization, whose electric field component is in the x-axis direction, according to axis definition on Fig. 1. Using two dimensional finite difference time domain simulations (2D-FDTD), we calculate optimum grating period and etching depth for getting maximum coupling efficiency for TE polarization at $\lambda=1550$nm, taking into account a resonant angle of $\theta_{in}=10°$, optimum fiber position, desired coupling length, and a filling factor of $ff=w/\Lambda=50\%$. Wafer silicon thickness used is $t=250$nm and SiO$_2$ thickness is 3µm. Maximum 50% coupling efficiency is theoretically obtained, for a 20 period long grating with a period of $\Lambda=600$nm and an etching depth of $a=70$nm. For these optimum design parameters, spectrum in telecom band is also computed. Fig. 2 depicts theoretical spectral response of the designed grating coupler. It is obtained a 1dB bandwidth of about 40nm.

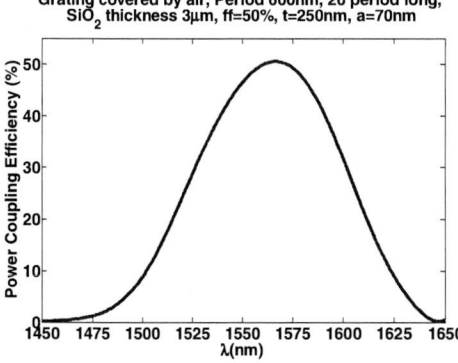

Figure 2: Simulation results for spectral response of designed grating coupler.

Theoretical fabrication and alignment tolerances

Sensitivity to fabrication and alignment tolerances of the structure has also been analyzed. To study fabrication tolerances, simulations were performed with different filling factors as well as different etching depths for the previously designed grating coupler. Fig. 3 shows coupling efficiency results when varying those parameters. It is obtained coupling efficiency higher than 45% for etching depth values of $a=70\pm10$ nm or filling factor values of $ff=50\pm10$ %.

To study alignment tolerances, the incident angles as well as the horizontal fiber positions were varied. Fig. 4 shows coupling efficiency results when varying those parameters. Coupling efficiency up to 45% is obtained for $\theta_{in}=10\pm2$ degrees or horizontal fiber positions of $d=4\pm2$ µm.

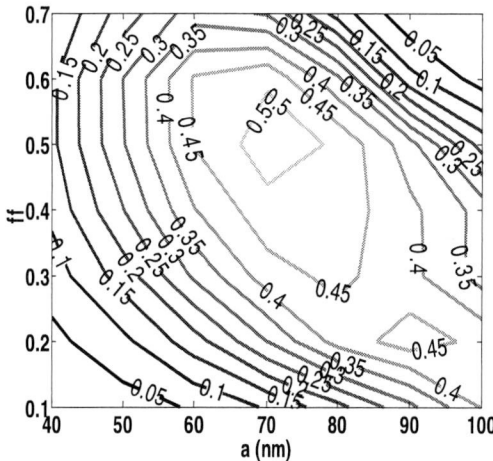

Figure 3: Grating fabrication tolerances.

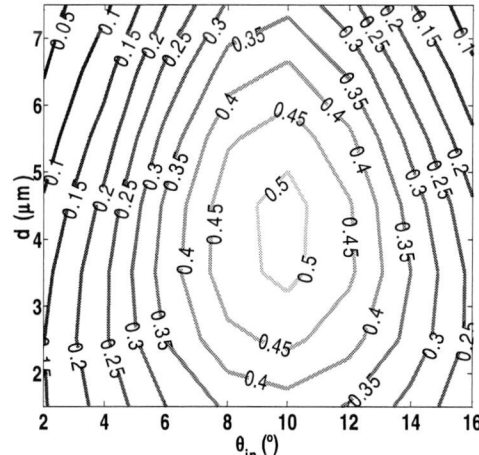

Figure 4: Grating alignment tolerances.

Fabrication and experimental measurements

A schematic of the layout for fabrication of the devices is depicted in Fig. 4. Basically, it consists on 500nm wide singlemode SOI waveguides, which are grating in/grating out attached. For adapting 12µm wide gratings to the waveguides, a 1mm long taper is used. This taper length is needed for minimazing higher order mode conversion in the singlemode waveguides. Waveguides with different lengths (L_w) will be fabricated for estimating propagation losses.

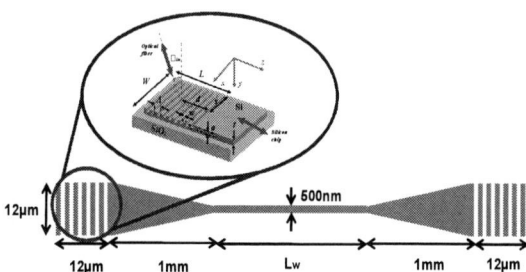

Figure 5: Schematic of the layout for fabrication of the devices.

The fabrication process of the grating structures was carried out by using e-beam lithography over PMMA resist. The electron dose was adjusted in order to achieve the optimized dimensions. After developing the sample, the patterned resist was employed as a mask in the following fabrication step consisting on dry etching by using an Inductive Coupled Plasma (ICP) system. This process was also optimized in order to reach grating design dimension (etching depth 70nm, period 600nm, and filling factor 50%). A Scanning Electron Microscope (SEM) image of fabricated gratings is depicted in Fig. 6. Actual etching depth of fabricated gratings was measured to be 67.5nm. In Fig. 6 it is also illustrated a SEM image detail when measuring grating period and filling factor after fabrication. Actual grating period is 611.7nm, and filling factor is ff=318.7/611.7=52%. For measured dimensions, coupling efficiency is expected to be 45%, according to theoretical fabrication tolerance graph depicted in Fig. 4.

Figure 6: SEM image of fabricated gratings.

Experimental transmission spectrum measurements for TE polarization of devices depicted in Fig. 5 are shown in Fig. 7. Minumum losses of 17dB are measured for λ=1550nm. Losses depicted in Fig. 7 are composed by input and output coupling losses, propagation losses in the waveguide, as well as losses in the input and output tapers. By measuring different waveguide length losses (L_w=1, 2, 4 mm), we estimated propagation losses of 7dB/cm in the waveguides for λ=1550 nm. So, we can estimate that experimental coupling losses of the gratings are around 4dB @ λ=1550 nm, and losses in each taper are around 1dB (extracted from a 1cm long device). This value corresponds to a coupling efficiency of about 40%. This experimental coupling efficiency value is in a very good agreement with 45% coupling efficiency previously estimated theoretically for grating dimension measured after fabrication. From Fig. 7 it can also be deduced that experimental 1dB bandwidth of the grating is about 30nm.

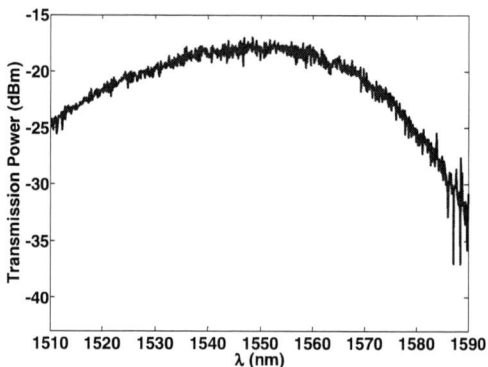

Figure 7: Experimental grating transmission spectrum measurements for TE polarization.

Highly efficient grating couplers

Although obtained grating coupling losses can be enough for lab prototypes, requeriments for assembled packaged devices in industry will require coupling losses even lower. One of the solutions for getting higher coupling efficiency relies on adding a bottom reflector on the SiO_2/Si substrate interface. This bottom reflector can take advantage of light which was previously lost by leakage to the substrate. Maximum amount of this light can be now reflected and constructively interfered with the light coupled to the waveguide, increasing coupling efficiency. According to simulation results of [5], coupling efficiency can be increased to about 79%.

Other solution to improve coupling efficiency is the use of nonuniform gratings. Output beam of a grating coupler with a uniform grating has an exponentially decaying power $P=P_0 exp(-2\alpha z)$ along the propagation direction (z). α is called the leakage factor or coupling strength of the grating. For a nonuniform grating, α becomes a function of z and the output beam can be shaped differently. Higher

978-1-4244-4722-0/09 $25.00
© 2009 IMAPS-ITALY

coupling efficiency to fiber can then obtained, by approximating the shape of the output beam to a Gaussian beam, matching better to the fiber optical mode [5].

By combining nonuniform gratings with the use of bottom reflectors in the SiO_2/Si substrate interface, maximum coupling efficieny up to 92% can be obtained [5]. This solution is being studied in the framework of IST-FP7 HELIOS project collaboration between IMEC (Gent University) and NTC (Technical University of Valencia) groups [6]. In Fig. 8 it is depicted a SEM image of fabrication optimization of first nonuniform grating prototypes by NTC. A very accurate resolution of e-Beam lithography is needed for the fabrication of the different filling factor periods of the grating. Optimization of e-beam lithography process is in progress.

Figure 7: SEM image of first prototype nonuniform grating fabrication by NTC.

Polarization diversity using grating couplers

In silicon photonics, diversity polarization schemes have to be implemented, as devices and components used in telecom networks are polarization sensitive. So, the use of polarization diversity approaches becomes mandatory in order to be sure proper polarization is injected to the devices and components. An example of polarization diversity approach in silicon photonics is depicted in Fig. 8. Light coming from the input fiber with random polarization is splitted in two orthogonal polarizations (TE and TM) using a polarization splitter. The TM arm is also TE rotated by using a polarization rotator. Then, by duplicating the same photonic component in both arms, only TE polarization interacts with the photonic component. At the output of the polarization diversity scheme, the same configuration as at the input is implemented for rotating the other arm of the polarization diversity approach, and combine them by using a polarization combiner, which is equal to the polarization splitter used at the input.

By means of 2D gratings it is possible to implement compact and efficient polarization diversity schemes. The advantage of using 2D gratings for the implementation is that no polarization rotator is needed anymore, as both polarizations coming from the input fiber are splitted properly in the right polarization by means of the 2D grating (see Fig. 9).

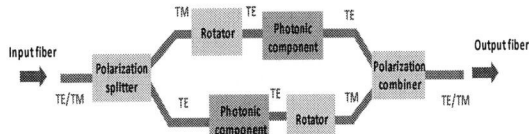

Figure 8: Example of polarization diversity scheme in silicon photonics.

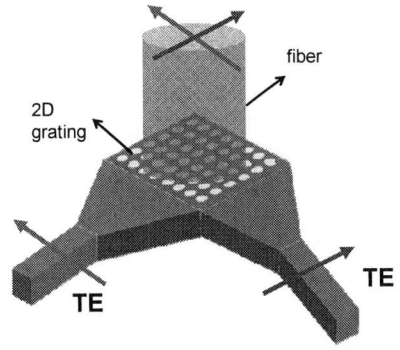

Figure 9: Polarization diversity scheme by means of a 2D grating coupler.

Packaging Consideration

As we have previously seen, grating couplers are a very attractive device for coupling fibers to SOI devices. Although coupling efficiency reachable with conventional designs is not very high, more sophisticated designs can be realized for improving coupling efficiency up to 90%. Furthermore, 1dB bandwidth inherent to grating couplers in SOI is still high for standard channel widths in WDM telecom networks. By implementing 2D gratings for polarization diversity schemes, polarization coming from the fiber can be splitted easily, so that there is no more need of polarization rotators which are so difficult to fabricate with standard fabrication technologies. Regarding alignment tolerances, multifiber vertical attachment for packaged assembled devices can be realized, as first prototypes have been already presented in literature [3]. Also, an advantage of the so obtained vertical orientation is that it is suitable for testing purpose at wafer level. It is possible in this case to measure the entire wafer (granted of course the proper tolerances) and, if necessary, also probe it on the pads in case of existing pads/lines. However, such vertical orientation is not easy to adapt to standard layouts for optoelectronics devices where, quite often, we have horizontal orientation. Also it is not really attractive in general to have such a loss of space on the top of the final device (we can have many mm of lost space in thickness).

978-1-4244-4722-0/09 $25.00
© 2009 IMAPS-ITALY

So, trying to overcome such limit, we recently started to design and develop a simple solution to allow the possibility to use such pigtailed components inside standard DIL or Butterfly Packages, a solution that could permit, where necessary, not to change the usual horizontal orientation of the completed packaged device as well to have a more rational footprint.

The simple concept arises from well known pin-diodes die-attachment applications where it is very usual to have substrates where to attach and connect the dies in the proper and most suitable way.

In our particular case we know that we have to change the fibers orientation from vertical, as requested by the grating structure, to horizontal, so we think that the best approach is to consider the SOI component as a simple die and to attach it, using UV epoxy, to the lateral side of a ceramic substrate, designed in such a way that the eventual transmission lines or electrical connection runs from the lateral side to the top side, as showed in figure 10 below.

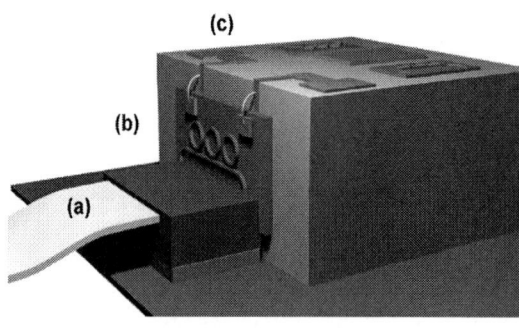

Figure 10: Subassembly for orientation change. Legenda: a) fiber-array with in/out fibers, b) SOI component, c) ceramic or silicon substrate with metalisation lines and pads

As a matter of fact we create a sub-assembly that can be, in a second time, easily mounted inside a package cavity, in the preferred position, and it is simple to imagine that the fibers can exit from the package throught a frontal cavity properly designed or a ferrule.

Looking at the draw in figure 10 it is clear that, with the use of such a submount, it is easy to connect, later on, the gold pads on the top of the same substrate directly to the package outputs. It is also possible, eventually, to integrate components on the submount itself, both on the lateral side and the top side. In figure 11 we show an example of design, placing our sub-assembly inside a cavity of a butterfly package. We started to work on such layout utilizing plastic packages, where this kind of cavity is quite easy to obtain, but of course it is also possible to use ceramic packages.

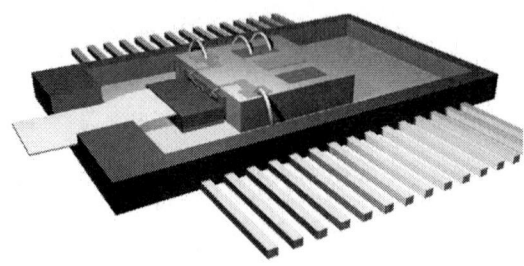

Figure 11: Package design concept with a butterfly package

Naturally, in case of a passive component, the only advantage of such solution lays in the final orientation and the optimized overall footprint (especially on the overall thickness).

But, in case of applications that need electrical connections on the SOI component (for example for thermal tuning utilizing integrated resistor), or in case that some active components have to be placed directly on the SOI component and have to be supplied, or, again, as mentioned, in case of necessity to place some kind of components on the ceramic submount, we think that such an approach can really bring good advantages in terms of integration and flexibility.

In fact, in these latter cases, we can imagine such kind of process flow:

1) Align, attach and mount the eventual components on the SOI chip, using any useful die-attachment technology from flip-chip with eutectic solder to standard epoxy die-mounting.
2) Proceed to the pigtail using standard fiber-arrays and themal curing epoxy.
3) Attach, mount and connect eventual components on the submount (top side but also, where possible, lateral side).
4) After attaching the chip to the substrate, connect the gold pads or the gold lines from the SOI chip to the gold pads or gold lines of the ceramic substrate, using standard wire-bonding technology.
5) Finally, after placing the obtained sub-assembly in the package, connect it to the package lids using the same wire-bonding technique (or ribbon–bonding if necessary).

Conclusion

SOI grating couplers have been presented as a promising approach for efficient packaging in silicon photonics. Around 40% coupling efficiency can be obtained by using conventional gratings for coupling to standard singlemode fibers. Coupling efficiency is almost constant for angle deviations of $\pm2°$ and fiber position deviations of about $\pm2\mu m$. Coupling efficiency can be improved above 90% by

978-1-4244-4722-0/09 $25.00
© 2009 IMAPS-ITALY

combining nonuniform gratings with the use of bottom reflectors. Furthermore, 2D gratings have also been shown to implement polarization diversity schemes. However, grating couplers are not easy to adapt to standard layouts for optoelectronics due to their vertical orientation. Therefore, a solution for allowing the use of standard DIL or Butterfly packages with horizontal orientation is also proposed.

Acknowledgements

Funding by EC under project HELIOS – FP7 – 224312 and Spanish MEC under contract TEC2008-06360 is acknowledged.

References

[1] H. Rong, S. Xu, O. Cohen, O. Raday, M. Lee, V. Sih and M. Paniccia, "A cascaded silicon Raman laser", Nature Photonics, Vol. 2, pp. 170-174, 2008.

[2] D. Taillaert, W. Bogaerts, P. Bienstman, T. F. Krauss, P. V. Daele, I. Moerman, S. Verstuyft, K. D. Mesel and R. Baets, "An Out-of-Plane Grating Coupler for Efficient Butt-Coupling Between compact Planar Waveguides and Single-Mode Fibers", IEEE J. Quantum Electronics, Vol 38, No. 7, pp. 949-955, 2002.

[3] L. Zimmermann, T. Tekin, H. Schroeder, P. Dumon, and W. Bogaerts, "How to bring nanophotonics to application – silicon photonics packaging", IEEE LEOS Newsletter, December 2008.

[4] L. Vivien, D. Pascal, S. Lardenois, D. Marris-Morini, E. Cassan, F. Grillot, S. Laval, J. M. Fedeli and L. El Melhaoui, "Light Injection in SOI Microwaveguides Using High Efficiency Grating Couplers", J. Ligthwave Technology, Vol. 24, No. 10, pp. 3810-3815, 2006.

[5] D. Taillaert, F. V. Laere, M. Ayre, W. Bogaerts, D. V. Thourhout, P. Bienstman and R. Baets, "Grating Couplers for Coupling between Optical Fibers and Nanophotonic Waveguides", Japanese Journal of Applied Physics, Vol. 45, No. 8A, pp. 6071-6077, 2006.

[6] http://www.helios-project.eu/

978-1-4244-4722-0/09 $25.00
© 2009 IMAPS-ITALY

Encapsulation Challenges for Wafer Level Packaging

[1]Eric Kuah TH*, JY Hao (PhD), JP Ding (PhD), QF Li , WL Chan & SC Ho

HM Huang[+] and   YJ Jiang[++]

[1]ASM Technology Singapore Pte Ltd
2 Yishun Avenue 7, Republic of Singapore 768924
*Email: eric.kuah@asmpt.com

[2]Siliconware Precision Industries Co. Ltd. (SPIL)
No. 153, Sec. 3, Chung Shan Road, Tan tzu, Taichung , Taiwan 427, R. O. C.
[+]Email: huiminghuang@spil.com.tw
[++]Email: jasejiang@SPIL.com.tw

Abstract

The interest of user for WLP has been raised because of benefits such as reduced package thickness, fan-out capability, high I/O, substrate-less process, integration of passives into structure, good thermal and electrical performance. The objective of this paper is to delineate technical challenges and issues that potential adopter of wafer level molding will face, technological solution availability and the broad application of WLP using granulated epoxy and liquid encapsulant such as epoxy, hybrid of epoxy-silicone, silicone. Result base on actual molding trial indicates among the different form of wafer level molding the challenges being faced are similar, including co-planarity, warpage, die shifting, coefficient of thermal expansion matching, incomplete filling, MBF and voiding

Key words: WLP (Wafer Level Package), Wafer Level Molding (WLM), Die Shift (DS), Warpage, Mold Bleed Flashing (MBF), Co-planarity, Voiding, Tape crinkle Incomplete Filling, PEMs (plastic encapsulated microelectronics)

Introduction and Cost

Using WLM to assemble packages allows the designer to reduce the package in z direction and parally minimize the lateral x-y dimension. At the same time optimizes the real estate for the number of I/O through minimum pitching. The electrical benefits of WLP for package include introduction of redistribute layer which increase functionality but not necessary I/O and increases in operating frequency, clock speed, which reducing parasitice effects and also the interconnect line length. Using RDL opens to designer the capability of functional integration of passives and system in package. In some instances such as the embedded wafer level package an increase fan-out capability such that the I/O increment can be dramatic [1]. Performance gains of package manufactured via WLM are electrical as discussed above and thermal. Thermal improvement comes from the fact that there is better board to chip coupling which results in efficient heat dissipation. Cost efficiency of WLM is from three angles. Firstly, a wafer-less process in case such as

embedded wafer level packaging [2] and also the encapsulation process does not result in material wastage as traditional transfer does. Secondly, the tooling use in WLM can be generic so long the range of wafer diameter is known in priori during design. As encapsulation of different mold cap thickness can be made using the same tool and the saving can be substantial in terms of the cost of tooling and the speed of execution to deliver just in time product to customer. Lastly, in WLM the cost of manufacturing is expected to be low in the long run, because lots of assembly steps can be done in batches, in parallel and testing of assemble package is at a wafer level format rather than an individual singulated units.

Like most package development over years from PDIP, SOIC, BGA, QFN and POP, encapsulation of WLP will certainly have challenges to meet. Encapsulation challenges in WLP can be broadly categorized into material selection, equipment mold, location of optima processes and moldability requirement. There are several types of

WLP package (Figure 1) in the market but this paper will discuss generically with reference to the embedded wafer level package [2] with fan out and the paper has four sections: materials, encapsulation, moldability, discussion and summary.

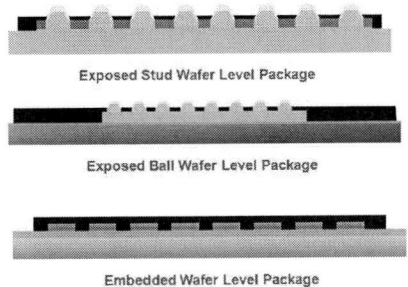

Figure 1: Types of WLP

Materials

The encapsulant used for WLM can be broadly classified into liquid and B-stage granular (powder) molding compound. This type of encapsulant provides protection against mechanical and hostile environmental damage for the assembled package when use. The encapsulant should have good flowability, moldability, well-controlled CTE and shrinkage. One additional desirable properties of encapsulant to meet such requirement is low temperature molding capability. Experimental data confirms the requirement of low encapsulation temperature indeed result in lower warpage as shown in Figure 2.

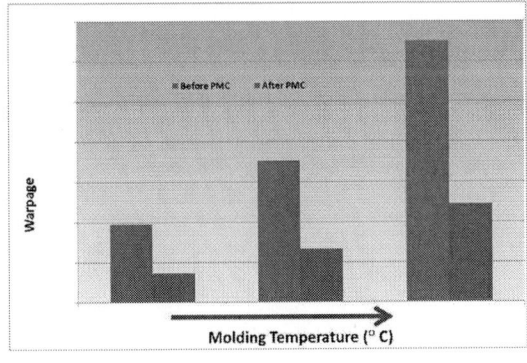

Figure 2: Warpage Variation with Temperature

Lower temperature molding also assist in mitigating better die shift which will be discussed further in following section.

Encapsulation

The design of the encapsulation mechanism should allow full control of the molding process. This include good parallelism to ensure good coplanarity, programmable clamping and packing profiles to serve optimum mold quality and minimum die shift, effective vacuum and venting design to achieve complete and void free molding. Figures 3 and 4 are typical commercial encapsulation equipment and the mold tool, respectively. The mold tools are generally design to be robust such that it can mold wafer or metal carrier ranges from 6 to 12 inches. Further it can encapsulated square substrate format. If the industry decide to move forward in this direction the gain of output can be substantial, refer to Figure 5.

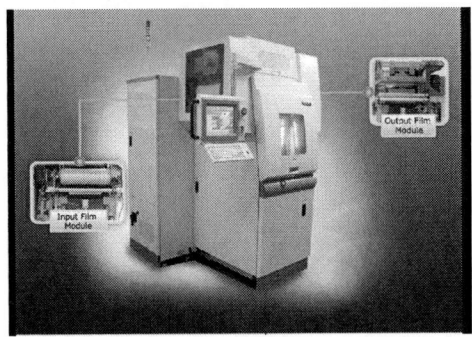

Figure 3: Encapsulation System

The challenges that manufacturer of wafer level molding faces are co-planarity, warpage, die movement, film wrinkle, voids, incomplete fill and mold bleeding. Co-planarity as defined here as the maintenance of thickness variation of $\pm$ 20 to 30 μm. It is achievable by designing mold that can be co-planar and balance by correct factoring of the manufacturing stack up tolerance and material thermal expansion. Figure 6 shows field data to support this claim of that such co-planarity is possible. Good co-planarity is required by downstream process such as RDL. A non-coplanar molded wafer will result in non-uniform layer of RDL.

Tight warpage requirement is critical because of the handling and alignment of the downstream process after encapsulation. Handling is because most assembly houses use front-end equipment are generally robotic handlers, thus warpage is a challenge for robot end-effector to handle, as these transport mechanism are initially design for moving fairly planar wafers. Base on current materials made up of the various components, warpage attainable after mold of 1000 μm is possible, see Figure 7.

978-1-4244-4722-0/09 $25.00
© 2009 IMAPS-ITALY

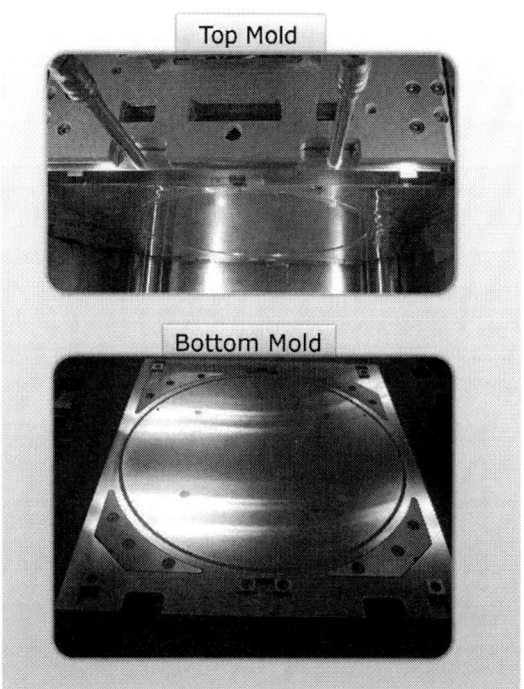

Figure 4: Molding Tool

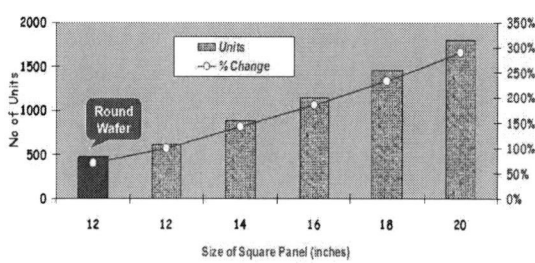

Figure 5: Round Vs Square Format Substrate Moldability

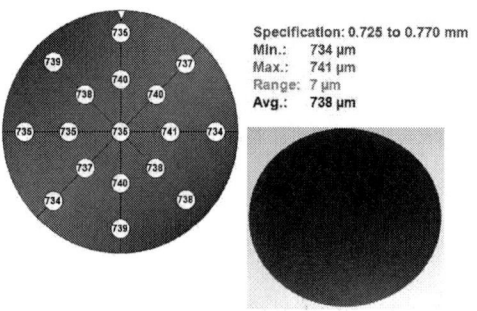

Figure 6: Co-planarity Measurement

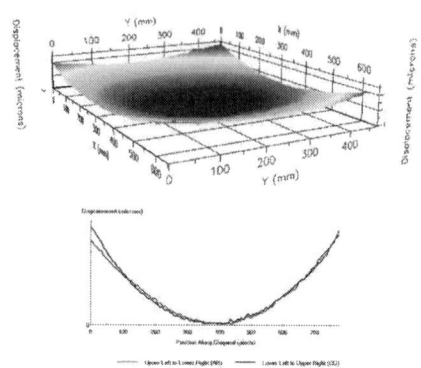

Figure 7: Warpage Using Shadow Moriè

A substantially warp wafer after molding will affect alignment during process such as RDL as it cannot target accurately for the layer redistribution to occur. Die movement can be of two categories namely, shifting and swimming (flying). The swimming of die meant that a die has displaced itself out of it original attached position drastically and general greater than 100 μm, Figure 8 illustrated this die swim defects. Reasons of such phenomenon include poor attachment of die to the double side tape, poor adhesion preparation, staging before molding, moisture, use of expire tape and too aggressive population of die toward edge of wafer. Die swimming can be actually avoided by choosing correct material, optimized taping as well as molding process.

Figure 8: Die Swim

Die shift with respect to wafer level packaging is referring to micro movement of less than 10 μm. Larger than that is generally not acceptable due to the existing process of RDL using "aligner" methodology. Offset UBM layer is the consequence of using encapsulated wafer with large die shift during RDL process.

978-1-4244-4722-0/09 $25.00
© 2009 IMAPS-ITALY

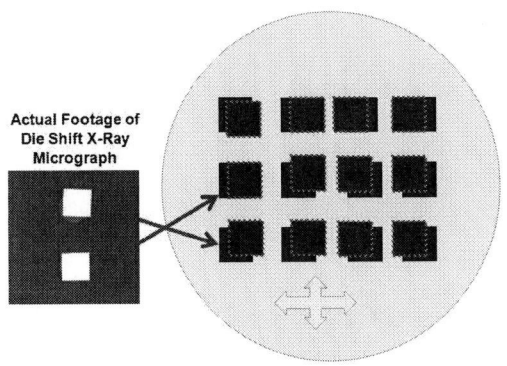

Figure 9: Die Shift

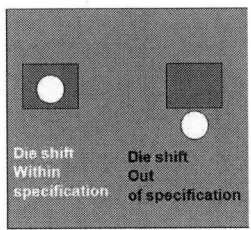

Figure 10: UBM Due to Die Shift

Unlike die swimming, die shifting is material related rather than molding process, as far as we know, die shifting is caused by either CTE mismatch, i.e. different thermal expansion of material, double side tape and carrier, or shrinkage of encapsulant during curing. Die shifting is one of the major obstacles for users to implement this technology and a lot of efforts are still ongoing to resolve this issue. Figure 10 shows the result of die shift after RDL process.

Film wrinkle imply the creation of imperfection on the mold piece as shown in Figure 11.

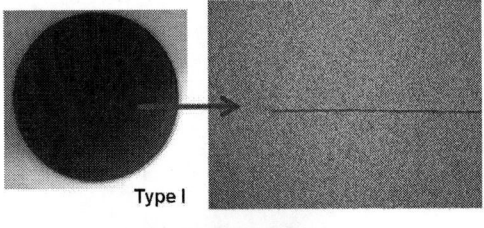

Figure 11: Film Wrinkle

There are two types namely Type I due to the tape using for molding and Type II due to the double sided taped use in embedded wafer package level to hold the die during assembly. There are a couple

reasons for such imperfection in Type I, that include sequencing of film attachment to the top mold tool, mold tool surface, and type of film. The film attachment is about sequence to apply air suction while type of film is about the property like elongation and surface finishing. Table 1 indicates test result of different film performance for film wrinkle. Typically, optimizing the vacuum efficiency and interface between film and metal piece can resolve issues such as wrinkle. Type II crinkle defect is primary due to poor preparation of the double side tape on the metal carrier. This defect can lead to mold flash on the die once exposed as shown in Figure 11.

Table 1: Film Type Influence on Wrinkle

Film Types	Type A	Type B	Type C
Film Sticky	O	O	O
Film Wrinkle	X	O	X
Outer Residue	O	O	O

Voiding in WLM can be external or internal. These defects are generally due to air entrapment during the encapsulation. One way to resolve these issues are using vacuum evacuation and design of strategic locate air venting. Another way is to optimize dispensing pattern to eliminate air entrapment inside B-stage compound. Figure 12 below shows the X-ray micrograph of internal void formation indicate by the white color spot.

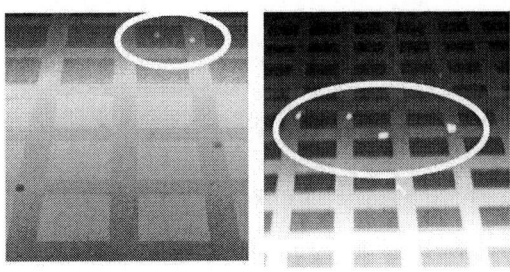

Figure 12: X-ray Micrograph of Internal Voids

The last moldability defects to discuss here is incomplete filling. It is defined here as the incomplete formation of the encapsulant to completely encapsulated around the die. Besides insufficient air evacuation and non-optimized process, e.g. temperature, preheat, filling time and balance between clamping and packing, some of the root causes are from material chemistry, such as gel time, catalyst, hardener etc. Sometimes bubbles under double side tape caused by poor taping or outgassing also induce incomplete fill. Figure 13 is an illustration of an incomplete fill defect.

978-1-4244-4722-0/09 $25.00
© 2009 IMAPS-ITALY

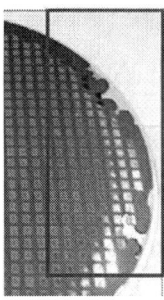

Figure 13: Incomplete Fill

Discussion and Summary

In sections above, discussion was focused on three main areas for WLM, namely, materials, encapsulation and moldability. They can be well summarized with the concept map shown in Figure 14.

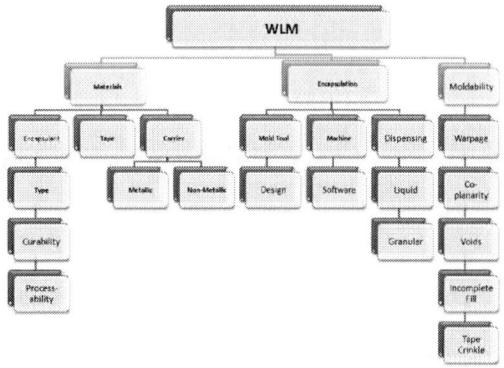

Figure 14: Concept Map of Wafer Level Molding

If due diligence is exercised during the assembly and/or packaging of a wafer level package, the chance of obtained process-able wafer for the next step process such as the RDL, laser dicing and finishing process shown in Figure 15, is absolutely possible without much hurdle during research and development phases. However, what are the factors and/or circumstances that will prevent the proliferation of this package through the assembly and packaging segment of the semiconductor industry? In our view, there are several - knowledge, manufacturability and cost structure. Firstly, the knowledge of WLM lies in the hand of several users (assembler and suppliers) which are deemed as intellectual property thus potential adopter of the technology would be either deter or have little incentive to pursue them vigorously. Secondly, the manufacturability of WLP involve a number of process steps (as much as 70 in most cases) with very stringent specification requirement from process to process partly due to the reverse of packaging the package starting from backend assembly follow by frontend assembly. It requires adopter to have good manufacturing capability from

both the assembly machines to processable materials. Lastly, the cost of manufacturing this fan out wafer level package should be substantial lower than existing alternative as such singulated flip chip, even if the better performance index are better. If this cannot happen, it would not be an incentive for the industry to adopt, as the economic value and the return of investment cannot be justified, furthermore, we are now in the age of high technology, which must also accompanied by substantially low cost to produce.

In sum, our perspective on success for the fan out wafer level packing is shown in Figure 17. The themes are about of sharing of knowhow with a win-win mindset; close collaboration among adopters and supplier; and teamwork with the customer who is the adopter and sellers of the final product, material and equipment suppliers that are inputs to the assembly of this package. Only through these efforts of collaboratively spinning of the gears, then the industry can realize the successful proliferation of fan out WLP, as an alternative to current methods of producing PEMs.

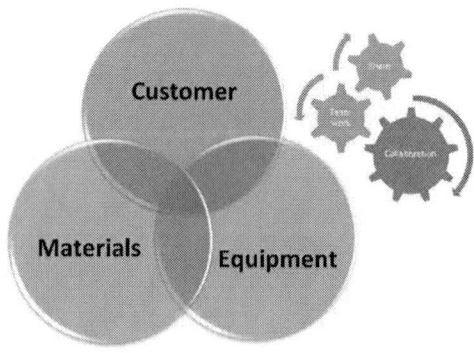

Figure 16: Assembly and Packaging of fan out WLP

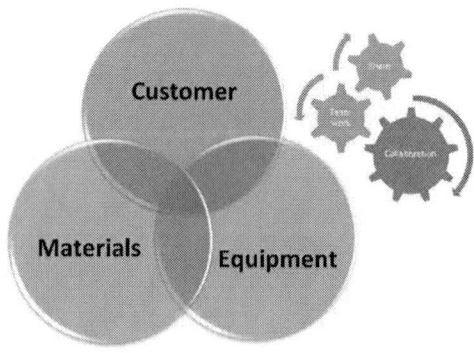

Figure 17: Wheels of Successful Proliferation of Wafer Level Package

Acknowledgements

The authors wish to thanks SPIL and ASM Pacific Technology higher management for permission to work and collaborated on this project and the publication of the work. In addition, our colleagues that help in this work, appreciation also

978-1-4244-4722-0/09 $25.00
© 2009 IMAPS-ITALY

goes to the material vendors that collaborate with us and our respectively subcontractors.

References

[1] S. C. Johnson, "Fan-out wafer-level packages catch on", 3D Packaging News Letter on 3D IC, TSV, WLP and embedded Technologies, Issue 10, March 2009.

[2] M. Brunnbauer , T. Meyer, G. Ofner , K. Mueller, R. Hagen, "Embedded Wafer Level Ball Grid Array (eWLB)", 33rd International Electronics Manufacturing Technology Conference 2008.

Mechanical Behaviour of SAC-Lead Free Solder Alloys with Regard to the Size Effect and the Crystal Orientation

Villain, Juergen; Mueller, Wolfgang*; Saeed, Usman; Weippert, Christina; Corradi, Ulrike; Svetly, Artur

University of Applied Sciences Augsburg, Germany;

An der Fachhochschule 1, 86161 Augsburg, Germany

juergen.villain@hs-augsburg.de, 0049(0)821 5586 3386

*Technical University of Berlin, Germany, Strasse des 17. Juni 135, 10623 Berlin, Germany

Abstract

Material parameters has to be determined with regard to the influence of the solder volume and the crystal orientation to understand the thermo-mechanical properties and therefore the reliability of small lead free solder joints in microelectronics. The influence of crystal orientation to the reliability is very important because a small solder joint of a 0201 or a 01005 electronic device consists of three to one tin dendrites only.

The results of creep and tensile tests were obtained with with very small test specimens (diameter 1 mm) of SAC-alloys with Ag-contents between 2 to 4 wt-% and Cu-contents with 0,5 to 1.2 wt-% and SAC-alloys with Bi, Sb and Ni components at room temperature, 80 and 150 °C. The highest strength and creep resistance show alloys with high Ag and Cu content combined with Bi, Sb and Ni. In solder joints the Cu content depends on the solder volume and results in diffusion and intermixture of Cu in the solder material during soldering. Nano hardness measurements in ß-tin crystals and intermetallic components to determine the hardness, the Young`s Moduli and the Yield strength give information to the size effect on mechanical parameters and the strong influence of the crystal orientation, which was determined using EBSD measurements, and the metallographic structure of the solder joints after reflow soldering.

The experimental results and their differences based on differential test methods will be compared with literature date and hints will be given to choose the real material parameters for thermo-mechanical simulations.

Introduction

Material parameters has to be determined with regard to the influence of the solder volume and the crystal orientation to understand the thermo-mechanical properties and therefore the reliability of small lead free solder joints in microelectronics. The influence of crystal orientation to the reliability is very important because a small solder joint of a 0201 or a 01005 electronic device consists of three to one tin dendrites only [1].

Under thermal-mechanical stresses normally the cracks run in the solder material of a solder joint. So it is one way to study the mechanical behaviour – tensile behaviour, creep - of solder alloys with small casted test specimens of different solder alloys. On the other hand it is necessary to used strain rates which are near by the strain rates in real solder joints.

Tested specimens and alloys

To determine the mechanical parameters of solder alloys small casted tensile specimens were used (Fig. 1).

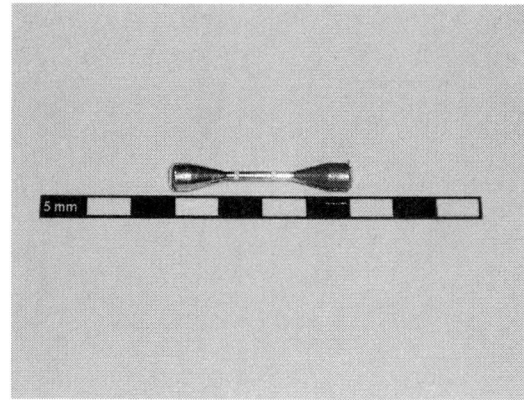

Fig. 1 Casted tensile specimen

The tensile specimens with a diameter of 1 mm were used to investigate nearly the same dimensions of solder volume as we found in small solder joints. Stress-strain diagrams were determined at room temperature using a strain velocity of 3×10^{-3} 1/s. A laser extensometer was used to determine the elongation.

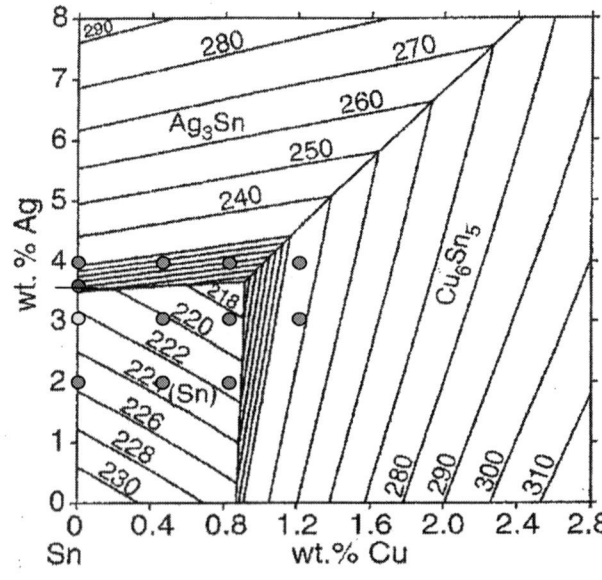

Fig. 2 Sn rich side of the ternary composition diagram [1]

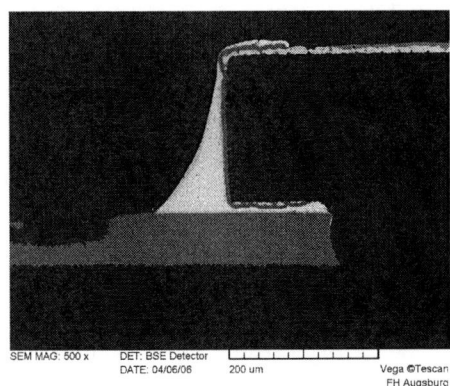

Fig. 23 Cross section of a small solder joint (0201)

SAC-alloys with Ag-contents between 2 to 4 wt-% and Cu-contents with 0,5 to 1.2 wt-% and SAC-alloys with Bi, Sb and Ni components (Innolot SnAg3.8Cu0.7Bi3Sb1,5Ni0.15) at room temperature, 80 and 150 °C were used. Fig. 2 shows the tin rich edge of the SnAgCu constitution diagram. The grenn points indicate some tested solder alloys.

To determine the real Cu content in solder joints the Cu content was determined in the meniscus area (Fig. 3), without the intermetallic phases of different electronic devices (0201, 1206, 2512, BGA 345).

Nanoindentation

The hardness was determined by Vickers indentation. An apparatus to indent small areas with a lateral reproducibility lower than 0.5 μm was used. It is a combination of a micro indenter (max load 20 mN, min load 50 μN) and an optical distance sensor (solution 2 nm). This apparatus could be used in a microscope or in a SEM (patented). The force and the indentation depth were measured on line, to get force indentation depth curves. These curves were used to determine the hardness, the Yield strength and the Young`s Modul.

Electron Back-Scattering Diffraction

The most attractive feature of the EBSD technique is its unique capability to perform concurrently rapid and automatic diffraction analysis to give crystallographic data and imaging with a spatial resolution of less than 0.5 μm. Normally, it is combined with a SEM (Scanning Electron Microscope) for chemical analysis and imaging of rough surfaces. A small angle is adjusted between the electron beam and the specimen surface, typically between 30° and 20°, to make sure that more electrons are diffracted and "escape" toward the detector. The diffraction is generated by the interaction of the primary backscattered electrons with the lattice plains (Bragg's law) close to the specimen surface (some nm deep). This means that some electrons must always arrive at the Bragg angle θ at every set of lattice planes. The locus of the diffracted

radiation is the surface of a cone (Kossel cone) which extends about the normal of the reflecting atomic planes. Due to the very small wavelength of the electron beam one half of the apex angle of this diffraction cone is close to 180°. If these diffraction cones were intercepted by a phosphor screen, a pair of parallel conic sections results (Fig. 4). These lines appear to be parallel lines and they are called as Kikuchi lines. From this data information about the crystallographic orientation of the scanned surface and, therefore, about the micro-texture can be constructed serially [2]. This technique was used to determine the orientation of the β-tin crystals which were investigated.

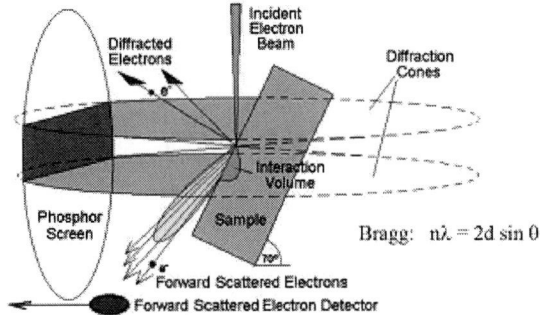

Fig. 4 EBSD technique schematically (Source: Albert-Ludwigs-Universität Freiburg, Germany).

Results

The influence of Ag or Cu on the mechanical behaviour of SAC alloys is shown exemplarily in Fig. 5 based on bulk material. More Cu raises the tensile strength for all tested temperatures and alloys even though the strength decreases with higher temperature. The highest increase of the strength

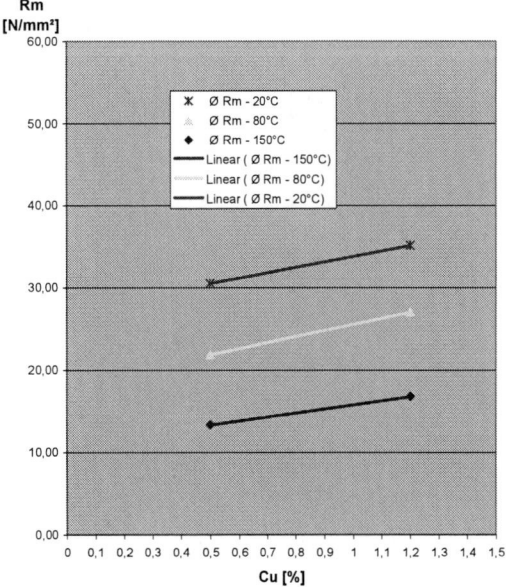

Fig. 5 Influence of Cu content on the tensile strength of SnAg2Cux

978-1-4244-4722-0/09 $25.00
© 2009 IMAPS-ITALY

could be observed for Cu content more than 1 wt.-% due to more Cu6Sn5 phase in the solder volume. For Ag a similar trend can be observed, but the influence on the strength is a little bit smaller and no critical Ag content could be seen. If we add Sb, Bi and Ni to SAC solder alloys (Innolot) a big increase of the strength can be observed due to many small intermetallic compounds and a mixture crystal hardening by Sb and Bi (Fig. 6) [???live]. The ultimate tensile strength decreases with lower strain rate.

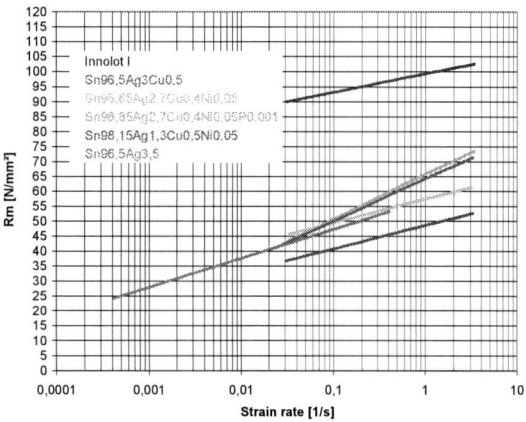

Fig. 6 Influence of strain rate on ultimate tensile strength for SAC alloys (lower curves) and Innolot

If we want to know the mechanical behaviour of solder joints, we need information about the chemical composition of a SAC solder joint depending on the used electronic devices, that means of the solder volume. Normally the metallization of electronic devices, R or CC, is Ni or NiPd with a layer of Sn and the metallization of the substrate is Cu OFC, Cu/Sn or Cu/Ni/Au. Due to the reaction of the SAC solder alloy with these elements, which diffuse into the liquid solder, SnNi- and

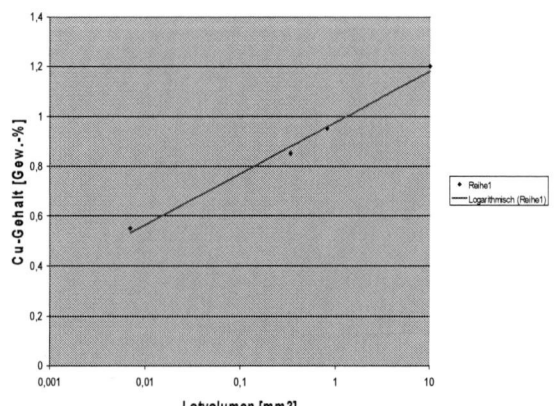

Fig. 7 Cu content in bulk solder volume of solder joints

SnCu intermetallic compounds arise at the boundaries to the electronic device and the pad of the substrate. Primary

Cu6Sn5 phases occure due to the higher Cu content in the solder volume.

Fig. 7 shows the Cu content in the meniscus area with out the intermetallic layers using a SAC solder alloy SnAg3Cu0.5. It shows that the Cu content in solder alloys is not determined by the used solder paste. Their composition depends on the solder volume that means the size effect of the solder volume. The smaller the solder volume, the lower the Cu content. This could be the result of different fluctuation behaviour in different solder volumes so the solidification of Cu in the bulk material area is smaller with decreasing solder volume. This indicates that different solder compositions on a board exists with different mechanical parameters as tensile strength, Yield strength, Young`s modules and thermal expansion coefficients and therefore the reliability of the solder joints built up by the same solder paste differs.

Fig. 8 shows a typical force-indentation curve of a Vickers indenter in tin crystals. The difference of the curves indicates different mechanical behaviour of pure tin crystal and ß-tin crystal. The mechanical parameters are summarized in Tab. 1

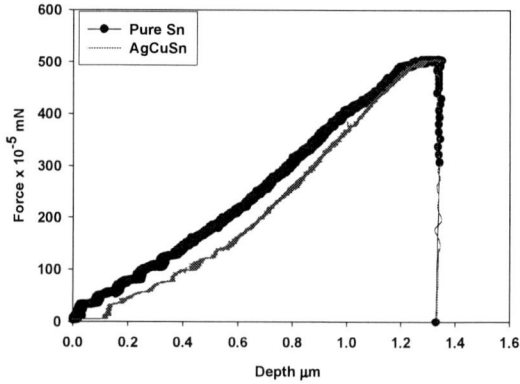

Fig. 8 Force-indentation curve of a Vickers indenter in tin crystals (pure tin, ß-tin in SAC-alloy)

The hardness of the pure tin crystals is tendencially lower than of the ß-tin crystals due to a small amount of solved elements as Cu or Ni in Sn. Two techniques were used to determine the hardness: AFM profile measurement (calculate AFM) and the theory of Oliver and Pfarr (experimental) [3,4]. The differences in the Young`s modulus are not significant and give a hint that the orientation of the tested crystals is comparable. Normally the anisotropy of tin has to be taken into account. The Young`s modulus of Sn varies from 28 to 64 GPa depending on the crystal orientation. These results compare well with literature data.

EBSD measurements were done to determine the crystal orientation of tin crystals. Before a nanohardness measurement starts, the orientation of the tin crystal was determined. In Tab. 2 a comparison of hardness and crystal orientation is done.

978-1-4244-4722-0/09 $25.00
© 2009 IMAPS-ITALY

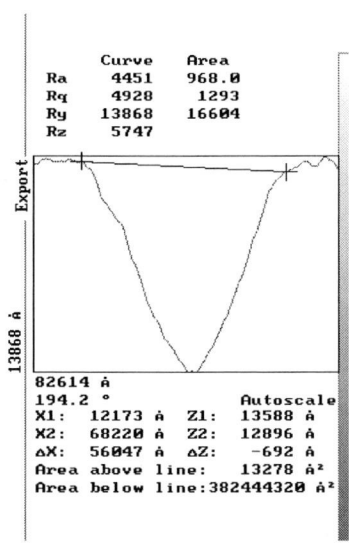

```
           Curve    Area
   Ra      4451     968.0
   Rq      4928     1293
   Ry     13868    16604
   Rz      5747
```

```
82614 Å
194.2 °                    Autoscale
X1:   12173 Å   Z1:   13588 Å
X2:   68220 Å   Z2:   12896 Å
ΔX:   56047 Å   ΔZ:    -692 Å
Area above line:      13278 Å²
Area below line:382444320 Å²
```

Fig. 9 Depth profile of a nanoindent (AFM measurement)

Tab. 1 Hardness and Young`s Modulus of tin and ß-tin crystals

Specimen	Hardness (calculated AFM) HV	Hardness (Experimental) HV	Indent. Modulus GPa	Indent. Modulus (lit.) GPa
Sn (Pure)				
1	9.84	10.43	46.8	50**
2	8.44	9.76	42.8	
3	10.43	10.62	48.92	
4	10.82	11.31	45.97	
5	10.32	11.02	45.31	
6	9.44	9.93	45.06	
Sn-9Zn	11.58	12.39	45.9	
Sn-1Ag-0.5Cu	16.21	17.34	45.5	
Sn-3Ag-0.5Cu	17.55	18.31	46.3	
Sn-3Ag-0.9Cu	18.53	19.50	46.0	
Sn-3Ag-1.2Cu	18.45	19.24	48	

Tab. 2 Influence of grain orientation on the hardness

Orientation (hkl) / tilt	Hardness [HV] / standard deviation
(212) / 6°	12 / 0.4
(001) / 12°	20 / 6
(320) / 2°	16 / 2

Discussion

To determine the mechanical behaviour of small solder joints many influences have to be taken into account:

- number of crystals in a solder joint
- orientation of the tin crystals which determine mainly the reliability of the solder joint
- chemical composition with regard to the Cu content
- mechanical behaviour of solder joints with different Cu content (size effect)

Real thermo-mechanical simulations can only be done if these parameters are considered. Otherwise wrong results lead to wrong reliability data and reliability models.

Conclusions

Material parameters small lead free solder joints in microelectronics material parameters have to be determined with regard to the influence of the solder volume and the crystal orientation in order to get valid results of the reliability. The combination of tensile tests, nanoindentation measurements and determination of the crystal orientation by EBSD measurements lead to understanding of the mechanical behaviour and therefore the reliability of SAC solder joints.

Acknowledgments

We would like to thank for the support by the BMBF project "LIVE", the AiF project "PIW" and the European COST action MP0602.

References

1. Materialmodifikation für geometrisch und stofflich limitierte Verbindungsstrukturen hochintegrierter Elektronikbaugruppen "LiVe"; Aufbau und Verbindungstechnik in der Elektronik – Aktuelle Berichte, Band 8, Verlag Dr. Markus A. Detert, Templin, 2009

2. Randle, V., Engler, O., Introduction to Texture Analysis, Taylor & Francis (2000)

3. Fischer-Cripp, A. C., Nanoindentation, Mechanical Engineering Series, , Springer Verlag, (New York, 2002)

4. W.C. Oliver and G.M. Pharr, J. Mater. Res. 7, 1564 (1992).

NanoBond® Assembly –
A Rapid, Room Temperature Soldering Process

Presented by:
Greg Caswell
VP-Engineering
Reactive NanoTechnologies
180 Lake Front Drive
Hunt Valley, MD 21030
443-834-9284
gcaswell@rntfoil.com

Abstract - Reactive NanoTechnologies (RNT) has commercialized a new technology that will revolutionize how manufacturers join components using solder materials. (See Figure 1) The joining process is based on the use of reactive multilayer foils as local heat sources. The foils are a new class of nano-engineered materials, in which self-propagating exothermic reactions can be ignited at room temperature through an ignition process. By inserting a multilayer foil between two solder layers and two components, heat generated by the reaction in the foil melts the solder and consequently bonds the components. The joining process can be completed at room temperature in air, argon or vacuum in approximately one second. The resulting metallic joints exhibit thermal conductivities two orders of magnitude higher, and thermal resistivities an order of magnitude lower, than current commercial TIMs.

The use of reactive foils as a local heat source eliminates the need for torches, furnaces, or lasers, speeds the soldering processes, and dramatically reduces the total heat that is needed. Thus, temperature-sensitive or small components can be joined without thermal damage or excessive heating. In addition, mismatches in thermal contraction on cooling can be avoided because components see very small increases in temperature. This is particularly beneficial for joining metals to ceramics. The fabrication and characterization of the reactive foils is described, and the value proposition for NanoBonding is presented. This paper also shows the applicability of this platform technology to many areas of packaging including Thermal Interface Materials, microelectronics, optoelectronics, and Light Emitting Diodes (LEDs)

Key words: Thermal transfer, TIM, NanoFoil, NanoBond, solder bonding

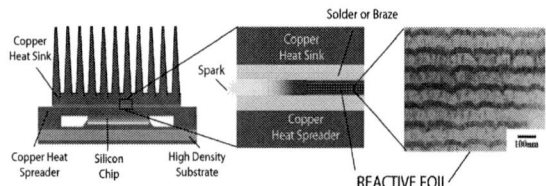

Figure 1-Cross section of NanoFoil and Application

Background

There are numerous instances where it is necessary to provide a thermal path from one element of a product to another, e.g. circuit board to heatsink, power amplifier to heatsink, High Brightness LED to printed wiring board, etc. Most of these applications have utilized some form of thermal adhesive or reflow soldering technique to provide the bond. Often it has been found that the thermal stresses

inherent in the product require a better methodology for the heat transfer. There are also situations where the materials to be joined together have Coefficient of Thermal Expansion (CTE) mismatches which then result in a stress being applied to the solder joint itself. The use of Nanofoil® and the process of Nanobonding® provide such a solution. This paper will delineate several applications where this process provided significant improvement in both the thermal and electrical performance of the product.

Nanofoil® Fabrication and Characterization

Nanostructured, Ni/Al reactive multilayer foils were fabricated using a large-area, magnetron sputter system. The foils were deposited onto substrates and then were peeled from their substrates for use as free-standing samples. The reactive foils incorporate thousands of alternating layers of Ni and Al that are approximately 25nm thick. Total foil thicknesses can range from 40 to 200µm and are varied by simply changing the total number of Ni and Al layers.

978-1-4244-4722-0/09 $25.00
© 2009 IMAPS-ITALY

To create a bond, the NanoFoil is placed between the two components being joined, along with a solder layer or Sn plated foil. An energy impulse such as a battery, power supply, laser, or soldering iron is then applied to the foil. This initiates a self-propagating reaction in the NanoFoil material. The like-like bonds of the atoms of each layer in the foil are exchanged for more stable unlike bonds between atoms from neighboring layers. As the atoms of each layer mix, additional heat is generated, creating a self-sustaining reaction traveling the length of the foil. Temperatures at the joining surface are raised above the temperature necessary to melt the solder or braze, and thus create a bond. With most of the heat focused only on the joining surface, the temperature of the components does not rise, and their microstructure is not compromised. Joining dissimilar materials with different coefficients of thermal expansion, such as metals and ceramics, does not present a problem because heat is applied only to the surface.

When choosing a solder for the reactive bonding process, several criteria were considered including a desire to: (1) minimize solder cost; (2) maximize the thermal conductivity of the solder, (3) ensure compatibility with reactive foil and component surfaces, and (4) maintain a moderate solder melting temperature so as to minimize thermal exposure of the components. This process has been found to work equally well with SnAg, SAC305, SAC405, SnPb or matte tin interfaces.

Figure 2 is a model of the actual reaction of the NanoFoil once ignited and shows the limited depth of penetration of the temperature increase with regard to the heat sink side as well as the die side of this structure.

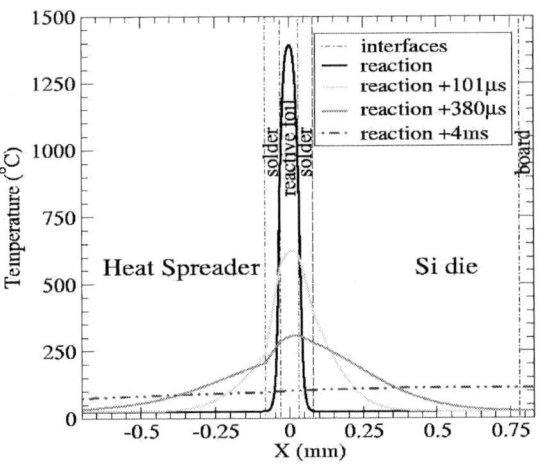

Figure 2 –Model representation of foil ignition
The data from this model is summarized in Table 1 which illustrates the maximum temperatures anticipated at each interface.

Die Thickness (µm)	700
Foil Thickness (µm)	60
Top of Spreader (oC)	60
HS/Solder (oC)	243
Solder to Foil (oC)	943
Foil to Solder (oC)	943
Solder to Die (oC)	389
Top of Die (oC)	113

Table 1- Maximum Temperatures at Interfaces

Applications

One of the byproducts of increasing power density in high power electronic components is higher heat dissipation and the consequent need to more effectively cool the component to ensure long term reliability. Minimizing the thermal resistance at the device/heat sink interface is critical for the development of new generations of high power electronic components. Similarly, reducing the junction temperature of LEDs is known to increase their lifetime significantly.

High Brightness LEDs

Bonds were fabricated between OSRAM Dragon-series LED packages and Sn-plated MCPCBs. The strength and thermal performance of the reactive multilayer bond was optimized by varying the reactive multilayer foil thickness and bond process parameters such as bonding pressure. The resulting optimized bonds exhibit average measured thermal conductivities of 30W/Km and shear strengths of 35 MPa (5130 PSI) with less than 5% void content. There is no significant degradation observed in the thermal performance and structural performance after a series of reliability tests.

The comparison tests between reactive multilayer bonds and epoxy thermal interfaces were conducted on both 1.6W Golden Dragon and 4.6W Platinum Dragon LED's attached to MCPCB's. The resulting reactive multilayer bonds have an average thermal

978-1-4244-4722-0/09 $25.00
© 2009 IMAPS-ITALY

resistance from the heat slug on the LED to ambient R_{ca} of 4.3°C/W and a slug temperature $T_{c(1A\ SS)}$ of 44°C at a steady state current of 1A, whereas the thermal epoxy bonds exhibit a thermal resistance of 5.2°C/W with a $T_{c(1A\ SS)}$ of 47°C on Golden Dragon LEDs. Platinum Dragon LEDs exhibit an average $R_{ca(2.5A\ SS)}$ values of 3.6°C/W for reactive multilayer bonds at 2.5A and 8.9°C/W for thermal epoxy. For all tests ambient was maintained at 25°C. Clearly the reactive multilayer bonds dramatically outperform thermal epoxy, especially at high operating currents. With further optimization it is believed that the gap between the reactive bonds and thermal epoxy will be even larger. It is well documented that this improvement in thermal resistance can more than double the anticipated lifetime of an LED device.

Heatsink to Printed Wiring Board

The Nanobonding process is extremely amenable to the attachment of circuit boards to heatsinks. Figure 3 shows two RF circuit boards and corresponding heatsinks. The circuit boards have an Electroless-Nickel Immersion-Gold (ENIG) finish as do the heatsinks (one copper and one aluminum). 60u thick tin plated foil was used to successfully bond these surfaces

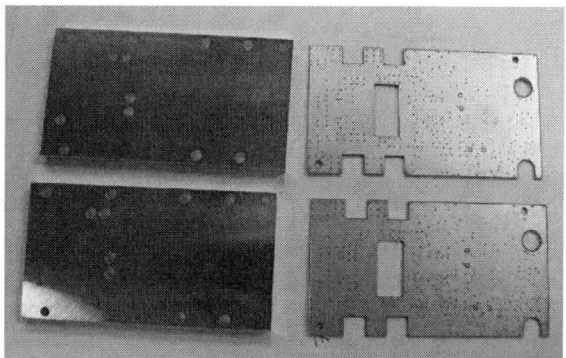

Figure 3 – RF circuit boards and heatsinks

For comparison purposes, two of the heatsink assemblies were tin plated. A tin plated NanoFoil preform was placed between the PCB or power amplifier and heat sink. A joining pressure of 100 psi was applied on the top of PCB or power amplifier. The NanoFoil was then ignited electrically; the reacting foil melted the solder and joined the components. RNT has also successfully bonded silver backed PWBs to silver heatsinks, tin plated interfaces and hot-air solder leveled (HASL) interfaces.

Scanning Acoustic Microscopy (SAM): Scanning acoustic microscopy images (C-scans) were used to determine the quality of the NanoBonded samples. Acoustic microscopy is an imaging technique that utilizes changes in material properties across interface boundaries to detect the presence of internal flaws and anomalies. Differences in acoustic impedance at a material interface result in a change in the response to an incident ultrasound wave. A portion of the ultrasound wave is reflected back to the transducer from the material boundary, while the remainder propagates through the boundary. The larger the change in impedance, the larger the fraction of the signal reflected. The technique is particularly useful for revealing voids or a lack of wetting in a bond; both appear as regions with white contrast in the acoustic image.

A typical SAM image of the bonded samples is shown in Figure 4. The black areas are either openings or holes in the printed wiring board for power amplifier attach at a subsequent time or are tooling holes. The gray contrast areas in the image indicate good bond quality at the interface between the circuit board and the heatsink.

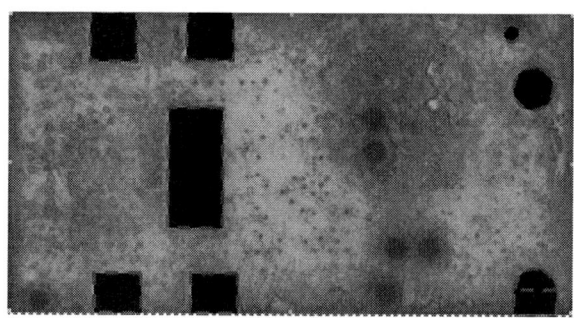

Figure 4 –C-Scan Image of PWB bonded to a Heatsink

The processed assemblies were then subjected to 500 cycles of -40 to +125C temperature cycling, (2 chamber system), following the JESD22-A104C standard. Each cycle consisted of a 15-minute dwell at each extreme temperature, and less than one minute transfer time from one extreme temperature to the other.

The samples were C-scanned after cycling; there was no evidence of an increase in voids or delamination of the bonds. Figure 5 is a cross section of the subject bond after temperature cycling. The stackup from top to bottom is circuit board layer, copper layer, tin plating, solder, Nanofoil, solder, tinned heatsink, copper heatsink.

978-1-4244-4722-0/09 $25.00
© 2009 IMAPS-ITALY

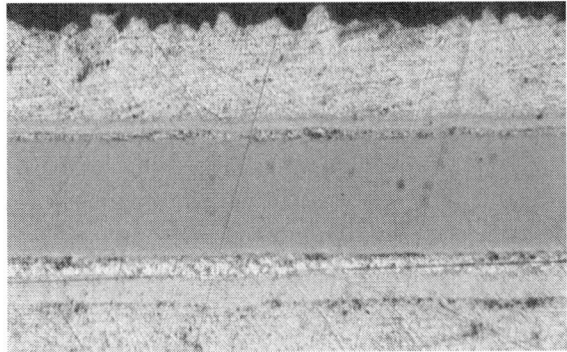

Figure 5 – Cross section of PWB to heatsink attach

Power Amplifier to Heatsink

Power Amplifiers (PA) are being utilized in several applications where the component must be attached to the heatsink through an opening in the circuit board. One of the byproducts of this trend is higher heat dissipation and the consequent need to more effectively cool the PA to ensure long term reliability. Typically, the largest contributor of thermal resistance along the heat conduction path from the die to ambient arises from the thermal interface between the PA and the adjacent heat sink. Minimizing the thermal resistance at this interface is critical for the development higher power amplifier installations. Figure 6 shows 3 power amplifier devices bonded to a copper heatsink and Figure 7 illustrates the C-Scan results obtained. Tin plated foil was used for this interface as the base of the power amplifier device had a thick gold layer. Utilizing tin plated foil enhanced the solder bond and the subsequent temperature cycling tests of 500 cycles from -40 to+125C were easily accomplished.

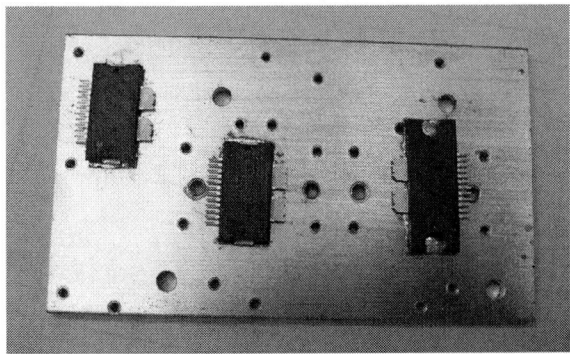

Figure 6 – Power Amplifiers bonded to heatsink

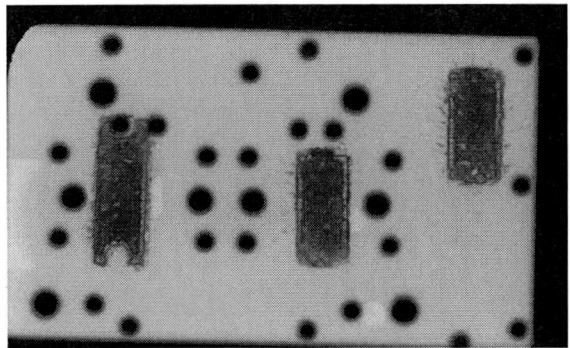

Figure 7 – C-Scan of Power Amplifiers

Figure 8 is a cross section of the bond of the power amplifier and illustrates the stability of the bond structure after completion of 500 temperature cycles from -40 to+125C.

Figure 8 – Cross Section of Power Amp to Heatsink

The stackup in the figure (from top to bottom) is the power amplifier base, the gold finish on the part, solder, NanoFoil, solder, tin plating and copper heatsink.

RF Filter Components

Nanofoil and the Nanobonding process has also been used to successfully bond RF filter components to silver plated chassis.

The bonding configuration used is shown in Figure 9. A piece of tin plated NanoFoil® was placed on the top of silver plated aluminum cavity, followed by silver plated ceramic parts. A bonding pressure (600 psi) was applied on the top of ceramic parts with a load cell, and the NanoFoil was ignited electrically. The foil melted the tin solder layers and joined the components.

978-1-4244-4722-0/09 $25.00
© 2009 IMAPS-ITALY

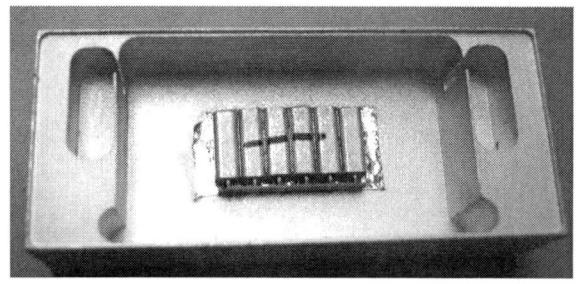

Figure 9 –RF Filter parts in Ag plated chassis

Concentrated PhotoVoltaics

RNT has also successfully bonded concentrated photovoltaic receivers to both aluminum and copper heatsinks. The receivers are Direct Bond Copper (DBC) construction with either an Alumina or BeO core. Solder was screen printed onto the base of the receiver to a thickness of 100u after reflow. Similarly, the heatsink was prepped with 250u of solder. The combination shown in the cross section in Figure 10 illustrates the reliability and stability of the bond after 540 cycles of -40 to +125 temperature cycles.

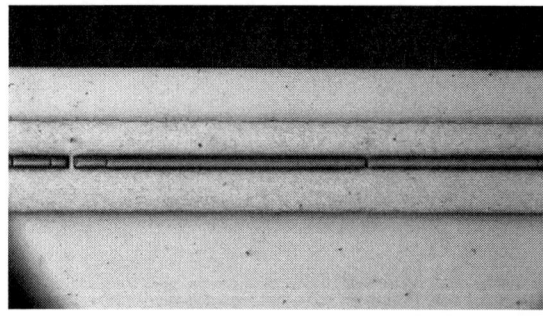

The stackup from top to bottom is alumina, copper with ENIG finish, 100u solder screen printed onto DBC, Nanofoil, 250u solder on heatsink and copper heatsink.

In addition, laser flash measurements were made on pieces subjected to this test sequence. Also, units were populated, for comparison purposes, using a thermal epoxy material for bonding the receiver. After 540 cycles the units NanoBonded showed no change in thermal resistance while those bonded with the epoxy showed a 20% increase.

Reliability and Stability

As previously noted one parameter that is used to determine the stability of a solder bond is the thermal resistance of the completed joint structures. Consistency of this resistance is paramount in determining the long term viability of an assembly from a thermal interface perspective. RNT has performed multiple tests of this parameter to assure stability over 50 lots of material which is illustrated in Figure 10.

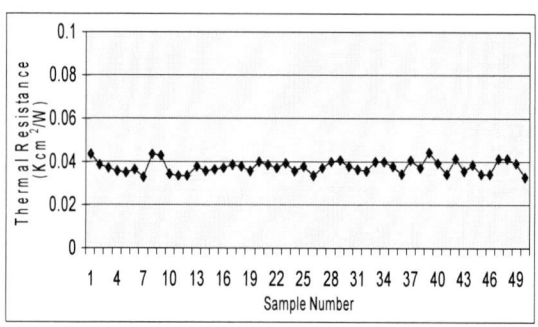

- Number of samples in this study: 50
- Bonding pressure: 50 psi
- Mean Thermal Resistance: 0.04 °C cm2/W (0.006 °C in2/W)
- Standard Deviation: 0.003 °C cm2/W
- Die Size: 17.5x17.5 mm
- R = 0.01°C/W

Conclusions

It has been demonstrated that a viable solder attachment technology where the resultant bond is formed in less than a millisecond is adaptable to numerous applications. It has also been shown that the resultant solder joints possess a high degree of reliability and stability when implemented in these applications. The nature of the bond permits materials that have dissimilar CTE's to be effectively bonded without incurring any stress to the resultant solder joint due to the bond being formed at room temperature.

Bibliography

1. A. Duckham, S. J. Spey, J. Wang, M. E. Reiss, and T. P. Weihs, E. Besnoin and O. M. Knio, "Reactive nanostructured foil used as a heat source for joining titanium," Journal of Applied Physics, 96, 4 (2004).
2. J. Wang, E. Besnoin, O.M. Knio, T.P. Weihs "Investigating the effect of applied pressure on reactive multilayer foil joining," Acta Materialia, 52, 5265–5274 (2004).

978-1-4244-4722-0/09 $25.00
© 2009 IMAPS-ITALY

3. J. Wang, E. Besnoin, A. Duckham, S. J. Spey, and M. E. Reiss, O. M. Knio, M. Powers, M. Whitener and T. P. Weihs, "Room-temperature soldering with nanostructured foils," Applied Physics Letters, 83, 19 (2003).

4. J. Wang, E. Besnoin, A. Duckham, S. J. Spey, M. E. Reiss, O. M. Knio,and T. P. Weihs, "Joining of stainless-steel specimens with nanostructured Al/Ni foils," Journal of Applied. Physics, 95, 248 (2004).

5. J. Subramanian et al, "Direct Die Attach With Indium Using A Room Temperature Soldering Process," Paper presented at the 37th IMAPS International Symposium on Microelectronics, Long Beach, California, November 14-18, 2004.

6. D. Van Heerden et al, "Thermal Behavior of a Soldered Cu-Si Interface," Paper presented at the 20th SEMITHERM Semiconductor Thermal Measurement and Management Symposium, San Jose, California, 11 March, 2004.

7. J. Levin et al, "Room Temperature Lead-Free Soldering of Microelectronic Components Using a Local Heat Source," Paper presented at the 2004 ASM Materials Solutions Conference & Exposition, Columbus, Ohio, October 18-21, 2004.

8. J. Wang, E. Besnoin and O. M. Knio, T. P. Weihs, "Effects of physical properties of components on reactive nanolayer joining," J. Appl. Phys. 97, 114307-1 to -8 (2005)

Characterization of oxidation of electroplated Sn for advanced flip-chip bonding

W. Zhang, and W. Ruythooren

IMEC

Kapeldreef 75, B-3001 Leuven, Belgium

Tel:0032-16-288219, wenqi.zhang@imec.be

Abstract

Sn based Pb-free solder material is often used for flip-chip bonding. However, Sn is prone to be oxidized in ambient due to its low standard Gibbs free energy. In this paper, we investigate the oxidation of electroplated Sn at room temperature and the temperature ramp-up during flip-chip bonding by XPS. It is found that the intial oxide of our electroplated Sn is about 1.43 nm thick, and the oxide thickness increases with time at room temperature. Acid wet clean can reduce the oxide thickness. After cleaning, the oxide thickness is reduced to less than 1.4 nm within 10 minutes, but after that the growth of oxide follows a similar trend as the as-deposited Sn. Moreover, oxide grows fast during the temperature ramp-up. It is found that about 1.0 nm oxide is grown when the temperature reaches 250ºC in less than 30 seconds. Therefore, it is important to find a solution to control the total amount of oxide for fluxless soldering.

Key words: Electroplated Sn, oxidation, oxide removal, flip-chip

Introduction

Tin (Sn) based Pb-free solder material is often used for flip-chip bonding [1]. However, it is prone to be oxidized in ambient due to its low standard Gibbs free energy. Sn has two oxidation states, SnO and SnO_2. The standard Gibbs free energy of SnO and SnO_2 at room temperature is -509 and -509.1 kJ/mol, respectively, indicating that the oxidation of Sn goes spontaneously [2]. In fact the situation is even worse when Sn is bonded to cooper (Cu), which is also known to be easily oxidized and the standard Gibbs free energy of Cu_2O and CuO are -170.6 and -127 kJ/mol, respectively [3].

It is well known that the oxidation of solder material can degrade its wettingability and solderability [4, 5]. However, the growth of Sn oxides is seldom studied. In this paper, we investigate the oxidation of electroplated Sn at room temperature and the temperature ramp-up during flip-chip bonding. We also studied the efficiency of oxide removal by dipping the Sn films into dilute aqueous acids, and the post-growth of Sn oxide is monitered after being re-exposed to air.

Experimental

Eight inch Si(001) wafers were used as base material for all experiments. First a 30 nm-thick Ti and 150 nm-thick Cu were sputtered on SiO_2 as the seed layer. Then, a 3 µm-thick Sn was deposited by electroplating. The growth of Sn oxides was then studied by X-ray Photoelectron Spectroscopy (XPS). High resolution XPS measurements were carried out by an SSX-100 photoelectron spectrometer with a monochromatic Al Kα X-ray source and concentric hemispherical electron energy analyzer. The base pressure in the spectrometer was in the low 10^{-10} Torr range. Details for XPS measurements can be found elsewhere [6].

Dilute HCl and HF were used to clean the oxidized Sn films, which were then sent to XPS measurement to study the cleaning efficiency. The post-growth of Sn oxide after cleaning was also monitored by XPS.

978-1-4244-4722-0/09 $25.00
© 2009 IMAPS-ITALY

In order to investigate the growth of Sn oxides during the temperature ramp-up, a piece of 8x8 mm² Si with above mentioned Sn film is bonded to another piece of Si covered with Ta. This simulated the real bonding process and prevented extra oxygen from diffusing to Sn surface. The flip-chip bonding process was carried out at a FC-6 bonder in ambient. The bonding temperatures were 130 and 250°C, and ramping speed was 10°C/sec. The bonding pressure was 1 MPa. Once the temperature reached the tagert value, the bonding was terminated immediately. The Si wafers were put in water and blow dried with N_2 to cool quickly. Since there was no reaction between Sn and Ta, the two chips were easily separated. Then the chip with Sn film was sent to the XPS measurement.

Results

Fig. 1 shows the XPS spectra obtained from Sn films that are exposed to air after plating. The Sn3d$_{5/2}$ peak is fitted with 3 sub-peaks at 485, 486.4, and 487.3 eV, which is attributed to Sn, SnO, and SnO$_2$, respectively. The difference in binding energy of SnO and SnO2 is only 0.9 eV, hence their peaks overlap with eachother. Therefore, unless otherwise indicated, the Sn peak is simply distinguished from the oxide peak, and hence the thickness of oxide is the sum of SnO and SnO$_2$. In general, the intensity of the Sn oxide peak increases with the exposed time in air (15 mins to 2 days), while the intensity of Sn is normalized to equality. It is different for the Sn film which has been stored for 2 days, where its oxide peak is the highest but the Sn peak is the lowest. Consequently, the thickness of Sn oxide increases with time, which is shown in Fig. 2.

oxided immediately. This is consistent with the observation from Peneva, et.al, where they found a very large initial rate of oxidation and the oxidation rate slows down [7].

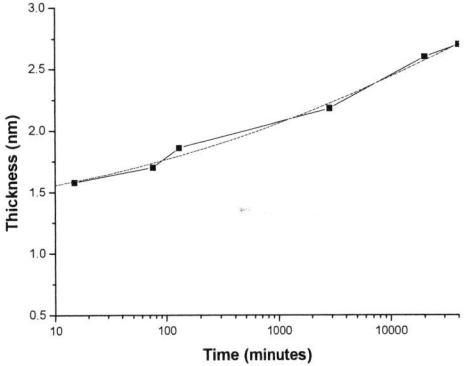

Figure 2. Oxide thickness of Sn vs time after plating. Sn is exposed to air at room temperature. The intial oxide thickness, 1.43 nm, is extroplated from this curve.

It has been reported that the oxide thickness of SnPb is about 3.5 nm regardless of the storge time in ambient within three monthes [5], which is different from observation on the electroplated Sn films. In fact, the absolute thickness of Sn oxide depends on the characterization technique and sample preparation. Therefore, it is quite normal that the Sn oxide thickness varies from each other in literature.

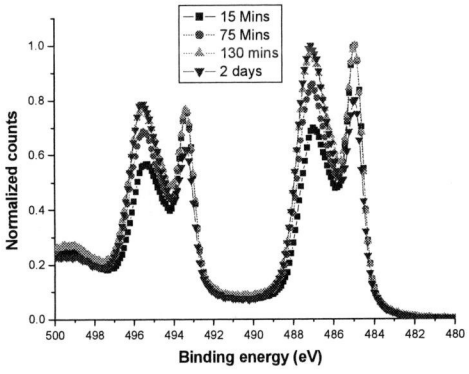

Figure 1. XPS spectra of Sn films after plating. The Sn3d $_{5/2}$ peak at 485, 486.4, and 487.3 eV represents Sn, SnO, and SnO$_2$, respectively. The difference in binding energy of SnO and SnO$_2$ is only 0.9 eV, hence their peaks overlap with eachother.

The intial oxide thickness (time zero) is about 1.43 nm by fitting Fig. 2. This indicates that once the Sn wafer is taken out from the plating bath, Sn is

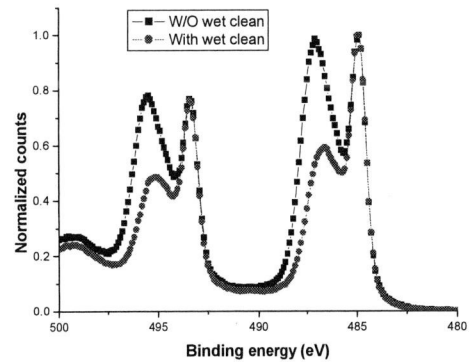

Figure 3. XPS spectra of Sn films before and after a wet cleaning. The intensity of Sn oxide is largely reduced, indicating less oxide after cleaning.

In order to remove Sn oxide, an acid wet cleaning step is introduced. Indeed, this cleaning can effectively reduce the thickness of Sn oxide. As shown in Fig. 3, the Sn intensity of the one with a wet clean is almost the same as the one without a wet clean, but the oxide intensity is much lower after

978-1-4244-4722-0/09 $25.00
© 2009 IMAPS-ITALY

cleaning. This suggests that the oxide layer becomes thinner after cleaning.

It is worth mentiong that both dilute HCl and HF show similar cleaning effect. In addition, the remaining oxide thickness is the same for Sn films cleaned for 20 secs and 70 secs. This may be due to the fact that the minnium time interval between cleaning and the XPS measurement is ~ 5 mins, during which the cleaned Sn is slightly re-oxidized. The post-growth of Sn oxide after cleaning is shown in Fig. 4. As expected, the oxidized Sn moves back to its initial stage after cleaning, and then the oxide thickness increases with time after being re-exposed to air. The oxide post-growth curve is quite similar with the one after plating.

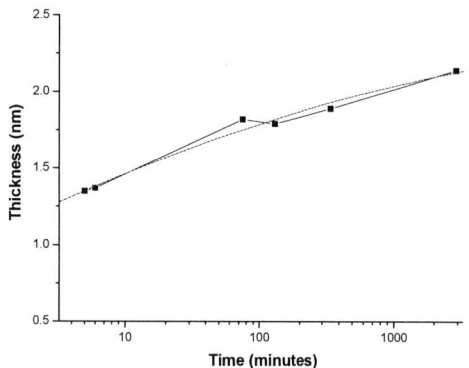

Figure 4. Oxide thickness of Sn vs time after a wet cleaning. The oxidized Sn is cleaned and re-exposed to air at room temperature.

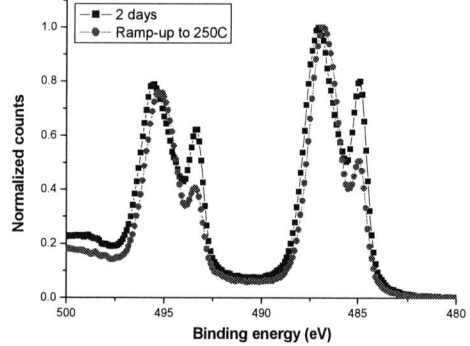

Figure 5. XPS spectra of Sn films. One of the films is exposed to air for two days after plating, while the other is only exposed to air for less than 3 hours after plating and then heated up to 250°C in 22 secs in a flip-chip bonder.

Oxide grows quickly during the temperature ramp-up in flip-chip bonding, as shown in Fig. 5. The sample used for the temperature ramp-up is only exposed to air for less than three hours after plating. After ramping to 250°C in 22 secs, its oxide intensity is almost the same as the one exposed to air

for 2 days, but the Sn intensity is much less. This indicates that the growth of oxide during the temperature ramp-up is very fast. Nevertheless, the difference is getting smaller when it is only ramped to a lower temperature for example 130°C with the same ramping speed.

The oxide growth rate during the temperature ramp-up is estimated to be 0.05 nm/sec, which results in about 1.0 nm-thick Sn oxide after ramping to 250°C. It is worth menting that the oxide growth at elevated temperatures is normally studied by heating up a sample in ambient for a certain time and then measured the oxide thickness. This in fact is different from the flip-chip bonding process, where two chips are placed together under a pressure and much less oxygn diffuses to the solder material surface. Especial, when the bonding temperature is higher than the melting point of solder materials such as Sn, 232°C, liquid Sn enhances intimate contact between the two chips and hence further prevents the diffusion of oxygen. Therefore, the total amout of Sn oxide growed during flip-chip bonding will be less than that growed during the normal heating up in ambient. However, the Sn oxide growth during flip-chip bonding is still significant.

An acid wet clean still helps to reduce the thickness of oxide after the temperature ramp-up. Here we compare two cases: one is directly heated up to 250°C, while the other is subjected to a wet clean step prior to the temperature ramp-up. The oxide intensity is almost the same for the two samples, but the Sn intensity is higher for the one with a wet clean. However, the difference is not much since the majority of oxide grows during the tempearture ramp-up.

Discussions

The time interval between bump formation for example by plating and flip-chip bonding could be very long, which depends on wafer dicing and die sorting, etc. This means that Sn can be oxidized heavily. Flux is often used to remove oxide, but for some applications such as MEMS sealing, fluxless soldering is required. An alternative approach is an acid wet clean prior to bonding, but it would be a time-critical process since Sn can be oxidized quickly again. In addition, immersion bonding, during which a dilute acqueous acid whose decomposition temperature is much lower than the peak temperature is applied as a flux, would be another option [8].

Conclusions

In this paper, we investigate the oxidation of the electroplated Sn at room temperature and the temperature ramp-up during flip-chip bonding by XPS. It is found that Sn can be easily oxidized in ambient at room temperature and the oxide thickness increases with time. Acid wet clean can reduce the oxide thickness. The oxidized Sn moves back to its

978-1-4244-4722-0/09 $25.00
© 2009 IMAPS-ITALY

initial stage after cleaning. However, once it is re-exposed to air, the growth of oxide follows a similar trend as the as-deposited Sn. Moreover, oxide grows fast during the temperature ramp-up of flip-chip bonding. Therefore, it is important to find a solution to control the total amount of oxide for fluxless soldering.

Acknoledgements

The authors would like to acknowledge the contribution from I. Hoflijk and T. Conard for their XPS measurements.

References

[1] R. Labie, T. Webers, B. Swinnen, and E. Beyne, "Electromigration behavior of Pb-free flip chip bumps", J. Microelectronics and Electronic Packaging, Vol. 3, pp. 32-36, 2006.

[2] Holmes RD, Kersting AB, and Arculus AJ, Standard molar Gibbs energy of formation for Cu_2O: High resolution electrochemical measurements from 900 to 1300 K, J. Chemical. Thermodynamics, Vol.21, pp. 351-361, 1989.

[3] C. Mallika, A. M. Edwin Suresh Raj, K. S. Nagaraja, and O. M. Screedharan, Use of SnO for the determination of standard Gibbs energy of formation of SnO_2 by oxide electrolyte e.m.f. measurements", Thermochimica Acta, Vol. 371, pp. 95-101, 2001.

[4] J. Kim, H. Schoeller, J. Cho, and S. Park, "Effect of oxidation on indium solderability", J. Electron. Mater., Vol. 37, pp. 483-489, 2008.

[5] J.F. Kuhman, A. Preuss, B. Adolphi, K. Maly, T. Wirth, W. Oesterle, W. Pittroff, G. Weyer, and M. Fanciulli, "Oxidation and reduction kinetics of eutectic SnPb, InSn, and AuSn: A knowledge base for fluxless solder bonding applications", IEEE Trans. Comp., Hybrids Manufact. Technol., Vol. 21, pp. 134-141, 1998.

[6] C.M. Whelan, M. Kinsella, H.M. Ho, and K. Maex, "Aorrosion inhibition by thiol-derived SAMs for enhanced wire bonding on Cu surface", J. Electro-chem. Soc., Vol. 151, pp. B33-B38, 2004.

[7] SK. Peneva, NS. Neykov, V. Rusanov, and DD. Chakarov, "Mossbauer spectroscopy investigations of the oxidation of α–Sn and β–Sn types of structure", J. Phys.:Condens. Mater., Vol. 6, pp. 2083-2094, 1994.

[8] R. Agarwal, W. Ruythooren, "Low Temperature Direct Cu-Cu Immersion Bonding for 3D Integration", to be published in MRS spring meeting, April 13-17, 2009.

978-1-4244-4722-0/09 $25.00
© 2009 IMAPS-ITALY

Miniaturisation of a LTCC High-Frequency Rat-Race-Ring by Using 3-Dimensional Integrated Passives and Embedded High-K Capacitors

Rubén Perrone[1], Polina Kapitanova[2], Dmitry Kholodnyak[2], Irina Vendik[2],
Stefan Humbla[1], Matthias Hein[1], Jens Müller[1]

[1] Institute for Micro- and Nanotechnologies, Ilmenau University of Technology, P O. Box 100565, 98684 Ilmenau, Germany, Tel.: +49 (0) 3677 691360, Fax: +49 (0) 3677 693379, e-mail: ruben.perrone@tu-ilmenau.de

[2] Department of Microelectronics & Radio Engineering, Saint-Petersburg Electrotechnical University, 5, Prof. Popov str., 197376, Saint-Petersburg, Russia, Tel.: +7 (812) 346 0867, e-mail: IBVendik@eltech.ru

Abstract

Some of the trends in the communication industry consist in increasing the variety and/or number of integrated functions in portable devices as well as reducing their size. Because of these trends, forthcoming portable wireless devices have to comply with more demanding hardware requirements, especially regarding higher wiring densities and miniaturisation level.

A suitable combination of system architecture, substrate technology, integrated materials with improved functionality and substrate design is crucial for enhancing the level of miniaturisation in RF-electronic circuits.

In this paper the miniaturisation of a rat-race-ring for 2.45 GHz (free-space wavelength of about 122 mm, Bluetooth and WLAN applications) with embedded lumped-elements in low-temperature co-fired ceramic technology is discussed and the results are presented.

The rat-race-ring was realised in three different topologies. The first version was realised using a quasi-planar architecture. Using this architecture a device size of 7.5 x 7.5 mm^2 was achieved. The second version of the rat-race-ring was designed as a module with three-dimensional fully integrated passives and land-grid-array (LGA) solder pads. Using this topology the size of the device was reduced to 3.8 x 3.8 mm^2. In the third design the capacitors were implemented using locally integrated thin high-k dielectric patches. The whole size of this LGA-module (third device) amounted to 2.4 x 2.4 mm^2, which implies an area miniaturisation to ten percent of the size of the first design.

In this paper technological and design issues as well as numerical simulations and measurements of the microwave scattering parameters are considered and discussed.

Key words: LTCC, passive integration, high-K capacitors, rat-race-ring, RF- and microwave applications

Introduction

Forthcoming portable wireless devices have to comply with more demanding hardware requirements, especially regarding higher wiring densities and miniaturisation level.

The level of miniaturisation of electronic RF-circuits can be substantially improved by integrating/embedding the passive components, such as resistors and decoupling capacitors, and passive structures needed to support the active components into the carrier substrate. Examples of passive RF-structures are filters, power dividers, matching networks, etc.. These passive structures can be realised whether using transmission line elements or lumped elements. For Bluetooth and WLAN applications around 2.5 GHz the size of the transmission line elements may become significant as it is related to the wave length (free-space wavelength of about 122 mm at 2.5 GHz). Therefore, a combination of a three-dimensional

(3D) passive integration with the use of a lumped element concept offer in many cases much more miniaturisation potential than the transmission line approach, even at frequencies above 10 GHz [1]. However, through the 3D-integration of lumped RF-elements their parasitic effects and their mutual interaction become stronger [2], [3] and should be carefully taken into account in the design.

By using an appropriate layout and fine line patterning technologies to realise passive elements their parasitic effects can be reduced or optimised to meet particular requirements regarding their RF-properties such as a high self resonant frequency or, in the case of inductors, a constant inductivity value over the frequency for example [1], [4], [5].

Applying the already mentioned technological concepts and a suitable design a low-pass filter with lumped-elements for 10 GHz with a stop-band of 40 GHz was previously realised as a multilayer module in low-temperature co-fired ceramic technology (LTCC) [6]. Furthermore, through the integration of materials with improved properties into the carrier substrate, such as high-quality dielectric materials with high dielectric constant, embedded capacitors with a high capacitance density and therefore small dimensions in LTCC substrates can be achieved [7], [8].

In this paper the miniaturisation of a 4 port rat-race-ring (RRR) for 2.45 GHz with embedded lumped-elements in LTCC technology is discussed. The RRR was realised in three different topologies. The first version was realised using a quasi-planar architecture. The second version of the RRR was designed as a module with three-dimensional fully integrated passives and land-grid-array (LGA) solder pads. In the last design the capacitors were implemented using locally integrated thin high-k dielectric patches. From the first to the last design the device area was miniaturised to ten percent of that of the first design.

In this paper technological and design issues as well as numerical simulations and measurements of the microwave scattering parameters are considered and discussed.

Device description

Rat-race-rings are four-port directional couplers, which are widely used as power dividers/combiners [9]. A conventional RRR consists of three $\lambda/4$ and one $3\lambda/4$ transmission line (TL) sections (see Fig.1.a). The power applied to port 1 is divided between port 2 and port 4. The power division ratio is determined by the wave impedance of the transmission lines. Port 3 is isolated from port 1 and usually terminated with a 50 Ohm load. For applications at frequencies up to 10 GHz and depending on the substrate material the size of these elements becomes usually rather large. Therefore, in order to be able to use RRRs in modules, where a high level of miniaturisation is required, alternative ways to reduce the size of these devices have to be explored.

Using special techniques the RRR based on transmission lines from Fig.1.a) was transformed to the equivalent lumped-element circuit shown in Fig.1.b) [10]. The device was designed to operate at a central frequency of 2.45 GHz.

Quasi-planar RRR-Design

The circuit of Fig.1.b) was implemented in 8 layers of LTCC material DP951PT (fired thickness of 95µm) using standard LTCC technology and therefore conservative design rules regarding line width and via diameter (line space and width of 200 µm, via diameter of 170 µm). For conductor-printing and via-filling silver ink was used.

In Fig.2.a) a 3D-view of the module design is shown. The capacitances C_1 and C_2 were realised

a)

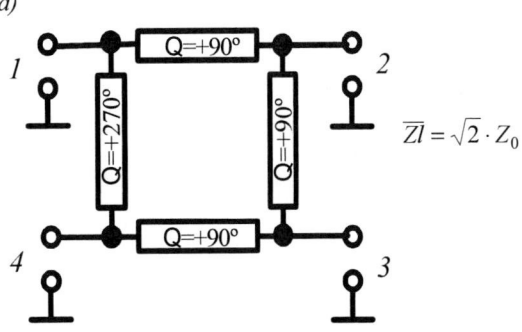

b)

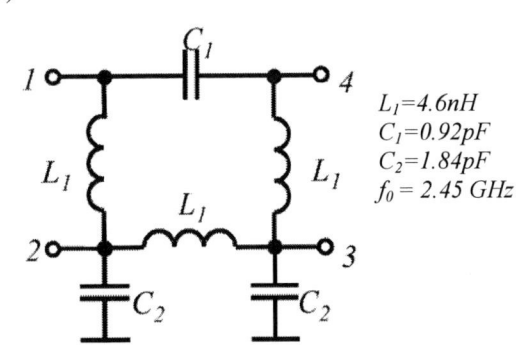

$L_1=4.6nH$
$C_1=0.92pF$
$C_2=1.84pF$
$f_0 = 2.45\ GHz$

Fig.1: RRR; a) line transmission model; b) equivalent circuit with lumped elements

as parallel plate capacitors with one dielectric layer in between the electrodes. The bottom electrode of both capacitors C_2 were connected to the general ground planes using vias. The inductors were implemented as turned stacked inductors with the individual turns situated on the top and first inner layer. These details can be clearly seen in Fig.2.a) and Fig.2.b). Due to the use of the 8-layer stack, the distance to ground of the inductors and capacitors amounted 570 µm, which caused the parasitic capacitance to ground of the inductors and capacitor C_1 to be negligible small.

The size of the structure containing the passive arrangement is 7.5x7.5 mm^2, thus about $0.17\lambda_g$ x $0.17\lambda_g$ at 2.45 GHz for the material LTCC DP951 (relative permittivity of 7.8). The size of the whole device is in this case slightly larger, as the ports for the RF-characterisation had to be arranged according to our measurement equipment.

This structure was optimised using the commercial 2.5D field simulator Sonnet EM$^®$.

The microwave scattering parameters of the four-port device were investigated using an Agilent E8631A vector network analyzer and microwave GSG coplanar probes from SÜSS with a 650 µm pitch. For the measurements a four-port SOLT (short, open, load, thru) calibration was used. For that purpose a cal substrate containing the

978-1-4244-4722-0/09 $25.00
© 2009 IMAPS-ITALY

a)

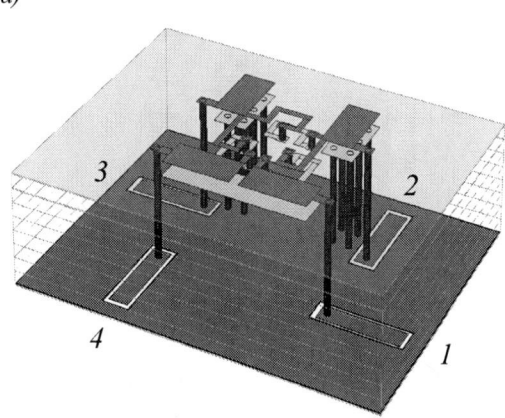

b)

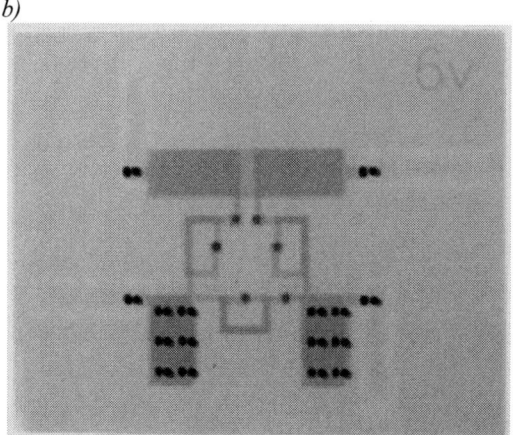

Fig.2: a) 3D view of the RRR module with quasi-planar design; b) X-ray image of the module from the top side

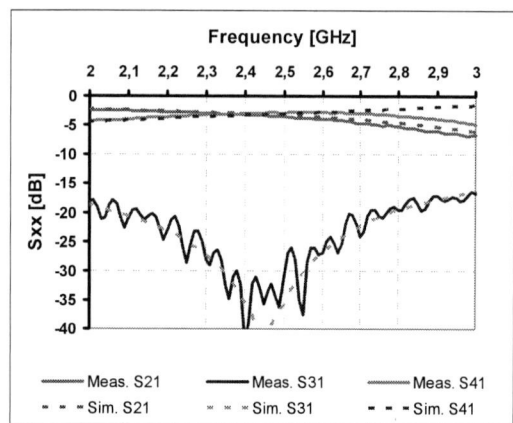

Fig.3: Results of the EM-simulation (dashed lines) and experimental investigation (solid lines) of the RRR with quasi-planar design

calibration elements was used. To calibrate adjacent probes special through calibration standards with 90° rotated ports were manufactured.

The measured and simulated scattering parameters of the device are compared in Fig.3. The measured data reproduce the simulated results well.

At the central frequency 2.45 GHz the measured insertion loss was less than 0.25 dB. In the frequency range from 2.2 to 2.6 GHz the amplitude unbalance was below than ±1 dB. The return loss in port 1 was better than 20 dB and in port 2 and port 4 were better than 14 dB. The isolation between port 1 and port 3 was better than 20 dB.

RRR-Module with 3D fully integrated passives

The boundaries of this design regarding the material for the dielectric layers and their disposition were defined by the technology used for a multi-project substrate. The module consists of a 7-layer stack of LTCC material DP951.

To further miniaturise the previous structure different modifications were introduced in the design. The first modification was applied to the coil geometry. These elements were designed with a round helical geometry. They consist of 3.5 turns distributed in 4 different layers and are fully embedded in the 7-layer LTCC stack. The coils were also placed between two shielding ground planes. To minimise their parasitic capacitance to ground, the coil turns were designed to have a width of 70 μm. Nevertheless, their parasitic capacitance could not be neglected and was taken into account in the design of the capacitor C_2. Using this arrangement for the coils their area was reduced to about 29 % of that of the coils of the first design.

In order to achieve fine-line inductors with a high quality factor, Silver ink was used for printing these elements as well for via-filling. Fine-line printing screens were also needed to obtain an accurate edge quality.

Regarding the plate capacitors, they were realised using one layer of LTCC DP951C2 with a fired thickness of 43 μm as dielectric layer. The remaining 6 LTCC layers have a sintered thickness of 210 μm. In this way the area of the capacitor could reduced to about 28 % compared to the capacitors used in the previous design. Moreover, the shape of the capacitors C_2 (see Fig.1.b)) was adapted to the design to allow a compact component arrangement. These details can be recognised in the design view shown in Fig.4. Both ground planes were electrically connected by means of a via arrangement, which also serves as an electromagnetic shielding. The ground via arrangement can be seen in the X-ray image of the module shown in Fig.5. The module also includes 4 LGA ports. The size of this structure is 3.8 x 3.8 mm². Thus, using fully embedded passives arranged in the LTCC module the RRR area was reduced by about 75 % compared to the previous design.

This structure was simulated and optimised using the full-wave field simulation programm "CST Microwave Studio®".

978-1-4244-4722-0/09 $25.00
© 2009 IMAPS-ITALY

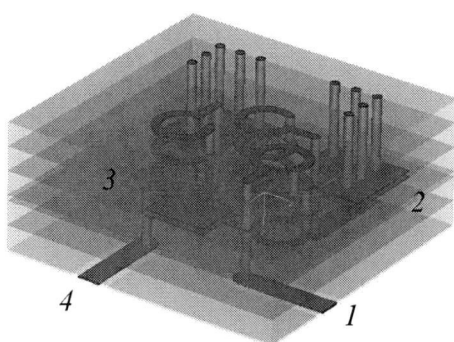

Fig.4: Spatial view of the RRR module design with fully integrated passives

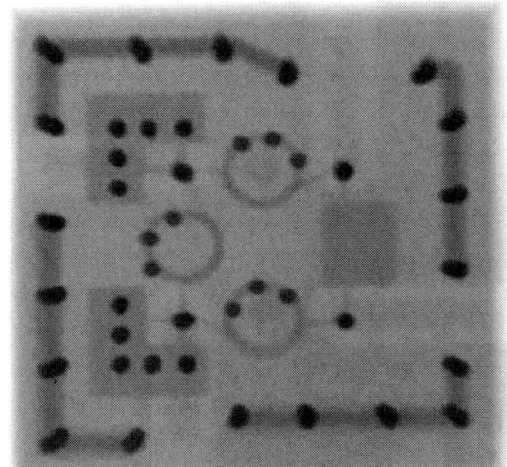

Fig.5: X-Ray image of the module with fully integrated passives

Fig.6: Measurement set-up with four coplanar wave probes and a two-port network analyzer

The RF-characterisation of the manufactured modules was carried out using a two-port network analyzer HP8753C (300 kHz – 6 GHz) and 650 µm-pitch Suess coplanar test probes. The network analyzer was calibrated using SOLT standards. Three measurements steps were required

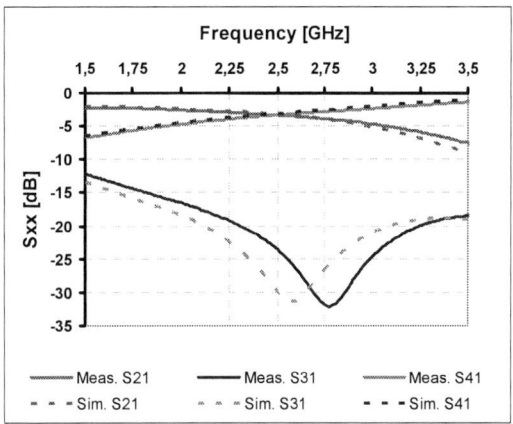

Fig.7: Results of the EM-simulation (dashed lines) and experimental investigation (solid lines) of the RRR with 3D fully integrated passives

to obtain all RRR parameters with the two-port analyzer. The two remaining ports of the device under test were terminated with precision 50 Ohm load standards through the coplanar probes. The whole measurement arrangement is shown in Fig.6.

The measured and simulated data are compared in Fig.7. This device shows an equal power division at 2.46 GHz with an insertion loss of 0.28 dB. In the frequency range from 2.2 to 2.6 GHz the amplitude unbalance was less than ±1.1dB. The return loss in port 1 was better than 17dB and in port 2 and port 4 better than 15 dB. The isolation from port 1 to port 3 was better than 18 dB. The isolation and return losses of the corresponding ports show a slightly frequency shift compared to the simulation results. Nevertheless, the simulated and measured results show a very good agreement.

RRR-Module with embedded high-k capacitors

Paying attention to the module design of Fig.5 it can be seen that the size of the capacitors constitutes the limiting factor for the further miniaturisation of this circuit. Therefore, for the next design all capacitors were designed and manufactured using locally inserted patches of high-k tape as dielectric layer between the capacitor plates.

The high-k patches are pre-structured by laser on the carrier tape and transferred to the LTCC tape by lamination. A special design of the electrodes minimises the interface area between the foreign high-k-tape and the base LTCC to prevent excessive material interactions [8]. The recently developed local tape insertion method makes it possible to achieve capacitors with low tolerance (< 5 %) and high capacitance density [7].

NAMICs K30-tape was used as the high-k material for capacitor miniaturization. The capacitor plate dimensions were determined based on the parallel plate capacitor equation and the dielectric

978-1-4244-4722-0/09 $25.00
© 2009 IMAPS-ITALY

material parameters measured earlier [8]. Using this technology the capacitor C_2 was reduced to a size of 0.322 mm² (see Fig.1.b)). This represents an area reduction of approximately 85 % and 45 % compared to the same capacitor of the first and second design respectively. For the capacitor C_1 similar size reduction values were achieved.

The design boundaries of this design were defined by the technology used for a multi-project substrate as well. This substrate is based on 6 layers of Du Pont LTCC tape DP951 with 210 µm single layer thickness. Silver paste is used for conductor printing and via fill. The low sheet resistance is necessary for the implementation of high quality inductors.

The full structure of the rat-race hybrid is depicted in Fig.8. Two ground planes were placed on the top and bottom layers of the LTCC module for shielding purposes and to provide the passives with defined electrical conditions. Both ground planes were electrically connected by means of four via stacks which were added in the corners of the design (see Fig.9). The lower electrodes of all three capacitors are located on top of tape 1 (counted from bottom). The upper electrode was printed on the back side of tape 2 (see Fig.8). The inductors used in this design have similar geometrical properties as the ones used in the previous RRR-module.

The x-y dimensions of this device are 2.4 x 2.4 mm², which represents a miniaturization of about 90% and 60% related to the first and second design respectively. In terms of wave length related to the dielectric material DP951 and a frequency of 2.45 GHz the size of this device represents $0.055\lambda_g \times 0.055\lambda_g$. The extremely compact component arrangement of this module can be appreciated in the X-ray image shown in Fig.9.

All individual components as well as the whole module were designed, simulated and optimised using the 2.5D simulator Sonnet EM®. The design methodology is thoroughly described in [11].

The S-parameter characterisation of this device was carried out in the same was as for the previous design. The simulation and measurement results are compared in Fig.10. The highly miniaturised device shows an equal power division at 2.42 GHz with an insertion loss of 0.24 dB. In the frequency range from 2.2 to 2.6 GHz the device provided an amplitude unbalance of less than ±1.3 dB. The return loss in port 1 was better than 15 dB and in port 2 and port 4 better than 12 dB. The isolation from port 1 to port 3 was better than 19 dB. The isolation curve (S31) show a frequency shift of about 300 MHz compared to the simulation results. Nevertheless, a good agreement between simulation and measurement results was achieved.

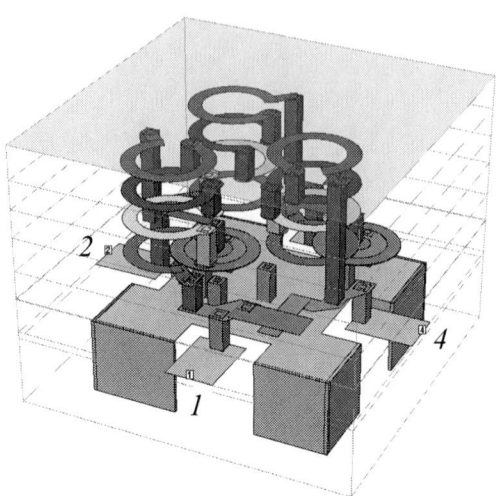

Fig.8: Spatial view of the RRR module with embedded high-k capacitors

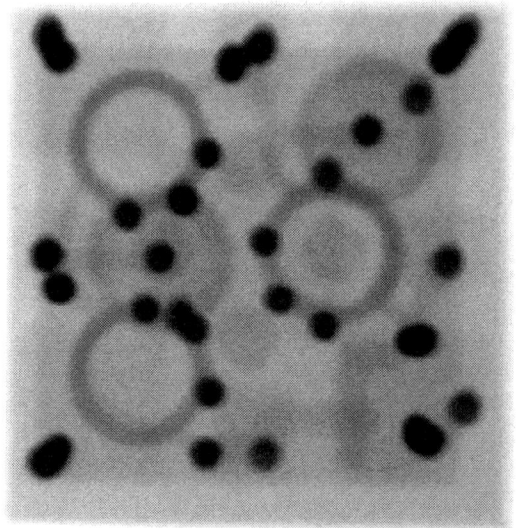

Fig.9: X-ray image of the module with high-k capacitors

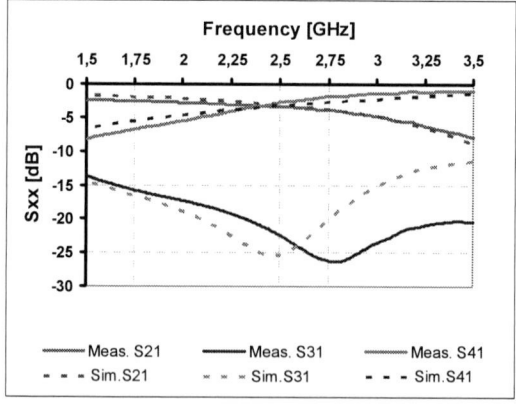

Fig.10: Simulated (dashed lines) and measured (solid lines) coupling and isolation behaviour of the RRR module with high-k capacitors

978-1-4244-4722-0/09 $25.00
© 2009 IMAPS-ITALY

In Fig.11 a photograph of all devices is shown. The significant miniaturisation improvement achieved for the RRR across the different design stages can be clearly recognised.

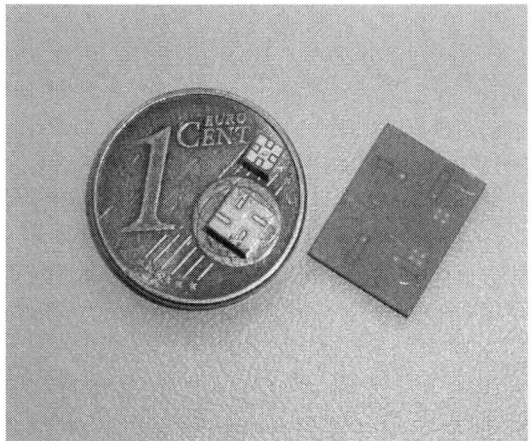

Fig.11: Photograph of the manufactured RRR-modules

Summary and Outlook

The examples presented in this paper demonstrate the flexibility and the miniaturisation potential of the multilayer LTCC technology used in combination with design approaches based on lumped-components and their three-dimensional integration.

Furthermore, the benefits yielded by the integration of functional materials into multilayer LTCC structures as well as by the development of innovative technological processes to allow that integration were clearly shown.

Finally, the measured RF-performances of the manufactured devices indicate that it is not necessary to resign an excellent performance in highly miniaturised devices.

Acknowledgements

The work carried out in the project *MultiSysTeM* is funded by the BMBF under the initiative "Centres for Innovation Competence". The work was also supported by the Network of Excellence "Metamorphose" of the 6-th Framework Program of the European Commission (Project No. 500252). P. K. appreciates very much a scholarship received from the President of the Russian Federation.

The authors would like to thank Namics Corp. for providing the tape samples.

References

[1] Perrone, R.: "Erweiterung des Frequenzbereichs und der Integrationsdichte von LTCC-Modulen mittels Photostrukturierung und Designoptimierung", ISLE, 2007, ISBN 978-3-938843-30-7.

[2] Müller J., Perrone R.: "Physical Model Based Design Optimization for LTCC RF Inductors", IMAPS/ACerS International Conference on Ceramic Interconnect and Ceramic Microsystems Technologies, Denver, Colorado, April 24-27, 2006

[3] Hartung J.: "Entwurf, Modellierung und Optimierung monolithisch integrierter Spulen und Koppler", Shaker Verlag, 2000.

[4] Perrone, R., Thust, H., Drüe, K.-H.: "Progress in the Integration of planar and 3D Coils on LTCC by using Photoimageable Inks", Journal of Microelectronics and Electronic Packaging, Volume 2, No.2, Second Quarter, pp. 155-161, 2005.

[5] Müller J., Perrone R., Rentsch, S., Hintz, M., Stephan, R.: "Improved RF Performance for Embedded Passives in LTCC by Fine Line Structuring Methods", XXX International Conference of IMAPS Poland Chapter Kraków, Karaków, Polen, 24.-27. September, 2006.

[6] Perrone, R., Müller, J.: "Compact RF-Filter-Modules with Lumped Elements in LTCC for Applications up to 10GHz", Proc. 2nd IEEE Electronic System Integration Technology Conference (ESTC), Greenwich/UK, Sept. 1-4, 2008.

[7] Müller, J., Mach, M., Perrone, R., Novel Method for Embedding High Quality LTCC RF-Capacitors Using mid-k Tapes, Proc. 4th Ceramic Interconnect Ceramic Microsystems Technology Conference (CICMT), Munich/Germany, April 21-24, 2008.

[8] Müller J., Perrone R.: "Technology and Design of Precise Embedded Capacitors in LTCC", Proceedings of IMAPS Nordic Conference 2008, Hotel Marienlyst, Helsingør, September 14 – 16, 2008.

[9] Pozar, David M.: "Microwave Engineering", 2nd ed., John Wiley & Sons, Inc., New York, 1998, ISBN 0-471-17096-8.

[10] Vendik I., Kholodnyak D., Kapitanova P., Hein M.A., Humbla S., Perrone R., Mueller J. "Tunable dual-band microwave devices based on a combination of left/right-handed transmission lines", Proceedings of 38th European Microwave Conference, Amsterdam, The Netherlands, October 27-31, 2008, pp. 273–276.

[11] Müller, J., et.al: "Highly integrated Passive LTCC Device with Embedded High-k Capacitors", IMAPS/ACerS 5th International Conference on Ceramic Interconnect and Ceramic Microsystems Technologies (*CICMT 2009*), Denver, Colorado, April 24-27, 2009

DreamPAK – Small Form Factor Package

L. A. Lim / Ramkumar. M / Charles J. Vath, III

ASM Technology Singapore Pte. Ltd.

2, Yishun Avenue 7, Singapore 768924.

Abstract

Recently there have been several developments in the areas of small form factor and high-density lead frame based IC packages. These packages have the benefits of occupying smaller PC board area, lower assembly material consumption due to their smaller volume package construction, and good thermal dissipation properties.

Another area of recent development in packaging integration is system in package SIP. A SIP package can integrate various ICs in a package. For example; CPU, logic, analog, digital, discrete and passive components can be integrated in a single package. SIP provides a high level of integration capability of several functional chips. It provides good miniaturization capability for hand held portable electronic communication devices. High-density capability and flexibility in design using a substrate like chip carrier will be covered. Design flexibility covering several main stream packages equivalents will be presented.

However there can be several challenges faced by the end user in assembly of these similar packages. This paper will cover the benefits of a newly developed package and how the challenges faced in the assembly downstream processes are addressed. Main assembly material consumption reduction includes wire and mold compound. Details of benefits of the relevant assembly processes would be covered including actual comparison.

Introduction

Packaging Engineers are in constant pressure for decades to accommodate the never ending increase of IC package I/O count and at the same time to cut the package cost. Right from the DIP package era, the industry had passed through various ways of accommodating the higher I/O count by innovating peripheral leaded packages like SOIC, QFP, TQFP etc. Later, the industry believed the new invention, namely, BGA will be the future. Several versions of BGAs like, FBGA, CSP came up and stayed in the semiconductor field. However the higher cost and reliability short comings put some break on the usage of laminate based packages.

In the recent decade, just package cost alone made the packaging engineers to take a relook at the leadframe based packaging technology. New etch techniques were developed and the QFN was successfully born. We think QFN is one successful technology that has driven the market as good as or better than the QFP or the early PDIP technologies.

For these many years, QFN has been a major industry driver due to the smaller size, lower cost, and ease of high volume manufacturing using higher unit density matrix array designed, etched lead frames. However since QFN is also a peripheral leaded package which is small in size, the I/O feasibility was limited to a little over 100. Recent developments in chemical etch technology, could stretch the QFN to have dual row of leads, that raised the I/O counts to >150. However this technique always had difficulty in sustaining the etching process capability and became a challenge to packaging processes like wire bond due to its fragile leads and inconsistency in etch process. Also QFN dual row technique failed to gain PBGA level I/O counts, though it can cross over 150 leads.

Furthermore, attempts to improve the etch technology were done and few methods of reaching the multiple row leads emerged. Using partial etch techniques, most of these methods required wet end processes, like a chemical etch line at the assembly house, in order to separate the leads. Wet end processes are additional investments and many companies are not willing to go in that direction.

The new DreamPAK technology is also similar to this half etch technique but unique on its own. This new technology is capable of achieving multiple rows of leads with small form factor but does not need any wet etch processes during the assembly. Thus DreamPAK is expected to provide the benefit of QFN like small form factor, but BGA type higher I/O at lower cost. This paper describes about this new packaging technique and explores its above said benefits as well as other unique new advantages.

DreamPAK vs. QFN Leadframe Design

QFN leadframe, is an all joined, partially etched leadframe, where the adjacent leads are connected typically by a half etched connecting bar. The die attach pad is also connected to the structure by QFP type tie bars at the corners. The resulting leadframe design, typically, is a matrix array with all leads and units connected.

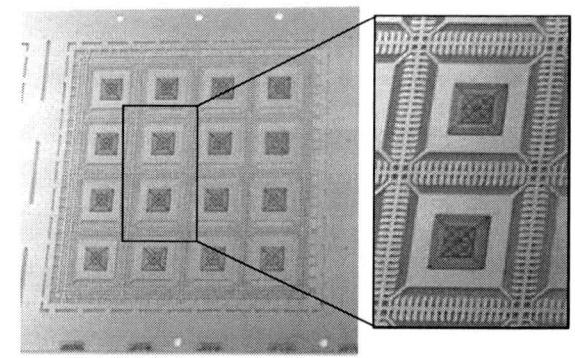

Fig.1 - QFN lead frame with die

Typically, it uses an adhesive tape at the bottom to hold all the leads and die attach pads in the same plane. This tape also serves a flash limiting barrier for the molding process also, so that an exposed pad is achieved after molding. The tape is removed after the molding before saw singulation. The half etched features ensure the good mold adhesion and aid in better reliability performance.

The DreamPAK leadframe has multiple rows of leads and a die attach pad, arranged in matrix array form. However, the

EMPC2009

978-1-4244-4722-0/09 $25.00
© 2009 IMAPS-ITALY

leads and die attach pads are not connected with each other by any metal links, like connecting bar or tie bars as in QFN.

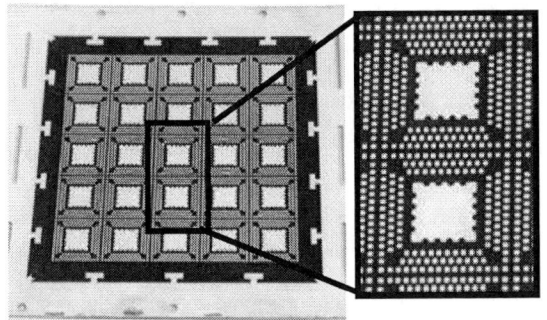

Fig.2 - DreamPAK lead frame

Also the tape we find in the QFN leadframe is also not needed. All the components of the DreamPAK leadframe are embedded in an insulative epoxy matrix and held together with the outer supportive copper frame. Since the leads and pads are just embedded in an epoxy matrix, the design layout of leads and pads, or the number of I/O are fully flexible and only limited to the PCB routing design rules. Thus DreamPAK, uniquely, is able to meet the packaging demand of multiple row fully flexible layout design feasibility and at the same time (since all leads are separated already) no special wet processes required when compared to some similar techniques.

DreamPAK and Assembly Process Flow

The process flow of DreamPAK assembly is very simple and uses existing machines of a typical assembly house. As indicated in the Fig.3, DreamPAK can be used as we handle the normal QFN lead frames. Starting from die attach, the strips can move to wire bond followed by typical QFN molding process. Any requirements for plasma cleaning are also possible as per the user needs.

DreamPAK	"F"	"E"	"T"	"C"
Leadframe	Leadframe	Leadframe	Leadframe	Leadframe
Die Attach	Die Attach	Die Attach	Die Attach	Die Attach
Wire Bond	Wire Bond	Wire Bond	Wire Bond	Wire Bond
Mold	Mold	Mold	Mold	Mold
Testing	Etching	Etching	Etching	Etching
Saw Singulation	Plating	Ag Strip	Testing	Testing
	Testing	Plating	Saw Singulation	Saw Singulation
Main stream process flow	Saw Singulation	Testing		
		Saw Singulation		
Lower assembly cost	Requirement of Etching / and Plating Line			
	Requirements of water treatment facilities for etchant / and plating discharge			

Fig.3 - Process flow comparison

After the PMC process, the strips can be sent to electrical testing in "strip form", since the leads are already electrically isolated. Later, the unit singulation can be performed by a standard dicing singulation process. As it is illustrated in Fig.3, the DreamPAK process uses the mainstream processes

only and does not require any etching/plating stage thus cutting down investment cost.

DreamPAK and Form Factor

DreamPAK has the freedom of flexible footprint layout. Though it can cater for high pin count packaging requirement, it can also successfully used for very low I/O (like 2L, 3L etc) packages like transistors and diodes. Higher pin count range can be as high as 300 leads as we see in the BGA applications. The biggest benefit of DreamPAK comes from the form factor it provides for the high pin count packages. A typical 0.5 mm lead pitch for a 12x12 mm package with 4 rows of leads can yield over 300L. In PBGA package, for this much pin counts the package size could be easily 27x27mm. In QFP package, for similar I/O it shall be 40 x 40mm.

If calculated, 12x12 mm DreamPAK will occupy only 9% of area, in comparison to QFP (40x40 mm) of similar I/O and when compared to a 27x27 mm PBGA, DreamPAK will yield 80% board space reduction. Please see illustration Fig 4.

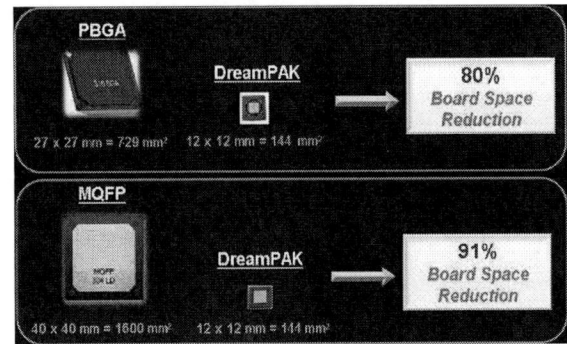

Fig.4 - Board space by DreamPAK

DreamPAK is very flexible in its pad layout since odd shapes can be etched out without any worry of structural complications. Due to its flexibility in the layout design, DreamPAK offers benefits like having a leadframe based package with power rings and (or) ground rings, which are otherwise not possible before and only can be seen so far with laminate based packages.

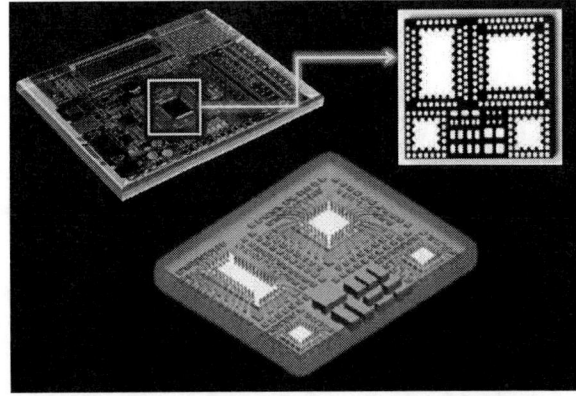

Fig.5 - SIP application demo with DreamPAK.

Another important capability arising from the flexibility in the layout design, is the achievement of SIP (System in Package). Any number (or any shape) of die attach pads and

EMPC 2009

978-1-4244-4722-0/09 $25.00
© 2009 IMAPS-ITALY

flexible lead layout / lead count can be planned together with mounted passive IC packages (if required). This is also another very unique capability of DreamPAK, which was so far achieved in laminate based packages only. Also this gives a great cost benefit since this is achieved in leadframe based substrate which is much cheaper than a multilayer laminate. Since DreamPAK is typically and exposed pad package, the thermal performance for the SIP package is expected to be much better than the laminate based package. An illustration of conceptual SIP is in Fig. 5

Many of the low pin count packages choose to have no die attach pad, due to either the characteristics of the IC chip or due to no space for it. DreamPAK fully satisfy this "no die attach pad" intention, since the user can attach the die on the epoxy matrix itself, which is rigid enough to serve to support the die with normal die attach process. Fig. 6 is an example.

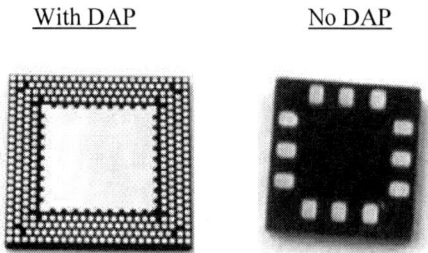

Fig.6 - DreamPAK applications without DAP.

Assembly Processing

Wire bonding in DreamPAK assembly can be a smooth sail. Tests show that good stitch pull strength of typically 7.5 gms can be achieved on 25 micron gold wire with simple process setup.

Stitch Pull :	(Wire size : 1mil)	
Minimum	7.09 gm	
Average	7.45 gm	
Maximum	8.20 gm	
St Dev	0.36	

Fig.7 - DreamPAK - Stitch pull test data.

The major problem with QFN wire bond process is the problem of stressed /cracked ball neck when higher ultrasonic power parameter is used. The connecting bar linked floating leads bounce and stress the wire when ultrasonic power is used. This stressing of wire typically cause thinning or crack in the ball neck. This is illustrated in Fig.8

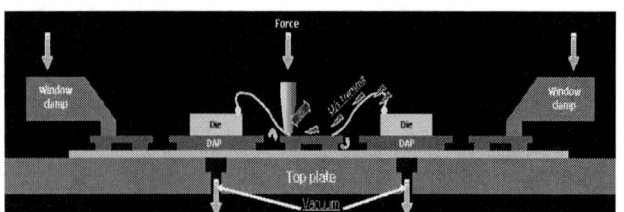

Fig.8 - Neck stress in QFN wire bonding

Due to this issue, QFN wire bonding uses lower ultrasonic power setting. Also to compensate this lower U/S power, typically, higher bonding time and special scrub motions are needed to successfully wire bond the QFN package. At times, one has to use virtually thermo-compression wire bonding process for QFN.

This drawback no longer valid for DreamPAK wire bond. Since the leads are not linked to each other, the flow of ultrasonic stress is interrupted and does not stress the neck. So, the special thermo compression bonding/scrub motions as used in QFN wire bond process, are no longer needed. Bonding can be achieved by simple parameters resulting in wire bonding UPH gain. Also, since the wire bond process is thermo-sonic, lower force is involved and higher power can be applied without necking problem. Fig 9 shows the neck stress on the wire bonding.

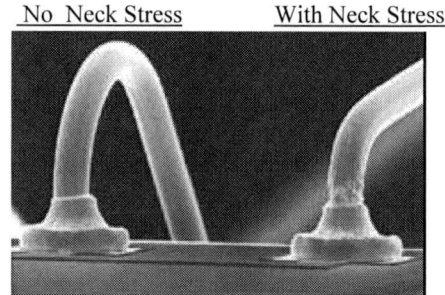

Fig.9 - Neck stress in thermosonic wire bond.

For the molding process, due to the small form factor for higher I/O count, lower EMC volume consumption is seen. In comparison to PBGA (~300L), an equivalent DreamPAK package will use considerably lower EMC (about 90% lower) as a direct comparison.

Fig.10 - DreamPAK - Tapeless molding.

Also, typically QFN molding is affected by tape residue since most of the QFN molding processes require tape below the leadframe for flash free molding. However for DreamPAK, molding will not be affected by any tape residue, as no tape will be involved during assembly process.

For saw singulation process, as the leads are just embedded in an epoxy matrix, during sawing, the blade is not cutting metal connecting bars. Thus the stress for the saw blade will be much lower and higher saw speed without metal smear can be realized. This helps in gaining increased UPH as well as longer blade life. Fig 11 shows the absence of copper smear concern and realization of higher UPH clearly.

EMPC 2009

978-1-4244-4722-0/09 $25.00
© 2009 IMAPS-ITALY

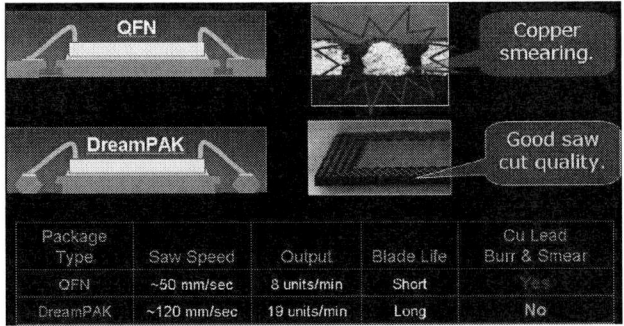

Fig.11 - Singulation speed advantage

All over the world, industries are going "Green". DreamPAK automatically support this cause, since these leadframes are pre-plated with the proven standard Ni-Pd-Au plating. Also the epoxy matrix material is Pb-free. Together with a green EMC at the final mold after wire bond, DreamPAK offers a "complete Pb-free solution". Also pre-plating removes the solder plating requirement of leads which results in one less process step in IC assembly.

Fig.12 - Pre-isolated leads aid strip testing

One of the major processes after package assembly is the testing of singulated packages. Handling and testing of packages, one by one, incur significant cost. Hence many assembly houses have looked into the possibilities of doing electrical testing of the units in the strip form itself. This is known as "Strip testing".

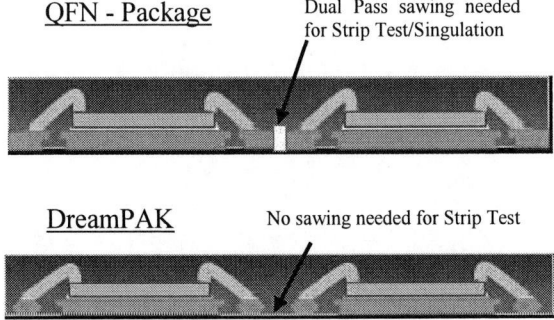

Fig.13 - No sawing for strip test

Since leads in DreamPAK are already isolated, it provides this freedom of strip testing and significantly reduces packaging & testing cost.

Thermal Performance

DreamPAK is expected to provide very good thermal performance. Thermal resistance simulation was done for different air velocity, in comparison to different packages. Results show that DreamPAK packages follow the thermal performance similar to QFN packages. Fig 15 shows this comparison.

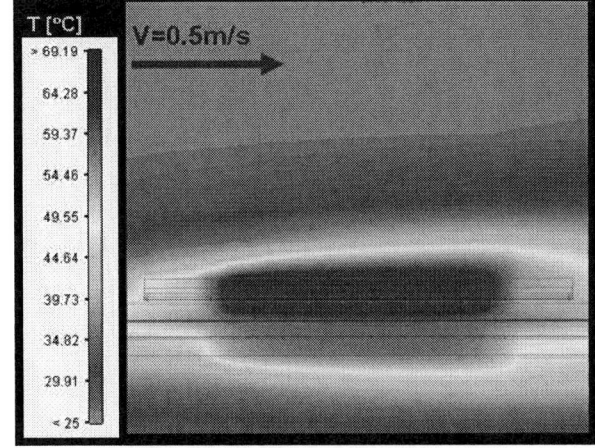

Fig.14 - Temperature distribution under air flow.

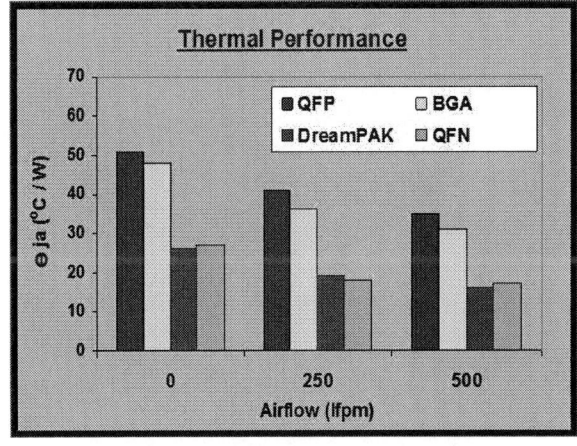

Fig.15 - Thermal resistance simulation

Lead Integrity

Many similar half etch based technologies that requires wet etch processes after assembly, typically had one significant problem. That is the lead integrity with the package. Instances were observed, where in such technologies, that the leads can come off easily from the package due to less EMC locking of leads.

This concern, due to suspected less molding locking, was tested for the DreamPAK, (based on the lead integrity test guidelines from MIL STD 883 for leadless packages). The method involves soldering a joint to the lead, allowing it to

EMPC 2009

978-1-4244-4722-0/09 $25.00
© 2009 IMAPS-ITALY

cool and subsequently pulling vertically upwards to test the strength of the solder bond with the lead. The criteria will be,

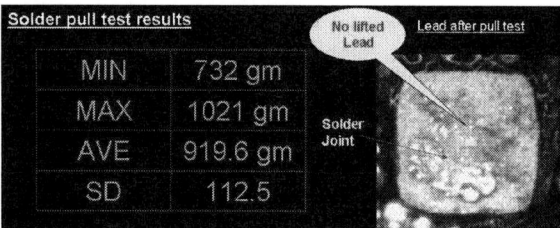

Fig.16 - Lead integrity (solder pull test)

the joint should withstand at least 227gm-f in tensile pull. DreamPAK is found to have performed very well to this test and observed that the lead could withstand an average 900 gm-f of load "without leads being pulled out" of the package. Fig.16 illustrates this.

Package Reliability

DreamPAK, by its design, has many highly elevated features created by etching technology. These features were found to greatly enhance the package integrity by acting as locks. The unique locking mechanism in DreamPAK not only provides good reliability but also good lead integrity.

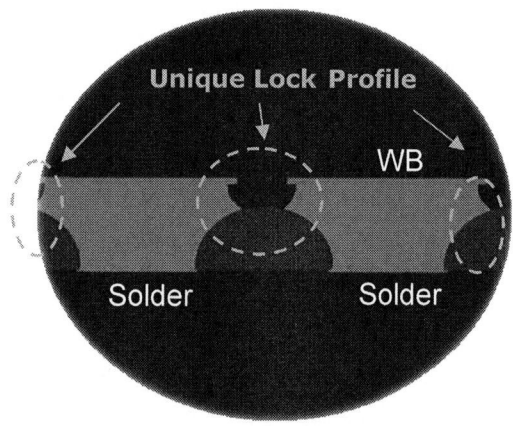

Fig.17 - DreamPAK - Lock features

Tests show that the package is able to withstand MSL precon level 1. Further to the preconditioning, other reliability tests are on being conducted and the results are awaited.

Fig.18 - T-SCAN after MSL1 precon

Board Level Package Reliability

DreamPAK is being tested for board level reliability tests like package pull test, board level thermal cycling, and so on.

The package pull test was done for DreamPAK 12x12 mm package size. Firstly, the DreamPAK package was solder mounted on the PCB test board. After solder reflow, the units were put on the special pull test machine to apply a pull load on to the package. Since 12x12mm package is a high pin count package, only the inner two rows of leads (124 leads only) were solder joint mounted on the PCB. Also the exposed DAP was not soldered too. This is done with an objective of testing only the lead joint reliability. Package pull test result is as shown in Fig. 19.

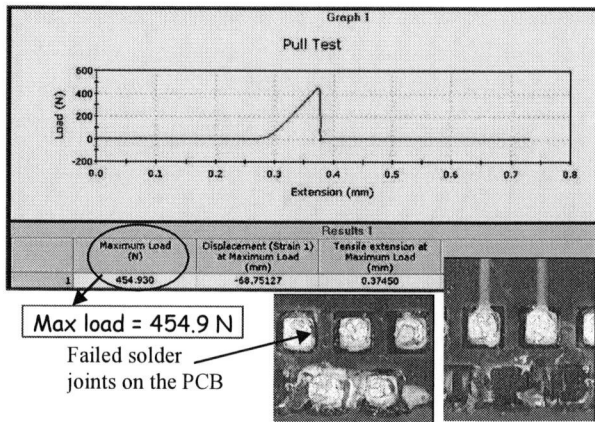

Fig.19 - Package pull test results

As one can see, the failure load for this package (with 124 leads soldered) is very high (454.9N). This value is comfortably passing the lead integrity requirements as per MIL-STD-833. Thus the solder joint reliability is very good.

Similarly, DreamPAK packages were mounted on Daisy chain PCB test board for thermal cycling reliability tests. These DreamPAK mounted PCB boards were then put inside the thermal cycle chamber for thermal cycles from -45 to 125 degC. At regular intervals, the units were taken out of the chamber and electrical open/short test is conducted. So far, our test results show no open or short failures even after 500 cycles. This is an ongoing test and so further thermal cycling is continuously being done to monitor the package robustness.

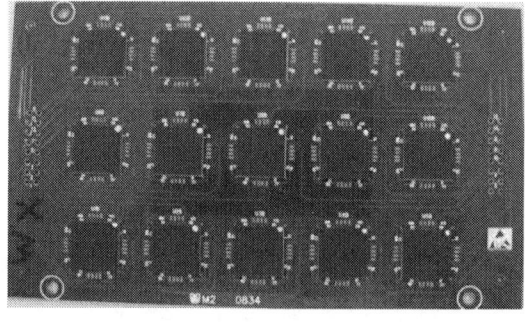

Fig.20 - Test board for DreamPAK TC test

EMPC 2009

978-1-4244-4722-0/09 $25.00
© 2009 IMAPS-ITALY

Assembly Challenges

Every technology has its own merits and demerits. DreamPAK is not immune to that.

DreamPAK offers a few challenges particularly when dealing with high I/O and thin packages. Typically high I/O packages are involved with many rows of leads and require multi-tier wire bonding. When package thickness is small, the loop height to wire length ratio will also be very small between wire layers. Proper selection and optimization of low loop profile might help to resolve the issue.

As the epoxy matrix is very thin and fragile, handling related epoxy matrix cracks is another challenge that can be encountered. These are typically caused by improper handling of the DreamPAK leadframe strip. Thus a careful handling will help to reduce the instances.

Wire sweep is another issue with multi-row high I/O packages, and DreamPAK is no exception. However, current EMCs are able of meeting the requirements. Further adjustment can also help to keep the wire sweep under control.

Conclusions

DreamPAK packaging technology addresses all the requirements of the Packaging Engineering. It offers low to high I/O with small form factor. The design flexible package / leads layout empowers the designer with more freedom to design the package. DreamPAK is environmentally friendly and offers a complete Pb-free packaging.

DreamPAK is fully compatible with mainstream assembly processes and does not require any wet processes like etching / plating etc., to complete the IC assembly processes. Thus, it lowers overall assembly costs.

Since the leads are already isolated, DreamPAK supports strip testing and improves productivity. Package singulation is easier and faster since no metal smear is encountered due to metal-free cut line at the saw operation.

Finally, the lead integrity as well as package reliability including the board level reliability are outstanding; thanks to the special etch profile lock mechanism design. Thus, despite some typical challenges with any high I/O packages, DreamPAK is truly offering the dream options to packaging engineers to reach optimal solutions for complex packaging challenges.

ADVANTAGES	DreamPAK	QFP	QFN	PBGA
√High I/O	●	●		●
√Small form factor	●		●	
√Flexible in package design layout	●			●
√Environmental friendly (Pb free)	●	●	●	●
√Compatible to mainstream assembly process	●	●	●	●
√Support in-strip testing	●	●	●	●
√High reliability	●		●	
√Good thermal performance	●			

Fig.21 DreamPAK package features.

Acknowledgments

The authors would like to thank the team members of Packaging Technology, Leadframe Development, Encapsulation Development and the management of ASM for the contributions and various supports to the DreamPAK Project. Last but not least the authors would like to thank the various material suppliers, customers and research institutes for their contributions to the program.

References

1. JEDEC Package Outline Specifications MO-220, MO-247
2. JEDEC JESD22-A104C
3 IPC-2226, 2221
4. IPC-2226, 2221

978-1-4244-4722-0/09 $25.00

Fatigue life prediction of plated through holes(PTH) under thermal cycling

Nochang Park[*], Jiho Kim, Chulmin Oh, Changwoon Han, Byungsuk Song, and Wonsik Hong

Physics-of-Failure Research Center, Korea Electronic Technology Institute, Seongnam, 463-816, Republic of Korea
*82-31-789-7285, 82-31-789-7059 and ncpark@keti.re.kr

Abstract

PTH(Plated Through hole) integrity becomes primary PCB(Printed circuit board) reliability concern during component mounting process or subsequent field condition. Main cause of PTH failures is traced to significant differences in the amount of thermal expansions between copper barrel and surrounding glass epoxy. When the different expansion due to thermal stresses makes strain that exceeds the fracture toughness of copper barrel or interconnection, ruptures develop in the latter. It results in open circuits or intermittent contacts in FR-4 PCB which leads to system failures. This paper presents a prediction results of PTH thermal fatigue life on field condition based on coffin-manson life model. Experiments are conducted with 0.3mm dia. PTH under 2 thermal cycle conditions (-55□~125□, -35□~125□ with 10 min. dwell time). The thickness of PCB is 1.6mm with 4 multilayers. The resistance of PTH is monitored one minute interval. Material constant are obtained after thermal cycling test. Crack propagation at failed PTHs is examined with optical microscope. For each test, thermal cycling failure data has been analysed using two parameter Weibull models to predict B_{10} life of PTH. Acceleration factor of PTH under thermal cycling test is obtained. Life Prediction of PTH thermal fatigue on field condition is conducted.

Key words: Plate through hole, PCB, Thermal fatigue life, Activation energy, Coffin-manson equation , B10 life

Introduction

A plated through hole(PTH) in multi-layer printed circuit board(PCB) is defined as a hole in which electrical connection is made between internal or external conductive patterns by plating of metal on the wall of the hole[1]. Recent studies have indicated that thermal stress testing at the lead free temperature level of 260℃ or close can possibly lead to a failure mode shift from barrel cracking to barrel-inner-layer interconnection failure[2-4] integrity becomes the primary PCB reliability concern either during the component assembly process or during subsequent field usage, where PCBs are subjected to thermal cycling. The main cause of PTH failures may be traced to significant differences in thermal expansions between the copper barrel and the surrounding glass epoxy. When, due to thermal stresses, the differential expansion causes strain that exceeds the fracture toughness of the copper barrel or interconnect, ruptures develop in the latter[5]. Corner cracks or barrel in near layer crack are the failure mode and the failure mechanism relates to exceeding the fracture strain of diurnal temperature variations experienced by the end product during its field life. Failure occur when the fatigue life of the plated barrel is exceed. This results in open circuits or intermittent contacts which may lead to system

failures. In general, Thermal cycling tests are employed to determine the capability of the PTH's to withstand cyclic temperature variations encountered in actual field life. In order to satisfy the reliability requirements, the PTH's must meet two criteria : a solder float test and temperature cycling tests such as MIL-P-55110D or MIL-STD-202E. Temperature cycling test simulates the temperature variations during its filed life. Failures occur when the fatigue life of the plated barrel or the foil pad its exceed.

In this paper, Experiments are carried out under 2 thermal cycling conditions of -55℃~125℃, -35℃~125℃ with 10 min. dwell time. Material constant is obtained after thermal cycling. Physical analysis shows the cracking of PTHs. Thermal cycling failure data has been analysed using two parameter Weibull models to predict B_{10} life of PTH. Acceleration factor of PTH under thermal cycling test is obtained. Life Prediction of PTH thermal fatigue on field condition is conducted.

Experimental procedures

All PTHs are connected through a daisy chained pattern of Fig 1. Total resistance of a series of 600 daisy chained holes is about 4 ohm. The thickness of is 1.6mm with 4 multi layer. The hole

978-1-4244-4722-0/09 $25.00
© 2009 IMAPS-ITALY

size is 0.3 mm of diameter, pad size is 1.5mm. The glass transition temperature of FR-4 is about 125 ℃. This is consonance with MIL-STD thermal/shock test temperature, where the high temperature limit is 125℃. So the test sample is heated to 120℃ and cooled to -55℃(condition 1), -35℃(Condition 2) in order to Z-axis expansion mismatch. The dwell time is 10 minutes.

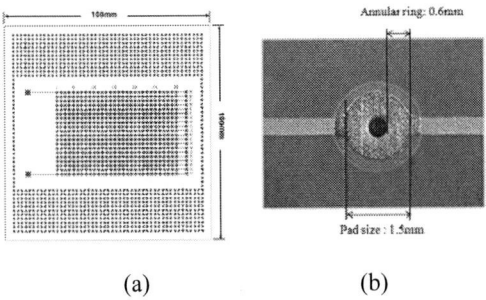

(a)　　　　　　　　　(b)

Figure 1 Test board for Thermal cycling test: (a) front side of board, and (b) front side of PTH hole.

Total 25 test coupons is used for thermal cycling test. Test is stopped at the time of 25th PTH fail occurred. The temperature of PCB is monitored by T-type thermocouple (Fig 2) and ETAC NT1200W chamber for thermal cycling is used. . The resistance of PTH is measured one minute interval by FLUKE 2680A. Failure criterion is 20% changing of initial resistance.

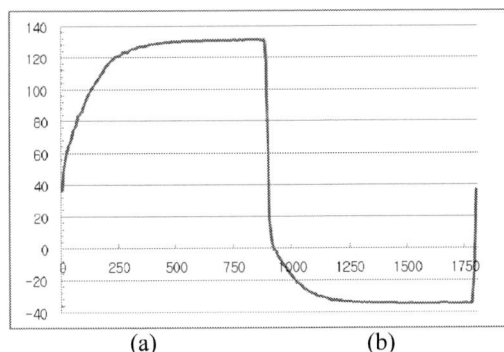

(a)　　　　　　　　　(b)

Figure 2 The temperature of PCB surface (Condition 2)

Results and discussions

The cycle to failure is listed in table 1. The A-D(Anderson-Darling) value is calculated for finding a valid life distribution(Table 2). Proper life distribution is selected in low A-D value distributions. A-D value of all distribution is little variation compared to minimum A-D value distribution. So weibull distribution can be assumed for this test. B_{10} life estimate of PTH is performed with statistics package (Minitab). B_{10} life of

Table 1 The thermal cycle of PTH failure

	Condition 1	Condition 2			Condition 1	Condition 2
1	475	903		14	958	1,644
2	564	922		15	975	1,676
3	570	1,085		16	1,023	1,707
4	621	1,091		17	1,047	1,729
5	695	1,095		18	1,050	1,733
6	724	1,156		19	1,051	1,764
7	735	1,224		20	1,054	1,786
8	767	1,224		21	1,082	1,792
9	854	1,246		22	1,106	1,870
10	860	1,292		23	1,233	1,877
11	863	1,380		24	1,310	1,928
12	920	1,390		25	1,393	1,939
13	951	1,639				

Table 2 A-D value on Life distribution

Distribution	Condition 1	Condition 2
Weibull	0.743	1.378
Log normal	0.938	1.454
Exponential	6.775	7.202
Loglogistic	0.795	1.335
Smallest Extreme Value	0.940	1.305
Normal	0.755	1.398
Logistic	0.736	1.320
B_{10} Life(Cycle)	606	1,065

condition 1 and 2 is 606 cycle and 1065 cycle respectively. Shape parameter and scale parameter is calculated by appling maximum likelihood estimation. Shape and scale parameter of condition 1 is 4.46949, 1003.57 and Condition 2 is 5.40435, 1615.04 respectively (Fig 3). P value is 0.402 for equal shape parameter of condition 1, 2. Accordingly this acceleration test is valid in significane level 5%.

Crack propagation at failed PTH is examined with optical microscope. Fig 4 show the crack on the Cu barrel.

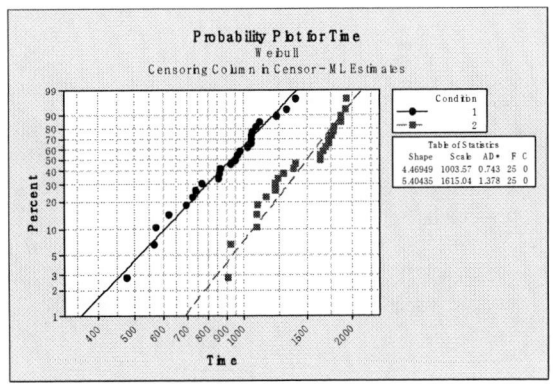

Figure 3 Probability plot of condition 1, 2

978-1-4244-4722-0/09 $25.00
© 2009 IMAPS-ITALY

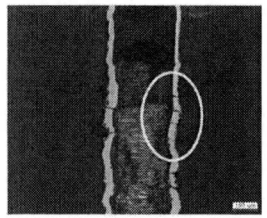 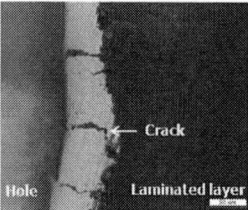

Figure 4 Cross sectional Image of failed specimen

Assuming N_f is PTH life or cycle to failure and ΔT is the stress level or the temperature range in thermal stress test, a acceleration relationship can generally given as

$$N_f = f(\Delta T) \qquad (1)$$

Then acceleration factor, AF, between B_{10} life correspondent to condition 1 temperature(ΔT_1) and B_{10} life correspondent to condition 2 temperature(ΔT_2) is given by

$$AF = \frac{N_{f1}}{N_{f2}} = \frac{f(\Delta T_1)}{f(\Delta T_2)} \qquad (2)$$

The IPL relationship is given by [6,7]

$$N_f = \frac{1}{k\Delta T^n} \qquad (3)$$

Where k > 0, n > 0 are empirical parameters that need to be determined from testing. Cycle to failure results that obstained using this model is in Fig 5. k and n is 4.787 and 2.6×10^{-14} respectively. The following definition of acceleration factor, which is consistent with the definition of (2)

$$AF \equiv \frac{N_{field}}{N_{accelerated}} \qquad (4)$$

For known $N_{accelerated}$, N_{field}, AF is then estimated by equation (4). AF can be seen from table. Using the result in AF, Field life can be predicted from accelerated test result. In case that the $\triangle T$ is 60, an predicted B_{10} life is 116,600 cycles to failure. For example, Assuming that the number of thermal cycle is 10 cycles per one day, B10 life of this sample equates to a period of approximately 31 years.

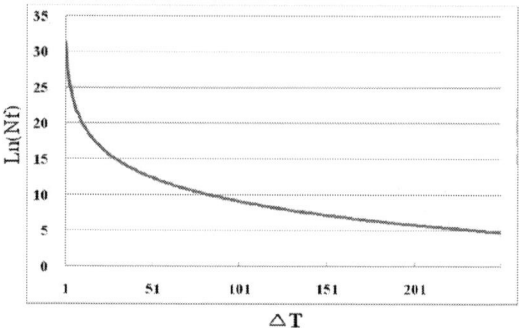

Figure 5 Cycle to failure of IPL Model according to $\triangle$T

Table 3 Acceleration factor estimated by experimental model

	Field condition ($\triangle$T)	AF	
		Condition 1 ($\triangle$T = 180)	Condition 1 ($\triangle$T = 160)
1	40 (0℃~40℃)	1,340	762
2	60 (-20℃~40℃)	192	109
3	80 (-40℃~40℃)	49	28

Conclusion

It was confirmed that the failure mode of PTH is copper barrel crack. This failure mechanism, which is related to material, structure, and the manufacturing process of PCB has more effect on reliability under actual service conditions.

A inverse power model was used to estimated the time to failure of PTH copper barrel crack subjected to TC test. The weibull parameters for copper barrel crack of PTH were determined based on the result of experiment. AF of barrel crack on PTH is determined experimentally. Shape and scale parameter of condition 1 is 4.46949, 1003.57 and Condition 2 is 5.40435, 1615.04 respectively.

References

[1] IPC-T-50, 1988, "Terms and definitions for interconnecting and packaging electronic circuits," The institute for interconnecting and packaging electronic circuits, Lincolnwood, IL.

[2] J. Xie, Y.J Huo, Y. Zhang, and M. Freda, "Effect of PWB design factors and glass transition temperature on PTH reliability," in Proc. 39th int. symp. Microelectron. (IMAPS 2006), San Diego, CA, Oct. 8-12, 2006, 891-898

[3] M. Freda and D. Barker, "Predicting plated through hole life ant assembly and in the field from thermal stress data," in Proc. IPC Printed circuits Expo, APEX and the Designers Summit, 2006, S36-01-1-S36-01-373

978-1-4244-4722-0/09 $25.00
© 2009 IMAPS-ITALY

[4] M. Freda and J. Furlong, "Application of reliability/Survival statistics to analyze interconnect stress test data to make life prediction on complex, lead free printed board assemblies," in Proc. EPC2004, Cologne, Germany, Oct. 5, 2004 CD-ROM

[5] "Round robin reliability evaluation of small diameter plated through holes in printed wiring boards," IPC-TR-579, pp. 19, 41, Sept. 1988

[6] M. Freda and D. Baker, "Predicting plated through hole life at assembly and in the field from thermal stress test data" in Proc. IPC Printed Circuits Expo, APEX and the Designers Summit, 2006, pp. S36-01-1-S36-01-37

[7] M Freda and J. Furlong, " Application of reliability/survival statistics to analyze interconnect stress test data to make life predictions on complex, lead-free printed circuit board assemblies," In Proc. EPC 2004, Cologne, Germany, Oct. 5, 2004. CD-ROM

[8] "IPC-TR-579 round robin reliability evaluation of small diameter plated through holes in printed wiring boards," IPC Tech. Rep., Sep. 1988

Thin Hermetic Borosilicate Glass Layers for Highly Reliable Chip-Passivations in Wafer-Level-Packaging

Ulli Hansen[1], Jürgen Leib[1], Simon Maus[1], Oliver Gyenge[1], Michael Töpper[2]

[1]MSG Lithoglas AG, Gustav-Meyer-Allee 25, 13355 Berlin, Germany
[2]Fraunhofer IZM, Gustav-Meyer-Allee 25, 13355 Berlin, Germany

+49 30 46403 618, ulli.hansen@lithoglas.de

Abstract

A technology yielding thin, hermetic borosilicate glass layers at high deposition rates and low substrate temperatures and its potentials for a novel approach on wafer-level passivation is described. The benefits of this CMOS-compatible technology are highlighted, comparing the achievable film characteristics to polymers commonly used for these purposes. The glass layer is deposited at low temperatures (T < 100°C) using a plasma-enhanced e-beam deposition and can be structured by a lift-off process using a standard photo resist process for masking. The process flow is fully compatible with standard CMOS post processing and is integrated in a state-of-the-art production environment.

Key words: borosilicate glass deposition, low temperature process, e-beam evaporation, passivation, wafer-level-packaging

Introduction

With advancing miniaturization in micro system technologies the challenges posed by industry demands for robust and hermetic packaging grow steadily more complex. To serve these needs – especially in combination with cost-critical high-volume applications – wafer-level-packaging technologies are becoming increasingly established. This overall trend calls for reliable and cost effective technologies for applying and structuring long-term stable materials.

Initially, the core functionality of materials used in wafer-level-passivation could be reduced to yielding a protection layer against environmental influences like humidity combined with a certain insulating capability against passivated conductor lines.

However, with the increasing demands for performance also the requirements for the passivation layers get more differentiated. In summary the most common requirements are [1]:

- sufficient adhesion to all surface materials
- low water absorption / moisture diffusion → protection against corrosion of metals and to serve as a diffusion barrier in electronic circuits and redistributions
- electrical properties including insulation resistance, dielectric constant, dissipation for DC to Gigahertz frequencies and breakdown voltage.
- ideally low intrinsic stress and a matched coefficient of thermal expansion to the substrate
- limited change of material characteristics within the range of device usage including temperature and humidity exposure as well as aging of the device.
- manufacturability by standard technologies and moderate costs of the base material, storing and processing

Driven by the increasing demand and advanced requirements of high density, high performance devices, also the requirements for dielectric materials in wafer-level-packages are becoming more critical.

Elevated Reliability Requirements

The influence of humidity on the reliability of electronic components is often underestimated. Moisture from atmospheric humidity may enter the packaging materials by diffusion and collects at the interfaces of dissimilar materials. The moisture sensitivity of a particular layer or package is influenced by different factors, including its internal dimensions and the physical properties of the used materials.

How much moisture a particular layer or package actually absorbs depends on factors such as temperature, the relative humidity of the ambient atmosphere, and how much time the component is exposed to those conditions. The higher the temperature, the faster the surrounding moisture will penetrate. The absorption process will continue until the internal moisture concentration reaches equilibrium with the ambient relative humidity. The higher the relative humidity, the greater the amount of absorbed moisture within the package.

978-1-4244-4722-0/09 $25.00
© 2009 IMAPS-ITALY

As applications in packaging and especially wafer-level-packaging spread into harsher environments like under-hood in automotive, industrial as well as avionic and defense, reliability requirements are getting more challenging. Thermal cycling, humidity and aging are posing the key challenges here.

Whereas in consumer markets testing from 0 °C to 85 °C for 300 cycles might be considered sufficient, Thermal Cycling Tests (TCT) for automotive application may be specified to be -55 °C to +125 °C (JEDEC JESD22-A104 Condition B) or -55 °C to +150 °C (JEDEC JESD22-A104 Condition H) for 1000 cycles. Even more challenging requirements can be seen in space industry, where -143 °C to +125°C for 2000 cycles needs to be met.

Temperature Humidity Storage (THS) tests are usually requiring devices and systems to pass 85 °C and 85 % humidity exposure for 1000 h. In visual testing or simple DC-electrical testing most polymers learned to pass these conditions. However in Moisture Sensitivity Level (MSL) test, where after a soak time of 168 h at 85 % / 85 °C solder reflow is simulated for three times, it might remain a challenge to achieve MSL-1 (JEDEC J-STD-20).

The main reason for failure is excessive moisture absorption in the polymer during the soak, which causes the redistribution layers to crack and/or delaminate, when the moisture evaporates during the exposure to reflow temperatures of up to 260 °C.

In the case of integrated passive devices, RF-devices or optical applications the passivating materials are also used as dielectric materials. However, the stability of the insulation properties and the mechanical integrity of the final device may change due to water absorption or aging [2].

For manufacturability in mass markets the electrical characteristics should not have any or only low dependence on process conditions, incl. post processes like molding and reflow, as well as there should be no or very low change due to the conditions and duration of use during the designed life time. This includes exposure to high or low temperature, humidity and (UV-)light as well as the aging of the device. Long time stability at elevated temperatures (>150 °C) is a limiting factor for automotive applications.

Polymers as Chip Passivations

Mostly common polymers are used as passivation layers [1, 3]. These layers are deposited by spin-coating from a liquid precursor and are cross-linked directly on the wafer by temperature. Depending on the type of the polymer the layers may be structured directly by lithography prior to curing (e.g. BCB, photosensitized polyimide or photosensitized epoxy). In case the material itself is not photosensitive, it can be structured by subsequent masked plasma etching.

Very specialized polymers (e.g. BCB, Polyimide or PBO) have been successfully used for highly reliable, high performance redistributions and can be processed on wafer level [4, 5, 6, 7]. Their glass temperatures (Tg) or temperatures of decomposition in case of thermoset materials can be as high as 400°C – depending on the polymer backbone. However, also these high performance materials show water absorption and swelling in humidity testing, which impacts the reliability and may impair the electrical performance of the materials, esp. looking at the long term stability of the devices.

A draw-back for most organic materials is the mismatch of the thermo-mechanic properties with regard to the substrate material, mostly silicon. Stress in the material interfaces as well as resulting wafer warp and delamination may cause failures in the designated product life time. Main reasons are the large differences in thermal expansion of the substrate materials and the polymers (CTE Si: 2.6 ppm/K; polymers: 12 – 60 ppm/K) as well as shrinkage of the polymers during processing.

The Benefit of Glass

Glass – especially borosilicate glass – is due to its excellent physical properties a very fitting material to be used as chip passivation. It is chemical inert yielding excellent durability when exposed to humidity and water as well as acids, bases and solvents. It is a bio-compatible material and suitable medical or micro fluidic applications. Furthermore it is hard and scratch resistant leading to mechanically robust surfaces. Its good electrical performance as well as its optical qualities allow for a wide range of applications in opto semiconductors and for embedded optical functions.

In its composition according to ISO 3585 it is among the most robust glasses available. It has a high hydrolytic resistance (ISO 719 - HGB 1 and ISO 720 HGA 1) and is resistant to most of the acids (ISO 1776 – Class 1) and bases (ISO 695 – Class A2). It is also resistant to solvents like acetone, kerosene and IPA.

Glass-to-Metal-Seals are used as a hermetic packaging technology since the early days of semiconductor industry and special solder glasses are used on wafer-level for passivation of diodes, power transistors and thyristors in larger volumes.

Borosilicate glass wafers like Borofloat 33 or Pyrex 7740 are well known for hermetic wafer-level-packaging of MEMS. They are used for anodic bonding, where the bond is performed at a temperature of about 400 °C. It is therefore essential, that the

978-1-4244-4722-0/09 $25.00
© 2009 IMAPS-ITALY

CTE of the cover material is matched to the silicon device wafer in order to avoid cracking, delamination, warp and excessive stresses in the bonded devices.

Although being successful in those special applications, the use of glass for mainstream packaging of semiconductors was limited by lacking CMOS process compatibility and other manufacturing limitations.

The Lithoglas Process –
Microstructured Glass as a Thin-Film Process

The novel deposition and microstructuring of glass allows the formation of thin films of dense borosilicate glass on a broad range of substrates at substrate temperatures below 100 °C [2, 8].

The deposition of the glass is done by a plasma-assisted e-beam evaporation process. It is a high rate deposition process with deposition rates above 200 nm/min and at the same time low substrates temperature. With the high deposition rate typical film thicknesses of 3 – 10 μm can be achieved easily and production can be run as a cost effective batch process depending on the required uniformity. Thicker layers as thick as 100 μm have been demonstrated.

As the evaporation process ensures low substrate temperatures standard photo resists can be used for masking and the necessary processing to yield a structured borosilicate glass layer consists of merely three steps (see figure 1).

Step 1: Lithography

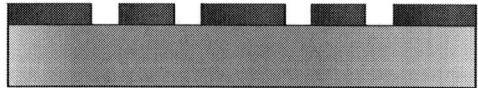

Step 2: Glass Deposition

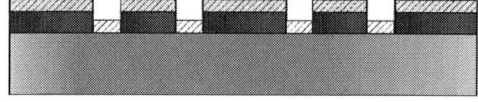

Step 3: Lift-Off

Figure 1: Lithoglas (hatched) is deposited at low temperatures on a substrate (grey) furnished with a patterned standard photo resist (dark grey). The glass is structured by lift-off-technique.

In the first step the necessary lithography is performed to define the intended structuring of the coating. With respect to the surface topography it is usually added by spin-on. After exposure and development the photo resist carries the negative image of the target structures in the glass layer.

As the second step the actual glass deposition is done by plasma-assisted e-beam evaporation. The exposed wafers are fully coated at substrate temperatures well below 100 °C.

The third and final step is the lift-off: the glass microstructures are developed by dissolving the photo resist mask and with this removing the glass on top of it. A structured glass layer remains at the locations which were not covered by the resist; areas covered by photo resist were protected troughout the process and reveal their original surface finish.

Depending on the layer thickness an aspect ratio of 1.8 can be achieved yielding 1.3 μm fine structures in a 3 μm glass layer by using a mask aligner for lithography (see figure 2).

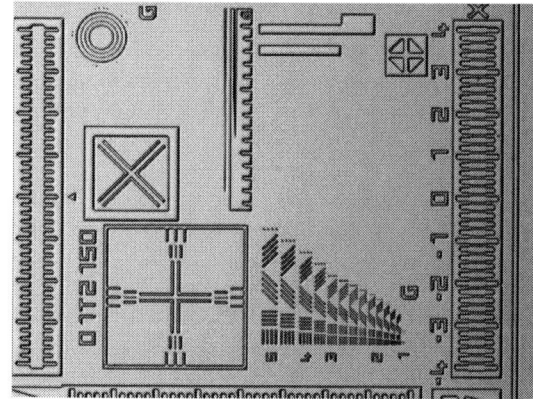

Figure 2: Microstructured glass on silicon with a smallest feature size of 1.3μm at an aspect ratio of 1.8. The deposited glass is structured by photo resist lift-off.

It is possible to structure very large areas by this technique as frequently used in the production of optical cavity windows for image sensors and optical MEMS. Throughout the whole process sensitive areas where the glass is not to be deposited are covered and protected by the lift-off photo resist. This is especially useful for MEMS and optical sensors, when e.g. microstructured glass bond frames are formed on a glass cap wafer using an antireflective or other coating in order enhance the optical performance of the final device. The optical quality of these surfaces is unchanged since the Lithoglas process is an additive structuring method, whereas most other technologies for structuring

978-1-4244-4722-0/09 $25.00
© 2009 IMAPS-ITALY

glass like etching, ultrasound milling or sandblasting are subtractive and are removing material from the glass surface thus changing its quality or finish.

Characteristics of Lithoglas layers

The characteristics of a Lithoglas layer are very close to the properties of the bulk borosilicate materials – comparable to Duran, as it is used in laboratory equipment or B33 used in wafer-level-packaging applications. Used as passivation and dielectric layer within wafer-level packaging it comes pretty close to the ideal case.

Excellent Adhesion: The e-beam deposited and microstructured glass thin films show excellent adhesion on a broad range of substrates and surface materials. This includes – but is not limited to – e.g. silicon, glass, SiO_2, Si_3N_4, GaAs, SiC, $LiNbO_3$ or LTCC, HTCC and Al, Cu, Ni, Steel as well as different plastics.

Hermetic: As mentioned borosilicate glass is well known for its low water uptake and vapor transmission. The glass exhibits several orders of magnitude better barrier properties to humidity and gases compared to polymers [9]. It was shown, that an 8 µm thick Lithoglas layer is hermetic in a Helium leak test according to method 1014 of MIL-STD 883 [8].

Low stress: The internal stresses in the glass thin film can be influenced by the choice of the process parameters. In principle, stress free layers are feasible; however, it is beneficial to have a moderate compressive stress designed in the glass layer to increase resistance to thermal and mechanical shock. Since the CTE of the glass layer is matched to the thermal expansion of silicon, the engineered stress conditions can be maintained over a wide range of temperatures. Lithoglas layers are used as anodic bond layer at temperatures beyond 400 °C as well as cooling down of the Lithoglas passivation to temperatures of liquid nitrogen (-196 °C) has been demonstrated on different substrate materials. Qualification of thermal cycling according to the specifications of space applications (-143 °C to +125°C for 2000 cycles) is on the way.

Chemically inert and stable, biocompatible: As mentioned above borosilicate glass is chemically very stable and inert. As the glass composition of the produced layers is closely matched to the bulk material, already thin films with thicknesses of a few 100nm have proven excellent stability. Being biocompatible it is an ideal material for medical and biochemical applications.

Electrical properties: High frequency optimized Lithoglas dielectric layers show better electrical performance compared to most of the polymers, where as they are similar to pure SiO_2 or Si_3N_4 coatings manufactured at similar conditions. The effective dielectric constant of a Lithoglas layer is 4.3 with a dissipation factor lower than $6·10-4$ measured up to at 50 GHz [8, 10]. Breakdown voltage is well above 200 V/µm. It appears to be important to note, that these properties only slightly depend on temperature, humidity and aging of the glass dielectric layer.

Optical properties: The produced thin films are fully transparent and – having thicknesses in the micrometer range - have an improved transmission range over the bulk material, which opens at about 270nm and closes at 2800nm.

Manufacturability and low costs: The Lithoglas process is CMOS back-end compatible and has been developed to fit into standard process flows. The number of process steps as shown in figure 1 are similar to a deposit & etch flow process for polymer or Oxide/Nitride-CVD. However batch processing is feasible.

Lithoglas as Final Chip Passivation

Its excellent hydrolytic and chemical performance makes borosilicate glass layers ideal as final chip passivations for applications with elevated reliability requirements. Especially in harsh-environment applications glass maintaining a constant low intrinsic stress may serve as a replacement for polymers, oxides or nitrides. In fact, using a highly hermetic material like borosilicate glass may reduce overall packaging costs as for certain applications ceramic packages may be replaced by more cost-efficient packaging solutions.

As an application example Lithoglas layers were used as a final passivation of avalanche photo diodes, both silicon and III/V-based. Apart from extending the device life-time by hermetic sealing the impact of the passivation layer on the electrical performance of the devices, which in some designs work with high voltages at the surface metalizations

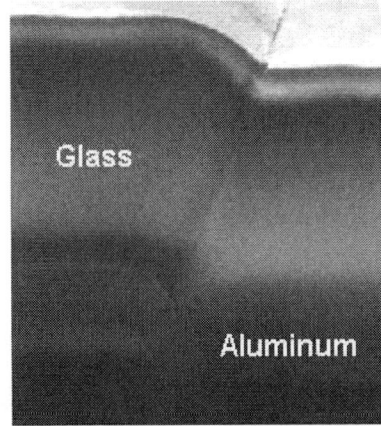

Figure 3: SEM of a cross-cut by focussed ion beam (FIB) through a metalization structure (aluminum, 800nm) covered by a glass passivation (3µm).

978-1-4244-4722-0/09 $25.00
© 2009 IMAPS-ITALY

of some 100 Volts, is of high importance. Consequently critical factors for device performance like e.g. the surface dark current can be influenced only within tight limits of typically several nA.

Additionally a layer thickness needs to be engineered, which guarantees a leak-proof sealing of the surface topography of the device. Lithoglas borosilicate glass layers of 500nm already show their excellent protective properties. However, to guarantee sufficient step coverage on surface topographies usually higher layer thicknesses become necessary. Figure 3 shows a cross-cut through a 800 nm thick aluminum metalization with 3μm glass layer as sealing. As tapered slopes as shown are not critical, 90° angles or undercuts may cause unwanted cavities and require adapted processing.
The devices were put through extensive reliability testing and showed no signs of impairment after temperature-humidity-storage of 8000h THS 85°C/85%.

Lithoglas as Interdielectric and Final Passivation in Flip-Chip-Packaging

As reported earlier microstructured glass has been used as a thin dielectric layer in integrated capacitors and micro strip line resonators as well as passivation on thin-film resistors improving the long term performance of these devices significantly [10, 11].

In combination with the promising results in passivating chip surfaces for avalanche photo diodes the feasibility of the microstructured glass for the use as an interdielectric and passivation layer was studied. The target was to process a standard redistribution design on wafer-level using the Lithoglas layer as a replacement for the dielectric as well as the final passivation. The schematic of the design is given in figure 4.

The design was made anonymous by producing costumized test wafers with a simulated top metal layout including contact pads (1μm aluminum covered by 700nm silicon nitride). The process flow for the formation of the flip-chip redistribution only needed minor modifications in the lithography with regard to standard polymer-based packages and the additional glass deposition processes were inserted.

As a first step in forming the flip-chip redistribution a glass layer with a thickness of 2 μm is deposited as the initial dielectric using the process flow described earlier. The contact pads of the test wafers remain free of glass.

The deposited glass is structured on the contact pads of the test wafer using the discussed lift-off-technique. The wafers show good adhesion of the glass to both silicon nitride and aluminum as relevant surfaces. The sidewalls of the glass openings are sloped with an angle of approx. 55° due to the lift-off-processing employed. As the glass deposition is a conformal process the topography of underlying structures is preserved.

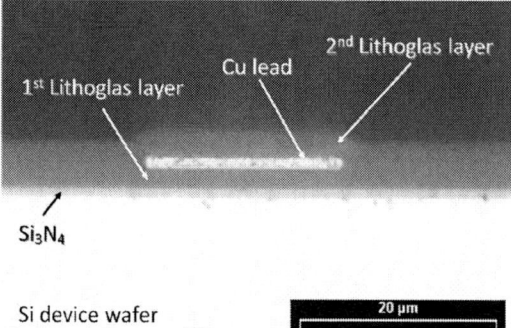

Figure 5: Cross-section through an embedded copper redistribution lead. The lead runs on top of the 2 μm thick 1st Lithoglas layer and is covered by the 3 μm thick 2nd Lithoglas coating.

The redistribution layer is composed of a 1.2μm thick electroplated copper on a Ti:W/Cu seed layer.

The second Lithoglas layer has a target thickness of 3 μm and is opened at the solder pad area to expose the copper lands formed by the Cu leads. The same process as for the first glass layer is used and the copper leads are conformally covered. This second Lithoglas layer acts as a final passivation of the chip surface. Figure 5 shows a cross-cut through the copper redistribution layer on top of the silicon device wafer.

To finalize the redistribution a nickel underbump-metalization (UBM) is formed by electroplating over the exposed copper lands at the locations where the solder balls are to be placed. The UBM

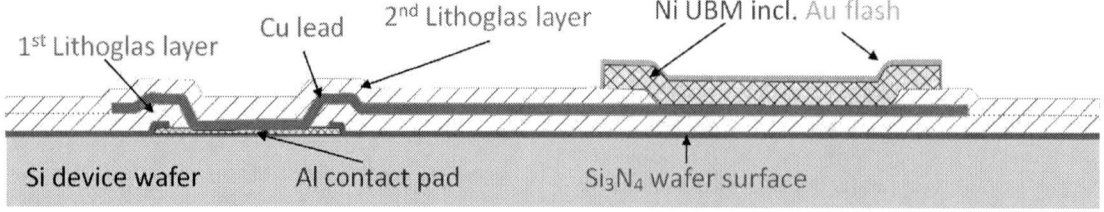

Figure 4: Schematic of a redistribution from aluminum contact pad via copper leads to Ni-UBM insulated and passivated by two borosilicate glass layers deposited with the Lithoglas process.

978-1-4244-4722-0/09 $25.00
© 2009 IMAPS-ITALY

opening is larger than the passivation openings on the copper lands allowing the Nickel to seal the interface to the passivation. It also allows for a more aggressive routing. A 5 μm thick nickel is electroplated to form the UBM including a thin layer of electro-less plated Au, which improves the wetting of the solder in subsequent processes (see Figure 6 and 7).

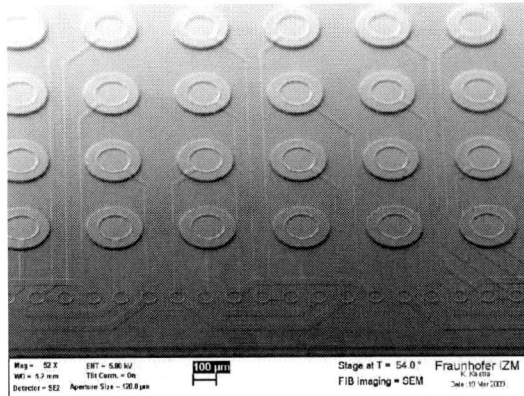

Figure 6: Top view on the complete redistribution including an array of UBMs for solder ball placement. The metalizations can be observed embossed in the glass.

Standard bumping technologies like stencil printing, preformed solder balls or plating can be used as board level interconnect.

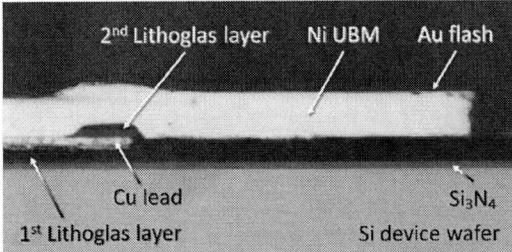

Figure 7: Partial cross-section of a Ni-UBM. The copper land is overlapped by the second glass layer. On top the Ni-UBM is formed in a larger diameter than the copper land.

Results

The Lithoglas process allowed the processing of standard device test wafers without any change in the design rules. The two borosilicate glass layers were deposited and structured with a high yield and within tight specifications esp. with regard to the critical dimensions of the pad openings. The achievable resolution is about 2 μm in the layers by using mask aligner lithography. It showed to be better compared to most direct photo-sensitive polymers. The Lithoglas layer does not exhibit residues or

debris on the contact pads nor swelling or deformation at the edges of structures. Due to the stress controlled deposition and the match of the CTE of the glass to silicon the device wafer is not warped during and after processing. The maximum process temperature throughout the Lithoglas redistribution process stays below 100 °C.

The cycle time of a single microstructured passivation layer can be less than an hour and does not require extensive curing steps in controlled atmosphere. The Lithoglas process is in production for 4", 6" and 8" wafers. The capability on other formats has been demonstrated.

Reliability testing according to automotive specifications of the redistribution technology is ongoing.

Summary

The feasibility of using thin borosilicate glass layers deposited by plasma-assisted e-beam evaporation as a CMOS-compatible technology has been demonstrated in two differing packaging applications. As a result a novel wafer-level packaging technology emerged using Lithoglas layers as interlayer dielectric and final passivation for redistribution. The glass films acted as a drop-in within a standard redistribution process flow – replacing polymers.

The microstructured glass layer has proven its resistance to temperature and humidity testing. While the mechanical, chemical, electrical and optical performance of most polymers will decrease when exposed to temperature, intensive light or humidity for a longer time the properties of the borosilicate glass have proven to be much more stable.

The low temperature load, high compatibility with other materials as well as the simplicity of the Lithoglas process has proven it to be a serious alternative to established passivation materials. This holds especially true for applications with elevated reliability requirements or need for additional functional properties such as optical performance or bio compatibility. It is a cost effective, high yielding technology.

Acknowledgments

MSG Lithoglas AG wishes to thank Prof. Herbert Reichl and his team at the Fraunhofer Institute for Reliability and Microintegration in Berlin for their ongoing support.

References

[1] James J. Licari, "Coating Materials for Electronic Applications", Noyes Publications New York, pp. 1-63, 2003.

978-1-4244-4722-0/09 $25.00
© 2009 IMAPS-ITALY

[2] K. Zoschke, J. Wolf, M. Töpper, O. Ehrmann, Th. Fritzsch, K. Scherpinski, F.-J. Schmückle, H. Reichl, ''Thin Film Integration of Passives – Single Components, Filters, Integrated Passive Devices'', Proceedings of 54th ECTC Conference, Las Vegas, June 2004, p. 294.

[3] Michael Töpper, "Wafer Level Chip Scale Packaging" in D. Lu and C.P. Wong, "Materials for Advanced Packaging", Springer, New York, pp. 547-600, 2009.

[4] M. Töpper, J. Simon, H. Reichl, ''Redistribution Technology for CSP using Photo-BCB'', Future Fab International, p. 363, 1996.

[5] J.H. Lau, S.W. Ricky Lee, "Chip Scale Package: Design, Materials, Process, Reliability, and Applications", McGraw-Hill, 1999

[6] M. Töpper, J. Auersperg, V. Glaw, K. Kaskoun, E. Prack, B. Keser, P. Coskina, D. Jäger, D. Petter, O. Ehrmann K. Samulewicz, C. Meinherz, C. Karduck, S. Fehlberg, H. Reichl "Fab Integrated Packaging (FIP) A New Concept for High Reliable Wafer-Level Chip Size Packaging", Proceedings of 50th ECTC Conference, Las Vegas, 2000, pp. 74-80, 2000.

[7] M. Töpper, M. Schaldach, S. Fehlberg, C. Karduck, C. Meinherz, K. Heinricht, V. Bader, L. Hoster, P. Coskina, A. Klöser, O. Ehrmann, H. Reichl „Chip Size Package - The Option of Choice for Miniaturized Medical Devices" Proceedings IMAPS Conference, San Diego, 1998

[8] D. Mund, J. Leib, "Novel Microstructuring Technology for Glass on Silicon and Glass-Substrates," Proceedings of 54th ECTC Conference, Las Vegas, June. 2004, pp. 939-942.

[9] Rao Tummala, "Fundamentals of Microsystems Packaging", McGraw-Hill, p. 586, 2001.

[10] K. Zoschke, C. Feige, J. Wolf, D. Mund, M. Töpper, "Evaluation of Micro Structured Glass Layers as Dielectric- and Passivation Material for Wafer Level Integrated Thin Film Capacitors and Resistors", Proceedings of 57th ECTC Conference, Reno, 2007, pp. 566-573.

[11] D. Mund, J. Leib, M. Töpper, "Novel Hermetic Wafer-Level-Packaging Technology Using Low-Temperature Passivation," Proceedings of 55th ECTC Conference, Orlando, June. 2005, pp. 562-565.

Modeling and Quantification of Conventional and Coax-TSVs for RF Applications

Ivan Ndip[1], Brian Curran[1], Stephan Guttowski[1], Herbert Reichl[1,2]

[1]Fraunhofer Institute for Reliability and Microintegration, IZM, Gustav-Meyer-Allee 25, 13355 Berlin, Germany

[2]Technische Universität Berlin, Straße des 17. Juni 135, 10623 Berlin, Germany

ivan.ndip@izm.fraunhofer.de

Abstract — **In this work we modeled and simulated Through Silicon Vias (TSV) in low, medium and high resistivity silicon (LRS, MRS and HRS) for frequencies up to 80 GHz. We then quantified the electromagnetic reliability (EMR) problems caused by conventional TSVs, in which silicon is used entirely as the medium for wave propagation. Our results revealed that using these conventional structures leads to high insertion loss, lack of impedance control, cross-talk and strong EMI. For example, for TSVs having a diameter 40 μm and depth of 200 μm, approximately 30 % of the power is lost through a conventional TSV in LRS at about 5 GHz if a SiO$_2$ thickness of 1 μm is considered. We then proposed three different configurations of TSVs, based on the concept of the coaxial transmission line, namely Coax-TSV (SF), Coax-TSV (MDF) and Coax-TSV (LDF) to overcome all the limitations of conventional TSVs. This enables LRS to be used for the development of low-cost silicon-based system modules.**

Index Terms —**Coax-TSV (SF), Coax-TSV (MDF), Coax-TSV (LDF), Conventional TSV, RF/high-speed.**

I. INTRODUCTION

Due to the multitude of advantages offered by TSVs over conventional chip-interconnection methods such as wire-bonds (e.g., by enabling the stacking of identical memory chips, presenting shorter interconnect paths, hence reduced interconnect delay), they have been considered as one of the key technologies needed for the development of future high-performance and miniaturized silicon system modules, based on 3D integration. Consequently, research projects are currently being developed at research institutions and industries worldwide to explore the advantages of TSVs.

In conventional TSV configurations, silicon serves as the medium for wave propagation. Due to the lossy nature of silicon, RF/high-speed signals propagating through these conventional TSVs may suffer severe electromagnetic reliability (EMR) problems such as cross-talk, electromagnetic interference, high insertion loss and delay, which may limit the performance of the entire system. So far, much of the published work regarding electrical design of TSVs has concentrated largely on using analytical approximations, equivalent circuit and 3D full-wave models of TSVs to extract their parasitic elements [1], study the propagation delay they cause [2], manage their losses by properly defining the return-

current paths [3] and also to perform comparative analysis of TSVs and other chip interconnection methods in ensuring low-impedance profile in chip packages [4]. Recent contributions on electrical modeling of TSVs have also focused on proposing novel TSV configurations, based on the coaxial transmission line concept [5], [6].

Despite all these outstanding contributions, there is still a need for thorough quantification of the EMR problems caused by TSVs in LRS, MRS and HRS. Such quantification would provide information on the amount of signal degradation caused by different configurations of TSVs in different silicon substrates at different frequency ranges. This information helps the designer to know the limit of application of conventional TSVs and the parameters that can be modified to enhance their performance, especially at microwave frequencies.

In this work, we modeled and simulated conventional TSVs in LRS, MRS and HRS for frequencies up to 80 GHz. We then quantified some of the EMR problems caused by conventional TSVs when used to transport power to components for WLAN and UWB applications. Furthermore, we propose two new coaxial TSV configurations. Finally, we quantified the power lost from the coaxial TSV configurations in LRS.

The rest of this paper is divided into two sections: In section two, EMR problems caused by conventional TSVs will be quantified, using the insertion loss as an example. This will be followed in section three by modeling and analysis of the coaxial TSV configurations, proposed to overcome the limitations of conventional TSVs.

II. QUANTIFICATION OF EMR PROBLEMS CAUSED BY CONVENTIONAL TSVs

The severity of EMR problems caused by conventional TSVs depends on a number of factors, amongst which includes the conductivity of silicon, dimensions of TSV, the thickness of SiO$_2$ layer as well as the frequency of operation. In this section, we'll examine the impact of these factors on the RF performance of conventional TSVs with the help of 3D full-wave electromagnetic field simulations.

978-1-4244-4722-0/09 $25.00
© 2009 IMAPS-ITALY

A. Impact of LRS, MRS and HRS

Although there is currently no industry standard which gives a distinct demarcation of the exact range of resistivities that can be termed LRS, MRS and HRS, there is however, a general consensus that the higher the resistivity, the more expensive it is. So, LRS is cheaper than HRS. In this work, 1• cm, 10 • cm and 500 • cm were used to represent LRS, MRS and HRS, respectively.

In order to study the impact of these silicon resistivities, we developed a 3D model for the TSV, whose cross-sectional view is shown in **Figure 1**.

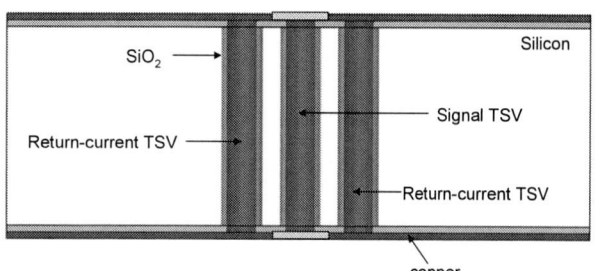

Figure 1: 3D model used to study the impact of LRS, MRS and HRS on the RF performance of conventional TSVs.

The dimensions of the TSV normally depend on the application. In this work, we considered application areas where silicon is used to replace traditional organic and ceramic substrates as interposer or carrier for the development of high-performance and ultra-miniaturized silicon-based system modules, as proposed in [7]. For this case, typical TSV dimensions include the following: diameter=40 μm, thickness=200 μm.

The 3D full-wave simulations were performed using Ansoft HFSS, a frequency-domain field solver that employs the finite element method to solve Maxwell's equations. In the simulation model, an air box surrounds the structure and PML-boundary conditions were used. The excitations were done using wave ports. The simulations were performed from 100 MHz to 80 GHz, so that most of the frequencies used for commercial communication and radar applications can be covered. For these simulations, a 1 μm thick SiO_2 (with $\varepsilon_r = 4$ and $\tan(\delta) = 0.0001$) was used to isolate the copper from silicon. In **Figure 2**, a comparison of the insertion loss of the conventional TSVs in LRS, MRS and HRS is shown. As expected, the insertion loss of the TSV in LRS is higher than in MRS and HRS. Now, to better understand the severity of the signal degradations, let's consider typical frequency ranges used by WLAN/WPAN and UWB systems. 2.4 GHz, 5 GHz and 60 GHz were chosen for WLAN/WPAN and 10 GHz for UWB, as examples. From the curves in **Figure 2**, the power lost through the TSV used for the above mentioned communication applications, is given in the **Table 1**.

It can be seen from this table that for dimensions of TSVs considered in this section, signal degradation and hence, power lost through TSVs in HRS for WLAN and UWB

applications is within acceptable limits. For both MRS and LRS, more than 10% of the power is lost even for the 2.4 GHz applications. Approximately 30% of power is lost through the TSV when LRS is used at about 5 GHz. Hence for applications above this frequency, it becomes very inefficient to use TSVs in LRS. Therefore, alternative techniques must be applied to minimize these losses.

Two possible ways to reduce these losses include; 1) the use of TSVs with smaller dimensions, 2) increase in the thickness of the SiO_2 layer on the TSV. In the section B, the amount of performance enhancement that can be achieved using these methods will be discussed.

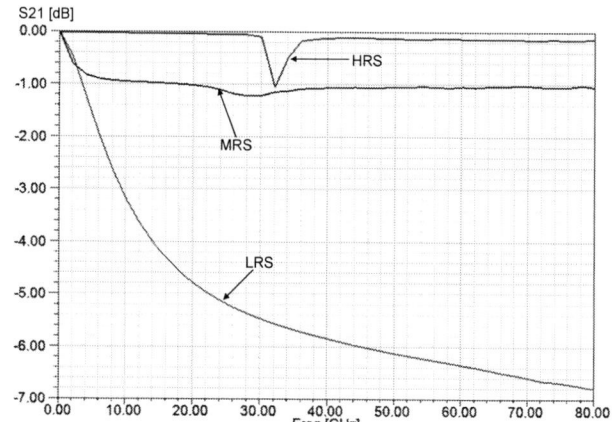

Figure 2: Impact of LRS, HRS and MRS on insertion loss of conventional TSVs (diameter=40 μm, length=200 μm, pitch between signal and return-current TSVs=40 μm).

Type of silicon	Approximate values of power lost through TSV for WLAN/WPAN applications [%]			Approximate values of power lost through TSV for UWB applications [%]
	2.4 GHz	5 GHz	60 GHz	10 GHz
LRS	13	29	76	52
MRS	13	17	24	19
HRS	2	2	4	2

Table 1: Comparison of power lost through TSV used for WLAN/WPAN and UWB applications.

B. Impact of Thickness of SiO_2 and Via Dimensions

During the fabrication process, SiO_2 and silicon nitride are deposited by plasma-enhanced chemical vapor deposition to line the via and the substrate. They serve two purposes: first, they act as an electrical insulation between Cu and silicon and secondly, as a barrier to prevent Cu migration to the substrate [5]. The thickness of silicon nitride is usually very small (in the nanometer range) and was not considered in the simulation models, because of the discretization errors that could occur as a result of the aspect ratio problem.

978-1-4244-4722-0/09 $25.00
© 2009 IMAPS-ITALY

To capture the performance enhancement achieved by increasing the thickness of SiO$_2$ and using smaller via dimensions, the model shown in **Figure 1** was used. LRS was chosen for these investigations because of its low-cost nature. The SiO$_2$ thickness was varied from 0.5 µm to 4 µm, while the TSV length was varied from 160 µm to 280 µm, with a constant diameter of 40 µm. Both parameters were investigated independently. The simulations were performed using the same boundary and excitation conditions as the one in section A. The results obtained are presented in **Figure 3** and **Figure 4**. For a better understanding, a comparison of the power lost is also given in **Table 2** and **Table 3**.

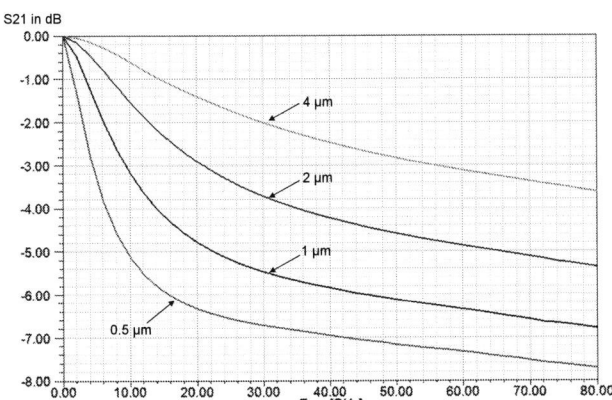

Figure 3: Impact of thickness of SiO$_2$ on insertion of TSV in LRS.

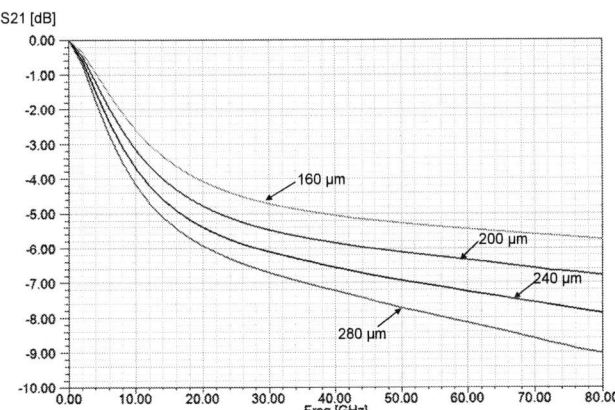

Figure 4: Impact of via length (with constant via diameter of 40 µm) on insertion loss of TSV in LRS.

From **Table 3**, it can be seen that there is not much improvement in the power lost when the TSV length is reduced, at least considering the dimensions used in this work). However, there is significant reduction in power loss when the thickness of SiO$_2$ is increased. For example, doubling the thickness from 1 µm to 2 µm leads to a reduction in the power lost by at least 50% for frequencies less than 10 GHz. However, despite this improvement, it must be noted that for frequencies greater than 5 GHz, more than 10% of the power is lost. Further more, in most of the technological process, a SiO$_2$ thickness of less than 1 µm is used (typically 0.5 µm). As seen in **Table 3**, about 30% of power is lost already at 2.4 GHz, when SiO$_2$ with a thickness of 0.5 µm is

used. In addition to the insertion loss problem discussed, the lossy nature of LRS also causes undesired electromagnetic coupling, leading to EMI problems. Furthermore, since the inductance of the return-current is not controlled, it becomes difficult to achieve impedance controlled signal paths using these conventional TSVs in LRS.

Thickness of SiO2 [µm]	Approximate values of power lost through TSV for WLAN/WPAN applications [%]			Approximate values of power lost through TSV for UWB applications [%]
	2.4 GHz	5 GHz	60 GHz	10 GHz
0.5	30	52	81	69
1	13	29	76	52
2	5	10	67	29
4	2	4	51	13

Table 2: Impact of SiO$_2$ thickness on power lost through TSV in LRS.

TSV length [µm]	Approximate values of power lost through TSV for WLAN/WPAN applications [%]			Approximate values of power lost through TSV for UWB applications [%]
	2.4 GHz	5 GHz	60 GHz	10GHz
160	11	24	71	45
200	13	29	76	52
240	17	33	80	57
280	20	39	84	61

Table 3: Impact of TSV length on power lost through TSV in LRS.

All of these EMR issues make it impractical to use these conventional TSVs for RF/high-speed applications. So, there is always the temptation to use HRS. However, using HRS makes the development of silicon-based systems extremely expensive. To overcome this problem, novel TSV configurations in LRS are needed.

II. COAX-TSVs FOR IMPROVED SIGNAL AND POWER INTEGRITY

In [5] and [6], a coaxial TSV configuration was proposed and analyzed for frequencies up to 40 GHz. In this section, we propose two new configurations and characterize them up to 80 GHz. The aim of these configurations is to overcome all the limitations of conventional TSVs in LRS and hence, tremendously reduce the cost of development of silicon-based system modules. Schematic top views of the three coaxial TSV configurations are shown in **Figure 5**.

We call the configuration on the top LHS in **Figure 5**, Coax-TSV (SF), where SF stands for silicon filled. This configuration can be applied to all the three classes of silicon substrates discussed in this paper. The aim of this structure is

978-1-4244-4722-0/09 $25.00
© 2009 IMAPS-ITALY

to control the inductance/capacitance of the TSV and hence, prevent any fluctuation in the characteristics impedance of the signal path or impedance profile of the power distribution network. Consequently, signal integrity and power integrity are ensured. Furthermore, since the outer ring is much thicker than the skin depth at RF/microwave frequencies, there is no leakage of power. Hence, cross talk and EMI are also eliminated. However, because silicon is the medium of propagation, there is still power lost. Hence, the insertion loss is still high.

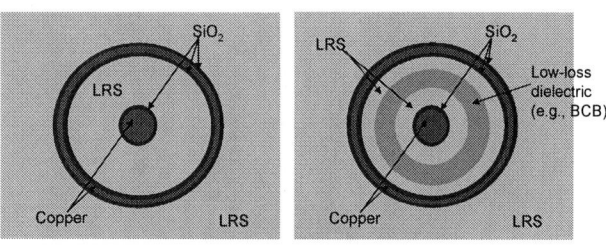

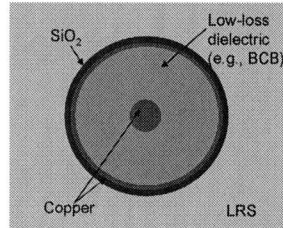

Figure 5: Three Coax-TSV configurations used to enhance the RF performance of TSVs in LRS. Coax-TSV (SF) – Top LHS; Coax-TSV (MDF) – Top RHS; Coax-TSV (LDF) – bottom.

By replacing part of the silicon between the inner and outer conductor with a low-loss dielectric, such as BCB, the insertion loss problem can be greatly minimized. We call the coaxial TSV configuration used for this purpose, Coax-TSV (MDF), where MDF stands for mixed-dielectric filled, since silicon and BCB are used as the medium for wave propagation. This configuration is shown on the top RHS in **Figure 5**. However, as can be seen in **Figure 6**, the improvement achieved depends on the ratio of silicon to BCB used. For example, consider a Coax-TSV (MDF) configuration with the following dimensions: an inner conductor radius of 2.5 μm, outer conductor radius of 22 μm, which consists of the following silicon to BCB distribution: 7.5 μm (silicon) – 5 μm (BCB) – 7.5 μm (silicon), and a SiO_2 thickness of 2 μm (i.e., 1 μm between the inner conductor and silicon and 1 μm between the outer conductor and silicon). For this configuration, only about 14 % of the power is lost at 60 GHz, as can be seen in **Figure 6**. This power lost reduces to 5%, when the thickness of BCB is doubled and that of silicon is reduced by 1/3. By further increasing the thickness of BCB by 50% and reducing the SiO_2 by same amount reduces the power lost to approximately 2% for frequencies up to 60 GHz. For all three cases of Coax-TSV (MDF) studied, the power lost for frequencies less than 10 GHz is about 2 %.

The final configuration of coaxial TSV shown on the bottom in Figure 5 is called Coax-TSV (LDF), where LDF stands for low-loss dielectric filled. Using BCB as the dielectric in this case leads to a negligible loss of power for frequencies up to 60 GHz as seen in **Figure 6**.

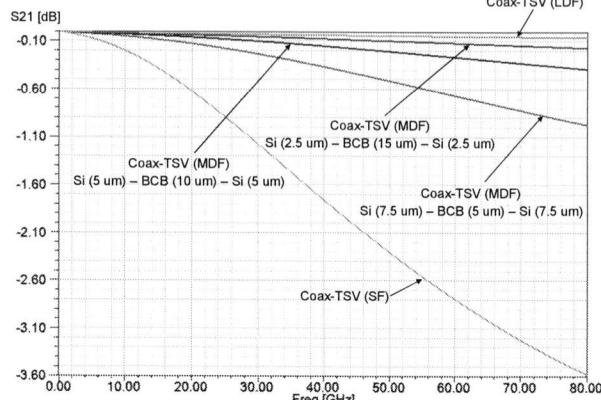

Figure 6: S12 for all three Coax-TSV configurations.

III. Conclusion

In this work we modeled and quantified TSVs in LRS, MRS and HRS for frequencies up to 80 GHz. Due to the very high conductivity of LRS, it causes more signal degradation than MRS and HRS. However, since it is far cheaper than MRS and HRS, we proposed TSV configurations based on the coaxial transmission line concept to overcome the limitations of conventional TSVs in LRS. This enables low-cost design of silicon-based system modules.

REFERENCES

[1] Jang D. M, et al. "Development and evaluation of 3D SiP with vertically interconnected through silicon vias," *2007 ECTC Conference, page 847-852.*

[2] Khalil E et al., "Analytical model for the propagation delay of through silicon vias", *9th international symposium on quality electronic design, pp. 553-556.*

[3] Curran B., Ndip I., et al, "Managing losses in through silicon vias with different return current path configurations," *2008 EPTC conference, Singapore.*

[4] Pak J. S. et al., "Wideband low power distribution network impedance of high chip density package using 3D stacked through silicon vias," *2008 APS on EMC and 19th International Symposium on EMC, May 19-22 2008, Singapore, pp. 351-354.*

[5] Ho S. W. et al., "Development of coaxial shield via in silicon carrier for high frequency application," *2006 EPTC Conference, Singapore, pp. 825-830.*

[6] Ho S. W. et al., "High RF performance TSV silicon carrier for high frequency application," *2008 ECTC Conference, Lake Buena Vista, FL, pp. 1946-1952.*

[7] Chatterjee R. et al.,, "3D technology and beyond: 3D All Silicon System Module", *Advanced Packaging Magazine, 2008.*

Stacking of Full Rebuilt Wafers For SiP and Abandoned Sensors/Applications

By Dr Christian Val, Dr Pascal Couderc and Pierre Lartigues

3D PLUS
641, Rue Hélène Boucher
78532 BUC – France
Tel. + 33 1 30 83 26 50 –
email : cval@3d-Plus.com

1. INTRODUCTION

The fantastic development of the components interconnection allows to imagine «Systems in Package» of around a few mm3, which opens new perspectives for industrial applications.

Since the use of hybrid modules, then Multichip Modules « MCM » which were identical to the previous ones but named differently, the coming of the interconnection in 3-D allowed to divide both the volume and weight of a module by a factor comprised between 25 and more than 100.

We can distinguish several technologies among the technologies used for the 3-D interconnection, (table figure 1)

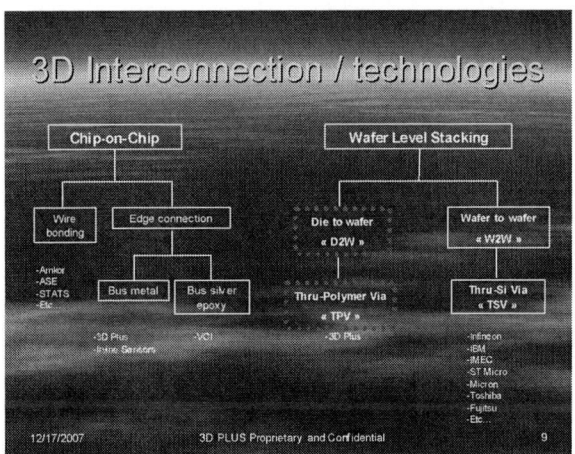

Figure 1

. The so-called Chip on Chip technology, with the chips interconnection carried out by wire bonding
. The Wafer Level Packaging technology with an interconnection carried out without wire bonding ; These lead to modules even denser than the above ones.

Among this last category, two families can be distinguished :

- The technology called Wafer to Wafer studied and developed by almost all the semiconductors manufacturers and the large laboratories worldwide, and which is based on the use of Thru Silicon Via « TS » ; This technology presents the following disadvantages :
. Use of non standard wafer with the need for smallest possible Thru Silicon Vias. For instance, vias of 2µm diameter lead to a thickness of 20µm or less for the 8 or 10inches wafer
. Copper's TSV inside the silicon leads to a composite material which modifies its mechanical properties
. Almost impossible to make "System in Package" since die of different size
. Impossible to have 100% good wafer (very low global yield)

978-1-4244-4722-0/09 $25.00
© 2009 IMAPS-ITALY

- The technology named Stacking of Known Good Rebuilt Wafer which is based on the Edge Connection and the Thru Polymer Via « TPV ». This technology avoids these disadvantages and adapts perfectly to the Systems In Package as well as to small and large volumes.

We have registered a trade mark : Wirefree Die on Die « WDoD » :
- Use of standard wafers without TSV
- Each layer testing prior to stacking, only "Known Good Rebuilt Wafer" are stacked
- Parallel processing from A to Z with the standard semi-conductor equipment
- Very small form factor:
- Thickness of each layer: 100µm (10 levels perm)
- Size: 100 µm around the larger die

ULTRA LOW PROFILE 3-D TECHNOLOGY

This technology resulted in the registration of more than 23 patents worldwide. The flowchart is represented on Figure 2.

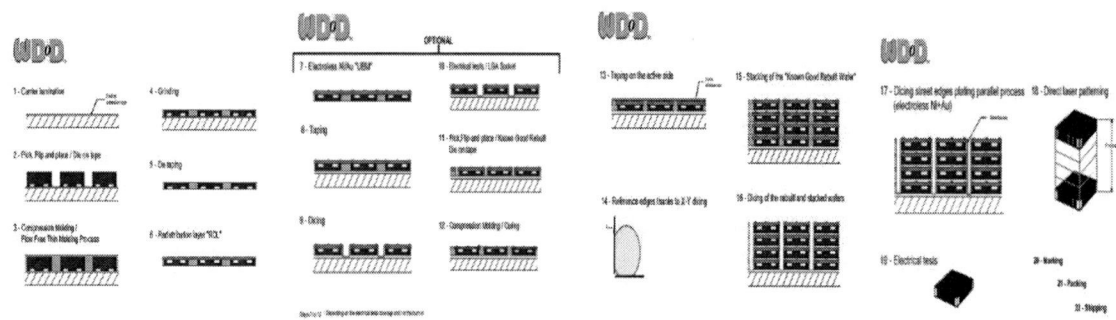

Figure 2

It can be observed that steps from 7 to 12 are optional, they are used only when the Burn-In is necessary, this is the case with some memories such as DDR or some analogic components or ASIC.

Figure 3 shows a « Known Good Rebuilt Wafer » of 6" diameter

Figure 3

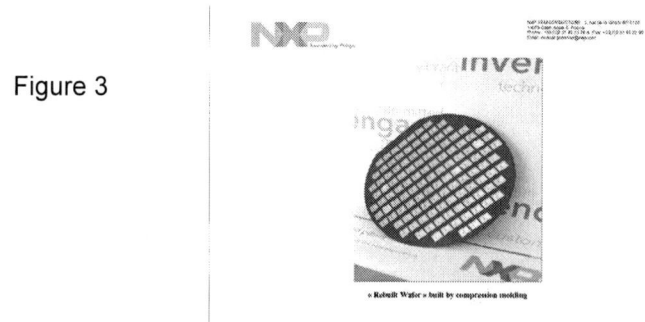

The status of this « WDoD » technology is the following :
- Proof of concept completed (2002 to 2005) with an European funding of 20M€ with several companies: CEA/LETI, Thales, ST Microelectronics, Axalto/Gemalto and 3D Plus
- Process development and optimization, based on Daisy chains die, with NXP/Philips (4Q 2008)
- Functional prototypes/Qualifications based on SRAM die (2Q 2009)
- Ramp up for the manufacturing (4Q 2009).

The main applications concern memories, SiP and Abandoned Sensors.
The most important production so far is carried out as per the Package on Package (PoP) technology.
A comparison is presented hereafter between the PoP, the Wafer to Wafer (W2W) and the WDoD.

Comparison of PoP / W2W and WDoD Technologies

The PoP usually contains two subsystems
(Figure 4)
. The top PoP usually integrates stacked memory devices in a fine-pitch ball grid array (FBGA); we call it : subsystem 1,
. The bottom PoP typically contains a logic device of some sort; we call it: subsystem 2 .

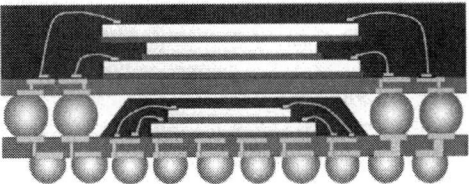

Figure 4 : Assembled Package-on-Package

Among the PoP main advantages presented hereafter, test and burn-in are considered essential.

Wirefree Die on Die (WDoD)

The 3D Plus wafer level package (WDoD) combines the main advantages of PoP technology with those of a smaller full wafer level package.

As can be seen on Figure 5, the two subsystems are the same
Than in the PoP :

. Top package with 2 memory devices such as Flash and DDR.
. Bottom package with digital logic lus passive devices, such as PICS from NXP. The WDoD technology enables the same electrical test and burn-in strategy as the PoP however, the WDoD package is:
. Smaller (die size of the largest device plus 100µm around 2 or 4 sides, 100µm between each level)
. Wirefree
. Solder ball free for internal subsystems connections (the handset makers must surface- mount the 2 subsystems on the PC board simultaneously, the same way as PoP).

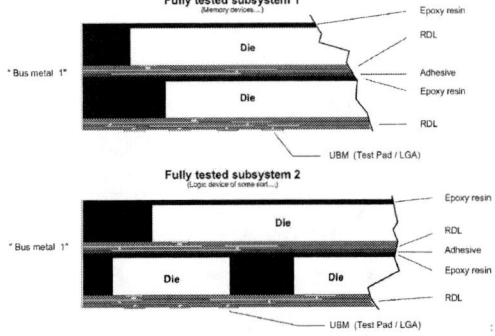

Figure 5: PoP and WDoD package

Figure 6 shows the 2 subsystems:

. Subsystem 1 contains two or more memories such as Flash, SDRAM, DDR, SRAM etc. The thickness of each level is 100μm, therefore if the subsystem contains 3 devices, the total thickness, including the external layer with the UBM pads, will be 350μm. The UBM pads will be used to make the electrical test and burn-in contact via a LGA socket.

. Subsystem 2 contains single or multiple devices to integrate high density digital logic with, or without, passive components (R,L,C). Here again, the external UBM pads enable testing and burn-in via a LGA socket.

The process to build subsystems 1 and 2 is illustrated in Figure 2.

The first 6 steps are similar to those utilized for new wirefree manufacturing processes announced by Infineon (in 8 inches diameter) and Freescale (in 12 inches diameter), etc.

Step 4 can be avoided if thin wafers are available (thickness between 50 and 70μm). In this case, the epoxy resin stays over the back of the die and no grinding is needed. If thin wafers are not available, step 4 is required. Once the die have been thinned down to 80μm, a CMP process is used to release the stress inside the silicon.

Once the redistribution layer has been applied, there are two process options.

Option 1 presents the process for the test and burn-in of one device. When the dicing step has been completed, we have a micro-package which can be placed in a standard LGA socket.
After test, we will have a Known Good Rebuilt Die (KGRD). This (KGRD) is now treated like a die and the same process is used:
. Pick, flip and place the (KGRD) on to a lamination film (step 11); note that the accuracy of this pick and place is not critical because there is no RDL process requirement.
. Once the active side has been mounted on the lamination film, the devices are molded and steps 13 up to 18 can be carried out.

Option 2 enables the use of 2 or more stacked rebuilt wafers in order to form a Subsystem Stacked Rebuilt Wafer (SSRW).The manufacturing flow for the SSRW is exactly the same as the standard WDoD process from steps 13 to18.

As with option 1, step 10 enables subsystem 1 to be tested and burned-in in a micro-package format, courtesy of the UBM pads which have a pitch used in a standard socket.
Subsystem 2 is manufactured exactly the same way as subsystem 1 utilizing the full wafer level WDoD process.

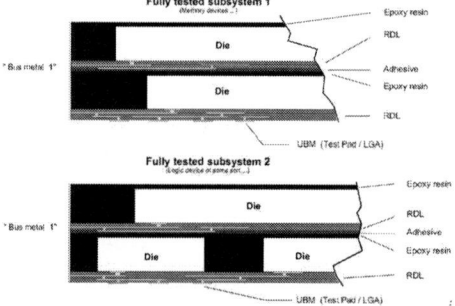

Figure 6 : Subsystem 1 and subsystem 2 after testing and burned-in
and before stacking

Figure 7 illustrates the two stacked subsystems. Again, the standard WDoD process is used, specifically steps 13 to 18. The two stacked rebuilt wafers subsystems are stacked (step 15) and diced (step 16). This

second dicing (cutting line N° 2) destroys "Bus metal 1" and again exposes each RDL copper trace. The layout of
the external RDL used to test subsystem 2 remains compatible with the test and the surface mounting of this PoP.

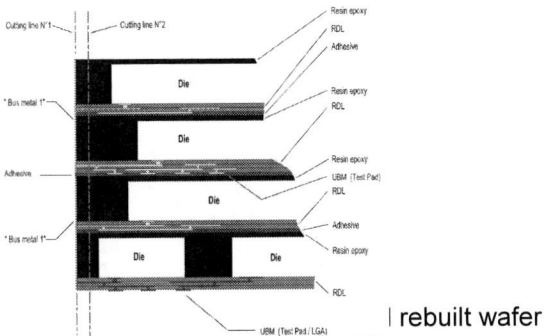

| rebuilt wafer

Figure 8 shows the finished PoP, with the second "Bus metal". The package can now be tested before shipping. The total thickness for this LGA example is 450μm. If a BGA format is required , the diameter of the solder balls has to be added. Compared to PoP, *the epoxy resin around the die is well-balanced thereby minimizing warpage.*

The surface mounting process is comparable to that of the micro package or LGA package. It can be seen that only one package, instead of two, needs to be surface mounted. *The warpage of the bottom package of the PoP is a critical factor in determining the yield for successful package stacking. "If the warpage is too large, open solder joints may appear between bottom package and motherboard, as well as between bottom package and top package, making the stacking fail," Wei Lin from Amkor said.*

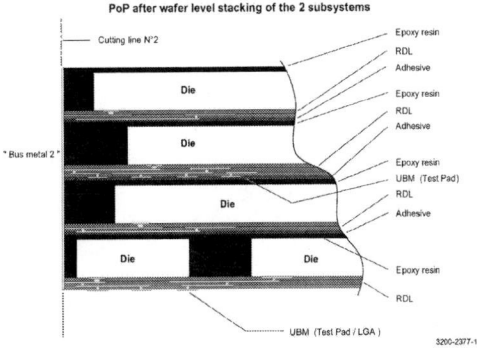

Figure 8: WDoD/PoP after wafer level stacking of the two subsystems

Comparison of PoP, Wafer to Wafer, and WDoD

In the table shown in Figure 9 the three major 3-D technologies are compared. It can be seen that:
. Wafer to Wafer(W2W) is not compatible with the System in Package,
. PoP and WDoD are on the same position for "***components*** sourcing" and "test and burn-in"

. Cost is now lower with PoP, as the volumes scale for WDoD, the efficiencies of wafer level process (vs. non-parallel processing) will enable it to be equally or more cost effective .

. W2W offers the smallest size, with WDoD being only slightly larger with (Die size plus 100μm). The PoP is significantly larger due to the need to fan out the ball pattern on around the mold cap of the lower package in order to enable the package stack.

978-1-4244-4722-0/09 $25.00
© 2009 IMAPS-ITALY

. The total height of WDoD and W2W is, and will remain, significantly lower than PoP due to the solder ball interconnect methodology utilized for PoP.

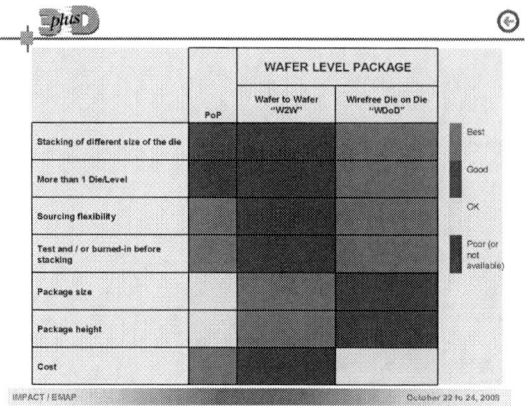

Figure 9 : Comparison of PoP, W2W and WDoD
(1 cross : bad – 3 crosses : good)

SiP and Abandoned Sensors applications

For the small volumes 3D Plus carries out the design and the manufacturing. For larger volumes of the industrial products a Joint Development Agreement was signed in 2007 with NXP and a second source with larger diameter of Rebuilt Wafer will be activated. Several applications are presented:

SiP with memories

Memories and memory-based SiP naturally represent a considerable market. On Figure10 we present a stacking of 8 SRAM memories on a 1mm thickness.

Figure 10

The Flash memories represent an important part of the applications for the nomad products. For instance the smart cards could have the need of a dense SiP to put it inside the SIM card. We are evaluate the possibility to integrate 4 Flash + 1 µcontroller + 1 oscillator + Capacitors in 5 levels within 550µm.

Medical applications

- **Microstimulators** in a 3mm diameter cylinder with 5 die on a 0.6mm thickness and a total volume of 3 mm3 Figure 11).

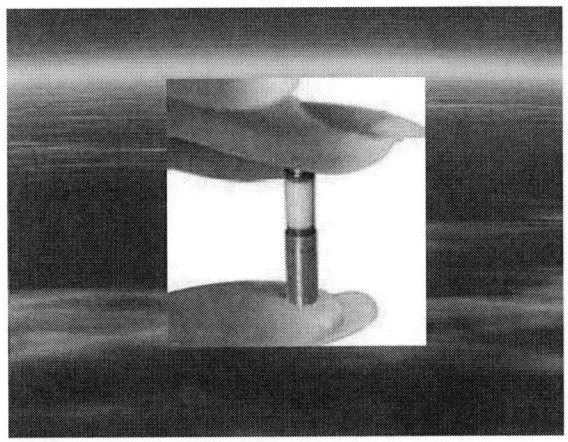

Figure11

-**Endoscopy** / Camera Head (Fig.11), Three CCD sensors can be used: 1/6 in, 1/10 in and 1/15 in the 2,6 x 2,6 mm (Figure 12).

Figure12

-**Module for pacemaker**. A very dense module is developing with our WDoD technology in order to decrease the size and the volume of the electronic function by 10 times.

-**X ray camera for Medical Scanner** is developing with Philips Medical

Security/Domain

-Multisensor module to locate the persons at risk (Fireman etc …).
- This module (10 x 5 x 6 mm) integrated: -1 triaxial accelerometer with associated electronics (6 ICs , 9 passive components).

-3 channels to provide a 3 axes magnetometer; each channel includes:

-1 magnetometer in bare die (1,5 x 1,5 mm)

-1 ASIC in bare die (2,2 x 2 mm)

-6 passive components.

Industrial Domain

-Gyroscope inside a watch (Figure 13)

        o   X-Y Axis Magnetic Sensor with 2 ASICs for digital processing and Passive components in a 3 x 3 x 3 mm Module.

Figure13: Magnetic SensorModule

-Gyroscope for avionics application

-A 3 axes module (10 x 6 x 5 mm) with accelerometer and magnetometer sensors (6) and associated electronics with 3 ASICs, 1 FPGA and passive components is developing.

-Abandoned Sensors for Avionics
-

-With an European program "e-Cubes", 3D Plus is making prototypes of abandoned sensor modules (7x7x7 mm) with 30 dice and sensors to measure the mechanical aging of the structure of aircrafts 14

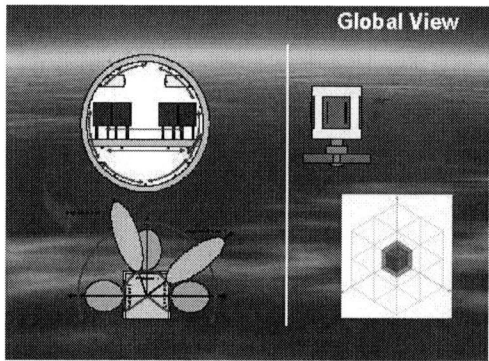

Figure 14

This module integrates :
-Sensors : Accelerometer, Temperature, Humidity and Pressure)
- signal treatment + memory
-Energy
-Antennas
-Communication This Module (Figure 15) is a full "Abandoned Sensors"

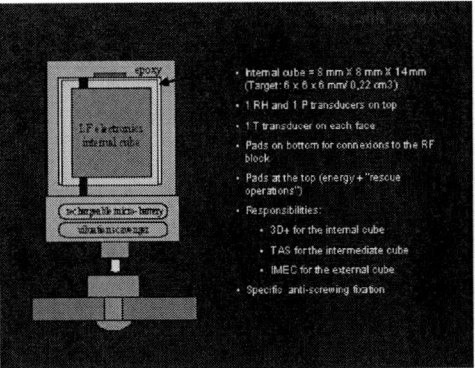

Figure 15

. Extremely dense micro-systems for medical applications are contemplated thanks to a future European project (7e PCRD) – see figure 16.

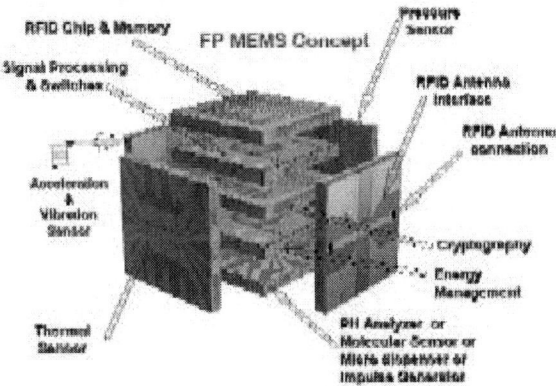

Figure 16 : Future comples : « Abandoned Senors »

CONCLUSION

Miniaturisation for commercial applications demands very high interconnection densities and low costs. Wiser for former experiences, multi-chip modules, wafer scale integration, 3-D modules for Space, Defense and professional (3D Plus) applications, we learned that the yield constituted an important part of the production costs. The WDoD process allows to stack Known Good Rebuilt Wafer (KGRW) only. Several applications in the Medical and industrial domains have been presented. This extremely important densification of 10 to 20 levels per mm with only 100 µm around the largest die allows to launch extremely ambitious applications with memories, Systems in Package and Abandoned Sensors.

978-1-4244-4722-0/09 $25.00
© 2009 IMAPS-ITALY

Creep Mechanism Fractography Analysis on SnPb Eutectic Solder Joint Failure

Chulmin Oh[*], Changwoon Han, Nochang Park, Byungsuk Song, and Wonsik Hong

Physics-of-Failure Research Center, Korea Electronic Technology Institute, Seongnam, 463-816, Republic of Korea
*82-31-789-7288, 82-31-789-7059 and cmoh@keti.re.kr

Abstract

Microstructural fracture mode observed in creep can be divided into intergranular and transgranular fracture. Depending on temperature and stress condition, creep fracture mode is decided. To design an accelerated life test, it should be confirmed that the failure mode in the accelerated test is identical to the mode in real field condition. Selecting optimal conditions of temperature and stress in accelerated creep rupture test requires extensive fractography analysis. In this study, SnPb eutectic solder joints for holding an anchor from heat sink system are subjected to creep rupture tests. After the test, failed solder joints are investigated and analyzed to identify creep fracture mode. Fracture microstructures of solder joints are analyzed using SEM and FIB. It is observed that transgranular fractures are predominant in the condition of low temperature and high stress and intergranular fractures are predominant in the condition of high temperature and low stress. Analysis results confirmed creep deformation mechanism map made by X.Q.Shi et al. and suggested optimal conditions of temperature and stress for accelerated creep rupture test with SnPb eutectic solder joints.

Key words: creep fracture mode, fractography analysis, accelerated life test, SnPb solder joint

Introduction

SnPb eutectic solder has been applied in electronics for a long time because the solder has several following properties;(i) good wettability with an aid of mildly active fluxes, (ii) no brittle intermetallic compound formation in solder itself, (iii) low melting point to permit the design of components that can endure the high temperature associated with the soldering process, and (iv) few problem occurrence of tin oxide film problems compared with the oxide films of other solder alloys[1]. SnPb eutectic solder is still used in specified area such as military, aviation and health due to an exception of the environmental regulations although Pb-free solder has been applied in electronic industry several years ago corresponding with environmental regulations.

SnPb eutectic solder shows a slow plastic strain(creep) under low permanent stresses at ambient temperature because the ambient operating temperature is above 0.5 times melting point of eutectic solder[1]. At high elevated temperature, the mobility of atom and dislocation increases and the concentration of vacancy increases with temperature. Creep deformation comes into play at evaluated temperature[2]. The deformation of SnPb eutectic solder at operating temperature has followed the creep deformation including the matrix diffusion, dislocation glide, dislocation climb and grain boundary sliding. The deformation of SnPb eutectic solder is increased by these thermal activitaed

processes and causes various failures of solder joint responsible for interconnecting the electronic components and making the robust structural reliability of the electronic package[3]. Therefore, it is necessary to understand the creep behavior of SnPb eutectic solder with stress and temperature in order to predict the lifetime of SnPb solder joint in electronic assembly.

A creep deformation map for SnPb solder alloy was established by X.Q.shi *et al.* and divided into two regions of dislocation-controlled creep and diffusion-controlled creep[4]. Diffusion–controlled creep is favored at high temperatures and low stresses, while dislocation-controlled creep is more dominant at low temperature and high stresses. In intermediate temperature regime(in between 0.4 and 0.6 Tm) the creep deformation mechanism can be a mix of those in the low temperature and high temperature regime[5]. The creep rate of this regime is plotted as a power(n) function of stress and an Arrhenius-type expression with characteristic activation energy(Q). The values of n and Q are variable with respect to temperature and stress. A break in the isothermal curve is shown between low stress and high stress[5]. The break in the curve occurs at stress at which the fracture mode changes from intergranular to transgranular facture[5]. The facture mechanism map is useful for indentifying the fracture mode at any conditions of temperature and stress.

Solder joints in electronics play role on not only making an electrical path by connecting the

978-1-4244-4722-0/09 $25.00
© 2009 IMAPS-ITALY

electronic components and board but also holding the anchor from heat sink pin of microprocessor package. The field temperature is above 0.5Tm and the applied stress in solder joints is constant so that the creep damage of solder joint would be accumulated during the usage and brings about the failure of solder joints. It is needed to establish an acceleration model based on creep deformation mechanism for SnPb eutectic solder joint in order to predict the lifetime for SnPb eutectic solder joint.

In our study, accelerated test was conducted on several conditions of temperature and stress corresponding to creep mechanism to acquire proper acceleration creep model for SnPb eutectic solder joint. The failure mode of failed solder joint is analyzed to indentify the failure mode in accelerated conditions. It should be confirmed that the failure mode in the accelerated test is identical to the mode in real field condition. We discussed the results for fracture analysis and provided a guideline to the accelerated test for creep rupture test.

Experimental procedures

The test board for accelerated creep rupture test is shown Fig. 1. Specimens were soldered with SnPb eutectic solder on specific coupon board with which was attached steel bar for preventing the board warpage during creep rupture test and involved Cu layer patterns for acquiring the failure time.

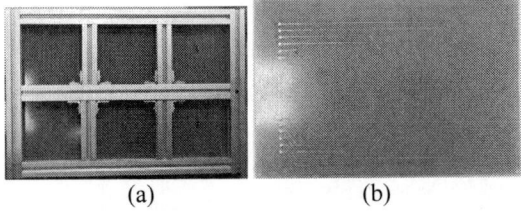

(a)   (b)

Figure 1 Test board for accelerated creep rupture test: (a) front side of board, and (b) back side of board.

Considering the operation environment for solder joint of an anchor to hold heat sink pin, it was found that the operation temperature is about ambient temperature (above 0.5Tm) and the applied load to anchor is around 1 kg. Therefore, the experimental conditions for creep rupture accelerated test are planned in Table 1. The specimen size is 10 units per each condition.

Table 1 Experimental conditions for accelerated creep rupture test

Temp.(°C) Load(kg)	35	55	75
4.0	10 ea	10 ea	10 ea
6.0g	10 ea	10 ea	10 ea
8.0	10 ea	10 ea	10 ea

As shown Fig. 2, the weight assigned to each condition was hung on every specimen and inserted into chambers capable of keeping constant temperature. It was able to acquire the failure time by measuring the resistance of solder joint continuously. When the creep fracture was happened during accelerated creep rupture test, microstructure fractography of solder joint was analyzed with scanning electron microscope(SEM) and focused ion beam(FIB) in order to identify the failure mode of fracture specimen.

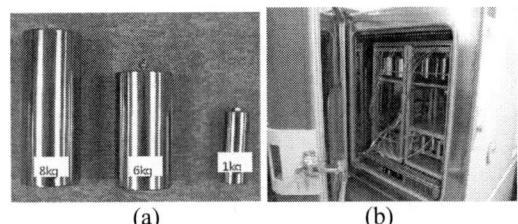

(a)   (b)

Figure 2 Experimental configurations with (a) weights used for creep rupture test, (b) test fixture with weights and specimens.

Results and discussions

Fig. 3 shows the SEM micrographs of microstructure of creep failure on solder joint of anchor at the condition of low stress(4kg) and high temperature(75°C). The microstructure of SnPb eutectic solder was composed of Pb-rich α phase and Sn-rich β phase. As indicated by arrows in Fig. 3(b), micro cracks were observed between Pb-rich α phase and Sn-rich β phase or along Sn-rich β phases with different grain orientation. Voids and cavities liked on grain boundary normal to principal stress were seen in Fig. 3(c),(d). It was reported that the voids and cavities could be formed at the early stage of creep deformation especially perpendicular to tensile stress and grow and coalesce to become micro cracks[3,5].

Although test temperature was decreased to minimum value under constant low stress, the failure mode was very similar with Fig. 3. Grain boundary sliding was prevalent and micro crack grows on the grain boundary under low applied stress. In this condition that applied load was relative low, it was revealed that intergranular fracture is predominant as creep fracture mode.

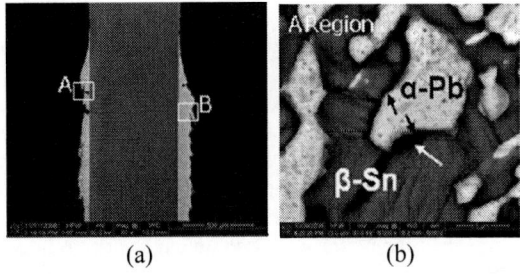

(a)   (b)

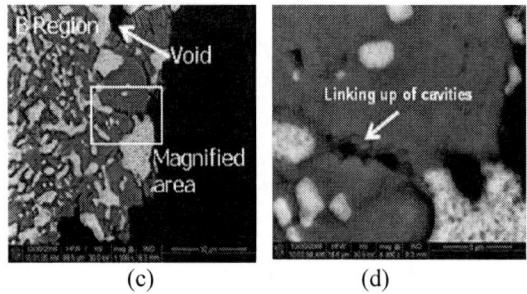

(c)   (d)

Figure 3 Cross-sectional SEM micrographs of (a) creep fracture on solder joint of anchor at the condition of low stress and high temperature, (b) magnified image of A region in (a), (c) magnified image of B region in (a) and (d) magnified image of interested area of (b)

Under the condition of medium load(6kg), microstructure of failed solder joint on board was shown in Fig. 4. As an applied load was increased, it was observed that a mircocrack grew along grain boundary and propagated inside grain at final stage. It was also observed that microcrack was penetrating into Sn-rich β phase. Voids were observed near crack tip as supplier on nucleation and growth of microcrack during creep process.

Considering the failure mode with increased applied load of creep test, experiment condition shown in Fig. 4 seemed to be a transition region from intergranular fracture to transgranular fracture.

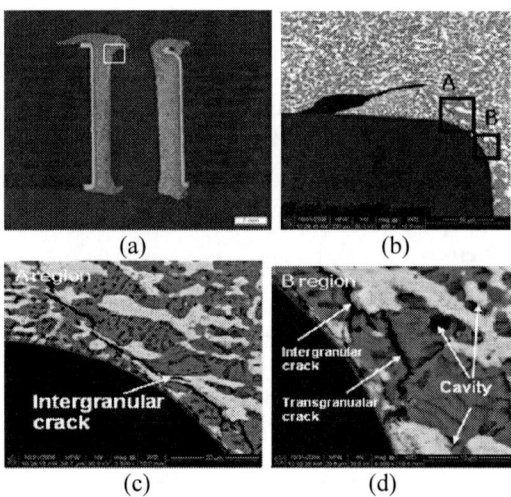

(a)   (b)

(c)   (d)

Figure 4 Cross-sectional optical micrograph of (a) solder joint on board at the condition of medium load, SEM micrographs of (b) interested area of (a), (c) magnified image of A region in (b) and (d) magnified image of B region in (b)

Micro crack penetrated into Pb-rich α phase and Sn-rich β phase along tensile stress axis under high stress(8kg), as shown in Fig. 5. With observation for the failure mode with temperatures under high constant stress it was revealed that a transglanular mode was predominant under relative high stress and the effect of temperature was more negligible than that of stress. According to previous studies, fracture mode was transgranular in limited range of creep where stress flow was stabilized and void coalescence was postponed due to the strain-rate dependence of creep[2]. As mentioned on previous clause, the break in the creep rate-stress curve represented in power-law creep occurred at stress at which the fracture mode changed from intergranular to transgranular fracture[5]. In power-law creep regime, there are different n and Q between high-temperature(HT) creep and low-temperature(LT) creep, that is, failure mode is changed with the shift of failure mechanism by increasing the applied stress. Therefore, it was recognized that the failure mode of high stress condition was different with that of low and medium stress condition.

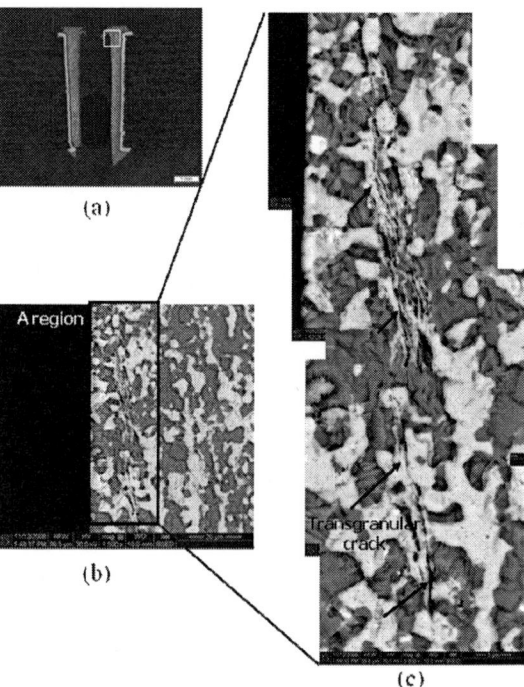

(a)

(b)

(c)

Figure 5 Cross-sectional optical micrograph of (a) solder joint on board at the condition of high stress, cross-sectional SEM micrographs of (b) interested area of (a), (c) magnified image of A region in (b).

Considering with overwriting our test conditions on SnPb eutectic solder creep mechanism map established by X. Q. shi et al,[4] as shown in Fig. 6, it was found that the diffusion creep was involved on condition of low and medium stress whereas dislocation creep was involved on the condition of high stress. Although creep mechanism boundary between each region might be changed by grain size or by other process, it was found that our results listed in Table 2 corresponded with previous work[4].

978-1-4244-4722-0/09 $25.00
© 2009 IMAPS-ITALY

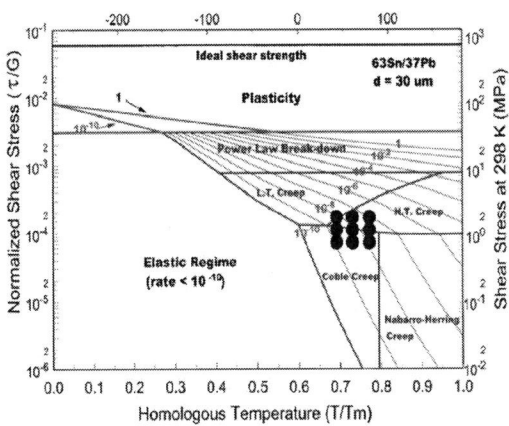

Figure 6 Creep deformation map for 63Sn/37Pb eutectic solder alloy by X.Q.Shi *et al.*[4] with tested conditions

Table 2 Results of fracture analysis with applied load and temperature

Applied Load	Temp. (℃)	Fracture Mode	Creep Mechanism
4kg	35	Intergranular	Diffusion
	55	Intergranular	Diffusion
	75	Intergranular	Diffusion
6kg	35	Intergranular & Transgranular	Diffusion & Dislocation
	55	Intergranular & Transgranular	Diffusion & Dislocation
	75	Intergranular & Transgranular	Diffusion & Dislocation
8kg	35	Transgranular	Dislocation
	55	Transgranular	Dislocation
	75	Transgranular	Dislocation

Conclusion

In order to build an accelerated life model for creep rupture of SnPb eutectic solder, the accelerated tests were conducted on several conditions of temperature and stress considering creep mechanism. As solder joints were failed, fractograpghy of solder joint was analyzed with SEM and FIB. We found that the failure mode was shifted from intergranular fracture to transgranular fracture with increased applied stress. We suggested optimal conditions of temperature and stress for accelerated creep rupture test on SnPb eutectic solder joints.

Acknowledgements

Authors would like to thank Heesun Lee, Jiho Kim for supporting failure analysis work.

References

[1] R.J.Klein Wassink, "Soldering in Electronics", electrochemical publication, second edition, Bristol, chapter 4, pp135, 1989.

[2] George E. Dieter, "Mechanical Metallurgy", McGraw-Hill, SI Metric edition, London, chapter13, pp.423, 1988

[3] Chin-Kuang Lin, De-You Chu, "Creep rupture of lead-free Sn-3.5Ag and Sn-3.5Ag-0.5Cu Solders", Journal of Materials Science: Materials in Electonics, Vol 16, pp355-365, 2005.

[4] X. Q. Shi, Z.P. Wang, Q,J, Yang, and H.L.Pang,"Creep behavior and Deformation Mechanism Map of Sn-Pb Eutectic Solder Alloy", Journal of Engineering Materials and Technology, Vol.125, pp 81-88, January, 2003.

[5] Alan Liu, "Mechanics and Mechanisms of Fracture", ASM international, first edition, chapter 2, pp105-111, 2005

Direct interconnection of chemical mechanical polishing (CMP)-Cu thin films at 150°C in ambient air

Akitsu Shigetou[1] and Tadatomo Suga[2]

[1] National Institute for Materials Science (NIMS)
1-1, Namiki, Tsukuba-shi, Ibaraki 305-0044, Japan
TEL & FAX +81-29-860-4669, SHIGETOU.Akitsu@nims.go.jp

[2] The University of Tokyo
7-3-1, Hongo, Bunkyo-ku, Tokyo 113-8656, Japan
TEL +81-3-5841-6491, FAX +81-3-5841-6485, suga@pe.t.u-tokyo.ac.jp

Abstract

This paper describes the feasibility of a low-temperature diffusion bonding process for Cu thin film electrodes in ambient air. After Cu thin film surfaces were bombarded by an Ar fast atom beam in vacuum to remove the initial thick adsorbate layer, O_2 gas was introduced into the vacuum chamber to prevent moisture-induced generation of thick $Cu(OH)_2$ layers, which was considered hydrated. Then the surfaces were contacted with each other at atmospheric pressure. Upon heating at 150°C for 600 s after the touchdown, high bonding strength, which was as high as that of Cu film breakage from the inside, was obtained through a CuO interfacial layer of around 10 nm thickness with considerably low electrical resistivity.

Key words: CMP-Cu, diffusion bonding, direct bonding, low temperature, Ar fast atom beam, and ambient air.

Introduction

A high-density Cu interconnection between layer structures, as achieved using assembly (that means bonding) technology, is expected to be important for future solution-oriented and three-dimensional integration of discrete electronic devices. Considering the potential requirement of mixed integration with non-heat-resistant substrates such as polymers and bio-inert materials typically used in the field of micro electro mechanical systems (MEMS), the Cu–Cu bonding process should be carried out at a temperature lower than 150°C. Moreover, interconnection through highly flattened layers including the surface of ultralow-profiled Cu electrodes, for instance by chemical mechanical polishing (CMP)-Cu films, is indispensable for the stacking of substrates. Some pioneering studies have achieved Cu–Cu wafer bonding at relatively low temperatures around 400°C using diffusion bonding methods assisted by chemical forming gas conditions [1, 2]. As for the ultrafine pitch interconnection, 8-μm-pitch bonding of thin Cu electrodes was achieved at around 350°C using oxide bonding technology to the dielectric surfaces [3, 4]. In such cases, heating times of several hours were necessary after the oxide bonding process to accelerate the diffusion of Cu atoms across a thick adsorbate (mainly oxide) layer to obtain sufficient bonding strength and electrical conductivity. Meanwhile, we performed a room temperature direct bonding of 6-μm-pitch bumpless Cu electrodes fabricated with the applied damascene process [5]. The attractive force between atomically clean metal surfaces, which was created using physical bombardment of an Ar fast atom beam (Ar-FAB) in a vacuum condition [6], was used to reduce the process temperature to ensure high alignment accuracy. In this case, it was indispensable to remove the initial thick adsorbate layer on the Cu surface for high bonding strength and low contact resistance at room temperature, whereas the use of high vacuum conditions increased the bonding process complexity, such as difficulty in sample handling and alignment procedure. Therefore, it was necessary to carry out the bonding process at atmospheric pressure, except for Ar-FAB bombardment; the use of ambient air was preferred for reasons of practicality. That is, the Cu-Cu interconnection should be obtained on the premise of adsorbate layer on clean surfaces. Consequently, to ensure high bonding strength and electrical conductivity at the same time at a temperature lower than 150°C in such an adsorptive condition, we had to modify the thickness and chemical composition of adsorbate after the Ar-FAB bombardment because sufficient volume diffusion of Cu atoms across the interface was necessary.

For this study, we carried out X-ray photoelectron spectroscopy (XPS) analyses on the growth behavior of adsorbate with the parameter of humidity in O_2 gas because O_2 and H_2O are

978-1-4244-4722-0/09 $25.00
© 2009 IMAPS-ITALY

considered the most critical molecules for the chemical binding status of an atomically clean Cu surface in ambient air. Subsequently, the bonding quality was evaluated based on results of die-shear testing, transmission electron microscopy (TEM), and electron energy-loss spectroscopy (EELS). Moreover, the changes in the interfacial resistivity were investigated by high temperature storage test using 4-wires patterns consisted of $10 \times 10 \ \mu m^2$ electrodes.

Adsorption Behavior of O_2 and H_2O Molecules onto Clean CMP-Cu Surface

As the test vehicle, CMP-Cu films with surface roughness (Ra) of 1–2 nm were prepared. This roughness was considered small enough to ensure good contact between Cu surfaces if the samples were placed parallel before the touchdown, assuming that the cross-sectional surface asperities formed a continuous elastic sinusoidal curve [7]. Therefore, the influence of morphology on the bondability could be ignored in this study. These samples were fabricated on thermally oxidized Si chips after sputter deposition of Cu/TaN layers. Figures 1(a) and 1(b) respectively show schematic representations of the bonding process and a flip-chip bonding apparatus. The sizes of upper and lower samples in the apparatus were 6×6 and $15 \times 15 \ mm^2$, respectively. The bonding process was performed as follows: 1) The initial adsorbate with thickness of more than 20 nm [8] was removed using Ar-FAB in vacuum; 2) the O_2 gas with controlled humidity was introduced at the pressure of 8.0×10^4 Pa;, then 3) the samples were mutually contacted and heating to 150°C was applied to them. In Step 2), the absolute humidity was controlled by changing the flow through a bubbling water bottle, as illustrated in Fig. 1(b).

Figures 2(a) and 2(b) present the XPS depth-profiling results of the sample shown in Step 2) of Fig. 1, respectively indicating the time evolution of adsorbate thickness and the main peaks of Cu and O with the parameter of humidity. Tested humidity conditions were dry O_2 (0), 6.3, 12.6, and 18.9 g/m^3; the adsorbate thicknesses in those conditions were derived from the Ar ion beam etching depth, where C 1s and O 1s peaks disappeared and Cu LMM peaks appeared at binding energies of metal Cu. Results showed that the growth behavior of adsorbate on clean Cu surface was apparently influenced by humidity. In every test condition in Fig. 2(a), the increments in adsorbate thickness approximately followed the cube law, although the reaching thickness at the same exposure time increased concomitantly with increased humidity. These results suggest that two factors limited the speed of adsorbate growth: a) the amount of volume diffusion of Cu atoms occurring from the inside of CMP-Cu film to the surface of the adsorbate layer,

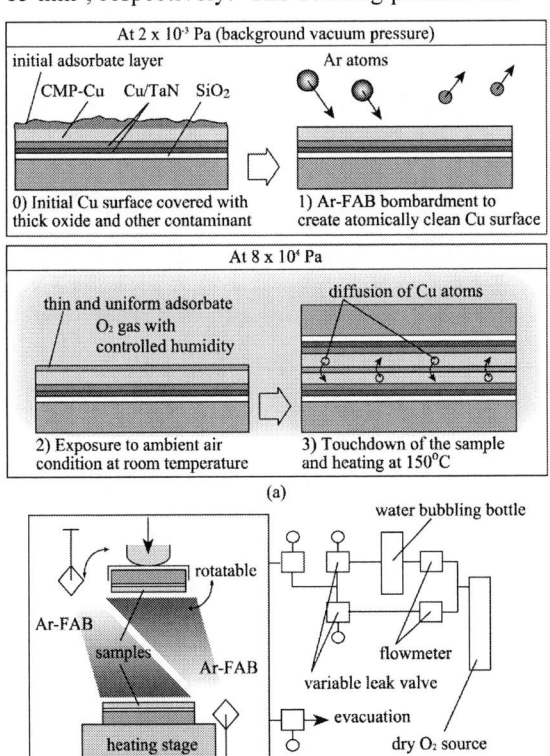

Figure 1. Schematic representations of: (a) the modified diffusion bonding process; and (b) the flip-chip bonding apparatus with an O_2 gas induction line.

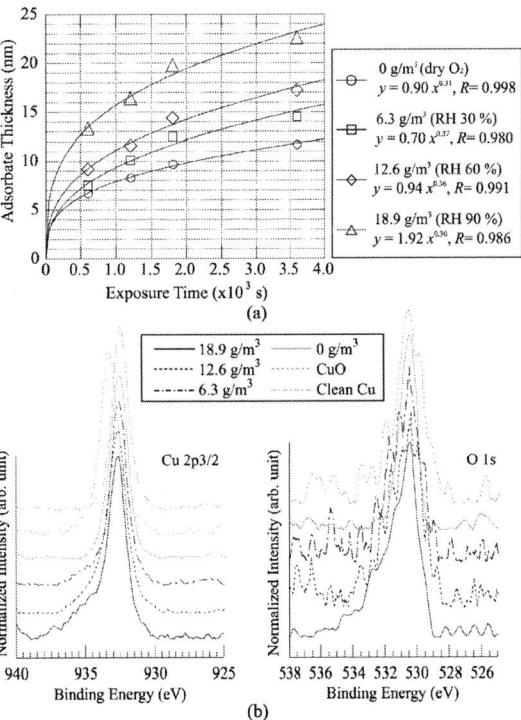

Figure 2. Results of XPS depth profiling with the parameter of humidity: (a) time evolution of the adsorbate thickness on a clean Cu surface; and (b) spectra of Cu 2p3/2 and O 1s at the Process 2) in Fig. 1(a).

978-1-4244-4722-0/09 $25.00
© 2009 IMAPS-ITALY

and b) the number of water molecules in collision with the surface in a unit time. In our previous studies of room temperature bonding of CMP-Cu films, low interfacial electrical resistance, which could be converted to the resistivity of 5×10^{-8} $\Omega \cdot m$, was obtained until an approximately 2-nm-thick oxide was generated on the clean surface [9]. Simply comparing the ratio of self-diffusion length of Cu, as derived from the one-dimensional solution of Fick's law, the diffusion length would increase by around seven times at 150°C. Therefore, assuming that the bondability was ruled mainly by diffusion of Cu atoms, the adsorbate thickness of around 15 nm, which was satisfied with the exposure time of 600 s in all tested humidity conditions, was considered allowable at 150°C. However, synthetic peaks of Cu 2p3/2, as presented in XPS spectra in Fig. 2(b), show that $Cu(OH)_2$ (935.0 eV) was detected in humidified conditions, whereas the peak of Cu_2O (932.7 eV) was dominant in the dry O_2 condition. Moreover, the component of $Cu(OH)_2$ was considered hydrated because peaks attributable to molecular H_2O (533.0 eV) were observed in O 1s spectra. Such a hydrate layer on the surface might promote undesirable formation of air bubbles and/or thick oxide formation at the interface when the applied temperature is increased. Therefore, we conducted bonding experiments mainly with samples exposed to the dry O_2 gas for 600 s.

Bonding Experiment in O_2 gas at 150°C – Bond Strength and Interfacial Structure

In bonding experiments, the heating temperature was set at 150°C, the heating time was varied: 300 (necessary to raise the temperature) – 3600 s. Figures 3(a) and 3(b) respectively represent the change in die-shear strength and the enlarged image of the debonded surface of dry O_2 sample heated for 600 s. In Fig. 3(a), the die-shear strength indicates the normalized values derived from the measured strength divided by the contact load. The contact load of 490 N, which was applied to the samples after the touchdown, was used to eliminate the influence of contact area on the bondability,

because this value had been proven to be sufficient to compensate for a tilt between upper and lower Cu surfaces and to make a nominally perfect contact using a rotatable bonding fixture. When the normalized strength was higer than 0.12, Cu film breakage wad observed. The testing results show that the die-shear strength of dry O_2 sample rose concomitantly with increased heating time, reaching the mechanical maximum strength at around 600 s. Since a kinetic increase was observed in the die-shear strength until 600 s, the major bond mechanisms was considered to be volume diffusion of Cu atoms across thin oxide layer. On the debonded surface shown in Fig. 3(b), the positions of apparent Cu film breakage were observed in a dot-like distribution. Because the distance between these dots was approximately equal to the wavelength of the surface asperities measured using an atomic force microscope (AFM), it is inferred that the deformation of surface summits at the sample touchdown might provoke the stress-induced acceleration of diffusion as well.

Figures 4(a) and 4(b) respectively represent the TEM images and EELS spectra of the interface created with the heating time of 600 s in dry O_2 gas. Figure 4(a) shows that a nominally void-free interconnection was obtained between CMP-Cu films through a polycrystalline interfacial layer with 10 –15 nm thickness. This interfacial layer had been partially broken mainly because of the deformation of surface asperities at the touchdown. Direct

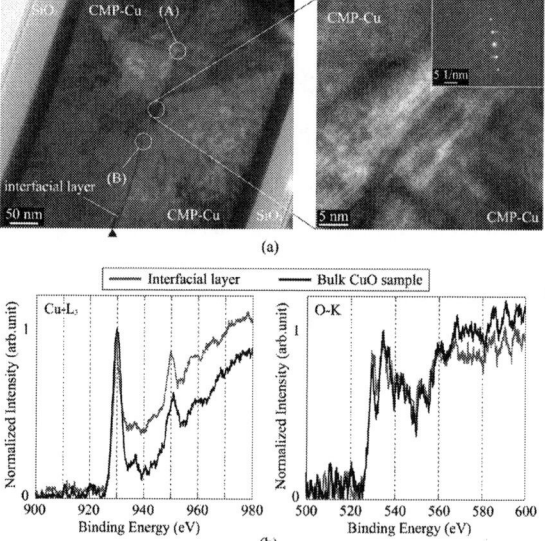

(a)

(b)

Figure 4. Results of interfacial analyses on the sample bonded with heating at 150°C for 600 s in a dry O_2 condition: (a) TEM cross sectional images; and (b) the EELS spectra of Cu-L$_3$ and O-K. In (a), no readily visible interfacial layer was found at positions marked as (A) and (B). In (b), the spectral intensity was normalized with that of the peak edges to emphasize the differences in peak positions.

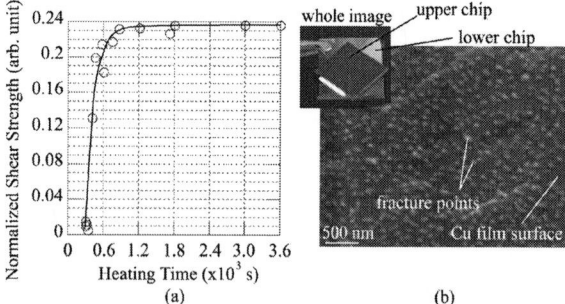

(a)

(b)

Figure 3. Results of Cu-Cu bonding experiments at 150°C in the dry O_2 condition: (a) heating-time evolution of the die-shear strength; and (b) scanning electron microscopy (SEM) image of the debonded surface at the heating for 600 s.

978-1-4244-4722-0/09 $25.00
© 2009 IMAPS-ITALY

Cu-Cu interconnections were obtained at those positions, where the electrical conduction across the interface would also be facilitated. From the EELS spectra of Cu-L_3 and O-K, as depicted in Fig. 4(b), the dominant component of this interfacial layer was proven to be CuO because the peaks appeared at the same binding energies as the bulk CuO sample. This CuO layer was inferred to be attributable to Cu_2O layer oxidation because the CuO thickness was almost equivalent to that of Cu_2O generated during exposure to dry O_2 gas.

Bonding Experiment in O_2 gas at 150°C – Electrical Conductivity

Figures 5(a) and 5(b) show the schematics of the patterned CMP-Cu film sample used in the electrical resistance test, respectively indicating the fabrication process and the enlarged image of the electrodes array. In the test vehicles, the electrodes with the contacting area of 10 x 10 μm^2 were fabricated in the line and space (L/S) of 20/5 μm using Ar ion beam etching process; they were arranged in the area of 4 x 4 mm^2 on 15 x 15 mm^2 chip. Some of the electrodes were designed to carry out the 4-wires measurements in the outer, middle, and center regions of the patterned area. The test vehicles were bonded to CMP-Cu simple film samples in the dimensions of 6 x 6 mm^2 using the same dry O_2 condition (that is, CuO interfacial layer) as previous section, then high temperature storage tests were carried out at 150°C for 1000 hours.

Figures 6(a) and 6(b) represent the positions of tested electrodes and the changes in electrical resistivity through the storage testing, respectively. In Fig. 6(b), the electrical resistivity denotes the value derived from the averaged resistance in each testing region, assuming that only the contacting area (10 x 10 μm^2) between the electrodes and the CMP-Cu film, with a thickness around 2 μm in total, contributed to the electrical conduction. In addition,

the values of electrical resistivity obtained from the samples bonded with conventional surface activated bonding (SAB) method at room temperature in high vacuum condition [6] were shown for comparison. Results show that considerably low electrical resistivity could be obtained in all testing regions of the sample, regardless of the CuO interfacial layer with around 10 nm thickness. Although the values obtained from dry O_2 samples were slightly higher than those of conventional SAB samples, the increment ratio was less than 10 % after the storage testing for 1000 hours.

Incidentally, the TEM and EELS analyses indicated that the CMP-Cu films could be bonded at 150°C in the humidified O_2 gas conditions as well, thorough the amorphous interfacial layers consisted of $Cu(OH)_2$ and hydro-carbonaceous compound (data not shown). The thickness of these layers showed a sudden increase at the temperatures around 100°C, with a concomitant rise in die-shear strength. The reaching strength and thickness at higher temperatures corresponded to humidity. For instance, the thickness of amorphous interfacial layer could be decreased to around 15 nm with humidity of 3.1 g/m^3, while the thickness had increased to over 250 nm with humidity of 18.9 g/m^3. Because the segregation of CuO grains was observed in these amorphous layers upon additional heating at higher than 100°C after the bonding, and the number and size of CuO grains increased concomitantly with humidity, it was suggested that the heating temperature facilitated the dissociation of water molecules, which might be coordinated with $Cu(OH)_2$, assisted the recrystallization of CuO, and increased bonding strength. Therefore, it is expected

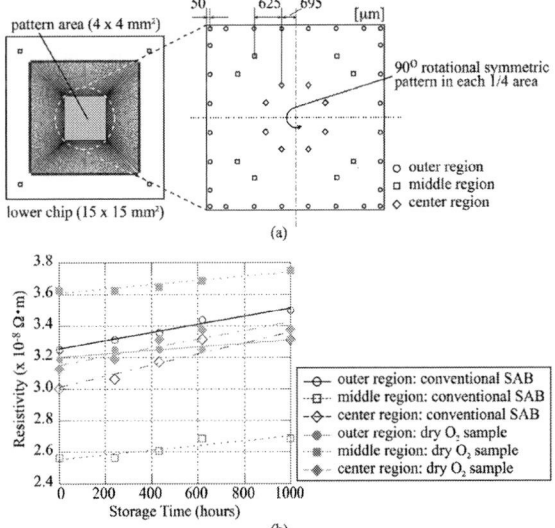

Figure 6. (a) Arrangement of the positions for 4-wires resistance measurement in the pattern area, and (b) time evolution of electrical resistivity through CuO interfacial layers at 150°C. The measured resistance in each region was averaged excluding the highest and lowest values.

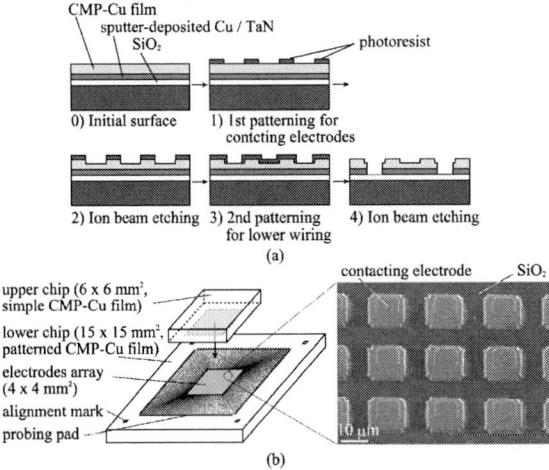

Figure 5. Schematics of the patterned CMP-Cu film sample: (a) outline of the fabrication process; and (b) the scanning electron microscope (SEM) image of the fabricated electrodes.

that the interfacial structure would be changed to a thin CuO layer similar to Fig. 4(a) using controlled humidity, and would result in high bonding strength and good electrical conductivity even in humidified conditions. The detailed investigations on these matters remain as a subject for future work.

Conclusions

In this study, the modified diffusion bonding of CMP-Cu films was realized at 150°C in an O_2 gas condition at atmospheric pressure, supposing the Cu-Cu interconnection in ambient air. The Cu surfaces were cleaned atomically with Ar-FAB bombardment in vacuum; then they were exposed to O_2 gas with controlled humidity at atmospheric pressure at room temperature. Subsequently, the samples came into mutual contact and heating was applied to them. Results show that humidity in the O_2 gas strongly influenced the growth speed and chemical composition of adsorbate at room temperature. A Cu_2O layer with thickness of around 7 nm, which was considered thin enough to ensure the bonding strength and electrical conductivity at 150°C, was created through dry O_2 exposure for 600 s. On the other hand, thick adsorbate layers, which might be composed of hydrated $Cu(OH)_2$, were generated in the humidified conditions. Upon heating at 150°C for 600 s in the dry O_2 gas condition, the CMP-Cu films were bonded successfully through a CuO layer with thickness of around 10 nm, which was considered to be an oxidized Cu_2O layer. This CuO layer was partially broken because of the surface deformation on the moment of surface touchdown, and direct interconnections were obtained at such positions. Moreover, high temperature storage tests at 150°C for 1000 hours showed that electrical resistivity as low as that of direct-bonded samples could be obtained through this oxide interfacial layer.

Acknowledgements

This study was partially supported by *Institute for Advanced Micro-System Integration (IMSI)* and by *"World Premier International Research Center Initiative on Material Nanoarchitectonics," MEXT, Japan.* The authors highly appreciate the financial and technical help from them.

References

[1] A. Fan, A. Rahman, and R. Reif, "Copper Wafer Bonding, *IEEE/ECS Electrochem. and Solid-State Lett.*, Vol. 2 (1999) pp. 534-536.

[2] K. N. Chen, C. S. Tan, A. Fan, R. Reif, "Morphology and Bond Strength of Copper Wafer Bonding," *IEEE/ECS Electrochem. and Solid-State Lett.*, Vol. 7 (2004), pp. G14-G16.

[3] C. W. C. Lin, S. C. L. Chiang, T. K. A. Yang, "Bumpless flip chip packages," *Proc. IEEE 4th Int'l. Conf. on Electron. Mater. and Packag. (EMAP)*, 2002, pp. 173-177.

[4] S. Towle, H. Braunisch, C. Hu, R. Emery, G. Vandentop, "Bumpless Build-up Layer Packaging," *Proc. ASME Int'l. Mech. Eng. Congress and Exposition (IMECE)*, 2001, EPP-24703.

[5] A. Shigetou, T. Itoh, K. Sawada, T. Suga, "Bumpless Interconnect of 6-μm-pitch Cu electrodes at room temperature," *IEEE. Trans. on Adv. Packg.*, Vol. 31 (2008), pp. 473-478.

[6] H. Takagi, K. Kikuchi, R. Maeda, T. R. Chung, T. Suga, "Surface Activated Bonding of Silicon Wafers at Room Temperature," *Appl. Phys. Lett.*, Vol. 68 (1996), pp. 2222-2224.

[7] K. Takahashi, T. Onzawa, "The Effect of Surface Roughness on The Low Energy Bonding," *J. High Pressure Inst. of Japan*, Vol. 35 (1997), pp. 159-164.

[8] A. Shigetou, T. Itoh, T. Suga, "Direct Bonding of CMP-Cu Films by Surface Activated Bonding (SAB) Method," *J. Mater. Sci.*, Vol. 40 (2005), pp. 3149-3154.

[9] A. Shigetou, T. Itoh, T. Suga, "Bumpless Interconnect of Cu Electrodes in Millions-Pins Level," *Proc. 56th Electronic Components and Technology Conf*, 2006, pp. 1223-1226.

Damage Risk Assessment of Under-Pad Structures in Vertical Wafer Probe Technology

Torsten Hauck*, Ilko Schmadlak*, Christopher Argento*, Wolfgang H. Müller**

* Freescale Halbleiter Deutschland GmbH
Schatzbogen 7, 81829 Munich, Germany
Phone: +49 89 92103 692, Fax: +49 89 92103 688

**Technische Universität Berlin, Fakultät V, LKM
Einsteinufer 5, 10587 Berlin, Germany
Phone: +49 30 314-27682, Fax: +49 30 31424499

Abstract

Due to the demand of the industry for an increase of the number of I/Os, while decreasing the die size, the bond pads had to shrink and design restrictions for the active structures underneath had to fall. This leads to new challenges for the electrical probing and the mechanical robustness of the under-pad structures. This paper presents analytical and numerical simulation approaches for predicting loading conditions, estimating stress states and assessing associated damage risks for the Back-End-Of-Line (BEOL) interconnect system underneath a probe pad. For this purpose we investigate, first, the elastic stability of the probe needle according to large deflection theory of buckled bars. Micro-spring and buckling beam probe technologies are compared. Second, we determine probe forces as functions of the probe card overdrive. By using finite element analysis we then determine the stress and deformation state in the probe pad and underneath. Various stress criteria are used to assess and rank fracture risk in brittle and ductile material members of the BEOL stack.

Key words: vertical probe technology, large deflection analysis

Vertical Probe Technology

As the industry demand for smaller probe marks and parallel multi-DUT (device under test) testing increases, so does the push toward finer pitch vertical probing solutions. To meet this demand, probe vendors have created a wide variety of vertical probe options that range from 25μm round probes to 100 μm rectangular probes. Some probes are stamped with fixed spring shapes while others are straight; some are simple extruded wires and others are grown using semiconductor process; and some are coated or plated or made of varying alloys. Because of these variations, probes of various technologies designed for the same spring rate will behave differently and thus produce varying effects on device test pads without. It is not uncommon that a probe with a low spring rate, often considered to be lower force, may result in drastic pad and under pad damage while a different type probe designed with a higher spring rate will produce much less pad and no under pad damage. For these reasons, it is desirable that a technique be created to help calculate the resulting applied force of the newer probe technologies and assist in low cost empirical prediction and analysis of various pad designs and technology stacks. The current study addresses this need by investigating the basic difference of buckling phenomena of vertical probes and derives a numeric model for calculating the probe force for use with FEM analysis and probe recipe comparison.

Elastic Buckling of a Vertical Probe Needle

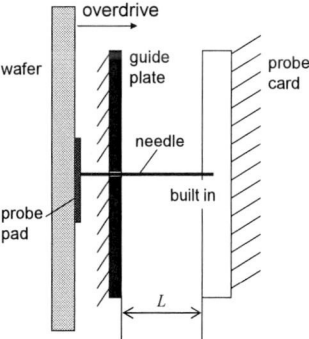

Fig. 1: Vertical probe setup

One of the leading edge probe contactor technologies is based on vertical buckling beam needles. The required elasticity for contacting the pad is achieved by buckling of a beam. Figure 1 shows the schematics of the probe card assembly. The lower end of each needle is held in place by an insulating guide plate. The upper end of the needle is built in. The wafer is moved toward the needle probe card and contacts the needle [1, 2].

978-1-4244-4722-0/09 $25.00
© 2009 IMAPS-ITALY

The contact load causes an axial compression force on the probe needle. If the compression force exceeds a critical load the needle will slightly bend or buckle, as shown in Figure 2. The critical load P can be calculated by using the differential equation of the deflection curve [3].

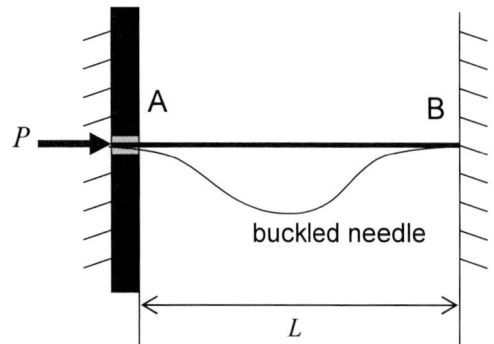

Fig 2: Mechanical model of the probe needle

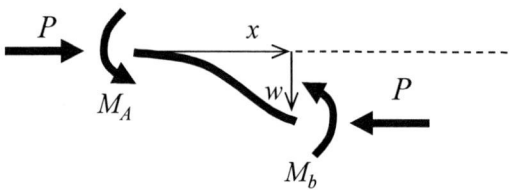

Fig. 3: Static equilibrium at the deformed needle

The coordinate axis x and the deflection w are defined as indicated in Figure 3. The reactive moment M_A prevents the end of the needle from rotating during buckling. By assuming a slightly deflected position of the needle, the bending moment M_b reads:

$$M_b(x) = P\,w(x) - M_A. \tag{1}$$

For small deformation values the differential equation of the deflection curve becomes:

$$EI\frac{d^2 w}{dx^2} = -M_b(x) = M_A - P\,w(x), \tag{2}$$

where E and I denote Young's modulus and the moment of inertia. Equation (2) can be rewritten in the form:

$$w''(x) + k^2\,w = k^2\frac{M_A}{P}, \tag{3}$$

with the notation:

$$k^2 = \frac{P}{EI}.$$

The general solution of this equation is:

$$w(x) = C_1\sin(kx) + C_2\cos(kx) + \frac{M_A}{P}. \tag{4}$$

The constants of integration C_1 and C_2 are determined from the boundary conditions at the guide plate in point A:

$$w = 0, \quad \frac{dw}{dx} = 0 \quad \text{at } x = 0, \tag{5}$$

$$C_1 = 0, \quad C_2 = -\frac{M_A}{P}.$$

The boundary conditions at the built in end of the needle in point B require that

$$w = 0, \quad \frac{dw}{dx} = 0 \quad \text{at } x = L.$$

A non-trivial solution to Equation (4) requires that $1 - \cos(kL) = 0$ and $\sin(kL) = 0$. Both conditions are fulfilled if:

$$kL = 2\pi n \quad n = 1,2,3,\dots \tag{6}$$

The critical probe force is obtained by taking $n = 1$:

$$P_{cr} = 4\pi^2\frac{EI}{L^2}, \tag{7}$$

and the buckling mode shape becomes:

$$w(x) = \frac{M_A}{P_{cr}}\left(1 - \cos\left(2\pi\frac{x}{L}\right)\right). \tag{8}$$

With maximum deflection:

$$\delta = 2\frac{M_A}{P_{cr}} \quad \text{at } x = \frac{L}{2},$$

the deflection curve, sloop, and the bending moment of the needle read:

$$w(x) = \delta\frac{1}{2}\left(1 - \cos\left(2\pi\frac{x}{L}\right)\right), \tag{9}$$

$$\frac{dw(x)}{dx} = \frac{\delta}{L}\pi\sin\left(2\pi\frac{x}{L}\right), \tag{10}$$

$$M_b(x) = M_A\cos\left(2\pi\frac{x}{L}\right), \tag{11}$$

where maximum deflection δ and reaction moment M_A are still indetermined. Equations (9) – (11) are plotted in Figures 4a)-c). Inspection of these plots indicate four equal regions of the needle, each of which represents the condition of a column with built-in lower end and free upper end, the first Euler buckling case.

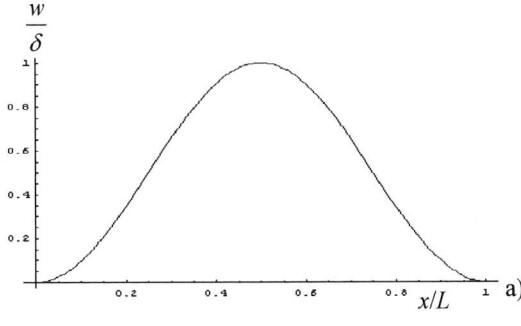

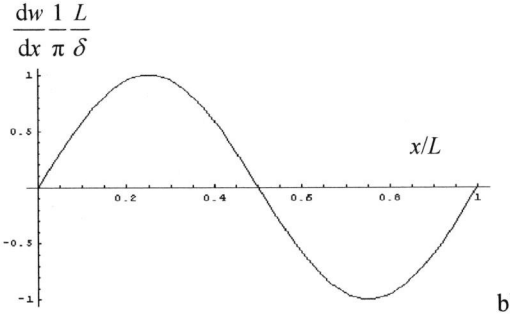

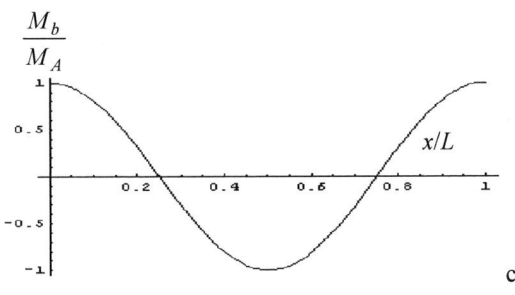

Figure 4: Deflection, slope and bending moments of the buckled needle

In what follows, we will make an attempt to determine the probe force as a function of the needle deformation at large deflections. It is sufficient to perform the analysis for the first Euler buckling case and then to transfer the results to the previously described needle buckling case by symmetric expansion. The calculation follows a solution from Timoshenko for the problem of large deflection of buckled bars, "The Elastica" [3].

Large Deflection of a Buckled Needle

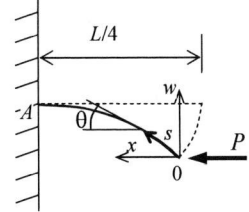

Fig. 5: Smallest repetitive buckling case

Figure 5 shows the mechanical model that is used. It represents the conditions of the first Euler buckling case with the left end A built-in and the free right end at point 0. It represents the conditions of the outermost left quarter of the needle. The coordinate s denotes the distance from the origin 0. The exact expression for the curvature of the needle reads $d\theta/ds$ and the differential equation of the deflection curve becomes

$$EI\frac{d\theta}{ds} = -P\,w\,. \tag{12}$$

Differentiation of (12) with respect to s and by using the relation

$$\frac{dw}{ds} = \sin(\theta),$$

as well as the previously defined notation for k gives:

$$\frac{d^2\theta}{ds^2} = -k^2\,\sin(\theta)\,. \tag{13}$$

Multiplication of (13) with $d\theta$ and integration leads to:

$$\int\frac{1}{ds}\frac{1}{2}\left(\frac{d\theta}{ds}\right)^2 ds = -k^2\int\sin(\theta)\,d\theta + C_3\,,$$

and:

$$\frac{1}{2}\left(\frac{d\theta}{ds}\right)^2 = k^2\cos(\theta) + C_3\,. \tag{14}$$

The integration constant C_3 is determined from the conditions at the origin 0:

$$\frac{d\theta}{ds} = 0 \quad \text{and} \quad \theta = \alpha \quad \text{at} \quad s = 0\,,$$

$$C_3 = -\cos\alpha\,.$$

Equation (15) can be rewritten to yield:

$$\frac{1}{2}\left(\frac{d\theta}{ds}\right)^2 = k^2\left(\cos(\theta) - \cos(\alpha)\right). \tag{15}$$

The evaluation of Equation (15) provides the probe force P and the overdrive u as functions of the sloop parameter α:

$$P(\alpha) = P_{cr}\frac{4}{\pi^2}K(p)^2\,, \tag{16}$$

$$u(\alpha) = 2L\left(1 - \frac{E(p)^2}{K(p)^2}\right), \tag{17}$$

where $p = \sin(\alpha/2)$, and the functions K and E denote the complete elliptic integrals of first and second kind, respectively.

Figure 6 shows deflection states of the buckling beam needle at increasing probe force. Figure 7 shows the probe force as a function of the wafer overdrive for a buckling beam probe needle with ∅2.5 mils. The comparison of the exact solution and a finite element solution shows a perfect match. The probe force vs. overdrive curve is also shown for a micro spring needle with a spring stiffness of 1.5 grams/mil. The difference between buckling beam and micro spring technologies is evident. The buckling beam reveals a constant probe force over a large range of overdrive. It guarantees a consistent contact pressure on every point tested, independently of the overdrive. This allows an optimal tolerance even under changing planarity conditions of the probe card. The micro-spring linearly increases the probe forces with the overdrive. Thus, the overdrive must be controlled within tight limits. Corresponding probe cards are much more sensitive to non-planarity issues.

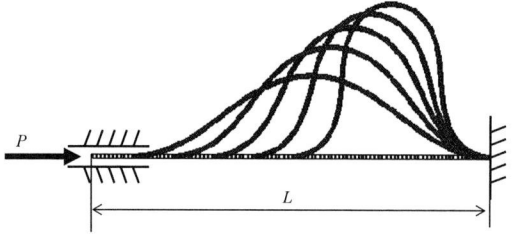

Fig. 6: Large deformation of the probe needle

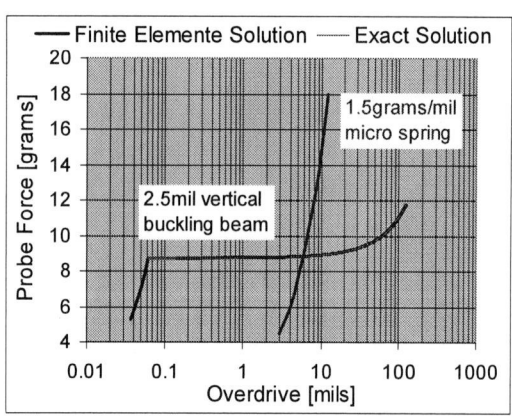

Fig. 7: Probe force versus overdrive

Stress State in Underpad Structure Subjected to Probe Force

The second part of this study had its focus on the under pad design, and in particular the design of the last via layer. The probe needle analysis had helped gaining a better understanding of the differences between the two vertical probe card concepts. It also revealed that the lateral force component can be neglected in comparative probe pad simulations. In both vertical probe card concepts, a guide plate limits the lateral scrub. Lateral stresses in the pad and the structures below are limited by the yield stress of the pad aluminum,

and should be of the same magnitude for the different needle options.

The assumption of a 6.5 grams vertical probe load was considered for all simulations in this comparison. This loading condition correlates with the buckling force of the investigated needle geometry. An array of buckling beam needles in a probe card can be seen in Figure 8, while Figure 9 shows the general approach of the probe event simulations. A probe contact pressure was applied to the top surface of a stack model. The Back End of Line (BEOL) stack was modeled in detail for the last layers. In order to simplify the model, the lower layers, which are of less importance, were neglected and a block of silicon represented the stiffness underneath the layers of interest.

Fig. 8: Vertical probe card with buckling beam probe needles

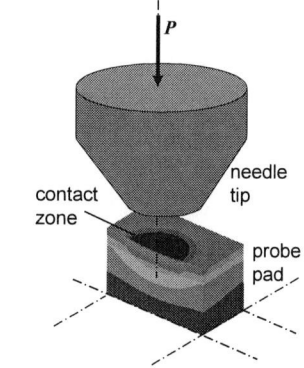

Fig. 9: Probe event and resulting probe pad deformation

A further simplification was accomplished by using a slice model as shown in Figure 10. The designs were compared by the stress states of the different materials in the stack. The results showed first of all that differences in the last via layer design had no effect on the Mises stress state in the metals of the stack. For all other materials in the stack, the first principal stress level was used to rank different design options. Only the inter layer dielectric (ILD) of the last via layer showed significant differences in this stress around the vias.

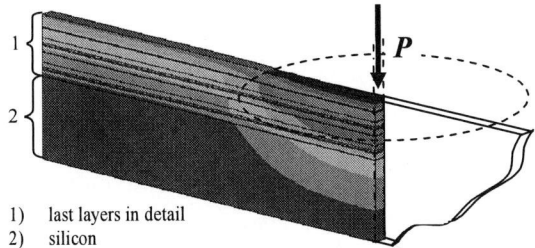

1) last layers in detail
2) silicon

Fig. 10: Displacement contour of slice model
considering half symmetry

Comparison of Underpad Designs

The simulations of different via patterns revealed that the maximum stress in the last via layer, occurs around the edges of vias which are located below the edge of the contact surface of the probe needle. Figure 11 shows that the stress below the probe needle is of compressive nature, which is not considered to be critical. Around this area of compression the stress is tensile, with a maximum in a thin ring very close to the projection of the contact surface edge.

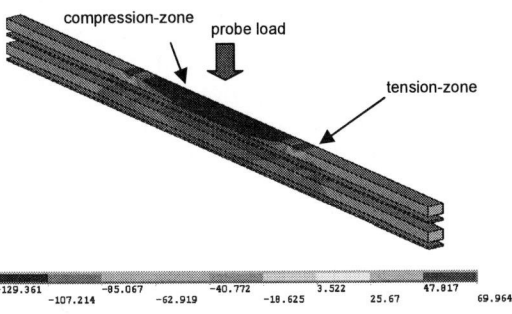

Fig. 11: First principal stress contour of ILD layers
(no vias considered in this simulation)

Different via patterns as well as via designs were simulated with the result that the maximum first principal stress was of the same magnitude if the vias had the same diameter and lay in the ring of maximum stress as shown in Figure 12.

Maximum stresses were observed at the periphery of the contact zone. Via structures in this region cause stress concentrations at the via edges inside the ILD material. Hence, each via that lies at the periphery of the contact zone is a potential crack initiation point.

As a conclusion, the number of vias that create the stress maxima in the ILD and their distance to each other seem to be critical parameters for the occurrence of a crack. The stress level seems to be of secondary importance.

Five different design options with varying via pattern were proposed for a design of experiments. An associated pizza-wafer was produced and is currently being tested.

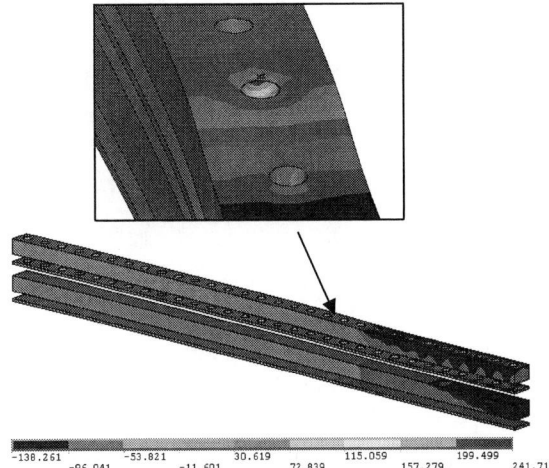

Fig. 12: Maximum first principal stress at top via
edge in tension ring

Summary

The analytical solution of the elastic buckling of a vertical probe needle lead to an exact equation for the probe forces that is acting onto the wafer probe pad. Large buckling deflection was investigated by means of nonlinear large deformation analysis. Probe forces and corresponding wafer overdrive were derived in parametric form. The resulting equation allows an easy prediction of the needle behavior and wafer probe parameter for vertical buckling beam technology.

The fundamental difference between spring and buckling probe technologies lies in the probe force versus wafer overdrive characteristics. Buckling probe needles provide a constant probe force over a large range of wafer overdrive. Spring probe contactors, such as micro-springs or cantilever beams, linearly increase the probe force with an increasing overdrive.

The stress state in the under pad structure on wafer was investigated by means of finite element analysis. The loading conditions that result from a probe event were taken from the deformation analysis of the probe needle technologies.

Design options of the last via layer were considered in detail. A stress analysis was carried out and potential fracture risks were predicted. A ranking of design options was provided to guide wafer redesigns and associated validation experiments.

References

[1] William R. Mann, Frederick L. Taber, Philip W. Seitzer, Jerry J. Broz, "The Leading Edge of Production Wafer Probe Test Technology", pp. 1168-1195, International Test Conference 2004 (ITC'04), 2004.

[2] www.feinmetall.com: ViProbe principle.

[3] Stephen P. Timoshenko: "Theory of Elastic Stability, Second Edition", McGraw-Hill, 1963, pp. 76-82.

978-1-4244-4722-0/09 $25.00
© 2009 IMAPS-ITALY

Impact of substrate coupling induced by 3D-IC architecture on advanced CMOS technology

Maxime Rousseau[1,3,4], *Student Member, IEEE*, Marie-Anne Jaud[2], Patrick Leduc[2],
Alexis Farcy[1] and Antoine Marty[3,4]

[1] STMicroelectronics - 850 rue Jean Monnet - 38926 Crolles cedex, France
[2] CEA-LETI/MINATEC - 17 rue des Martyrs - 38054 Grenoble cedex 9, France
[3] CNRS ; LAAS ; 7 avenue du colonel Roche, F-31077 Toulouse, France
[4] Université de Toulouse ; UPS, INSA, INP, ISAE ; LAAS ; F-31077 Toulouse, France

Phone: +33 4 38 78 06 35, fax: +33 4 38 78 50 12, mail: maxime.rousseau@cea.fr

Abstract

A TCAD-based simulation approach is proposed to study the impact of transient coupling that occurs within a generic 3D integration on 65 nm technology based CMOS devices. This coupling is mainly due to signals applied on redistribution layer (RDL) and through-silicon vias (TSV). These both 3D-inherent metal structures may cause variations on normal operating conditions of advanced devices. Influence of design and technology parameters such as keep-away zone, TSV/RDL isolation oxide thicknesses and remaining silicon thickness are investigated on NMOS transistors, in order to extract application-driven 3D-specific design rules. We also show that significant variations on saturation drain current and especially on leakage current appear each time TSV/RDL-applied signals switch. These current variations are strongly dependent on rise and fall potential ramp times applied on TSV or RDL. Shorter rise or fall ramp time induces a more aggressive coupling on devices. In certain cases, it may be destructive for advanced CMOS technology. Dynamic variations on saturation drain current can be tolerated under specific design rules and process options but those on leakage current are very important compared to static leakage current value, and are of the order of 10^{-6} A/µm.

Key words: 3D integration, through-silicon via, redistribution layer, substrate coupling, TCAD, device performance.

Introduction

As 2-dimensional IC scaling becomes more and more difficult to achieve for the next technological nodes, 3D integration technology is being considered as a real breakthrough approach. It seems to be an interesting solution in terms of IC design and manufacturing [1,2]. 3D integration doubtlessly brings significant benefits concerning circuit performance, density of integration, interconnect power consumption and heterogeneous technology integration capabilities [3].

Microelectronics worldwide actors tend to develop a through-silicon via technology (TSV) in order to interconnect stacked ICs. TSV appears to be one of the greatest technology challenges brought by 3D integration. But the redistribution layer (RDL) can also be considered as a brand new entity in conventional IC architectures. RDL features large metal lines implemented on the backside of the thinned active stratum. Many papers showed that 3D integration process has no impact on CMOS technology, or very limited [4-5]. Beyond process issues, the fact is that both structures are needed for power, I/Os and signal routing through all the thinned strata, leading to electrical parasitic coupling

and critical substrate noise on neighboring active devices [6]. This coupling might be restrictive for design capabilities and needs to be quantified as a function of layout and technology parameters, such as keep-away zone, TSV/RDL isolation oxide thicknesses and remaining silicon thickness. This will lead to define tunable 3D-specific design rules that will ensure reliable circuit design according to the application choice. In a first time, we propose to explain our 2-dimensional TCAD-based simulation methodology. Then, the impact of technology and layout parameters – in case of TSV or RDL induced coupling – is investigated on NMOS transistors. From that point, a specific structure for 3D integration is set and the impact of various TSV-applied potential ramp times is investigated.

Simulation methodology

Two dimensional TCAD (Technology Computer Aided Design) transient simulations [7] have been performed on a 3D integration structure (described on figure 1). Drift-diffusion transport model with usual mobility model of Lombardi and Shockley-Read-Hall generation rate are used.

978-1-4244-4722-0/09 $25.00
© 2009 IMAPS-ITALY

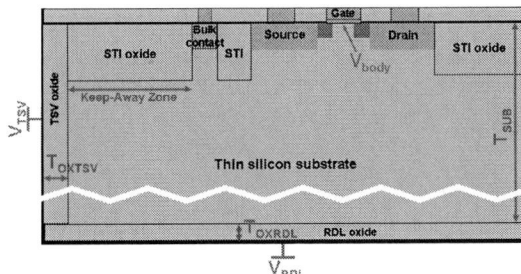

Figure 1. 2-D structure of the thinned active stratum.

3D integration can be implemented with various device technologies. This paper focuses on 65 nm bulk CMOS devices integrated with high-density TSV. Silicon substrate is thinned in the range of 5 to 30 μm (T_{SUB}). Thinned active stratum is bonded face to face with the bottom active stratum. As only electrical coupling phenomena occurring within the thinned stratum are considered, all other integration process steps do not impact our methodology. This study can be applied to a majority of case.

Simulated NMOS transistor features low power 65 nm bulk technology [8]. The channel length is 1 μm in order to avoid short-channel effects. The doping profile of the P-doped well is based on SIMS measurement. Intrinsic doping level for bulk silicon is 10^{+15} cm^{-3}. Table 1 refers to the electrical characteristics of the simulated transistor. The electrical output characteristics $I_{DS}(V_{GS})$ calculated at V_{DS}=1.2V is plotted on Figure 2. The saturation drain and leakage currents demonstrate rather accurate performance for this range of channel length.

TSV and RDL are respectively considered as intra bulk and backside electrodes – independent from each other and isolated from the thinned substrate with an appropriate oxide layer. TSV and RDL isolation oxide thicknesses are two important technology parameters. They are respectively noted as T_{OXTSV} and T_{OXRDL}.

In this work, we propose to observe how a square signal applied independently on these two electrodes may impact the electrical characteristics of the NMOS transistor. In order to understand how the substrate coupling occurs within the structure, the three technology parameters previously described as T_{SUB}, T_{OXTSV} and T_{OXRDL}, and a layout parameter, namely "keep-away zone", are investigated in the simulation methodology. The range of dimensions for each parameters investigated hereafter are listed in Table 2.

The MOS transistor is plugged on its static mode ($V_{GS} = V_{DS} = 1.2$ V). A square voltage (f = 200 MHz, 50 ps-long rise and fall ramp time, 1.2 V peak voltage) is then applied on the TSV or on the RDL. The saturation drain current I_{DSAT} and the local body potential V_{body}, located in between the conduction channel at 5 nm under the gate oxide-

silicon interface, are extracted during the transient analysis. This position enables to quantify the finest electrical disturbances that may modify the electrical behavior of the transistor.

Table 1. Electrical characteristics of the simulated Low Power NMOS transistor with a channel length of 1 μm.

	NMOS
Threshold Voltage V_T (V)	0.40
Saturation drain current I_{DSAT} (μA/μm) extracted at $V_{GS}=V_{DS}$=1.2 V	50
Leakage current $I_{D\ leak}$ (pA/μm) extracted at V_{DS}=1.2 V and V_{GS}=0 V	0.72

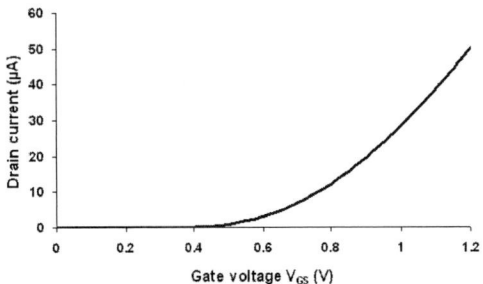

Figure 2. $I_{DS}(V_{GS})$ characteristics at V_{DS}=1.2V of the simulated NMOS transistor.

Table 2. Range of dimension of the parameters investigated in this work.

	Dimension range (μm)
Substrate thickness (T_{SUB})	5 - 30
TSV isolation oxide thickness (T_{OXTSV})	0.05 - 0.5
RDL isolation oxide thickness (T_{OXRDL})	0.4 - 1.4
Keep-away zone	2.5 - 5

TSV-induced substrate coupling

Following the methodology described above, variations on body potential V_{body} and saturation drain current I_{DSAT} are extracted during transient analysis. Maximum dynamic variations for both are reported on Figure 3 as a function of remaining substrate thickness T_{SUB} after grinding for different TSV isolation oxide thicknesses T_{OXTSV}. Potential and current variations follow the same behaviour. It is shown that a thicker oxide for TSV isolation reduces significantly variations on channel potential (3 % to 1,5 %) and saturation current (5,5 % to 2,5 %) because of the decrease of TSV oxide capacitance. In all cases, increasing the oxide thickness for TSV isolation from 50 nm to 500 nm makes decreasing the coupling in a range of 50 %. Silicon substrate thickness has a rather low impact on dynamic variations of current for thickness in the range of 10 to 30 μm. For thickness lower than 10 μm, the bulk resistance between TSV and active

978-1-4244-4722-0/09 $25.00
© 2009 IMAPS-ITALY

area increases and the TSV oxide capacitance decreases. The consequence is a significant decrease of TSV-induced coupling.

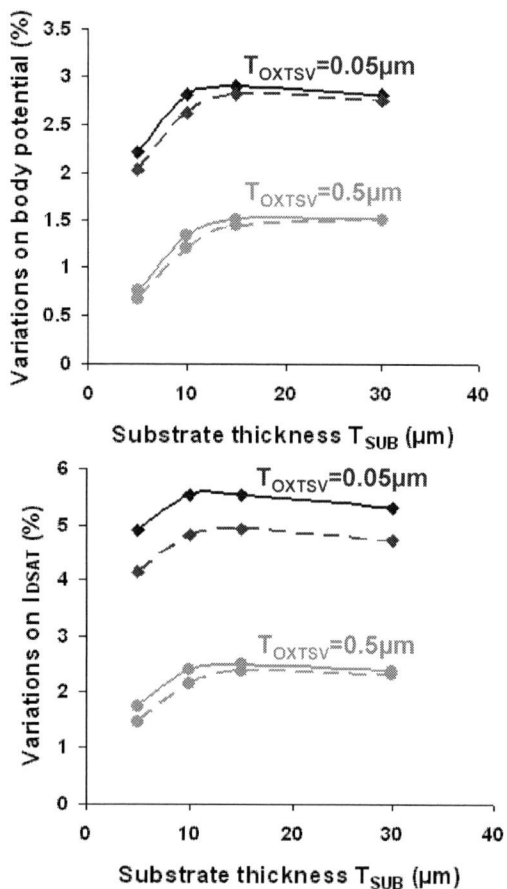

Figure 3. Maximum dynamic variations on body potential (top) and I_{DSAT} (bottom) for NMOS transistor as a function of substrate thickness T_{SUB} for two T_{OXTSV} values: {0.05 µm; 0.5 µm} and two keep-away zones: 2.5 µm (full lines) and 5 µm (dotted lines).

As well as technology parameters, layout design may impact the coupling on CMOS devices. For instance, the position of bulk contact, that helps to control the body potential, and the position of the closest active area beside the TSV edge (the so-called keep-away zone), may control more or less substrate coupling. As the bulk contact is always implemented close to a single transistor, the dimension of the STI located between the bulk contact and the source area (cf. Figure1) is set at 200 nm for all simulations. Only the impact of keep-away zone is investigated and reported on Figure 3.

The first keep-away zone at 2.5 µm is the minimum distance allowed because of the 2 µm guard for TSV alignment. The second keep-away zone at 5 µm is the typical distance not to be exceeded to keep density of device integration as high as possible. If referring to Figure 3, it seems obvious that keep-away zone, taken in this range of

dimensions, does not have enough impact to decrease coupling, compared to the influence of technology parameters. The bulk contact seems to be unsufficient to control accurately the body potential, almost independently of its position. The fact is TSV-induced coupling occurs all along the TSV, so that substrate noise propagates at the same time through the low-resistivity doped well, and through the high-resistivity bulk silicon. The lower resistivity of the active area (because of its high doping level compared to bulk silicon) makes easier the noise propagation. As the TSV has a thin isolation thickness, it can be assumed that TSV behaves like a parasitic oxide capacitance that bulk contact cannot control because of the large dimensions of the structure (bulk contact has a very localized effect). For the following, we set the keep-away zone at 2.5 µm.

By referring to Figure 3, thinning down the substrate to 5 µm - or below if possible - decreases the substrate coupling. But this will place the RDL line closer to active devices, so that RDL-induced coupling may increase.

RDL-induced substrate coupling

Impact of RDL-induced substrate coupling is reported on Figure 4.

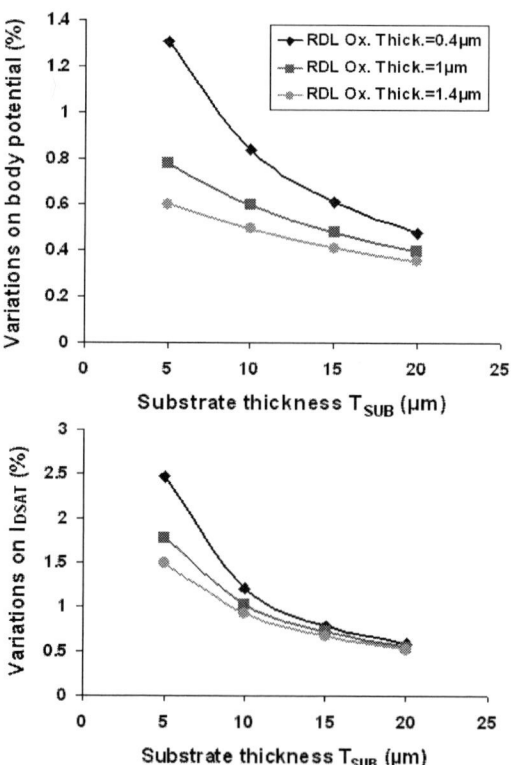

Figure 4. Maximum dynamic variations on body potential (top) and I_{DSAT} (bottom) for NMOS as a function of substrate thickness T_{SUB} for various T_{OXRDL} values: {0.4, 1, 1.4 µm}.

As expected, maximum variations on body potential and saturation drain current are observed for the thinnest substrates. Compared to previous TSV study, RDL-induced noise coupling seems to be less substantial. Maximum variation on I_{DSAT} is lower than 2.5 % for minimum RDL isolation oxide thickness (400 nm). As for TSV isolation, a thicker RDL isolation makes decreasing the coupling. Moreover, from 15 µm-thick substrate and above, there is no more effect of RDL isolation thickness on variations of saturation drain current. It may be due to the fact that RDL isolation behaves like an oxide capacitance whose effect decreases when the bulk resistance increases with thicker silicon substrate.

Coupled impact of TSV and RDL

It was explained how TSV and RDL structures produce substrate noise independently and what typical response on active devices is. Considering the TSV is connected to the RDL line – meaning that the same signal is applied on both structures - their respective coupling noise can be added to obtain the global substrate noise produced by 3D architecture. The global response on NMOS saturation drain current is depicted on Figure 5, for equivalent thickness of TSV and RDL isolation (400 nm). The highest coupling response on I_{DSAT} is obtained for the thinnest substrate (5 µm). At a glance, the global behaviour of the curve is driven by RDL coupling behaviour, mainly because the TSV coupling remains rather constant around 2 and 3 %. Beyond that, the global coupling jointly produced by TSV and RDL may be considered as mainly driven by TSV from 15 µm thick and more because of the significant decrease of RDL coupling for this range of substrate thickness.

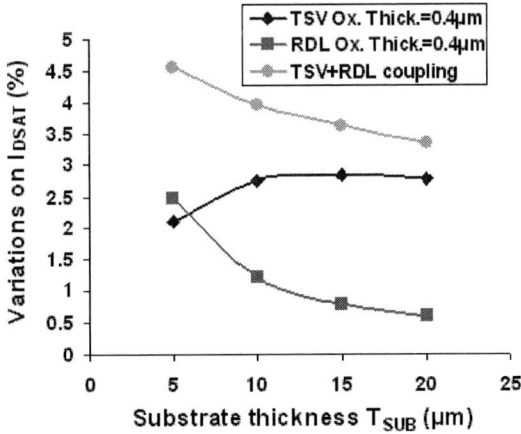

Figure 5. Contributions of TSV and RDL induced coupling on saturation drain current.

Impact of TSV potential ramp times on coupling

It was demonstrated that substrate noise is produced by 3D integration, *i.e.* TSV and RDL structures, when a square signal is applied.

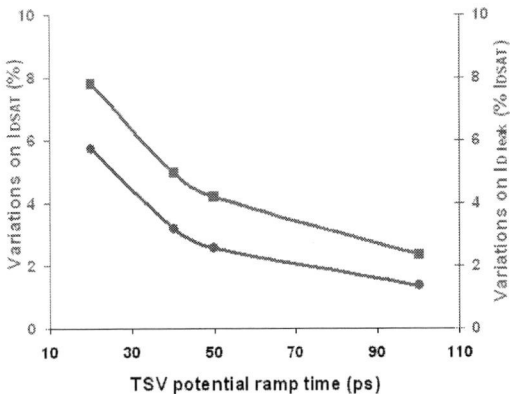

Figure 6. Maximum dynamic variations on I_{DSAT} (green squares) and $I_{D\,leak}$ (blue circles expressed in terms of % I_{DSAT}) for NMOS as a function of various rise and fall ramp times applied on TSV (from 20 ps to 100 ps).

As variations on I_{DSAT} and channel potential are only detected during TSV/RDL potential ramp time, parasitic potential may have a scalable impact depending on its ramp time. In this part, similar results as previously described are reported for increasing TSV potential ramp times, from 20 ps (most aggressive ramp time in CMOS65 technology) to 100 ps. Moreover, leakage current (I_{DS} for $V_{DS} = 1.2$ V and $V_{GS} = 0$ V) is also investigated. These results are summarized in Figure 6. For this study, substrate thickness is 5 µm, keep-away zone is 2.5 µm and oxide thickness for TSV isolation is 50 nm (worst case).

NMOS currents seem to be affected quite similarly. Variations on saturation drain current and variations on leakage current are expressed in the same unit (% I_{DSAT}) so that they can be compared to each other. Variations on leakage current appear dramatically important because in the range of those on I_{DSAT} (see Table 1 for static current values).

In all cases, and independently of the ramp time, the dynamic variation on leakage current is really critical for logic circuits. It means that the transistors could be turned on temporarily while the off state is established.

The strongest impact on drain currents is shown to appear with the shortest ramp time applied on TSV potential, here 20 ps. In that case, maximum I_{DSAT} variation is 8 %. For instance, 10 % is the maximum accepted value for static I_{DSAT} shift after 5-year-long normal operation (transistor aging).

The effect of TSV isolation oxide thickness T_{OXTSV} is also investigated regarding the results on drain leakage current for various TSV potential ramp times. These results are reported on Figure 7. It is shown that an increase of TSV isolation oxide thickness from 50nm to 300nm decreases the impact on leakage current of only 50 %. That makes typical shifts on leakage current in the range of 10^{-7} to 10^{-6} A/µm compared to its static value at 10^{-13} A/µm.

978-1-4244-4722-0/09 $25.00
© 2009 IMAPS-ITALY

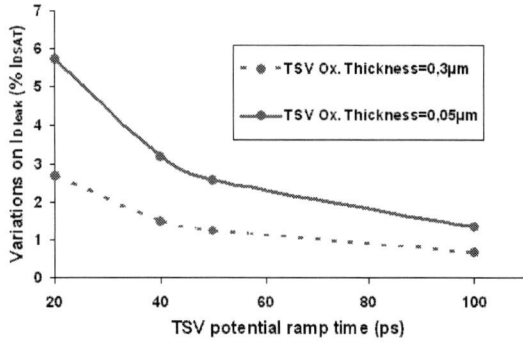

Figure 7. Maximum dynamic variations of leakage current for NMOS transistor at various TSV potential ramp times for two values of T_{OXTSV}: 50 nm (solid lines) and 300 nm (dotted lines).

In all cases, the variations on both currents only occur during the parasitic signal ramp time. Once the TSV potential remains static in either its state 'on' (1.2 V) or 'off' (0 V), no more electrical coupling occurs through the substrate. Leakage current or saturation drain current go back to their initial static values.

Conclusion

Through-silicon vias and redistribution layers, respectively considered as intra bulk and backside electrodes, generate dynamic parasitic coupling within the thinned silicon substrate in a high density 3D integration. This dynamic noise propagates through the thin silicon, mainly in the active area where the resistivity is lower than in the bulk silicon. It was also shown that bulk contact is not efficient enough to control accurately the variation of substrate potential (the so called body potential) through all the depth of silicon. Moreover, technology parameters like oxide thickness for TSV/RDL isolation and silicon thickness have a significative impact on control of the coupling, unlike the layout parameter (keep-away zone). Electrical impact of this dynamic coupling on NMOS transistors has been observed on drain saturation current, leakage current and body potential, with an equivalent behaviour of the electrical response on each of them. Concerning saturation drain current, dynamic variations are less than 8 % that is possibly critical for transistor aging and may disrupt normal operating conditions of CMOS logic circuits. Parasitic potential ramp time seems to be a critical parameter that makes increasing coupling when shorter. What is the most worrisome fact is probably the variations of leakage current. Coupling increases the static steady-state value of 10^{-13} A/μm for NMOS to around 10^{-6} A/μm, *i.e.* the range of saturation drain current variations. Transistors under their 'off' state may then move to their 'on' state temporarily.

Acknowledgements

This work has been carried out in the frame of CEA-LETI/MINATEC and STMicroelectronics collaboration. The authors would like to thank Simon Deleonibus, Olga Cueto and Jean-Charles Barbé from CEA-LETI/MINATEC for fruitful discussions about this work.

References

[1] K. Banerjee *et al.*, "3D ICs: a novel chip design for improving deep-submicrometer interconnect performance and Systems-on-Chip integration", Proceedings of the IEEE, vol. 89, No. 5, May 2001.

[2] W.R. Davis *et al.*, "Demystifying 3D ICs: The pros and cons of going vertical", IEEE Design & Test of Computers, vol. 22, No. 6, pp. 498-510, November/December 2005.

[3] S.J. Souri *et al.*, "Multiple Si layer ICs: motivation, performance analysis and design implications", Proceedings of the 37th Conference on Design Automation (DAC'00), pp. 213-220, 2000.

[4] P.R. Morrow *et al.*, "Three-dimensional wafer stacking via Cu-Cu bonding integrated with 65-nm strained-Si/Low-k CMOS technology", IEEE Electron Device Letters, vol. 27, No. 5, pp. 335-337, May, 2006.

[5] N. Tanaka *et al.*, "Characterization of MOS transistors after TSV fabrication and 3D-assembly", Proceedings of the 2nd Electronics Systemintegration Technology Conference (ESTC'08), Greenwich, UK, September 1-4, pp. 131-134, 2008.

[6] M. Rousseau *et al.*, "Through-silicon via based 3D IC technology: electrostatic simulations for design methodology", Proceedings of the 2008 IMAPS Device Packaging Conference, Phoenix, Arizona, March 17-20, 2008.

[7] Silvaco International, Atlas User's Manual, 2008, USA, www.silvaco.com

[8] ITRS roadmap, www.itrs.net

978-1-4244-4722-0/09 $25.00
© 2009 IMAPS-ITALY

Versatile MEMS and MEMS integration technology platforms for cost effective MEMS development

Philip Pieters

IMEC
Kapeldreef 75, 3001 Leuven, Belgium
Phone: +32 16 281259; Fax: +32 16 281812; E-mail: Philip.Pieters@imec.be
www.imec.be

Abstract

For fast and cost effective development of novel MEMS devices, it is advantageous to start from advanced and stable MEMS technology platforms. IMEC offers such platforms for post-CMOS MEMS integration, RF-MEMS, MEMS interconnection and packaging and for Si-photonics. In this paper these versatile MEMS technology platforms together with various examples are described. In IMEC's CMORE offering, not only technology development but also prototyping and small volume production becomes possible.

Key words: MEMS integration, post-CMOS MEMS, RF MEMS, MEMS packaging, photonics

1. Introduction

MEMS or MEMS-based devices can offer a lot of added value for many applications. This may be for (potentially) high volume applications for telecom or consumer electronics, but this may also be for small volume applications in niche markets. In the latter case MEMS can be the key enabling factor for such dedicated application.

Developing a new MEMS process or new MEMS application involves a lot of money. If the market is estimated to be (initially) small, the risk to come to a profitable product is high. One way to reduce the development cost and risk is to try to get a "head start" by starting the development from a technology level that is as mature and already as industrial as possible. For this reason, IMEC has been working towards establishing stable and versatile MEMS and MEMS integration base-line technologies platforms. These base-line technology platforms are the ideal starting point for, on the one hand the creation of new devices with minimal development effort, and on the other hand also allow doing small volume production of such novel devices. Because of the advanced level of these base-line technology platforms, because of the use of industrial equipment and because these tools are shared for different developments or applications, the cost per new development and the related time to market can be minimized.

At IMEC base-line technology platforms exist for above-IC MEMS integration (based on poly-SiGe), for RF MEMS applications, and Si photonics. Additionally, IMEC has base-line technology platforms to package, interconnect or integrate MEMS into systems. These platforms comprise multilayer thin film interconnection technologies with integrated passive devices, wafer level MEMS capping technologies, and 3D stacking technologies.

In the paper, we will give an overview of the different MEMS and MEMS integration technology platforms and show results of recently developed innovative devices built from these platforms.

.

2. Post-CMOS MEMS intergration

Post-processing, integrating and packaging MEMS devices directly above their CMOS control circuitry allows a significant reduction in size of the total system function and provides a very short and hence very low loss interconnection.

In this field of MEMS on CMOS integration, IMEC has developed a low temperature poly-SiGe MEMS post-IC processing technology. In the next sections this approach is further described.

2.1 MEMS-CMOS integration

The majority of current MEMS products on the market still use a hybrid approach for the MEMS and the processing circuitry. Such an approach is modular and, as a consequence, has a shorter development time as compared to the monolithic approach. Also it allows for an independent optimization of the integrated circuit (IC) and the MEMS technology. Disadvantage of the hybrid approach are on the one hand the higher assembly and packaging cost and on the other hand the fact that the interconnections between the MEMS and the processing circuitry induce additional parasitic effects that may limit the system performance. Monolithic integration of MEMS and processing circuitry yields simpler assembly and packaging with minimum interconnection parasitics.

978-1-4244-4722-0/09 $25.00
© 2009 IMAPS-ITALY

Post-CMOS MEMS integration is, in our view, the most promising approach for monolithic integration as it enables integrating MEMS without introducing any changes in the standard CMOS fabrication process. In fact, this type of MEMS back-end-integration keeps a modular approach to a large extent: MEMS and IC can first be developed and optimized separately. It is only in a later stage of the development path, when a certain level of optimization is already reached, that MEMS are processed on top of the IC surface. In addition, a new generation of circuitry can easily replace the older one without affecting the MEMS on top of it. However, post-processing imposes very stringent requirements towards the allowed MEMS processes. It restricts for instance the chemicals that can be used and, most importantly it limits the maximum fabrication temperature of the MEMS because of the risk of damaging the existing electronics or degrading its performance.

2.2 IMEC's Poly-SiGe technology for the realization of above-IC MEMS devices

Polycrystalline silicon (Poly-Si) has been widely used for MEMS applications (e.g. accelerometers, gyroscopes, ...). This material requires a high processing temperature (>800°C) to achieve a low tensile stress and to activate dopants. Poly-SiGe is an attractive alternative to poly-Si, as it has similar properties, while the desired electrical and mechanical properties can be realized at a temperature suitable for post-processing MEMS on top of standard CMOS wafers with Al on-chip interconnects and W plugs [1,2]. Most CMOS processes are capable of withstanding post-processing temperatures of 450–520ºC, possibly with a slight increase in the on-chip interconnect sheet resistance which can be taken into account during the design phase [3].

IMEC has been working for many years on a above-IC MEMS integration technology using poly-SiGe. Various applications and processing details are presented in [1-9]. Recently IMEC published CMOS integrated 11 MPixel SiGe based micro mirror array for high-end industrial applications [10]. This device was realized on top of standard 0.18μm analog-CMOS wafers fabricated by NXP featuring 6 interconnect levels, 8μm pitch micro-mirrors, spaced 0.3 μm apart. A photograph of the completed die is shown in Figure 1. Cross-sections are shown in Figure 2.

3. RF MEMS

The RF MEMS activities at IMEC focus on the development of generic technology platforms integrating switches as well as tunable and fixed passives components (see also [11-13]).

The reliability of RF-MEMS switching devices is very important. The lifetime of capacitive switches is mainly limited by dielectric charging. By realizing a capacitive switch with only air as dielectric its lifetime can be significantly improved [14].

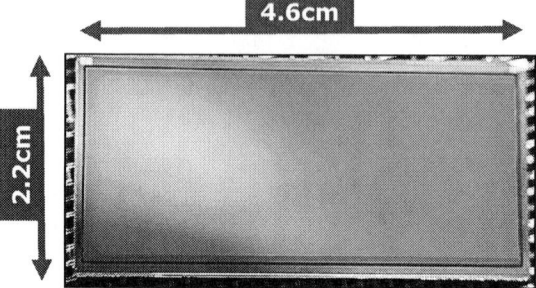

Figure 1. Photograph of the 10cm² large die with the 11 Mpixel micromirror array integrated on top.

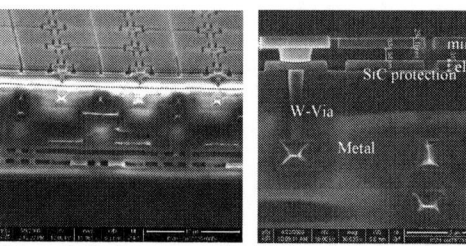

Figure 2. Left: cross-sectional view of integrated micro-mirror array, showing the mirrors on top of the 6 layers of Al interconnect. Right: cross-sectional of mirror showing W-via, SiC protection layer, SiGe electrodes, SiGe mirror layer.

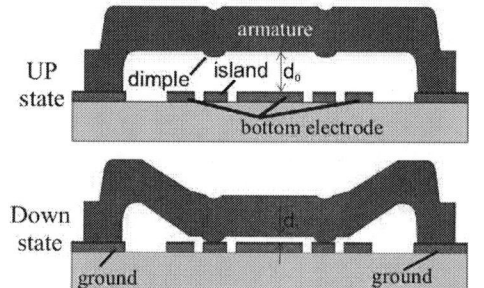

Figure 3. Conceptual cross-section of the air-gap based capacitive switch [14].

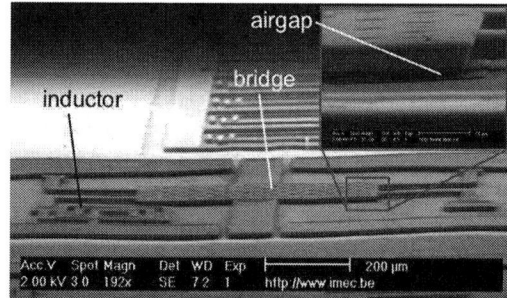

Figure 4. Top view SEM image of the air-gap based capacitive switch [14].

The switch shown in figures 3 and 4 consists of a thick membrane that defines an airgap capacitor with 2 different states. By switching between the 2 capacitor states (even though having a low capacitance ratio) adequate switching and RF circuits can be

978-1-4244-4722-0/09 $25.00
© 2009 IMAPS-ITALY

realized by proper design and combining these devices with high-Q inductors and transmission lines. Lifetimes of more than 1×10^8 cycles with unipolar actuation are observed.

4. RF MEMS circuit integration

The RF-MEMS technology as e.g. described above offers the possibility to define tunable passive devices and circuits. These components are however characterized by lower quality factors than the fixed passives realized in the multilayer thin film (MCM-D) integrated passive device (IPD) technology available at IMEC [15-16] (see also Figure 6). The realization of fixed and tunable fully integrated bandpass filters in MCM-D/RF-MEMS technologies for GPS-Galileo bands/band-switching confirmed this conclusion. However, by proposer selection of a filter architecture, the amount of switchable capacitors may be limited. They can be isolated and in this way a hybrid filter with a flip-chipped a RF-MEMS chip (constituing of 1 shunt and 2 series MEMS capacitors) on a MCM-D substrate, (incorporating high-Q inductors) could be realised. Figure 5 shows a picture of this hybrid filter as well as the transmission measurement results. The results are in good agreement with the design goals, i.e. 1.2 GHz and 1.6 GHz center frequencies for 100 MHz bandwidth. The insertion loss is better than 6 dB, very close to the values realizable in MCM-D only version for such narrow filters.

A schematic view of the MCM-D/RF-MEMS technology platform with IPD's is shown in Fig. 6.

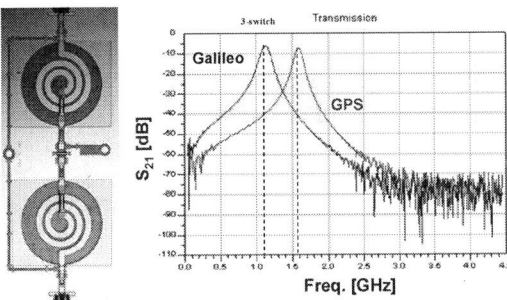

Figure 5. Photograph (left) and measured S_{21} parameter (right) of the tunable Galileo-GPS bandpass filter using MEMS tunable capacitors.

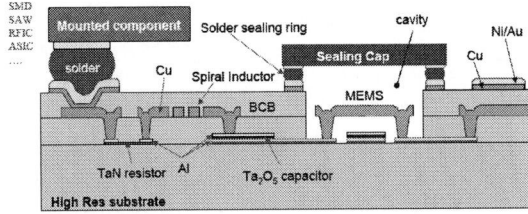

Figure 6. Schematic view of IMEC's MCM-D/RF-MEMS technology platform for high density integration of RF-MEMS components [16].

5. MEMS wafer-level packaging

The performance of MEMS devices is direct-ly influenced by the package containing the device. So, during the development phase of the device, the packaging and interconnection has to be developed at the same time. IMEC is very much involved in this "MEMS-package co-design" approach in order to realize functional and highly reliable MEMS devices. There are different approaches

5.1. Thin film wafer level poly-SiGe MEMS packaging

Next to integrating MEMS devices directly above CMOS, the poly-SiGe technology presented in section 2 can also be used for wafer level encapsulation of the devices. These thin film MEMS caps are typically thinner and consume less lateral area than traditional wafer bonded caps. In Figure 7 a cross-section of thin film poly-SiGe capping of MEMS capacitive fingers is shown [17-18].

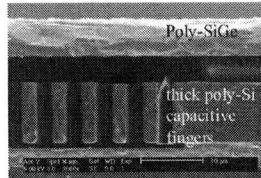

Figure 7. Thin film poly-SiGe cap on capacitive fingers [17-18].

5.2. Wafer level MEMS packaging using flip-chip or wafer bonded caps

Another way to protect MEMS devices is to mount caps on top of the devices using flip-chip or wafer bonding techniques. The caps can be mounted above the MEMS device using a polymer glue layer ring (e.g. BCB) that is cured, or by applying a solder ring that is reflown. The polymer solution is easier to realize and allows easier signal feedthrough realization, but is typically not fully hermetic. The solder based solution can potentially yield hermetic packages but makes it more difficult to electrically connect to the devices. In Figure 8 an RF MEMS switch with a glas cap made by wafer level processing is shown.

Figure 8. Wafer level glas cap on top of an RF MEMS switch.

6. Si-Photonics

The advanced CMOS processing equipment at IMEC can be used for photonics applications. Especially the lithography equipment (e.g. the 193 nm scanner) is important with respect to critical dimensions and reproducibility. Not only all kinds of passive components (splitters, resonators, couplers, ...) can be realized,

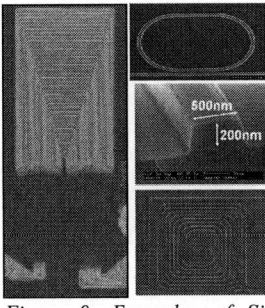

Figure 9. Examples of Si-photonic devices.

but also integration of active components together with the photonic passives by using e.g. 3D technology is possible.

7. CMORE

From the above, it is clear that IMEC has built-up extensive know-how and expertise in the field of MEMS. IMEC has a strategy to develop stable technology platforms and for thereon further extent its know-how by performing new research projects. In this way a base-line technology becomes available for application oriented projects (which aim to realized device prototypes and products), while innovative research activities remain possible.

The availability of stable MEMS, photonics and high density integration and packaging technology platforms have led to the setting up IMEC's so-called CMORE program. "CMORE" refers to "more than CMOS". It covers the More than Moore activities at IMEC where we aim to combine CMOS processing capacities together with MEMS, packaging and interconnect capabilities in order to realize new application and small series production for customers.

IMEC's capabilities include 8000 m² of state of the art cleanroom, a 200 mm wafer size CMOS pilot line covering 130 nm and 90nm CMOS nodes, the in this paper described MEMS and packaging technology platforms and various other capabilities in the field of RF, high voltage, sensors and detectors. This is a unique combination of processing infrastructure and processing know-how that can be devoted for the realizing of innovative product development.

8. Summary and conclusion

In this paper we have given an overview of the MEMS activities at IMEC. We presented the poly-SiGe technology for efficient integration of MEMS on CMOS, we discussed IMEC's RF MEMS technologies and we showed Si-photonics examples. Next to this, we presented packaging, integration and MEMS capping technologies. It is essential that the packaging and interconnection approaches are taken into account in the MEMS device development as soon as possible as the package is an essential part of the total MEMS system and may considerably influence the device performance and reliability.

The presented MEMS technology platforms are very versatile and can be used as starting point for fast and cost effective innovative MEMS developments. Because of the use of production oriented equipment, these developments can easily be followed by small volume production at IMEC in IMEC's 'CMORE' program.

Acknowledgements

The author acknowledges the MEMS teams of IMEC for their input for this paper and for their continuous effort and wit to develop innovative MEMS solutions.

References

[1] S. Sedky, et al., 'Poly SiGe, a promising material for MEMS post-processing on top of standard CMOS wafers', in Proc. Transducers 2001, pp. 988-991.

[2] S. Sedky, et al., 'Poly SiGe, a Promising Material for Monolithic Integration With the Driving Electronics', Sensors and Actuators A, Vol.97-98, pp. 503-511, 2002.

[3] S. Sedky, et al., "Experimental determination of the maximum post-process annealing temperature for standard CMOS wafers", IEEE Trans. on Electron Devices, Vol. 48(2), pp. 377-385, 2001.

[4] S. Sedky, et al., 'IR bolometers made of polycrystalline silicon germanium', Sensors and actuators A 66, pp. 193-199, 1998.

[5] A. Mehta, et al., 'Novel High Growth Rate Processes for Depositing Poly-SiGe Structural Layers at CMOS Compatible Temperatures', Proc. IEEE MEMS 2004, pp. 721-724.

[6] M. Gromova, et al., 'The Novel Use of Low Temperature Hydrogenated Microcrystalline Silicon Germanium (μcSiGe:H) for MEMS Applications', J. of Microelectr. Eng., vol. 76, pp. 266-271, 2004.

[7] M. Gromova, et al., 'Highly reliable and extremely stable SiGe micro-mirrors", Technical Digest of 20th IEEE International MEMSConference, MEMS 2007, Kobe, Japan, pp. 759-762, 2007.

[8] http://www.imec.be/SiGeM/SiGeM_web.pdf

[9] A. scheurle, et al., 'A 10μm thick poly-SiGe gyroscope processed above 0.35μm CMOS", Technical Digest of 20th IEEE International MEMS Conference, MEMS 2007, Kobe, Japan, pp. 39-42.

[10] L. Haspeslagh, et al., "Highly reliable CMOS-integrated 11MPixel SiGe-based micro-mirror arrays for high-end industrial applications", Proceedings IEDM 2008, San-Francisco, December 2008.

[11] T. Lisec and B. Wagner, "Integration aspects of RF-MEMS technologies", Proc. of the European Microwave Conference, pp. 297-300, 2005.

[12] O. Vendier, et al., "Main achievements to date towards the use of RF MEMS into space satellite payloads", Proc. of the European Microwave Conference, pp. 285-288, 2005.

[13] D. Dubuc, et al., "MEMS-IC integration for RF and millimeterwave applications", Proc. of the European Microwave Conference, pp. 529-532, 2005.

[14] P. Ekkels, et al, "Simple and robust air gap-based MEMS technology for RF-applications", Proc. 9th Int. Symp. on RF MEMS and RF Microsystems, MEMSWAVE 2008, Heraklion, Greece, 2008.

[15] X. Rottenberg, et al., "Filter-Through device: A distributed RF MEMS capacitive series switch", Journal of Micromechanics and Microengineering, vol. 15, 2005, S97-S102

[16] H. Tilmans, et al., "MEMS for wireless communications: 'from RF-MEMS components to RF-MEMS-SiP'", J. of Micromech. and Microeng., vol 13 (2003), S139–S163

[17] C. Rusu, et al., "MEMS 0-level packaging using thin film poly-SiGe caps", Proc. IMAPS ATW on Packaging of MEMS and Related Micro Integrated Nano Systems, Denver, Colorado, Sept 6-8, 2002.

978-1-4244-4722-0/09 $25.00
© 2009 IMAPS-ITALY

[18] H. Stahl, et al., 'Thin Film Encapsulation of Acceleration Sensors Using Polysilicon Sacrificial Layers' Proc. Transd. 2003, pp. 1899- 1902, 2003.

Reliability Testing of Frequency Converters with Salt Spray and Temperature Humidity Tests

J. Kiilunen and L. Frisk

Department of Electronics, Tampere University of Technology, P.O. Box 692, FIN-33101 Tampere, Finland

Phone: +358 40 849 0621, Fax: +358 3 3115 3394, E-mail: janne.kiilunen@tut.fi

Abstract

In this study an accelerated corrosion test was developed for a frequency converter with markedly high reliability. Constant humidity and salt spray testing were used, as they are commonly used methods for accelerated corrosion testing. The purpose of the test was to study the persistence of the device against corroding atmosphere and to examine failure locations and mechanisms. Testing included 24 hour exposure to neutral salt spray in accordance with ISO 9227, after which the excess salt residues were removed with deionized water. This was followed by a 24 hour drying period at 23 °C and 50 %RH. After drying the functionality of frequency converters was tested. The corrosion effect was then precipitated by placing the devices in 85 °C and 85 %RH temperature and humidity test for 500 hours. The aforementioned drying period was then repeated before final functional testing. The functional testing was performed by running the frequency converters with overvoltage in a test rig. In order to study the effect of the salt spray, another eight converters were tested using only the 85 °C and 85 %RH temperature and humidity test. No failures were seen in these converters during the test. However, five of the converters exposed to the salt spray failed during or after the test. Using salt spray testing, quick results could be obtained, as normally the test times required to fail the frequency converter can be very long. The interface between the printed circuit board and the power module of the device was noticed to be the most susceptible to corrosion damage.

Key words: Corrosion, Frequency Converter, Reliability, Salt Spray

Introduction

Accelerated environmental testing is commonly used to investigate the reliability of various electronics products. The aim of these tests is to predict the future performance of a product for a shorter period of time than the service life of the product [1]. Accelerated life tests can also be used to detect failure mechanisms occurring in products under different conditions of use. The acceleration is accomplished by using elevated stress levels or higher stress cycle frequency during testing compared to those under normal operational conditions of the product [2]. The use of elevated environmental stresses is especially important in applications, which have very high reliability. Testing of high reliability electronics can take considerable time and using well designed accelerated tests this testing time can be decreased significantly.

Depending on the testing conditions failures may occur through several mechanisms. It is essential that the testing conditions are determined so that the failure mechanisms during testing are similar to those occurring under normal conditions of use. The test conditions depend significantly on the application of the product. For example, the test conditions for products used in military and space applications are much more rigorous than those for consumer electronics.

The aim of this study was to investigate the reliability of a frequency converter with high reliability demands. A frequency converter, or commonly inverter, is a device that can convert the frequency of alternating current and they are used to control the speed, direction of rotation, and torque of AC motors. In comparison to driving an AC motor with constant speed, significant energy and power savings can be achieved with frequency converters. Currently, frequency converters are used all around the world in all kinds of service conditions. The environmental stresses these devices may face during operation can include high and low ambient temperatures, high humidity, voltage transients and interruptions, and corrosion. Because of the large variations in use conditions, the reliability of the devices has to be tested with several different accelerated environmental tests.

In order to study the corrosion resistance of a three-phase inverter a salt spray and constant temperature/humidity test was used. The object was to examine failure locations and mechanisms, but also the applicability of the test method. The electrical functionality of the devices was tested during and after the environmetal tests in a purpose-built test rig. The failed devices were examined with the help of an optical microscope.

Experimental

Frequency converters were investigated in this study. The studied device was a 0.75 kW three-phase frequency converter with input and output voltage ranges of 380-480 V and 0-480 V, respectively. The input frequency (50/60 Hz) can be altered between 0-120 Hz. The device consisted of a single printed circuit board (PCB) inside of an aluminium casing. The rectifier stage and the insulated gate bipolar transistor (IGBT) semiconductor switches were housed in a separate module between the PCB and the casing. The operability of the devices was tested in a test rig where the frequency converters drove an 80 Hz AC motor. An attached weight plate was used as the load of the motor. With this test setup, the three-phase supply voltage and its frequency could be altered, supply voltage interruptions of desired lengths could be produced, and the input current of the device could be measured in real time. The devices were controlled with a PC running a National Instruments LabView coded program. The frequency converters drove the motors in drive cycles comprising of "on" and "off" periods, during which the motors accelerated and decelerated cyclically. The test setup is described in more detail in [3].

To test the reliability of the device, two environmental stress tests were used. A salt spray test based on ISO 9227 standard [4] was conducted with Ascott S450XP salt spray corrosion chamber. The test temperature was 35 °C, relative humidity was 98 %, and concentration of the salt water sprayed on to the devices was 5 %. The duration of the test was 24 hours. The second test was a constant temperature/humidity test in accordance with JEDEC's JESD22-A101-B [5] and it was done with Espec Global-N EGNX12-6CAL chamber. The test temperature was 85 °C and the relative humidity was 85 %. The test lasted for 500 hours.

Sixteen frequency converters were tested. In order to study the effect of salt fog testing eight of the tested converters were first placed inside the salt spray chamber with their plastic covers removed for a 24 hour period. After the 24 hours, the devices were removed and accumulated salt residue was rinsed of with deionized (DI) water. Inverters were then dried with paper towels and a warm-air heater. After that, the devices were also disassembled, i.e. the printed circuit board (PCB) was removed for a more thorough 24 hour drying period. This procedure was done to avoid possible short circuits during operational testing because of residuary moisture inside e.g. the transformer coil structure or IGBT module. After the drying process the functionality of the devices was tested by placing them to the test rig for three complete drive cycles with nominal supply voltage.

After the salt fog test all sixteen inverters were put to the constant temperature and humidity test for 500 hours. This test was conducted without an operating voltage. The objective was to accelerate the corrosion process with the elevated temperature and humidity. After this test the inverters were again left drying for a couple of days but without disassembly. The condition of the inverters was then checked visually and with a fluke multimeter before a final operational test in the test rig. In the preceding check, the condition of the components on the PCB, the functionality of the IGBTs, and the connectivity of the PCB and the IGBT module were inspected. In the final test, the devices were subjected one at a time to an overvoltage of 520 V (50 Hz) for three drive cycles during which the input phase current was measured in real time.

Results and Discussion

After the constant temperature and humidity test no failures were seen in the converters without exposure to the salt fog. All eight samples passed the operational test that was performed after the drying period. On the other hand, five of the converters exposed to the salt fog failed during the test. This indicates that the exposure to the salt markedly increases the occurrence of failures in these types of devices. Results and failure types of the eight converters subjected to the salt fog and constant temperature and humidity testing are presented in table 1, below.

Table 1. Failure modes of tested devices.

Device:	Status:	Failure mode:
1	Failed	Transformer coil wire damaged
2	Failed	IGBT module short circuit
3	Intact	-
4	Failed	Inoperable IGBT module and corroded PCB
5	Failed	Corroded PCB and IGBT module interface
6	Failed	Abnormal drive current during operation
7	Intact	-
8	Intact	-

As can be seen from table 1, three devices remained intact after the operational test (devices 3, 7, and 8). Based on the measurements done before the final functional test, two test samples, devices 1 and 5, were noticed to have failed. The remaining three devices (2, 4, and 6) either failed during the electrical testing or their inoperability was then discovered. Findings from the analysis done to the failed devices are presented next in more detail.

A picture of the upper PCB side of device 1 is presented in figure 1.

978-1-4244-4722-0/09 $25.00
© 2009 IMAPS-ITALY

Figure 1. Top of the PCB of device 1 after salt spray and temperature/humidity tests.

As can be noticed from the picture above, salt residues were found on top of the PCB and especially on the joints of the components despite the cleansing with DI water after the salt spray test. This was common for all the tested samples. With device 1, measurements from the secondary winding of the voltage transformer showed unusually large resistance values. In other words, the connection between the secondary winding wire and one of the contact pins was damaged, which inhibited the operation of the device.

Device 2 suffered from a large current spike during the first drive cycle, which caused the safety contactor to trip in the test rig. Failure analysis showed that three gate resistors and the current measurement resistor had been damaged on the PCB of the sample (fig. 2).

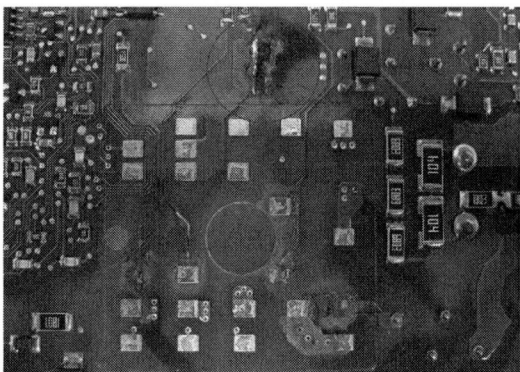

Figure 2. PCB underside of device 2. Broken current measurement resistor circled.

The contacts between the PCB and the IGBT module were also found to be corroded (fig. 2). In addition, the corrosion could be noticed to have spread under the solder resist, which was seen as discolouration of the conductor line. This was also observed with other sample devices.

IGBT module measurements indicated that two IGBTs were short circuited between the gate and output phase and one between the gate and negative DC bus voltage. Disassembly of the device

also revealed imprints of the PCB solder resist coating on top of the IGBT module casing (fig. 3). It seems that salt residue and moisture between the PCB and IGBT module interface caused the shorting of the device.

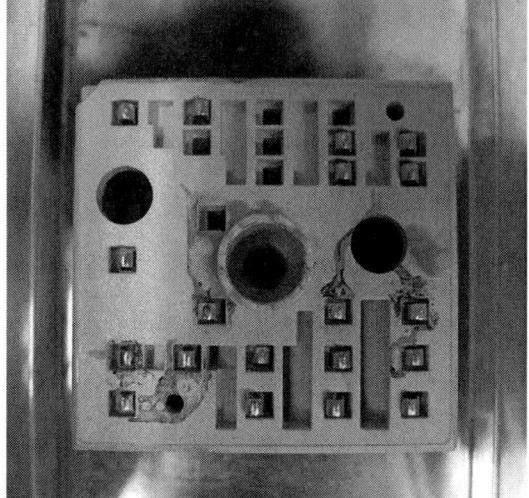

Figure 3. IGBT module casing of device 2.

When put to the test rig, device 4 switched on but signal leds on the PCB indicated a fault situation immediately. Because of this, the device did not drive the motor during the drive cycle. Inspection of the sample showed that on the PCB side, the gate contact of one of the IGBTs had clearly corroded (fig. 4) but it was still functional.

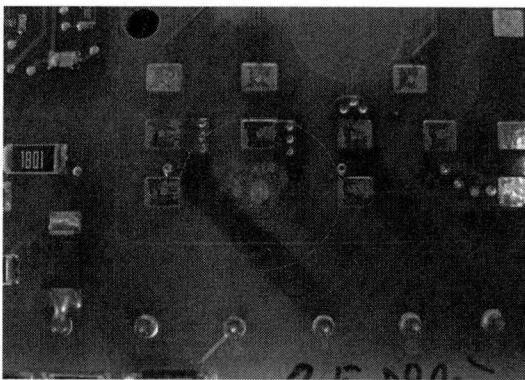

Figure 4. PCB underside of device 4. Corroded contact circled.

The contact pins on the top side of the IGBT module also seemed to be intact at first. However, when the module casing was opened, two contact pins were noticed to have rusted almost completely through at the middle section (fig. 5).

978-1-4244-4722-0/09 $25.00
© 2009 IMAPS-ITALY

663

Figure 5. Corroded IGBT module contact pin of device 4.

The PCB of device 4 was tested with a new IGBT module but the same fault situation occurred as previously. The corroded contact pins of this device's IGBT module were also replaced and it was then tested with and working PCB. At the start of the first drive cycle, the device failed to operate as it indicated a short circuit error with its signal leds. Measurements with a multimeter did not reveal the reason for the fault situation and visually the IGBTs appeared to be undamaged. However, traces of corrosion debris from the contact pins were found on top of the silicone gel, which protects the IGBTs. It is possible that impurities and moisture can get into contact with the IGBTs via the contact pins.

Device 5 suffered from corroded connections between the PCB and the IGBT module, which prevented the operational testing. The contact pins on the IGBT module had not suffered much, but on the PCB side, two input phase contacts, one DC bus contact, and one IGBT gate contact had notably corroded (fig. 6). These contacts were tried to be cleaned but with no success.

Figure 6. PCB underside and corroded contacts of device 5.

During the operational testing of device 6, large current spikes were noticed during both "on" and "off" periods of the drive cycle. During the "on" period the measured phase current was clearly above normal values and also during the "off" period the device tried to drive the motor intermittently. Despite of this abnormal operation, the safety contactor did not trip but the test was aborted after the two first drive cycles.

Inspection of the device did not reveal any clear reasons for the unusual operation. Components on the PCB and the IGBT module were measured to be within operational limits. There were either no significant signs of solder resist residue on top of the IGBT module, like with device 2. Nevertheless, corroded conductor lines were found on the underside of the PCB as pictured in figure 7.

Figure 7. PCB underside and corroded conductor lines of device 6.

Conclusions

In this paper, an accelerated corrosion test for frequency converters was presented. The aim was to find a test method capable of producing information on failure modes and mechanisms in short amount of time. A 24 hour exposure to salt spray was performed to eight devices. This was followed by 500 hours in 85 °C and 85 %RH temperature and humidity test. In order to investige the effect of the salt spray, other eigth devices were tested at the same time using only the 85 °C and 85 %RH test. No failures were seen in these devices during testing. However, five of the converters exposed to the salt spray failed during the test. Two of the devices were found to be inoperable after the environmental tests and three failed during operational testing. Three samples passed the tests intact. The results show that with exposure to salt spray occurrence of failures during 85 °C and 85 %RH temperature and humidity testing can be considerably accelerated. Additionally, it was noticed that the inverter PCB - IGBT module interface was quite susceptible to corrosion damage. Furthermore, it could be seen from the results that to

978-1-4244-4722-0/09 $25.00
© 2009 IMAPS-ITALY

avoid short circuits during operational testing, the samples have to be dried thoroughly after exposure to moisture.

The results indicate that the salt spray test is an interesting option when high reliability products are tested. However, a difficulty with the salt spray test is its poor reproducibility, i.e. it is quite difficult to remove the excess salt from the samples equally after testing. Because of this, there will be differences between the places and rate that corrosion occurs.

Acknowledgements

We would like to thank TEKES (Finnish Funding Agency for Technology and Innovation) and the following companies which supported this work: Vacon, Kone, and Konecranes.

References

[1] W. Nelson, "Accelerated Testing: Statistical Models, Test Plans, and Data Analysis", John Wiley & Sons, Inc., New Jersey, USA, p.601, 2004.

[2] E. Suhir, "Accelerated Life Testing (ALT) in Microelectronics and Photonics: Its role, Attributes, Challenges, Pitfalls, and Interaction With Qualification Tests", Journal of Electronic Packaging, Vol. 124, No. 3, pp. 281-91, 2002.

[3] S. Kuusiluoma, "System Level Reliability Testing for High Reliability Devices", Proceedings of the 2007 Electronics Packaging Technology Conference (EPTC), Singapore, December 10-12, pp. 897-901, 2007.

[4] Finnish Standards Association SFS, "SFS-ISO 9227 Corrosion Tests in Artificial Atmospheres. Salt Spray Tests", Federation of the Finnish Metal and Engineering Industries, Standards Department, 21 p., 2001.

[5] EIA/JEDEC, "JESD22-A101-B Steady State Temperature Humidity Bias Life Test", JEDEC Solid State Technology Association, 6 p., 1997.

978-1-4244-4722-0/09 $25.00
© 2009 IMAPS-ITALY

Screen-Printed Polymer-Based Microfluidic and Micromechanical Devices Based on Evaporable Compounds

N. Serra, T. Maeder, C. Jacq, Y. Fournier and P. Ryser

Laboratoire de Production Microtechnique, Ecole Polytechnique Fédérale de Lausanne (EPFL), Station 17, BM 0.141, CH-1015 Lausanne, Switzerland

Phone: +41.21.693.77.58, Fax: +41.21.693.38.91, nathalie.serra@epfl.ch; http://lpm.epfl.ch

Abstract

We investigate in this work the fabrication of polymer fluidic and mechanical devices based on evaporable compounds as sacrificial layers using the screen-printing process.

The combination of thermosetting polymer resins with evaporable compounds as sacrificial layers allows straightforward fabrication of polymer microfluidic and micromechanical devices. Channel, cavities and spacings are first defined by sacrificial material layers. This is followed by polymer resin deposition and polymerization / solvent removal. Removal of the sacrificial layers is then accomplished by heating at their decomposition / evaporation temperature. This process does not require an ulterior dissolution step, and systems exist where the sacrificial layer molecules escape by diffusion through the polymer at high temperature: we are therefore not limited to open structures. Furthermore, the use of screen-printed thick-film compositions is very advantageous regarding process cost and flexibility.

The selected sacrificial layer must be stable at the polymer processing temperature, yet decompose / evaporate cleanly at a temperature low enough to avoid degradation of the polymer. In this work, we used polymers such as ethylcellulose and silicone resins filled with graphite to impart electrical conductivity and improve mechanical stability. The selected sacrificial materials were polyol-based and were formulated to sublime rather than evaporate, in order to preserve the structures. They were formulated as thick-film pastes using a suitable cyclohexanol-based solvent mix.

Key words: sacrificial layer, evaporable compounds, screen-printing inks, micromechanics, fluidics.

Introduction

In the micro-electromechanical systems (MEMS) field, extensive use is made of sacrificial layers for fabrication of free-standing structures by surface [1] or volume micromachining [2]. However, many applications, for size or cost reasons, do not require the rather involved clean-room processes, and/or are just as well or better fabricated using other substrates than silicon. The related thick-film and low-temperature co-fired ceramic (LTCC) technologies, for instance, principally based on glass/ceramic materials, are widely used for sensors [3], and have recently seen considerable developments in structuration techniques [4-7] for fabrication of elements such as cantilevers, bridges, membranes and fluidic structures.

Another important recent development is the move towards polymer structures, whose properties such as low cost, transparency and good biocompatibility make them particularly suitable for disposable biomedical applications. Microstructuration using sacrificial layers is somewhat limited, by their thermal stability. Nevertheless, the combination of a stable photoresist

epoxy resins such as SU-8 and a sacrificial material such as polypropylene carbonate (PPC), which decomposes below 300°C, allows to fabricate all-polymer fluidic structures [8], which can be closed (i.e. channels, membranes,...), as the propylene carbonate monomer evolved during decomposition and can diffuse through the SU-8.

The concept of an organic sacrificial material thermally removed (which can be combined with polymerisation of the resin used for the structures) is attractive. However, 300°C is still much too high for many polymer materials, and formulation as paste or ink materials instead of photolithographic processing would be more advantageous in terms of cost and versatility. Nitrocellulose is an attractive alternative, as it decomposes at low temperature (<200°C) [9]; this has led to its use for Ag conductive inks printable on polymer substrates [10]. While this is very promising for mechanical structures, these studies show that nitrocellulose does leave significant residue, which could especially be an issue for biological applications.

Therefore, we have endeavoured in this work to explore mixes of non-polymeric polyol-type organic materials (so-called plastic crystals).

978-1-4244-4722-0/09 $25.00
© 2009 IMAPS-ITALY

Table 1. Compounds used in this work.

CAS = Chemical Abstracts Service Nr. ; M.P. / B.P. = melting / boiling point (supplier information).

Code	CAS	Name [vol.%]	Source	M.P. [°C]	B.P. [°C]
CH	108-93-0	Cyclohexanol	Sigma-Aldrich 105899	24	160
H	111-27-30	1-hexanol	Sigma-Aldrich 52840	-52	157
PG	57-55-6	Propylene glycol	Sigma-Aldrich	-60	187
PC	99-89-8	4-isopropylphenol	Sigma-Aldrich 175404	60	212
H_2O		Water	Deionised water	0	100
NPG	126-30-7	Neopentyl glycol	Sigma-Aldrich 538256	130	207
TME	77-85-0	Trimethylolethane / pentaglycerine	Sigma-Aldrich 93340	200	293
TMP	77-99-6	Trimethylolpropane	Sigma-Aldrich 93370	58	295
Epoxy		2-component epoxy resin	EpoTek 354T or 377		
EC-46-48	9004-57-3	Ethylcellulose, 46 cps grade, 48% ethoxyl content	Sigma-Aldrich 433837	170	
Silicone		2-component silicone resin	Dow Corning Q5-8401		

Polyols are widely used in industry for paints and coatings formulation, due to the viscosity and fast drying properties they give to the final paste. They have also attracted considerable attention due to their solid-solid order-disorder phase transitions, which make them interesting for thermal energy storage [11]. Solid solutions of these polyols allow tuning of the temperature of this phase transition [12], which is important for the present paper.

Three polyols were investigated for the formulation of sacrificial layers: trimethylolethane (TME), trimethylolpropane (TMP) and neopentyl glycol (NPG). Table 1 gives their properties and that of the other used materials. TME, NPG and TMP are all solid at room temperature and sublime / evaporate well below 200°C (Fig. 1, data from [13], [14]), based on their vapor pressure. Sublimation is preferable to evaporation, as it obviates surface tension effects. Therefore, NPG and TME, which have appreciable vapor pressures (ca. 5 kPa) at their melting point, are particularly attractive.

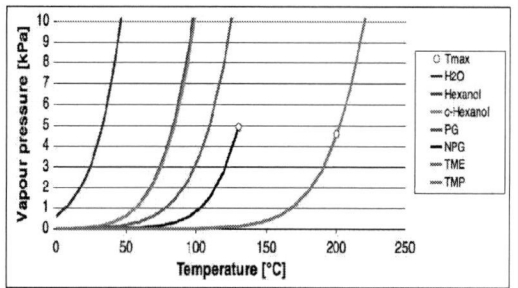

Fig. 1. Vapour pressure vs. temperature for the compounds used in this study (almost perfect overlap between CH & H, and TME & TMP).

Experimental

Formulation of low-molecular-weight sacrificial materials as screen-printable pastes is significantly more difficult than with polymeric materials, as there is no polymer to impart high viscosity to the solution, and the solvent thus tends to separate out of the solid material. To mitigate this issue, the solvents mixes were based on cyclohexanol (CH), one of the rare easily evaporable solvents having a high viscosity. Its melting point is 24°C, but can be depressed easily well below room temperature by only a small amount of additions such as water, 1-hexanol (H), propylene glycol (PG) or even dissolved TMP.

First tests of preparing the paste by mixing plastic crystal powders were not successful, giving too coarse pastes that tended to separate out. All pastes were therefore prepared with the method giving the most homogeneous materials, which involved dissolving the plastic crystals in a 100 ml beaker into the solvent at high temperature (ca. 100°C), followed by reprecipitation under agitation and cooling by immersing the beaker into cold water, yielding fine-grained suspensions. Using this process, we prepared three different pastes: the first two based on a mix of TME and TMP and the last one based on NPG and TME. It should be noticed that instead of using pure compounds, we only used mixes, which may combine the following advantages: limit grain growth during precipitation but also introducing a kind of binder between the different parts composing the paste. Moreover, in each case, solvents based on CH with PG, water and/or 4-isopropylphenol were adapted in order to get the required viscosity. After preparation, the sacrificial pastes were printed and dried (around 60°C), and then overprinted using three different polymers: 1) epoxy resins – EpoTek 354T and 377, 2) ethylcellulose (EC, "46 cps" grade – 48% ethoxyl content [15]), and 3) silicone resin.

978-1-4244-4722-0/09 $25.00
© 2009 IMAPS-ITALY

Graphite (TIMCAL® KS 15 or KS4) was added as filler in order to impart electrical conductivity and additional mechanical stability. The polymerization of this layer is then performed around 80°C. Finally, the sacrificial material is removed at 150°C by diffusion through the polymer layer.

The substrate used for all our experiments here is alumina.

Results

Preliminary evaporation tests consisting in dispensing bridges of conductive paste above a sacrificial layer strip were first performed to check compatibility of sacrificial paste – polymer combinations, which was verified for ethylcellulose polymer and silicone resin. For epoxy resins, however, interactions between the sacrificial layer and the polymer coating occur, as acid anhydride groups can react with the polyols contain in the sacrificial paste, preventing proper polymerisation. The resin goes thus inside the sacrificial layer preventing it from evaporation. This is evidenced from Fig. 2a and 2b, which show a comparison between sacrificial layers coated with epoxy and ethylcellulose respectively.

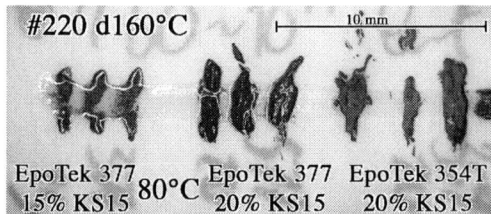

Fig. 2a: TME-TMP sacrificial layer coated with epoxy resin-graphite composites, showing strong reaction, especially with EpoTek 377.

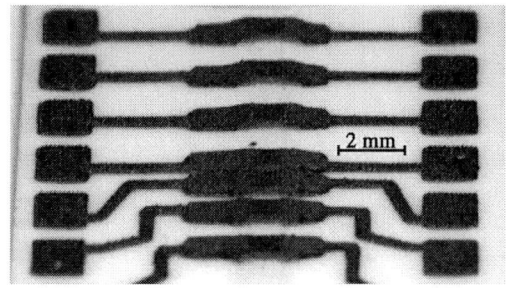

Fig. 2b: Ethylcellulose bridges after evaporation of the underlying TME-TMP sacrificial layer (yellow colour is remaining colorant).

From these results, we decided to forgo epoxy and focus on silicone and ethylcellulose for the rest of our study. As far as the sacrificial material is concerned, when appropriately coated, it sublimed easily at 150°C in all three formulated pastes.

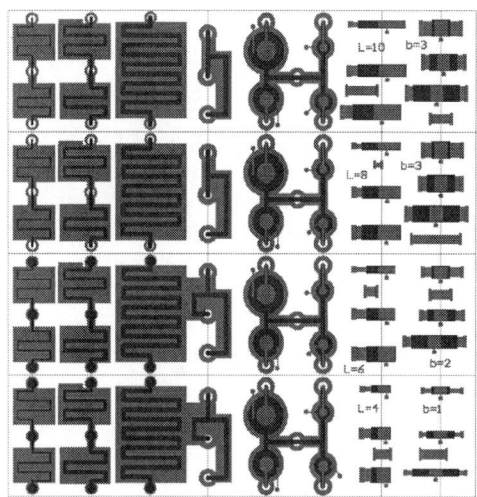

Fig.3: Layout of the structures (silver layer in red, sacrificial layer in brown and polymer layer in green; 101.6 mm square total substrate size).

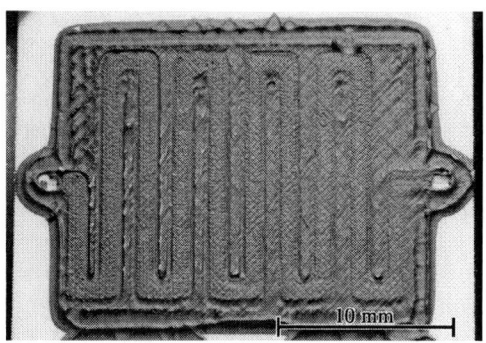

Fig.4a: Microchannels covered with silicone - graphite

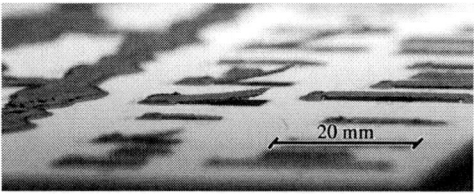

Fig. 4b: Cantilevers (left) and bridges (right) in silicone - graphite

Different structures were then tested in order to see how the sacrificial materials evaporate and behave depending on the geometry: channels, bridges, cantilevers and membranes. For each type, several sizes were tested with different sacrificial layers. Fig. 3 show the layout used.

After the 150°C cycle, complete removal of the sacrificial layer can be observed on the substrate, leaving clean and relatively well preserved structures (see Fig. 4a and 4b). In order to prove the good functionality of the process, cantilevers were qualitatively tested. This was only

978-1-4244-4722-0/09 $25.00
© 2009 IMAPS-ITALY

possible with samples coated with silicone resin; ethylcellulose, being thermoplastic, tended to sag if large spans were used for cantilever and bridge structures. The principle of the set-up is based on capacitive measurement, performed by an Analog Devices AD7746/7745 integrated circuit that allows differential capacitance measurements between -4 and +4 pF. A force is then applied at the end of the cantilever with a tip, which changes the distance between the electrodes, and thereby the electrical capacitance using the equation:

$$C = \frac{\varepsilon_0 . \varepsilon_r . A}{d}$$

with C the capacitance, ε_0 the electric constant, ε_r the relative permittivity, A the electrodes area and d the distance between the electrodes.

Fig. 5 shows an example: the first graph (a) corresponds to the unloaded sample and the second one when there is excitation of the cantilever (b). While only noise is present in (a), the second graph exhibits some peaks corresponding to the moment when the cantilever was loaded, i.e. when d was smaller. Note that the larger peak at the beginning of this graph corresponds to the moment when we approach the tip to the cantilever, showing that it is really sensitive. Therefore we can conclude that the process works for micromechanical devices.

An example of a fluidic structure is given in Fig. 6, where the ethylcellulose-graphite "roof" of a fluidic channel was removed using a scalpel: the TME-TMP sacrificial layer has completely sublimed through the ethylcellulose.

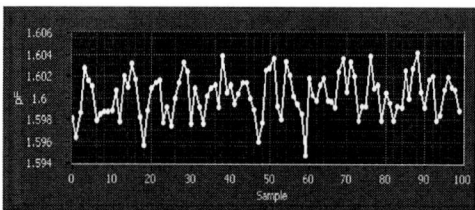

Fig. 5a: Reference capacitance measurement.

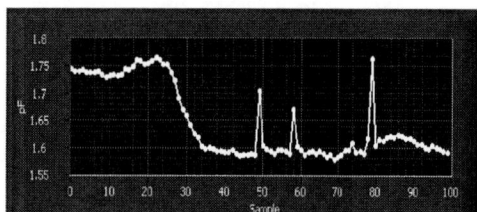

Fig. 5b: Capacitance measurement with excitation of the silicone cantilever.

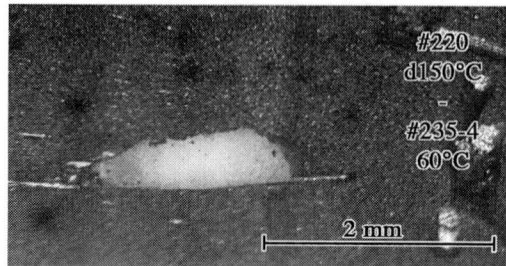

Fig. 6. Ethylcellulose-graphite removed with a scalpel, showing complete sublimation of sacrificial layer.

Conclusion

Sacrificial pastes removed by evaporation or sublimation were successfully formulated with mixes of polyhydric alcohols, TME/TMP and TME/NPG in a suitable cyclohexanol-based solvent mixture in order to obtain the required viscosity for screen-printing. Based on their interesting physical properties, we succeeded in designing non-toxic sacrificial layers that disappear cleanly around 150°C, thus considerably below the temperatures that were previously obtained with polypropylene carbonate in [8]. Our results showed that these polyol materials were compatible with silicone and ethylcellulose polymers, which allowed the manufacture of microfluidic and micromechanical devices. Epoxy resins were incompatible with these materials, but it should be also possible to formulate adequate sacrificial layers. Camphor and also camphor-isoborneol mixes appear as good candidates. Indeed, camphor evaporates at ca. 200°C (lit) but it presents a ketone group that may be more inert towards the reactive groups of the epoxy resins and thus not disturb their polymerisation. Mixes of camphor and isoborneol may exhibit a similar behavior as TME/TMP and prevent formation of aggregates, giving thus fine grain dispersions and pastes with a good viscosity.

Copyright

By submitting the paper to the conference, the authors transfer the copyright to the EMPC2009 conference committee.

Acknowledgements

The authors are indebted to M. Garcin for the fabrication of the samples.

References

[1] Kotani-T Nakanishi-T Nomura-K, "Fabrication of a new pyroelectric infrared sensor using MgO surface micromachining", Japanese Journal of Applied Physics Vol. 32, 6297-6300, 1993.

[2] Luo-J He-J Flewitt-A Moore-DF Spearing-SM Fleck-NA Milne-WI, "Development of all

978-1-4244-4722-0/09 $25.00
© 2009 IMAPS-ITALY

metal electrothermal actuator and its applications", Journal of Microlithography, Microfabrication, and Microsystems Vol. 4 (2), 023012, 2005.

[3] White-NM Turner-JD, "Thick-film sensors: past, present and future", Proceedings, 14th International Conference on Solid-State Sensors, Actuators and Microsystems - Transducers / Eurosensors'07, Lyon, France, 107-111, 2007.

[4] Lucat-C Ginet-P Ménil-F, "New sacrificial layer based screen-printing process for free-standing thick-films applied to MEMS", Journal of Microelectronics and Electronic Packaging Vol. 4 (3), 86-92, 2007.

[5] H. Birol, T. Maeder, P. Ryser, "Application of graphite-based sacrificial layers for fabrication of LTCC (low temperature co-fired ceramic) membranes and micro-channels", Journal of Micromechanics and Microengineering Vol. 17, 50-60, 2007.

[6] Fournier-Y Triverio-O Maeder-T Ryser-P, "LTCC free-standing structures with mineral sacrificial paste", Proceedings, International Conference on Ceramic Interconnect and Ceramic Microsystems Technologies (CICMT), Munich (DE), 11-18 (TA12), 2008.

[7] Maeder-T Fournier-Y Wiedmer-S Birol-H Jacq-C Ryser-P, "3D structuration of LTCC / thick-film sensors and fluidic devices", Proceedings, 3rd International Conference on Ceramic Interconnect and Ceramic Microsystems Technologies (CICMT), Denver, USA, THA13, 2007.

[8] Metz-S Jiguet-S Bertsch-A Renaud-P, "Polyimide and SU-8 microfluidic devices manufactured by heat-depolymerizable sacrificial material technique", Lab on a Chip Vol. 4, 114-120, 2004.

[9] Phillips-RW Orlick-CA Steinberger-R, "The kinetics of the thermal decomposition of nitrocellulose", Vol. 59, 1034-1039, 1955.

[10] Nguyen-BT Gautrot-JE Nguyen-MT Zhu-XX, "Nitrocellulose-stabilized silver nanoparticles as low conversion temperature precursors useful for inkjet printed electronics", Journal of Materials Chemistry Vol. 17, 1725-1730, 2007.

[11] William N. Hunter, "Alcohols, Polyhydric" in Kirk-Othmer Encyclopedia of Chemical Technology, John Wiley & Sons (USA), Vol.2, pp 46-58, 2000.

[12] Font-J Muntasell-J Navarro-J Tamarit-JL, "Mélanges pentaglycérine / néopentylglycol : formation d'une solution solide [pentaglycerine / neopentylglycol mixes: formation of a solid solution]", Thermochimica Acta, No. 136, pp 55-71, 1988.

[13] Font-J Muntasell-J, "Comparative study on solid crystalline–plastic–vapour equilibrium in plastic crystals from pentaerythritol series", Journal of Materials Chemistry Vol. 5, 1137-1140, 1995.

[14] Dykyj-J Svoboda-J Wilhoit-RC Frenkel-M Hall-KR, "Vapor pressure and Antoine constants for oxygen containing organic compounds - Organic compounds - C1 to C57 - part 1", Landolt-Börnstein, Group IV - Physical Chemistry, Vol. 20B, Advances in Chemistry Series - Numerical Data and Functional Relationships in Science and Technology, Springer-Verlag, DE, 2000.

[15] "ETHOCEL ethylcellulose polymers technical handbook", Dow Chemical, product information No. 192-00818-0905 X AMS, 2005.

3D integration of ultra-thin functional devices inside standard multilayer flex laminates

W. Christiaens, T. Torfs, W. Huwel, C. Van Hoof, J. Vanfleteren

CMST (Ghent University - IMEC)
Technologiepark 914A, 9052 Zwijnaarde, Belgium

Phone: +32-9-2645371
Fax: +32-9-2645374
E-mail: Wim.Christiaens@elis.ugent.be

Abstract

Nowadays, more and more wearable electronic systems are being realized on flexible substrates. Main limiting factor for the mechanical flexibility of those wearable systems are typically the rigid components - especially the relatively large active components - mounted on top and bottom of the flex substrates. Integration of these active devices inside the flex multilayers will not only enable for a high degree of miniaturization but can also improve the total flexibility of the system. This paper now presents a technology for the 3D embedding of ultra-thin active components inside standard flex laminates.

Active components are first thinned down to 20-25 μm, and packaged as an Ultra-Thin Chip Package (UTCP). These UTCP packages will serve as flexible interposer: all layers are so thin, that the whole package is even bendable. The limited total package thickness of only 60 μm makes them also suitable for lamination in between commercial flex panels, replacing for example the direct die integration. A fan-out metallization on the package facilitates easy testing before integration, solving the KGD issue, and can also relax the chip contact pitch, excluding the need for very precise placement and the use of expensive, fine-pitch flex substrates.

The technology is successfully demonstrated for the 3D-integration of a Texas Instrument MSP430 low-power microcontroller, inside the conventional double sided flex laminate of a wireless ECG system. The microcontrollers are first thinned down and UTCP packaged. These packages are then laminated in between the large panels of the flex multilayer stack and finally connected to the different layers of the flex board by metallized through-hole interconnects.

The thinning down, the UTCP packaging and the 3D-integration inside the commercial flex panels did not have any affect on the functionality of the TI microcontroller. Smaller SMD´s were finally mounted on top and bottom of the integrated device.

Key words: 3D integration, flex, ultra-thin chip, chip package

Introduction

One of the main challenges in the electronics manufacturing and packaging development is how to integrate more functions inside the same, or even smaller, size. The electrical performance and the number of functions of every new product generation are increasing while the size and the weight of the products are decreasing. To meet this, the semiconductor industry is integrating more and more transistors into the same silicon area. This allows either to reduce the size of the IC or to increase the functionality and performance of the same size IC components. At the same time there is also an interest to increase the packaging density. While silicon chips continue integrating more functionality as per Moore's law, the packaging is challenged to also integrate and shrink.

A high degree of miniaturization can be achieved by removing the assembled active devices from the surface and integrate them inside the inner layers of multilayer FCB or PCB boards.

On flex substrates typically only pure electronic assembly is performed up to flip-chip components. Embedding of passive or active components in conventional flexible printed circuits is at least not state of the art. This limitation to two component layers (front and backside of the flex laminate) limits of course the compactness of the circuit, and the presence of relatively large rigid components also limits the flexibility of the FCB.

This paper now presents the embedding of an ultra-thin, functional microcontroller device inside a double sided flex board, by means of a unique concept for packaging ultra-thin chips: the Ultra-Thin Chip Package (UTCP).

The UTCP serves as flex interposer and can be used for the embedding inside the substrate, replacing for example the direct integration of bare dies. The UTCP allows for easy testing of the chip

978-1-4244-4722-0/09 $25.00
© 2009 IMAPS-ITALY

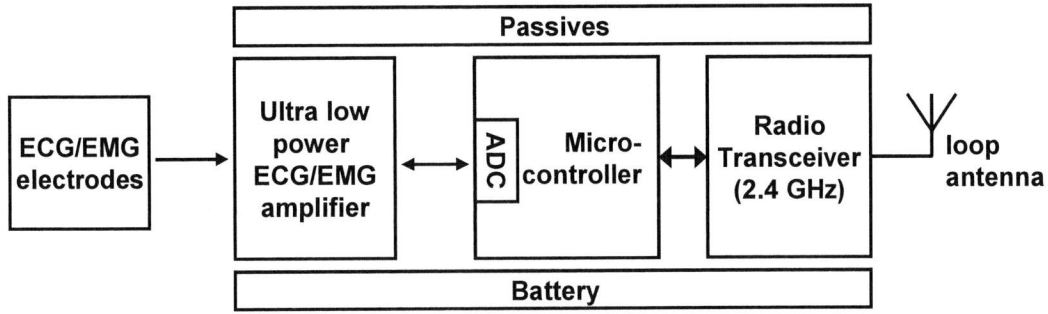

Figure 1 : Block scheme of the ECG circuit

before embedding, solving the KGD issue, and provides a contact fan out with more relaxed pitches, eliminating the need for very fine pitch PCB or FPC compatible with the chip contact pad pitch.

A Texas Instrument MSP430 low-power microcontroller is first UTCP packaged and next laminated inside the conventional double sided flex laminate of a wireless ECG system. This 3D integration of the microcontroller leads to high density integration, since SMD components can be mounted on top and bottom of the integrated devices. In addition, the UTCP packaged thin silicon devices are mechanically flexible itself, leading to an increased total flexibility of the resulting system.

ECG demonstrator

The demonstrator for this 3D integration technology consists of a high-density flexible ExG circuit for personal health or wellness applications. IMEC's ExG system is a bio-potential wireless sensor node, able to monitor the vital body signs provided by portable electrocardiogram (ECG, which monitors the heart activity), electromyogram (EMG, which monitors muscle contraction) and electroencephalogram (EEG, which monitors brain waves). The system collects and processes data from (external) human body sensors and wirelessly transmits the data to a central monitoring system. Small size and low power consumption of the system enables non-invasive and ambulatory monitoring of vital body parameters.

The block scheme of e.g. an ECG/EMG circuit is given in Figure 1. The system uses IMEC's proprietary ultra-low-power bio-potential readout ASIC (application specific integrated circuit) to extract the bio-potential signals produced during the ECG, EMG or EEG measurements. The low power microcontroller MSP430F149 drives the bio-potential chip, also digitalizes the samples with its A/D converter. The signals are finally sent by the 2.4GHz low power radio Nordic NRF2401A using a coplanar antenna. Information can be transmitted to a pc by a 'USB stick' receiver for visualization and recording.

This ECG system was already realized on standard double sided flex laminates. This is shown in Figure 2: the use of flex substrates ensures a certain degree of mechanical flexibility for the wearable systems. However, an important factor limiting miniaturization and flexibility is typically the presence of rigid components, especially the relatively large active devices.

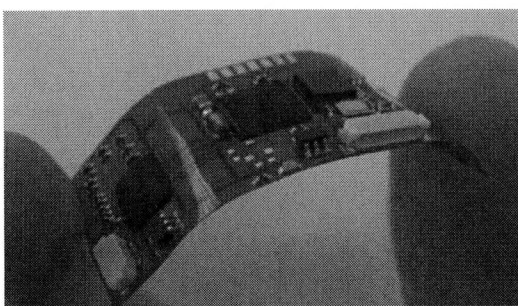

Figure 2 : ExG system realized on standard double sided flex substrate. Flexibility of system is limited by the presence of the large IC's.

In this paper, one of those 3 used actives, the TI microcontroller device, is removed from the surface and embedded as UTCP inside a commercial standard flex substrate. Smaller SMD components can be mounted above and below the embedded chip. This leads to a high density of integration. In addition, the embedded ultra-thin silicon chip becomes mechanically flexible itself. This leads to an increase in total flexibility of the resulting system.

UTCP packaging of ultra-thin microcontroller

Before integration inside the substrate the Texas Instruments microcontroller MSP430F149 is first embedded as Ultra-Thin Chip Package. This UTCP will serve as interposer during the 3D integration, as an alternative for direct die integration.

The UTCP interposers can provide a fan-out of the chip contact pads, enabling for easy testing before integration (compared with naked die integration). Moreover this fan-out metallization can also bridge the gap between the chip contact pitch

978-1-4244-4722-0/09 $25.00
© 2009 IMAPS-ITALY

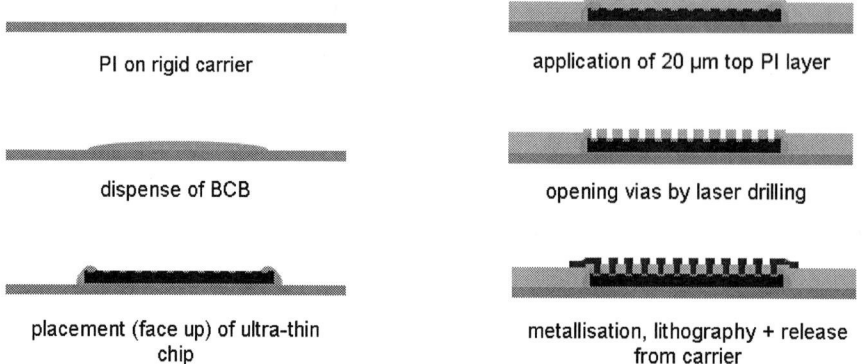

Figure 3 : Overview of UTCP process flow.

and the minimum available substrate pitch, or can exclude the need for expensive fine pitch interconnection substrates.

As the UTCP technology is developed for using ultra-thin chips, the functional TI microcontrollers were also first thinned down to only 20-25 μm.

UTCP technology

After thinning, the active devices are integrated as a UTCP package. An overview of the individual process steps of the Ultra-Thin Chip Package production is depicted in Figure 3. More details on this technology can also be found in [1].

The base substrates for this technology are a 20 μm polyimide layer spincoated on a rigid glass carrier. For the fixation and the placement of the chips a benzocyclobutene (BCB) is used as adhesive. The chip is covered with a next 20 μm thick polyimide layer. For the contacting to the chip, contact openings to the bumps of the chips are laser drilled and a 1 μm TiW/Cu layer is sputtered and photolithographically patterned. This metal layer provides a fan out to the contacts of the chips.

After processing, the whole package can be released easily from the rigid carrier. The result is a very thin, even flexible, chip package, with a total package thickness of only 50–60 μm.

Design

The top metallization on the UTCP package can provide a fan-out of the chip contact pads, with smaller pitch. This is shown in Figure 4: the metallization provides a small fan-out of the 64 die contacts to 400 x 400 μm² contacts with pitch of 500 μm, and a larger fan-out to 650 x 1300 μm contacts. The dimensions of this design are: 9 x 9 mm² for the small fan-out and 3 x 3cm² for the large fan-out. The design also includes solder pads to solder mount an SMD resistor and a LED component on top of the package for visual functionality demonstration after processing.

The outer, larger contact pads on the package can be used to connect the integrated microcontroller for programming and testing. This solves the KGD issue: a UTCP packaged device can easily be tested before integration, compared with a bare die.

Figure 4 : Fan-out metallization to the UTCP packaged microcontroller device.

The inner contacts match the design rules of conventional flex board manufacturers and will be used for embedding the UTCP package inside multilayer flex boards. This relaxes the chip contact pad pitch to pitches compatible with conventional, low cost flex substrates.

Functionality testing

Functionality of the UTCP integrated microcontroller devices was compared with the bare, unthinned dies.

The fan-out metallization of the UTCP package allows for easy testing of the thinned microcontroller devices, after processing and before the integration inside the flex substrates. First the microcontroller was programmed to generate a blinking sequence on the LED to demonstrate its functionality: see Figure 4. Also more extensive tests of the analog-to-digital converter of the embedded microcontrollers are performed after

978-1-4244-4722-0/09 $25.00
© 2009 IMAPS-ITALY

UTCP integration. The linearity and also the AC specs (including signal to noise ratio (SNR), effective number of bits (ENOB), total harmonic distortion, spurious free dynamic range) were compared, between a bare, unthinned microcontroller die and UTCP packaged thinned microcontroller. The UTCP performs slightly worse than the naked die (most likely due to parasitics introduced by the package interconnection), but the small deviations are limited and will have only little effect on the usability of the packaged microcontroller.

Flex circuit production

After testing, the UTCP package can be embedded in turn inside a double-layer flex printed circuit board (PCB) using standard flex PCB production techniques. This was done at ACB, a Belgian flex manufacturer.

The principle is as follows: the very thin UTCP packages can be laminated inside the flex multilayer stacks. The packages are first fixed on a flex inner layer sheet, before the multilayer flex is laminated together and the packages are interconnected afterword to the multilayer substrates by conventional through hole interconnections. All these processing steps (lamination, through hole interconnections) are exactly the same as for conventional multilayer board production.

The UTCP packages are first aligned and fixed on a patterned inner layer, having already a prelaminated adhesive sheet. The metal pattern of this first flex inner layer contains a copy of the contact pad layout of the UTCP package, which is used as alignment marks during placement of the UTCP. For the ECG demonstrator only a double sided flex is needed, so the UTCP packaged microcontroller will laminated in between only two flex sheets. For the ECG demonstrator the UTCP packages are aligned manually on the prepatterned alignment marks of a first panel and are fixed by heat tack (using a solder iron) on an adhesive sheet, prelaminated on this panel. Once the package(s) are fixed on this adhesive layer, the second panel of the flex multilayer is laminated on top of these packages. Next, the through holes are drilled, for the interconnection between both sides of the panel, but also for the interconnections to the UTCP contact pads. Finally the through holes are metallized, as for a standard multilayer flex.

After substrate production, the other components are assembled on the flex board. The result after assembly is shown in Figure 5: the microcontroller device is removed from the surface and integrated inside a standard flex substrate. Small SMD components are even mounted on top and/or bottom of the integrated component.

Measurements

The functionality of this wireless biopotential system is tested after assembly. The final ExG demonstrator with integrated microcontroller device is completely functional, and able to monitor ECG signals. Figure 6 shows an example ECG waveform recorded with the wireless system using standard disposable stick-on Ag/AgCl foam electrodes.

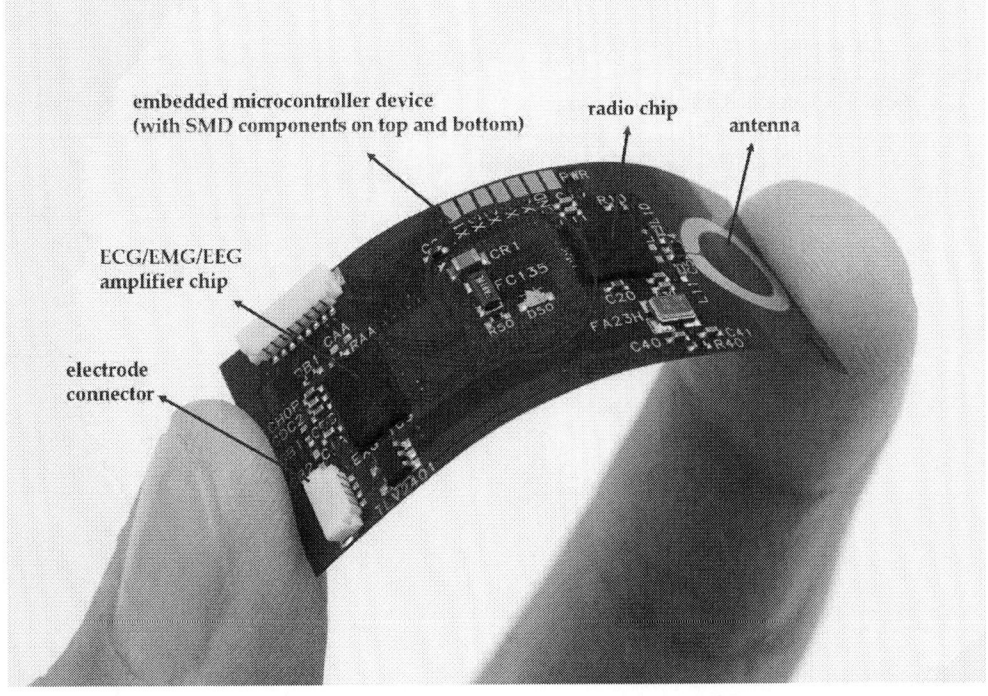

Figure 5 : The flexible wireless biopotential system with integrated microcontroller device (excluding battery).

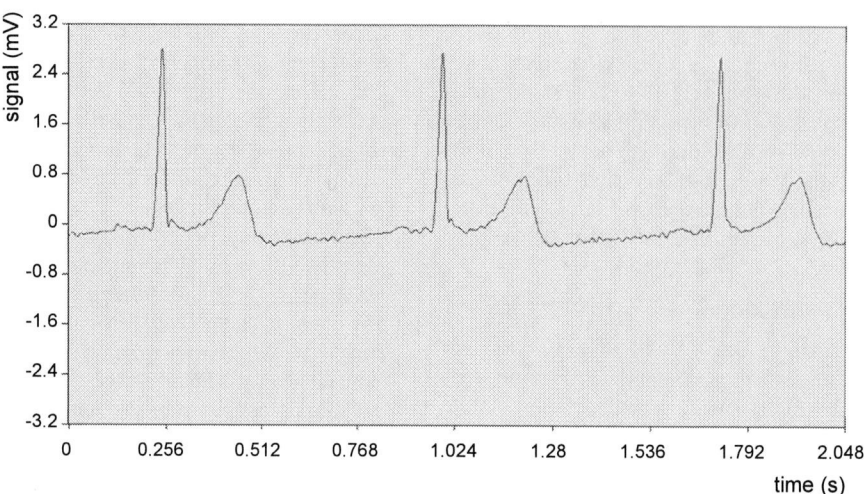

Figure 6 : An example ECG waveform recorded with the wireless system.

Conclusions

The main limiting factor for the total mechanical flexibility of typical electronic assemblies on standard flex substrates are the presence of relatively large rigid components. This paper presents a technology for removing the large IC's from the surface of the flex substrates and integrates them inside the flex multilayer boards.

The components are first embedded as UTCP package. These UTCP's are used as flex interposer during integration inside the multilayer FCB. The packages are only 60 μm thin, and are mechanically flexible. These packages can also provide a fan-out of the chip contacts, enabling easy testing before integration, and relaxing the contact pitch matching the minimum pitch on the flex substrates.

The technology is demonstrated for the integration of a microcontroller device of a wearable, wireless ECG system. The 3D integration of such relatively large components, not only enhances the mechanical flexibility of the wearable system, but can also lead to an increased miniaturization since smaller SMD components are mounted on top and bottom of the integrated device.

The other active components in the system are an ultra-low power biopotential amplifier chip developed by IMEC and a commercial off-the-shelf 2.4GHz low power radio transceiver (Nordic nRF2401A). These components can in principle also be embedded inside the flex PCB just like the microcontroller. This will increase the flexibility and integration density of the device even more.

Acknowledgements

The work was carried out under the EC IST funded project IP-SHIFT (contract number 507745).

References

[1] W. Christiaens, B. Vandevelde, E. Bosman, and J. Vanfleteren, "UTCP: 60 mm thick bendable chip package", Proceedings of IWLPS, San Jose, November 1-3, pp. 114-119, 2006.

978-1-4244-4722-0/09 $25.00
© 2009 IMAPS-ITALY

3D integration process flow for set-top box application: description of technology and electrical results

S. Cheramy (1), J. Charbonnier(1), D. Henry(1), A. Astier(1), P. Chausse(1),
M. Neyret(1), C. Brunet-Manquat(1), S. Verrun(1), N. Sillon(1).
L. Bonnot(2), X. Gagnard(2), J. Vittu(3).

(1) CEA Léti – MINATEC ; 17 rue des Martyrs ; F-38054 GRENOBLE - France

(2) STMicroelectronics ; 850 rue Jean MONNET ; F-38926 CROLLES - France

(3) STMicroelectronics; 12 rue Jules Horowittz ; F-38000 Grenoble - France

Abstract

In this paper, the technological steps specifically developed for 3D integration of a set top box demonstrator will be presented (figure 1). The integration flow is based on a 45nm node technology top chip stacked on a 130nm node technology active bottom wafer [1]. This flow needed to develop specific wafer level packaging technologies such as:

- *Top & bottom chips interconnections*
- *Temporary bonding and debonding of bottom wafer*
- *High aspect ratio TSV's designed into the bottom wafer*
- *Backside interconnections for subsequent packaging step*
- *Top chip stacking on bottom wafer*

In this paper, the complete process flow will be presented. Then, a technical focus will be done on the most important process steps for the 3D integration. The preliminary electrical results of the demonstrator will be discussed. Finally, some prospects for 3D integration technologies and applications will be proposed.

Key words: Trough Silicon Vias (TSV) – Wafer level Packaging – Stacking

Introduction

Today, a new trend in wafer level packaging is to add more than one dies in the same package. To minimize the package size and to increase the system performances, stacking of components is one solution, using TSV in the bottom die, and specific interconnection between each component. If 2 levels of components will satisfy lots of applications, we can easily imagine future applications with more than 2 levels. The figure 1 shows the integration scheme for set-top box application, which in the same manner could be used for other applications and can be extended to several layers of components.

Figure 1: Set-top box application integration scheme

This type of integration allows using different types of technology for the top and the bottom chips (heterogeneous integration). For example, it could be interesting to select the top chips in order to improve the yield in the case of very advanced technology. In this particular case, the advanced technology is used for the top chip (Known Good Die concept) and the mature technology in which TSV process will be done is used as the interposer. This scheme implies TSV process in the interposer, as well as small pitch back-side connection for the report on the plastic board. All of those processes are done on thin wafers (via last approach), using a temporary carrier [2][5].

Concerning the face to face stacking of components, copper pillar technology has been chosen for its possibility of low pitch.

After a short description of the flow chart and major explanation on our technological choices, technical focuses are proposed step by step. Finally, some preliminary electrical tests will be presented.

978-1-4244-4722-0/09 $25.00
© 2009 IMAPS-ITALY

Process flow

The complete integration scheme proposed for the active silicon interposer, for the top chip as well as for the die to wafer stacking is described in figure 2.

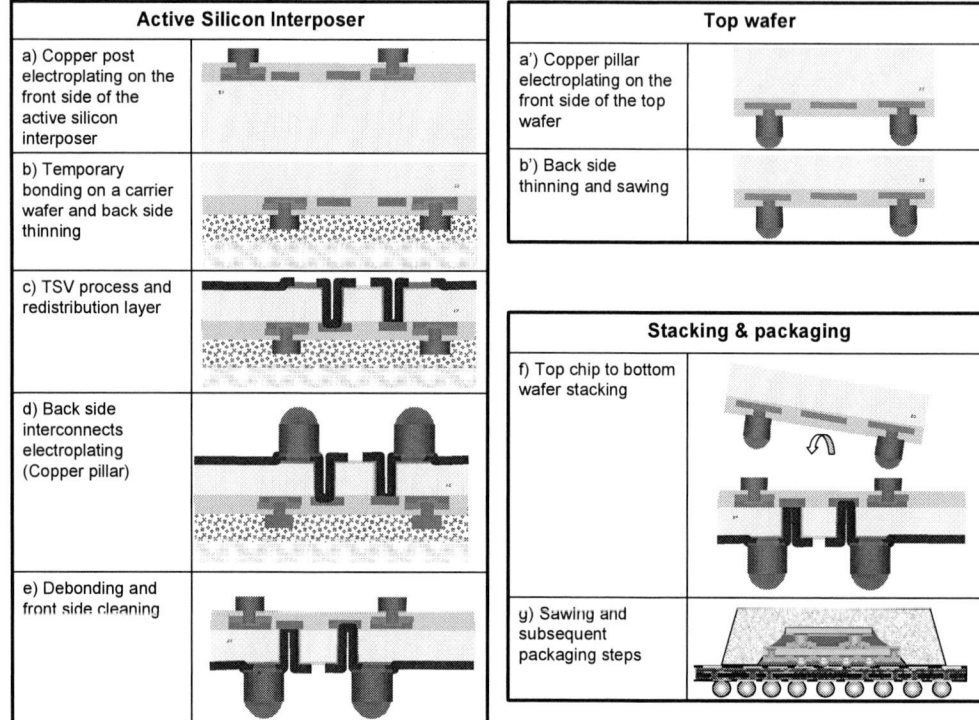

Figure 2: Process flow for the active silicon interposer, for the top chip and the die to wafer stacking

The die to wafer (face to face) interconnection solution, using copper pillar, has been chosen for its possibility of fine pitch (20-25µm) and consequently high density of interconnections.

The thickness of the interposer is defined in function of the desired via aspect ratio as well as via diameter. Benefiting from Leti's previous studies and test vehicles (TSV for CMOS Image Sensor [4]) and taking into account the density of TSV required, our development has been focused on 60µm via diameter with 2:1 aspect ratio leading to a 120µm silicium thicknesses).

To handle and process the thin wafers during the process flow of the active interposer, a temporary bonding & debonding process has been developed [5] with a specific focus to the debonding of wafer with back side high topology (developed by EVG [3]), allowing the final packaging.

The final packaging, here PBGA, also brings some technical challenges to overcome which will be the subject to further publications.

Technology description

Face to face interconnection
Firstly, an interconnection solution has been performed in order to connect the top chip onto the bottom wafer. The pitch for those interconnections was 50µm and the solution was based on pillars, formed by a copper post on the bottom wafer (only Cu) and a copper pillar (Cu and lead free alloy) on the top chip (Figure 3).

For the copper post step, a titanium nitride diffusion barrier and copper seed layer are deposited by sputtering (PVD) technique. The photo-lithography is done with 20µm of MicroChemicals AZ4562 resist on a MA8 SUSS MicroTech mask aligner. The electrochemical deposition (ECD) of copper is then done in a Raider Semitool equipment. The copper post obtained have a 25µm diameter for 12µm height with a lower than 5% uniformity.

For the copper pillar, the photolithography is done with the same resist or with a new solution of thick resist (40µm), allowing a larger gap between chips and easiest die to die underfilling process. After the same copper electrodeposition step as described for

978-1-4244-4722-0/09 $25.00
© 2009 IMAPS-ITALY

copper post, the wafers are dipped in a dilute HF bath and wetted in a Quick Dump Rinser (QDR) in order to have a reactive copper and hydrophilic resist surface. The Tin-Silver-Copper (TSC) alloy is then straightly deposited without any drying of the wafers in between by ECD process.

For both case (copper pillar and copper post), the resist is stripped in a Semitool equipment with EKCLE stripping solution. An Advantec Ecofrec202 no clean flux is spin coated on the wafers and the reflow is then operated in an infrared furnace (tungsten halogen lamps) with a TSC reflow peak at 230°C. On figure 3, the spherical shape of the reflowed TSC alloy is shown. The complete copper pillar including the copper and the alloy shows a height uniformityof 6,4%.

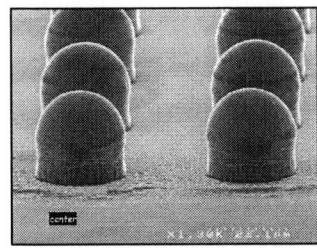

Figure 3: SEM pictures of Copper pillar (Cu and reflowed TSC) on the top chip

The barrier and seed layer are lastly etched by subsequent HF and dilute sulfuric peroxide (DSP) chemistry in both case of copper post and pillar.

Bonding process

In order to achieve TSV into the bottom wafer, a temporary bonded carrier has been used [2] [5].

The device wafers are coated with a spin-on adhesive on the active surface where the technology is embedded. The adhesive thickness is adjusted depending on the technology topography. The adhesive was then cured with two successive proximity bakes: 120°C and 160°C for 2 minutes each so as to remove the solvent. Those wafers are then bonded to borosilicate carriers at 180°C under vacuum. After the through silicon vias (TSV) backside processes (thinning, etching, metallization, etc.) the thin device wafers are then debonded by slide off from the carrier wafer at medium temperature (from 150°C to 220°C).

TSV

The TSV process used for this project was based on an existing process developed for CMOS image Sensors packaging, previously described [4]. Nevertheless, this process has been adapted for higher aspect ratio (2:1) than for CMOS Image Sensors packaging.

- Thinning

The flow starts with a mechanical thinning at 120µm and edge grinding of the bottom wafer bonded to the temporary carrier. A first low temperature PECVD SiH₄ based oxide is then deposited as hard mask.

- Vias etching

For the vias etching, a double side lithography step is done to align the vias lithography on the back side with the metal 1 pad on the front side [4]. For the further deep reactive ion etching (DRIE) the same Bosch process has not been possible as previously. Indeed, due to AR increase, the silicon notching at the via bottom was too pronounced (several hundreds of nanometers). Compare to the 70µm height TSV, inside the 120µm height TSV the bottom oxide is loading very quickly inducing this notching after only 5% of overetch. This phenomenon is critical if the etch uniformity is not below 2%. In order to avoid this trouble, another DRIE equipement has been used, the Alcatel AMS3200 with an electrostatic clamping system. A specific Bosch type recipe has been developed using a duty cycle base time of 10ms for the pulsed BIAS power for a etch uniformity of 2.5%. This process shows no micromasking and a possible overetch of 30% with a notching below 150nm in the vias bottom. The silicon etch rate obtained with this recipe is 8.8µm/min with a very low undercut and slightly open profile (Figure 4).

Figure 4: SEM cross section of the bottom side (notching), global view, top corner (undercut) of a 60µm diameter, 120µm height TSV inside the interposer wafer

The last step of etching is a standard RIE CHF₃/SF₆ process for removing the SiO₂ layer at the bottom of the cavity [4].

- Via Insulation

Due to the higher temperature tolerance of the device compare to the CIS, the temperature limit was set at 300°C to avoid any degradation of the Brewer HT10.10 temporary adhesive. This increase in temperature tolerance allowed us to use a TEOS precursor instead of the SiH₄ for the insulation PECVD layer. We characterized the conformality of our deposit by using SEM cross section after sawing into the cavities (Fig. 5a 5b). These cross sections showed that for AR 2.4:1 vias of 65µm diameter:

-The thickness on the side wall at the bottom of the cavity is 0.84µm for a deposition of 3.07µm on the top (3.07µm of TEOS based oxide

978-1-4244-4722-0/09 $25.00
© 2009 IMAPS-ITALY

and 1μm of hard mask on Fig. 5a). The conformality top to bottom side wall [4] is so 27%.

-The thickness at the bottom of the cavity is 0.96μm for a deposition of 3.07 μm on the top. The conformality top to bottom is so 31%.

No difference has been observed between vias at the center of the wafer compare to vias at the edge of the wafer in the conformality point of view, confirming the good deposit uniformity. The conformality obtained with this process is enough to provide a good electrical insulation of AR 2:1 vias.

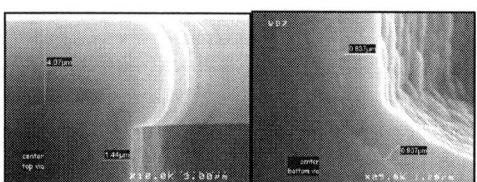

Figure 5a,b: SEM cross sections: a- top step coverage of a AR 2.4:1 via with insulation TEOS based oxide layer. b- bottom side of the same via

The following steps of the TSV process flow (via metallization, metal passivation) are similar to the previous CIS processes [4].

Backside bumping process

In this specific test vehicle, a backside bumping process has been developed close to an upscalling of the copper pillar process. A titanium nitride diffusion barrier and copper seed layer are deposited by sputtering (PVD) technique on the BCB metal passivation layer after an outgasing bake. The photolithography is done with the 50μm thick MX5050 dry film from Dupont. The copper and TSC ECD are done in a Raider Semitool as for the copper pillar. The dry film stripping is done in a Surface Strip 419 bath. The reflow, barrier and seed etching are identical as those of the copper pillar. Figure 6 shows the morphological shape of the bump interconnection.

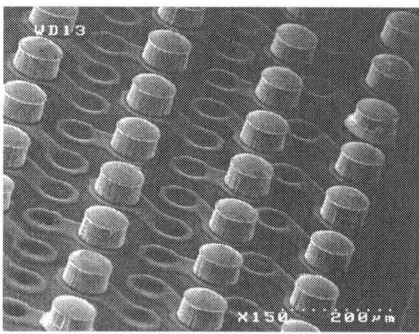

Figure 6: SEM picture of back side bumping

Top chip stacking and hybridation reflow

Finally, the top chip is stacked using a Datacon pick and place tool. This stacking needs some specific recognition patterns on top & bottom chips, to be able to stack with a precision lower than +/-5μm. To avoid any oxidation during reflow, the top chip is partially dipped into a specific flux just before the stacking. A cavity plate designed in accordance to the height of the copper pillar is used.

After stacking and reflow of the component, this 3D stack is ready for subsequent packaging steps. The step of dicing, pick & place, underfilling & overmolding require also some technical fine tuning that will be detailed in further publication.

Electrical results and discussions

2 batches of demonstrator, on vehicule test and on functional parts, have been manufactured and characterized. We have processed the complete process flow, from interposer manufacturing, to stacking and finally packaging (figure7).

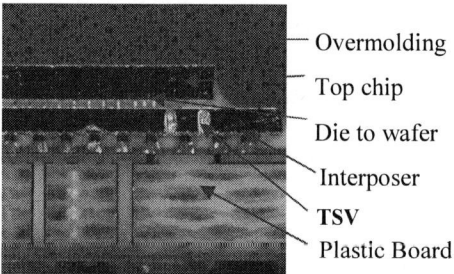

— Overmolding

Top chip

Die to wafer

Interposer

TSV

Plastic Board

Figure7: cross section of the final package (courtesy of ST Microelectronics)

To simplify failure analysis and subsequent corrective actions, we have electrically tested the products at 3 different steps: On the interposer, after TSV and back-side connection process, after die to wafer stacking, and after final packaging.

1/ TSV resistance on the interposer

To characterize the TSV process and back side redistribution line, electrical tests can be done first of all at the end of the interposer process, either before or after debonding. The comparison has proven that the debonding does not impact the TSV behavior [5].

To obtain this value, some specific daisy chains have been implemented into the lay-out (so called Kelvin measure), measuring directly the resistance of the vias, thanks to a four-points probe technique (Figure 8).

The following figure 9 shows the resistance per via AR2, in ohms. 67% of the chips have a resistance below 20mOhms per via, which is in the range of the analytical value estimated.

80% of the chips show a resistance below 50mOhms.

Considering those wafers are sampled in the very first batch of TSV AR2 manufactured at CEA - LETI, the results are very encouraging. We can see a small edge / center effect, process induced.

978-1-4244-4722-0/09 $25.00
© 2009 IMAPS-ITALY

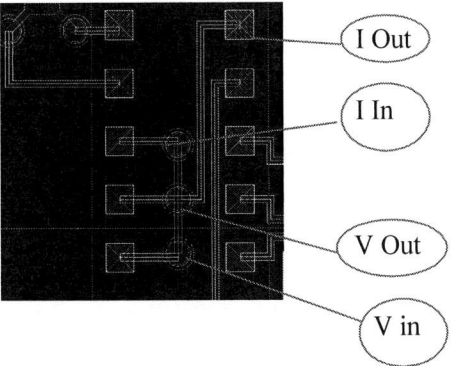

Figure 8: Electrical testing of the via (yellow: frontside rerouting, green: back side RDL)

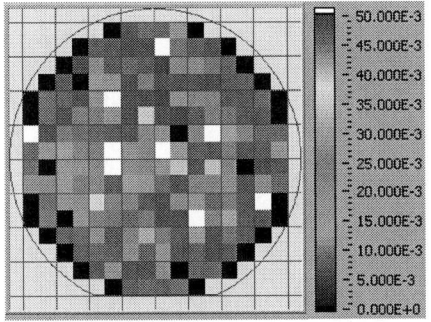

Figure 9: wafer mapping of TSVs electrical resistance in the interposer (ohms).

The black areas (which are in fact negative values) are points on which the measure has not been possible at standard voltage. It can be a TTV too important between the four points of the pseudo Kelvin.

The value superior to 50mOhms can be understood as a loss of contact inside the via: probably an issue of continuity in the RDL ECD.

Both of those defects can be improved by adjusting parameters of dielectric deposition inside the via or during RDL copper electroplating. Some improved batches are currently under process.

2/ Electrical contact top chip / bottom chip

Electrical tests are also done after stacking and reflow, at wafer level.

The very first tests have been done on a specific vehicule tests whose design is given above. Depending on the line tested, the quantity of interconnections top chip / bottom chip is different. Furthermore, for this demonstrator the line of interconnection are in aluminium, this gives intrinsically a high value of resistance. We do not have to take into account this resistance value itself, but the continuity of the resistance. First of all, 100% of the stacking is functional, and 100% of the lines per chips are functional, which proves the right alignment between top & bottom chip, as well as a correct interconnection thanks to the copper pillar, even in a dense matrix of copper pillar.

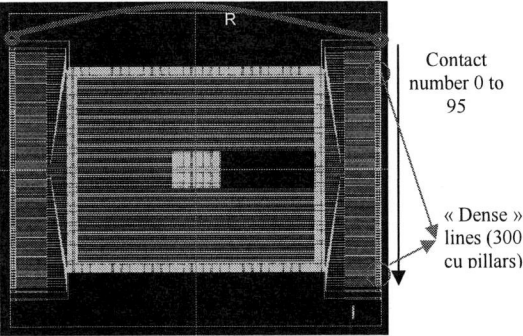

Figure 10: lay-out of the bottom die before stacking

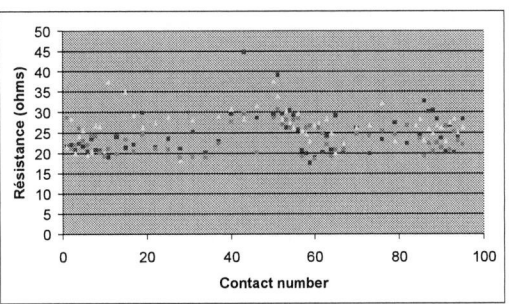

Figure 11: Value of resistance on different stacked chips

The resistance of a line with a high density of interconnects (contact number 1 to 4 and 91 to 95) is similar to the resistance of a line with smaller quantity of interconnects. This means that the resistance of the cu pillar interconnection is completely masked by the resistance of the aluminium line. In addition to these parametric tests, we have also proven thanks to functional test on devices that some functionalities of the top die (memory blocks) are operational after stacking.

3/ Electrical tests after packaging

Finally, each piece can be also functionally tested after final packaging. We have used a PBGA package for 15 chips. Some parts are showing some functional face to face daisy chains as well as ring oscillators. Top chip is also proven to be alive on some parts.

Conclusion & Prospects

In the frame of this work, a lot of technological elementary bricks for 3D integration have been developed, such as temporary bonding and debonding, low pitch interconnections based on Cu pillars technology, high aspect ratio TSV, thin wafer handling and chips on wafer alignment and stacking. Those bricks have been implemented on both test vehicle and functional demonstrator (Fig. 12) and even if the yield of the technology is not still acceptable in an industrial point of view, the

978-1-4244-4722-0/09 $25.00
© 2009 IMAPS-ITALY

electrical results obtained on the functional demonstrator are very encouraging. That means that the 3D integration technological feasibility has been proven thanks to this project.

Figure 12: SEM global cross section of the backside bump, RDL and TSV of a functional demonstrator

Nevertheless, this work was a first step for the 3D integration technologies development in LETI and a lot of projects are currently on going in which we are developing new technological bricks for 3D integration. The main developments we are leading are concerning the following aspects:

- *Low pitch C2W interconnections*: we are currently focusing on pitch decreasing for the connection between the top chip and the interposer. The current pitch is aroud 50μm but the applications roadmaps show that we need to reach less than 30μm pitch.

- *High aspect ratio TSV:* In another hand, we also need to decrease the TSV diameter faster than the Si interposer thickness. That means that the TSV aspect ratio will increase in a very near future from 2:1 to 5:1. Meantime, we will have to improve the electrical performances of the TSV: resistance and parasitic capacitance reduction, in order to be compatible with the high frequency applications.

- *Temporary bonding & debonding:* the strategy for the carrier implementation has to be improved. Actually, our current process flow needs to have a debonding with interconnections on the backside and to stack the top chips on the thin Si interposer after debonding. This approach is probably not optimized in terms of throughput and industrial risks. It's the reason why we are developing alternative process flows in order to secure the complete process.

- *Underfilling:* The current process uses a classical capillary underfill between the top chip and the Si interposer. Nevertheless, the diminution of the gap between the top and the bottom chip implies to develop new underfill technology based on pre-applied underfill products.

- *Components stacking:* last but not least, we are also developing new techniques for high accuracy alignment and components stacking, using pick and place tools. The objectif is obviously to be able to stack more than two components on the same Si interposer. Note that the stacking will also be impated in case of pre-applied underfill technology.

We are currently working on all these topics in LETI, thanks to different projects with industrial partners and suppliers. The results of those studies will be presented in a very near future through other publications.

Acknowledgements

We would like to acknowledge all the co-authors of this paper for their hard work in the projects. We also would like to acknowledge ST Micoelectronics and the European community (Seventh Framework Programme FP7-ICT-2007-1) for their funding.

References

[1] L. Bonnot & Al, "3D integration program overview", DATE 2009, April 2009.

[2] A. Jouve & Al, "Facilitating Ultrathin Wafer Handling for TSV Processing", EPTC 2008, 9-12 december 2008, Singapour.

[3] S. Pargfrieder & Al, "3D Integration by Through-Silicon-Via (TSV) processing enabled by Temporary Bonding and Debonding Technology", Advanced Packaging, April 2009 issue.

[4] D.Henry & Al, "Through silicon vias technology for cmos image sensors packaging: presentation of technology and electrical results", EPTC 2008, 9-12 december 2008, Singapour.

[5] J. Charbonnier & Al, "Integration of a temporary carrier in a TSV Process Flow", ECTC 2009, 26-29 May 2009

978-1-4244-4722-0/09 $25.00
© 2009 IMAPS-ITALY

Advanced Failure Analysis Methods and Microstructural Investigations of Wire Bond Contacts for Current Microelectronic System Integration

R. Klengel, S. Bennemann, J. Schischka, C. Grosse, M. Petzold

Fraunhofer Institute for Mechanics of Materials, Walter-Huelse-Strasse 1, 06120 Halle, Germany

E-Mail: robert.klengel@iwmh.fraunhofer.de, Phone: +49 345 5589 159, Fax: +49 345 5589 101

Abstract

The paper demonstrates that new failure modes can be analyzed and understood if an improved comprehensive flow in diagnostics involving non-destructive failure localization, ion-beam-supported target preparation, high-resolution electron microscopy and ultra-sensitive surface analytics are implemented in the physical failure analysis chain.

For illustration, the potential of combining new non-destructive Lock-In-Thermography (LIT) and, 2D/3D-X-Ray inspection to localize fine pitch wire bond failures inside moulded packages will be shown.

In addition to non-destructive methods, the increasing demands related to current "physics of failure" approaches in reliability aspects require ultra-high-resolution microstructure investigations. Corrosive failing of the intermetallic Au_4Al formed in the Au bond/Al pad contact is discussed. Results are presented based on preparation with Focused Ion Beam (FIB) device followed by Scanning Electron Microscopy (SEM) and Transmission Electron Microscopy (TEM) allowing analysing the corrosion process. Further application of Time-of-Flight Secondary Ion Mass Spectroscopy (ToF-SIMS) verified the presence of low concentrated contaminations as a root cause.

Key words: wire bonding, reliability, failure analysis

Introduction

Wire bonding is still the most common interconnect technology for semiconductors. Many efforts have been continuing to increase the integration density and to minimize the packaging size while, the reliability requirements such as in automotive or power applications were rapidly increasing. In addition to shrinked bond pad sizes and smaller wire diameters, higher process temperatures due to lead-free soldering, the introduction of new materials such as green mold compounds and harsh application conditions lead to specific intensified loading situations, to enhanced failure risks and new defect modes such as cracking in the bond pad/ ball interface, corrosion of the formed intermetallic, swaying effects during the moulding process caused by lower loop stability or lift off failures induced by alternating mechanical stress caused by high temperature variations. To meet these new requirements it is indispensable to improve and to adapt the test and physical failure analysis methods and equipment.

Fig. 1: SEM image of ball-wedge wire bond contact

Analytical Methods and Equipment

In many cases, X-Ray inspection can be used for non-destructive investigations of electrical boards, devices and components. Unfortunately several materials in electronic packaging like Silicon or Aluminum have very low or similar x-ray absorption coefficients so that the method has several limitations. In 2D x-ray imaging, all planes penetrated by the radiation form a superposed image. Thus, it is sometime difficult to resolve close-set details. In these cases, the 3D-Computer-Tomography can be useful. The object will be rotated step wise in the x-ray beam for 360°. In every single step an image is generated. After the full scan, it is possible to reconstruct a 3D model of the object. Every single part of the component can now be separately analysed.

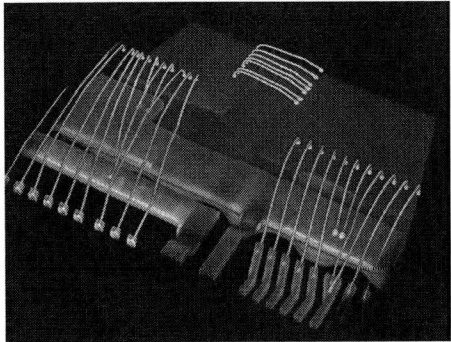

Fig. 2: 3D X-ray image of a Multi Chip Module

978-1-4244-4722-0/09 $25.00
© 2009 IMAPS-ITALY

While X-ray analysis and SAM have been industrially commercialized already for a considerable time, the Lock-In-Thermography (LIT) is a rather new sensitive method in non-destructive diagnostics for microelectronic components [3]. The defect is stimulated electrically by a periodically pulsed supply voltage which produces a periodically thermal response (hot spot) detected by a free-running IR camera. The amplitude but also the phase signal of the thermal response with respect to the electric current can be analyzed which is the main advantage compared to the steady state thermography. Due to the periodically pulsed excitation the heat spreading and therefore the thermal blurring is limited depending on the used lock-in-frequency. In addition, due to the lock in amplifier principle, the detection limit is improved by the 3-4 orders compared to the steady-state-thermography. In particular, the phase signal contains information about the time delay between electrical stimulation and thermal response. Thus, the method is not limited to IC surfaces but can also be applied to molded packages [3]. The phase shift is influenced by thermal spreading inside the mold compound or adhesives and can also be used for three-dimensional defect localisation, e.g. in System in Package or stacked die applications [3].

The Backside-Preparation-Tool (BPT) is a development of Fraunhofer IWM Halle to locally prepare electrical components mechanically. It allows highly accurate local grinding of package or IC material for following preparation steps or investigations. The software driven preparation routine is freely programmable and has an accuracy of about 1μm in x-, y- and z-direction.

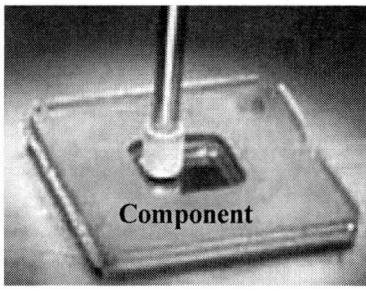

Fig. 3: Grinding of mold compound with the BPT

A cross beam combined Scanning Electron Microscope (SEM) / Focussed Ion Beam (FIB) device was used for preparation and analytical tasks. With the ion beam it is possible to remove sample material highly accurately until the exact analysis place is reached. This is very important for extremely small points of interest or preparing electron transparent lamellas for TEM analyzes (thickness about 60-100nm). Simultaneously the SEM can be used as process control for the FIB preparation. Performing FIB imaging the grain structure of the sample material can be visualized due to the different orientations of the grains. SEM is also used for high resolution imaging and for element identification with Energy Dispersive X-ray Spectroscopy (EDX).

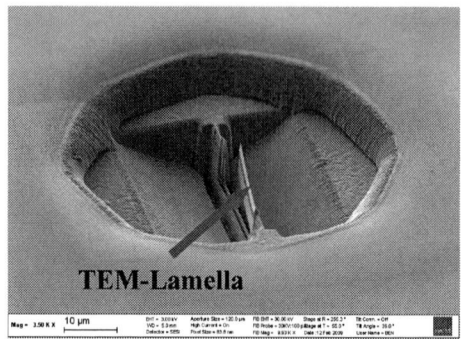

Fig. 4: Electron transparent lamella in a solder pad for TEM analysis prepared with FIB

Transmission Electron Microscopy (TEM) is used for ultra high resolution imaging and nano-spot EDX analyses. An electron transparent lamella prepared by FIB is ex situ manipulated on a sample grid and afterwards transferred in the TEM. Performing 200kV through radiation imaging of atomistic structures can be realized. The extremely small sample thickness of about 60-100nm allows a very small acquisition spot for EDX analyzes (< 5nm) and so the possibility of a chemical element characterization in thin structures or small particles.

Time of Flight–Secondary Ion Mass Spectroscopy (ToF-SIMS) is a very high sensitive surface analytical method (sensitivity 1-10ppm). From the sample surface bombarded by an ion gun (acquisition gun) secondary ions were released and after deflection using an ion mirror emitted by a detector. The flight time of the ions is measured and referred to the specific ion masses. Using the acquisition gun in a pulsed modus during the stand time a second sputter gun can remove the already analyzed layer. So also measuring of depth profiles is possible.

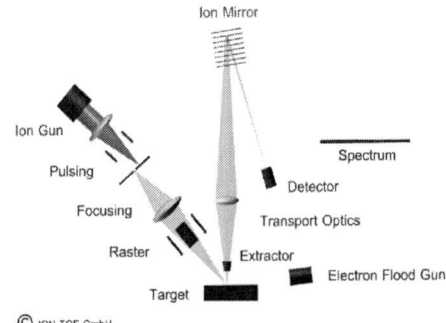

Fig. 5: Functional principe of ToF-SIMS (source: ION-TOF GmbH)

Application example for Systematic Flow of Causal Research

Fig.6 shows an overview of a multi chip component that failed during electrical testing. A 2D X-ray inspection showed no shorts of bond wires or lead fingers as well as any other hints of failure.

Fig. 6: Overview of the Multi Chip Module, 2D X-ray image

Thus, a mechanical cross section preparation or chemical decapsulation could not be applied due to the unknown failure location and the risk to destroy or remove the decisive defect detail.

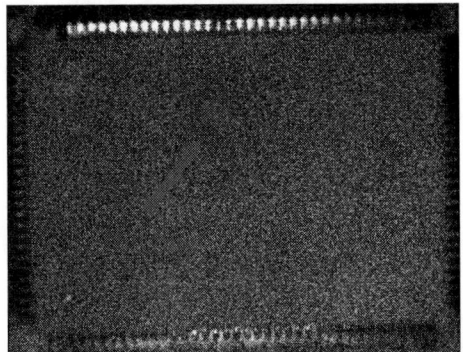

Fig. 7: IR camera image during LIT measuring, arrow marks the local hot spot

In this case, the performed LIT inspection was applied revealing a local hot spot in the center of the package. Fig. 7 pictures the overview with the region of increased temperature marked by the red arrow. Still, the accuracy of the failure localization is relatively low caused by the strong scattering effect by the mold compound. Thinning the mold compound over the chip down to close above the bond wires using the BPT improved the sharpness of the hot spot detection considerable. Fig. 8 shows the related image of the IR camera with the visible edge of the mold thinning and the point of failure. Now, the overview image from the 2D X-ray inspection and the new IR camera image from the LIT were overlaid. Thus it is clearly visible that the failure is located on the upper left corner of the right chip in the region of the bond wires (see red arrow in Fig. 9).

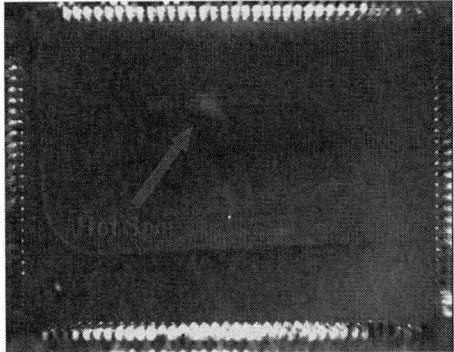

Fig. 8: IR camera image during LIT measuring after thinning the mold compound using the BPT, arrow marks the local hot spot

With the knowledge of the failure position, it is possible to start a mechanical target preparation. The component was potted in epoxy and afterwards grinded to the destination area. Then the further preparation was performed stepwise very carefully with fine grinding paper under permanent control using light microscopy. When the failure occurred the grinding was stopped immediately. Finally the cross section was polished using clothes moistened with diamond suspension.

Fig. 9: Overlaid 2D X-ray image and LIT image allows exact failure localisation

Light Microscopy and SEM inspection clarified that an aluminium splint in the mold created a short between several wire bond contacts.

Fig. 10: Aluminum splint created a short between wire bond contacts

Application example for Analyzes of contact corrosion in Au-ball/ Al-pad interface

The process of gold wire thermosonic ball-wedge bonding is a well-established technology [1, 2]. For the ball bonding process, it is known that specific Au/Al intermetallics are susceptible to corrosion effects [4, 5], particularly affecting the gold-rich intermetallic Au_4Al. The activators can be halides like Iodine, Chlorine, Bromine or Fluorine, each in combination with moisture and temperature. Halide contaminations may occur in mold compounds, Silgels, package materials, may be due to bond pad residues from the semiconductor patterning process (etch resist, photo lack, rinsing baths) or may result from environmental conditions.

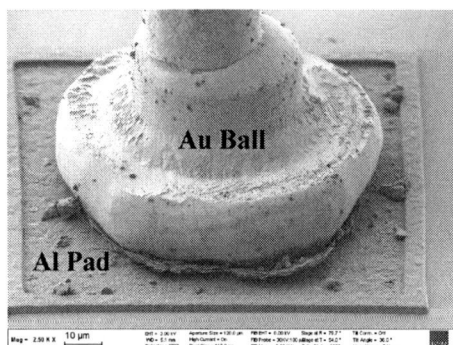

Fig. 11: SEM image of Au ball bond contact

The sample given in Figs. 11 and 12 represents an example of the dramatic failure mode caused by reaction effects with the intermetallics. Fig. 11 shows a typical thermosonic bonded Au ball contact on an Al pad metallisation. Standard pull tests resulted in ball lift off for all contacts at very low load values. After cross sectioning using FIB preparation and following SEM inspection the reason for the bad test results became apparent.

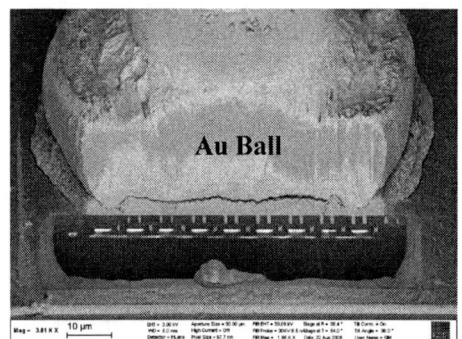

Fig. 12: SEM image of FIB cut, contact interface is nearly completely delaminated

In Fig. 12, it is visible that the contact interface is nearly completely delaminated between the intermetallic layer and the gold bond ball reducing the mechanical stability distinctly. However, as long as the bond is not mechanically stressed the electrical interconnect is still working. Fig. 13 shows a detail of the delamination in the edge region of the contact.

It becomes obvious that the failing interface is between the top layer of the intermetallic system and the Au ball.

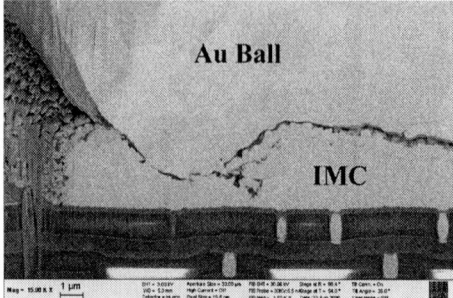

Fig. 13: Detail of corroded contact interface

For failure analysis, the bond pad of a defective contact was analyzed using SEM/ EDX to detect possible corrosion activators. Fig. 14 shows a bond pad with the intermetallic area of a lifted ball bond contact. The colored frame symbolizes the acquisition area for the EDX measurement. The recorded spectrum pictured in Fig. 15 shows clearly a significant Fluorine signal.

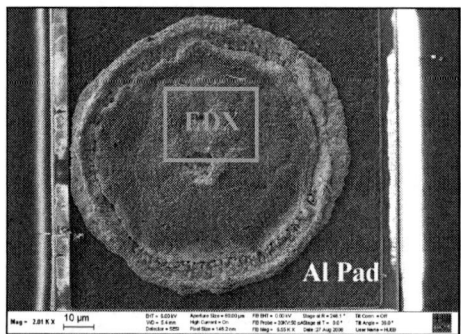

Fig. 14: Failed bond pad region with EDX analysis position in the IMC

Although for the Au_4Al/Au ball interface a high gold signal and only a low aluminum signal have to be expected, a relatively high aluminum peak combined with a high oxygen peak was found by EDX. This is a typical effect for the oxidation of Au_4Al induced by Fluorine as was also found in TEM analyzes. Fluorine contaminations can be formed due to insufficient cleaning after the chip structuring process.

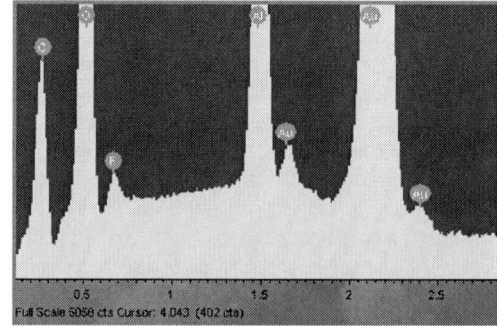

Fig. 15: EDX spectrum of failed bond interface, arrow marks significant Fluorine peak

Application example for Analyzes of contact corrosion in Al-wedge/ Flash-Au interface

A similar failure case can be detected if aluminum wedge contacts bonded to gold surfaces are considered. Again, aggressive substances like halides that may be due to harsh environments can cause corrosion effects of intermetallic compounds [4, 5]. Due to the very small dimensions of the formed intermetallics and the extremely low concentrations of contaminations, the application of high resolution analytical methods is required.

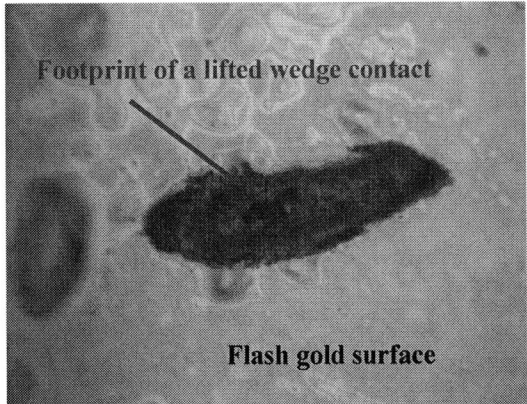

Fig. 16: Footprint of a lifted wedge contact caused by corrosion

Fig. 16 shows reflected light microscopy image of a footprint of a corroded aluminum bonded wedge contact after the wedge contact was lifted off . The SEM image of the opposite wire surface is given in Fig. 17.

Fig. 17: Undersurface of a corroded and lifted aluminum wedge contact, SEM image

To analyze the failed interface, cross sections of partially corroded bond contacts are required. The cross section could be done both by metallographic procedures (grinding and polishing with diamond suspensions) and Focused Ion Beam Technique (FIB). Fig. 18 shows an aluminum bonded wedge contact cross sectioned by FIB. The interface region was inspected by SEM and a porous morphology could be observed (see Fig. 19). Thus the interface region that leads to the failure could be identified.

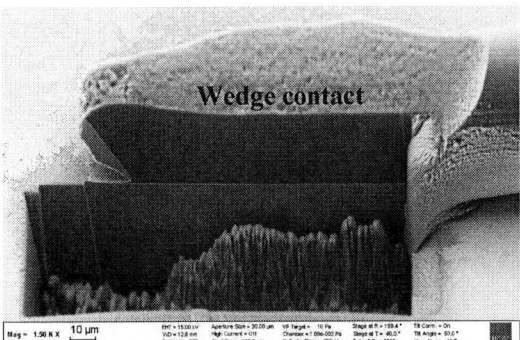

Fig. 18: FIB cut aluminum wedge contact, SEM image

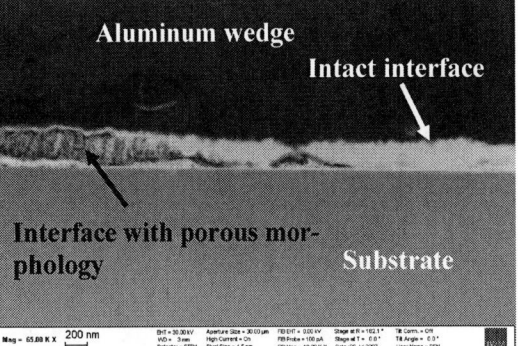

Fig. 19: Interface area of a corroded aluminum wedge contact, SEM image

Now high resolution analyzes like TEM with Nanospot-EDX are necessary for further investigations on the area with the critical morphology. Thus an electron transparent lamella was prepared and manipulated by FIB lift-out-technique. Fig. 20 shows the TEM image of the region with the critical morphology. The area consists of two parts, one dark colored without pores and one with strong defects.

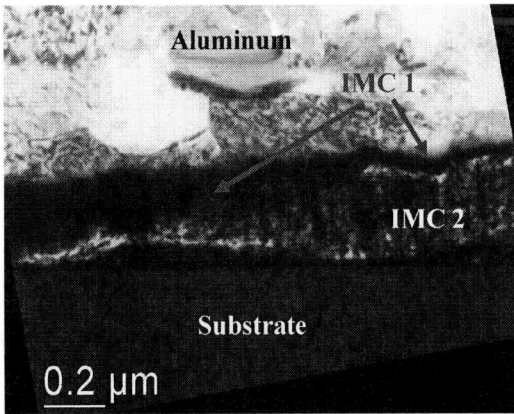

Fig. 20: TEM image shows corroded interface of a aluminum wedge contact with two different intermetallic areas

Nanospot-EDX analyzes show that the dark coloured region consists of gold and aluminum (see Fig. 21) and the region with the porous morphology consists of gold, aluminum and oxide. Thus it is assumed that the dark region is an intact intermetallic compound, and the area with the critical morphology a corroded intermetallic compound. Again, it has to be checked whether the presence of halides was responsible to form such aluminum-oxide containing corrosion products.

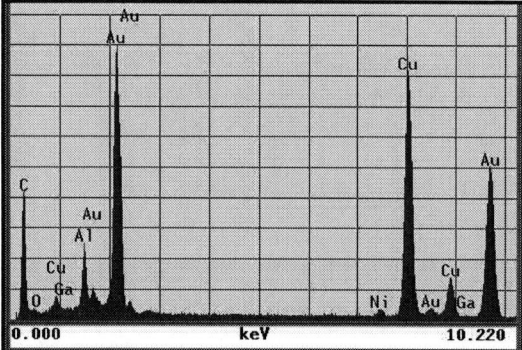

Fig. 21: TEM/EDX spectrum of IMC1 in Fig. 20, no significant oxide signal

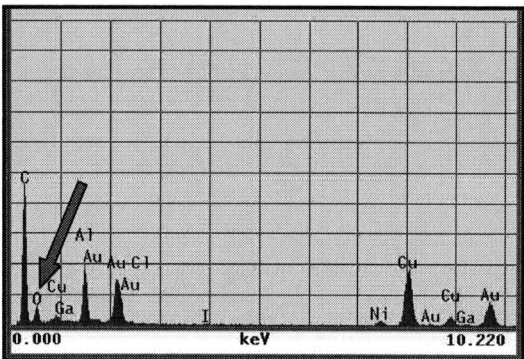

Fig. 22: TEM/EDX spectrum of IMC2 in Fig. 20, significant oxide signal and a remarkable higher aluminum part detectable

Because of the volatility and typical small concentrations of halides it is very difficult to determine these elements by TEM/EDX.

Thus, analytical methods with a ultra high resolution like TOF-SIMS are required. The TOF-SIMS analysis performed at the contact area of a lifted wedge contact shows significant Iodine signals (see Fig. 23). The presence of Iodine approves the assumption of the halide induced corrosion and allowed to identify the root cause of the failure.

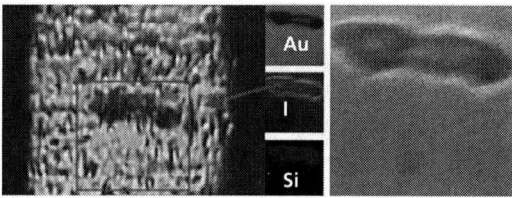

Fig. 23: Tof-SIMS mapping shows iodine accumulation in the region of the lifted wedge contact

Summary

The examples presented in this paper showed the absolute neccessity to level analytical methods and equipment concerning the increasing requirements of present packaging systems due to smalle dimensions, new material combinations and heavier operation terms. Thus its very important to combine different high resolution analysis methods to detect the miscellaneous possibilities of root causes. Especially for power and automotive applications the increasing demands in harsh operation environments lead to new and unknown failure modes. Thus fundamental background regarding material properties, binding mechanisms, contact degradation and physics of failure is required.

References

[1] Harman: Wire Bonding in Microelectronics: Materials, Processes, Reliability, and Yield; Mcgraw-Hill Publ.Comp.; Edition 2. (Aug. 1997)

[2] Zschech: Bondkontakte; Edition 1, Akademie-Verlag Berlin, 1990

[3] C. Schmidt et. al.: Lock-in Thermography for 3-dimensional localization of electrical defects inside complex package devices; 34th International Symposium for Testing and Failure Analysis 2008 (ISTFA2008)

[4] Klengel et. al.: Micro structure analysis for thermosonic bonded Au wire and ultrasonic bonded Al coated Au wire; IMAPS and SEMI Wire Bonding Workshop, San Francisco 2008

[5] Lue et. al.: Bromine- and Chlorine-Induced Degradation of Gold-Aluminum Bonds; Journal of Electronic Materials , Oct. 2004

978-1-4244-4722-0/09 $25.00
© 2009 IMAPS-ITALY

Electrical Modeling and Analysis of the Impact of Slits on Microstrip Lines in Thin Film Polymer Layers

Ivan Ndip[1], Michael Töpper[1], Karl-Friedrich Becker[1], Matthias Hirte[1], Ines Eidner[1], Thorsten Fischer[1], Brian Curran[1], Jörg Bauer[1], Wolfgang Scheel[1], Stephan Guttowski[1], Herbert Reichl[1,2]

[1]Fraunhofer Institute for Reliability and Microintegration, IZM, Gustav-Meyer-Allee 25, 13355 Berlin, Germany

[2]Technische Universität Berlin, Straße des 17. Juni 135, 10623 Berlin, Germany

ivan.ndip@izm.fraunhofer.de

Abstract—In this contribution, the impact of slits on reference planes of microstrip lines in thin-film polymer layers is studied. The return-currents of the microstrip lines excite the slits, causing them to behave like slot transmission lines that are short-circuited at both ends. Consequently, strong electromagnetic coupling occurs, especially when the slits are resonant within the frequency range of interest. We realized that the intensity of this coupling and hence, the amount of power lost from the microstrip lines depends on whether the fields that are coupled away from the lines through the slits penetrate into the silicon on which the thin film polymer layers are developed. For example, a resonant 5 mm long slit causes approximately 55% of power to be lost from a non-resonant 2.5 mm long microstrip line, when the fields penetrate into silicon. When the fields are shielded by a copper layer, the power lost from the line reduces to 15%. To experimentally validate the modeling approach, test structures were designed, fabricated and measured. A very good correlation was obtained between measurement and simulation results. Finally, guidelines to minimize power lost from microstrip lines in the presence of slits were deduced.

I. INTRODUCTION

In most wafer level packages, thin-film polymer layers are used to re-route the peripheral bonding pads of each chip to UBM pads arranged in an area array configuration. Since these re-distribution layers carry signal to and from the devices, they have a huge impact on the system performance. To isolate noisy devices, reference planes in these re-distribution layers are segmented (i.e., they are not continuous). The gaps/slits in these planes create a discontinuity for return-currents of signal lines routed above and/or below the slits (Return Path Discontinuity – RPD). This forces the return-current to flow as displacement current across the slit and/or as uncontrolled conduction current around it. Consequently, electromagnetic reliability (EMR) problems such as huge insertion loss, delay, electromagnetic interference (EMI) may occur, especially at RF/microwave frequencies. Hence, the impact of slits must be properly analyzed and minimized right at the beginning of the design cycle.

So far, slits on reference planes in different substrate technologies have been extensively studied and outstanding

results published in the open literature. The focus in most of the published work has been on studying the undesired electromagnetic (EM) radiation caused by very large slits [1], using full-wave simulation and lumped-element models to capture the interactions between slits, transmission lines and power-ground plane pairs [2]-[5], and proposing components/structures (e.g., RF chokes and corrugated slits) to suppress the impact of slits [6]. However, the impact of slits depends on whether they are resonant or non-resonant within the frequency range of interest. Furthermore, in a multilayered stack-up, the dielectric layer beneath the metal layer on which the slit is, plays an important role, especially if the dielectric is lossy. So far, none of these issues have been extensively addressed in the published literature. Hence, the goal of this work is to fill that gap.

For fixed lengths of microstrip lines in thin-film polymer layers on silicon (Si), the impact of resonant and non-resonant slits on the RF characteristics of the lines was modeled, simulated and analyzed for frequencies up to 20 GHz. The impact of Si beneath the slits on the performance of the microstrip lines was also investigated. In order to validate the simulation results, test structures were designed, fabricated and measured. A very good correlation was obtained between the measurement and simulation results. Finally, guidelines to minimize the impact of slits were then developed.

In section two, the build-up of the multilayered configuration studied will be presented. The impact of resonant and non-resonant slits on resonant and non-resonant microstrip lines will be discussed in section three. In section four, the comparison between measurement and simulation results will be presented.

II. LAYER STACK-UP

The build-up we developed and studied in this work is shown in Figure 1. It consist of three copper (Cu) and four Benzocyclobutene (BCB) layers developed on Si and arranged as follows: BCB-Cu-BCB-Cu-BCB-Cu-BCB-Silicon. Each Cu layer is 5 μm thick. The first and last BCB layers are 10 μm and 5 μm thick, respectively. Each of the other BCB layer

978-1-4244-4722-0/09 $25.00
© 2009 IMAPS-ITALY

is 15 μm thick. The first and second Cu layers were used as signal and reference layers of the microstrip line, respectively. Actually, the transmission line structure is an "embedded" microstrip line. However, throughout this work, it will be referred simply to as a microstrip line. The slits were formed on the reference layer (Cu-layer 2). The third Cu layer shields signals that are coupled away from the microstrip line through the slit, from Si. However, to understand the impact of Si, this third Cu-layer was removed in some investigations. The Si used in this work has a relative dielectric constant of 11.9 and conductivity of 10 S/m.

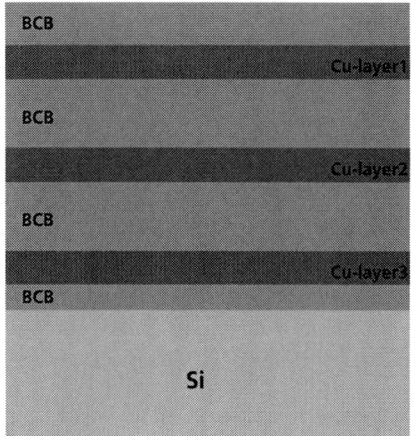

Figure 1: Layer stack-up.

III. INTERACTIONS BETWEEN SLITS AND MICROSTRIP LINES IN THE PRESENCE AND ABSENCE OF CU-LAYER 3

The top and cross-sectional views of the geometrical model that was used to study the impact of the interactions between slits and microstrip lines are shown in Figure 2. The dimensions used are also given in this figure. When the microstrip line is excited, its return-current flows on Cu-layer 2. Consequently, the slit is excited and it behaves like a slot transmission line that is short-circuited at both ends. This leads to an interaction between the electromagnetic (EM) fields of the slit and the microstrip line. Hence, power to be transported along the line is coupled away and lost. The intensity of this interaction depends on a number of issues, amongst which includes the impedance matching between the slit and the line, the length of the microstrip line that extends beyond the slit and whether the microstrip line and/or the slit is resonant.

In this section, the impact of resonant and non-resonant slits on resonant and non-resonant microstrip lines is extensively studied.

A. Impact of Resonant Slits on Non-Resonant Microstrip Lines in the Presence and Absence of Cu-Layer 3

For the investigations in this section, a microstrip line of constant length of 2.5 mm (non-resonant within the frequency

range of interest – up to 20 GHz) and two resonant slits of lengths 5 mm and 7.5 mm were used. The width of each slit was held kept constant at a value of 0.15 mm in all experiments in this work.

In order to accurately capture the effects of the interaction, 3D full-wave EM modeling was used. For this purpose, Ansoft HFSS (a frequency-domain field solver that employs the finite element method to solve Maxwell's equations in the frequency domain) was employed. The simulation model consists of the geometrical model shown in Figure 2, enclosed by an air box on which PML boundary conditions are defined. Wave ports (shown as P1 and P2 in Figure 2) were used for the excitations and the simulations were performed from 100 MHz to 20 GHz. In Figure 3, a comparison of the S-parameters of the insertion loss of the 2.5 mm long microstrip line for the two different slit lengths, in the presence and absence of Cu- layer 3, is shown. Note: If Cu-layer 3 (see Figure 1) is absent, then Si will have an impact on the signals coupled away from the microstrip line. However, in the presence of Cu-layer 3, the effects of Si can be neglected.

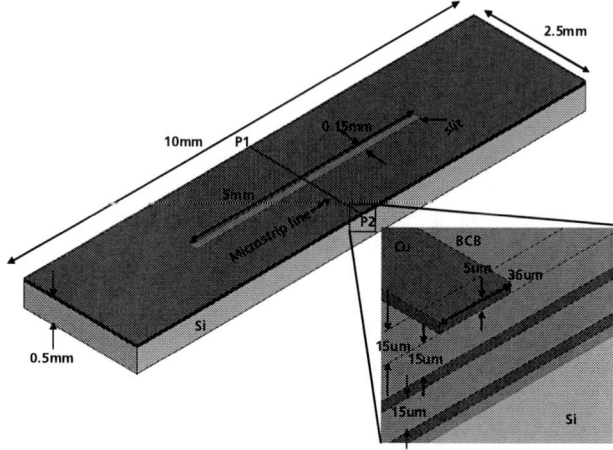

Figure 2: Geometrical model used for the EM field simulations.

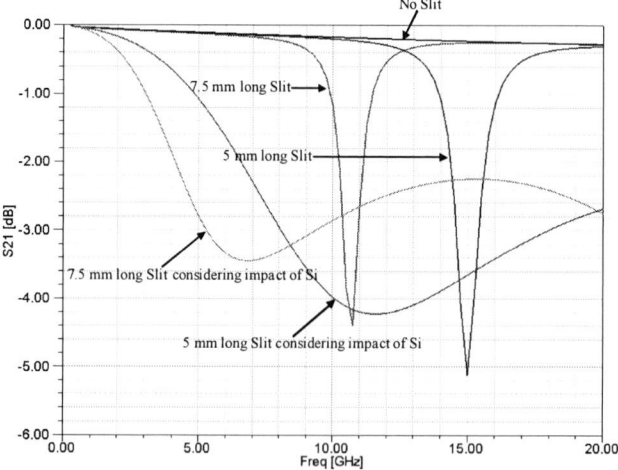

Figure 3: Impact of resonant slits on the insertion loss of 2.5 mm long microstrip line in the presence and absence of Cu-layer 3.

978-1-4244-4722-0/09 $25.00
© 2009 IMAPS-ITALY

As seen in Figure 3, in the absence of the slit, the insertion loss of the 2.5 mm long microstrip line is free of resonances. Besides, it is very small (less than 7% of power is lost from the line at frequencies up to 20 GHz) because of the good dielectric properties of BCB (ε_r=2.65, tan δ=0.008). For frequencies up to 1 GHz, the impact of both slit lengths considered (5 mm and 7.5 mm) in the presence and absence of Cu-layer 3 can be neglected, because less than 3% of power is lost from the microstrip line as a result of the interaction between the slit and the microstrip line. Up to 5 GHz, this power lost stays below 5% if Cu-layer 3 is present. However, in the absence of Cu-layer 3, at least 15% (considering the 5 mm long slit) and 40% (considering the 7.5 mm long slit) of power is lost as a result of the lossy nature of Si. The greatest interaction between the slits and the microstrip line occurs around the resonance frequency of the slit. For example, considering the 5 mm long slit, it can be seen from Figure 3 that approximately 55% of power is lost from the line over a frequency bandwidth of approximately 500 MHz in the presence of Cu-layer 3. In the absence of this Cu-layer, the situation gets worst. At least 55% of power is lost from the line over a 3.5 GHz bandwidth. This may lead to system failure. Furthermore, the signal is strongly attenuated at higher frequencies. The shift in resonance frequency when Cu-layer 3 is absent is because of the increase in effective dielectric constant as a result of Si.

B. Impact of Non-Resonant Slits on Resonant Microstrip Lines in the Presence and Absence of Cu-Layer 3

For the investigations in this section, a constant slit length of 2.5mm (non-resonant within the frequency range of interest) was chosen to study the impact of the interaction in the case when the microstrip line is resonant. 5 mm and 7.5 mm long microstrip lines (resonant lengths within frequency range of interest) were used. The simulations were performed using the same tool and boundary condition as those in section A. A comparison of the S-parameters obtained is shown in the figure below.

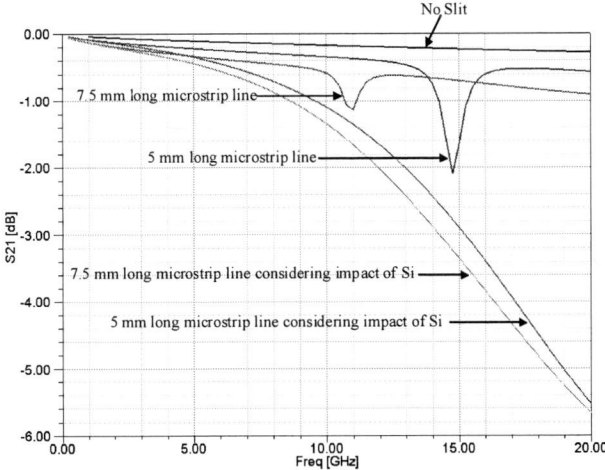

Figure 4: Impact of non-resonant slits on insertion loss of microstrip lines in the presence and absence of Cu-layer 3.

In this case, the impact of the slit can also be neglected, considering both microstrip lines for frequencies up to 1 GHz, because less than 3 % of power is lost as a result of the interaction between the non-resonant slit and the resonant lines. In the presence of Cu-layer 3, the signal degradation caused by the resonant lines is not as high as that caused by the resonant slits. For example, considering the 5mm long microstrip line, approximately 22 % of power is lost around its resonance frequency over a 500 MHz bandwidth, as compared to 55% of power lost around the resonance frequency of the slit over also over a 500 MHz bandwidth. In the absence of Cu-layer 3, the weak microstrip line resonances are not visible, most probably because of the strong attenuation as a result of the lossy nature of Si. However, the insertion losses of both lines increase steadily with frequency. For example, approximately 20% of power is lost from the 7 mm long line at 10 GHz. Above 15 GHz, it becomes almost impractical to use line lengths greater than 5mm, when Cu-layer 3 is absent, because more than 45% of power is lost.

C. Impact of Non-Resonant Slits on Non-Resonant Microstrip Lines in the Presence and Absence of Cu-Layer 3

When both the slit and the microstrip line are non-resonant, not much power is coupled away from the microstrip line in the presence of Cu-layer 3. As seen in Figure 5, less than 5% and 10% of power is lost from the line at 10 GHz and 20 GHz, respectively. However, in the absence of Cu-layer 3, the fields penetrate into Si and much power is lost, although neither the microstrip line nor the slit is resonant. For example, at 20 GHz, approximately 65% of the power is lost.

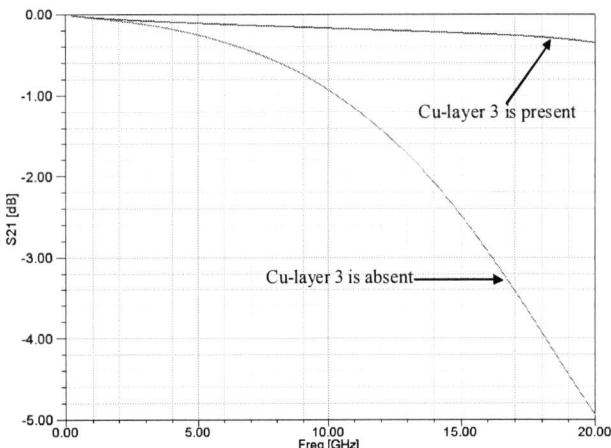

Figure 5: Impact of the insertion loss of a microstrip line when both the microstrip line and the slit are non-resonant.

When both the microstrip and the slit are resonant, their interaction is very strong around their resonance frequencies. The power lost from the line is similar to that discussed in sections A and B.

IV. EXPERIMENTAL VALIDATION

In order to experimentally validate the simulation results, test structures were designed, fabricated and measured. The measurements were performed with a vector network analyzer from 100 MHz to 20 GHz. As can be seen in Figure 6 where the insertion loss of the microstrip line is shown, a very good correlation was obtained between the measurement and simulation results. The microstrip line considered in this case has a length of 5.5 mm and width of 18 m. The slit has a length of 2.5 mm (along the microstrip line) and width of 0.6 mm (perpendicular to the line). Cu-layer 3 was not considered.

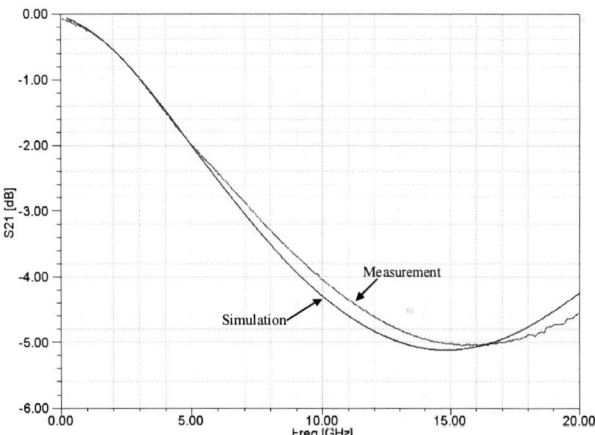

Figure 6: Comparison between measurement and simulation

V. SUMMARY AND DESIGN GUIDELINES

In this paper, we studied the impact of slits on reference planes of microstrip lines in thin-film polymer layers. Our results reveal that the return-currents of the microstrip lines excite the slits, causing them to behave like slot transmission lines that are short-circuited at both ends. Hence, strong EM interaction occurs, especially when the slits are resonant within the frequency range of interest. The intensity of this coupling and hence, the amount of power lost from the microstrip lines depends on whether the fields that are coupled away from the line through the slits penetrate into the Si on which the thin-film polymer layers are developed. For experimental validation, test structures were designed, fabricated and measured. A very good correlation was obtained between measurement and simulation results. Some guidelines to minimize power lost from the microstrip lines in the presence of slits were deduced. Examples of which include the following:

1. The dimensions of a slit should be chosen such that it is non-resonant within the frequency range of interest.

2. If the slit is resonant, then the layer stack-up should be chosen such that a metal layer shields the signals that are coupled away from the microstrip line, thus preventing them from penetrating into the lossy Si.

3. In the presence of such a shielding layer, the impact of resonant and non-resonant slits on the insertion loss of microstrip lines can be neglected for frequencies up to 1 GHz.

REFERENCES

[1] M. Kahrizi et al., "Analysis of a Wide Radiating Slot in the Ground Plane of a Microstrip Line" *IEEE Transaction on MTT*, Vol. 41, No. 1, January 1993, pp. 29-37.

[2] J. Kim et al., "Slot Transmission Line Model of Interconnections Crossing Split Power/Ground Plane on High-speed Multi-layer Board", *6th IEEE Workshop on Signal Propagation on Interconnects*, May 12-15, 2002, pp. 23-26.

[3] Y. Mizuguchi et al., A study of the model of the slit on the ground plane", *2000 IEEE International Symposium on EMC*, pp. 575-580, vol. 2.

[4] Liaw H-J., "Signal Integrity Issues at Split Ground and Power Planes", 1996 *ECTC Conference*, pp. 752 – 755.

[5] Wu, C-T et al., "Composite Effects of Reflections and Ground Bounce for Signal Line through a Split Power Plane", *IEEE Transactions on Advanced Packaging*, Vol. 25, No. 2, May 2002, pp. 297-301.

[6] Moran T.E. et al., "Methods to Reduce Radiation From Split Ground Planes in RF and Mixed Signal Packaging Structures", *IEEE Transactions on Advanced Packaging*, vol. 25, No. 3, August 2002, pp. 409 – 416.

3D Integration Technologies for Ceramic Substrates in a SHM Application

Samuel Hildebrandt, Klaus-Jürgen Wolter

Dresden University of Technology, Electronics Packaging Laboratory, 01062 Dresden, Germany

Phone: +49 (0)351 463-36426; Fax: +49 (0)351 463-37035; E-mail: hildebrandt@iavt. de

Abstract

Many applications restrict the lateral dimensions of a substrate. LTCC is one way to overcome this restriction by integrating passive components into the substrate. LTCC is used for building structural health monitoring (SHM) sensors in aerospace. One goal is to obtain compact and reliable sensors that are as intelligent as possible, resulting in a high number of passive and active components. This leads to a demand for more than one or two component layers, requiring substrates to be arranged 3-dimensionally. Research goals were to build a stack of sintered LTCC substrates, being reliable, high enough to give room for components and easy to manufacture. The paper analyzes several ways to build such a stack. The first part describes stacking technologies known from literature. In the second part, several technologies to build frames for stacking are analyzed. Technologies to connect the frame to ceramic substrates are also examined, including conductive paths through the frame. Finally, a SHM system built from a piezoelectric sensor (LTCC substrate one) and its signal preamplifiers (LTCC substrate two) is shown.

Key words: 3D Integration, Ceramic, LTCC, Stacking

Introduction

A structural health monitoring (SHM) system in an airplane needs to be a very robust piece of electronics. The lifetime of such a plane can reach 30 to 40 years. Over the whole time the system has to work without failure, since SHM modules usually are mounted permanently onto the observed part. Ceramic circuits offer the necessary properties regarding lifetime and mechanical robustness. Especially LTCC make highly, partly 3D integrated sensor constructions possible. For special applications even more integration is needed, for example if certain surfaces cannot be populated with components due to system mounting requirements. In those scenarios the stacking of several ceramic substrates helps to offer enough space for components. 3D integration technologies permit the construction of very high integrated intelligent sensor nodes.

Literature research

A well-known technology to stack ceramic substrates is building a ceramic frame. Low Temperature Cofired Ceramics (LTCC) offer an elegant way to construct such a frame: simply use the existing process, cut frame structures from the tape, stack and laminate the tape, and finally sinter the object. Of course, electrical connections can be integrated the same way it is done for LTCC circuits: realize a via hole (by laser drilling or punching), fill it with thick film paste, eventually print catch pads. Everything looks quite manageable,

but a closer look reveals some difficulties. One of the biggest drawbacks is the huge waste of material from cutting the form. If conductive paths are required, there are many stencil or screen printing steps to absolve, even more for staggered vias which may be needed in narrow frames to avoid mechanical weakening of the structure. High wall aspect ratios (height vs. width) require special lamination techniques to keep their shape. An idea to overcome the high material consumption is using L-shaped pieces of LTCC tape to stack the frame. However, alignment and stacking of these parts is even more difficult.

Another way to connect several ceramic substrates is by soldering an area array connection field to a flexible printed circuit (FPC). In theory, this is a very convenient way to obtain an 3D substrate stack. First, a FPC is constructed, for example a long stripe with several area array contacts on it, corresponding to the planned position of the ceramic substrates. The already built up circuits are then placed on the FPC and soldered again to form a ball-grid-array-like connection. Finally, the FPC can be folded, resulting in a package with high component and interconnection density. All these processes are quite difficult to manage, for example keeping the FPC waveless during the area array contacting or applying an underfiller to strengthen the connection. Furthermore, the array structure consumes space on the substrate, and it cannot be positioned freely on the ceramic. Another drawback is the reduced

978-1-4244-4722-0/09 $25.00
© 2009 IMAPS-ITALY

reliability in thermal cycling due to the different thermal coefficients of expansion of the involved materials. As most ceramic materials can handle compressive loads much better than tensile stress, the assembly usually starts failing at low temperatures where the polymer shrinks more than the inorganic substrate. Research is done on this field of interconnection to increase the reliability [1,2], but at the moment the way looks long and challenging.

When talking about stacked (ceramic) substrates for microsystems, the Match-X System [3] cannot be ignored. This keyword is hard to position into the context of this paper, since Match-X defines much more than a way to stack substrates. The idea behind it is to supply a modular construction system to the developer. During the early experimental phase detachable modules can be interconnected with spring-loaded pins, allowing maximum flexibility in combining different functional blocks. In the validation stage, the bottom of one module is connected permanently to the top of the next one by an BGA structure. (The modules are the same as in the experimental phase, only the interface changes.) The final goal is to engineer an customized highly integrated microsystem. The modules mentioned are made from FR4 or from LTCC, Al₂O₃ is also possible. Unfortunately, the whole interconnection system is not designed for high reliability. In the described concept this fact does not matter since the stacked substrates prototypes are intended to serve only for testing and validation purposes.

Ideas and Implementations

(1) Casting a frame from LTCC slurry. Compared to the lamination of LTCC sheets, the number of process steps can be reduced by casting a frame from LTCC slurry. Although the required casting forms can be expensive, cost reduction is possible: no special single sheet processes (cutting, printing, stacking, laminating) are necessary. The shape of the frames is variable, but complicated ones may need in-depth knowledge of form construction. The height/width aspect ratio of the walls is limited by the drying and sintering processes. All the solvent in the slurry needs to exhaust over the upper surface during drying, and the organic binder needs to be driven out before the sintering starts. This results in long temperature profiles with slow temperature rises. For common LTCC tapes the manufacturers recommend sintering cycles of 3 to 12 hours, so for thicker frames even longer profiles may be required. To obtain electrical vias through the frame, the holes can either be laser-drilled after drying the frames (cold laser ablation with short wavelength radiation recommended) or casted (by implementing pins in the form). Fig. 1 shows a casting form for four frames, consisting of three types of parts: one ground plate, one outer border and four inlays. The white parts are made from PTFE (Teflon) and bolted

down to the aluminum plate. The corners are rounded to avoid cracks.

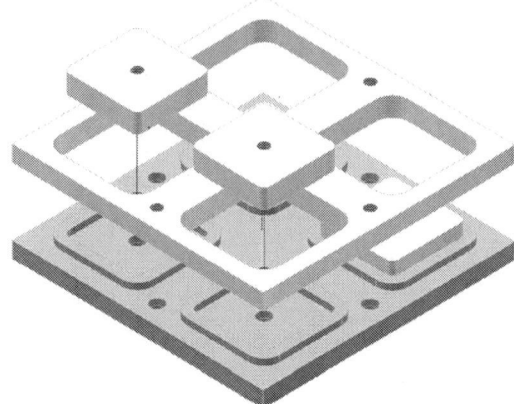

Fig. 1: Casting form for LTCC slurry

The advantages of the casting technology results from the reduced number of processing steps and LTCC sheets. Furthermore, only the casting form is required additional to a common LTCC process line. The TCE of the frame will be very close to that of the substrate material. However, slurry casting is not regarded as the best possibility to make ceramic shaped parts. The process is normally used to cast tape (LTCC, HTCC, alumina, PZT, ...). It still provides an opportunity to build the whole stack with nearly constant TCE. That will not be the case for pressure-sintered Al₂O₃ frames.

(2) Sintering several alumina frames together. It is possible to sinter Al₂O₃ parts together using standard isolation thick film pastes as a glass solder. In our case, an overglaze sintering at 520 °C was tested successfully. The process is quite simple. First, the paste is screen printed and sintered on both contact partners separately. Then the frames are cut (inner side) and scribed (outer edges) to prepare the separation. The process continues with stacking and sintering (eventually at a 20 K higher peak temperature than the paste's datasheet recommends). Finally, the frames are separated. If the substrates or ceramic parts are warped, an additional print of paste may be required before stacking. Generally, an increased film thickness is recommended, for example by using a coarse-mesh screen. Experiments have shown that a SD 125/65 stainless steel screen (133 mesh) results in an adequate film of isolation paste. The connection is expected to be hermetic and stable. Integrating electrical connections would be difficult and is not recommended due to mechanical weakening of the frame.

This technique is quite expensive (laser cutting, printing), but the tool costs are very low, only one screen is needed. Electrical connections require additional work (see (4)).

978-1-4244-4722-0/09 $25.00
© 2009 IMAPS-ITALY

Fig. 2: Laser-cutted and -scribed Al2O3 frames with pre-sintered isolation paste

The process makes sense in development or in a laboratory environment, where the needed machines are on-hand and fast availability and low tool costs are more important than the higher price per unit. Glass soldering is also used in the industry to connect alumina formed parts to substrates. One example is a pressure sensor where a cap (with pressure connection) is sintered onto a thin alumina substrate with thick film resistors and conductors (Fig. 4).

Fig. 3: Al$_2$O$_3$ base plate with 5 frames (walls 2mm wide, height 3.2 mm, whole part 25x25 mm²)

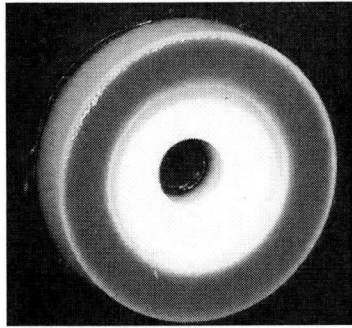

Fig. 4: Al$_2$O$_3$ pressure sensor cap prepared for sintering onto a substrate

(3) passive SMT components as a spacer.

The basic idea is to solder SMT components between two (not necessarily ceramic, but rigid) substrates. Depending on the number of adequate sized passives in the circuit, dummy components are required. If the technology is chosen early in the design process, it should be possible to consider

choosing e.g. capacitors with the desired height. If there are no constraints due to vibration loads, three or four components may be enough to stack the substrates. Adding more components rises the stability, also directed vibration reduction is possible by placing "stabilizing capacitors" under heavy or mechanically stressed components (e.g. connectors) on the top substrate. No additional machines are needed compared to a standard SMT line (stencil printer for solder paste, component placer, reflow oven). Fig. 5 shows one possible process flow.

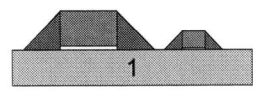

1. Print solder paste on top of substrate 1
2. Place components (circuit + spacer caps)
3. Solder substrate 1

4. Print solder paste on top of substrate 2
5. Place components
6. Solder substrate 2

7. Print solder paste on bottom of substr. 2
8. Place components

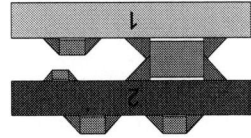

9. Place substrate 1 (bottom up)
10. Solder stack

Fig. 5: Process flow for connecting two ceramic substrates by soldering large components

(4) Separate mechanical and electrical functions. The first three ideas were driven by the idea to combine mechanical functions (stacking, hermetic shielding) with electrical ones (implementing vias). Especially point 2 shows the difficulties of implementing conductive paths in a alumina frame. Additionally, via filling requires quite some printing steps. To overcome these drawbacks, the fourth approach intentional separates the mechanical from the electrical tasks. For the mechanics, the principles illustrated in (1) and (2) can be used. Since no vias are necessary, the frame can be made thinner. High volume applications make tools for compression molding profitable. The difficult stacking process for LTCC tape or alumina frames is eliminated. The ceramic frames should be sintered from LTCC or alumina powder to ensure maximum stability. Frames with standardized dimensions can be offered like other parts made from structural ceramics. To connect the frame to the substrates, glass soldering, soldering (after metallization) or adhesive bonding are feasible.

Fig. 6 illustrates a process where the frame is soldered to both substrates, resulting in a hermetic sealing of the components between the ceramic

978-1-4244-4722-0/09 $25.00
© 2009 IMAPS-ITALY

substrates. An inert atmosphere can be realized by flushing the package with nitrogen before soldering the wires to the top substrate (step 10). Of course this involves the final soldering to be carried out under an inert atmosphere, too, e.g. in a glove box.

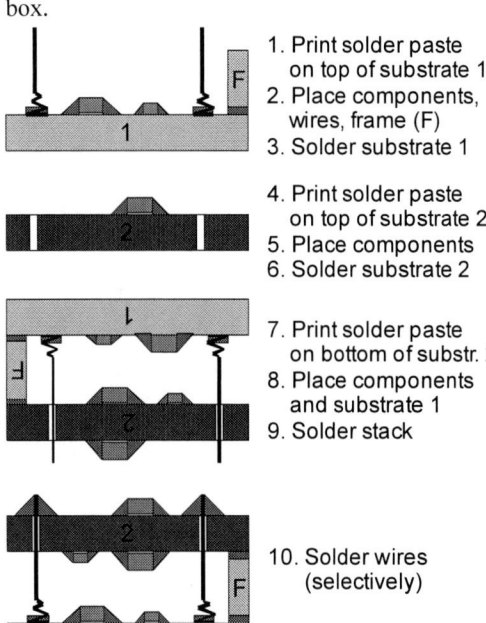

1. Print solder paste on top of substrate 1
2. Place components, wires, frame (F)
3. Solder substrate 1

4. Print solder paste on top of substrate 2
5. Place components
6. Solder substrate 2

7. Print solder paste on bottom of substr. 2
8. Place components and substrate 1
9. Solder stack

10. Solder wires (selectively)

Fig. 6: Sample process for connecting two substrates with elastic wires and a rigid frame

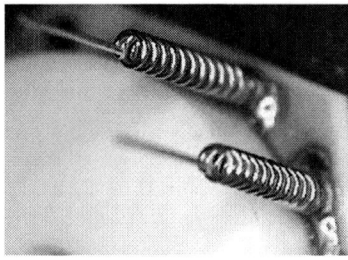

Fig. 7: Elastic, spring-like wire

If a through-hole solder joint is not acceptable, the elastic wires can be replaced with springs that are soldered to substrate 1 and force-fit to substrate 2. The last process step (10) can be saved, but the soldering of the stack (step 9) has to be carried out under a well-defined mechanical load: If the force is to low, the springs will lift the top; if it is too high, the solder will be squeezed out between frame and substrates. The frame can also be soldered selectively after the rest of the circuit is built. This allows a detachable connection for development or test purposes. For uncritical applications a clamping may be sufficient. This method is widely used to contact CPUs in computers (known as Land Grid Array, LGA; electrical contact applied by several hundred micro springs or spring-loaded pins).

Example: Piezoelectric Sensor for SHM

All the considerations were done in the context of a project that will result in a ultrasound-based sensor system for structural health monitoring in an airplane. Carbon fiber reinforced polymers (CFRP) are widely used wherever weight has to be reduced without compromising the mechanical properties of parts. One popular application is in aerospace. If a CFRP structure is damaged, it is usually not possible to analyze nature, scale and seriousness of the damage visual. Ultrasonic measurements can help to identify, locate and evaluate hidden defects. At the moment, these time-consuming measurements are done within the regular inspections of a plane by scanning critical parts manually with an ultrasound transducer. To reduce maintenance time and eventually detect impacts even during flight it is planned to mount an array of miniaturized sensors on CFRP elements. LTCC-based ultrasound transducers allow the integration of reliable electronics into the ultrasound sensor. There are several techniques how to realize a piezoelectric transducer on LTCC. Both sides of a substrate are needed – the bottom side as acoustic interface to the CFRP part, the top side to mount a piezo onto the LTCC. The footprint of the whole sensor has a strong influence on its acoustic behavior, so the circuit cannot be placed next to the transducer. Since the piezo transducer and circuit are still under development, it makes sense to build them on two substrates. Furthermore, the circuit might require two component layers at the size given by the acoustic properties of the system. A hermetic enclosure is preferred to reduce environmental influences. Ideally, the sensor shows only sintered LTCC and a robust wire connector to the outside world.

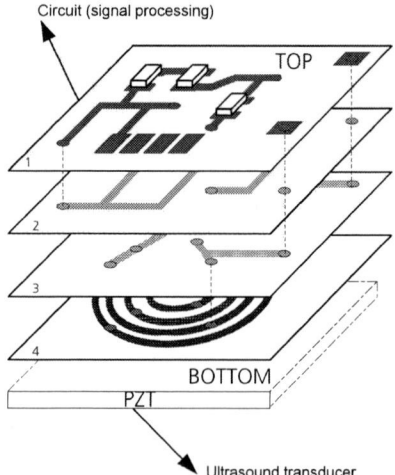

Fig. 8: Concept of a piezo-based SHM sensor

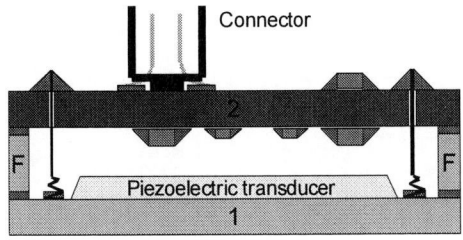

Fig. 9: Implementation of a 3D stack to connect a piezoelectric transducer to its signal processing circuit board

Conclusion and Outlook

Several alternative ways to stack ceramic substrates were shown. Some of the principles can also be transferred to other rigid substrates like FR4, although the main advantage of an all-ceramic system is its robustness and durability. Furthermore the stacking of substrates allows the combination of different process levels (e.g. one expensive fine-line substrate for high frequency circuits combined with a standard substrate for electrical power supply) as well as a (limited) modular building system (e.g. use one driving circuit for several types of LTCC-based piezoelectric transducers).

Unfortunately, the focus of the project in which the research was started changed recently. Therefore, some of the planned investigations could not be done. The next steps are to build a sufficient number of test circuits from the different construction versions and to carry out reliability investigations.

Acknowledgements

The author would like to thank Dr. Markus Detert for many fruitful discussions. Furthermore many thanks to all colleagues working in the MiFaLu project.

References

[1] Detert, M., Rebenklau, L., Schröder, S., Wolter, K.-J.: "Thick film modules assembled to flexible printed circuits", 2nd Electronics System-Integration Technology Conference (ESTC), London, 2008, pp. 873-876

[2] Schröder, S.: "Entwurf und Bewertung von Verbindungstechniken für die Kontaktierung von Area-Array-Strukturen (Dickschicht, LTCC) auf flexiblen Verdrahtungsträgern", Diploma Thesis, TU Dresden, 2007.

[3] Fraunhofer IZM: Match-X. Website: http://www.pb.izm.fhg.de/match-x/index.html, April 2009.

978-1-4244-4722-0/09 $25.00
© 2009 IMAPS-ITALY

A New Low Cost, Elastic and Conformable Electronics Technology for Soft and Stretchable Electronic Devices by use of a Stretchable Substrate

F. Bossuyt, T. Vervust, F. Axisa and J. Vanfleteren

CMST, University of Ghent, Technologiepark 914A, 9052 Zwijnaarde

Phone: +32 (0) 9 264 53 54, Fax: +32 (0) 9 264 53 74 and Frederick.Bossuyt@elis.ugent.be

Abstract

A growing need for ambient electronics in our daily life leads to higher demands from the user in the view of comfort of the electronic devices. Those devices should become invisible to the user, especially when they are embedded in clothes (e.g. in smart textiles). They should be soft, conformable and to a certain degree stretchable. Electronics for implantation on the other hand should ideally be soft and conformable in relation to the body tissue, in order to minimize the rejecting nature of the body to unknown implanted rigid objects. Conformable and elastic circuitry is an emerging topic in the electronics and packaging domain. In this contribution a new low cost, elastic and stretchable electronic device technology will be presented, based on the use of a stretchable substrate. The process steps used are standard PCB fabrication processes, resulting in a fast technology transfer to the industry. This new developed technology is based on the combination of rigid standard SMD components which are connected with 2-D spring-shaped metallic interconnections. Embedding is done by moulding the electronic device in a stretchable polymer. The reliability of the overall system is improved by varying the thickness of the embedding polymer, wherever the presence and type of components requires to. Manufacturability issues are discussed together with the need for good reliability of the stretchable interconnections when stress is applied during stretching.

Key words: Ambient electronics, biomedical electronic implants, moulding, PDMS, polymers, smart textiles, stretchable electronics, stretchable interconnections

I. Introduction

Stretchable, elastic electronic interconnection technologies will be a major improvement in the development of biomedical implantable electronics and smart textiles. User comfort expressed in softness and elasticity of the electronic device is a major issue.

Our main philosophy of stretchable electronic devices is that standard SMD electronic components are used, typically being non-stretchable. They are grouped in non-stretchable functional islands. To make stretchability happen, the different islands are connected by 2-D spring-shaped copper connections. This principle is shown in Figure 1.

Copper connectors are preferred above conductive polymers, due to their high conductivity, reliability and low cost. Some research groups [1]-[6] reported already their activities on the development of stretchable metallic interconnections on or in elastic, stretchable substrates. In [7,8,9] our technology based on plating gold meanders was presented together with optimizations of stretchability by use of finite element analysis (FEA) to obtain the optimal shape of the conductor shape. The shape of the copper conductors used in this paper is based on these results.

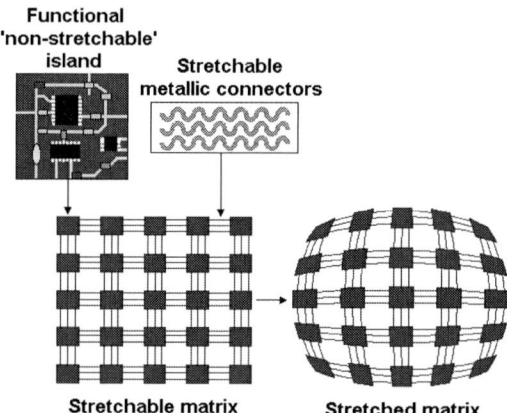

Figure 1: Stretchable electronics' architecture principle

II. Method

In [7-14] a technology based on plating gold metal tracks on copper foil was presented. This approach has some major disadvantages. While processing, all metal tracks are short circuited. This gives problems when embedding batteries. Moreover, etching of a copper substrate poses environmental issues.

A new approach for embedding metal tracks and standard electronic components in elastic substrates is shown in Figure 2.

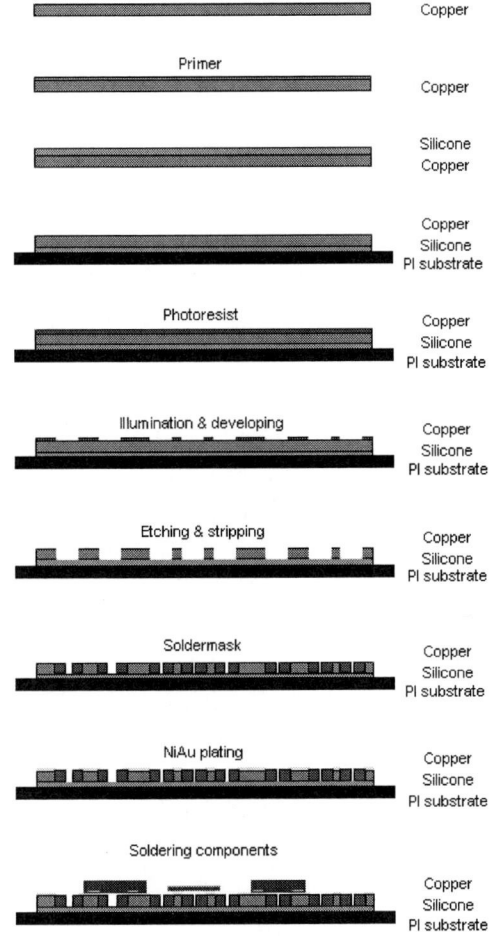

Figure 2: Process sequence

An 18 μm TW/YE copperfoil (Circuitfoil) is treated on the rough side with adhesion primer OS1200 (Dow Corning). A thin layer of silicone Sylgard 186 (Dow Corning) is spun or casted on it and cured at 50° C during 2 hours in air. The overall thickness of the silicone is ~100μm. The low curing temperature is needed to avoid thermal stresses gradients in between the copper and the PDMS. These give rise to problems after etching the copper: when the copper is patterned, silicone cured at high temperature would shrink and curl, making it impossible to align e.g. the soldermask on it.

The substrate is placed, with the copper on top, on a (perforated) Cirlex polyimide foil (Dupont)

with thickness of 300μm. Perforation holes are recommended in order to avoid entrapment of air between the temporary PI carrier and the silicone: air bubbles would cause problems because of thermal expansion during reflow soldering at elevated temperatures.

The physical adhesion between the silicone and the polyimide keeps our stretchable substrate on the polyimide temporary carrier during processing; no temporary adhesive is needed here.

Before application of a photoresist, the surface of copper has to be prepared for cleanliness and good adhesion. Preposit-Etch E25/29 (Shipley) is used as a micro-etchant for our surface preparation. The substrates are subsequently etched in 10% HCl solution, followed by a rinse in DI water. Next, photoresist AZ4562 is spun and a lithography step takes place in order to define the copper tracks. Etching is done by use of a spray-etcher resulting in copper tracks laying on the silicone substrate.

Before application, the silicone is treated with air-based plasma, in order to make the surface hydrophilic.

After stripping the photoresist, a 25μm layer of soldermask ELPEMER SD2463 FLEX HF is applied on the substrate by screenprinting Finally an electroless Ni, finished with an electroless Au flash is deposited on the copper, to improve the solder connection reliability. Typical thickness for the Ni deposition is a few (2-3) microns; thickness of the Au is 150nm.

Components are soldered by vapour phase soldering using SAC305 alloy. Due to the high soldering temperature of 260° C, the silicone Sylgard 186 expands. The linear coefficient of thermal expansion is 330 μm/m.ΔT. The small thickness of the silicone and the adhesion to the polyimide substrate prevents the thin silicone layer to move very drastically. Soldering at lower temperatures is recommended using low-temperature solderpastes or conductive epoxy glues. Silicones with a lower Young's modulus or a lower CTE are also advisable, leading to less deformation during soldering at high temperatures.

Before embedding in a polymer, the functionality of the circuit can be tested. This is a unique feature of this technology. Components can be replaced and resoldered if necessary.

Adhesion of the metal tracks, soldermask and components to the embedding elastic material is improved by using the same adhesion primer OS1200 (Dow Corning).

Finally, the substrate is embedded using the same elastic material (Sylgard 186 by Dow Corning). This can be done by casting or by using MID technology being shown in Fig. 3. In the MID technology, the stretchable electronic circuit on the carrier is put into a PMMA or ULTEM® 1000 mould. In the mould, cavities are made in order to have thicker layers of silicone on places where components are and thinner layers on places where the stretchable interconnections are defined. During stretching, the system will stretch more in the thinner parts. The upper layer of PDMS is injected and cured.

Next, the bottom mould is removed and the flexible polyimide substrate is peeled off. Another injection mould is mounted to inject the final layer of PDMS. After curing this layer the system is unmoulded leading to a stretchable system.

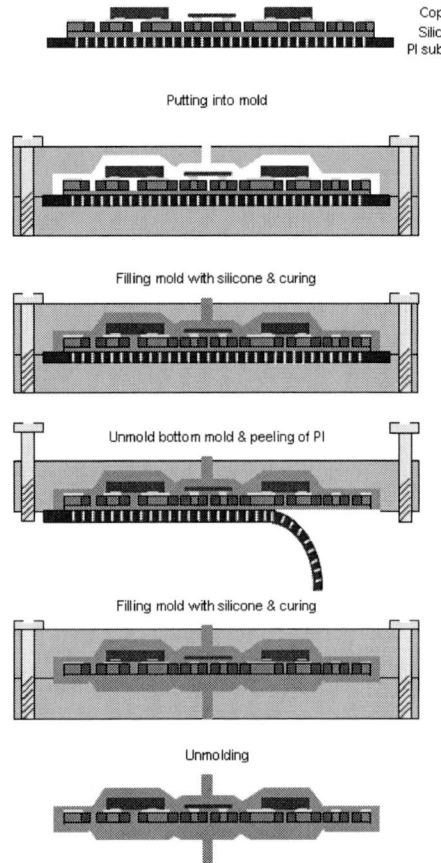

Figure 3: Moulding process sequence

III. Results

In Figure 4 copper meanders are shown made on Sylgard 186. They can be used as stretchable cables between 2 non-stretchable circuits and can be soldered by use of SAC or glued by use of a conductive glue. Completely embedding of the overall circuit in an elastomer can be done.

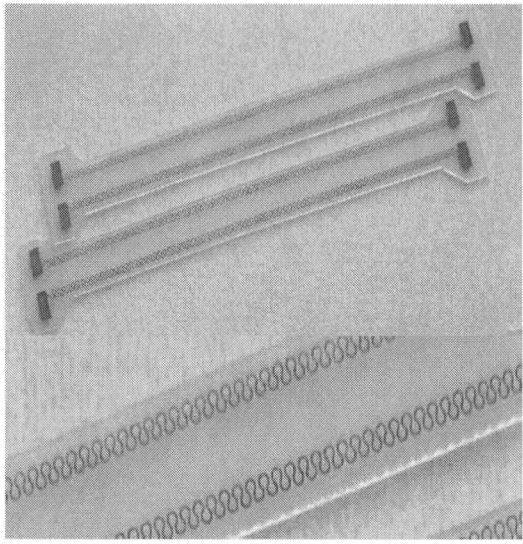

Figure 4: Stretchable copper meanders on Sylgard 186

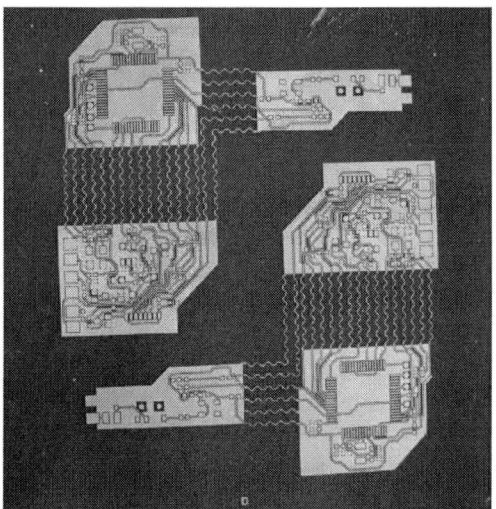

Figure 5: Stretchable electronic circuit with stretchable interconnetions and non-stretchable functional component islands

In Figure 5, a stretchable electronic circuit is shown where the copper is defined on a thin layer (~100μm) of Sylgard 186. In the copper, stretchable parts are defined by use of meander shaped tracks and non-stretchable parts are defined where component pads and straight tracks are surrounded with an electrical ground plane. The whole stretchable circuit is lying on a (perforated) Cirlex polyimide foil (Dupont) with a thickness of 300μm.

NiAu plating of the copper is followed and the result is shown in Figure 6.

978-1-4244-4722-0/09 $25.00
© 2009 IMAPS-ITALY

Figure 6: Stretchable electronic circuit on a 100μm thick Sylgard 186 layer

The following step is the mounting of components by vapour phase soldering them with SAC305 alloy. In Figure 7, the result is shown after mounting SMD components.

Figure 7: Stretchable electronic circuit with soldered components

As can be seen in Figure 7, due to the high temperature during vapour phase soldering, the Sylgard 186 has expanded and some air bubbles are created under the substrate. This can lead to soldering problems, especially for components with a lot of I/O pins. We have noticed that an air bubble can lift a component, leading to a badly soldered component. This problem can be solved by using a silicone with a lower Young's modulus and a lower CTE. Some tests were performed with Sylgard 527 (Dow Corning) which is a silicone gel and hasn't a

noticeable expansion during vapour phase soldering. In the end, a different encapsulating silicone material can be used to embed the whole stretchable circuit.

In Figure 8, the good result is shown of vapour phase soldering a component with a high number of I/O pins (TQFP44 package, 0.8mm pitch) on the Sylgard 527 substrate.

Figure 8: Feasibility of vapour phase soldering high I/O components on PDMS elastomer

In Figure 9, a completely embedded stretchable electronic circuit is shown. Not all components have been implemented and the embedding was done by casting a layer of Sylgard 186 on top, curing it and removing the temporary polyimide carrier and adding another Sylgard 186 layer at the back. The overall thickness is ~3mm.

978-1-4244-4722-0/09 $25.00
© 2009 IMAPS-ITALY

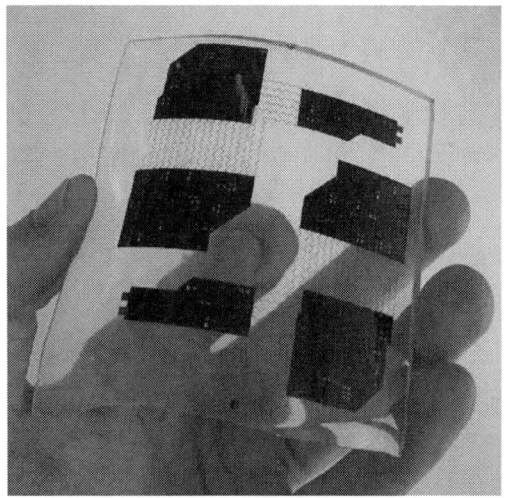

Figure 9: Feasibility of vapour phase soldering high I/O components on PDMS elastomer

Reliability tests have been performed for this technology by embedding meander shaped copper conductors in a 2mm Sylgard 186 substrate. An Instron 5543 has been used for this purpose in order to subject those samples to cyclic stretching (Figure 10-11) until failure (no conductivity of the meandered copper tracks). Depending on the meander shape, ~2000 cycles of cyclic stretching can be achieved for an elongation of 10% and a strain rate of 1%/s, without an increase in resistivity. This result gives an indication of the life-time of such embedded meander shaped copper conductors.

Figure 10: Instron 5543 for cyclic stretching of the embedded meandered copper tracks

Figure 11: Testing the reliability of the copper meanders embedded in Sylgard 186 by application of a cyclic strain of 10% at 1%/s strain rate

IV. Discussion

An advantage of this technology is that copper of several thicknesses can be used. The minimal copper thickness we used was 9μm, to obtain small features. Use of 18μm thick copper makes pitches of 100μm possible.

This technology is ideal for making stretchable connectors embedded in an elastomer, where stretchability and life-time till failure will be determined by the type of meander shape and the elastomer.

The problems arising during vapour phase soldering can be avoided by use of low temperature soldering pastes (based on Sn, Bi and In which have melting temperatures less than 183°C). Also, conductive epoxy glues can be used to overcome the problem of thermal expansion at 260°C. Another kind of elastomer, with lower Young's modulus and lower CTE can overcome the soldering problems for components with a high number of I/O pins, as has been shown demonstrated.

Besides the further work needed to optimize the soldering steps in order to obtain reliable soldered connections - especially for components with a high number of I/O pins – this technology has a high potential to realize stretchable, conformable electronic devices.

V. Conclusion

In this contribution a new low cost, elastic and stretchable electronic device technology has been presented, starting from a stretchable substrate. All used processes are standard PCB fabrication processes, leading to a less complicated transfer to the industry of this technology. The combination of rigid standard SMD components and 2-D spring-shaped metallic interconnections leads to a stretchable electronic system. Embedding can be done by moulding the electronic device in a stretchable substrate polymer in a way that the reliability of the overall system is improved by varying the thickness of the embedding polymer depending on the presence and type of components.

Different kinds of polymers can be used (able to withstand soldering temperatures) for other types of applications like implantable biomedical systems, smart textiles, sensors, actuators, robotic skins, etc…

VI. Acknowledgements

The work is supported by the Institute for the Promotion of Innovation by Science and Technology in Flanders (IWT) through the SBO-Bioflex project (contract number 04101). This work is also supported by European Commission Research programme STELLA (contract number 028026). The work is also supported by the Belgian Science Policy (Belspo) through the SWEET project (contract number P2/00/08).

VII. References

[1] D. S. Gray, J. Tien, and C. S. Chen, "High-conductivity elastomeric electronics," Adv. Mater., vol. 16, no. 5, pp. 393-397, Mar. 2004.

[2] S.P. Lacour, J. Jones, S. Wagner, T. Li, S. Suo, "Stretchable Interconnects for Elastic Electronic Surfaces", Proceedings of IEEE, Vol. 93, N°. 8, august 2005.

[3] Z. Yu, O. Graudejus, C. Tsay, SP Lacour, S. Wagner and B. Morrison, Stretchable microelectrode array: A potential tool for monitoring neuroelectrical activity during brain tissue deformation, J. Neurotrauma, Vol. 24 (2007), p. 1278 P200.

[4] M.N. Maghribi, PhD Thesis LLNL, Preprint UCRL-LR-153347, 2003

[5] T.A. Green, M.J. Liew, S. Roy, "Electrodeposition of Gold from a Thiosulfate-Sulfite Bath for Microelectronic Applications", Journal of the electrochemical society, 150(3),C104-C110,2003

[6] C.P Wong, "High Performance Screen Printable Silicone as Selective Hybrid IC Encapsulant", IEEE Transactions on components, hybrids and manufacturing technology, Vol. 13, N°4, December 1990

[7] D. Brosteaux, F. Axisa, J. Vanfleteren, N. Carchon, M. Gonzalez, "Elastic Interconnects for Stretchable Electronic Circuits using MID (Moulded Interconnect Device) Technology", MRS spring 2006.

[8] D. Brosteaux, F. Axisa, M. Gonzalez, and J. Vanfleteren, "Design and fabrication of elastic interconnections for stretchable electronic

circuits, IEEE Electron Device Letters, Vol. 28 (2007), pp. 552-554.

[9] M. Gonzalez, F. Axisa, M. Vanden Bulcke, D. Brosteaux, B. Vandevelde and J. Vanfleteren, "Design of Metal Interconnects for Stretchable Electronic Circuits using Finite Element Analysis", Proc. of 8th. Int. Conf. on Thermal, Mechanical and Multiphysics Simulation and Experiments in Micro-Electronics and Micro-Systems, EuroSimE 2007.

[10] F. Axisa, D. Brosteaux, E. De Leersnyder, F. Bossuyt, J. Vanfleteren, B. Hermans, and R. Puers, "Biomedical stretchable systems using MID based stretchable electronics technology", Proc. of IEEE EMBS 2007, pp 5687-5690.

[11] R. Carta, P. Jourand, B. Hermans, J.Thoné, D. Brosteaux, F. Axisa, J. Vanfleteren and R. Puers, "Design and implementation of complex systems on flexible-stretchable technology towards embedding in textile", Proc. of Eurosensors XXII, pp. 1384-1387.

[12] F. Axisa et al. Engineering Technologies for the development of advanced stretchable polymeric systems, Proc. of Polytronic 2008.

[13] F. Axisa, D. Brosteaux, E. De Leersnyder, F. Bossuyt, M. Gonzalez, N. De Smet, E. Schacht, M. Rymarczyk-Machal and J. Vanfleteren, "Low cost, biocompatible elastic and conformable electronic technologies using MID in stretchable polymer", Proc. of IEEE-EMBS 2007, pp. 6592-6595.

[14] F. Axisa, D. Brosteaux, E. De Leersnyder, F. Bossuyt, M. Gonzalez, M. Vanden Bulcke, J. Vanfleteren, "Elastic and Conformable Electronic Circuits and Assemblies Using MID in Polymer", Proc. of Polytronics 2007, p280-286.

[15] J. Vanfleteren, D. Brosteaux, F. Axisa "Methods for embedding of conducting material and devices resulting from the methods", US patent application, #US2006/0231288 A1, October 19, 2006; EP patent application EP 1 746 869 A1, January 24, 2007.

978-1-4244-4722-0/09 $25.00
© 2009 IMAPS-ITALY

Comparison between Die Attach Film (DAF) and Film over Wire (FOW) on Stack-die CSP Application

C.L. Chung*, C.W. Ku, H.C. Hsu and S. L. Fu

I-SHOU UNIVERSITY

Department of Materials Science and Engineer,

No.1, section 1, shiuecheng Rd., Dashu Shiang,

Kaohsiung Country, Taiwan, 84008, R.O.C.

Tel: 886-7-6577711 Ext.3121, Fax: 886-7-6578444

E-Mail: markchun@isu.edu.tw

Abstract

Due to the strong requirement of the miniaturization of micro electronic products, the development of IC package has been pushed toward to a smaller, thinner, lighter and higher density package structures. For the purpose of mass production, the designs of stack die-attached process should be simplified and wire-penetrated. In this paper, the characteristics of the Die Attach Film (DAF) and Film over Wire (FOW) as well as the following process to stack dies were defined. The cure kinetics and thermal resistances of the DAF and FOW materials were analyzed by Differential Scanning Calorimetry (DSC) and Thermo gravimetric Analysis (TGA). The geometry and distribution of filler of attached materials were analyzed by were examined by Scanning Electronic Microscopy (SEM) and Optical Microscope (OM). The thermo-deformation and pressure-induced flow behaviors of the attached materials were evaluated by the dynamic mode in rheology test. The results revealed that the FOW materials exhibit optimum and wider lowest-viscosity working window than DAF that obviously would contribute enough time to flow over the wires and easily to control bond line thickness, that can enhance the quality of Stack-die CSP. Besides, the geometry influences of filler were fully discussed in this paper.

Key words--- Film Over Wire (FOW), Stack-die CSP (Chip Scale Package).

INTRODUCTION

For the current advanced packages, Bismaleimide Triazine (BT) substrates are widely used for the purpose of PBGA, MCM with high-density I/O applications[1,2]. Stacked Chip Scale Package (CSP) are developed under the concept not only for the above advantages but also multifunctional applications[3]. There are some important papers discussed about triple-Chip stacked CSP[4], mold flow simulation[5], failure and fatigue life[6] and stress analysis of spacer paste[7]. Base on the mass product requirements, the die mount process should be simplified and improved. Cross-section of stacked CSP package can be seen in Figure 1. There are 2 dice in the package and some anther information shown in this table.

Within the limited space/thickness, to carefully evaluate the attached film materials were vital for the quality controls of the stacked CSP package. Reported here were our updated studies about the roles played by attached film characters and microstructure during stacked die mount processes.

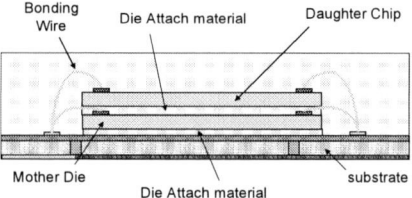

Figure 1 Detailed structure and cross-section of 2-layer stacked CSP package.

Experiments

Figure 2 shows the schematic illustrate of the process flow of stacked CSP package with FOW. Noteworthy, the support film using in wafer saw process also as the dice mount material in the following die bond process[8].

978-1-4244-4722-0/09 $25.00
© 2009 IMAPS-ITALY

* Process Flow:

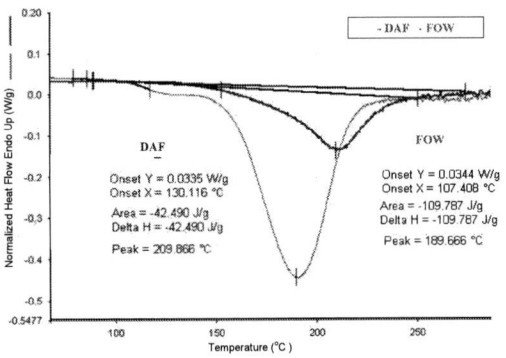

Figure 2 Assembly process flow of stacked CSP package with Film over Wire (FOW).

Rheology Test.

A routinely calibrated ARES strain control type (Rheometric Scientific) equipped with an ice or LN$_2$ bath cooling accessories is used. Heating rate of normal run is 5°C/min. Scanning range was from 25°C to 230 $^{\circ}$C for the attached materials. The dynamic test for G' and G'' were examined by parallel plate disk tool with 1 Hz frequency.

RESULTS AND DISCUSSION
Curing Kinetic Analysis.

Shown in Figure 3 were the DSC thermograms of DAF and FOW. The FOW revealed an earlier and larger exothermic heat than DAF. The initial cure point of FOW and DAF were at 107°C and 130°C, and the exothermic heats are -109.8 J/g and -42.5 J/g for FOW and DAF system. It is mean that the DAF system is more stable than FOW system during the ambit storage and followed assembly process.

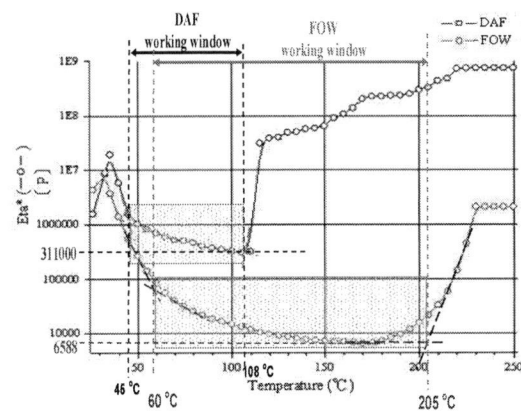

Figure 3 The DSC heating thermogram of DAF and FOW.
Thermal Resistance Analysis.

Given in Figure 4 were the TGA thermograms of DAF and FOW. Both film materials revealed very stable and without serious degradation before 300 $^{\circ}$C. The thermogram curves show the filler contents of DAF and FOW are 7.4 wt% and 40.1 wt%, respectively.

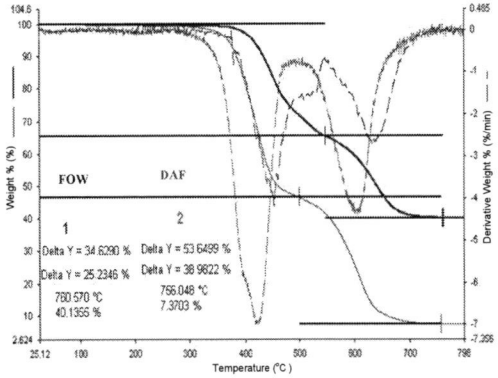

Figure 4 TGA heating thermogram of DAF and FOW.
Rheology Test.

Combination of Rheology behavior and cure kinetic are very important information for die mount condition designs. Shown in Figure 5 were the viscosity curves of DAF and FOW. The viscosity curve increase point identified the initial point of crosslinkage that are almost the start point of crosslinking reaction. The viscosity of DAF and FOW were increased located ca. 108 and 205°C, respectively. The FOW with wider lowest part of the viscosity curve that revealed wider process window than DAF system. Noteworthy, FOW exhibited lower value of the viscosity after curing reaction that indicates the packaging has a lower stress level inside after process.

978-1-4244-4722-0/09 $25.00
© 2009 IMAPS-ITALY

Figure 5 The viscosity curve of DAF and FOW.

Microstructrue Examination in Both Film

Figure 6 exhibited the SEM examination on the cross-section of (a) DAF and (b) FOW. The filler distribution of ADF and FOW were bi-model (0.5-0.8um and 2.5-3.6um) and mono-model (filler diameter ca. 2.5~4.7 um), respectively. There is a concern about die surface damage during the die mount process even a few special big filler in the DAF system. Due to 40.1 wt% fine filler uniform distributed inside FOW film, the FOW film with optimum viscosity can provide better bond line thickness control during the die mount process. The Figure 7(b) showed almost no deformations of gold wire in the FOW layer of the crossection of final package.

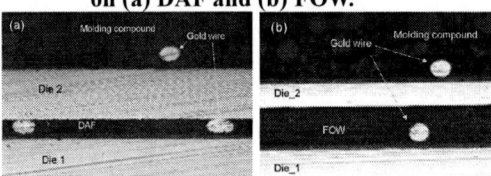

Figure 6 The cross-section examination by SEM on (a) DAF and (b) FOW.

Figure 7 The crossection of (a) DAF and (b) FOW by using optical microscopy.

Conclusions

Base on cure kinetic study, it is clear that the DAF system is more stable than FOW system for the ambit storage and whole assembly process. Rheology test results also indicate The FOW with relative wider working window in die mount process than DAF system. The FOW system with fine and uniform distribution filler about 40.1 wt%, could provide optimum viscosity and better bond line thickness control to protect gold wire during the die mount process. Noteworthy, FOW showed lower the viscosity after curing process that indicates low stress conditions in stacked CSP package.

Acknowledgements

The authors would like to thank Dr. Tsai-Sha Huang for his great helps in the study on materials supply and discussion in detail. Also thanks are due to MANALAB at ISU in experimental work.

References

[1] A.T., Cheung, "Dicing Die Attach Films for High Volume Stacked Die Application", Electronic Components and Technology Conference, Proceedings. 56th, 2006, pp.1312-1316.

[2] I., Ahmad, N.N., Bachok, N.C., Chiang, M. Z.M., Talib, M.F., Rosle, F.L.A., Latip, Z.A., Aziz, "Evaluation of Different Die Attach Film and Epoxy Pastes for Stacked Die QFN Package", IEEE 9th Electronics Packaging Technology Conference, 2007, pp.869-873.

[3] S.N., Song, and H.H., Tan, P.L., Ong, "Die Attach Film Application in Multi Die Stack Package", IEEE Electronics Packaging Technology Conference, Vol. 2, 2005, pp.848-852.

[4] X.L., Cheng, and F., Qian, "Analysis on the Effects of Packaging Materials on Structural Thermal Stress in Stacked Chip Scale Package", Electronic Packaging Technology, ICEPT 2007. 8th International Conference on, 2007, p.1.

[5] H.C., Hsu, Y.C., Hsu, H.Y., Lee, C.L., Yeh, and Y.S., Lai, "Application of Submodeling Technique to Transient Drop Impact Analysis of Board-level Stacked Die Packages", IEEE Electronics Packaging Technology Conference, 2006, pp.412-418.

[6] Y.M., Cheung, A.C.M., Chong and B., Huang, "Determination of the Interfacial Fracture Toughness of Laminated Silicon Die on Adhesive Dicing Tape from Stud Pull Measurement", Electronic Materials and Packaging, EMAP 2006, International Conference on, 2006, pp.1-10.

[7] M., Todd, "Material Systems Enable High Density Packaging", IEEE Proceedings of HDP 07, 2007, pp.1-5.

[8] E.J., Vardaman, and L., Matthew, "New Developments in Stacked Die CSPs", IEEE Proceedings of HDP 04, 2004, pp.29-30.

978-1-4244-4722-0/09 $25.00
© 2009 IMAPS-ITALY

A Novel Thermo-Mechanical Test Method of Fatigue Characterization of Real Solder Joints

R. Metasch*, M. Roellig** and K.-J. Wolter*

* Technische Universität Dresden, Electronics Packaging Laboratory, Dresden, Germany/Saxony

Tel.: +49351 463 36416, Email: metasch@avt.et.tu-dresden.de

** Fraunhofer Institut für Zerstörungsfreie Prüfverfahren, Dresden, Germany/Saxony

Tel.: +49 351 888 15557, Email: mike.roellig@izfp-d.fraunhofer.de

Abstract

The paper presents a novel thermo-mechanical test method for in-situ force measurement on real solder joints during thermal cycling test. Due to plastic deformation in solder joints the solder-material-matrix is damaged and micro cracks occur. This degradation behaviour can be determined by direct force measurement under thermal cycling loads. After the overall characterization of the machine, first successful results were generated based on 400µm solder balls on micro BGA-package stressed under -40°C and +125°C conditions. A force drop down to approx. 60% of initial force at the first cycles was measured. The measured forces kept in a steady state ranged over approx. 80% of lifetime until the material mechanically failed. The main tasks planned to be solved are the direct measurement of degradation of solder alloys for comparison purpose, the determination of solder material fatigue behaviour and derived material fatigue laws.

Key words: solder, material, degradation, fatigue, LCF, force drop

1. Motivation

The comparison of the solder fatigue behaviour is very important in order to choose the right solder alloy for specific application. The solder creep measurements [1,3] help to understand the time-dependent mechanical material behaviour, the micro structure analysis of solder alloys as function of composition, the solder volume, the solder cooling rate as well as the undercooling which all influence the formation of micro structure under specific boundary conditions [2].

For the determination of the degradation process further investigations are necessary to understand the fatigue behaviour of solder. A common way to compare solder alloys according to their fatigue resistance is the thermal cycle test of soldered electronic components on printed circuit boards. The main advantage is the very high number of tested solder joints to provide good statistical dataset. Disadvantages are the only roughly known mechanical conditions at the solder joint, the only stepwise analysed micro structure changes and the high mechanical influence of the substrate degradation on solder fatigue behaviour. Dudek presents a good experimental method to track the degradation of one solder joint over thermal cycles [3]. Using a visual microscopic analysis the joint deformation, the plastic strain and finally the crack

growth could be shown. So the micro structure changes could be observed.

If the continual force load on the solder joint is known the material degradation could be tracked over thermal cycle. The presented and developed fatigue machine allows the in-situ measurement of the force continuously during thermal cycles. So the force drop behaviour of different solder joints (alloys) can be observed. This allows a comparison of the solder fatigue behaviour. The stepwise reducing of the inducted force load on solder joints allows controlling the fatigue speed of the solder joints to derive the material fatigue behaviour.

2. Experimental setup

The load frame is a thermal controlled loading machine and allows the induction of shear force on solder joints. It consists of three symmetrical constructed main columns, which are connected by two aluminium arms. Inside the aluminium arms there are two flexure-hinges which can be used as force sensors (see Fig. 1).

The outer columns material is copper and the inner one is aluminium. The copper columns have holes inside to reduce their thermal mass.

The specimen is located in the lap shear shaped gap in the centre of the inner column because the inner aluminium column is divided and offers

978-1-4244-4722-0/09 $25.00
© 2009 IMAPS-ITALY

the opportunity to clamp two aluminium cylinders on a level with each other that the base areas of the cylinders are coplanar.

The base areas of the cylinders are adjusted arbitrarily towards each other. One of these is connected with the upper arm while the other is connected with the lower. Therefore the change in length which results from the temperature differences only appears between the cylinders.

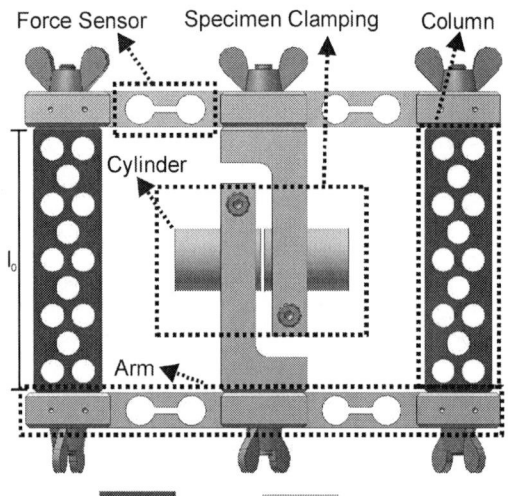

Fig. 1: Load frame construction (CAD)

At room temperature all three columns have the same length l_0. When the temperature of the load frame is changed (ΔT), a length difference Δl between the columns appears due to the materials different coefficient of thermal expansion CTE (see Tab. 1):

$$\Delta l = l_0 \cdot \Delta T \cdot (CTE_{Cu} - CTE_{Al}) . \qquad (1)$$

Material	CTE	E-Modul (20°C) in GPa
Al 2017	22,9	72.5
Cu	17,1	118.6

Tab. 1: Material data of load frame components

If a temperature change causes a length difference Δl, the specimen is deformed (shear length Δs) and the force sensor deformed. This displacement of the force sensor can be calculated from the acting force F and the spring stiffness k of the force sensor. The sum of shear length Δs and displacement result in the length difference:

$$\Delta l = \Delta s + \frac{F}{k} . \qquad (2)$$

As Δl and k are known, the shear strain of the specimen can be calculated from equation (3).

$$\Delta s = \Delta l - \frac{F}{k} \qquad (3)$$

The force F is measured by the force sensor. It is constructed symmetrically and equipped with two flexure-hinges which are recording the displacement. At one of the flexure-hinges there are arranged resistance strain gauges (DMS) that are connected up to a full bridge (see Fig. 2). The bridge voltage U_B amplified by $v = 500$. From this signal U_M there can be measured the acting force via the sensitivity a. This DMS full bridge compensates the thermal deformation in a temperature cycle.

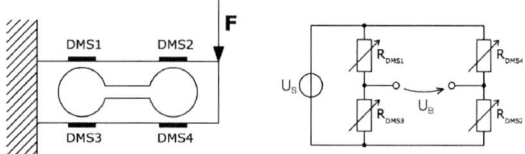

Fig. 2: Model of the force sensor (left), full bridge circuit (right)

The linear coherence between force and strain is explicitly recognizable (see Fig. 3). In order to be able to apply a force $F < 0$ when calibrating, the force sensor was turned around. Because of its self-weight the displacement was slightly changed which explains the switch at $F = 0N$.

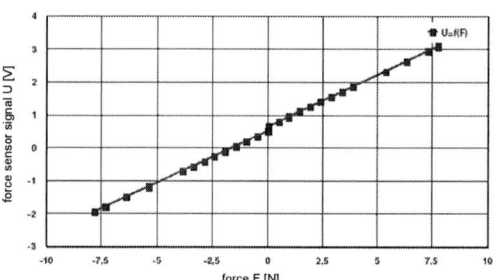

Fig. 3: Calibration curve of the force sensors

The complete measurement system is shown in Fig. 4.

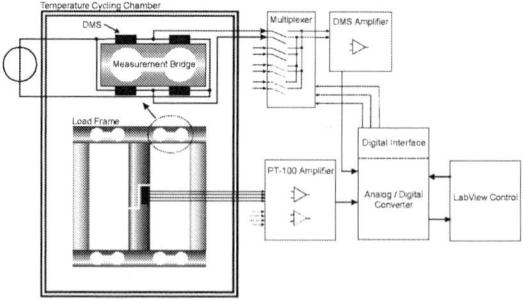

Fig. 4: Schematic of the measurement system

978-1-4244-4722-0/09 $25.00
© 2009 IMAPS-ITALY

It contains the development of the measurement signal amplifier for the force sensors and temperature sensor (PT-100), the assembly of a multiplexer unit for containing several force sensors, the development of the already introduced force sensors basing on the work in [1] and the programming of a Lab View based operating surface.

3. Specimen

In order to be able to examine the solder contacts of the load frame these have to be clamped appropriately. Especially there should not appear any stresses and strains due to clamping; otherwise the measurement would be affected from the beginning, which could adulterate the reproducibility of the measurement.

The specimen consists of two Al_2O_3 substrates with four solder pads and four solder joints. Hence, the solder contacts are situated in a sandwich of these two Al_2O_3 substrates, whereas the outer areas are plane parallel towards each other (see Fig. 5).

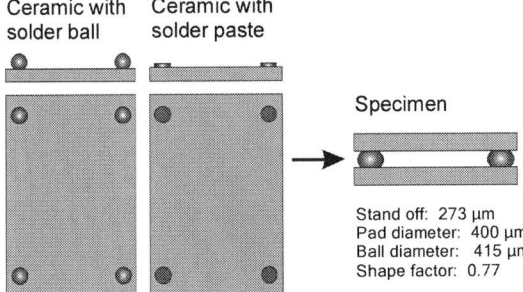

Fig. 5: Specimen assembling and specifications

The specimen is bonded to the base area of the cylinder (see Fig. 6). Then a high temperature curing adhesive is applied to both: back side of the specimen and the second cylinder. Both parts are put in contact when the cylinders are assembled. The hardening of the adhesive will take place only during the first temperature cycle, what ensures no stress from the clamping will be transferred to the specimen. A temperature sensor PT-100 was attached to the specimen.

Fig. 6: Specimen bonded to aluminium bolt (left), load frame (middle), the aluminium bolt with adhesive (right)

4. Characterization of the load frame

Three characterization tests were applied to access the quality of the measurement system. On the 1st test the force sensor alone was temperature cycled. On the 2nd test the load frame with the force sensor was temperature cycled as well but without a specimen. For the 3rd test the same configuration as for 2nd test was used with the difference that now instead of 2 cylinders only one was used.

The applied temperature cycling test ranges from -40°C to +125°C with a changing rate of 5K/min and a cycle time of 2h. The temperature measurements were accomplished at the flexure-hinges of the force sensor.

At the 1st measurement the force sensor, which was not built into the load frame, was exposed to the temperature cycling test. Fig. 7 shows the progress of the first four cycles of this measurement.

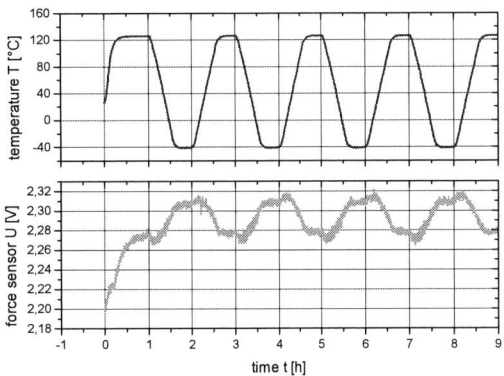

Fig. 7: Thermal drift characterization of force sensor

In the 1st cycle the sensor signal moved about 0.07V. The following cycles show a slight temperature dependence which is to be lead back to production tolerances of the force sensor as well as the DMS. These have to be compensated with a linear term (see Fig. 8).

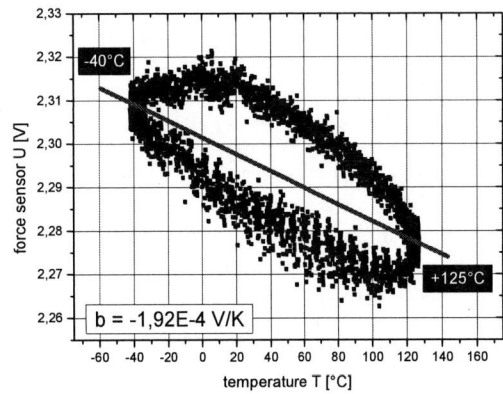

Fig. 8: Temperature dependency of the force sensor

978-1-4244-4722-0/09 $25.00
© 2009 IMAPS-ITALY

The apparent hysteresis behaviour of the force sensors is caused because the temperature can only be determined at the surface of the force sensor. For this reason the straight line was only fitted above the plateau data of the temperature cycles, where one can presuppose the homogeneous temperature distribution within the force sensor.

In the 2nd test the force sensor was build into the load frame (the middle bifid aluminium core was open). The intention was to be able to judge the restraint of the load frame during the temperature cycle. The progression of the 1st four cycles of the measurement is shown in Fig. 9.

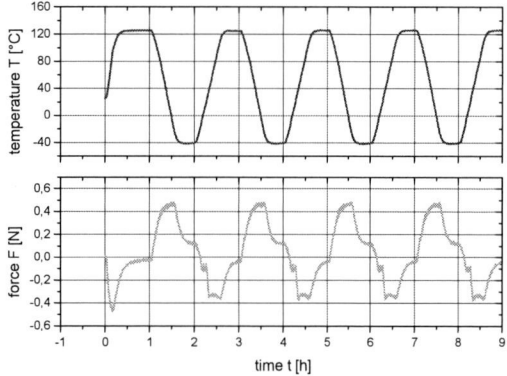

Fig. 9: Force sensor signal under thermal cycling (force sensor was build in of the load frame, the specimen clamping was open)

The displacement of the force sensor signal (as observed at the beginning of the 1st test) is hardly to be recognised. Hence, one can assume that this might have been a unique effect, which was caused by a change of the DMS features. The attachment of the DMS to the force sensor joints was realized using special adhesive that hardens at a temperature of 240°C. Nevertheless, the first cycle is slightly different from the following, which especially at the change of the nominal temperature show a continuous fluctuation of the force signal of about ±0.5N. This is caused by a fluctuation in the temperature cycling chamber which flushes the chamber with a strong air flow when changing to a new nominal temperature. Therefore supposedly two effects are interfering with each other which influence the force sensor signal and let it rotate as a kind of hysteresis (see Fig. 10).

Firstly, when cooling down or heating up, mechanical strains in the load frame occure that dispand when reaching the high or low temperature plateau. Secondly, the temperature gradient in the force sensor joint, which is not compensated by the DMS Bridge, increases because of the air-stream. After reaching the nominal temperature decreases or increases the measurement signal of the force sensor again and the effective fluctuation between the plateaus differs only about ±0.1N.

Furthermore, the cooling and heating profile clearly differ from each other. This is to be lead back to the thermal mass of the load frame and the superficial freezing of the DMS after exceeding the 0°C air temperature. However, this effect only appears for a short time and will be observed in the last test again.

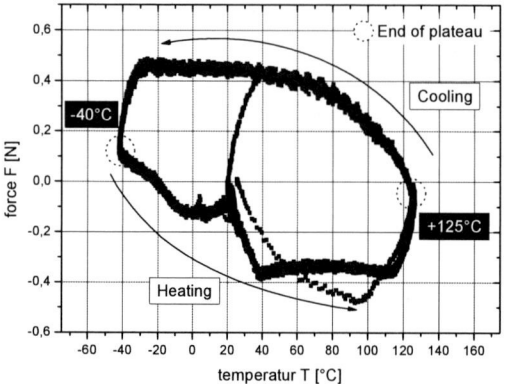

Fig. 10: Force measurement of the restraint inside the load frame

In this last test the bifid middle aluminium core was fastly braced by contrary to the 2nd test. Hence, the displacement of the force sensor equals the one in the temperature cycle generated length difference Δl. That means the shear length Δs is 0 in the temperature cycle. The progress of the 1st four cycles of this measurement is shown in Fig. 11. The intension of the last measurement was, to find out the maximum measurement range and how the qualitative progress might be changing, because of the mentioned effects (freezing, restraint, temperature gradient in the load frame).

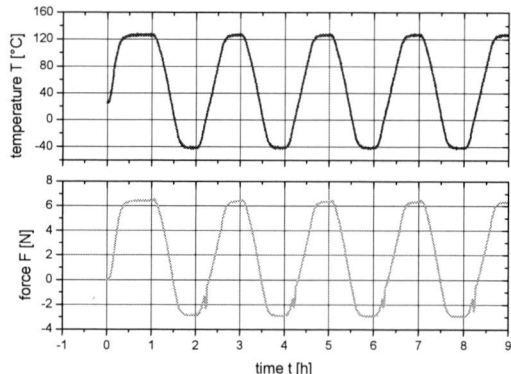

Fig. 11: Measurement of the load frame under thermal cycling (the specimen clamping was fixed)

Qualitatively the force sensor signal follows the one of the temperature of the temperature change chamber. The hysteresis behaviour of this measurement is shown on Fig. 12. The thermal mass of the load frame which stores heat and the superficial temperature measurement which never mirrors the temperature of the whole load frame, are the reasons for this. The effect of the freezing can be seen, however it is only shortly dominant.

These three tests show some conclusions that have to be taken into account for the following tests. The forces measured with the load frame are reproducible after the first half cycle. The measurement difference on the plateau is about 0.1N, whereas the difference of the temperature change is up to 0.5N. Before every test it is recommended to record the behaviour of the load frame (restraint effect) at least for three cycles.

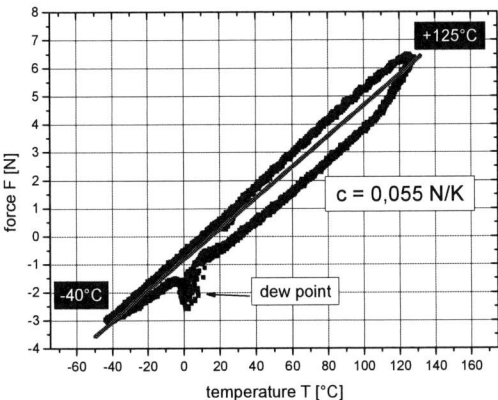

Fig. 12: Force vs. temperature

5. Results

For this experiment a specimen (SnAgCu alloy) was clamped stress free into the load frame, as described in chapter 3 and loaded with the temperature cycle described at the beginning of chapter 4. The progress of the first 30 temperature cycles (TC) of the experiment is shown in Fig. 13.

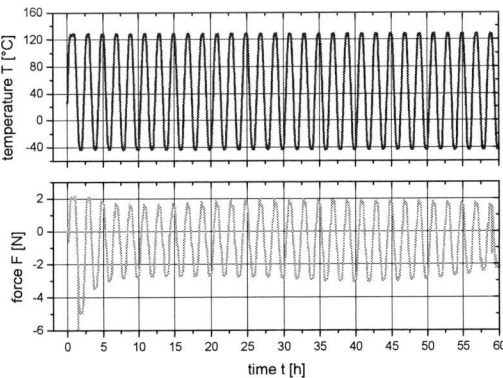

Fig. 13: Temperature and force signal of the experiment

The only possible way to get the real stress and strain inside the solder contact is to use numeric simulation systems like finite elements method because of the multiaxial stress conditions inside the solder contact. A uniaxial stress condition model

like a deformed cylinder (see Fig. 14) leads to a good approximation.

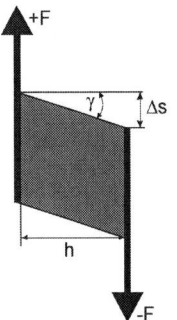

Fig. 14: Uniaxial stress model

The rough conversion of the temperature and the force signal into the shear strain of the solder contact is possible to be calculated with the equations 3, 4 and 6.

The shear length Δs from equation 3 and the known specimen stand off h determined the shear angle γ as shown in equation 4.

$$\gamma = \frac{\Delta s}{h} \qquad (4)$$

With the equation 5 the effective strain after Mises can be calculated. Finally, as the specimen in the load frame is only stressed by one shear component, the equation simplifies to equation 6.

$$\varepsilon = \frac{\sqrt{2}}{3} \sqrt{ \begin{array}{l} \left(\varepsilon_x + \varepsilon_y\right)^2 + \left(\varepsilon_y + \varepsilon_z\right)^2 + \left(\varepsilon_z + \varepsilon_x\right)^2 \\ + \frac{3}{2}\left(\gamma_{xy}^2 + \gamma_{yz}^2 + \gamma_{xz}^2\right) \end{array} } \qquad (5)$$

$$\varepsilon = \frac{\Delta s}{h \cdot \sqrt{3}} \qquad (6)$$

The shear stress in the specimen is to be calculated from the effective force F and the specimen surface. In the specimen there are several solder contacts, so that its base area A needs to be multiplied with the number of contacts n.

$$\tau = \frac{F}{n \cdot A} \qquad (7)$$

According to the shape modification hypotheses by Mises the effective stress is calculated in equation 8 and simplified in 9.

$$\sigma = \sqrt{ \begin{array}{l} \frac{1}{2}\left[\left(\sigma_{xx}-\sigma_{yy}\right)^2 + \left(\sigma_{yy}-\sigma_{zz}\right)^2 + \left(\sigma_{zz}-\sigma_{xx}\right)^2\right] \\ + 3\left(\tau_{xy}^2 + \tau_{yz}^2 + \tau_{xz}^2\right) \end{array} } \qquad (8)$$

$$\sigma = \sqrt{3} \cdot \tau . \qquad (9)$$

With the won parameters the progress of the strain and stress of the solder material can be described (see Fig. 15). This progress shows that the largest changings take place already at the beginning of the measurement. Hence, the strain of the solder is much smaller at the beginning by contrary to the stress which is clearly higher. Already after a few cycles a nearly static condition is shown in which the strain or the stress is hardly changed.

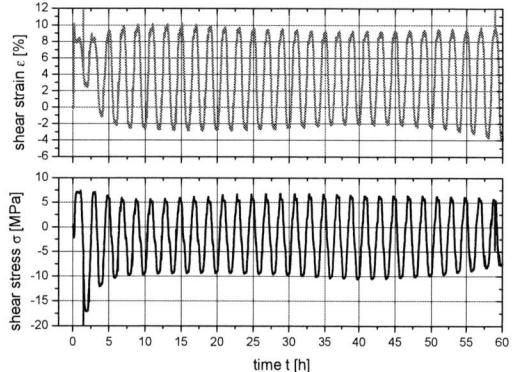

Fig. 15: Calculated shear strain and shear stress vs. the time

With a directed analysis the differences between the plateau data of a temperature cycle the strain and the stress of a temperature cycle is collected (see Fig. 16).

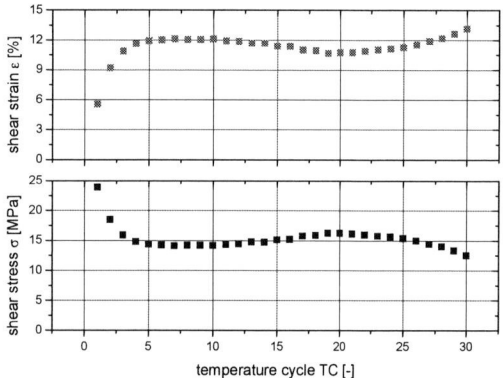

Fig. 16: Determined shear strain and shear stress vs. the temperature cycles

The figure shows a force drop from 24MPa to 14MPa stress and simultaneously a rise of the strain from 5% to 12% strain at the first 5TC. After that the experiment shows quasi constant stress and strain conditions.

The demonstration of the stress-strain-hysteresis of this experiment takes place in a summary of 5 TC each (see Fig. 17).

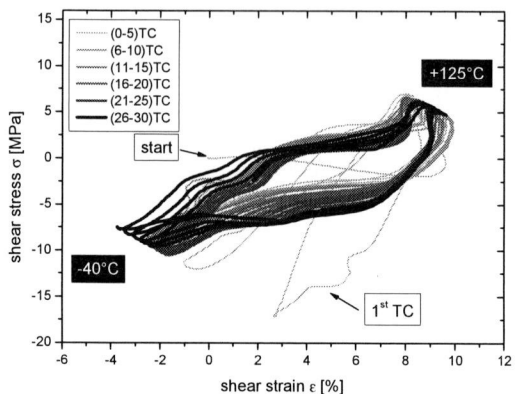

Fig. 17: Shear stress vs. shear strain

It shows that the shape of the hysterese curves changed. Through the determination of the surface area A of the single hysterese curves their relative progress can be determined (see Fig. 18). Because there was no metrological dividing between the elastic and the plastic material behaviour, there is no quantitative evaluation possible about the progress of the dissipated energy of the plastic deformation.

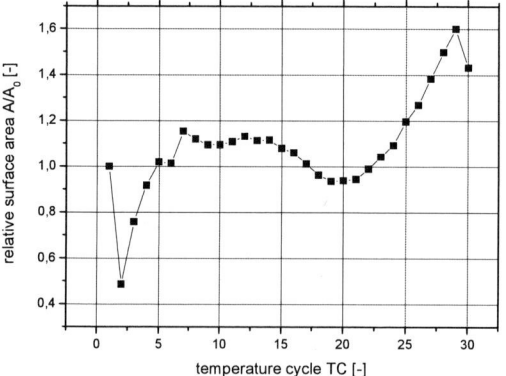

Fig. 18: Relative dissipaded strain EnergyShear

The figure shows that the progress of the surface area correlated from the 2nd temperature cycle with the progress of the shear strain and the shear stress (see Fig. 16).

6. Conclusion

Points discussed in this paper:

- an approved thermal controlled experimental setup
  - for force drop measurement on real solder joints
  - for in-situ force oberservation at solder joints
- load frame is suitable for solder fatigue compensation based on in-situ force drop measurement
- variable load frame columns configurations allow variable inducted forces for adjustment of fatigue speed
- different stress-strain amplitudes and different fatigue speeds – creation of material fatigue laws

Further activities planed:

- investigation of different solder alloys to compare the fatigue
- generation of material fatigue laws by variation of stress amplitude
- determination of precise stress, strain and dissipated strain energy by FEM calculation

References

[1] Roellig, M.; Wiese, S.; Meier, K. Wolter, K.-J.: Creep Measurements of 200 μm - 400 μm Solder Joints, International Conference on Thermal, Mechanical and Multi-Physics Simulation Experiments in Microelectronics and Micro-Systems, EUROSIME 2007, pp. 255 - 263

[2] Mueller, M.; Wiese, S.; Roellig, M.; Wolter, K.-J.: The Dependence of Composition, Cooling Rate and Size on the Solidification Behaviour of SnAgCu Solders, International Conference on Thermal, Mechanical and Multi-Physics Simulation Experiments in Microelectronics and Micro-Systems, 2007. EUROSIME 2007, London, Great Britain, pp. 446 - 455

[3] Dudek, R.; Faust, W.; Wiese, S.; Roellig, M.; Michel, B.: Low-Cycle-Fatigue of First and Second Generation Leadfree Solders, Electronics Packaging Technology Conference, EPTC2007, Singapore, in press

[4] Wiese, S.; Roellig, M.; Wolter, K.-J.: Creep of eutectic SnAgCu in thermally treated solder joints, 55th Electronic Components and Technology Conference, 2005, Orlando, United States of America, pp. 1272 - 1281

Thermo Mechanical Characterization of Packaging Polymers

Bjoern Boehme*, K.M.B. Jansen**, Sven Rzepka***, Klaus-Juergen Wolter*
* Technische Universität Dresden, Electronics Packaging Laboratory, Dresden, Germany
** Delft University of Technology, Mechanics of Materials, Delft, The Netherlands
*** Qimonda Dresden GmbH & Co. OHG, Dresden, Germany
bboehme@avt.et.tu-dresden.de

Abstract

In this study, two highly filled molding compounds were used as example to demonstrate the characterization scheme. In addition, two low filled packaging polymers are included for comparison. The characterization scheme consists of the steps sample preparation, measurement of the material data, and modeling the material behavior. The 'sample preparation' step included a DSC analysis to understand the cure reaction and to establish the cure kinetics model. In the 'measurement' step, two different sets of equipment were applied. The elongation modulus is determined by dynamic mechanical analysis (equipment: DMA 'Q800') in a wide range of temperatures and frequencies. The other parameters are measured by pressure-volume-temperature experiments (equipment: PVT 'Gnomix'). Conducting these characterization tests, the bulk modulus (K), the coefficient of thermal expansion (CTE), and the cure shrinkage was determined.

The paper describes this comprehensive characterization with the measurement setups and parameter selection. E(T,t), K(T,t), CTE(T), Tg and cure shrinkage are determined to define a complete and consistent material model. Subsequently, the characterization results are presented, discussed and further work of implementing the complete material model into FEM simulation tools like $ANSYS^{TM}$ is outlined.

Key words: glass transition, bulk modulus, viscoelastic

Introduction

Organic packaging materials gain a steady increase in importance for electronics packages. Most parts of the package consist of them in different ways, e.g., low filled adhesives and solder masks, highly filled encapsulates and underfills, and epoxy-glass laminates (interposers and substrates). This paper addresses the influence of the consistency of thermo-mechanical characterization on accuracy and efficiency of finite element modeling (FEM), which is the key tool in virtual prototyping to speed up the packaging development cycle.

In general, organic packaging materials or polymers behave visco-elastically, which is much more complex than elastic behavior. Hence, creep may increase the deformation and/or relaxation processes may decrease the stress of the polymer parts with time. These processes are most effective in the temperature range of glass transition (Tg), at which the material behavior switches between glassy and rubbery states and all material parameters change drastically.

Most of the organic packaging materials have their Tg well within the temperature range of manufacturing and/or service. Thus, FEM models must consider the visco-elastic behavior in order to provide for reliable simulation results. In addition, these models need to account for the chemical shrinkage that occurs during polymer curing and for the temperature dependent coefficient of thermal expansion (CTE).

The viscoelasticity is described by the temperature and time dependencies of the engineering parameters elongation modulus (E), shear modulus (G), bulk modulus (K), and Poisson's ration (μ). In case of isotropic materials, the full set of the engineering parameters has only two independent contributors, e.g. E and K. That means, the remaining parameters can be computed based on characterization experiments providing values for two of them. The chemical shrinkage determines the initial stress state in the organic parts after their curing. The thermal expansion is a key factor in heterogeneous electronics packages due to the thermal mismatch between the materials involved.

Altogether, these material properties need to be modeled in a comprehensive way fitting to each other. The common practice of just compiling data from different sources has been found to fail yielding in reliable and accurate results. The conditions under which the data were determined may cause mismatches between them and cause inconsistencies within the model. If a convergent solution was obtained at all, much simulation time would be needed as many iterations with small time steps were needed. In order to avoid this, the paper reports an approach of characterizing the temperature and time dependent mechanical material properties in one comprehensive scheme.

Procedures

978-1-4244-4722-0/09 $25.00
© 2009 IMAPS-ITALY

CURE KINETICS

The cure kinetics of a material describes the conversion rate as a function of time and temperature. In thermosetting resins, the conversion rate (α) can be defined as the number of reacted epoxy groups divided by the total number of epoxy groups. It ranges from 0% to 100%. The reaction rate depends on how much of the unreacted epoxy and hardener is present at that moment and thus depends on the degree of cure itself [JAN07]. In order to conduct cure shrinkage experiments, the cure kinetics need to be understood for choosing the proper temperatures in the cure shrinkage experiments using the PVT setup. DSC experiments allow developing a cure kinetics model for thermosetting resins. The degree of cure (α) is defined as

$$\alpha(t) = \frac{H(t)}{(H_u)} \qquad (1)$$

where $H(t)$ is the reaction heat released until time t and H_u is the ultimately released heat of the reaction. The concentration dependent conversion rate can be described with a principle equation

$$\frac{d\alpha}{dt} = k \cdot f(\alpha) \qquad (2)$$

$d\alpha/dt$ is the change in conversion rate over time, k is the rate constant and $f(\alpha)$ is a function which depends on reaction mechanism.

For thermosetting resins, a convenient model, which is relatively simple and sufficiently accurate is the Kamal-Sourour equation proposed by Kamal and Sourour (1976)

$$\frac{d\alpha}{dt} = k_T \, \alpha^m (1-\alpha)^n \qquad (3)$$

with m and n as reaction orders and k_T as the reaction rate constant that follows the Arrhenius relation

$$k_T = A \, exp\left(-\frac{Q}{RT}\right) \qquad (4)$$

For the relationship between Tg and conversion rate, the DiBenedetto equation can be used [YAN07a].

$$\frac{T_g - T_{g0}}{T_{g\infty} - T_{g0}} = \frac{\lambda \alpha}{1 - (1-\lambda)\alpha} \qquad (5)$$

The equipment DSC 2920 (TA Instruments) was used to analyze the reaction during the curing of the materials. The materials were studied using dynamic cure with scan rates between 1K/min and 15 K/min. In addition, isothermal curing experiments were carried out at RT/100/120/140 °C to gain the information about the cure kinetics. The range of the temperature scan was between -50°C and 350°C - dependent on the material. Each sample was scanned twice (at least) in order to see the effect of dynamic cure on Tg.

To have a good contact between the sample material and the aluminum pan, the pellet shaped samples (molding compound) were grinded to powder.

Afterwards, 5…10 mg of the sample powder was inserted into the sample pan and closed hermetically. For the two liquid samples (WPR1, SM1), 5…10 mg were hermetically enclosed in the aluminum pan.

PVT MEASUREMNENTS

The measurement of the bulk modulus is not a very common. In literature, a few reports refer to PVT measurements [ZOL89, SAR08]. The GNOMIX high pressure dilatometer allows measuring the pressure and temperature dependent volume change of polymeric materials. It consists of a pressure vessel filled with silicone oil, in which the sample cell is located and by which its temperature and pressure can be controlled. The sample is inserted in the sample cell and surrounded by a confining fluid like mercury. The temperature and pressure dependent volume change of the sample can be monitored by sensing the deformation of the bellows. This allows determining the volume change and calculating the specific volume (v), bulk modulus (K), and the volumetric coefficient of thermal expansion (CTE$_V$). Additional volumetric effects like cure shrinkage can also be analyzed very accurately. The pressure, temperature, and displacement (change of volume) is monitored. The useable temperature range spans from 25 to 400°C with pressures from 10 to 200 MPa (100...2000 bar) to be applied. In Figure 1 the PVT setup (left) and its schematic of the setup (right) is shown.

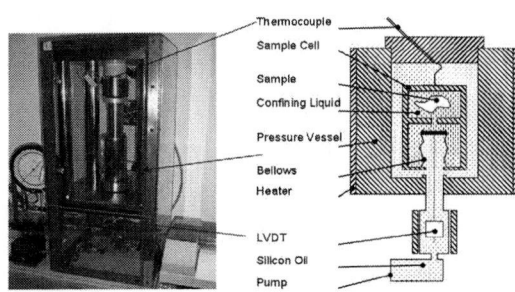

Figure 1: GNOMIX PVT setup for experiments

CURE SHRINKAGE MEASUREMENTS

Chemical shrinkage (ε_{chem}) is typical for polymerization processes and causes residual stresses. The residual stress (σ_R) can be expressed as follows [WAN08]:

$$\sigma_R = \int E d\varepsilon_{chem} = \int_{t_{gel}}^{t_{final}} (E(t) \, \varepsilon_{chem}(t)) dt \qquad (6)$$

The cure shrinkage quantifies the decrease of the specific volume due to crosslinking in the polymer matrix. Here, it is defined as the decrease of the specific volume during isothermal cure at a moderate reaction temperature. The cure shrinkage experiment with the Gnomix PVT apparatus consists of three main sections as shown in figure 2.

978-1-4244-4722-0/09 $25.00
© 2009 IMAPS-ITALY

The first section (1 → 2 → 3) is a fast heating to the desired cure temperature, which is about 120°C for the tested molding compounds with an ultimate (fully cured) Tg of 100-120°C. The cure temperature must be chosen properly to be able to monitor most of the reaction shrinkage and to allow a complete cure of the material. Therefore, the preliminary DSC studies must be conducted to understand the cure kinetics of the material. For the low filled materials, higher cure temperatures (150°C, 200°C) were needed due to the higher ultimate Tg values. As seen in figure 2 at point 2, a change in the slope of the specific volume vs. temperature plot indicates the Tg of the uncured material.

The second part of the cure experiment is the actual cure shrinkage measurement (3 → 4) for several hours. Depending on the polymer and the filler content, the specific volume will decrease for increasing curing times. This change in the specific volume is the cure shrinkage and can be monitored by the measurement system.

The third section (4 → 5 → 6) is the cooling down to room temperature. At point 5, the Tg of the cured material is reached. Its shift to higher temperatures can also be evaluated (points 2 vs. 5) and indicates the cure reaction of the thermosetting resins. As the result of the experiment, the uncured and cured Tg as well as the cure shrinkage are known. In addition, the CTE of the uncured and cured material can be estimated by the slope of the specific volume vs. temperature curve.

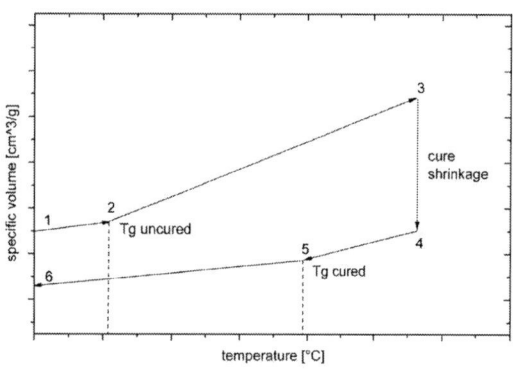

Figure 2: Schematic of the cure shrinkage experiment in the Gnomix PVT setup

TIME DEPENDENCE OF BULK MODULUS

The bulk modulus of a material can be seen as the resistance against a compression load. If a material with a constant pressure loading changes its specific volume over time under constant pressure, its bulk modulus is time dependent. Therefore, pressure stepping experiments are a convenient method for studying the time dependency of the bulk modulus of materials. As seen in figure 3, load steps between 10 MPa and 50 MPa (10-30 MPa, 10-40 MPa) were applied with dwell times up to 3 hours (10800 s). The typical rate of the pressure change was

0.25 MPa/s. To account for the temperature dependency of the viscoelastic material behavior, the pressure stepping experiments were repeated at different temperature levels between 23°C and 200°C. Depending on the Tg of the materials, the glassy, the transition, and the rubbery state were covered by the experiments.

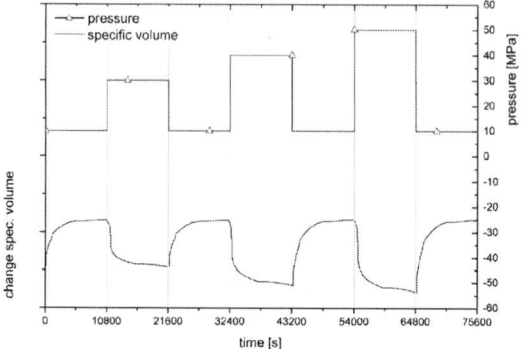

Figure 3: Schematic of the pressure step experiment for time dependent bulk modulus measurements

TEMPERATURE DEPENDENCE OF VOLUMETRIC MATERIAL BEHAVIOR

For this kind of PVT experiments, the temperature and pressure was increased stepwise (10 K, 10 MPa) and the specific volume was monitored (temperature range: 30-250°C, pressure range: 10-100 MPa). To ease the data analysis, the measurement data was than fitted to the Tait-Equation [ZOL98] to describe the temperature and pressure dependent behavior.

$$v(T,P) = v_0(T)\left[1 - C\,ln\left(1 + \frac{P}{B(T)}\right)\right], \qquad (7)$$

$$B(T) = b_1\,exp(-b_2 T) \qquad (8)$$

Instead of the bilinear function for describing the volume change vs. temperature for the glassy and the rubbery state (next equation)

$$v_0^{bilinear}(T) = \begin{cases} v_{00}\left[1 + k_1'\left(T - T_{gp}\right)\right] & \text{for} \quad T \le T_{gp} \\ v_{00}\left[1 + k_2'\left(T - T_{gp}\right)\right] & \text{for} \quad T \ge T_{gp} \end{cases} \qquad (9)$$

$$T_{gp} = T_{g0} + s_0 P \qquad (10)$$

a smoother function is used, which is taking into account that the slope will not change instaneously.

$$v_0(T) = v_{00}\left\{ \begin{array}{l} 1 + k_1\left(T - T_{gp}\right) \\ + \frac{1}{2}k_2\left[T - T_{gp} + ln\left(cosh\left[c_1\left(T - T_{gp}\right)\right]\right)/c_1\right] \end{array} \right\}$$

(11)

Differentiation with respect to temperature then results in a function for the volumetric CTE while differentiation with respect to pressure yields the compressibility (bulk modulus).

$$CTE_V(T,0) = k_1 + \frac{1}{2}k_2\left(1 + tanh\left[C_1\left(T - T_{gp}\right)\right]\right) \quad (12)$$

$$\beta(T,P) = k_1 s_0 + \frac{1}{2} k_2 s_0 \left(1 + tanh\left[C_1\left(T - T_{gp}\right)\right]\right) \quad \textbf{(13)}$$
$$+ C_1\left(1 - b_2 s_0 P\right)/\left(B(T) + P\right)$$
$$K(T,P) = 1/\beta(T,P) \quad \textbf{(14)}$$

The parameters $b_{1,2}$, c_1, $k_{1,2}$, s_0, and v_{00} were determined by non-linear curve fitting using MatLab®.

ELONGATION MODULUS

The investigation of the elongation modulus of the materials was done on rectangular strips in tension mode using a DMA Q800 (Fa. TA Instruments). To evaluate the time and temperature dependency of the material behavior, a heat scan (1 K/min) in the multi-frequency mode is the preferred and a very efficient method. The frequency range should be as large as possible, e.g., between 0.5 Hz and 60 Hz. The deformation amplitude depends on the material studied. It was 2 μm for the very stiff molding compounds (25 mm sample), 12.4 μm for the solder mask (25 mm sample), and 2 μm for the wafer photo resist (8 mm sample).

Investigated Materials

Two commercially available highly filled molding compounds (MC4, MC5) were used as example for demonstrating the characterization scheme. Table 1 lists the composition data of the molding compounds taken from the supplier's datasheet. For molding compounds a very high (82-94wt.%) content of inorganic fillers are typical to decrease the thermal expansion and to increase the thermal conductivity of the material. Additionally, proprietary additives such as color pigments, flame retardants, release agent, and others are added to the epoxy resin and to the hardener to optimize the material behavior. Several studies have been done on model compounds [YAN07a]. Here we focus on the study of commercially available materials. In Figure 4 a cross section of the molding compound material is shown. The filler particles (silica, spheres) are about 5-50μm in diameter.

In addition, two low-level filled packaging polymers are included for comparison. One was a wa-fer photo resist (WPR1) and the other was a solder mask material (SM1). The WPR1 is an organic polymer, which may stay permanently on the dies as insulation layer between silicon and copper traces in electronic packages. The composition is taken from the supplier's datasheet and is given in Table 3.

SM1 is a two-component, heat curable solder mask material. The data of the raw material constituencies given by the supplier's datasheet is listed in Table 2.

Results and Discussion

CURE KINETICS

The DSC dynamic cure results of all four pack-

Table 1: Comparison of molding compound (MC) composition of the experiments

Constituent	MC4	MC5	Function
	wt %		
Silica, vitreous	82-94	84-94	Filler
2,2'-((3,5',5,5'-tetramethyl-(1,1-biphenyl)-4,4'-diyl)-bis(oxymethylene))-bis-oxirane	1-5	3-7	Hardener
Carbon Black	0.2	0.2	Filler
Phenolic Resin 1	3-6	1-5	Resin
Phenolic Resin 2		1-5	Resin
Epoxy Resin	1-5		Resin

Table 2: Composition of the solder mask material (SM1)

Constituent	wt %	Function
Epoxy resin 1	<13	Resin
Epoxy resin 2	<9	Resin
Barium Sulfate	<24	Filler,
Dipropylene Glycol monomethyl ether	<16	Regulator, solvent
Talc	<4	Filler
Morpholinederivative	<4	
Silica, amorphous	<1	Filler

Table 3: Composition of the wafer photo resist material (WPR1)

Constituent	wt %	Function
Phenolic resin	20-30	Resin
Trisphenolderivative	1…3	Hardener
Melanin compound	1…5	
Epoxy Resin	1…5	Resin
Silan Part	0.5…3	
Rubber Particles	0…1	Filler
Surfacetant.	0.05…0.1	
Photo sensitizer	5-15	Energy Absor-
Ethyl lactate	45-65	Solvent

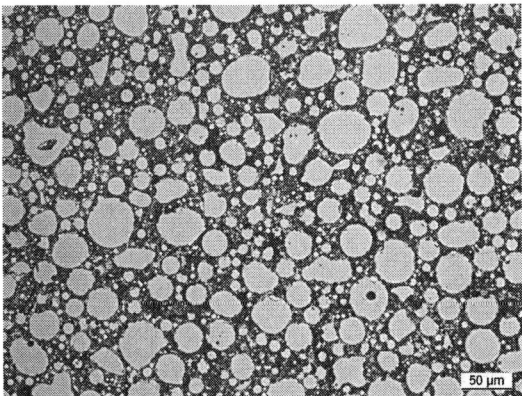

Figure 4: Cross section of a highly filled molding compound material

978-1-4244-4722-0/09 $25.00
© 2009 IMAPS-ITALY

aging materials are compared in Figure 5. The graph shows the heat flow (between sample and reference) during a temperature scan with 10 K/min (temperature ramp). Very similar reaction curves have been obtained for both molding compounds. An exothermic cure reaction peak occurs between 100°C and 180°C. The initial Tg_{ini} of MC4 was about 25°C. At 90°C, the exothermic cure reaction starts. The total heat released per gram material H_u is about 21 J/g. Rescanning of the cured sample showed a final Tg_u of 112°C that was reproducible (Figure 5, Figure 6).

For MC5, the released heat H_u is about 21.6 J/g. The initial Tg_{ini} is difficult to determine but ranges somewhere between 20°C and 40°C. The final Tg_u was about 119°C.

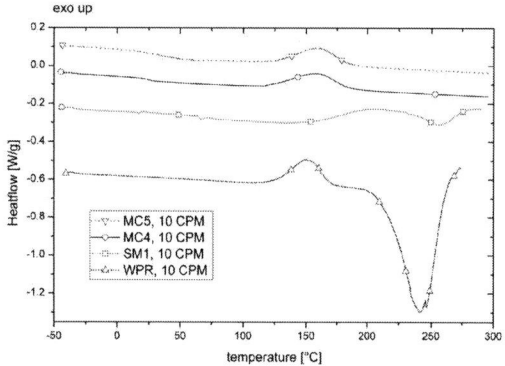

Figure 5: Comparison of heat scan signature for all tested materials (DSC, heat rate 10K/min)

For SM1, the cure reaction was more complex than for the other materials. In addition to the exothermic peak between 150°C and 250°C (Figure 5), there is a second exothermic peak occurring above 250°C. In the second scan of the sample, the first peak has disappeared (Figure 6). Hence, it is seen as the cure reaction signal with a released heat H_u of about 20 J/g. The second peak may indicate an additional high temperature reaction (post cure reaction of secondary groups) or the onset of material decomposition (destruction of bonds). For WPR1, the exothermic cure reaction peak starts at about 100°C and reaches its maximum at about 150°C. At 200°C, a second but endothermic peak starts reaching its minimum at 240°C. In the second scan, the exothermic peak is no longer present. Hence, it is concluded that this is the cure reaction. The endothermic peak could be a sign of material decomposition, the release of low molecular weight volatiles or even a measurement artifact.

In Figure 6 the comparison of the DSC scans of the uncured samples (MC4, SM1) and the fully cured samples (MC4 rescan, SM1 rescan) is shown.

For MC4 the single exothermic reaction peak is disappeared and the glass transition is increased.

For the SM1 material the complex reaction is visible.

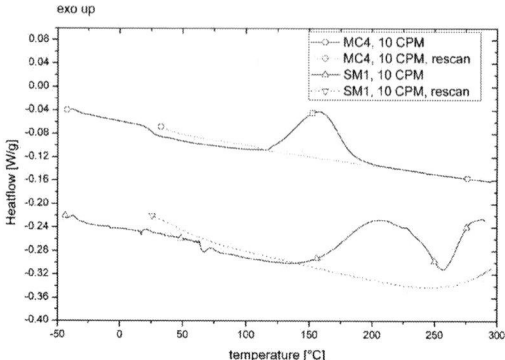

Figure 6: Scan and rescan results for MC4 and SM1 material

In Figure 7 the results for the dynamic cure with different scan rates (1 CPM to 15 CPM) is shown for the MC5 material. This data and isothermal experiment results were used to fit the cure kinetics model applying Matlab® routines.

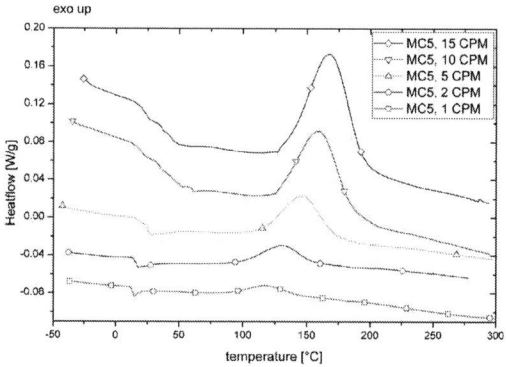

Figure 7: Experimental results for the dynamic cure experiments of MC5 material (scan rate 1-15 CPM)

Table 4 compiles the coefficients fitted by the Kamal-Sourour equation. In Table 5 the fit parameters for the DiBenedetto equation for the molding compounds are given. For the low filled materials SM1 and WPR1, the complex cure reaction has not been fully understood, yet. Therefore, the kinetics model is not given here. It will be the focus of further studies.

Table 4: Cure kinetics parameters according to the Kamal-Sourour equation from the DSC experiments of the molding compounds

Parameter	Unit	MC 4	MC5
m	-	0.31	0.4
n	-	1.1	1.1
A	1/s	2.0e6	1.77e6
Q	KJ/mol	68000	67800

Table 5: Cure kinetics parameters for the DiBenedetto equations from the DSC experiments of the molding compounds

Parameter	Unit	MC 4	MC5
λ	-	0.668	0.57
Tg_{ini}	°C	25	24.5
Tg_u	°C	112	121.2
H_u	J/g	21	21.6

CURE SHRINKAGE

For the molding compounds, the cure shrinkage experiments were carried out at 120°C for 820 min to ensure a complete cure. This temperature avoids a vitrification to be reached prior to the fully cured state and leaves the reaction rate at a moderate level to allow the measurement. A cure shrinkage of 0.71 vol.% (MC4, Figure 8) and 0.66 vol.% (MC5), respectively, was seen for the two molding compounds. This is the expected behavior as the materials are highly filled by 82-94 wt.% of vitreous silica and 0.2 wt.% carbon black. The cure shrinkage of the unfilled epoxy matrix should be about 1-2 vol.% [YAN07a].

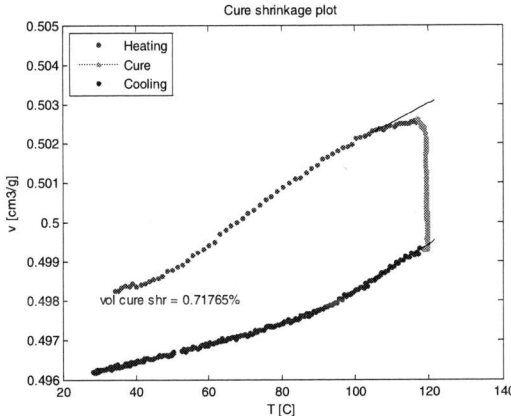

Figure 8: Cure shrinkage of MC4 material

For the thin film materials SM1 and WPR1, a much higher cure shrinkage of several percent was expected due to the low filler level (SM1 about 30 wt.%, WPR1 below 10 wt.%) and a high solvent content to adopt the viscosity. The measurements in the PVT cell at 150°C and 200°C (due to cure kinetics data) showed cure shrinkages of more than 1 vol.%. However, they also showed the limits of the cure experiments in the PVT. The solvents, which evaporate freely in the manufacturing process, were trapped in the sample and caused errors. Pretreatments to evaporate the solvents by vacuum and heat improved the results.

Still, it was found that accurate measurements of packaging polymers, which are filled at low level and contain much solvent, require a method applying open systems. In [WEN05], a method is presented for a epoxy materials that show a cure shrinkage of 5 vol.%. Although the inaccuracy of this method was 0.3 vol.%, this ε_{chem} is much higher than that measured by PVT. In [WAN08], an alternative method is suggested. The use of a Fiber Bragg Grating surrounded by a cylindrical shaped polymer is suggested for monitoring the cure process by the Bragg wavelength and converting the signals into cure shrinkage. Unfortunately, this method seems to be not applicable for materials like the WPR1 and SM1 due to the low viscosity and high solvent content. A single plunger experiment to measure the volume change during cure in a mold cavity for molding compounds is described in [HWA06]. This setup might be critical due to the very low viscosity of the SM1 and WPR1 material. Finally, [Yan07a] and [BLU03] introduced a method for estimating the cure dependent density by measuring the weight and volume of a silicone oil (at cure temperature) immersed sample (Archimedes principle). Hence, the characterization of SM1 and WPR1 is to be repeated applying a more suitable measurement technique to measure the complete cure shrinkage with very high accuracy.

BULK MODULUS - TIME DEPENDENCY

Figure 9 shows the bulk modulus vs. time after a pressure step up (10 MPa to 50 MPa) for the SM1 and MC4 material. As expected based on the high filler fraction, the MC4 material has a much higher bulk modulus in the glassy (50 °C) and rubbery state (150 °C) than the solder mask material. The bulk modulus of the studied materials showed almost no time dependent effects during the pressure step experiments. At higher temperatures (100°C, 150°C), the bulk modulus values decreased and showed that the temperature influence is much higher than the influence of time. In fact, the time dependency of the bulk modulus may even be neglected completely in the material models.

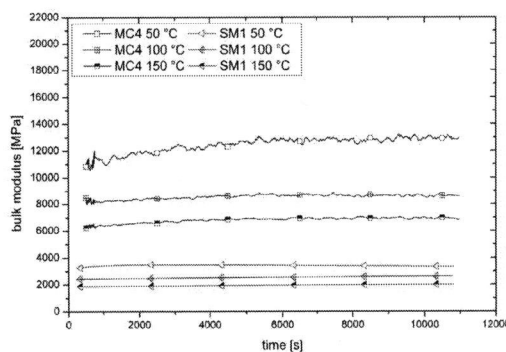

Figure 9: Time dependent bulk modulus K(t) for pressure step up experiments

BULK MODULUS - TEMPERATURE DEPENDENCY

As described before, the pressure and temperature dependent specific volume was determined by PVT scans of the fully cured sample. Subsequently, the dataset was fitted to the modified Tait equation. Figure 10 plots the measurement results (symbols) and the Tait model (lines). The model fits the data points very well. A clear glass transition in the range of 80-110 °C was seen for both molding compounds. For the low filled materials, no real transition was seen.

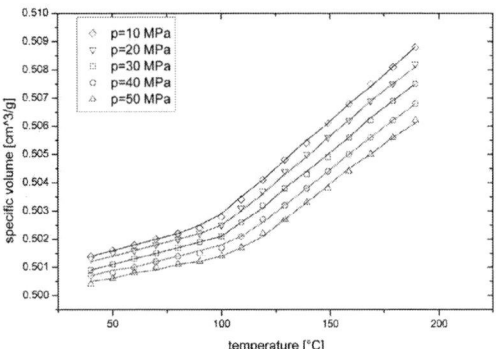

Figure 10: Specific volume for MC4 material (time and pressure dependent)

Table 6 gives the coefficients of the Tait model for all materials describing the temperature and pressure dependent specific volume in the temperature range from 40°C to 190°C.

Table 6: Fit parameters for modified Tait equation

Parameter	Units	MC4 Post cured @ 175°C	MC5 Post cured @ 175°C	SM1 Cured @ 200°C, 600 min	WPR1 Cured @ 200°C, 600 min
v_{00}	[cm^3/g]	0.503	0.496	1.094	1.260
k_1	[1/°C]	4.426e-5	3.920e-5	2.727e-4	4.183e-4
k_2	[1/°C]	9.558e-5	1.055e-4	6.666e-4	1.626e-3
c_1	[1/°C]	9.088e-2	8.439e-2	3.345e-3	5.025e-3
b_1	[MPa]	3496	2054.6	178.55	202.94
b_2	[1/°C]	6.75e-3	5.09e-3	4.38e-3	6.46e-3
Tg_0	[°C]	94.4	88.6	193.3	292.6
s_0	[K/MPa]	0.33954	0.35445	-0.4814	-0.1404

Figure 11 compares the temperature dependent bulk moduli of the four materials. There is a remarkable difference between the highly filled molding compounds (K ≈ 14-21 GPa at 50°C) and the low filled materials (K ≈ 2-3 GPa at 50°C). The molding compounds show a clear glass transition Tg between 80°C and 110°C, whereas the low filled materials (SM1, WPR1) do not.

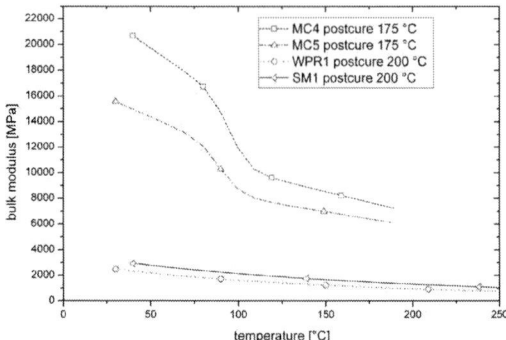

Figure 11: Bulk modulus K(T) of the materials

COEFFICIENT OF THERMAL EXPANSION (CTE)

From the Tait equation, the volumetric CTE can be derived by deviation versus temperature. Figure 12 shows the temperature dependent CTE for the four packaging polymers in a temperature range from 30 to 250 °C. The highly filled molding compounds have a glassy volumetric CTE of about 45-50 ppm/K, which increases to about 135-140 ppm/K in the rubbery state. A clear Tg is visible. The volumetric CTE of the low filled materials is much higher (above 400 ppm/K) and a Tg is not visible in the tested range.

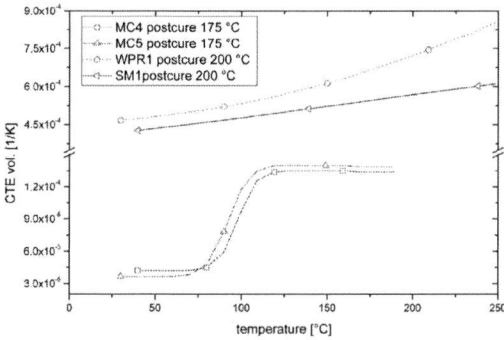

Figure 12: Volumetric CTE of the materials

POSTCURE EFFECTS ON VOLUMETRIC RESPOND IN MOLDING COMPOUNDS

The effect of post cure on the bulk modulus K(T,P) and CTE(T,P) of both mold compounds was investigated by several post cure steps. After the initial cure at 120°C for 820 min, post cure steps were performed at 175°C, 200°C, and 250°C, respectively, for 240 min in each case. It is expected that during post cure, free reactants remaining in the resin may finally crosslink so that the Tg increases and/or the CTE decreases. In addition, remaining solvents in the polymer network may be removed by post cure. Both processes can change the modulus and increase the Tg. In Figure 13 two post cure steps of the MC5 are compared to the initial cure state. A decrease in the glassy bulk modulus was seen due to post cure at 175 °C.

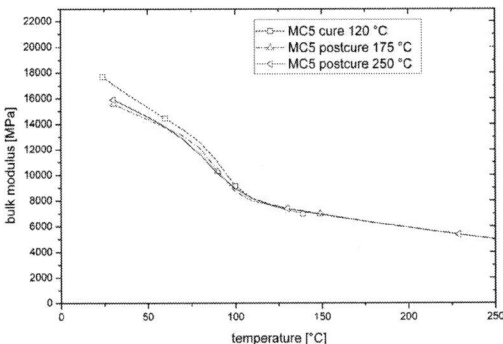

Figure 13: Post cure induced variation of temperature dependent bulk modulus of MC5

The CTE graph showed a broader Tg range was seen for both molding compounds. For the MC4 material, the variation of the glassy volumetric CTE for the different post cure steps was small (<10 ppm/K) in all post cure steps, while the MC5 showed a systematic decrease of up to 20 ppm/K (Figure 14). The rubbery volumetric CTE decrease after the first post cure step and stayed constant.

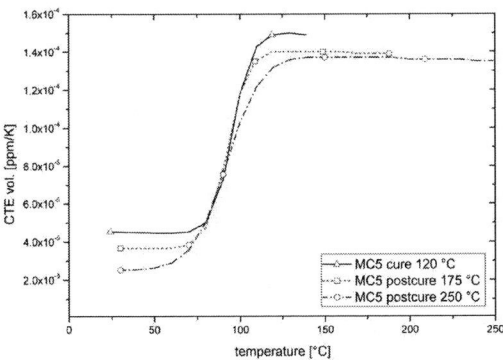

Figure 14: Volumetric CTE change due to post cure in the mold material MC5

ELONGATION MODULUS

As mentioned in the "Procedures" section, the visco-elastic elongation modulus was determined by DMA metrology in tension film mode. Figure 15 plots the temperature dependent elongation modulus E(T, 1Hz) of the four materials tested. The molding compounds show a high storage modulus of 25 GPa (MC4) and 23.5 GPa (MC5) at 50°C. The Tg DMA (peak of tan δ at 1Hz) can be determined as 107°C (MC4) and 114°C (MC5). The rubbery modulus is about 0.65 GPa, i.e., approximately 2.5% of the glassy value, for both molding compounds.

At 50°C, the low filled materials SM1 and WPR1 had a glassy modulus of 3 GPa (SM1) and 2 GPa (WPR1), respectively. The Tg values range from 70-135°C (SM1) and 220-230°C (WPR1). For these materials, the thermal history needs to be considered for obtaining the correct Tg due to post cure

effects. In Figure 17 a significant change in the viscoelastic elongation properties can be seen.

A dynamic postcure was used to show the Tg and rubbery modulus increase. During the first heatscan the material continued its cure after reaching the glass transition Tg and the rubbery modulus increased from about 300 MPa to 550 MPa.

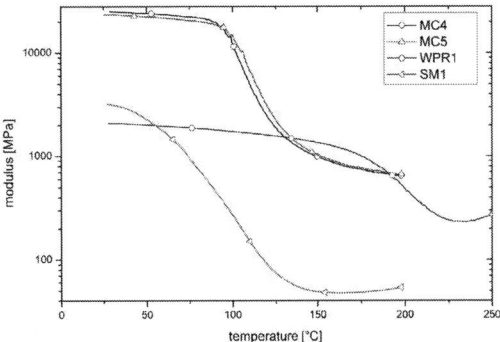

Figure 15: Comparison of temperature dependent elongation modulus E(T) from DMA experiments (1Hz)

POSTCURE EFFECTS ON ELONGATION RESPONSE

In Figure 16 the changes in the temperature and the frequency dependent elongation modulus are shown for the MC5 molding compound for three dynamic post cure steps (DMA heat scan, heat ramp 1 K/min). The post cure to 200°C (MC5 scan 2 curve) causes a Tg increase of about 5 K. The (extreme) dynamic post cure to 280°C (1K/min) has a substantial effect and increases the Tg (about 10K) and the rubbery modulus (0.6 to 3 GPa) significantly.

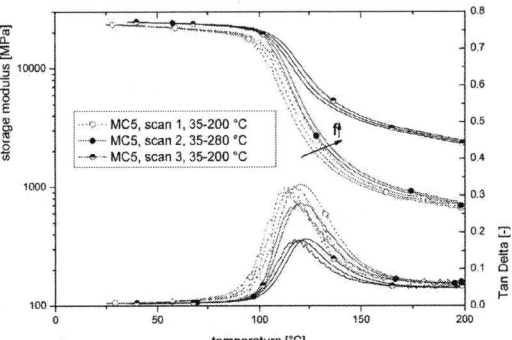

Figure 16: Significant postcure for molding compound material (MC5, f=1, 3, 10 Hz)

For comparison post cure behavior for the low filled WPR1 material is given in Figure 17. For this material an extreme postcure was seen during DMA as well as PVT experiments. The Tg shifts from about 225°C to more than 300°C (peak of tanδ) and the rubbery elongation modulus 0.3 to 0.8 GPa. Remarkable is the visible cure during the heat scan (rubbery modulus increases in WPR scan 1 graph).

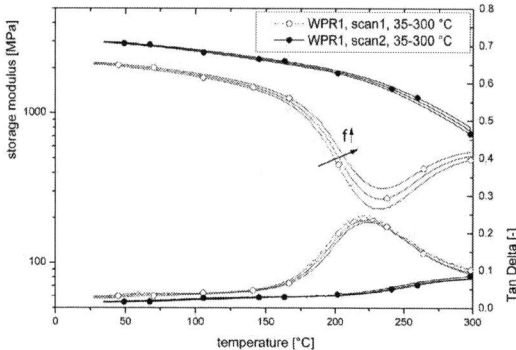

Figure 17: Significant postcure for wafer photo resist material (WPR)

Material Model
CONSISTENCY CHECK

As already mentioned in the first section, the full set of the engineering parameters of isotropic materials has only two independent contributors. That means, the remaining parameters can be computed based on characterization experiments providing E and K by use of the equations:

$$\mu(T) = \frac{1}{2} - \frac{E(T)}{6K(T)} \tag{15}$$

$$G(T) = \frac{3E(T)K(T)}{9K(T) - E(T)} \tag{16}$$

They are most valid outside the Tg region only since they do not account for abrupt changes in the moduli. G and μ can be computed for the consistency check of the material parameters in the glassy state. For MC4 [$\mu(50°C) = 0.30$, $G(50°C) = 9.64$ GPa] and MC5 [$\mu(50°C) = 0.25$, $G(50°C) = 9.42$ GPa] the glassy values are consistent. The moduli E(t,T) and G(t,T) are time dependent.

IMPLEMENTATION IN ANSYS

ANSYS™ supports the implementation of temperature dependency on elongation modulus (E), shear modulus (G), and Poisson's ratio (μ) via temperature tables. Applying the consistency relations, the temperature dependency of the bulk modulus (K) can also be considered. The time dependency (relaxation data) can be introduced by a Prony series for both, shear and bulk, moduli. In addition, a shift function needs to be provided. In PVT experiments (Gnomix), both mold compounds did not show significant time dependency of the bulk moduls while a clear visco-elastic reaction was seen in the DMA. Therefore, a Prony series is needed for G only but not for K. In order to still keep the full set of material data consistent, the use the user programmable feature USRMAT seems advisable.

Conclusions

The paper presents the material characterization results for two commercial molding compound as well as two low filled organic materials typical to electronics packaging (solder mask, wafer photo resist). The full data set consists of a cure kinetics model, the temperature and time dependent elongation E and bulk modulus K, the temperature dependent CTE and the cure shrinkage. In addition, post cure effects were found to influence the bulk modulus, the CTE and the elongation modulus.

The data of the full set has been obtained for the molding compounds and has been shown to be consistent. In contrast, the behavior of the low filled materials is not fully understood, yet. The high solvent content makes preparation, handling, and testing of the thin samples quite difficult. Further studies are necessary for completing characterization. Still, the results achieved to provide an adequate set of preliminary material data that is applicable to reliability analysis. The implementation of post cure effects requires new routines in the ANSYS® FEM code which can be developed applying the user subroutines available.

Acknowledgments

The presented work was a joint effort of TU Dresden, TU Delft, and Qimonda Dresden. The author would like to thank Leo Ernst and his group for the support during the study. Also thanks to the FEM simulation group at Qimonda Dresden.

References

[BLU03] Blumenstock, Tobias, Analyse der Eigenspannungen während der Aushärtung von Epoxidharzmassen, PhD thesis, Universität Stuttgart, 2003

[HWA06] Hwang, Sheng-Jye, P-V-T-C Equation for Epoxy Molding Compound, IEEE Transactions on Components and Packaging Technologies, vol. 29, 2006

[JAN07] Jansen, K.M.B., Kinetic Characterisation of Molding Compounds Proceeding EuroSimE 2007, London, 2007

[RZE07] Rzepka, S. & Müller, A., The Effect of Visco-elasticity on the Result Accuracy of FEM Panel Warpage Simulations Supporting Industrial Microelectronics Packaging, Proceeding EuroSimE 2007,London, 2007

[SAR08] Saraswat, M.K., A Characterization Method for Viscoelastic Bulk Modulus of Molding Compounds, Proceedings EuroSimE 2008

978-1-4244-4722-0/09 $25.00
© 2009 IMAPS-ITALY

[SCH90]   Schwarzl, Polymermechanik, Springer-Verlag, 1990

[WAN08]   Wang, Yong, Simultaneous Measurement of Effective Chemical Shrinkage and Modulus Evolution During Polymerization, Proceedings ECTC 2008, 2008

[WEN05]   Wenzel, Mirko, Spannungsbildung und Relaxationsverhalten bei der Aushärtung von Epoxidharzen, Darmstadt, 2005

[YAN07a] Yang, D., Cure-Dependent Viscoelastic Behaviour of Electronic Packaging Polymers, PhD thesis, TU Delft, 2007

[ZOL89]   Zoller, Paul, in Polymer Handbook, PVT Relationships and Equations of State of Polymers, 1998 in Polymer Handbook, 3rd ed., edited by J. Brandrup and H. Immergut Wiley, New York, 1989

Au–Sn SLID Bonding:
Fluxless Bonding with High Temperature Stability,
to Above 350 °C

Knut E. Aasmundtveit[1][*], Kaiying Wang[1], Nils Hoivik[1], Joachim M. Graff[2],
and Anders Elfving[3]

[1]Vestfold University College, PO Box 2243, N-3103 Tonsberg, Norway

[2]SINTEF Materials and Chemistry, PO Box 124 Blindern, N-0314 Oslo, Norway

[3]SensoNor Technologies, PO Box 196, N-3192 Horten, Norway

[*]Phone: +47-3303 7726, Fax: +47-3303 1103 and E-mail Address: knut.aasmundtveit@hive.no

Abstract

A fluxless SLID (Solid-Liquid Inter Diffusion) bonding process based on Au and Sn, where the final bond consists of intermetallics with high melting point, is presented. The decomposition temperature of the bond was tested by applying shear force while heating bonded samples. No bond delamination was observed for temperatures up to 350–400 °C, which is 100 °C higher than the melting temperature of the commonly used eutectic Au–Sn bonds (80 wt% Au, melting at 278 °C). The Au–Sn metal system is of great interest since it is oxidation resistant, allowing fluxless bonding. The high temperature stability of the presented process opens the possibility to use Au–Sn bonding for true high-temperature applications.

The bonded samples had electroplated Au–Sn layers, with an overall composition of 8 wt% Sn (13 at% Sn), thus being a surplus of Au relative to the eutectic point. The Sn layer was converted to an intermetallic compound prior to bonding. No flux agent or chemical surface treatment was used.

SEM/ EDS analysis of cross-sections shows uniform bond lines consisting of a layered structure: Au / Au–Sn-alloy / Au. The bonding alloy, being rich in Au, was identified as the ζ/ζ' phase (Au_5Sn). This phase, with a melting point up to 519 °C, explains the elevated delamination temperature of the bonded samples. Since the Au–Sn phase diagram does not contain room-temperature phases between the ζ' phase (Au_5Sn) and the Au phase, the bond is expected to be stable over time.

Key words: Fluxless bonding, 3D integration, High-temperature applications, Electroplating, Au, Sn

Introduction

Au–Sn bonding is a commonly performed bonding method, for instance for optoelectronic devices, 3D integration, and for hermetic sealing of cavities [1-4]. One major advantage of this metal system is the high stability against oxidation, making fluxless bonding possible [5]. Standard Au–Sn bonding makes use of the eutectic composition (80 wt% Au), with a melting point at 278 °C. This gives a soldering type process, with the solidification temperature and the melting/ re-melting temperature equal.

Eutectic Au–Sn bonding thus have a significantly higher thermal stability than the commonly used Sn–Ag–Cu lead-free solders (melting temperatures around 220 °C), or the traditional eutectic Sn–Pb solder (183 °C). This enhanced melting temperature is an important property for applications which will be subject to high temperatures.

Stability at high temperatures is important for several applications: Engine control, oil and gas extraction, and geothermal energy, to mention a few examples. For such applications, stable bondings at even higher temperatures than what eutectic Au–Sn can provide, is desired.

Also high temperatures during subsequent processing of the bonded parts may pose a challenge to the bond integrity: For example when the bonding represents the sealing frame for a vacuum cavity where a getter is needed to obtain the required vacuum. Such getters may require activation at high temperature (for instance 350 °C). Other examples are when chips are to be stacked and bonded in successive processes, as in 3D integration, or if interconnections and sealing rings are being bonded in different process steps. In these latter cases, there is need for the bond to be stable at higher temperatures than the actual process temperature, so that interconnections or sealing rings bonded in one

978-1-4244-4722-0/09 $25.00
© 2009 IMAPS-ITALY

process step do not melt when the bonding process is repeated.

SLID (Solid-Liquid InterDiffusion) is a novel interconnection and bonding technique that has received much interest, giving bonds that are stable at temperatures higher than the processing temperature. This bonding technique relies on IMC (intermetallic compound) formation, using a two-metal system: a high-temperature melting metal and a low-temperature melting metal [6, 7]. At a process temperature above the lower melting point, interdiffusion causes IMC to form. A typical SLID process uses Cu and Sn (melting points 1083 °C and 232 °C), heading for a bond comprised of a Cu / Cu_3Sn / Cu layer structure. Cu_3Sn has a melting point around 700 °C, much higher than the processing temperature (typically 250–300 °C), making the bond very robust against high-temperature excursions. Because Cu is easily oxidized, gaseous flux is generally required in a Cu–Sn SLID process.

Fluxless, SLID-like bonding using a Ag–Sn metal system has also been reported [8], with a theoretical thermal stability to 700 °C. Experimental data for the thermal stability is not provided in [8].

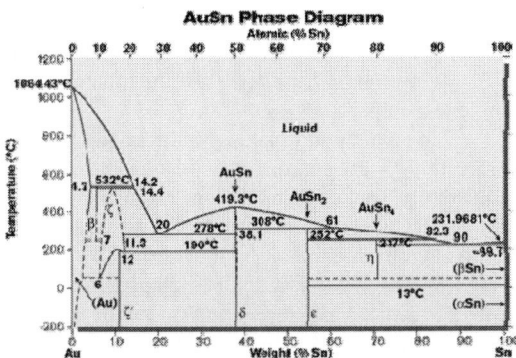

Figure 1: Au–Sn phase diagam. Note the double horizontal line to the left in the diagram: The upper line is at 532 °C (given), the lower line is at 519 °C.

The Au–Sn phase diagram in figure 1 shows the existence of several Au–Sn intermetallic compounds (IMC). All of these have a higher melting point than pure Sn, and the δ and ζ/ ζ' phases have melting points higher than the eutectic composition (419.3 °C and up to 519 °C, respectively). For applications where the bond is to resist higher temperatures, either during later processing or during applications, a bond made of one of these IMCs may be appropriate. For a system where there is surplus of pure gold after formation of the IMC bond line, a bond made of the δ-phase may be susceptible to be converted into a eutectic or near-eutectic structure over time, due to Au–Sn interdiffusion, thus lowering the melting point. A bond made of the ζ/ ζ' phases, in the vicinity of surplus Au, is not expected to convert into lower-melting phases over time. Note that the eutectic composition itself is actually composed of the δ and ζ/ ζ' phases, in a ratio corresponding to 80 wt% Au.

"Off-eutectic" Au–Sn bonding, using such a metal system with surplus of Au has been reported, [9]. These workers demonstrate a bond that survives long-term aging at 400 °C. However, bond characterization at high temperature has not been performed.

We have previously reported fluxless bonding of Au–Sn electroplated layers (using a thin Au layer on top of Sn), with a resulting bondline of the IMC $AuSn_2$ [10]. The present work investigates the possibility to use a Au–Sn system, rich in Au, for a SLID-type bonding. The potential of such a process is to have a flux-free process, with a high temperature stability. Bond strength is characterized at elevated temperatures.

Fabrication of test vehicles

Oxidized (300 nm oxide) silicon wafers with sputtered TiW/ Au adhesion/ seed layers were purchased from Reinhardt Microtech AG, Switzerland. The thickness of the TiW adhesion layer is 60 nm, and the Au seed layer is 100 nm. These metalized wafers are patterned with photoresist AZ4562 for electroplating metallic layers in the shape of rectangular bonding frames. Gold electroplating is performed in gold cyanide solution at a temperature range of 60–65 °C, with a current density 5.4 mA/cm^2. Tin electroplating (on top of the gold layer) is performed in tin sulphate solution at room temperature, with current density 10 mA/cm^2.

Two different types of samples were made:
- Single Au layer: 4.0 μm Au layer thickness.
- Multilayer Au / Sn / Au: Layer thickness 4.0 μm / 2.0 μm / 0.1 μm respectively.

None of the samples have exposed Sn. This minimizes oxidation, and makes fluxless bonding possible [10].

The wafer was diced as 4.3 x 6.6 mm^2 chips for die bonding and testing. The total electroplated area is 3.6 mm^2 per chip.

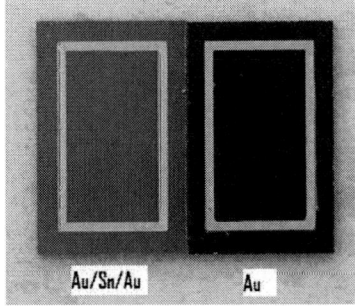

Figure 2: Top view images of electroplated Au (4.0 μm) and Au (4.0 μm) / Sn (2.0 μm) / Au (0.1 μm) chips. The bonding frames are 200 μm wide.

The samples were stored in ambient condition for prolonged time (12 months) prior to bonding. For the multilayer (Au / Sn / Au) samples, reference samples were cross-sectioned (by Struers, Denmark) and investigated by scanning electron microscopy (SEM) and energy-dispersive X-ray spectroscopy (EDS) to identify the phases present prior to bonding.

During the storing time, Au–Sn interdiffusion is expected to convert the Sn layer into an alloy consisting of intermetallics.

Pairs of samples, consisting of one Au-layered chip and one Au / Sn / Au-layered chip were bonded. The bonding was carried out in two steps. First, a flip chip bonder (MAT-6400), was used for pick and place at room temperature in air (applying a force of 30 N for 30 seconds), then the positioned sample pair was bonded, using a hotplate in a vacuum chamber.

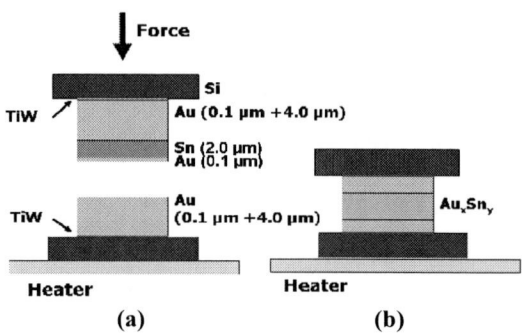

Figure 3: Sketch of layer structure of samples for bonding. a): Layers as plated. b): Expected structure after bonding.

The layer structure of the samples gives an overall composition of 8 wt% Sn (13 at% Sn). This gives a surplus of Au relative to the eutectic point, and also a small surplus of Au relative to the ζ' phase (Au_5Sn).

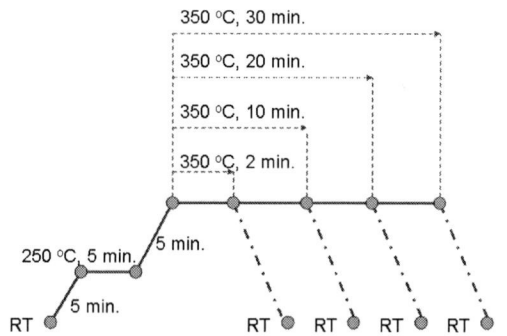

Figure 4: Bonding temperature profiles.

A bonding temperature of 350 °C was selected. This is a realistic temperature for getter activation, a required process for obtaining low pressure in a vacuum cavity. Using such a bonding

temperature, there may not be need for an additional process step for getter activation. Samples were bonded using different bonding times (2 min, 10 min, 20 min, 30 min). The bonding temperature profiles are shown in figure 4.

Several samples (3 pairs) were bonded for each choice of bonding time. This allows testing of high-temperature shear strength, and microscopic investigations of cross-sections of samples bonded under identical conditions.

Bond Integrity Experiments

The bond strength at room temperature (for a 10 minutes bonded sample) was verifed by standard die shear testing, using a Delvotec 5000. The chip was diced in 4 pieces prior to die shear testing, to allow the max. force (50 N) to be sufficient to test the bond destructively.

To verify the temperature stability of the bonded samples, the resistance to shear force at elevated temperatures was investigated. Two independent experiments were set up:

Experiment a): Hot plate

Recesses were machined in an aluminium plate, with dimensions allowing a close fit of a bonded sample in a recess. The recess depth was (0.3 ± 0.02) mm, a little less than the thickness of a single test chip.

The aluminium plate with the bonded sample for testing, was put on a hotplate. The hotplate temperature was controlled from room temperature up to 400 °C. A shear force (around 2 N) was applied to the uppermost chip during the heating, by pushing with a glass slide. The actual temperature in the bond was checked by measuring on top of a bonded sample in a recess; using a thermocouple, with a molten solder ball as thermal interface.

Experiment a) was carried out for one sample of each bonding time (four samples in total).

Experiment b): Oven

The bonded pair was mounted in a fixture giving a constant shear force in the order of 5 N to the bond, as sketched in figure 5. The fixture was put in a thermal chamber. The oven temperature was set to 300 °C, 325 °C, 350 °C, 375 °C and 400 °C, with a 2 hour thermal stabilization time at each temperature step. Thus, the temperature at the bond frame is expected to be equal to the chamber temperature.

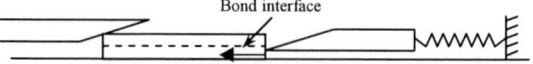

Figure 5: Sketch of fixture giving shear force to bonded sample at elevated temperatures.

978-1-4244-4722-0/09 $25.00
© 2009 IMAPS-ITALY

Experiment b) was carried out for two samples: one sample bonded 20 minutes and one sample bonded 30 minutes.

The experiments a) and b) were designed to reveal melting of the bonding layer: The topmost chip would be easily pushed off if the bonding layer melts.

Method for Microscopic Studies of Cross-sections

The samples were embedded in epoxy resin prior to grinding. They were grinded on SiC paper, grade 320 through 4000, using water cooling. During the fine grinding, soap water was used for lubrication. The grinded samples were polished using 6 µm diamond particles and an alcohol based lubricant prior to fine polishing using 3 µm and 1 µm diamond particles with a water and oil based lubricant.

The cross-sectioned samples were investigated by optical microscopy, scanning electron microscopy (SEM) and energy-dispersive spectroscopy (EDS).

Structure of Multilayered Au–Sn Samples Prior to Bonding

Figure 6 shows a cross-section of an aged, multilayered (Au / Sn / Au) sample. EDS measurements show clearly that the bottom layer is Au and the topmost layer is a uniform AuSn (δ-phase) layer.

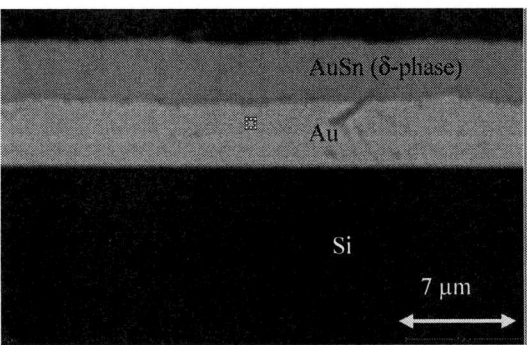

Figure 6: SEM image (backscattered electrons) of cross-section of an aged multilayer (Au / Sn / Au) sample.

Bond Integrity Results

The alignment of the bonding frame and possible smearing of the bonding frame was examined by transmission infrared imaging. Figure 7 shows the image of a chip pair bonded at 350 °C for 20 minutes. It can be seen that the width of the bonding frame is almost the same for top, right and left side, which indicates good alignment for the bonded chips.

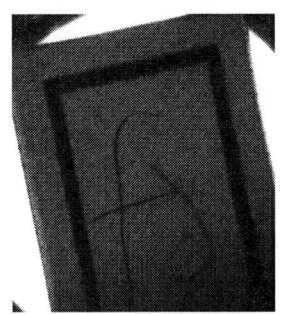

Figure 7: Infrared image (transmission mode) of a pair of chips bonded at 350 °C for 20 minutes. The figures inside the bonding rings are scratches on the substrate backsides for identification purposes.

The bond strength at room temperature for the sample with 10 minutes bonding time, was measured to exceed 60 MPa.

Experiment a): Hot plate

The four bonded samples were shear tested for workholder temperatures up to 400 °C. 400 °C workholder temperature was measured to correspond to 378 °C on top of the bonded pair, implying a temperature at the metal bond frame around 380 °C. With shear forces up to 2 N, no delamination or movement of the uppermost chip occurred within the temperature testing range. This is valid for all samples, independent of the bonding time at 350 °C.

Experiment b): Oven

The sample bonded for 20 minutes survived oven temperatures up to 350 °C, but delaminated at oven temperature of 375 °C. The sample bonded for 30 minutes showed no delamination for temperatures up to 400 °C. Higher temperatures were not tested.

Microscopy of Cross-sections

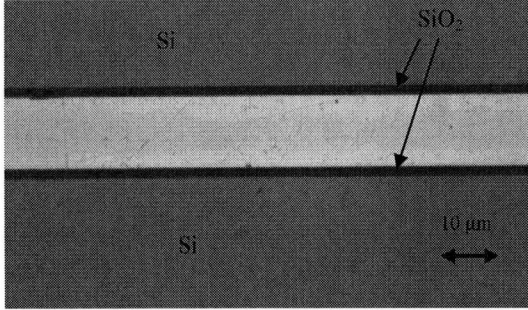

Figure 8: Optical micrograph of cross-sectioned bonded sample (2 minutes bonding time).

Figure 8 shows an optical micrograph of a cross-section of a sample bonded for 2 minutes at 350 °C. The bond line is seen to be uniform. A yellow/ golden phase is found in the vicinity of the Si surfaces, and a uniform, greyish phase makes up the bonding layer. All cross-sectioned samples show a similar phase structure, independent of the bonding time at 350 °C.

For SEM images, the cross-sectioned samples turn out to have a poor phase contrast, as shown in figure 9. This applies to imaging either by secondary electrons or by backscattered electrons. This indicates that the difference in composition between the two phases is relatively small.

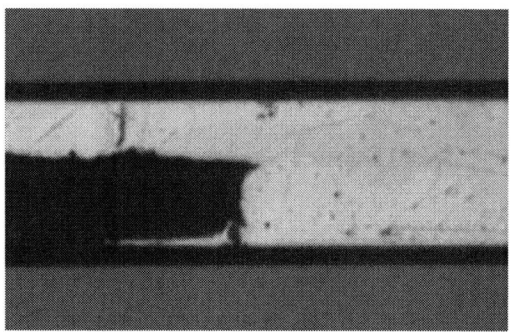

(a)

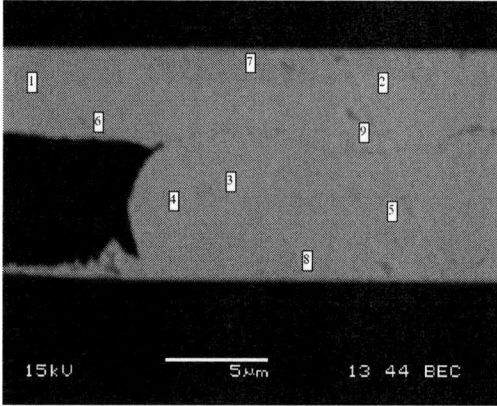

(b)

Figure 9: Optical microscopy (a) and SEM image (backscattered electrons) (b) of the same region of the sample bonded for 10 minutes. The phase contrast seen in optical microscopy is not discernable in SEM imaging. The numbers in (b) indicates positions for EDS analysis. Note that the pictures are selected at the edge of the bond line, showing some μm misalignment. This ensures unique identification of the same regions in optical and SEM images.

A more detailed analysis was done using EDS, the positions for analysis are given in figure 9 b). For all positions, Au and Sn are the elements found, as expected. For analysis done within the

greyish phase (position 2, 6 & 9), EDS gives compositions in the range 8–10 at% Sn. For analysis done within the golden phase (positions 4, 5 & 8), EDS gives compositions in the range 0–2 at% Sn. Positions 1 & 3, being close to the phase border, give intermediate values (4–5 at% Sn). Since EDS probes a finite volume, measurements close to the phase border may have contributions from both phases.

Discussion

The bonded samples resist shear forces in the temperature range from room temperature up to well above 350 °C. One of the tested samples delaminates at 375 °C, whereas the other five tested samples resist shear forces also for temperatures higher than 380 °C. This shows that the obtained bondline does not melt at these temperatures. Bonding of AuSn δ-phase to Au at 350 °C therefore has potential as a bonding method for high-temperature applications, as well as to tolerate high-temperature processes such as getter activation after bonding.

The obtained bond is uniform, and the actual bondline consists of a single phase, being a Au-rich Au–Sn compound. According to EDS analysis, this consists of 8–10 at% Sn. However, for these rather low Sn concentrations, the quantitative EDS analysis is questionable, due to the small spectral peaks compared to the background to be subtracted in the data analysis. Hence, the absolute values of Sn concentration may not be correctly calibrated, whereas the relative comparison between different measurements is valid.

According to the Au–Sn phase diagram (figure 1), two Au-rich phases exist at room temperature: Au with maximum 3 at% Sn (2 wt% Sn) in solid solution, and the ζ' phase (Au_5Sn) consisting of 17 at% Sn (11 wt% Sn).

The optical micrographs cleary indicate that the phase remaining at the Si surfaces is Au (possibly with a small amount of Sn in solid solution). The bonding layer is obviously not Au (from the optical micrographs). Nor can it be δ-phase (AuSn), since this phase is known to give both a far better contrast to Au in SEM (ref. figure 6) and a more reliable EDS value of Sn concentration. The conclusion is therefore that the greyish bonding phase consists of the ζ' phase (Au_5Sn). This phase has a melting temperature that may reach 519 °C, which explains the high-temperature stability of the bond.

The overall composition of the as-plated metal layers is 13 at% Sn. This is consistent with a resulting structure of ζ' phase (Au_5Sn, 17 at% Sn) and Au phase (0–3 at% Sn), with the ζ' phase occupying the larger volume fraction (as seen in figure 8).

The total bond structure is a layered Au / ζ' phase (Au_5Sn) / Au structure. The phase diagram does not show phases with composition between these two phases. Au–Sn interdiffusion during aging of the bond is therefore not expected to result in further phase transformations. The obtained bond is therefore expected to be reliable and stable over time.

The bond structure, with an intermetallic layer sandwiched between layers of pure metal, is similar to the desired bond structure for a Cu–Sn SLID bond [6]. The bonding mechanism in our case is also similar to SLID bonding, with Au–Sn interdiffusion giving rise to the bonding ζ' (Au$_5$Sn) phase. Liquid phase may exist during the initial part of the bonding process, as long as δ (AuSn) and ζ/ζ' (Au$_5$Sn) phases coexist.

Conclusion

We have performed bonding of Au to AuSn δ-phase at 350 °C. The AuSn δ-phase was obtained by aging a multilayered Au / Sn / Au structure. The obtained bonding structure is a layered structure: Au / ζ' phase (Au$_5$Sn) / Au. This bonding structure has a theoretical temperature stability up to 519 °C. We have shown that the actual bond is stable up to 375 °C or higher, being around 100 °C higher than the eutectic point. Varying the bonding time (in the range 2–30 minutes) does not have significant effect on the result. The bonding structure is expected to be stable over time, and not to change composition due to interdiffusion of Au and Sn.

The investigated bonding method therefore has potential for high-temperature applications, and as a bonding method to tolerate high-temperature processes, such as getter activation, after bonding.

Acknowledgements

This work was funded by the RCN (Research Council of Norway) funded BIA project No. 174320, "3DHMNS – 3D Heterogeneous Micro Nano Systems".

Assistance with laboratory work from Tormod Vinsand and Finn M. Reinhardtsen, both at VUC, is greatly acknowledged.

References

[1] R.S. Forman and G. Minogue, "The Basics of Wafer-Level AuSn Soldering," Chip Scale Review, Vol. 8, pp. 55-59, 2004.

[2] C.C. Lee and C.Y. Wang, "A low temperature bonding process using deposited gold tin composites", Thin Solid Films, 208, pp 202-209, 1992.

[3] D.Q. Yu, H. Oppermann, J. Kleff and M. Hutter, "Interfacial metallurgical reaction between small flip-chip Sn/Au bumps and thin film Au/TiW metallizatgion under multiple reflow", Scripta Materials, 58, pp 606-609, 2008.

[4] L. Dietrich, G. Engelmann, O. Ehrmann and H. Reichl, "Gold and gold-tin wafer bumping by electrricchemical deposition for flip chip and TAB", EuPac'98, Nürnberg, pp 28-31, 1998.

[5] J.S. Kim, W.S. Choi, D. Kim, A. Shkel, C.C. Lee, "Fluxless silicon to alumina bonding using electroplated Au–Sn–Au structure at eutectic composition," Materials Science & Engineering A, Vol. 458, pp. 101-107, 2007.

[6] H. Huebner, S. Penka, B. Barchmann, M. Eigner, W. Gruber, M. Nobis, S. Janka, G. Kristen and M. Schneegans, "Microcontacts with sub-30 µm pitch for 3D chip-on-chip integration", Microelectronic Engineering 83, pp 2155-2162, 2006.

[7] L. Li, J. Jiao, L. Luo and Y. Wang, "Cu/Sn Isothermal Solidification Technology for Hermetic Packaging of MEMS", Proceedings of the 1st IEEE International Conference on Nano/Micro Engineered and Molecular Systems, Zhuhai, China, pp 1133-1137, 2006.

[8] J.S. Kim, T. Yokozuka, and C. C. Lee, "Fluxless bonding of silicon to Ag-cladded copper using Sn-based alloys", Materials Science & Engineering A, vol. 458, pp. 116-122, 2007.

[9] R. W. Johnson, C. Wang, Y. Liu, and J. D. Scofield, "Power device packaging technologies for extreme environments", IEEE Transactions on Electronics Packaging Manufacturing, vol. 30, pp. 182-193, 2007.

[10] K. Wang, K. Aasmundtveit and H. Jakobsen: "Surface Evolution and Bonding Properties of Electroplated Au/Sn/Au", Proceedings of the 2nd Electronic System-Integration Technology Conference, London, UK, pp 1131-1133, 2008.

978-1-4244-4722-0/09 $25.00
© 2009 IMAPS-ITALY

Optimization of Flip-chip Laser Soldering for Low Temperature Stability Substrate

Tamás Hurtony, Bálint Balogh, Péter Gordon

Budapest University of Technology and Economics
Department of Electronics Technology
Goldmann tér 3, 1111 Budapest, Hungary

Phone: +36-1463-2748, Fax: +36-1463-4118, Email: hurtony@ett.bme.hu

Abstract

Interconnect technology is becoming increasingly more complex due to miniaturization of surface mount devices. These trends in electronics industry have led to the demand for new, highly controllable selective soldering technologies. Laser soldering methods can be an adequate answer for the mentioned demands. Optimization of laser soldering process is extremely important especially when there are more than 100°C difference between the temperature limit of the substrate and the melting point of the solder eg. PMMA substrate (T_g~105°C) and SnAgCu solder (MP=217°C). Temperature distribution of the laser soldered structure can be simulated by our model which also considers reflection, absorption and transmission as well as the Gaussian energy distribution of the beam, beyond the thermal properties of the sample. Simulations and experiments were carried out at frequency trippled Nd:YAG laser wavelength (355 nm) by direct heating of the flip-chip. Soldering process parameters (pulse energy, average power, soldering time, beam intensity) were optimized based on both the simulation and experimental results. The solder joints were qualified by resistance measurements, X-ray micrographs, micro-sections and shear tests.

Key words: Laser soldering, flip-chip, thermal simulation, low temperature stability substrate

Introduction

New packaging concepts for electronic components and assemblies are essential bases for the continuing miniaturization [1, 2]. The successful use of flip-chip, chip scale packages, and microball-grid arrays depends on the availability of suitable repair techniques [[3]]. Although the majority of the solder joints are created by conventional reflow or wave soldering processes, in certain cases it is necessary to form some joints individually, which can be achieved by selective soldering methods [[4]].

One of the most commonly used selective joining technologies is laser beam soldering. The energy transfer from the laser to the material is achieved by means of irradiation, which has couple of advantages over conventional selective soldering methods, ie. soldering iron or hot bar. In case of laser soldering no contamination can be transferred between consecutive samples, no mechanical deformation is caused by the heating tool, no extra space is required around the to be soldered component [5]. The soldering temperature profile can be controlled by varying the laser processing parameters such as input energy and processing speed. Heating by high intensity laser beams, such as CO_2 and Nd:YAG lasers, solder joints can be formed within a few microseconds to a few seconds, which is much shorter than by conventional

soldering methods [6, 7]. With this process the heating is localized only to the component to be soldered and unnecessary heating of other components or the substrate can be avoided. In addition, the short reflow time minimizes the dissolution of the pad material and the rapid cooling results in fine grain structure of the solder [5]. At the same time optimization of laser soldering process is extremely important especially when there are more than 100°C difference between the temperature limit of the substrate and the melting point of the solder eg. PMMA substrate (T_g~105°C) and SnAgCu solder (MP=217°C).

Thermal simulation of the laser soldering process –in case of applying low temperature stability substrates (LTSS) – can be applied in order to reduce the number of experiments needed for gaining the optimal parameters of the soldering process.

This paper describes how the laser soldering parameters were optimized in order to achieve the highest shear strength values. A thermal model of the silicon flip-chip and the laser beam has been created, which takes reflection, absorption, transmission as well as the thermal properties of the sample into consideration. A laser source with Gaussian distribution is considered moving with constant velocity along an elliptic trajectory. The thermal properties are temperature independent.

978-1-4244-4722-0/09 $25.00
© 2009 IMAPS-ITALY

Surface heat losses toward the ambient are taken into account. Simulations had been verified with experimental results and had been reconsidered to reach better accuracy.

Experimental

A special layout has been formed on a LTSS, which can be completed to a daisy chain, when the flip-chip and the substrate are placed together. The 12 contact points allow us to measure the resistance between each two pads by using four wire measuring method. This is important, because we have to be able to validate the electrical quality of the soldered joints. Twenty 80 μm diameter SnAgCu solder (MP=217 °C) bumps are placed on the bottom of the flip chip in a 4 by 5 matrix with 400 μm raster. Type 4 tin-bismuth solder paste (58Bi/42Sn, MP=139 °C), was printed onto the LTSS manually with 70 μm thick nickel-iron stencil through laser cut 120·120 μm^2 apertures.

Pilot laser soldering experiments were carried out at two Nd:YAG laser wavelengths (1064 nm and 355 nm) by transmission soldering through transparent substrate and also by direct heating of the flip-chip. It was found that heating the flip-chip directly with the 355 nm wavelength Coherent Avia 355-4500 laser resulted in better reproducibility and wider process window than with 1064 nm wavelength thus all further experiments were carried out by this equipment. The properties and parameters influencing the laser soldering process are collected in Table 1.

Table 1. Parameters of the laser soldering process and typical values

ts	N/f	soldering time, s
tf	1/f	time between pulses, s
tp	2.00E-6	pulse length, s
f	5.00E+4	pulse repetition rate, Hz
w	2*π*40	angular frequency, rad/s
v	2.00E-1	scan velocity, m/s
N	round(K*n/v*f)	number of shots
n	60	number of multiple scans
σ	d/4	sigma of the Gaussian beam, m
d	0.2E-3	spot diameter, m
As	(d/2)^2*π	spot area, m^2
A	1.25E-3/2	A-semiaxis of the trajectory, m
B	1.00E-3/2	B-semiaxis of the trajectory, m
K	3.545E-3	scanned length, m
Ep	6.00E-05	pulse energy, J
Pp	Ep/tp	peak power, W
Pa	Ep*f	average power, W
Qa	Pa/As	heat flux average, W/ m^2
Qina	(1-ref)*C* Qa	coupled average heat flux
Alfa	107	absorption coefficient um^{-1}
Ref	0.58	reflection coefficient
C	0.38	coupling in constant

The parameters above can be separated into two major groups. The first one is the group of the parameters which are characterized by the physical properties of the laser and the beam delivery system so they cannot be varied. These fix parameters are: t_p, σ, d, As. The rest of the parameters can be freely choosen to optimize the soldering process. Seven of them (w, v, N, n, A, B and K) describe how the beam is scanned during the soldering process. They have to be optimized to obtain as homogenouos lateral heat distribution as possible. The scanning parameters determine the soldering time (t_s) as well. The energy absorbed by the flip-chip depends on the surface properties of the silicon die, which depends on the manufacturing technology, thus it can not be found in the literature. This is considered in C (coupling in constant) factor, which can only be determined experimentally by harmonizing experimental and simulation results. We found that the two most important process parameters influencing the soldering process and the quality of the solder joints are

1. P_a: average power, W
2. t_s: soldering time.

Thus during the optimazitaion we focused on these two values.

Mathematical description

We applied a thermal based macroscopic model which is detailed in [8]. The primary state variable is the temperature. The initial temperature distribution is generated by the laser exposure. Once the initial boundary condition is given the whole process can be traced back to a heat conduction problem, where the solution comes from the solution of the partial differential equation of heat conducton:

$$Q + \lambda \cdot \left[\frac{\partial^2 T}{\partial x^2} + \frac{\partial^2 T}{\partial y^2} + \frac{\partial^2 T}{\partial z^2} \right] = \rho \cdot c \cdot \frac{\partial T}{\partial t}, \qquad (1)$$

where Q is the inward heat flux, λ is the coefficient of thermal conduction, ρ is the density, and c is the specific heat. A frequency tripled, Q-switched Nd:YAG laser (Coherent Avia 355-4500) with Gaussian energy distribution was applied. According to the Lambert-Beer's law the (2) equation describes the absorbtion of the Gaussian laser beam in a material with an absorbtion coefficient of α

$$F(x,y,z) = F_0 \cdot e^{-(\frac{x^2+y^2}{2\sigma^2}+\alpha z)}, \qquad (2)$$

where F_0 is the fluence of the laser beam in the axis of symmetry, and σ is the parameter of width of the beam. The absorption coefficient of the silicon flip-chip is 107 μm^{-1} at λ =355 nm wavelength, which means that the total energy is absorbed in the upper 9 nm of the surface, which is comparable to the size of the applied meshgrid so in this case it can be neglected.

978-1-4244-4722-0/09 $25.00
© 2009 IMAPS-ITALY

Simulation model and results

The partial differencial equation was calculated by Finite Element Method (FEM) in Comsol Multhiphysics 3.5. We used the laser in pulsed mode, but as it has been shown in [9] the laser is delivering the same amount of energy to the substrate despite using pulsed mode with peak power or countinous mode with average power, thus the temperature converges to a somewhat constant value far enough downstream from the heat affected zone [9]. In this frequency range the flip-chip acts as a thermal integrator whose integrating time is RC, where R and C are the thermal resistance and the heat capacitance of the flip-chip respectively.

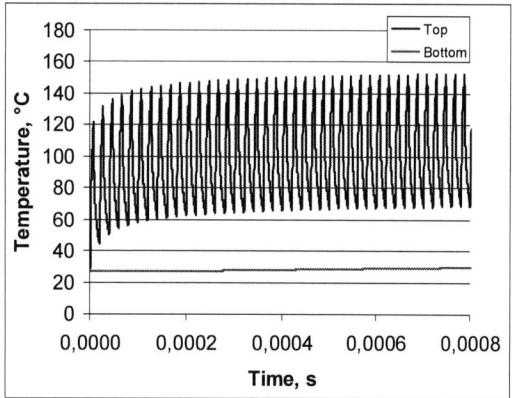

Figure 1: Temperature at the top (heated by the laser) and bottom (bump) side of the flip-chip when heated by 50 kHz repetition rate pulses.

Thus far enough from the impact zone, there is no difference between using the average power or the impulse mode as it is shown in Fig. 1 and 2. That is why we applied countinous mode with averaged input power in order to reduce the simulation runtime. The amount of input energy is proportional to the area under the temperature profile graph.

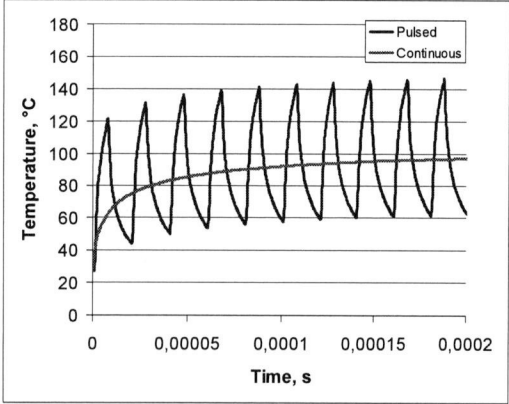

Figure 2: Temperature vs time diagram simulated with continuous and pulsed laser energy input.

In Comsol we should use periodic boundary condition to produce inward heat flux pulses. It can be created in the following way:

$$Q(t) = \begin{cases} Q_{peak} \cdot e^{-\left(\frac{x^2+y^2}{2\sigma^2}\right)}, & if \sin(2 \cdot \pi \cdot f \cdot t) > 0.9999 \\ 0, & else \end{cases} \quad (3)$$

where t is the runnig time index of the simulator f is the frequency of the impulses as in Table 1. If we use this boundary condition the duty cycle of the switching signal will be 0.47, thus in case of 50 kHz pulse repetition frequency the pulse length is 94 ns.

In practice 5 mm defocusing of the laser beam was applied, in order to reduce laser fluence below ablation threshold on the surface of the flip-chip not to damage the silicon die. The trajectory, along which the beam was deflected was an elliptic trace, which can be implemented as the addition of two periodic signal, with 90° phase shift between them and with different amplitude:

$$Q(x,y,t) = Q_0 \cdot e^{-\frac{(x-(A \cdot \sin(\varpi \cdot t))^2 + (y-(B \cdot \cos(\varpi \cdot t))^2}{2\sigma^2}}, \quad (4)$$

This elliptic trajectory can be observed in Fig. 3. where the mesh of the finite element model and the temperature distribution during laser irradiation are also shown. Accuracy of the simulation is expected to converge, until a mesh-independent state is reached. Increasing the mesh density beyond this point would not yield significantly increased accuracy in observed results. The applied mesh is fine, where the temperature gradient tends to be high (see Fig. 4.), thus around the ellipse and the mesh can be coarser where the temperature gradient is lower.

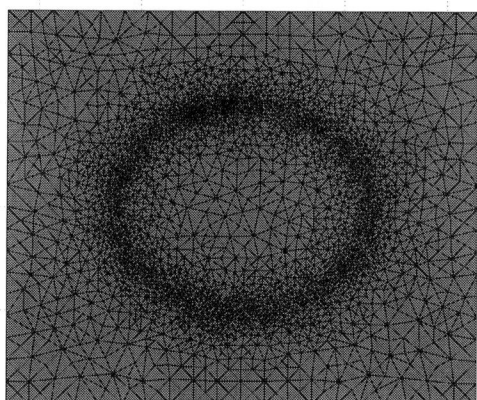

Figure 3: Applied mesh for finite element modeling of the flip-chip.

Fundamentally two parameters influence the energy input during soldering: the average power and the soldering time. The latter can be varied by changing the number of multiple scans (how many times the ellipse is scanned). The higher the number of multiple scans, the longer the soldering time and the higher the input energy is. At a given average power it has both lower and upper limits, ie. the temperature has to reach the melting point of the solder but should not exceed the temperature limit of the flip-chip or the LTSS. Fig. 5. shows the temperature of the coldest bump as the function of

978-1-4244-4722-0/09 $25.00
© 2009 IMAPS-ITALY

the number of multiple scans. It can be seen that more than 40 scans are needed to reach the melting point of the eutectic tin-bismuth solder.

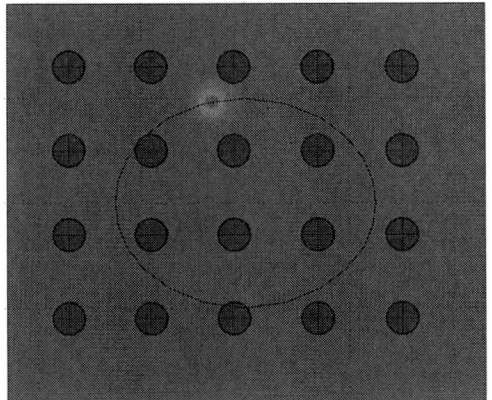

Figure 4: Simulated temperature distribution of the flip-chip. The hottest spot shows where the laser beam was at the screenshot while traveling around the ellipse.

Fundamentally two parameters influence the energy input during soldering: the average power and the soldering time. The latter can be varied by changing the number of multiple scans (how many times the ellipse is scanned). The higher the number of multiple scans, the longer the soldering time and the higher the input energy is. At a given average power it has both lower and upper limits, ie. the temperature has to reach the melting point of the solder but should not exceed the temperature limit of the flip-chip or the LTSS. Fig. 5. shows the temperature of the coldest bump as the function of the number of multiple scans. It can be seen that more than 40 scans are needed to reach the melting point of the eutectic tin-bismuth solder.

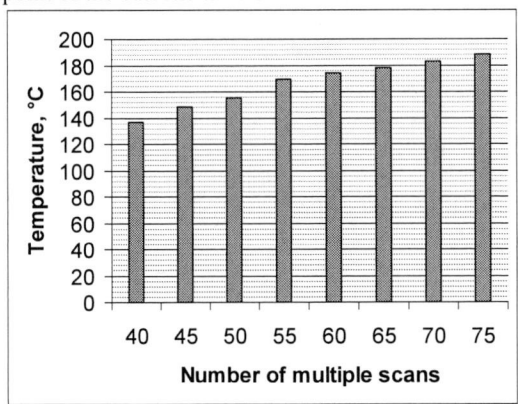

Figure 5: Simulated temperature of the coldest bump vs number of multiple scans.

Experimental results

Laser soldering experiments were carried out with the same parameters as the simulation shown in Fig. 5. The shear strength of the soldered flip-chips was measured by a Dage BT2400 shear tester. The

maximal shear strength was obtained at 60 scans, applying longer or shorter soldering times resulted in lower strength, as it can be seen in Fig. 6.

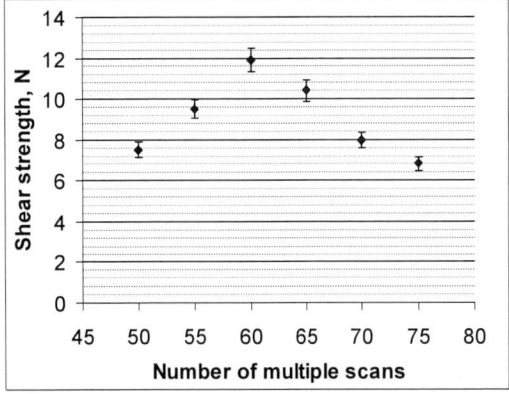

Figure 6: Shear strength of the soldered flip-chips was maximal at 60 scans.

The temperature distribution of the flip-chip can be simulated by our model. In order to be able to verify the simulation results it is important to monitor the temperature of the flip-chip during soldering. This can be achieved by a pyrometer.

The experimental set-up is shown in Figure 7. Meanwhile the flip-chip was about to be soldered onto the substrate the temperature of the surface of the chip was monitored by a pyrometer. The measurement was repeated at different number of multiple scans. Experimentally 2 well definied temperature values can be detected, the melting point of the solder paste (138°C) and the solder bumps (217°C). The shorter the heat affect time the lower the possibility of having delamination on the chip under the bump, or on the substrate. That is is why the termperature, where the SnBi solder paste reflows but the SnAgCu does not is a useful point, such situation can be seen in Fig. 12.

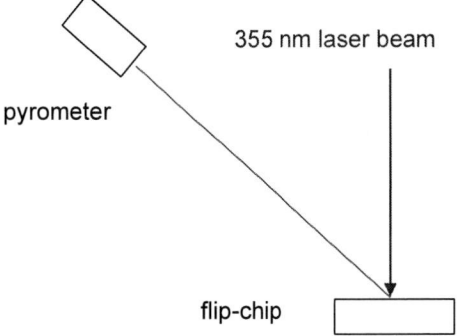

Figure 7: In-situ temperature measurement set up

The test chips have been placed onto a golden thin film coated glass substrate without solder paste. Since the 150 nm thin film layer had negligible heat capacitance and the glass substrate is a good thermal insulator the chips could be considered as they were placed onto thermal insulator. But at the same time

978-1-4244-4722-0/09 $25.00
© 2009 IMAPS-ITALY

due to the golden surface the wetting of the substrace can be examined. We applied various numbers of multiple scans to find out at which point the solder bumps are molten. Lower temperature values were measured than expected from the simulation results (Fig. 8.).

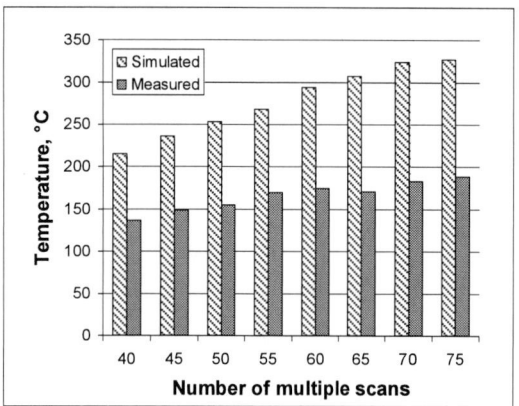

Figure 8: Simulated and measured temperature values.

The pyrometer measures the total intensity of radiation, which is coming from the surface area under its focal spot. When measuring with pyrometers the intensity of radiation is changed into an electric signal. The intensity of radiation across the whole band of all wavelengths is formed by the integral of spectral intensity between 0 μm and infinity, at a given temperature [10]. Thus the total intensity is proportional to the area under the black body radiation curves. According to Stefan-Boltzmann law a given temperature can be determined by the total amount of intensity. This also implies that if the temperature of an object has to be measured which has smaller area than the focal spot of the pyrometer, then the measured values will be lower. The measured data have to be corrected with the area ratio of the pyrometer spot size and the object to be measured.

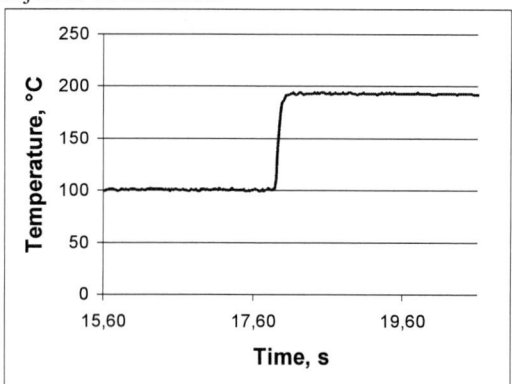

Figure 9: pyrometer control exp: the measured temperature is lower when the chip-size sheet is under the focal spot.

Unfortunately in our case the spot size of the pyrometer was not known, so in order to be able to

determine what correction factor we have to apply to our temperature measurement results, we carried out the following experiment. The same geometry of measurement setting has been set up as in Fig. 7. We placed a $100*100$ mm^2 plate which had a relatively small reflection coefficient (black oxide-coated aluminium) on the top of a hotplate. After a given temperature had been set up the measured value of the pyrometer was recorded. The same has been repeated with a higher reflectance nickel-iron stencil sheet. Then a flip-chip sized nickel-iron block had been fixed under the focal spot of the pyrometer. As we removed the tiny block a rapid temperature increase was recorded (Fig. 9.). From the measured temperature difference the area ratio of the focal spot ($t=t_1+t_2$) and the chip (t_1) can be calculated

$$t_2 \cdot I_2 + (t-t_2) \cdot I_1 \cong 55.54 °C \qquad (5)$$

We measured the total intensity which was radiated from t_1 and t_2. According to the Boltzmann's law this gives us a certain temperature.

$$t \cdot I_1 \cong 33.43 °C @ 100 °C \qquad (6)$$

Where $t*I_1$ is proportional to the temperature that would be radiated from the total area on nickel-iron surface, and it is known from previous experiments. The difference of the reflaction coefficient for t_2 can be described as a temperature value.

$$t_2*(I_2 - I_1) \cong 55.54 - t*I_1 = 22.41 \qquad (7)$$

The same calculation can be applied for t_1

$$t*I_2 \cong 99.21 °C @ 100 °C \qquad (8)$$

$$t_1*(I_2 - I_1) \cong t*I_2 - 55.54 = 43.67 \qquad (9)$$

$$\frac{t_2*(I_2 - I_1)}{t_1*(I_2 - I_1)} \cong \frac{22.41}{43.67} = 0.5131 \qquad (10)$$

The same process was repeated at different temperatures and in conlclusion the average area ratio between the flip-chip and the focal spot turned out to be $t_2/t_1=0.588$. Using this correction factor the simulation and experimental results are in good correlation, see Fig. 10.

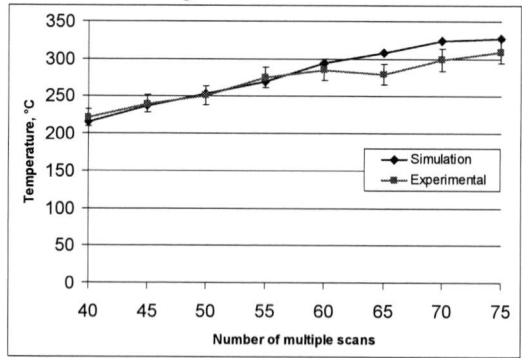

Figure 10: applied number of multiple scans vs T

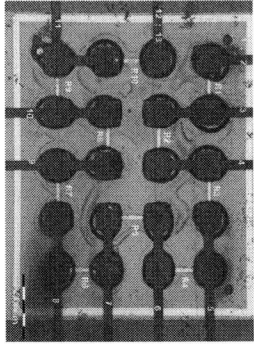

	Ohm
R1	0,651
R2	0,736
R3	0,65
R4	0,535
R5	0,629
R6	0,506
R7	0,631
R8	0,643
R9	0,495
R10	0,984
R1-	3,299
R6-	3,559

Figure 11: Optical microscopic image of a soldered flip-chip through the PMMA substrate and measured resistance values.

Figure 12: SEM micrograph of a microsectioned soldered flip-chip. The SnBi solder was molten and wetted the bump.

Conclusion

A numerical model was developed to simulate the laser soldering process of a flip-chip on low temperature stability substrate with 355 nm Nd:YAG UV laser. The optimization of the manufacturing processes can be supported by using our simulation model. We have validated the model with experiments, making sure that the model is calibrated to give suitably accurate results.

We found that the significant process parameters of soldering chip-on-LTSS are soldering time and average power, the laser source can be considered continuous at 50 kHz pulse repetition frequency.

In our model the amount of energy that is heating up the flip-chip is so charachterised that we were able to maintain the temperature which was above the melting point of the tin-bismuth eutectic alloy but below the MP of the SnAgCu solder bumps. At 50 kHz pulse repeptition frequency, 60 µJ pulse energy, and 0.2 m/s scan velocity this point was at 37 multiple scans

Mechanically and electrically acceptable solder joints were produced as the shear tests and the electrical measurements (Fig. 11.) proved. These results show the possibility of applying laser soldering for chip-on-LTSS.

References

[1] A. J. Flanagan, G. Lowe, T. J. Glynn, Finite element thermal modeling of the laser soldering process, Key Engineering Materials, vol. 86-87, pp 329-336, 1993.

[2] Tony Hoult, Laser Solutions for Soldering, Circuits Assembly, February 2004, pp. 52-56

[3] G. Hanreich, L. Musiejovsky, K.-J. Wolter, M. Fasching, J. Nicolics, H. Kök, Optimisation of a Laser Soldering / Desoldering Process for Flip-Chips Using a New Thermal Simulation Tool, Electronic Packaging, pp. 68-69

[4] John Vivari, Alex Kasman, Laser Solder Reflow: A Process Solution Part I, http://www.efd-inc.com/NR/rdonlyres/ 41EE0746-E444-4CB3-944D-7E0D063C5B48 /0/EFDLaserReflowSolderingI.pdf

[5] Jong-Hyun Lee, Won-Yong Kim, Dong-Hoon Ahn, Yong-Ho Lee, Yong-Seog Kim, Laser Soldering for Chip-on-Glass Mounting in Flat Panel Display Application, Journal of Electronics Materials, vol. 30., No. 9., 2001.

[6] J.S. Hwang, Solder Paste in Electronics Packaging (Technology and Applications in Surface Mount, Hybrid Circuits, and component Assembly), New York: Van Nostrand Reinhold, 1992, p. 221.

[7] E. Semerad, L. Musiejovsky, and J. Nicolics, J. Mater. Sci. 28, 5065 1993.

[8] Péter Gordon, Bálint Balogh, Bálint Sinkovics: Thermal simulation of UV laser ablation of polyimide Microelectronics Reliability (47), 2006., pp. 347-353

[9] Weixue Tian, Wilson K.S. Chiu, Temperature prediction for CO2 laser heating of moving glass rods, Optics & Laser Technology 36. 2004, pp. 131-137.

[10] IMPAC Infrared GmbH, Pyrometer-Handbook , 2004,http://www.impacinfrared.com/Bedanl_be schraenkt/IN5plus_d_e.pdf (2009.03.25)

978-1-4244-4722-0/09 $25.00
© 2009 IMAPS-ITALY

Low energy consumption thick-film pressure sensors

Darko Belavič[1], Marina Santo Zarnik[1], Matej Možek[2], Sandi Kocjan[3],
Marko Hrovat[4], Janez Holc[4], Mitja Jerlah[3], Srečko Maček[4]

[1] HIPOT-RR, Trubarjeva 7, 8310 Šentjernej, Slovenia
2 Faculty of Electrical Engineering, Tržaška 25, 1000 Ljubljana, Slovenia
3 HYB, Levičnikova 34, 8310 Šentjernej, Slovenia
4 Jožef Stefan Institute, Jamova 39, 1000 Ljubljana, Slovenia

Phone: ++386 1 4773479   Fax: ++386 1 4773887   E-mail: darko.belavic@ijs.si

Abstract

This paper is focused on three different types of ceramic pressure sensors for the use in low-energy-consumption applications. We investigated the design issues for low energy consumption of sensing elements and compared the results and other sensors' characteristics for three different types of thick-film pressure sensors. The first type is the capacitive sensor, which is based on changes to the capacitance values between two electrodes: one electrode is fixed and the other is movable. The displacement of the movable electrode depends on the applied pressure. The energy consumption depends mostly on the values of operating frequency and the capacitance of the capacitor. The second type is the piezoelectric, resonant pressure sensor, which is based on the shifting of the resonant frequency of the diaphragm. The thick-film piezoelectric actuator structure on the diaphragm generates the vibration of the diaphragm in the resonant-frequency mode. This resonant frequency then shifts due to the static deflection of the diaphragm caused by the applied pressure. The energy consumption depends mostly on the values of operating frequency and the capacitance of the actuator. The third type is the piezoresistive pressure sensor, which is made with four thick-film resistors on the diaphragm. Each thick-film resistor acts as a strain gauge, which is capable of translating the strain into an electrical signal. The energy consumption depends mostly on the value of resistance of thick-film resistors and the operating voltage. All three types of pressure sensors were made with low-temperature cofired ceramic (LTCC) and designed as ceramic capsules consisting of a circular edge-clamped deformable diaphragm that is bonded to a rigid ring and the base substrate. This construction forms the cavity of the pressure sensor. The diaphragm has a radius of 4.5 mm and a thickness of 200 µm. The depth of the cavity is from 100 to 250 µm.

Key words: sensors, pressure sensors, ceramic pressure sensors, low energy consumption, LTCC substrate

Introduction

Pressure sensors, like other sensors, become more and more common as a part of the Intelligent and Integrated Micro-Systems (IMS). The typical IMS combine some or all of the functions of sensing, actuation, processing information, communication and power management within a single integrated package. The needs for such system are not only to combine mentioned functions, but also to realize efficient systems with reduced size, weight, power and cost. The IMS applications often need to be energy autonomous which requires wisely managed energy, storage energy, to use independent energy sources or to obtain energy from the environment (energy harvesting), and the systems must be fabricated with the low energy consumption components.

The applications of wireless-sensor networks, which are typical intelligent integrated microsystem, are rapidly growing. A large number of sensors in the network situated at remote locations require,

beside other requirements, very low power consumption. Therefore the additional energy saving is possible with the use of sensing element with low power consumption.

The pressure sensor market is dominated by the silicon pressure sensors. On the other hand, complex sensor systems combine different materials (silicon, ceramic, metal, polymer, etc.) and technologies (semiconductor, thin and thick film, etc.). In some demanding applications thick-film technology and ceramic materials are a very useful alternative for a fabrication sensor systems i.e. ceramic or thick-film pressure sensors. In comparison with semiconductor sensors they are larger, more robust and have a lower sensitivity, but they have higher resistance against harsh environment and they can operate over a wider operating-temperature range [1-6].

Most ceramic or thick-film pressure sensors are made with flexible and deformable diaphragms. The applied pressure deforms the diaphragm and regarding to the sensing principle changes one of the sensor's characteristics: resistance, capacitance, impendence, phase, frequency [5,6].

Low-temperature cofired ceramic (LTCC) technology and materials are suitable for forming a three-dimensional (3D) construction. The LTCC technology has a great deal of flexibility when designing a 3D construction. This technology is, therefore, very promising in sensor applications, and the number of applications in this technology is growing. The advantage of LTCC material is also lower modulus of elasticity in comparison with alumina. Therefore, by using the LTCC material for diaphragm sensors have higher sensitivity and the operating range can be extended, or alternatively, a thinner, and because of this a more flexible, diaphragm can be used. However, a disadvantage is the low flexural strength and some problems with fatigue [5-9].

In this contribution we investigated and compared the design issues of sensing element for low energy consumption applications of three different types of thick-film pressure sensors. The types of sensors are based on capacitance (thick-film electrode), resonance (with a thick-film piezoelectric actuator), and piezoresistance (thick-film resistors), principles.

Ceramic Body of a Pressure Sensor

The LTCC technology and materials are suitable for made the ceramic structure of a thick-film pressure sensor. This structure consists of a circular edge-clamped deformable diaphragm that is bonded to a rigid ring and the base substrate. In the base substrate is the hole for applied reference or differential pressure. These elements form the cavity of the pressure sensor. The depth of the cavity depends on the thickness of the rigid ring. The top-view and the cross-section of this type of thick-film pressure sensor are schematically shown in Figure 1.

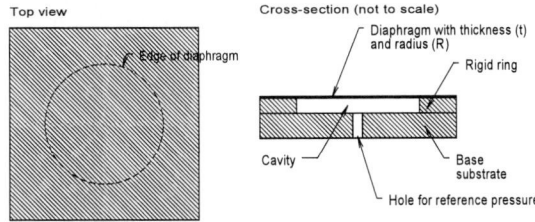

Figure 1: The top-view and the cross-section of the ceramic pressure sensor body (3D LTCC structure). (Schematic – not to scale)

The structure in Figure 1 was constructed for measuring pressures in the range from 0 to 100 kPa, and for the burst pressure of about 400 kPa. For these characteristics we designed the diaphragm with the thickness of 200 μm and with the radius of 4.5 mm. The depth of the cavity is about 100μm when the structure is designed for capacitive type of

pressure sensor, and about 200μm when the structure is designed for piezoresistance or resonant type of pressure sensor. The cross-section of the cavity within the 3D LTCC structure fabricated with LTCC materials is presented in Figure 2.

Figure 2: The cross-section of the ceramic pressure sensor body (3D LTCC structure).

The applied pressure on ceramic pressure sensors with flexible and deformable diaphragms deforms the diaphragm. The deformation is presented in Figure 3.

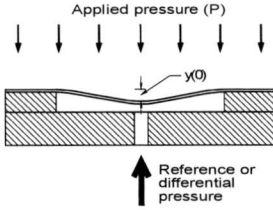

Figure 3: The cross-section of the ceramic pressure sensor body (3D LTCC structure) under applied pressure. (Schematic – not to scale)

The construction, the dimensions and the material properties of the sensor structure influence on the sensors' characteristics of all three types of thick-film pressure sensors. The dependence of the geometry and the material properties of the LTCC construction on the deflection of an edge-clamped deformable diaphragm under an applied pressure is described by equation (1)

$$y(r) = \frac{3P\left(1-v^2\right)\left(R^2-r^2\right)^2}{16\,E\,t^3} \quad (1)$$

where the deflection y at the position r from the centre of the diaphragm is a function of the applied pressure, P, the material characteristics (elasticity, E, and Poisson's ratio, v) of the diaphragm, and the dimensions (thickness, t, and radius, R) of the diaphragm (Figures 1 and 3).

Capacitive Ceramic Pressure Sensor

The capacitive ceramic pressure sensor is based on the fractional change in capacitance ($\Delta C/C$) induced by the applied pressure. The capacitance change is due to the changing distance between the electrodes of the capacitor [10-13]. These electrodes are within the cavity of the LTCC structure (Figures 1 and 4). The bottom electrode (Electrode 2) of the capacitor is on the rigid substrate and the upper electrode (Electrode 1) is on the deformable diaphragm. The areas of the electrodes and the

978-1-4244-4722-0/09 $25.00
© 2009 IMAPS-ITALY

distance between them define the value of the initial capacitance(C_0) of the pressure sensor. The distance between the electrodes (D) is calculated from cavity depth and the thickness of both electrodes.

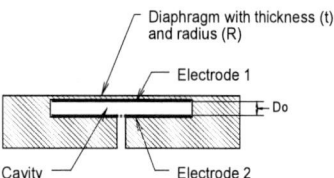

Figure 4: The cross-section of a capacitive ceramic pressure sensor without applied pressure. (Schematic – not to scale)

When the applied pressure is within the nominal range and when the deflection of the diaphragm y(r) is much smaller than both the thickness of the diaphragm and the distance between the electrodes then the capacitance between electrodes is given by equation (2) [12].

$$C(P) = \varepsilon_0 \cdot \varepsilon_r \cdot \int_0^R \frac{2 \cdot \pi \cdot r \cdot dr}{D_0 - y(r)} \qquad (2)$$

where C is the capacitance under an applied pressure P, ε_0 is the permittivity in vacuum, ε_r is the relative permittivity, R is the radius of the electrode, r is the current radius, D_0 is the distance between the electrodes at zero applied pressure and $y(r)$ is the deflection (described by equation (1)) at the current radius r when the pressure P is applied.

The fabricated test capacitive ceramic pressure sensor is shown in Figure 5. The construction and dimensions are presented in Figures 1 and 4. The air-gap capacitor has distance between the electrodes about 80 μm and the radius of the upper and bottom electrodes are 4.3 mm.

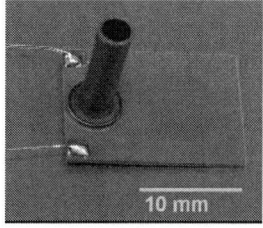

Figure 5: The sensing element made within a 3D-LTCC structure for capacitive ceramic pressure sensors.

The fabricated test samples have capacitances around 10 pF and the changes in this capacitance are of the order of a few fF. The calculated pressure sensitivities from the measured data (in the range from 0 to 100 kPa) are between 10 and 20 fF/kPa. The capacitive ceramic pressure sensor is the part of the electronic conditioning circuit with the frequency output (Figure 6). The typical output frequency is between 10 and 14 kHz, and depends on the applied pressure [13].

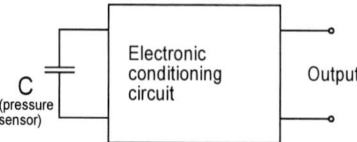

Figure 6: The capacitive sensing element is the part of the electronic conditioning circuit.

The power consumption of the sensing element for the capacitive ceramic pressure sensor is very low and depends mostly on the values of the operating frequency and voltage, and the capacitance of the sensing element. In our case the capacitance was 8 pF, the operating voltage is 1 V with the frequency of 10 kHz. These parameters resulting in power consumption of about 0.5 μW. If we reduce the values of any mentioned parameters the power consumption of the sensing element is lower. But in the same time the pressure sensitivity is lower and the problems with the quality of the output signal and also with the parasitic capacitance are more serious. On the other hand, if we design more stabile sensing element with higher pressure sensitivity and lower sensitivity on different interferences, than the values of mentioned parameters must be higher and the power consumption of the sensing element is also higher.

Piezoelectric Resonant Ceramic Pressure Sensor

The thick-film piezoelectric resonant pressure sensor is based on the piezoelectric properties of ferroelectric thick films that act as actuator/sensor on a deformable diaphragm [20]. The basic construction of the sensor is shown in Figures 1 and 7. The resonant sensor has an additional integrated device (circular thick-film piezoelectric actuator/sensor) for stimulating and sensing the oscillation of the diaphragm.

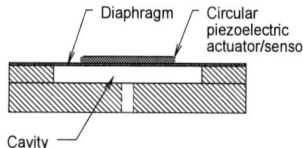

Figure 7: The cross-section structure of a piezoelectric resonant ceramic pressure sensor. (Schematic – not to scale)

The piezoelectric structure works as an actuator and generates stresses in the diaphragm, which induces the vibration of the diaphragm at its resonant frequency. In the same time this piezoelectric structure detects the vibration frequency. This resonant frequency of the circular edge-clamped diaphragm is given by equation (3).

978-1-4244-4722-0/09 $25.00
© 2009 IMAPS-ITALY

$$f_r = \frac{\lambda \cdot t}{R^2} \sqrt{\frac{E}{\rho(1-\nu^2)}} \qquad (3)$$

where f_r is the resonant frequency of the diaphragm; it is a function of the material characteristics (elasticity, E; density, ρ; and Poisson's ratio, ν) of the diaphragm, and the dimensions (thickness, t, and radius, r) of the diaphragm. The constant λ has value 0.412 for circular edge-clamped diaphragm [14,15].

The applied pressure bends the diaphragm and induces additional stresses in the diaphragm. By changing the mechanical tensile condition in the diaphragm the resonant frequency is shifted, which can be used as the output signal of the piezoelectric resonant pressure sensor.

The piezoelectric actuator/sensor is fabricated on the deformable diaphragm as a vertical thick-film structure, and consists of a bottom electrode, an active PZT (Pb(Zr,Ti)O$_3$) layer, and an upper electrode. The PZT material for this structure is a ferroelectric thick-film paste based on PZT 53/47 powder (PbZr$_{0.53}$Ti$_{0.47}$O$_3$). One of our previous investigations [14] of printed thick-film PZT layer on the alumina and LTCC substrates indicated that due to the interaction between the PZT layers and the LTCC substrates during firing the electrical and the piezoelectrical characteristics deteriorate significantly. In Table 1 the electrical parameters (ε' and tg δ) and the piezoelectric constants d$_{33}$ on different substrates are presented. For a comparison, the dielectric constant and the piezoelectric constant d$_{33}$ of a bulk PZT 53/47 ceramic are around 1000 and 200 pC/N, respectively [14].

Table 1: Electrical characteristics (dielectric constant ε'; dielectric loss tg δ; and piezoelectric constant d$_{33}$) of the PZT layers on alumina and LTCC substrates.

Characteristic	LTCC	Alumina
ε'	200	500
tg δ	0.015	0.020
d$_{33}$ (pC/N)	70	140

The fabricated test thick-film piezoelectric resonant pressure sensor is shown in Figure 8. The test samples have on the diaphragm a planar thick-film piezoelectric actuator/sensor with a radius of 2.3 mm and a thickness of about 50 μm.

The test sensing elements for piezoelectric resonant ceramic pressure sensors were fabricated and then tested in the pressure range from 0 to 100 kPa. The impedance spectrum, Z, and the phase, Θ, around the resonant frequency were measured with an HP4192A LF impedance analyser.

The sensing element is the part of the electronic conditioning circuit (Figure 9). The resonant frequencies of the test samples are shifted around 26 kHz, depending on the pressure. The

pressure sensitivities from the measured data are about 2.6 Hz/kPa [15].

Figure 8: The sensing element made with a 3D-LTCC structure for piezoelectric resonant ceramic pressure sensors.

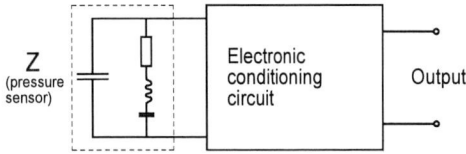

Figure 9: The piezoelectric resonant sensing element is the part of the electronic conditioning circuit.

The power consumption of the sensing element for the piezoelectric resonant ceramic pressure sensor is relatively low and depends mostly on the values of the operating frequency and voltage, and the impedance of the actuator/sensor device. The operating frequency of sensor is defined by the geometry and the material properties of the diaphragm. On the other hand the electrical characteristics of the actuator/sensor device depend on its dimensions and material properties, which are given by thick-film PZT material. The values of the dielectric constant ε' are relative low in comparison with bulk material and the layer on alumina substrate, while the dielectric loss tg δ are comparable (Table 1).

The test samples have the values of resonant frequencies about 26 kHz and the values of impedance about 16 kOhm. The value of the operating voltage is 1 V. These three parameters resulting in power consumption of about 70 μW. The resonant frequency and impedance are determined by the geometry and the material properties. Therefore only with the reducing value of the operating voltage we can reduce the power consumption of the sensing element. On the other hand, with the reducing this value the quality of the sensing element is lower and the sensitivity on different interferences is higher.

Piezoresistive Ceramic Pressure Sensor

A piezoresistive ceramic pressure sensor is based on the piezoresistive properties of the thick-film resistors that are screen-printed and fired onto the deformable diaphragm [5,6,16-18]. The piezoresistive ceramic pressure sensor has four thick-film resistors, which acts as strain gauges and translate a strain into an electrical signal. The sensing resistors are located on the diaphragm so that two are under tensile strain, and two are under compressive strain. These four resistors are electrically connected in a Wheatstone-bridge configuration and excited with a stabilised bridge voltage. The Wheatstone-bridge is the part of the electronic conditioning circuit (Figure 10).

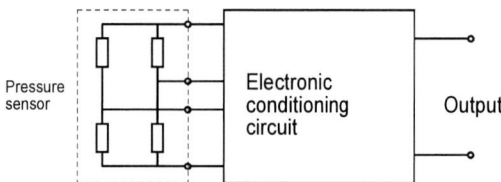

Figure 10: The piezoresistive sensing element is the part of the electronic conditioning circuit.

The test piezoresistive ceramic pressure sensor is presented in Figure 11.

Figure 11: The sensing element made with a 3D-LTCC structure for piezoresistive ceramic pressure sensor

For the fabrication of the sensing resistors for piezoresistive pressure sensors can be used different thick-film resistor materials with different properties e.g. sheet resistivities (R_{sh}), gauge factors (GF) and current noise. Some results of our previous investigations [17,18] of thick-film resistors for strain-gauge applications are presented in Table 2.

Table 2: Resistivities, gauge factors, and current noise of some thick-film resistors with dimensions 1.6×1.6 mm^2 [17].

Denotation of resistor material	R_{sh} [Ω/sq.]	GF	Noise [μV/V]
C2	100	5	0.04
C3	1k	7	0.06
C4	10k	10	0.15
C5	100k	13	0.55
C6	1M	15	1.80
E4	10k	22	1.25

In the Table 2 the results of five resistor materials of resistor series denoted C and one resistor material denoted E, which is specially designed for strain-gauge application are presented. The gauge factors from the same resistor series increase with increasing sheet resistivity. The current noise increases with increasing resistance values of the resistors and decreases with increasing dimensions. The highest gauge factor has resistors made with resistor material denoted E. Unfortunately the current noise of these resistors is also relatively high. This resistor material is also a few times more expensive than resistor material denoted C.

The resistances of the sensing resistors connected in a Wheatstone-bridge and the bridge voltage have direct influence on the power consumption of the sensing element for the piezoresistive ceramic pressure sensor. The lower bridge voltage causes lower current and therefore lower power consumption, but the value of the output signal is also lower. The power consumption can be also reduced by the use resistor with higher resistance (higher sheet resistivity). In this case the sensitivity is slightly higher but the signal/noise ratio is lower. The lower bridge voltage and/or the higher resistance cause another problem – sensor becomes more susceptible to the parasitic effects and other interferences.

Typical value of resistances of the sensing resistors is 10 kOhm and typical value of the bridge voltage is 5 V. These parameters resulting in power consumption of about 2.5 mW.

Conclusions

We investigated three different types of ceramic pressure sensors: a capacitive sensor, a piezoelectric resonant sensor, and a piezoresistive sensor. All the pressure sensors were made on LTCC structures consisting of a circular edge-clamped deformable diaphragm. We studied and compared the design issues for low energy consumption applications only of sensing elements. The power consumption of the signal conditioning and the data processing is usually more important for low-energy design, but they were not the subject of this work.

The most suitable ceramic pressure sensor for low energy consumption applications is capacitive sensor followed by piezoelectric resonant sensor. In both cases the power consumption is low. The power consumption is mainly defines by sensor's construction and applied materials. The power consumption can be slightly reduced by the construction and the materials approach but it is not an easy task because there are many interdependent relationships between power consumption and other sensor's characteristics.

In the case of piezoresistive ceramic pressure sensor the energy consumption can be easy controlled by the values of Wheatstone-bridge

978-1-4244-4722-0/09 $25.00
© 2009 IMAPS-ITALY

resistances. These values can be from few hundred ohms to few hundred kOhms. For a comparison, the Wheatstone-bridge resistances of silicon pressure sensors are typically between few hundred and few thousand ohms.

Acknowledgement

The financial support of the Slovenian Research Agency in the frame of the project L2-0186 is gratefully acknowledged.

References

[1] N. Maluf, K. Williams, "An Introduction to Microelectromechanical System Engineering", Artech House, Inc., Norwood, 2004.

[2] D. Belavič, M. Hrovat, M. Pavlin, M. Santo Zarnik, "Thick-film technology for Sensor Applications", Informacije MIDEM, Vol. 33, pp. 45-48, 2003.

[3] M. Pavlin, D. Belavič, M. Santo Zarnik, M. Hrovat, M. Možek, "Packaging technologies for pressure-sensors". Microelectron. int., Vol. 19, pp. 9-13, 2002.

[4] M. Pavlin, F. Novak, "Yield enhancement of piezoresistive pressure sensors for automotive applications", Sensors and Actuators, A, Physical, Vol. 141, No. 1, pp. 34-42, January 2007.

[5] D. Belavič, M. Santo Zarnik, M. Hrovat, S. Maček, M. Pavlin, M. Jerlah, J. Holc, S. Drnovšek, J. Cilenšek, M. Kosec, "Benchmarking different types of thick-film pressure sensors", Proceedings of the IMAPS/ACerS 2007, 3rd International Conference and Exhibition on Ceramic Interconnect and Ceramic Microsystems Technologies (CICMT), Denver, Colorado, USA, pp. 278-285 April 23-26, 2007.

[6] D. Belavič, M. Hrovat, J. Holc, M. Santo Zarnik, M. Kosec, M. Pavlin. "The application of thick-film technology in C-MEMS", Journal of electroceramics, vol. 19, no. 4, pp. 363-368, 2007

[7] L.J. Golonka, A. Dziedzic, J. Kita, T. Zawada, "LTCC in microsystem application", Informacije MIDEM, Vol. 32, No.4, pp. 272-279, 2002.

[8] T. Maeder, C. Jacq, H. Briol, P. Ryser, "High-strength Ceramic Substrates for Thick-film Sensor Applications", Proceedings of the 14th European Microelectronics and Packaging Conference, Friedrichshafen, Germany, pp. 133-104, 2003.

[9] U. Partsch, D. Arndt, H. Georgi, "A new concept for LTCC-based pressure sensors", Proceedings of the IMAPS/ACerS 2007, 3rd International Conference and Exhibition on Ceramic Interconnect and Ceramic Microsystems Technologies (CICMT), Denver, Colorado, USA, pp. 367-372, April 23-26, 2007.

[10] R. Puers, "Capacitive sensors; when and how to use them", Sensors and Actuators A, Vol. 37, pp. 93-105, 1993.

[11] C. B. Sippola, C. H. Ahn, "A thick film screen-printed ceramic capacitive pressure microsensor for high temperature applications", Journal of Micromechanics and Microengineering, Vol.16, pp. 1086-1091, 2006.

[12] D. Belavič, M. Santo Zarnik, M. Jerlah, M. Pavlin, M. Hrovat, S. Maček, "Capacitive thick-film pressure sensor : material and construction investigation", Proceedings of the XXXI International Conference of IMAPS Poland 2007, Rzeszóv, Krasiczyn, Poland, September 23-26, pp. 249-253, 2007.

[13] D. Belavič, M. Santo Zarnik, S. Maček, M. Jerlah, M. Hrovat, M. Pavlin, "Capacitive pressure sensors realized with LTCC technology", Proceedings of the 31st International Spring Seminar on Electronics Technology (ISSE 2008), Reliability and life-time prediction, Piscataway: IEEE, Budapest, Hungary pp. 271-274, 7-11 May, 2008.

[14] D. Belavič, M. Santo Zarnik, J. Holc, M. Hrovat, M. Kosec, S. Drnovšek, J. Cilenšek, S. Maček, "Properties of lead zirconate titanate thick-film piezoelectric actuators on ceramic substrates", International journal of applied ceramic technology, vol. 3, no. 6, pp. 448-454, 2006.

[15] M. Santo Zarnik, D. Belavič, S. Maček, J. Holc, "Feasibility study of a thick-film PZT resonant pressure sensor made on a prefired 3D LTCC structure", International journal of applied ceramic technology, Vol. 6, No. 1, pp 9-17, 2009.

[16] C. Grimaldi, T. Maeder, P. Ryser, S. Strässler, "Critical behaviour of the piezoresistive response in RuO2-glass composites", Journal of Physics D: Applied Physics, Vol. 36, pp. 1341-1348, 2003.

[17] D. Belavič, M. Hrovat, M. Pavlin, S. Gramc, "Low-cost thick-film strain gauge applications", Proceedings of the 13th European Microelectronics and Packaging Conference, Strasbourg, France, pp. 103-108, 2001.

[18] M. Hrovat, D. Belavič, Z. Samardzija, J. Holc, "A characterisation of thick film resistors for strain gauge applications", Journal of Materials Science, Vol. 36, No. 11, pp. 2679-2689, 2001.

Reliability Comparison of Aluminum Redistribution based WLCSP Designs

Umesh Sharma, Ph.D., Harry Gee, and Phil Holland

California Micro Devices, Inc., Milpitas, CA

umeshs@cmd.com, harryg@cmd.com, and philh@cmd.com

Abstract

Redistribution layer (RDL) WLCSP technology is often used in area array ICs where both high performance and low cost are important considerations. RDL can be implemented in two ways: a) as topmost level metal in the wafer fab (Wafer fab–RDL), and b) as an additional metal layer during bumping operation (Assembly-RDL). Selection of an appropriate RDL technology and its optimization are necessary for building reliable products. In this paper, we describe the structural differences between the two technologies and compare their thermal reliability performance. Aluminum is used as the redistribution layer material for both technologies. To explain the experimental findings, we construct a finite element model (FEM) of the experimental IC which is packaged as CSP area array and mounted on a FR4 board. We investigate the critical role of process parameters such as UBM stack composition, and thicknesses of various PBO layers in determining the overall device reliability. Finally, we also explore the reliability improvements obtained as a result of using an underfill layer during board assembly.

Key words: WLCSP, RDL, UBM, Thermal Shock, Reliability

Introduction

WLCSP technology is being rapidly adopted as the package technology of choice for ICs in portable personal electronic products. Recently, nanometer scale CMOS products have utilized area array WLCSP packaging for both performance and cost reasons. However, selection of appropriate WLCSP technology and its optimization are necessary to build reliable products. Nanometer scale CMOS technology has multiple levels of metal layers interconnects that are separated by low temperature deposited oxide layers. This fragile interconnect structure is susceptible to delamination, cracks, and even physical rupture during routine thermal reliability testing such as thermal shock and temperature cycling. The level of thermo-mechanical stress and subsequent damage introduced in the interconnect layers depends on the process integration scheme chosen for implementing the RDL-CSP technology.

In this paper we discuss the reliability of RDL technologies based on Aluminum interconnects. After a brief description of the experimental methods and devices, we discuss two methods of implementing RDL in CSP devices. For each RDL method, we numerically model the entire device structure using ANSYS modeling software. Simulation results are discussed and experiments are designed to improve the device reliability.

Description of RDL Technologies

The two RDL technologies are shown schematically in Fig. 1. In both cases, the silicon wafer processing is the same up to the last metal layer. For Wafer fab-RDL, the topmost metal layer (Aluminum) is also used as the RDL interconnect layer. After metallization etch, a passivation layer is deposited and I/O pads are lithographically defined. A re-passivation layer of polyimide is applied to seal the device surface. UBM is then deposited, patterned, and followed by the solder ball drop CSP process sequences. This process integration scheme is schematically shown in the diagram on the left side of Fig. 1.

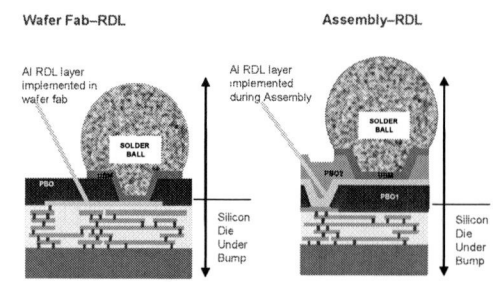

Figure 1: Structural Comparison of Wafer-Fab RDL and "Assembly-RDL" CSP devices.

978-1-4244-4722-0/09 $25.00
© 2009 IMAPS-ITALY

The Assembly-RDL integration approach uses one less metal layer in the wafer fab. As shown in the right side of Fig. 1, after pad definition, a re-passivation layer labeled PBO1 is applied followed by Aluminum RDL layer deposition and definition. The RDL layer is followed by a second re-passivation layer labeled PBO2. PBO2 is lithographically defined to open areas where UBM is deposited. Solder spheres are then dropped through a stencil mask and the wafers go through the conventional reflow cycle to complete the CSP process.

In essence, the ball structure in the Assembly-RDL process is situated higher from the wafer surface and is cushioned by an extra layer of PBO under the Aluminum interconnect and the periphery.

Test Vehicle and Experimental Procedures

ICs are fabricated in state of the art 0.18um CMOS, single poly, 5/6 metal technology followed by either Wafer Fab-RDL or Assembly-RDL process steps. In both cases, we use Aluminum as the redistribution interconnection material and the I/Os are arranged as 10 row x 10 column area array with a 0.5mm pitch. The ICs are assembled on FR4 boards and then subjected to thermal shock (-40°C to 75 °C; 45 min dwell time at each temperature) and temperature-cycle tests (-40 °C to 125 °C; 5min ramp and 10 min dwell time). Electrical tests are performed at several intermediate cycles.

Several devices exhibit functional electrical failure after thermal tests. I/O failures are observed predominantly around the chip periphery. Emission microscopy reveals hot spots related to the circuits physically located near the I/O structures as shown in Fig. 2. Systematic de-processing of the silicon shows cracks in the hot spot regions, underneath the ball structure.

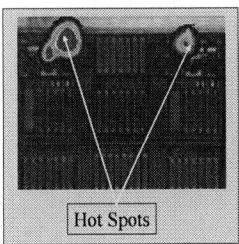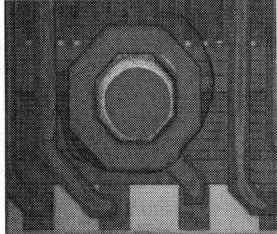

Figure 2: Emission Microscopy of a failed unit. The failed or hot spot regions are directly connected to the I/O balls. After de-processing, the failed region is examined carefully and a crack around the periphery of the I/O pad can be seen. The crack is in the region where the ball existed before de-processing.

Cracks are observed in the inter-metal oxide layers around the periphery of failed solder balls. Fig. 2 (right) is a top level microphotograph of a failed I/O section. The crack is highlighted by red arrows. Fig. 3 shows the SEM of a failed device after thermal shock (left) and the SEM corresponding to another device that failed after temperature cycling (right). Both photos show fracture of the inter-metal insulator layer and thus the failure signatures appears to be the same in both cases.

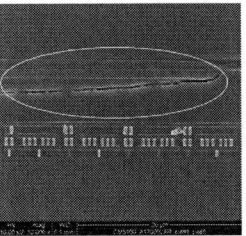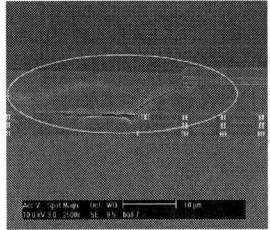

Figure 3: SEM X-section of a failed unit after thermal shock (left). SEM of another failed unit after temperature cycling (right). Notice that the cracks are deep in both SEMs and extend through several layers of inter-metal dielectric.

Theoretical Analysis:

Numerical Modeling with ANSYS

To understand the failure mechanism, we construct a finite element model (FEM) of the IC packaged as a (10 x 10) area array and mounted on a FR4 board. Figure 4 shows a unit cell of the WLCSP structure. Finite Element Model is generated using APDL (Ansys Parametric Design Language) parametrically such that the dimensions, material properties, and loading conditions can be changed for subsequent simulations. Periodic boundary conditions are applied on the finite element model boundaries to model physics with the representative structure. Details of the device mounted on a FR-4 board are shown in the ANSYS model diagram below (Fig. 5). It is assumed that for a specified thermal test, the stress free state is achieved at the highest temperature value. Stress vectors are calculated for the lowest temperature value and plotted as contours or path plots.

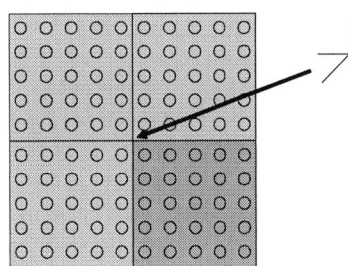

978-1-4244-4722-0/09 $25.00
© 2009 IMAPS-ITALY

Figure 4: A schematic model of the 10 x 10 array IC, partitioned with quad symmetry around the origin.

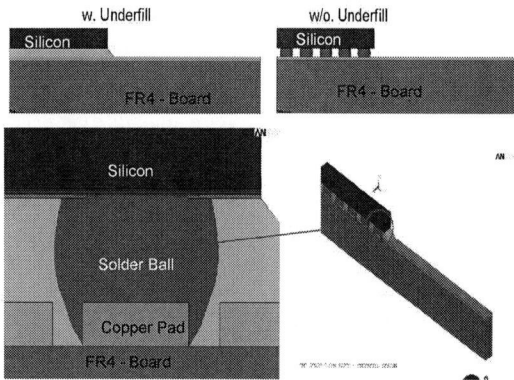

Figure 5: Details of the solder ball construction highlighting various materials involved with different colors. Two cases of special interest – ICs with and without underfill are shown in the figure.

The simulations assume that at 75°C the device is in a completely stress free state. As the temperature is reduced from 75°C to -40°C, the material mismatch between the IC and the FR4 board results in varying amounts of stress in each material layer. Stress and deformation vectors can be plotted in the two dimensional plane of the IC. If desired, stress can be computed for each layer of the device structure. In Fig. 6 we summarize the stress calculations for the condition after temperature shock test.

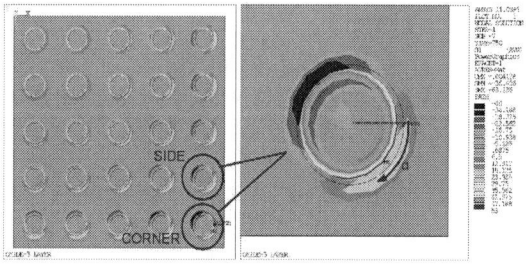

Figure 6: The calculated stress magnitude in oxide layer 5 (between Metal 5 and Metal 6 layers) increases from center to the edge balls. A magnification of the corner ball stress plot is presented on the right. The color scale corresponds to the magnitude of stress vector ranging from Blue (compressive) to Red (Tensile). Stress contour plots can be generated for any given distance from the center of the ball. α in the figure is the angle extended from the horizontal for constructing path plots.

In the next few paragraphs, we present path plots around the edge to extract stresses along this

critical path and compare results for various simulation conditions. In particular, the role of an underfill layer underneath the CSP device is examined in detail. As illustrated by the simulation results in Fig. 7, the tensile stress value changes along the periphery of the solder ball. The stress initially increases, becoming more tensile and then reduces until the polarity changes and it becomes compressive instead of tensile. The behavior of the device with underfill is remarkably different. Regions of compressive stress are replaced by tensile stress and vice versa. The stress magnitude is also lower for the underfill case. It is quite interesting to examine the stress magnitude and deformation produced in the solder balls for the two cases. As the temperature is lowered, the stress increases in magnitude across the entire IC. The stress is clearly non uniform from the edge to the center of the chip. The edge balls are severely deformed for the "no underfill" case. The solder balls are also twisted and elongated from the attachment to the FR4 board. This observation suggests that this "twist and elongation" mechanism will create a physical separation at the weakest point in the device structure. The weakest point for products manufactured using advanced CMOS processes is often the inter-metal dielectric.

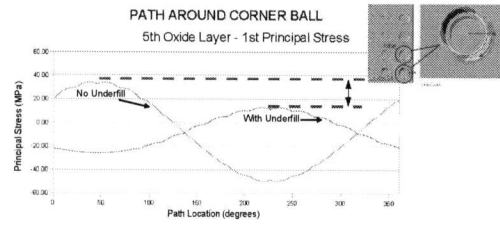

Figure 7: Path plot of 1ˢᵗ principal stress around the corner solder ball as shown in the inset. The red curve is the simulation result for a device structure that has no underfill and the blue curve is for a device structure with underfill.

Finally, we show the path of maximum stress along the edge solder balls in Fig. 8 and compare it with the observed crack in the device structure after thermal tests. Clearly, the crack location and length coincides with the location and length of the high stress contour region confirming our theoretical simulation model.

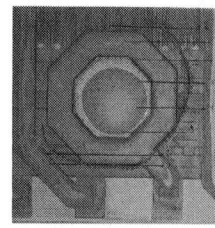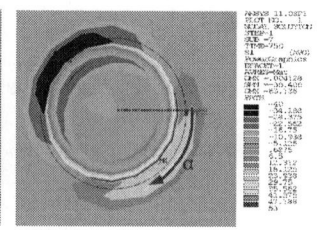

978-1-4244-4722-0/09 $25.00
© 2009 IMAPS-ITALY

Figure 8: Crack observed in the inter-metal oxide layers under the solder ball and the contours of maximum stress around the solder ball region in Oxide layer 5. The maximum tensile stress is roughly at the same location as the fracture.

Comparison of Assembly-RDL and Wafer fab-RDL

Numerical simulations of the Assembly-RDL process are carried out in a manner similar to that described in the previous paragraphs. The structure described previously is inserted in the ANSYS modeling software. All other parameters and experimental conditions are kept the same as before.

In the next few paragraphs, we compare the simulation results of the two RDL processes. Path plots around the edge of the corner ball are shown in Figures 9 and 10. In general, we can conclude the following from our simulation studies:

- Assembly-RDL structure provides a thick layer of PBO between solder ball structure and silicon die and is expected to be more robust than Wafer fab-RDL. PBO has a higher CTE and lower Young's modulus vs. deposited silicon dioxide/nitride and therefore expands/contracts during thermal shocks

- Increasing the thickness of PBO reduces the stress in the dielectric layers underneath. Devices with thicker PBO layers should perform better.

- Underfill improves the thermal reliability of both RDL technologies.

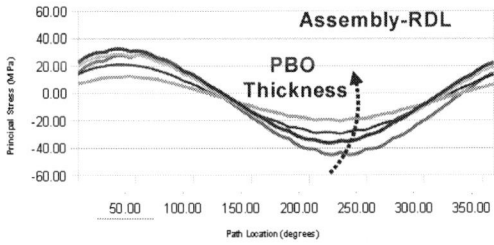

Figure 9: Stress levels reduce with increasing Repassivation-2 (PBO2) thickness

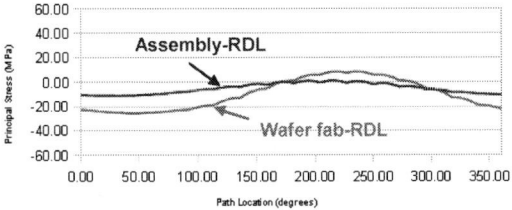

Figure 10: Underfill reduces the oxide stress levels for both Assembly-RDL and Wafer fab-RDL processes

Experimental Results

We performed extensive temperature cycle tests and thermal shock tests on the WLCSP product test vehicle described earlier. This product is designed in 0.18um CMOS technology with 1 poly and 6 metal layers. The sixth layer of metal (Aluminum) is also used as RDL for this product. After passivation opening, a re-passivation PBO layer is applied followed by conventional WLCSP processing. Devices are mounted on FR4 board and a few devices are assembled with underfill applied between the CSP and the board.

Fig. 11 summarizes the temperature cycle performance of the assembled parts, with and without underfill. As expected, the parts with underfill are more robust and fail after several hundreds of cycles. The first failure is noticed at 617 cycles for the parts with underfill.

The results for thermal shock were similar. Without underfill, we noticed device failures at less than 12 thermal shock cycles. The failures increased to almost 2% after 75 cycles. The parts with underfill showed no failures for several hundreds of thermal shock cycles.

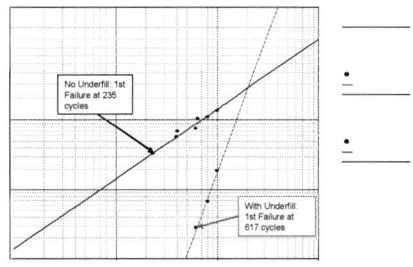

Figure 11: Probability Weibull plot of temperature-cycle failures for the 10 x 10 array WLCSP product. Underfill dramatically improves the reliability of the device.

In Tables 1 and 2, we compare the thermal shock and temperature cycle performance of various processes discussed in the paper. Thermal shock tests were conducted up to 90 cycles. Parts assembled without underfill start failing within a few

978-1-4244-4722-0/09 $25.00
© 2009 IMAPS-ITALY

cycles and approximately 2% parts fail after 75 cycles. The Assembly-RDL parts on the other hand did not fail up to the tested limits. The temperature cycle behavior is quite similar. Samples with underfill are able to withstand higher number of temp-cycle failures. The Assembly-RDL samples are very robust and not a single failed up to 1000 cycles.

Table 1: Thermal shock testing results

Temp: -40C to 75C, 45m dwell time

Samples	Sample Size	Cycles/Failure %			
		12	25	35	75
Fab-RDL (no underfill)	346	0.6%	1.4%	1.4%	2.0%
Fab-RDL (w underfill)	246	No failures for > 90 cycles			
Assembly-RDL	90	No failures for > 90 cycles			

Table 2: Temperature cycling test results

Temp: -40C to 125C, 5 min ramp; 10min dwell time

Samples	Sample Size	Parts Failed @ # of Cycles				
Fab-RDL No underfill	147	6@235	4@397 2@407	1@616 5@642	1@804	4@1000
Fab-RDL W Underfill	245			1@617	1@800	3@1000
Assembly-RDL	120	No failures > 1000 cycles				

Conclusions

In this paper we discussed the thermal reliability performance of large area WLCSP arrays. Thermal stress applied to devices packaged as WLCSP products can result in fracture of the inter-metal dielectric layers. For large arrays, underfill is required to minimize thermal shock and temperature-cycling induced failures. We have also demonstrated that implementing the RDL as part of the WLCSP operation instead of wafer fab operation results in improved thermal reliability performance.

Furthermore, the Assembly-RDL process is very robust and does not require underfill to meet even the most stringent reliability criteria. Thicker PBO films and UBM layers result in further improvements of device reliability.

Simulation results agree with experiments, confirming our hypothesis and validating the model. The model is also successful in explaining the improved reliability results for the Assembly-RDL process.

Acknowledgments

The authors would like to thank CMD management for their support of this work.

978-1-4244-4722-0/09 $25.00
© 2009 IMAPS-ITALY

Miniaturization of Printed Wiring Board Assemblies into System in a Package (SiP)

Steven G. Rosser, Irving Memis and Harry Von Hofen
Endicott Interconnect Technologies
1701 North St.
Endicott, NY 13760
Ph: 607-755-2229, Fax: 607-755-6082, E-Mail: rossersg@eitny.com

Abstract

The demand for system miniaturization in many applications has lead to efforts to put all or most of the functions on a single chip. However, there are many situations where this is not possible or cost prohibitive. Memory uses large amounts of chip area and several different memory types may be needed to fulfill the functional requirements. In many cases, the need for analog and digital functions may make consolidation on a single chip impossible. An alternate approach is to preserve the proven functional design and miniaturize at the package level to achieve the desired space savings. The approaches explored in this publication include eliminating active chip packages by directly attaching the chip to the SiP with flip chip technology. Additionally, the area devoted to passive components can be greatly reduced by embedding many of the capacitors and resistors. In some instances, the connector systems that were consuming large amounts of space in the traditional Printed Wiring Board (PWB) assembly can be reduced with a small pitch connector system. This PWB assembly can then be transformed into a much smaller SiP with the full surface area on both sides of the package effectively utilized by active and passive components. A further benefit of the SiP is a major reduction in total height. Two specific cases will be detailed and the size reductions shown. The concept of a SiP index will be introduced to show how the SiP area compares to active die area. The miniaturized SiP with its reduced package size and demand for passives requires a high wireability package with embedded passives and excellent communication from top to bottom. Endicott Interconnect Technologies has a Core EZ^{TM} package [1] that meets these requirements. The details of the package design parameters and package electrical performance are demonstrated.

Key words: package, miniaturization, assemblies, system, MCM, embedded passives

Introduction – PWB Assemblies – Candidates to Migrate to SiP

Traditional PWB assemblies logically integrate a variety of functions to achieve an end result that will communicate to the rest of the system. These components are complemented by passives, mainly resistors and capacitors, to achieve desired waveforms and filter out electrical noise due to circuit switching. Since the advent of Surface Mount Technology (SMT), most components can be assembled directly to pads on the surfaces of the PWB. In order to keep the total height down and ease assembly, typical designs have most active components on one surface which we will call the top surface and passive components on the opposite surface which will be called the bottom surface. The assembled PWB top surface is typically more completely populated than the bottom and this unbalance leads to a larger area for the product

than would have been achieved with a totally balanced placement of components on the two sides. However, the totally balanced placement would result in a much thicker assembled PWB and have significant system impact. Therefore, our base point for this study is a PWB with active components and connectors on the top and passives on the bottom. Two different designs are quantified; Example A and Example B. Both have a mixture of analog and digital active circuits which makes integration into a single chip extremely difficult. In both cases, a variety of individual functions are represented in the digital active components further discouraging chip integration. The distribution of active components, passive components and connectors for these two cases is shown in Table 1.

978-1-4244-4722-0/09 $25.00
© 2009 IMAPS-ITALY

Table 1 - Component Counts for Examples A and B

	Total Components	Active Components	Passive Components	Connectors
Example A	310	17	277	16
Example B	705	49	655	1

Example B is a more efficient PWB assembly because the connectors have already been integrated and the passives in Example B have been miniaturized. In Example A, there was little motivation to miniaturize the passives because the back side had more than enough room to handle the passives without miniaturization.

When these assemblies are redesigned into a SiP, the component count drops significantly with Example A dropping to 99 components and Example B dropping to 136 components; a 3x to 5x reduction. The area reduction is even more dramatic; 8x to 9x. Since the SiP has eliminated all the very tall components, the total height is reduced and the volume reduction is 17x to 21x. The next sections will show the technologies that enabled these improvements and quantify the gain achieved with each element.

SiP Features for Size Reduction

The SiP has four features that can significantly reduce the size of the final assembly. The first is to remove the active components from their package and assemble them directly to the SiP. The most efficient approach is to use Flip Chip Assembly (FCA). Many components have not been designed for FCA but if desired they can be converted to make FCA possible. The approaches to achieve this will be described. The second key item is to embed into the PWB as many of the passive components as possible. Resistors with values ranging from 15 ohms to 30,000 ohms can be efficiently embedded. Capacitors that are used for noise suppression (bypass / decoupling) and have values up to 0.1 microfarads (uf) can also be embedded. The SiPs described here will use a Passive Core that incorporates both the resistors and capacitors. The third key item is balancing the component area on both the top and bottom layers. With many passives eliminated and the active components in FCA format, components can be freely placed on top or bottom with little impact on the total height of the SiP assembly. The fourth item is efficiently using connector area. A fine pitch connector that is placed near the edge of the part minimizes the loss of functional area.

FCA

Components that use Flip Chip (FC) Packaging are the easiest to accommodate and can be shipped to the SiP assembly site after dicing. If the chip was meant to be assembled to the package with wire bonding, there are two approaches that can make them compatible with FCA. Stud Bumping [2] [3] with solder reflow can be utilized if the pad pitch on the chip is large. A good rule of thumb for Stud Bumping with solder reflow is a minimum pitch of 115 um between pads. The stud bump is equivalent to applying a wire bond (gold or copper) to the chip pad and then shearing off the wire. Typical size of the bump is 50 um diameter and 75 um high. Figure 1 shows a top view of chip pads with stud bumps applied and ready for assembly.

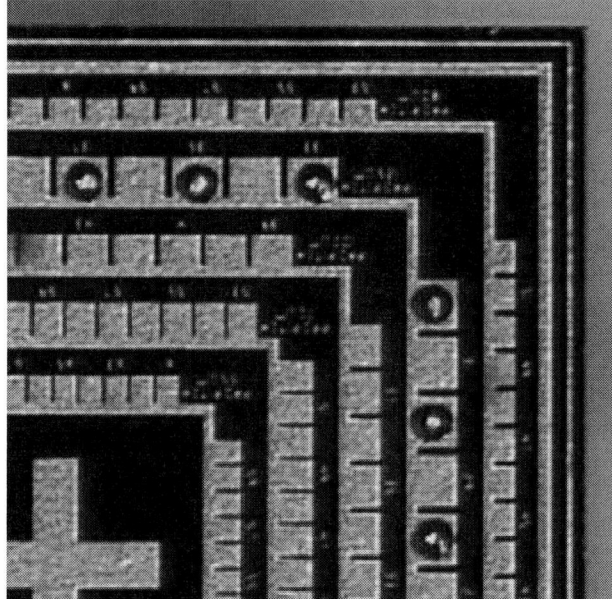

Figure 1 - Stud Bumps on Chip

The solder supply on the substrate is used to connect to the stud bump and concern for solder bridging limits the minimum pitch for this approach. The second approach is to redistribute the pads on the top surface of the chip into a pattern that uses typical Flip Chip technology. This is achieved with one or two thin film layers using semiconductor materials such as polyimide insulation and aluminum-copper conductors. The top surface then has a solder bump applied in the same manner as a chip intended for Flip Chip assembly. This approach is not limited by pad pitch and is the preferred approach for high volume production. However, the investment in design, wafer masks and wafer purchase makes the cost prohibitive for prototypes. Therefore, the wire

978-1-4244-4722-0/09 $25.00
© 2009 IMAPS-ITALY

bond chips for the prototypes may have a mixture of stud bump and redistributed chips with the intention to have all the chips redistributed for the production phase. The significant reduction in area consumed by the active components, about 7x, is achieved by converting from Packaged parts to Flip Chip as noted in Table 2. In all cases, assembly required spacing was included in the component areas.

Table 2 - Area Comparison for Active Components- Packaged vs. Flip Chip

	PWB Packaged Component Area (sq mm)	SiP Flip Chip Die Area (sq mm)	Area Reduction Ratio
Example A	3598	495	7.3
Example B	13,222	1,989	6.6

Passives

Capacitors, resistors and inductors represent the majority of the component count on these assemblies.

The inductors were left in component format since there are a small number of them and integrating them into a package would not be efficient. Many of the capacitors are used for bypass or power supply decoupling and are fairly low in value. Capacitance material capable of providing 50 nanofarads (nf) per square inch has been proven feasible [4]. When the capacitor is embedded in the substrate, the impedance from the active device to the supporting capacitor can be much lower than with a discrete SMT capacitor. Therefore, a much lower capacitor value can provide the required filtering. The designs shown here use a 40x factor for effectiveness of embedded capacitance vs. discrete SMT capacitors based upon embedded capacitance results reported to the industry [5]. Example A requires 4.6 embedded layers to replace 9.1 uf of discrete capacitance. Example B requires 2.9 layers to replace 25.7 uf of discrete capacitance. The passive core technology (Fig. 2) developed by Endicott Interconnect Technologies is capable of providing up to 6 layers of embedded capacitance and could be extended further. Table 3 shows the impact on component count and area by embedding capacitors up to a value of 0.1 uf in the SiP.

Via Structures			Layer	CoreEZ Thin Build-Up 3-8-3	Estimated Thickness (um)	Potential Function
			Top		12	Mounting / Signal / Power
				Driclad Filled RCC	32	
Stacked			2T		12	Signal / Power
(if needed)				Driclad Filled RCC	32	
			1T		12	Signal / Power
				Driclad Filled RCC	34	
			CL1		18	Power / Signal
				Embedded Capacitance	12	
			CL2		6	Power
				Embedded Capacitance	12	
			CL3		6	Power
				Embedded Capacitance	12	
PTH			CL4	Embedded Resistance	6	Power / Resistance
				Thermount	110	
			CL5	Embedded Resistance	6	Power / Resistance
				Embedded Capacitance	12	
			CL6		6	Power
				Embedded Capacitance	12	
			CL7		6	Power
				Embedded Capacitance	12	
			CL8		18	Power / Signal
				Driclad Filled RCC	34	
			1B		12	Signal / Power
				Driclad Filled RCC	32	
			2B		12	Signal / Power
				Driclad Filled RCC	32	
			Bottom		12	Mounting / Signal / Power
				Total	522	
				Core (Incl cu)	254	

Figure 2 - Cross - Section of SiP with Passive Core

Table 3 - Capacitor Count and Area Impact by Embedding Capacitors in SiP

	PWB Capacitor Count	SiP Capacitor Count	Capacitor Count Ratio	PWB Capacitor Area (sq mm)	SiP Capacitor Area (sq mm)	Capacitor Area Reduction Ratio
Example A	200	71	2.8	467	269	1.7
Example B	490	68	7.2	647	251	2.6

One notes that the area reduction ratio is lower than the component count ratio because the larger capacitors cannot be embedded.

The 40x factor was experimentally developed based on reconfigured systems performing acceptably in functional mode. It is a good starting point for assessing miniaturization feasibility. However, it is often necessary to characterize the impedance profile of the power delivery network (PDN) as a function of frequency. Required impedance and bandwidth are often known.

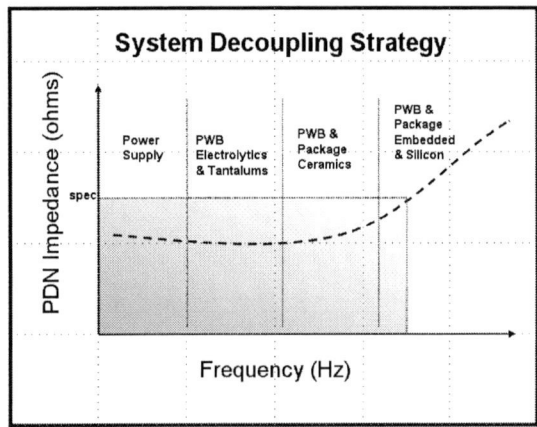

Figure 3 - System Decoupling Strategy

System decoupling, in the form of bypass capacitors on the PWB, package, and silicon, is then utilized to bring the PDN impedance within specification. Electrolytic bulk capacitors (high C) are often employed on the PWB to extend the bandwidth into the MHz range. Tantalum capacitors provide a mid-frequency boost and high frequency (low ESL) ceramic capacitors further extend the PDN bandwidth. The equivalent series inductance (ESL) of even these "high-frequency" SMT capacitors is still ineffective in the bandwidth beyond 1 GHz, even when many capacitors are placed in parallel to reduce the effective impedance.

Given the speeds of today's IC's, a PDN must supply more than voltage at DC. Figure 3 depicts a common system decoupling strategy. While a power supply will typically deliver voltage at low impedance, its bandwidth usually only extends to the KHz range.

The PDN can be characterized in the design phase by modeling with simulators such as Ansoft's SIWave. It is possible to extract the complete PDN from the physical design database. This extraction can include all internal power planes, power vias, traces, embedded components, discrete components, and interconnect arrays for both die and PWB attach. Figure 4 shows a side view of a typical PDN extraction done in SIWave. Once this extraction is accomplished, s-parameters and impedance can be calculated directly and compared to the specification. This technique provides a more accurate determination of the potential to embed surface components and realize miniaturization.

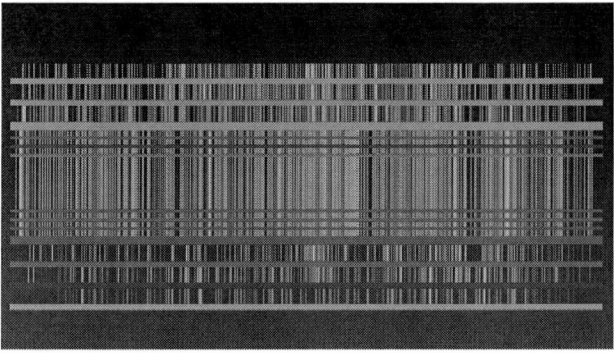

Figure 4 - PDN Extraction

Embedded capacitance tends to be most effective at higher frequencies. Given the constraints of the currently available embedded materials, the maximum capacitance density is usually not sufficient to provide significant bulk values of capacitance. Discrete SMT capacitors are still necessary for low and mid-frequency impedance reduction. Figure 5 shows simulated PDN impedance as a function of frequency for a one inch package. Four cases are examined; 1) with no decoupling capacitors, 2) with one decoupling capacitor, 3) with 201 decoupling capacitors and 4) with embedded capacitance and no decoupling capacitors. At higher frequencies, the inductive parasitics associated with SMT devices and their

978-1-4244-4722-0/09 $25.00
© 2009 IMAPS-ITALY

termination become limiting. For this reason, large quantities of SMT decaps are often populated to reduce the effective inductance. This is counter-productive to system miniaturization. Due to the distributed nature of embedded capacitance, the inherent impedance and that associated with connection are significantly lower than SMT devices. Therefore, replacing SMT decaps with embedded capacitance at 40:1 or a greater capacitor value ratio improves high frequency performance and miniaturizes the part at the same time.

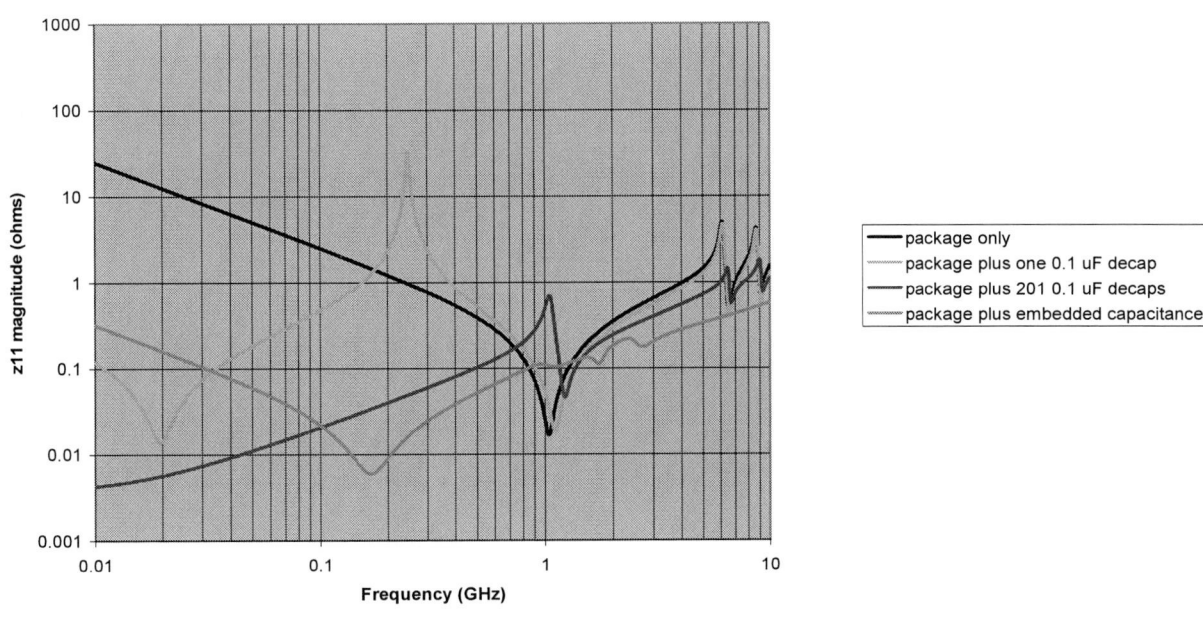

Figure 5 - PDN Impedance vs. Frequency as a Function of Capacitor Configuration

Embedded resistor technology can use either thin film materials that are applied on the copper foil or screened resistor material that can be applied at any level. The approach chosen in these examples is thin film material and it is incorporated into the center layer of the core. The top of the core uses 25 ohm per square material and the bottom of the core uses 250 ohm per square material. This combination enables resistor ranges from 15 ohms through 30,000 ohms with efficient sizes for the embedded resistors. The embedded resistors can be laser trimmed to a targeted value which is derived based upon the final desired value and performance of the embedded resistors through the entire process exposure including assembly. Power is also a consideration and the embedded resistor area needs to be matched to the power requirement. In these examples, this is only a concern for low value resistors because the maximum voltage in these applications is about 3 volts and a resistor of 100 ohms or higher can only draw 0.09 watts which can be dissipated in a small resistor area. Figure 6 shows a serpentine 10,000 ohm resistor using 250 ohm per square material. The large rectangular area on the right side is the area that will be used by the laser trimmer to reach the targeted value. One of the features of the passive core is the use of Core EZ's fine pitch 50 um via diameter on a pitch as small as 200 um. This allows vias to be placed within the legs of the serpentine resistors as shown in Figure 6. The ability to provide a high density of vias through the resistor network is critical for both signal wiring density and the effectiveness of the capacitor layers. Table 4 shows the component count reduction and area reduction attributed to embedding resistors.

978-1-4244-4722-0/09 $25.00
© 2009 IMAPS-ITALY

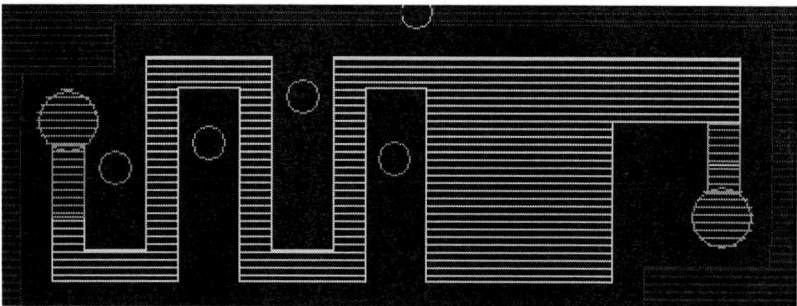

Figure 6 – Embedded Serpentine 10,000 Ohm Resistor with Vias in Legs – 1.0 mm x 3.3 mm

Table 4 - Resistor Count and Area Impact by Embedding Resistors in SiP

	PWB Resistor Count	SiP Resistor Count	Resistor Count Ratio	PWB Resistor Area (sq mm)	SiP Resistor Area (sq mm)	Resistor Area Reduction Ratio
Example A	70	3	23.3	137	8	16.7
Example B	151	18	8.4	143	17	8.5

Connectors

In example A, the use of multiple connectors contributed to significant area consumption which is not consistent with miniaturization. A single connector at 1 mm pitch between connections replaced the multiple connectors and reduced the connector area from 1614 sq mm to 308 sq mm; a 5.2x reduction in area. The single connector will most likely result in higher wiring demands and a SiP with good wiring capacity is required.

Double Sided Effectiveness

The elimination of large components through FCA and reduction in components through embedded passives eliminates concerns about total height of the assembled SiP and allows flexibility to fully use both sides of the SiP. Full double sided utilization saved 2510 sq mm in example A and saved 8265 sq mm in example B. Good double sided utilization can place a large demand on z-axis communication and a SiP with small holes in the core layer and ability to thread holes through the resistor legs enables an efficient double sided product.

SiP Index

One of the measures of the effectiveness of a package is the package area vs. the area that is consumed by active chips on the package. A SiP Index of 1.0 is achieved by a single chip. In Example A, the PWB version had a SiP Index of 11.5 which indicates a package significantly larger than the sum of the active chip area. The redesign of Example A into a SiP reduced the SiP Index to 1.3, very close to the ideal of 1.0. Example B has a similar result. Table 5

summarizes the SiP Index as a function of the two examples.

Table 5 - SiP Index for PWB vs. SiP

	PWB SiP Index	SiP SiP Index
Example A	11.5	1.3
Example B	8.8	1.1

Summary

Effective miniaturization requires a combination of changes which all interact to achieve the desired goals for a SiP. Large active components must be removed from their packages and assembled directly to the SiP. As many passive components as possible need to be embedded inside the SiP thereby relieving real estate on the surfaces and in many cases improving electrical performance. Connectors need to be consolidated. SiP wiring capacity must be high because component placement will be more concerned with real estate than with easy routing. The SiP must also have an ability to communicate efficiently in the z-direction to handle the double sided demands. The substrate shown in Figure 2 has both high wireability and high z-direction communication by use of 25 um lines, 50 um diameter Microvias and 50 um diameter PTHs. An overall summary of the final achievement of an area reduction of 8x to 9x and volume reduction of 17x to 25x is summarized in Table 6. This is illustrated to scale in Figures 7 and 8.

978-1-4244-4722-0/09 $25.00
© 2009 IMAPS-ITALY

Table 6 – Summary of Area, Height and Volume Reduction with SiP vs. PWB

	PWB Area (sq mm)	SiP Area (sq mm)	Area Ratio	PWB Ht (mm)	SiP Ht (mm)	Ht Ratio	PWB Volume (cu mm)	SiP Volume (cu mm)	Volume Ratio
Example A	5715	632	9.0	6.8	2.5	2.7	38,863	1581	24.6
Example B	17,346	2253	7.7	6.5	3.0	2.2	113,332	6758	16.8

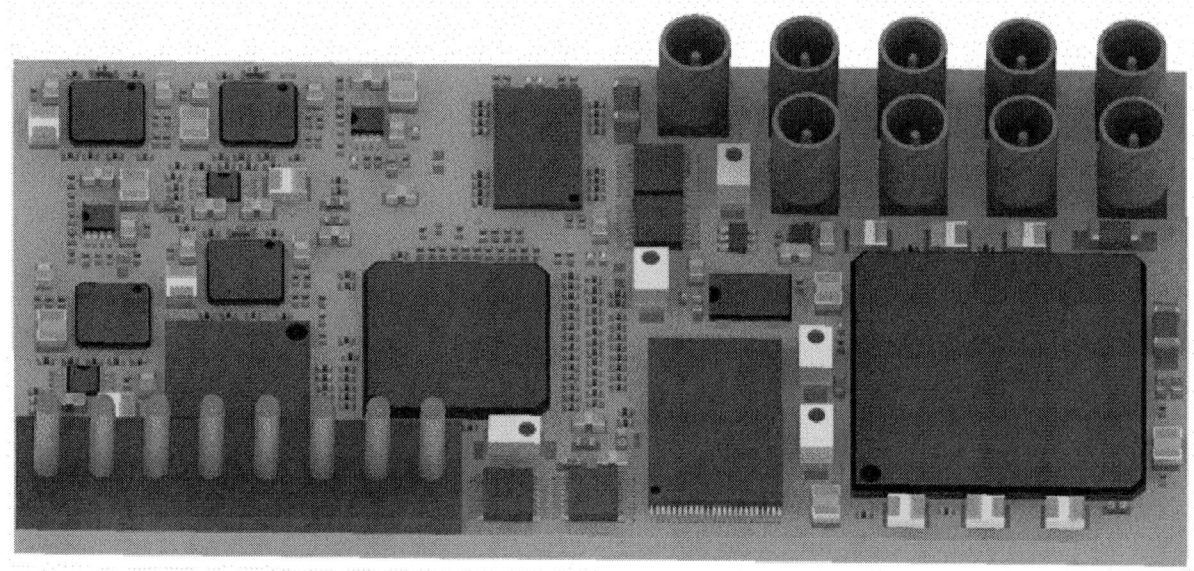

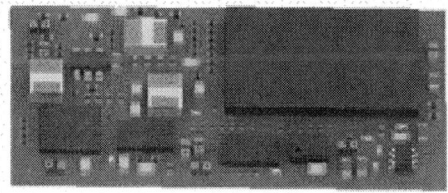

Figure 7 - Example A - PWB vs SiP

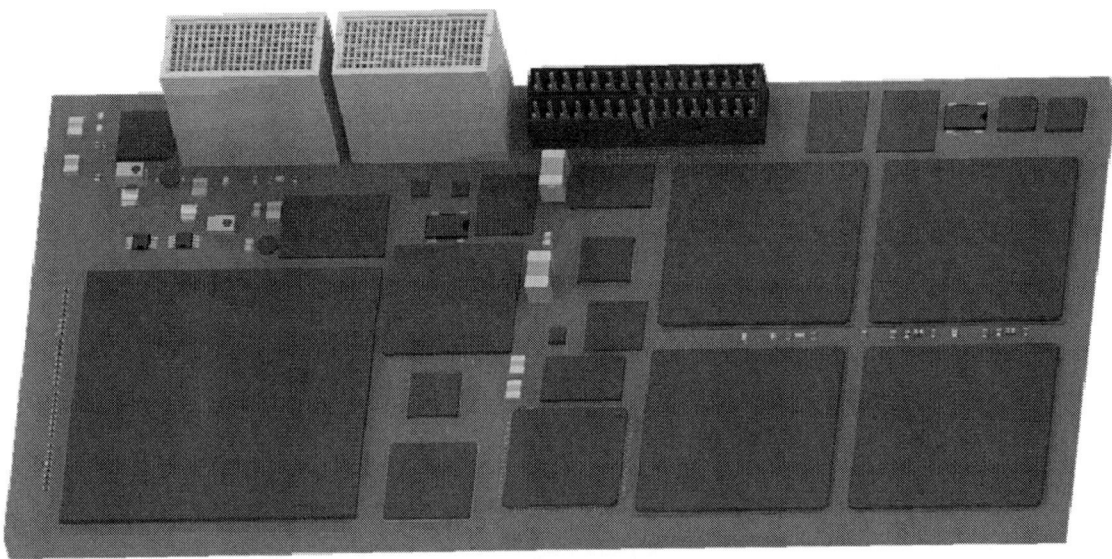

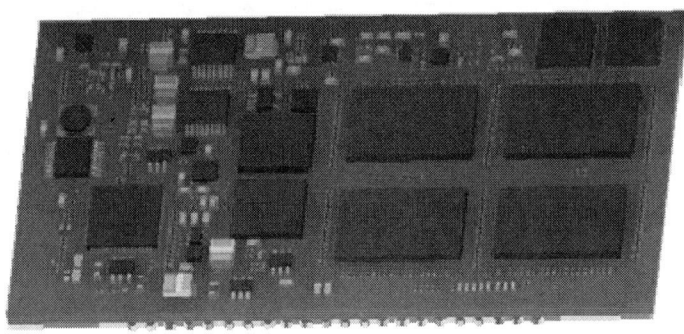

Figure 8 - Example B - PWB vs. SiP

References

[1] Michael J. Rowlands, Steven G. Rosser, "Simulation and Measurement of High Speed Serial Link Performance in a Dense, Thin Core Flip Chip Package", 2006 Electronic Components and Technology Conference, San Diego, CA, May 2006.

[2] Max Osborne, Jamin Ling, Cuong Huynh, Ivy Qin and Vincent McTaggart, "Stud Bumping for Flip Chip – An Alternate Strategy", IMAPS Device Packaging Conference, March 2005, Session 3.

[3] Laurie S. Roth and Vince McTaggart, "Stud Bump Bonding", Advanced Packaging Magazine, Feb. 2005.

[4] Rabindra N. Das, Steven Rosser, Konstantinos I. Papathomas, Mark D. Poliks, John M. Lauffer and Voya R. Markovich, "Resin Coated Copper Capacitive (RC3) Nanocomposites for Multilayer Embedded Capacitors", 2008 Electronic Components and Technology Conference, Lake Buena Vista, Florida, May 2008.

[5] Joel S. Peiffer, "Ultra-Thin Embedded Passive Laminates", CircuiTree, May 2007, pp. 22 – 33.

Long Term Stability of Polymer Based Resistors Tested by Noise, Non-Linearity and Electro-Ultrasonic Spectroscopy

Vlasta Sedlakova, Pavel Tofel, Josef Sikula

Brno University of Technology

Physics Department of FEEC, Technicka 8, 616 00 Brno, Czech Republic

Tel/Fax: +420 54114 3398, E-mail: sedlaka@feec.vutbr.cz

Abstract

The long term stability of polymer based thick film resistors was correlated with the results of standard testing methods: the low frequency noise measurements, non-linearity and new proposed method: Electro-Ultrasonic Spectroscopy. The samples were made using different resistive and conducting pastes. The resistive pastes were made using one type of conducting grains - carbon and graphite particles, suspended in different polymer vehicles. Contacts were made by polymer based dipping silver. The results of standard measuring methods are compared with those of the electro-ultrasonic spectroscopy. This method is based on the phonon interaction with conducting electrons and on the change of the contact area among the conducting particles. The ultrasonic signal changes the contact area between the conducting grains in the resistor structure and then the resistance is modulated by the frequency of ultrasonic excitation. Resultant intermodulation voltage appearing on the sample depends on the value of AC current varying with frequency f_E and on the ultrasonic excited resistance change ΔR varying with frequency f_U. Ultrasound-excited resistance change in polymer based thick film structures is in the range 0.1 to 1.1 mΩ, which corresponds to the relative resistance change of the order from 10^{-6}.

Key words: Low frequency noise, 1/f noise, non-linearity, polymer based TFR, electro-ultrasonic spectroscopy

Introduction

Accurate method for the resistor stability prediction and quality evaluation is required by the component producers. Low frequency noise and resistance non-linearity measurements are the methods, which are frequently used for the resistor quality assessment. New measuring method – Electro-Ultrasonic Spectroscopy – is based on the interaction of two exciting signals with the granular structure of measured samples. We give the comparison between the resistance stability and the results of these three methods.

Low frequency noise

1/f noise can be used as a diagnostic tool in resistance type devices. For stationary and ergodic stochastic process the noise spectral density is proportional to the square of voltage or current $S_U \propto U^2$ or $S_I \propto I^2$. Voltage noise spectral density is than given by:

$$S_U(f) = C_Q \frac{U_{Rx}^2}{f^a} \qquad (1)$$

Where S_U is voltage noise spectral density, U_{Rx} is DC voltage applied to the measured sample R_X, f is frequency, and a is frequency exponent.

It is very convenient to normalize the measured noise spectral density for applied voltage and frequency and to use noise quality indicator C_Q for the resistor quality evaluation:

$$C_Q = S_U(f) \cdot \frac{f}{U_{Rx}^2} \qquad (2)$$

C_Q is a dimensionless parameter with a value dependent on sample quality and reliability.

Non-linearity

Non-linearity of thick film resistors is proportional to the distortion of pure harmonic signal applied to the sample [1-3]. It can be shown, that the number of harmonics can approximate any kind of voltage time dependence with different amplitudes superimposed upon the fundamental frequency. If an AC voltage is applied to a component, where the current paths consist of perfect elements the corresponding current will exhibit a true picture of the applied signal. In this case the transmission is linear. If the elements on the other hand are imperfect, the current will be distorted and generate a voltage inside the component that will produce a correspondingly distorted signal.

978-1-4244-4722-0/09 $25.00
© 2009 IMAPS-ITALY

The third harmonic voltage U_3 measured for the same value of the first harmonic voltage U_1 can be used for the resistors quality evaluation.

We apply the first harmonic voltage with frequency 10 kHz, and we measure the amplitude of the third harmonic voltage. The measurements were performed by non-linearity meter RADIOMETER COPENHAGEN, type CLT1.

Electro-ultrasonic spectroscopy

The electro-ultrasonic spectroscopy was used as a non-destructive testing method for the polymer based thick film resistors evaluation. Proposed method is exploiting two different signal sources – ultrasonic wave and alternating electric current. Mechanical vibrations affect the defects in the sample structure and it influences the electric charge transport through the measured structure. Resulting information is measured on the differential frequency given by the superposition or subtraction of exciting signals frequencies. Measured sample together with the ultrasonic transducer creates the resonant system. For the thick film resistor samples prepared on the alumina substrate we achieved high sensitivity measuring on the resonant frequencies determined by the alumina substrate size. The intermodulation signal was measured on the differential frequency 2 kHz. The intermodulation signal amplitude increases linearly with the amplitude of alternating electric current, and with the square of increasing voltage on the ultrasonic transducer. The intermodulation voltage is influenced by the AC current value and by the resistance change of contacts among the conducting grains in the thick film resistor structure. Ultrasonic signal changes the area of the contact between the conducting grains in the resistor structure, hence the square-law between the voltage on the ultrasonic transducer and the intermodulation voltage is observed. The method sensitivity is influenced by the measuring set-up noise background. In our case the background noise voltage spectral density is about 1.6×10^{-17} V^2/Hz, which corresponds to the equivalent noise resistance 1 kΩ. The intermodulation signal amplitude was measured to be about one order above the measuring set-up background noise voltage for the current 1 mA flowing through the structure. In this case no sample heating is observed even through the long term measurements. The relative resistance change is of the order of 10^{-6}.

The electro-ultrasonic measurement setup consists of two parts, the electric and the ultrasonic one.

The ultrasonic part consists of the generator Agilent and the power amplifier WPD 100. The measured sample was fixed (using beeswax) on the power piezoceramic transmitter (HTP05) which is used for ultrasonic signal generation.

Electric part consists of an AC voltage source. This signal is led to the measured sample over the protective resistor.

The measured signal is amplified by the low noise amplifier with frequency filters with adjustable input and output gain. All parameters are programmed over GPIB or the front panel of the amplifier. The amplified signal is led to the A/D converter. The digital oscilloscope Agilent 54624A with sampling rate 200 Msa/s is used as the A/D converter. The digitized signal is stored in the computer and signal spectral density frequency dependence is evaluated using discrete FFT.

The voltage U_S can be calculated from following equation:

$$U_S = \sqrt{S_U \Delta f} \qquad (3)$$

Where S_U is the peak value of the signal spectral density measured on the differential frequency f_E - f_U, and Δf is the band with or the distance between two successive lines in the signal spectra for pure harmonic voltage.

From the measured voltage U_S we can find the resistance change ΔR as:

$$\Delta R = U_s / I_{AC} \qquad (4)$$

Our measurements were performed for the exciting ultrasonic signal of frequency 31,8 kHz and AC electrical signal of frequency 33,8 kHz.

In order to evaluate the influence of the ultrasonic wave on the measured sample we compared the signal measured on the sample fixed on the piezoceramic transmitter using beeswax with the signal measured on the sample just laying on the top of the transmitter. The results of sample 40 measured for $f_E - f_U$ = 2 kHz, U_E = 10 V and U_U = 10 V are shown in Fig. 1. For the sample non-fixed on the transmitter no peak is observed on the differential frequency.

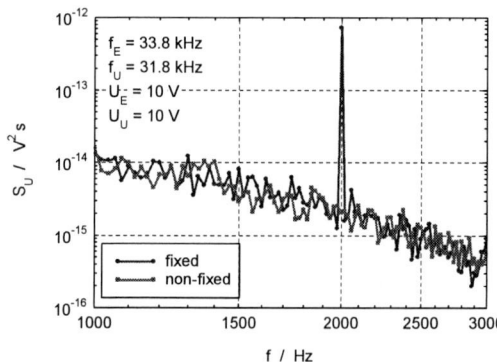

Figure 1: Comparison between the signal measured on the sample fixed on the piezoceramic transmitter using beeswax (blue line) and the signal measured on the sample just laying on the top of the transmitter (red line)

Experimental

Two sets of polymer based thick film resistors were evaluated. The resistive pastes were made from carbon (C) spherical particles and graphite (Gr) flakes suspended in polymer. One type of C/Gr conducting particles and two different polymers were used for our samples. The sets are denoted as Tech 4 and Tech 5. Ten samples were evaluated within each technology. Resistive pastes were applied on the alumina substrate of dimensions 5 by 40 mm. Resistive layer thickness was about 20 μm. The contacts were made by dipping silver (DiAg) – polymer based paste with Ag filling. Different DiAg was used for each set of samples to obtain optimal resistor – contact system. The resistance of our samples was measured three times: (i) after the resistive layer application and drying - denoted later as R_a; (ii) after the resistor curing – denoted as R_0, and (iii) after the storing at the room temperature for 4000 hours - denoted as R_{4000}. The mean value of R_a was 165 Ω for Tech 4, and 142 Ω for Tech 5. After the curing at the elevated temperature the sample resistances increased to about 200 Ω for good samples and up to 6times for unstable samples in both technologies (see Figs. 3 to 6). The correlation between the resistance drift during the curing R_0/R_a and the resistance drift during the storing R_{4000}/R_0 is shown in Fig. 2. We can see that the resistance increases during the curing process and this slightly decreases during the storing.

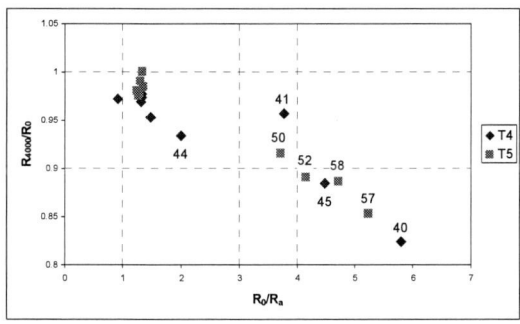

Figure 2: Correlation between the resistance drift during the curing R_0/R_a and the resistance drift during the storing R_{4000}/R_0

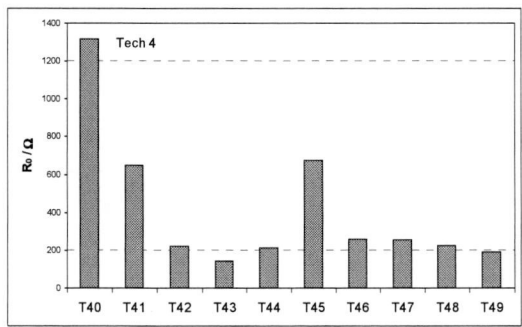

Figure 3: Resistance R_0 for Technology 4

From the resistance drift during the storing we have calculated the resistance change during the storing (see Figs. 4,6) as

$$\frac{R_0 - R_{4000}}{R_0} \cdot 100\% \qquad (5)$$

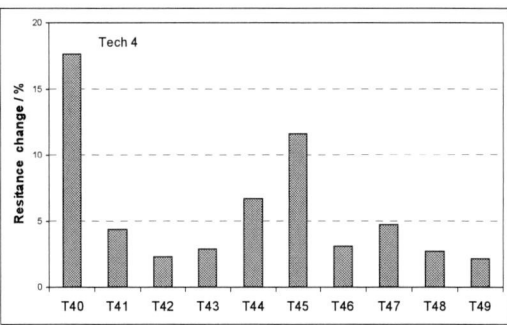

Figure 4: Resistance change during the storing for Technology 4

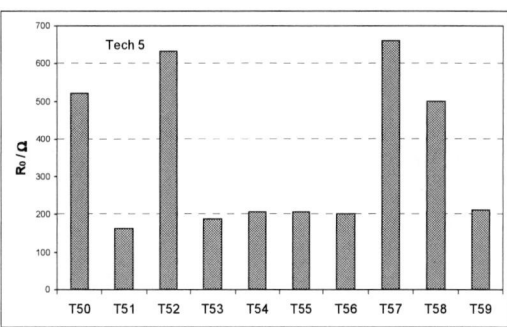

Figure 5: Resistance R_0 for Technology 5

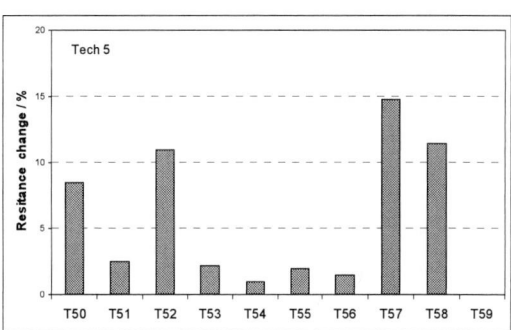

Figure 6: Resistance change during the storing for Technology 5

The dependence between the noise quality indicator C_Q and resistance change for both technologies is shown in Fig. 7. Resistance noise measured for the samples of Tech 5 is two orders of magnitude higher comparing to the results measured for Tech 4. For both technologies good correlation between the resistance change during the storing and C_Q is observed.

The dependence between the third harmonic voltage U_3 measured for the first harmonic voltage $U_1 = 7$ V and resistance change for all the samples is shown in Fig. 8. We can see, that all the samples showing higher resistance change during the storing exhibit also higher value of the third harmonic voltage. From the results measured for sample 40 we can see that the sources of noise and non-linearity are not exactly the same.

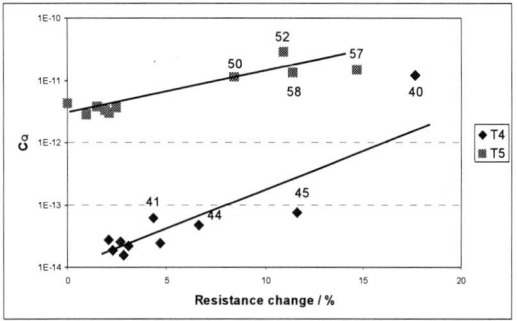

Figure 7: Noise quality indicator C_Q vs. resistance change during the storing for samples of both Tech 4 and Tech 5

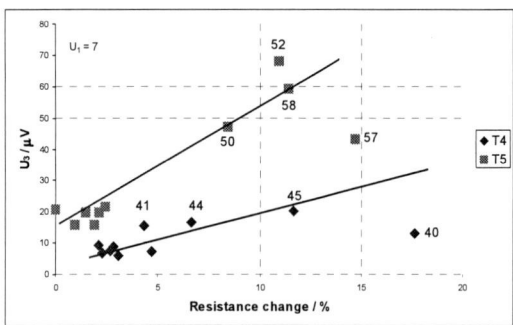

Figure 8: The third harmonic voltage U_3 measured for $U_1 = 7$ V vs. resistance change during the storing for samples of both Tech 5

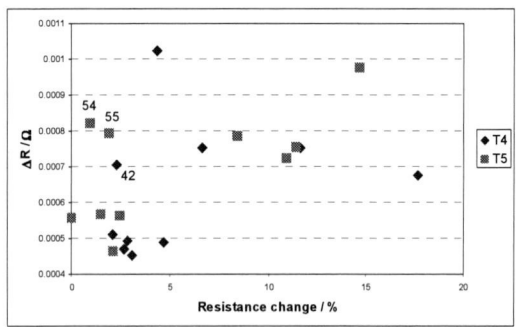

Figure 9: Ultrasound induced resistance change ΔR vs. resistance change during the storing for samples of both Tech 4 and Tech 5

The dependence between the ultrasound induced resistance change ΔR and resistance change

during the storing for all the samples is shown in Fig. 9. We have observed higher value of ΔR not only for all the samples with higher resistance change during the storing, but also for samples 42, 54 and 55. The explanation of this behavior is a subject for next study.

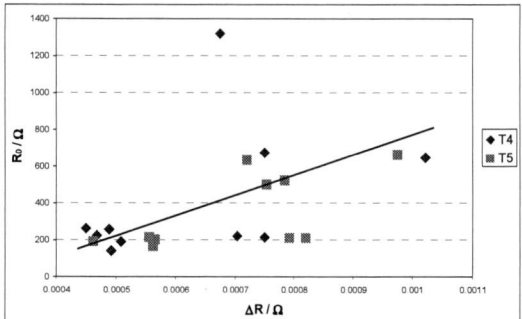

Figure 10: Correlation between the resistance value R_0 and ultrasound induced resistance change ΔR

There is a correlation between the resistance value R_0 and the ultrasound induced resistance change ΔR (see Fig. 10). At the same time there exists the correlation between the resistance change ΔR and the value of the noise quality indicator C_Q and the value of the third harmonic signal amplitude U_3, respectively (see Figs. 11 and 12).

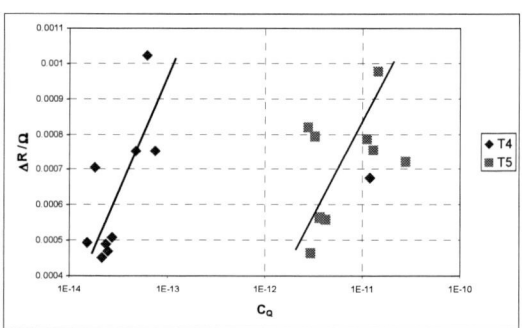

Figure 11: Correlation between C_Q and ultrasound induced resistance change ΔR

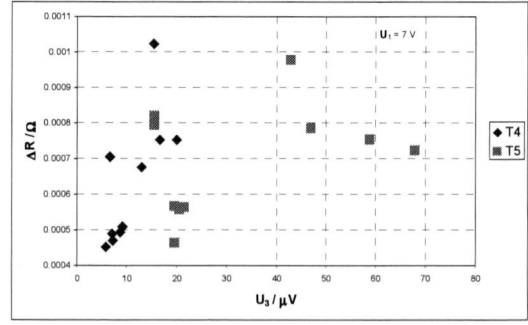

Figure 12: Correlation between U_3 (for $U_1 = 7$ V) and ultrasound induced resistance change ΔR

978-1-4244-4722-0/09 $25.00
© 2009 IMAPS-ITALY

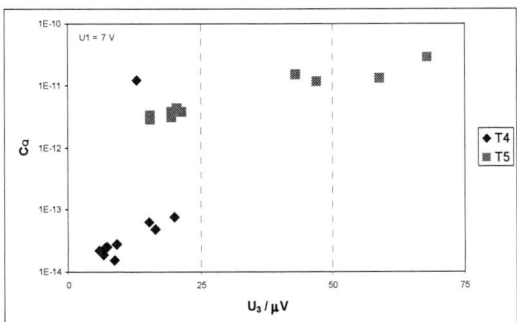

Figure 13: Correlation between U_3 (for $U_1 = 7$ V) and noise quality indicator C_Q

The correlation between the noise and non-linearity results is shown in Fig. 13. As it was mentioned before there should be an important noise source in sample 40, which is not pronounced as a source of resistance non-linearity.

It happens in all cases that with the increasing number of contacts among the conducting grains in the sample there are decreasing the ultrasound induced resistance change ΔR, as well as the value of the noise quality indicator, the value of the third harmonic signal amplitude, and the resistance for the samples with identical geometry.

Conclusions

The object of our work was to find the correlation between the quality tests rejects and the long term stability. We could see, that all the samples showing the resistance change during the storing above 5 % exhibit increased values of noise quality indicator, the third harmonic voltage and ultrasound induced resistance change, respectively. In addition will be rejected also the sample 41 showing increased values of all indicators, and samples 42, 54, and 55 on the base of electro-ultrasonic spectroscopy results. The last 3 samples are quite stable and they are showing good results in other testing methods. We recommend using the combination of two testing methods to increase the test reliability.

Acknowledgements

This research has been supported by the Czech Ministry of Education in the frame of MSM 0021630503 and by grants GACR 106/07/1393 and GACR 102/09/H074.

References

[1] Zhigal'skii, G. P., Russian Journal of Physical Chemistry, 1995, vol. 69, No. 8, pp. 1218–1220.

[2] Kirby P. L. "The non-linearity of fixed resistors", Electronic Engineering, Vol. 32 (1960), pp.722.

[3] Fagerholt, P.O., Ewell, G. J., "Third harmonic testing: An Initial Review", Proceedings of CARTS - EUROPE 2001. Copenhagen, (Denmark), 2001, October 15 – 19, pp. 221-231, ISSN 0887-7491.

System packaging & integration for a swallowable capsule using a direct access sensor

Pio Jesudoss[1], Alan Mathewson[1], William Wright[2], Colm McCaffrey[1], Vladimir Ogurtsov[1], Karen Twomey[1] and Frank Stam[1]

[1]Tyndall National Institute, Lee Maltings, Prospect Row, Cork, Ireland

[2] Department of Electrical and Electronic Engineering, University College Cork, Cork, Ireland

Phone: +353214904441, and pio.jesudoss@tyndall.ie

Abstract

Technological developments in biomedical microsystems are opening up new opportunities to improve healthcare procedures. Swallowable diagnostic capsules are an example of this. In this paper, a diagnostic capsule technology is described based on direct-access sensing of the Gastro Intestinal (GI) fluids throughout the GI tract.

The objective of this paper is two-fold: i) develop a packaging method for a direct access sensor, ii) develop an encapsulation method to protect the system electronics. The integrity of the interconnection after sensor packaging and encapsulation is correlated to its reliability and thus of importance. The zero level packaging of the sensor was achieved by using a so called Flip Chip Over Hole (FCOH) method. This allowed the fluidic sensing media to interface with the sensor, while the rest of the chip including the electrical connections can be insulated effectively. Initial tests using Anisotropic Conductive Adhesive (ACA) interconnect for the FCOH demonstrated good electrical connections and functionality of the sensor chip. Also a preliminary encapsulation trial of the flip chipped sensor on a flexible test substrate has been carried out and showed that silicone encapsulation of the system is a viable option.

Key words: Flip chip over hole, direct-access sensor, flexible substrate, ACA, polysiloxane.

Introduction

The Human gastro-intestinal (GI) tract is prone to various distressing and fatal disorders which have dramatic effect on health and quality of life of people. As an example each year, around 3 million people in the US are hospitalized from gastrointestinal (GI) related disease [1]. No suitable reason or cause was ever found in more than one third of the cases.

The conventional method used to investigate suspected pathology employs an endoscope which is inserted through patient's mouth, nose (gastroscopy) or anus (colonoscopy). These procedures provide some information on the state of the GI tract: gastroscopy provides data about the oesophagus and the stomach while colonoscopy helps investigate the large intestine. These procedures are not only unpleasant for the patients but are also unable to provide any information on the condition of the small intestine.

With recent advances in electronics, wireless communication and microelectronic miniaturisation the limitation of endoscopy can be overcome by using a swallowable electronic capsule. The swallowable capsules can be classified into families of imaging (PillCam [2], Olympus Optical [1]), drug delivery systems (Enterion, ipill) and sensing capsules (Smart [3], Bravo device, IDEA [4], [5], ipill [6]. In none of these sensing capsules is the sensor interconnection achieved through Flip Chip (FC) and in particular Flip Chip Over Hole (FCOH) technology.

FCOH interconnection involves attaching a sensor chip's bond pads face down on to a substrate with an opening in it. This allows interaction between a sensor die and the medium to be sensed. FCOH is particularly suitable for low I/O count applications such as few I/O sensors [7]-[9] because it provides:

- Rugged connections;
- Requires low processing temperature (which results in low thermal stress during processing);
- Dual function, ACA interconnect vertical electrical conduction and liquid insulation around the substrate hole perimeter;
- Mask free process; potentially no post clean step.

This paper presents the zero level packaging and priliminary encapsulation trials of the flip chipped sensor. In the next section, a detailed review of the chip, test substrate, the FCOH packaging and the preliminary encapsulation trials will be presented with the results and discussion.

System Schematic

The electronic system is complex consisting of many elements, see figure 1. The sensor is interfaced to analogue circuitry for signal conditioning. A power supply is placed centrally along with a single lithium ion cell. A processing unit, or microcontroller, controls the measurement and communication process. An ultra low power wireless communication system is added, which provides the transfer of measured data to an external receiver module. This receiver has been developed to acquire the data and interface it to a PC end-user.

The latest system prototype implements the instrumentation in a modular fashion on circular PCB disks which are interconnected with a flexible polyimide core. In this way the disks can be folded for encapsulation and the flexible core provides a reliable interconnect between them.

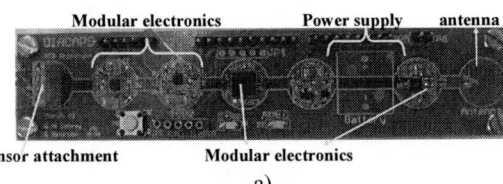

a)

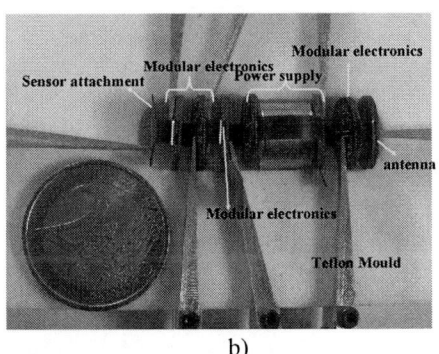

b)

Fig 1. a) PCB electronic system and b) system in serpentine form awaiting encapsulation.

System Components

Sensor Chip

The sensing chip was fabricated using silicon multi-layer process and photolithography techniques. Gold and platinum were deposited on the chip sensing area. The sensing chip had an I/O count of 5 300μm square pads, which were positioned on the periphery of a 6x6mm^2 die as shown in figure 2. The microelectronic sensor comprised of four gold working electrodes (WE) of 1 mm diameter and a platinum counter electrode (CE) of 2mm diameter. The distance between centers of counter electrode and the working electrode was set at 0.5mm.

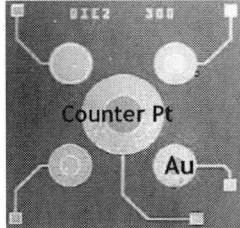

Fig 2. The Sensing Chip

Test Substrate

A thin subsrate with thickness 0.025mm was fabricated[1]. The board pad metallization scheme consists of 15μm Cu, 5μm Ni and 0.05μm of electroplated flash gold. Using a laser, a square window of 4.4mm was cut from the centre of the board to expose the chip to external conditions, see figure 3

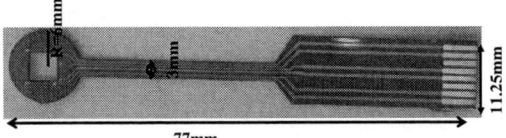

Fig 3. Flexible Polyimide FCOH sensor test substrate.

Assembly Process

A pre-cleaning procedure was carried out separately on both the chip and the substrate. This involved placing the chips and the substrates into a barrel type chamber of a March Plasmod system and exposing them to an oxygen plasma for 40sec at 150 watts. This was followed by IPA immersion in a bench top ultraware ultrasonic precision cleaner for 5 min followed by a DI water rinse. The samples were then dried in a conventional Heraeus vacuum oven at 150°C for an hour.

Gold stud bumps were formed on the die pads using a Kulicke and Soffa ball wedge gold bonder. The bumps had a mean diameter of 103 microns and a mean height of 108 microns, see figure 4a. This was followed by coining the gold stud bump on to a glass substrate using Finetech Fineplace 96 Lambda flip chip bonder. The main purpose of the gold bumps was to provide an under bump metallurgy so that a standoff would be provided when assembling the device onto to the substrate. The gold stud bump was coined at 11.7N at a coining temperature of 180°C for 20 sec, see figure 4b.

Bonding

ACA material from Loctite was dispensed on the test board using a CAM/ALOT 1414 liquid dispense system. A brown viscous epoxy paste with gold coated nickel filler particles of 7μm was used.

---

[1] Trulon printed circuits – UK.

978-1-4244-4722-0/09 $25.00
© 2009 IMAPS-ITALY

The alignment of the chip/substrate was done using the Finetech Flip-Chip bonder. Bonding was carried out at 180°C and a bonding pressure of 10N for 8 sec. Figure 4c provides an overview of the assembly process.

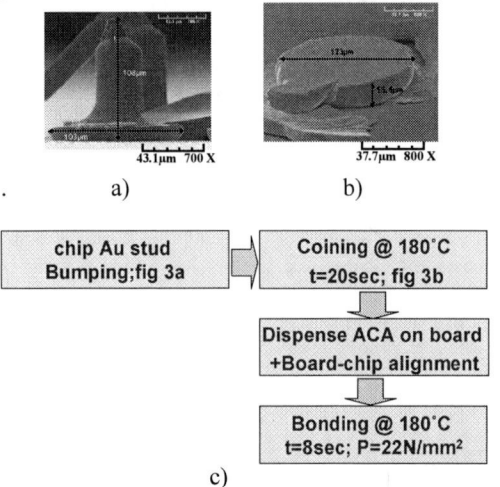

c)

Fig 4. (a) SEM image of a gold stud bump; (b) SEM image of a coined bump and (c) Flow diagram of assembly process.

Direct Access sensor testing

The electrical connection and the robustness of the packaging as well as the functionality of the sensor were tested using cyclic voltammetry of the three electrode sell comprising of WE and CE on the designed sensing chip and a standard Ag/AgCl electrode which was used as the reference electrode. Elecrochemical reactions occurring at the interface between the WE and the solution were monitored by a CH instruments 620B computer controlled potentiostat The fabricated test assemblies were dipped into a solution of 0.5M of H_2SO_4 and cyclic voltammetry test at a scan rate of 0.2V/sec was applied to the electrode system. The chemical reactions that occured at the gold WE in this solution is well documented [10] so any change in the performance of ACA or the component will be identified at this stage.

The measured voltammograms for different assemblies are presented in figure 5. Each voltammogram showed a similar response. A peak was obtained at 1.4V during the positive voltage sweep from 0 to 1.5V, and a corresponding peak was obtained at 0.9V during the negative voltage sweep. These gold peaks were due to gold oxide formation and reduction, and they illustrated the correct function of the sensor and interconnect.

Encapsulation of the direct access sensor.

The test assemblies were then encapsulated with silicone. The encapsulation process consisted from a number of steps: the first step involved dispensing silicone on the perimeter of the window using CAM/A LOT and cured in the oven at 80° for 3hrs. The cured silicone acted as a dam around the window. Then the protection of the sensor was carried out via AZ photoresist – Diazonnaphthoquinones (AZ Electronic Materials GmbH). The photoresist was applied using the pendant drop method - 6 drops of AZ on the sensor area - and cured at 80°C for 1 hr. It had a height of around 596.7μm and acted as a plug covering the exposed area of 19.36mm². Once the dam and the plug were ready, the assembly was inserted into a gelatin glycerin capsule (33mm*13mm) and secured in place. The fixed assembly was then filled with silicone and cured at room temperature for 24-48 hrs. This was followed by immersing the capsule in warm water (50°) for 10-15 min. to dissolve the glycerin capsule. The sensor was exposed by dissolving the AZ photoresist in acetone for 5-10 min, as shown in figure 6.

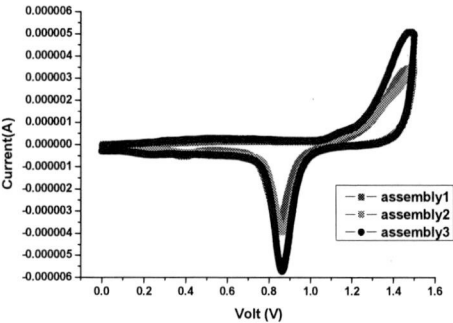

Fig 5. Cyclic current-voltage curve of the Au WE of 3 different chips with the same assembly method

To prove a quality of the encapsulation process the cyclic voltammetry test was carried out on both an unencapsulated sensor and an encapsulated sensor. The derived voltammograms are shown in figure 7. A similar response was obtained for both of the tested sensors. Oxidation and reduction peaks were obtained at 1.4V and 0.9V for both sensors; minor changes in the shape of the voltammogramms can be related to a standard voltagramm dispersion as a result of difference between surface of the WE on the sensing chips and decreasing of the leakage current in case of encapsulated sensor. Thus this test results allows the conclusion that the assembly process for the sensor encapsulation did not affect the sensor operation.

Conclusion

These preliminary results show that FCOH can be used for direct-access sensor packaging and integration. Futrhermore ACA can be used as a suitable material for applications with few relatively large bond pads and particularly in relation to measurements in the fluidic environment when the sensing area needs to be sealed off from the electronics.

Electrical tests after system encapsulation with silicone showed that electronic functionality

978-1-4244-4722-0/09 $25.00
© 2009 IMAPS-ITALY

and chemical sensing performance wasn't compromised.

Future work will incorporate the study of leakage and impedance of the encapsulated assembly.

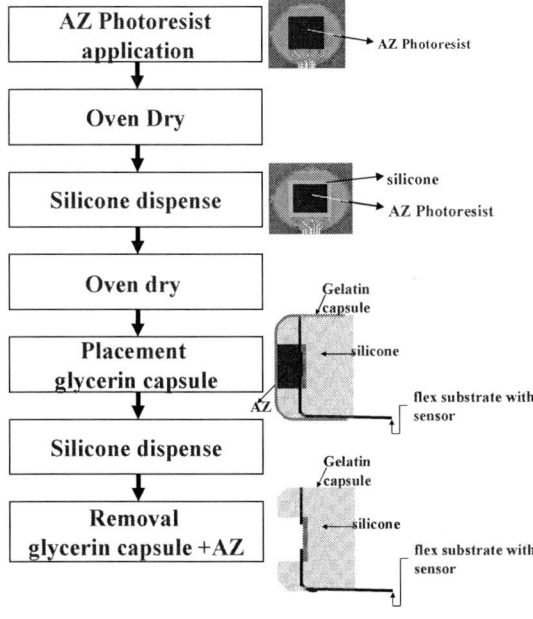

a)

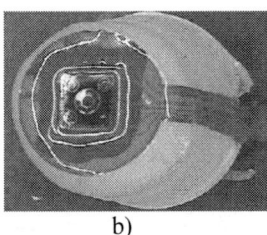

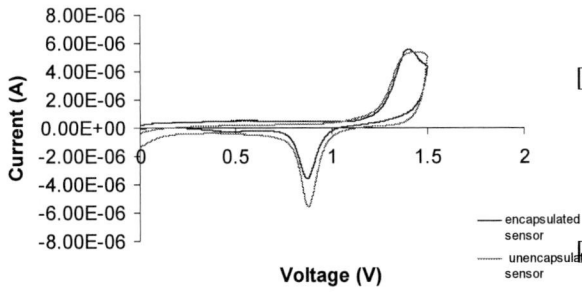

b)

Fig 6. (a) Encapsulation assembly process and (b) picture of an encapsulated assembly.

Fig 7. Voltammetric responses of an unencapsulated sensor and an encapsulated sensor

Acknowledgement

The author would like to acknowledge Enterprise Ireland CFTD /05 / 122 and HEA PRTLI-IV project NEMBES for providing the opportunity to carry out the work presented in this paper

Tyndall's Ken Rodgers is also acknowledged for his generous assistance in the lab

References

[1] Colm Mc Caffrey, Olivier Chevalerias, Cian O'Mathuna, and Karen Twomey, "Swallowable-Capsule Technology", IEEE Pervasive Computing, Vol. 7, No. 1, pp. 23-29, January-March, 2008.

[2] Guido Costamagna, Saumil K. Shah, Maria Elena Riccioni, Francesca Foschia, Massimiliano Mutignani, Vincenzo Perri, Amorino Vecchioli, Maria Gabriella Brizi, Aurelio Picciocchi, and Pasquale Marano, "A Prospective Trial Comparing Small Bowel Radiographs and Video Capsule Endoscopy for Suspected Small Bowel Disease", Gastroentrology, pp. 999-1005, 2002.

[3] Roger Allan, "Smart Pill goes on a Fantastic Voyage", Electronic Design, Vol. 54, No. 27, pp. 66, December 2006.

[4] Erik A. Johannessen, Lei Wang, Li Cui, Tong Boon Tang, Mansour Ahmadian, Alexander Astaras, Stuart W. J. Reid, Philippa S. Yam, Alan F. Murray, Brian W. Flynn, Steve P. Beaumont, David R. S. Cumming, and Jonathan M. Cooper, "Implementation of Multichannel Senosrs for Remote Biomedical Measurements in a Microsystems Format", IEEE Transsactions on Biomedical Engineering, Vol. 51, No. 3, pp. 525-535, March 2004.

[5] K. Twomey, J. Marchesi, "Swallowable Capsule Technology: current perspectives and future directions", Endoscopy, vol. 41, pp. 357-361, 2009.

[6] http://technologyreview.coverleaf.com/technologyreview/200902/?pg=80.

[7] Francesca Campabadal, Josep Lluís Carreras, Enric Cabruja, "Flip-Chip packaging of piezoresistive pressure sensors", Sensors and Actuators, A 132, pp. 415-419, 2006.

[8] Jun Karasawa, Masao Segawa, Yasukazu Kishimoto, Makoto Aoki, Tomoyuki Sasaki, "Flip Chip Interconnection Method Applied to Small Camera Module", IEEE Electronic Components and Technology Conference, pp. 1024-1028, 2001.

[9] R. Briegel, M. Ashauer, H. Ashauer, H. Sandmaier, W. Lang, "Anisotropic conductive adhesion of microsensors applied in the instance of a low pressure sensor", Sensors and Actuators A: Physical, Vol. 97-98, pp. 323-328, April 2002.

[10] L. D. Burke, B. H. Lee, "An investigation of the electrocatalytic behaviour of gold in aqueous media", Journal of electroanalytical chemistry, pp. 637-661, 1992.

978-1-4244-4722-0/09 $25.00
© 2009 IMAPS-ITALY

Interface Resistance between Polymer Based Conducting and Resistive Layers

Pavel Tofel, Vlasta Sedlakova, Milos Chvatal, Jiri Majzner

Brno University of Technology

Physics Department of FEEC, Technicka 8, 616 00 Brno, Czech Republic

Tel/Fax: +420 54114 3254, E-mail: xtofel01@stud.feec.vutbr.cz

Abstract

We have studied the interface resistance between the polymer based conducting and resistive thick film layers. The samples were made using different resistive pastes and dipping silvers. The composite of carbon and graphite (C/Gr) conducting particles suspended in different polymer vehicles were used for the thick film resistive layers preparation. Interface resistance R_I created between the contact layer made from dipping silver (DiAg) and resistive layer was determined from the surface potential distribution measurements and its value was less than 1% of total sample resistance. Measuring apparatus DISPOT® designed in our laboratory provides the measuring of a surface potential distribution. The measuring probe is sliding on the surface of measured structure and potential change between the successive steps is normalized by the total current flowing through the structure. Elementary step (the shortest distance between two measurements) is 1.25 µm. The equipment is arranged for current and voltage four-point measurement. Jump in potential on the interface between DiAg and C/Gr layer corresponds to the contact resistance R_I which is created between these layers. Interface resistance plays important role in the capacitors technology thus the analysis of the charge carrier transport through the interface between different layers can be used to improve the low ESR capacitors technology.

Key words: interface resistance, 1/f noise, non-linearity, polymer based TFR

Introduction

Measuring apparatus DISPOT® designed in our laboratory (Fig. 1) provides the measuring of a surface potential distribution. The measuring probe is sliding on the surface of measured structure and potential change between the successive steps is normalized by the total current flowing through the structure.

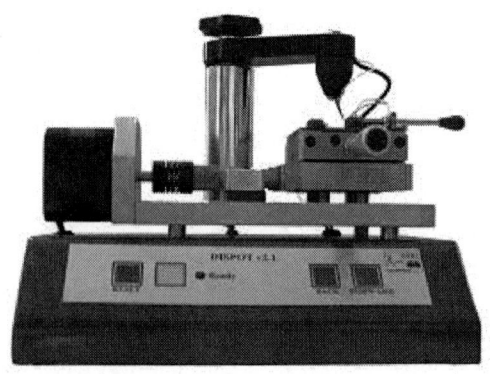

Figure 1: DISPOT® designed in our laboratory

Elementary step (the shortest distance between two measurements) is 1.25 µm. The equipment is arranged for current and voltage four-point measurement. The slope *s* (see Fig. 2 and Fig. 3) corresponds to normalized potential change on distance 1 mm.

Samples of resistors

Two sets of polymer based thick film resistors were evaluated. The resistive pastes were made from carbon (C) spherical particles and graphite (Gr) flakes suspended in polymer. One type of C/Gr conducting particles and two different polymers were used for our samples. The sets are denoted as Tech 4 and Tech 5. Ten samples were evaluated within each technology. Resistive pastes were applied on the alumina substrate of dimensions 5 by 40 mm. Resistive layer thickness was about 20 µm. The contacts were made by dipping silver (DiAg) – polymer based paste with Ag filling. Different DiAg was used for each set of samples to obtain optimal resistor – contact system. The resistance of our samples was measured three times: (i) after the resistive layer application and drying -

denoted later as R_a; (ii) after the resistor curing – denoted as R_0, and (iii) after the storing at the room temperature for 4000 hours - denoted as R_{4000}. The mean value of R_a was 165 Ω for Tech 4, and 142 Ω for Tech 5. In Fig. 2 you can see the resistance of the samples before storing from Tech 4 and in Fig. 3 the resistance of the samples before storing from Tech 5.

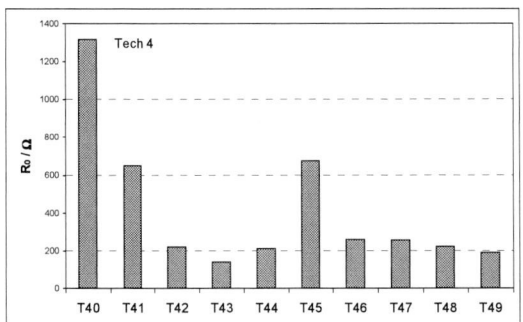

Figure 2: The resistance of the samples before storing from Tech 4

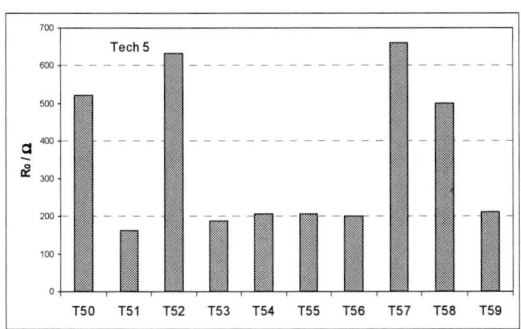

Figure 3: The resistance of the samples before storing from Tech 5

Jump in potential on the interface between DiAg and C/Gr layer corresponds to the contact resistance R_c which is created between these layers (see Fig. 4).

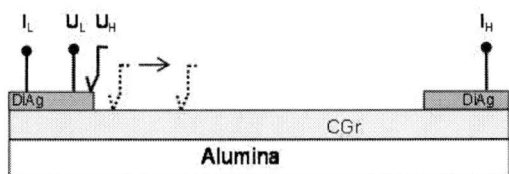

Figure 4: The sample and 4-point contacts measuring by DISPOT®

Measuring with DISPOT®

The DISPOT® is equipment which can measure the distribution of surface potential. The interface resistance R_I is jump in potential on the

interface between two layers and designates the compatibility of the layers (contact and resistive paste). For example the interface resistance R_I is 6.8 Ω on the sample T57 from Tech 5 (Fig. 5) and 0.65 Ω on the sample T58 from Tech 5 (Fig. 6).

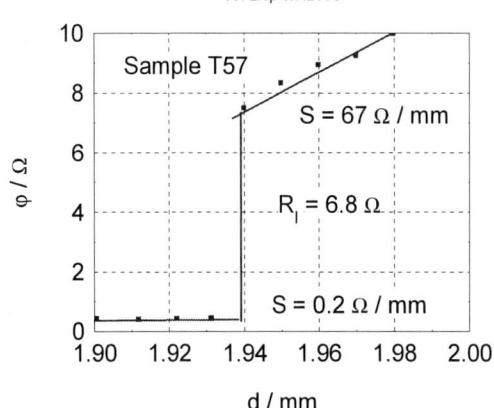

Figure 5: Potential distribution measured on sample T57

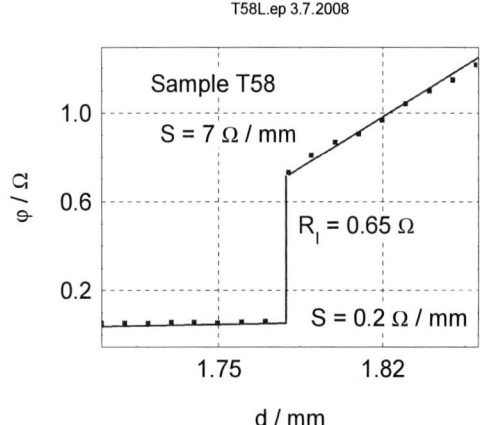

Figure 6: Potential distribution measured on sample T58

Experimental

We have stored our samples for a period 4000 hours at room temperature. The resistance stability is defined as:

$$S_T = \left(1 - \frac{R_{4000}}{R_0}\right) \cdot 100\% \qquad (1)$$

Where R_{4000} is resistance after storing 4000 hour and R_0 is resistance of the sample before storing.

The resistance change for samples from Tech 4 is shown in Fig. 7.

978-1-4244-4722-0/09 $25.00
© 2009 IMAPS-ITALY

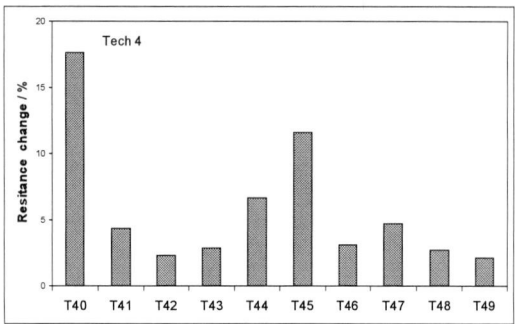

Figure 7: The resistance change of Tech 4 after 4000 hours

The maximum value of the resistance change was found for the sample T40. We supposed that the maximum value of the resistance change occurs in the sample of the lowest quality. The resistance change for the samples from Tech 5 is shown in Fig. 8. We can see that the sample T59 have the same resistance before and after 4000 hours. The maximum value of the resistance change has occurred in the sample T57.

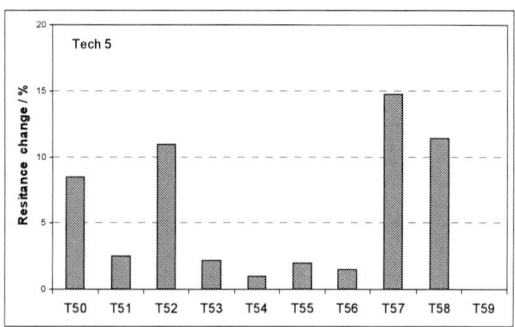

Figure 8: The resistance change of Tech 5 after 4000 hours

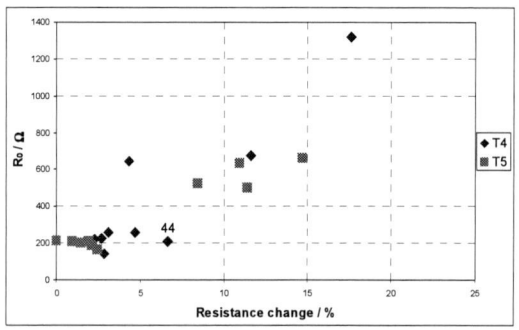

Figure 9: The resistance R_0 vs. the resistance change for the samples from Tech 4 and Tech 5

The dependence of the sample resistance after the curing R_0 on the resistance change for both technologies is shown in Fig. 9. After the curing the resistance increased to about 200 Ω for stable samples for both technologies. For these samples the resistance decreases for less than 5 % with 4000 hour storing.

After the curing at the elevated temperature the sample resistances increased to about 200 Ω for good samples and up to for unstable samples 1 kΩ in both technologies. The correlation between the resistance drift during the curing R_0/R_a and the resistance drift during the storing R_{4000}/R_0 is shown in Fig. 10. We can see that the resistance increases during the curing process and this one slightly decreases during the storing.

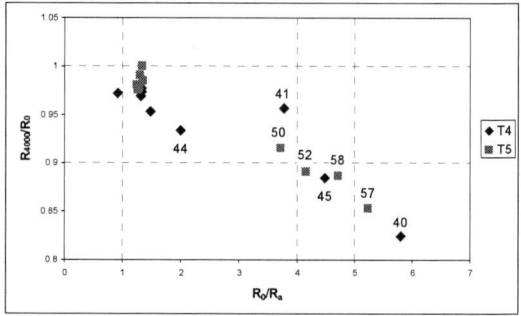

Figure 10: Correlation between the resistance drift during the curing R_0/R_a and the resistance drift during the storing R_{4000}/R

The interface resistance R_I is jump in potential on the interface between contact and resistive paste measured by DISPOT®. The samples of Tech 4 have the same quality of the interface resistance between contact and resistive paste as the samples Tech 5 (Fig. 11). The samples with the resistance change less than 5% have the interface resistance below 2 Ω. Only the sample T41 has higher interface resistance and resistance change is less then 5 %.

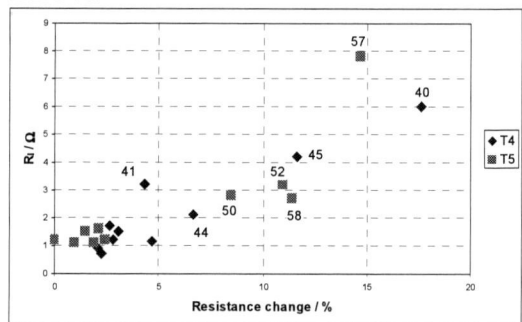

Figure 11: The interface resistance R_I vs. the resistance change for the samples from Tech 4 and Tech 5

1/f noise was used as a diagnostic tool in the thick film resistors. Voltage noise spectral density is given by:

$$S_U(f) = C_Q \frac{U_{Rx}^2}{f^a} \qquad (2)$$

978-1-4244-4722-0/09 $25.00
© 2009 IMAPS-ITALY

Where S_U is voltage noise spectral density, U_{Rx} is DC voltage applied to the measured sample R_X, f is frequency, and a is frequency exponent.

It is very convenient to normalize the measured noise spectral density for applied voltage and frequency and to use noise quality indicator C_Q for the resistor quality evaluation:

$$C_Q = S_U(f) \cdot \frac{f}{U_{R_x}^2} \qquad (3)$$

C_Q is a dimensionless parameter with a value dependent on sample quality and reliability. The dependence of the interface resistance R_I on the noise quality indicator C_Q is shown in Fig. 12.

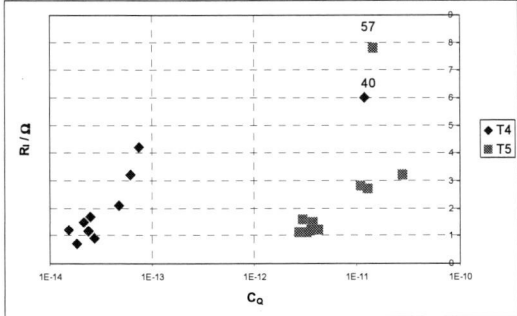

Figure 12: The interface resistance R_I vs. the noise quality indicator C_Q for the samples from Tech 4 and Tech 5

It was found that the sample with lower value of the interface resistance has lower noise quality indicator.

The third harmonic voltage U_3 measured for the same value of the first harmonic voltage U_1 can be used for the resistors quality evaluation.The third harmonic voltage U_3 measured for all the samples of polymer based resistors for the first harmonic voltage $U_1 = 7$ V is shown in Fig. 13.

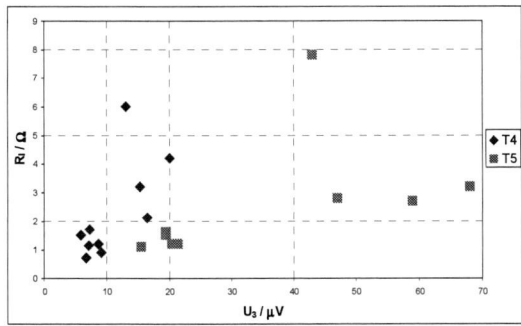

Figure 13: The interface resistance vs. the third harmonic signal for the samples from Tech 4 and Tech 5

In this case the samples with lower value of the interface resistance have the lower value of the third harmonic signal again.

Conclusions

1. The lower interface resistance:
   (i)   the higher stability.
   (ii)   the lower noise quality indicator C_Q.
   (iii)   the lower the third harmonic signal.

2. The important noise and non-linearity sources are on the interface between (or in the vicinity of) carbon-graphite layer and silver contact layer.

3. Measurement of potential distribution by DISPOT® can be used for sample quality and stability assessment.

Acknowledgements

This research has been supported by the Czech Ministry of Education in the frame of MSM 0021630503 and by grants GACR 102/07/1393 and GACR 102/09/H074

References

[1] Dziedzic A., Kolek A. "1/f Noise in Polymer Thick-Film Resistors". Journal of Physics D: Applied Physics, 31 (1998), p. 2091-2097.

[2] Soliman L.I., Sayed W.M. "Some Physical Properties of Vinylpyridine Carbon-Black Composites". Eyptian Journal of Solids, **25** (2002), No.1, p. 103-113.

[3] Sedlakova V., Brustlova J., Sikula J., Hlavka J., Coocker J., Adams K., Greenhill D.A. "Noise of Carbon/Graphite Thick Conducting Films:. Proceedings of the 17th International Conference on Noise and Fluctuations, 2003, Prague, Czech Republic, p. 201-204.

[4] Sedlakova V., Sikula J. "Charge Carrier Transport and Noise in Polymer Based Thick Films". Proccedings of the 4th Europeand Microelectronics and Packaging Symposium, 2006, Terme Catez, Slovinia, p. 15-20.

New packaging technology enabling integration of Magnetics and Semiconductors in one component

Abel Pot, Application Development Manager, DSM Engineering Plastics, Geleen, The Netherlands

Dr. Horst Roehm, Technical Marketing Manager, NXP Semiconductors, Hamburg, Germany

Rinus v.d. Berg, Architectural and Industrial Designer, DSM Engineering Plastics, Geleen, The Netherlands

Shanmugam T, Assitant General Manager Technology, Nano Technology Mfg. Pte. Ltd., Singapore

See-Wee Ong, Senior Application Development Support Engineer, DSM Engineering Plastics, Singapore

Frank van der Burgt, Product Development Engineer, DSM Engineering Plastics, Geleen, The Netherlands

Dr. Tamim P. Sidiki, Innovation Program Manager, DSM Engineering Plastics, Sittard, The Netherlands

Abstract

The continuous trend towards convergence and miniaturization is recently generating significant interest in new technologies for Electronics. This requires the integration of Semiconductors and Magnetics, two entirely different industries with different players in the value chain. In this paper, we demonstrate, a packaging technology which allows three dimensional stacking of Magnets and Semiconductors. We realized the integration of a Semiconductor chip - which provides protection against electro static discharge (ESD) – and a common mode filter (CMF) into one thermoplastic package. For the first time ever, this filter is integrated directly into the thermoplastic part which is used as the substrate, filter and housing at the same time.

Laser direct structuring in combination with Stanyl® ForTii™ as an ultra high performance, entirely halogen high temperature thermoplastic does omit any wires for the realized coil, and also facilitates high flexibility in design and manufacturing, allowing ultra small footprints and the realization of components suitable for surface mount technology.

As an example of this new technology, we demonstrate a component which can provide full ESD protection and common mode filtering for a high speed USB2.0 interface.

ESD Protection

Impact of Moore's Law on ESD protection of advanced CMOS ICs

The continuous trend of feature-size miniaturization has enabled semiconductor manufacturers over decades to improve chip performance, reduce power consumption, and drive cost down by squeezing billions of transistors into a single IC. Despite all obvious advantages, there is one major disadvantage in miniaturization of sub- circuits: integration of sufficiently robust ESD protection.

978-1-4244-4722-0/09 $25.00
© 2009 IMAPS-ITALY

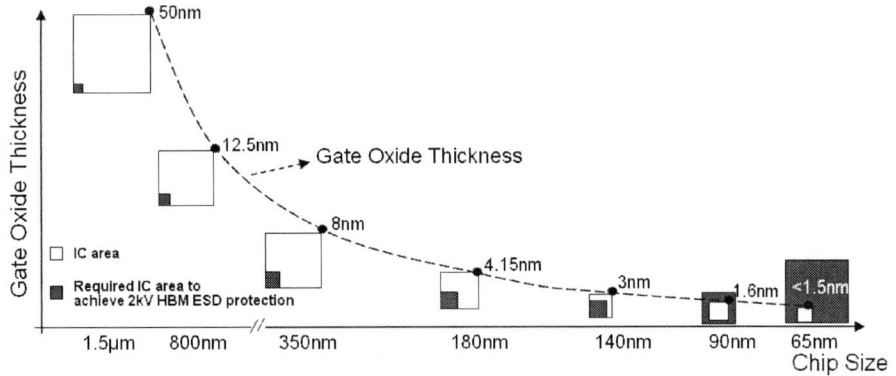

Figure 1: ESD considerations for advanced CMOS ICs

Figure 1 shows the reduction of the total IC area for various technology nodes. The red boxes within each of these ICs indicate schematically the required area to implement a minimum 2-kV ESD protection into the IC. With each technology node the relative area required for ESD protection increases. The reason is that ESD protection scales with the area of the diodes and these diodes can not be shrunk at the same scale as transistors required for logic functions. It is obvious that for very advanced technology nodes there is a physical and economical limitation to integrate robust enough ESD protection. Advanced ICs are optimized for power consumption and speed, not for ESD protection. An optimization for ESD protection would blow up the chip above any acceptable limit.

Smaller feature sizes (channel length) related with thinner and smaller gate oxides drive down the maximum gate (e.g. for CMOS90 below 1.5V) and drain-source voltages (e.g. for CMOS90 <1.6V). Such ICs are very sensitive to over voltage and therefore especially sensitive to ESD discharges, which destroy sub- circuits already at very low ESD levels. As such, external board-level ESD protection becomes a must if developers of consumer/computer appliances want to build "CE"-compliant devices and furthermore want to prevent high field return rates due to ESD and other discharge issues. In general, one can say that today's ESD issue will become tomorrow's nightmare when even smaller feature sizes are applied.

External interfaces to other appliances are subject to ESD damage. In specific, higher speed, hot-plug interfaces such as HDMI, USB or Display port are most critical. Users can connect any sink or source equipment while at least one of the applications is still running, i.e. there is a supply voltage at the port. Needless to say, such a powered port will be affected by serious ESD issues. The question is not if, but only when the related transceivers (standalone or integrated) will be seriously damaged.

The high interface speed in conjunction with "hot plug" characteristics implies stringent requirements for an ESD protection solution, including:

- very high diode switching speed (nsec) and ultra low line capacitance (<1pF) can ensure signal integrity
- robust ESD protection without degradation after several ESD strikes
- low leakage even after several hundred ESD discharges

Based on main stream monolithic silicon technology, NXP Semiconductors provides ESD protection ICs fulfilling highest performance and meeting today's and tomorrow's requirements of OEMs in Electronics like:

- the required low-cost solution for the mass consumer and computer market
- ultra-low total line capacitance of below 0.5 pF (Silicon chip incl. bonding wires, package and any existing parasitic)
- no degradation even after thousands of high-level ESD strikes (IEC61000-4-2)

978-1-4244-4722-0/09 $25.00
© 2009 IMAPS-ITALY

- a fast diode reaction time (nsec range) to ESD pulses
- highest integration
- full compliance with high speed interfaces such as HDMI 1.3, Display Port or USB3.0

EMI Filtering

High speed digital interfaces like HDMI, USB or Display Port are widely used in the mobile as well as in the computer and TV area. As base for data exchange, all these interfaces use differential signals to exchange data which means, that two complementary signals were sent on two separate wires. As long as ideal differential signals are transmitted, no electromagnetic interference (EMI) will occur. Unfortunately, in a real electronic system phase lag between differential signals, potential differences between differential signals and rise (fall) time lag between differential signals leads to common mode signals and therefore to EMI. This affects other electrical circuits by electromagnetic conduction or electromagnetic radiation from external antennas. It is obvious that EMI has to be suppressed in electronic systems.

In systems with differential signals common mode filter (CMF) are widely used to suppress the unwanted EMI generated by common mode signals. Especially, if unshielded twisted pair (UTP) cables – which acts as an antenna for common mode signals - are used as interconnects between devices, the use of CMF is a must.

High Speed Interface Protection

Today, state of the art solutions are using separated devices for ESD protection and EMI filtering to protect the highly integrated silicon chips and to suppress unwanted EMI (Figure 2).

To overcome disadvantages of the discrete solution like space requirement, performance mismatch, inventory costs, etc. the integration of ESD protection and EMI filtering in one device is the next step, a straight forward approach.

Because CMF are build in principle with wires winded around a ferrite core but on the other hand the ESD protection devices consists of diodes diffused in a block of silicon connected to a lead frame and covered by plastic the main challenge was to combine two different technologies in one package.

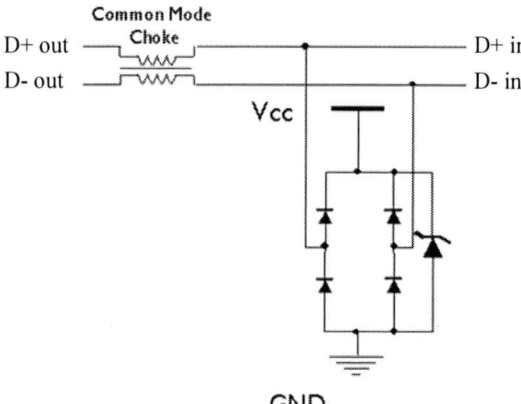

Figure 2: Schematic of a differential signal interface ESD & EMI protection

Global, cross-industry collaboration

Since this project involves entirely new technology, four companies have been working closely together to make it happen. The design of the package concept was proposed by DSM Engineering Plastics in The Netherlands, where a package was crafted enabling the integration of Semiconductors and Magnetics into one thermoplastic package based on injection molding. The injection molding of the package was done at NTM (NanoTechnology Mfg. Pte. Ltd.) in Singapore, the transfer of the EMI filter was achieved by laser Direct Structuring (LDS) at laser Micronics in Germany. In order to meet the high requirements of this technology to the thermoplastic, a new high temperature polyamide called Stanyl ForTii has been selected.

Market need for new package concept

OEMs in Electronics industry with their strong drive of application conversion seek for increased functionality integration and reduction of form factor to focus on PCB space and component count reduction. Since all external interfaces, in specific those operating at higher speeds such as HDMI, USB or Display Port, do require EMI filtering as well as ESD protection, two different components populate such interfaces and eat up valuable real estate on the PCB: ESD protection devices and EMI filters. From various discussions with leading OEMs, it is clear that a component which can integrate both these functionalities and at same time offers a space and component count reduction is highly appreciated and can solve some of the existing issues of OEMs realizing easier and denser application designs.

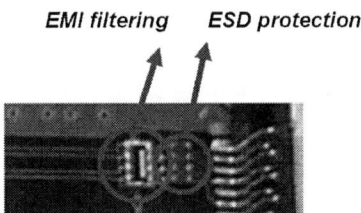

Figure 3: Optical microscopy of a typical external interface in Consumer Electronics using two individual components for ESD protection (first part after the interface pins) and EMI filtering (second component) in the electrical path on its way from the interface to the transceiver IC

We have designed and realized a package which integrates both these components into one package. This is a breakthrough technology which involves many industry first actions.

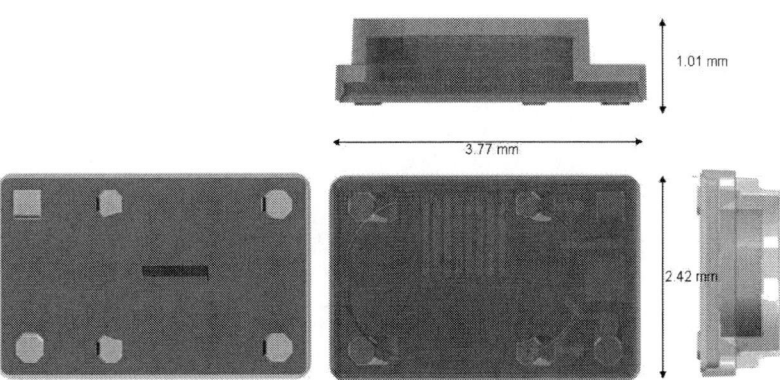

Figure 4: View of the package from the top, bottom and side showing total package dimensions

Figure 4 shows the sizes of the new package. With a total footprint of 3.77mm x 2,42mm x 1,01mm this is currently the world's smallest package integrating a fully EMI filters and an ESD protection for a high speed interface such as e.g. USB. The equivalent space reduction on the PCB is >75% by integration of these two components into one package. The small footprint in combination with low height enables PCB space reduction for OEMs.

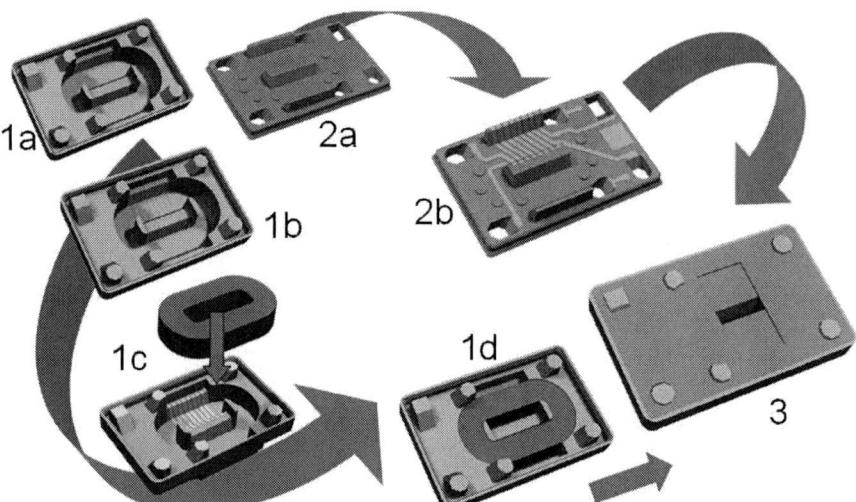

Figure 5: Schematics of manufacturing steps to realize a package integrating IC and magnet

Figure 5 shows the various manufacturing steps. In step 1a and 2a the top and cap layer of the package are molded in Stanyl ForTii, a high temperature polyamide suitable for lead free surface mount assembly due to high melting temperature Tm=320°C, a high glass transition temperature of Tg=135°C and a stiffness across a broad temperature range. From similar work in air cavity packages which are used for e.g. MEMS sensors it is well known that co-planarity is a key issue since it can lead to delamination of ICs mounted on lead frames. Due to the high stiffness and a comparable CTE (Coefficient of Thermal Expansion) values in the parallel and vertical flow direction, Stanyl ForTii was selected for this application. In specific the high stiffness at lead free reflow temperature range between 260-288°C makes Stanyl ForTii an unbeatable solution for such applications. Although reflow temperature is typically 260°C, we have also looked into a higher range up to 288°C in order to account for potential hot spots during assembly. Due to its high toughness before and after reflow (flexural strength), Stanyl polyamide family is one of few materials enabling such designs as applied in this concept. Any other halogen free material which would fit the temperature requirements of reflow soldering such as Liquid Chrystal Polymers (LCPs) are commonly known to be very brittle and would hence fail during later stages of package assembly. After the top and bottom part are molded, the parts will be exposed to laser Direct Structuring (LDS) to transfer all electrical tracks (Figure 5: steps 1b and 2b). Later in this paper the process is described in more detail. At a next stage, chip and magnet are inserted (Figure 5: steps 1c and 1d) and finally the two parts of package are put together (Figure 5: step 3).

Package Molding
The specific design expertise at NTM allows design and building of tools and concepts with the help of mold simulation to seek application approval before starting with tool fabrication. 2D & 3D drawings are drafted to tooling specialist's to start with micro tool fabrication. The micro tools are fabricated using state-of-art machines like Kern Pyramid Nano, Charmilles Robform Die Sinker & Hauser Jig Grinder for highly accuracy and finishing. Every process is precisely machined and quality controlled before the tool are being assembled.

The micro tool is set on the Battenfeld microsystem injection molding machine in the 10k clean room facility for better control and cleanness. The raw material is dried according to DSM material recommendation before injection molding in the micro tool starts. Micro molding process Engineer carefully process control the micro molding machine to ensure the parts are produced with high precision and stringent quality. Final inspection is supported by state of art measurement equipment and methodology to meet customer satisfaction.

978-1-4244-4722-0/09 $25.00
© 2009 IMAPS-ITALY

Making use of the good rheological properties of the high temperature polyamide Stanyl ForTii, the cover and base parts could easily be molded in an injection molding machine fit for micro molding. Even smallest details of the mold were easily transferred into the plastic.

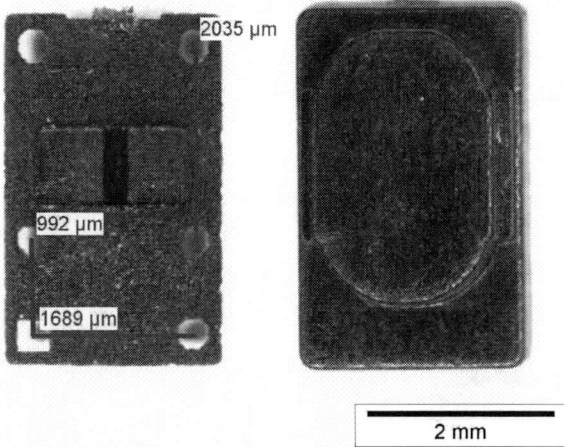

Figure 6: Top and bottom cap of the package after being micro molded at NTM

Figure 6 shows the outcome of the micro molding. Despite the ultra small feature sizes, all features have been transferred perfectly into Stanyl ForTii. Process setup is easy and fast and no flashing occurs with Stanyl ForTii.

Stanyl ForTii
The thermoplastic materials for micro molding also should be carefully selected to fit the process and required part design. Here the brand new high temperature polyamide Stanyl ForTii was selected for its good combination of thermal, mechanical and rheological properties. Parts of this material can withstand lead-free soldering conditions without degradation. Therefore the material is suitable for manufacturing miniaturized, small footprint SMT components. Stanyl ForTii is currently the only available high temperature polyamide enabling such a high flexibility in design and full compliance to lead free reflow SMT assembly. Stanyl ForTii furthermore is entirely halogen free, fully meeting OEM specifications.
With Stanyl (polyamide 46), Stanyl ForTii (polyamide 4T) and Xantar (PC/ABS) DSM Engineering Plastics is offering the broadest available portfolio of LDS grades and covers the entire temperature range in Electronics applications. A global presence of DSM Engineering Plastics, a presence since more than 20 years in the Electronics industry as well as a high application and design support level makes DSM Engineering Plastics a strong partner to OEMs and connector and component manufacturer.

LDS to selectively create tracks on thermoplastic parts
Molded parts with integrated tracks are called MID's (Molded Interconnected Devices). There are several technologies available to create MID's. Laser Direct Structuring (LDS) is the technology with the most design freedom to selectively plate plastic parts. The process to create a MID with LDS technology consists of three basic steps: molding of the thermoplastic part, selective laser ablation of the track-layout on the part and plating of the track layout.

Plastic parts of all shapes and materials fulfilling all kinds of functions are daily practice. Examples are housing parts of household appliances, mobile phones, interior parts of cars and even lunch boxes. A common and widely available process to manufacture plastics parts is injection molding. This process allows a lot of design freedom to come to 3D design solutions. This can also be translated in miniaturization. In order to do so, tooling and molding machines need to be adapted to the situation. Additional design rules should be considered. Not every molder has the capability for micro molding. The parts used in this study are molded in Nano Technology Manufacturing; they are specialized in ultra-precision- manufacturing.

After molding, the next step is laser ablation. During the selective laser ablation, the track pattern is written in the surface top layer of the plastic parts. The LDS additive in modified polymer is transformed into micro metal cores they appear on the surface of the tracks and are fixed in the polymer matrix. This allows electroless Cu plating of the tracks.

Design solutions

The design in our study demonstrates several design solutions. Between two parts a spool around a ferrite core is created, contact or soldering pads are created with only one sided lasering lasing and mechanical fixation between the cover and base part is created.

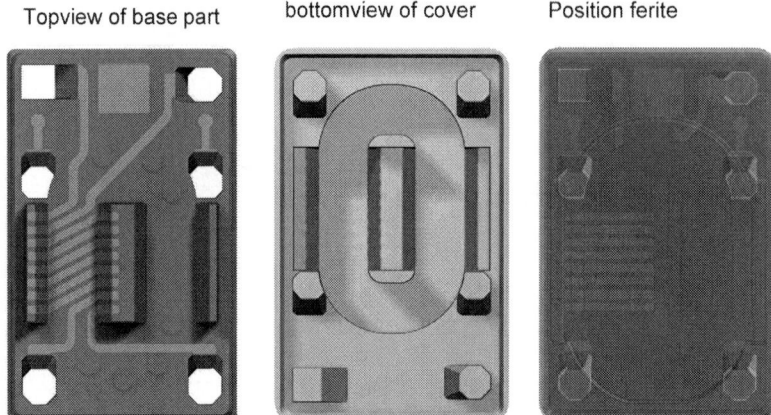

Figure 7: Top and bottom view schematics of the package as well as position of the ferrite

Compared to stitching, the windings of the spool can be more close to the ferrite core. There is no need to compensate the design for the use of stitching the wires. Hindering of the stitching head does not occur is simply no hindering. Apart form the used spool design in this study other design solutions are possible.

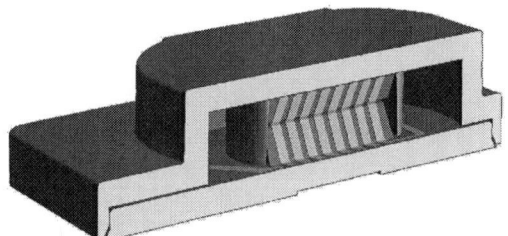

Figure 8: Assembly of top and bottom part closes the Cu tracks around the magnet and puts together both parts of the EMI filter

For ease of manufacturing it is chosen to use only one sided laser structuring to create the 3D track. Of coarse multi sided laser structuring is also possible. This allows even greater design freedom to create additional functions.

The solder pads for mounting on the PCB are part of the cover design. Due to through contacting of the legs of the cover and the tracks in the receiving holes interconnection with the tracks on the base is established. It demonstrates interconnection between tracks on 3D surfaces of different parts. In this way stacks of layers and via's are easily created. Again, only with one sided laser structuring.

Apart from fixation of the assembly on the PCB the cover and base should also be mechanically fixated. In micro-molded parts the standard snap fit solutions in plastics is not possible. There is no design space available leading to any mechanical strength. The demonstrated solution is a track, isolated from the electronic circuit, all around the inside of the cover and outside of the base. The soldering seals the parts into one assembly.

978-1-4244-4722-0/09 $25.00
© 2009 IMAPS-ITALY

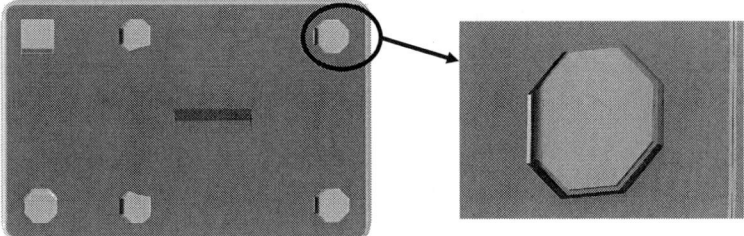

Figure 9: Realization of the solder pads by LDS

The last step in LDS process is the plating of the tracks. Due to the laser exposure the LDS additive in the polymer is chanced into Cu-cores. Those metal cores are sensitive for electroless plating. For this plating, 15 µm standard commercial chemical processes are available, like Mc Dermid bath and Roehm &Haas. Once the tracks are covered with Cu additional layers can be added by galvanic plating.

Figure 10: Optical microscopy of the two assembled LDS parts including the Semiconductor chip and magnetic coil

Stanyl ForTii

Stanyl ForTii is the very first entirely new high temperature thermoplastic introduced by any Chemical company in this millennium. Stanyl ForTii is a polyamide 4T, which DSM is marketing under the Stanyl brand family and which enlarges the DSM portfolio into the ultra high performance polymers best suitable for lead free reflow soldering applications. Stanyl ForTii is entirely halogen free meeting latest industry requirements.

Stanyl ForTii is an ultra high performance polymer meeting the highest requirements of lead free reflow soldering with a unique balance of thermal, mechanical and electrical properties. Stanyl ForTii is most suitable for demanding applications in the electronics industry/lighting/automotive/aerospace. In this project, we have selected this material due to its best fit to the required properties. In addition to the regular high performance requirements demanded by lead-free reflow soldering, the integration of

978-1-4244-4722-0/09 $25.00
© 2009 IMAPS-ITALY

components inside the package do require an excellent co-planarity of the package in order to avoid possible delamination of the chip from the polymer. Stanyl ForTii with its very high stiffness at reflow temperatures does enable this.

Conclusion

In summary, we have successfully shown the integration of Semiconductor ICs and Magnets in one thermoplastic package. LDS and micro molding technologies have enabled aggressive space reduction which can be used by OEMs to add additional functionality onto their PCBs or to simply reduce PCB size. Furthermore, OEMs can omit standalone components and hence also reduce component count which directly reduced their assembly cost.

The availability of an ultrahigh performance thermoplastic such as Stanyl ForTii from DSM Engineering Plastics has opened the door to realize LDS concepts on 3D designs in air cavity packages with high toughness, high co-planarity and excellent fit to lead free reflow soldering temperatures fulfilling JEDEC MSL 1 standard. The ease of processing of Stanyl ForTii enables an excellent material fit to the stringent requirements of micro molding.

Large panel, highly flexible multilayer thin film boards

H. Burkard, W. Kapischke, J. Link

Hightec MC AG, CH-5600 Lenzburg, Switzerland

Phone: +41 62 885 85 85; E-mail: hans.burkard@hightec.ch

Abstract

A new production line for large panel multilayer thin film boards has been installed and is now on the way to get functional. The line allows the manufacturing of flexible HiCoFlex multilayers in the format of 24" x 24". Thus larger sized panels and greater production capacity are now enabled.

HiCoFlex is a technology for producing flexible multilayer circuits using the conventional thin film techniques. The multilayer is built-up on temporary rigid glass substrates by repetitive application of polyimide layers by a spin-on process and metal layers by a sputtering and if needed enforced by galvanic deposition. Via's to underneath conductor levels are opened by laser cutting. The steps (metal layer by sputtering - photolithographic and galvanic processes - polyimide layer by spin-on, drying and curing - via's opening) can be repeated several times, resulting in a multilayer structure. Assembling and bonding of the components and tests of the circuits are possible while the film is still sticking on the glass, avoiding handling problems of conventional flexprints. At the end the flex multilayer is released from the substrate making use of a special release layer.

Applications of HiCoFlex multilayers are in the fields of high-density interconnect (HDI) technologies for sensors, industrial and medical micro systems. In addition, the unique combination of a flexible material with the electrical properties of a MCM-D offers new applications in the field of 3D-packaging. Application fields of such highly integrated modules are strongly growing areas like medical and health monitoring, both for implanted and non-implanted medical devices, sensors, portable and wearable electronic systems.

In this paper the properties and features of these flex as well as some applications are discussed.

Key words: Multilayer flex, HiCoFlex, Thin Film, 3D-package, Medical device

HiCoFlex Technology and Properties

The HiCoFlex process [1] starts with rigid substrates, alumina or glass plates, which are used as a carrier during the multilayer build-up process and the assembly of components. First a thin 'release layer' is applied on these substrates. The multilayer is built-up by repetitive application of polyimide layers using a spin-on process from the liquid solution and curing and metal layers by a sputtering and if needed enforced by electroplating. Vias between conductor levels are opened by laser or plasma processing.

Handling is easy as all process steps including flip chip assembly of thin IC's and other components, reflow soldering and testing of the circuits are done while the film is still fixed to the rigid carrier substrate. After that, the flex multilayer can be released from the substrate. The structuring technique allows line widths of 15 μm, spacing of 10 μm and vias of 30 μm. Actually circuits with up to 4 metal layers are realized. The total thickness of such a foil is about 50 μm, resulting in highly flexible foil-like circuits with excellent mechanical and electrical properties. Folding of the substrates does not impede the electrical properties of the circuitry. The minimum bending radius is smaller than 0.5

mm. The resulting 3D-packages have excellent mechanical and electrical properties.

Application fields of such highly integrated modules are strongly growing areas like medical and health monitoring, both for implanted and non-implanted medical devices, sensors, portable and wearable electronic systems.

A folded thin flex module containing a hearing aid flip-chip-set has been realised as a demonstrator.

Figure 1: Highly flexible Multilayer assembled with SMD

Integration of components

Integration of passive components, mainly RF structures and RF lines, thin film resistors, and embedding of active chips into HiCoFlex multilayers have recently been investigated. The advantages are higher miniaturizations, new ways of 3D packaging and more design possibilities.

RF lines and RF structures - Narrow and well-defined lines and gaps enabled by the thin film technology ensure a perfect high frequency performance. This allows the realization of very thin, highly flexible microstrips, stripeline and waveguide structures for RF cables and interconnections.

Different polymers for RF applications, e.g. spin-on polyimides, BCB, Kapton, LCP and similar have been studied. The handling of the foil-type polymers is described in the section 'Thin Film on Foils'. LCP and BCB have the advantage of a low water uptake, an important point for high frequencies. Losses of coplanar waveguides were measured to verify the performance until 20 GHz. Polyimides and LCP showed to be acceptable, at least to 20 GHz with a bandwidth ≥ 20 dB.

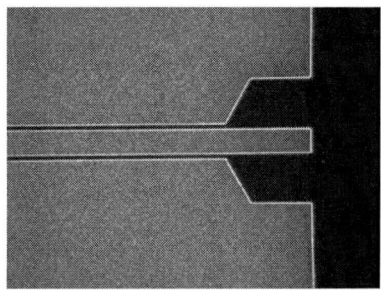

Figure 2: Contact part of a flexible co-planar waveguide test samples

PI-BCB multilayer sandwiches were analysed with a microstrip test pattern up to 40 GHz. The results indicate a good RF performance for this material combinations which is not susceptible to moisture and still very flexible [2].

Thin film NiCr resistors – Integrated resistor on polyimide (PI) were produced by standard thin film methods: Sputtering NiCr, photolithography, Ti/Cu/Ni contacts, annealing, laser trimming, protection by a further PI, release from the carrier substrate [3]. Test structures were analyzed by measurement of the resistance drift during PI curing (380°C peak temp), the temperature coefficient of resistance TCR, by a temperature life test for 1000h at 125°C, a humidity test for 1000h at 85% r.H. / 85°C and a bias of 18 VDC and a simple bending test. The evaluation showed that NiCr thin film resistors in a range of 10 Ω to 100 kΩ can be integrated into HiCoFlex, that laser trimming is possible and that the properties are nearly as on rigid alumina substrates (figures 3 and 4). The test results are: TCR: -18 ± 8 ppm/°C; the temperature life test

at 125°C, 1000h showed a drift < 0.1%; drift in the humidity tests after 1000h with 18 VDC is typically < 1%; the bending effect is typically < 1% for bending radius of 1.25 mm.

Figure 3: Resistor foil elements

Figure 4: 3500 Ω thin film resistor

Thin film Ni resistors – Integrated Ni resistors are made in a similar way. The high temperature coefficient of resistance TCR of Ni enables temperature and mass-flow sensors applications. Low thermal mass of the integrated Ni resistors results in fast response.

Embedding active chips - First attempts to embed chips into HiCoFlex were encouraging. Embedding technique has been published by IMEC [4].

Thin Film on Foils - Alternative solutions are by applying thin film techniques on commercial foils, e.g. Kapton, Liquid Crystalline Polymer (LCP) or other polymers with suitable RF properties. Methods for temporary attachment of the foils on rigid carriers during the thin film coating and their detachment afterwards have been used.

Lamination of foils with integrated thin film components into PCB

Small foil elements with integrated thin film components like resistors, RF lines and RF structures can be laminated into conventional printed flex-boards as local high-resolution part and connected to the wiring of the print. By this method expensive high-resolution features such as precision resistor arrays or RF structures in high frequency circuits, can be limited to the areas where they are

978-1-4244-4722-0/09 $25.00
© 2009 IMAPS-ITALY

needed. The concept has been proven by lamination of integrated chip foils into PCB [5].

3D Packaging of Medical Devices using Flip Chip on foldable Flex

Medical devices need a maximum of miniaturization for highest functionality in smallest volumes. An approach using ultra-thin highly flexible substrate technology, flip chip reflow soldering of thinned IC dies thereon and 3D-folding has been tested as a demonstrator in the framework of the European project SHIFT. The concept and results will be presented here.

3D packaging concept with folded thin flex for hearing aids - Besides the packaging and size aspects the folded thin flex modules are anticipated to ease the development and production of hearing aids – even at reduced costs. The reason for this is the concentration of all complex fine pitch routing onto a substrate which has very high panel utilization. This means that the main substrate in the hearing aid can be realized on standard flex or rigid boards at much lower cost. Furthermore, the compact module once designed and qualified, will allow a shorter development time and thus time to market.

Even though the CMOS technology keeps reducing the die size an increasing number of dies is expected to be found in future hearing aids due to the split up of different features/functions on dedicated dies. Dense packaging methods will therefore remain in focus. The folded thin flex seems to be an excellent solution to this demand. It is kept flat throughout the full assembly line enabling very low profile solder bumps – a technology that would otherwise call for advanced carriers. It is – obviously – thin, and allows very narrow bending radiuses taking up only a minimum extra foot print space as opposed to wire-bonded die stacks which requires space for the bonds onto the substrate. The thin flex is thus a very efficient way of die stacking especially for high-tech and medical products where size is a truly critical factor.

In hearing aids the increasing number of dies is becoming a challenge in terms of space on the substrate. For this reason it becomes beneficial to focus more on 3D packaging methods in order to increase the packaging density. One solution to this problem is the folded thin flex. A further reduction in volume can be obtained by thinning the dies in the folded stack.

The current concept for the folded hearing aid thin flex module is shown in the figure 5. The first assembly of the module has been realized using standard bumped hearing aid flip chips with a thickness of 180 μm and 90 μm solder bumps. The next step is to use even thinner dies of 50 μm and low profile bumps, 20 μm. Thinning is offered on commercial basis from several wafer service providers, but low profile bumping is still somewhat

exotic, maybe because it only makes sense on very flat substrates like the thin flex on glass carrier.

As the thin flex technology at the present stage offers one side assembly only two 180° bends are needed in order to fold the module. However, as described previously, this is no problem to the thin flex technology.

Figure 5: 3D packaging concept with folded thin flex

Ultra-thin flex technology realisation - The hearing aid thin flex has been laid out with a relatively coarse BGA array (to the right in the figure 6). The two main dies, analogue and digital CMOS, are placed in the centre being folded back-to-back. One the left side, being the top layer when folded, a memory die and a sensor die have been placed. The routing and the BGA interconnections allow cutting away this last "wing" (the top layer) in case the module needs to be even lower and optional placement of the dies can be found elsewhere in the hearing instrument.

The whole flip chip module will be populated with solder balls on the bottom side to realize a BGA structure for soldering onto a motherboard with SMD components and passives.

Bending zones of the flex substrate and bending radius can be kept very small due to the high flexibility of the ultra-thin substrate technology. The IC's are all thinned to 180μm silicon thickness. Bumps are applied by electroplating of SnAgCu onto redistributed pads for optimal flip chip design rules.

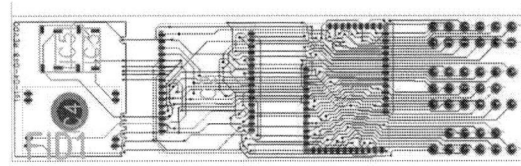

Figure 6: Substrate layout for flip chip set with

The module design was miniaturized onto a 4" square ceramic carrier with 100 single substrates on top. A processed flexible substrate is shown in fig. 7.

978-1-4244-4722-0/09 $25.00
© 2009 IMAPS-ITALY

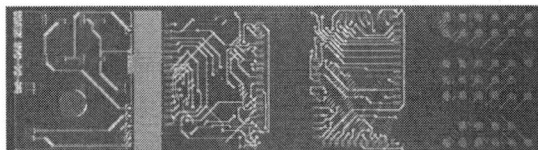

Figure 7: Thin film flexes substrate for single module after flex processing

Flip chip assembly - Multilayer substrates have to be well dried before next assembly steps to eliminate water and humidity inside the material. Wafer level assembly of 4 flip chips per module is realized on the ceramic carrier with thin film flex on top. 4 different chips have to be placed. The minimum bump pitch is 180µm. The solder bumps are deposited by galvanic processes and reflowed first. As the chips are bumped with SnAgCu solder flux application before placement is advisable. A liquid flux type was dispensed onto the pads of the substrate. The chips were placed into the wet flux depot using fully automated placing equipment.

The reflow of the solder interconnection was performed at T(max)=260°C in nitrogen atmosphere.

After reflow the assembled modules onto the carrier can easily be inspected using x-ray-microscopy [7].

Underfilling of the soldered flip chips is necessary to protect the chip surfaces and to stabilize the chip at the bonded areas before 3D-folding of the flex. Acoustic microscopy using sonsoscan was performed during underfilling setup process to check for voids within the underfiller. An acoustic microscopy image of a single assembled substrate is depicted in figure 8.

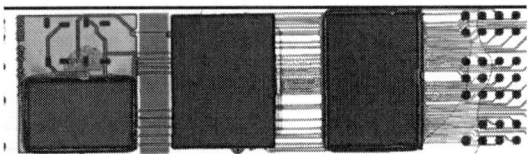

Figure 8: Acoustic microscopy image after underfilling and curing

The BGA solder balls have been attached for the first test samples after release and singulation. In production solder balling can be performed before or after flip chip soldering using preferably the same reflow process. (BGA array with solder balls see figure 9)

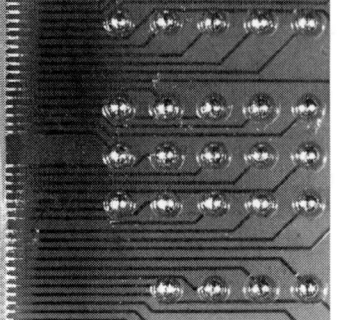

Figure 9: Solder balls on BGA pads

Figure 10: Singulated and released assembled modules on Gel-Pack

As the modules are still fixed onto the carrier the thin film substrate can be cutted using a UV laser. The release of the modules is performed afterwards.

After singulation the hearing aid modules are folded carefully. Folding is a crucial and rather difficult task as the thin flex will easily be misplaced (figure 11). A dedicated mechanical fixture is thus needed in order to guide the thin flex throughout the folding process. After folding the modules can be electrically tested before continuing through the SMD production line.

Figure 11: Folded thin film flex module with flip chips

978-1-4244-4722-0/09 $25.00
© 2009 IMAPS-ITALY

Applications of thin film flex technology

HiCoFlex multilayers are used in miscellaneous fields:

- High-density interconnect (HDI) technologies for sensors, industrial and medical micro systems,
- Ultra-thin, bendable flex boards used for 3D packaging, e.g. for hearing aids, pill cams and other, strongly miniaturized electronic applications,
- The integration of passive components, as thin film resistors, inductors and RF lines and structures,
- The resulting very thin and flexible multilayer foils can be laminated into conventional rigid and flexible PCB as a local high-resolution part,
- Embedding of thinned active chips into HiCoFlex multilayers,
- HiCoFlex on steel or other metal foils for special applications,
- Long Micro Cables (typical length ≥1.5m, width ≤100-200µm, 2-128 lines) for catheter connection, which are under study (figure 12).

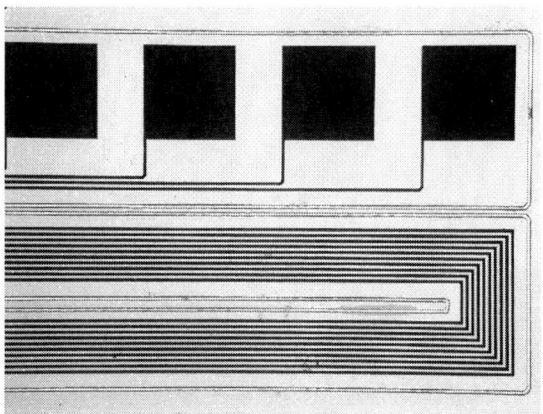

Figure 12: Long micro cables, test sample, 10 lines, length 850 mm, right end and turning point

Large area panel processing

Until recently HiCoFlex has been produced on 4" and 6" square carrier substrates. Now a new production line has been installed which allows the manufacturing of flexible HiCoFlex multilayers on the extremely large 24" x 24" glass panels. The machinery includes all process steps like substrate cleaning, vapour deposition, sputtering (figure 13), spin coating, laser direct imaging (LDI), resist develop and strip, chemical etch, electroplate, drying on hotplates and in furnaces, curing, plasma treatments, optical inspection by AOI, layer thickness measurement and electrical flying prober test.

Figure 13: Loading 24" substrate into load look of sputter tool

Thus larger sized panels and a greater production capacity are now enabled.

Conclusion and outlook

The discussion of the hearing aid module demonstrator shows that extremely high routing densities on flexible substrates can be realized using the HiCoFlex technology. The portion of the electronic system with the highest functionality is thereby interconnected and further condensed by mechanical folding. The high density module can subsequently be connected a low density periphery like power supply or in the presented case to microphones and speakers. The future integration of passive components (resistors/capacitors) into the HiCoFlex substrate provides additional potential for miniaturization. Efficient partitioning, modularization, and the combination of high and low density parts will in future enable versatile electronic system assembly.

The potential in using thin film flexible substrates is furthermore to be seen in flip chip applications with ultra-thin IC's (below 50µm thickness) and subsequent thin flip chip interconnections. Reliability and process details have been investigated using standard flex substrates and thermode soldering process with thin solder caps as well as adhesive bonding technologies using ACA [6].

The new large panel production facilities allow the manufacturing of flexible HiCoFlex multilayers on 24" x 24" susbstrates. This opens the way for larger sized parts, greater production capacity at competitive conditions to meet the increasing demand mainly from the medical and sensor field.

Acknowledgements

This work has been supported by the EU 6th framework program under contract 507352 SHIFT and funded by the Swiss State Secretariat for Education and Research SER, under project 03.0233.

978-1-4244-4722-0/09 $25.00
© 2009 IMAPS-ITALY

Special thanks to the SHIFT project partners for contributions in the fields of large area panel processing (ACREO, Norrköping, Sweden) and hearing aid demonstrator design and testing (Oticon, Smorum, Denmark) and assembly (Fraunhofer Institute of Reliability and Microintegration IZM, Berlin, Germany).

References

1) A. Fach et al., "Multilayer polyimide film substrate for interconnections in Microsystems", Microsystem Technologies, Volume 5, pp. 166-168, 1999

2) H. Burkard et al., "Ultra-Thin, Highly Flexible Cables and Interconnections for Low and High Frequencies", MicroTech 2006, Cambridge UK, 7-8 March 2006

3) H. Burkard, "Thin Film Resistor Integration into Flex-Boards", 5th International Workshop 'Flexible Electronic Systems', November 29, 2006, Munich

4) Maarten Cauwe et al., "Embedding active components as a 3D packaging solution", Advancing Microelectronics, May/June 2006, p. 15-19

5) W. Christiaens, H. Burkard, J. Link, J. Vanfleteren, "Integration of Thin Flexible RF Structures into Flexible PCB", EMPC 2007, Oulu, Finland, June 17-20

6) J. Haberland, B. Pahl et al., "Super Thin Flip Chip Assemblies on Flex Substrates Adhesive Bonding and Soldering Technology – Reliability Investigations and Applications", Proc. of IMAPS 2006, San Jose

7) B. Pahl, T. Loeher, H. Burkard, J. Link, A. Petersen, R. Aschenbrenner, "Flex Technology for Foldable Medical Flip Chip Devices", IMAPS Device Packaging Conference, 3D Workshop, March 17-20, 2008

Closing Technology Knowledge Gaps
Projects arising from the iNEMI Technology Roadmap

Bob Pfahl*, Jim Arnold* and Grace O'Malley^

*iNEMI, Virginia, USA and ^iNEMI, Limerick, Ireland.

Tel: +353 87 9040363   Fax: +353 61 351935   Email: gomalley@inemi.org

Abstract

iNEMI has been publishing its Technology Roadmap biannually for the last 14 years. The strength of these roadmaps has been the depth of the knowledge from experts worldwide combined with the breath of view taken across the whole electronic manufacturing supply chain. Every new roadmap helps identify the technology gaps in various aspects of the supply chain. These gaps then stimulate and encourage dialogue to develop consensus-based strategies resulting in industry-based joint project work or promoting research areas to the academic communities. Since 2007, there has been a focus on three main technology areas for iNEMI industry-based projects: Energy and the Environment, Miniaturization and Medical Electronics. This paper will cover the goals of these three areas and the strategies being adopted to address the present and future technology gaps, as identified in the most recent roadmaps. Results from key projects in the three areas will be presented. These will include projects on Pb-free alloy alternatives and HFR-free substrates; higher density testing and nanosolders, and Medical Components Reliability.

Key words: Environmental, Pb-free; HFR-free, Nano-solder, Boundary Scan Test, Functional Test, Medical Electronics, Miniaturization

Introduction

The International Electronics Manufacturing Initiative (iNEMI), an industry-led consortium, is advancing electronics manufacturing technology through identification of technology needs, development of industry infrastructure, accelerated deployment of new technologies, dissemination of efficient business practices and stimulation of standards. It is the only corporate membership organization focused exclusively on electronics manufacturing technology from IC to system.

iNEMI forms industry-wide projects to address technology and infrastructure gaps identified through the consortium's roadmapping and gap analysis activities. iNEMI identifies areas: (i) that are not currently being addressed by other industry efforts, and (ii) where members can collectively have an impact. Using a proven methodology projects aim to eliminate gaps through:

- accelerated deployment of new technology
- development of industry infrastructure
- dissemination of efficient business practices
- stimulation of standards

Projects are organized within Technology Integration Groups (TIGs) and iNEMI membership is usually required for project participation. Member companies invest their respective skills and resources in these deployment projects that will improve the global industry supply infrastructure. Our current collaborative efforts emphasize three areas: Energy and the Environment, Miniaturization and Medical electronics.

Energy and the Environment

The electronics industry's interest in and concerns with environmental issues continues to grow. To remain competitive the industry must keep pace with the continuing emergence of material restrictions, end-of-life requirements, customer preferences and legal requirements for the development of energy efficient products as seen in Table 1. Over the last number of years iNEMI members have driven a number of projects and held workshops that address many of these areas [1, 2].

Table 1: Major Focus Areas

Materials	Continuing emergence of material restrictions
Energy	Energy efficiency requirements and renewable energy
Recycling/Reuse	End-of life requirements
Eco-Design	Holistic Eco-design requirements
Sustainability	Sustainable business practice

978-1-4244-4722-0/09 $25.00
© 2009 IMAPS-ITALY

For example iNEMI has a long and successful history in developing the area of Pb-free solder alloys. Since 1999 it has coordinated projects developing the supply chain and the manufacturing capabilities for Pb–free products as summarized in Table 2. These projects have helped increase communication and cooperation in the industry as well as contributing to the development of standardization of the industry supply chain [3]. In particular, recommendations from the Pb-free assembly and rework project provided manufacturing data to IPC and JEDC to develop the J-STD-020C; Recommendations from both the Materials Composition Data Exchange project and the RoHS Materials Declarations project formed the basis of the IPC-1725, Materials Declaration Management [4, 5].

This Pb-free work still continues as alternative alloys are now being considered in the industry. More recently new projects in the areas of Halogen Flame Retardant (HFR) free have also begun.

Table 2: Table of Completed Environmental Projects

Project Name	Status
HFR-free PCB Material Evaluation	Completed
Halogen-Free – Phase I	Completed
Tin Whisker User Group	Completed
Tin Whisker Accelerated Test	Completed
Tin Whisker Modeling	Completed
Lead Free Assembly & Rework – Phase 1 & 2	Completed
RoHS Transition Task Group	Completed
Lead-Free Assembly	Completed

Pb-free Alloys Alternatives: The impact of new alloys with lower levels of Silver (Ag) on the entire PCA system of materials and on the long term reliability is not well understood. The goal of this project work is to develop the framework for testing new Pb-free alloys. The first phase of this project focused on analysis of existing knowledge and assessment of critical gaps and on driving standards to help manage supply change complexity and risk. Team members included Agilent Technologies, Alcatel-Lucent, Celestica, Cisco, Cookson Electronics, Delphi HP, Huawei, Indium, Intel, Jabil, Nihon Superior, Plexus, Sanmina-SCI, Senju Comtek and Vitronics Soltec. Results of phase 1 of this project have been published [6]. Phase 2 of this work which will characterize Pb-free alloy alternatives is now underway. The first thrust of this work is to characterize the thermal fatigue an acceleration behavior of alternative alloys through accelerated thermal cycle testing. The second thrust is to develop a set of test data requirements that will allow OEMs and other to evaluate alloy properties against their requirements [7]. These requirements and recommended test methods will be proposed to the IPC Solder Product Value Council to consider for standardization.

HFR-Free Technology Leadership Program: The electronics industry is moving towards the elimination of Halogenated Flame Retardants (HFRs) from Printed Circuit Board (PCB) materials. While mobile phone manufacturers are well along the way, the next area of impact will likely be driven by the high volume consumer computer applications. This industry wide conversion to HFR-free materials faces numerous challenges including:

- Reliability of materials with alternative flame retardants has not been fully qualified
- Complete technical specifications have not been established for various product applications
- Incomplete design knowledge in segments of the supply chain increase risk of conversion issues
- A rapid complete conversion of computer products will have a major impact on the supply chain and needs to be coordinated.

Thus, the industry needs to test processes and product performance to optimize product quality for a smooth transition. This project has also being subdivided into two parts.

The HFR-Free Signal Integrity group, co-chaired by Intel and Cisco, is focused on the critical electrical properties of HF dielectrics ensuring suitability for high-speed digital designs parameters (Figure 1). The team will define key performance characteristics and test criteria, design test vehicles and test methodologies and identify candidate materials leveraging standards where possible. The project will also assess technology readiness and identity gaps plus address manufacturing capability and supply capacity [8].

978-1-4244-4722-0/09 $25.00
© 2009 IMAPS-ITALY

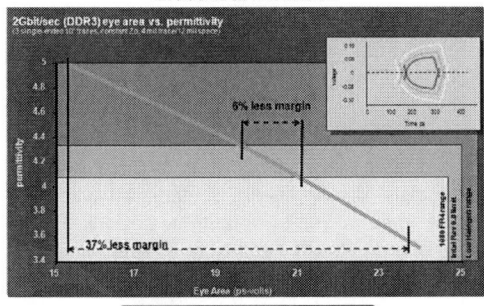

Figure 1: Example of Electrical Concern with HFR-free materials

The second HFR-free group, chaired by Intel, is studying the key mechanical properties of HFR free materials, such as outlined in Figure 2. In particular they will initially look at issues such as solder joint reliability, via and Plated Through Hole (PTH) reliability, pad cratering, and warpage. The project will identify technical risks, and flag unexplored issues related to transitioning to HFR-free PCB materials in high volume manufacturing. [9].

Basic Materials Properties

Micro and macro hardness
Glass transition temperature (Tg)
Decomposition temperature (Td)
Moisture absorption
Fracture Toughness of Resin / Resin Cohesive Strength
Stiffness
Dk & Df
Coefficient of thermal expansion (z-axis and x-, y-axes)
Flexural strength

· Basic Material Properties have been measured and shown to be different from the Brominated Epoxy baselines.

· Test methods adopted from IPC/ASTM/other sources may be refined or modified to give a quantifiable value.

Figure 2: Example of Properties to be Studied

Miniaturization

Miniaturization affects everything from design and test to materials and packaging. As the size of electronic components and circuitry continues to shrink there are an increasing number of issues that have to be addressed. Several iNEMI projects addressing the challenges related to miniaturization have already been completed as shown in Table 3. For the purposes of this paper three ongoing projects which address different aspects of the drive to miniaturization will be discussed.

Table 3: List of Completed Projects in the Area of Miniaturization

Project (Complete)	Status
Pb-Free Nano-solder – Phase 1	Completed
Nano-Attach – Phase 1	Completed
DPMO (Defective Parts per Million Opportunities)	Completed
Pb-free Component & Board Finish Reliability	Completed
Advanced Embedded Passives Technology	Completed
High Frequency Material Effects on HDI Formation	Completed
Optoelectronics for Substrates	Completed
Fiber Optic Splice Improvement	Completed
Fiber Optic Signal Performance	Completed
Fiber Connector End-Face Inspection,	Completed

Boundary Scan Adoption: Increasing circuit densities and speeds are quickly reducing electrical test point access for the printed circuit assembly test. Boundary scan is a technology that will allow continued testability of Printed Circuit Assemblies (PCA), but its use requires that it be designed into semiconductors. Currently not all semiconductor vendors support boundary scan. Wider availability of complying devices is necessary to enable cost-efficient and effective board test of future designs. In addition tools to support boundary scan based test need to be developed and integrated into manufacturing test equipment. The project, co-chaired by Intel and Cisco, was organized to promote wider industry adoption of boundary scan (IEEE 1149.1, 1149.6, P1581 and others). Efforts will focus on encouraging semiconductor vendors to include technology in their products, promoting the development of tools by ATE vendors to support boundary scan-based board test and promoting the development, refinement and adoption of synthesis and verification tools to assist implementation. Team members include Dell, Cisco Systems, Agilent Technologies, Corelis, TRI, Huawei, HP and Asset.

The goal for the project is to enhance current IPC/JEDEC standards to promote the use of spherical bend test methods and the use of a common type of strain for reliability risk assessment. The project has kicked off with a survey of the suppliers and users to identify the current levels of Boundary Scan implementation in the industry today and the projected short term use [10].

Functional Test Coverage: This project is working to develop a standardized functional test coverage assessment method that will allow reliable comparisons of test coverage among different test environment, test conditions and different assessors. Functional test equipment is more customized than structural test equipment, including hardware, software and test generation process. This variability in the test environment and the fact that the functional test requires the board to perform its native functions during testing, make it difficult to determine whether a resistor value is correct in two

978-1-4244-4722-0/09 $25.00
© 2009 IMAPS-ITALY

differing designs. A coverage assessment method for functional test must be able to accommodate differences in the test environment and yet still offer information that allows the coverage of the test to be comprehended and compared to other test stages. Creating more consistency in functional test coverage assessment will open opportunities to automate reports which enable informed decision making on issues pertaining to test.

To date the team has complied usage models of function test coverage and developed and distributed a survey to gauge industry use of functional test coverage. Work is focusing on the second phase including:

- Compiling a list of defects that encompass structural faults referencing existing categorization methods
- Add functional test specific defects to structural defects list
- Define assessment methods
- Create confidence margin and weighting factors that allow emphasis for important assessment items
- Develop guidelines for assessing coverage and assigning confidence margin

Pb-free Nano Solder: Nanotechnology encompasses many diverse disciplines to allow the manipulation of matter at the atomic level, enabling radical new approaches to material property enhancement and synthesis. Nano–material solutions have the potential to augment and enhance the traditional reliability solution and enabling new product concepts [11]. The first phase of this project investigated the application of nano technology to surpress Pb-free solder reflow temperature. Pb-free materials and products require the use of solders that have higher melting points and, therefore, require higher processing temperatures than SnPb solder. These higher reflow temperatures can negatively affect product reliability, require tougher qualification requirements for components, and sometimes result in significant changes in manufacturing processes. The project successfully demonstrated production of nanoscale particles of oxide free tin and confirmed significant melting point depression The project also made strides in improving coalescence of nano particles. Phase 2 of the project will focus on identifying compatible fluxes that enable homogenous melting of the nanometal particles and demonstrating the ability to reflow solder and form homogenous solder joints.

Medical Electronics

The medical electronics equipment market was estimated to reach $66B in 2007, growing to $84B by 2013 [12]. This constant expanding market of an aging population with higher expectations for continued full life is providing an opportunity for the development of specialized medical systems. These medical products increasingly rely on electronics for their functionality. The range of electronics components covers a wide spectrum; everything from stand-alone large diagnostic systems to portable units to small implantable devices. Each has unique reliability and operating requirements. One of the common technology gaps identified in roadmaps over the years has been the lack of minimum guidelines or standards to ensure the reliability of medical electronics. Component suppliers must satisfy many different customer requirements and the lack of standardization can lead to duplication of effort and delays in adopting new technologies. For example the whole medical electronics industry must prepare for the future Pb-Free requirements. Thus development of guidelines and test standards which could be used by component suppliers to assure the reliability quality and consistency of electronic components used in medical applications and speed up the adoption of new technology requirements are very desirable.

Medical Components Reliability Specifications Project: This project was set up to address the unique performance requirements and use environments that are found in the medical products sector. The goal of this project was to develop guidelines and methods to assess component reliability as related to implantable medical or other life critical applications. Team members included Boston Scientific, Dyconex, Micro Systems Engineering, Medtronics, Texas Instruments, Celestica, and NIST. The focus was on the four commodity categories: discrete, array packages, substrates and interconnects, and hybrids. The teams approach was to develop an understanding of the expected failure modes and mechanisms, use conditions and comparatives rate of each type of failure. Once established, test methods could then developed to accelerate failures and could then be used to improve the reliability of those component types where a correlation between failure and acceleration methods can be established. Initial work established the data used to create a working Design of Experiments (DOE) matrix.

As part of this project a review forum was held with US Food and Drug Administration (FDA) in November 2008. The goal of this iNEMI Medical TIG Forum was to present to the industry the Technology Integration Group's (TIG) process for the establishment of a minimum set of requirements for electronics components for application in life critical applications developed in the TIG's Medical Component Reliability Project Review Project. In addition, there was a review of the TIG's validation process, demonstrated in the Medical Reliability for

MLCC Project, which is determining accelerated life test methods of long-term leakage and breakdown failures of Multi-Layer Chip Capacitors (MLCC) [13].

Medical Reliability for MLCC Project:

The goal of this project is to determine the accelerated life test methods of long term leakage and breakdown failure of Multi Layer Ceramic Capacitors (MLCCs). The construction of the MLCC (multi-layer ceramic capacitor) for example in Figure 2a is ideal. It is free of voids, has good wetting and fillet, there is no detachment at the ceramic, and coverage of Ag/Cu layers is uniform. In comparison, the photo in Figure 2b shows blemishes on the case of the MLCC. Currently, it is unknown whether these blemishes cause reliability concerns. iNEMI's Medical Reliability for MLCCs Project is working to develop an accelerated test method that will help anticipate long-term leakage and break-down failures.

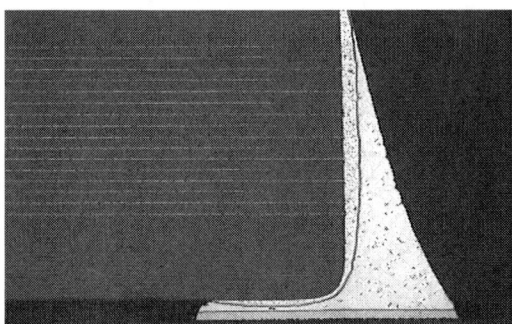

Figure 2a: Ideal

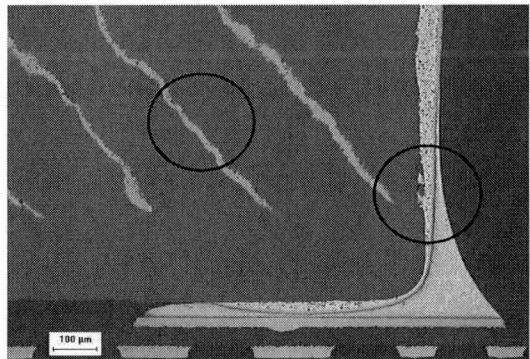

Figure 2b: Highlight Areas of Potential Concern

The Design of Experiment (DOE) matrix will be used to establish correlation between accelerated tests and recommended test methods. This project has designed and fabricated a test vehicle populated with MLCC components from multiple suppliers using industry standard methods and testing is being done at a NIST facility. Initial electrical test has been completed including insulation resistance, capacitance and dissipation factor. Thermal testing is underway and failure analysis has started. Remaining DOE cells are planned for thermal and vibration testing.

Acknowledgements

The authors would like to acknowledge the efforts of all the project team members and their organizations in conducting these projects.

References

[1] N. Grayeli et al, iNEMI Sustainability Summit, Schaumburg IL 2009.

[2] R. Pfahl et al. "Future Initiatives for Sustainability" Proceedings of Electronics Goes Green 2008, Berlin Germany September 2008.

[3] E. Bradley et al , "Lead Free Electronics", Wiley Interscience Publishers, New Jersey, Chapter 1, pp. 9-43, 2007.

[4] M. Kelly et "Pb-free Reflow and Rework", Circuits Assembly, pp 32-35, November 2004.

[5] R. Kubin "Roadmap to Compliance" proceedings APEX 2005, February 24 2005.

[6] G. Henshall et al, "iNEMI Pb-Free Alloy Alternatives Project Report: State of the Industry", Proceedings SMTA International 2008, Orlando, Florida, August 2008.

[7] G. Henshall et al, "Addressing Opportunities and Risks of Pb-free Solder Alloy Alternatives", Proceedings of EMPC 2009, Rimini, Italy, June 2009.

[8] "HFR-free Signal Integrity Initiative", iNEMI and ITRI HFR-Free leadership meeting, Taipei, Taiwan, April 15, 2009.

[9] "HFR-free PCB Materials Initiative", iNEMI and ITRI HFR-Free leadership meeting, Taipei, Taiwan, April 15, 2009.

[10] "iNEMI Boundary Scan Adoption Survey", www.inemi.org

[11] A. Rae et al. "Emerging Nanotechnology and its Effect on Electronics Manufacturing", Proceedings of SMTA International, September 26, 2005.

[12] "Executive Summary", iNEMI 2009 Technology Roadmap, pp15-17, January 2009.

[13] A. Primavera, "Medical Components Reliability Specifications Meeting – MLCC Phase I Wrap Up and Phase II Introduction", Proceedings of iNEMI Medical TIG Forum, November 14, 2008.

978-1-4244-4722-0/09 $25.00
© 2009 IMAPS-ITALY

Addressing Opportunities and Risks of Pb-Free Solder Alloy Alternatives

Gregory Henshall, Ph.D.
Hewlett-Packard Co.
Palo Alto, CA, USA
greg.henshall@hp.com

Robert Healey and Ranjit S. Pandher, Ph.D.
Cookson Electronics South Plainfield, NJ, USA

Keith Sweatman and Keith Howell
Nihon Superior Co., Ltd. Osaka, Japan

Richard Coyle, Ph.D.
Alcatel-Lucent Murray Hill, NJ, USA

Thilo Sack and Polina Snugovsky, Ph.D.
Celestica Inc. Toronto, ON, Canada

Stephen Tisdale and Fay Hua, Ph.D.
Intel Corporation Chandler, AZ and Santa Clara, CA, USA

and

Grace O'Malley
iNEMI, Limerick, Ireland

Abstract

Significant innovations in Pb-free solder alloy compositions are being driven by volume manufacturing and field experiences. As a result, the industry has seen an increase in the number of Pb-free solder alloy choices beyond the common near-eutectic Sn-Ag-Cu (SAC) alloys. The increasing number of Pb-free alloys provides opportunities to address shortcomings of near-eutectic SAC, such as the poor mechanical shock performance, alloy cost, copper dissolution, and poor mechanical behavior of joints in bending. At the same time, the increase in alloy choice presents challenges in managing the supply chain and introduces a variety of technical and logistical risks, such as a potential decrease in thermal fatigue resistance and the complexity of managing process parameters given the variability of alloy compositions.

This paper summarizes the results of an iNEMI project to address the opportunities and risks of new Pb-free alloy alternatives. The results of our analysis of the state of industry knowledge on Sn-Ag-Cu alloy alternatives are provided, and focus areas for closing key gaps are identified. Progress in updating or creating industry standards to manage the introduction and use of new alloys is also presented. Finally, our plans to investigate thermal fatigue reliability of new alloys are described.

Key words: Pb-free solder alloy, microalloying, low silver alloys, lead free, reliability

1. Introduction

The electronics industry recently has seen an increase in the number of Pb-free solder alloy choices beyond the near-eutectic Sn-Ag-Cu (SAC) favored during the initial adoption phase of RoHS. These developments have been driven by both processing and reliability problems experienced by different segments of the industry. In particular, new wave solder alloys were developed with the intent of addressing concerns with copper dissolution, barrel fill, wave solder defects, and the high cost of alloys containing significant amounts of silver. The poor mechanical shock performance of near-eutectic SAC alloys has driven the handheld product segment to consider the development of low Ag or low Cu ball alloys to improve the mechanical strength of BGA and CSP solder joints, especially under dynamic loading conditions. Most recently, investigations into new solder paste alloys for mass reflow have begun. The full impact of these materials on Printed Circuit Assembly (PCA) reliability though has yet to be determined.

The aforementioned increasing number of Pb-free alloys available provides opportunities to address the issues described above. At the same time, these alloy choices present challenges in managing the supply chain and introduce a variety of risks. For example, high melting point, low Ag alloys represent a risk in the reflow process if not managed properly. They may also present risks for thermal fatigue failure of solder joints in some circumstances, though much appears to be unknown about the impacts of Ag, Cu, and dopant concentration on thermal fatigue resistance of these alloys. Further, many "high reliability" OEMs have not switched to Pb-free technology and have rigorous requirements for evaluation and qualification of Pb-free materials and processes. The introduction of new alloys creates a "moving target" for these companies in making their transition (or partial transition) to Pb-free technology. Also, experience with near-eutectic SAC alloys is small relative to Sn-Pb, but new alloys have even less of a database and track record. Overall, there is a general lack of knowledge throughout the supply chain regarding new Pb-free alloys, their properties, advantages, and risks.

Thus, the wide variety of Pb-free alloy choices is both an opportunity and a risk for the electronics industry. In order to take advantage of the former while minimizing the impact of the latter, much still needs to be learned and the visibility to existing knowledge needs to be improved. In addition, an assessment of critical knowledge gaps needs to be performed so that industry efforts can be focused on them and not on repeating investigations into issues already resolved. Further, standards need to be updated and improved to account for the new alloys and to better manage the risks they present.

Addressing these concerns is the goal of the recently formed iNEMI Alloy Alternatives Team. This team is comprised of representatives from 16 companies spanning the entire supply chain: solder suppliers, component suppliers, EMS providers, and OEMs.

Overall, the goals of this project are to: (1) help manage the supply chain complexity created by alloy choices, (2) address reliability concerns and, (3) highlight the opportunities created by the new Pb-free alloy alternatives. Specific goals in Phase 1 include:

- Assess existing knowledge and identify critical gaps related to new Pb-free alloys. Provide technical information to the industry that will make selection and management of alloys easier.
- Raise awareness of this information through publication and presentation of findings.

- Propose a methodology and set of test requirements for assessing new alloys.
- Work with industry standards bodies to address standards that require updating to account for new alloys.
- Use findings to drive follow-on work in Phase 2.

2. Considerations in Alloy Selection

The Evolution of Tin-Silver-Copper Alloys: The first phase of the transition to lead-free solder alloys was based around the Sn-Ag-Cu (SAC) eutectic. This was based initially on various industry consortia projects, such as the National Center for Manufacturing Center (NCMS) alloy down-selection study, and later strengthened by the original iNEMI Pb-free reliability study. Although the initial motivation for the addition of silver to the tin-copper eutectic was the approximately 10°C reduction in melting point, the SAC alloys also exhibited a substantial increase in flow stress and improved thermal fatigue life. Because of concern about the cost of silver and in the hope of a avoiding a patent held by Iowa State University, the Japanese Electronics Industry Association (JEITA) and then the IPC recommended use of the hypoeutectic alloy commonly known as SAC305 (Sn-3.0Ag-0.5Cu). However, many companies, particularly in Europe, chose to stay with the higher silver SAC405 because of the advantages that a eutectic alloy offers, in particular a lower incidence of shrinkage cavities and a lower melting temperature and pasty range.

The transition to Pb-free solder happened to coincide with a dramatic increase in the popularity of hand-held devices, such as cell phones, in which the disadvantage of the high flow stress of the near eutectic SAC alloys quickly became apparent. The high flow stress was accompanied by high elastic stiffness (modulus), which led to brittle solder joint failures when these hand-held devices were accidentally dropped. These failures occurred in the intermetallic compound (IMC) layers between the bulk solder and the pad or by printed circuit board (PCB) "cratering." The reason for these failures is that high stresses are transmitted to the IMC layers or to the underlying laminate rather than being absorbed by strain in the solder itself, as happened with the soft, elastically compliant Sn-Pb eutectic solder. This topic is discussed in more detail later in the paper.

Microalloy Fundamentals: Microalloying is the addition of an element, other than the major constituents, that has the effect of modifying the behavior of the alloy in a way advantageous to its performance. The level of the microalloying addition is typically 0.1% or lower. Thus, in parallel with the evolution of SAC alloys to lower levels of silver, the impact of microalloying additions to the

978-1-4244-4722-0/09 $25.00
© 2009 IMAPS-ITALY

properties of Sn-Cu eutectic has been explored. Such additions, e.g. Ni, have been shown to benefit the high strain-rate performance of Sn-Cu and SAC alloys. For example, Sweatman et al. [3] demonstrated improved fracture toughness at high strain rates through microalloy additions of Ni and Ge to eutectic Sn-Cu, Fig. 1.

The performance issues with SAC305 have forced the industry to investigate (and offer) modified alloys with reduced or no silver, and with microalloy additions. Some problems potentially addressed by reduced Ag content and microalloy additions are set out in Table I.

Table I. Problems with high silver SAC alloys and possible solutions.

Problem	Reduce or Remove Silver	Micro-alloy
High Flow Stress	X	X
Brittle Joint Failure	X	X
Low Impact Strength	X	X
Shrinkage Defects	X	X
Copper Erosion		X
Cost	X	

Some commonly investigated elements for microalloying are: nickel (Ni), bismuth (Bi), phosphorus (P), germanium (Ge), cobalt (Co), indium (In), and chromium (Cr), with several already being used commercially. Some elements selectively incorporate into the interfacial intermetallic layer to: (1) control the IMC thickness, (2) slow the growth of the IMC in service, (3) modify its morphology, (4) prevent disruptive phase changes, and (5) increase toughness. Some microalloy additions go into solid solution within the tin matrix to increase both strength and ductility, and thus reliability, while others control oxidation. For example, the properties and behavior of the tin-copper-nickel alloy have been enhanced by additions of Ge and P as antioxidants [4]. The effect of an antioxidant, such as Ge, is to produce less dross and to provide resistance to tarnishing during exposure to elevated temperature.

Nickel is perhaps the most common microalloy additive. The beneficial effect of microalloying with nickel, which was first identified in the tin-copper eutectic, was found to extend to the low silver SAC alloy joints.

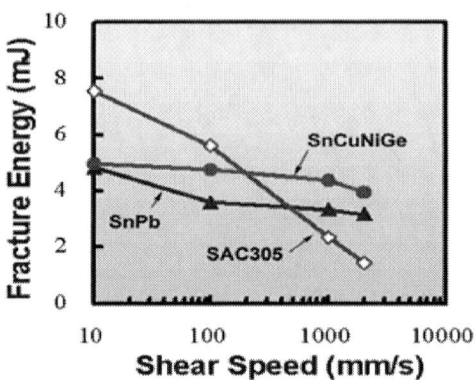

Figure 1. Fracture energy of 0.5mm BGA spheres as a function of shear impact speed [3].

3. Mechanical Shock Reliability

The major motivation for component and solder suppliers to develop new alloys for BGA/CSP solder balls was to improve the mechanical shock performance relative to SAC 305/405. In the past few years, a significant number of studies have been performed to assess the mechanical shock performance of new Pb-free alloy interconnections on area-array packages. These studies consistently show that low Ag (<3%) SAC alloys perform better in mechanical shock (drop) testing than alloys with Ag content of 3% or more [4,19-22]. Fig. 2 illustrates this finding for two low Ag alloys relative to SAC405.

Figure 2 also shows the positive impact that many studies have demonstrated on copper surfaces for the addition of microalloy additions, specifically the addition of Ni to the SAC125 base alloy for LF35. Another example is provided in Fig. 3 for the addition of 0.03% Cr to a SAC105 alloy with 0.1% Ni. Other elements that have been added to improve properties include Bi, Co, In, and Ge.

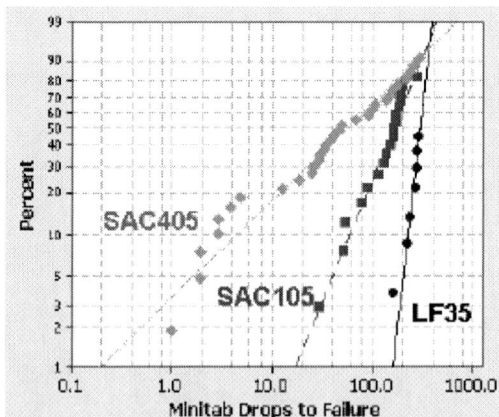

Figure 2. Mechanical shock performance for SAC405, SAC105, and LF35 (Sn-1.2Ag-0.5Cu+Ni). After Kim et al. [21].

978-1-4244-4722-0/09 $25.00
© 2009 IMAPS-ITALY

Various reasons have been cited for the improved mechanical shock performance of low Ag and doped SAC alloys, particularly on copper surfaces. For example, Pandher et al. [4] provided data showing that microalloying additions slow inter-diffusion, thus reducing IMC thickness or propensity for void formation. Further, they showed that small amounts of Ni can decrease Cu_3Sn IMC growth, improving reliability. Finally, these authors noted that low Ag content will decrease the strength and elastic modulus of the solder, transferring less stress to the solder/substrate interface compared to a high Ag alloy. Similarly, work at Intel emphasized that low elastic modulus and low yield strength improve the mechanical shock resistance of low Ag alloys, and that optimization of these properties requires increasing the amount of primary Sn relative to the Ag_3Sn and Cu_6Sn_5 phases in the alloy [19,21]. H. Kim et al. [19] also found that the majority of cracking in SAC405 solders was through the IMC layer (package side). Cracking in the SAC105 joints was more complex, with cracks going through the bulk solder near the IMC layer and in the IMC. Pandher et al. [4] found that when small amounts of Cr and Ni are combined in low Ag alloys the occurrence of flat, brittle interfacial fractures (Mode 4) are reduced 80% compared to the base alloy without additives.

Syed et al. [20, 23] provided one note of caution when making conclusions regarding mechanical shock performance of various alloys: the effect of solder pad surface finish. They found that SAC125 + Ni did not produce a significant drop/shock performance improvement over SAC305 for Ni/Au package finish (with Cu-OSP PCB finish). However, this alloy was the best performer for Cu-OSP finish on both the package and the PCB. Other literature data also indicate a strong dependence of mechanical shock performance on pad finish [24, 25].

Still, it now appears clear that reduced Ag content solder ball alloys improve mechanical shock performance over near eutectic alloys with Ag content ≥3%.

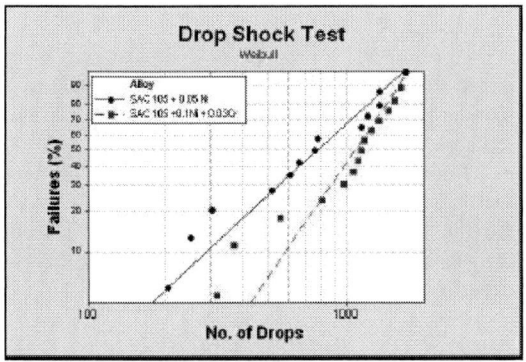

Figure 3. Addition of 0.03% Cr improves the mechanical shock performance of SAC 105 + 0.1Ni alloy. After Pandher et al. [4].

4. Thermal Fatigue – SAC Alloy Alternatives

Development of Thermal Cycling Data: The combination of thermal fatigue and solder joint creep is considered a major source of failure of surface mount (SMT) components [6]. The standard technique for assessing susceptibility to low cycle fatigue failure commonly is referred to as thermal cycling or accelerated thermal cycling (ATC). The thermal fatigue reliability of eutectic Sn-Pb solder has received thorough treatment in the literature and generally is well understood. Reliability of Pb-free solders, however, continues to be a topic of intense study and debate as the conversion to Pb-free solder and processes proceeds throughout the electronics industry [7-11].

Due to resource limitations, the costs, and time required, only a few thermal fatigue studies have been performed on new Pb-free alloys. Although solder suppliers have played an active role in Pb-free solder development, there is little precedent for these companies to serve as a primary source for ATC fatigue data. OEMs and consortia, from which ATC data traditionally have originated, have yet to publish much ATC data on the new alloys.

Challenges Evaluating ATC Test Data: Little thermal fatigue data exist for lower Ag alloys, such as SAC105. This poses a potential problem because the lower Ag alloys are being implemented in components used in products with long life, high reliability requirements. Unpublished Unovis consortium test data on a custom area array substrate indicate that SAC205 has slightly better reliability than either SAC405 or 305 [13]. One of the most detailed studies on a commercial area array component was by Kang et al. [14]. Their results suggest that a low Ag alloy will have better thermal fatigue reliability than a high Ag alloy.

A primary objective of the Kang study was to determine if low Ag content improved thermal fatigue resistance by minimizing formation and growth of Ag_3Sn intermetallic platelets. Kang also examined cooling rate and thermal cycling profile using a CBGA package with a nominal characteristic lifetime of 1000 cycles. The alloy comparison of interest in this study was SAC387 to SAC219.

The ATC data from the Kang study are summarized in Table II. The lowest lifetimes using the 0/100°C temperature cycle were recorded with the longest dwell time (120 min cycle). This confirms the hypothesis that SAC alloys have reduced reliability at longer dwell times. Slow cooling produced the best reliability regardless of ATC conditions, which the authors attributed to the improved microstructure. Slow cooling produces a coarser β-Sn microstructure that is more fatigue resistant, more ductile, and has lower residual stresses from SMT assembly. For the 120 min. cycle, the low Ag alloy (SAC219) recorded the best reliability. However, a systematic and consistent

978-1-4244-4722-0/09 $25.00
© 2009 IMAPS-ITALY

impact of Ag content on ATC life across all test conditions was not established, as indicated in Table II.

Table II. ATC test data from the Kang CBGA study [14].

Alloy Composition (cooling rate)	ATC Stress Conditions		
	0 to 100°C, 30 min	0 to 100°C, 120 min	-40-125°C, 42 min
Sn-3.8Ag-0.7Cu (Slow)	1,408 (7.25)*	1,012 (6.92)*	455 (6.12)*
Sn-3.8Ag-0.7Cu (Fast)	1,164 (7.06)	982 (6.89)	392 (5.97)
Sn-2.5Ag-0.9Cu (Slow)	1,212 (7.10)	1,108 (7.01)	446 (6.10)
Sn-2.5Ag-0.9Cu (Fast)	1,200 (7.09)	1,054 (6.96)	384 (5.95)
Sn-2.3Ag-0.5Cu-0.2Bi (Slow)	1,212 (7.10)	953 (6.86)	407 (6.01)
Sn-2.3Ag-0.5Cu-0.2Bi (Fast)	1,130 (7.03)	934 (6.84)	376 (5.93)
Sn-2.1Ag-0.9Cu (Slow)	1,224 (7.11)	1,188 (7.08)	372 (5.92)
Sn-2.1Ag-0.9Cu (Fast)	1,064 (6.97)	1,012 (6.92)	384 (5.95)

In contrast to Kang's results, a study by Terashima et al. [15] found that increasing Ag content increases the thermal fatigue resistance of SAC solder. Their results, summarized in Fig. 4, show that: (1) the 1% Ag alloy had the fastest failure rate and (2) the 4% Ag alloy had twice the cycles to first failure (N_0) as the 1% Ag alloy. Similar to the findings of Terashima et al., the more recent data of Henshall et al. shown in Figure 5 for a 676 PBGA package suggest that low Ag alloys perform worse in accelerated thermal cycling than alloys with high Ag content.

To date, there are very limited published studies on the impact of microalloy additions on thermal fatigue performance. Recently, solder suppliers have begun generating such data. Pandher et al. recently published data that suggest the addition of bismuth (Bi) to low Ag alloys significantly improves temperature cycling performance, while other additives such as nickel have little to no effect in improving temperature cycling performance [5]. A full understanding of how microalloy additions affect thermal fatigue performance represents a major gap in our knowledge as an industry.

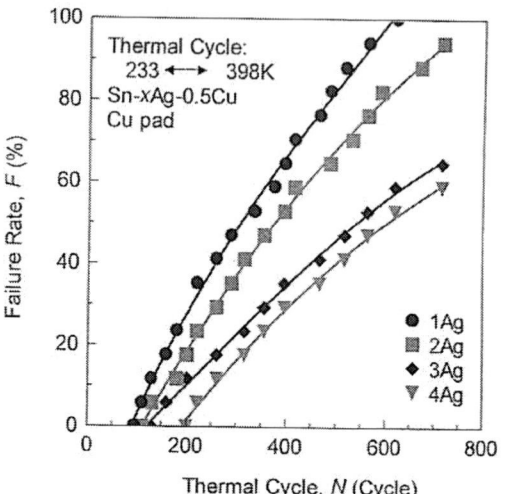

Figure 4. ATC test data from Terashima showing the direct relationship between Ag content and thermal fatigue life [15].

Finally, the impact of significant alloy changes on the acceleration factor that relates field life to accelerated test life is unknown. Darveaux and Reichman [31] performed hysteresis loop predictions for various Pb-free alloys based on measured mechanical property data. The shape and area of the loops, and their dependence on thermal cycle parameters, varied significantly for different alloys. This led the authors to suggest that the acceleration factor will also vary by alloy. However, direct measurements of the acceleration factors from accelerated thermal cycle testing has yet to be published for any of the new alloys.

Summary: The following conclusions can be drawn from this brief review of thermal fatigue of SAC alternative alloys:
- Thermal cycling studies for low Ag content and microalloyed SAC solders are extremely limited.
- Some ATC results show better performance at low Ag contents, while others show better performance at high Ag contents. Thermal fatigue reliability appears to be dependent on process, microstructure, and microalloy content, and those dependencies have yet to be characterized fully and understood.
- A number of proprietary industry studies are either underway or in the advanced stages of planning. The results from these studies are expected to become public eventually.
- Prior to launching additional, independent thermal fatigue studies, researchers should review the literature and should also consider the scope, technical details, and timetables of existing experimental programs within the industry.

978-1-4244-4722-0/09 $25.00
© 2009 IMAPS-ITALY

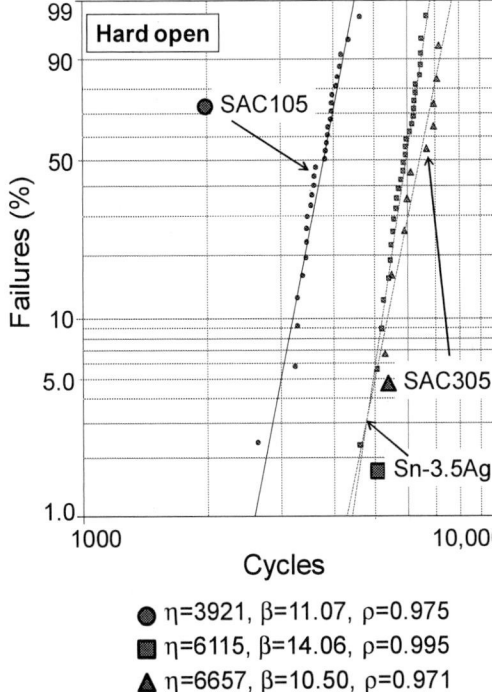

● η=3921, β=11.07, ρ=0.975

■ η=6115, β=14.06, ρ=0.995

▲ η=6657, β=10.50, ρ=0.971

Figure 5. Weibull plots of failure life for three different ball alloy joints using an electrical failure criterion of a hard open. Data of Henshall et al. [35].

5. Impact of Low Ag Balled BGA Components on PCA Manufacturing

Although low Ag BGAs have been successfully integrated into the many products today, issues have surfaced when attempting to use them in either more thermally challenging assemblies or if having to use them in a backwards compatibility scenario where they are being soldered to the PCB using Sn-Pb solder.

Thermally Challenging Assemblies: With the potential benefits of alternate alloy balled BGA components come some associated impacts to the manufacturing process, specifically if having to incorporate them into a thermally challenging assembly. To understand the issue better, the impact of alloy composition on melting point needs to be understood. Figure 6 depicts the melting point of several common SAC alloy compositions (note: the temperatures depicted on the graph represent the points at which the full liquid phase exists).

The addition of other alloying elements meant to affect undercooling, formation of various intermetallics, matrix properties and microstructure may also impact the melting behavior of the alloy. Such changes in composition can increase the melting point of the solder ball by as much as 10°C over that for SAC305 or SAC405 ball compositions. In many cases, either not all suppliers on the approved vendor list (AVL) for a package have a

consistent ball composition or the supplier has made a change in the composition and not changed either the marking or part number of the package to indicate such. This situation can impact the assembly yields or worse yet, create unacceptable solder joints because the assembly was soldered at too low a temperature. Improperly assembled components, such as those shown in Fig. 7, are a significant reliability risk, since they may pass electrical test but fail more rapidly in the field than a properly formed solder joint. The obvious solution would be to simply raise the assembly temperature of all Pb-free products from the current minimum peak temperature of 230-232°C by approximately 5-7°C. This might be practical for simple, less thermally challenging products where the range of package types drives a low thermal gradient and none of the parts exceed the maximum allowable body temperatures as specified by J-STD-20. However, raising the soldering temperature on more thermally challenging products can be next to impossible without running the risk of overheating certain packages on the board [36]. Increasing temperatures will also put more strain on the PCB laminate, as well resulting in potentially more warpage or increasing the likelihood for pad cratering. Profiling studies appear to suggest that 1% Ag alloys are incompatible with any current industry Pb-free assembly specification that requires a minimum reflow peak temperature and time above liquidus (TAL) of 230°C/60 sec. This could preclude their use on thermally massive components on thermally challenging assemblies.

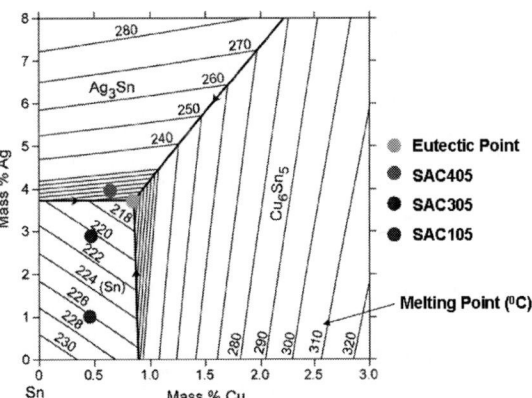

Figure 6. SAC Phase Diagram

Backwards Compatibility: Not all products produced today are manufactured using Pb-free solders. Those OEMs that qualify for exemptions specified in the RoHS Directive are still assembling many of their products using Sn-Pb solder pastes. The challenge for these OEMs has been a shrinking supply of Sn-Pb balled BGAs, especially if the part is also used by OEMs building products for consumer applications where no exemption for Sn-Pb solder exists. In some cases, the use of Pb-free balled BGAs is the only available option. The only recourse is to use the Pb-free BGA in a Sn-Pb soldering process. Previous reliability studies on backwards compatibility conducted using SAC305 and SAC405 BGAs soldered using Sn/Pb paste have demonstrated that by soldering at peak temperatures in excess of 217°C, Sn-Pb solder paste and SAC BGA balls mix completely to form a homogenous microstructure with adequate reliability performance for most electronic applications. For more information on this issue, the reader is referred to the recent report by the iNEMI Mixed Alloy BGA project team [32].

However, the introduction of lower Ag content balled BGA components has changed this situation, particularly for rework operations. Unlike SAC305 or SAC405 BGAs, when SAC105 BGAs are reworked with Sn-Pb solder paste, severe voiding can occur at the component to ball interface, as shown in Fig. 8 and discussed in more details elsewhere [33, 34].

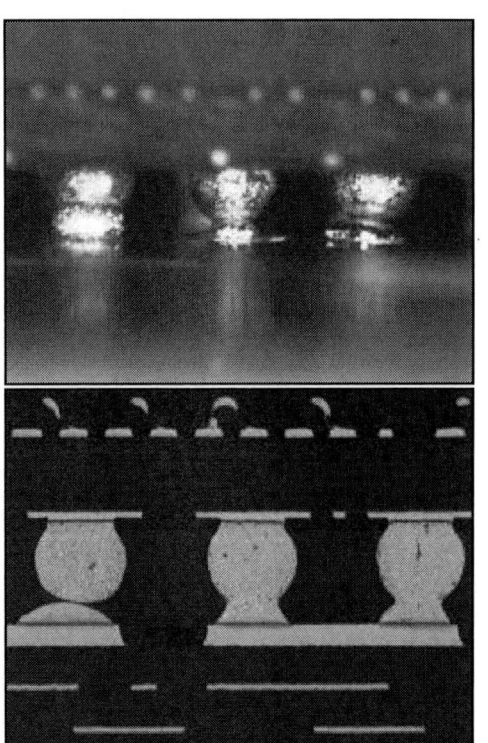

Figure 7. Unmelted Solder Balls and Unacceptable Solder Joints [22].

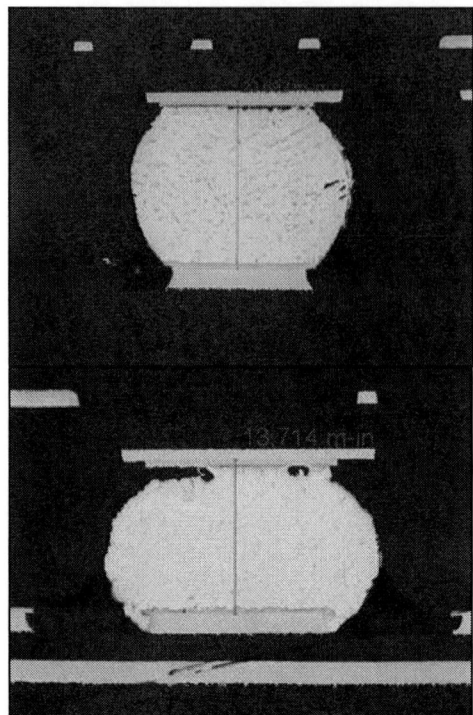

Figure 8. Voiding Observed after Rework of a SAC105 BGA using Sn/Pb Paste.

When a SAC105 ball is mixed with Sn-Pb eutectic solder paste, the resulting alloy has a very wide "pasty" range of approx. 45°C (177°C to 224°C) compared to that of the SAC305/Sn-Pb combination (approx. 30°C). The extra 15°C are the result of the additional Sn in the SAC105 versus the SAC305, driving up the overall melting point of the alloy and resulting in more time being required for the solidification of certain constituents. For example, Sn will need more time to solidify in the SAC105 case than the SAC305 case, resulting in much larger Sn dendrites. In real life soldering situations, the presence of additional alloy dopants (e.g., Ni, Mn, Bi, Ce etc.) or constituents resulting from the dissolution of substrate material, such as organic and inorganic additives that were co-deposited with the PCB surface finish, can lower the final solidification temperature even further.

6. Standards

There are a number of key industry standards that require updating or modification to address the new Pb-free alloys. The iNEMI Alloy Alternatives team is helping to drive these changes, where possible.

First, the iNEMI team has provided input to the IPC/JEDEC committee to provide guidance on J-STD-609, "Marking and Labeling of Components, PCBs and PCBAs to Identify Lead (Pb), Pb-Free and Other Attributes." There is confusion regarding how to label low Ag and microalloyed materials. The

978-1-4244-4722-0/09 $25.00
© 2009 IMAPS-ITALY

committee is currently considering our proposal to provide clarification on labeling for the new alloys.

Second, the iNEMI team presented to the JEDEC JC-14 committee our concerns about part numbers and customer notification when BGA/CSP suppliers change ball alloys. We noted that a Pb-free BGA ball alloy change may have an impact on printed circuit assembly (PCA) manufacturing due to the higher melting point of some alloys, as discussed in the previous section. In particular, the change to low Ag ball alloys may require a change to PCA manufacturing processes. Our request was that the committee consider mandating, or at least recommending, that new part numbers be issued when a BGA supplier changes the solder ball alloy such that a manufacturing process change is needed. A new task group was formed to consider this issue, our recommendation, and other related topics for standardization by JEDEC.

Another standard being discussed by the iNEMI team is J-STD-006, "Requirements for Electronic Grade Solder Alloys and Fluxed and Non-Fluxed Solid Solders for Electronic Soldering Applications." Our goal is to update J-STD-006 to account for new alloys, particularly those with microalloy additions. Presently, some microalloying elements are present in amounts that would normally be considered an impurity. We have communicated our concerns to the committee responsible for this standard, and will be working with them to update the document.

One situation that creates uncertainty in the industry regarding new alloys, and which may slow the adoption of improved materials, is the lack of defined information requirements for alloy acceptance. The acceptability of any alloy may vary from product class to product class, and possibly from company to company. However, the methodology and data requirements may be largely same, regardless of product requirements or company.

Table III. Areas where knowledge is complete/adequate

Sufficient Knowledge
Low Ag alloys improve drop/shock resistance
Micro alloy additions significantly improve drop/shock performance on Cu surfaces but not on Ni surfaces
Decreasing Ag content decreases elastic modulus, yield and tensile strength of SAC
Decreasing Ag content decreases creep strength of SAC
Alloy additions can increase the creep strength of low Ag SAC alloys
SAC alloys are not inherently brittle (needs to be better communicated, however)

As described in Section 8, the iNEMI Alloy Alternatives team is examining the possibility of establishing a set of data requirements and the methods to generate these data. Ultimately, the goal is to take this approach to the relevant standards bodies for industry acceptance and standardization.

7. Knowledge Assessment

Phase 1 efforts of the iNEMI Alloy Alternatives team were focused on establishing and communicating the industry state of knowledge regarding new Pb-free solder alloys. Table III summarizes the major areas where industry understanding is relatively complete, or at least adequate. A summary of key knowledge gaps is provided in Table IV.

The iNEMI team is now focused on addressing the key knowledge gaps to further the industry's understanding of the benefits and possible risks of new Pb-free solder alloys. Follow-on activities are being planned that will add value to the industry and avoid redundancy with other efforts. In addition, the team continues to engage standards bodies in the creation or updating of standards that will help the industry to manage the increasing number of alloys choices. The goal is to drive standards that will help companies take advantage of the benefits of new alloys without suffering unintended negative consequences.

8. Phase 2 Activities

The iNEMI Alloy Alternatives project team is nearing completion of planning efforts for further work on second generation Pb-free solder alloys. In some cases, activity has already begun or is continuing from Phase 1, such as driving updates to key standards. Planning and activity in Phase 2 centers around two focus areas: driving industry standards, in particular alloy test methods, and thermal fatigue behavior. As shown in Table IV, thermal fatigue and alloy test methods were identified as high priority knowledge gaps to be filled.

Development of standard test requirements and methods has begun with a review of the approach being developed by Hewlett-Packard [28]. The iNEMI team is discussing the merits of this approach and possible modifications that may be necessary in order to meet the needs of the broader electronics industry. A dialogue also is taking place between the Alloy Alternatives team and the Solder Products Value Council (SPVC), with the ultimate goal of providing a formal starting point for the development of an IPC standard, or set of standards, addressing testing of new Pb-free alloys.

Table IV. Key knowledge gaps regarding the performance and impact of new Pb-free alloys.

Gap or Concern
High Priority
Advantages and disadvantages of specific alloys
Composition limits for microalloy additions; ranges of effectiveness
Standard method to assess new alloys; standard data requirements
Consistency of testing methods, including test vehicles & assembly, test parameters, etc.
Establish the microstructural characteristics of specific alloys
Long term reliability data for new alloys, particularly low Ag & microalloyed
Lack of thermal cycle data for evaluating new alloys; benchmark to Sn-Pb and SAC 305/405
Medium Priority
Assessment of new alloys for use in "mission critical, long life" products
Impact of rework on microstructure and properties
Mixed Sn-Pb/Pb-free assembly, including rework
Impact of alloy composition on work hardening rates & other flow properties; effect of strain rate and temperature
Impact of alloy composition on bend/flex limits (moderate strain rate; ICT, handling, card insertion, etc.)
Thermal fatigue accelerations factors (not yet fully established for SAC 305/405)
Impact of aging on microstructure and mechanical properties
Low Priority
Solder process margins required for new alloys used in various product classifications
Mixing of different BGA ball alloys and paste alloys for various component and board designs

The lack of information on the thermal fatigue performance for many new Pb-free alloys has motivated the Alloy Alternatives team to plan accelerated thermal cycle experiments. The project team has considered many possible sets of experiments in order to answer a variety of questions. In the end, the team has decided to:

- Validate the impact of Ag concentration in the range of 0 to 4% on thermal fatigue resistance.

- Evaluate the impact of commercially common dopants, such as Ni, on thermal fatigue performance.

- Assess how alloy composition affects the acceleration behavior.

- Provide basic thermal fatigue data for several of the most common alternate alloys on the market today, benchmarking them against eutectic Sn-Pb and SAC305.

Currently, we are in the process of finalizing details of the accelerated thermal cycle test plan, including which alloys to test and which thermal cycle profiles to investigate such that meaningful acceleration factor data will result. If we are successful in executing these plans, the data could be of major benefit to the industry, especially since such large studies are nearly impossible for a single company to undertake.

9. Summary and Conclusions

The knowledge assessment efforts of the iNEMI Alloy Alternatives team have been described. This multi-company, multi-sector team has assessed the recent literature regarding new Pb-free solder alloys alternatives and come to the following conclusions.

1. Considerable progress has been made in understanding the fundamental relationships between alloying elements and properties for the SAC family of new Pb free solders. Additional work is needed to fully characterize the complex microstructures and their influence on physical and mechanical properties.

2. Areas where the performance of new alloys is reasonably well established have been identified. Some of these include: (i) impact of Ag content and microalloy additions on mechanical shock reliability; (ii) impact of Ag content on elastic stiffness, plastic flow and creep behavior of SAC alloys.

3. Areas where more knowledge is needed in order to properly assess the benefits and potential risks of new alloys also have been identified. Some of these include: (i) thermal fatigue performance, including the impact of microalloy additions and development of acceleration models; (ii) the impact of alloy composition on the full range of solder processes; (iii) impact of thermal aging on microstructure and properties; (iv) impact of composition on bend/flex limits related to PCA manufacturing, test, board handling, etc.

4. Standardized data requirements for assessment of new alloys are needed so that each company can compare alloy performance with product requirements over the full range of relevant properties. The iNEMI Alloy Alternatives team is currently considering the HP approach as a starting point for such standardization.

5. The iNEMI Alloy Alternatives team is actively engaged with relevant standards bodies to create or update industry standards related to new Pb-free solder alloys.

Acknowledgements

The authors would like to thank Jim McElroy and Jim Arnold of iNEMI for their assistance in launching this project and for many valuable discussions.

References

1. Yoshiharu Kariya et al., J. of Elect. Mat, 33, No. 4, 2004.

2. C.M. Gourlay, J. Read, K. Nogita, and A.K. Dahle, "The Maximum Fluidity Length of Solidifying Sn-Cu-Ag-Ni Solder Alloys", Journal of Electronic Materials, Special Issue Paper DOI: 10.1007/s11664-007-0248-8

3. K Sweatman, S. Suenaga and T. Nishimura, "Strength of Lead-free BGA Spheres in High Speed Loading" Proceedings Pan Pacific, 2008.

4. Ranjit S Pandher, Brian G Lewis, Raghasudha Vangaveti and Bawa Singh, "Drop Shock Reliability of Lead-Free Alloys – Effect of Micro-Additives," Proceedings 57th Electronic Components and Packaging Technology (ECTC), Reno, May 29-June 1, 2007.

5. Ranjit S Pandher, Robert Healey, "Reliability of Pb-Free Solder Alloys in Demanding BGA and CSP Applications," Proceedings 58th Electronic Components and Packaging Technology (ECTC), Orlando, May 27-30, 2008.

6. Werner Engelmaier "Surface Mount Solder Joint Long-Term Reliability: Design, Testing, Prediction," *Soldering and Surface Mount Technology*, vol 1, no. 1, 14-22, February, 1989.

7. J. Bartelo, et al., "Thermomechanical Fatigue Behavior of Selected Pb-Free Solders, Proceedings IPC APEX 2001, LF2-2, January 14-18, 2001.

8. N. Pan, et al., "An Acceleration Model for Sn-Ag-Cu Solder Joint Reliability Under Various Thermal cycle Conditions," Proceedings of SMTAI 2005, 876-883, Chicago, IL, September 2006.

9. J. Bath, et al., "Reliability Evaluations of Lead-Free SnAgCu PBGA676 Components Using Tin-Lead and Lead-Free SnAgCu Solder Paste," Proceedings of SMTAI, 891-901, Chicago, IL, September 25-29, 2005.

10. John Manock, et al., "Effect of Temperature Cycling Parameters on the Solder Joint Reliability of a Pb-free PBGA Package," Proceedings of SMTAI 2007, 564-573, Orlando, FL, October 2007.

11. B. Nandagopal, et al., "Study on Assembly, Rework Process, Microstructures and Mechanical Strength of Backward Compatible Assembly," Proceedings of SMTAI, 861-870, Chicago, IL, September 25-29, 2005.

12. H. McCormick et al. ,"The Great Debate: Comparing the Reliability of SAC305 and SAC405 Solders in a Variety of Applications," Proceedings Pan Pacific Symposium, January 31, 2007.

13. Unpublished results, Peter Borgeson, Unovis Consortium, November 2007.

14. S.K. Kang et al., "Evaluation of Thermal Fatigue Life and Failure Mechanisms of Sn-Ag-Cu Solder Joints with Reduced Ag Contents," Proceedings ECTC 2004, June , 2007.

15. S. Terashima et al., "Effect of Silver Content on Thermal Fatigue Life of Sn-xAg-0.5Cu Flip-Chip Interconnects," J. Electronic Materials, Vol 32, no. 12, 2003.

16. J. Liang, N. Dariavich, and D. Shangguan, "Solidification Condition Effects on Microstructure and Creep Resistance of Sn-3.8Ag-0.7Cu Lead-Free Solder," Metallurgical and Materials Transactions A, Vol. 38A, 1530-1538, July 2007.

17. C. Shea, et al., "Low-Silver BGA Assembly Phase I – Reflow Considerations and Joint Homogeneity Initial Report," Proceedings IPC APEX, April 2008.

18. C. Shea, et al., "Low-Silver BGA Assembly Phase I – Reflow Considerations and Joint Homogeneity Second Report: SAC105 Spheres with Tin-Lead Paste," Proceedings SMTAI, August 2008.

19. H. Kim, et al., "Improved Drop Reliability Performance with Lead Free Solders of Low Ag Content and Their Failure Modes," Proceedings ECTC, p. 962, 2007.

20. A. Syed, et al., "Effect of Pb free Alloy Composition on Drop/Impact Reliability of 0.4, 0.5 & 0.8mm Pitch Chip Scale Packages with NiAu Pad Finish," Proceedings ECTC p. 951, 2007.

21. D. Kim, et al., "Evaluation of High Compliant Low Ag Solder Alloys on OSP as a Drop Solution for the 2nd Level Pb-Free Interconnection," Proceedings ECTC p. 1614, 2007.

22. G. Henshall, et al., "Manufacturability and Reliability Impacts of Pb-Free BGA Ball Alloys," Unpublished research, Hewlett-Packard, 2007.

23. A. Syed, T. Kim, S Cha, "Alternate Solder Balls for Improving Drop/Shock Reliability," Proceedings SMTAI, p. 390, 2007.

24. Tanaka et al., Proceedings ECTC, p. 78 2006.

25. Y-S Lai et al., Microelectronics Reliability, 46, p. 645-650, 2006.

26. P. Snugovsky, et al., "Microstructure, Defects, and Reliability of Mixed Pb Free / SnPb Assemblies," Proceedings TMS, V 1: Materials Processing and Properties p.p. 631- 642, 2008.

27. B. Smith, P. Snugovsky, M. Brizoux, A. Grivon., "Industrial Backward Solution for Lead Free Exempted AHP Electronic Products: Process Technology Fundamentals and Failure Analysis," Proceedings IPC APEX, April 2008.

28. H. Holder, et al., "Test Data Requirements for Assessment of Alternative Pb-Free Solder Alloys" Proceedings SMTAI, p. 109, 2008.

29. S. Athavale, MS Dissertation, SUNY Binghamton, 2005.

30. K. Nogita & T. Nishimura, Scripta Materialia 59, 2 (2008) 191-194.

31. R. Darveaux and C. Reichman, "Mechanical Properties of Lead-Free Solders," Proceedings ECTC,. 695, 2007. p

32. R. Kinyanjui, et al., "Solder Joint Reliability of Pb-free Sn-Ag-Cu Ball Grid Array (BGA) Components in Sn-Pb Assembly Process," proceedings APEX, 2008.

33. G. Henshall et al, "iNEMI Pb-Free Alloy Alternatives Project Report: State of the Industry", Proceedings SMTA International 2008, Orlando, Florida, August 2008

34. G. Henshall et al., "Addressing Industry Knowledge Gaps Regarding New Pb-Free Solder Alloy Alternatives" Proceedings of 33rd International Electronics Manufacturing Technology Conference 2008

35. G. Henshall et al., "Comparison of Thermal Fatigue Performance of SAC105 (Sn-1.0Ag-0.5Cu), Sn-3.5Ag, and SAC305 (Sn-3.0Ag-0.5Cu) BGA Components with SAC305 Solder Paste," Proceedings APEX, p. S05-03, 2009

36. L. G. Pymento, et al., "Lead Free Process Development with Thick Multipayer PCBA Density in Server Applications," Proceedings APEX, p. S05-02, 2009

A Comprehensive Overview on Today's Ceramic Substrate Technologies

Franz Bechtold

VIA electronic GmbH, Robert-Friese-Strasse 3, D-07629 Hermsdorf

Phone +49 36601 81529, Fax +49 36601 81530, E-mail bechtold@via-electronic.de

Abstract

Ceramic packaging solutions offer superior reliability performance compared to all organic technologies and are only a fraction of the development cost required for monolithic semiconductor integration. System in package integration capabilities, thermal management, temperature resistivity and heterogeneous system integration (matching different TCE's) are the driving forces for the utilisation of ceramic substrate technologies in microelectronics.

This presentation gives a wide and comprehensive overview of today's ceramic substrate technologies used in microelectronic packaging. It deals with double sided and multilayer ceramics, with low temperature and high temperature manufacturing processes and different material systems. The current state of the art is described by an overview of the main technical and economical characteristics, relevant market sections and the driving forces of these markets to use ceramic substrates. Typical applications, manufactured in volumes for the actual market will be presented to highlight the key benefits for the technology chosen.

The focus of the contribution will be on advanced ceramic technologies and LTCC in particular. LTCC material systems, widely used in the automotive and the telecommunication business, will be highlighted and some market penetrating applications as well as low volume applications in the field of sensors will be presented to demonstrate the advantages of ceramic multilayer substrates. An overview of European sources and market shares will complete the picture of the current state of ceramic substrate technologies. Future market needs, technical trends and actual developments in current R&D programmes will be reviewed. This will show perspectives and challenges for the use of ceramic substrate technologies in new applications and emerging markets like MEMS, Biosensors, MOEMS and others.

Key words: Ceramic substrates, Alumina, Aluminum Nitrid, LTCC, HTCC, RF capability, integrated components, power dissipation, system in package, reliability performance, Microsystem Technologies, costs

Introduction

Ceramic substrate technologies in the electronic business is always considered in the sense of "ceramic printed boards" or more precisely as film integrated circuits. Ceramic substrates are widely used in the field of microelectronic packaging, sensors and actuators and passive components. The most frequently used kind of substrate is the plane and rectangular standard 96% Alumina thickfilm substrate, manufactured by tape casting and high temperature firing at about 1600°. Beside 96% Alumina thickfilm substrates, 99% Alumina thinfilm substrates, ZrO_2 stabilized Alumina substrates, Aluminum Nitride in thickfilm and thinfilm quality, are used for thickfilm and thinfilm integration.

But also other very specific forms and shapes are well known for special applications like pressure sensors, heating tubes, minicoolers and others. They are usually dry pressed or extruded and fired at 1600 °C. Thickfilm processes have to be adapted to the forms and shapes.

Very different from single and double sided standard substrates is the processing of ceramic multilayer Substrates and packages. Most of the thickfilm processes are applied on unfired ceramic sheets, the so called "green" tape which is produced by tape casting. After metallisation with refractory metals inks in the case of HTCC or noble metal inks in the case of LTCC, the sheets are stacked together, laminated and cofired at high temperature (1600°C), which resuslts in **High Temperature Cofired Ceramics HTCC** or cofired at low temperature (850°C) which results in **Low Temperature Cofired Ceramics LTCC.** Manufacturing processes have to be adapted to the special needs of green ceramic. The integration density respectively the high degree of miniaturisation together with the excellent reliability performance and the high frequency capability are the driving forces to apply this technology for very complex solutions in the field of automotive and telecommunication.

An **overview of today's fired ceramic substrates** used in microelectronic packaging is given by the product portfolio of Ceramtec/Marktredwitz.

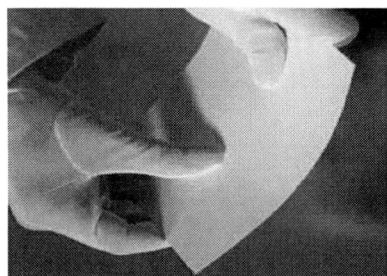

Fig. 1: ZrO_2 stabilised Al_2O_3 Substrate

Beside the typical 96% and 99% standard Alumina products Ceramtec is offering also ZrO_2 stabilised Al_2O_3 Substrate, which shows outstanding reliability performance with respect to thermomechanical stress, flexural strength and elasticity. These substrates, metallised by direct copper bonding, are replacing standard Al_2O_3 direct copper substrates for power applications.

AlN substrates, rarely used in the past, are more and more important as far as power and energy management are concerned. They show extremely good thermal conductivity and together with direct bonded or active brazed copper metallisations they play a very important roll for high power electronics.

Fig. 2:   AlN substrates

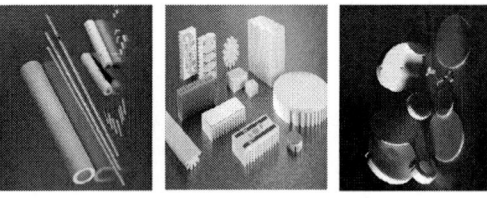

Fig. 3:   Extruded and dry pressed devices

Extruded ceramic tubes for heater elements, dry pressed alumina pressure sensor membranes and micro cooler components are applied in special applications in harsh environment.

Multilayer Substrates HTCC and LTCC

Single and double sided ceramic substrates show limited performance with respect to integration density, miniaturisation RF performance and system in package capabilities. Whereas standard multilayer HTCC packages are on the market since more then 20 years, mainly sourced by the big Japanese companies Kyocera and NTK, customer tailored HTCC packaging solutions start to become more and more attractive in the field of RF housings and micro system integration. Egide in France is the only European supplier of HTCC. The main material used for HTCC is Alumina, white or black coloured and the metal inks are made from refractory metal powders such as Tungsten, Manganese and Molybdenum. Wire bonding and solder surfaces need to be plated for interconnecting to the chip and to the board. LTCC multilayer packages for RF are also available from the Japanes suppliers since a few years and most recently also AlN multilayer packages are offered. In all cases the package/substrate supplier works on the bases of proprietary material receipts, doing all the processes from powder preparation to tape casting and plating. Due to that, high volumes are mandatory to fill a complete manufacturing line.

The situation is different with LTCC substrates. More or less mature "green tape" material systems are available from the shelf, provided by well reputed thickfilm suppliers like DuPont, Heraeus, Ferro, ESL and partially CeramTec. The expertise in the technology together with the relevant application know how are the backbone of the small and medium sized European LTCC Foundries like MSE, Selmic, C-Mac and VIA to serve the free market with high flexibility and medium to low volumes. Free sintering, co- and postfired metallisations ready for all types of post processes (soldering, brazing and wire bonding), a full range of resistor inks and different dielectric inks are providing the full system capability without any additional process.

Large entities like the big Japanese three are dealing mainly with high volume markets whereas Bosch and Epcos cover mainly the captive market in the automotive and telecom business. They all have their proprietary material systems either provided by external partners or made themselves. Constrained sintering and plating are key process for cost and performance reasons. All the processes are dedicated to automated high volume production.

The material compounds differ from manufacturer to manufacturer and from material supplier to material supplier. In most of the cases the LTCC material compounds result into a feldsparoid recrystallised mixed oxide compound CaO.

978-1-4244-4722-0/09 $25.00
© 2009 IMAPS-ITALY

$Al_2O_3.SiO_2.B_2O_3$ with different content of Al_2O_3 grains and a remaining glassy phase. The sintering behaviour of the materials has to be matched to the noble metal metallisations of Ag, Au, Pt, Pd and their alloys, to buried, Cofired and postfired resistor inks and in some cases to high k and low k dielectric inks for capacitors and insulators.

Advantages of LTCC solutions

LTCC Packaging solutions distinguish from all organic solutions by it`s superior reliability characteristic, from standard thickfim technology by a much better degree of miniaturisation, from HTCC by increased functionality and from semiconductor integration by the fact that LTCC asks only a fraction of the development and tooling costs. The line and via resolution today is below 100µm line width and line distance, 100µm via diameter and 150µm via distance. Passive components like resistors, capacitors and inductors can be integrated in a limited range of values monolithically inside the board and the 3d-integration of fluidic structures such as reaction chambers and channels is possible and realised.

Ceramic substrates between Silicon and PCB

System in package integration capabilities, thermal management, temperature resistivity and heterogeneous system integration (matching different TCE's) are the driving forces for the utilisation of ceramic substrate technologies in microelectronics.

The position of ceramics between Silicon and PCB is mainly destinated by costs and performance. In the following chapter the **main economical characteristics** are carried out and shown. They present a cross cost comparison of the technologies for medium to high volumes of customer specified industrial applications.

Cost intensive low volume manufacturing, full Au bearing systems and hermetic variants of the technologies are expressively excluded. All values refer to a finished substrate/wafer assuming a total area of 150cm2.

Typical NRE costs for product development including Design verification, prototype run, masks, tools and test range from 1.000€ for PCB, 5.000 € for thickfilm board, 10.000 to 15.000 € for HTCC and LTCC and up to 250.000 € for a Application specified integrated circuit on Silicon.

The material content of the technologies is different as well: almost pure silicon for Asics, a few sheets of glassfibre inforced low priced plastics together with about 5g of Copper for PCB's, expensive green ceramic sheets and about 3g of noble metals for conductors and resistors in the case of LTCC and a few less expensive Alumina sheets together with about 3g of refractory metals for the conductor. On these assumptions, the average material costs based on different volumes were

calculated and compared. In a similar way the process costs were estimated and calculated.

Table 1: Average Material costs in €					
	PCB	Thickfilm	HTCC	LTCC	Silicon
1000 panels	2,0	15	20	40	20
10.000 Panels	1,5	12	16	32	16
100.000 Panels	1,0	10	14	26	13

Table 2: Average Process cost					
	PCB	Thickfilm	HTCC	LTCC	Silicon
1000 panels	1,50	15	40	40	500
10.000 Panels	1,00	9	32	32	300
100.000 Panels	0,50	5	26	26	200

For completeness, it was considered, that for HTCC and LTCC at a certain volume soft tools will be replaced by **hard tools**. The cost is in the range of 150.000 € and the cost saving is about 15 € per panel. Hard tooling seems feasible at a volume above 100.000 panels.

Adding the material costs and the process costs, the pure manufacturing costs were estimated. Yield losses and overhead costs were not included.

Table 4: Total costs at different volumes					
	PCB	Thickfilm	HTCC	LTCC	Silicon
1000 panels	3,50	30	60	80	520
10.000 Panels	2,50	21	48	64	316
100.000 Panels	1,50	15	40	52	213

In order to get a complete picture, the integration density of the technologies was reflected. It was assumed that the PCB Panel contains 25 curcuits. Thickfilm will have a similar integration density with the same 25 circuits per panel, whereas HTCC and LTCC provide a improved integration density with a factor of 6 and resulting at 150 panels per circuit.

Table 5: Total cost per circuit					
	PCB	Thickfilm	HTCC	LTCC	Silicon
1000 panels	0,18	1,40	0,47	0,63	1,28
10.000 Panels	0,10	0,86	0,33	0,44	0,57
100.000 Panels	0.06	0,60	0,27	0,35	0,36

Moving from multilayer ceramic to Silicon integration another factor of 4 was calculated having then 600 circuits per panel. Thus, if substrate costs per area are compared, it is mandatory to know the number of circuits to achieve these costs.

Table 6: Minimum number of circuits to be realised (yield losses not included)

	PCB	Thickfilm	HTCC	LTCC	Silicon
1000 panels	25k	25k	150k	150k	600k
10.000 Panels	250k	250k	1.500k	1.500k	6.000k
100.000 Panels	2.500k	2.500k	15.000k	15000k	60.000k

Conclusion

Comparing the costs of any kind of ceramic to PCB, ceramic cannot be competitive, independent from volumes and miniaturisation. Comparing ceramic with monolithic semiconductor integration, there is a clear overlap in the range of 1 Mio circuits, where hybrid integration may be competitive to monolithic integration.

Comparing the main different ceramic technologies, HTCC and LTCC are very close together whereas standard thickfilm is significantly behind. Dependend from the technical requirements of the application, each of the ceramic technologies may be the most appropriate one.

Complementary to the economical analysis, the **technical performance** and it`s key characteristics are analysed in the following table, giving a comparison of ceramic packages compared to other technologies. The figures for HTCC are based on Al_2O_3. Thickfilm is not explicitly included.

Table 7: Technical characteristics	PCB	LTCC	HTCC	Si
Guaranteed Life time in years	3	>20	>20	>20
Volume resistivity In Ohm*m (RT)	$\geq 10^{11}$	$>10^{12}$	$>10^{12}$	10^6
Break down voltage In V/μm	40	>40	15	-
Thermal Coefficient of Expansion(TCE) x,y	16	6	7	3
TCE z < tg / TCE z > tg	60 / 290	6	7	3
Flexural strength In Mpa	-	210	350	100
Thermal Conductivity In W/m*K (RT to 400°C)	0,1	3	20	125
Conductor Resistance In mOhm/square	<1	<5	<15	<30
Dielectric Const. 1MHz / Dielectric Const. 10GHz	4,7 / -	7,85 / 7.83	9,8 / 9,0	12,0 / -
Loss Tangent 1MHz / Loss Tnangent 10GHz	0,025 / 0,020	0,002 / 0,005	0,0004 / 0,001	- / -
Line pitch	200	200	200	2
Integration of Passive components	L	L, R, C	L	L, R, C
Water absorption In %	$\leq$0,3	0	0	0
Hermeticity in Torr	$>10^{-3}$	$<10^{-8}$	$<10^{-8}$	$<10^{-8}$

The most important technical feature is the reliability of the integration technology respectively the life time. And one of the most critical parameters in that realtion is the mismatch between the TCE of the different interconnection interfaces, i.E. silicon and board material for Chip and Wire.

Conclusion

The weaknesses of organic PCBs are the high TCE in x, y which leads to thermo-mechanical fatigue defects of chip on board components during thermal cycling and thermal shock, and the much higher TCE in z direction which leads, together with the absorbtion of humidity, to changes in the mechanical and electrical properties of the board (thickness, flatness, dielectric constant, break through voltage..), the low mechanical strength, which impacts the behaviour of sensitive semiconductor devices and the overall parasitic effects in the circuit due to low distances in x, y and Z direction.

Based on these disadvantages of PCB, ceramic solutions are well justified for applications in harsh environment like under the hood in automotive, long life outdoor applications in the telecommunication like base stations, highly miniaturised devices in the mobile phone business like front end modules and for a large variety of sensors and microsystems having specific reliability requirements.

World market for Hybrid Circuits

The recent yearly report 2008 of the ZVEI gives an excellent overview of the market situation.

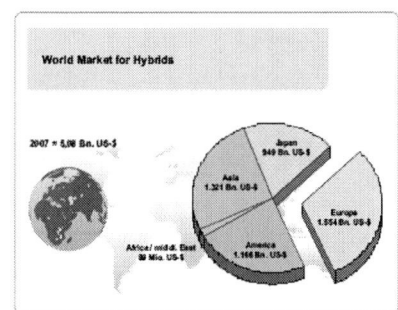

Fig. 4: World market for Hybrids, ZVEI Yearly Report 2008

The European market is 1.200 Bn € and the regional distribution is Germany 48%, France 14%, UK 14%, Benelux 9%, Italien 5%, Scandinavia 5%, Others 5%. From a technological viewpoint, Thick film with 38% and LTCC with 36,5% are the main sales drivers in the ceramic microcircuits segment in Germany. DCB technology arriving at 21% is gaining in importance, while thin film technology with 4,5% still has virtually the same market

978-1-4244-4722-0/09 $25.00
© 2009 IMAPS-ITALY

significance it had in the past. This situation in Germany should be similar all over Europe.

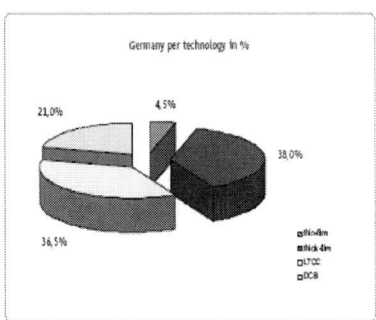

Fig. 5:   Market per technology, ZVEI

According to the "Handbuch der Leiterplattentechnik" (2003) and confirmed by the reports of ZVEI, the market sections are distributed as follows: Automotive: 40%, Telecom: 30%, Industrial: 15%, Military, Avionic: 10%, others: 5% The captive market is about 2/3 of the total market.

Standard Thickfilm applications and key benefits

Fig. 6:   Courtesy of Siegert electronic: Dual Inline Hybrid Circuit for power stations

Key features are the high degree of miniaturisation with printed resistors on top of the dielectric and on both sides of the substrate. High reliability application in harsh environment.

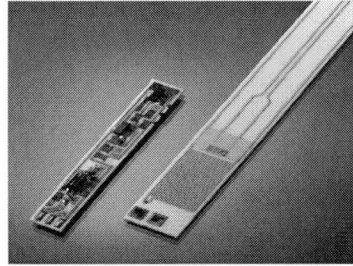

Fig. 7:   Courtesy of Siegert electronic: Oil sensor

Industrial application for measurements in aggressive environment: hot oil.

Fig. 8:   Courtesy of Siegert electronic: Smart power Gear Controll Unit

Key benefits are the Chip on board technology for high power requiring thick wire alumina bonding, the high currency due to thick silver lines and good heat dissipation for high reliability and (-40/+125) in automotive under the hood application.

Micro-Hybrid has developed inhouse a thickfilm pressure sensor system for the measurement and control of yarn tension in industrial seewing machines.

Fig. 9:   Courtesy of Micro Hybrid: Yarn tension controlling sensor system

The matched TCE between the mechanical component and the ceramic board, together with a long term stable Young's modulus of 320 GPa of the ceramic can provide a thermo mechanically stress free and hermetic packaging construction for pressure sensors.

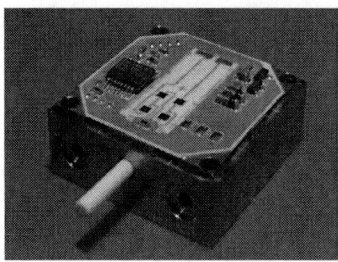

Fig. 10:   Courtesy of Micro Hybrid: Sensor system, open frame

978-1-4244-4722-0/09 $25.00
© 2009 IMAPS-ITALY

The pressure sensor is thickfilm integrated and connected to the force introducing mechanical device. Evaluation electronic is hybrid integrated close to the sensor. The construction principle is demonstrated in the following picture.

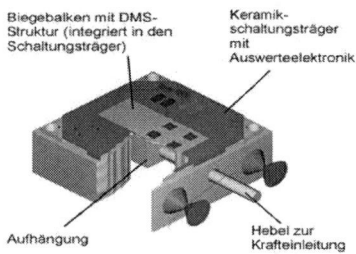

Fig. 11:   Courtesy of Micro Hybrid: Sensor system, principle

Thickfilm on Aluminum Nitrid and their key benefits

The following figures demonstrate the benefit of AlN for power dissipation.

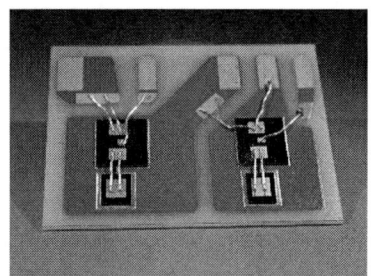

Fig. 12:   Courtesy of Lust Hybrid: H-Bridge hybrid circuit

A high power H-Bridge is possible to be thickfilm integrated on a 0,5 mm AlN substrate. This power circuit works up to 1600 V and 24 A. It is sealed by Glob Top, encapsulated by Silcone and plastic packed. Leads are soldered to the ceramic.

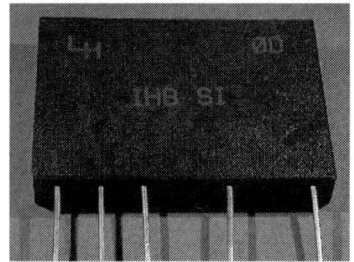

Fig. 13:   Courtesy of Lust Hybrid: H-Bridge sealed and plastic packed

CCD Camera board on AlN for space application

Together with Fairchild Imaging, Anceram developed a so called „Header Bar" based on AlN ceramic, which went to series production 2008. The AlN ceramic was the electric board for a satellite supported CCD camera systems with dimensions of 260 x 15 x 7mm. 384 I/Os were realised int thickfilm technology around the corners of the lapped ceramic bar. Tolerances < 50µm could be realised. Further advantages of AlN are the stiffness, the very well matched TCE and the low weight.

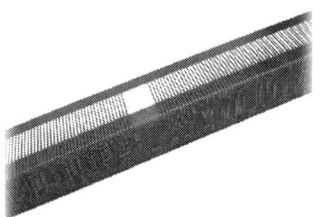

Fig. 14:   Courtesy of Anceram: AlN Camera Board

Pins are Alloy 42, brazed with AuSn to the AuPt solder pads. A ring of AlN was mounted on top for hermetic sealing.

High power applications on direct copper bonded substrates and their key benefits

Direct bonded copper (DBC) substrates are composed of a ceramic insulator, Al_2O_3 (Aluminum Axide) or ZrO_2 inforced Al_2O_3 or AlN (Aluminum Nitride) onto which pure copper metal is attached by a 1065 °C eutectic melting process and thus tightly and firmly joined to the ceramic.

Fig. 15:   Courtesy of Electrovac Curamik: DBC substrates AlN and Al_2O_3

Key features of the ceramic are the high thermal conductivity of 24 W/m*K in the case of Al_2O_3 and 180W/m*K in the case of AlN. Together with Cu as conductor and heat sink material with a thickness of 100 to 300µm, single or double sided (electrically connected by Cu Vias!) high currency interconnections, heat spreaders and heat sinks can

978-1-4244-4722-0/09 $25.00
© 2009 IMAPS-ITALY

be realised at substrate level. A remarkable cost effective approach to reduce the thermal resistance by the reduction of the thickness of the substrates. With respect to mechanical stability, the ZrO_2 inforced Alumina enables a thickness of 250µm without loosing the mechanical stability needed for the DBC process.

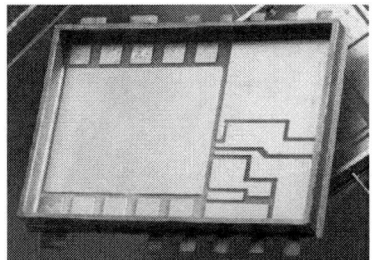

Fig. 16: Courtesy of Electrovac Curamik: hermetically sealed DBC/AlN

DBC substrates are widely used for of power electronic integration for automotive and industrial applications.

HTCC and LTCC Processes

The processes of making ceramic multilayer circuits do not differ very much between LTCC and HTCC as long as green processes are concerned.

Tape blanking

Via punching

Via filling

conductor printing

Stacking

Lamination

scribing/cerving

Burnout,

Cofiring

Singulation

Fig. 17: LTCC/HTCC green processes

For HTCC, the dielectric tape is based on Alumina, the inks based on tungsten. Whereas powder preparation, slurry formulation, tape casting and ink formulation is a proprietary part of the HTCC provider, LTCC semifinnished dielectric tape and inks are available on the shelf and offered by qulified suppliers like DuPont, Heraeus and others. Cofiring is different, in the case of HTCC it is up to

1600°C under reducing atmosphere, whereas LTCC is fired at 875°C under oxidising atmosphere. HTCC metallisation need to have a plated Ni or NiAu finnising, LTCC material systems provide full soldering and bonding capability. Also plating of Ag lines for soldering and bonding compatibility is state of the art. Passive integration of resistors is restricted to LTCC.

HTCC applications and their key benefits

HTCC substrates are mainly used as submounts in metal packages i.e. for fibre optic transmission but also as MCM modules for high power and high reliability applications. The only European suppliere is Egide in France. All informations and applications presented ware provided by Egide.

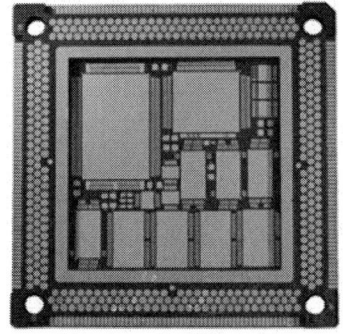

Fig. 18: Courtesy of MDBA –France and Egide: MCM for aerospace

The key features are the high integration density together with a good thermal conductivity for power management. The size of the circuit is 50x50mm, it contains 15 internal layers, 75µm lines and spaces and > 20.000 vias with a diameter of 75µm.

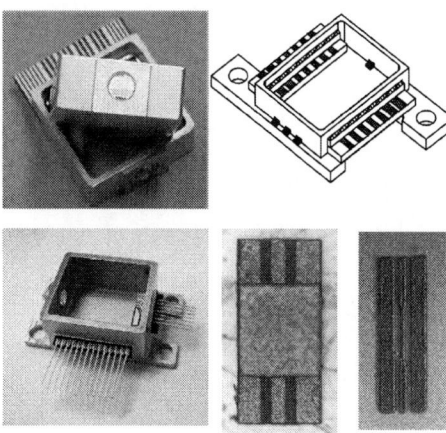

Fig. 19: Courtesy of Egide: Tosa/Rosa fibre optics transmissions

Another field of HTCC applications are feed throughs in packages for fibe optic transmissions. The key features are Hermeticity and robustness, high density, RF transmissions and high temperature package assembly for subsequent brazing technologies.

Coplanar transmissions have a very small dimension of 0.7 x 1.6 x 0,26 mm and a minimum internal line with of 40 µm. It is applied in a package for a 100 Gb/s multiplexer module using 4 feedthroughs brazed in a cover frame.

Fig. 20: Courtesy of Egide: circular feedthrough for infrared sensors

Flexibility to the package construction, good long term hermeticity under high vacuum and I/O pins robustness are the key features of this feedthrough.

Application field of LTCC substrates and their key benefits

LTCC substrates and packages are widely used in the automotive and the telecommunication business. They are made from different material systems either from the free market (DuPont, Heraeus, ESL, Ferro) or proprietary formulations (Epcos, Bosch, Murata). An overview of different suppliers, types and characteristics is given in the table below.

Table 8	Commercial LTCC materials				
	951 DuPont	943 DuPont	A6 Ferro	41110 ESL	CT 2000 Heraeus
Permittivity ε					
≤ 1 GHz	7,4	7,5	5,9	4,7	9,1
1-20 GHz	-	7,4			9,1
20-40 GHz	-	7,4	5,9	4,6	
Dielectric losses tan δ [10⁻³]					
≤ 1 GHz	< 2	< 1	< 2	≤ 4	≤ 1
1-20 GHz	5	1			≤ 2
20-40 GHz	<15	< 2	< 2		
Insertion loss [dB inch⁻¹]					
Conductor	Ag	Ag	Ag	Ag	
1- 20 GHz	≤1,4	≤ 0,3	≤ 0,4	≤ 0,4	
TCE					
α [ppm K⁻¹] (25-300°C)	5,8	4,5	7,0	6,4	5,6
Roughness as fired					
R$_a$ [µm]	0,3	≤ 0,3	≤ 0,38		≤ 0,22

The datas are focussing on RF capabilities, which are most important for many applications. One of the most mature systems used is DuPont 951. The following table gives an overview of the complete material system:

Table 9: Material System 951 DuPont

Material	Process	Function	properties	Tol.%
Au/AuPt	cofire	Wire/solder	3/30 mOhm	20
Ag/AgPd	cofire	Wire/solder	3/30 mOhm	20
Ag/AgPd/Au	cofire	Wire/solder/ bond	3/30/3 mOhm	20
Au, AuPtPd	postfire	Wire/solder	3/30 mOhm	20
Ag, AgPd	Postfire	Wire/solder	3/30 mOhm	20
Dielectric glass	Cofire	Solder stop		
Dielectric glass	Postfire	Insulator		
RuO2-Bi2O3	Cofire	Resistor	Not qualified	30
RuO2-Bi2O3	Buried	Resistor	10 to 10kOhm	50
RuO2-Bi2O3	postfire	Resistor	10 to 10MOhm	1 (tr.)
Metallisation	Co/post	Inductor, capac.	E = 7	20

High volume application: automotive

The biggest market for LTCC in Europe is the automotive business. It is estimated to be 400 to 500 Bn €. Most of it is captive and covered by Bosch and Conti-Temic. Whereas Bosch has its own high volume LTCC lines, Conti Temic is sourced with LTCC substrates from the biggest Japanese and European suppliers. The main applications on this market are Antiblocking breaking systems, Engine control units, Gear control units, Injection Systems, and Speed Control The reason for LTCC is its high integrations density and it`s reliability performance which makes it suitable for under the hood applications..

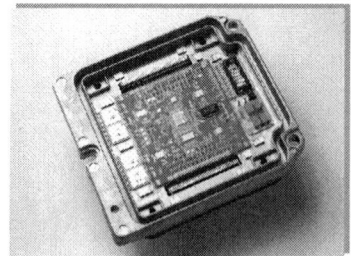

Fig. 21: Courtesy of Siemens-VDO: Gear Box Control (Source: W.C. Heraeus)

Vibration is up to 50g, and operating temperature ranges from -40 up to 155°C. The electronic equipment is directly exposed to the

978-1-4244-4722-0/09 $25.00
© 2009 IMAPS-ITALY

working environment at the brake, in the gear box and on the motor block

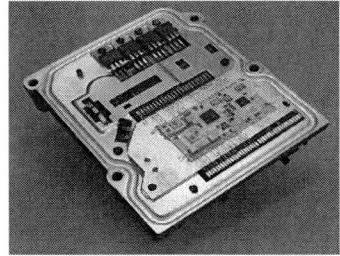

Fig. 22: Courtesy of Bosch: ABS/ESP 8 Antiblocking and Stability Controller

RF electronic has been introduced into LTCC technology for automotive for communication (Navigation, WLan, Bluetooth, GSM/UMTS, Keyless entry and security (Cruise Control, Collision Warning. Side Radar, Distance Radar.

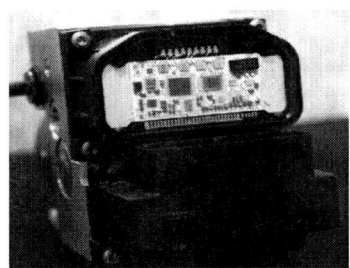

Fig. 23: Courtesy of Bosch: ABS 5.3 Old Genereration Antiblocking System

New applications are coming up, introduced by Bosch, the main player in that field and the biggest European automotive supplier. Steer by wire is one of the key word and LTCC is already introduced into this application by Bosch in high end automobiles.

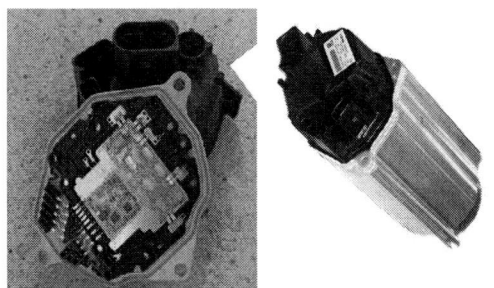

Fig. 24: Courtesy of Bosch: Steer by wire

Another very recent and important application is transmission control.

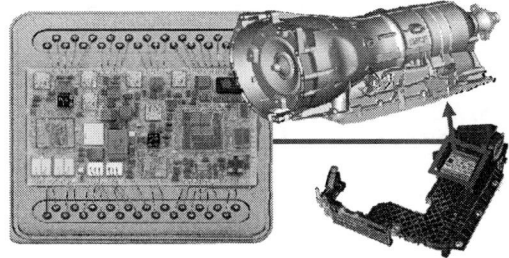

Fig. 25: Courtesy of Bosch: Transmission control unit

High volume application: telecommunication

Telecommunication has been one of the technology drivers until the big market decrease in 2001. Nevertheless the mobil phone business is still the most relevant high volume market of LTCC for RF applications. Complete front are realized today in LTCC.

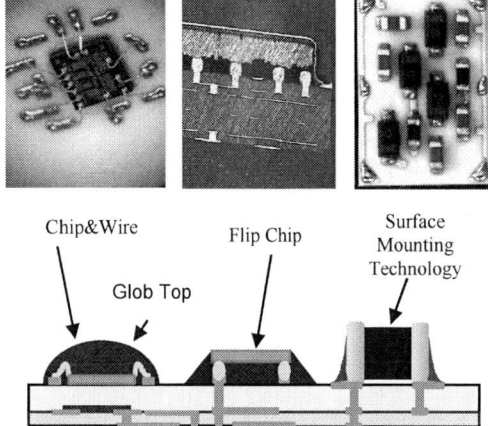

Fig. 26: Courtesy of Epcos: Front End for Mobile Phones

The key benefit of LTCC is the possibility to integrate passive RF functions at substrate level. Using a material system which consists of two different dielectric materials, one for signals with K=8 and the other with K=20 (the so called electronic ceramic) enables to integrate a number of such functions like filters, baluns, matching networks and others.

Low volume application: Telecommunication

Not only high volumes are requested in the telecommunication business. One example are radio links for point to point RF transmission. LTCC is offering again numerous advantages. One is the RF performance and the capability of passive

978-1-4244-4722-0/09 $25.00
© 2009 IMAPS-ITALY

integration, another one is the design flexibility combined with the approach of modularity.

A number of different T/R modules which work at different frequencies from 15 to 39 GHz have been designed, using the same or similar RF building blocks for the realisation of the circuit.

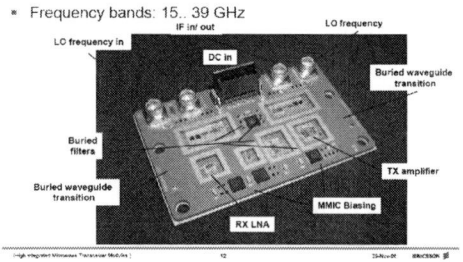

Fig. 27: Courtesy of Ericsson: T/R modules Source: Workshop MakroNano, Ilmenau 2006

Filters, waveguide transition and other RF functions were successfully integrated at substrate level together with buried resistors. The substrates are sourced on the free market in Europe.

Low volume application: X-Ray Detector

For industrial applications like X-ray detection LTCC has proven as a proper and competitive solution. Concerning costs, modularity plays again a significant role, in this case because similar electrical designs can easily be adapted to different housing requirements. The key benefit is the flexibility of LTCC with respect to forms and shapes together with the high integration density and the possibilities of thermal management. Reliability is an important issue as well and the reason to use LTCC as packaging technology were: the matched TCE between the silicon sensor and the ceramic package, the high insulation resistivity of the ceramics and good wire bonding and solder properties of the metallisation.

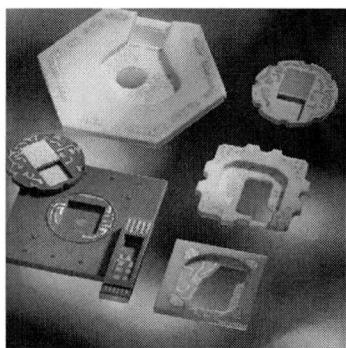

Fig. 28: Courtesy of VIA: Customized LTCC Packages for Si based sensors

For the application it was important to be spectroscopical neutral, that only few x-ray spectral lines are coming from impurities and that full vacuum compatibility (no outgassing) for x-ray microscopy is guaranteed. For the thermal management, heat dissipation of the semiconductor is efficiently guided through thermal vias in the LTCC to the cooling plate.

Fig. 29: Courtesy of Ketek and VIA: Fully integrated X-Ray Detector System

LTCC in these low volume applications is offering a sole solution for the user and plays an enabling roll to bring new system developments onto the market.

Conclusions for LTCC

LTCC is providing manifold advantages for packaging tasks with respect to the material properties, the superior reliability performance and the flexibility of package structures. These are mainly the low TCE of (4ppm to 7ppm), the multilayer construction and the hermeticity of multilayer ceramic. Additionally there are all the „System in Package" functionalities which may be integrated with standard micro techniques i.e. for fluidics or in thickfilm technology i.e. for passive components. For sensor applications the LTCC technology is particularly attractive in the case of special applications with high reliability and high miniaturisation requirements asking for very specific solutions for low and medium volumes. The following.

European Ceramic board sources

An overview of European material and substrate sources, technology providers and technology users will complete the picture of the current state of ceramic substrate technologies. Dealing with the sources for ceramic substrates needs some differentiation with respect to processes and materials along the value chain as well as with respect to markets. Following the value chain we have material and technology suppliers.

The main material suppliers are listed in the following table. This summary relates to fired ceramic substrates, thickfilm materials and complex LTCC material systems.

Table 10 Supplier	Materials and services
CeramTec	Al_2O_3 substrates 96%, 99%, Al_2O_3 ZrO_2 stabilised, Al_2O_3 dry pressed and extruded parts AlN Substrates thinfilm/thickfilm LTCC dielectric tape Laserscribing and cutting Single sided cofired metallisation
AnCeram	AlN substrates thinfilm/thickfilm Active brazed Copper
Heraeus	Thickfilm inks LTCC material systems
DuPont	Thickfilm inks LTCC material systems

Of course, the big Japanese NTK, Kyocera and Murata are offering their services for the European market as well.

The market itself is distinguished between the open and the captive market. On the **open market** we have the following technology providers for ceramic board technologies in Europe:

Table 11 Technology Provider	Technology services
AB Mikroelektronik	Thickfilm
Advanced Microelectronics	Thickfilm, Thinfilm
C-Mac	Thickfilm, LTCC
Curamik	Direct bonded Copper
Egide	HTCC
Epcos	LTCC
Hightec	Thinfilm
Hybrid Electronic	Thickfilm
Lewicki	Thickfilm
Lust	Thickfilm
Microdul	Thickfilm
Microtel	Thickfilm
Micro-Hybrid	Thickfilm
Micro Circuit Engineering	Thickfilm
Micro System Engineering	Thickfilm, LTCC
Radeberger	Thickfilm, Thinfilm
Reinhardt Microtech	Thinfilm
Selmic	Thickfilm, LTCC
Siegert Electronic	Thickfilm
Siegert TFT	Thinfilm
VIA electronic	LTCC

The **captive market** is dominated by

Table 12 Technology User	Technology
Bosch	Thickfilm,
Conti-Temic	Thickfilm
Magneti Marelli	Thickfilm
Sagem	Thickfilm

Future market needs

Future market needs, technical trends and actual developments in current R&D programmes were reviewed and summarized in the following table. This demonstrates the perspectives and challenges for ceramic substrates as enabler from the technical point of view.

Table 13   Perspectives and Challenges

Opportunity	Challange	Technical Requirements
Biotechnology, chemical sensors	Lab on Chip, integrated fluidic, integrated reactors and sensors	Chemical performance Biocompatibility Fluidic interfaces
Heterogeneous Integration	High pin count, QFP, LGA, BGA	Flatness, TCE matching of Ceramic to PCB,
System in Package	3D interconnection, passive integration, thermal management	New and adapted materials, new and adapted processe
MEMS Packaging	3D integration, hermeticity, Wafer level packaging	TCE matching of Ceramic to Si, no XY shrinkage
MOEMS Packaging	Integration of optical and electrical conductors	no XY shrinkage, excellent planarity and accuracy
Energy efficiency	High power/high complexity integration	High currency conductor, micro cooler, high temp.

It is obvious, that standard ceramic substrates for thickfilm and thinfilm integration are at their limits and no significant R&D efforts are envisaged. Complexity, temperature and power are permanently increasing and will drive the future developments of the technology with respect to cost and performance.

Microsystems and sensors are constantly increasing and require new and adapted packaging solutions. The major R&D efforts will be put into innovative LTCC technology to cope with increasing complexity of microelectronic and microsystem devices and into DBC and comparable technologies to cope with increasing power and operating temperatures requirements.

The R&D road maps for LTCC ceramic are dominated by efforts for **cost improvements at process level** like self constrained sintering, high resolution screen printing, micro via laser drilling, wafer level packaging, for **cost improvements at integration level** with new functional materials like high k, low k, high permeability and TCE graduation and cost improvements by combining and optimising different technologies for specific applications.

A significant improvement with respect to high power high complexity requirements is the development of an AlN multilayer technology at the Fraunhofer IKTS in Dresden.

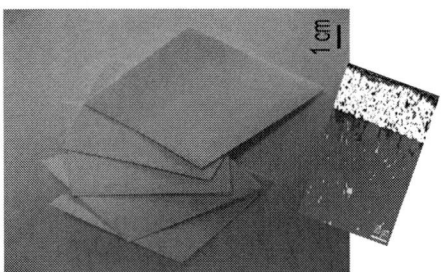

Fig. 29: Courtesy of Fraunhofer IKTS: AlN Multuilayer System (W ink Cofired)

Conclusions

All ceramic substrates provide high reliability for applications in Harsh environment. Ceramic substrates provide economic and ecologic solutions at low, medium and high volumes for high reliability applications.

Increasing power of the electronic devices needs well adjusted thermal properties of the ceramic board technology selected.

Increasing frequency of the electronic devices arrives to the request for passive integration of RF functions and thermal management for high power dissipating GaAs RF semiconductors together with increased integration density for complexity and RF functionality and needs a ceramic board technology fitting all of these requirements for the different applications at the best.

Acknowledgements

This presentation was supported by:

Dr. Knuth Baumgärtel, Micro-Hybrid, Hermsdorf
Dr. Dieter Brunner, Anceram, Bindlach,
Walter Distler, Siegert electronic, Cadolzburg
Dr. Erwin Effenberger, ZVEI, Frankfurt
Dr. Marco Fritsch, Fraunhofer IKTS, Dresden
Thomas Heise, Ceramtec, Marktredwitz
Dr. Sebastian Brunner, Epcos, Deutschlandsberg
Michel. Massiot, Egide, Bollene
Christina Modes, W.C. Heraeus, Hanau
Walter Röthlingshöfer, Bosch, Reutlingen
Dr. Jürgen Schulz-Harder, Electrovac Curamik,
Ralf Stötter, Ketek, München
Thomas. Walther, Lust Hybrid, Hermsdorf

References

[1] J. John, "A Fine Presentation", Proceedings of the 1993 Electronic Components and Technology Conference (ECTC), Atlanta, Georgia, August 12-15, pp. 21-33, 1993.

[2] F. Bechtold, "Innovative Integrationslösungen für Sensoren mit LTCC", Sensor Magazin Vol.2 pp 16-20, Juni,2008

[3] M.Kallenbach e.a., "High-Inductive Small-Size Microcoils with High Ampacity", Proceedings of the 2007 Smart System Integration Conference, Paris,27-28.3., pp.505ff, 2007

[4] J. Töpfer e.a., "Soft Ferrites for multilayer inductors", Int. J. Appl. Ceram. Technol., 3[6], pp 455-462, 2006

[5] F.Bechtold, "A Specialist Manufacturer`s Outlook on the Future of Ceramics and Ceramic Microsystems", Keynote Presentation at the 2nd International Conference on Ceramic Interconnect and Ceramic Microsystems Technologies Denver, April 2006

[6] T. Kraft, H. Riedel, D. Schwanke und E. Müller: „ Simulation von Rissbildung und Verzug in der Mikroelektronik", PLUS, Bd.7 (2005) 883-886

[7] T. Bartnitzek, "MATCH-X-LTCC-Packages for Modular Microsystems", Proceedings of teh 14th European Microelectronics and Packaging Conference, Friedrichshafen, 23.-25. Juni pp. 2003

[8] Dr.W. Schiller e.a., C. Modes, W. Brode, F. Bechtold, „Neue Zero Shrinkage LTCC für Mikrosysteme hoher Präzision" DKG/BAM Sypposium Berlin, 2005

[9] Prof. Dr.-Ing. W.Jillek, G.Keller,"Handbuch der Leiterplattentechnik Band 4", 2003

[10]ZVEI (German Electrical and Electronic Manufacturers Association) Annual Report 2008, Frankfurt

Compression molding solutions for various high end Package and cost savings for standard Package applications

Hideaki Matsutani

Towa Corporation, Kyoto, Japan

Tel. 81-75-692-0263

h_matsutani@towajapan.co.jp

Introduction

There have been substantial changes on high density IC packages.

Packages are becoming smaller and thinner, but, more numbers of chips are installed and stacked and wires are becoming longer and finer.

Towa has been proving mold solutions for IC packages based on their original innovative Multi-Plunger molding method.

Recently, Transfer Molding method is facing difficulty to mold advanced packages such as multi-stack die and fine wire bonded packages.

Transfer molding requires resin flow path with high injection pressure to fill in the cavity using thermoset compound. The advanced packages have limited space for resin flow and more numbers of longer and finer wires are inside.

Also, it becomes more difficult to mold large wafer and large area size substrate by Transfer molding method.

To slove these issues, Compression FFT (Flow-Free and Thin) Molding[Pat.P] was introduced which has become new Defacto standard.

Compression Molding

Compression Mold which was developed by Towa is to mold packages with minimum compound flow.

Basic process is as below:

1. Supply substrate on top Mold.

   After release film is sucked on bottom Mold, Substrate is loaded onto top Mold by mechnical handler.

2. Supply granular compound

   Granular compound is prepared in the box after measuring required amount by detecting number of chip attached on the substrate and the compound is dropped into bottom Mold Chase.

3. Clamp substrate and vacuum starts

   Mold is beginning to close and vacuum is on to suck up air, gas and moisture coming out from the compound.

4. Close mold and cure starts

Mold is completely closed by applying required pressure and compound starts the curing process.

5. Mold open and release

   After cure is complete, mold starts to open and molded substrate is released by ejecting release film.

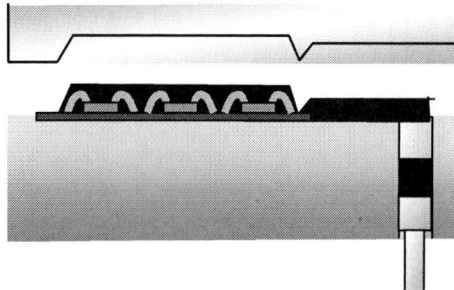

Figure 1: Tranfer Molding

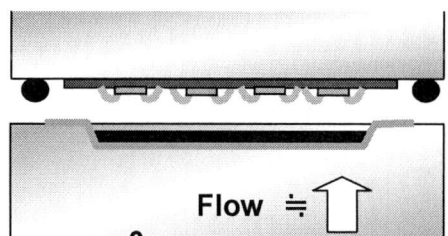

Figure 2: Compression Molding

There is almost no resin flow.and compound can go in small area.

The release film helps to mold without cleaning mold after shot and to release from the mold

Towa's unique high vacuum method "FM(fine mold)" helps to mold any open area to fill in.

The correct amount of compound is supplied after counting number of chip attached on actual

978-1-4244-4722-0/09 $25.00
© 2009 IMAPS-ITALY

substrate and the precise amount of the compound is loaded into the mold cavity.

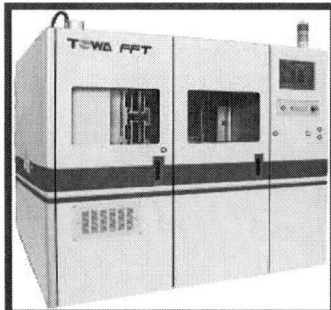

Picture1:Compression Mold System:FFT1030

Advantages by Compression FFT Mold

We have found that there are the following significant advantages by compression FFT Molding.
1. No mold cleaning

Every after shot, no mold cleaning process is required because compound never touches mold cavity surface. Melamine mold cleaning is not required which will reduce scheduled Down Time.
2. High effective usages of mold compound

There is no wastage of mold compound as the required amount of compound is supplied and molded without any cull and runners.
3. High quality molding

This method is quite suitable for the following advanced packages without having any internal defects:

a. Multi-stack die package

When chip is stacked in thin package, there is limited space between top surface of package and top of chip. But, Compression mold can make sure that the limited space can be filled with molding compound.

b. Long or fine wire bonded package

In case of Transfer molding, the injection pressure may push wires when wire becomes finer or longer. Also, the velocity of package through out the gate is very high, so have wire sweep. But, Compression Mold can make sure that there is no X or Y direction resin flow which secures no wire sweep condition. The Picture2 shows no wire sweep by compression FFT molding method.

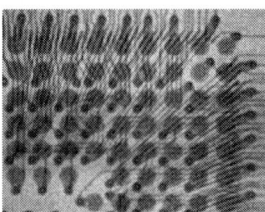

Picture2:Package after molding

The Picture2: Mold evaluation example of wires; ϕ 15 μ m,5mm length by compression FFT

c. Low-K die package

By the soft close and no resin flow, Low-K die is not damaged.The high pressure isn't needed because of no resin flow and no void is created
d. Wafer molding

200 mm or 300 mm wafer can be molded without creating any internal defects.

Figure3:Wafer Level Package

e. Large area size substrate molding

Transfer molding may create different filler distribution and air may be trapped due to long resin flow by injection pressure. But, Compression mold can make sure the symmetric mold condition on any place of the substrate.Compression mold can also keep the resin temperature changes equally on any place of the substrate.
4. Common use of Mold

It is not necessary to change mold cavity for each unique package type. There is up to 0.3 mm different package thickness without changing any mold parts.

Different thickness Mold Cavity is not required when the supplied amount of compound is changed. This will also reduce down time by the conversion if case different thickness molding is required.

Chip

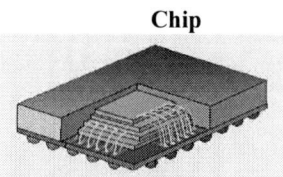

Package

Figure 4: Stacked die package

Package cost reduction with high productivitiy

Recenlty,.we have realized that this method allows significant cost reduction in current standard package assembly process.
1. Gold wire saving

It is possible to mold finer wires such as $\phi 15$ μ m which is approx. 64 % reduction of gold consumption from $\phi 25$ μ m wires. Gold cost is one of most expensive material in package assembly.

2. CMP process elimination

There is no limitation between the top surface of package and top of IC Chip for Compression Molding. Because transfer molding needs minimum mold gap flowing resin compound, but compression FFT molding was no flow. Therefore, thicker chip which reduces CMP or back grinding process can be used without increasing total package height.

3. Large size substrate with larger number of packages

Tranfer mold faces unbalanced mold injection flow which results in possible yield loss. But, Compression Mold has no such limitation. Therefore, larger area size of substrate can be designed which helps to increase more number of packages per strip and higher productivity.

4. The other items

As described in Compression mold advantages, Compound usages as well as no scheduled down time, the CoO is decreased.

In addition, System conversion time and cost can be saved within 0.3 mm package thickness variation.

Compression Molding method is now receiving high attention in the molding market as an important method of CoO reduction.

The only problem was that there was no mass production system for Compression Molding in the marketplace.

This is now solved.

Figure 3: PMC System

Mass Production System, PMC

Towa has developed PMC which is Modular press type, fully automatic Compression Molding System. This System can mold maximum 4 strips in one system.

Single Press Compression Molding System, FFT1030 was developed in 2004 which has been used for R&D and initial production.

In 2008, Towa finished the development of PMC in response to the high demand from the market.

This System adopts unique press structure with "Hold Frame" which could reduce System depth to 1,250 mm which helps to reduce floor space in the clean room.

Main features as as below:

1. Maximun 4 press

Press Module Unit can be added in the site when the production volume is increased.

2. Modular design

Towa's patented Modular Press design makes sure that each press is independently controlled and it is possible to skip the press when something happens.

3. Small foot print

The hold frame press design makes it possible to reduce the machine depth.

4. Pre-cut release film method reduces film usage

Release film has been used in Transfer Molding System by installing indexing film roller which has certain width and every after shot, the film is indexed into used film roller. There was wide area of unused film which was one of higher cost factor. But, Towa invented the "pre-cut" method which can reduce film usages by 30 %.Moreover,we take measures against littering granular compound and contaminations because of separating for the substrate supply and the compound supply and carrying compound on the film.

5. Accurate granular compound control

The required amount of compound is counted after measuring actual number of chip on substrate and is supplied by accurate compound distribution mechanism and correctly supplied into Mold Cavity.

6. Max.100 x 260 mm substrate

This size is 30 % larger than the current popular substrate size. The system is already prepared for the next capital investment by larger substrate size to reduce CoO.

7. Possible to detect number of stacked die

Special Laser scan sensor is equipped to detect number of stacked die on each position on the substrate.

Varioud applications

There are various possible applications for Compression FFT Molding.

One of the best applications is for LED.

LED application

Figure5:LED Package by Compression mold

978-1-4244-4722-0/09 $25.00
© 2009 IMAPS-ITALY

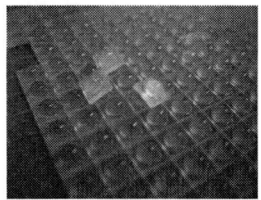

Basic process is the same with IC packages. Compression Molding is now becoming most popular by the following reasons:

1. Lens shape can be molded by high precision mold tooling capability.
2. LED package size can be significantly reduced.
3. Silicone liquid resin can be molded by 1-liquid or 2-liquid mix type Dispenser.
4. High yield
5. Fragile Ceramic substrate can be molded without any crack.
6. Wider light emmiting scope can be achieved.

Towa has also developed mass production System, LCM for LED.

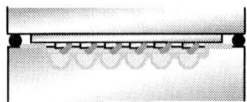

Figure7:Compression Mold

The other possible applications

1. Solar Cell protection

Various kind of plastic is used to protect Solar Cell. The solar needs to guarantee its life for more than 30 years. Compression mold receives high interests to encapsulate the cell.

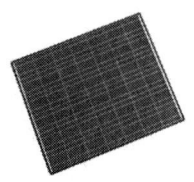

2. Electric Component module

Recenlty, the need to encapsulate the Module by plastic compound becomes high as high reliability is achieved and the cost is economic.

3. MEMS Package

This is also new possible application.

Conclusion

There has been great progress on Compression Molding Technology and Equipment.

This Technology is used not only for most advanced semiconductor package, but also regular BGA type substrate molding to reduce CoO.

There is wide range of applications including LED production.

The system for mass production is fully developed and is ready to support the production.

Compression Molding is the next Defacto standard product in mold industry.

Reference:

Articles from conference proceedings

[1] K. Kawakubo, " Molding System ", Semiconductor Fab/Equipment/Facilities 2009 by Electronic Journal , September 5 , 2008.

[2] H. Matsutani, "LED Package Molding Technology", Proceedings of Electric Journal No.202 Technical Symposium, Tokyo, Japan, March 17, 2009.

978-1-4244-4722-0/09 $25.00
© 2009 IMAPS-ITALY

Advanced Solutions for
Ultra-Thin Wafers and Packaging

Gerald Klug

DISCO HI-TEC EUROPEGmbH
Liebigstrasse 8
D-85551 Kirchheim b. München

Tel.: +49 89 9090 3204
Fax.: +49 89 9090 3298
Email: g.klug@discoeurope.com

Abstract

DISCO Corporation is a leading manufacturer for equipment and tools for wafer thinning and dicing. "Bringing science to comfortable living by Kiru (Dicing), Kezuru (Grinding) and Migaku (Polishing)" is DISCO's mission. By combining these three core technologies, DISCO provides total solutions to meet the more and more demanding requirements of the Semiconductor industry in terms of manufacturing thin dies with high die-strengths and several new approaches for advanced packaging. When developing such processes, circumstances for the total process flow from front-end to packaging are actively taken into consideration.

This article describes various process flows which on the one hand are implemented in the industry as state of the art processes (Conventional process, DBG), but on the other hand also processes which are considered for the future to meet the upcoming requirements for the packaging industry, such as TSV (Through Silicon Via) for die-stacking, thin power devices and the combination of MEMS- and logic devices in one package.

Key words: Conventional process, DBG, TAIKO, Stealth Dicing, Cool expansion, TSV

Conventional Process

The process of Grinding-Polishing-Dicing is the so called standard process. Grinding and polishing takes place in a 3-axis-grinder with the 3rd axis utilizing a dry polishing pad for stress relief.

Dry polishing has become a standard over the recent years having the advantages of slurry free processing, long pad lifetime, easy operation and low cost of ownership. Over the recent years DISCO has further developed new grinding wheels and polishing pads which allow a dedicated selection to meet the best Cost of Ownership (CoO) with regards to the required die thicknesses and die strengths.

Figure 1: Conventional Process

DBG

The process of Dicing-Before-Grinding (DBG) was developed for smart card and memory devices to meet the requirements of thin devices with high die-strengths. The procedure from the conventional process was therefore turned around.

Firstly the wafer is diced partially by Half-Cut-Dicing. Afterwards the grinder separates the pre-cut dies on the grinding tape. This process enables grinding to any die thickness at zero risk of wafer breakage and is reducing the amount of backside chipping to less than 5 μm.
A double dicing speed at dual cut mode increases the UPH of the dicing process and the high die-strength just from grinding can skip the need for stress relief on 3rd grind axis.

Figure 2: DBG-Process

978-1-4244-4722-0/09 $25.00
© 2009 IMAPS-ITALY

TAIKO

TAIKO is the Japanese word for "drum". The wafer edge remains unground so that the ring around the wafer thereby stabilizes the thin inner area. The wafer can then go through further process steps such as implant, back metal or others.

In contrast to the DBG-process, TAIKO enables to manufacture thin dies which need backside processing steps after grinding, e. g. power devices.

Figure 3: TAIKO wafer vs. conventional wafer

TAIKO wafers do not require carrier wafers for handling sequences which makes TAIKO very competitive in terms of CoO.

Different solutions for the removal of the ring are available - such as ring grinding, circle cut, direct dicing - and need to be selected with respect to the customer's requirements.

Laser Dicing

Laser Dicing was originally developed for the removal of low-k materials from the dicing street. Nowadays laser dicing is also being used in a wide range of applications and materials beside Silicon, e. g. GaAs, Germanium, SiC, LiNb.

Furthermore new applications - such as separation of Die-Attach-Films (DAF-tapes) in the Conventional Process and in the DBG-Process, scribe and break for hard and brittle materials and removal of TEG-structures from dicing streets - are currently under development.

A wide range of laser sources is available to the meet the required laser characteristics (wave lengths, frequency, power, beam shape) for the various materials.

Optics for beam shaping and process improvement are completely designed in-house.

Laser dicing is split up in two separate sections:
1) Ablation Process
2) Stealth Dicing

Ablation processing

During the ablation processing material is removed from the wafer and a kerf of app. 10 μm width and 15 – 20 μm depth is cut into the wafer surface. Several passes can be programmed so that also wafers with thicknesses of 100 μm can be completely cut by the ablation process.

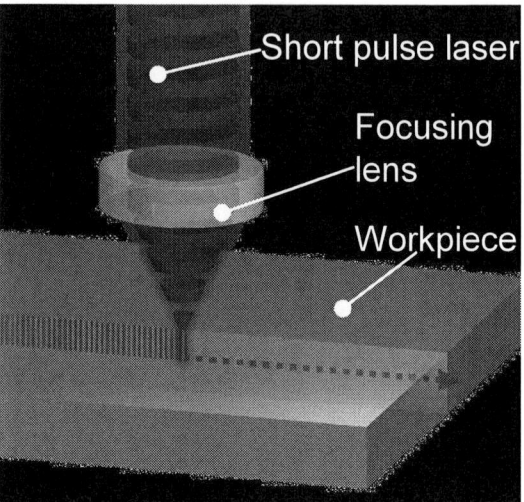

Figure 4: Ablation processing by Laser

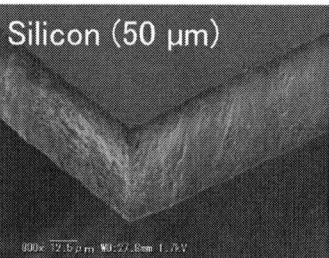

Figure 5: Silicon die cut by ablation laser

During the ablation process the wafer surface can be protected against residues. A water solvable layer (Hogomax) is applied to the wafer front side fore the cutting. Later this layer is removed during the washing sequence inside the spinner section.

This significantly reduces debris adhesion and contributes to an improvement in wafer tolerance and device reliability.

A two headed fully automatic laser saw broadens the range of applications, because it can be equipped with two laser oscillators.

Stealth Dicing

"Stealth dicing" is a dicing method that forms a modified layer in the work piece by focusing a laser inside the work piece, and then a tape expander is used to separate the die. It is a dry process that does not require cleaning. Therefore it is suitable to process wafers such as MEMS that are vulnerable to load.

978-1-4244-4722-0/09 $25.00
© 2009 IMAPS-ITALY

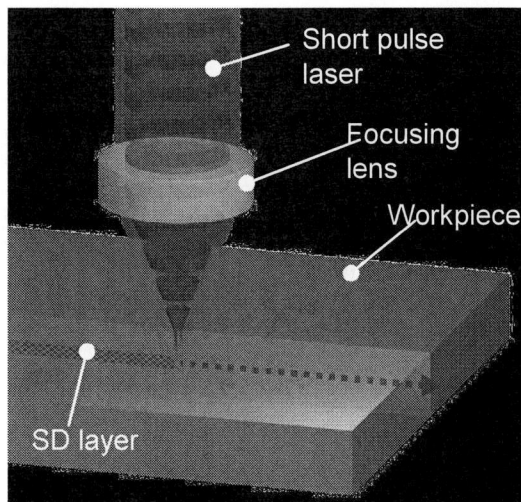

Figure 6: Stealth Dicing

The profile of a MEMS die, such as one with a hollowed structure, or one that is already embedded with complicated minute elements, is generally not strong enough for cleaning water or the dicing load. Stealth dicing can be expected to result in high-quality processing of MEMS because it does not use water for processing or cleaning, and there is little or no damage to the die front or back surfaces.

Figure 7: MEMS photograph after stealth dicing

Another application for Stealth Dicing is the singulation of multi-reticle device wafers. By using the Hasen Cut, it is possible to process a wafer with combinations of irregular die shapes that could not be realized with the existing laser full cut processing or a blade dicer.

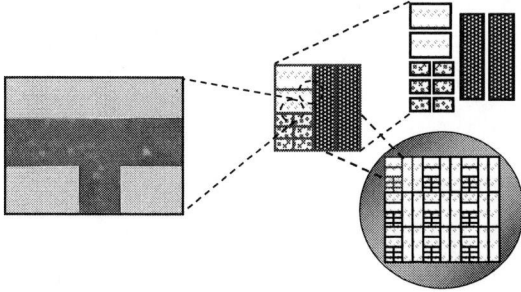

Figure 8: Systematic complex die size wafer cut

Finally the processing of irregular shaped die can efficiently process work pieces like polygon shaped die, such as hexagons and octagons.

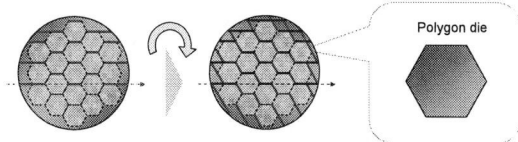

Figure 9: Processing of non rectangular die

Fully automatic die separator

For the expansion of dies DISCO has developed the so called fully automatic die separator DDS2300 which covers following processes:
- Die separation after Stealth Dicing
- Separation of DAF-tapes in combination with the above mentioned DBG-process
- Separation of DAF-tapes in combination with Stealth Dicing of MEMS-wafers

Stealth Dicing in combination with DAF-expansion on DDS2300 enables MEMS-customers to design packages where the MEMS-device is placed sitting on DAF-tape. Thus such sensitive devices can easily and accurately be implemented in compact packages.

The separation of DAF-tapes is achieved by so called "Cool expansion". For this purpose the DDS2300 is equipped with chamber to cool down the DAF-tape to 0°C which improves the rate and quality of DAF-separation tremendously.

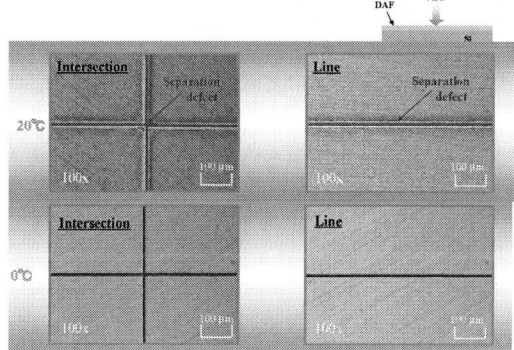

Figure 10: DAF separation at 20°C and 0°C

The combination of DBG with DAF-separation and the combination of Stealth Dicing with DAF- separation are examples how Disco combines its various technologies to for providing total solutions and meeting new requirements.

Further examples are:
- TAIKO and Plasma Etching for TSV on non-bonded wafers
- DBG and Stealth Dicing for TSV on bonded wafers
- Edge trimming in combination with grinding for bonded wafers

978-1-4244-4722-0/09 $25.00
© 2009 IMAPS-ITALY

TAIKO and Plasma Etching for TSV

Grinding wheels which can continuously grind into Cu-posts were developed for TSV. Thus TAIKO followed by Plasma Etching can be used to create a TSV wafer as shown on hereafter.

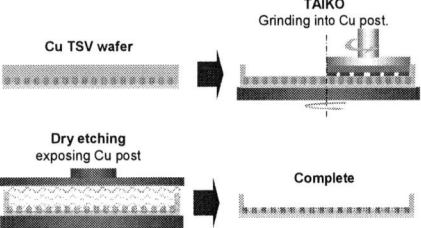

Figure 11:
TSV- wafer after TAIKO grind and Plasma Etching

By grinding into the CU-posts it is assured that all posts have the same height.
The height of exposure of the Cu-post is then controlled by the time that Plasma is applied to the surface.

Figure 12: Via exposed by Plasma Etching

Finally a laser saw can cut the TAIKO wafer from the backside whilst applying a circle cut to the TAIKO ring. Thus the TAIKO ring can easily be removed and the singulated dies are ready for pick up.

DBG and Stealth Dicing for TSV on bonded wafers

The combination of DBG, Plasma Etching, Stealth Dicing, Grinding and Die Separation by Expansion enables customers to create following package when using bonded wafers:

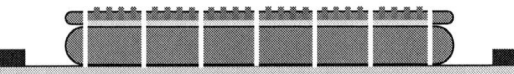

Figure 13: Thin TSV die with exposed vias on top of carrier wafer dies

With this process grinding tapes are not applied at all so that the TTV of TSV-dies is very accurate. The utilization of DBG ensures to have a very little backside chipping on the TSV die and to have no wafer breakage during thinning of the TSV-wafer. Finally the separation of the bottom wafer by Stealth Dicing helps to shrink the kerfs between the TSV-dies to a minimum.

Edge trimming

Further on the topic of bonded wafers so called edge trimming can be conducted for the CAP-wafer prior to grinding or prior to wafer bonding.

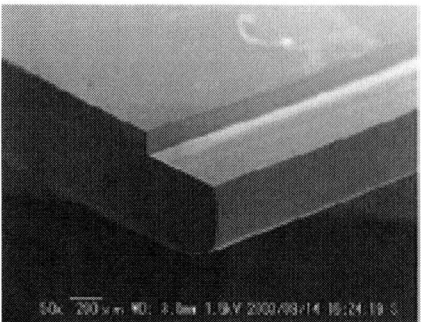

Figure 14: Edge trimmed wafer

Then after thinning of the CAP-wafer there are no sharp wafer edges remaining which can cause wafer breakage easily due to edge chipping from grinding process.

Grinding with Non-Contact-IR-Gauge to thickness of <10 μm

When thinning the top wafer of bonded wafers down to a very small thickness like <10 μm customers request to control the thickness of this top wafer very accurately.
DISCO has therefore implemented a Non-Contact-Gauge which is measuring the thickness of the top wafer whilst grinding by means of a IR-laser.

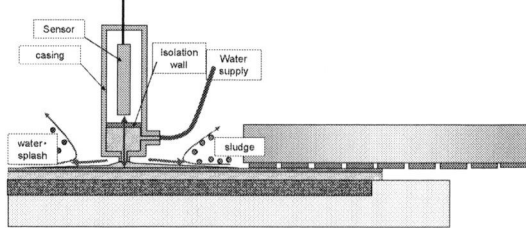

Figure 15: Non-Contact-IR-Gauge

Beside the accurate thinning of top wafers on bonded wafers also the top-layer of SOI-wafers can be reduced to thicknesses of <10 μm very precisely.

978-1-4244-4722-0/09 $25.00
© 2009 IMAPS-ITALY

AUTHOR INDEX

Aasmundtveit, K. 92
Aasmundtveit, K. E. 723
Ahr, A. 367
Aikio, J. 1
Aikio, M. 217
Alajoki, T. 1
Alderman, J. 447
Ali, W. 559
Ansorge, F. 497, 502
Argento, C. 646
Arnold, J. 782
Astier, A. 676
Auchere, D. 78
Avenas, Y. 425
Axisa, F. 697
Azzopardi, C. 167
Azzopardi, M. 58
Bailini, A. 180
Baldo, L. 58
Balogh, B. 729
Balut, C. E. 367
Baraton, X. 321
Barras, A. 338
Barthelmes, J. 290
Batut, N. 87
Bauer, J. 688
Baumgartner, T. 7
Bechtold, F. 798
Becker, K. 688
Beelen-Hendrikx, C. 12
Begbie, M. 232
Belavic, D. 266, 735
Belharet, D. 20
Belmonte, M. 35
Bembnowicz, P. 27
Bennemann, S. 682
Berg, R. V. D. 767
Bhattacharjee, A. 371
Bhatti, N. S. 31
Bjorklund, N. 391
Bock, K. 403
Boehme, B. 713
Boettcher, L. 451
Bonazzoli, M. 35
Bonfert, D. 270, 403
Bonino, S. 35
Bonnot, L. 676
Bos, A. 221
Bosman, E. 275
Bossuyt, F. 697
Bougataya, M. 383
Brannen, C. 69
Brun, J. 40
Brunet, P. 199, 248
Brunet-Manquat, C. 676

Brunke, O. 163
Brunner, S. 159
Burkard, H. 776
Bursik, M. 513
Byun, K. Y. 413
Caccioli, D. 128
Campaniello, M. 180
Campos, D. 203
Canegallo, R. 487
Cardu, R. 487
Caswell, G. 591
Cavallaro, A. 191
Chaehoi, A. 232
Chan, W. L. 581
Chang, L. 555
Chang, Y. 555
Charbonnier, J. 676
Chau, H. 563
Chausse, P. 676
Che, F. X. 315
Chen, W. 555
Cheramy, S. 676
Chiang, K. 280, 528
Chin, K. 555
Chin, L. W. 321
Chou, C. 280, 528
Christiaens, W. 671
Chung, C. 354, 703
Chung, Q. H. 413
Chvatal, M. 763
Cobussen, H. 151
Codreanu, N. D. 270
Colin, D. 20
Collander, P. 286
Connors, M. 569
Conway, P. P. 548
Copeland, D. 569
Corradi, U. 587
Couderc, P. 628
Coyle, R. 787
Crema, P. 290
Cristaldi, G. 45
Curran, B. 624, 688
Debono, J. 78
Dekker, J. 7
Delaney, K. 109
Demosthenous, A. 447
Dietzel, A. 534
Dijkstra, P. 226
Ding, J. P. 581
Dohle, R. 98
Donaldson, N. 447
Doriol, P. J. 52
Dreiza, M. 203
Dresbach, C. 299, 307

AUTHOR INDEX

Drost, A. 403
Dubreuil, P. 20
Dunn, G. 243
Dunne, T. 98
Dutron, A. 361
Eidner, I. 688
Elfving, A. 723
Endrinal, L. 151
Farcy, A. 651
Fenner, M. 327
Fiori, F. 437
Fischer, T. 688
Fledderus, H. 534
Flossel, M. 333
Fontana, F. 58
Fornes, T. D. 74
Forzan, C. 52
Fournier, Y. 122, 338, 666
Fowkes, C. R. 252
Frassati, F. 40
Frisk, L. 466, 661
Fu, S. L. 703
Gabriel, M. 211
Gacusan, R. 139
Gagnard, X. 676
Galan, J. V. 575
Galeotti, R. 35
Garrou, P. 345
Gatt, S. 58
Gebhardt, S. 333
Gee, H. 741
Gerrinck, P. 275
Gindy, N. 252
Giry, J. P. 191
Gobbi, L. 35
Golonka, L. 27
Gonthier, L. 87
Gordon, P. 729
Goßler, J. 98
Graff, J. M. 723
Granier, H. 20
Graziosi, G. 52, 78, 139
Griol, A. 575
Gromala, P. 474
Grosse, C. 682
Grubl, W. 491
Gualandris, D. 65
Guedon, S. 425
Guillou, Y. 361
Guttowski, S. 624, 688
Gyenge, O. 617
Hakansson, A. 575
Halonen, E. 391
Hamoui, A. 383
Han, C. 613, 637

Hansen, U. 617
Hao, J. Y. 581
Harjunpaa, H. 103
Hast, J. 217
Hauck, T. 646
Healey, R. 787
Hegde, P. 171
Heikkinen, M. 1
Hein, M. 601
Heino, P. 103
Hejatkova, E. 513
Helfenstein, M. 457
Hennemeyer, M. 211
Henry, D. 676
Henshall, G. 787
Hertl, M. 378
Heyes, D. 399
Hiitola-Keinanen, J. 217
Hildebrandt, S. 692
Hirte, M. 688
Ho, L. N. 443
Ho, S. C. 581
Hoffmann, C. 159
Hoivik, N. 723
Holc, J. 735
Holland, P. 741
Hong, W. 613, 637
Hough, P. 69, 74
Howell, K. 787
Hraiz, W. 122
Hrovat, M. 735
Hsu, H. C. 703
Hu, G. 321
Hua, F. 787
Huang, C. J. 280
Huang, H. M. 581
Huffman, A. 345, 367
Humbla, S. 601
Hung, T. 528
Hurtado, J. 575
Hurtony, T. 729
Hutt, D. A. 548
Huwel, W. 671
Hyun, C. 294
Imbs, Y. 78
Imran, M. 31
Indelli, G. F. 191
Innocenti, M. 487
Ionescu, C. 266, 270
Ishikawa, T. 237
Izquierdo, R. 383
Jacq, C. 116, 122, 666
Jacques, S. 87
Jang, J. 259
Jang, K. 259

AUTHOR INDEX

Jansen, K. M. B. 713
Jarvinen, P. 203
Jaud, M. 651
Jerlah, M. 735
Jesudoss, P. 759
Jiang, Y. J. 581
Jin, H. 217
Johannessen, R. 92
Juntunen, E. 1
Kaija, K. 391
Kaiser, S. 425
Kang, S. 569
Kapischke, W. 776
Kapitanova, P. 601
Karaszkiewicz, S. 451
Kashiwagi, Y. 443
Kattelus, H. 7
Kellomaki, M. 103
Kemethmuller, S. 98
Kemppainen, A. 391
Kengen, M. 399
Kholodnyak, D. 601
Kiilunen, J. 661
Kim, J. 203, 294, 613
Kim, S. C. 413
Kim, Y. 413
Kimura, K. 518
Kittel, H. 497
Klein, M. 7
Klengel, R. 682
Klug, G. 814
Klumpp, A. 403
Knauf, B. J. 548
Knodler, D. 7
Kocjan, S. 735
Kokko, K. 103
Kopola, P. 217
Koponen, M. 1
Kosonen, T. 1
Kramer, N. 226
Kristiansen, H. 92
Ku, C. W. 703
Kuah, E. 581
Kuhlkamp, P. 290
Kuisma, H. 7
Kurtz, O. 290
Kusters, R. 534
Kuusiluoma, S. 466
Lakhssassi, A. 383
Landesberger, C. 403
Lang, F. 408
Lartigues, P. 628
Lecomte, J. 378
Leduc, P. 651
Lee, C. 509

Lee, J. 294, 413
Lee, S. 259, 354, 509, 563
Leib, J. 617
Lepine, B. 40
Leroy, R. 87
Li, Q. F. 581
Liebsch, W. 367
Lim, L. A. 607
Link, J. 776
Lishchynska, M. 109
Liu, C. 548, 555
Lorenz, G. 299
Luan, J. 315, 321
Luebbehuesen, J. 163
Macek, S. 735
Mach, M. 159, 418
Mackie, A. 327
Maeder, T. 116, 122, 338, 666
Maggi, L. 128, 133
Magni, P. 139
Majzner, J. 763
Makinen, J. T. 1
Malgioglio, G. 45
Manessis, D. 145, 451
Manivannan 371
Mantysalo, M. 391
Marechal, L. 78
Marechal, Y. 425
Marghescu, C. 266
Marsala, J. 569
Marti, J. 575
Martins, O. 425
Marty, A. 651
Masto, A. 569
Mathewson, A. 759
Matsutani, H. 810
Maus, S. 617
Mavinkurve, A. 151
Mazenq, L. 20
McCaffrey, C. 759
Mehta, K. 371
Memis, I. 746
Merlin, M. 437
Metasch, R. 706
Michaelis, A. 333
Milke, E. 299, 307
Min, C. 559
Mittag, M. 299, 307
Mourey, B. 40
Mozek, M. 735
Mueller, W. 587
Muller, J. 159, 418, 601
Muller, T. 299, 307
Muller, W. H. 646
Nakamoto, M. 443

AUTHOR INDEX

Ndip, I. 624, 688
Neubert, B. 211
Neubrand, T. 163
Neyret, M. 676
Nguyen, H. 92
Niehoff, K. 497
Nishikawa, H. 443
Nitin, G. 371
Nonomura, M. 243
Noren, M. 159
Nowakowska, D. 27
Nurmi, S. 7
O'Connell, D. 232
Ogurtsov, V. 759
Oh, C. 613, 637
Ohashi, H. 408
Oldervoll, F. 92
Ollila, J. 1
O'Malley, G. 782, 787
Ong, S. 767
Osterbacka, R. 391
Ostmann, A. 145, 451
Owzar, A. 457
Paik, K. 259, 354
Pandher, R. S. 787
Pandini, D. 52
Park, J. 259, 461
Park, N. 613, 637
Parviainen, A. 466
Passagrilli, C. 167
Patzelt, R. 145
Peels, W. 399
Peltier, N. 425
Perala, J. 466
Perrone, R. 601
Perugini, L. 487
Petaja, J. 1
Petersen, W. 457
Petzold, M. 299, 307, 682
Pfahl, B. 782
Phung, A. 367
Piascik, J. 345
Pieters, P. 656
Pignataro, S. 191
Podprocky, T. 534
Pohlner, J. 98
Pot, A. 767
Preve, G. B. 575
Radji, M. 383
Ramanan 371
Ramkumar, M. 607
Ramm, P. 403
Ray, S. 232
Rebholz, C. 502
Reichelt, J. 474

Reichl, H. 145, 451, 624, 688
Reznicek, M. 482
Reznicek, Z. 482
Reznicek Jr., Z. 482
Roehm, H. 767
Roellig, M. 706
Rosser, S. G. 746
Roth, H. 163
Rotigni, M. 52
Roubion, J. 87
Rouelle, G. B. 338
Rousseau, M. 651
Rubingh, E. 534
Russo, S. 191
Ruythooren, W. 597
Ryser, P. 116, 122, 338, 666
Rzepka, S. 474, 713
Sack, T. 787
Saeed, U. 587
Saeidi, N. 447
Sala, S. A. 180
Sanchis, P. 575
Sandera, J. 513
Sano, M. 280
Santospirito, G. 187
Sarma, G. H. 371
Saugier, E. 203
Scandiuzzo, M. 487
Scandurra, A. 191
Schaber, U. 403
Schachler, R. 7
Scheel, W. 688
Scheithauer, U. 333
Schindler-Saefkow, F. 497
Schischka, J. 682
Schmadlak, I. 646
Schmid, B. 7
Schmitz, S. 491
Schneider-Ramelow, M. 491
Schonecker, A. 333
Schreier-Alt, T. 497, 502
Schuch, B. 491
Scrofani, E. 45, 191
Sedlakova, V. 754, 763
Serra, N. 666
Shanmugam, T. 767
Sharma, U. 741
Sheach, K. 199, 248
Shigetou, A. 641
Shin, Y. 509
Sidiki, T. P. 767
Sikula, J. 754
Silberschmidt, V. V. 171
Sillon, N. 676
Sim, G. 354

AUTHOR INDEX

Sitomaniemi, A.	1
Smetana, W.	543
Smith, L.	203
Snugovsky, P.	787
Song, B.	613, 637
Stam, F.	759
Stapleton, R.	69
Starkey, D.	563
Stary, J.	513
Stephan, R.	457
Suga, T.	641
Suh, M. S.	413
Sun, Y.	252
Svasta, P.	266, 270
Svetly, A.	587
Sweatman, K.	787
Szendiuch, I.	482, 513, 543
Takemoto, T.	443
Tanimoto, S.	408
Terzoli, A.	187
Ticozzi, G.	133
Tisdale, S.	787
Tiziani, R.	139, 167
Tofel, P.	754, 763
Tonnies, D.	211
Topper, M.	7, 617, 688
Torfs, T.	671
Tsai, C.	555
Tuomikoski, M.	217
Twomey, K.	759
Tyldum, H.	92
Uno, T.	518
Upadhya, G.	569
Val, C.	628
Valimaki, M.	217
Van Daele, P.	275
Van Den Brand, J.	534
Van Der Burgt, F.	767
Van Dort, M.	151
Van Driel, W. D.	151
Van Hoof, C.	671
Van Steenberge, G.	275
Van Weelden, T.	221
Vanek, J.	543
Vanfleteren, J.	275, 671, 697
Vath III, C. J.	607
Vendik, I.	601
Verrun, S.	676
Verspeek, J.	226
Vervust, T.	697
Vicard, D.	40
Villa, C.	65, 139
Villain, J.	587
Villavicencio, Y.	52
Vitali, B.	167

Vittu, J.	676
Vogel, M.	569
Von Hofen, H.	746
Walczyk, S.	226
Wang, C.	232
Wang, K.	723
Wang, L.	221
Webb, D. P.	548
Weidmann, D.	378
Weiland, D.	232
Weilguni, M.	543
Weippert, C.	587
Whalley, D. C.	171
Whitney, B.	569
Wieland, R.	403
Wilson, G.	327
Windemuth, R.	237, 243
Wolter, K.	692, 713
Wolter, K. J.	706
Wright, W.	759
Xiang, G.	199, 248
Yamada, T.	518
Yamaguchi, H.	408
Yamamoto, M.	443
Yew, M.	528
Ylikunnari, M.	217
Yong, G. K.	321
Yoo, S.	509
Yun, H.	367, 563
Zafarana, R.	191
Zarnik, M. S.	266, 735
Zhang, J.	563
Zhang, M.	563
Zhang, W.	597
Zoba, D.	69
Zoberbier, M.	211
Zoberbier, R.	211